S0-ABC-613

BIG IDEAS MATH®

Advanced 2

A Common Core Curriculum

Ron Larson
Laurie Boswell

Erie, Pennsylvania
BigIdeasLearning.com

Big Ideas Learning, LLC
1762 Norcross Road
Erie, PA 16510-3838
USA

For product information and customer support, contact Big Ideas Learning at **1-877-552-7766** or visit us at ***BigIdeasLearning.com***.

Cover Image
Pavelk/Shutterstock.com, © Ivan Cholakov | Dreamstime.com, valdis torms/Shutterstock.com

About the Cover
The cover images on the *Big Ideas Math* series illustrate the advancements in aviation from the hot-air balloon to spacecraft. This progression symbolizes the launch of a student's successful journey in mathematics. The sunrise in the background is representative of the dawn of the Common Core era in math education, while the cradle signifies the balanced instruction that is a pillar of the *Big Ideas Math* series.

Printed in the U.S.A.

ISBN 13: 978-1-60840-527-5
ISBN 10: 1-60840-527-3

4 5 6 7 8 9 10 WEB 17 16 15

AUTHORS

Ron Larson is a professor of mathematics at Penn State Erie, The Behrend College, where he has taught since receiving his Ph.D. in mathematics from the University of Colorado. Dr. Larson is well known as the lead author of a comprehensive program for mathematics that spans middle school, high school, and college courses. His high school and Advanced Placement books are published by Holt McDougal. Ron's numerous professional activities keep him in constant touch with the needs of students, teachers, and supervisors. Ron and Laurie Boswell began writing together in 1992. Since that time, they have authored over two dozen textbooks. In their collaboration, Ron is primarily responsible for the pupil edition and Laurie is primarily responsible for the teaching edition of the text.

Laurie Boswell is the Head of School and a mathematics teacher at the Riverside School in Lyndonville, Vermont. Dr. Boswell received her Ed.D. from the University of Vermont in 2010. She is a recipient of the Presidential Award for Excellence in Mathematics Teaching. Laurie has taught math to students at all levels, elementary through college. In addition, Laurie was a Tandy Technology Scholar, and served on the NCTM Board of Directors from 2002 to 2005. She currently serves on the board of NCSM, and is a popular national speaker. Along with Ron, Laurie has co-authored numerous math programs.

ABOUT THE BOOK

The *Big Ideas Math Advanced* series allows students to complete the Common Core State Standards for grades 6, 7, and 8 in two years. After completing this series, students will be ready for Algebra 1 in the eighth grade. The *Big Ideas Math Advanced* series uses the same research-based strategy of a balanced approach to instruction that made the *Big Ideas Math* series so successful. This approach opens doors to abstract thought, reasoning, and inquiry as students persevere to answer the Essential Questions that introduce each section. The foundation of the program is the Common Core Standards for Mathematical Content and Standards for Mathematical Practice. Students are subtly introduced to "Habits of Mind" that help them internalize concepts for a greater depth of understanding. These habits serve students well not only in mathematics, but across all curricula throughout their academic careers.

Big Ideas Math exposes students to highly motivating and relevant problems. Woven throughout the series are the depth and rigor students need to prepare for career-readiness and other college-level courses. In addition, *Big Ideas Math* prepares students to meet the challenge of PARCC and Smarter Balanced testing.

We consider *Big Ideas Math* to be the crowning jewel of 30 years of achievement in writing educational materials.

Ron Larson Laurie Boswell

TEACHER REVIEWERS

- Lisa Amspacher
  Milton Hershey School
  Hershey, PA
- Mary Ballerina
  Orange County Public Schools
  Orlando, FL
- Lisa Bubello
  School District of Palm Beach County
  Lake Worth, FL
- Sam Coffman
  North East School District
  North East, PA
- Kristen Karbon
  Troy School District
  Rochester Hills, MI
- Laurie Mallis
  Westglades Middle School
  Coral Springs, FL
- Dave Morris
  Union City Area School District
  Union City, PA
- Bonnie Pendergast
  Tolleson Union High School District
  Tolleson, AZ
- Valerie Sullivan
  Lamoille South Supervisory Union
  Morrisville, VT
- Becky Walker
  Appleton Area School District
  Appleton, WI
- Zena Wiltshire
  Dade County Public Schools
  Miami, FL

STUDENT REVIEWERS

- Mike Carter
- Matthew Cauley
- Amelia Davis
- Wisdom Dowds
- John Flatley
- Nick Ganger
- Hannah Iadeluca
- Paige Lavine
- Emma Louie
- David Nichols
- Mikala Parnell
- Jordan Pashupathi
- Stephen Piglowski
- Robby Quinn
- Michael Rawlings
- Garrett Sample
- Andrew Samuels
- Addie Sedelmyer
- Tyler Steffy
- Erin Taylor
- Reid Wilson

CONSULTANTS

- Patsy Davis
  Educational Consultant
  Knoxville, Tennessee
- Bob Fulenwider
  Mathematics Consultant
  Bakersfield, California
- Linda Hall
  Mathematics Assessment Consultant
  Norman, Oklahoma
- Ryan Keating
  Special Education Advisor
  Gilbert, Arizona
- Michael McDowell
  Project-Based Instruction Specialist
  Fairfax, California
- Sean McKeighan
  Interdisciplinary Advisor
  Norman, Oklahoma
- Bonnie Spence
  Differentiated Instruction Consultant
  Missoula, Montana

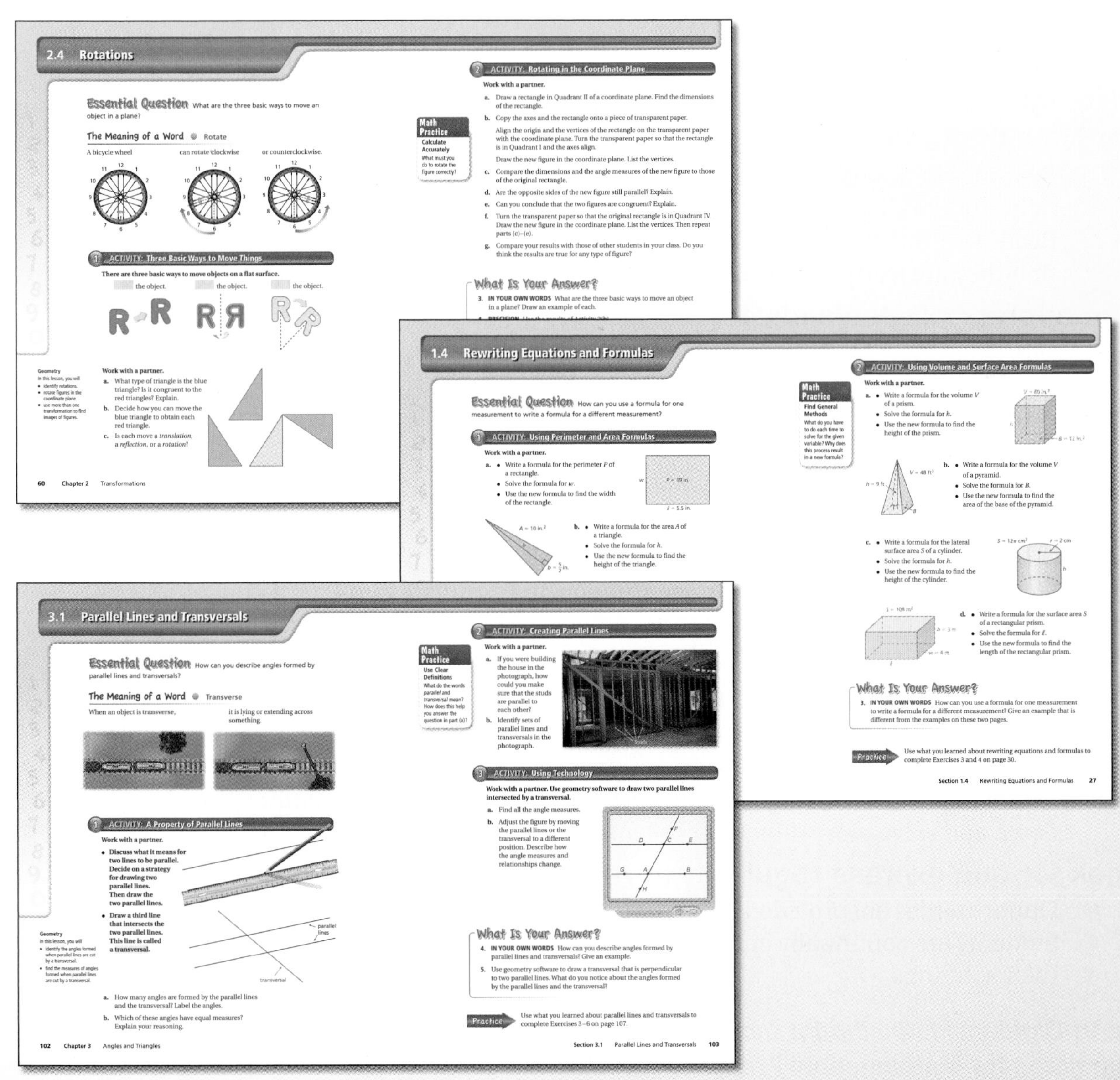

Common Core State Standards for Mathematical Practice

Make sense of problems and persevere in solving them.

- Multiple representations are presented to help students move from concrete to representative and into abstract thinking
- *Essential Questions* help students focus and analyze
- *In Your Own Words* provide opportunities for students to look for meaning and entry points to a problem

Reason abstractly and quantitatively.

- Visual problem solving models help students create a coherent representation of the problem
- Opportunities for students to decontextualize and contextualize problems are presented in every lesson

Construct viable arguments and critique the reasoning of others.

- *Error Analysis*; *Different Words, Same Question*; and *Which One Doesn't Belong* features provide students the opportunity to construct arguments and critique the reasoning of others
- *Inductive Reasoning* activities help students make conjectures and build a logical progression of statements to explore their conjecture

Model with mathematics.

- Real-life situations are translated into diagrams, tables, equations, and graphs to help students analyze relations and to draw conclusions
- Real-life problems are provided to help students learn to apply the mathematics that they are learning to everyday life

Use appropriate tools strategically.

- *Graphic Organizers* support the thought process of what, when, and how to solve problems
- A variety of tool papers, such as graph paper, number lines, and manipulatives, are available as students consider how to approach a problem
- Opportunities to use the web, graphing calculators, and spreadsheets support student learning

Attend to precision.

- *On Your Own* questions encourage students to formulate consistent and appropriate reasoning
- Cooperative learning opportunities support precise communication

Look for and make use of structure.

- *Inductive Reasoning* activities provide students the opportunity to see patterns and structure in mathematics
- Real-world problems help students use the structure of mathematics to break down and solve more difficult problems

Look for and express regularity in repeated reasoning.

- Opportunities are provided to help students make generalizations
- Students are continually encouraged to check for reasonableness in their solutions

Go to *BigIdeasMath.com* for more information on the Common Core State Standards for Mathematical Practice.

Common Core State Standards for Mathematical Content for Grade 7 Advanced

Chapter Coverage for Standards

1 · 2 · 3 · 4 · 5 · 6 · **7** · 8 · 9 · 10 · 11 · 12 · 13 · 14 · 15

Domain The Number System

- Know that there are numbers that are not rational, and approximate them by rational numbers.

1 · 2 · 3 · **4** · **5** · 6 · **7** · 8 · 9 · **10** · **11** · 12 · 13 · 14 · 15

Domain Expressions and Equations

- Use properties of operations to generate equivalent expressions.
- Solve real-life and mathematical problems using numerical and algebraic expressions and equations.
- Work with radicals and integer exponents.
- Understand the connections between proportional relationships, lines, and linear equations.
- Analyze and solve linear equations and pairs of simultaneous equations.

1 · 2 · 3 · 4 · 5 · **6** · 7 · 8 · 9 · 10 · 11 · 12 · 13 · 14 · 15

Domain Functions

- Define, evaluate, and compare functions.
- Use functions to model relationships between quantities.

1 · **2** · **3** · 4 · 5 · 6 · **7** · **8** · 9 · 10 · 11 · **12** · **13** · **14** · 15

Domain Geometry

- Draw, construct, and describe geometrical figures and describe the relationships between them.
- Solve real-life and mathematical problems involving angle measure, area, surface area, and volume.
- Understand congruence and similarity using physical models, transparencies, or geometry software.
- Understand and apply the Pythagorean Theorem.
- Solve real-world and mathematical problems involving volume of cylinders, cones, and spheres.

1 · 2 · 3 · 4 · 5 · 6 · 7 · 8 · **9** · 10 · 11 · 12 · 13 · 14 · **15**

Domain Statistics and Probability

- Use random sampling to draw inferences about a population.
- Draw informal comparative inferences about two populations.
- Investigate chance processes and develop, use, and evaluate probability models.
- Investigate patterns of association in bivariate data.

Go to *BigIdeasMath.com* for more information on the Common Core State Standards for Mathematical Content.

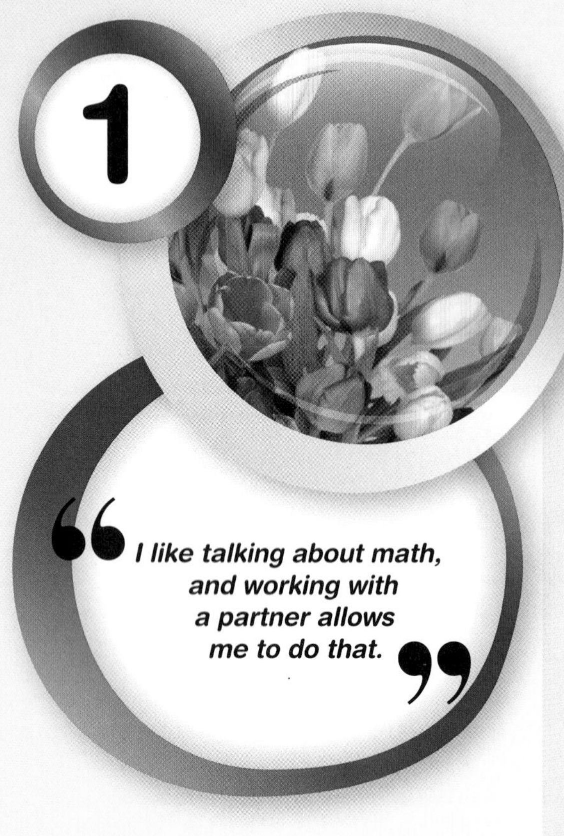
1
“I like talking about math, and working with a partner allows me to do that.”

Equations

Transformations

“With my eBook, I get to decide when I use technology and when I use print.”

3 Angles and Triangles

“I like that the Essential Question helps me begin thinking about the lesson.”

Graphing and Writing Linear Equations

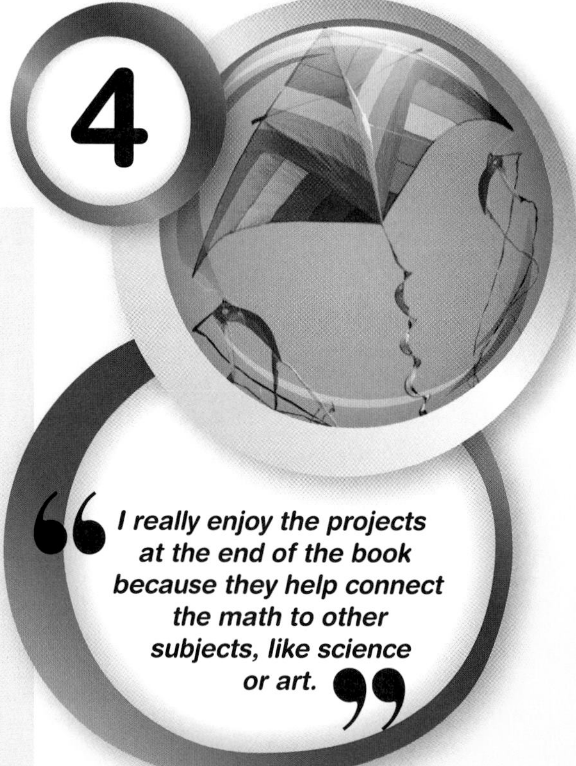

"I really enjoy the projects at the end of the book because they help connect the math to other subjects, like science or art."

5
I like Newton and Descartes! The cartoons are funny and I like that they model the math that we are learning.

Systems of Linear Equations

Functions

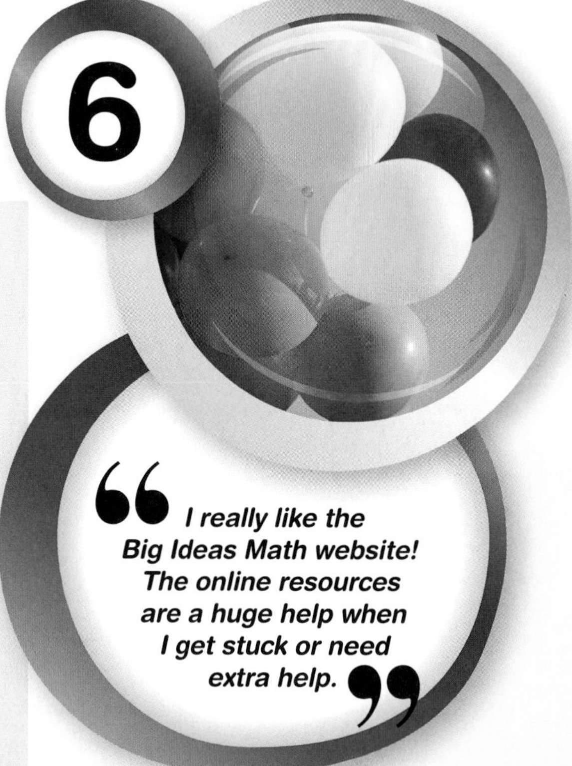

"I really like the Big Ideas Math website! The online resources are a huge help when I get stuck or need extra help."

7

Real Numbers and the Pythagorean Theorem

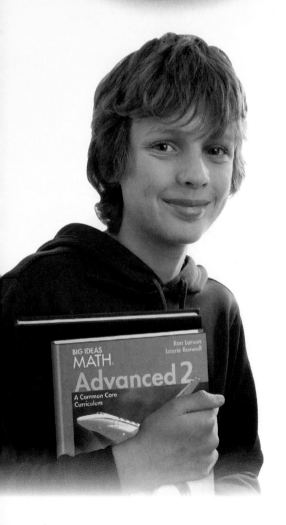

“I like the real-life application exercises because they show me how I can use the math in my own life.”

Volume and Similar Solids

8
I like playing the games in the Game Closet! They are a fun way to practice concepts we are learning in class.

9
With the BigIdeasMath.com website I don't have to worry if I forget my book or my workbook at school.

Data Analysis and Displays

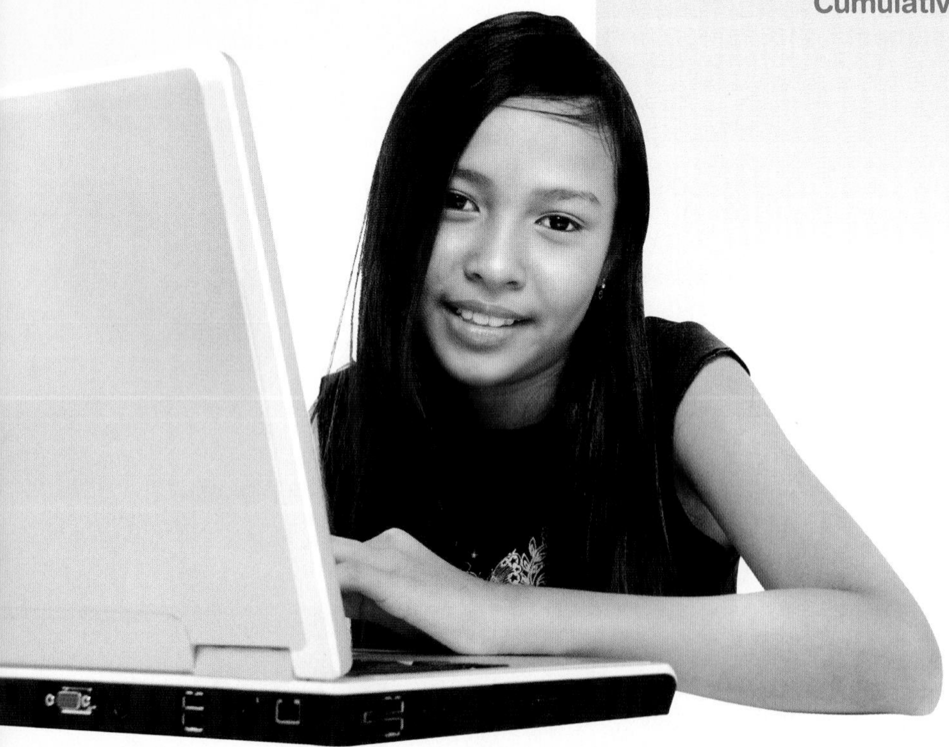

Exponents and Scientific Notation

“I like that I can print the Glossary Flashcards to review before a test.”

11 Inequalities

Before my school had Big Ideas Math I would always lose test points because I left units off my answers. Now I see why they are so important.

Constructions and Scale Drawings

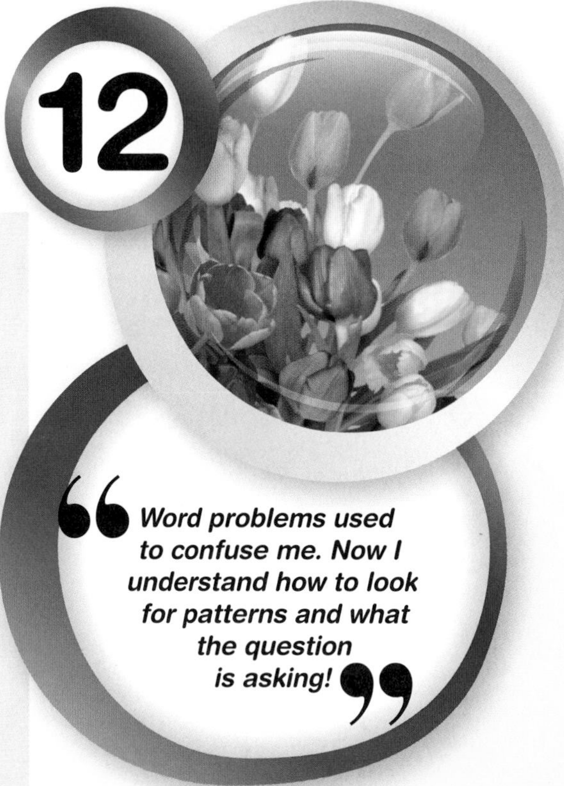

12
“Word problems used to confuse me. Now I understand how to look for patterns and what the question is asking!”

13 Circles and Area

"I like the Big Ideas Math Tutorials because they help explain the math when I am at home."

Surface Area and Volume

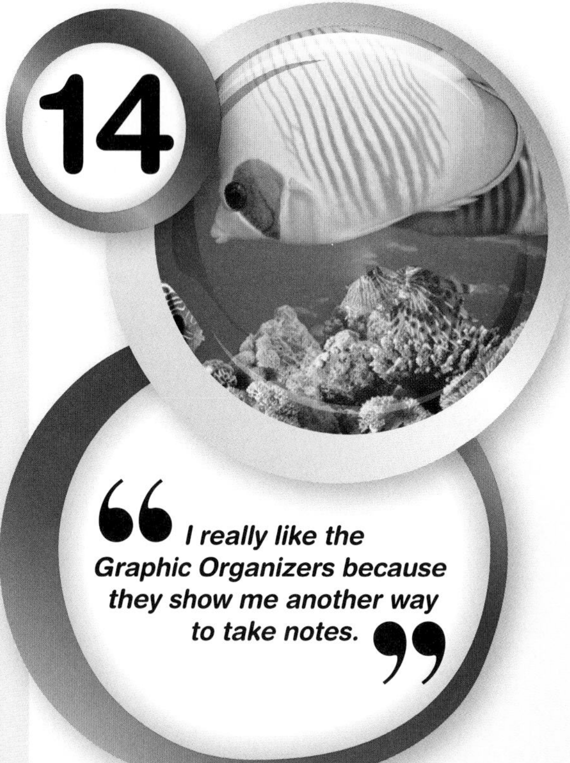
14
I really like the Graphic Organizers because they show me another way to take notes.

BIG IDEAS MATH
Ron Larson
Laurie Boswell
Advanced 2
A Common Core Curriculum

Probability and Statistics

"Using the Interactive Manipulatives from the Dynamic Student Edition helps me to see the mathematics that I am learning."

Appendix A: My Big Ideas Projects

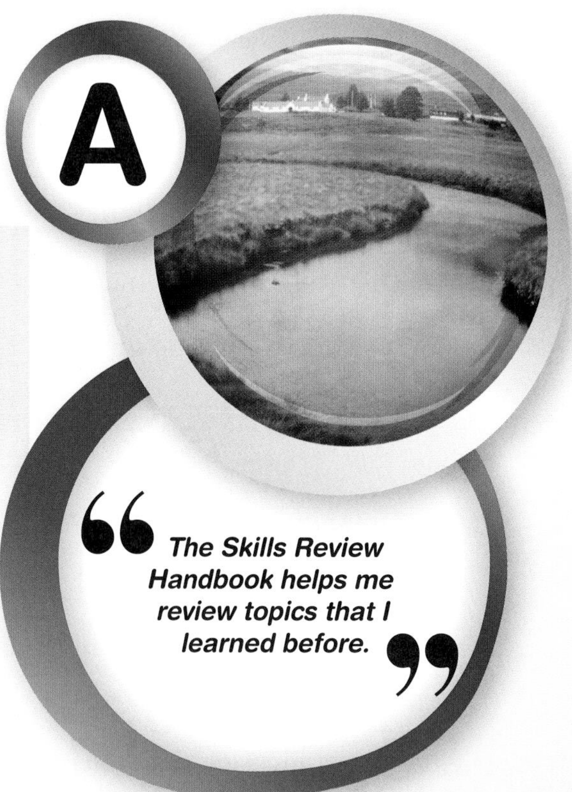

"The Skills Review Handbook helps me review topics that I learned before."

How to Use Your Math Book

- Read the **Essential Question** in the activity.

  Discuss the **Math Practice** question with your partner.

  Work with a partner to decide **What Is Your Answer?**

  Now you are ready to do the **Practice** problems.

- Find the **Key Vocabulary** words, **highlighted in yellow**.

  Read their definitions. Study the concepts in each **Key Idea**.

  If you forget a definition, you can look it up online in the

  Multi-Language Glossary at BigIdeasMath.com.

- After you study each **EXAMPLE**, do the exercises in the **On Your Own**.

  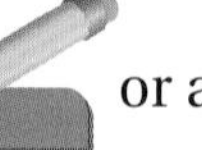

  to do the exercises that correspond to the example.

  As you study, look for a **Study Tip** or a **Common Error** .

- The exercises are divided into 3 parts.

  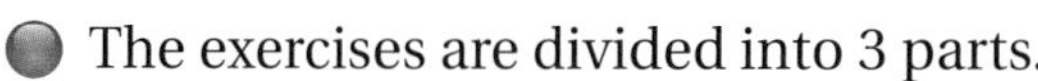

  If an exercise has a 1 next to it, look back at Example 1 for help with that exercise.

  More help is available at 

- To help study for your test, use the following.

  Quiz **Study Help**

  Chapter Review **Chapter Test**

SCAVENGER HUNT

Use this *Scavenger Hunt* to find where things are in **Chapter 1**.

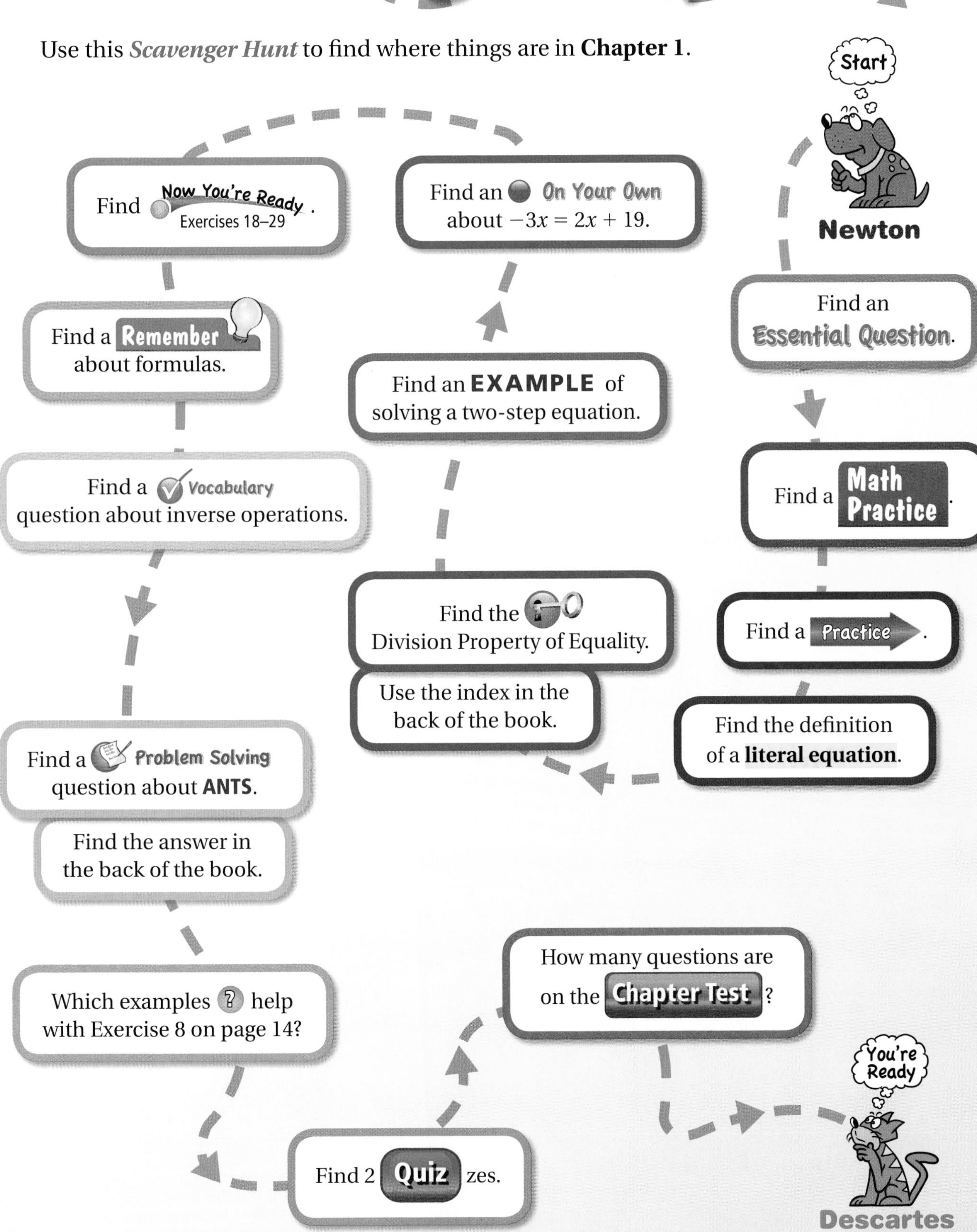

Start
Newton
Find an Essential Question.
Find a Math Practice.
Find a Practice.
Find the definition of a **literal equation**.
Find the Division Property of Equality.
Use the index in the back of the book.
Find an EXAMPLE of solving a two-step equation.
Find an On Your Own about $-3x = 2x + 19$.
Find Now You're Ready Exercises 18–29.
Find a Remember about formulas.
Find a Vocabulary question about inverse operations.
Find a Problem Solving question about ANTS.
Find the answer in the back of the book.
Which examples help with Exercise 8 on page 14?
Find 2 Quizzes.
How many questions are on the Chapter Test?
You're Ready
Descartes

1 Equations

1.1 Solving Simple Equations

1.2 Solving Multi-Step Equations

1.3 Solving Equations with Variables on Both Sides

1.4 Rewriting Equations and Formulas

"Dear Sir: Here is my suggestion for a good math problem."

"A box contains a total of 30 dog and cat treats. There are 5 times more dog treats than cat treats."

I need to learn to type so that I can write the story problems.

"How many of each type of treat are there?"

I think D=RT stands for Descartes is Really Tired.

"Push faster, Descartes! According to the formula $R = D \div T$, the time needs to be 10 minutes or less to break our all-time speed record!"

What You Learned Before

I've heard this story many times.
27. Writing Write a story problem that uses the Addition Property of Equality.

"Once upon a time, there lived the most handsome dog who just happened to be a genius at math. He..."

Simplifying Algebraic Expressions

Example 1 **Simplify $10b + 13 - 6b + 4$.**

$10b + 13 - 6b + 4 = 10b - 6b + 13 + 4$ Commutative Property of Addition

$= (10 - 6)b + 13 + 4$ Distributive Property

$= 4b + 17$ Simplify.

Example 2 **Simplify $5(x + 4) + 2x$.**

$5(x + 4) + 2x = 5(x) + 5(4) + 2x$ Distributive Property

$= 5x + 20 + 2x$ Multiply.

$= 5x + 2x + 20$ Commutative Property of Addition

$= 7x + 20$ Combine like terms.

Try It Yourself

Simplify the expression.

1. $9m - 7m + 2m$ **2.** $3g - 9 + 11g - 21$ **3.** $6(3 - y)$

4. $12(a - 4)$ **5.** $22.5 + 7(n - 3.4)$ **6.** $15k + 8(11 - k)$

Adding and Subtracting Integers

Example 3 **Find $4 + (-12)$.**

$4 + (-12) = -8$

$|-12| > |4|$. So, subtract $|4|$ from $|-12|$.

Use the sign of -12.

Example 4 **Find $-7 - (-16)$.**

$-7 - (-16) = -7 + 16$ Add the opposite of -16.

$= 9$ Add.

Try It Yourself

Add or subtract.

7. $-5 + (-2)$ **8.** $0 + (-13)$

9. $-6 + 14$ **10.** $19 - (-13)$

1.1 Solving Simple Equations

Essential Question How can you use inductive reasoning to discover rules in mathematics? How can you test a rule?

1 ACTIVITY: Sum of the Angles of a Triangle

Work with a partner. Use a protractor to measure the angles of each triangle. Copy and complete the table to organize your results.

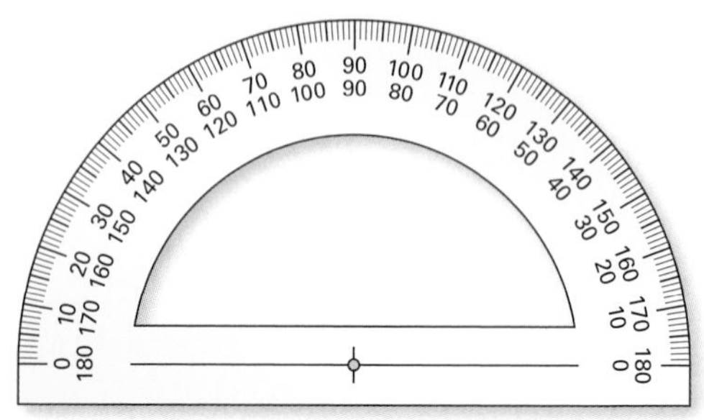

a.

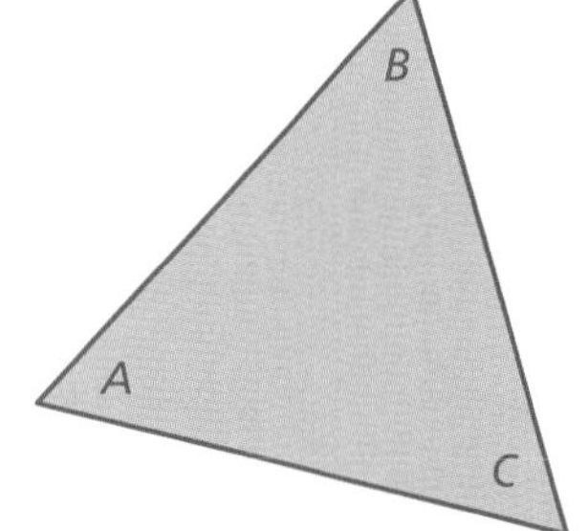

b.

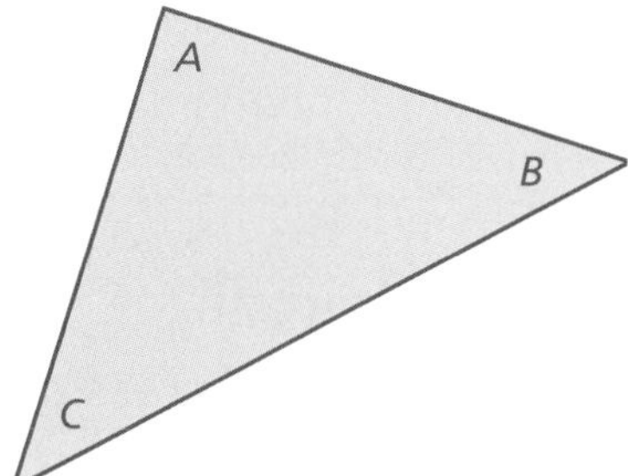

c.

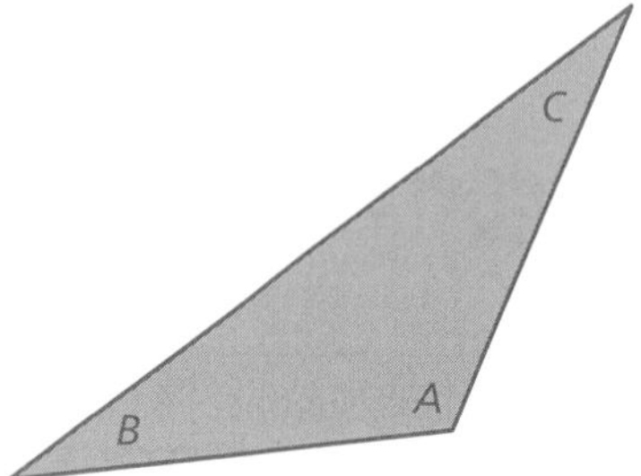

d.

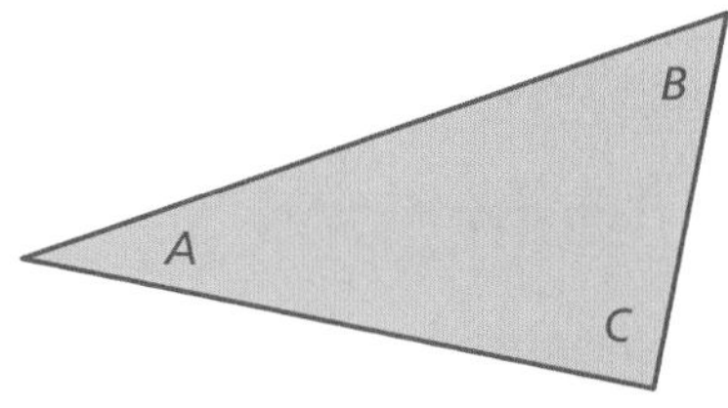

Solving Equations
In this lesson, you will
- solve simple equations using addition, subtraction, multiplication, or division.

Triangle	Angle A (degrees)	Angle B (degrees)	Angle C (degrees)	$A + B + C$
a.				
b.				
c.				
d.				

2 ACTIVITY: Writing a Rule

Work with a partner. Use inductive reasoning to write and test a rule.

a. **STRUCTURE** Use the completed table in Activity 1 to write a rule about the sum of the angle measures of a triangle.

b. **TEST YOUR RULE** Draw four triangles that are different from those in Activity 1. Measure the angles of each triangle. Organize your results in a table. Find the sum of the angle measures of each triangle.

3 ACTIVITY: Applying Your Rule

Math Practice

Analyze Conjectures

Do your results support the rule you wrote in Activity 2? Explain.

Work with a partner. Use the rule you wrote in Activity 2 to write an equation for each triangle. Then solve the equation to find the value of *x*. Use a protractor to check the reasonableness of your answer.

a.

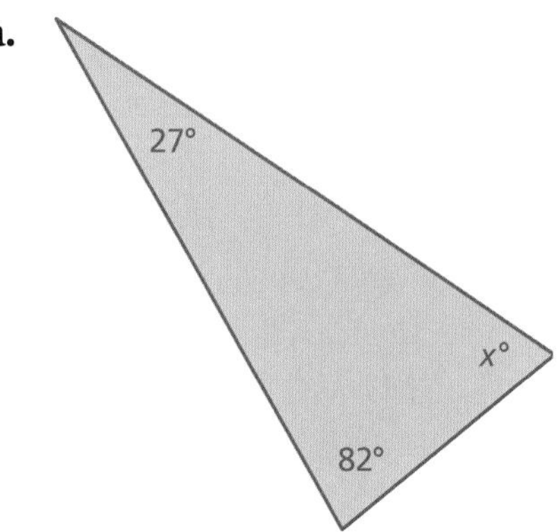

b.

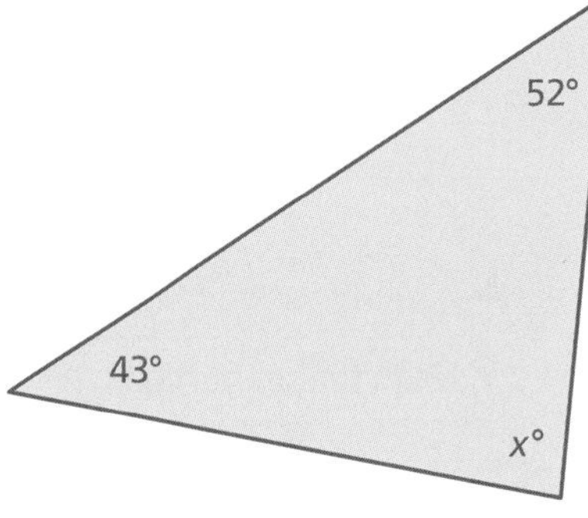

c.

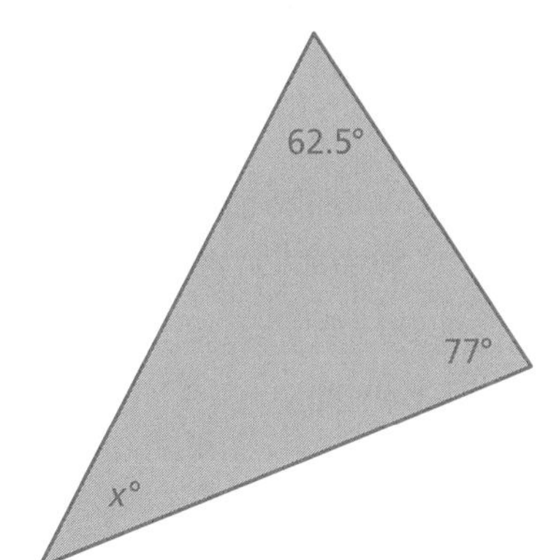

d.

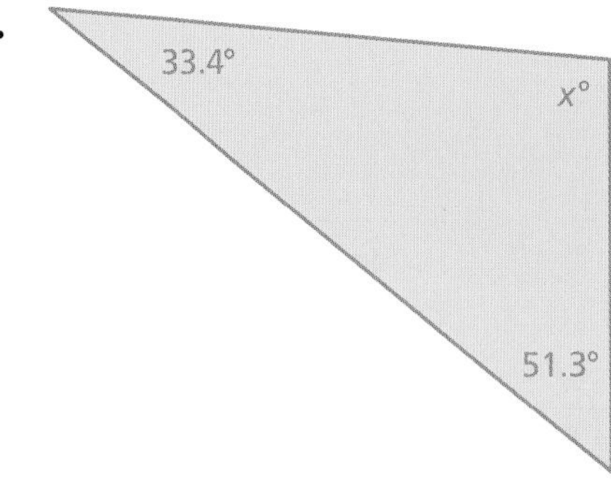

What Is Your Answer?

4. **IN YOUR OWN WORDS** How can you use inductive reasoning to discover rules in mathematics? How can you test a rule? How can you use a rule to solve problems in mathematics?

Practice Use what you learned about solving simple equations to complete Exercises 4–6 on page 7.

Remember

Addition and subtraction are inverse operations.

Addition Property of Equality

Words Adding the same number to each side of an equation produces an equivalent equation.

Algebra If $a = b$, then $a + c = b + c$.

Subtraction Property of Equality

Words Subtracting the same number from each side of an equation produces an equivalent equation.

Algebra If $a = b$, then $a - c = b - c$.

EXAMPLE 1 Solving Equations Using Addition or Subtraction

a. Solve $x - 7 = -6$.

	$x - 7 = -6$	Write the equation.
Undo the subtraction. →	$\underline{+7} \quad \underline{+7}$	Addition Property of Equality
	$x = 1$	Simplify.

The solution is $x = 1$.

Check

$x - 7 = -6$

$1 - 7 \stackrel{?}{=} -6$

$-6 = -6$ ✓

b. Solve $y + 3.4 = 0.5$.

	$y + 3.4 = 0.5$	Write the equation.
Undo the addition. →	$\underline{-3.4} \quad \underline{-3.4}$	Subtraction Property of Equality
	$y = -2.9$	Simplify.

The solution is $y = -2.9$.

Check

$y + 3.4 = 0.5$

$-2.9 + 3.4 \stackrel{?}{=} 0.5$

$0.5 = 0.5$ ✓

c. Solve $h + 2\pi = 3\pi$.

	$h + 2\pi = 3\pi$	Write the equation.
Undo the addition. →	$\underline{-2\pi} \quad \underline{-2\pi}$	Subtraction Property of Equality
	$h = \pi$	Simplify.

The solution is $h = \pi$.

On Your Own

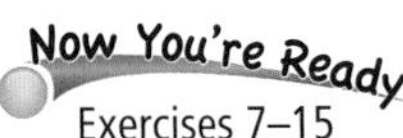

Now You're Ready
Exercises 7–15

Solve the equation. Check your solution.

1. $b + 2 = -5$ **2.** $g - 1.7 = -0.9$ **3.** $-3 = k + 3$

4. $r - \pi = \pi$ **5.** $t - \frac{1}{4} = -\frac{3}{4}$ **6.** $5.6 + z = -8$

Key Ideas

Remember

Multiplication and division are inverse operations.

Multiplication Property of Equality

Words Multiplying each side of an equation by the same number produces an equivalent equation.

Algebra If $a = b$, then $a \cdot c = b \cdot c$.

Division Property of Equality

Words Dividing each side of an equation by the same number produces an equivalent equation.

Algebra If $a = b$, then $a \div c = b \div c$, $c \neq 0$.

EXAMPLE 2 Solving Equations Using Multiplication or Division

a. Solve $-\frac{3}{4}n = -2$.

$-\frac{3}{4}n = -2$ Write the equation.

Use the reciprocal. → $-\frac{4}{3} \cdot \left(-\frac{3}{4}n\right) = -\frac{4}{3} \cdot (-2)$ Multiplication Property of Equality

$n = \frac{8}{3}$ Simplify.

∴ The solution is $n = \frac{8}{3}$.

b. Solve $\pi x = 3\pi$.

$\pi x = 3\pi$ Write the equation.

Undo the multiplication. → $\frac{\pi x}{\pi} = \frac{3\pi}{\pi}$ Division Property of Equality

$x = 3$ Simplify.

∴ The solution is $x = 3$.

Check

$\pi x = 3\pi$

$\pi(3) \stackrel{?}{=} 3\pi$

$3\pi = 3\pi$ ✓

On Your Own

Now You're Ready
Exercises 18–26

Solve the equation. Check your solution.

7. $\frac{y}{4} = -7$ **8.** $6\pi = \pi x$ **9.** $0.09w = 1.8$

EXAMPLE 3 Identifying the Solution of an Equation

What value of k makes the equation $k + 4 \div 0.2 = 5$ true?

Ⓐ -15 Ⓑ -5 Ⓒ -3 Ⓓ 1.5

$$k + 4 \div 0.2 = 5$$ Write the equation.

$$k + 20 = 5$$ Divide 4 by 0.2.

$$\underline{-20} \quad \underline{-20}$$ Subtraction Property of Equality

$$k = -15$$ Simplify.

The correct answer is Ⓐ.

EXAMPLE 4 Real-Life Application

The melting point of bromine is $-7°C$.

The *melting point* of a solid is the temperature at which the solid becomes a liquid. The melting point of bromine is $\frac{1}{30}$ of the melting point of nitrogen. Write and solve an equation to find the melting point of nitrogen.

Words The melting point of bromine is $\frac{1}{30}$ of the melting point of nitrogen.

Variable Let n be the melting point of nitrogen.

Equation $-7 = \frac{1}{30} \cdot n$

$$-7 = \frac{1}{30}n$$ Write the equation.

$$30 \cdot (-7) = 30 \cdot \left(\frac{1}{30}n\right)$$ Multiplication Property of Equality

$$-210 = n$$ Simplify.

So, the melting point of nitrogen is $-210°C$.

On Your Own

10. Solve $p - 8 \div \frac{1}{2} = -3$.

11. Solve $q + |-10| = 2$.

12. The melting point of mercury is about $\frac{1}{4}$ of the melting point of krypton. The melting point of mercury is $-39°C$. Write and solve an equation to find the melting point of krypton.

1.1 Exercises

Vocabulary and Concept Check

1. **VOCABULARY** Which of the operations $+$, $-$, $\times$, and $\div$ are inverses of each other?
2. **VOCABULARY** Are the equations $3x = -9$ and $4x = -12$ equivalent? Explain.
3. **WHICH ONE DOESN'T BELONG?** Which equation does *not* belong with the other three? Explain your reasoning.

   $x - 2 = 4$ $\quad$ $x - 3 = 6$ $\quad$ $x - 5 = 1$ $\quad$ $x - 6 = 0$

Practice and Problem Solving

CHOOSE TOOLS **Find the value of x. Check the reasonableness of your answer.**

4. 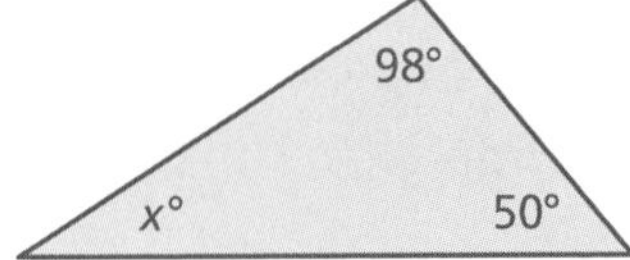

5. 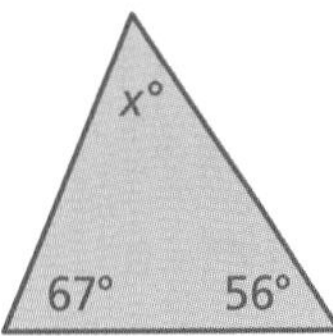

6. 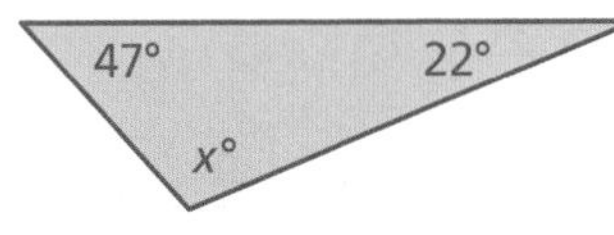

Solve the equation. Check your solution.

1 7. $x + 12 = 7$

8. $g - 16 = 8$

9. $-9 + p = 12$

10. $0.7 + y = -1.34$

11. $x - 8\pi = \pi$

12. $4\pi = w - 6\pi$

13. $\frac{5}{6} = \frac{1}{3} + d$

14. $\frac{3}{8} = r + \frac{2}{3}$

15. $n - 1.4 = -6.3$

16. **CONCERT** A discounted concert ticket costs \$14.50 less than the original price p. You pay \$53 for a discounted ticket. Write and solve an equation to find the original price.

17. **BOWLING** Your friend's final bowling score is 105. Your final bowling score is 14 pins less than your friend's final score.

    a. Write and solve an equation to find your final score.

    b. Your friend made a spare in the 10th frame. Did you? Explain.

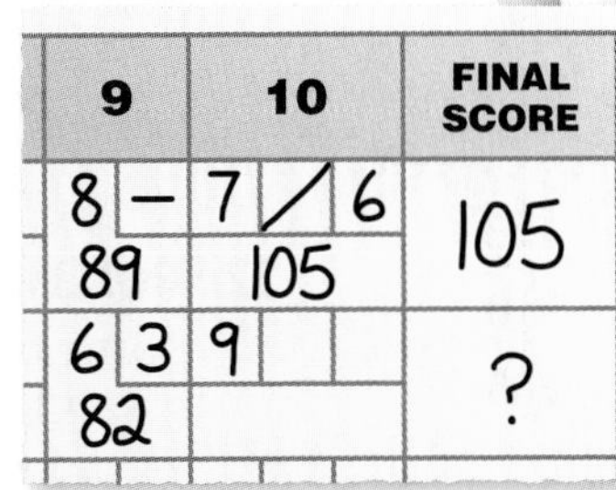

9	10	FINAL SCORE
8 – 89	7 / 6 105	105
6 3 82	9	?

Solve the equation. Check your solution.

② **18.** $7x = 35$

19. $4 = -0.8n$

20. $6 = -\frac{w}{8}$

21. $\frac{m}{\pi} = 7.3$

22. $-4.3g = 25.8$

23. $\frac{3}{2} = \frac{9}{10}k$

24. $-7.8x = -1.56$

25. $-2 = \frac{6}{7}p$

26. $3\pi d = 12\pi$

27. ERROR ANALYSIS Describe and correct the error in solving the equation.

> ✗
> $-1.5 + k = 8.2$
> $k = 8.2 + (-1.5)$
> $k = 6.7$

28. TENNIS A gym teacher orders 42 tennis balls. Each package contains 3 tennis balls. Which of the following equations represents the number x of packages?

$x + 3 = 42$ $\qquad$ $3x = 42$ $\qquad$ $\frac{x}{3} = 42$ $\qquad$ $x = \frac{3}{42}$

MODELING In Exercises 29–32, write and solve an equation to answer the question.

29. PARK You clean a community park for 6.5 hours. You earn $42.25. How much do you earn per hour?

30. ROCKET LAUNCH A rocket is scheduled to launch from a command center in 3.75 hours. What time is it now?

Launch Time
11:20 A.M.

31. BANKING After earning interest, the balance of an account is $420. The new balance is $\frac{7}{6}$ of the original balance. How much interest did it earn?

Roller Coasters at Cedar Point	
Coaster	**Height (feet)**
Top Thrill Dragster	420
Millennium Force	310
Magnum XL-200	205
Mantis	?

32. ROLLER COASTER Cedar Point amusement park has some of the tallest roller coasters in the United States. The Mantis is 165 feet shorter than the Millennium Force. What is the height of the Mantis?

Solve the equation. Check your solution.

3 **33.** $-3 = h + 8 \div 2$

34. $12 = w - |-7|$

35. $q + |6.4| = 9.6$

36. $d - 2.8 \div 0.2 = -14$

37. $\frac{8}{9} = x + \frac{1}{3}(7)$

38. $p - \frac{1}{4} \cdot 3 = -\frac{5}{6}$

39. **LOGIC** Without solving, determine whether the solution of $-2x = -15$ is *greater than* or *less than* -15. Explain.

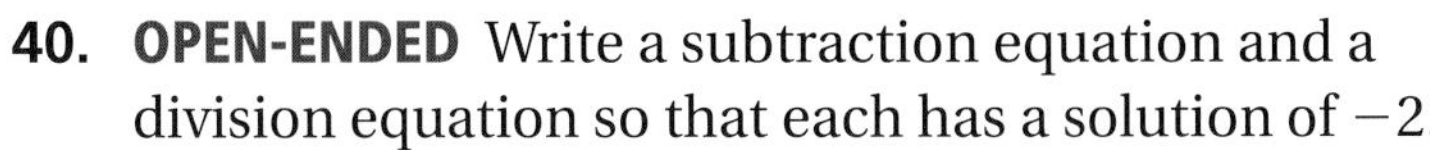

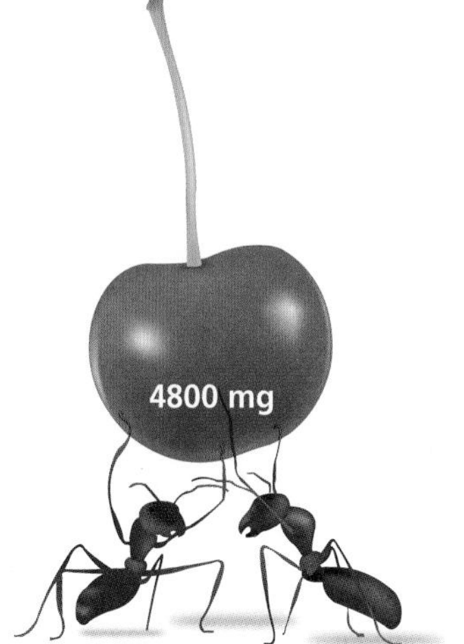

40. **OPEN-ENDED** Write a subtraction equation and a division equation so that each has a solution of -2.

41. **ANTS** Some ant species can carry 50 times their body weight. It takes 32 ants to carry the cherry. About how much does each ant weigh?

42. **REASONING** One-fourth of the girls and one-eighth of the boys in a class retake their school pictures. The photographer retakes pictures for 16 girls and 7 boys. How many students are in the class?

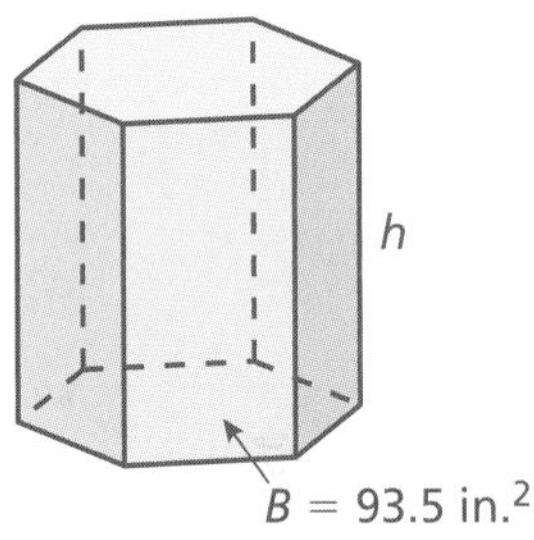

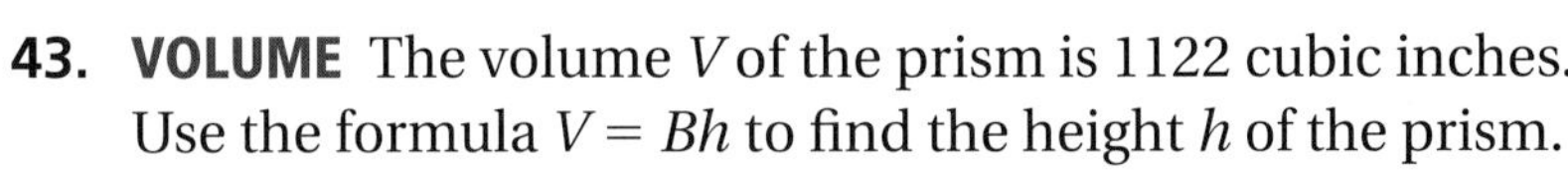

43. **VOLUME** The volume V of the prism is 1122 cubic inches. Use the formula $V = Bh$ to find the height h of the prism.

44. Critical Thinking A neighbor pays you and two friends \$90 to paint her garage. You divide the money three ways in the ratio 2 : 3 : 5.

a. How much does each person receive?

b. What is one possible reason the money is not divided evenly?

Fair Game Review What you learned in previous grades & lessons

Simplify the expression. *(Skills Review Handbook)*

45. $2(x - 2) + 5x$

46. $0.4b - 3.2 + 1.2b$

47. $\frac{1}{4}g + 6g - \frac{2}{3}$

48. **MULTIPLE CHOICE** The temperature at 4:00 P.M. was -12 °C. By 11:00 P.M., the temperature had dropped 14 °C. What was the temperature at 11:00 P.M.? *(Skills Review Handbook)*

Ⓐ -26 °C Ⓑ -2 °C Ⓒ 2 °C Ⓓ 26 °C

1.2 Solving Multi-Step Equations

Essential Question How can you solve a multi-step equation? How can you check the reasonableness of your solution?

1 ACTIVITY: Solving for the Angles of a Triangle

Work with a partner. Write an equation for each triangle. Solve the equation to find the value of the variable. Then find the angle measures of each triangle. Use a protractor to check the reasonableness of your answer.

a.

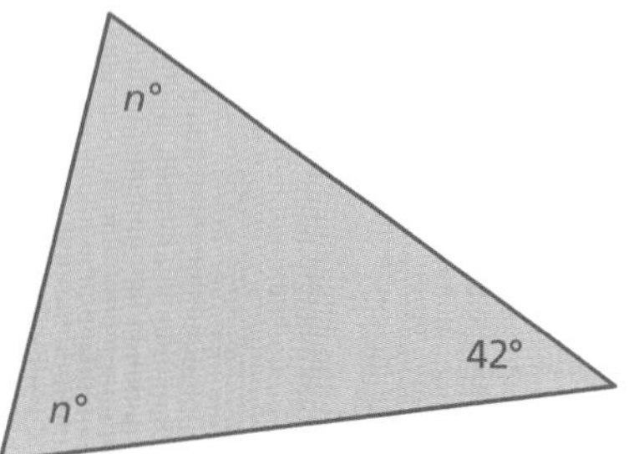

b.

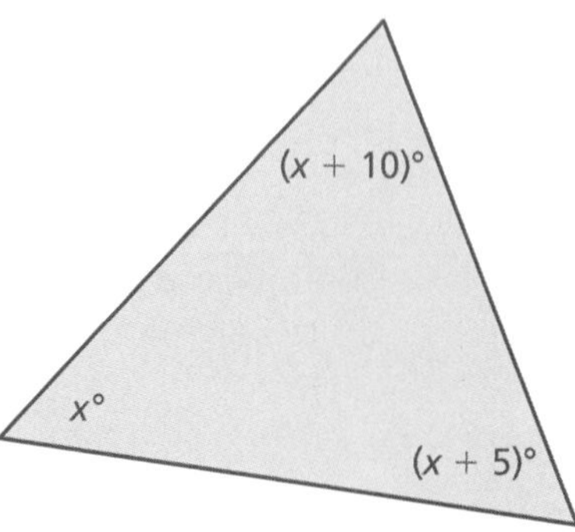

c.

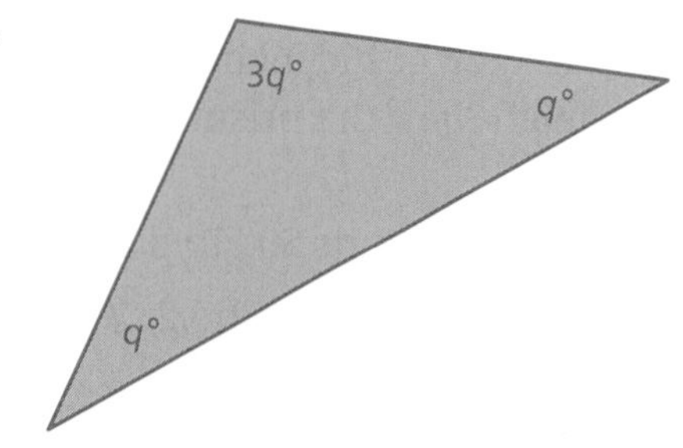

d.

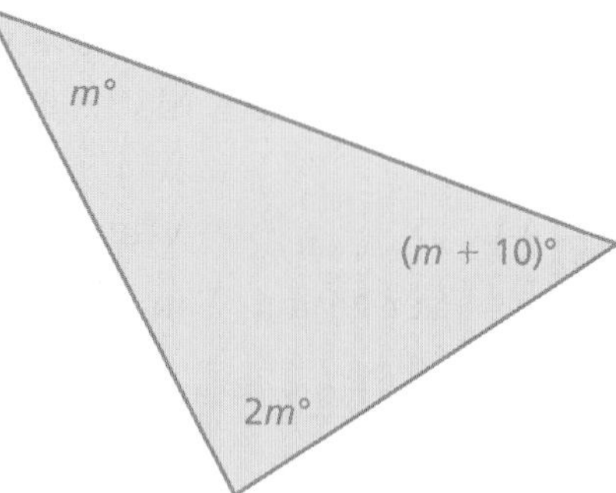

Solving Equations

In this lesson, you will

- use inverse operations to solve multi-step equations.
- use the Distributive Property to solve multi-step equations.

e.

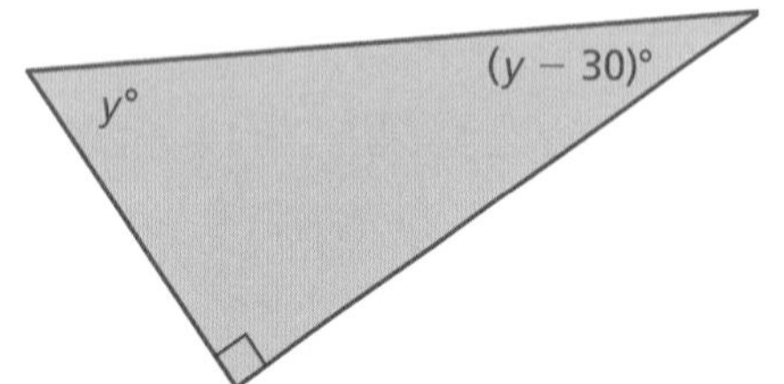

f.

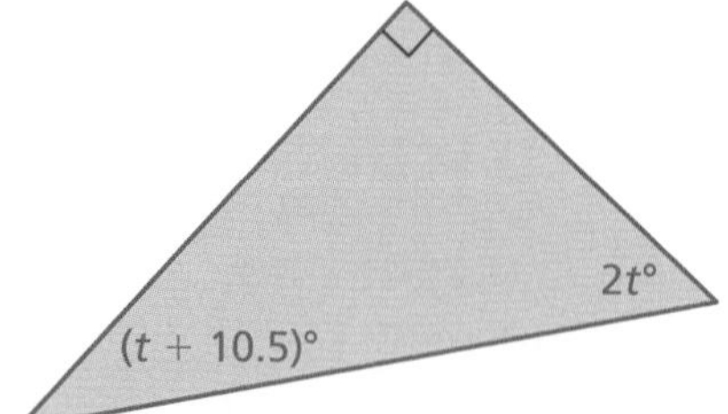

2 ACTIVITY: Problem-Solving Strategy

Math Practice

Find Entry Points

How do you decide which triangle to solve first? Explain.

Work with a partner.

The six triangles form a rectangle.

Find the angle measures of each triangle. Use a protractor to check the reasonableness of your answers.

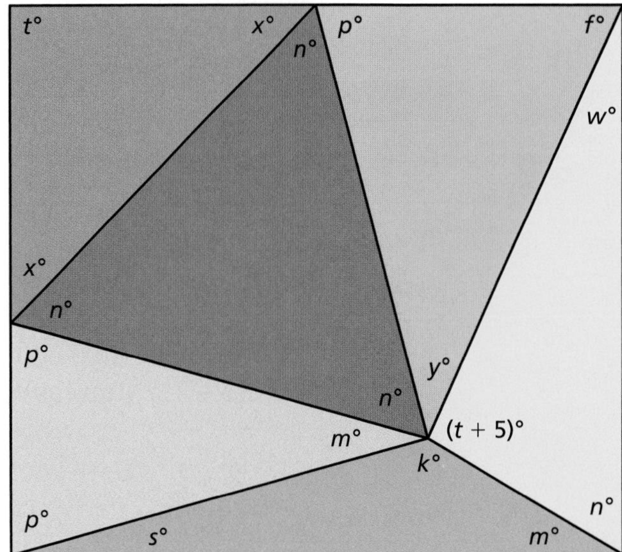

3 ACTIVITY: Puzzle

Work with a partner. A survey asked 200 people to name their favorite weekday. The results are shown in the circle graph.

a. How many degrees are in each part of the circle graph?

b. What percent of the people chose each day?

c. How many people chose each day?

d. Organize your results in a table.

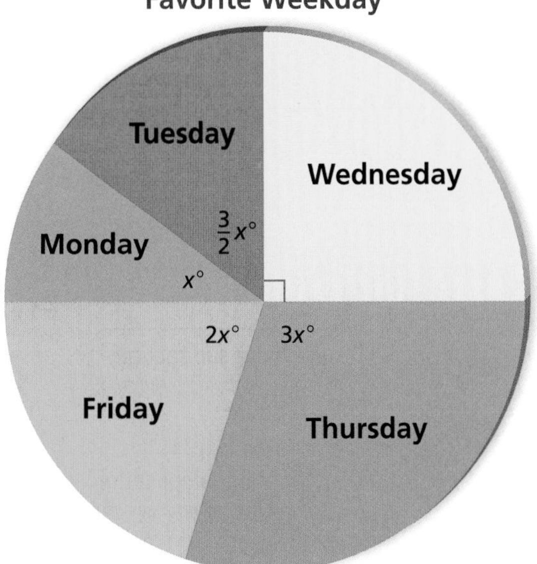

What Is Your Answer?

4. IN YOUR OWN WORDS How can you solve a multi-step equation? How can you check the reasonableness of your solution?

Practice Use what you learned about solving multi-step equations to complete Exercises 3–5 on page 14.

Solving Multi-Step Equations

To solve multi-step equations, use inverse operations to isolate the variable.

EXAMPLE 1 Solving a Two-Step Equation

The height (in feet) of a tree after x years is $1.5x + 15$. After how many years is the tree 24 feet tall?

	$1.5x + 15 = 24$	Write an equation.
Undo the addition. →	$-15 \quad -15$	Subtraction Property of Equality
	$1.5x = 9$	Simplify.
Undo the multiplication. →	$\dfrac{1.5x}{1.5} = \dfrac{9}{1.5}$	Division Property of Equality
	$x = 6$	Simplify.

So, the tree is 24 feet tall after 6 years.

EXAMPLE 2 Combining Like Terms to Solve an Equation

Solve $8x - 6x - 25 = -35$.

	$8x - 6x - 25 = -35$	Write the equation.
	$2x - 25 = -35$	Combine like terms.
Undo the subtraction. →	$+25 \quad +25$	Addition Property of Equality
	$2x = -10$	Simplify.
Undo the multiplication. →	$\dfrac{2x}{2} = \dfrac{-10}{2}$	Division Property of Equality
	$x = -5$	Simplify.

The solution is $x = -5$.

On Your Own

Now You're Ready
Exercises 6–9

Solve the equation. Check your solution.

1. $-3z + 1 = 7$
2. $\frac{1}{2}x - 9 = -25$
3. $-4n - 8n + 17 = 23$

EXAMPLE 3 Using the Distributive Property to Solve an Equation

Solve $2(1 - 5x) + 4 = -8$.

$2(1 - 5x) + 4 = -8$	Write the equation.
$2(1) - 2(5x) + 4 = -8$	Distributive Property
$2 - 10x + 4 = -8$	Multiply.
$-10x + 6 = -8$	Combine like terms.
$\underline{-6} \quad \underline{-6}$	Subtraction Property of Equality
$-10x = -14$	Simplify.
$\dfrac{-10x}{-10} = \dfrac{-14}{-10}$	Division Property of Equality
$x = 1.4$	Simplify.

Study Tip

Here is another way to solve the equation in Example 3.

$$2(1 - 5x) + 4 = -8$$
$$2(1 - 5x) = -12$$
$$1 - 5x = -6$$
$$-5x = -7$$
$$x = 1.4$$

EXAMPLE 4 Real-Life Application

Use the table to find the number of miles x you need to run on Friday so that the mean number of miles run per day is 1.5.

Day	Miles
Monday	2
Tuesday	0
Wednesday	1.5
Thursday	0
Friday	x

Write an equation using the definition of *mean*.

sum of the data → number of values →	$\dfrac{2 + 0 + 1.5 + 0 + x}{5} = 1.5$	Write the equation.
	$\dfrac{3.5 + x}{5} = 1.5$	Combine like terms.
Undo the division. →	$5 \cdot \dfrac{3.5 + x}{5} = 5 \cdot 1.5$	Multiplication Property of Equality
	$3.5 + x = 7.5$	Simplify.
Undo the addition. →	$\underline{-3.5} \qquad \underline{-3.5}$	Subtraction Property of Equality
	$x = 4$	Simplify.

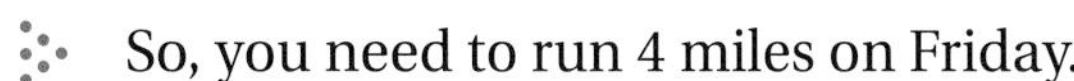

So, you need to run 4 miles on Friday.

On Your Own

Exercises 10 and 11

Solve the equation. Check your solution.

4. $-3(x + 2) + 5x = -9$
5. $5 + 1.5(2d - 1) = 0.5$
6. You scored 88, 92, and 87 on three tests. Write and solve an equation to find the score you need on the fourth test so that your mean test score is 90.

1.2 Exercises

Vocabulary and Concept Check

1. **WRITING** Write the verbal statement as an equation. Then solve.

   2 more than 3 times a number is 17.

2. **OPEN-ENDED** Explain how to solve the equation $2(4x - 11) + 9 = 19$.

Practice and Problem Solving

CHOOSE TOOLS Find the value of the variable. Then find the angle measures of the polygon. Use a protractor to check the reasonableness of your answer.

3. 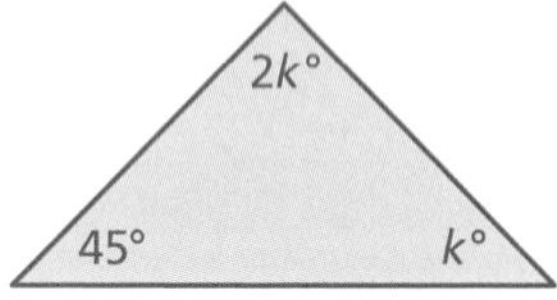

   Sum of angle measures: 180°

4. 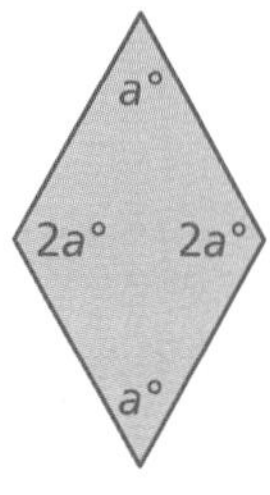

   Sum of angle measures: 360°

5. 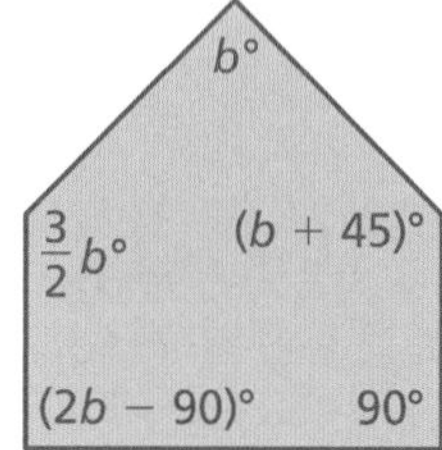

   Sum of angle measures: 540°

Solve the equation. Check your solution.

① ② 6. $10x + 2 = 32$

7. $19 - 4c = 17$

8. $1.1x + 1.2x - 5.4 = -10$

9. $\frac{2}{3}h - \frac{1}{3}h + 11 = 8$

③ 10. $6(5 - 8v) + 12 = -54$

11. $21(2 - x) + 12x = 44$

12. **ERROR ANALYSIS** Describe and correct the error in solving the equation.

   ✗
   $$-2(7 - y) + 4 = -4$$
   $$-14 - 2y + 4 = -4$$
   $$-10 - 2y = -4$$
   $$-2y = 6$$
   $$y = -3$$

13. **WATCHES** The cost C (in dollars) of making n watches is represented by $C = 15n + 85$. How many watches are made when the cost is $385?

14. **HOUSE** The height of the house is 26 feet. What is the height x of each story?

In Exercises 15–17, write and solve an equation to answer the question.

15. **POSTCARD** The area of the postcard is 24 square inches. What is the width b of the message (in inches)?

PARIS
PARIS
NOV 09 2013
FRANCE
Dear Miguel,
I'm having a great
time in Paris.
Yesterday I saw the
Eiffel Tower.
See you soon!
Gloria
Miguel Martinez
123 Any Street
Any Town, USA
4 in.
b
3 in.

16. **BREAKFAST** You order two servings of pancakes and a fruit cup. The cost of the fruit cup is \$1.50. You leave a 15% tip. Your total bill is \$11.50. How much does one serving of pancakes cost?

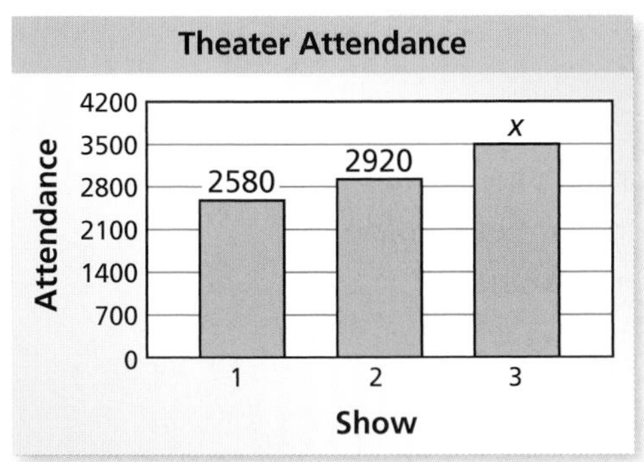

17. **THEATER** How many people must attend the third show so that the average attendance per show is 3000?

18. **DIVING** Divers in a competition are scored by an international panel of judges. The highest and the lowest scores are dropped. The total of the remaining scores is multiplied by the degree of difficulty of the dive. This product is multiplied by 0.6 to determine the final score.

a. A diver's final score is 77.7. What is the degree of difficulty of the dive?

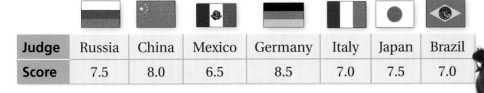

Judge	Russia	China	Mexico	Germany	Italy	Japan	Brazil
Score	7.5	8.0	6.5	8.5	7.0	7.5	7.0

b. Critical Thinking The degree of difficulty of a dive is 4.0. The diver's final score is 97.2. Judges award half or whole points from 0 to 10. What scores could the judges have given the diver?

Fair Game Review What you learned in previous grades & lessons

Let $a = 3$ and $b = -2$. Copy and complete the statement using <, >, or =.
(Skills Review Handbook)

19. $-5a$ ☐ 4

20. 5 ☐ $b + 7$

21. $a - 4$ ☐ $10b + 8$

22. **MULTIPLE CHOICE** What value of x makes the equation $x + 5 = 2x$ true?
(Skills Review Handbook)

Ⓐ -1 Ⓑ 0 Ⓒ 3 Ⓓ 5

1 Study Help

You can use a **Y chart** to compare two topics. List differences in the branches and similarities in the base of the Y. Here is an example of a Y chart that compares solving simple equations using addition to solving simple equations using subtraction.

Solving Simple Equations Using Addition	Solving Simple Equations Using Subtraction
• Add the same number to each side of the equation.	• Subtract the same number from each side of the equation.

- You can solve the equation in one step.
- You produce an equivalent equation.
- The variable can be on either side of the equation.
- It is always a good idea to check your solution.

On Your Own

Make Y charts to help you study and compare these topics.

1. solving simple equations using multiplication and solving simple equations using division
2. solving simple equations and solving multi-step equations

After you complete this chapter, make Y charts for the following topics.

3. solving equations with the variable on one side and solving equations with variables on both sides
4. solving multi-step equations and solving equations with variables on both sides
5. solving multi-step equations and rewriting literal equations

"I made a Y chart to compare and contrast Fluffy's characteristics with yours."

1.1–1.2 Quiz

Solve the equation. Check your solution. *(Section 1.1)*

1. $-\frac{1}{2} = y - 1$

2. $-3\pi + w = 2\pi$

3. $1.2m = 0.6$

4. $q + 2.7 = -0.9$

Solve the equation. Check your solution. *(Section 1.2)*

5. $-4k + 17 = 1$

6. $\frac{1}{4}z + 8 = 12$

7. $-3(2n + 1) + 7 = -5$

8. $2.5(t - 2) - 6 = 9$

Find the value of x. Then find the angle measures of the polygon. *(Section 1.2)*

9.

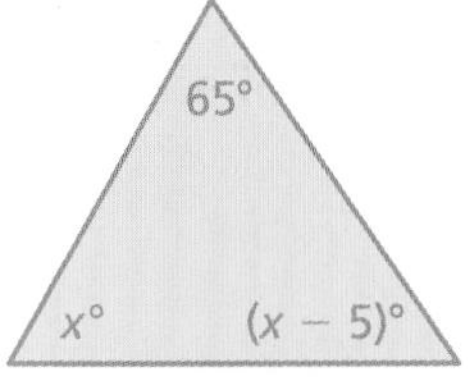

Sum of angle measures: 180°

10.

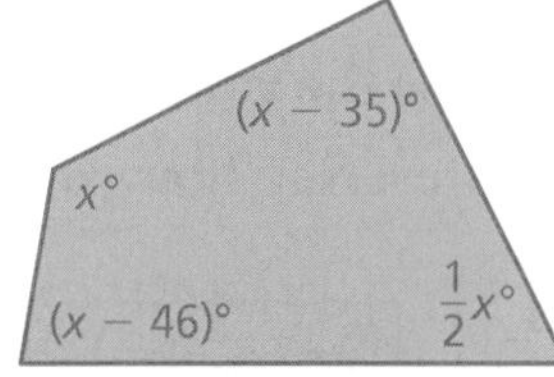

Sum of angle measures: 360°

11. JEWELER The equation $P = 2.5m + 35$ represents the price P (in dollars) of a bracelet, where m is the cost of the materials (in dollars). The price of a bracelet is \$115. What is the cost of the materials? *(Section 1.2)*

12. PASTURE A 455-foot fence encloses a pasture. What is the length of each side of the pasture? *(Section 1.2)*

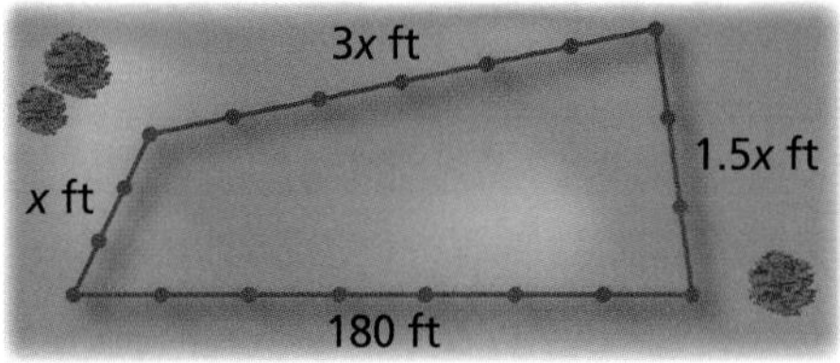

13. POSTERS A machine prints 230 movie posters each hour. Write and solve an equation to find the number of hours it takes the machine to print 1265 posters. *(Section 1.1)*

14. BASKETBALL Use the table to write and solve an equation to find the number of points p you need to score in the fourth game so that the mean number of points is 20. *(Section 1.2)*

Game	Points
1	25
2	15
3	18
4	p

1.3 Solving Equations with Variables on Both Sides

Essential Question How can you solve an equation that has variables on both sides?

1 ACTIVITY: Perimeter and Area

Work with a partner.

- **Each figure has the unusual property that the value of its perimeter (in feet) is equal to the value of its area (in square feet). Write an equation for each figure.**
- **Solve each equation for x.**
- **Use the value of x to find the perimeter and the area of each figure.**
- **Describe how you can check your solution.**

a.

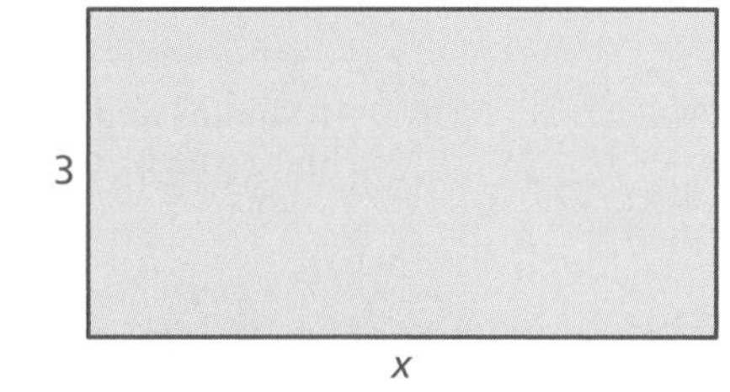

b.

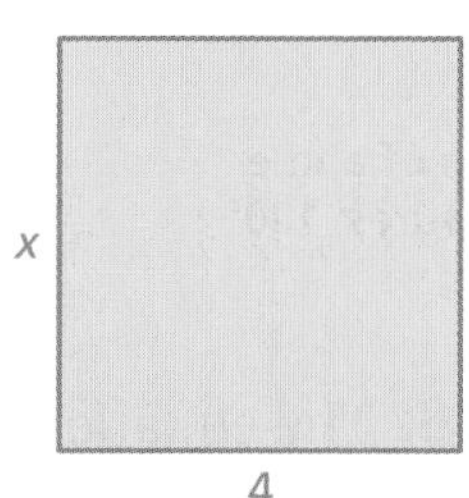

c.

d.

e.

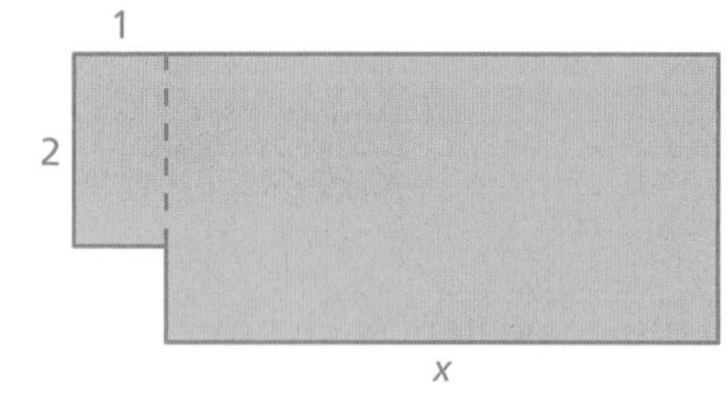

f.

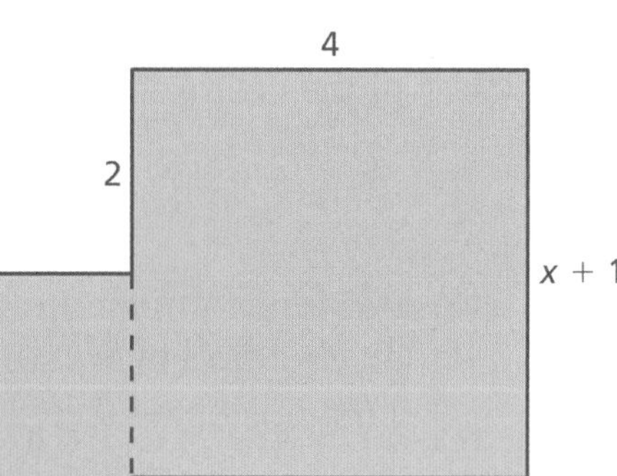

g.

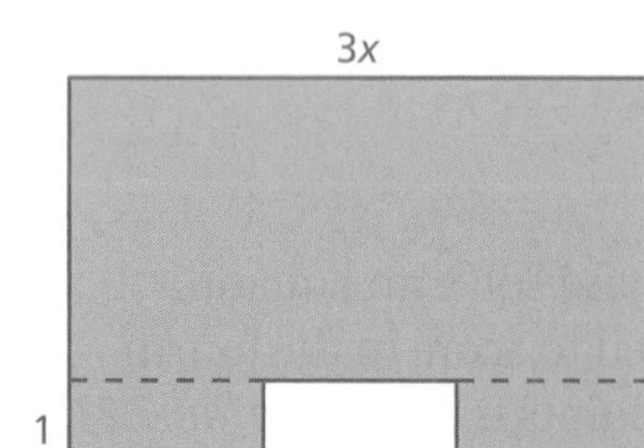

Solving Equations

In this lesson, you will

- solve equations with variables on both sides.
- determine whether equations have no solution or infinitely many solutions.

2 ACTIVITY: Surface Area and Volume

Math Practice

Use Operations

What properties of operations do you need to use in order to find the value of x?

Work with a partner.

- **Each solid has the unusual property that the value of its surface area (in square inches) is equal to the value of its volume (in cubic inches). Write an equation for each solid.**
- **Solve each equation for x.**
- **Use the value of x to find the surface area and the volume of each solid.**
- **Describe how you can check your solution.**

a.

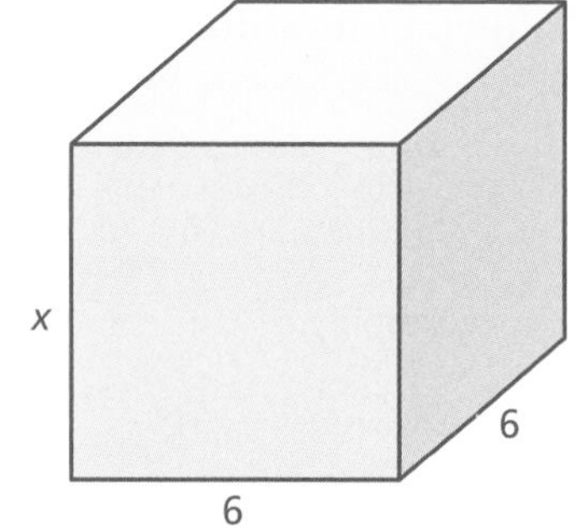

b.

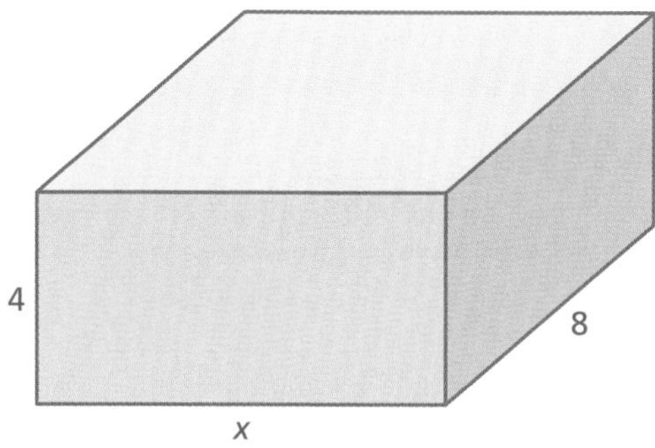

3 ACTIVITY: Puzzle

Work with a partner. The perimeter of the larger triangle is 150% of the perimeter of the smaller triangle. Find the dimensions of each triangle.

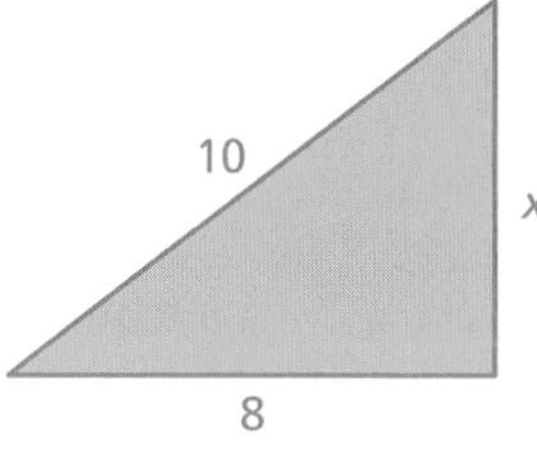

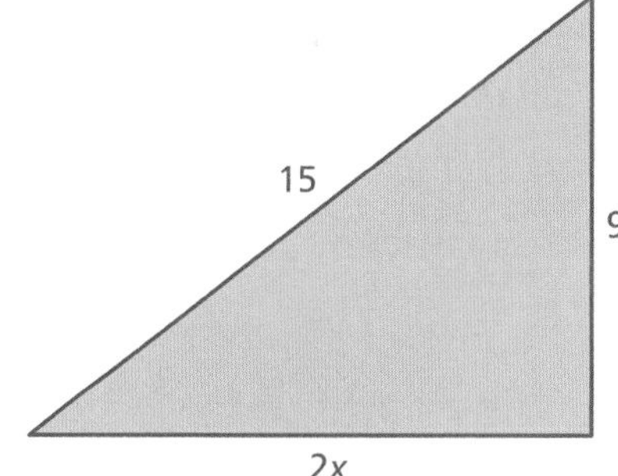

What Is Your Answer?

4. **IN YOUR OWN WORDS** How can you solve an equation that has variables on both sides? How do you move a variable term from one side of the equation to the other?

5. Write an equation that has variables on both sides. Solve the equation.

Practice Use what you learned about solving equations with variables on both sides to complete Exercises 3–5 on page 23.

1.3 Lesson

Solving Equations with Variables on Both Sides

To solve equations with variables on both sides, collect the variable terms on one side and the constant terms on the other side.

EXAMPLE 1 Solving an Equation with Variables on Both Sides

Solve $15 - 2x = -7x$. Check your solution.

	$15 - 2x = -7x$	Write the equation.
Undo the subtraction. →	$+2x \quad +2x$	Addition Property of Equality
	$15 = -5x$	Simplify.
Undo the multiplication. →	$\frac{15}{-5} = \frac{-5x}{-5}$	Division Property of Equality
	$-3 = x$	Simplify.

The solution is $x = -3$.

Check

$15 - 2x = -7x$

$15 - 2(-3) \stackrel{?}{=} -7(-3)$

$21 = 21$ ✓

EXAMPLE 2 Using the Distributive Property to Solve an Equation

Solve $-2(x - 5) = 6\left(2 - \frac{1}{2}x\right)$.

	$-2(x - 5) = 6\left(2 - \frac{1}{2}x\right)$	Write the equation.
	$-2x + 10 = 12 - 3x$	Distributive Property
Undo the subtraction. →	$+3x \quad +3x$	Addition Property of Equality
	$x + 10 = 12$	Simplify.
Undo the addition. →	$-10 \quad -10$	Subtraction Property of Equality
	$x = 2$	Simplify.

The solution is $x = 2$.

On Your Own

Now You're Ready
Exercises 6–14

Solve the equation. Check your solution.

1. $-3x = 2x + 19$
2. $2.5y + 6 = 4.5y - 1$
3. $6(4 - z) = 2z$

Some equations do not have one solution. Equations can also have no solution or infinitely many solutions.

When solving an equation that has no solution, you will obtain an equivalent equation that is not true for any value of the variable, such as $0 = 2$.

EXAMPLE 3 Solving Equations with No Solution

Solve $3 - 4x = -7 - 4x$.

$$3 - 4x = -7 - 4x$$ Write the equation.

Undo the subtraction. → $$\underline{+\,4x} \qquad \underline{+\,4x}$$ Addition Property of Equality

$$3 = -7 \quad ✗$$ Simplify.

The equation $3 = -7$ is never true. So, the equation has no solution.

When solving an equation that has infinitely many solutions, you will obtain an equivalent equation that is true for all values of the variable, such as $-5 = -5$.

EXAMPLE 4 Solving Equations with Infinitely Many Solutions

Solve $6x + 4 = 4\left(\frac{3}{2}x + 1\right)$.

$$6x + 4 = 4\left(\frac{3}{2}x + 1\right)$$ Write the equation.

$$6x + 4 = 6x + 4$$ Distributive Property

Undo the addition. → $$\underline{-\,6x} \qquad \underline{-\,6x}$$ Subtraction Property of Equality

$$4 = 4$$ Simplify.

The equation $4 = 4$ is always true. So, the equation has infinitely many solutions.

On Your Own

Exercises 18–29

Solve the equation.

4. $2x + 1 = 2x - 1$

5. $\frac{1}{2}(6t - 4) = 3t - 2$

6. $\frac{1}{3}(2b + 9) = \frac{2}{3}\left(b + \frac{9}{2}\right)$

7. $6(5 - 2v) = -4(3v + 1)$

EXAMPLE 5 Writing and Solving an Equation

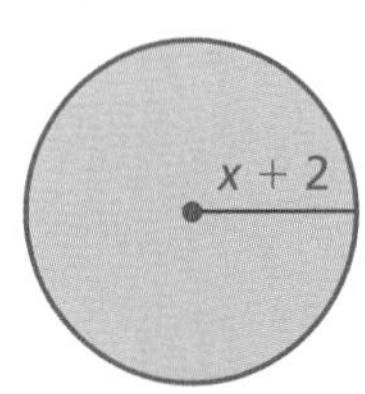

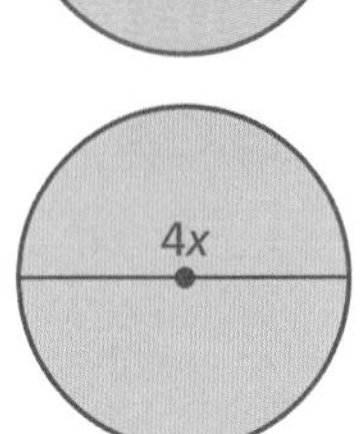

The circles are identical. What is the area of each circle?

Ⓐ 2 Ⓑ 4 Ⓒ 16π Ⓓ 64π

The circles are identical, so the radius of each circle is the same.

$x + 2 = 2x$	Write an equation. The radius of the purple circle is $\frac{4x}{2} = 2x$.
$\underline{-x} \quad \underline{-x}$	Subtraction Property of Equality
$2 = x$	Simplify.

Because the radius of each circle is 4, the area of each circle is $\pi r^2 = \pi(4)^2 = 16\pi$.

So, the correct answer is Ⓒ.

EXAMPLE 6 Real-Life Application

A boat travels x miles per hour upstream on the Mississippi River. On the return trip, the boat travels 2 miles per hour faster. How far does the boat travel upstream?

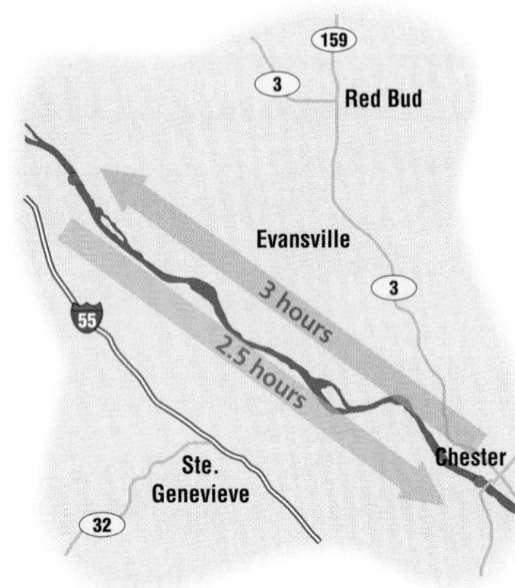

The speed of the boat on the return trip is $(x + 2)$ miles per hour.

Distance upstream = Distance of return trip

$3x = 2.5(x + 2)$	Write an equation.
$3x = 2.5x + 5$	Distributive Property
$\underline{-2.5x} \quad \underline{-2.5x}$	Subtraction Property of Equality
$0.5x = 5$	Simplify.
$\frac{0.5x}{0.5} = \frac{5}{0.5}$	Division Property of Equality
$x = 10$	Simplify.

The boat travels 10 miles per hour for 3 hours upstream. So, it travels 30 miles upstream.

On Your Own

8. **WHAT IF?** In Example 5, the diameter of the purple circle is $3x$. What is the area of each circle?

9. A boat travels x miles per hour from one island to another island in 2.5 hours. The boat travels 5 miles per hour faster on the return trip of 2 hours. What is the distance between the islands?

1.3 Exercises

Vocabulary and Concept Check

1. **WRITING** Is $x = 3$ a solution of the equation $3x - 5 = 4x - 9$? Explain.
2. **OPEN-ENDED** Write an equation that has variables on both sides and has a solution of -3.

Practice and Problem Solving

The value of the solid's surface area is equal to the value of the solid's volume. Find the value of x.

3. 

4. 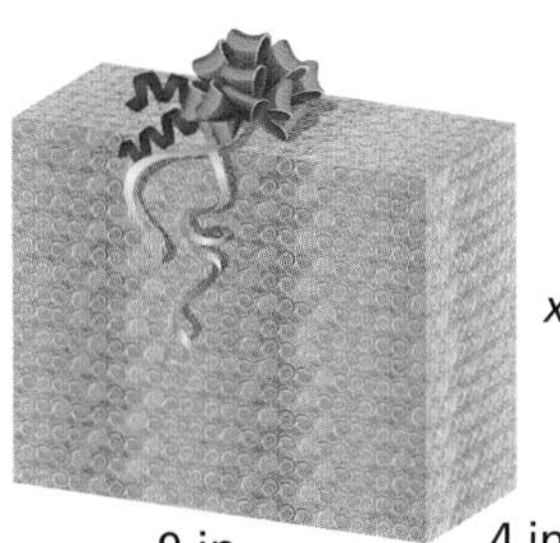

5. 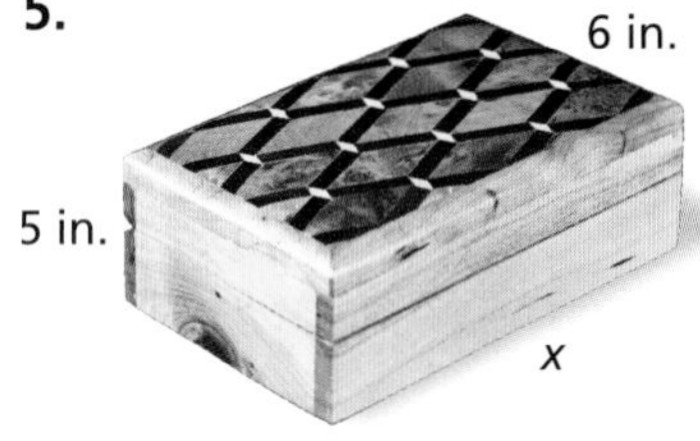

Solve the equation. Check your solution.

① ② 6. $m - 4 = 2m$ 7. $3k - 1 = 7k + 2$ 8. $6.7x = 5.2x + 12.3$

9. $-24 - \frac{1}{8}p = \frac{3}{8}p$ 10. $12(2w - 3) = 6w$ 11. $2(n - 3) = 4n + 1$

12. $2(4z - 1) = 3(z + 2)$ 13. $0.1x = 0.2(x + 2)$ 14. $\frac{1}{6}d + \frac{2}{3} = \frac{1}{4}(d - 2)$

15. **ERROR ANALYSIS** Describe and correct the error in solving the equation.

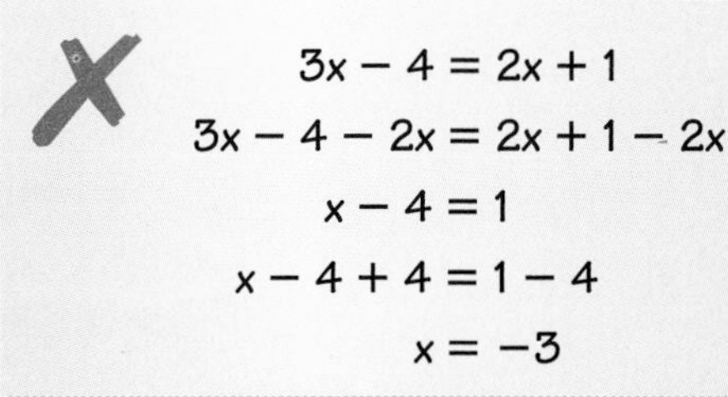

16. **TRAIL MIX** The equation $4.05p + 14.40 = 4.50(p + 3)$ represents the number p of pounds of peanuts you need to make trail mix. How many pounds of peanuts do you need for the trail mix?

17. **CARS** Write and solve an equation to find the number of miles you must drive to have the same cost for each of the car rentals.

$15 plus $0.50 per mile

$25 plus $0.25 per mile

Solve the equation. Check your solution, if possible.

③ ④ **18.** $x + 6 = x$ **19.** $3x - 1 = 1 - 3x$ **20.** $4x - 9 = 3.5x - 9$

21. $\frac{1}{2}x + \frac{1}{2}x = x + 1$ **22.** $3x + 15 = 3(x + 5)$ **23.** $\frac{1}{3}(9x + 3) = 3x + 1$

24. $5x - 7 = 4x - 1$ **25.** $2x + 4 = -(-7x + 6)$ **26.** $5.5 - x = -4.5 - x$

27. $10x - \frac{8}{3} - 4x = 6x$ **28.** $-3(2x - 3) = -6x + 9$ **29.** $6(7x + 7) = 7(6x + 6)$

30. ERROR ANALYSIS Describe and correct the error in solving the equation.

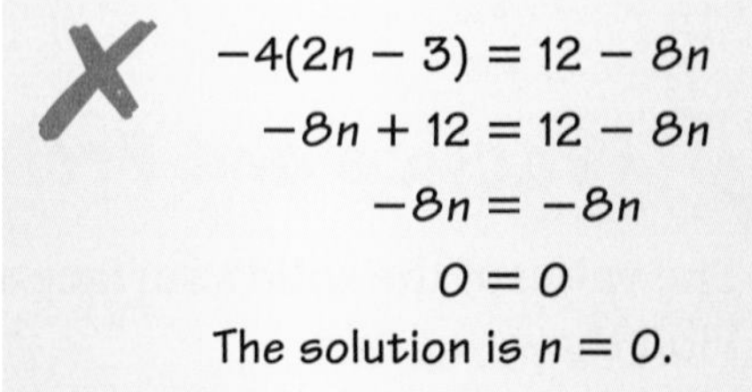

31. OPEN-ENDED Write an equation with variables on both sides that has no solution. Explain why it has no solution.

32. GEOMETRY Are there any values of x for which the areas of the figures are the same? Explain.

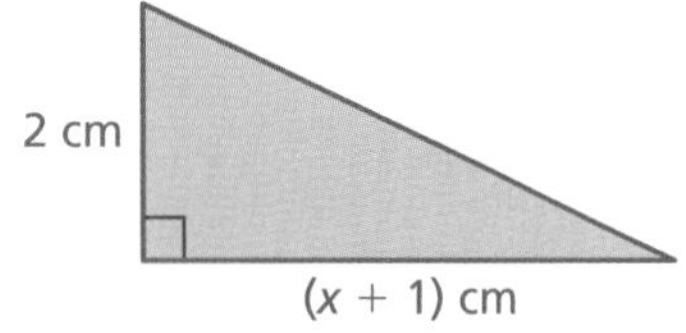

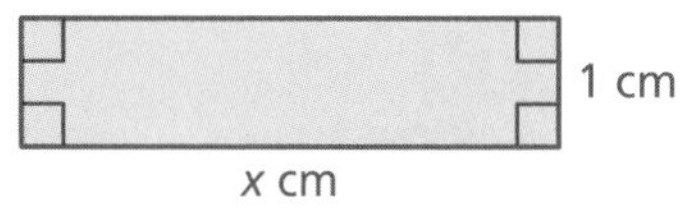

33. SATELLITE TV Provider A charges \$75 for installation and charges \$39.95 per month for the basic package. Provider B offers free installation and charges \$39.95 per month for the basic package. Your neighbor subscribes to Provider A the same month you subscribe to Provider B. After how many months is your neighbor's total cost the same as your total cost for satellite TV?

34. PIZZA CRUST Pepe's Pizza makes 52 pizza crusts the first week and 180 pizza crusts each subsequent week. Dianne's Delicatessen makes 26 pizza crusts the first week and 90 pizza crusts each subsequent week. In how many weeks will the total number of pizza crusts made by Pepe's Pizza equal twice the total number of pizza crusts made by Dianne's Delicatessen?

35. PRECISION Is the triangle an equilateral triangle? Explain.

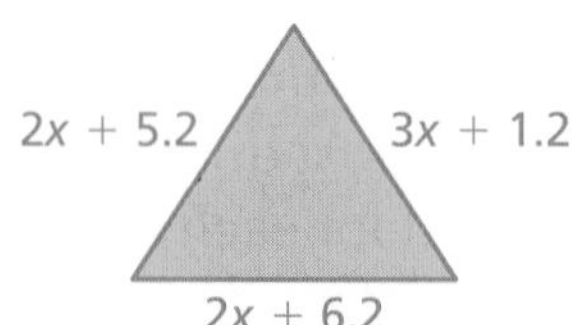

A polygon is *regular* if each of its sides has the same length. Find the perimeter of the regular polygon.

36.

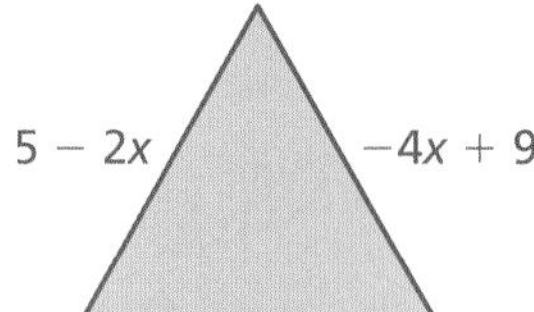

37.

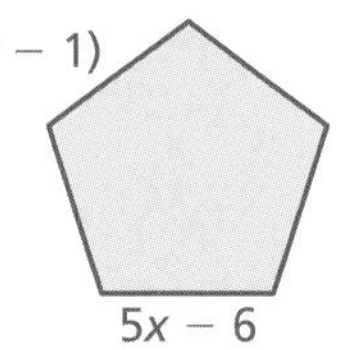

38.

$x + 7$

$\frac{4}{3}x - \frac{1}{3}$

39. PRECISION The cost of mailing a DVD in an envelope by Express Mail® is equal to the cost of mailing a DVD in a box by Priority Mail®. What is the weight of the DVD with its packing material? Round your answer to the nearest hundredth.

	Packing Material	Priority Mail®	Express Mail®
Box	\$2.25	\$2.50 per lb	\$8.50 per lb
Envelope	\$1.10	\$2.50 per lb	\$8.50 per lb

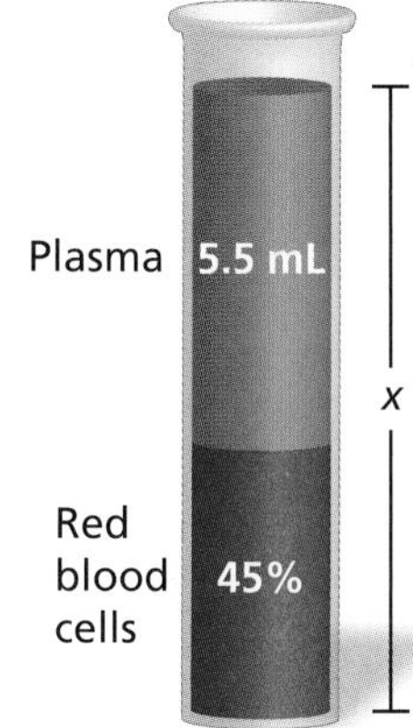

40. PROBLEM SOLVING Would you solve the equation $0.25x + 7 = \frac{1}{3}x - 8$ using fractions or decimals? Explain.

41. BLOOD SAMPLE The amount of red blood cells in a blood sample is equal to the total amount in the sample minus the amount of plasma. What is the total amount x of blood drawn?

42. NUTRITION One serving of oatmeal provides 16% of the fiber you need daily. You must get the remaining 21 grams of fiber from other sources. How many grams of fiber should you consume daily?

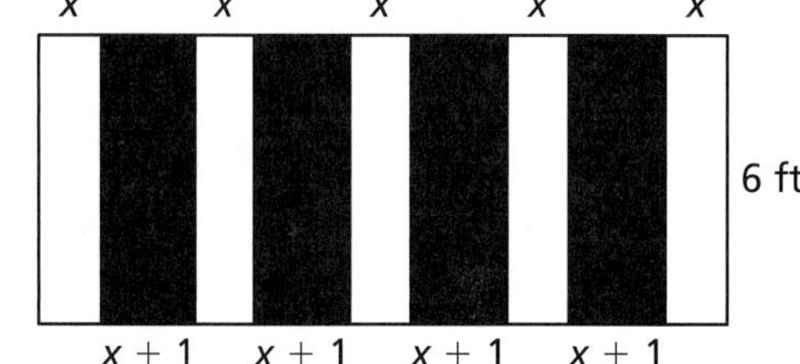

43. Geometry A 6-foot-wide hallway is painted as shown, using equal amounts of white and black paint.

a. How long is the hallway?

b. Can this same hallway be painted with the same pattern, but using twice as much black paint as white paint? Explain.

Fair Game Review What you learned in previous grades & lessons

Find the volume of the solid. *(Skills Review Handbook)*

44.

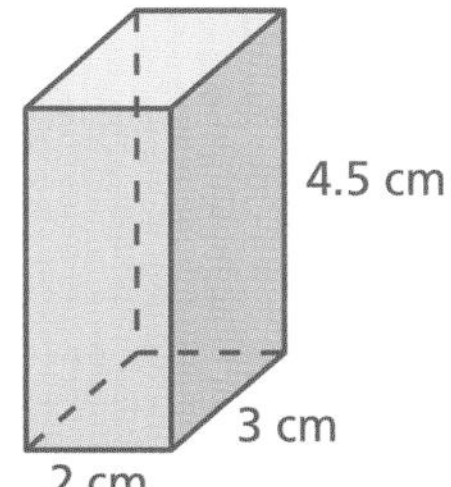

45.

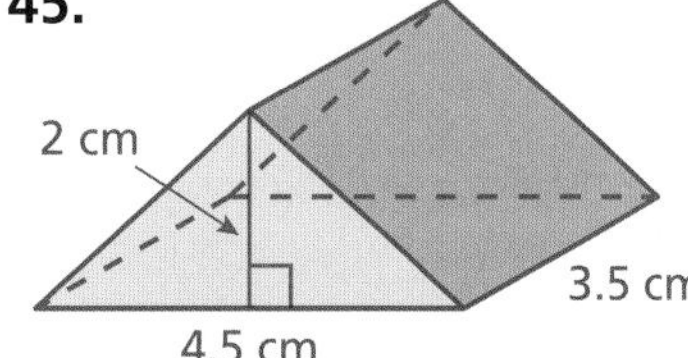

46.

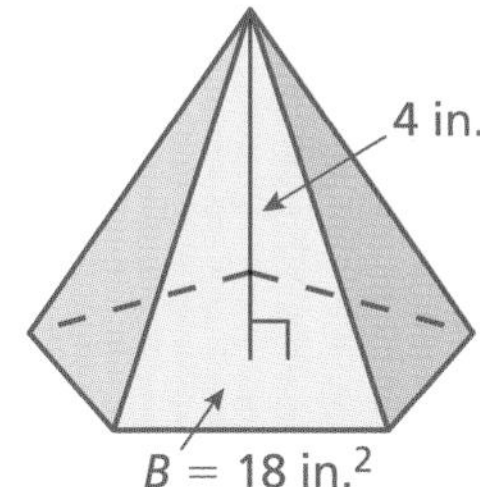

47. MULTIPLE CHOICE A car travels 480 miles on 15 gallons of gasoline. How many miles does the car travel per gallon? *(Skills Review Handbook)*

Ⓐ 28 mi/gal Ⓑ 30 mi/gal Ⓒ 32 mi/gal Ⓓ 35 mi/gal

1.4 Rewriting Equations and Formulas

Essential Question How can you use a formula for one measurement to write a formula for a different measurement?

1 ACTIVITY: Using Perimeter and Area Formulas

Work with a partner.

a.
- Write a formula for the perimeter P of a rectangle.
- Solve the formula for w.
- Use the new formula to find the width of the rectangle.

w

$P = 19$ in.

$\ell = 5.5$ in.

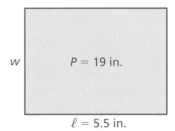

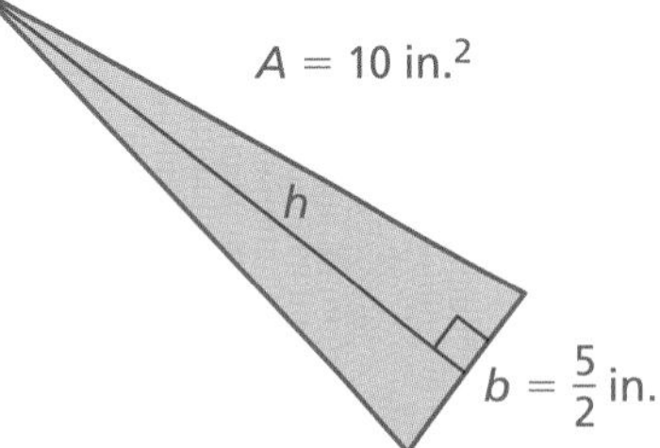

b.
- Write a formula for the area A of a triangle.
- Solve the formula for h.
- Use the new formula to find the height of the triangle.

c.
- Write a formula for the circumference C of a circle.
- Solve the formula for r.
- Use the new formula to find the radius of the circle.

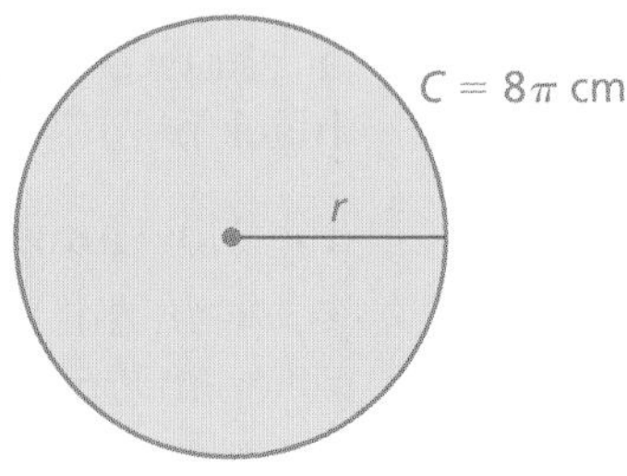

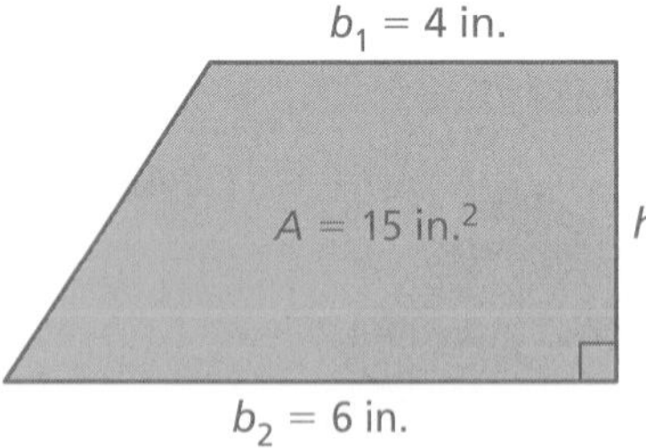

d.
- Write a formula for the area A of a trapezoid.
- Solve the formula for h.
- Use the new formula to find the height of the trapezoid.

Solving Equations

In this lesson, you will
- rewrite equations to solve for one variable in terms of the other variable(s).

e.
- Write a formula for the area A of a parallelogram.
- Solve the formula for h.
- Use the new formula to find the height of the parallelogram.

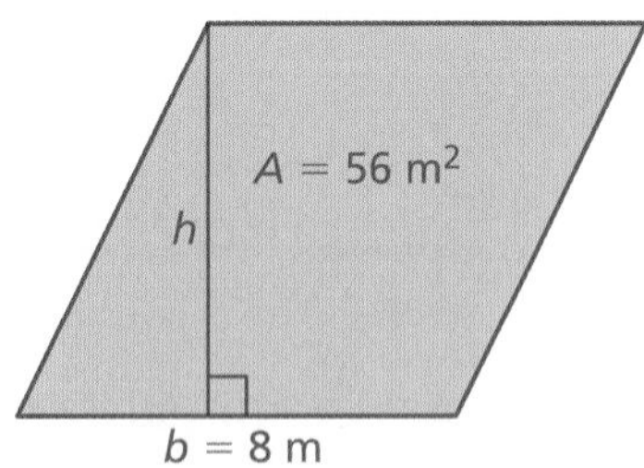

2 ACTIVITY: Using Volume and Surface Area Formulas

Math Practice

Find General Methods

What do you have to do each time to solve for the given variable? Why does this process result in a new formula?

Work with a partner.

a.
- Write a formula for the volume V of a prism.
- Solve the formula for h.
- Use the new formula to find the height of the prism.

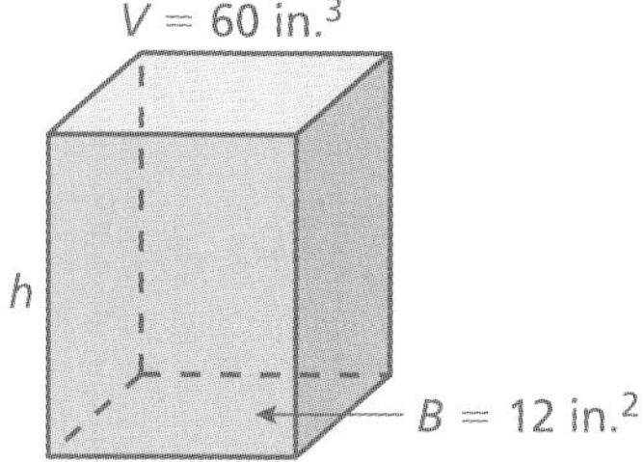

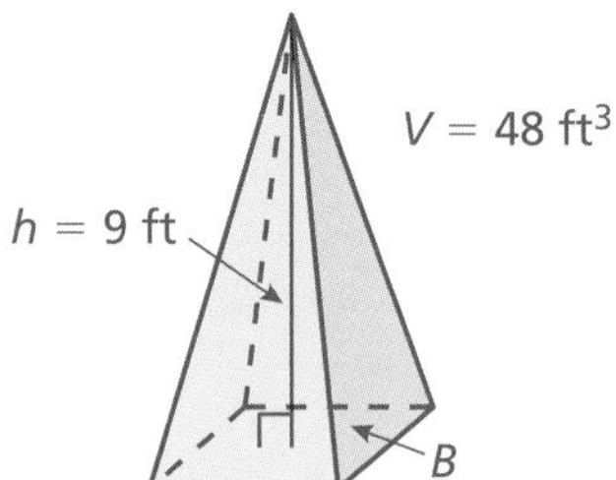

b.
- Write a formula for the volume V of a pyramid.
- Solve the formula for B.
- Use the new formula to find the area of the base of the pyramid.

c.
- Write a formula for the lateral surface area S of a cylinder.
- Solve the formula for h.
- Use the new formula to find the height of the cylinder.

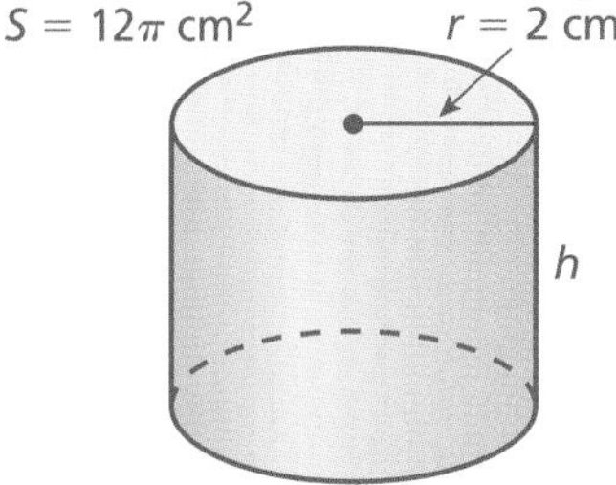

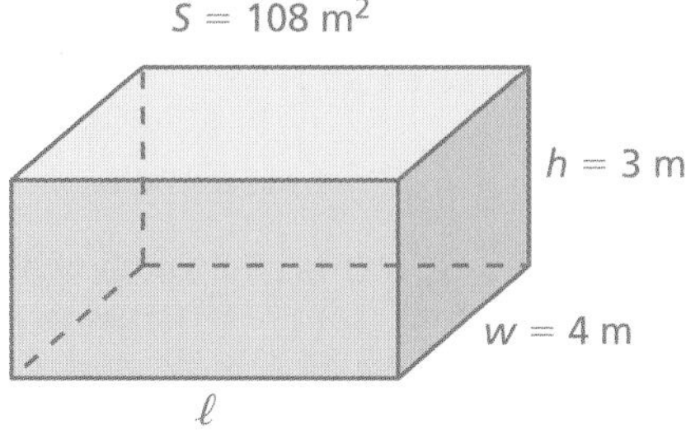

d.
- Write a formula for the surface area S of a rectangular prism.
- Solve the formula for ℓ.
- Use the new formula to find the length of the rectangular prism.

What Is Your Answer?

3. IN YOUR OWN WORDS How can you use a formula for one measurement to write a formula for a different measurement? Give an example that is different from the examples on these two pages.

Practice Use what you learned about rewriting equations and formulas to complete Exercises 3 and 4 on page 30.

1.4 Lesson

Key Vocabulary
literal equation, *p. 28*

An equation that has two or more variables is called a **literal equation**. To rewrite a literal equation, solve for one variable in terms of the other variable(s).

EXAMPLE 1 Rewriting an Equation

Solve the equation $2y + 5x = 6$ for y.

$2y + 5x = 6$ — Write the equation.

Undo the addition. → $2y + 5x - 5x = 6 - 5x$ — Subtraction Property of Equality

$2y = 6 - 5x$ — Simplify.

Undo the multiplication. → $\dfrac{2y}{2} = \dfrac{6 - 5x}{2}$ — Division Property of Equality

$y = 3 - \dfrac{5}{2}x$ — Simplify.

On Your Own

Exercises 5–10

Solve the equation for y.

1. $5y - x = 10$
2. $4x - 4y = 1$
3. $12 = 6x + 3y$

EXAMPLE 2 Rewriting a Formula

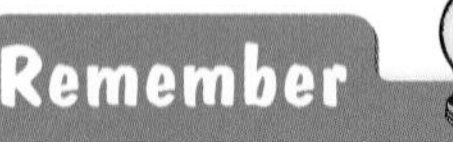

A *formula* shows how one variable is related to one or more other variables. A formula is a type of literal equation.

The formula for the surface area S of a cone is $S = \pi r^2 + \pi r\ell$. Solve the formula for the slant height ℓ.

$S = \pi r^2 + \pi r\ell$ — Write the formula.

$S - \pi r^2 = \pi r^2 - \pi r^2 + \pi r\ell$ — Subtraction Property of Equality

$S - \pi r^2 = \pi r\ell$ — Simplify.

$\dfrac{S - \pi r^2}{\pi r} = \dfrac{\pi r\ell}{\pi r}$ — Division Property of Equality

$\dfrac{S - \pi r^2}{\pi r} = \ell$ — Simplify.

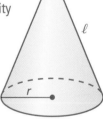

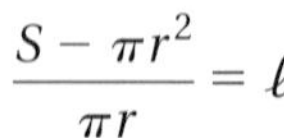

On Your Own

Exercises 14–19

Solve the formula for the red variable.

4. Area of rectangle: $A = bh$
5. Simple interest: $I = Prt$
6. Surface area of cylinder: $S = 2\pi r^2 + 2\pi rh$

Multi-Language Glossary at BigIdeasMath.com

Key Idea

Temperature Conversion

A formula for converting from degrees Fahrenheit F to degrees Celsius C is

$$C = \frac{5}{9}(F - 32).$$

EXAMPLE 3 Rewriting the Temperature Formula

Solve the temperature formula for F.

$C = \frac{5}{9}(F - 32)$ Write the temperature formula.

Use the reciprocal. → $\frac{9}{5} \cdot C = \frac{9}{5} \cdot \frac{5}{9}(F - 32)$ Multiplication Property of Equality

$\frac{9}{5}C = F - 32$ Simplify.

Undo the subtraction. → $\frac{9}{5}C + 32 = F - 32 + 32$ Addition Property of Equality

$\frac{9}{5}C + 32 = F$ Simplify.

The rewritten formula is $F = \frac{9}{5}C + 32$.

EXAMPLE 4 Real-Life Application

Which has the greater temperature?

Convert the Celsius temperature of lightning to Fahrenheit.

$F = \frac{9}{5}C + 32$ Write the rewritten formula from Example 3.

$= \frac{9}{5}(30{,}000) + 32$ Substitute 30,000 for C.

$= 54{,}032$ Simplify.

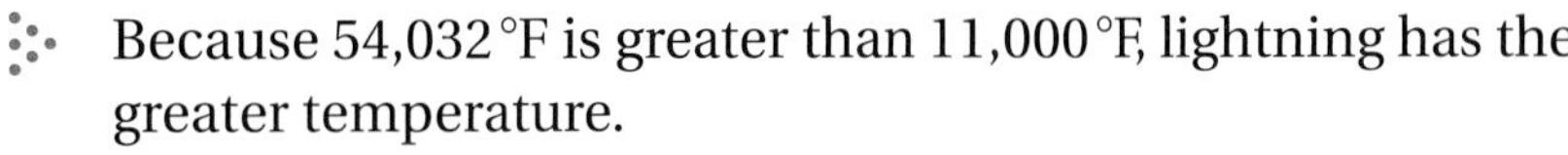

Because 54,032 °F is greater than 11,000 °F, lightning has the greater temperature.

On Your Own

7. Room temperature is considered to be 70 °F. Suppose the temperature is 23 °C. Is this greater than or less than room temperature?

1.4 Exercises

Vocabulary and Concept Check

1. **VOCABULARY** Is $-2x = \frac{3}{8}$ a literal equation? Explain.

2. **DIFFERENT WORDS, SAME QUESTION** Which is different? Find "both" answers.

Solve $4x - 2y = 6$ for y.	Solve $6 = 4x - 2y$ for y.
Solve $4x - 2y = 6$ for y in terms of x.	Solve $4x - 2y = 6$ for x in terms of y.

Practice and Problem Solving

3. a. Write a formula for the area A of a triangle.
   b. Solve the formula for b.
   c. Use the new formula to find the base of the triangle.

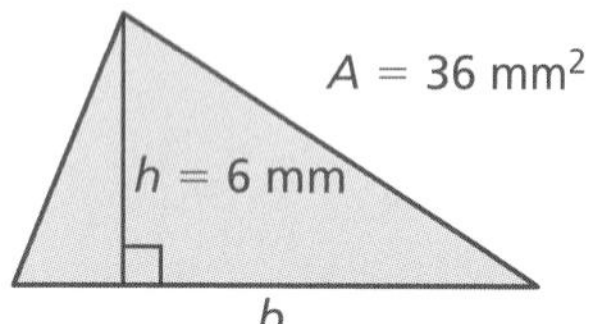

4. a. Write a formula for the volume V of a prism.
   b. Solve the formula for B.
   c. Use the new formula to find the area of the base of the prism.

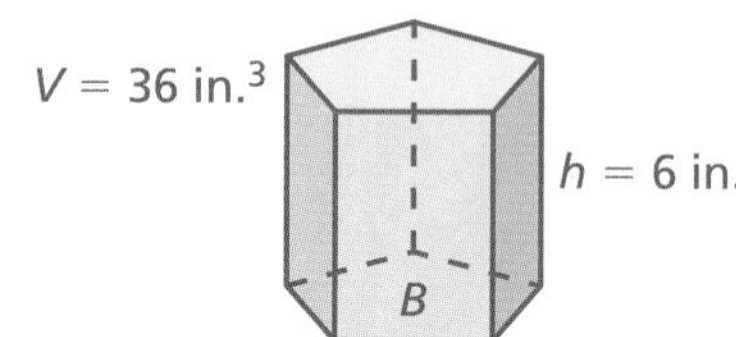

Solve the equation for y.

1 5. $\frac{1}{3}x + y = 4$

6. $3x + \frac{1}{5}y = 7$

7. $6 = 4x + 9y$

8. $\pi = 7x - 2y$

9. $4.2x - 1.4y = 2.1$

10. $6y - 1.5x = 8$

11. **ERROR ANALYSIS** Describe and correct the error in rewriting the equation.

✗
$2x - y = 5$
$y = -2x + 5$

12. **TEMPERATURE** The formula $K = C + 273.15$ converts temperatures from Celsius C to Kelvin K.
   a. Solve the formula for C.
   b. Convert 300 Kelvin to Celsius.

13. **INTEREST** The formula for simple interest is $I = Prt$.
   a. Solve the formula for t.
   b. Use the new formula to find the value of t in the table.

I	\$75
P	\$500
r	5%
t	

Solve the equation for the red variable.

2 **14.** $d = rt$ **15.** $e = mc^2$ **16.** $R - C = P$

17. $A = \frac{1}{2}\pi w^2 + 2\ell w$ **18.** $B = 3\frac{V}{h}$ **19.** $g = \frac{1}{6}(w + 40)$

20. LOGIC Why is it useful to rewrite a formula in terms of another variable?

21. REASONING The formula $K = \frac{5}{9}(F - 32) + 273.15$ converts temperatures from Fahrenheit F to Kelvin K.

a. Solve the formula for F.

b. The freezing point of water is 273.15 Kelvin. What is this temperature in Fahrenheit?

c. The temperature of dry ice is −78.5 °C. Which is colder, dry ice or liquid nitrogen?

Navy Pier Ferris Wheel

22. FERRIS WHEEL The Navy Pier Ferris Wheel in Chicago has a circumference that is 56% of the circumference of the first Ferris wheel built in 1893.

a. What is the radius of the Navy Pier Ferris Wheel?

b. What was the radius of the first Ferris wheel?

c. The first Ferris wheel took 9 minutes to make a complete revolution. How fast was the wheel moving?

23. Repeated Reasoning The formula for the volume of a sphere is $V = \frac{4}{3}\pi r^3$. Solve the formula for r^3. Use Guess, Check, and Revise to find the radius of the sphere.

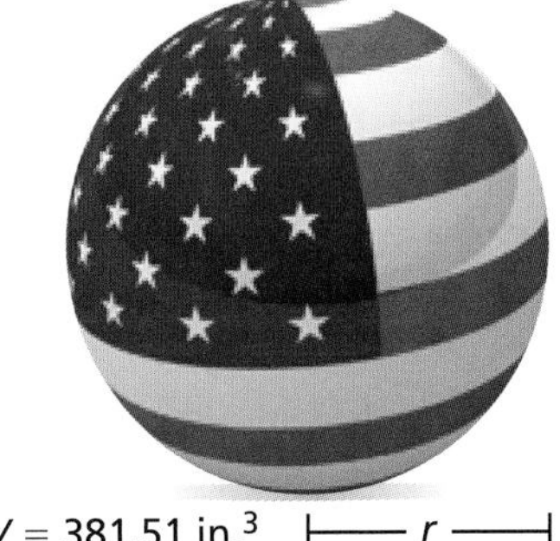

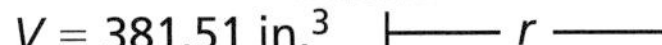

$V = 381.51$ in.^{3} r

Fair Game Review What you learned in previous grades & lessons

Multiply. *(Skills Review Handbook)*

24. $5 \times \frac{3}{4}$ **25.** $-2 \times \frac{8}{3}$ **26.** $\frac{1}{4} \times \frac{3}{2} \times \frac{8}{9}$ **27.** $25 \times \frac{3}{5} \times \frac{1}{12}$

28. MULTIPLE CHOICE Which of the following is not equivalent to $\frac{3}{4}$? *(Skills Review Handbook)*

Ⓐ 0.75 Ⓑ 3 : 4 Ⓒ 75% Ⓓ 4 : 3

1.3–1.4 Quiz

Solve the equation. Check your solution, if possible. *(Section 1.3)*

1. $2(x + 4) = -5x + 1$

2. $\frac{1}{2}s = 4s - 21$

3. $8.3z = 4.1z + 10.5$

4. $3(b + 5) = 4(2b - 5)$

5. $n + 7 - n = 4$

6. $\frac{1}{4}(4r - 8) = r - 2$

Solve the equation for *y*. *(Section 1.4)*

7. $6x - 3y = 9$

8. $8 = 2y - 10x$

Solve the formula for the red variable. *(Section 1.4)*

9. Volume of a cylinder: $V = \pi r^2 h$

10. Area of a trapezoid: $A = \frac{1}{2}h(b_1 + b_2)$

11. TEMPERATURE In which city is the water temperature higher? *(Section 1.4)*

12. SAVINGS ACCOUNT You begin with $25 in a savings account and $50 in a checking account. Each week you deposit $5 into savings and $10 into checking. After how many weeks is the amount in checking twice the amount in savings? *(Section 1.3)*

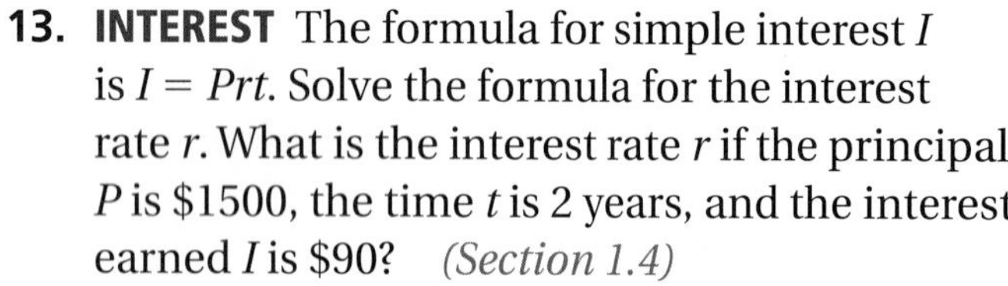

13. INTEREST The formula for simple interest I is $I = Prt$. Solve the formula for the interest rate r. What is the interest rate r if the principal P is $1500, the time t is 2 years, and the interest earned I is $90? *(Section 1.4)*

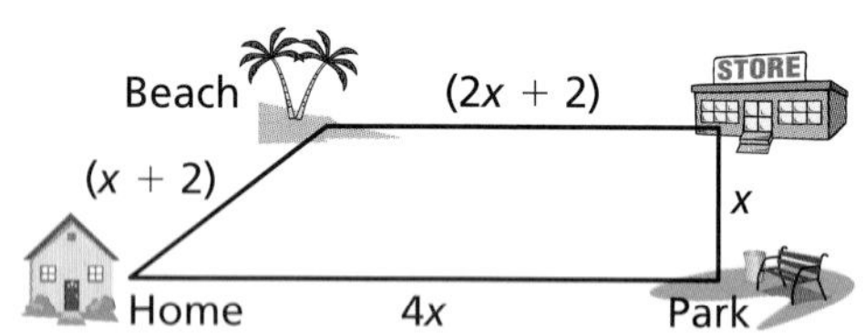

14. ROUTES From your home, the route to the store that passes the beach is 2 miles shorter than the route to the store that passes the park. What is the length of each route? *(Section 1.3)*

15. PERIMETER Use the triangle shown. *(Section 1.4)*

a. Write a formula for the perimeter P of the triangle.

b. Solve the formula for b.

c. Use the new formula to find b when a is 10 feet and c is 17 feet.

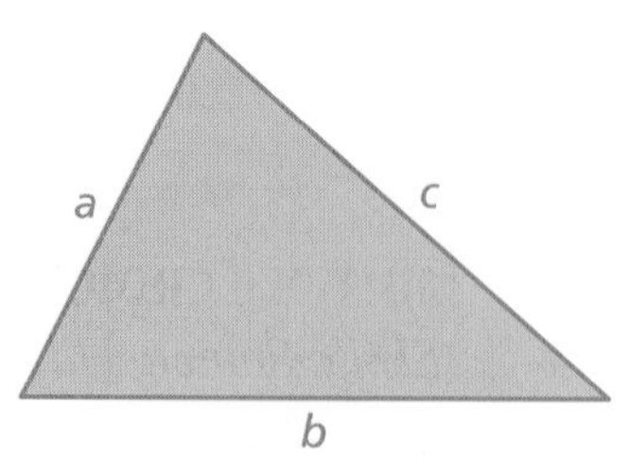

1 Chapter Review

Review Key Vocabulary

literal equation, *p. 28*

Review Examples and Exercises

1.1 Solving Simple Equations (pp. 2–9)

The *boiling point* of a liquid is the temperature at which the liquid becomes a gas. The boiling point of mercury is about $\frac{41}{200}$ of the boiling point of lead. Write and solve an equation to find the boiling point of lead.

Let x be the boiling point of lead.

$\frac{41}{200}x = 357$ Write the equation.

$\frac{200}{41} \cdot \left(\frac{41}{200}x\right) = \frac{200}{41} \cdot 357$ Multiplication Property of Equality

$x \approx 1741$ Simplify.

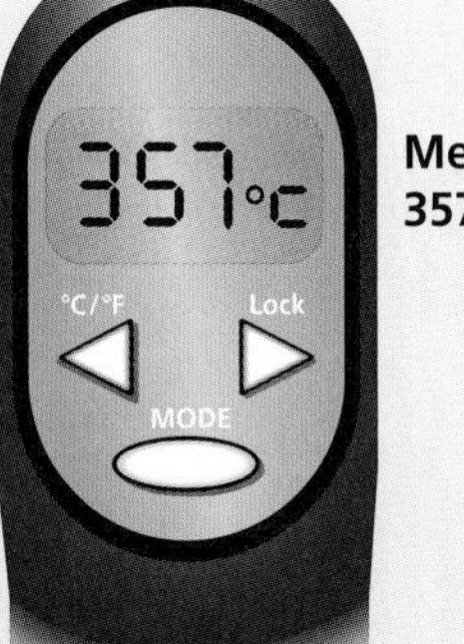

Mercury 357°C

The boiling point of lead is about 1741°C.

Exercises

Solve the equation. Check your solution.

1. $y + 8 = -11$
2. $3.2 = -0.4n$
3. $-\frac{t}{4} = -3\pi$

1.2 Solving Multi-Step Equations (pp. 10–15)

Solve $-14x + 28 + 6x = -44$.

$-14x + 28 + 6x = -44$ Write the equation.

$-8x + 28 = -44$ Combine like terms.

$\underline{-28} \quad \underline{-28}$ Subtraction Property of Equality

$-8x = -72$ Simplify.

$\frac{-8x}{-8} = \frac{-72}{-8}$ Division Property of Equality

$x = 9$ Simplify.

The solution is $x = 9$.

Exercises

Find the value of x. Then find the angle measures of the polygon.

4.

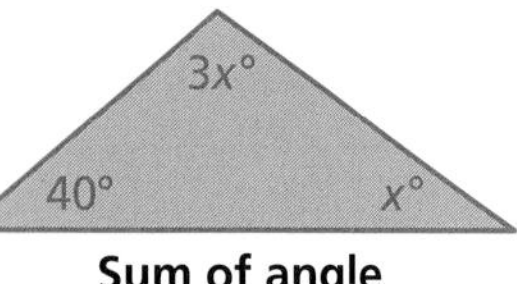

Sum of angle measures: 180°

5.

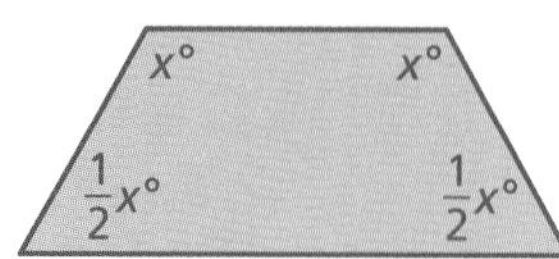

Sum of angle measures: 360°

6.

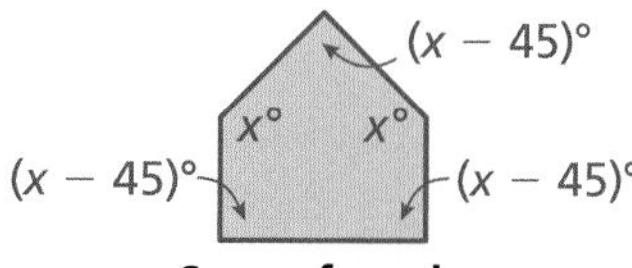

Sum of angle measures: 540°

1.3 Solving Equations with Variables on Both Sides (pp. 18–25)

a. Solve $3(x - 4) = -2(4 - x)$.

$3(x - 4) = -2(4 - x)$	Write the equation.
$3x - 12 = -8 + 2x$	Distributive Property
$\underline{-2x} \qquad \underline{-2x}$	Subtraction Property of Equality
$x - 12 = -8$	Simplify.
$\underline{+12} \qquad \underline{+12}$	Addition Property of Equality
$x = 4$	Simplify.

∴ The solution is $x = 4$.

b. Solve $4 - 5k = -8 - 5k$.

$4 - 5k = -8 - 5k$	Write the equation.
$\underline{+5k} \qquad \underline{+5k}$	Addition Property of Equality
$4 = -8$ ✗	Simplify.

∴ The equation $4 = -8$ is never true. So, the equation has no solution.

c. Solve $2\left(7g + \frac{2}{3}\right) = 14g + \frac{4}{3}$.

$2\left(7g + \frac{2}{3}\right) = 14g + \frac{4}{3}$	Write the equation.
$14g + \frac{4}{3} = 14g + \frac{4}{3}$	Distributive Property
$\underline{-14g} \qquad \underline{-14g}$	Subtraction Property of Equality
$\frac{4}{3} = \frac{4}{3}$	Simplify.

∴ The equation $\frac{4}{3} = \frac{4}{3}$ is always true. So, the equation has infinitely many solutions.

Exercises

Solve the equation. Check your solution, if possible.

7. $5m - 1 = 4m + 5$ **8.** $3(5p - 3) = 5(p - 1)$ **9.** $\frac{2}{5}n + \frac{1}{10} = \frac{1}{2}(n + 4)$

10. $7t + 3 = 8 + 7t$ **11.** $\frac{1}{5}(15b - 7) = 3b - 9$ **12.** $\frac{1}{6}(12z - 18) = 2z - 3$

1.4 Rewriting Equations and Formulas (pp. 26–31)

a. Solve $7y + 6x = 4$ for y.

$7y + 6x = 4$	Write the equation.
$7y + 6x - 6x = 4 - 6x$	Subtraction Property of Equality
$7y = 4 - 6x$	Simplify.
$\frac{7y}{7} = \frac{4 - 6x}{7}$	Division Property of Equality
$y = \frac{4}{7} - \frac{6}{7}x$	Simplify.

b. The equation for a line in slope-intercept form is $y = mx + b$. Solve the equation for x.

$y = mx + b$	Write the equation.
$y - b = mx + b - b$	Subtraction Property of Equality
$y - b = mx$	Simplify.
$\frac{y - b}{m} = \frac{mx}{m}$	Division Property of Equality
$\frac{y - b}{m} = x$	Simplify.

Exercises

Solve the equation for y.

13. $6y + x = 8$ **14.** $10x - 5y = 15$ **15.** $20 = 5x + 10y$

16. a. The formula $F = \frac{9}{5}(K - 273.15) + 32$ converts a temperature from Kelvin K to Fahrenheit F. Solve the formula for K.

b. Convert 240°F to Kelvin K. Round your answer to the nearest hundredth.

17. a. Write the formula for the area A of a trapezoid.

b. Solve the formula for h.

c. Use the new formula to find the height h of the trapezoid.

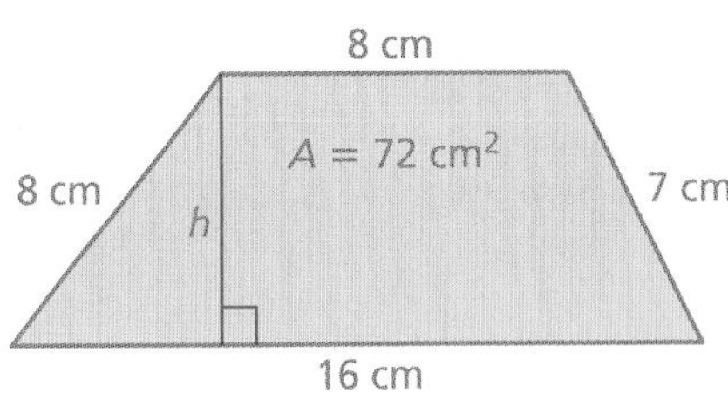

1 Chapter Test

Check It Out
Test Practice
BigIdeasMath.com

Solve the equation. Check your solution, if possible.

1. $4 + y = 9.5$
2. $-\frac{x}{9} = -8$
3. $z - \frac{2}{3} = \frac{1}{8}$
4. $3.8n - 13 = 1.4n + 5$
5. $9(8d - 5) + 13 = 12d - 2$
6. $9j - 8 = 8 + 9j$
7. $2.5(2p + 5) = 5p + 12.5$
8. $\frac{3}{4}t + \frac{1}{8} = \frac{3}{4}(t + 8)$
9. $\frac{1}{7}(14r + 28) = 2(r + 2)$

Find the value of *x*. Then find the angle measures of the polygon.

10. 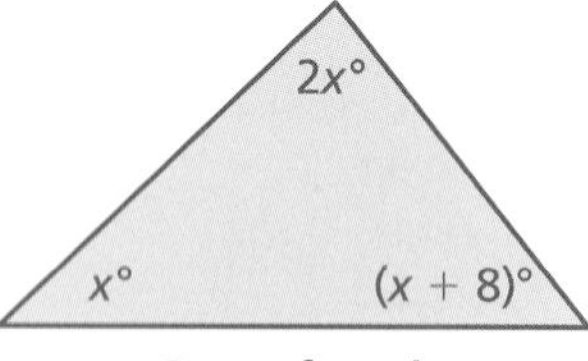

Sum of angle measures: 180°

11. 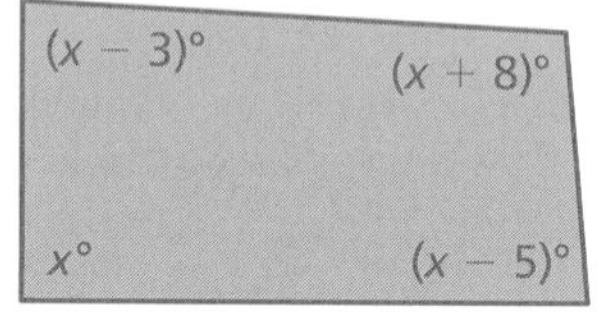

Sum of angle measures: 360°

Solve the equation for *y*.

12. $1.2x - 4y = 28$
13. $0.5 = 0.4y - 0.25x$

Solve the formula for the red variable.

14. Perimeter of a rectangle: $P = 2\ell + 2w$
15. Distance formula: $d = rt$

16. **BASKETBALL** Your basketball team wins a game by 13 points. The opposing team scores 72 points. Explain how to find your team's score.

17. **CYCLING** You are biking at a speed of 18 miles per hour. You are 3 miles behind your friend, who is biking at a speed of 12 miles per hour. Write and solve an equation to find the amount of time it takes for you to catch up to your friend.

18. **VOLCANOES** Two scientists are measuring lava temperatures. One scientist records a temperature of 1725°F. The other scientist records a temperature of 950°C. Which is the greater temperature? $\left(\text{Use } C = \frac{5}{9}(F - 32).\right)$

19. **JOBS** Your profit for mowing lawns this week is \$24. You are paid \$8 per hour and you paid \$40 for gas for the lawn mower. How many hours did you work this week?

1 Cumulative Assessment

Test-Taking Strategy
Solve Directly or Eliminate Choices

When a cat wakes up, it's grumpy for x hours, where 2x-5x=x-4. What's x?
(A) 0 (B) 1 (C) 2 (D) -3

Don't talk to me until I've had my morning milk.

"You can eliminate A and D. Then, solve directly to determine that the correct answer is B."

1. Which value of x makes the equation true?

$$4x = 32$$

A. 8 **C.** 36

B. 28 **D.** 128

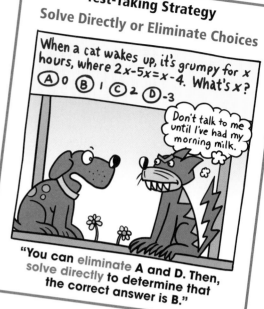

2. A taxi ride costs \$3 plus \$2 for each mile driven. When you rode in a taxi, the total cost was \$39. This can be modeled by the equation below, where m represents the number of miles driven.

$$2m + 3 = 39$$

How long was your taxi ride?

F. 72 mi **H.** 21 mi

G. 34 mi **I.** 18 mi

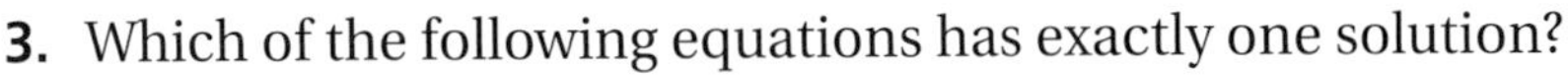

3. Which of the following equations has exactly one solution?

A. $\frac{2}{3}(x + 6) = \frac{2}{3}x + 4$ **C.** $\frac{4}{5}\left(n + \frac{1}{3}\right) = \frac{4}{5}n + \frac{1}{3}$

B. $\frac{3}{7}y + 13 = 13 - \frac{3}{7}y$ **D.** $\frac{7}{8}\left(2t + \frac{1}{8}\right) = \frac{7}{4}t$

4. The perimeter of the square is equal to the perimeter of the triangle. What are the side lengths of the square?

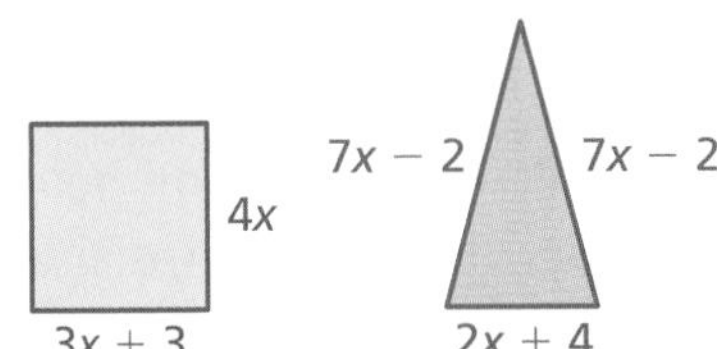

5. The formula below relates distance, rate, and time.

$$d = rt$$

Solve this formula for t.

F. $t = dr$ **H.** $t = d - r$

G. $t = \frac{d}{r}$ **I.** $t = \frac{r}{d}$

6. What could be the first step to solve the equation shown below?

$$3x + 5 = 2(x + 7)$$

A. Combine $3x$ and 5.

B. Multiply x by 2 and 7 by 2.

C. Subtract x from $3x$.

D. Subtract 5 from 7.

7. You work as a sales representative. You earn \$400 per week plus 5% of your total sales for the week.

Part A Last week, you had total sales of \$5000. Find your total earnings. Show your work.

Part B One week, you earned \$1350. Let s represent your total sales that week. Write an equation that you could use to find s.

Part C Using your equation from Part B, find s. Show all steps clearly.

8. In 10 years, Maria will be 39 years old. Let m represent Maria's age today. Which equation can you use to find m?

F. $m = 39 + 10$

G. $m - 10 = 39$

H. $m + 10 = 39$

I. $10m = 39$

9. Which value of y makes the equation below true?

$$3y + 8 = 7y + 11$$

A. -4.75

B. -0.75

C. 0.75

D. 4.75

10. The equation below is used to convert a Fahrenheit temperature F to its equivalent Celsius temperature C.

$$C = \frac{5}{9}(F - 32)$$

Which formula can be used to convert a Celsius temperature to its equivalent Fahrenheit temperature?

F. $F = \frac{5}{9}(C - 32)$

G. $F = \frac{9}{5}(C + 32)$

H. $F = \frac{9}{5}C + \frac{32}{5}$

I. $F = \frac{9}{5}C + 32$

11. You have already saved \$35 for a new cell phone. You need \$175 in all. You think you can save \$10 per week. At this rate, how many more weeks will you need to save money before you can buy the new cell phone?

12. What is the greatest angle measure in the triangle below?

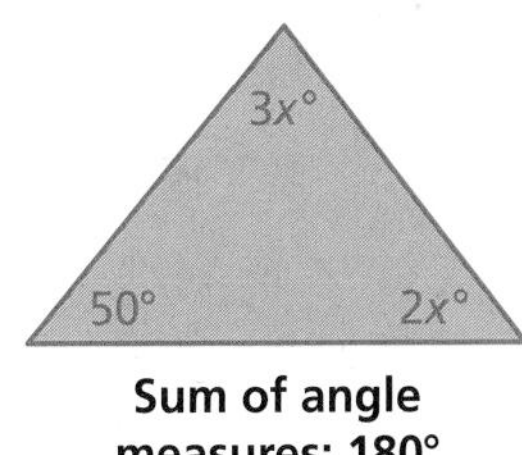

Sum of angle measures: 180°

A. 26°

B. 78°

C. 108°

D. 138°

13. Which value of x makes the equation below true?

$$6(x - 3) = 4x - 7$$

F. -5.5

G. -2

H. 1.1

I. 5.5

14. The drawing below shows equal weights on two sides of a balance scale.

What can you conclude from the drawing?

A. A mug weighs one-third as much as a trophy.

B. A mug weighs one-half as much as a trophy.

C. A mug weighs twice as much as a trophy.

D. A mug weighs three times as much as a trophy.

2 Transformations

"Just 2 more minutes. I'm almost done with my 'cat tessellation' painting."

"If you hold perfectly still..."

"...each frame becomes a horizontal..."

"...translation of the previous frame..."

What You Learned Before

Does that mean that my mirror image is better at music than I am?

"Did you know that when you look at yourself in the mirror, your left and right get switched?"

Reflecting Points

Example 1 **Reflect (3, −4) in the x-axis.**

Plot (3, −4).

To reflect (3, −4) in the x-axis, use the same x-coordinate, 3, and take the opposite of the y-coordinate. The opposite of −4 is 4.

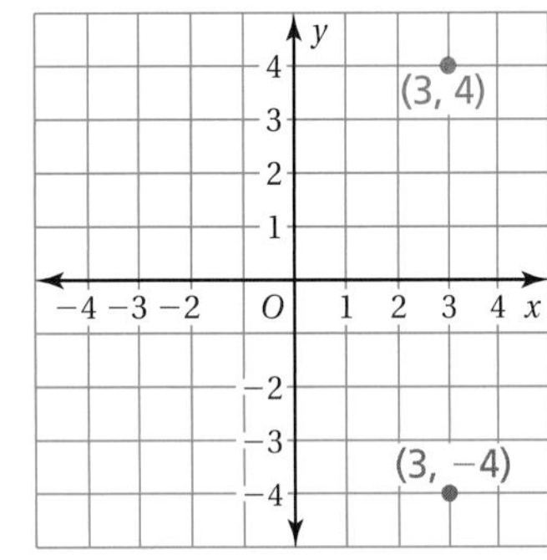
(3, 4)
(3, −4)

So, the reflection of (3, −4) in the x-axis is (3, 4).

Try It Yourself

Reflect the point in (a) the x-axis and (b) the y-axis.

1. (7, 3)
2. (−4, 6)
3. (5, −5)
4. (−8, −3)
5. (0, 1)
6. (−5, 0)
7. (4, −6.5)
8. $\left(-3\frac{1}{2}, -4\right)$

Drawing a Polygon in a Coordinate Plane

Example 2 **The vertices of a quadrilateral are A(1, 5), B(2, 9), C(6, 8), and D(8, 1). Draw the quadrilateral in a coordinate plane.**

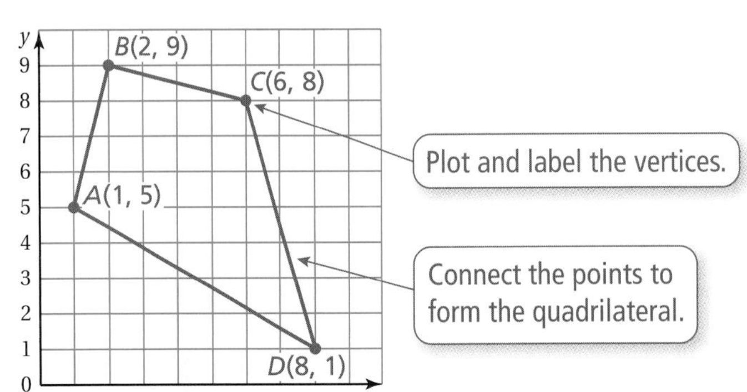
B(2, 9)
C(6, 8)
A(1, 5)
D(8, 1)
Plot and label the vertices.
Connect the points to form the quadrilateral.

Try It Yourself

Draw the polygon with the given vertices in a coordinate plane.

9. $J(1, 1)$, $K(5, 6)$, $M(9, 3)$

10. $Q(2, 3)$, $R(2, 8)$, $S(7, 8)$, $T(7, 3)$

2.1 Congruent Figures

Essential Question How can you identify congruent triangles?

Two figures are congruent when they have the same size and the same shape.

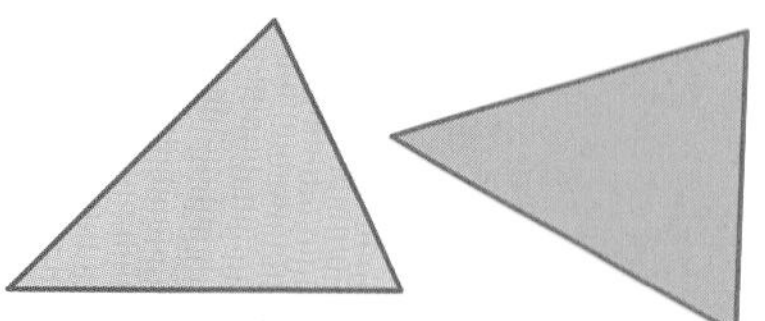

Congruent
Same size *and* shape

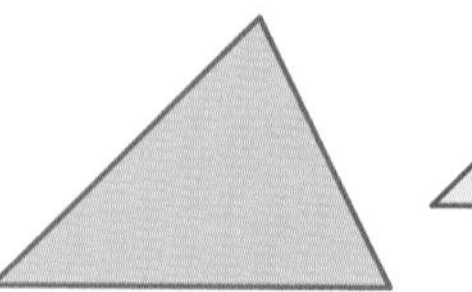

Not Congruent
Same shape, but not same size

1 ACTIVITY: Identifying Congruent Triangles

Work with a partner.

- **Which of the geoboard triangles below are congruent to the geoboard triangle at the right?**
- **Form each triangle on a geoboard.**
- **Measure each side with a ruler. Record your results in a table.**
- **Write a conclusion about the side lengths of triangles that are congruent.**

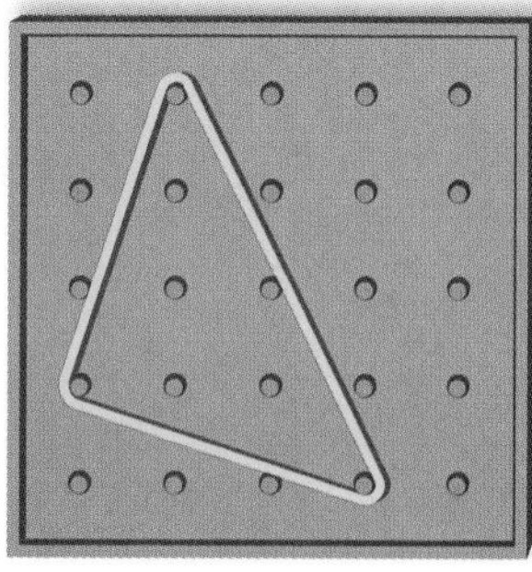

a.

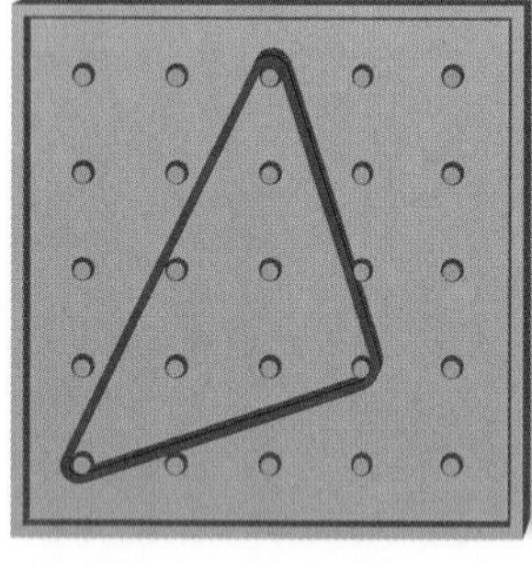

b.

c.

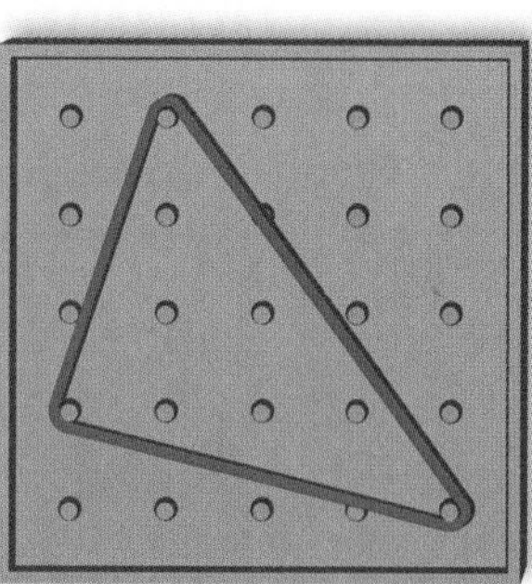

d.

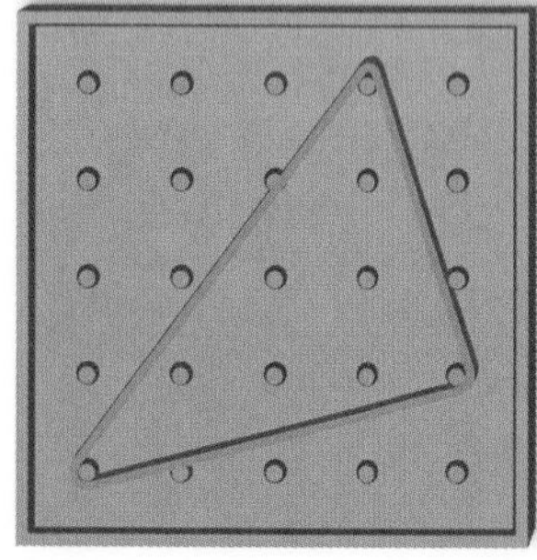

e.

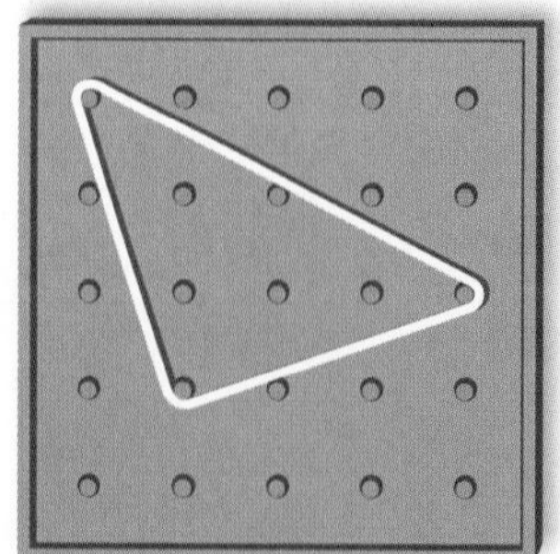

f.

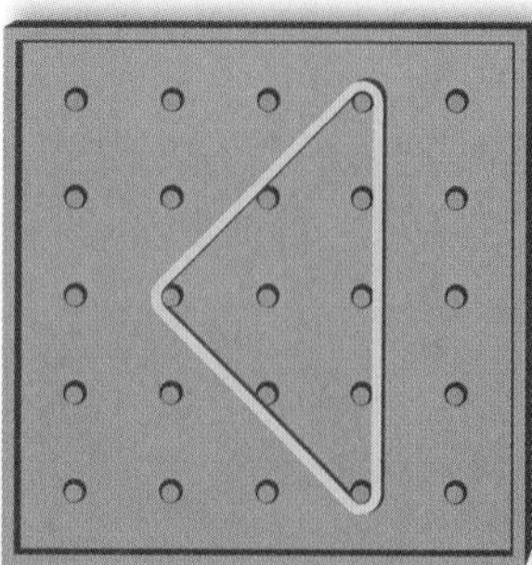

Geometry
In this lesson, you will
- name corresponding angles and corresponding sides of congruent figures.
- identify congruent figures.

Math Practice

Recognize Usefulness of Tools

What are some advantages and disadvantages of using a geoboard to construct congruent triangles?

The geoboard at the right shows three congruent triangles.

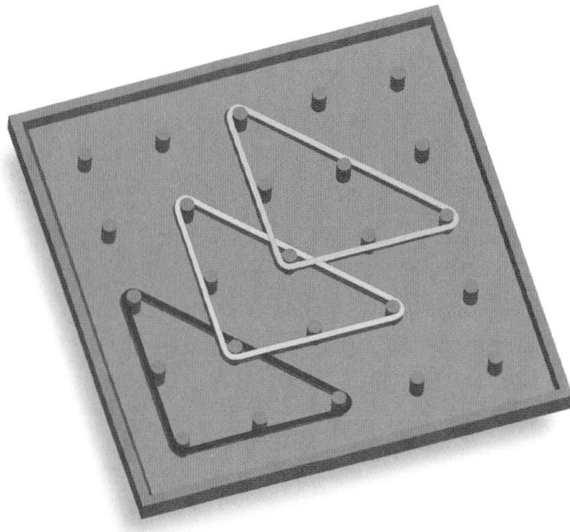

2 ACTIVITY: Forming Congruent Triangles

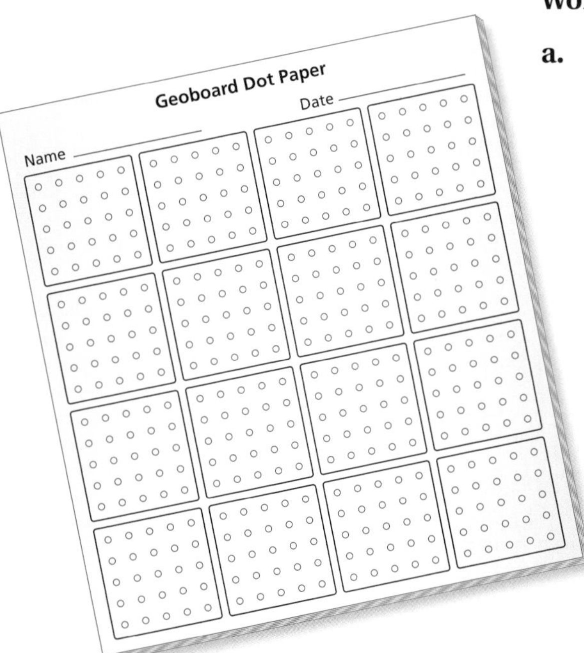

Work with a partner.

a. Form the yellow triangle in Activity 1 on your geoboard. Record the triangle on geoboard dot paper.

b. Move each vertex of the triangle one peg to the right. Is the new triangle congruent to the original triangle? How can you tell?

c. On a 5-by-5 geoboard, make as many different triangles as possible, each of which is congruent to the yellow triangle in Activity 1. Record each triangle on geoboard dot paper.

What Is Your Answer?

3. **IN YOUR OWN WORDS** How can you identify congruent triangles? Use the conclusion you wrote in Activity 1 as part of your answer.

4. Can you form a triangle on your geoboard whose side lengths are 3, 4, and 5 units? If so, draw such a triangle on geoboard dot paper.

Practice Use what you learned about congruent triangles to complete Exercises 4 and 5 on page 46.

Key Vocabulary
congruent figures, *p. 44*
corresponding angles, *p. 44*
corresponding sides, *p. 44*

Key Idea

Congruent Figures

Figures that have the same size and the same shape are called **congruent figures**. The triangles below are congruent.

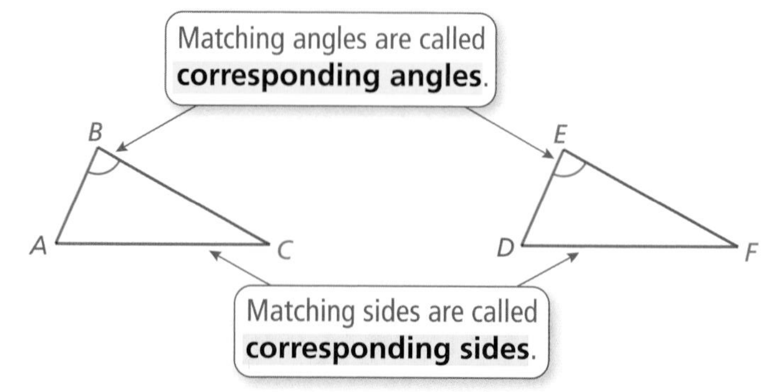

EXAMPLE 1 Naming Corresponding Parts

The figures are congruent. Name the corresponding angles and the corresponding sides.

Corresponding Angles	*Corresponding Sides*
$\angle A$ and $\angle W$	Side AB and Side WX
$\angle B$ and $\angle X$	Side BC and Side XY
$\angle C$ and $\angle Y$	Side CD and Side YZ
$\angle D$ and $\angle Z$	Side AD and Side WZ

On Your Own

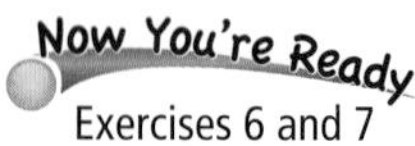

1. The figures are congruent. Name the corresponding angles and the corresponding sides.

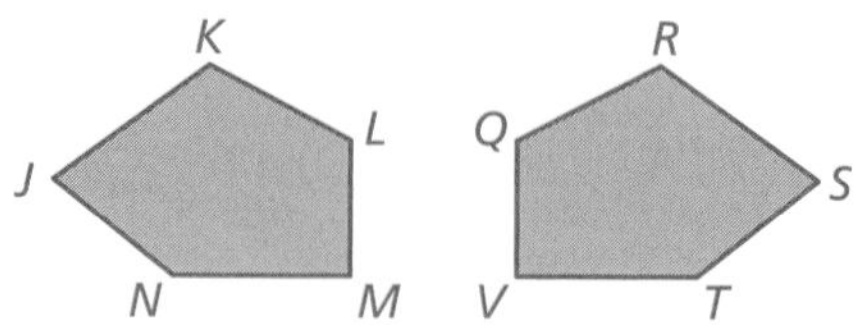

Key Idea

Identifying Congruent Figures

Two figures are congruent when corresponding angles and corresponding sides are congruent.

Triangle ABC is congruent to Triangle DEF.

$\triangle ABC \cong \triangle DEF$

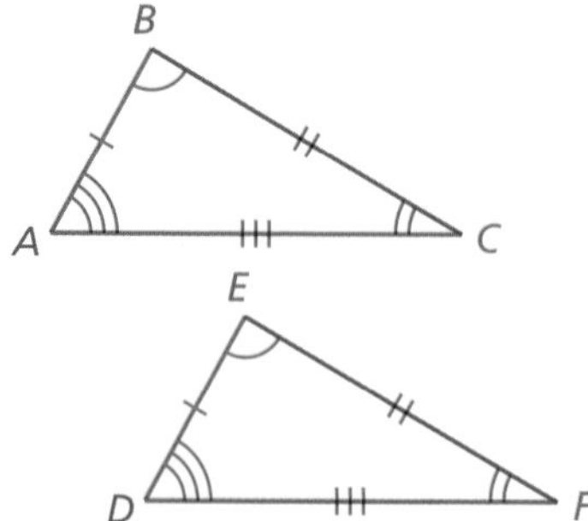

The symbol ≅ means *is congruent to.*

EXAMPLE 2 Identifying Congruent Figures

Which square is congruent to Square A?

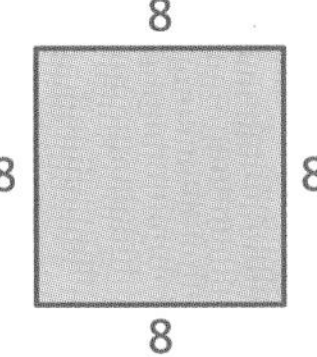

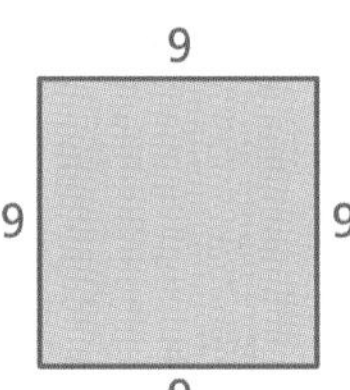

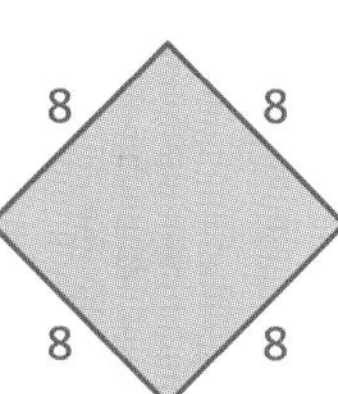

Each square has four right angles. So, corresponding angles are congruent. Check to see if corresponding sides are congruent.

Square A and Square B

Each side length of Square A is 8, and each side length of Square B is 9. So, corresponding sides are not congruent.

Square A and Square C

Each side length of Square A and Square C is 8. So, corresponding sides are congruent.

So, Square C is congruent to Square A.

EXAMPLE 3 Using Congruent Figures

Trapezoids *ABCD* and *JKLM* are congruent.

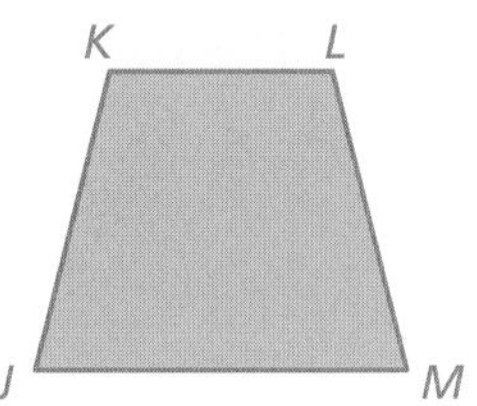

a. What is the length of side *JM*?

Side *JM* corresponds to side *AD*.

So, the length of side *JM* is 10 feet.

b. What is the perimeter of *JKLM*?

The perimeter of *ABCD* is $10 + 8 + 6 + 8 = 32$ feet. Because the trapezoids are congruent, their corresponding sides are congruent.

So, the perimeter of *JKLM* is also 32 feet.

On Your Own

Now You're Ready
Exercises 8, 9, and 12

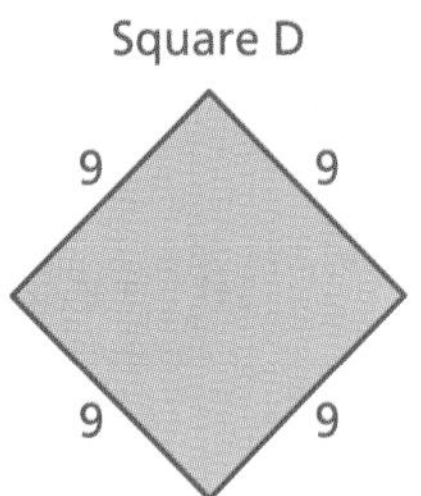

2. Which square in Example 2 is congruent to Square D?
3. In Example 3, which angle of *JKLM* corresponds to $\angle C$? What is the length of side *KJ*?

2.1 Exercises

Vocabulary and Concept Check

1. **VOCABULARY** $\triangle ABC$ is congruent to $\triangle DEF$.

   a. Identify the corresponding angles.

   b. Identify the corresponding sides.

2. **VOCABULARY** Explain how you can tell that two figures are congruent.

3. **WHICH ONE DOESN'T BELONG?** Which one does *not* belong with the other three? Explain your reasoning.

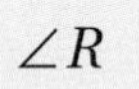

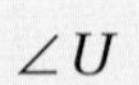

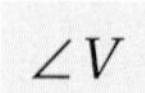

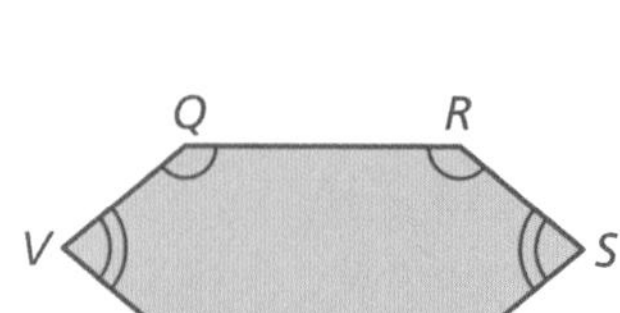

Practice and Problem Solving

Tell whether the triangles are *congruent* or *not congruent*.

4. 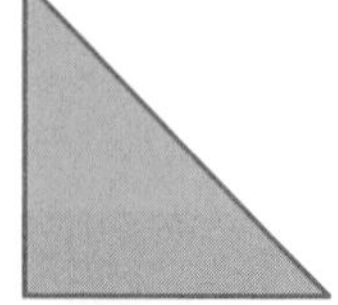 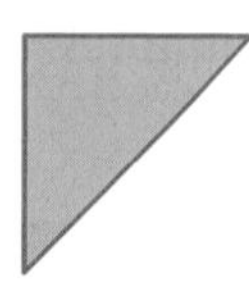

5. 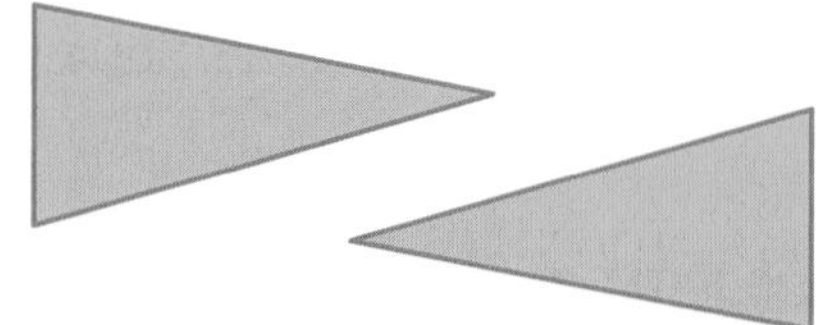

The figures are congruent. Name the corresponding angles and the corresponding sides.

1 6. 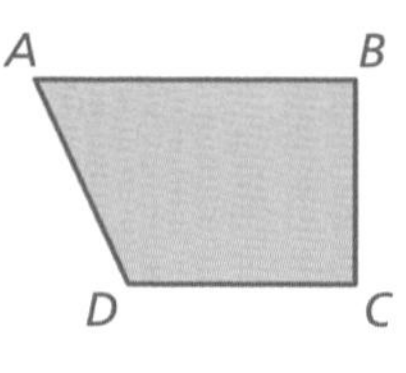

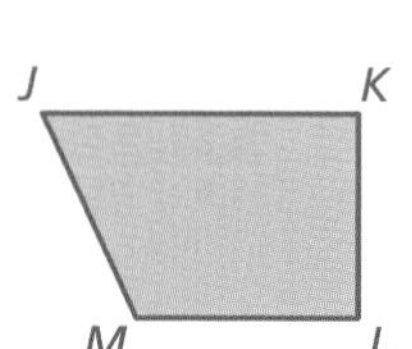

7. 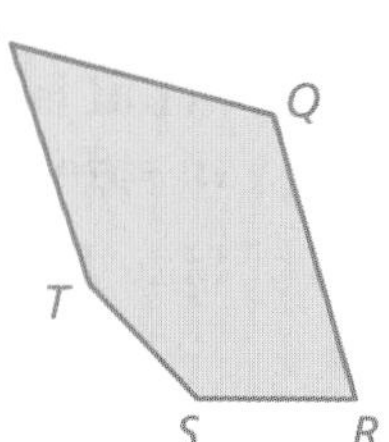

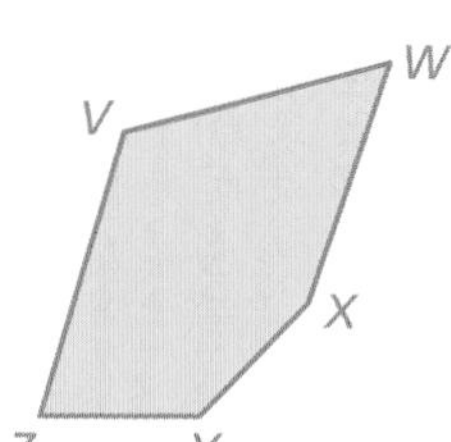

Tell whether the two figures are congruent. Explain your reasoning.

2 8. 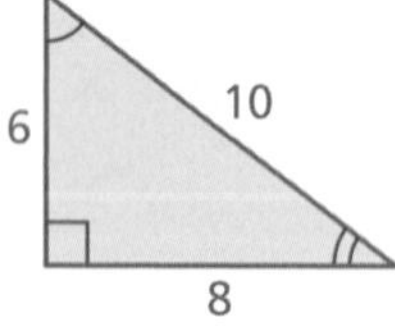

9. 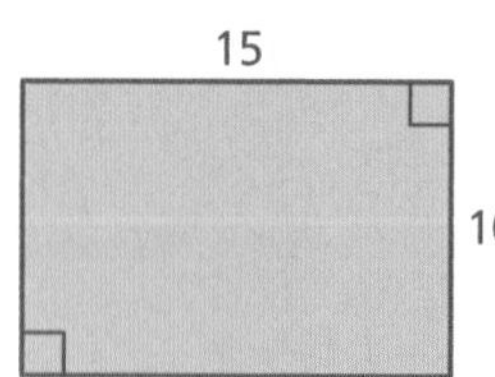

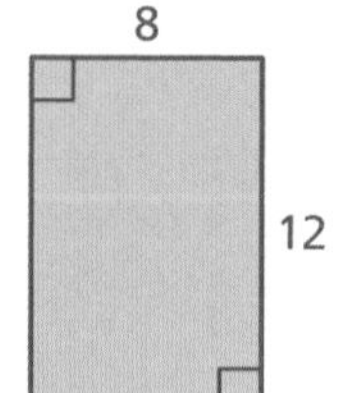

10. **PUZZLE** Describe the relationship between the unfinished puzzle and the missing piece.

11. **ERROR ANALYSIS** Describe and correct the error in telling whether the two figures are congruent.

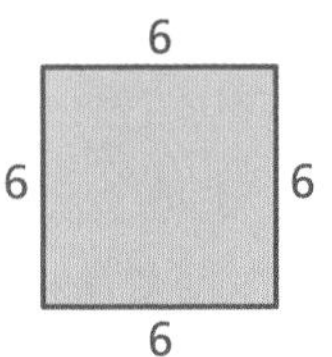

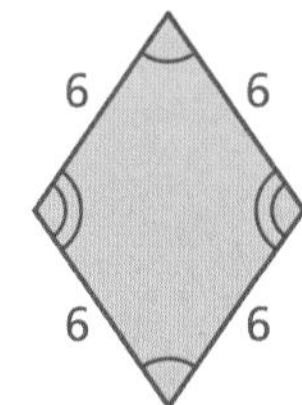

Both figures have four sides, and the corresponding side lengths are equal. So, they are congruent.

3 12. **HOUSES** The fronts of the houses are identical.

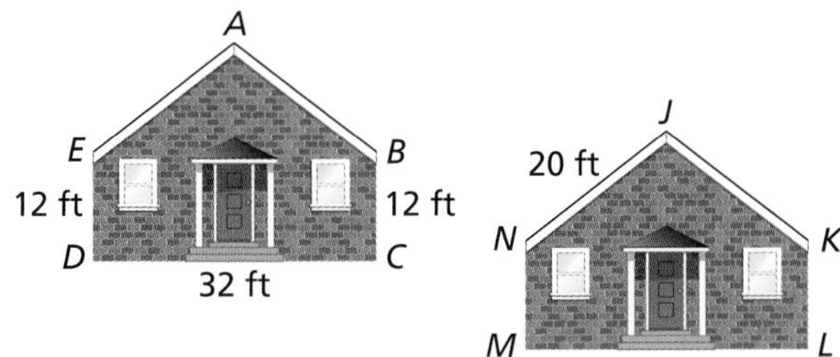

a. What is the length of side LM?

b. Which angle of $JKLMN$ corresponds to $\angle D$?

c. Side AB is congruent to side AE. What is the length of side AB?

d. What is the perimeter of $ABCDE$?

13. **REASONING** Here are two ways to draw *one* line to divide a rectangle into two congruent figures. Draw three other ways.

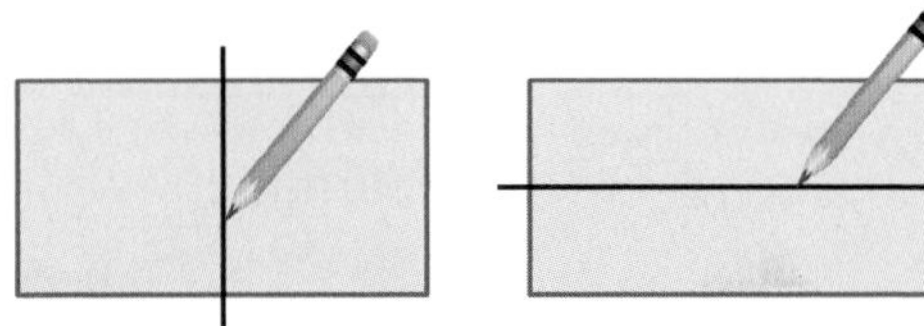

14. **CRITICAL THINKING** Are the areas of two congruent figures equal? Explain. Draw a diagram to support your answer.

15. True or False? The trapezoids are congruent. Determine whether the statement is *true* or *false*. Explain your reasoning.

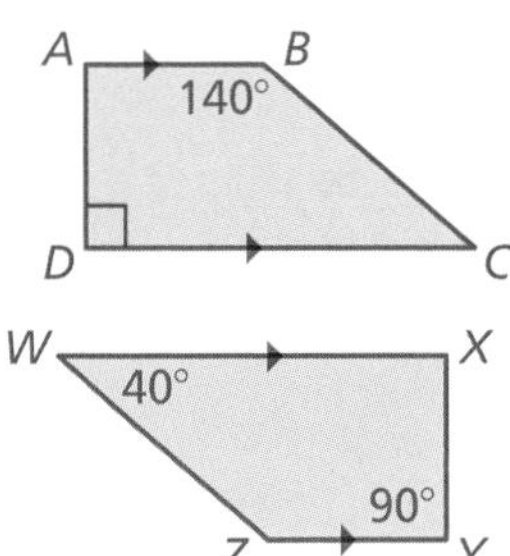

a. Side AB is congruent to side YZ.

b. $\angle A$ is congruent to $\angle X$.

c. $\angle A$ corresponds to $\angle X$.

d. The sum of the angle measures of $ABCD$ is 360°.

Fair Game Review What you learned in previous grades & lessons

Plot and label the ordered pair in a coordinate plane. *(Skills Review Handbook)*

16. $A(5, 3)$ 17. $B(4, -1)$ 18. $C(-2, 6)$ 19. $D(-4, -2)$

20. **MULTIPLE CHOICE** You have 2 quarters and 5 dimes in your pocket. Write the ratio of quarters to the total number of coins. *(Skills Review Handbook)*

(A) $\frac{2}{5}$ (B) 2 : 7 (C) 5 to 7 (D) $\frac{7}{2}$

2.2 Translations

Essential Question How can you arrange tiles to make a tessellation?

The Meaning of a Word ● Translate

When you **translate** a tile, you slide it from one place to another.

When tiles cover a floor with no empty spaces, the collection of tiles is called a *tessellation*.

1 ACTIVITY: Describing Tessellations

Work with a partner. Can you make the tessellation by translating single tiles that are all of the same shape and design? If so, show how.

a. Sample:

Tile Pattern

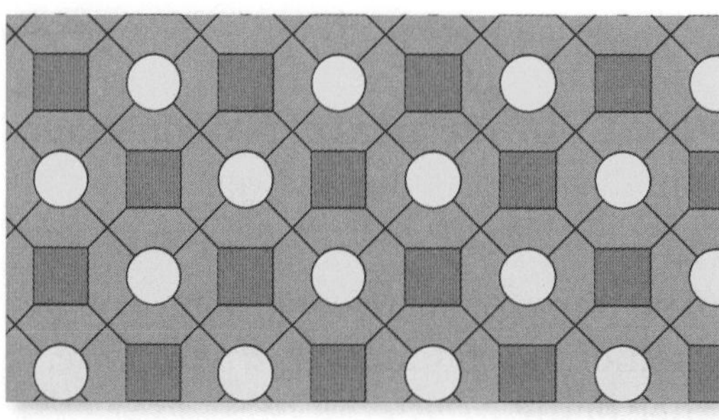

Single Tiles

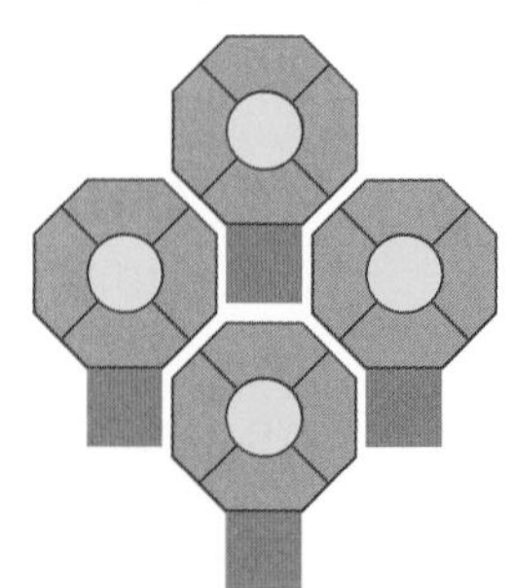

b.

c.

Geometry
In this lesson, you will
- identify translations.
- translate figures in the coordinate plane.

2 ACTIVITY: Tessellations and Basic Shapes

Work with a partner.

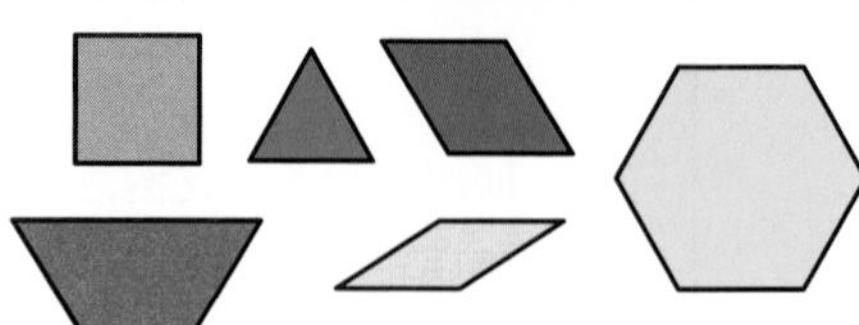

a. Which pattern blocks can you use to make a tessellation? For each one that works, draw the tessellation.

b. Can you make the tessellation by translating? Or do you have to rotate or flip the pattern blocks?

3 ACTIVITY: Designing Tessellations

Work with a partner. Design your own tessellation. Use one of the basic shapes from Activity 2.

Sample:

Step 1: Start with a square.

Step 2: Cut a design out of one side.

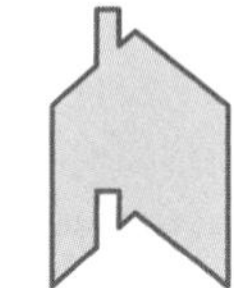

Step 3: Tape it to the other side to make your pattern.

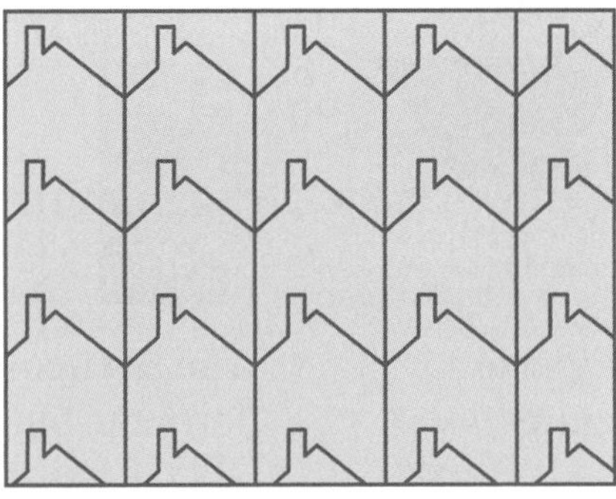

Step 4: Translate the pattern to make your tessellation.

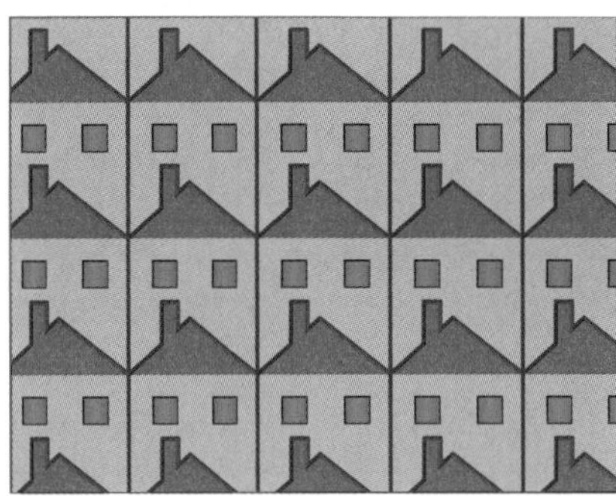

Step 5: Color the tessellation.

4 ACTIVITY: Translating in the Coordinate Plane

Work with a partner.

a. Draw a rectangle in a coordinate plane. Find the dimensions of the rectangle.

b. Move each vertex 3 units right and 4 units up. Draw the new figure. List the vertices.

c. Compare the dimensions and the angle measures of the new figure to those of the original rectangle.

d. Are the opposite sides of the new figure still parallel? Explain.

e. Can you conclude that the two figures are congruent? Explain.

f. Compare your results with those of other students in your class. Do you think the results are true for any type of figure?

Math Practice

Justify Conclusions

What information do you need to conclude that two figures are congruent?

What Is Your Answer?

5. **IN YOUR OWN WORDS** How can you arrange tiles to make a tessellation? Give an example.

6. **PRECISION** Explain why any parallelogram can be translated to make a tessellation.

Practice Use what you learned about translations to complete Exercises 4–6 on page 52.

2.2 Lesson

Key Vocabulary
transformation, *p. 50*
image, *p. 50*
translation, *p. 50*

A **transformation** changes a figure into another figure. The new figure is called the **image**.

A **translation** is a transformation in which a figure *slides* but does not turn. Every point of the figure moves the same distance and in the same direction.

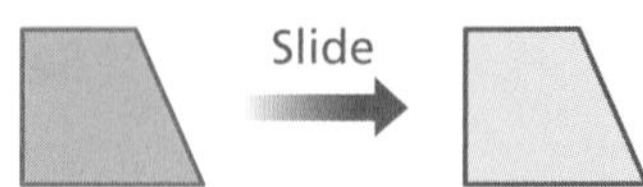

EXAMPLE 1 Identifying a Translation

Tell whether the blue figure is a translation of the red figure.

a.

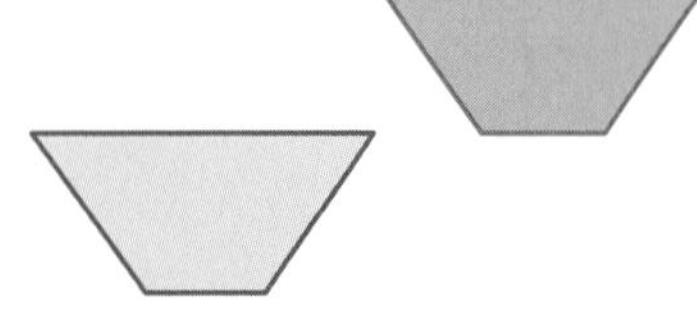

The red figure *slides* to form the blue figure.

So, the blue figure is a translation of the red figure.

b.

The red figure *turns* to form the blue figure.

So, the blue figure is *not* a translation of the red figure.

On Your Own

Now You're Ready
Exercises 4–9

Tell whether the blue figure is a translation of the red figure. Explain.

1.

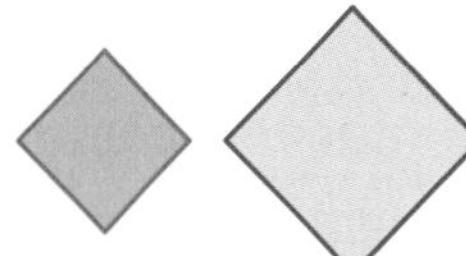

2.

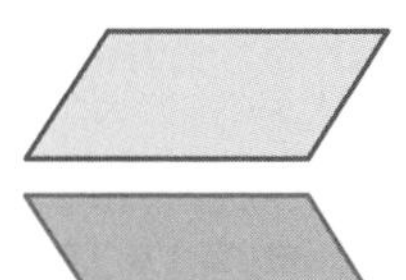

3.

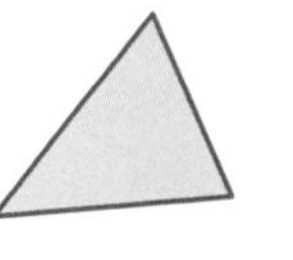

Key Idea

Translations in the Coordinate Plane

Words To translate a figure a units horizontally and b units vertically in a coordinate plane, add a to the x-coordinates and b to the y-coordinates of the vertices.

Positive values of a and b represent translations up and right. Negative values of a and b represent translations down and left.

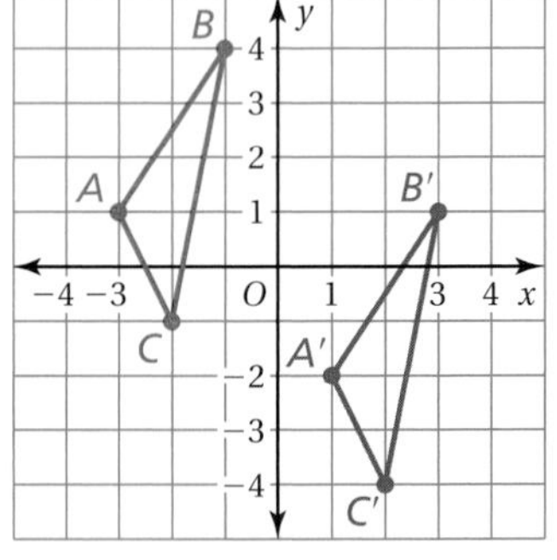

Algebra $(x, y) \rightarrow (x + a, y + b)$

In a translation, the original figure and its image are congruent.

Reading

A' is read "A prime." Use *prime* symbols when naming an image.

$A \rightarrow A'$
$B \rightarrow B'$
$C \rightarrow C'$

EXAMPLE 2 Translating a Figure in the Coordinate Plane

Translate the red triangle 3 units right and 3 units down. What are the coordinates of the image?

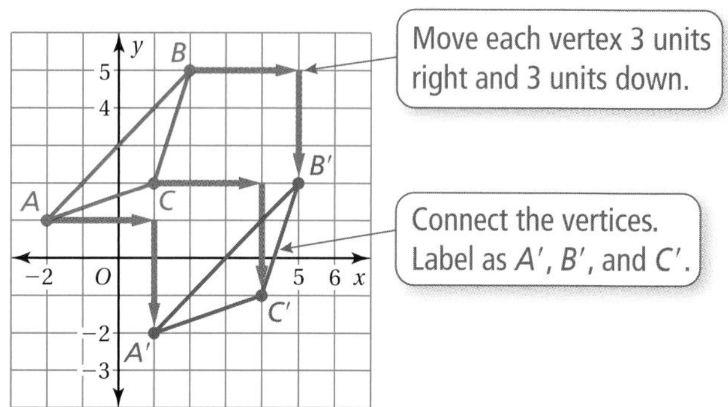

∴ The coordinates of the image are $A'(1, -2)$, $B'(5, 2)$, and $C'(4, -1)$.

On Your Own

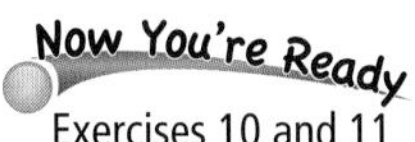

Now You're Ready
Exercises 10 and 11

4. **WHAT IF?** The red triangle is translated 4 units left and 2 units up. What are the coordinates of the image?

EXAMPLE 3 Translating a Figure Using Coordinates

The vertices of a square are $A(1, -2)$, $B(3, -2)$, $C(3, -4)$, and $D(1, -4)$. Draw the figure and its image after a translation 4 units left and 6 units up.

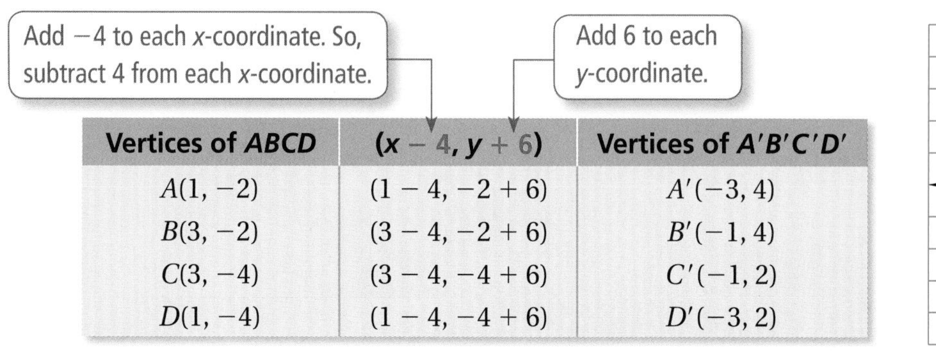

Add −4 to each x-coordinate. So, subtract 4 from each x-coordinate.

Vertices of ABCD	$(x - 4, y + 6)$	Vertices of A′B′C′D′
$A(1, -2)$	$(1 - 4, -2 + 6)$	$A'(-3, 4)$
$B(3, -2)$	$(3 - 4, -2 + 6)$	$B'(-1, 4)$
$C(3, -4)$	$(3 - 4, -4 + 6)$	$C'(-1, 2)$
$D(1, -4)$	$(1 - 4, -4 + 6)$	$D'(-3, 2)$

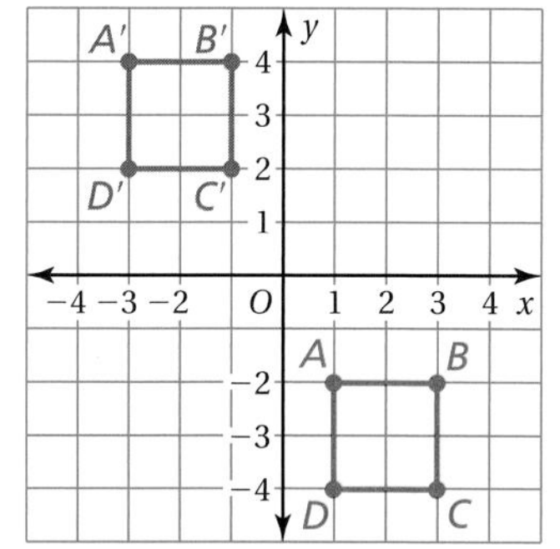

∴ The figure and its image are shown at the above right.

On Your Own

5. The vertices of a triangle are $A(-2, -2)$, $B(0, 2)$, and $C(3, 0)$. Draw the figure and its image after a translation 1 unit left and 2 units up.

2.2 Exercises

Vocabulary and Concept Check

1. **VOCABULARY** Which figure is the image?
2. **VOCABULARY** How do you translate a figure in a coordinate plane?

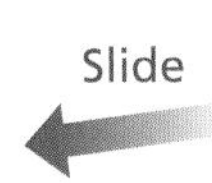

3. **WRITING** Can you translate the letters in the word TOKYO to form the word KYOTO? Explain.

Practice and Problem Solving

Tell whether the blue figure is a translation of the red figure.

① 4. 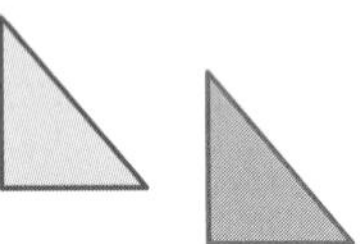

5. 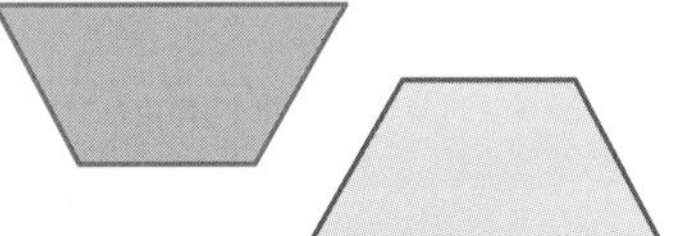

6. 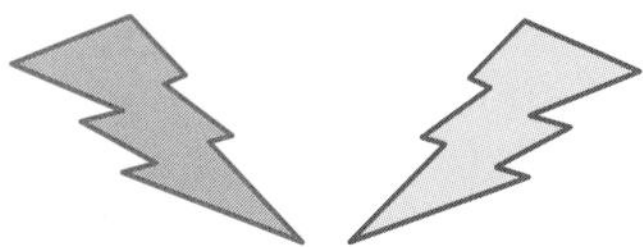

7. 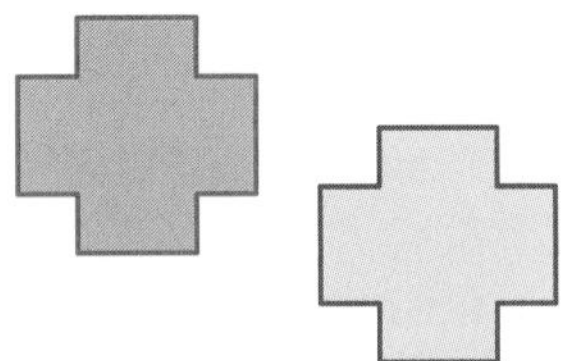

8. 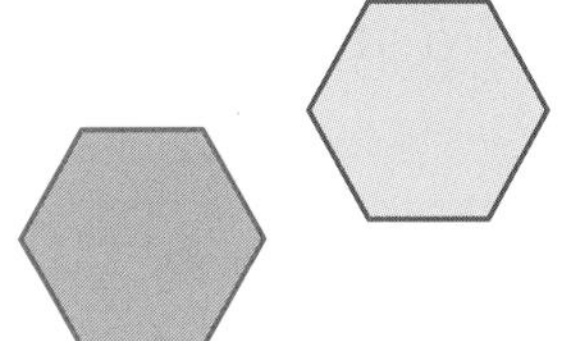

9. 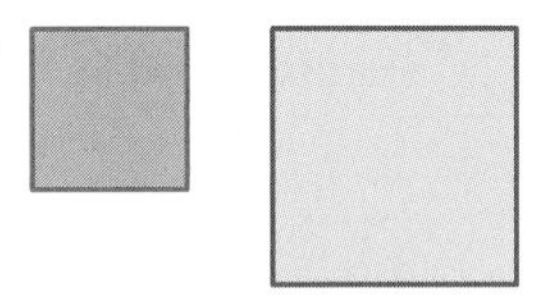

② 10. Translate the triangle 4 units right and 3 units down. What are the coordinates of the image?

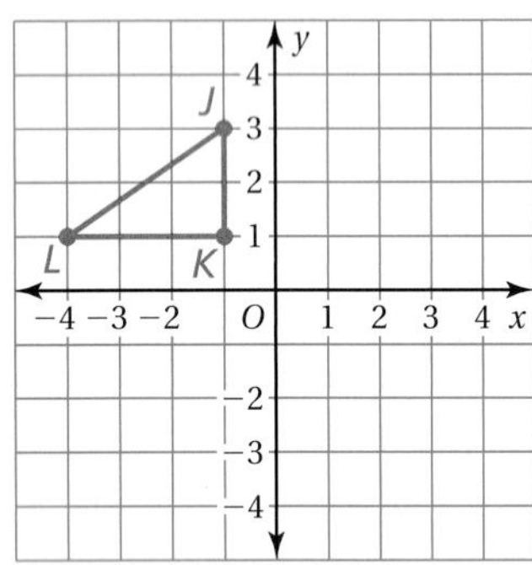

11. Translate the figure 2 units left and 4 units down. What are the coordinates of the image?

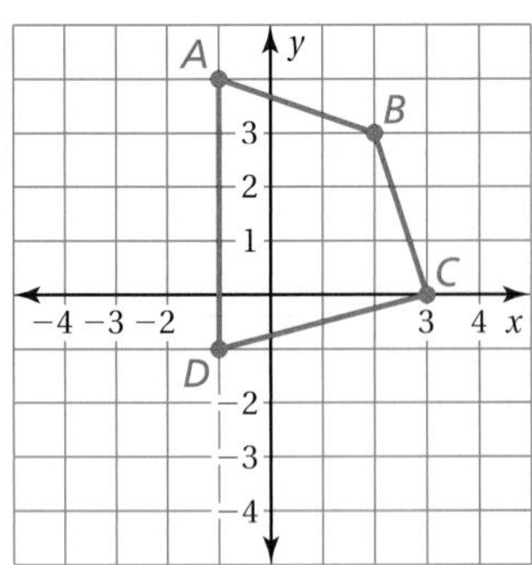

The vertices of a triangle are $L(0, 1)$, $M(1, -2)$, and $N(-2, 1)$. Draw the figure and its image after the translation.

③ 12. 1 unit left and 6 units up

13. 5 units right

14. $(x + 2, y + 3)$

15. $(x - 3, y - 4)$

16. **ICONS** You can click and drag an icon on a computer screen. Is this an example of a translation? Explain.

Describe the translation of the point to its image.

17. $(3, -2) \rightarrow (1, 0)$

18. $(-8, -4) \rightarrow (-3, 5)$

Describe the translation from the red figure to the blue figure.

19.

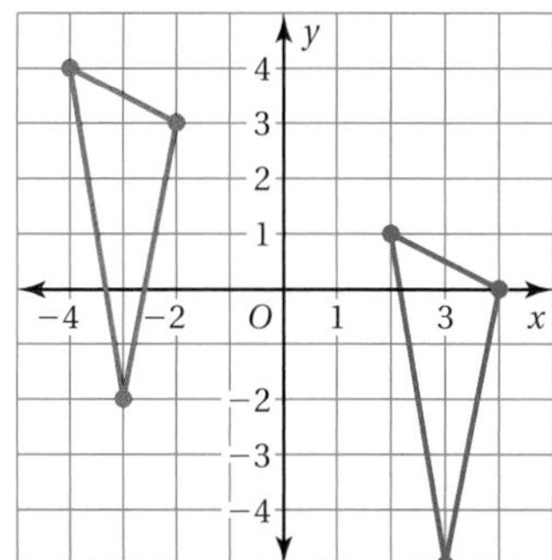

20.

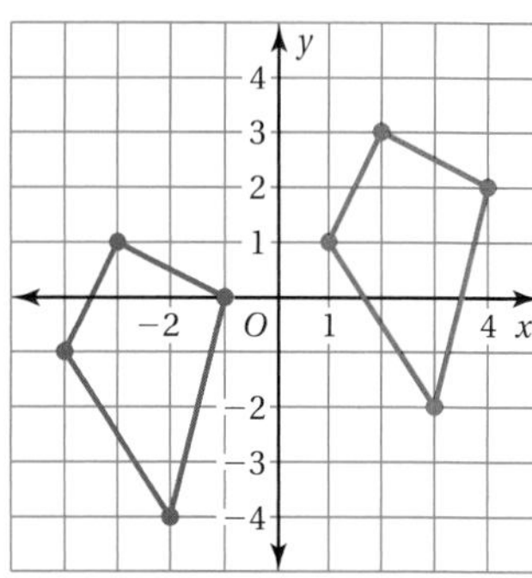

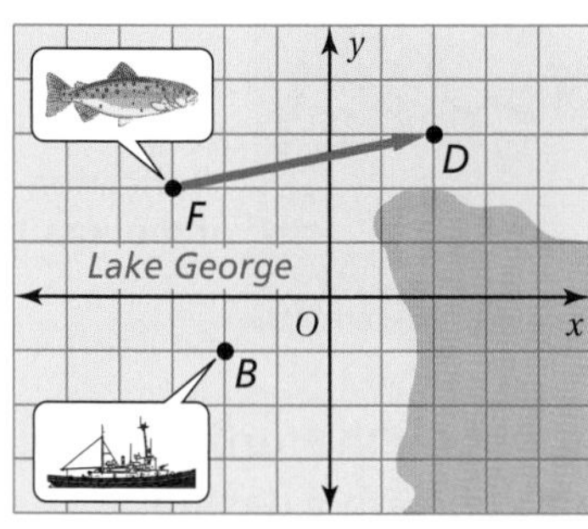

21. FISHING A school of fish translates from point F to point D.

a. Describe the translation of the school of fish.

b. Can the fishing boat make the same translation? Explain.

c. Describe a translation the fishing boat could make to get to point D.

22. REASONING The vertices of a triangle are $A(0, -3)$, $B(2, -1)$, and $C(3, -3)$. You translate the triangle 5 units right and 2 units down. Then you translate the image 3 units left and 8 units down. Is the original triangle congruent to the final image? If so, give two ways to show that they are congruent.

23. Problem Solving In chess, a knight can move only in an L-shaped pattern:

- *two* vertical squares, then *one* horizontal square;
- *two* horizontal squares, then *one* vertical square;
- *one* vertical square, then *two* horizontal squares; or
- *one* horizontal square, then *two* vertical squares.

Write a series of translations to move the knight from g8 to g5.

Fair Game Review What you learned in previous grades & lessons

Tell whether you can fold the figure in half so that one side matches the other. *(Skills Review Handbook)*

24. 

25. 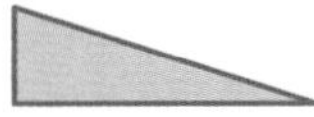

26. 

27. 

28. MULTIPLE CHOICE You put $550 in an account that earns 4.4% simple interest per year. How much interest do you earn in 6 months? *(Skills Review Handbook)*

Ⓐ $1.21 Ⓑ $12.10 Ⓒ $121.00 Ⓓ $145.20

2.3 Reflections

Essential Question How can you use reflections to classify a frieze pattern?

The Meaning of a Word ● Reflection

When you look at a mountain by a lake, you can see the **reflection**, or mirror image, of the mountain in the lake.

If you fold the photo on its axis, the mountain and its reflection will align.

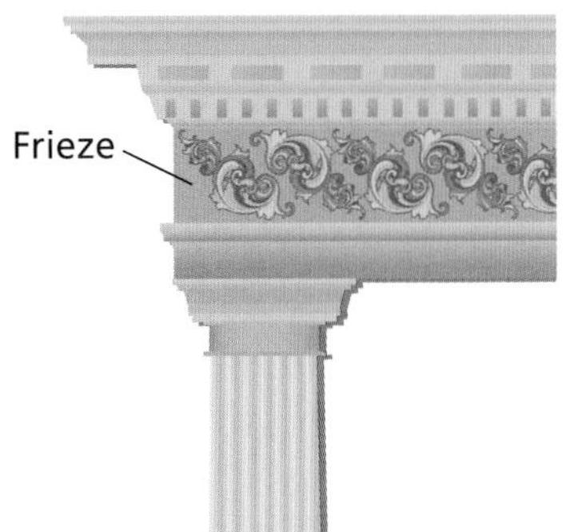

A *frieze* is a horizontal band that runs at the top of a building. A frieze is often decorated with a design that repeats.

- All frieze patterns are translations of themselves.
- Some frieze patterns are reflections of themselves.

1 ACTIVITY: Frieze Patterns and Reflections

Work with a partner. Consider the frieze pattern shown.

a. Is the frieze pattern a reflection of itself when folded horizontally? Explain.

b. Is the frieze pattern a reflection of itself when folded vertically? Explain.

Geometry
In this lesson, you will
- identify reflections.
- reflect figures in the *x*-axis or the *y*-axis of the coordinate plane.

2 ACTIVITY: Frieze Patterns and Reflections

Work with a partner. Is the frieze pattern a reflection of itself when folded *horizontally*, *vertically*, or *neither*?

a.

b.

3 ACTIVITY: Reflecting in the Coordinate Plane

Work with a partner.

a. Draw a rectangle in Quadrant I of a coordinate plane. Find the dimensions of the rectangle.

b. Copy the axes and the rectangle onto a piece of transparent paper.

Flip the transparent paper once so that the rectangle is in Quadrant IV. Then align the origin and the axes with the coordinate plane.

Draw the new figure in the coordinate plane. List the vertices.

c. Compare the dimensions and the angle measures of the new figure to those of the original rectangle.

d. Are the opposite sides of the new figure still parallel? Explain.

e. Can you conclude that the two figures are congruent? Explain.

f. Flip the transparent paper so that the original rectangle is in Quadrant II. Draw the new figure in the coordinate plane. List the vertices. Then repeat parts (c)–(e).

g. Compare your results with those of other students in your class. Do you think the results are true for any type of figure?

Math Practice

Look for Patterns

What do you notice about the vertices of the original figure and the image? How does this help you determine whether the figures are congruent?

What Is Your Answer?

4. **IN YOUR OWN WORDS** How can you use reflections to classify a frieze pattern?

Practice

Use what you learned about reflections to complete Exercises 4–6 on page 58.

2.3 Lesson

Key Vocabulary
reflection, *p. 56*
line of reflection, *p. 56*

A **reflection**, or *flip*, is a transformation in which a figure is reflected in a line called the **line of reflection**. A reflection creates a mirror image of the original figure.

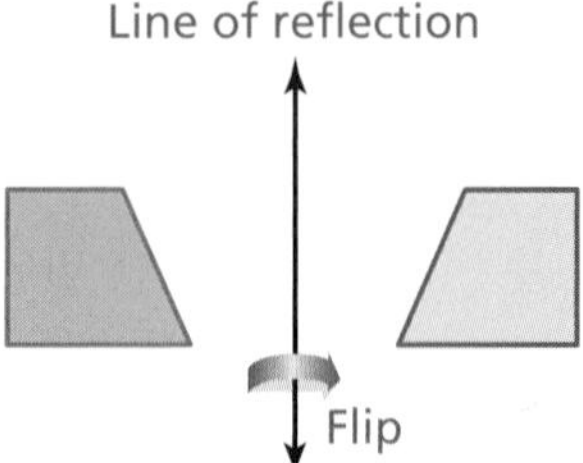

EXAMPLE 1 Identifying a Reflection

Tell whether the blue figure is a reflection of the red figure.

a.

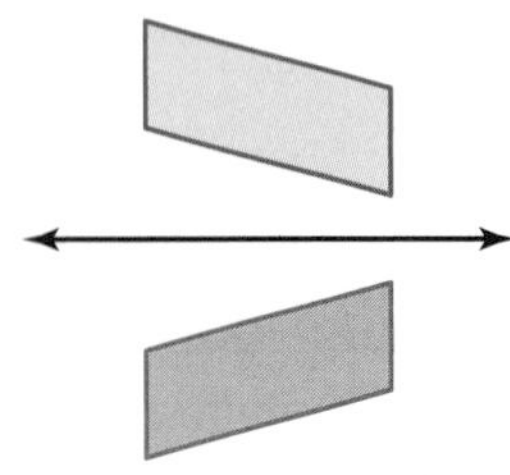

The red figure can be *flipped* to form the blue figure.

So, the blue figure is a reflection of the red figure.

b.

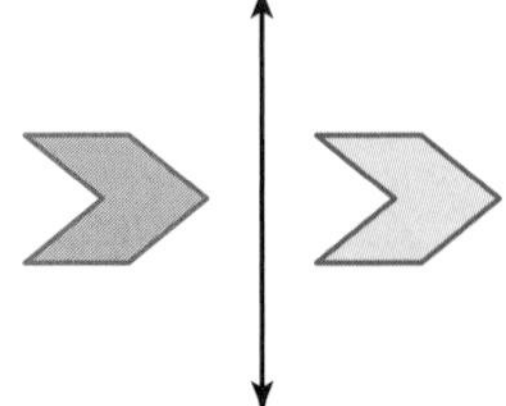

If the red figure were *flipped,* it would point to the left.

So, the blue figure is *not* a reflection of the red figure.

On Your Own

Now You're Ready
Exercises 4–9

Tell whether the blue figure is a reflection of the red figure. Explain.

1.

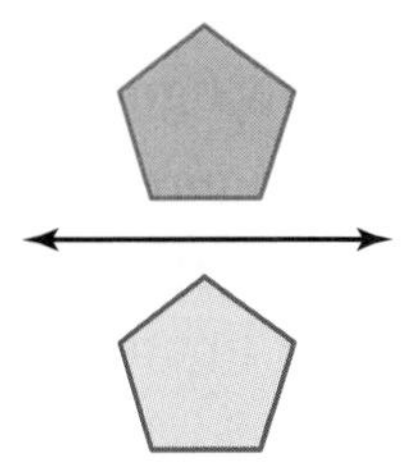

2.

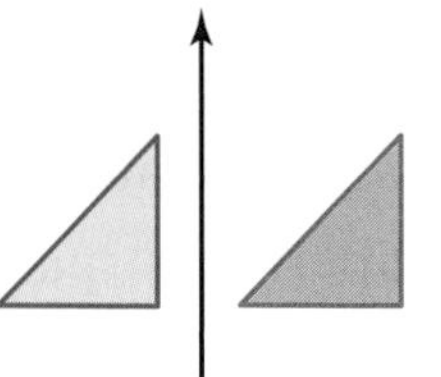

3.

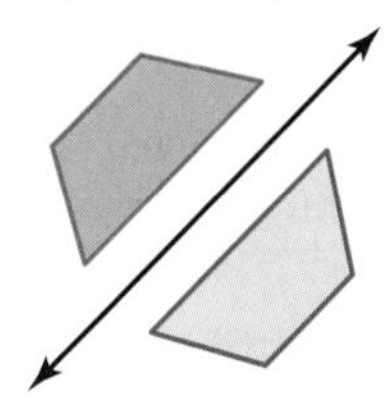

Reflections in the Coordinate Plane

Words To reflect a figure in the x-axis, take the opposite of the y-coordinate.

To reflect a figure in the y-axis, take the opposite of the x-coordinate.

Algebra Reflection in x-axis: $(x, y) \rightarrow (x, -y)$
Reflection in y-axis: $(x, y) \rightarrow (-x, y)$

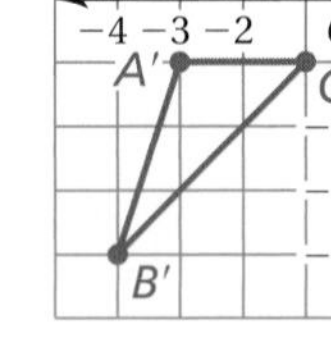

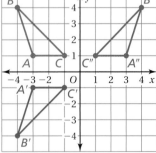

In a reflection, the original figure and its image are congruent.

Multi-Language Glossary at BigIdeasMath.com

EXAMPLE 2 Reflecting a Figure in the *x*-axis

The vertices of a triangle are $A(-1, 1)$, $B(-1, 3)$, and $C(6, 3)$. Draw the figure and its reflection in the x-axis. What are the coordinates of the image?

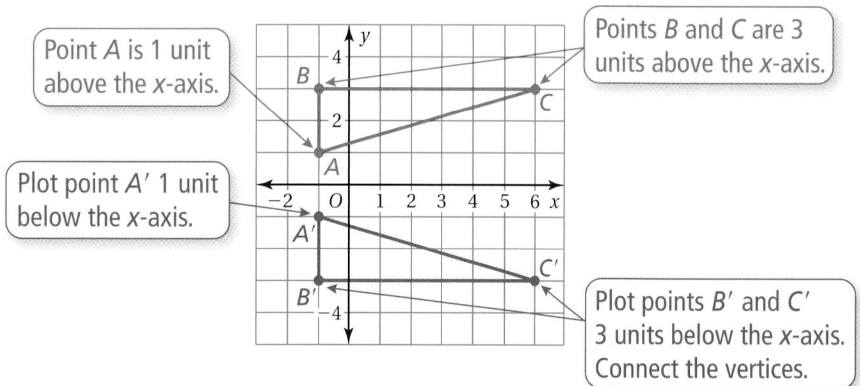

Point A is 1 unit above the x-axis.
Points B and C are 3 units above the x-axis.
Plot point A′ 1 unit below the x-axis.
Plot points B′ and C′ 3 units below the x-axis. Connect the vertices.

The coordinates of the image are $A'(-1, -1)$, $B'(-1, -3)$, and $C'(6, -3)$.

EXAMPLE 3 Reflecting a Figure in the *y*-axis

The vertices of a quadrilateral are $P(-2, 5)$, $Q(-1, -1)$, $R(-4, 2)$, and $S(-4, 4)$. Draw the figure and its reflection in the y-axis.

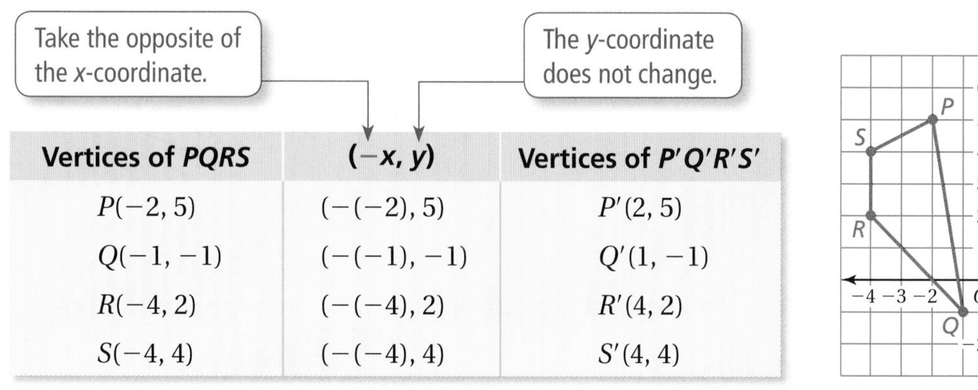

Take the opposite of the x-coordinate.
The y-coordinate does not change.

Vertices of $PQRS$	$(-x, y)$	Vertices of $P'Q'R'S'$
$P(-2, 5)$	$(-(-2), 5)$	$P'(2, 5)$
$Q(-1, -1)$	$(-(-1), -1)$	$Q'(1, -1)$
$R(-4, 2)$	$(-(-4), 2)$	$R'(4, 2)$
$S(-4, 4)$	$(-(-4), 4)$	$S'(4, 4)$

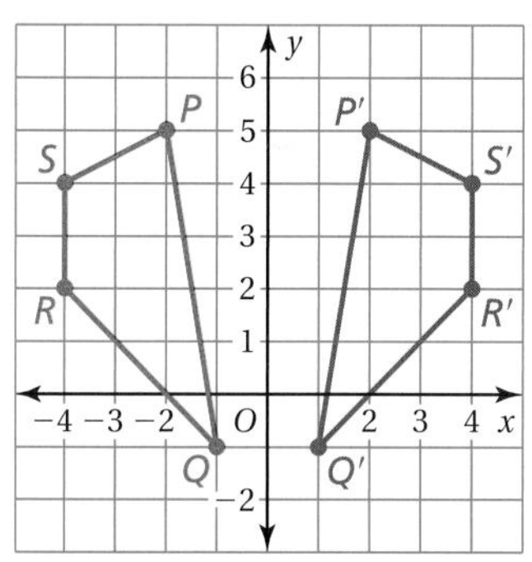

The figure and its image are shown at the above right.

On Your Own

Now You're Ready
Exercises 10–17

4. The vertices of a rectangle are $A(-4, -3)$, $B(-4, -1)$, $C(-1, -1)$, and $D(-1, -3)$.
   a. Draw the figure and its reflection in the x-axis.
   b. Draw the figure and its reflection in the y-axis.
   c. Are the images in parts (a) and (b) congruent? Explain.

2.3 Exercises

Vocabulary and Concept Check

1. **WHICH ONE DOESN'T BELONG?** Which transformation does *not* belong with the other three? Explain your reasoning.

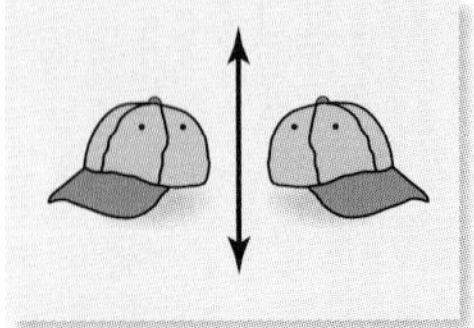 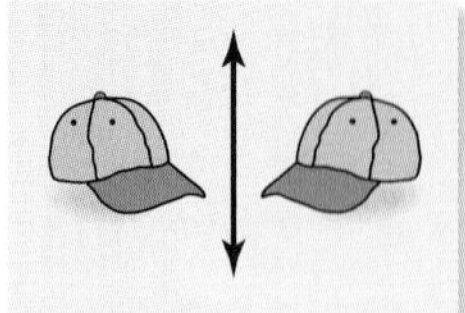 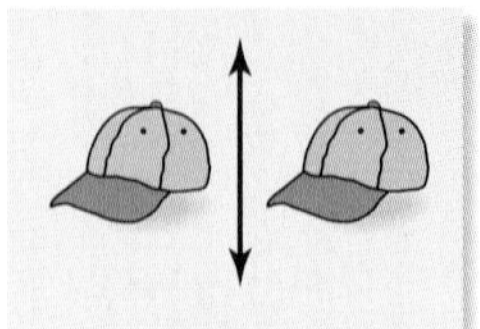 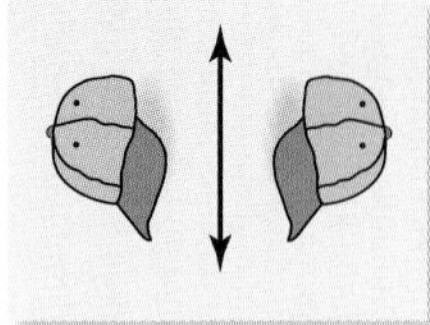

2. **WRITING** How can you tell when one figure is a reflection of another figure?

3. **REASONING** A figure lies entirely in Quadrant I. The figure is reflected in the x-axis. In which quadrant is the image?

Practice and Problem Solving

Tell whether the blue figure is a reflection of the red figure.

1 4. 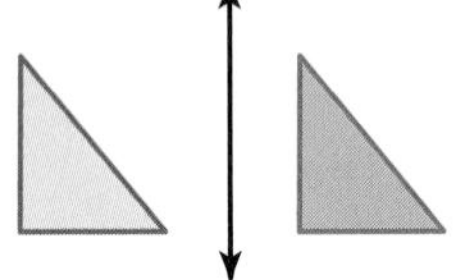

5. 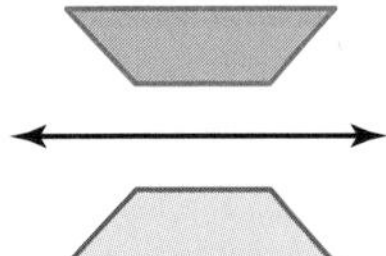

6. 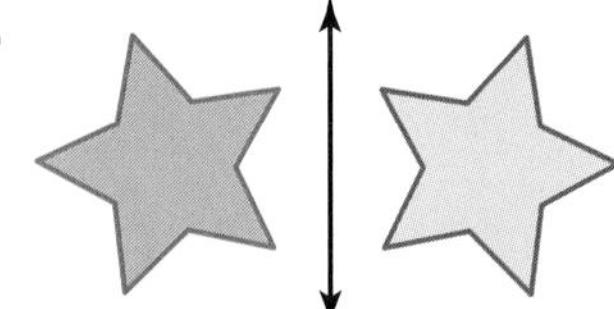

7. 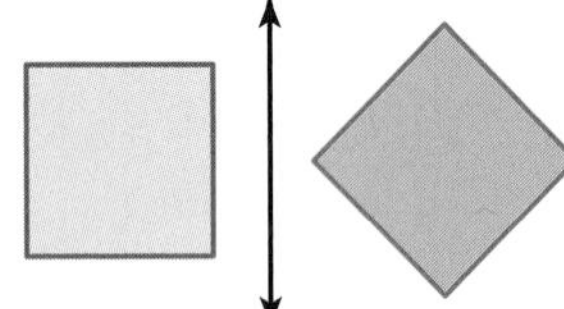

8. 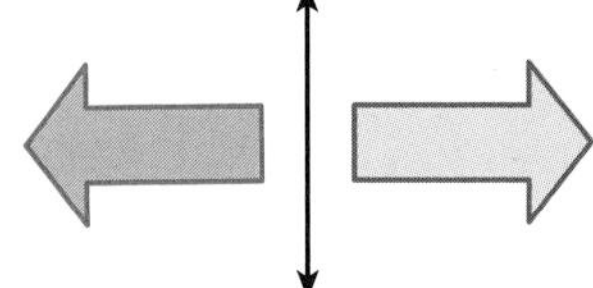

9. 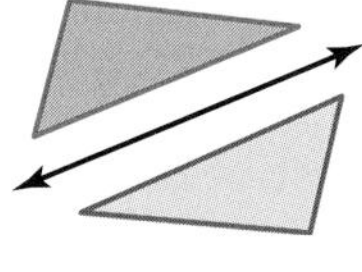

Draw the figure and its reflection in the x-axis. Identify the coordinates of the image.

2 10. $A(3, 2)$, $B(4, 4)$, $C(1, 3)$

11. $M(-2, 1)$, $N(0, 3)$, $P(2, 2)$

12. $H(2, -2)$, $J(4, -1)$, $K(6, -3)$, $L(5, -4)$

13. $D(-2, -1)$, $E(0, -1)$, $F(0, -5)$, $G(-2, -5)$

Draw the figure and its reflection in the y-axis. Identify the coordinates of the image.

3 14. $Q(-4, 2)$, $R(-2, 4)$, $S(-1, 1)$

15. $T(4, -2)$, $U(4, 2)$, $V(6, -2)$

16. $W(2, -1)$, $X(5, -2)$, $Y(5, -5)$, $Z(2, -4)$

17. $J(2, 2)$, $K(7, 4)$, $L(9, -2)$, $M(3, -1)$

18. **ALPHABET** Which letters look the same when reflected in the line?

A B C D E F G H I J K L M N O P Q R S T U V W X Y Z

The coordinates of a point and its image are given. Is the reflection in the *x-axis* or *y-axis*?

19. $(2, -2) \longrightarrow (2, 2)$

20. $(-4, 1) \longrightarrow (4, 1)$

21. $(-2, -5) \longrightarrow (2, -5)$

22. $(-3, -4) \longrightarrow (-3, 4)$

Find the coordinates of the figure after the transformations.

23. Translate the triangle 1 unit right and 5 units down. Then reflect the image in the y-axis.

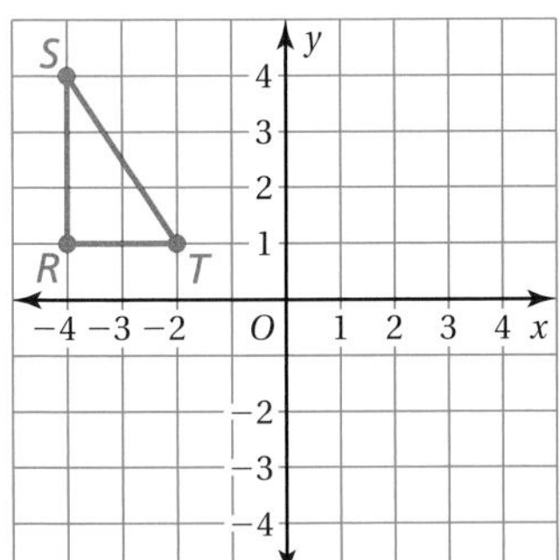

24. Reflect the trapezoid in the x-axis. Then translate the trapezoid 2 units left and 3 units up.

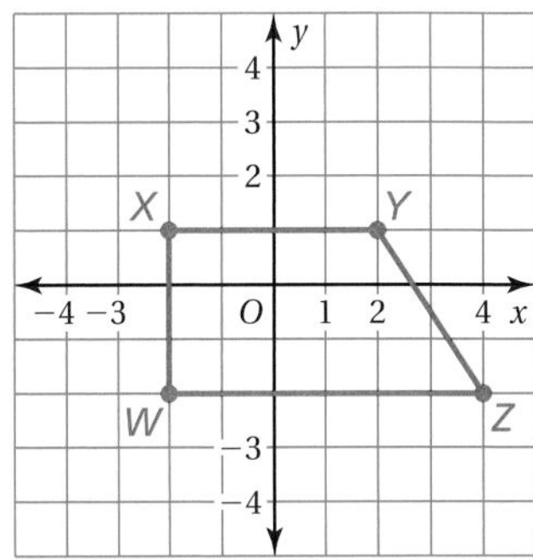

25. REASONING In Exercises 23 and 24, is the original figure congruent to the final image? Explain.

26. NUMBER SENSE You reflect a point (x, y) in the x-axis, and then in the y-axis. What are the coordinates of the final image?

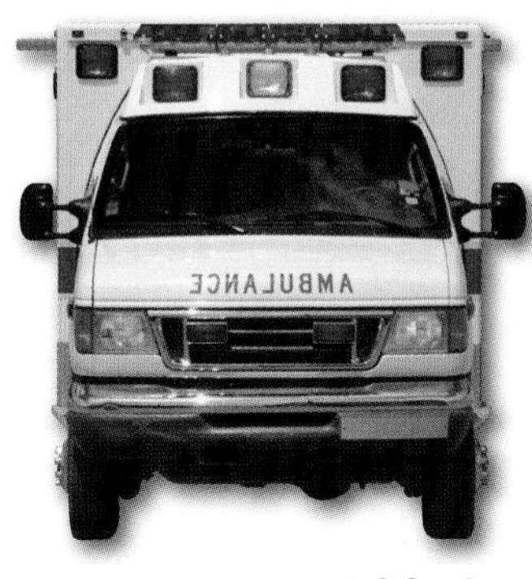

27. EMERGENCY VEHICLE Hold a mirror to the left side of the photo of the vehicle.

a. What word do you see in the mirror?

b. Why do you think it is written that way on the front of the vehicle?

28. Critical Thinking Reflect the triangle in the line $y = x$. How are the x- and y-coordinates of the image related to the x- and y-coordinates of the original triangle?

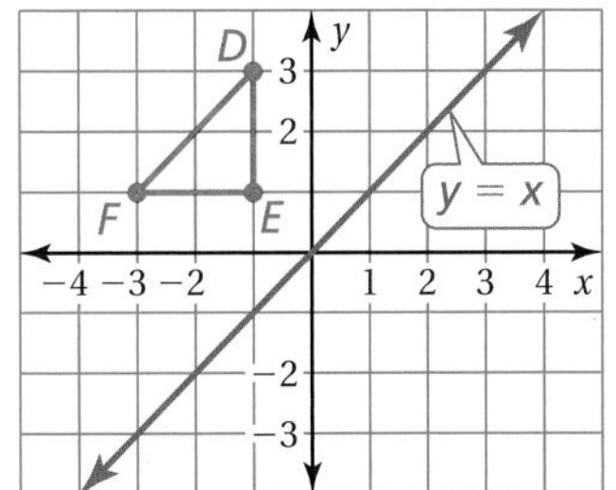

Fair Game Review What you learned in previous grades & lessons

Classify the angle as *acute, right, obtuse,* or *straight.* *(Skills Review Handbook)*

29.

30. 

31. 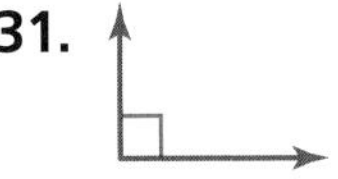

32. 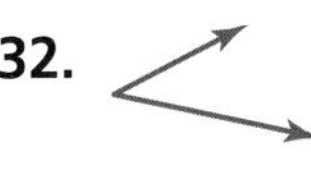

33. MULTIPLE CHOICE 36 is 75% of what number? *(Skills Review Handbook)*

Ⓐ 27 Ⓑ 48 Ⓒ 54 Ⓓ 63

2.4 Rotations

Essential Question What are the three basic ways to move an object in a plane?

The Meaning of a Word ● Rotate

A bicycle wheel can **rotate** clockwise or counterclockwise.

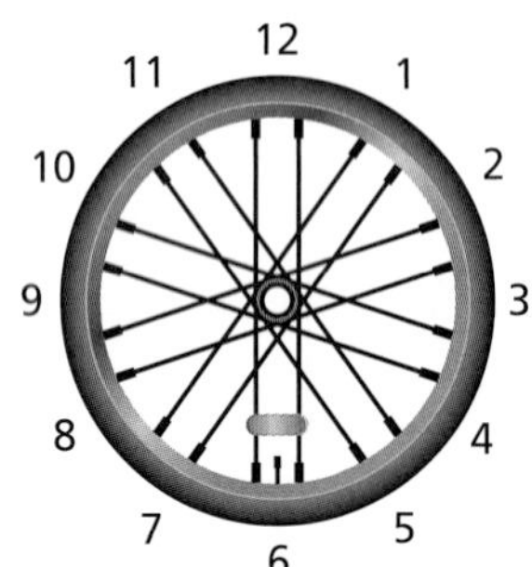

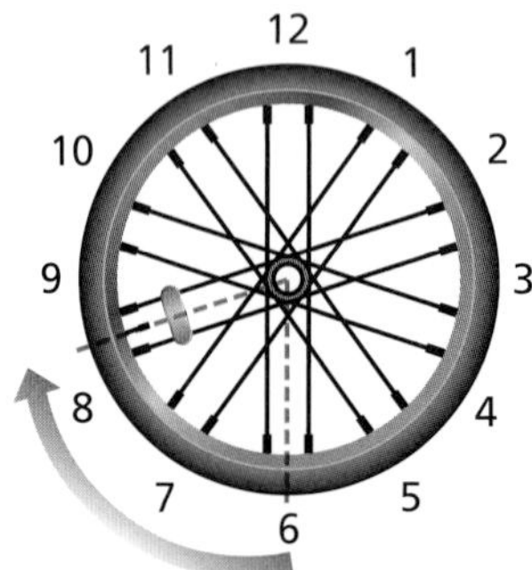

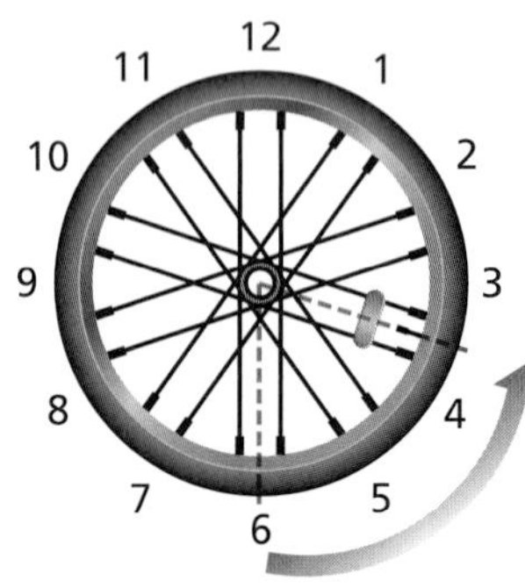

1 ACTIVITY: Three Basic Ways to Move Things

There are three basic ways to move objects on a flat surface.

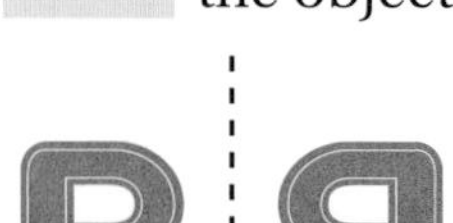

the object. the object. the object.

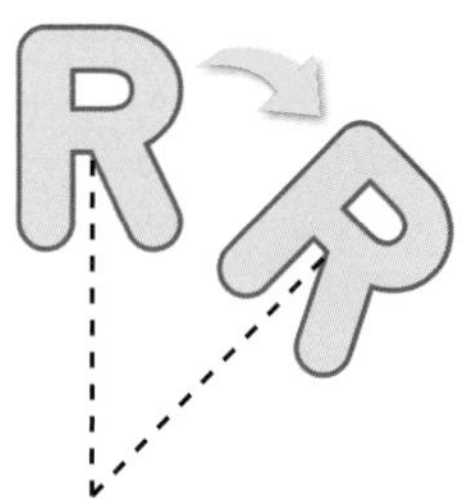

Geometry

In this lesson, you will

- identify rotations.
- rotate figures in the coordinate plane.
- use more than one transformation to find images of figures.

Work with a partner.

a. What type of triangle is the blue triangle? Is it congruent to the red triangles? Explain.

b. Decide how you can move the blue triangle to obtain each red triangle.

c. Is each move a *translation*, a *reflection*, or a *rotation*?

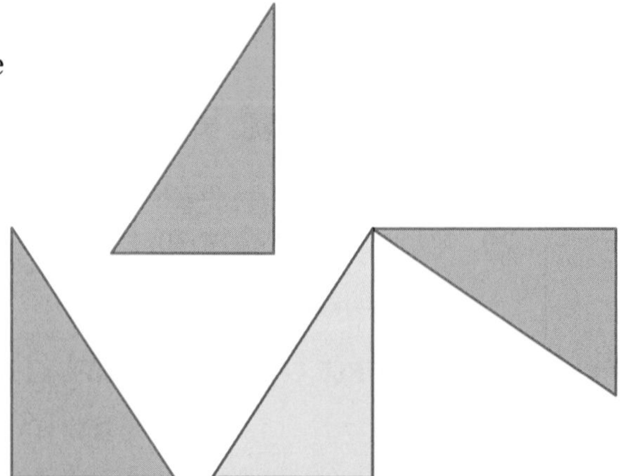

2 ACTIVITY: Rotating in the Coordinate Plane

Work with a partner.

a. Draw a rectangle in Quadrant II of a coordinate plane. Find the dimensions of the rectangle.

b. Copy the axes and the rectangle onto a piece of transparent paper.

Align the origin and the vertices of the rectangle on the transparent paper with the coordinate plane. Turn the transparent paper so that the rectangle is in Quadrant I and the axes align.

Draw the new figure in the coordinate plane. List the vertices.

c. Compare the dimensions and the angle measures of the new figure to those of the original rectangle.

d. Are the opposite sides of the new figure still parallel? Explain.

e. Can you conclude that the two figures are congruent? Explain.

f. Turn the transparent paper so that the original rectangle is in Quadrant IV. Draw the new figure in the coordinate plane. List the vertices. Then repeat parts (c)–(e).

g. Compare your results with those of other students in your class. Do you think the results are true for any type of figure?

> **Math Practice**
> **Calculate Accurately**
> What must you do to rotate the figure correctly?

What Is Your Answer?

3. **IN YOUR OWN WORDS** What are the three basic ways to move an object in a plane? Draw an example of each.

4. **PRECISION** Use the results of Activity 2(b).

   a. Draw four angles using the conditions below.

   - The origin is the vertex of each angle.
   - One side of each angle passes through a vertex of the original rectangle.
   - The other side of each angle passes through the corresponding vertex of the rotated rectangle.

   b. Measure each angle in part (a). For each angle, measure the distances between the origin and the vertices of the rectangles. What do you notice?

   c. How can the results of part (b) help you rotate a figure?

5. **PRECISION** Repeat the procedure in Question 4 using the results of Activity 2(f).

Practice Use what you learned about transformations to complete Exercises 7–9 on page 65.

2.4 Lesson

Key Vocabulary
rotation, *p. 62*
center of rotation, *p. 62*
angle of rotation, *p. 62*

Key Idea

Rotations

A **rotation**, or *turn*, is a transformation in which a figure is rotated about a point called the **center of rotation**. The number of degrees a figure rotates is the **angle of rotation**.

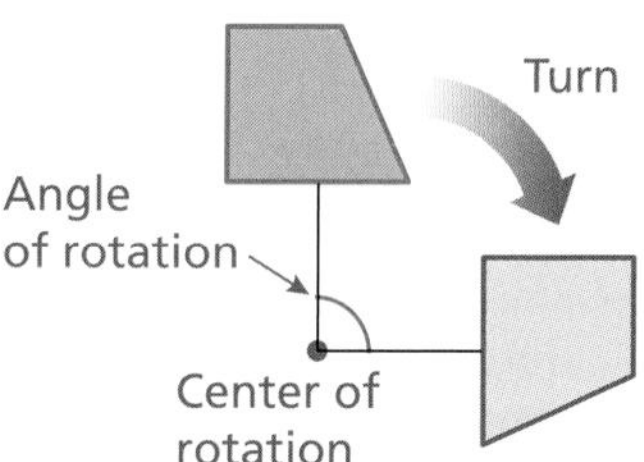

In a rotation, the original figure and its image are congruent.

EXAMPLE 1 Identifying a Rotation

You must rotate the puzzle piece 270° clockwise about point *P* to fit it into a puzzle. Which piece fits in the puzzle as shown?

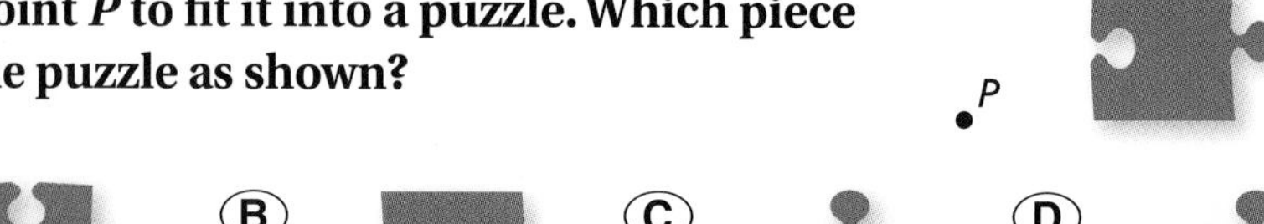

Ⓐ  Ⓑ  Ⓒ  Ⓓ 

Rotate the puzzle piece 270° clockwise about point *P*.

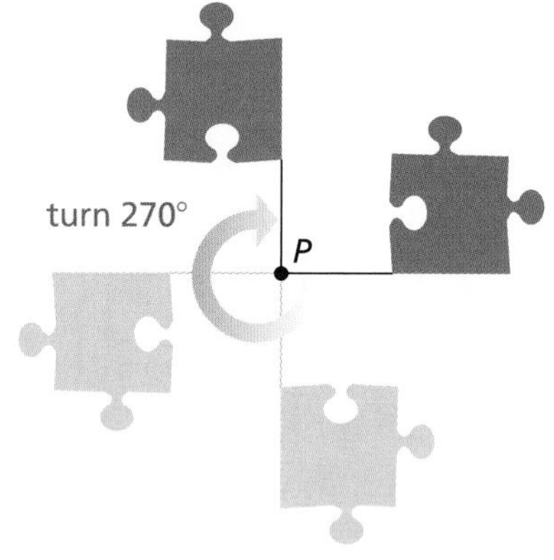

Study Tip
When rotating figures, it may help to sketch the rotation in several steps, as shown in Example 1.

So, the correct answer is Ⓒ.

On Your Own

Now You're Ready
Exercises 10–12

1. Which piece is a 90° counterclockwise rotation about point *P*?
2. Is Choice D a rotation of the original puzzle piece? If not, what kind of transformation does the image show?

Multi-Language Glossary at BigIdeasMath.com

EXAMPLE 2 Rotating a Figure

The vertices of a trapezoid are $W(-4, 2)$, $X(-3, 4)$, $Y(-1, 4)$, and $Z(-1, 2)$. Rotate the trapezoid 180° about the origin. What are the coordinates of the image?

> **Study Tip**
>
> A 180° clockwise rotation and a 180° counterclockwise rotation have the same image. So, you do not need to specify direction when rotating a figure 180°.

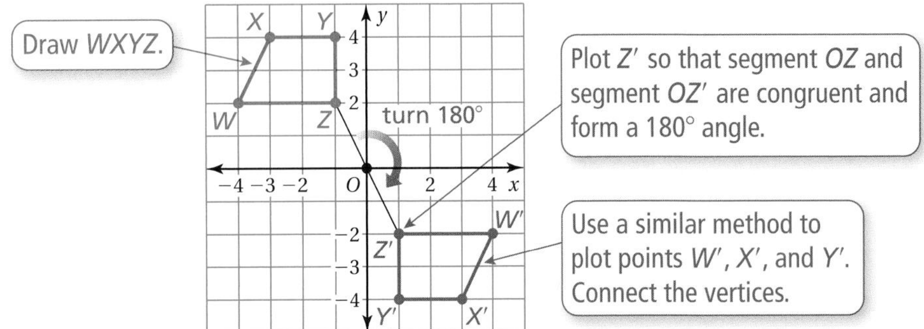

Draw WXYZ.
Plot Z′ so that segment OZ and segment OZ′ are congruent and form a 180° angle.
Use a similar method to plot points W′, X′, and Y′. Connect the vertices.
turn 180°
X
Y
W
Z
W′
Z′
Y′
X′
O
x
y

The coordinates of the image are $W'(4, -2)$, $X'(3, -4)$, $Y'(1, -4)$, and $Z'(1, -2)$.

EXAMPLE 3 Rotating a Figure

The vertices of a triangle are $J(1, 2)$, $K(4, 2)$, and $L(1, -3)$. Rotate the triangle 90° counterclockwise about vertex L. What are the coordinates of the image?

> **Common Error**
>
> Be sure to pay attention to whether a rotation is clockwise or counterclockwise.

Draw JKL.

Plot K' so that segment KL and segment $K'L'$ are congruent and form a 90° angle.

Use a similar method to plot point J'. Connect the vertices.

turn 90°

The coordinates of the image are $J'(-4, -3)$, $K'(-4, 0)$, and $L'(1, -3)$.

On Your Own

Now You're Ready
Exercises 13–18

3. A triangle has vertices $Q(4, 5)$, $R(4, 0)$, and $S(1, 0)$.
   a. Rotate the triangle 90° counterclockwise about the origin.
   b. Rotate the triangle 180° about vertex S.
   c. Are the images in parts (a) and (b) congruent? Explain.

EXAMPLE 4 Using More than One Transformation

The vertices of a rectangle are $A(-3, -3)$, $B(1, -3)$, $C(1, -5)$, and $D(-3, -5)$. Rotate the rectangle 90° clockwise about the origin, and then reflect it in the y-axis. What are the coordinates of the image?

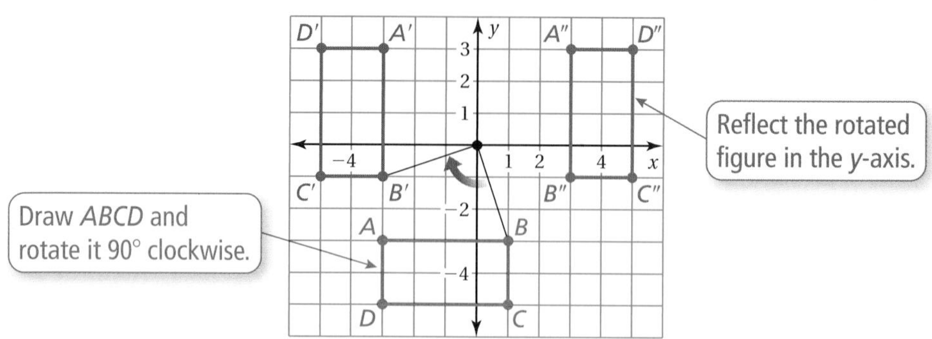

Reflect the rotated figure in the y-axis.
Draw ABCD and rotate it 90° clockwise.

∴ The coordinates of the image are $A''(3, 3)$, $B''(3, -1)$, $C''(5, -1)$ and $D''(5, 3)$.

The image of a translation, reflection, or rotation is congruent to the original figure. So, two figures are congruent when one can be obtained from the other by a sequence of translations, reflections, and rotations.

EXAMPLE 5 Describing a Sequence of Transformations

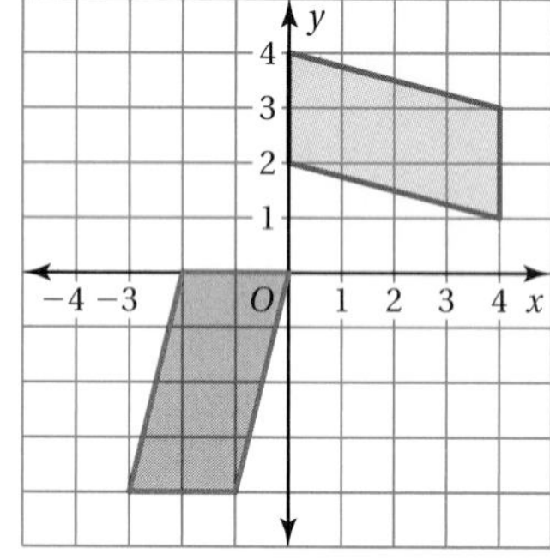

The red figure is congruent to the blue figure. Describe a sequence of transformations in which the blue figure is the image of the red figure.

You can turn the red figure 90° so that it has the same orientation as the blue figure. So, begin with a rotation.

After rotating, you need to slide the figure up.

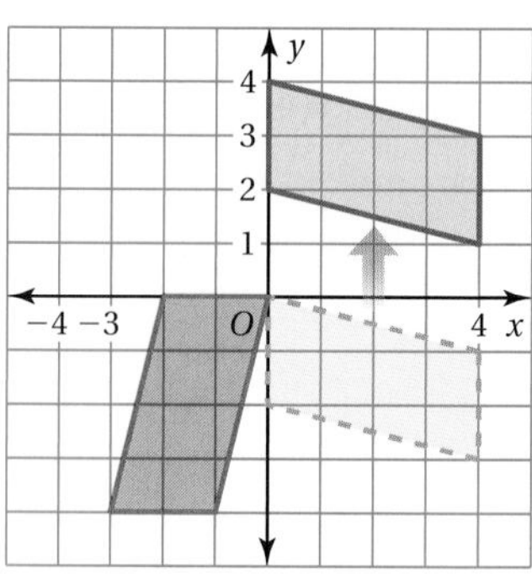

∴ So, one possible sequence of transformations is a 90° counterclockwise rotation about the origin followed by a translation 4 units up.

On Your Own

Now You're Ready
Exercises 22–25

4. The vertices of a triangle are $P(-1, 2)$, $Q(-1, 0)$, and $R(2, 0)$. Rotate the triangle 180° about vertex R, and then reflect it in the x-axis. What are the coordinates of the image?
5. In Example 5, describe a different sequence of transformations in which the blue figure is the image of the red figure.

2.4 Exercises

Vocabulary and Concept Check

1. **VOCABULARY** What are the coordinates of the center of rotation in Example 2? Example 3?

MENTAL MATH A figure lies entirely in Quadrant II. In which quadrant will the figure lie after the given clockwise rotation about the origin?

2. 90°
3. 180°
4. 270°
5. 360°

6. **DIFFERENT WORDS, SAME QUESTION** Which is different? Find "both" answers.

What are the coordinates of the figure after a 90° clockwise rotation about the origin?

What are the coordinates of the figure after a 270° clockwise rotation about the origin?

What are the coordinates of the figure after turning the figure 90° to the right about the origin?

What are the coordinates of the figure after a 270° counterclockwise rotation about the origin?

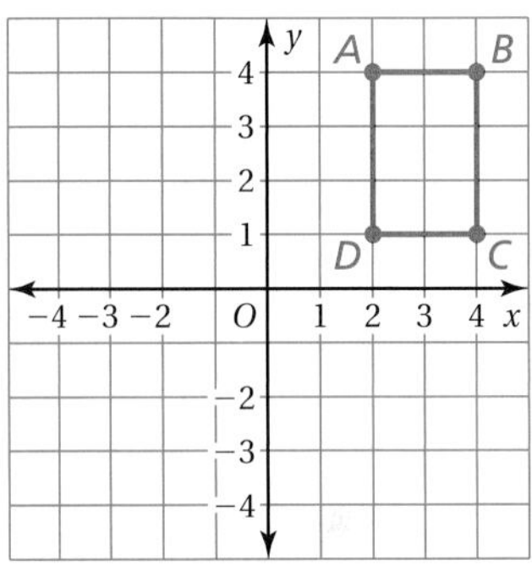

Practice and Problem Solving

Identify the transformation.

7.

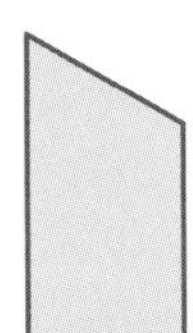

8.

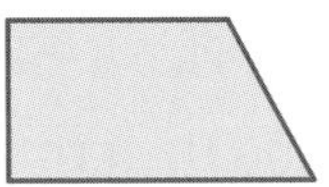

9.
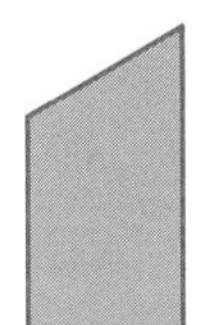
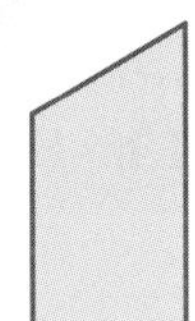

Tell whether the blue figure is a rotation of the red figure about the origin. If so, give the angle and direction of rotation.

1 10.
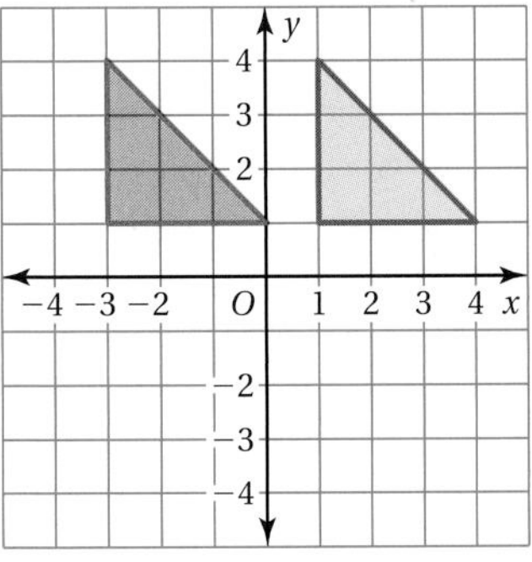

11.
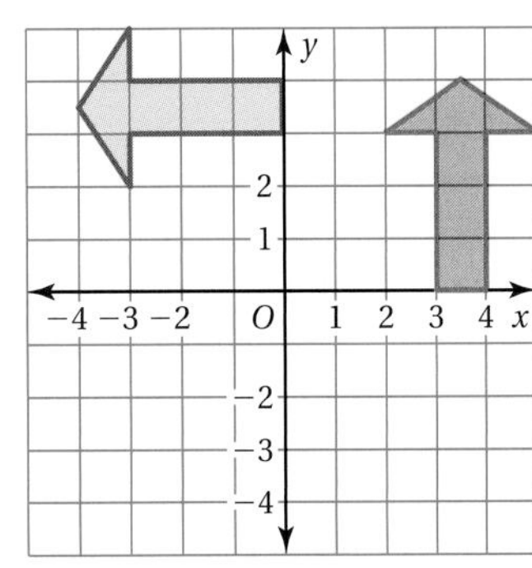

12.
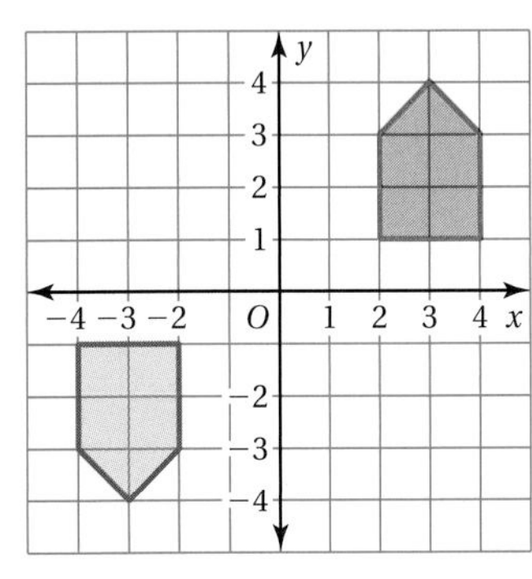

The vertices of a figure are given. Rotate the figure as described. Find the coordinates of the image.

② ③ **13.** $A(2, -2)$, $B(4, -1)$, $C(4, -3)$, $D(2, -4)$
90° counterclockwise about the origin

14. $F(1, 2)$, $G(3, 5)$, $H(3, 2)$
180° about the origin

15. $J(-4, 1)$, $K(-2, 1)$, $L(-4, -3)$
90° clockwise about vertex L

16. $P(-3, 4)$, $Q(-1, 4)$, $R(-2, 1)$, $S(-4, 1)$
180° about vertex R

17. $W(-6, -2)$, $X(-2, -2)$, $Y(-2, -6)$, $Z(-5, -6)$
270° counterclockwise about the origin

18. $A(1, -1)$, $B(5, -6)$, $C(1, -6)$
90° counterclockwise about vertex A

A figure has *rotational symmetry* if a rotation of 180° or less produces an image that fits exactly on the original figure. Explain why the figure has rotational symmetry.

19.

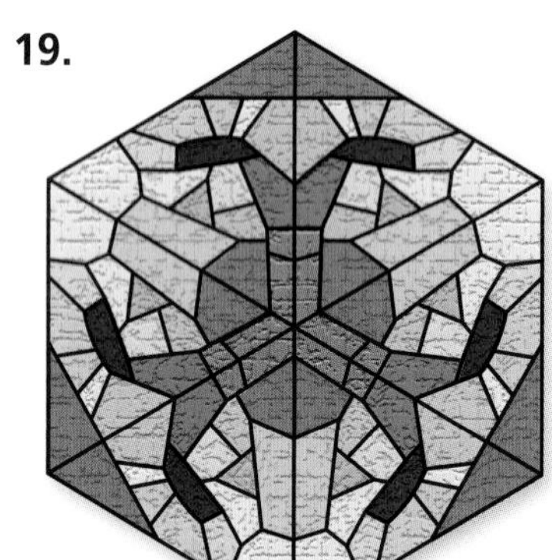

20.

21.

The vertices of a figure are given. Find the coordinates of the figure after the transformations given.

④ **22.** $R(-7, -5)$, $S(-1, -2)$, $T(-1, -5)$

Rotate 90° counterclockwise about the origin. Then translate 3 units left and 8 units up.

23. $J(-4, 4)$, $K(-3, 4)$, $L(-1, 1)$, $M(-4, 1)$

Reflect in the x-axis, and then rotate 180° about the origin.

The red figure is congruent to the blue figure. Describe two different sequences of transformations in which the blue figure is the image of the red figure.

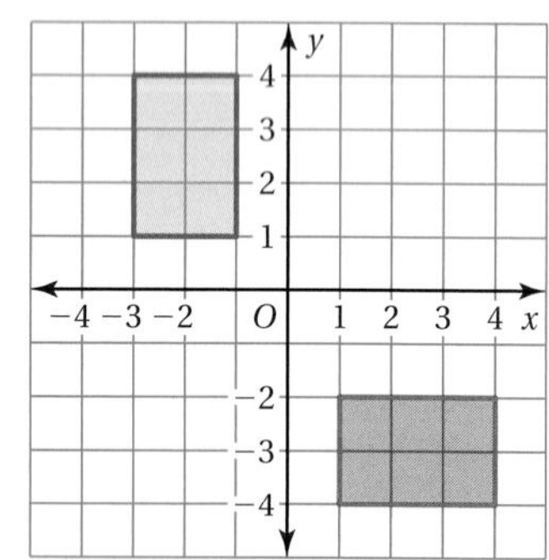

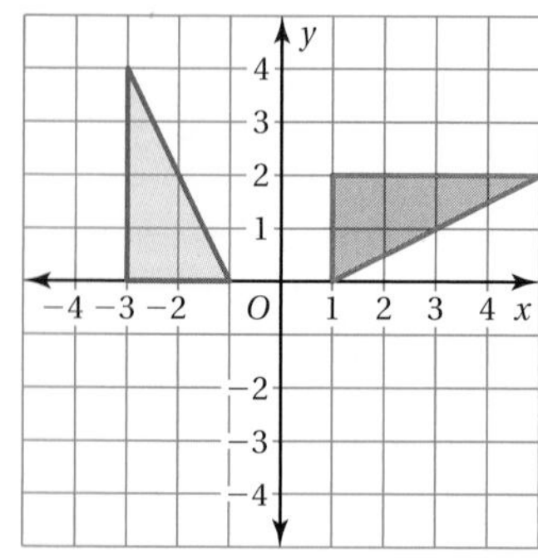

26. REASONING A trapezoid has vertices $A(-6, -2)$, $B(-3, -2)$, $C(-1, -4)$, and $D(-6, -4)$.

  a. Rotate the trapezoid 180° about the origin. What are the coordinates of the image?

  b. Describe a way to obtain the same image without using rotations.

27. TREASURE MAP You want to find the treasure located on the map at ×. You are located at ●. The following transformations will lead you to the treasure, but they are not in the correct order. Find the correct order. Use each transformation exactly once.

- Rotate 180° about the origin.
- Reflect in the y-axis.
- Rotate 90° counterclockwise about the origin.
- Translate 1 unit right and 1 unit up.

28. CRITICAL THINKING Consider $\triangle JKL$.

  a. Rotate $\triangle JKL$ 90° clockwise about the origin. How are the x- and y-coordinates of $\triangle J'K'L'$ related to the x- and y-coordinates of $\triangle JKL$?

  b. Rotate $\triangle JKL$ 180° about the origin. How are the x- and y-coordinates of $\triangle J'K'L'$ related to the x- and y-coordinates of $\triangle JKL$?

  c. Do you think your answers to parts (a) and (b) hold true for any figure? Explain.

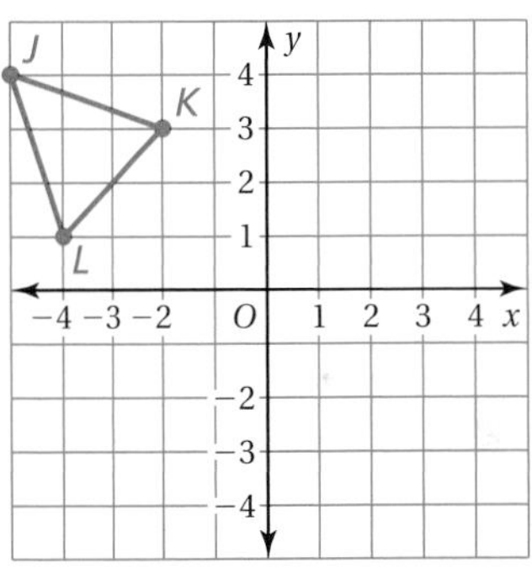

29. Reasoning You rotate a triangle 90° counterclockwise about the origin. Then you translate its image 1 unit left and 2 units down. The vertices of the final image are $(-5, 0)$, $(-2, 2)$, and $(-2, -1)$. What are the vertices of the original triangle?

Fair Game Review What you learned in previous grades & lessons

Tell whether the ratios form a proportion. *(Skills Review Handbook)*

30. $\frac{3}{5}, \frac{15}{20}$ **31.** $\frac{2}{3}, \frac{12}{18}$ **32.** $\frac{7}{28}, \frac{12}{48}$ **33.** $\frac{54}{72}, \frac{36}{45}$

34. MULTIPLE CHOICE What is the solution of the equation $x + 6 \div 2 = 5$? *(Section 1.1)*

Ⓐ $x = -16$ Ⓑ $x = 2$ Ⓒ $x = 4$ Ⓓ $x = 16$

2 Study Help

You can use a **summary triangle** to explain a concept. Here is an example of a summary triangle for translating a figure.

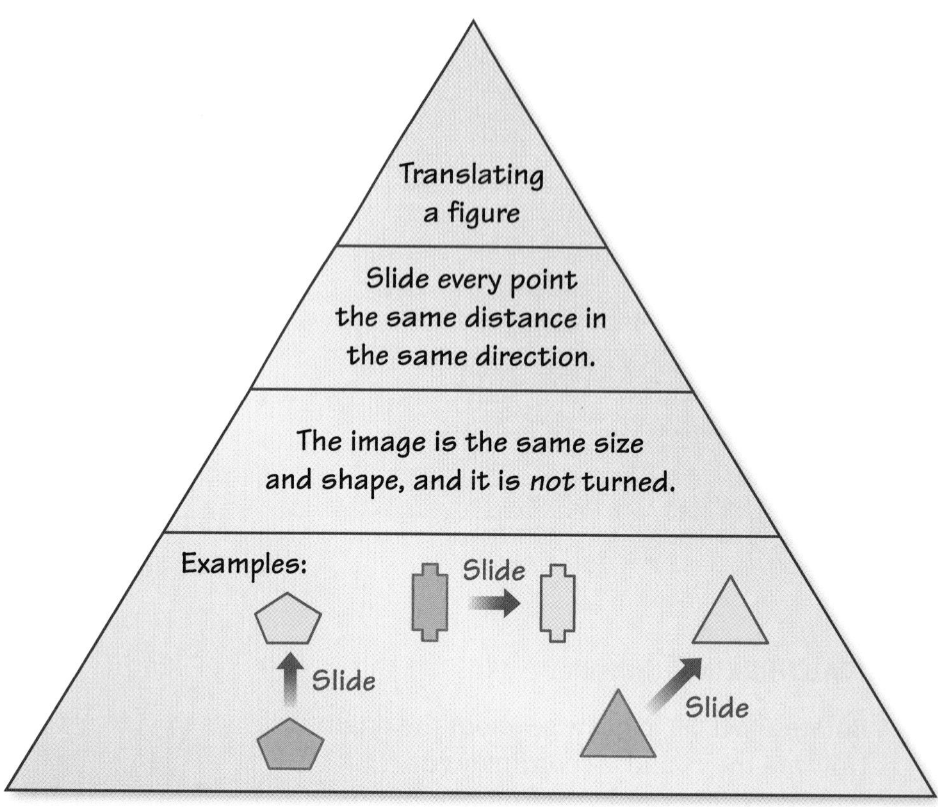

On Your Own

Make summary triangles to help you study these topics.

1. congruent figures
2. reflecting a figure
3. rotating a figure

After you complete this chapter, make summary triangles for the following topics.

4. similar figures
5. perimeters of similar figures
6. areas of similar figures
7. dilating a figure
8. transforming a figure

"I hope my owner sees my summary triangle. I just can't seem to learn 'roll over.' "

2.1–2.4 Quiz

Tell whether the two figures are congruent. Explain your reasoning. *(Section 2.1)*

1.

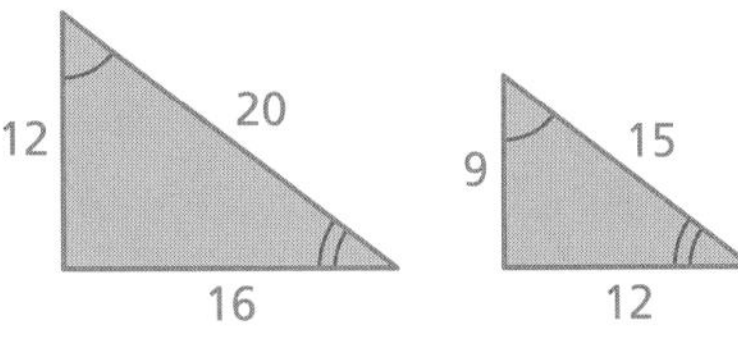

2.

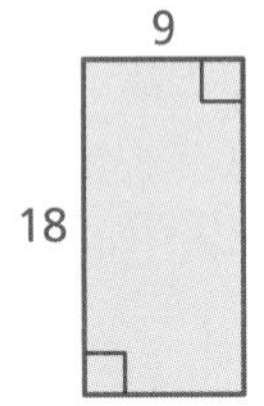

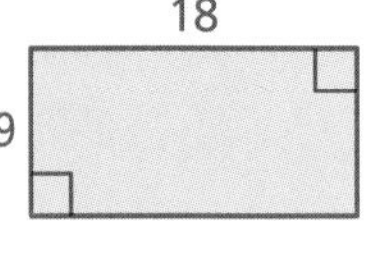

Tell whether the blue figure is a translation of the red figure. *(Section 2.2)*

3.

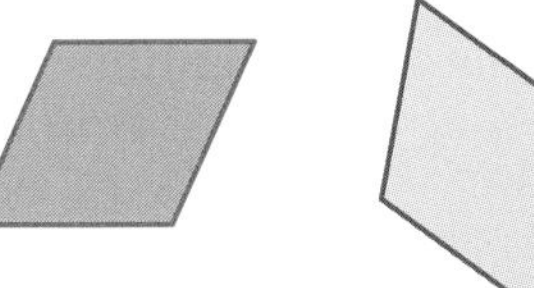

4.

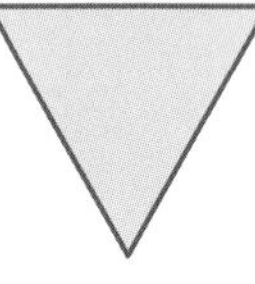

Tell whether the blue figure is a reflection of the red figure. *(Section 2.3)*

5.

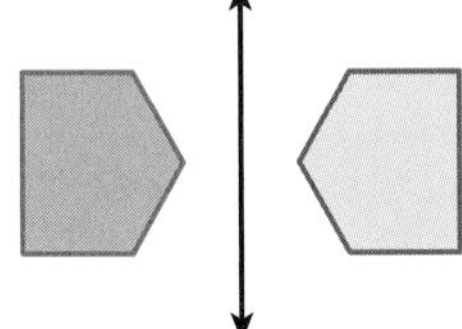

6.

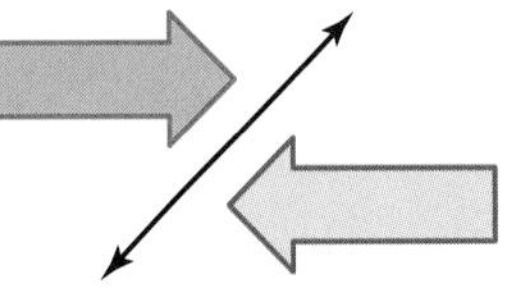

The red figure is congruent to the blue figure. Describe two different sequences of transformations in which the blue figure is the image of the red figure. *(Section 2.4)*

7.

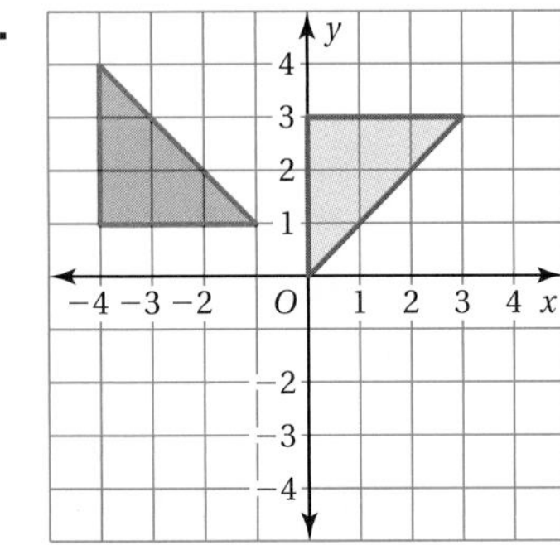

8.

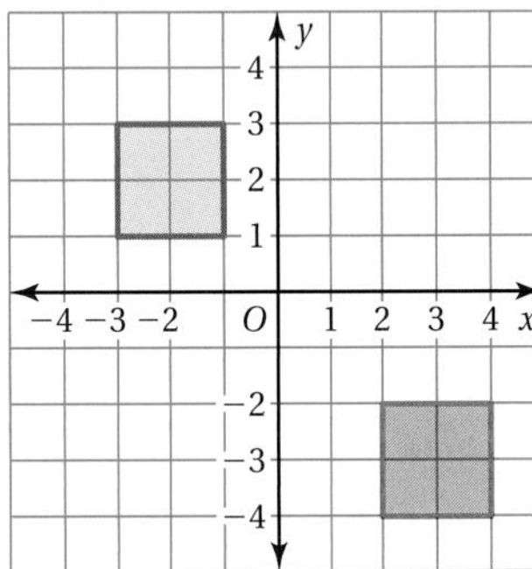

9. AIRPLANE Describe a translation of the airplane from point A to point B. *(Section 2.2)*

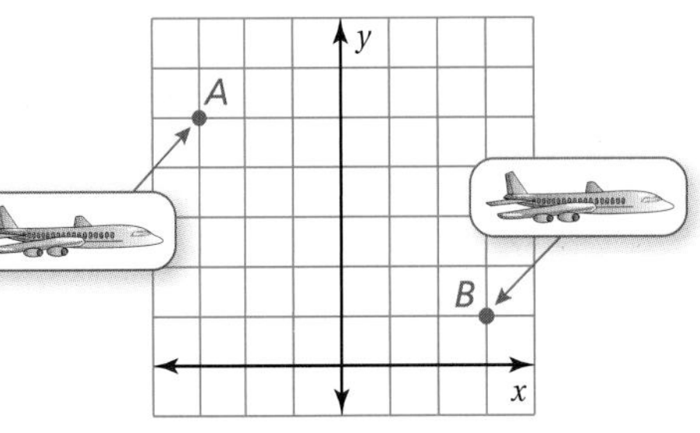

10. MINIGOLF You hit the golf ball along the red path so that its image will be a reflection in the y-axis. Does the golf ball land in the hole? Explain. *(Section 2.3)*

2.5 Similar Figures

Essential Question How can you use proportions to help make decisions in art, design, and magazine layouts?

Original photograph

In a computer art program, when you click and drag on a side of a photograph, you distort it.

But when you click and drag on a corner of the photograph, the dimensions remain proportional to the original.

Distorted

Distorted

Proportional

1 ACTIVITY: Reducing Photographs

Work with a partner. You are trying to reduce the photograph to the indicated size for a nature magazine. Can you reduce the photograph to the indicated size without distorting or cropping? Explain your reasoning.

a.

5 in.

6 in.

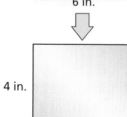

4 in.

5 in.

b.

6 in.

8 in.

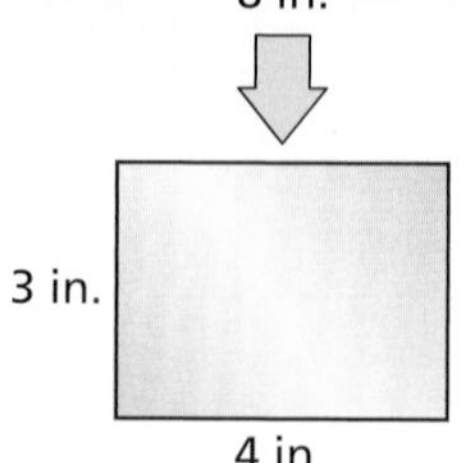

3 in.

4 in.

Geometry

In this lesson, you will

- name corresponding angles and corresponding sides of similar figures.
- identify similar figures.
- find unknown measures of similar figures.

2 ACTIVITY: Creating Designs

Work with a partner.

a. Tell whether the dimensions of the new designs are proportional to the dimensions of the original design. Explain your reasoning.

Math Practice

Analyze Relationships

How can you use mathematics to determine whether the dimensions are proportional?

Original

Design 1

Design 2

b. Draw two designs whose dimensions are proportional to the given design. Make one bigger and one smaller. Label the sides of the designs with their lengths.

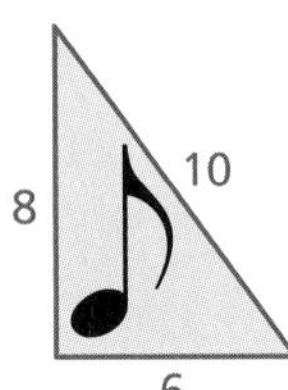

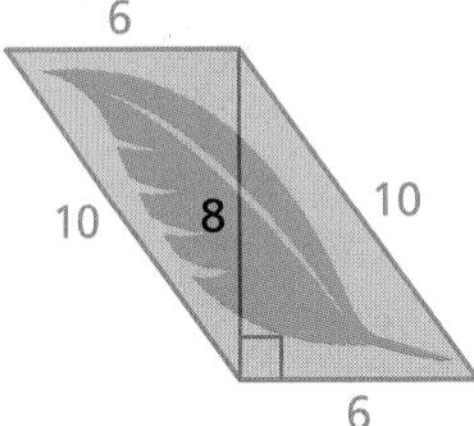

What Is Your Answer?

3. **IN YOUR OWN WORDS** How can you use proportions to help make decisions in art, design, and magazine layouts? Give two examples.

4. **a.** Use a computer art program to draw two rectangles whose dimensions are proportional to each other.

b. Print the two rectangles on the same piece of paper.

c. Use a centimeter ruler to measure the length and the width of each rectangle.

d. Find the following ratios. What can you conclude?

$$\frac{\text{Length of larger}}{\text{Length of smaller}} \qquad \frac{\text{Width of larger}}{\text{Width of smaller}}$$

You've got to hand it to him. He's right.

"I love this statue. It seems similar to a big statue I saw in New York."

Practice Use what you learned about similar figures to complete Exercises 4 and 5 on page 74.

2.5 Lesson

Key Vocabulary
similar figures, *p. 72*

Key Idea

Similar Figures

Figures that have the same shape but not necessarily the same size are called **similar figures**.

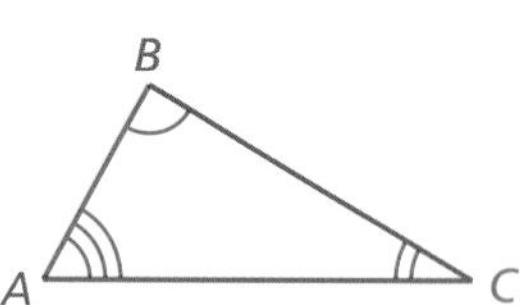

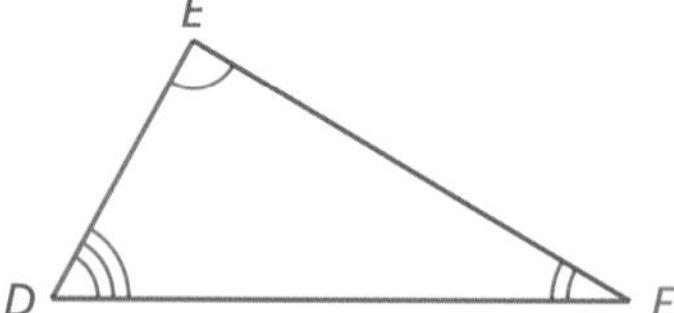

Triangle *ABC* is similar to Triangle *DEF*.

Words Two figures are similar when

- corresponding side lengths are proportional and
- corresponding angles are congruent.

Symbols

Side Lengths	*Angles*	*Figures*
$\frac{AB}{DE} = \frac{BC}{EF} = \frac{AC}{DF}$	$\angle A \cong \angle D$ $\angle B \cong \angle E$ $\angle C \cong \angle F$	$\triangle ABC \sim \triangle DEF$

The symbol ~ means *is similar to*.

Common Error

When writing a similarity statement, make sure to list the vertices of the figures in the correct order.

EXAMPLE 1 Identifying Similar Figures

Which rectangle is similar to Rectangle A?

Rectangle A

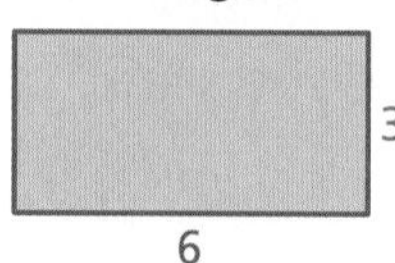

Rectangle B

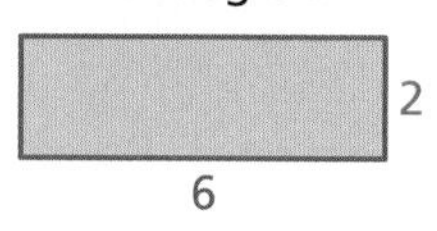

Rectangle C

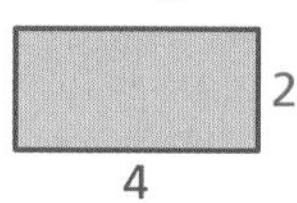

Each figure is a rectangle. So, corresponding angles are congruent. Check to see if corresponding side lengths are proportional.

Rectangle A and Rectangle B

$\frac{\text{Length of A}}{\text{Length of B}} = \frac{6}{6} = 1$ $\quad \frac{\text{Width of A}}{\text{Width of B}} = \frac{3}{2}$ Not proportional

Rectangle A and Rectangle C

$\frac{\text{Length of A}}{\text{Length of C}} = \frac{6}{4} = \frac{3}{2}$ $\quad \frac{\text{Width of A}}{\text{Width of C}} = \frac{3}{2}$ Proportional

So, Rectangle C is similar to Rectangle A.

On Your Own

1. Rectangle D is 3 units long and 1 unit wide. Which rectangle is similar to Rectangle D?

EXAMPLE 2 Finding an Unknown Measure in Similar Figures

The triangles are similar. Find *x*.

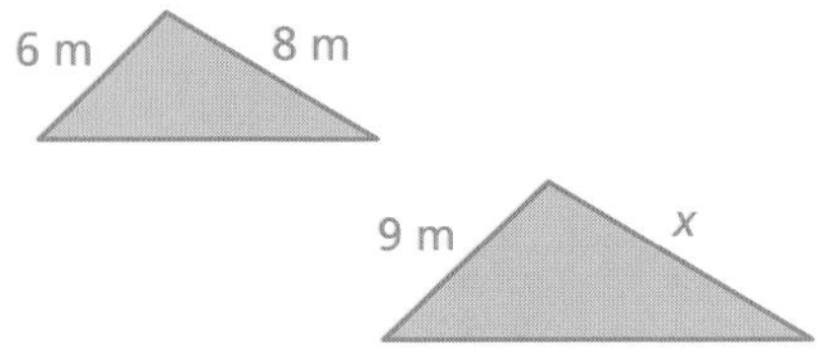

Because the triangles are similar, corresponding side lengths are proportional. So, write and solve a proportion to find x.

$\frac{6}{9} = \frac{8}{x}$ Write a proportion.

$6x = 72$ Cross Products Property

$x = 12$ Divide each side by 6.

So, x is 12 meters.

On Your Own

Now You're Ready
Exercises 8–11

The figures are similar. Find *x*.

2.

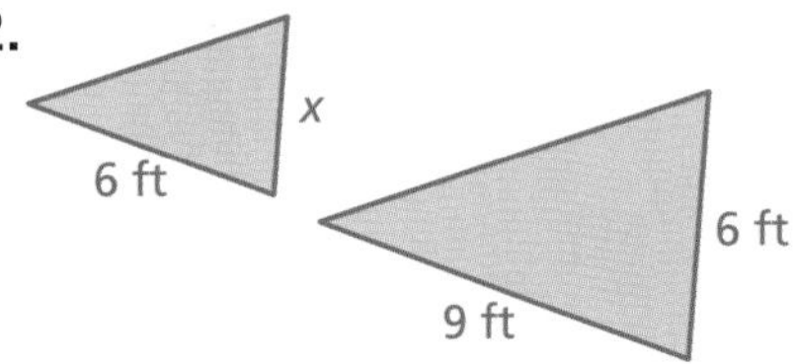

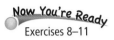

3.

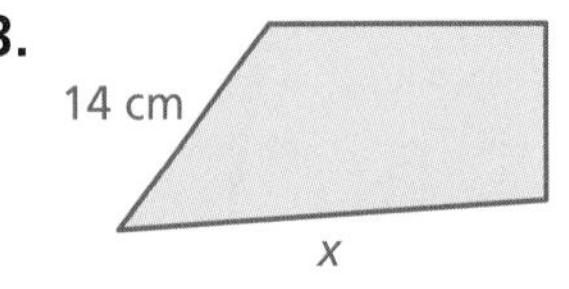

EXAMPLE 3 Real-Life Application

An artist draws a replica of a painting that is on the Berlin Wall. The painting includes a red trapezoid. The shorter base of the similar trapezoid in the replica is 3.75 inches. What is the height *h* of the trapezoid in the replica?

12 in.
15 in.
Painting

Because the trapezoids are similar, corresponding side lengths are proportional. So, write and solve a proportion to find h.

$\frac{3.75}{15} = \frac{h}{12}$ Write a proportion.

h
3.75 in.
Replica

$12 \cdot \frac{3.75}{15} = 12 \cdot \frac{h}{12}$ Multiplication Property of Equality

$3 = h$ Simplify.

So, the height of the trapezoid in the replica is 3 inches.

On Your Own

4. **WHAT IF?** The longer base in the replica is 4.5 inches. What is the length of the longer base in the painting?

2.5 Exercises

Vocabulary and Concept Check

1. **VOCABULARY** How are corresponding angles of two similar figures related?
2. **VOCABULARY** How are corresponding side lengths of two similar figures related?
3. **CRITICAL THINKING** Are two figures that have the same size and shape similar? Explain.

Practice and Problem Solving

Tell whether the two figures are similar. Explain your reasoning.

(1) 4.

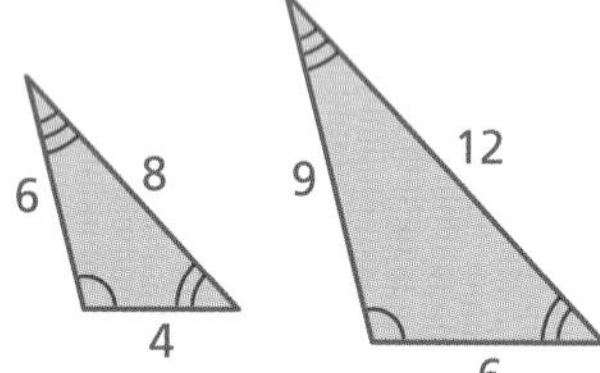

5.

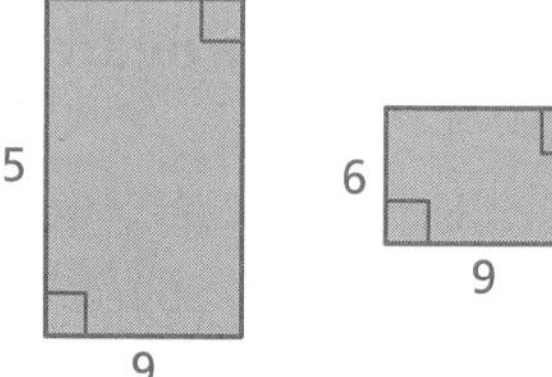

In a coordinate plane, draw the figures with the given vertices. Which figures are similar? Explain your reasoning.

6. Rectangle A: (0, 0), (4, 0), (4, 2), (0, 2)
   Rectangle B: (0, 0), (−6, 0), (−6, 3), (0, 3)
   Rectangle C: (0, 0), (4, 0), (4, 2), (0, 2)

7. Figure A: (−4, 2), (−2, 2), (−2, 0), (−4, 0)
   Figure B: (1, 4), (4, 4), (4, 1), (1, 1)
   Figure C: (2, −1), (5, −1), (5, −3), (2, −3)

The figures are similar. Find *x*.

(2) 8.

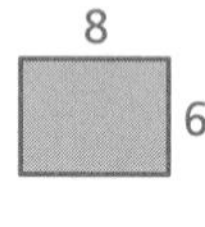

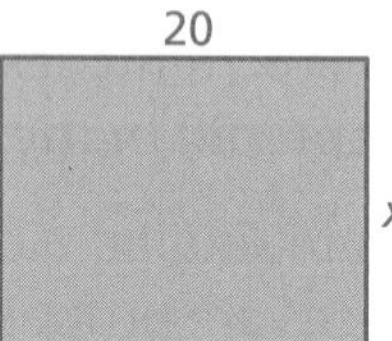

9.

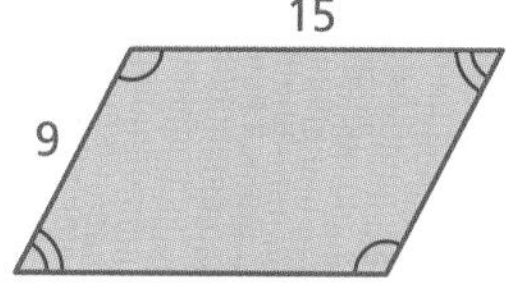

10.

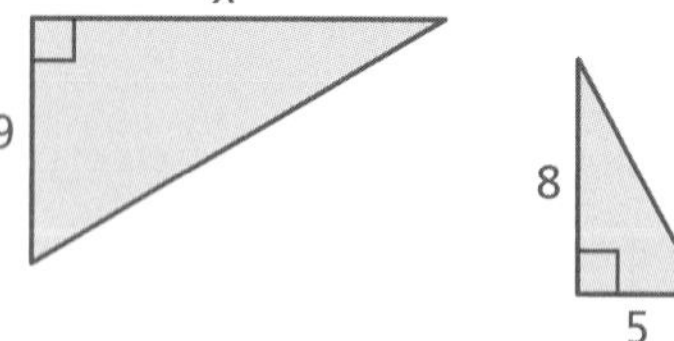

11.

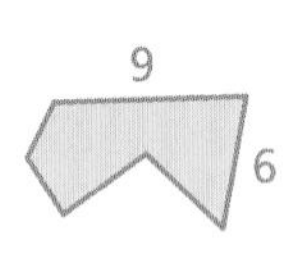

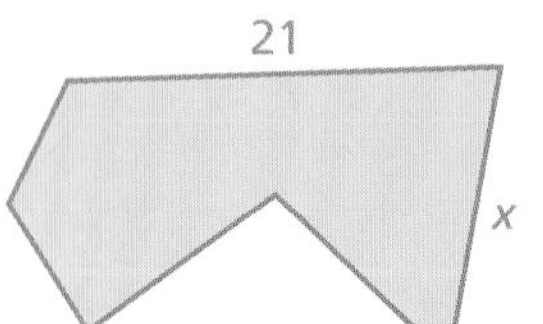

12. **MEXICO** A Mexican flag is 63 inches long and 36 inches wide. Is the drawing at the right similar to the Mexican flag?

13. **DESKS** A student's rectangular desk is 30 inches long and 18 inches wide. The teacher's desk is similar to the student's desk and has a length of 50 inches. What is the width of the teacher's desk?

14. LOGIC Are the following figures *always*, *sometimes*, or *never* similar? Explain.

a. two triangles
b. two squares
c. two rectangles
d. a square and a triangle

15. CRITICAL THINKING Can you draw two quadrilaterals each having two 130° angles and two 50° angles that are *not* similar? Justify your answer.

16. SIGN All the angle measures in the sign are 90°.

a. You increase each side length by 20%. Is the new sign similar to the original?

b. You increase each side length by 6 inches. Is the new sign similar to the original?

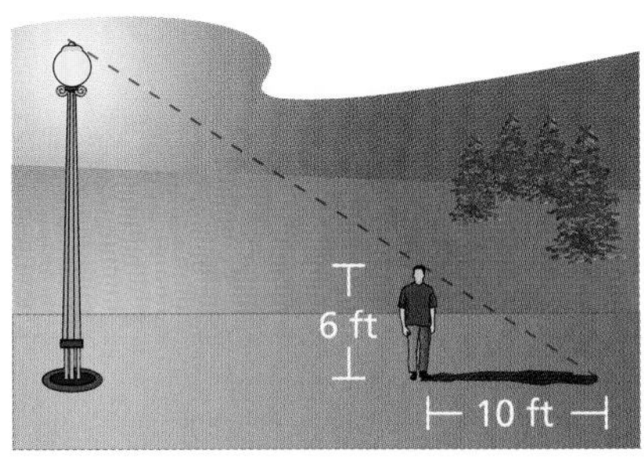

17. STREETLIGHT A person standing 20 feet from a streetlight casts a shadow as shown. How many times taller is the streetlight than the person? Assume the triangles are similar.

18. REASONING Is an object similar to a scale drawing of the object? Explain.

19. GEOMETRY Use a ruler to draw two different isosceles triangles similar to the one shown. Measure the heights of each triangle to the nearest centimeter.

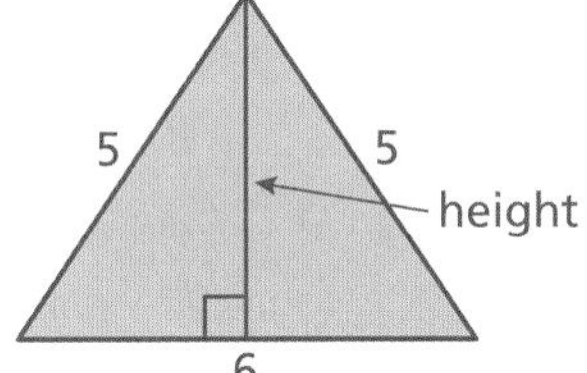

a. Is the ratio of the corresponding heights proportional to the ratio of the corresponding side lengths?

b. Do you think this is true for all similar triangles? Explain.

20. Critical Thinking Given $\triangle ABC \sim \triangle DEF$ and $\triangle DEF \sim \triangle JKL$, is $\triangle ABC \sim \triangle JKL$? Give an example or a non-example.

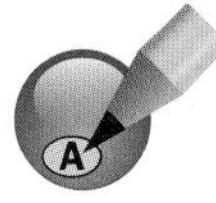

Fair Game Review What you learned in previous grades & lessons

Simplify. *(Skills Review Handbook)*

21. $\left(\frac{4}{9}\right)^2$ **22.** $\left(\frac{3}{8}\right)^2$ **23.** $\left(\frac{7}{4}\right)^2$ **24.** $\left(\frac{6.5}{2}\right)^2$

25. MULTIPLE CHOICE You solve the equation $S = \ell w + 2wh$ for w. Which equation is correct? *(Section 1.4)*

Ⓐ $w = \frac{S - \ell}{2h}$ Ⓑ $w = \frac{S - 2h}{\ell}$ Ⓒ $w = \frac{S}{\ell + 2h}$ Ⓓ $w = S - \ell - 2h$

2.6 Perimeters and Areas of Similar Figures

Essential Question How do changes in dimensions of similar geometric figures affect the perimeters and the areas of the figures?

1 ACTIVITY: Creating Similar Figures

Work with a partner. Use pattern blocks to make a figure whose dimensions are 2, 3, and 4 times greater than those of the original figure.

a. Square

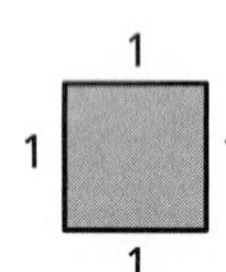

b. Rectangle

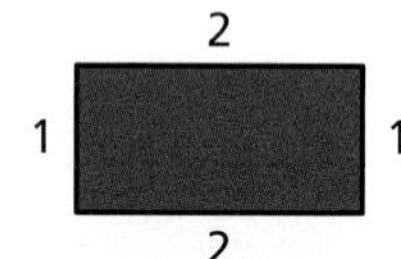

2 ACTIVITY: Finding Patterns for Perimeters

Work with a partner. Copy and complete the table for the perimeter *P* of each figure in Activity 1. Describe the pattern.

Figure	Original Side Lengths	Double Side Lengths	Triple Side Lengths	Quadruple Side Lengths
(square)	$P =$			
(rectangle)	$P =$			

3 ACTIVITY: Finding Patterns for Areas

Work with a partner. Copy and complete the table for the area *A* of each figure in Activity 1. Describe the pattern.

Figure	Original Side Lengths	Double Side Lengths	Triple Side Lengths	Quadruple Side Lengths
(square)	$A =$			
(rectangle)	$A =$			

Geometry

In this lesson, you will

- understand the relationship between perimeters of similar figures.
- understand the relationship between areas of similar figures.
- find ratios of perimeters and areas for similar figures.

4 ACTIVITY: Drawing and Labeling Similar Figures

Work with a partner.

a. Find a blue rectangle that is similar to the red rectangle and has one side from (−1, −6) to (5, −6). Label the vertices.

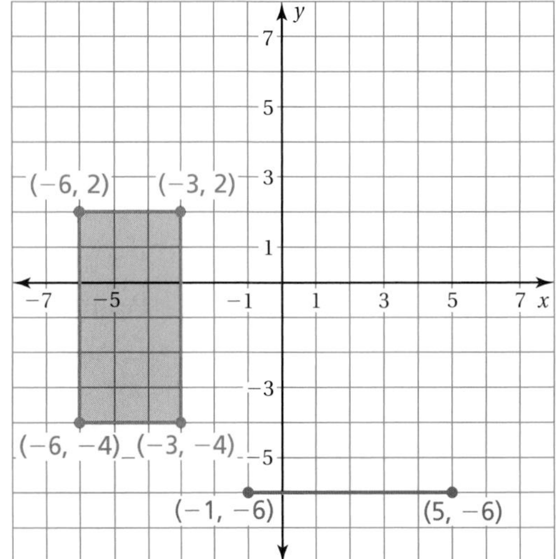

Check that the two rectangles are similar by showing that the ratios of corresponding sides are equal.

$$\frac{\text{Red Length}}{\text{Blue Length}} \stackrel{?}{=} \frac{\text{Red Width}}{\text{Blue Width}}$$

$$\frac{\text{change in } y}{\text{change in } y} \stackrel{?}{=} \frac{\text{change in } x}{\text{change in } x}$$

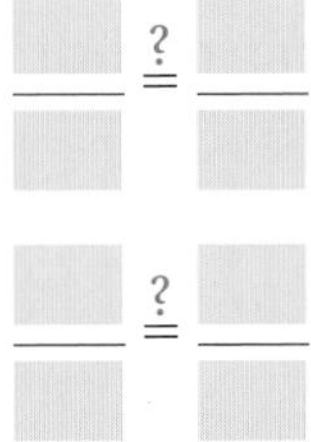

The ratios are equal. So, the rectangles are similar.

Math Practice

Analyze Givens

What values should you use to fill in the proportion? Does it matter where each value goes? Explain.

b. Compare the perimeters and the areas of the figures. Are the results the same as your results from Activities 2 and 3? Explain.

c. There are three other blue rectangles that are similar to the red rectangle and have the given side.

- Draw each one. Label the vertices of each.
- Show that each is similar to the original red rectangle.

What Is Your Answer?

5. **IN YOUR OWN WORDS** How do changes in dimensions of similar geometric figures affect the perimeters and the areas of the figures?

6. What information do you need to know to find the dimensions of a figure that is similar to another figure? Give examples to support your explanation.

Practice Use what you learned about perimeters and areas of similar figures to complete Exercises 8 and 9 on page 80.

Key Idea

Perimeters of Similar Figures

When two figures are similar, the ratio of their perimeters is equal to the ratio of their corresponding side lengths.

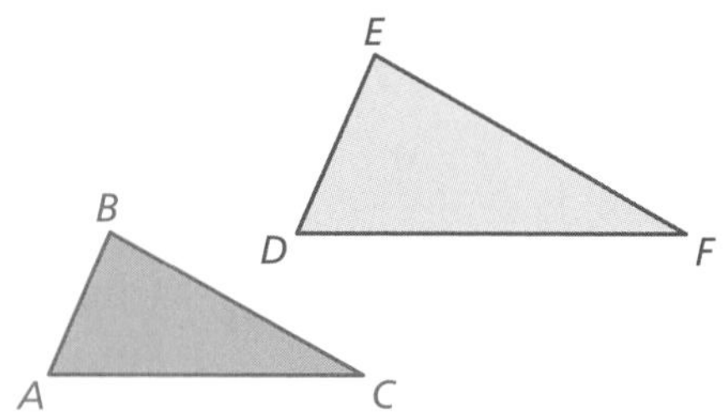

$$\frac{\text{Perimeter of } \triangle ABC}{\text{Perimeter of } \triangle DEF} = \frac{AB}{DE} = \frac{BC}{EF} = \frac{AC}{DF}$$

EXAMPLE 1 Finding Ratios of Perimeters

Find the ratio (red to blue) of the perimeters of the similar rectangles.

$$\frac{\text{Perimeter of red rectangle}}{\text{Perimeter of blue rectangle}} = \frac{4}{6} = \frac{2}{3}$$

∴ The ratio of the perimeters is $\frac{2}{3}$.

On Your Own

1. The height of Figure A is 9 feet. The height of a similar Figure B is 15 feet. What is the ratio of the perimeter of A to the perimeter of B?

Key Idea

Areas of Similar Figures

When two figures are similar, the ratio of their areas is equal to the *square* of the ratio of their corresponding side lengths.

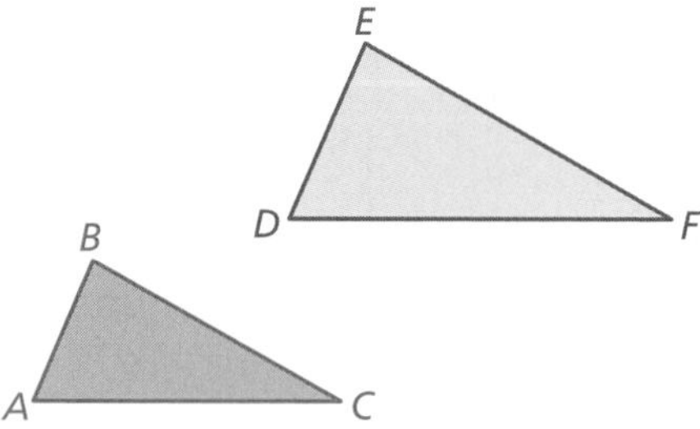

$$\frac{\text{Area of } \triangle ABC}{\text{Area of } \triangle DEF} = \left(\frac{AB}{DE}\right)^2 = \left(\frac{BC}{EF}\right)^2 = \left(\frac{AC}{DF}\right)^2$$

EXAMPLE 2 Finding Ratios of Areas

Find the ratio (red to blue) of the areas of the similar triangles.

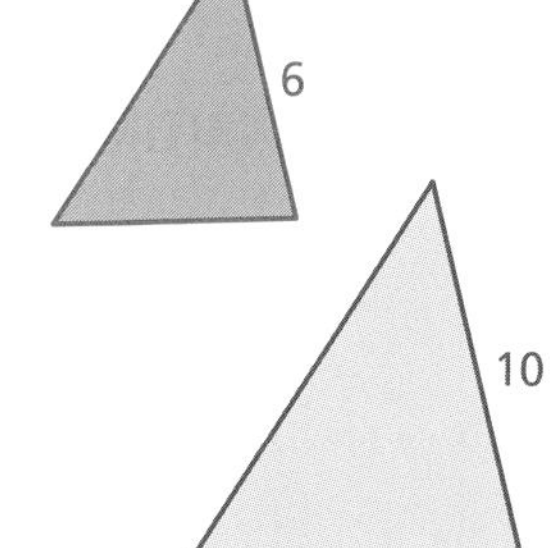

$$\frac{\text{Area of red triangle}}{\text{Area of blue triangle}} = \left(\frac{6}{10}\right)^2$$

$$= \left(\frac{3}{5}\right)^2 = \frac{9}{25}$$

∴ The ratio of the areas is $\frac{9}{25}$.

On Your Own

2. The base of Triangle P is 8 meters. The base of a similar Triangle Q is 7 meters. What is the ratio of the area of P to the area of Q?

EXAMPLE 3 Using Proportions to Find Perimeters and Areas

A swimming pool is similar in shape to a volleyball court. Find the perimeter P and the area A of the pool.

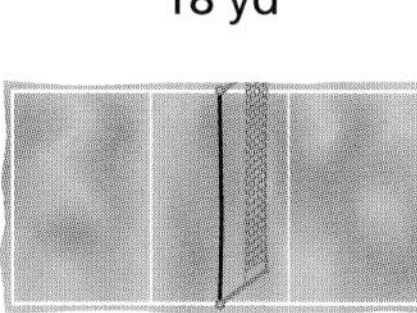

Area = 200 yd^2
Perimeter = 60 yd

The rectangular pool and the court are similar. So, use the ratio of corresponding side lengths to write and solve proportions to find the perimeter and the area of the pool.

Perimeter

$$\frac{\text{Perimeter of court}}{\text{Perimeter of pool}} = \frac{\text{Width of court}}{\text{Width of pool}}$$

$$\frac{60}{P} = \frac{10}{18}$$

$$1080 = 10P$$

$$108 = P$$

Area

$$\frac{\text{Area of court}}{\text{Area of pool}} = \left(\frac{\text{Width of court}}{\text{Width of pool}}\right)^2$$

$$\frac{200}{A} = \left(\frac{10}{18}\right)^2$$

$$\frac{200}{A} = \frac{100}{324}$$

$$64{,}800 = 100A$$

$$648 = A$$

∴ So, the perimeter of the pool is 108 yards, and the area is 648 square yards.

On Your Own

3. **WHAT IF?** The width of the pool is 16 yards. Find the perimeter P and the area A of the pool.

2.6 Exercises

Vocabulary and Concept Check

1. **WRITING** How are the perimeters of two similar figures related?

2. **WRITING** How are the areas of two similar figures related?

3. **NUMBER SENSE** Rectangle $ABCD$ is similar to Rectangle $WXYZ$. The area of $ABCD$ is 30 square inches. Explain how to find the area of $WXYZ$.

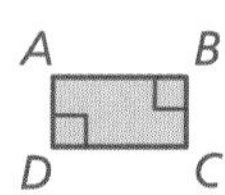

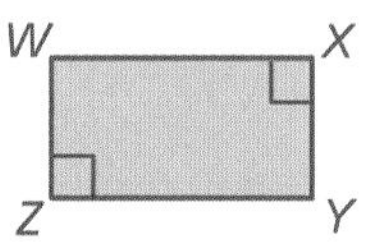

$\frac{AD}{WZ} = \frac{1}{2}$ $\frac{AB}{WX} = \frac{1}{2}$

Practice and Problem Solving

The two figures are similar. Find the ratios (red to blue) of the perimeters and of the areas.

4.

5.

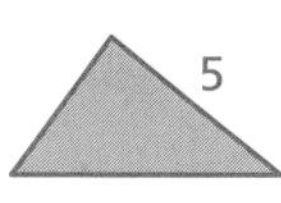

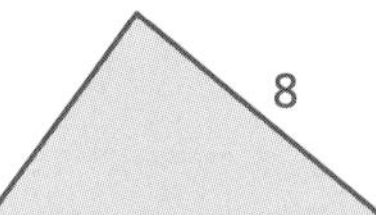

6.

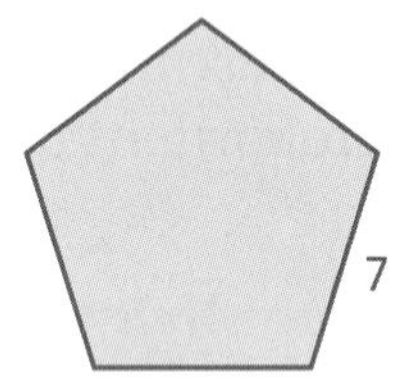

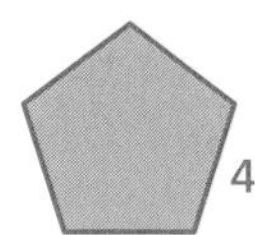

7.

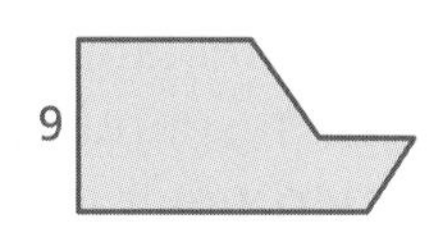

8. **PERIMETER** How does doubling the side lengths of a right triangle affect its perimeter?

9. **AREA** How does tripling the side lengths of a right triangle affect its area?

The figures are similar. Find x.

10. The ratio of the perimeters is 7 : 10.

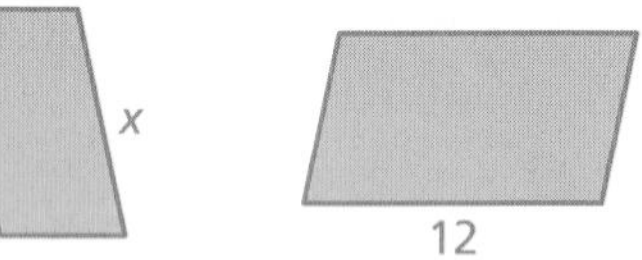

11. The ratio of the perimeters is 8 : 5.

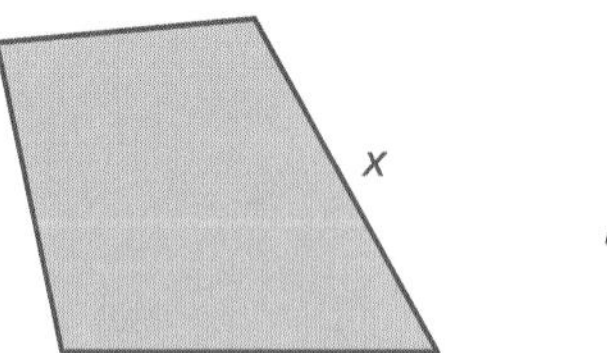

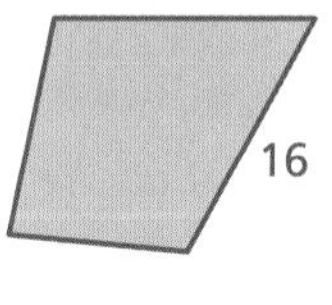

12. **FOOSBALL** The playing surfaces of two foosball tables are similar. The ratio of the corresponding side lengths is 10 : 7. What is the ratio of the areas?

13. **CHEERLEADING** A rectangular school banner has a length of 44 inches, a perimeter of 156 inches, and an area of 1496 square inches. The cheerleaders make signs similar to the banner. The length of a sign is 11 inches. What is its perimeter and its area?

14. REASONING The vertices of two rectangles are $A(-5, -1)$, $B(-1, -1)$, $C(-1, -4)$, $D(-5, -4)$ and $W(1, 6)$, $X(7, 6)$, $Y(7, -2)$, $Z(1, -2)$. Compare the perimeters and the areas of the rectangles. Are the rectangles similar? Explain.

15. SQUARE The ratio of the side length of Square A to the side length of Square B is 4 : 9. The side length of Square A is 12 yards. What is the perimeter of Square B?

16. FABRIC The cost of the fabric is $1.31. What would you expect to pay for a similar piece of fabric that is 18 inches by 42 inches?

17. AMUSEMENT PARK A scale model of a merry-go-round and the actual merry-go-round are similar.

a. How many times greater is the base area of the actual merry-go-round than the base area of the scale model? Explain.

b. What is the base area of the actual merry-go-round in square feet?

18. STRUCTURE The circumference of Circle K is π. The circumference of Circle L is 4π.

a. What is the ratio of their circumferences? of their radii? of their areas?

b. What do you notice?

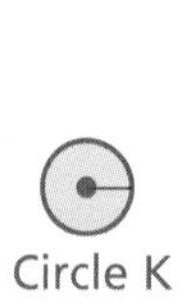

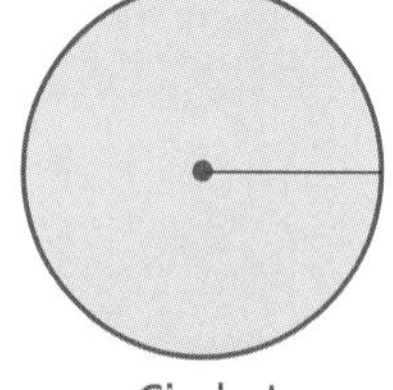

19. GEOMETRY A triangle with an area of 10 square meters has a base of 4 meters. A similar triangle has an area of 90 square meters. What is the *height* of the larger triangle?

20. Problem Solving You need two bottles of fertilizer to treat the flower garden shown. How many bottles do you need to treat a similar garden with a perimeter of 105 feet?

Fair Game Review What you learned in previous grades & lessons

Solve the equation. Check your solution. *(Section 1.3)*

21. $4x + 12 = -2x$ **22.** $2b + 6 = 7b - 2$ **23.** $8(4n + 13) = 6n$

24. MULTIPLE CHOICE Last week, you collected 20 pounds of cans for recycling. This week, you collect 25 pounds of cans for recycling. What is the percent of increase? *(Skills Review Handbook)*

Ⓐ 20% Ⓑ 25% Ⓒ 80% Ⓓ 125%

2.7 Dilations

Essential Question How can you enlarge or reduce a figure in the coordinate plane?

The Meaning of a Word ● Dilate

When you have your eyes checked, the optometrist sometimes **dilates** one or both of the pupils of your eyes.

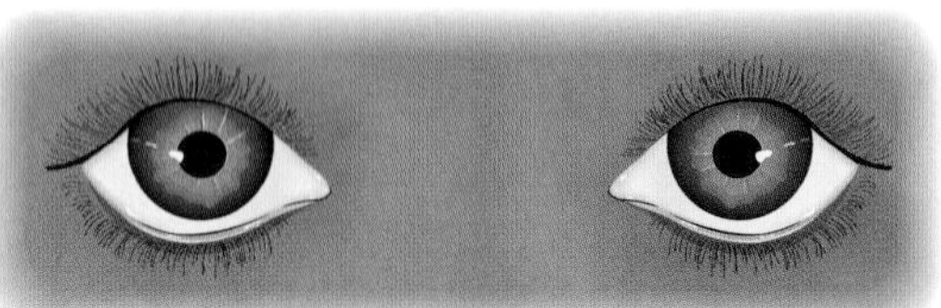

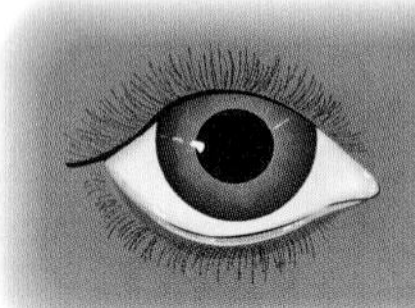

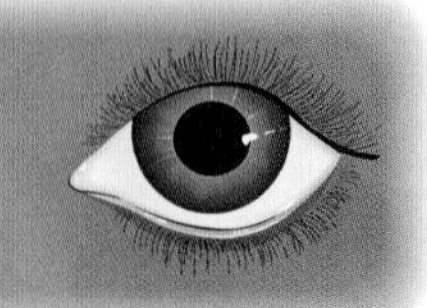

1 ACTIVITY: Comparing Triangles in a Coordinate Plane

Work with a partner. Write the coordinates of the vertices of the blue triangle. Then write the coordinates of the vertices of the red triangle.

a. How are the two sets of coordinates related?

b. How are the two triangles related? Explain your reasoning.

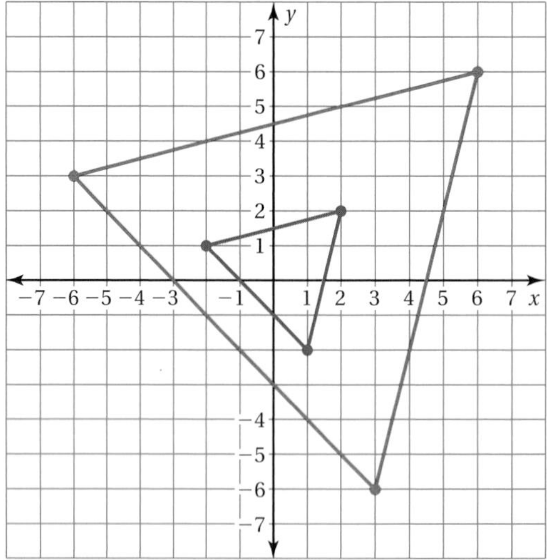

c. Draw a green triangle whose coordinates are twice the values of the corresponding coordinates of the blue triangle. How are the green and blue triangles related? Explain your reasoning.

d. How are the coordinates of the red and green triangles related? How are the two triangles related? Explain your reasoning.

Geometry

In this lesson, you will

- identify dilations.
- dilate figures in the coordinate plane.
- use more than one transformation to find images of figures.

2 ACTIVITY: Drawing Triangles in a Coordinate Plane

Work with a partner.

a. Draw the triangle whose vertices are (0, 2), (−2, 2), and (1, −2).

b. Multiply each coordinate of the vertices by 2 to obtain three new vertices. Draw the triangle given by the three new vertices. How are the two triangles related?

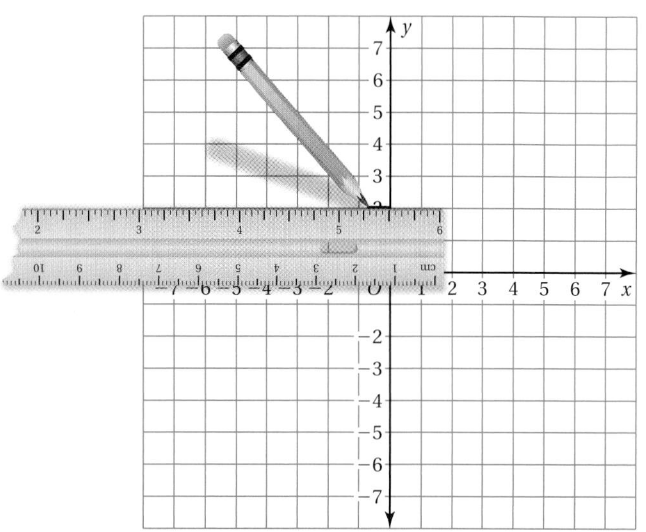

c. Repeat part (b) by multiplying by 3 instead of 2.

Math Practice

Use Prior Results

What are the four types of transformations you studied in this chapter? What information can you use to fill in your table?

3 ACTIVITY: Summarizing Transformations

Work with a partner. Make a table that summarizes the relationships between the original figure and its image for the four types of transformations you studied in this chapter.

What Is Your Answer?

4. **IN YOUR OWN WORDS** How can you enlarge or reduce a figure in the coordinate plane?

5. Describe how knowing how to enlarge or reduce figures in a technical drawing is important in a career such as drafting.

Practice Use what you learned about dilations to complete Exercises 4–6 on page 87.

2.7 Lesson

A **dilation** is a transformation in which a figure is made larger or smaller with respect to a point called the **center of dilation**.

EXAMPLE 1 Identifying a Dilation

Key Vocabulary 
dilation, *p. 84*
center of dilation, *p. 84*
scale factor, *p. 84*

Tell whether the blue figure is a dilation of the red figure.

a.

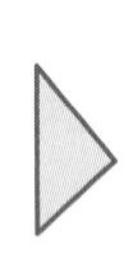

Lines connecting corresponding vertices meet at a point.

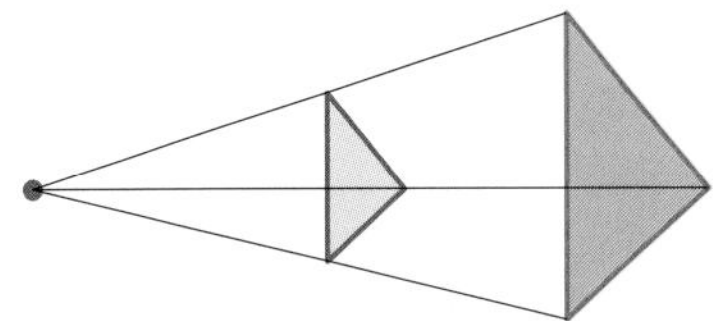

So, the blue figure is a dilation of the red figure.

b.

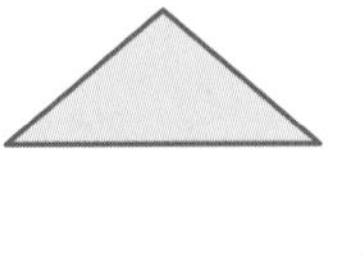

The figures have the same size and shape. The red figure *slides* to form the blue figure.

So, the blue figure is *not* a dilation of the red figure. It is a translation.

On Your Own

Now You're Ready
Exercises 7–12

Tell whether the blue figure is a dilation of the red figure. Explain.

1.

2.

In a dilation, the original figure and its image are similar. The ratio of the side lengths of the image to the corresponding side lengths of the original figure is the **scale factor** of the dilation.

Key Idea

Dilations in the Coordinate Plane

Words To dilate a figure with respect to the origin, multiply the coordinates of each vertex by the scale factor k.

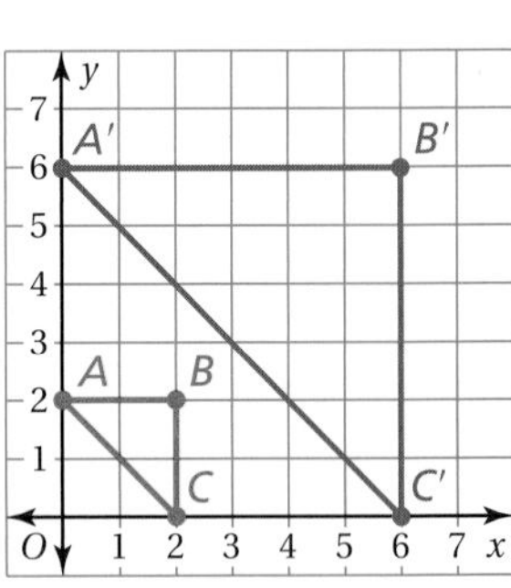

Algebra $(x, y) \rightarrow (kx, ky)$

- When $k > 1$, the dilation is an enlargement.
- When $k > 0$ and $k < 1$, the dilation is a reduction.

Multi-Language Glossary at BigIdeasMath.com

EXAMPLE 2 Dilating a Figure

Draw the image of Triangle *ABC* after a dilation with a scale factor of 3. Identify the type of dilation.

Multiply each x- and y-coordinate by the scale factor 3.

Vertices of *ABC*	$(3x, 3y)$	Vertices of *A′B′C′*
$A(1, 3)$	$(3 \cdot 1, 3 \cdot 3)$	$A'(3, 9)$
$B(2, 3)$	$(3 \cdot 2, 3 \cdot 3)$	$B'(6, 9)$
$C(2, 1)$	$(3 \cdot 2, 3 \cdot 1)$	$C'(6, 3)$

The image is shown at the right. The dilation is an *enlargement* because the scale factor is greater than 1.

Study Tip

You can check your answer by drawing a line from the origin through each vertex of the original figure. The vertices of the image should lie on these lines.

EXAMPLE 3 Dilating a Figure

Draw the image of Rectangle *WXYZ* after a dilation with a scale factor of 0.5. Identify the type of dilation.

Multiply each x- and y-coordinate by the scale factor 0.5.

Vertices of *WXYZ*	$(0.5x, 0.5y)$	Vertices of *W′X′Y′Z′*
$W(-4, -6)$	$(0.5 \cdot (-4), 0.5 \cdot (-6))$	$W'(-2, -3)$
$X(-4, 8)$	$(0.5 \cdot (-4), 0.5 \cdot 8)$	$X'(-2, 4)$
$Y(4, 8)$	$(0.5 \cdot 4, 0.5 \cdot 8)$	$Y'(2, 4)$
$Z(4, -6)$	$(0.5 \cdot 4, 0.5 \cdot (-6))$	$Z'(2, -3)$

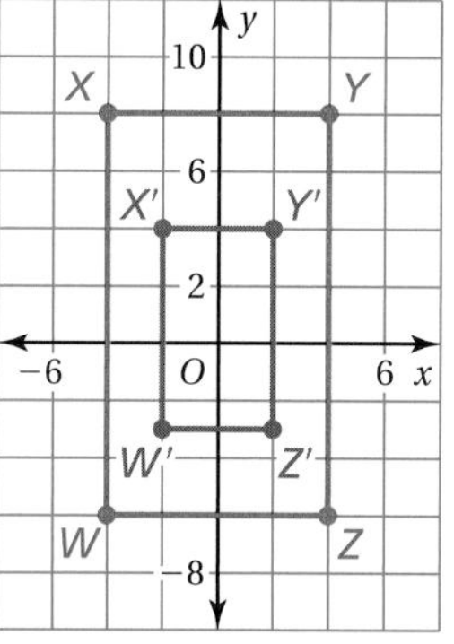

The image is shown at the right. The dilation is a *reduction* because the scale factor is greater than 0 and less than 1.

On Your Own

Now You're Ready
Exercises 13–18

3. **WHAT IF?** Triangle *ABC* in Example 2 is dilated by a scale factor of 2. What are the coordinates of the image?

4. **WHAT IF?** Rectangle *WXYZ* in Example 3 is dilated by a scale factor of $\frac{1}{4}$. What are the coordinates of the image?

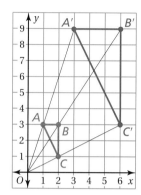

EXAMPLE 4 Using More than One Transformation

The vertices of a trapezoid are $A(-2, -1)$, $B(-1, 1)$, $C(0, 1)$, and $D(0, -1)$. Dilate the trapezoid with respect to the origin using a scale factor of 2. Then translate it 6 units right and 2 units up. What are the coordinates of the image?

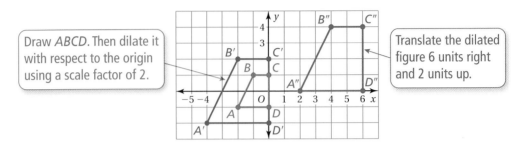

Draw *ABCD*. Then dilate it with respect to the origin using a scale factor of 2.

Translate the dilated figure 6 units right and 2 units up.

∴ The coordinates of the image are $A''(2, 0)$, $B''(4, 4)$, $C''(6, 4)$, and $D''(6, 0)$.

The image of a translation, reflection, or rotation is congruent to the original figure, and the image of a dilation is similar to the original figure. So, two figures are similar when one can be obtained from the other by a sequence of translations, reflections, rotations, and dilations.

EXAMPLE 5 Describing a Sequence of Transformations

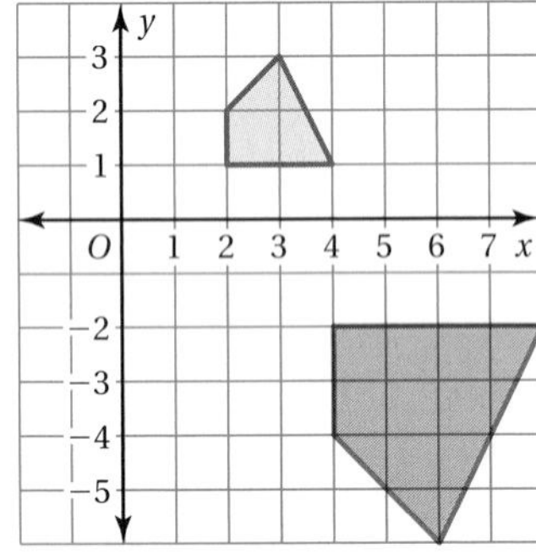

The red figure is similar to the blue figure. Describe a sequence of transformations in which the blue figure is the image of the red figure.

From the graph, you can see that the blue figure is one-half the size of the red figure. So, begin with a dilation with respect to the origin using a scale factor of $\frac{1}{2}$.

After dilating, you need to flip the figure in the x-axis.

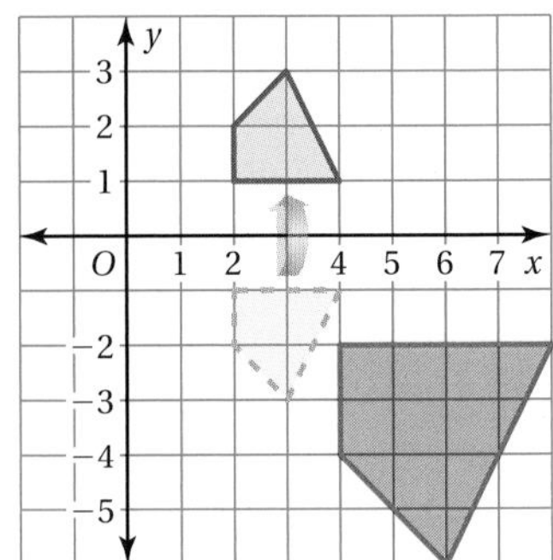

∴ So, one possible sequence of transformations is a dilation with respect to the origin using a scale factor of $\frac{1}{2}$ followed by a reflection in the x-axis.

On Your Own

5. In Example 4, use a scale factor of 3 in the dilation. Then rotate the figure 180° about the image of vertex *C*. What are the coordinates of the image?

6. In Example 5, can you reflect the red figure first, and then perform the dilation to obtain the blue figure? Explain.

2.7 Exercises

Vocabulary and Concept Check

1. **VOCABULARY** How is a dilation different from other transformations?
2. **VOCABULARY** For what values of scale factor k is a dilation called an *enlargement*? a *reduction*?
3. **REASONING** Which figure is *not* a dilation of the blue figure? Explain.

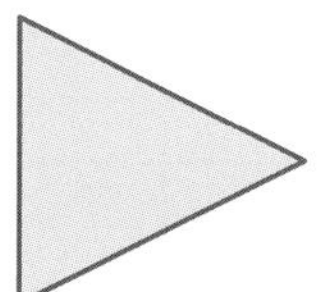 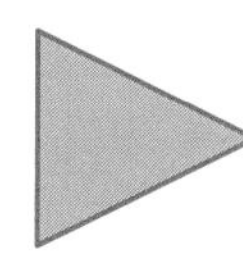  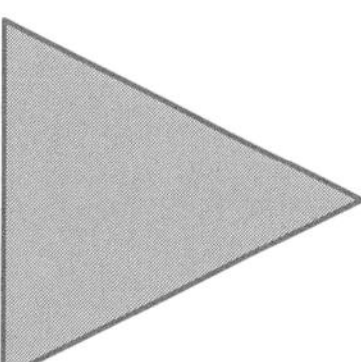

Practice and Problem Solving

Draw the triangle with the given vertices. Multiply each coordinate of the vertices by 3, and then draw the new triangle. How are the two triangles related?

4. (0, 2), (3, 2), (3, 0)
5. (−1, 1), (−1, −2), (2, −2)
6. (−3, 2), (1, 2), (1, −4)

Tell whether the blue figure is a dilation of the red figure.

7. 

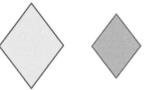

8. 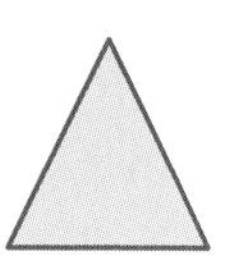 

9. 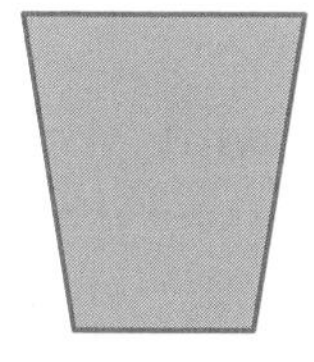 

10. 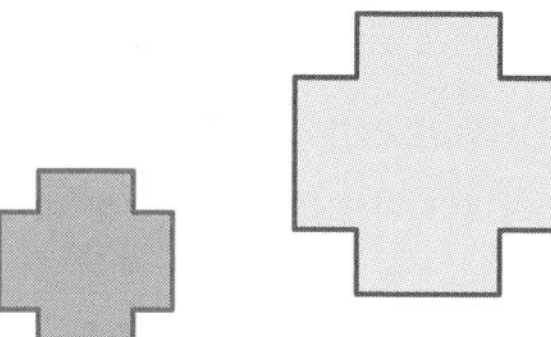

11. 

12. 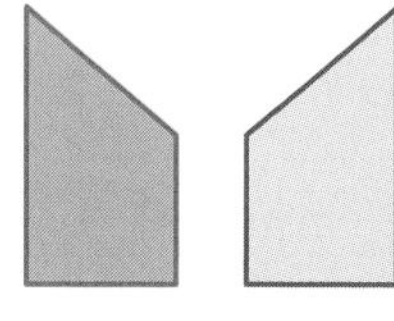

The vertices of a figure are given. Draw the figure and its image after a dilation with the given scale factor. Identify the type of dilation.

13. $A(1, 1)$, $B(1, 4)$, $C(3, 1)$; $k = 4$
14. $D(0, 2)$, $E(6, 2)$, $F(6, 4)$; $k = 0.5$
15. $G(-2, -2)$, $H(-2, 6)$, $J(2, 6)$; $k = 0.25$
16. $M(2, 3)$, $N(5, 3)$, $P(5, 1)$; $k = 3$
17. $Q(-3, 0)$, $R(-3, 6)$, $T(4, 6)$, $U(4, 0)$; $k = \frac{1}{3}$
18. $V(-2, -2)$, $W(-2, 3)$, $X(5, 3)$, $Y(5, -2)$; $k = 5$

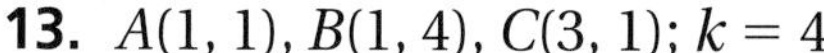

19. ERROR ANALYSIS Describe and correct the error in listing the coordinates of the image after a dilation with a scale factor of $\frac{1}{2}$.

✗

Vertices of ABC	$(2x, 2y)$	Vertices of $A'B'C'$
$A(2, 5)$	$(2 \cdot 2, 2 \cdot 5)$	$A'(4, 10)$
$B(2, 0)$	$(2 \cdot 2, 2 \cdot 0)$	$B'(4, 0)$
$C(4, 0)$	$(2 \cdot 4, 2 \cdot 0)$	$C'(8, 0)$

The blue figure is a dilation of the red figure. Identify the type of dilation and find the scale factor.

20. 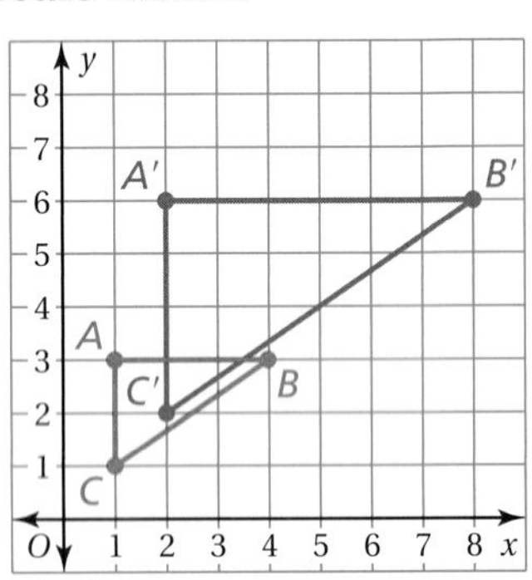

21. 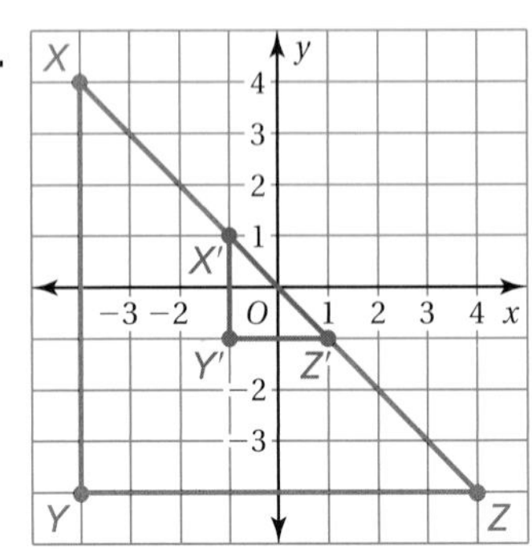

22. 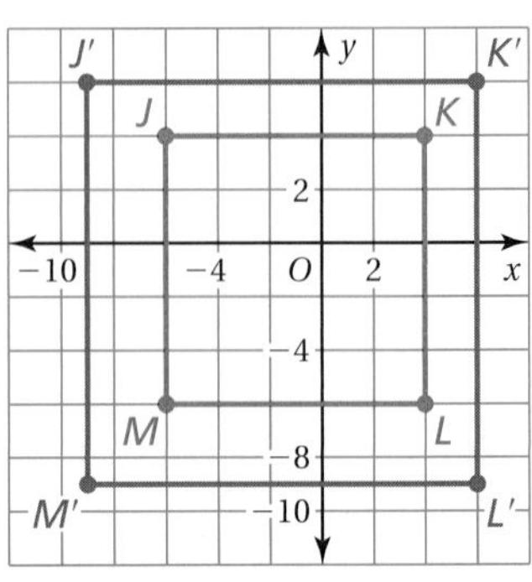

The vertices of a figure are given. Find the coordinates of the figure after the transformations given.

4 **23.** $A(-5, 3), B(-2, 3), C(-2, 1), D(-5, 1)$

Reflect in the y-axis. Then dilate with respect to the origin using a scale factor of 2.

24. $F(-9, -9), G(-3, -6), H(-3, -9)$

Dilate with respect to the origin using a scale factor of $\frac{2}{3}$. Then translate 6 units up.

25. $J(1, 1), K(3, 4), L(5, 1)$

Rotate 90° clockwise about the origin. Then dilate with respect to the origin using a scale factor of 3.

26. $P(-2, 2), Q(4, 2), R(2, -6), S(-4, -6)$

Dilate with respect to the origin using a scale factor of 5. Then dilate with respect to the origin using a scale factor of 0.5.

The red figure is similar to the blue figure. Describe a sequence of transformations in which the blue figure is the image of the red figure.

5 **27.** 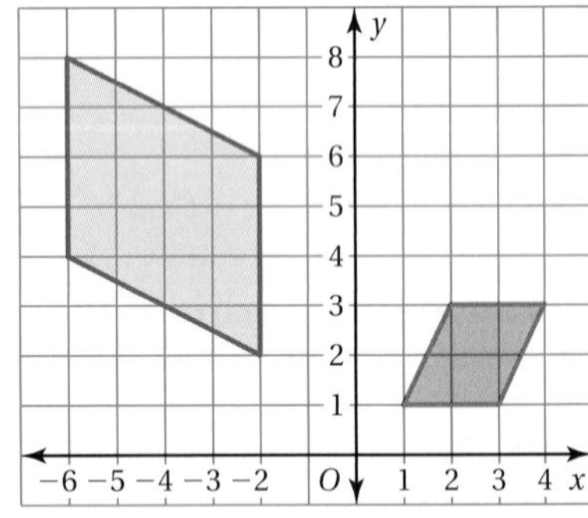

28. 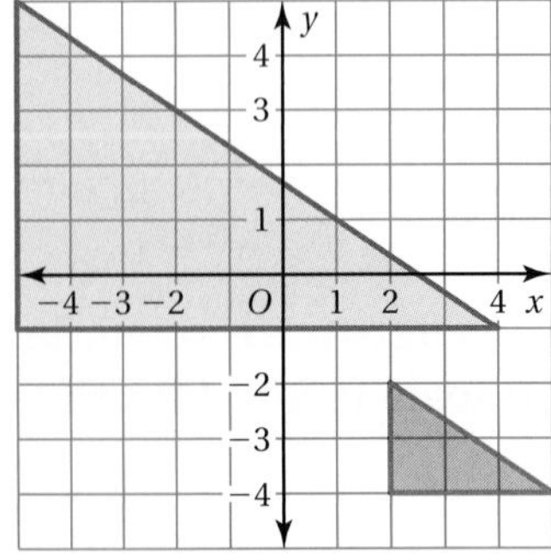

29. STRUCTURE In Exercises 27 and 28, is the blue figure still the image of the red figure when you perform the sequence in the opposite order? Explain.

30. **OPEN-ENDED** Draw a rectangle on a coordinate plane. Choose a scale factor of 2, 3, 4, or 5, and then dilate the rectangle. How many times greater is the area of the image than the area of the original rectangle?

31. **SHADOW PUPPET** You can use a flashlight and a shadow puppet (your hands) to project shadows on the wall.

   a. Identify the type of dilation.

   b. What does the flashlight represent?

   c. The length of the ears on the shadow puppet is 3 inches. The length of the ears on the shadow is 4 inches. What is the scale factor?

   d. Describe what happens as the shadow puppet moves closer to the flashlight. How does this affect the scale factor?

32. **REASONING** A triangle is dilated using a scale factor of 3. The image is then dilated using a scale factor of $\frac{1}{2}$. What scale factor could you use to dilate the original triangle to get the final image? Explain.

CRITICAL THINKING The coordinate notation shows how the coordinates of a figure are related to the coordinates of its image after transformations. What are the transformations? Are the figure and its image similar or congruent? Explain.

33. $(x, y) \rightarrow (2x + 4, 2y - 3)$

34. $(x, y) \rightarrow (-x - 1, y - 2)$

35. $(x, y) \rightarrow \left(\frac{1}{3}x, -\frac{1}{3}y\right)$

36. **STRUCTURE** How are the transformations $(2x + 3, 2y - 1)$ and $(2(x + 3), 2(y + 1))$ different?

37. 

The vertices of a trapezoid are $A(-2, 3)$, $B(2, 3)$, $C(5, -2)$, and $D(-2, -2)$. Dilate the trapezoid with respect to vertex A using a scale factor of 2. What are the coordinates of the image? Explain the method you used.

Fair Game Review What you learned in previous grades & lessons

Tell whether the angles are *complementary* or *supplementary*. Then find the value of *x*. *(Skills Review Handbook)*

38. 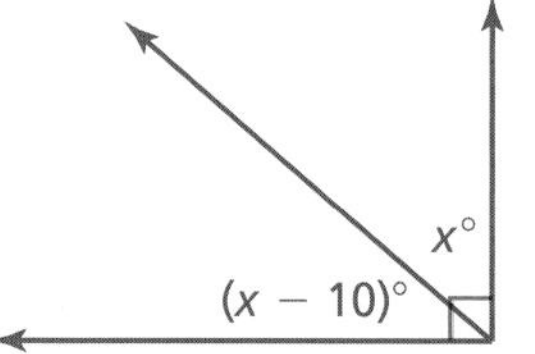

39. 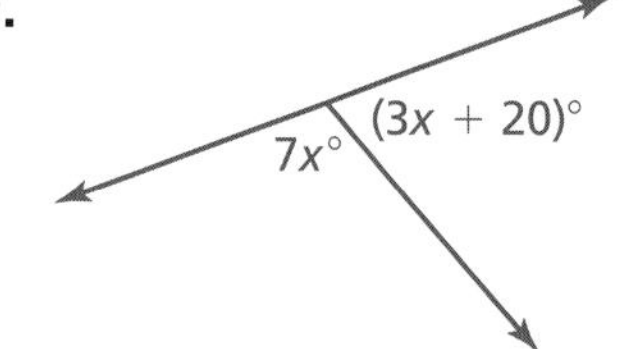

40. 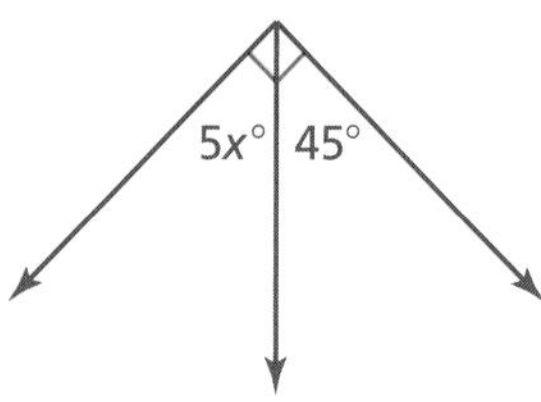

41. **MULTIPLE CHOICE** Which quadrilateral is *not* a parallelogram? *(Skills Review Handbook)*

   Ⓐ rhombus   Ⓑ trapezoid   Ⓒ square   Ⓓ rectangle

2.5–2.7 Quiz

Check It Out
Progress Check
BigIdeasMath.com

1. Tell whether the two rectangles are similar. Explain your reasoning. *(Section 2.5)*

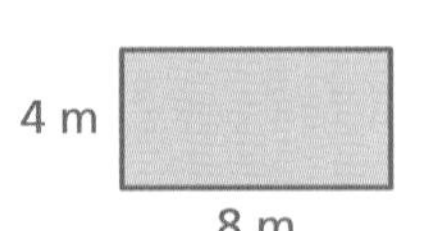

The figures are similar. Find *x*. *(Section 2.5)*

2. 

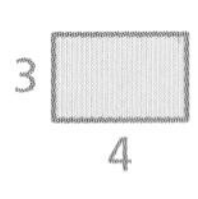

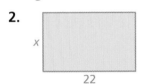

3. 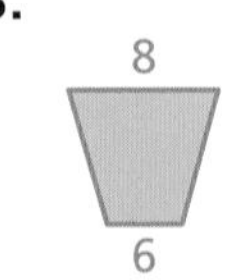

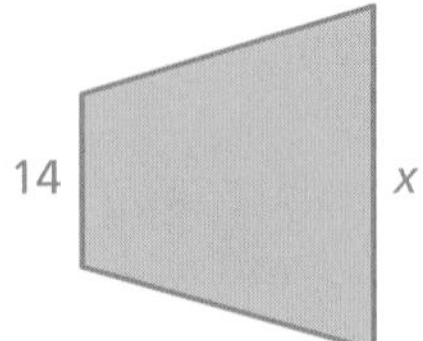

The two figures are similar. Find the ratios (red to blue) of the perimeters and of the areas. *(Section 2.6)*

4. 

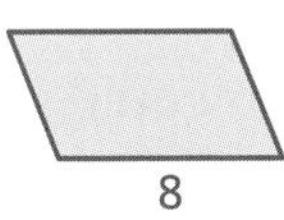

5. 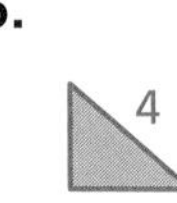

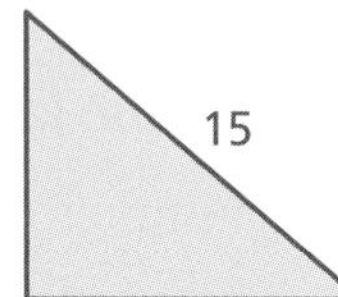

Tell whether the blue figure is a dilation of the red figure. *(Section 2.7)*

6. 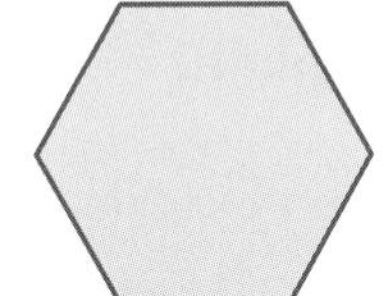

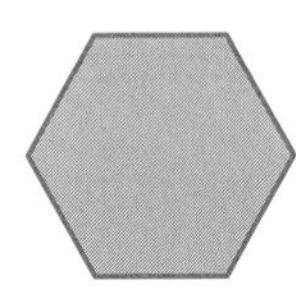

7. 

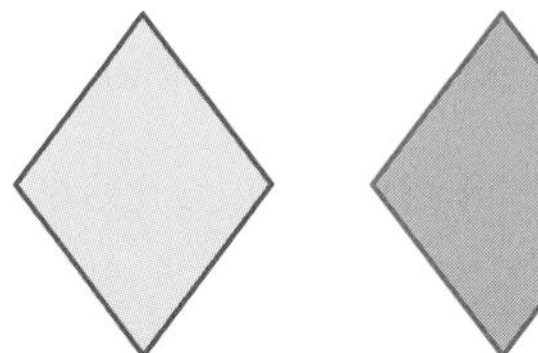

8. **SCREENS** The TV screen is similar to the computer screen. What is the area of the TV screen? *(Section 2.6)*

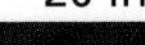

9. **GEOMETRY** The vertices of a rectangle are $A(2, 4)$, $B(5, 4)$, $C(5, -1)$, and $D(2, -1)$. Dilate the rectangle with respect to the origin using a scale factor of $\frac{1}{2}$. Then translate it 4 units left and 3 units down. What are the coordinates of the image? *(Section 2.7)*

10. **TENNIS COURT** The tennis courts for singles and doubles matches are different sizes. Are the courts similar? Explain. *(Section 2.5)*

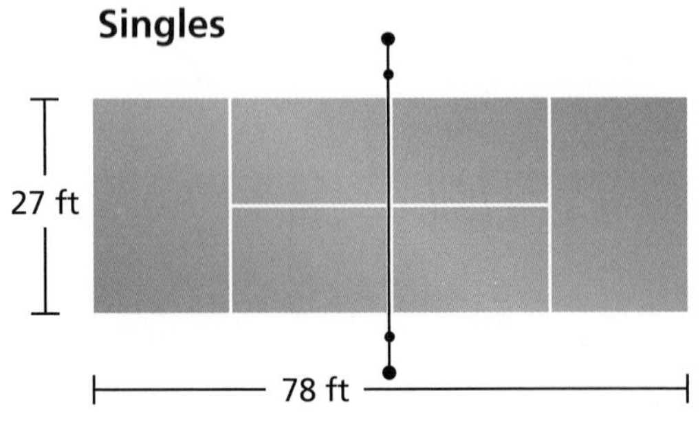

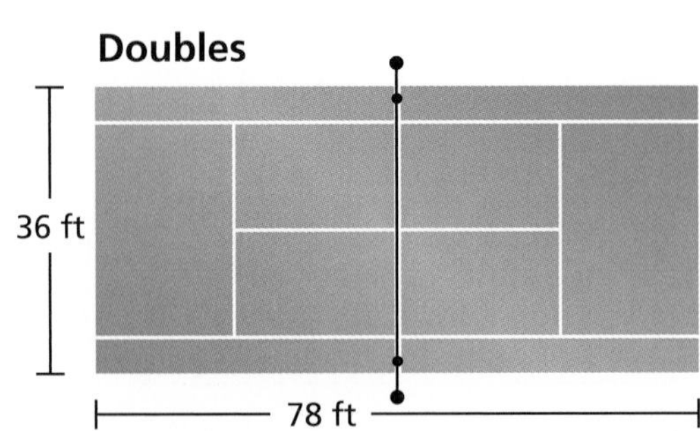

2 Chapter Review

Check It Out
Vocabulary Help
BigIdeasMath.com

Review Key Vocabulary

congruent figures, *p. 44*	translation, *p. 50*	angle of rotation, *p. 62*
corresponding angles, *p. 44*	reflection, *p. 56*	similar figures, *p. 72*
corresponding sides, *p. 44*	line of reflection, *p. 56*	dilation, *p. 84*
transformation, *p. 50*	rotation, *p. 62*	center of dilation, *p. 84*
image, *p. 50*	center of rotation, *p. 62*	scale factor, *p. 84*

Review Examples and Exercises

2.1 Congruent Figures *(pp. 42–47)*

Trapezoids *EFGH* and *QRST* are congruent.

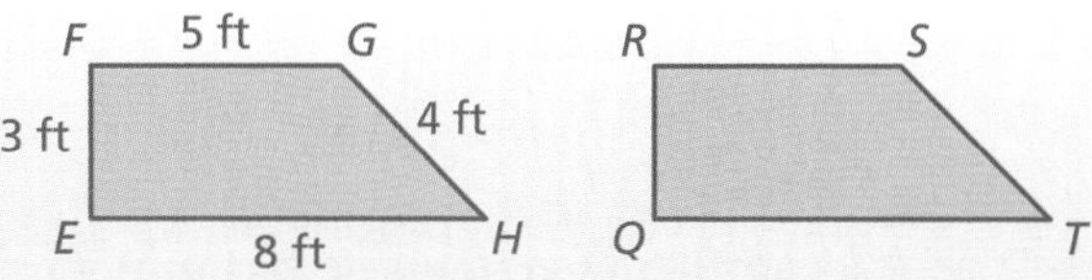

a. What is the length of side *QT*?

Side *QT* corresponds to side *EH*.

∴ So, the length of side *QT* is 8 feet.

b. Which angle of *QRST* corresponds to ∠*H*?

∴ ∠*T* corresponds to ∠*H*.

Exercises

Use the figures above.

1. What is the length of side *QR*?
2. What is the perimeter of *QRST*?

The figures are congruent. Name the corresponding angles and the corresponding sides.

3. 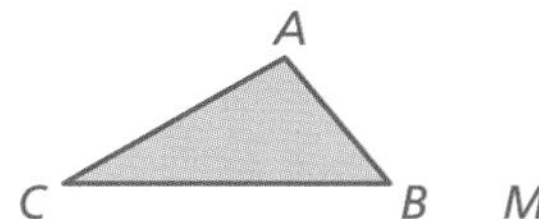

4. 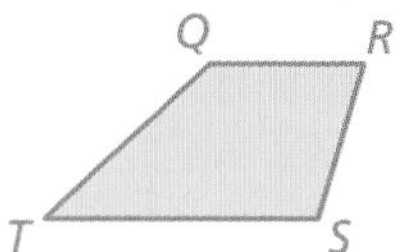

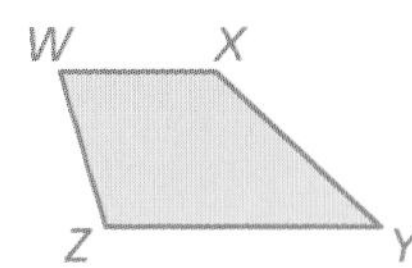

2.2 Translations *(pp. 48–53)*

Translate the red triangle 4 units left and 1 unit down. What are the coordinates of the image?

Move each vertex 4 units left and 1 unit down.

Connect the vertices. Label as A', B', and C'.

y, A, A′, B′, B, C, C′, 1 2 3 4 5 6 x

∴ The coordinates of the image are $A'(-1, 4)$, $B'(2, 2)$, and $C'(0, 0)$.

Exercises

Tell whether the blue figure is a translation of the red figure.

5.

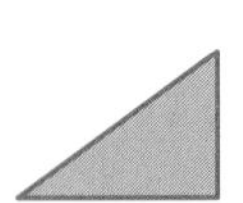
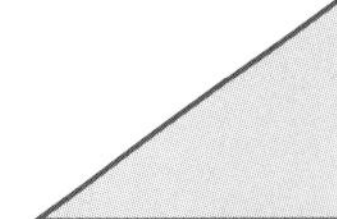

6.

7. The vertices of a quadrilateral are $W(1, 2)$, $X(1, 4)$, $Y(4, 4)$, and $Z(4, 2)$. Draw the figure and its image after a translation 3 units left and 2 units down.

8. The vertices of a triangle are $A(-1, -2)$, $B(-2, 2)$, and $C(-3, 0)$. Draw the figure and its image after a translation 5 units right and 1 unit up.

2.3 Reflections (pp. 54–59)

The vertices of a triangle are $A(-2, 1)$, $B(4, 1)$, and $C(4, 4)$. Draw the figure and its reflection in the x-axis. What are the coordinates of the image?

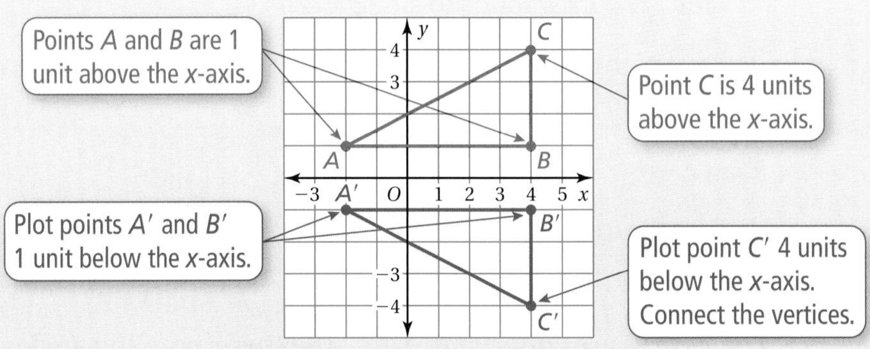
Points A and B are 1 unit above the x-axis.
Point C is 4 units above the x-axis.
Plot points A′ and B′ 1 unit below the x-axis.
Plot point C′ 4 units below the x-axis. Connect the vertices.
A
B
C
A′
B′
C′
x
y
O
-3
1
2
3
4
5
4
3
-3
-4

The coordinates of the image are $A'(-2, -1)$, $B'(4, -1)$, and $C'(4, -4)$.

Exercises

Tell whether the blue figure is a reflection of the red figure.

9.

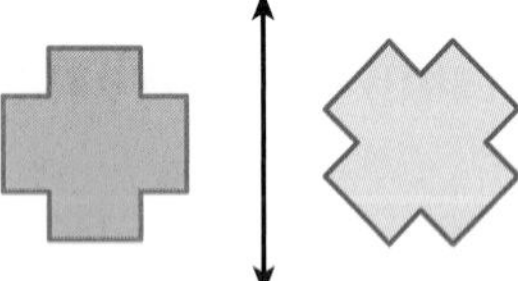

10.

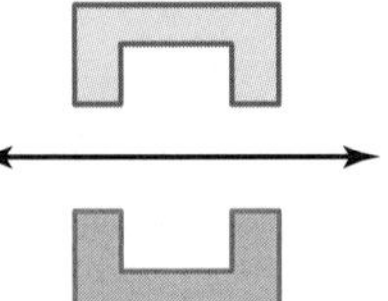

Draw the figure and its reflection in (a) the x-axis and (b) the y-axis.

11. $A(2, 0)$, $B(1, 5)$, $C(4, 3)$

12. $D(-5, -5)$, $E(-5, -1)$, $F(-2, -2)$, $G(-2, -5)$

13. The vertices of a rectangle are $E(-1, 1)$, $F(-1, 3)$, $G(-5, 3)$, and $H(-5, 1)$. Find the coordinates of the figure after reflecting in the x-axis, and then translating 3 units right.

2.4 Rotations (pp. 60–67)

The vertices of a triangle are $A(1, 1)$, $B(3, 2)$, and $C(2, 4)$. Rotate the triangle 90° counterclockwise about the origin. What are the coordinates of the image?

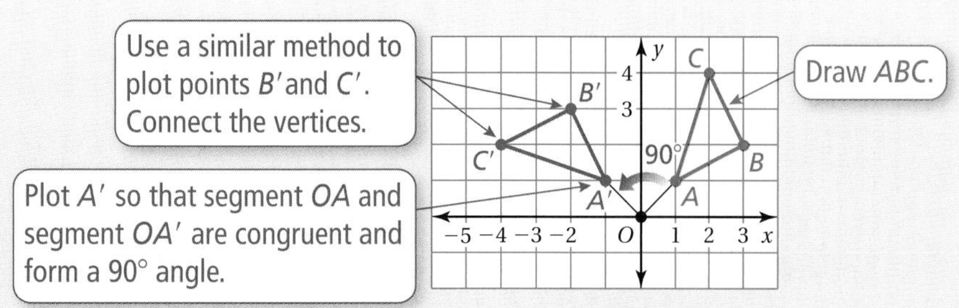

Use a similar method to plot points B' and C'. Connect the vertices.
Plot A' so that segment OA and segment OA' are congruent and form a 90° angle.
Draw ABC.

The coordinates of the image are $A'(-1, 1)$, $B'(-2, 3)$, and $C'(-4, 2)$.

Exercises

Tell whether the blue figure is a rotation of the red figure about the origin. If so, give the angle and the direction of rotation.

14.

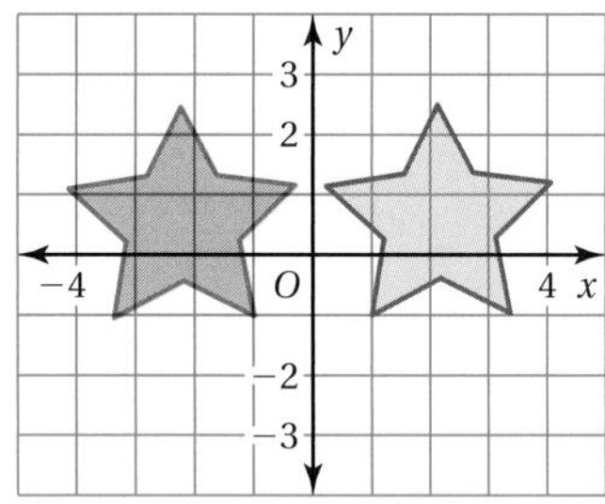

15.

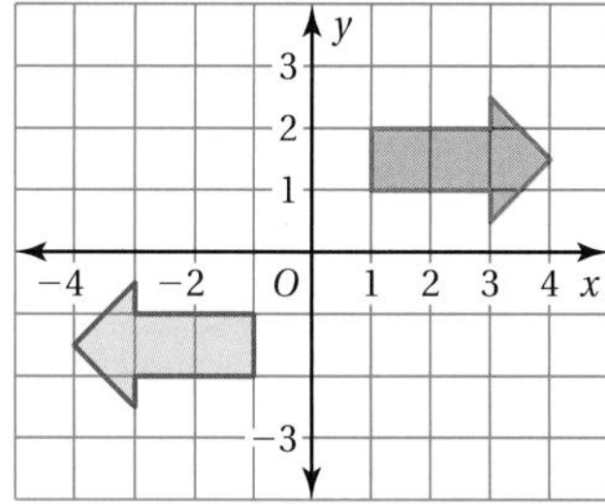

The vertices of a triangle are $A(-4, 2)$, $B(-2, 2)$, and $C(-3, 4)$. Rotate the triangle about the origin as described. Find the coordinates of the image.

16. 180°

17. 270° clockwise

2.5 Similar Figures (pp. 70–75)

a. Is Rectangle A similar to Rectangle B?

Each figure is a rectangle. So, corresponding angles are congruent. Check to see if corresponding side lengths are proportional.

Rectangle A: 10 by 4

Rectangle B: 5 by 2

$$\frac{\text{Length of A}}{\text{Length of B}} = \frac{10}{5} = 2 \qquad \frac{\text{Width of A}}{\text{Width of B}} = \frac{4}{2} = 2$$

Proportional

So, Rectangle A is similar to Rectangle B.

b. The two rectangles are similar. Find x.

Because the rectangles are similar, corresponding side lengths are proportional. So, write and solve a proportion to find x.

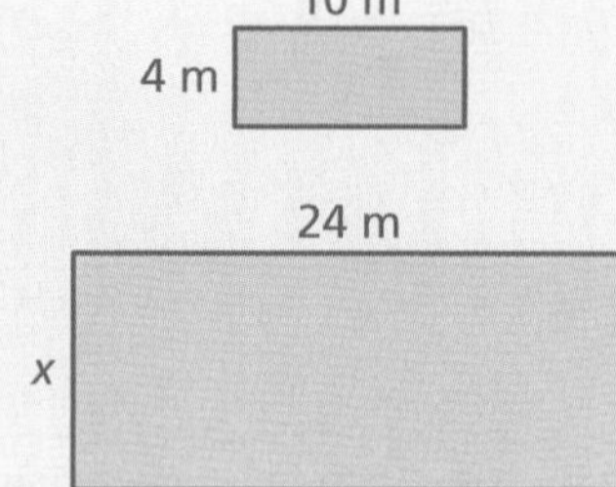

$\frac{10}{24} = \frac{4}{x}$ Write a proportion.

$10x = 96$ Cross Products Property

$x = 9.6$ Divide each side by 10.

So, x is 9.6 meters.

Exercises

Tell whether the two figures are similar. Explain your reasoning.

18.

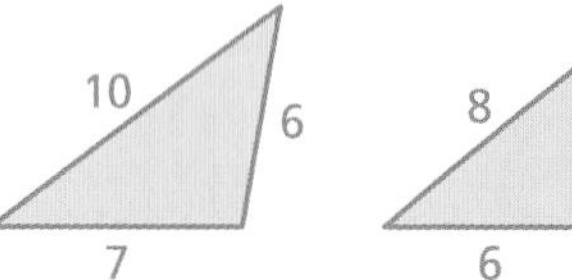

19.

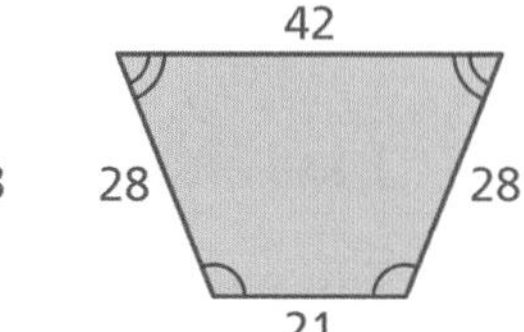

The figures are similar. Find x.

20.

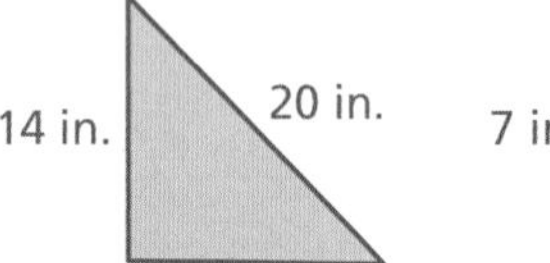

21.

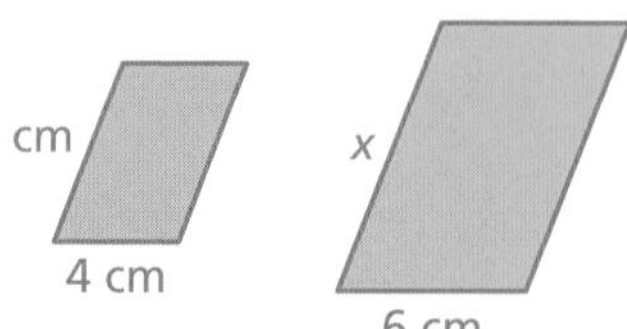

2.6 Perimeters and Areas of Similar Figures (pp. 76–81)

a. Find the ratio (red to blue) of the perimeters of the similar parallelograms.

$$\frac{\text{Perimeter of red parallelogram}}{\text{Perimeter of blue parallelogram}} = \frac{15}{9}$$

$$= \frac{5}{3}$$

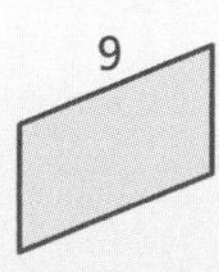

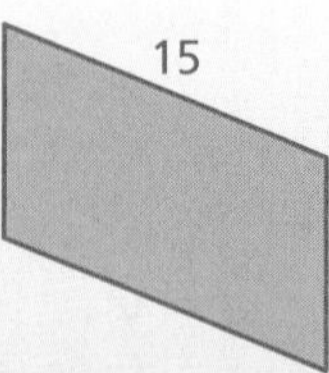

The ratio of the perimeters is $\frac{5}{3}$.

b. Find the ratio (red to blue) of the areas of the similar figures.

$$\frac{\text{Area of red figure}}{\text{Area of blue figure}} = \left(\frac{3}{4}\right)^2$$

$$= \frac{9}{16}$$

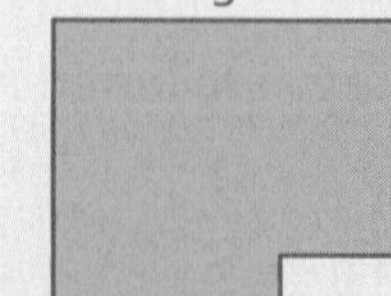

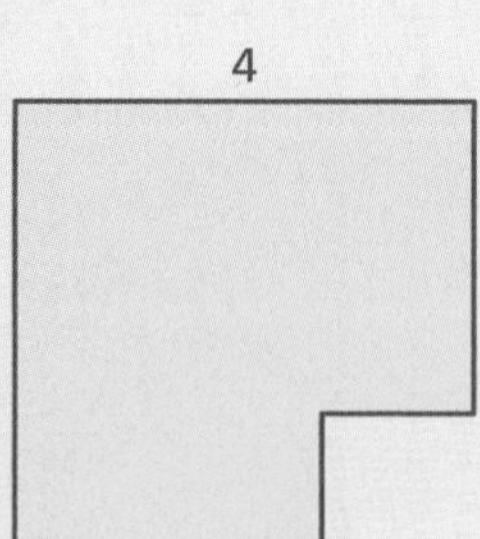

The ratio of the areas is $\frac{9}{16}$.

Exercises

The two figures are similar. Find the ratios (red to blue) of the perimeters and of the areas.

22. 

23. 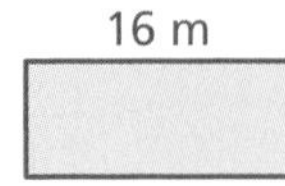

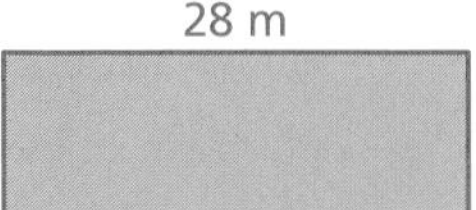

24. **PHOTOS** Two photos are similar. The ratio of the corresponding side lengths is 3 : 4. What is the ratio of the areas?

2.7 Dilations *(pp. 82–89)*

Draw the image of Triangle *ABC* after a dilation with a scale factor of 2. Identify the type of dilation.

Multiply each *x*- and *y*-coordinate by the scale factor 2.

Vertices of *ABC*	(2*x*, 2*y*)	Vertices of *A′B′C′*
$A(1, 1)$	$(2 \cdot 1, 2 \cdot 1)$	$A'(2, 2)$
$B(1, 2)$	$(2 \cdot 1, 2 \cdot 2)$	$B'(2, 4)$
$C(3, 2)$	$(2 \cdot 3, 2 \cdot 2)$	$C'(6, 4)$

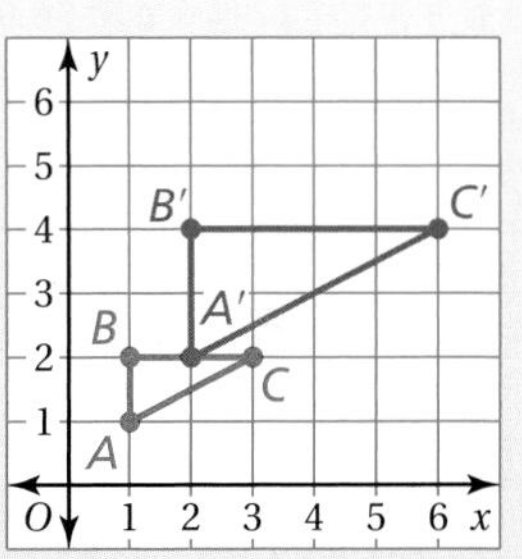

The image is shown at the above right. The dilation is an *enlargement* because the scale factor is greater than 1.

Exercises

Tell whether the blue figure is a dilation of the red figure.

25.  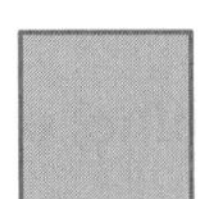

26. 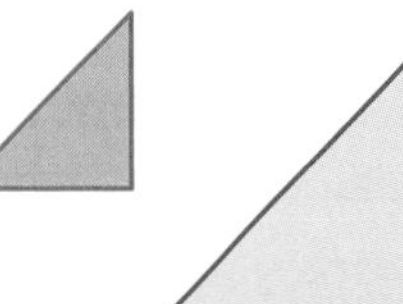

The vertices of a figure are given. Draw the figure and its image after a dilation with the given scale factor. Identify the type of dilation.

27. $P(-3, -2)$, $Q(-3, 0)$, $R(0, 0)$; $k = 4$

28. $B(3, 3)$, $C(3, 6)$, $D(6, 6)$, $E(6, 3)$; $k = \frac{1}{3}$

29. The vertices of a rectangle are $Q(-6, 2)$, $R(6, 2)$, $S(6, -4)$, and $T(-6, -4)$. Dilate the rectangle with respect to the origin using a scale factor of $\frac{3}{2}$. Then translate it 5 units right and 1 unit down. What are the coordinates of the image?

2 Chapter Test

Triangles *ABC* and *DEF* are congruent.

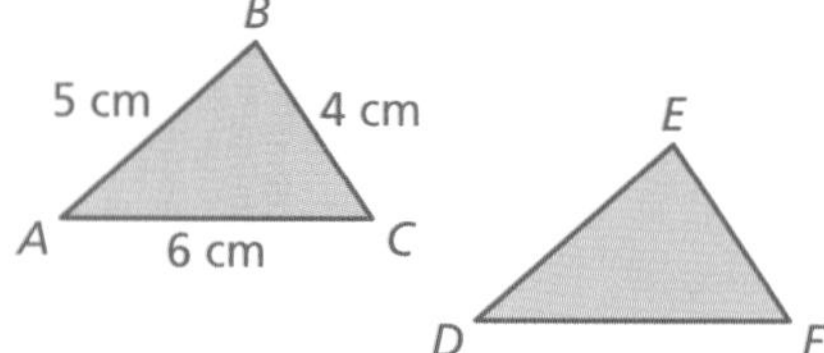

1. Which angle of *DEF* corresponds to $\angle C$?
2. What is the perimeter of *DEF*?

Tell whether the blue figure is a *translation, reflection, rotation,* or *dilation* of the red figure.

3.  

4.  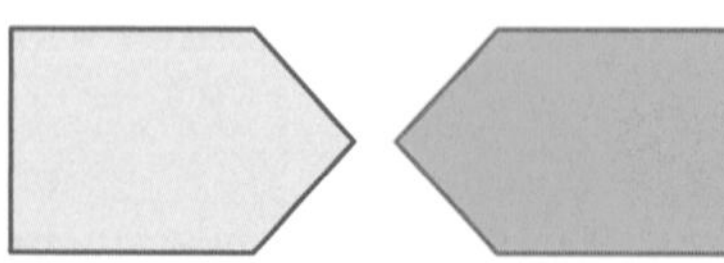

5. 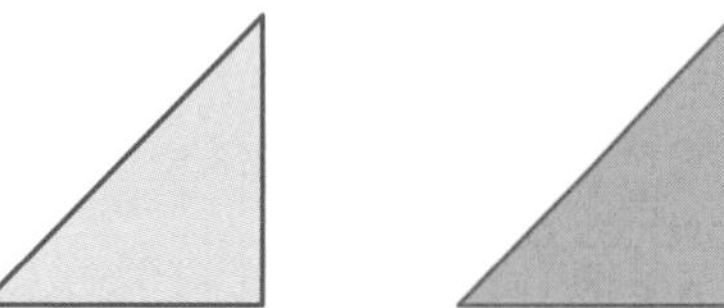

6.  

7. The vertices of a triangle are $A(2, 5)$, $B(1, 2)$, and $C(3, 1)$. Reflect the triangle in the x-axis, and then rotate the triangle 90° counterclockwise about the origin. What are the coordinates of the image?
8. The vertices of a triangle are $A(2, 4)$, $B(2, 1)$, and $C(5, 1)$. Dilate the triangle with respect to the origin using a scale factor of 2. Then translate the triangle 2 units left and 1 unit up. What are the coordinates of the image?
9. Tell whether the parallelograms are similar. Explain your reasoning.

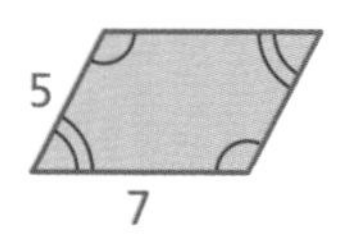

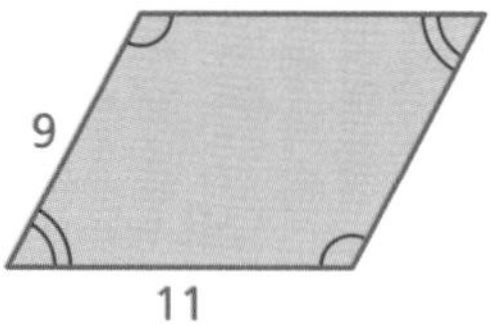

The two figures are similar. Find the ratios (red to blue) of the perimeters and of the areas.

10. 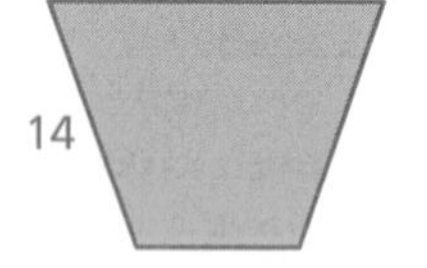

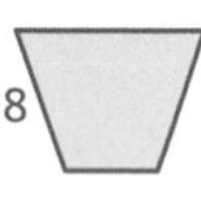

11. 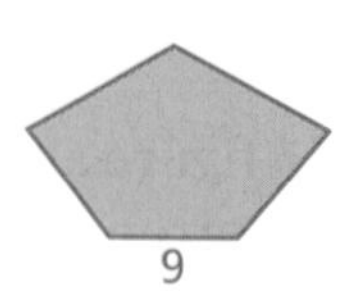

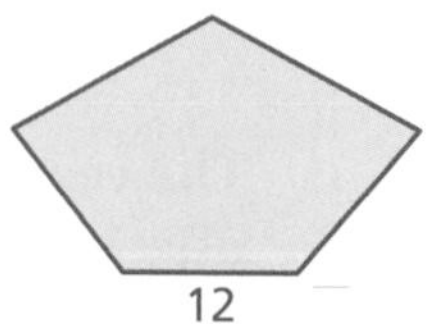

12. **SCREENS** A wide-screen television measures 36 inches by 54 inches. A movie theater screen measures 42 feet by 63 feet. Are the screens similar? Explain.
13. **CURTAINS** You want to use the rectangular piece of fabric shown to make a set of curtains for your window. Name the types of congruent shapes you can make with one straight cut. Draw an example of each type.

2 Cumulative Assessment

Test-Taking Strategy
After Answering Easy Questions, Relax
What type of transformation is shown?
Ⓐ rotation Ⓑ translation
Ⓒ dilation Ⓓ reflection
Lookin' good!
"After answering the easy questions, relax and try the harder ones. For this, the image is flipped. So, it's D."

1. A clockwise rotation of 90° is equivalent to a counterclockwise rotation of how many degrees?

2. The formula $K = C + 273.15$ converts temperatures from Celsius C to Kelvin K. Which of the following formulas is *not* correct?

   A. $K - C = 273.15$

   B. $C = K - 273.15$

   C. $C - K = -273.15$

   D. $C = K + 273.15$

3. Joe wants to solve the equation $-3(x + 2) = 12x$. What should he do first?

   F. Subtract 2 from each side.

   G. Add 3 to each side.

   H. Multiply each side by -3.

   I. Divide each side by -3.

4. Which transformation *turns* a figure?

   A. translation

   B. reflection

   C. rotation

   D. dilation

5. A triangle is graphed in the coordinate plane below.

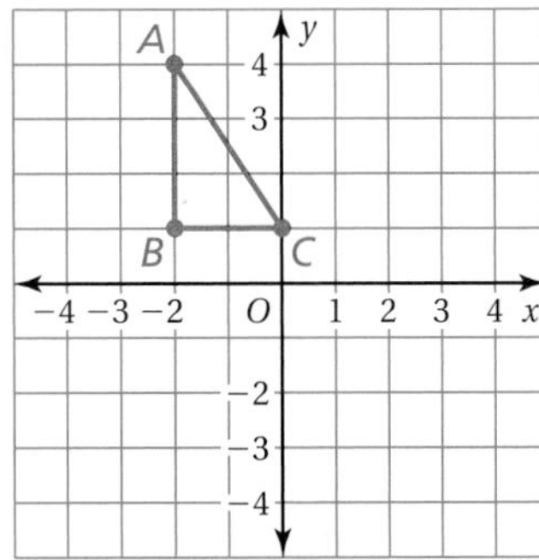

Translate the triangle 3 units right and 2 units down. What are the coordinates of the image?

   F. $A'(1, 4)$, $B'(1, 1)$, $C'(3, 1)$

   G. $A'(1, 2)$, $B'(1, -1)$, $C'(3, -1)$

   H. $A'(-2, 2)$, $B'(-2, -1)$, $C'(0, -1)$

   I. $A'(0, 1)$, $B'(0, -2)$, $C'(2, -2)$

6. Dale solved the equation in the box shown. What should Dale do to correct the error that he made?

$$-\frac{x}{3} + \frac{2}{5} = -\frac{7}{15}$$
$$-\frac{x}{3} + \frac{2}{5} - \frac{2}{5} = -\frac{7}{15} - \frac{2}{5}$$
$$-\frac{x}{3} = -\frac{13}{15}$$
$$3 \cdot \left(-\frac{x}{3}\right) = 3 \cdot \left(-\frac{13}{15}\right)$$
$$x = -2\frac{3}{5}$$

A. Add $\frac{2}{5}$ to each side to get $-\frac{x}{3} = -\frac{1}{15}$.

B. Multiply each side by -3 to get $x + \frac{2}{5} = \frac{7}{5}$.

C. Multiply each side by -3 to get $x = 2\frac{3}{5}$.

D. Subtract $\frac{2}{5}$ from each side to get $-\frac{x}{3} = -\frac{5}{10}$.

7. Jenny dilates the rectangle below using a scale factor of $\frac{1}{2}$.

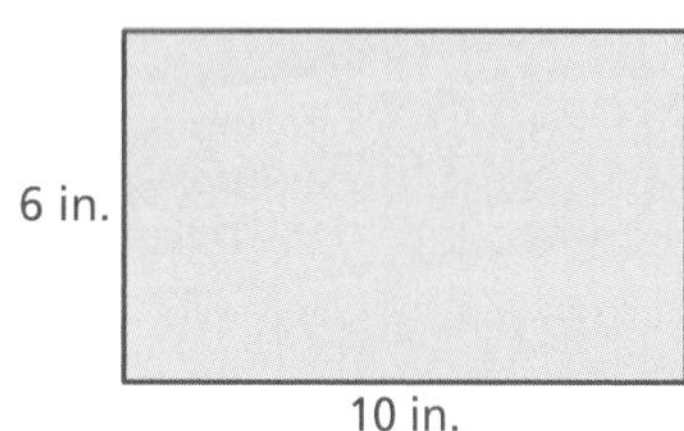

What is the area of the dilated rectangle in square inches?

8. The vertices of a rectangle are $A(-4, 2)$, $B(3, 2)$, $C(3, -5)$, and $D(-4, -5)$. If the rectangle is dilated by a scale factor of 3, what will be the coordinates of vertex C'?

F. $(9, -15)$

G. $(-12, 6)$

H. $(-12, -15)$

I. $(9, 6)$

9. In the figures, Triangle *EFG* is a dilation of Triangle *HIJ*.

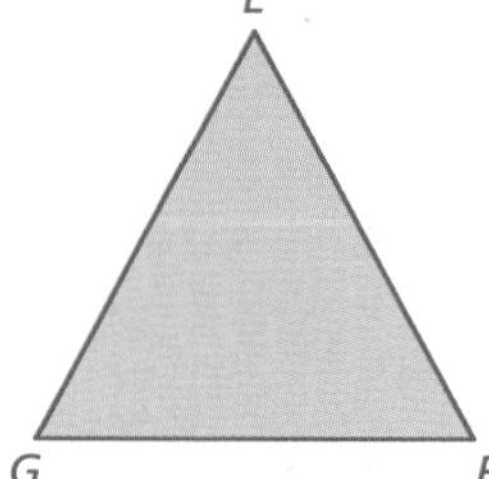

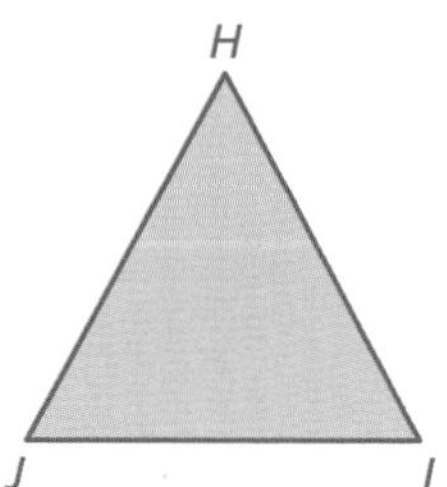

Which proportion is *not* necessarily correct for Triangle *EFG* and Triangle *HIJ*?

A. $\frac{EF}{FG} = \frac{HI}{IJ}$

B. $\frac{EG}{HI} = \frac{FG}{IJ}$

C. $\frac{GE}{EF} = \frac{JH}{HI}$

D. $\frac{EF}{HI} = \frac{GE}{JH}$

10. In the figures below, Rectangle *EFGH* is a dilation of Rectangle *IJKL*.

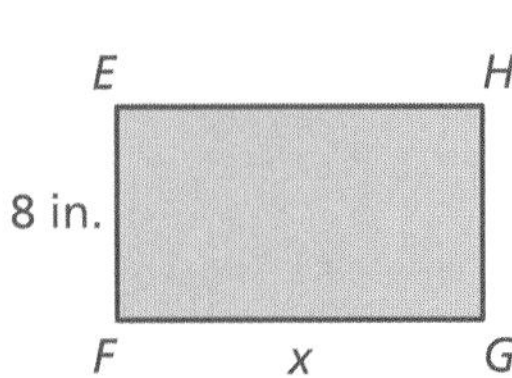

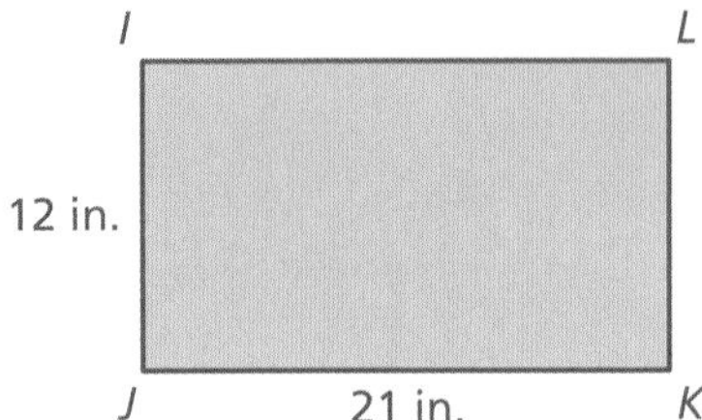

What is x?

F. 14 in.

G. 15 in.

H. 16 in.

I. 17 in.

11. Several transformations are used to create the pattern.

Part A Describe the transformation of Triangle *GLM* to Triangle *DGH*.

Part B Describe the transformation of Triangle *ALQ* to Triangle *GLM*.

Part C Triangle *DFN* is a dilation of Triangle *GHM*. Find the scale factor.

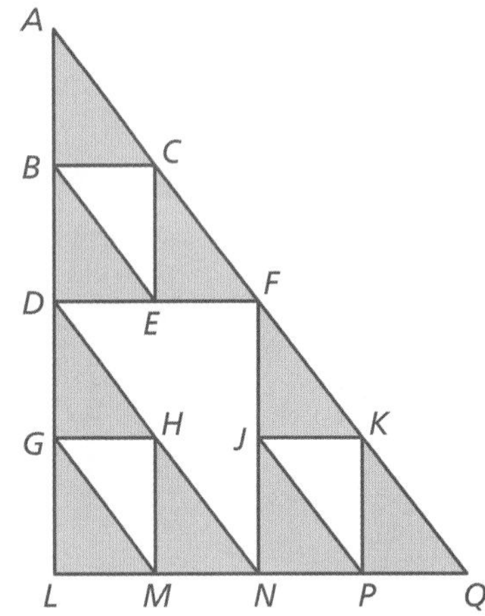

12. A rectangle is graphed in the coordinate plane below.

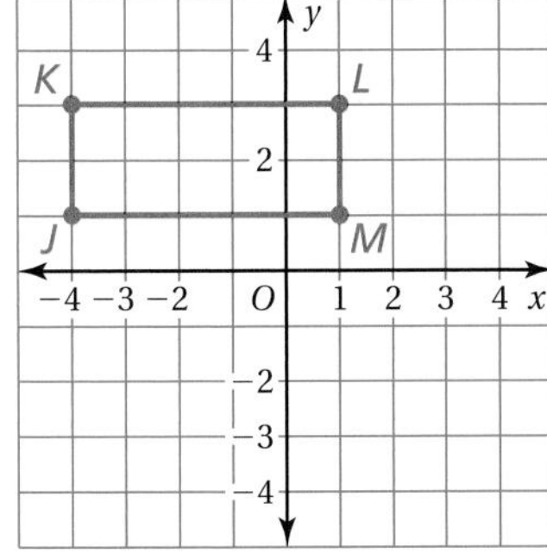

Rotate the rectangle 180° about the origin. What are the coordinates of the image?

A. $J'(4, -1)$, $K'(4, -3)$, $L'(-1, -3)$, $M'(-1, -1)$

B. $J'(-4, -1)$, $K'(-4, -3)$, $L'(1, -3)$, $M'(1, -1)$

C. $J'(1, 4)$, $K'(3, 4)$, $L'(3, -1)$, $M'(1, -1)$

D. $J'(-4, 1)$, $K'(-4, 3)$, $L'(1, 3)$, $M'(1, 1)$

3 Angles and Triangles

3.1 Parallel Lines and Transversals

3.2 Angles of Triangles

3.3 Angles of Polygons

3.4 Using Similar Triangles

"Start with any triangle."

"Tear off the angles. You can always rearrange the angles so that they form a straight line."

"What does that prove?"

"Let's use shadows and similar triangles to indirectly measure the height of the giant hyena standing right behind you."

What You Learned Before

Complementary Supplementary

30 degrees said to 60 degrees, "It's neat, you make me complete."

"I just remember that C comes before S and 90 comes before 180. That makes it easy."

Adjacent and Vertical Angles

Example 1 **Tell whether the angles are *adjacent* or *vertical*. Then find the value of *x*.**

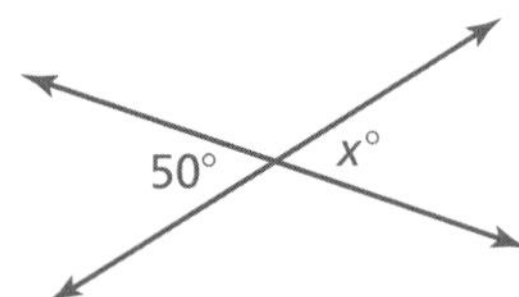

The angles are vertical angles. Because vertical angles are congruent, the angles have the same measure.

So, the value of x is 50.

Try It Yourself

Tell whether the angles are *adjacent* or *vertical*. Then find the value of *x*.

1. $(x + 8)°$ 120°

2.

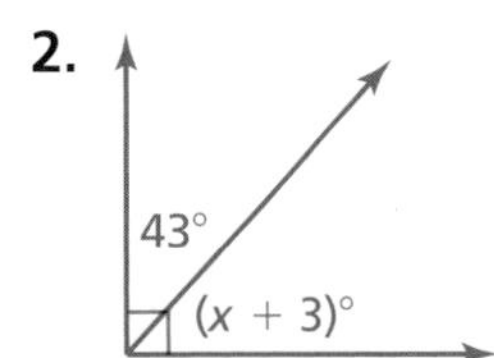

Complementary and Supplementary Angles

Example 2 **Tell whether the angles are *complementary* or *supplementary*. Then find the value of *x*.**

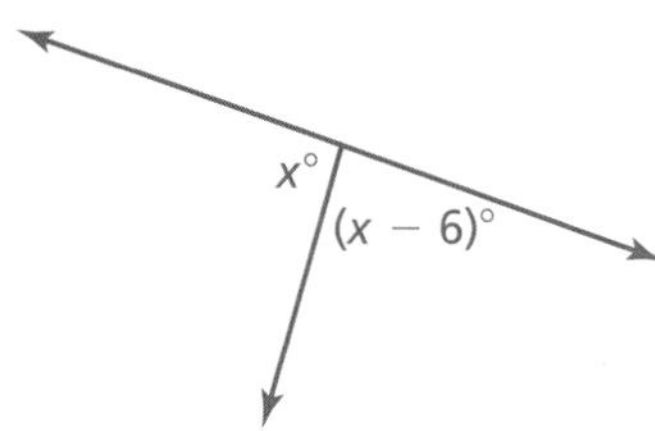

The two angles make up a straight angle. So, the angles are supplementary angles, and the sum of their measures is 180°.

$x + (x - 6) = 180$ Write equation.

$2x - 6 = 180$ Combine like terms.

$2x = 186$ Add 6 to each side.

$x = 93$ Divide each side by 2.

Try It Yourself

Tell whether the angles are *complementary* or *supplementary*. Then find the value of *x*.

3.

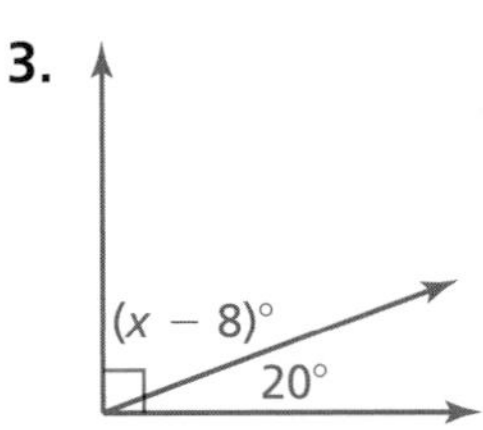

4.

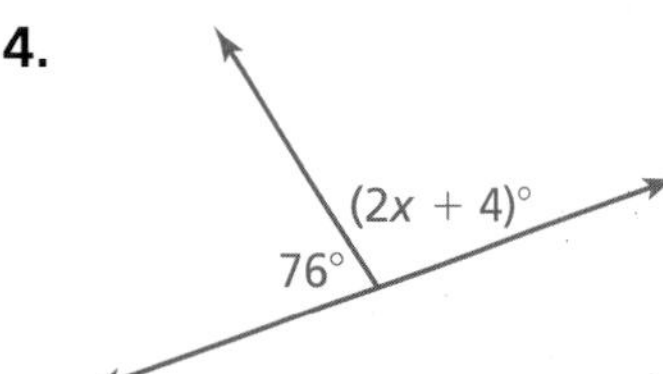

3.1 Parallel Lines and Transversals

Essential Question How can you describe angles formed by parallel lines and transversals?

The Meaning of a Word ● Transverse

When an object is **transverse**, it is lying or extending across something.

1 ACTIVITY: A Property of Parallel Lines

Work with a partner.

- **Discuss what it means for two lines to be parallel. Decide on a strategy for drawing two parallel lines. Then draw the two parallel lines.**
- **Draw a third line that intersects the two parallel lines. This line is called a transversal.**

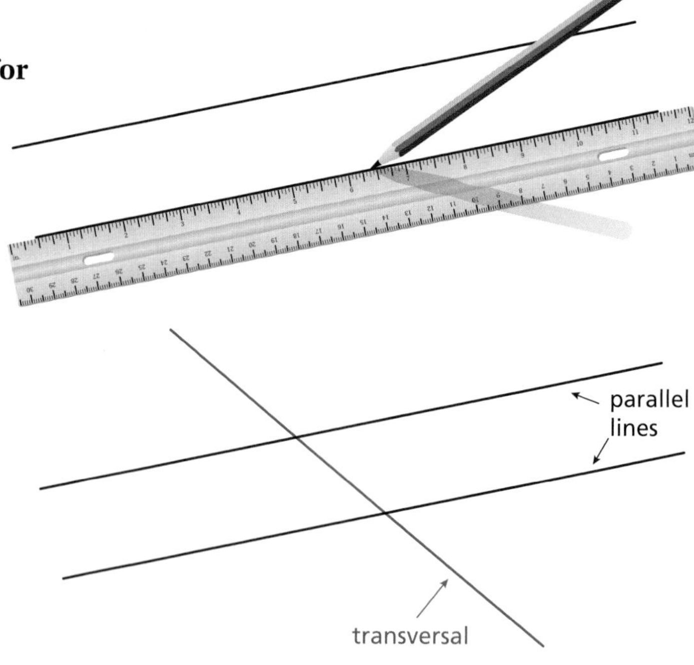

Geometry

In this lesson, you will

- identify the angles formed when parallel lines are cut by a transversal.
- find the measures of angles formed when parallel lines are cut by a transversal.

a. How many angles are formed by the parallel lines and the transversal? Label the angles.

b. Which of these angles have equal measures? Explain your reasoning.

2 ACTIVITY: Creating Parallel Lines

Math Practice

Use Clear Definitions

What do the words *parallel* and *transversal* mean? How does this help you answer the question in part (a)?

Work with a partner.

a. If you were building the house in the photograph, how could you make sure that the studs are parallel to each other?

b. Identify sets of parallel lines and transversals in the photograph.

3 ACTIVITY: Using Technology

Work with a partner. Use geometry software to draw two parallel lines intersected by a transversal.

a. Find all the angle measures.

b. Adjust the figure by moving the parallel lines or the transversal to a different position. Describe how the angle measures and relationships change.

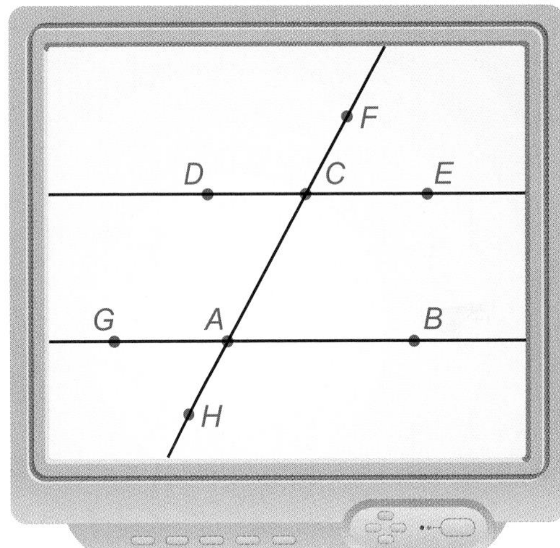

What Is Your Answer?

4. **IN YOUR OWN WORDS** How can you describe angles formed by parallel lines and transversals? Give an example.

5. Use geometry software to draw a transversal that is perpendicular to two parallel lines. What do you notice about the angles formed by the parallel lines and the transversal?

Practice Use what you learned about parallel lines and transversals to complete Exercises 3–6 on page 107.

3.1 Lesson

Key Vocabulary
transversal, *p. 104*
interior angles, *p. 105*
exterior angles, *p. 105*

Lines in the same plane that do not intersect are called *parallel lines*. Lines that intersect at right angles are called *perpendicular lines*.

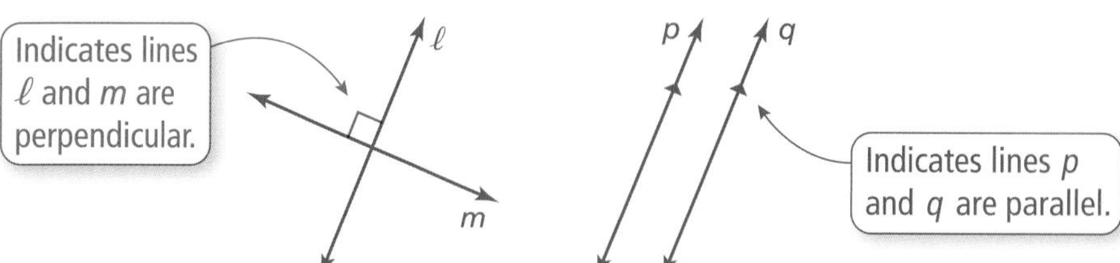

A line that intersects two or more lines is called a **transversal**. When parallel lines are cut by a transversal, several pairs of congruent angles are formed.

Key Idea

Study Tip
Corresponding angles lie on the same side of the transversal in corresponding positions.

Corresponding Angles

When a transversal intersects parallel lines, corresponding angles are congruent.

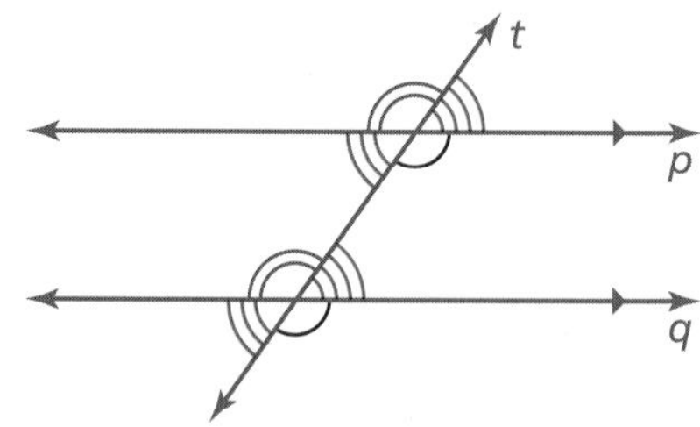

Corresponding angles

EXAMPLE 1 Finding Angle Measures

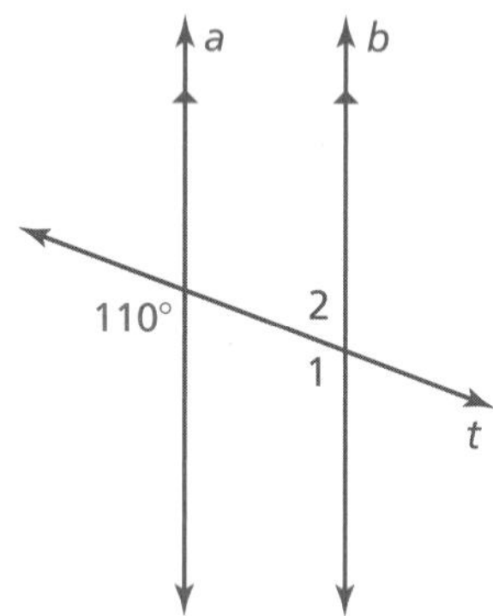

Use the figure to find the measures of (a) $\angle 1$ and (b) $\angle 2$.

a. $\angle 1$ and the 110° angle are corresponding angles. They are congruent.

So, the measure of $\angle 1$ is 110°.

b. $\angle 1$ and $\angle 2$ are supplementary.

$\angle 1 + \angle 2 = 180°$	Definition of supplementary angles
$110° + \angle 2 = 180°$	Substitute 110° for $\angle 1$.
$\angle 2 = 70°$	Subtract 110° from each side.

So, the measure of $\angle 2$ is 70°.

On Your Own

Now You're Ready
Exercises 7–9

Use the figure to find the measure of the angle. Explain your reasoning.

1. $\angle 1$ **2.** $\angle 2$

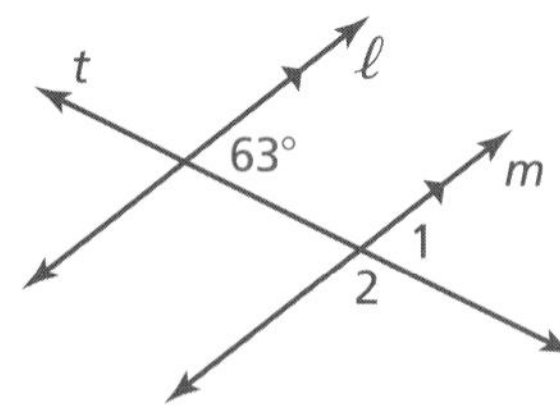

Multi-Language Glossary at BigIdeasMath.com

EXAMPLE 2 Using Corresponding Angles

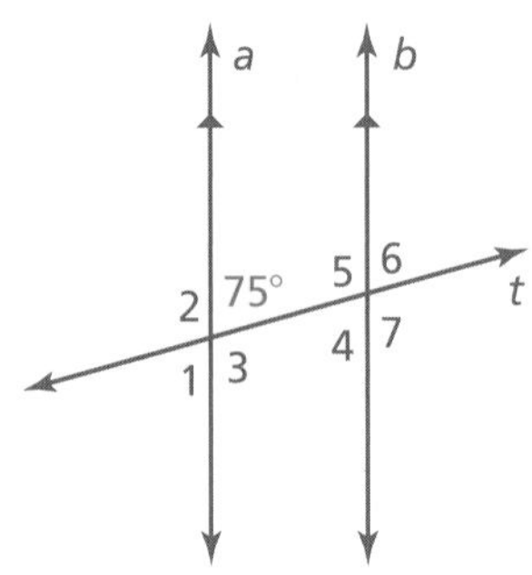

Use the figure to find the measures of the numbered angles.

$\angle 1$: $\angle 1$ and the 75° angle are vertical angles. They are congruent.

So, the measure of $\angle 1$ is 75°.

$\angle 2$ and $\angle 3$: The 75° angle is supplementary to both $\angle 2$ and $\angle 3$.

$75° + \angle 2 = 180°$ Definition of supplementary angles

$\angle 2 = 105°$ Subtract 75° from each side.

So, the measures of $\angle 2$ and $\angle 3$ are 105°.

$\angle 4$, $\angle 5$, $\angle 6$, and $\angle 7$: Using corresponding angles, the measures of $\angle 4$ and $\angle 6$ are 75°, and the measures of $\angle 5$ and $\angle 7$ are 105°.

On Your Own

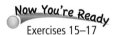

3. Use the figure to find the measures of the numbered angles.

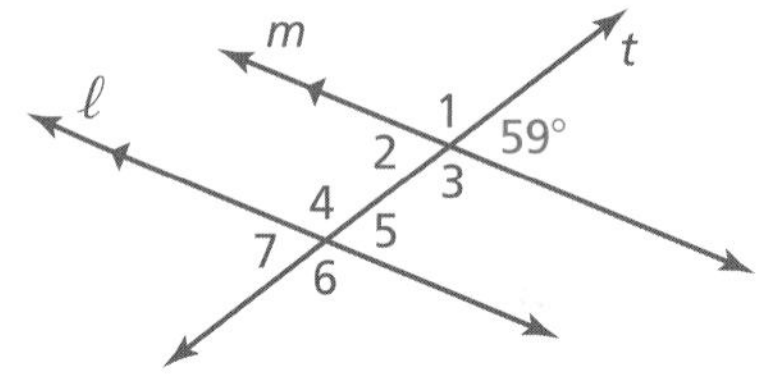

When two parallel lines are cut by a transversal, four **interior angles** are formed on the inside of the parallel lines and four **exterior angles** are formed on the outside of the parallel lines.

$\angle 3$, $\angle 4$, $\angle 5$, and $\angle 6$ are interior angles.
$\angle 1$, $\angle 2$, $\angle 7$, and $\angle 8$ are exterior angles.

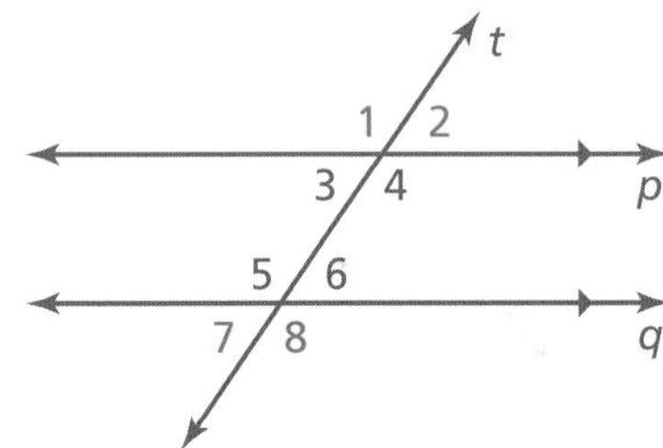

EXAMPLE 3 Using Corresponding Angles

A store owner uses pieces of tape to paint a window advertisement. The letters are slanted at an 80° angle. What is the measure of $\angle 1$?

Ⓐ 80° Ⓑ 100° Ⓒ 110° Ⓓ 120°

Because all the letters are slanted at an 80° angle, the dashed lines are parallel. The piece of tape is the transversal.

Using corresponding angles, the 80° angle is congruent to the angle that is supplementary to $\angle 1$, as shown.

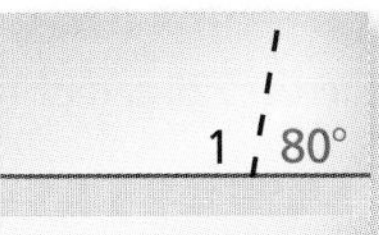

The measure of $\angle 1$ is $180° - 80° = 100°$. The correct answer is Ⓑ.

On Your Own

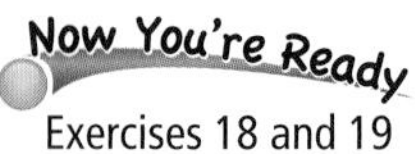

Exercises 18 and 19

4. **WHAT IF?** In Example 3, the letters are slanted at a 65° angle. What is the measure of $\angle 1$?

Key Idea

Alternate Interior Angles and Alternate Exterior Angles

When a transversal intersects parallel lines, alternate interior angles are congruent and alternate exterior angles are congruent.

Study Tip

Alternate interior angles and alternate exterior angles lie on opposite sides of the transversal.

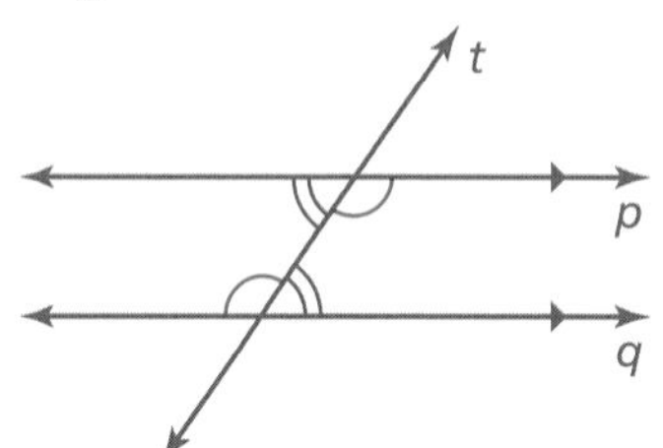

Alternate interior angles

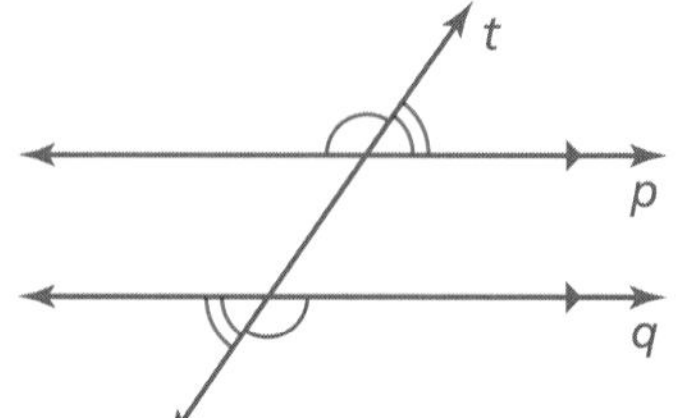

Alternate exterior angles

EXAMPLE 4 Identifying Alternate Interior and Alternate Exterior Angles

The photo shows a portion of an airport. Describe the relationship between each pair of angles.

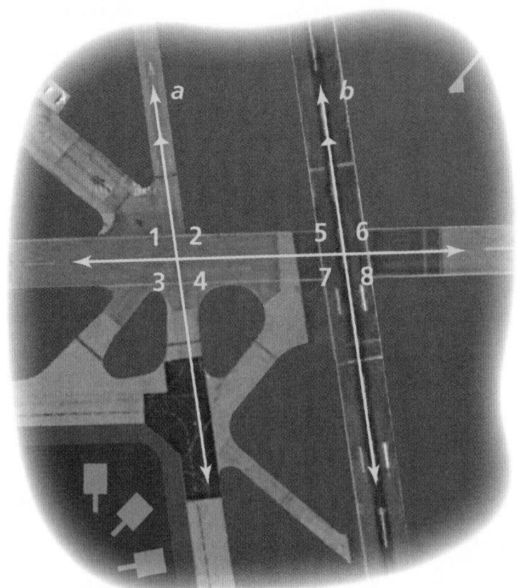

a. $\angle 3$ **and** $\angle 6$

$\angle 3$ and $\angle 6$ are alternate exterior angles.

So, $\angle 3$ is congruent to $\angle 6$.

b. $\angle 2$ **and** $\angle 7$

$\angle 2$ and $\angle 7$ are alternate interior angles.

So, $\angle 2$ is congruent to $\angle 7$.

On Your Own

Now You're Ready
Exercises 20 and 21

In Example 4, the measure of $\angle 4$ is 84°. Find the measure of the angle. Explain your reasoning.

5. $\angle 3$
6. $\angle 5$
7. $\angle 6$

3.1 Exercises

Vocabulary and Concept Check

1. **VOCABULARY** Draw two parallel lines and a transversal. Label a pair of corresponding angles.

2. **WHICH ONE DOESN'T BELONG?** Which statement does *not* belong with the other three? Explain your reasoning. Refer to the figure for Exercises 3–6.

The measure of $\angle 2$	The measure of $\angle 5$
The measure of $\angle 6$	The measure of $\angle 8$

Practice and Problem Solving

In Exercises 3–6, use the figure.

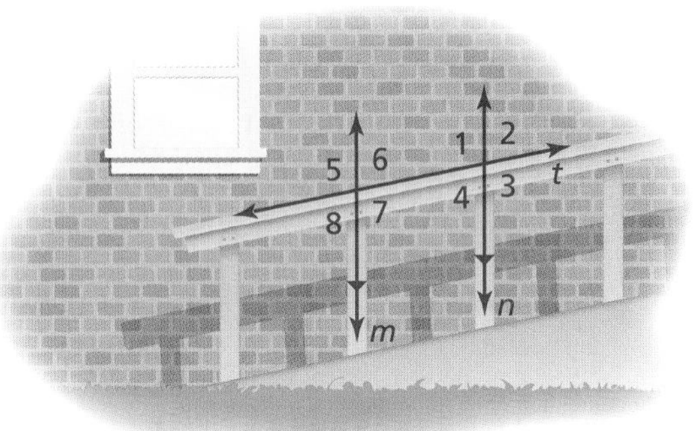

3. Identify the parallel lines.
4. Identify the transversal.
5. How many angles are formed by the transversal?
6. Which of the angles are congruent?

Use the figure to find the measures of the numbered angles.

1 7.

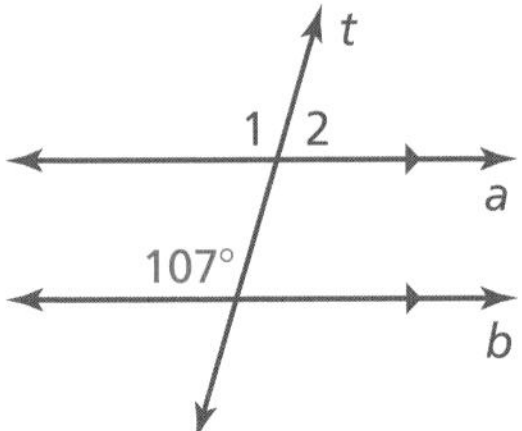

8.

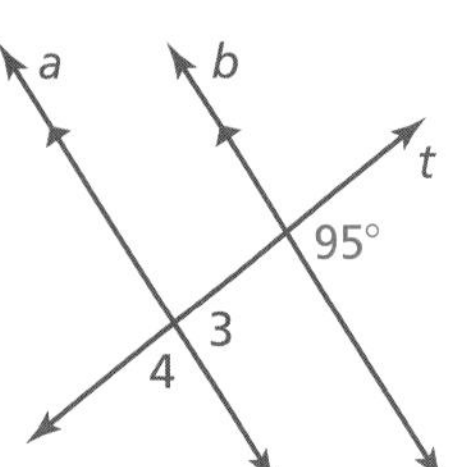

9.

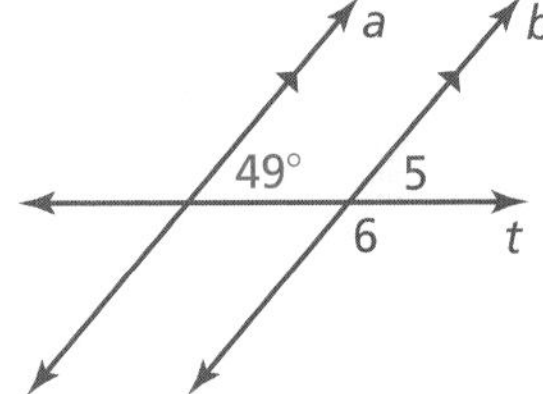

10. **ERROR ANALYSIS** Describe and correct the error in describing the relationship between the angles.

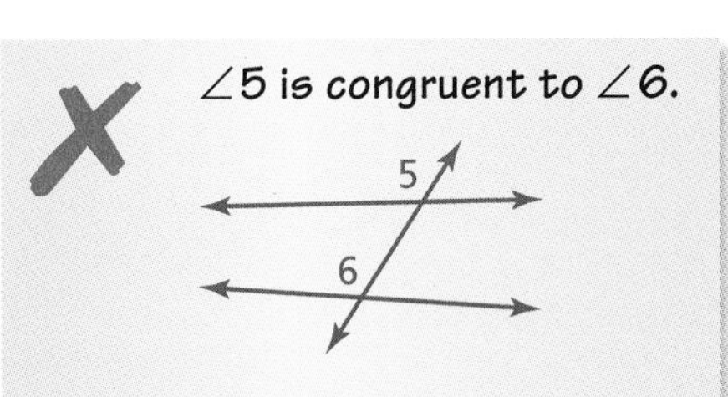

11. **PARKING** The painted lines that separate parking spaces are parallel. The measure of $\angle 1$ is 60°. What is the measure of $\angle 2$? Explain.

12. **OPEN-ENDED** Describe two real-life situations that use parallel lines.

13. **PROJECT** Trace line p and line t on a piece of paper. Label $\angle 1$. Move the paper so that $\angle 1$ aligns with $\angle 8$. Describe the transformations that you used to show that $\angle 1$ is congruent to $\angle 8$.

14. **REASONING** Two horizontal lines are cut by a transversal. What is the least number of angle measures you need to know in order to find the measure of every angle? Explain your reasoning.

Use the figure to find the measures of the numbered angles. Explain your reasoning.

② 15.

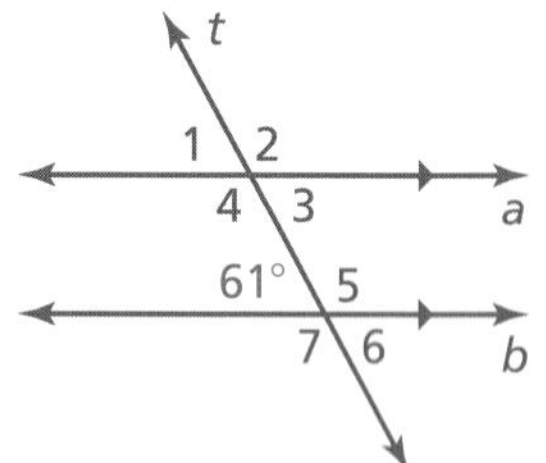

16.

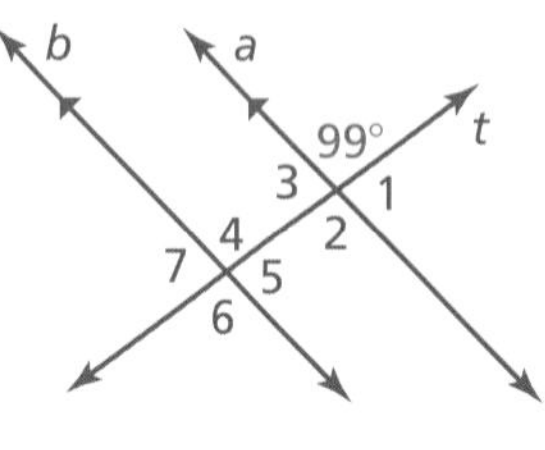

17.

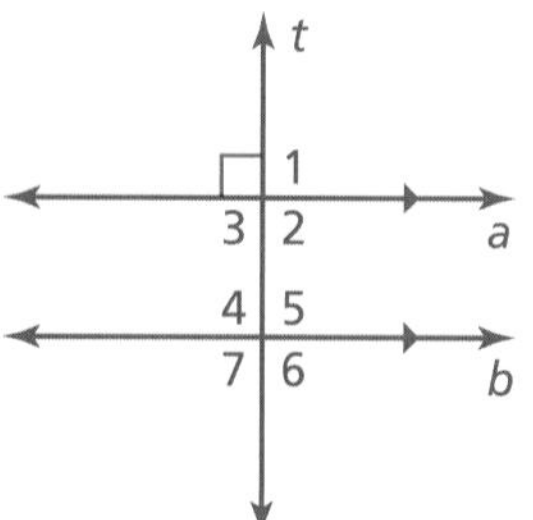

Complete the statement. Explain your reasoning.

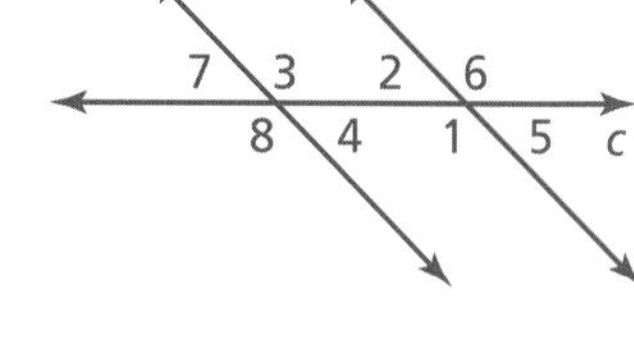

③ 18. If the measure of $\angle 1 = 124°$, then the measure of $\angle 4 =$ ___.

19. If the measure of $\angle 2 = 48°$, then the measure of $\angle 3 =$ ___.

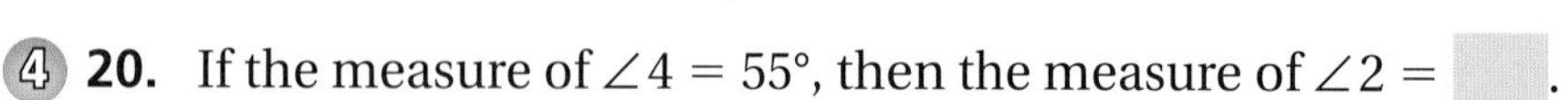

④ 20. If the measure of $\angle 4 = 55°$, then the measure of $\angle 2 =$ ___.

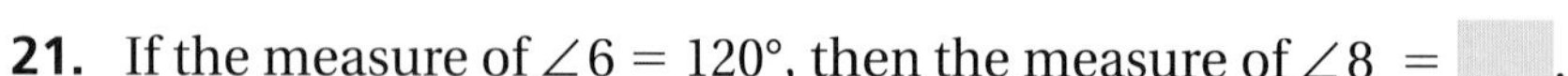

21. If the measure of $\angle 6 = 120°$, then the measure of $\angle 8 =$ ___.

22. If the measure of $\angle 7 = 50.5°$, then the measure of $\angle 6 =$ ___.

23. If the measure of $\angle 3 = 118.7°$, then the measure of $\angle 2 =$ ___.

24. **RAINBOW** A rainbow forms when sunlight reflects off raindrops at different angles. For blue light, the measure of $\angle 2$ is 40°. What is the measure of $\angle 1$?

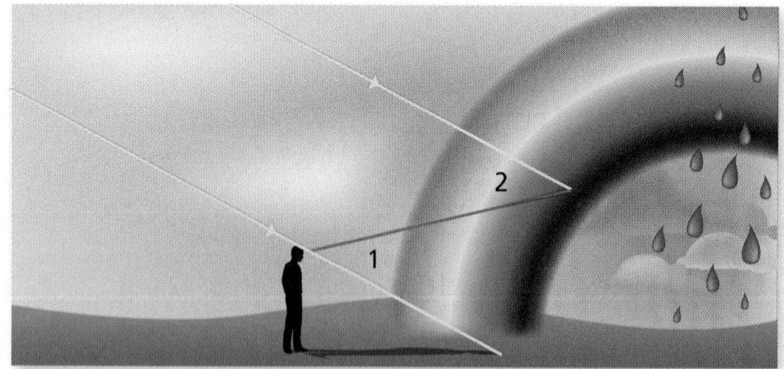

25. **REASONING** When a transversal is perpendicular to two parallel lines, all the angles formed measure 90°. Explain why.

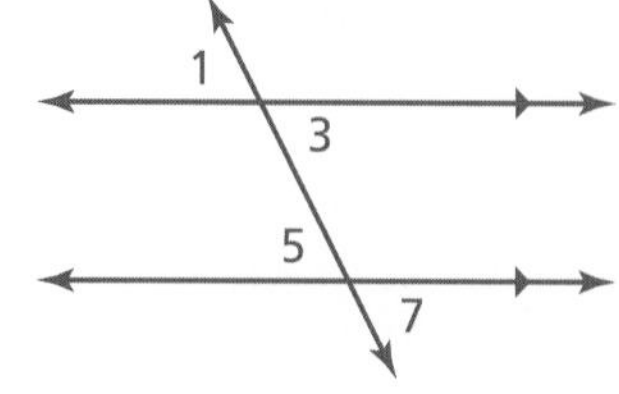

26. **LOGIC** Describe two ways you can show that $\angle 1$ is congruent to $\angle 7$.

CRITICAL THINKING **Find the value of x.**

27.

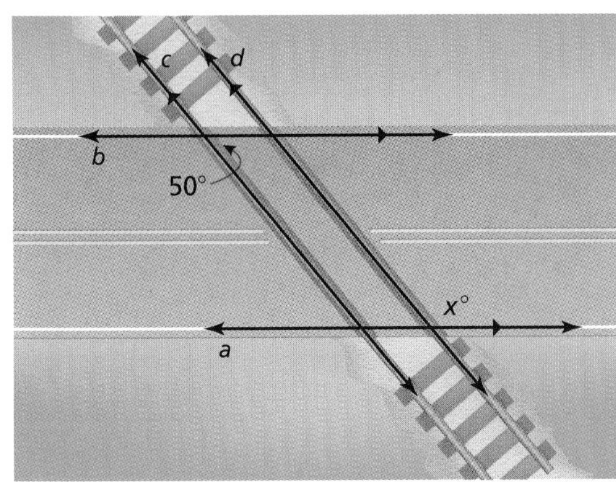

28.

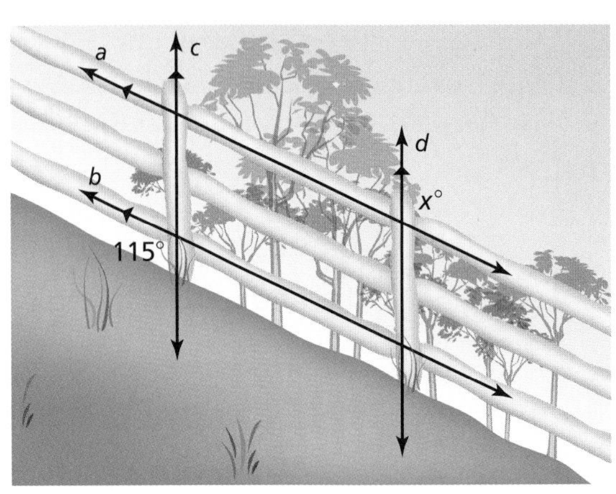

29. **OPTICAL ILLUSION** Refer to the figure.

a. Do the horizontal lines appear to be parallel? Explain.

b. Draw your own optical illusion using parallel lines.

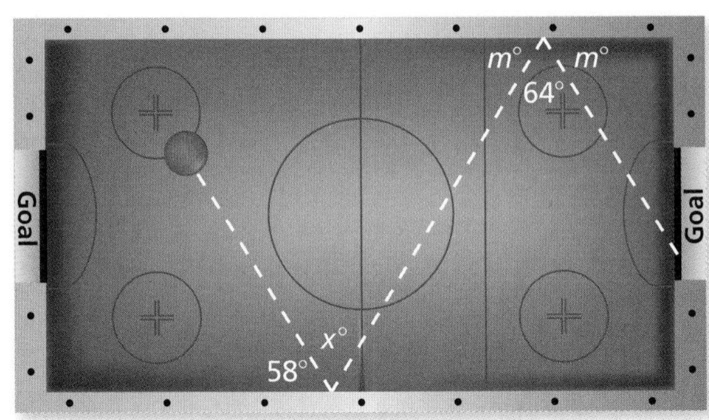

30. Geometry The figure shows the angles used to make a double bank shot in an air hockey game.

a. Find the value of x.

b. Can you still get the red puck in the goal when x is increased by a little? by a lot? Explain.

Fair Game Review What you learned in previous grades & lessons

Evaluate the expression. *(Skills Review Handbook)*

31. $4 + 3^2$

32. $5(2)^2 - 6$

33. $11 + (-7)^2 - 9$

34. $8 \div 2^2 + 1$

35. **MULTIPLE CHOICE** The triangles are similar. What length does x represent? *(Section 2.5)*

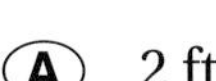 Ⓐ 2 ft Ⓑ 12 ft Ⓒ 15 ft Ⓓ 27 ft

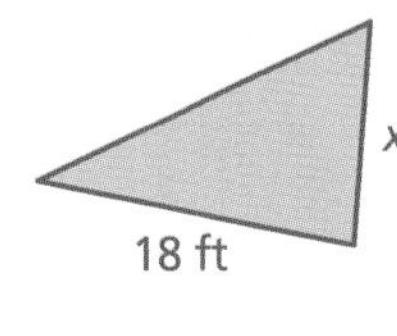

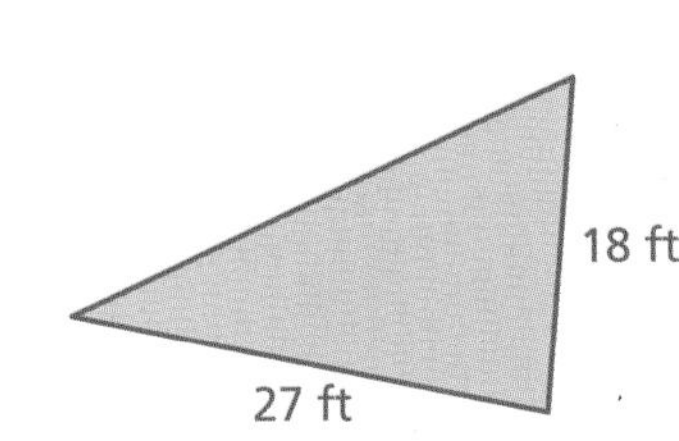

3.2 Angles of Triangles

Essential Question How can you describe the relationships among the angles of a triangle?

1 ACTIVITY: Exploring the Interior Angles of a Triangle

Work with a partner.

a. Draw a triangle. Label the interior angles *A*, *B*, and *C*.

b. Carefully cut out the triangle. Tear off the three corners of the triangle.

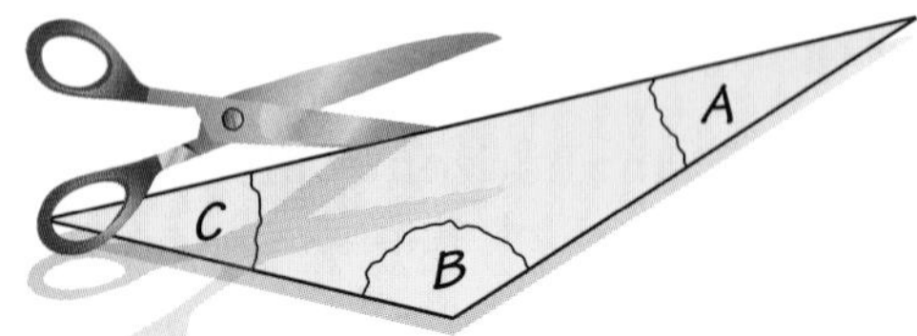

c. Arrange angles *A* and *B* so that they share a vertex and are adjacent.

d. How can you place the third angle to determine the sum of the measures of the interior angles? What is the sum?

e. Compare your results with those of others in your class.

f. **STRUCTURE** How does your result in part (d) compare to the rule you wrote in Lesson 1.1, Activity 2?

2 ACTIVITY: Exploring the Interior Angles of a Triangle

Work with a partner.

a. Describe the figure.

b. **LOGIC** Use what you know about parallel lines and transversals to justify your result in part (d) of Activity 1.

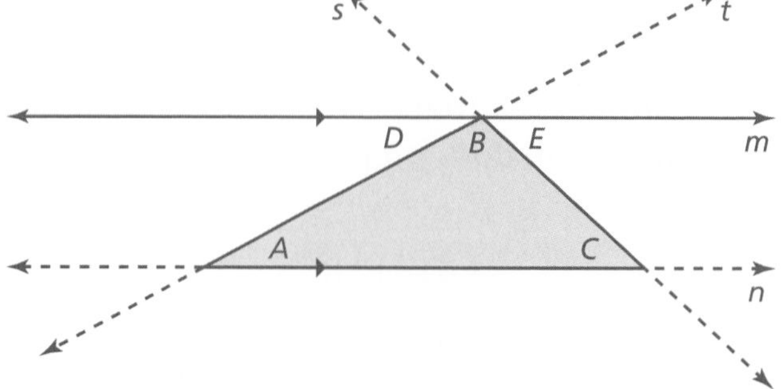

Geometry

In this lesson, you will

- understand that the sum of the interior angle measures of a triangle is 180°.
- find the measures of interior and exterior angles of triangles.

3 ACTIVITY: Exploring an Exterior Angle of a Triangle

Math Practice

Maintain Oversight

Do you think your conclusion will be true for the exterior angle of any triangle? Explain.

Work with a partner.

a. Draw a triangle. Label the interior angles *A*, *B*, and *C*.

b. Carefully cut out the triangle.

c. Place the triangle on a piece of paper and extend one side to form *exterior angle D*, as shown.

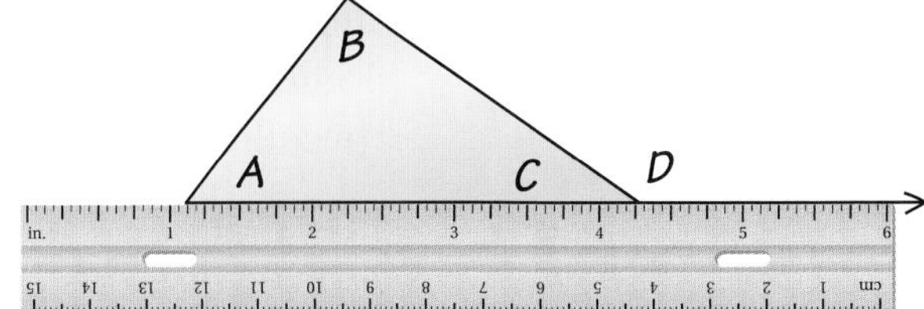

d. Tear off the corners that are not adjacent to the exterior angle. Arrange them to fill the exterior angle, as shown. What does this tell you about the measure of exterior angle *D*?

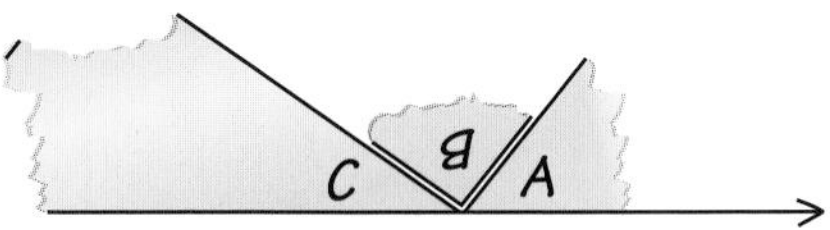

4 ACTIVITY: Measuring the Exterior Angles of a Triangle

Work with a partner.

a. Draw a triangle and label the interior and exterior angles, as shown.

b. Use a protractor to measure all six angles. Copy and complete the table to organize your results. What does the table tell you about the measure of an exterior angle of a triangle?

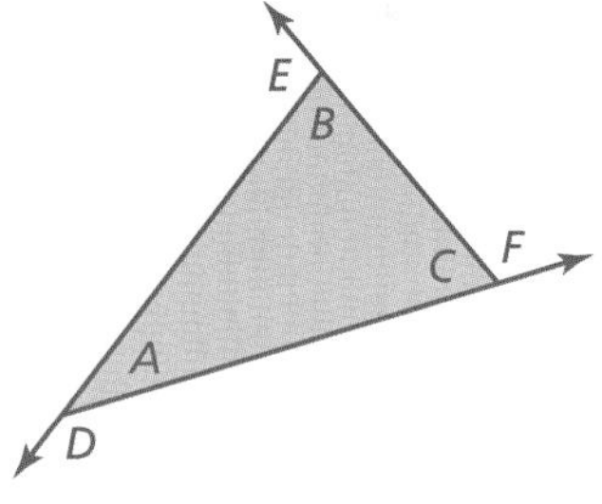

Exterior Angle	$D = \square°$	$E = \square°$	$F = \square°$
Interior Angle	$B = \square°$	$A = \square°$	$A = \square°$
Interior Angle	$C = \square°$	$C = \square°$	$B = \square°$

What Is Your Answer?

5. **REPEATED REASONING** Draw three triangles that have different shapes. Repeat parts (b)–(d) from Activity 1 for each triangle. Do you get the same results? Explain.

6. **IN YOUR OWN WORDS** How can you describe the relationships among angles of a triangle?

Practice Use what you learned about angles of a triangle to complete Exercises 4–6 on page 114.

3.2 Lesson

Key Vocabulary
interior angles of a polygon, *p. 112*
exterior angles of a polygon, *p. 112*

The angles inside a polygon are called **interior angles**. When the sides of a polygon are extended, other angles are formed. The angles outside the polygon that are adjacent to the interior angles are called **exterior angles**.

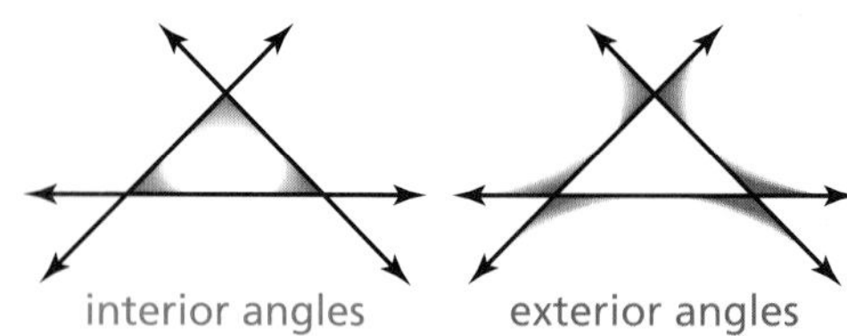

Key Idea

Interior Angle Measures of a Triangle

Words The sum of the interior angle measures of a triangle is 180°.

Algebra $x + y + z = 180$

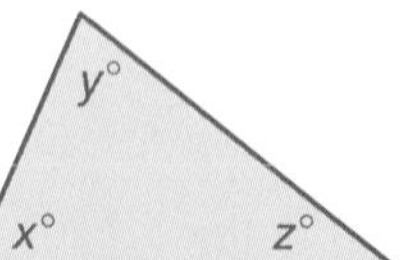

EXAMPLE 1 Using Interior Angle Measures

Find the value of *x*.

a.

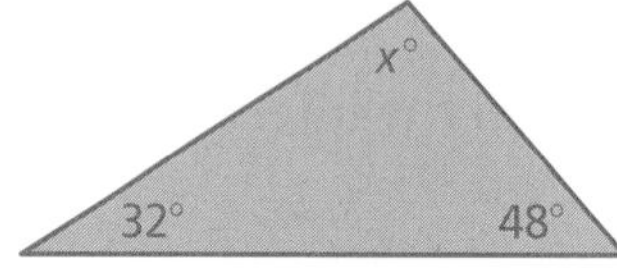

$$x + 32 + 48 = 180$$
$$x + 80 = 180$$
$$x = 100$$

b.

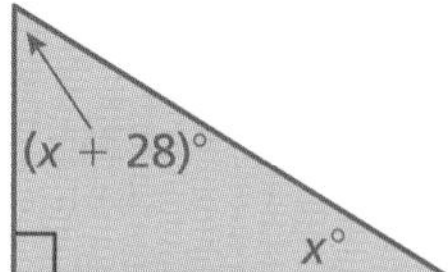

$$x + (x + 28) + 90 = 180$$
$$2x + 118 = 180$$
$$2x = 62$$
$$x = 31$$

On Your Own

Now You're Ready
Exercises 4–9

Find the value of *x*.

1.

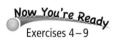

2.

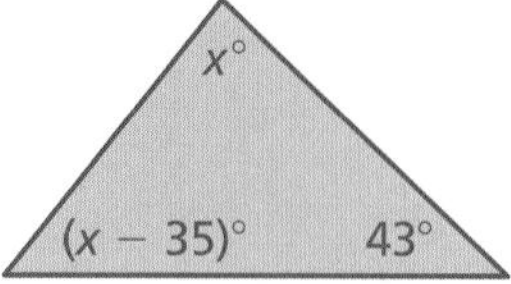

Key Idea

Exterior Angle Measures of a Triangle

Words The measure of an exterior angle of a triangle is equal to the sum of the measures of the two nonadjacent interior angles.

Algebra $z = x + y$

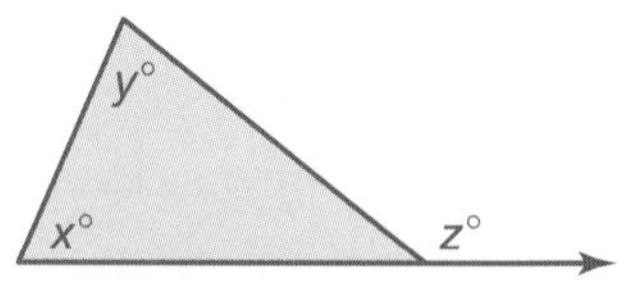

Multi-Language Glossary at BigIdeasMath.com

EXAMPLE 2 Finding Exterior Angle Measures

Study Tip

Each vertex has a pair of congruent exterior angles. However, it is common to show only one exterior angle at each vertex.

Find the measure of the exterior angle.

a.

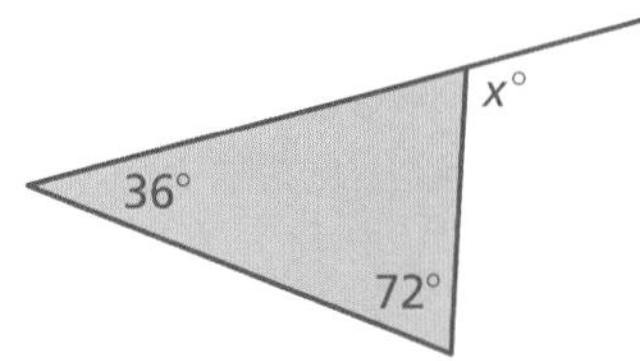

$x = 36 + 72$

$x = 108$

So, the measure of the exterior angle is 108°.

b.

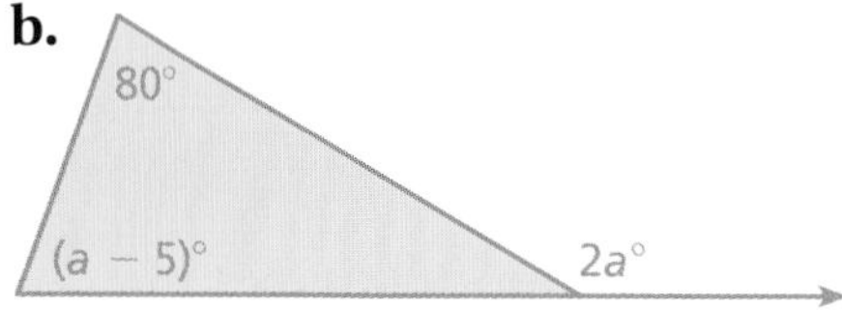

$2a = (a - 5) + 80$

$2a = a + 75$

$a = 75$

So, the measure of the exterior angle is $2(75)° = 150°$.

EXAMPLE 3 Real-Life Application

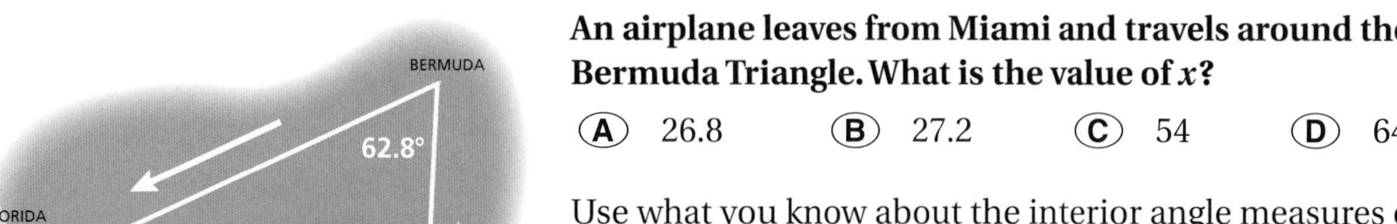

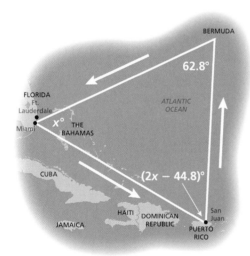

An airplane leaves from Miami and travels around the Bermuda Triangle. What is the value of x?

Ⓐ 26.8 Ⓑ 27.2 Ⓒ 54 Ⓓ 64

Use what you know about the interior angle measures of a triangle to write an equation.

$x + (2x - 44.8) + 62.8 = 180$	Write equation.
$3x + 18 = 180$	Combine like terms.
$3x = 162$	Subtract 18 from each side.
$x = 54$	Divide each side by 3.

The value of x is 54. The correct answer is Ⓒ.

On Your Own

Now You're Ready
Exercises 12–14

Find the measure of the exterior angle.

3.

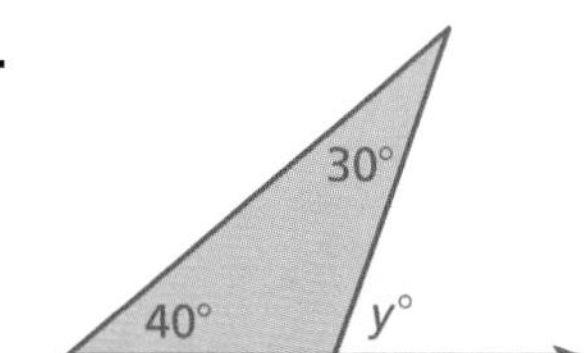

4.

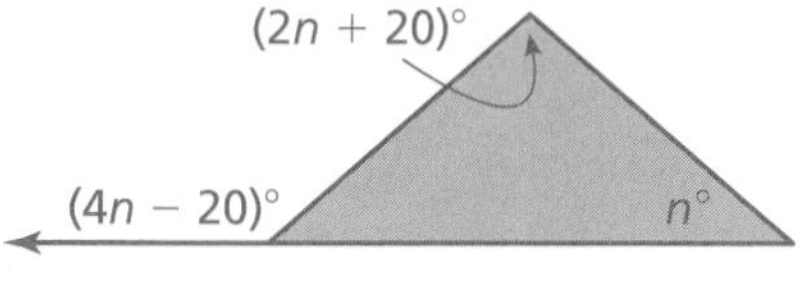

5. In Example 3, the airplane leaves from Fort Lauderdale. The interior angle measure at Bermuda is 63.9°. The interior angle measure at San Juan is $(x + 7.5)°$. Find the value of x.

3.2 Exercises

Vocabulary and Concept Check

1. **VOCABULARY** You know the measures of two interior angles of a triangle. How can you find the measure of the third interior angle?

2. **VOCABULARY** How many exterior angles does a triangle have at each vertex? Explain.

3. **NUMBER SENSE** List the measures of the exterior angles for the triangle shown at the right.

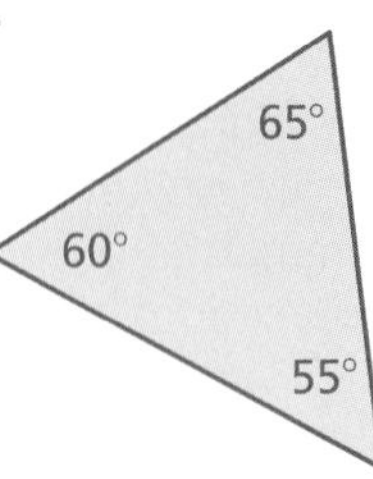

Practice and Problem Solving

Find the measures of the interior angles.

① 4.
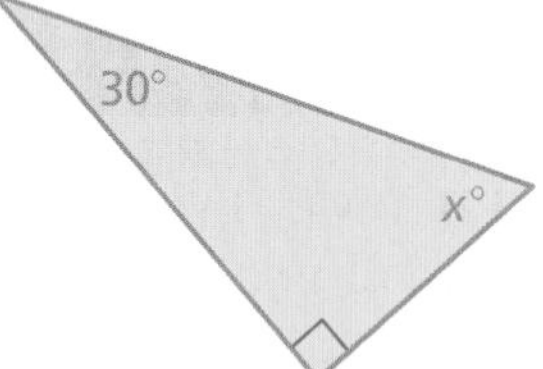

5.
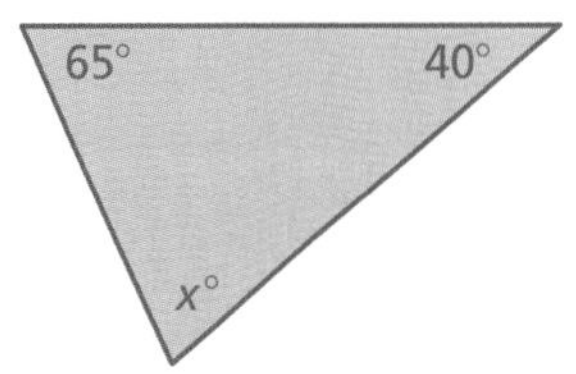

6.
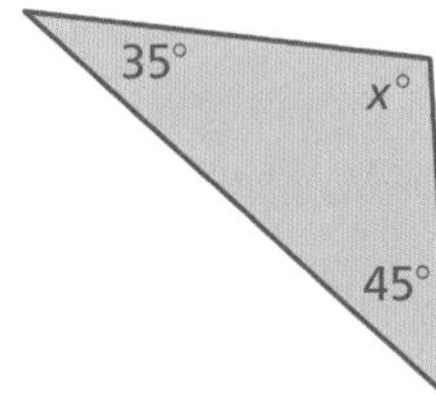

7.
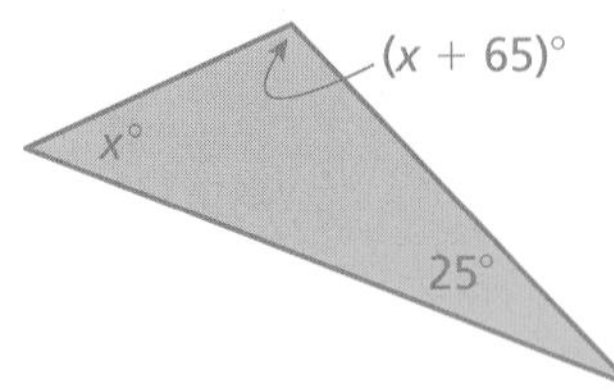

8.
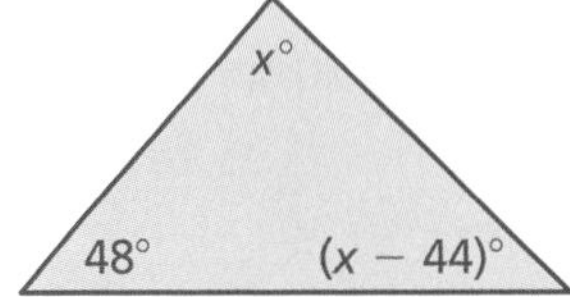

9.
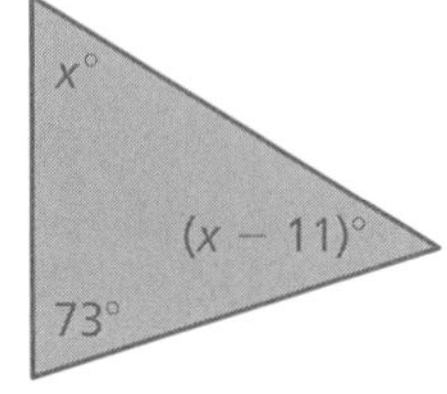

10. **BILLIARD RACK** Find the value of x in the billiard rack.

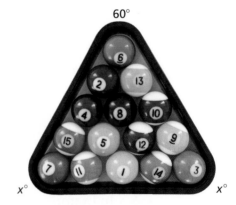

60°

x° x°

11. **NO PARKING** The triangle with lines through it designates a no parking zone. What is the value of x?

Find the measure of the exterior angle.

2 **12.**

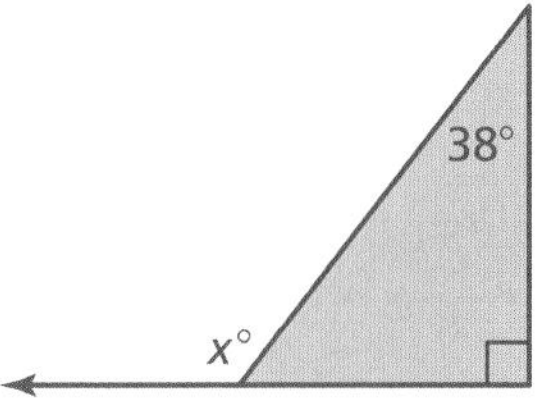

13.

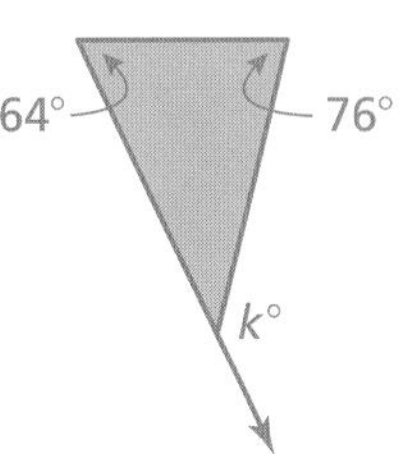

14.

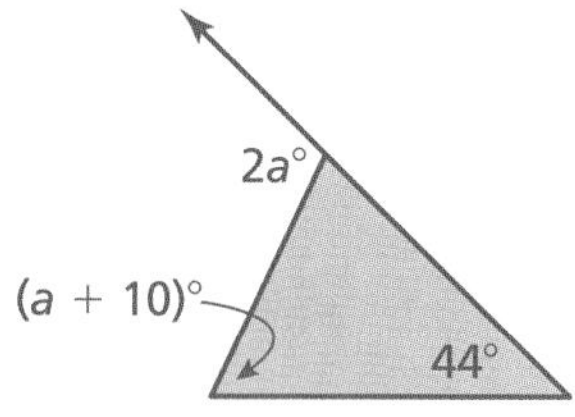

15. ERROR ANALYSIS Describe and correct the error in finding the measure of the exterior angle.

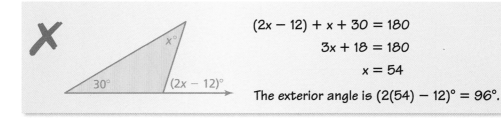

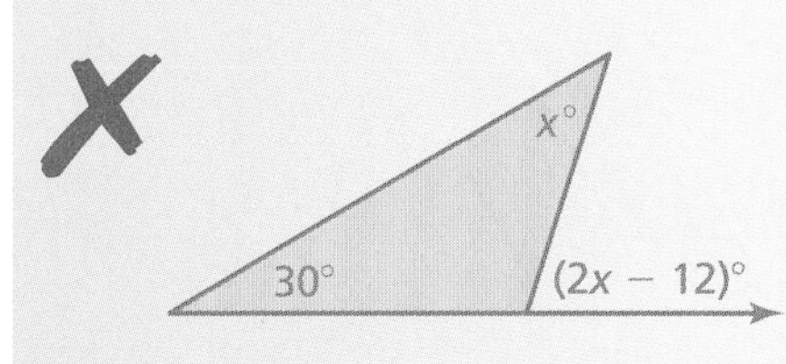

$(2x - 12) + x + 30 = 180$

$3x + 18 = 180$

$x = 54$

The exterior angle is $(2(54) - 12)° = 96°$.

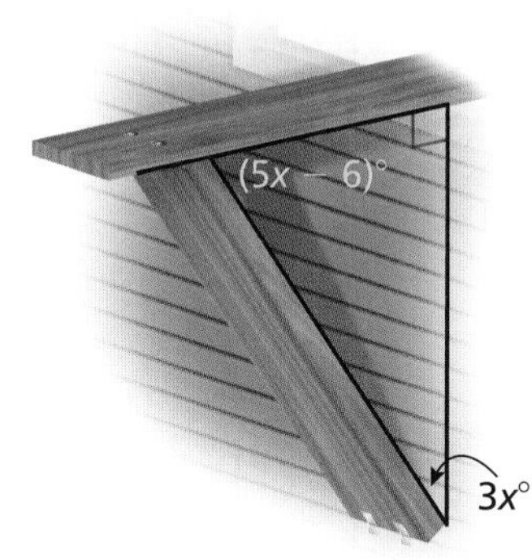

16. RATIO The ratio of the interior angle measures of a triangle is 2 : 3 : 5. What are the angle measures?

17. CONSTRUCTION The support for a window air-conditioning unit forms a triangle and an exterior angle. What is the measure of the exterior angle?

18. REASONING A triangle has an exterior angle with a measure of 120°. Can you determine the measures of the interior angles? Explain.

Determine whether the statement is *always, sometimes,* or *never* true. Explain your reasoning.

19. Given three angle measures, you can construct a triangle.

20. The acute interior angles of a right triangle are complementary.

21. A triangle has more than one vertex with an acute exterior angle.

22. Precision Using the figure at the right, show that $z = x + y$. (*Hint:* Find two equations involving w.)

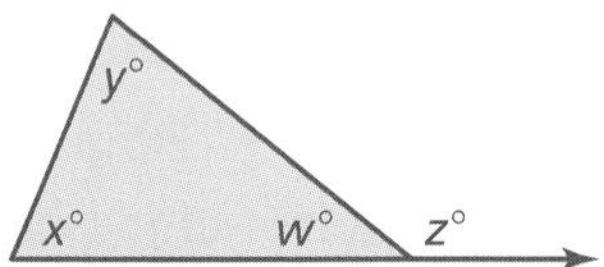

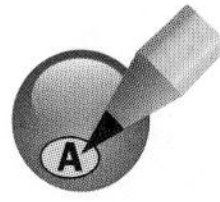

Fair Game Review What you learned in previous grades & lessons

Solve the equation. Check your solution. *(Section 1.2)*

23. $-4x + 3 = 19$

24. $2(y - 1) + 6y = -10$

25. $5 + 0.5(6n + 14) = 3$

26. MULTIPLE CHOICE Which transformation moves every point of a figure the same distance and in the same direction? *(Section 2.2)*

Ⓐ translation Ⓑ reflection Ⓒ rotation Ⓓ dilation

3 Study Help

You can use an **example and non-example chart** to list examples and non-examples of a vocabulary word or item. Here is an example and non-example chart for transversals.

Transversals

Examples	Non-Examples
line *p* line *q* line *r*	line *a* line *b* line *c*
line *a* line *b* line *c* line *d*	line *p* line *t*

On Your Own

Make example and non-example charts to help you study these topics.

1. interior angles formed by parallel lines and a transversal
2. exterior angles formed by parallel lines and a transversal

After you complete this chapter, make example and non-example charts for the following topics.

3. interior angles of a polygon
4. exterior angles of a polygon
5. regular polygons
6. similar triangles

"What do you think of my example & non-example chart for popular cat toys?"

3.1–3.2 Quiz

Check It Out
Progress Check
BigIdeasMath.com

Use the figure to find the measure of the angle. Explain your reasoning. *(Section 3.1)*

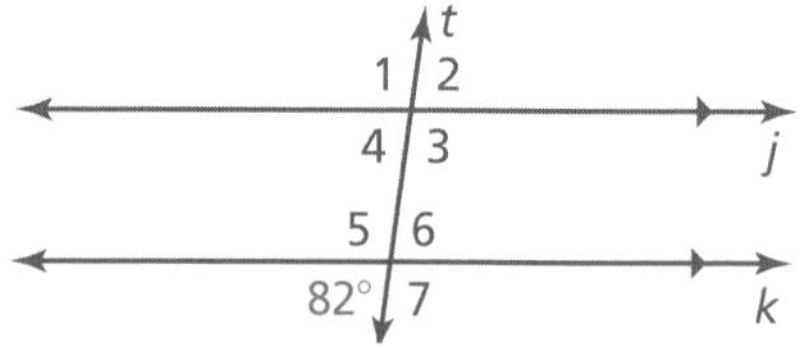

1. $\angle 2$
2. $\angle 6$
3. $\angle 4$
4. $\angle 1$

Complete the statement. Explain your reasoning. *(Section 3.1)*

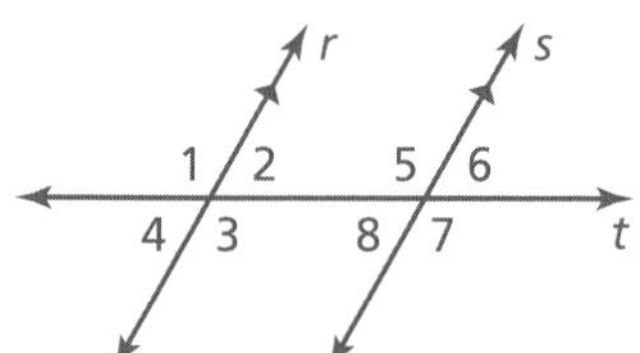

5. If the measure of $\angle 1 = 123°$, then the measure of $\angle 7 =$ ▭.
6. If the measure of $\angle 2 = 58°$, then the measure of $\angle 5 =$ ▭.
7. If the measure of $\angle 5 = 119°$, then the measure of $\angle 3 =$ ▭.
8. If the measure of $\angle 4 = 60°$, then the measure of $\angle 6 =$ ▭.

Find the measures of the interior angles. *(Section 3.2)*

9.

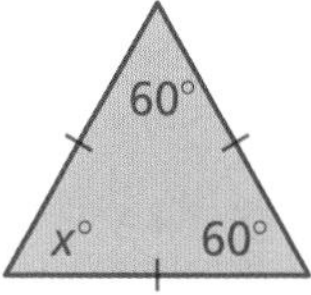

10.

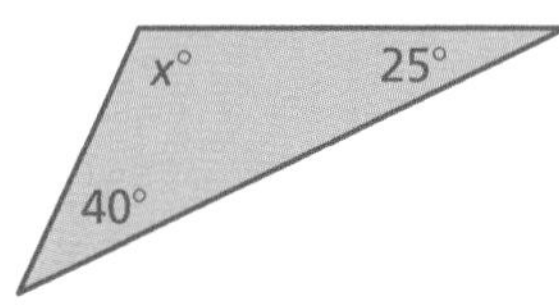

11.

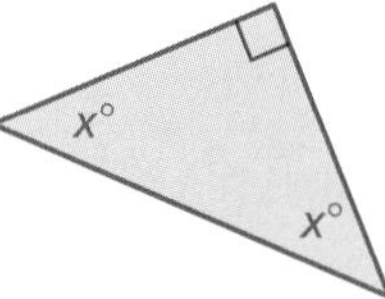

Find the measure of the exterior angle. *(Section 3.2)*

12.

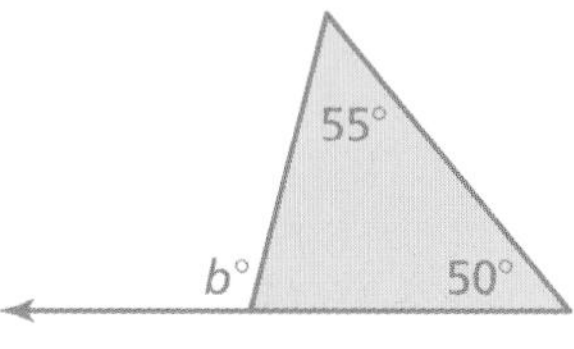

13.

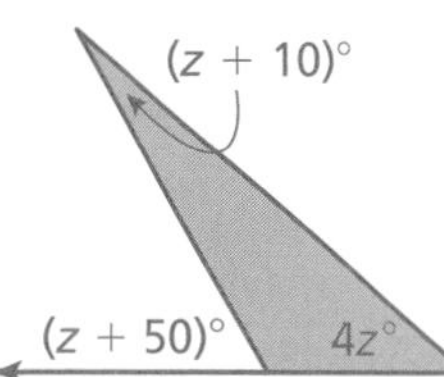

14. **PARK** In a park, a bike path and a horse riding path are parallel. In one part of the park, a hiking trail intersects the two paths. Find the measures of $\angle 1$ and $\angle 2$. Explain your reasoning. *(Section 3.1)*

15. **LADDER** A ladder leaning against a wall forms a triangle and exterior angles with the wall and the ground. What are the measures of the exterior angles? Justify your answer. *(Section 3.2)*

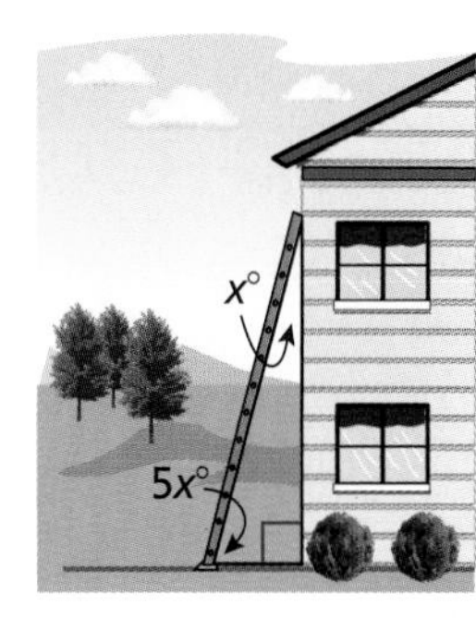

3.3 Angles of Polygons

Essential Question How can you find the sum of the interior angle measures and the sum of the exterior angle measures of a polygon?

1 ACTIVITY: Exploring the Interior Angles of a Polygon

Work with a partner. In parts (a)–(e), identify each polygon and the number of sides *n*. Then find the sum of the interior angle measures of the polygon.

a. Polygon: ______ Number of sides: $n =$ ______

Draw a line segment on the figure that divides it into two triangles. Is there more than one way to do this? Explain.

What is the sum of the interior angle measures of each triangle?

What is the sum of the interior angle measures of the figure?

b.

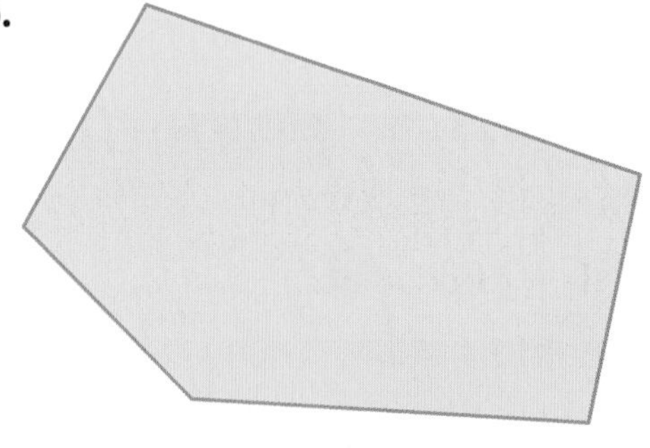

c.

d.

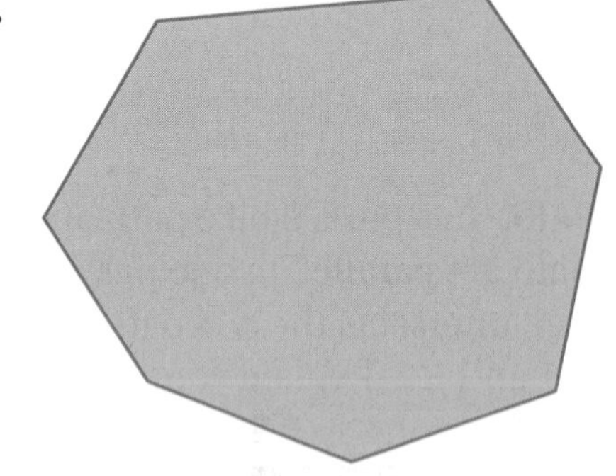

e.

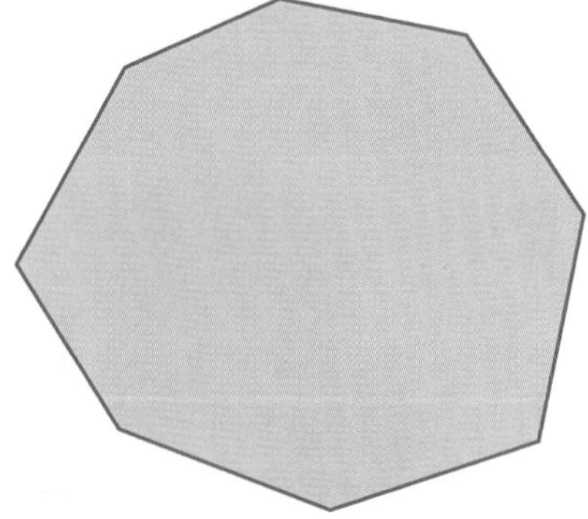

Geometry

In this lesson, you will

- find the sum of the interior angle measures of polygons.
- understand that the sum of the exterior angle measures of a polygon is 360°.
- find the measures of interior and exterior angles of polygons.

f. **REPEATED REASONING** Use your results to complete the table. Then find the sum of the interior angle measures of a polygon with 12 sides.

Number of Sides, *n*	3	4	5	6	7	8
Number of Triangles						
Angle Sum, *S*						

A polygon is **convex** when every line segment connecting any two vertices lies entirely inside the polygon. A polygon is **concave** when at least one line segment connecting any two vertices lies outside the polygon.

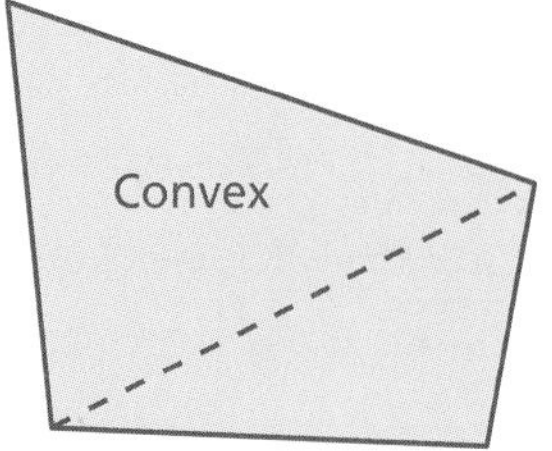

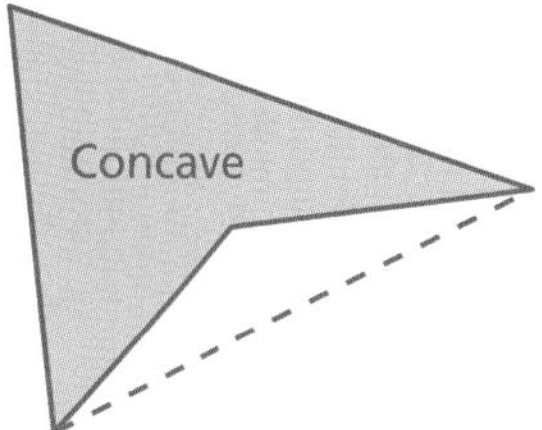

2 ACTIVITY: Exploring the Exterior Angles of a Polygon

Math Practice

Analyze Conjectures

Do your observations about the sum of the exterior angles make sense? Do you think they would hold true for any convex polygon? Explain.

Work with a partner.

a. Draw a convex pentagon. Extend the sides to form the exterior angles. Label one exterior angle at each vertex A, B, C, D, and E, as shown.

b. Cut out the exterior angles. How can you join the vertices to determine the sum of the angle measures? What do you notice?

c. **REPEATED REASONING** Repeat the procedure in parts (a) and (b) for each figure below.

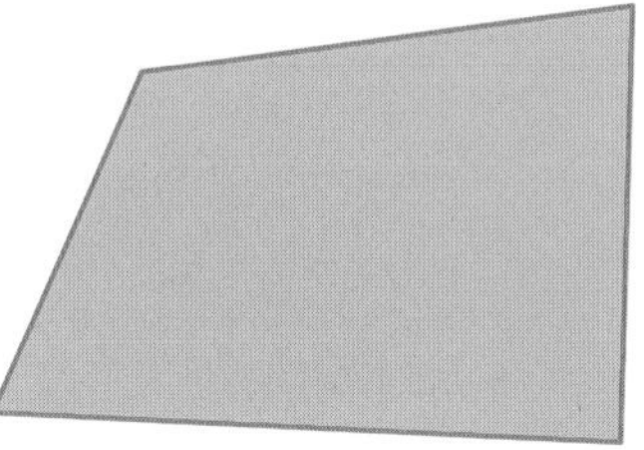

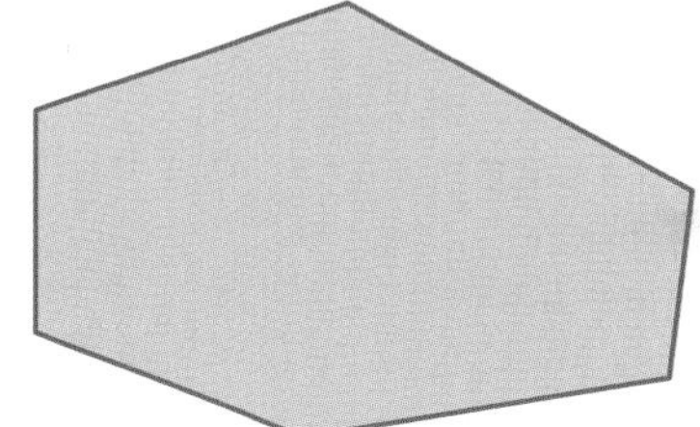

What can you conclude about the sum of the measures of the exterior angles of a convex polygon? Explain.

What Is Your Answer?

3. **STRUCTURE** Use your results from Activity 1 to write an expression that represents the sum of the interior angle measures of a polygon.

4. **IN YOUR OWN WORDS** How can you find the sum of the interior angle measures and the sum of the exterior angle measures of a polygon?

Practice Use what you learned about angles of polygons to complete Exercises 4–6 on page 123.

3.3 Lesson

Key Vocabulary
convex polygon, *p. 119*
concave polygon, *p. 119*
regular polygon, *p. 121*

A *polygon* is a closed plane figure made up of three or more line segments that intersect only at their endpoints.

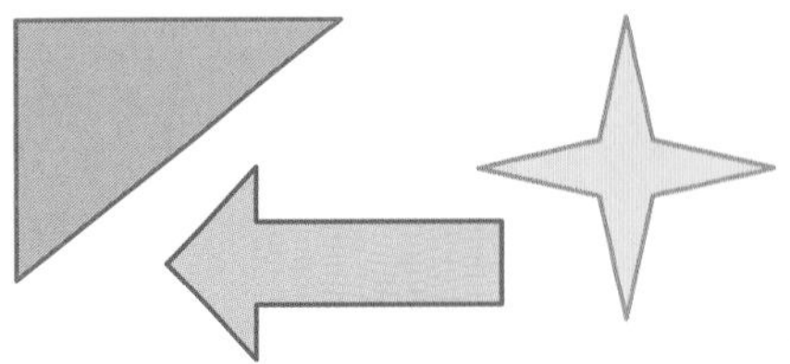

Polygons

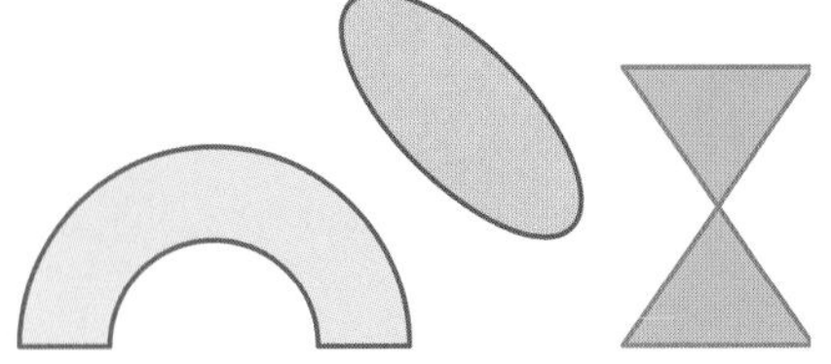

Not polygons

Interior Angle Measures of a Polygon

The sum S of the interior angle measures of a polygon with n sides is

$$S = (n - 2) \cdot 180^\circ.$$

EXAMPLE 1 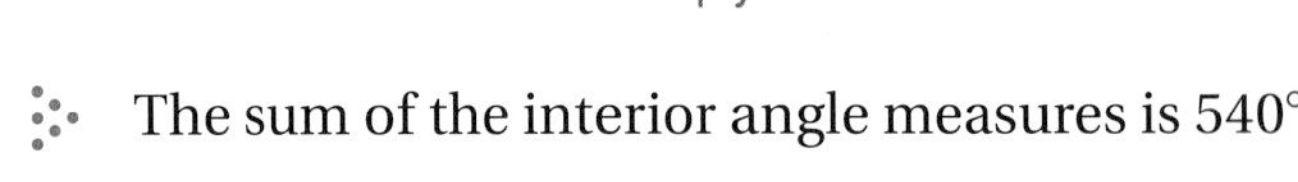

Find the sum of the interior angle measures of the school crossing sign.

The sign is in the shape of a pentagon. It has 5 sides.

$S = (n - 2) \cdot 180^\circ$	Write the formula.
$= (5 - 2) \cdot 180^\circ$	Substitute 5 for n.
$= 3 \cdot 180^\circ$	Subtract.
$= 540^\circ$	Multiply.

The sum of the interior angle measures is 540°.

Reading

For polygons whose names you have not learned, you can use the phrase "n-gon," where n is the number of sides. For example, a 15-gon is a polygon with 15 sides.

On Your Own

Now You're Ready
Exercises 7–9

Find the sum of the interior angle measures of the green polygon.

1. 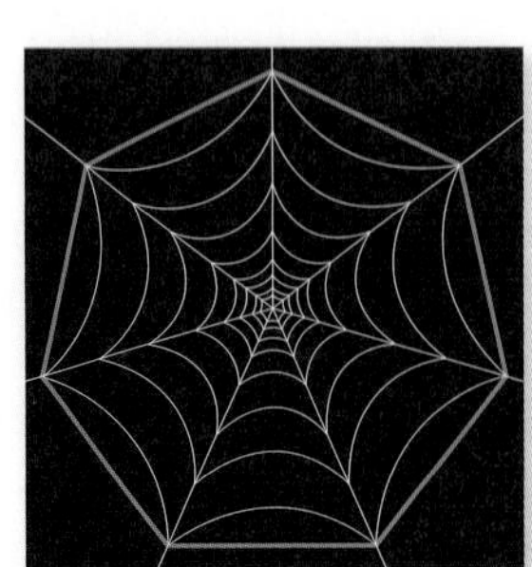

2. 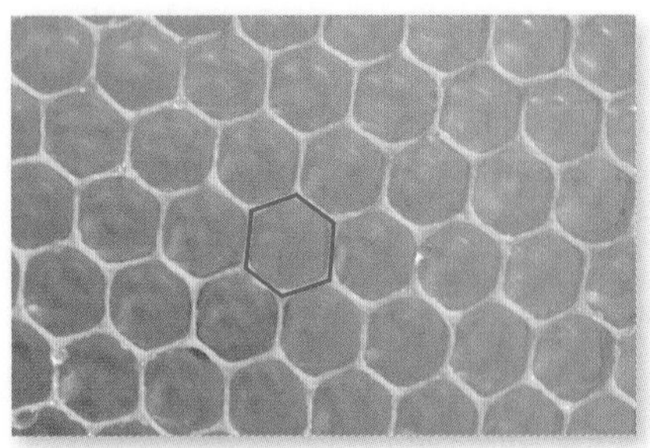

EXAMPLE 2 Finding an Interior Angle Measure of a Polygon

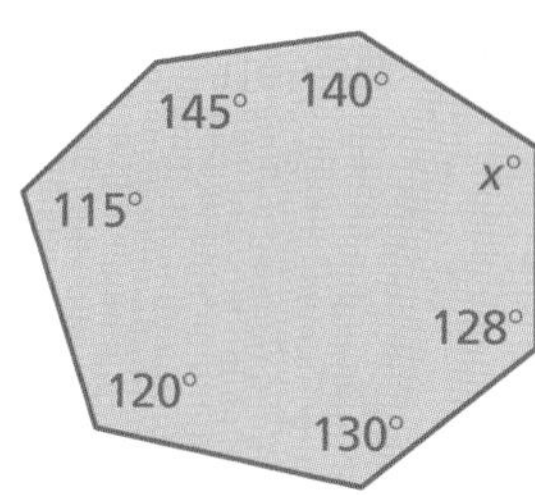

Find the value of x.

Step 1: The polygon has 7 sides. Find the sum of the interior angle measures.

$S = (n - 2) \cdot 180°$ Write the formula.

$= (7 - 2) \cdot 180°$ Substitute 7 for n.

$= 900°$ Simplify. The sum of the interior angle measures is 900°.

Step 2: Write and solve an equation.

$$140 + 145 + 115 + 120 + 130 + 128 + x = 900$$
$$778 + x = 900$$
$$x = 122$$

The value of x is 122.

On Your Own

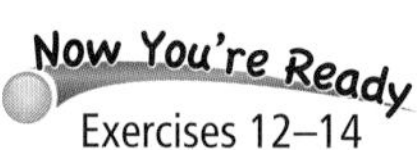

Find the value of x.

3. 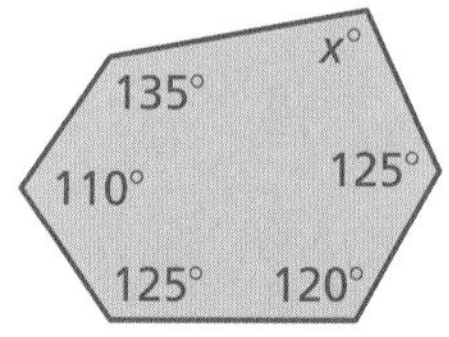

4. 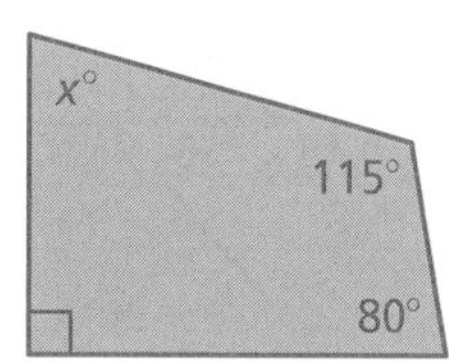

5. 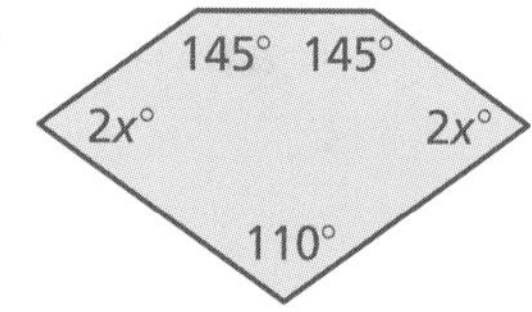

In a **regular polygon**, all the sides are congruent, and all the interior angles are congruent.

EXAMPLE 3 Real-Life Application

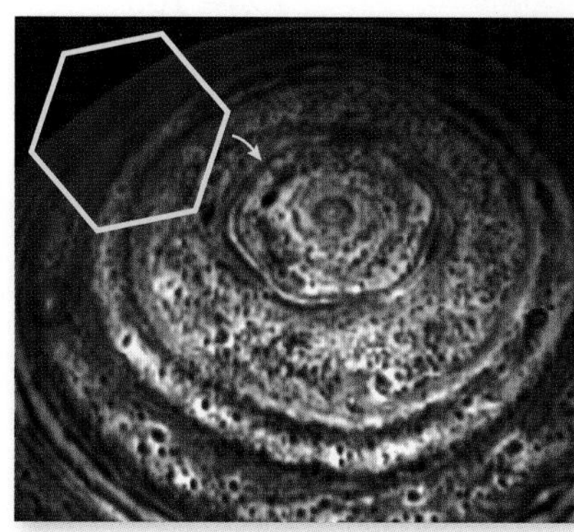

The hexagon is about 15,000 miles across. Approximately four Earths could fit inside it.

A cloud system discovered on Saturn is in the approximate shape of a regular hexagon. Find the measure of each interior angle of the hexagon.

Step 1: A hexagon has 6 sides. Find the sum of the interior angle measures.

$S = (n - 2) \cdot 180°$ Write the formula.

$= (6 - 2) \cdot 180°$ Substitute 6 for n.

$= 720°$ Simplify. The sum of the interior angle measures is 720°.

Step 2: Divide the sum by the number of interior angles, 6.

$720° \div 6 = 120°$

The measure of each interior angle is 120°.

Now You're Ready
Exercises 16–18

Find the measure of each interior angle of the regular polygon.

6. octagon **7.** decagon **8.** 18-gon

Exterior Angle Measures of a Polygon

Words The sum of the measures of the exterior angles of a convex polygon is 360°.

Algebra $w + x + y + z = 360$

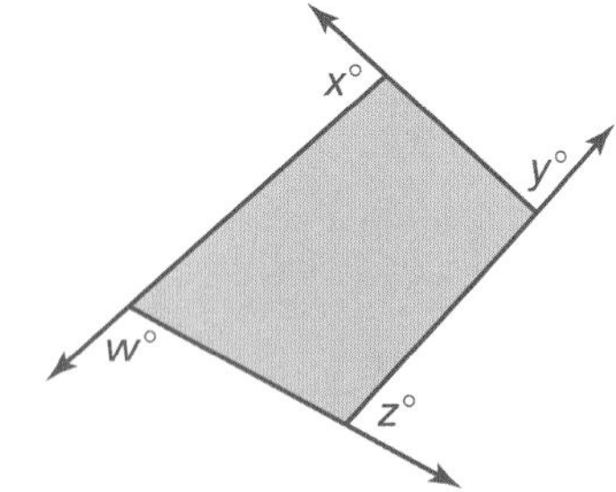

EXAMPLE 4 Finding Exterior Angle Measures

Find the measures of the exterior angles of each polygon.

a.

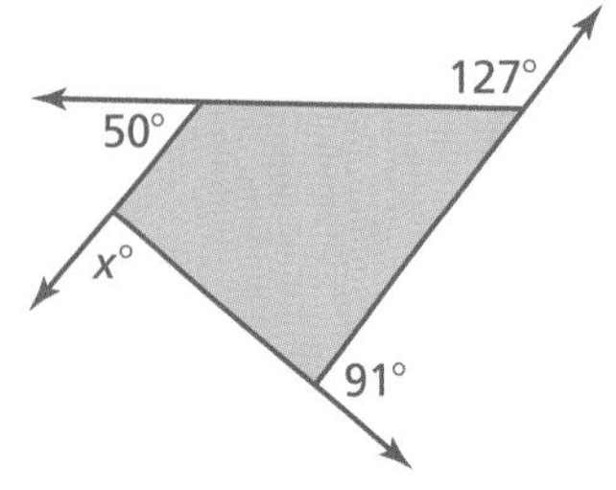

Write and solve an equation for x.

$$x + 50 + 127 + 91 = 360$$
$$x + 268 = 360$$
$$x = 92$$

So, the measures of the exterior angles are 92°, 50°, 127°, and 91°.

b.

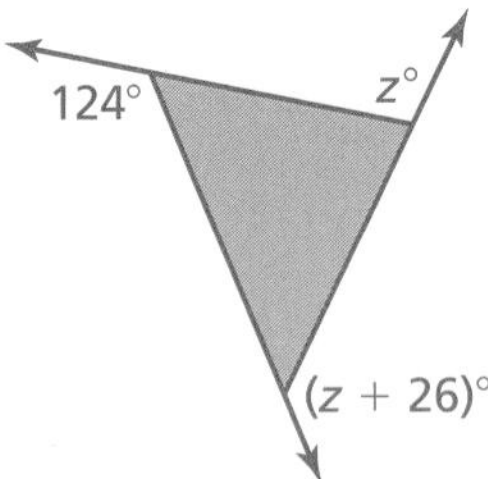

Write and solve an equation for z.

$$124 + z + (z + 26) = 360$$
$$2z + 150 = 360$$
$$z = 105$$

So, the measures of the exterior angles are 124°, 105°, and $(105 + 26)° = 131°$.

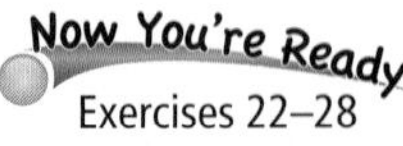

Now You're Ready
Exercises 22–28

9. Find the measures of the exterior angles of the polygon.

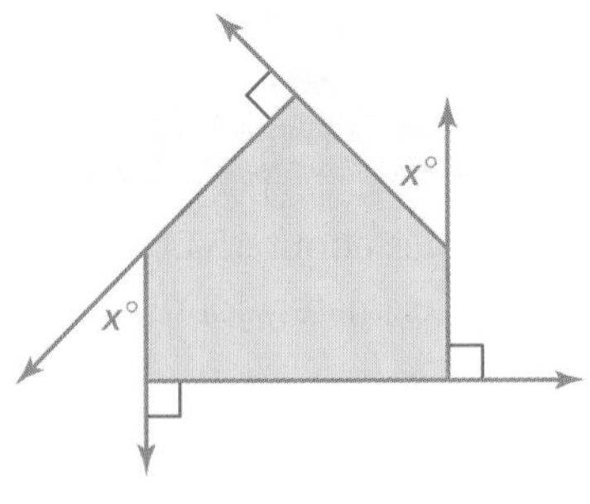

3.3 Exercises

Vocabulary and Concept Check

1. **VOCABULARY** Draw a regular polygon that has three sides.

2. **WHICH ONE DOESN'T BELONG?** Which figure does *not* belong with the other three? Explain your reasoning.

   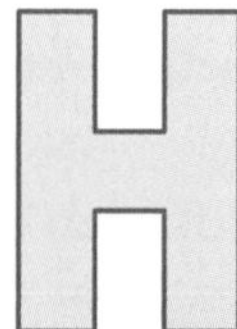

3. **DIFFERENT WORDS, SAME QUESTION** Which is different? Find "both" answers.

What is the measure of an interior angle of a regular pentagon?	What is the sum of the interior angle measures of a convex pentagon?
What is the sum of the interior angle measures of a regular pentagon?	What is the sum of the interior angle measures of a concave pentagon?

Practice and Problem Solving

Use triangles to find the sum of the interior angle measures of the polygon.

4. 

5. 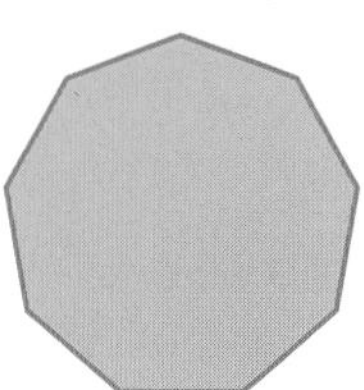

6. 

Find the sum of the interior angle measures of the polygon.

1 7. 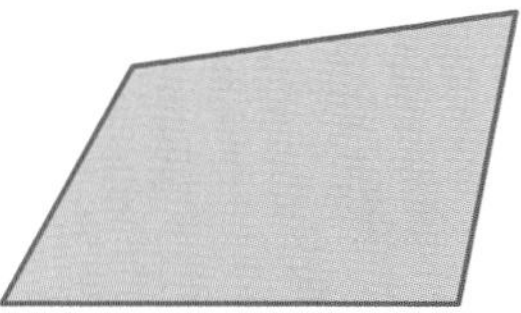

8. 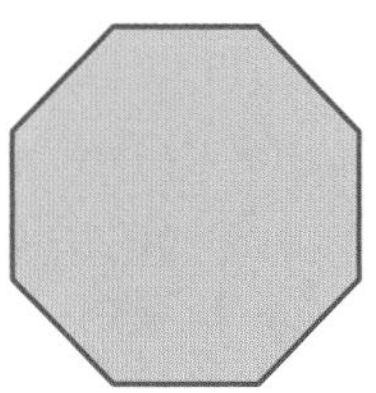

9. 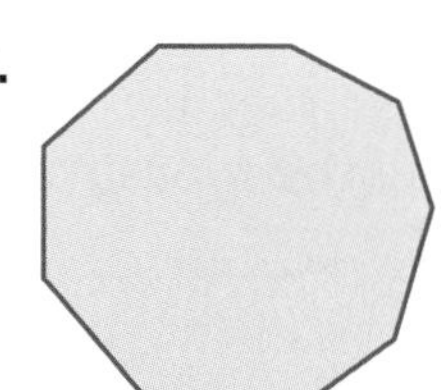

10. **ERROR ANALYSIS** Describe and correct the error in finding the sum of the interior angle measures of a 13-gon.

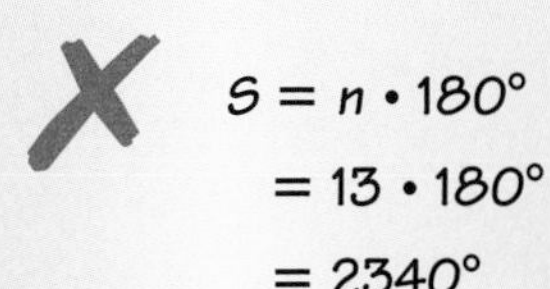

11. **NUMBER SENSE** Can a pentagon have interior angles that measure 120°, 105°, 65°, 150°, and 95°? Explain.

Find the measures of the interior angles.

2 **12.**

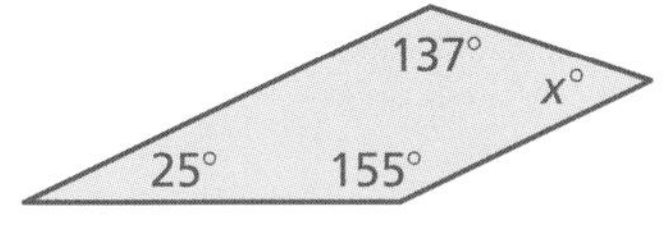

13.

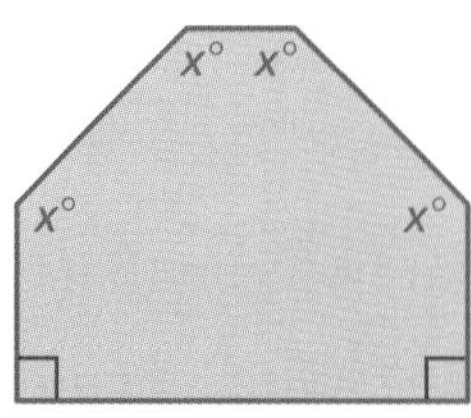

14.

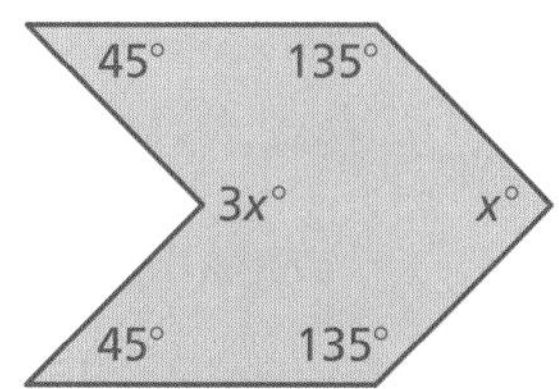

15. REASONING The sum of the interior angle measures in a regular polygon is 1260°. What is the measure of one of the interior angles of the polygon?

Find the measure of each interior angle of the regular polygon.

3 **16.**

17.

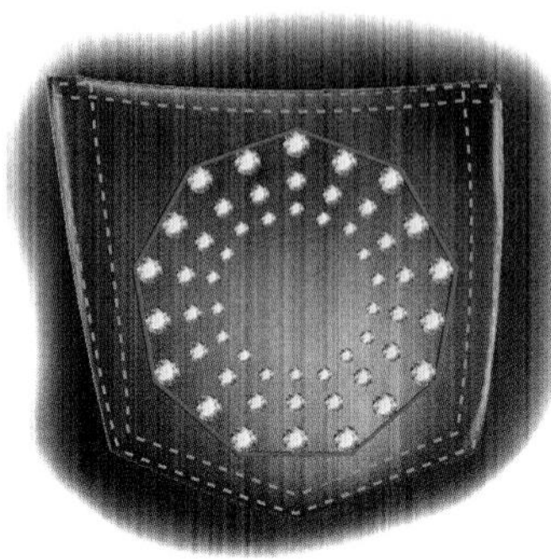

18.

19. ERROR ANALYSIS Describe and correct the error in finding the measure of each interior angle of a regular 20-gon.

✗ $S = (n - 2) \cdot 180°$
$= (20 - 2) \cdot 180°$
$= 18 \cdot 180°$
$= 3240°$
$3240° \div 18 = 180$
The measure of each interior angle is 180°.

20. FIRE HYDRANT A fire hydrant bolt is in the shape of a regular pentagon.

a. What is the measure of each interior angle?

b. Why are fire hydrants made this way?

21. PROBLEM SOLVING The interior angles of a regular polygon each measure 165°. How many sides does the polygon have?

Find the measures of the exterior angles of the polygon.

4 **22.**

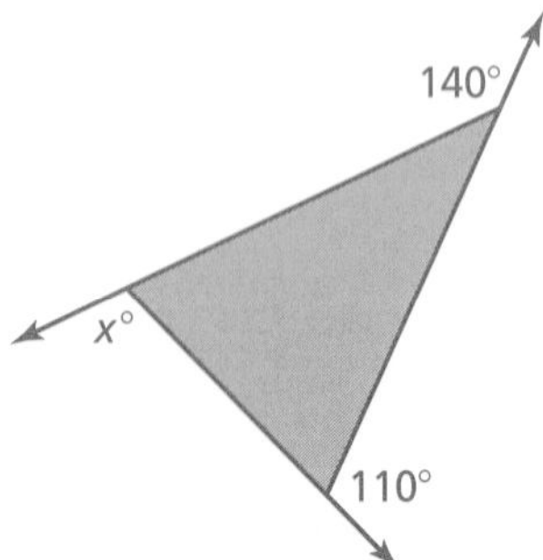

23.

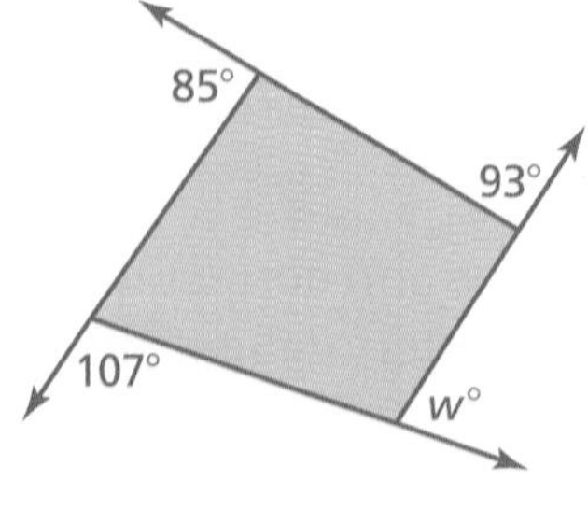

24.

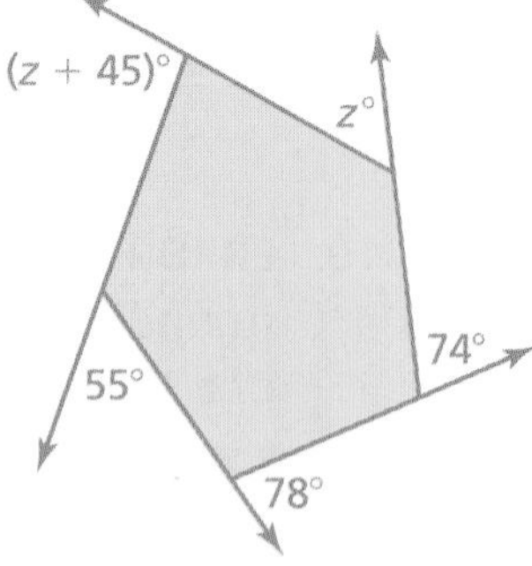

25. REASONING What is the measure of an exterior angle of a regular hexagon? Explain.

Find the measures of the exterior angles of the polygon.

26.

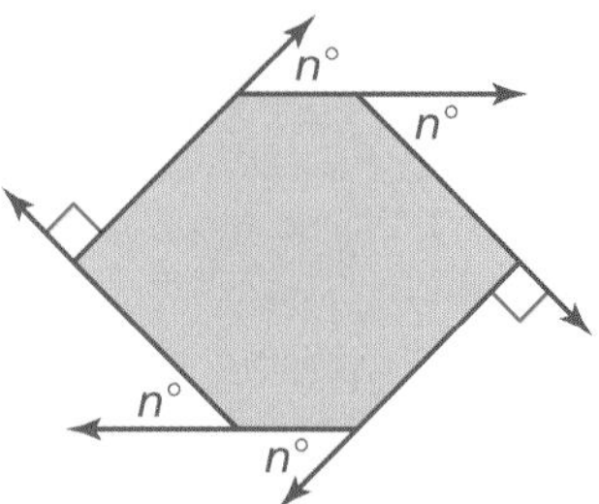

27.

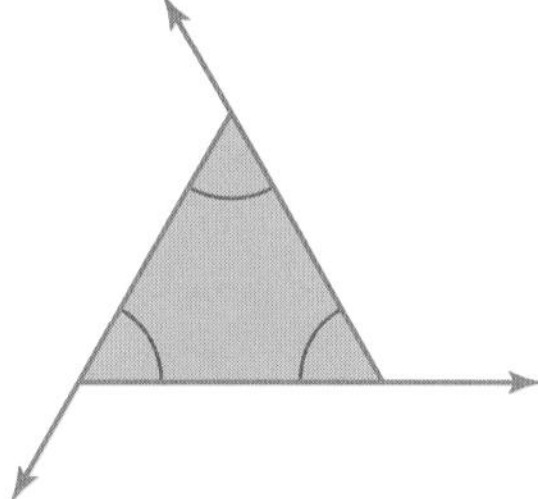

28.

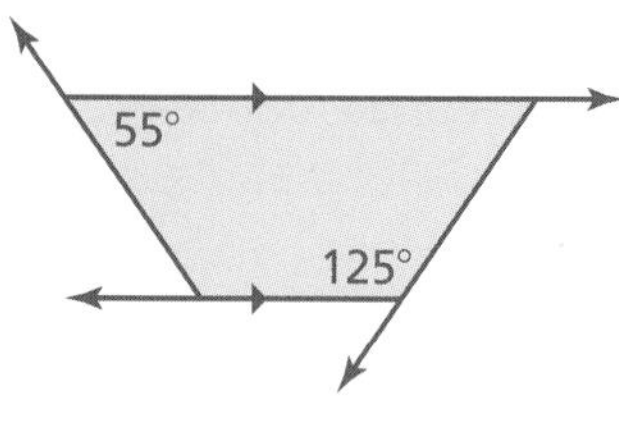

29. STAINED GLASS The center of the stained glass window is in the shape of a regular polygon. What is the measure of each interior angle of the polygon? What is the measure of each exterior angle?

30. PENTAGON Draw a pentagon that has two right interior angles, two 45° interior angles, and one 270° interior angle.

31. GAZEBO The floor of a gazebo is in the shape of a heptagon. Four of the interior angles measure 135°. The other interior angles have equal measures. Find their measures.

32. MONEY The border of a Susan B. Anthony dollar is in the shape of a regular polygon.

a. How many sides does the polygon have?

b. What is the measure of each interior angle of the border? Round your answer to the nearest degree.

33. Geometry When tiles can be used to cover a floor with no empty spaces, the collection of tiles is called a *tessellation*.

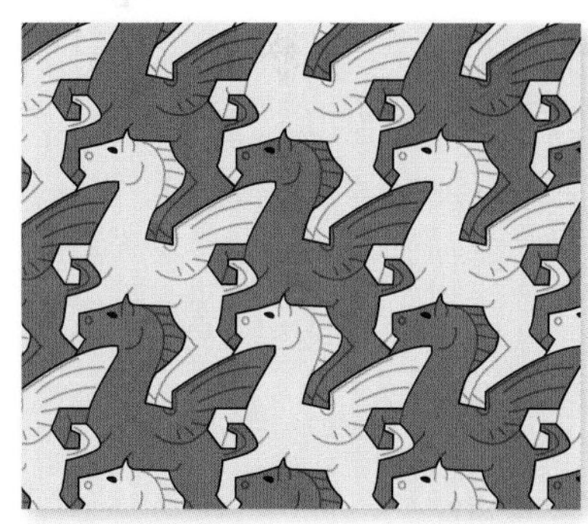

a. Create a tessellation using equilateral triangles.

b. Find two more regular polygons that form tessellations.

c. Create a tessellation that uses two different regular polygons.

d. Use what you know about interior and exterior angles to explain why the polygons in part (c) form a tessellation.

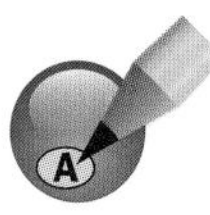

Fair Game Review What you learned in previous grades & lessons

Solve the proportion. *(Skills Review Handbook)*

34. $\frac{x}{12} = \frac{3}{4}$ **35.** $\frac{14}{21} = \frac{x}{3}$ **36.** $\frac{9}{x} = \frac{6}{2}$ **37.** $\frac{10}{4} = \frac{15}{x}$

38. MULTIPLE CHOICE The ratio of tulips to daisies is 3 : 5. Which of the following could be the total number of tulips and daisies? *(Skills Review Handbook)*

Ⓐ 6 Ⓑ 10 Ⓒ 15 Ⓓ 16

3.4 Using Similar Triangles

Essential Question How can you use angles to tell whether triangles are similar?

1 ACTIVITY: Constructing Similar Triangles

Work with a partner.

- **Use a straightedge to draw a line segment that is 4 centimeters long.**
- **Then use the line segment and a protractor to draw a triangle that has a 60° and a 40° angle, as shown. Label the triangle *ABC*.**

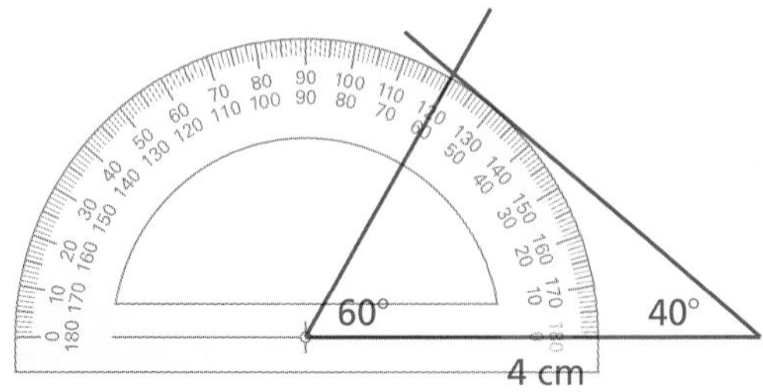

a. Explain how to draw a larger triangle that has the same two angle measures. Label the triangle *JKL*.

b. Explain how to draw a smaller triangle that has the same two angle measures. Label the triangle *PQR*.

c. Are all of the triangles similar? Explain.

2 ACTIVITY: Using Technology to Explore Triangles

Work with a partner. Use geometry software to draw the triangle below.

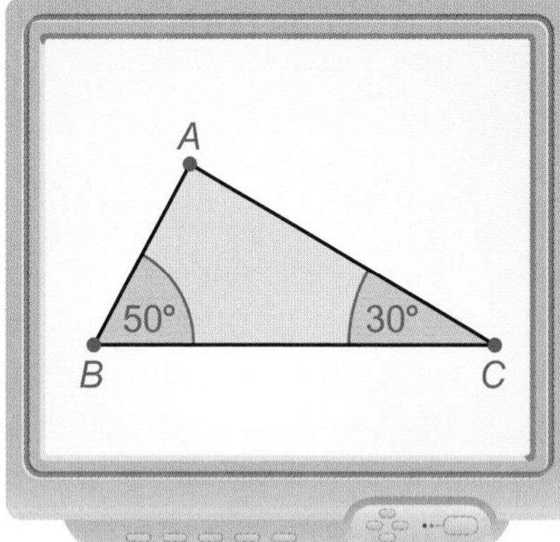

a. Dilate the triangle by the following scale factors.

$2 \qquad \frac{1}{2} \qquad \frac{1}{4} \qquad 2.5$

b. Measure the third angle in each triangle. What do you notice?

c. **REASONING** You have two triangles. Two angles in the first triangle are congruent to two angles in the second triangle. Can you conclude that the triangles are similar? Explain.

Geometry

In this lesson, you will

- understand the concept of similar triangles.
- identify similar triangles.
- use indirect measurement to find missing measures.

3 ACTIVITY: Indirect Measurement

Math Practice

Make Sense of Quantities

What do you know about the sides of the triangles when the triangles are similar?

Work with a partner.

a. Use the fact that two rays from the Sun are parallel to explain why $\triangle ABC$ and $\triangle DEF$ are similar.

b. Explain how to use similar triangles to find the height of the flagpole.

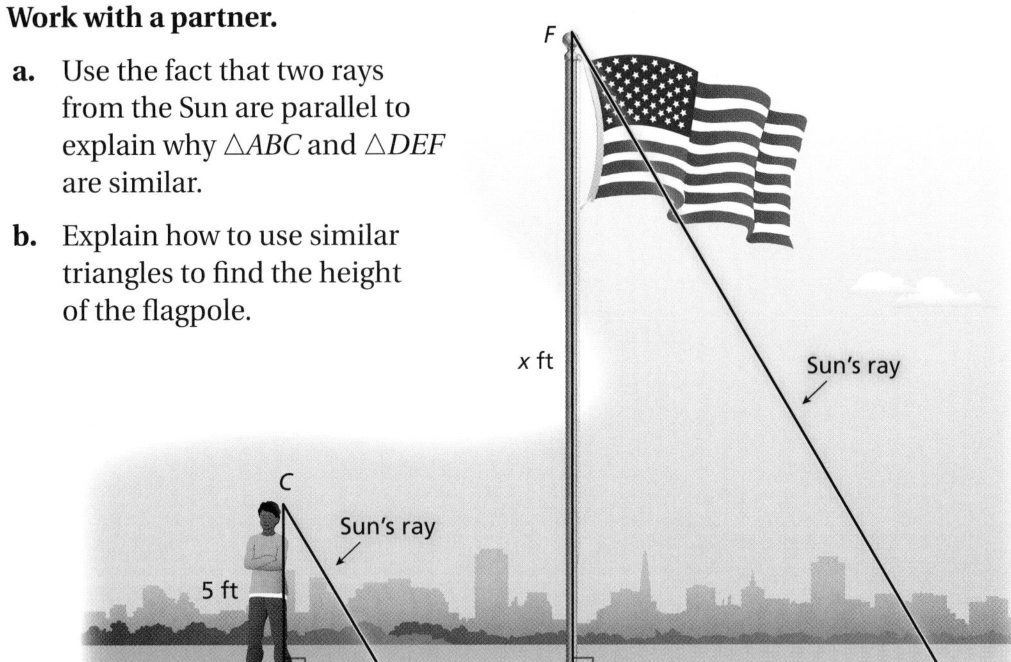

What Is Your Answer?

4. **IN YOUR OWN WORDS** How can you use angles to tell whether triangles are similar?
5. **PROJECT** Work with a partner or in a small group.
   - **a.** Explain why the process in Activity 3 is called "indirect" measurement.
   - **b.** **CHOOSE TOOLS** Use indirect measurement to measure the height of something outside your school (a tree, a building, a flagpole). Before going outside, decide what materials you need to take with you.
   - **c.** **MODELING** Draw a diagram of the indirect measurement process you used. In the diagram, label the lengths that you actually measured and also the lengths that you calculated.
6. **PRECISION** Look back at Exercise 17 in Section 2.5. Explain how you can show that the two triangles are similar.

Practice Use what you learned about similar triangles to complete Exercises 4 and 5 on page 130.

3.4 Lesson

Key Vocabulary
indirect measurement, p. 129

Angles of Similar Triangles

Words When two angles in one triangle are congruent to two angles in another triangle, the third angles are also congruent and the triangles are similar.

Example

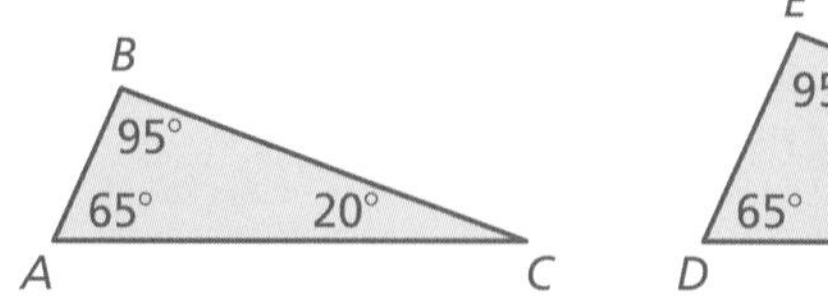

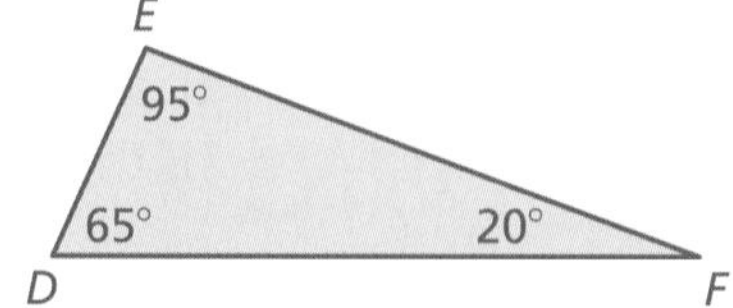

Triangle ABC is similar to Triangle DEF: $\triangle ABC \sim \triangle DEF$.

EXAMPLE 1 Identifying Similar Triangles

Tell whether the triangles are similar. Explain.

a.

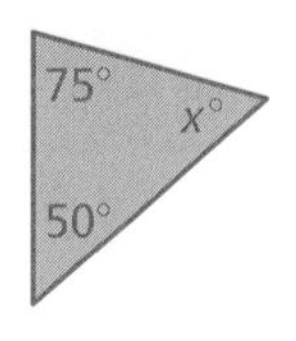

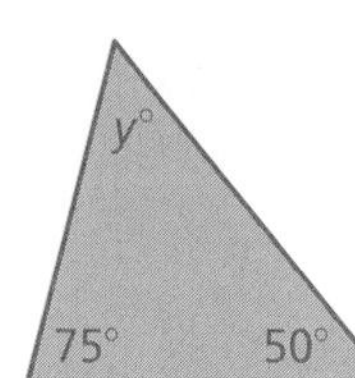

The triangles have two pairs of congruent angles.

So, the third angles are congruent, and the triangles are similar.

b.

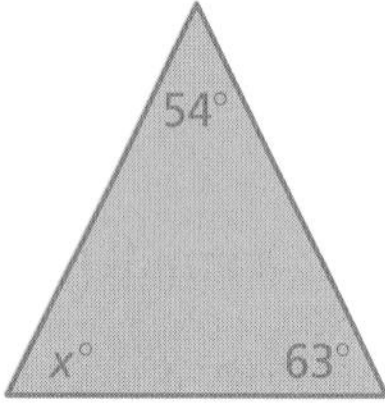

63° 63° y°

Write and solve an equation to find x.

$$x + 54 + 63 = 180$$

$$x + 117 = 180$$

$$x = 63$$

The triangles have two pairs of congruent angles.

So, the third angles are congruent, and the triangles are similar.

c.

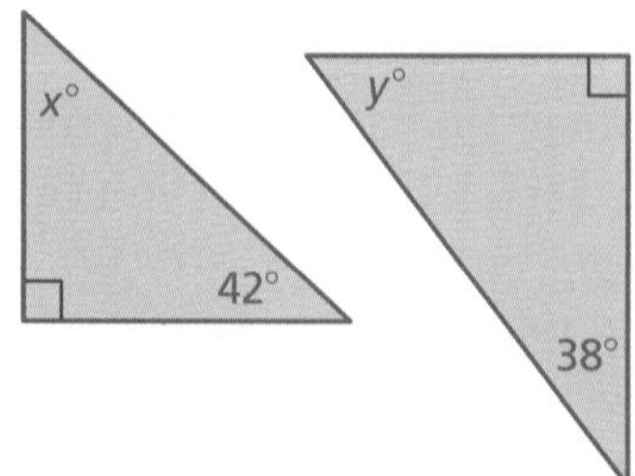

Write and solve an equation to find x.

$$x + 90 + 42 = 180$$

$$x + 132 = 180$$

$$x = 48$$

The triangles do not have two pairs of congruent angles.

So, the triangles are not similar.

On Your Own

Now You're Ready
Exercises 6–9

Tell whether the triangles are similar. Explain.

1.

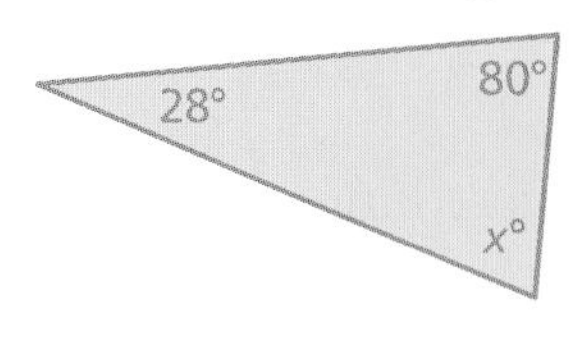

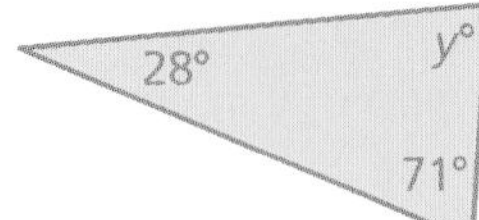

2.

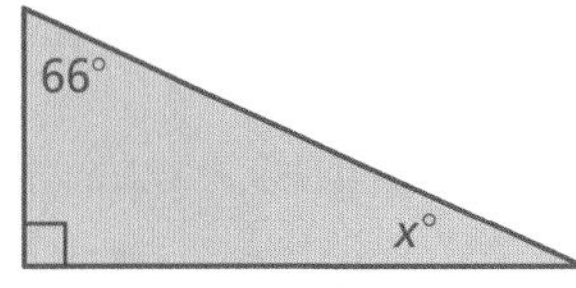

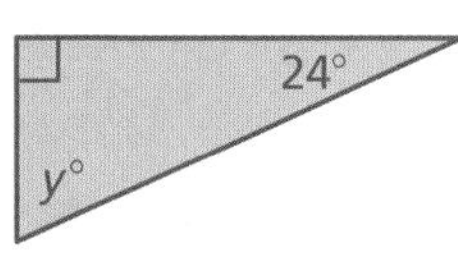

Indirect measurement uses similar figures to find a missing measure when it is difficult to find directly.

EXAMPLE 2 Using Indirect Measurement

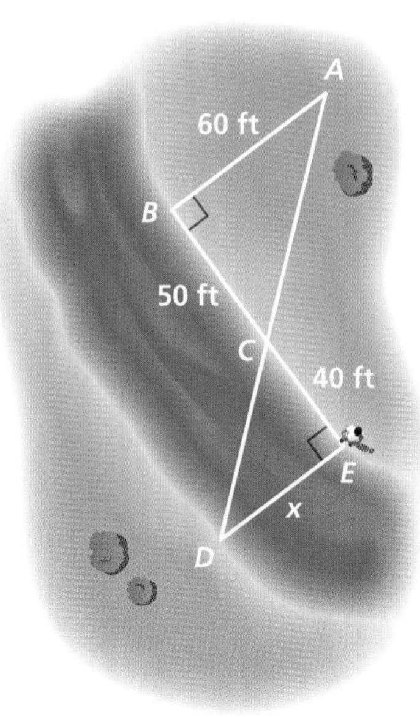

You plan to cross a river and want to know how far it is to the other side. You take measurements on your side of the river and make the drawing shown. (a) Explain why $\triangle ABC$ and $\triangle DEC$ are similar. (b) What is the distance x across the river?

a. $\angle B$ and $\angle E$ are right angles, so they are congruent. $\angle ACB$ and $\angle DCE$ are vertical angles, so they are congruent.

Because two angles in $\triangle ABC$ are congruent to two angles in $\triangle DEC$, the third angles are also congruent and the triangles are similar.

b. The ratios of the corresponding side lengths in similar triangles are equal. Write and solve a proportion to find x.

$$\frac{x}{60} = \frac{40}{50} \quad \text{Write a proportion.}$$

$$60 \cdot \frac{x}{60} = 60 \cdot \frac{40}{50} \quad \text{Multiplication Property of Equality}$$

$$x = 48 \quad \text{Simplify.}$$

So, the distance across the river is 48 feet.

On Your Own

3. **WHAT IF?** The distance from vertex A to vertex B is 55 feet. What is the distance across the river?

3.4 Exercises

Vocabulary and Concept Check

1. **REASONING** How can you use similar triangles to find a missing measurement?
2. **WHICH ONE DOESN'T BELONG?** Which triangle does *not* belong with the other three? Explain your reasoning.

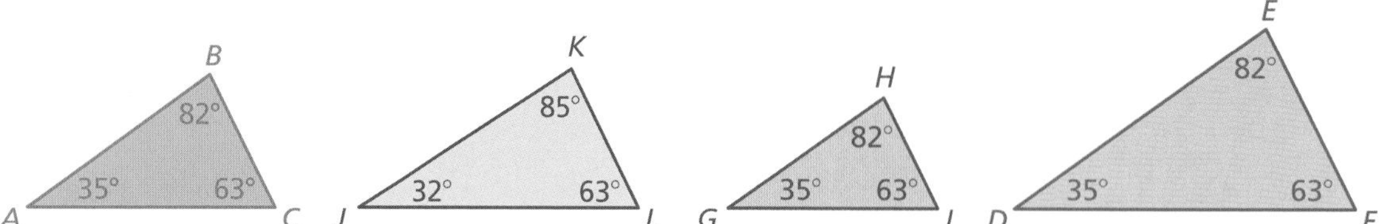

3. **WRITING** Two triangles have two pairs of congruent angles. In your own words, explain why you do not need to find the measures of the third pair of angles to determine that they are congruent.

Practice and Problem Solving

Make a triangle that is larger or smaller than the one given and has the same angle measures. Find the ratios of the corresponding side lengths.

4. 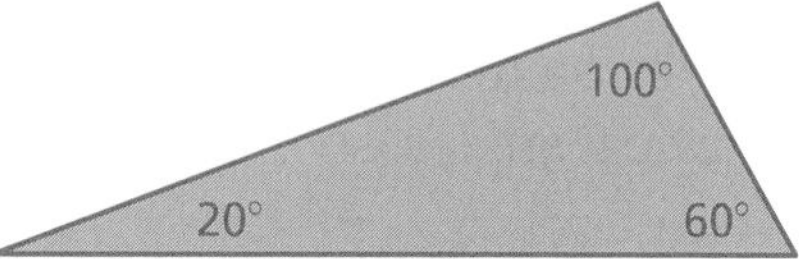

5. 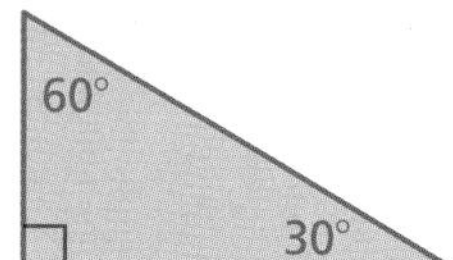

Tell whether the triangles are similar. Explain.

1 6. 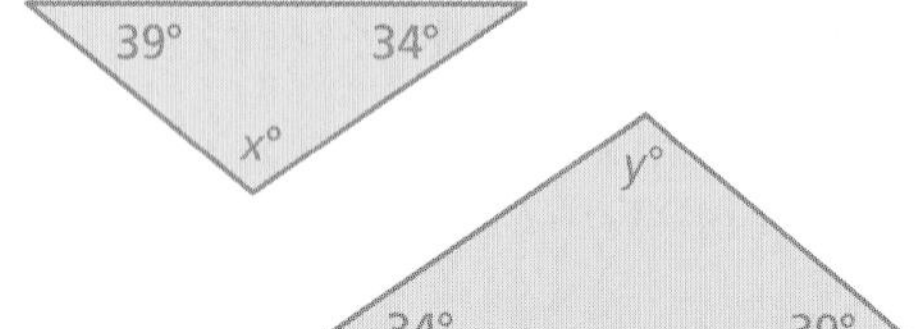

7. 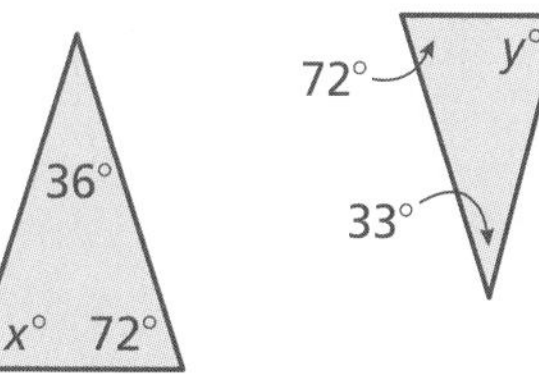

8. 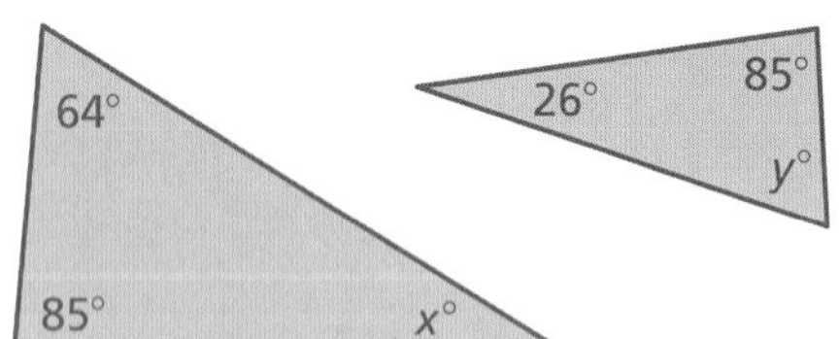

9. 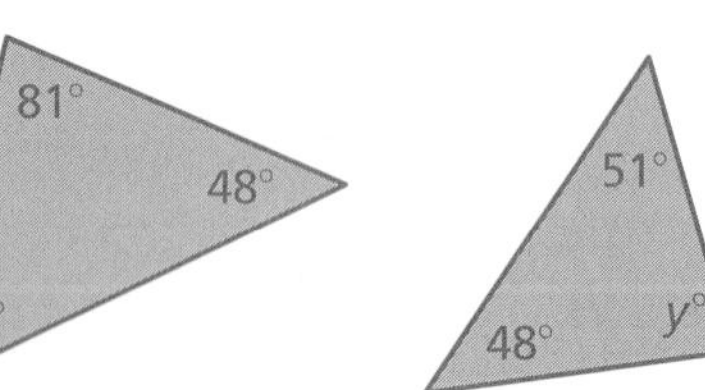

10. **RULERS** Which of the rulers are similar in shape? Explain.

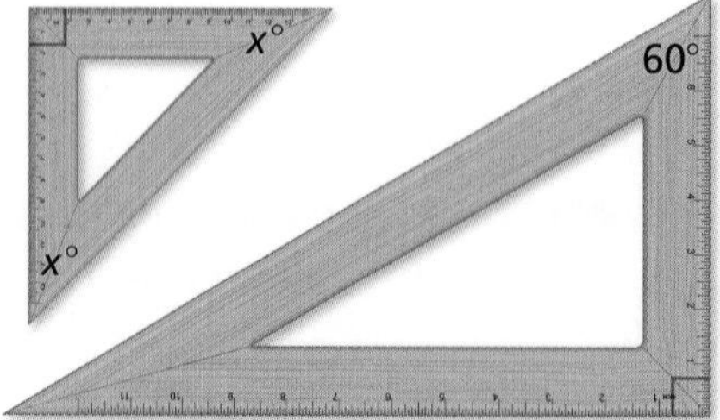

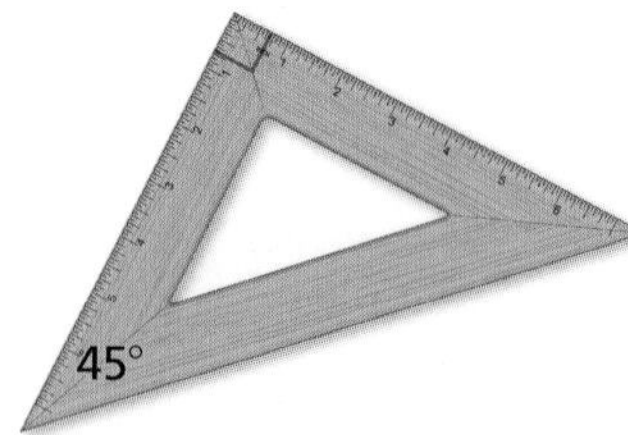

Tell whether the triangles are similar. Explain.

11.

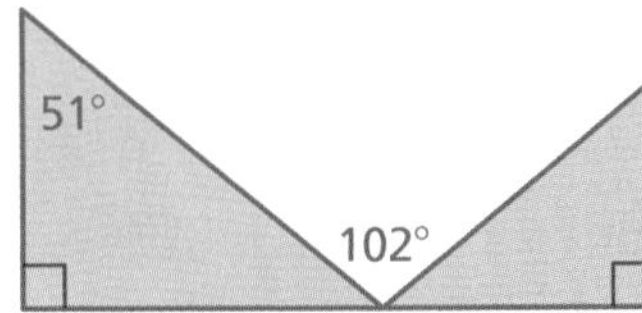

12.

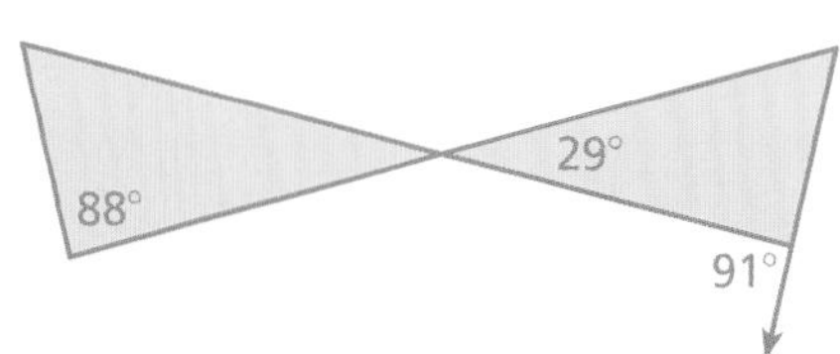

2 13. **TREASURE** The map shows the number of steps you must take to get to the treasure. However, the map is old, and the last dimension is unreadable. Explain why the triangles are similar. How many steps do you take from the pyramids to the treasure?

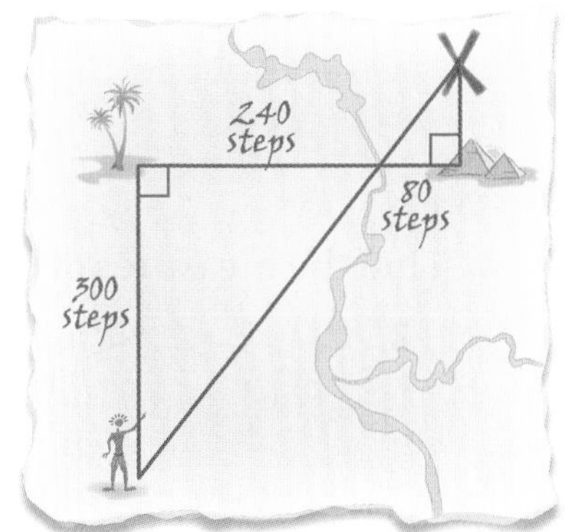

14. **CRITICAL THINKING** The side lengths of a triangle are increased by 50% to make a similar triangle. Does the area increase by 50% as well? Explain.

15. **PINE TREE** A person who is 6 feet tall casts a 3-foot-long shadow. A nearby pine tree casts a 15-foot-long shadow. What is the height h of the pine tree?

16. **OPEN-ENDED** You place a mirror on the ground 6 feet from the lamppost. You move back 3 feet and see the top of the lamppost in the mirror. What is the height of the lamppost?

17. **REASONING** In each of two right triangles, one angle measure is two times another angle measure. Are the triangles similar? Explain your reasoning.

18.  In the diagram, segments BG, CF, and DE are parallel. The length of segment BD is 6.32 feet, and the length of segment DE is 6 feet. Name all pairs of similar triangles in the diagram. Then find the lengths of segments BG and CF.

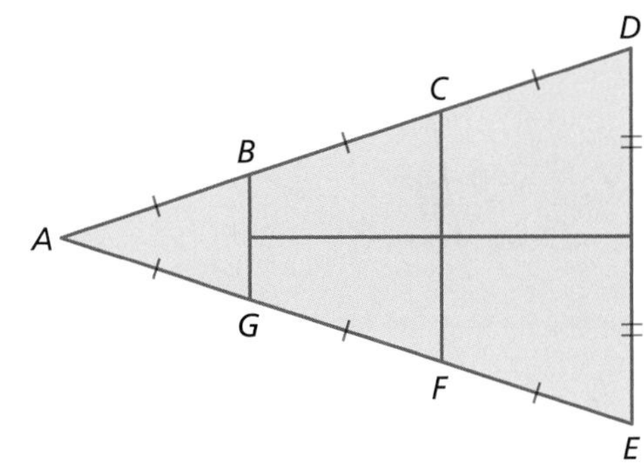

Fair Game Review What you learned in previous grades & lessons

Solve the equation for *y*. *(Section 1.4)*

19. $y - 5x = 3$

20. $4x + 6y = 12$

21. $2x - \frac{1}{4}y = 1$

22. **MULTIPLE CHOICE** What is the value of x? *(Section 3.2)*

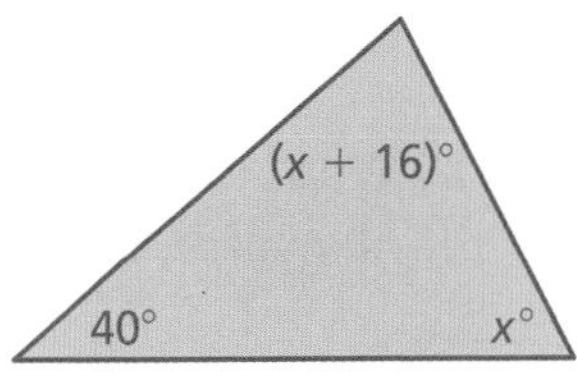

Ⓐ 17 Ⓑ 62

Ⓒ 118 Ⓓ 152

3.3–3.4 Quiz

Find the sum of the interior angle measures of the polygon. *(Section 3.3)*

1.

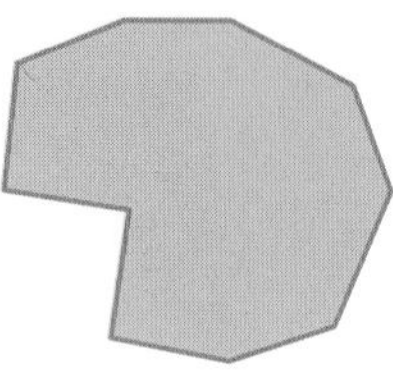

2.

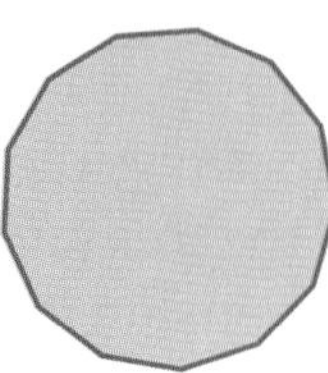

Find the measures of the interior angles of the polygon. *(Section 3.3)*

3.

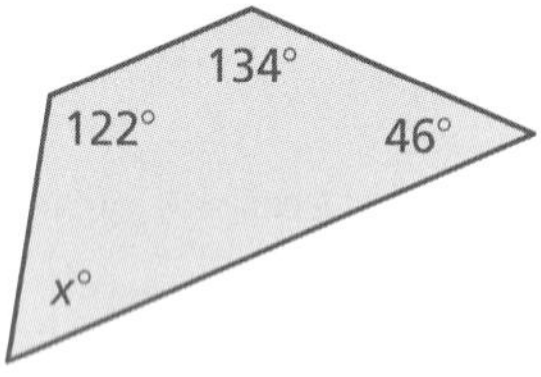

4.

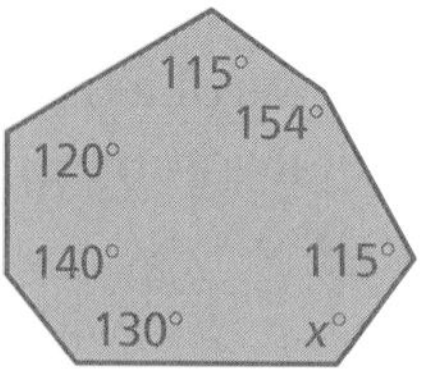

5.

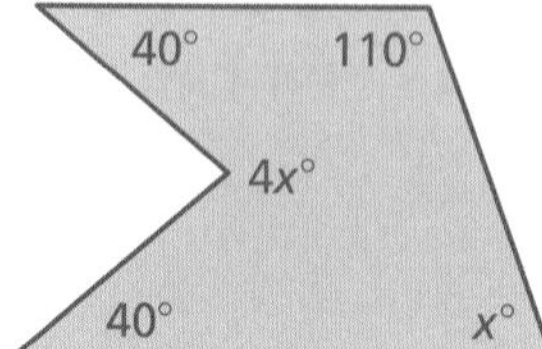

Find the measures of the exterior angles of the polygon. *(Section 3.3)*

6.

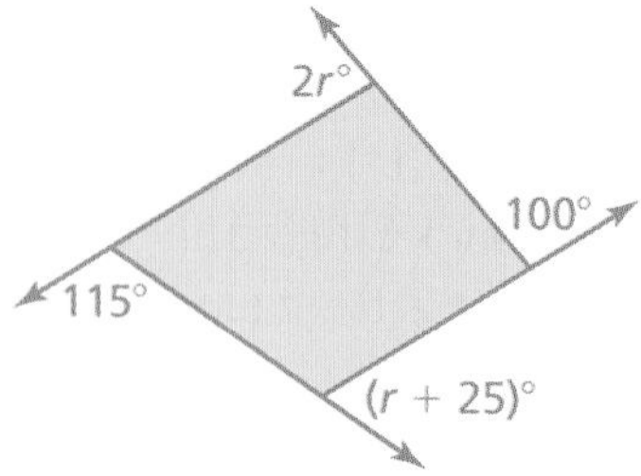

7.

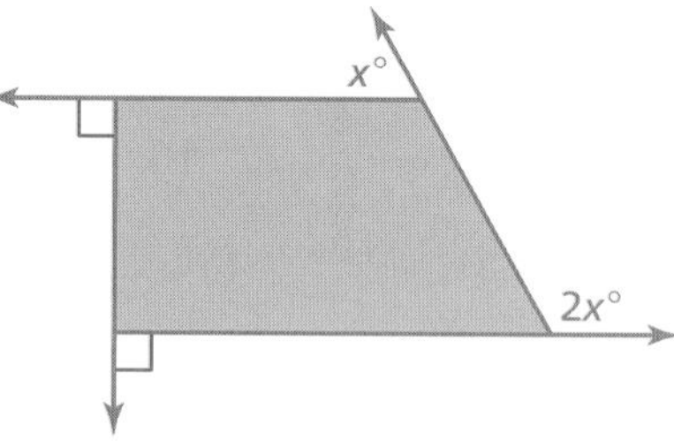

Tell whether the triangles are similar. Explain. *(Section 3.4)*

8.

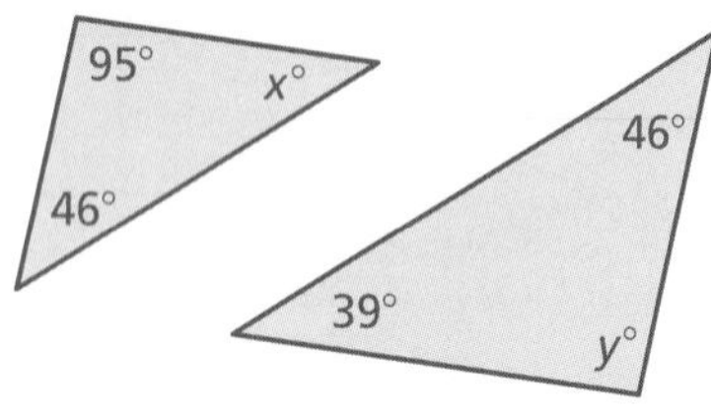

9.

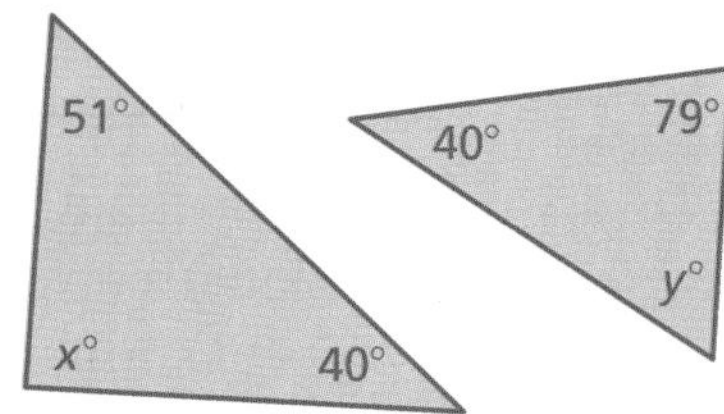

10. REASONING The sum of the interior angle measures of a polygon is 4140°. How many sides does the polygon have? *(Section 3.3)*

11. SWAMP You are trying to find the distance ℓ across a patch of swamp water. *(Section 3.4)*

a. Explain why $\triangle VWX$ and $\triangle YZX$ are similar.

b. What is the distance across the patch of swamp water?

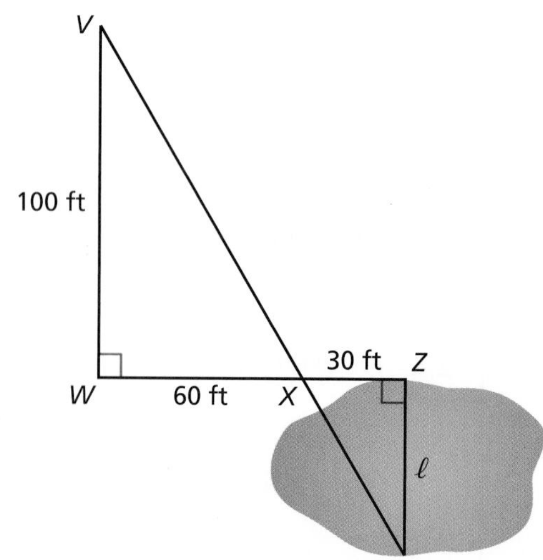

3 Chapter Review

Check It Out
Vocabulary Help
BigIdeasMath.com

Review Key Vocabulary

transversal, *p. 104*
interior angles, *p. 105*
exterior angles, *p. 105*
interior angles of a polygon, *p. 112*
exterior angles of a polygon, *p. 112*
convex polygon, *p. 119*
concave polygon, *p. 119*
regular polygon, *p. 121*
indirect measurement, *p. 129*

Review Examples and Exercises

3.1 Parallel Lines and Transversals *(pp. 102–109)*

Use the figure to find the measure of $\angle 6$.

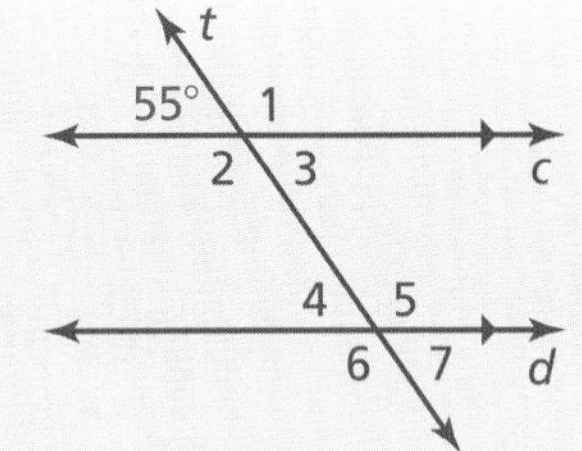

$\angle 2$ and the 55° angle are supplementary.
So, the measure of $\angle 2$ is $180^\circ - 55^\circ = 125^\circ$.

$\angle 2$ and $\angle 6$ are corresponding angles.
They are congruent.

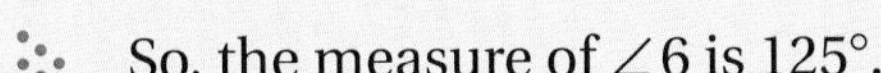

∴ So, the measure of $\angle 6$ is 125°.

Exercises

Use the figure to find the measure of the angle. Explain your reasoning.

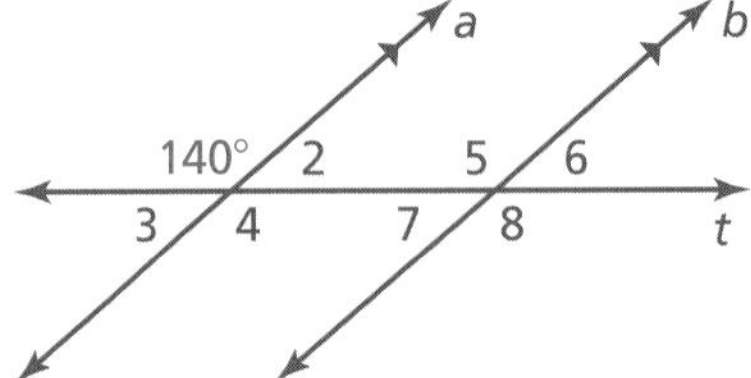

1. $\angle 8$

2. $\angle 5$

3. $\angle 7$

4. $\angle 2$

3.2 Angles of Triangles *(pp. 110–115)*

a. Find the value of x.

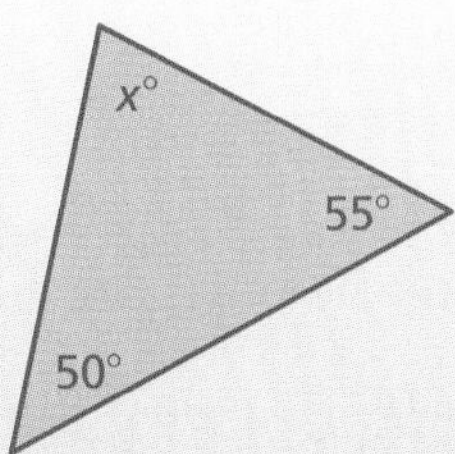

$$x + 50 + 55 = 180$$
$$x + 105 = 180$$
$$x = 75$$

∴ The value of x is 75.

b. Find the measure of the exterior angle.

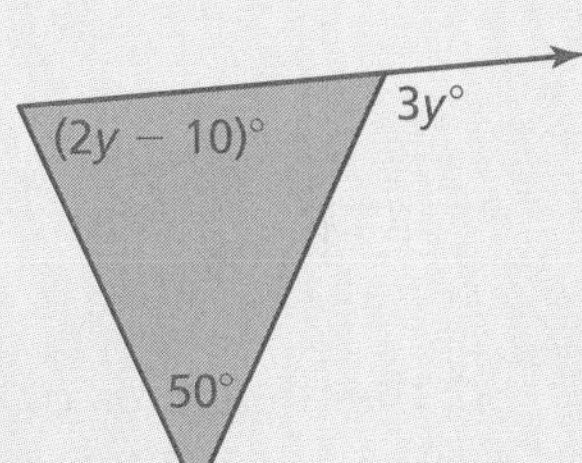

$$3y = (2y - 10) + 50$$
$$3y = 2y + 40$$
$$y = 40$$

∴ So, the measure of the exterior angle is $3(40)^\circ = 120^\circ$.

Exercises

Find the measures of the interior angles.

5.

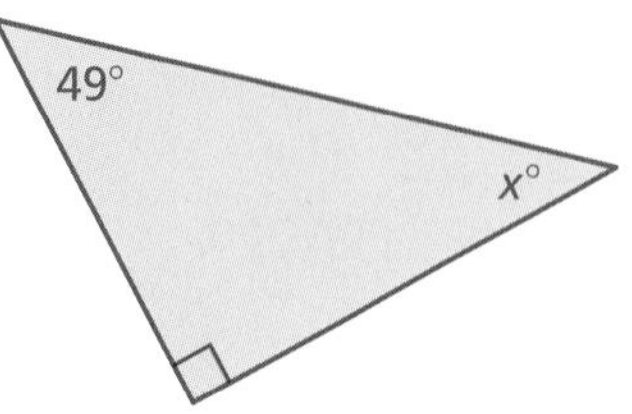

6.

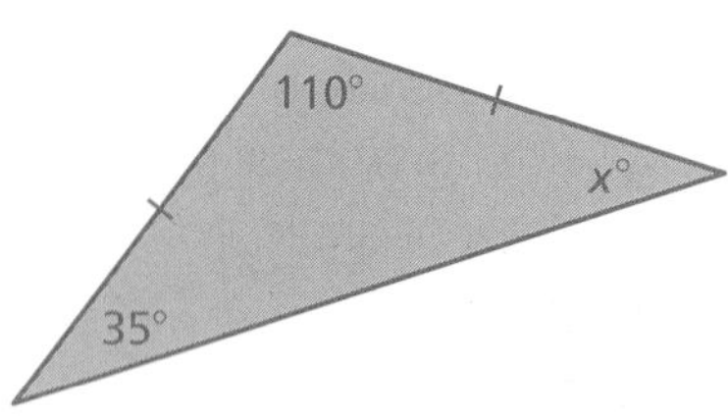

Find the measure of the exterior angle.

7.

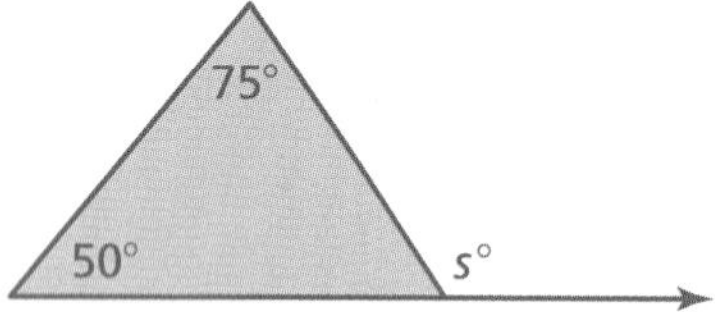

8.

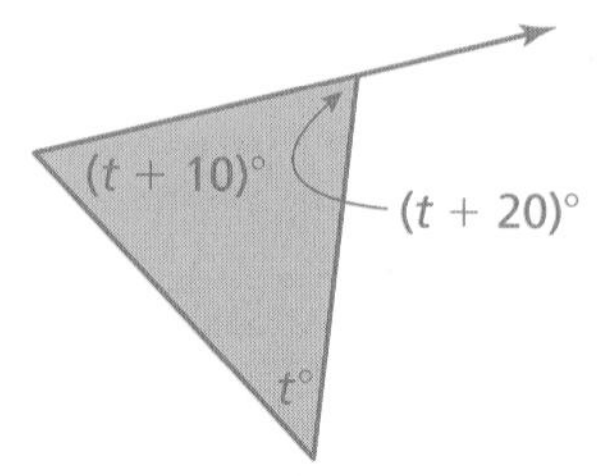

3.3 Angles of Polygons *(pp. 118–125)*

a. Find the value of x.

Step 1: The polygon has 6 sides. Find the sum of the interior angle measures.

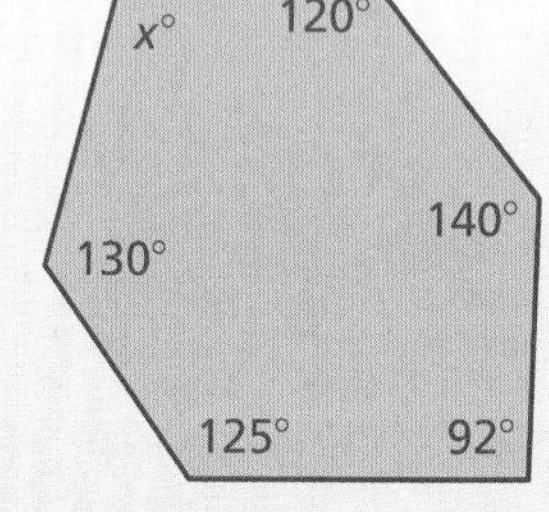

$S = (n - 2) \cdot 180°$ Write the formula.

$= (6 - 2) \cdot 180°$ Substitute 6 for n.

$= 720$ Simplify. The sum of the interior angle measures is 720°.

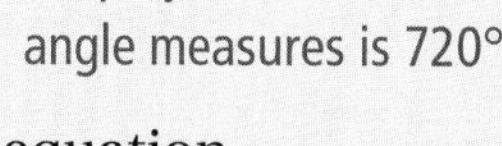

Step 2: Write and solve an equation.

$$130 + 125 + 92 + 140 + 120 + x = 720$$

$$607 + x = 720$$

$$x = 113$$

The value of x is 113.

b. Find the measures of the exterior angles of the polygon.

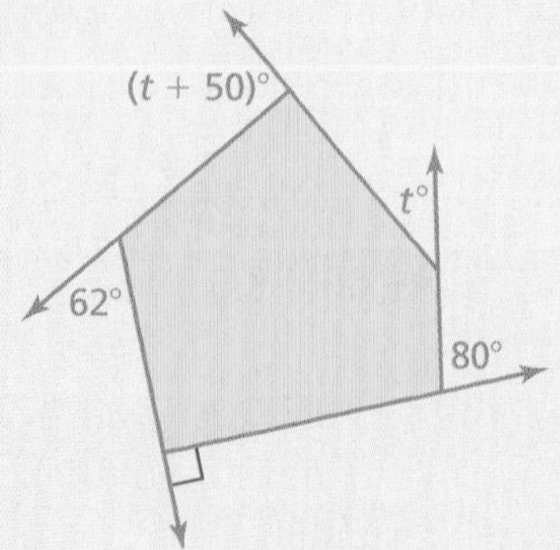

Write and solve an equation for t.

$$t + 80 + 90 + 62 + (t + 50) = 360$$

$$2t + 282 = 360$$

$$2t = 78$$

$$t = 39$$

So, the measures of the exterior angles are 39°, 80°, 90°, 62°, and $(39 + 50)° = 89°$.

Exercises

Find the measures of the interior angles of the polygon.

9.

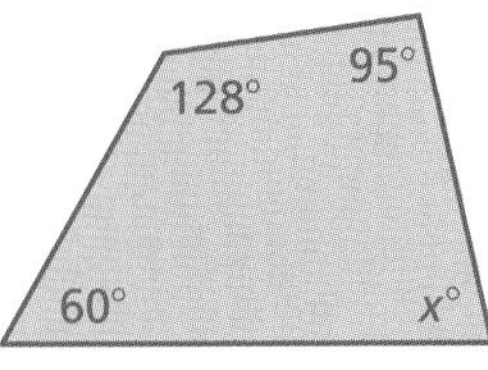

10.

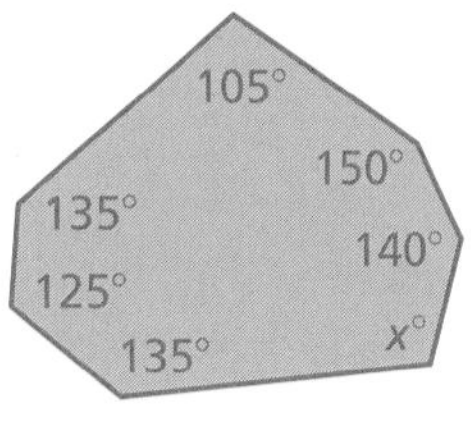

11.

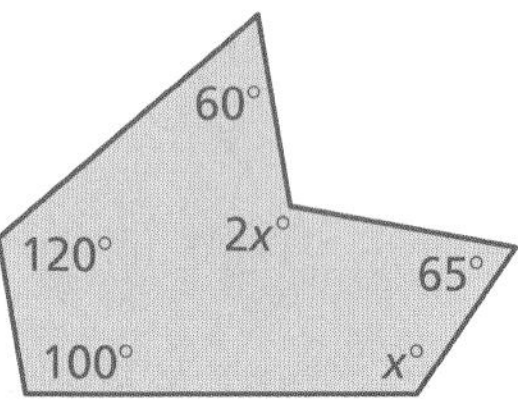

Find the measures of the exterior angles of the polygon.

12.

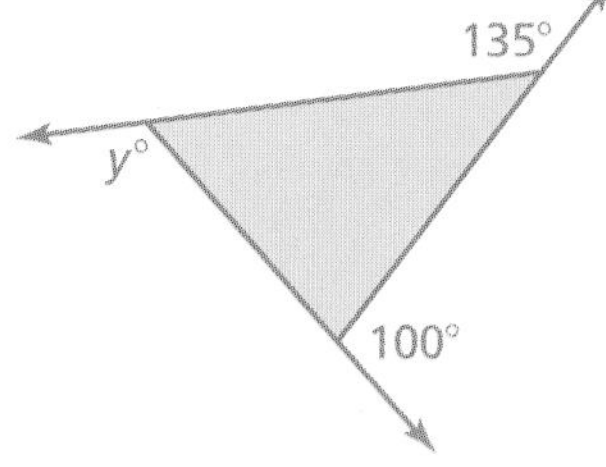

13.

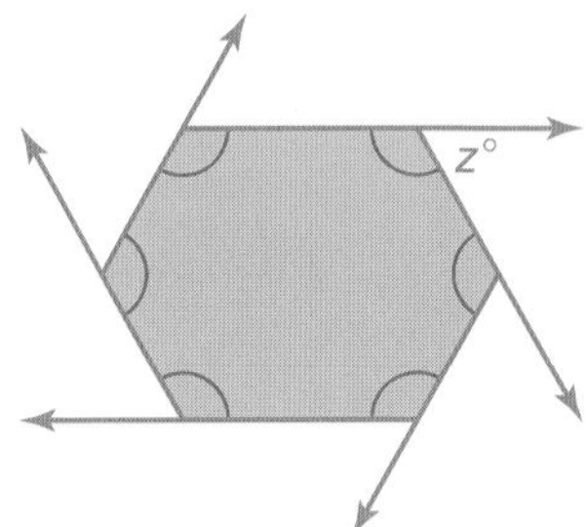

3.4 Using Similar Triangles *(pp. 126–131)*

Tell whether the triangles are similar. Explain.

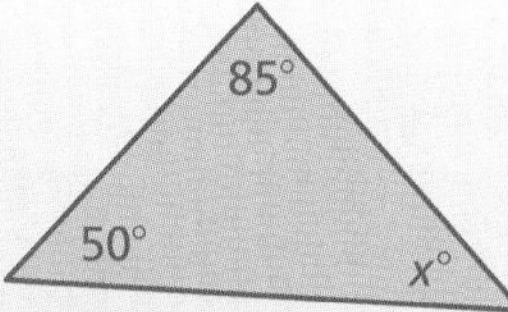

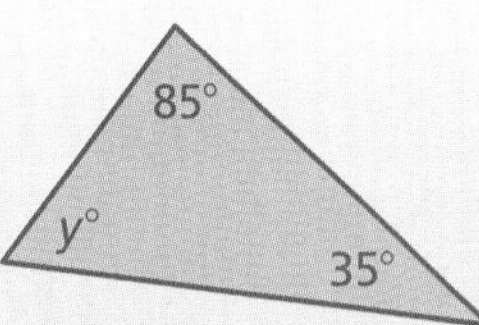

Write and solve an equation to find x.

$$50 + 85 + x = 180$$

$$135 + x = 180$$

$$x = 45$$

∴ The triangles do not have two pairs of congruent angles. So, the triangles are not similar.

Exercises

Tell whether the triangles are similar. Explain.

14.

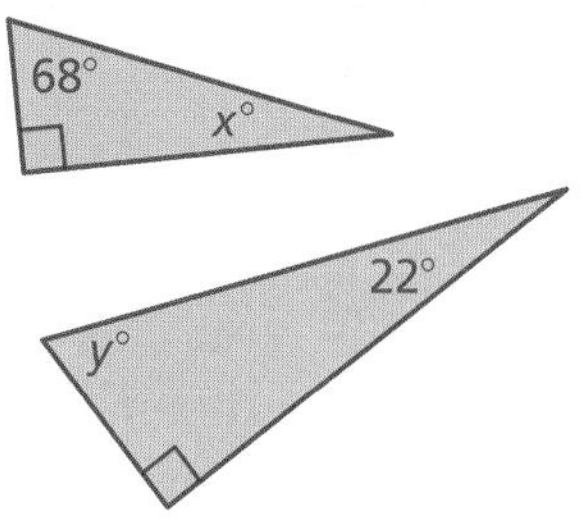

15.

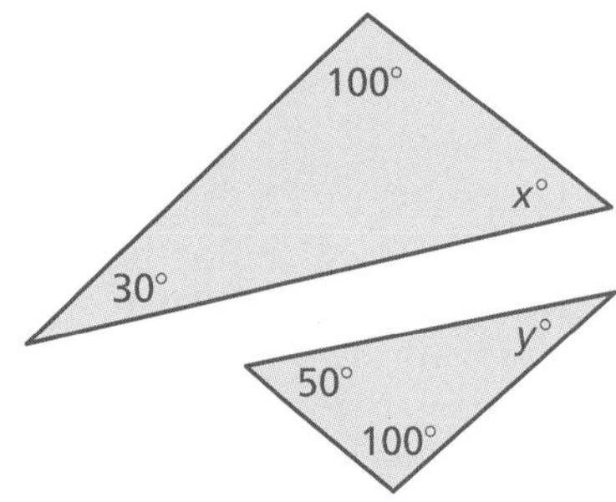

3 Chapter Test

Use the figure to find the measure of the angle. Explain your reasoning.

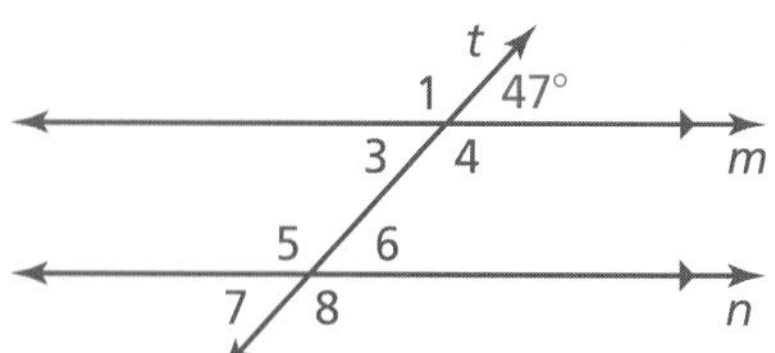

1. $\angle 1$
2. $\angle 8$
3. $\angle 4$
4. $\angle 5$

Find the measures of the interior angles.

5.

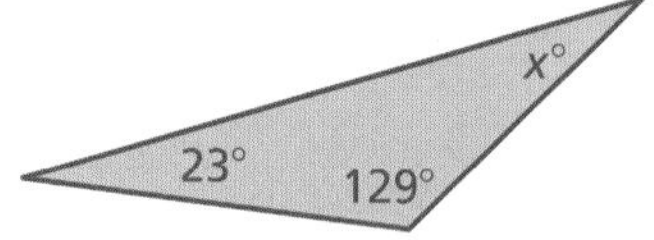

6.

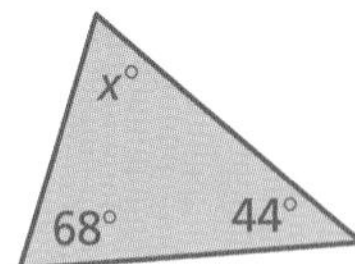

7.

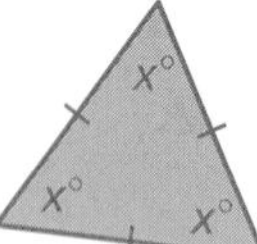

Find the measure of the exterior angle.

8.

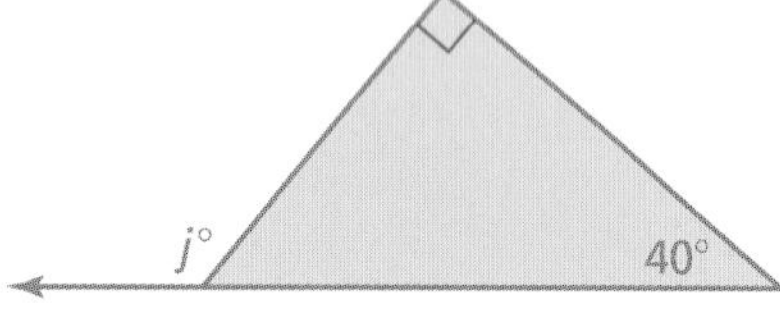

9.

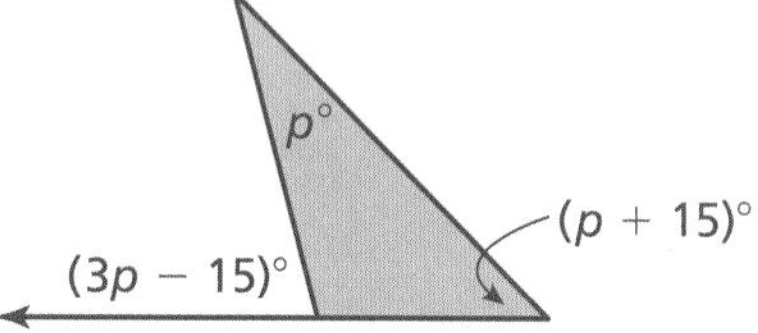

10. Find the measures of the interior angles of the polygon.

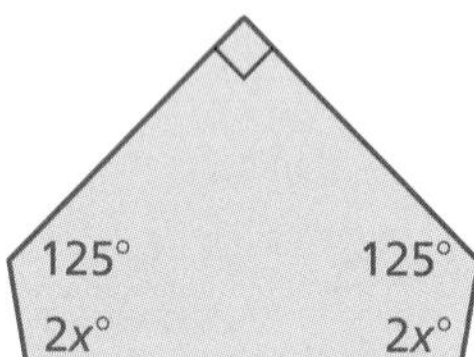

11. Find the measures of the exterior angles of the polygon.

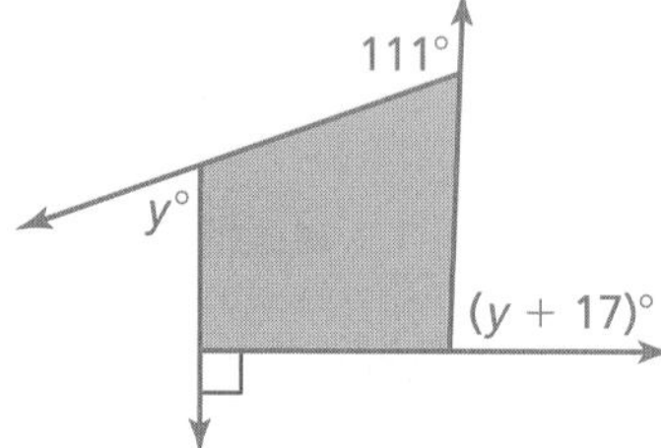

Tell whether the triangles are similar. Explain.

12.

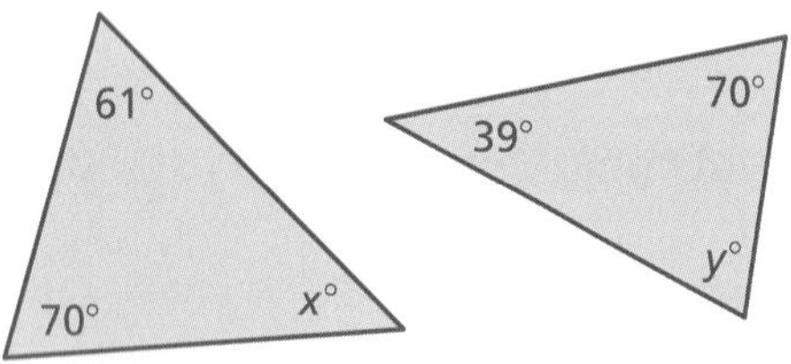

13.

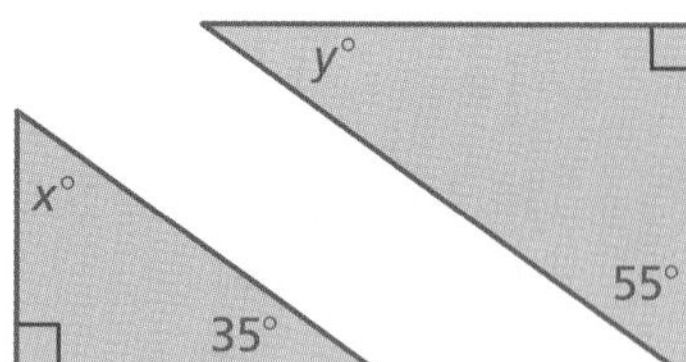

14. **WRITING** Describe two ways you can find the measure of $\angle 5$.

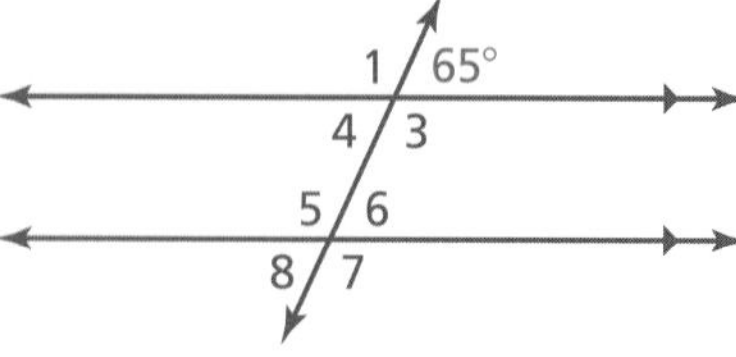

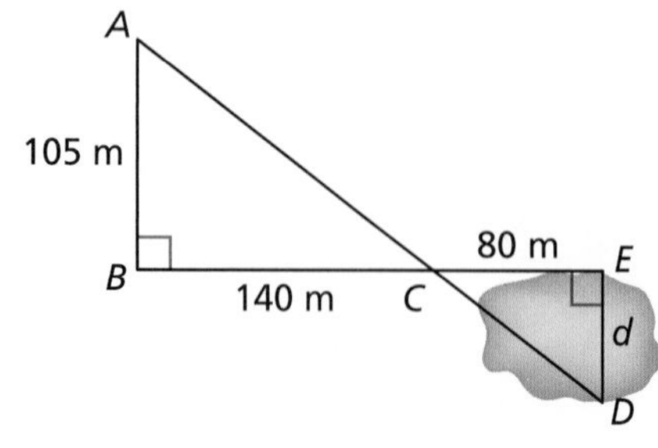

15. **POND** Use the given measurements to find the distance d across the pond.

3 Cumulative Assessment

Solve Problem Before Looking at Choices

"Solve the problem before looking at the choices. You know 180 −2(70) = 40. So, the answer is C."

1. The border of a Canadian one-dollar coin is shaped like an 11-sided regular polygon. The shape was chosen to help visually impaired people identify the coin. How many degrees are in each angle along the border? Round your answer to the nearest degree.

2. A public utility charges its residential customers for natural gas based on the number of therms used each month. The formula below shows how the monthly cost C in dollars is related to the number t of therms used.

$$C = 11 + 1.6t$$

Solve this formula for t.

A. $t = \dfrac{C}{12.6}$

B. $t = \dfrac{C - 11}{1.6}$

C. $t = \dfrac{C}{1.6} - 11$

D. $t = C - 12.6$

3. What is the value of x?

$$5(x - 4) = 3x$$

F. -10

G. 2

H. $2\frac{1}{2}$

I. 10

4. In the figures below, $\triangle PQR$ is a dilation of $\triangle STU$.

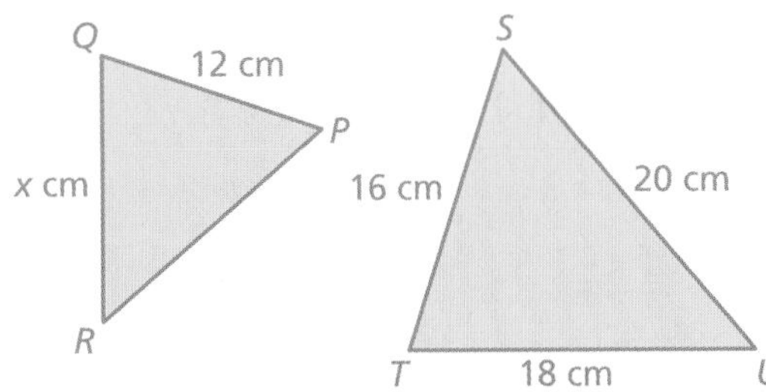

What is the value of x?

A. 9.6

B. $10\frac{2}{3}$

C. 13.5

D. 15

5. What is the value of x?

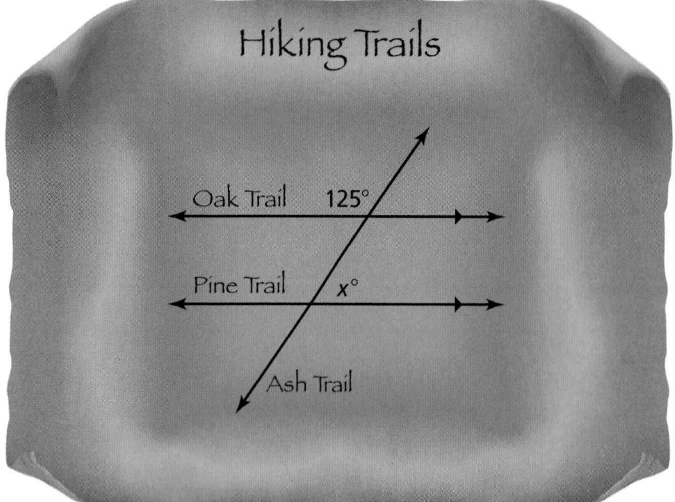

6. Olga was solving an equation in the box shown.

$$-\frac{2}{5}(10x - 15) = -30$$

$$10x - 15 = -30\left(-\frac{2}{5}\right)$$

$$10x - 15 = 12$$

$$10x - 15 + 15 = 12 + 15$$

$$10x = 27$$

$$\frac{10x}{10} = \frac{27}{10}$$

$$x = \frac{27}{10}$$

What should Olga do to correct the error that she made?

F. Multiply both sides by $-\frac{5}{2}$ instead of $-\frac{2}{5}$.

G. Multiply both sides by $\frac{2}{5}$ instead of $-\frac{2}{5}$.

H. Distribute $-\frac{2}{5}$ to get $-4x - 6$.

I. Add 15 to -30.

7. In the coordinate plane below, $\triangle XYZ$ is plotted and its vertices are labeled.

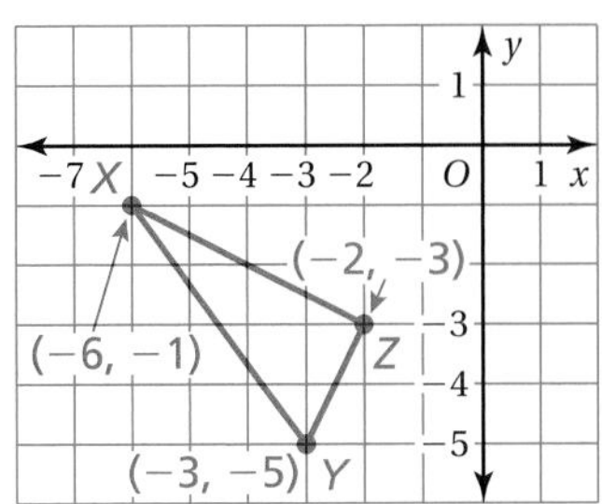

Which of the following shows $\triangle X'Y'Z'$, the image of $\triangle XYZ$ after it is reflected in the y-axis?

A.

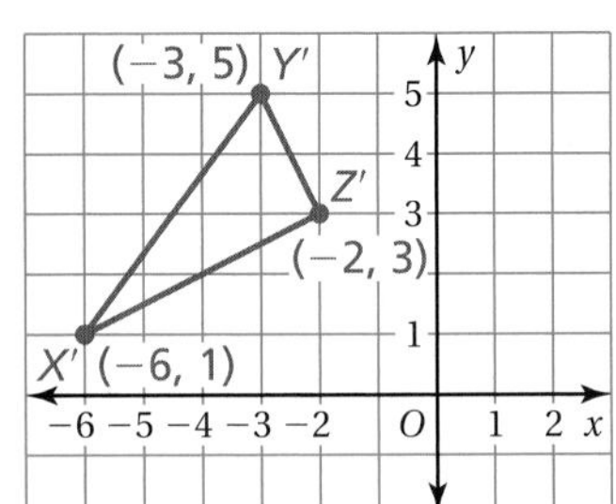

C.

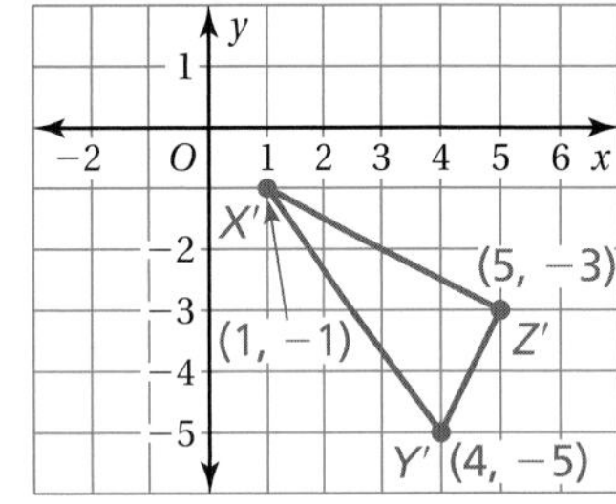

B.

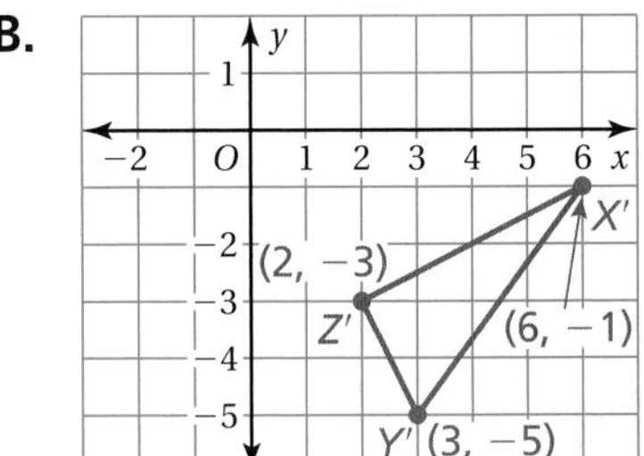

D.

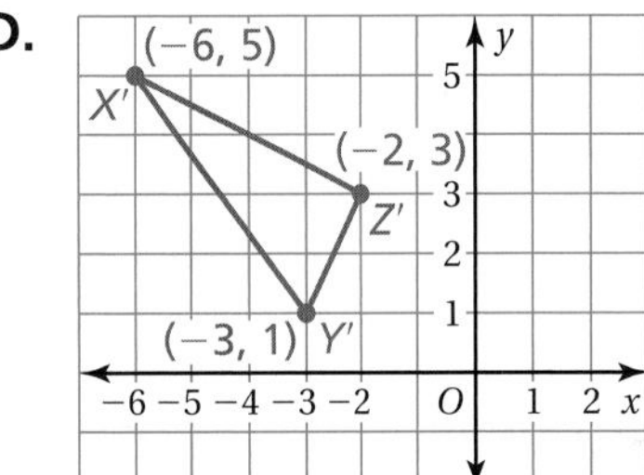

8. The sum S of the interior angle measures of a polygon with n sides can be found by using a formula.

Part A Write the formula.

Part B A quadrilateral has angles measuring 100°, 90°, and 90°. Find the measure of its fourth angle. Show your work and explain your reasoning.

Part C The sum of the measures of the angles of the pentagon shown is 540°. Divide the pentagon into triangles to show why this must be true. Show your work and explain your reasoning.

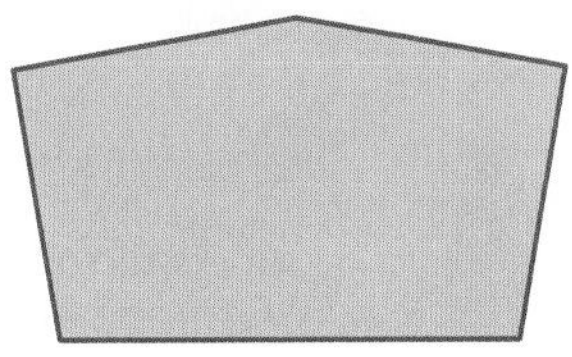

4 Graphing and Writing Linear Equations

"Okay Descartes, stand on the *y*-axis and try to intercept the pass when I throw."

"Here's an easy example of a line with a slope of 1."

"You eat one mouse treat the first day. Two treats the second day. And so on. Get it?"

What You Learned Before

"I estimate that we are on a slope of about −0.625. What do you think?"

Evaluating Expressions Using Order of Operations

Example 1 **Evaluate $2xy + 3(x + y)$ when $x = 4$ and $y = 7$.**

$2xy + 3(x + y) = 2(4)(7) + 3(4 + 7)$	Substitute 4 for x and 7 for y.
$= 8(7) + 3(4 + 7)$	Use order of operations.
$= 56 + 3(11)$	Simplify.
$= 56 + 33$	Multiply.
$= 89$	Add.

Try It Yourself

Evaluate the expression when $a = \frac{1}{4}$ and $b = 6$.

1. $-8ab$ **2.** $16a^2 - 4b$ **3.** $\frac{5b}{32a^2}$ **4.** $12a + (b - a - 4)$

Plotting Points

Example 2 **Write the ordered pair that corresponds to point U.**

Point U is 3 units to the left of the origin and 4 units down. So, the x-coordinate is -3, and the y-coordinate is -4.

∴ The ordered pair $(-3, -4)$ corresponds to point U.

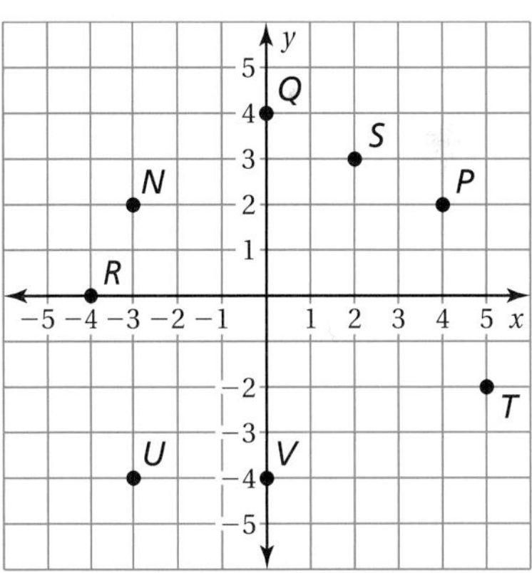

Example 3 **Which point is located at $(5, -2)$?**

Start at the origin. Move 5 units right and 2 units down.

∴ Point T is located at $(5, -2)$.

Try It Yourself

Use the graph to answer the question.

5. Write the ordered pair that corresponds to point Q.

6. Write the ordered pair that corresponds to point P.

7. Which point is located at $(-4, 0)$?

8. Which point is located in Quadrant II?

4.1 Graphing Linear Equations

Essential Question How can you recognize a linear equation? How can you draw its graph?

1 ACTIVITY: Graphing a Linear Equation

Work with a partner.

a. Use the equation $y = \frac{1}{2}x + 1$ to complete the table. (Choose any two x-values and find the y-values.)

	Solution Points	
x		
$y = \frac{1}{2}x + 1$		

b. Write the two ordered pairs given by the table. These are called *solution points* of the equation.

c. **PRECISION** Plot the two solution points. Draw a line *exactly* through the two points.

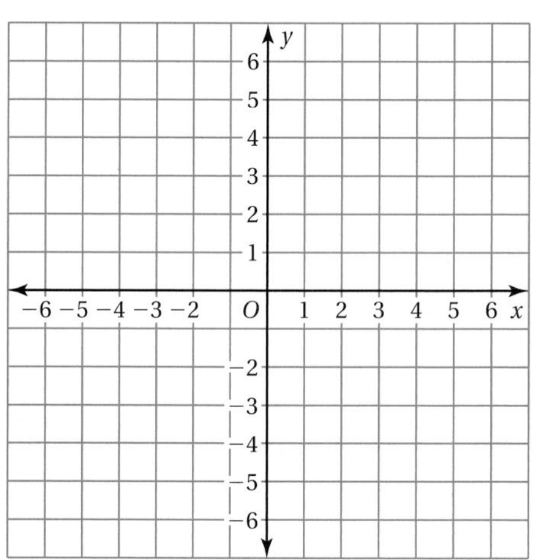

d. Find a different point on the line. Check that this point is a solution point of the equation $y = \frac{1}{2}x + 1$.

e. **LOGIC** Do you think it is true that *any* point on the line is a solution point of the equation $y = \frac{1}{2}x + 1$? Explain.

f. Choose five additional x-values for the table. (Choose positive and negative x-values.) Plot the five corresponding solution points. Does each point lie on the line?

	Solution Points				
x					
$y = \frac{1}{2}x + 1$					

g. **LOGIC** Do you think it is true that *any* solution point of the equation $y = \frac{1}{2}x + 1$ is a point on the line? Explain.

h. Why do you think $y = ax + b$ is called a *linear equation*?

Graphing Equations

In this lesson, you will

- understand that lines represent solutions of linear equations.
- graph linear equations.

2 ACTIVITY: Using a Graphing Calculator

Use a graphing calculator to graph $y = 2x + 5$.

a. Enter the equation $y = 2x + 5$ into your calculator.

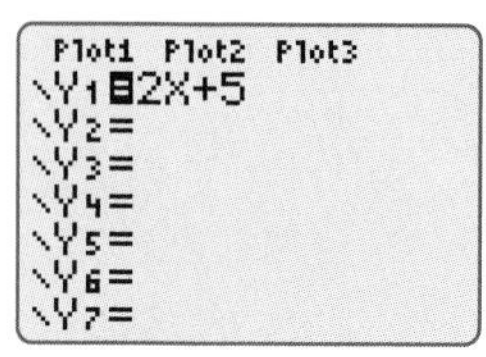

b. Check the settings of the *viewing window*. The boundaries of the graph are set by the minimum and the maximum x- and y-values. The numbers of units between the tick marks are set by the x- and y-scales.

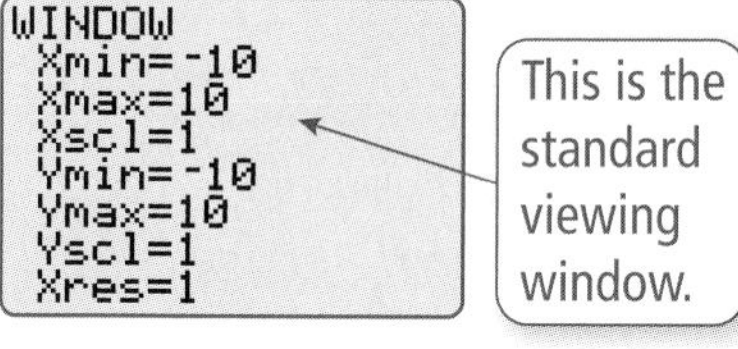

c. Graph $y = 2x + 5$ on your calculator.

d. Change the settings of the viewing window to match those shown.

Compare the two graphs.

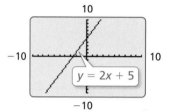

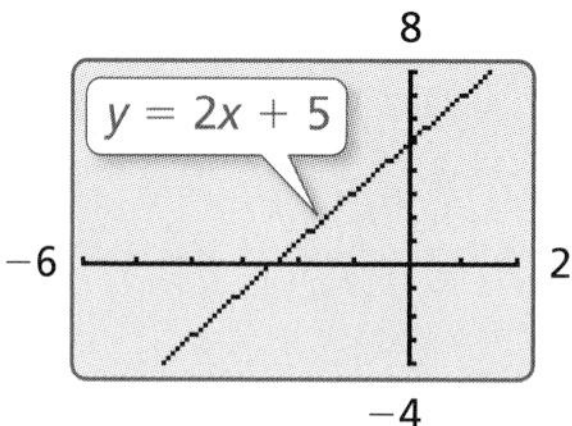

> **Math Practice**
>
> **Recognize Usefulness of Tools**
>
> What are some advantages and disadvantages of using a graphing calculator to graph a linear equation?

What Is Your Answer?

3. **IN YOUR OWN WORDS** How can you recognize a linear equation? How can you draw its graph? Write an equation that is linear. Write an equation that is *not* linear.

4. Use a graphing calculator to graph $y = 5x - 12$ in the standard viewing window.

   a. Can you tell where the line crosses the x-axis? Can you tell where the line crosses the y-axis?

   b. How can you adjust the viewing window so that you can determine where the line crosses the x- and y-axes?

5. **CHOOSE TOOLS** You want to graph $y = 2.5x - 3.8$. Would you graph it by hand or by using a graphing calculator? Why?

Practice Use what you learned about graphing linear equations to complete Exercises 3 and 4 on page 146.

4.1 Lesson

Key Vocabulary

linear equation, *p. 144*
solution of a linear equation, *p. 144*

Remember

An ordered pair (x, y) is used to locate a point in a coordinate plane.

Key Idea

Linear Equations

A **linear equation** is an equation whose graph is a line. The points on the line are **solutions** of the equation.

You can use a graph to show the solutions of a linear equation. The graph below represents the equation $y = x + 1$.

x	y	(x, y)
-1	0	$(-1, 0)$
0	1	$(0, 1)$
2	3	$(2, 3)$

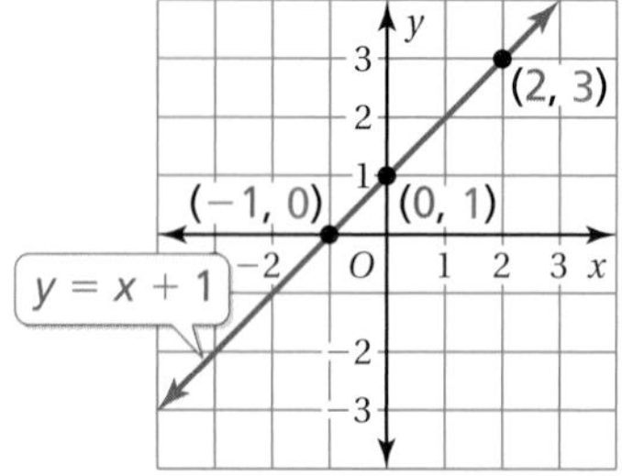

EXAMPLE 1 Graphing a Linear Equation

Graph $y = -2x + 1$.

Step 1: Make a table of values.

x	$y = -2x + 1$	y	(x, y)
-1	$y = -2(-1) + 1$	3	$(-1, 3)$
0	$y = -2(0) + 1$	1	$(0, 1)$
2	$y = -2(2) + 1$	-3	$(2, -3)$

Step 2: Plot the ordered pairs.

Step 3: Draw a line through the points.

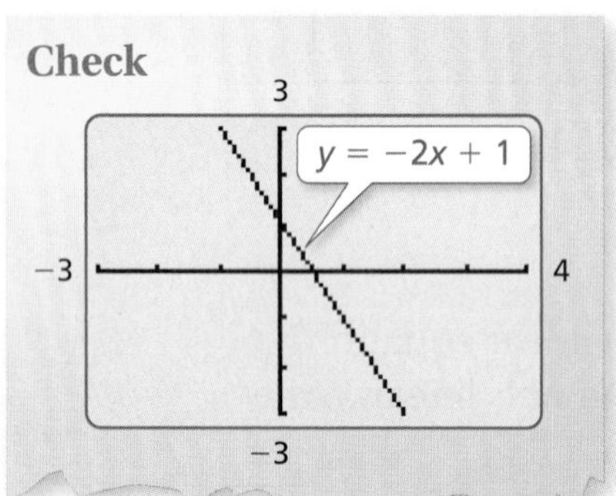

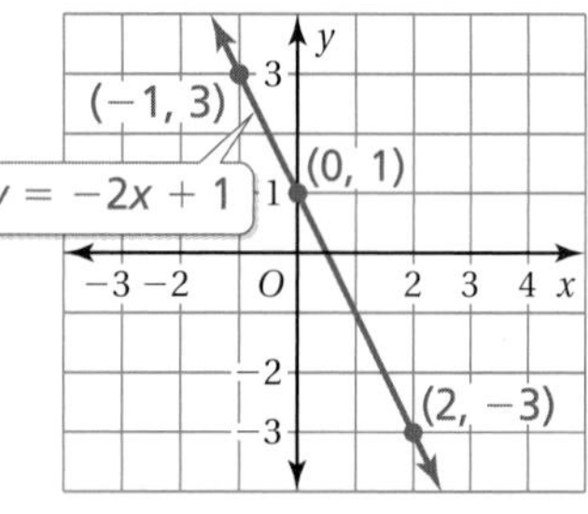

Key Idea

Graphing Horizontal and Vertical Lines

The graph of $y = b$ is a horizontal line passing through $(0, b)$.

The graph of $x = a$ is a vertical line passing through $(a, 0)$.

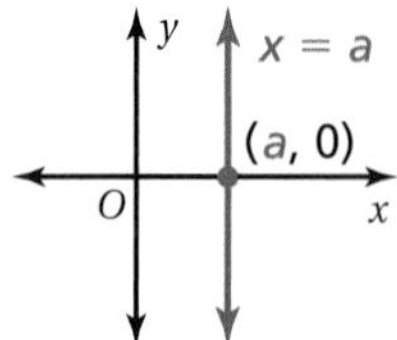

Multi-Language Glossary at BigIdeasMath.com

EXAMPLE 2 Graphing a Horizontal Line and a Vertical Line

a. Graph $y = -3$.

The graph of $y = -3$ is a horizontal line passing through $(0, -3)$. Draw a horizontal line through this point.

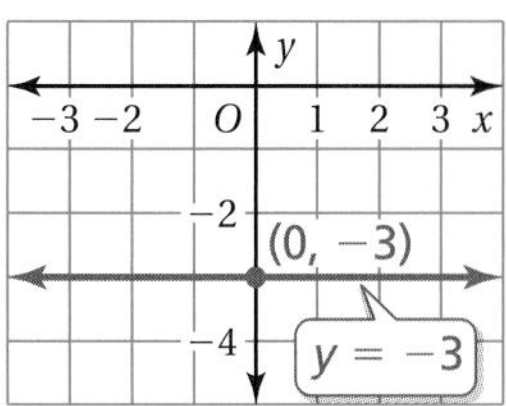

b. Graph $x = 2$.

The graph of $x = 2$ is a vertical line passing through $(2, 0)$. Draw a vertical line through this point.

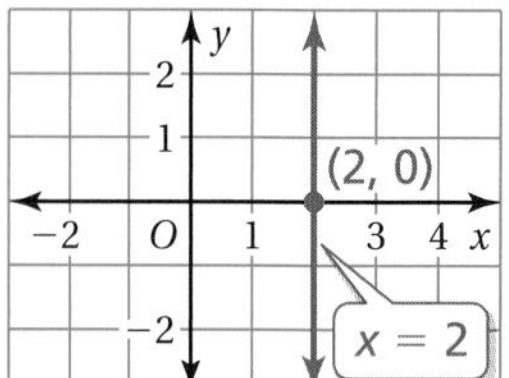

On Your Own

Now You're Ready
Exercises 5–16

Graph the linear equation. Use a graphing calculator to check your graph, if possible.

1. $y = 3x$
2. $y = -\frac{1}{2}x + 2$
3. $x = -4$
4. $y = -1.5$

EXAMPLE 3 Real-Life Application

The wind speed y (in miles per hour) of a tropical storm is $y = 2x + 66$, where x is the number of hours after the storm enters the Gulf of Mexico.

a. Graph the equation.

b. When does the storm become a hurricane?

A tropical storm becomes a hurricane when wind speeds are at least 74 miles per hour.

a. Make a table of values.

x	$y = 2x + 66$	y	(x, y)
0	$y = 2(0) + 66$	66	(0, 66)
1	$y = 2(1) + 66$	68	(1, 68)
2	$y = 2(2) + 66$	70	(2, 70)
3	$y = 2(3) + 66$	72	(3, 72)

Plot the ordered pairs and draw a line through the points.

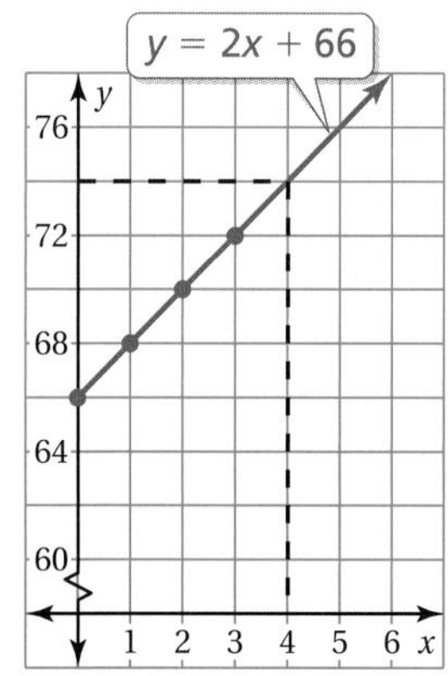

b. From the graph, you can see that $y = 74$ when $x = 4$. So, the storm becomes a hurricane 4 hours after it enters the Gulf of Mexico.

On Your Own

5. **WHAT IF?** The wind speed of the storm is $y = 1.5x + 62$. When does the storm become a hurricane?

4.1 Exercises

Vocabulary and Concept Check

1. **VOCABULARY** What type of graph represents the solutions of the equation $y = 2x + 4$?

2. **WHICH ONE DOESN'T BELONG?** Which equation does *not* belong with the other three? Explain your reasoning.

$y = 0.5x - 0.2$ | $4x + 3 = y$ | $y = x^2 + 6$ | $\frac{3}{4}x + \frac{1}{3} = y$

Practice and Problem Solving

PRECISION Copy and complete the table. Plot the two solution points and draw a line *exactly* through the two points. Find a different solution point on the line.

3.

x		
$y = 3x - 1$		

4.

x		
$y = \frac{1}{3}x + 2$		

Graph the linear equation. Use a graphing calculator to check your graph, if possible.

5. $y = -5x$
6. $y = \frac{1}{4}x$
7. $y = 5$
8. $x = -6$
9. $y = x - 3$
10. $y = -7x - 1$
11. $y = -\frac{x}{3} + 4$
12. $y = \frac{3}{4}x - \frac{1}{2}$
13. $y = -\frac{2}{3}$
14. $y = 6.75$
15. $x = -0.5$
16. $x = \frac{1}{4}$

17. **ERROR ANALYSIS** Describe and correct the error in graphing the equation.

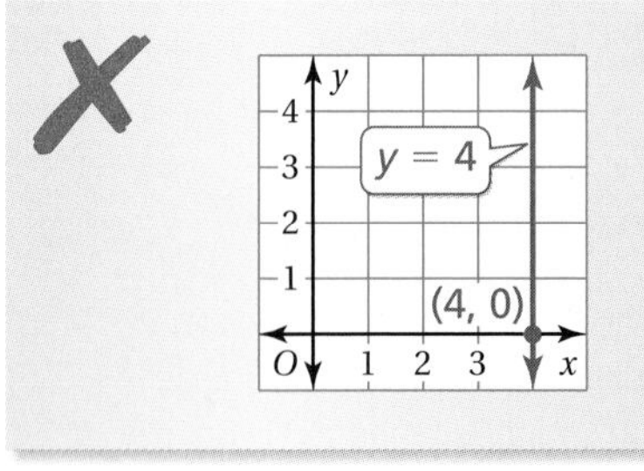

18. **MESSAGING** You sign up for an unlimited text-messaging plan for your cell phone. The equation $y = 20$ represents the cost y (in dollars) for sending x text messages. Graph the equation. What does the graph tell you?

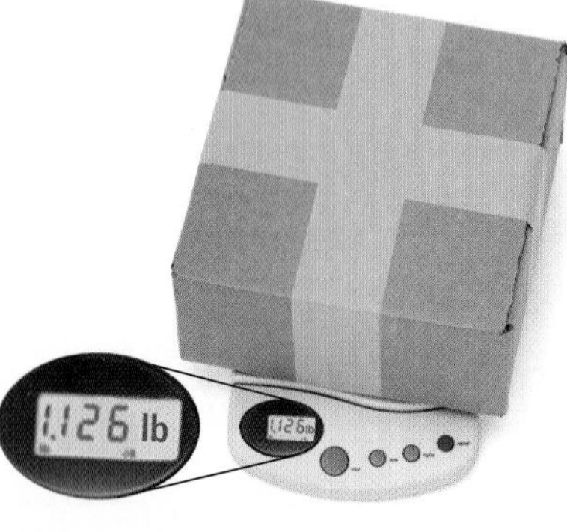

19. **MAIL** The equation $y = 2x + 3$ represents the cost y (in dollars) of mailing a package that weighs x pounds.

   a. Graph the equation.

   b. Use the graph to estimate how much it costs to mail the package.

   c. Use the equation to find exactly how much it costs to mail the package.

Solve for y. Then graph the equation. Use a graphing calculator to check your graph.

20. $y - 3x = 1$

21. $5x + 2y = 4$

22. $-\frac{1}{3}y + 4x = 3$

23. $x + 0.5y = 1.5$

24. SAVINGS You have \$100 in your savings account and plan to deposit \$12.50 each month.

a. Graph a linear equation that represents the balance in your account.

b. How many months will it take you to save enough money to buy 10 acres of land on Mars?

25. GEOMETRY The sum S of the interior angle measures of a polygon with n sides is $S = (n - 2) \cdot 180°$.

a. Plot four points (n, S) that satisfy the equation. Is the equation a linear equation? Explain your reasoning.

b. Does the value $n = 3.5$ make sense in the context of the problem? Explain your reasoning.

26. SEA LEVEL Along the U.S. Atlantic coast, the sea level is rising about 2 millimeters per year. How many millimeters has sea level risen since you were born? How do you know? Use a linear equation and a graph to justify your answer.

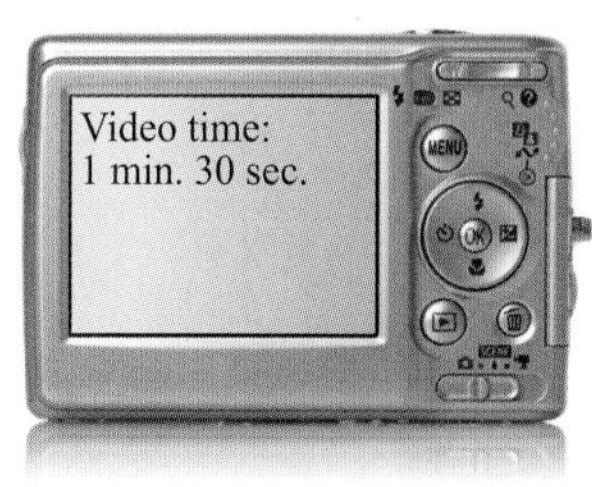

27. Problem Solving One second of video on your digital camera uses the same amount of memory as two pictures. Your camera can store 250 pictures.

a. Write and graph a linear equation that represents the number y of pictures your camera can store when you take x seconds of video.

b. How many pictures can your camera store in addition to the video shown?

Fair Game Review What you learned in previous grades & lessons

Write the ordered pair corresponding to the point. *(Skills Review Handbook)*

28. point A

29. point B

30. point C

31. point D

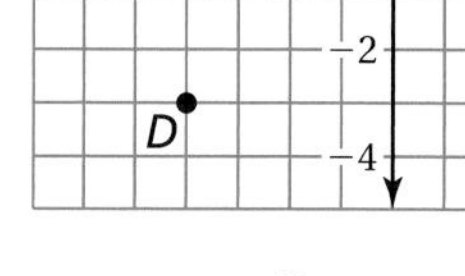

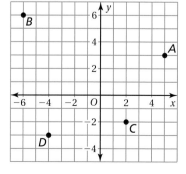

32. MULTIPLE CHOICE A debate team has 15 female members. The ratio of females to males is 3 : 2. How many males are on the debate team? *(Skills Review Handbook)*

Ⓐ 6 Ⓑ 10 Ⓒ 22 Ⓓ 25

4.2 Slope of a Line

Essential Question How can you use the slope of a line to describe the line?

Slope is the rate of change between any two points on a line. It is the measure of the *steepness* of the line.

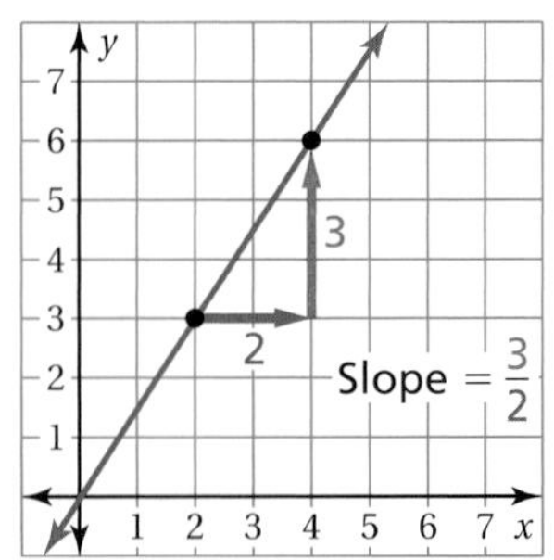

To find the slope of a line, find the ratio of the change in y (vertical change) to the change in x (horizontal change).

$$\text{slope} = \frac{\text{change in } y}{\text{change in } x}$$

1 ACTIVITY: Finding the Slope of a Line

Work with a partner. Find the slope of each line using two methods.

Method 1: Use the two black points. •

Method 2: Use the two pink points. •

Do you get the same slope using each method? Why do you think this happens?

a.

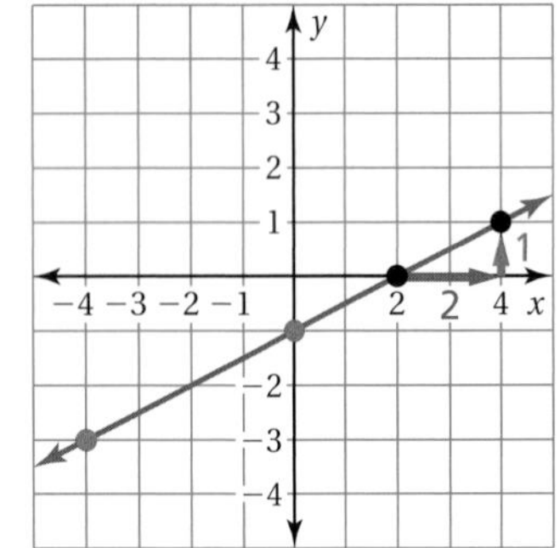

b.

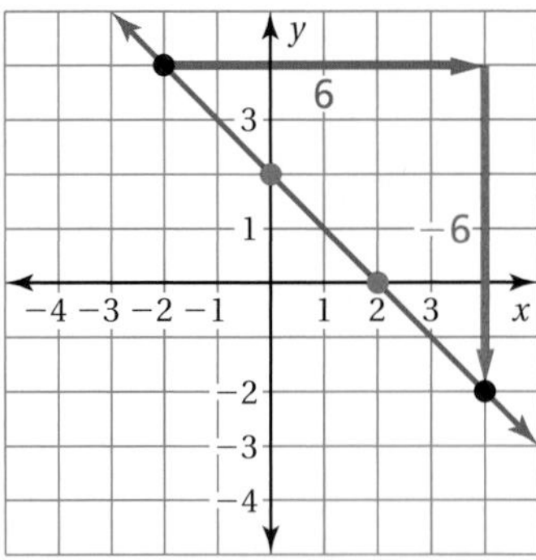

c.

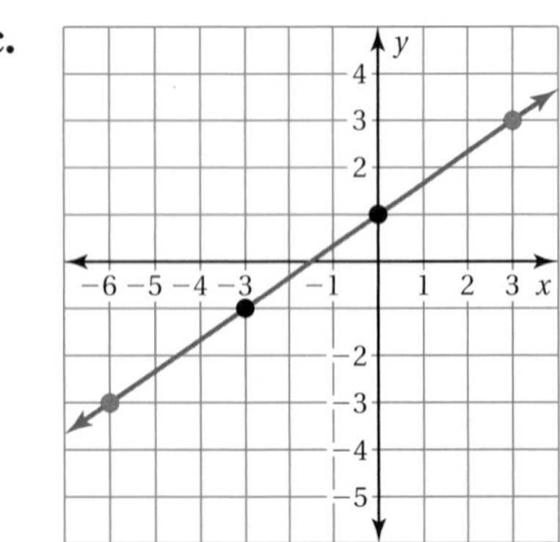

d.

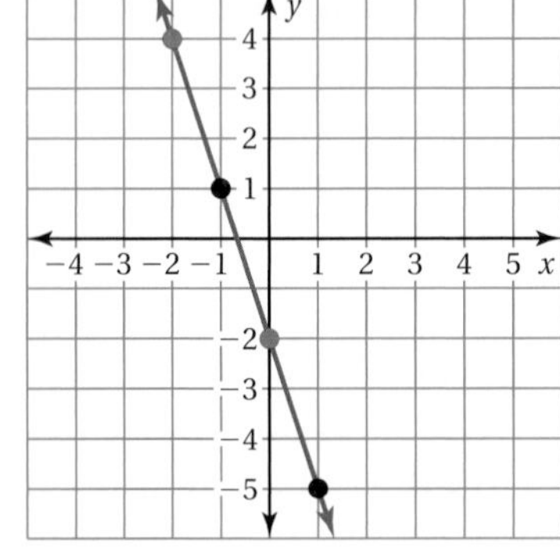

Graphing Equations

In this lesson, you will

- find slopes of lines by using two points.
- find slopes of lines from tables.

2 ACTIVITY: Using Similar Triangles

Work with a partner. Use the figure shown.

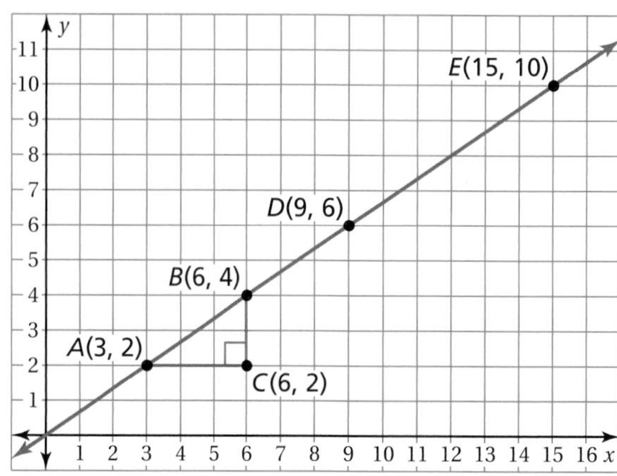

a. $\triangle ABC$ is a right triangle formed by drawing a horizontal line segment from point A and a vertical line segment from point B. Use this method to draw another right triangle, $\triangle DEF$.

b. What can you conclude about $\triangle ABC$ and $\triangle DEF$? Justify your conclusion.

c. For each triangle, find the ratio of the length of the vertical side to the length of the horizontal side. What do these ratios represent?

d. What can you conclude about the slope between any two points on the line?

3 ACTIVITY: Drawing Lines with Given Slopes

Work with a partner.

a. Draw two lines with slope $\frac{3}{4}$. One line passes through $(-4, 1)$, and the other line passes through $(4, 0)$. What do you notice about the two lines?

b. Draw two lines with slope $-\frac{4}{3}$. One line passes through $(2, 1)$, and the other line passes through $(-1, -1)$. What do you notice about the two lines?

c. **CONJECTURE** Make a conjecture about two different nonvertical lines in the same plane that have the same slope.

d. Graph one line from part (a) and one line from part (b) in the same coordinate plane. Describe the angle formed by the two lines. What do you notice about the product of the slopes of the two lines?

e. **REPEATED REASONING** Repeat part (d) for the two lines you did *not* choose. Based on your results, make a conjecture about two lines in the same plane whose slopes have a product of -1.

> **Math Practice**
> **Interpret a Solution**
> What does the slope tell you about the graph of the line? Explain.

What Is Your Answer?

4. **IN YOUR OWN WORDS** How can you use the slope of a line to describe the line?

Practice Use what you learned about the slope of a line to complete Exercises 4–6 on page 153.

4.2 Lesson

Key Vocabulary
slope, *p. 150*
rise, *p. 150*
run, *p. 150*

Key Idea

Slope

The **slope** m of a line is a ratio of the change in y (the **rise**) to the change in x (the **run**) between any two points, (x_1, y_1) and (x_2, y_2), on the line.

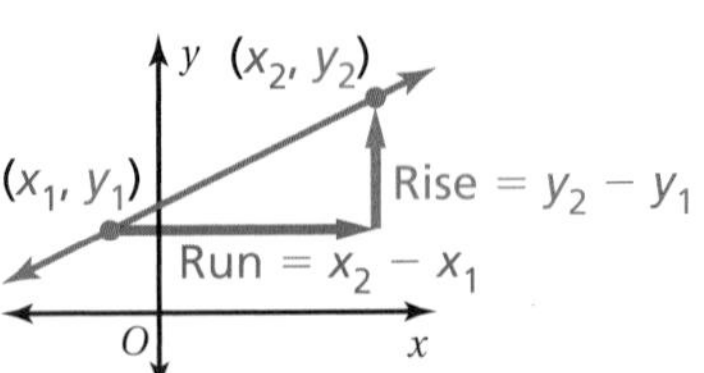

$$m = \frac{\text{rise}}{\text{run}} = \frac{\text{change in } y}{\text{change in } x} = \frac{y_2 - y_1}{x_2 - x_1}$$

Positive Slope

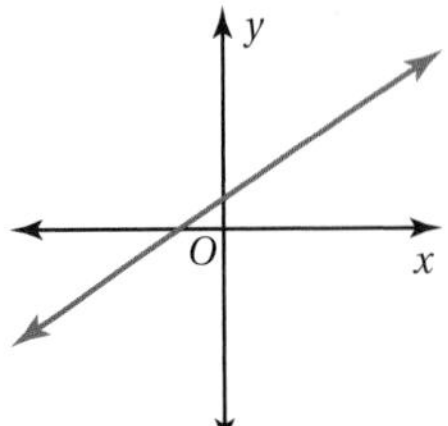

The line rises from left to right.

Negative Slope

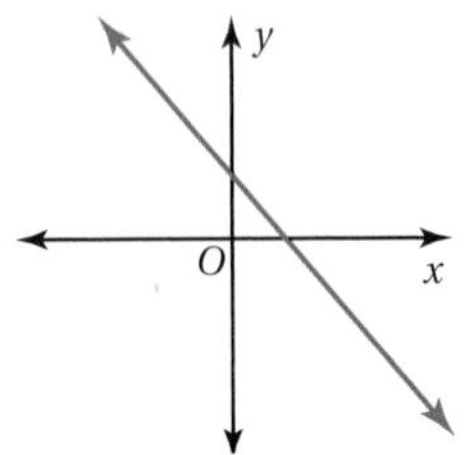

The line falls from left to right.

Reading

In the slope formula, x_1 is read as "x sub one," and y_2 is read as "y sub two." The numbers 1 and 2 in x_1 and y_2 are called *subscripts*.

EXAMPLE 1 Finding the Slope of a Line

Describe the slope of the line. Then find the slope.

a.

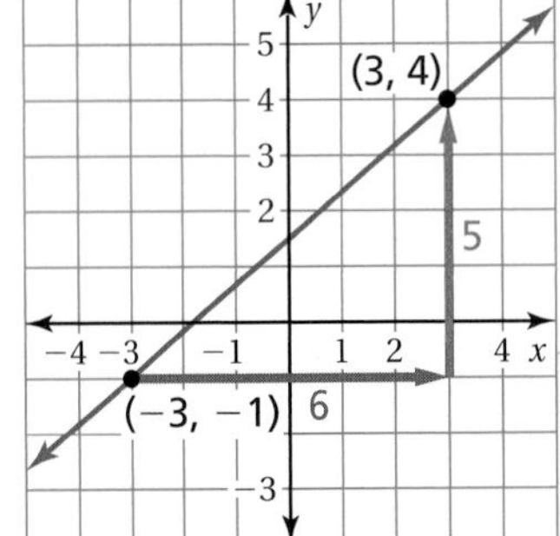

The line rises from left to right. So, the slope is positive. Let $(x_1, y_1) = (-3, -1)$ and $(x_2, y_2) = (3, 4)$.

$$m = \frac{y_2 - y_1}{x_2 - x_1}$$

$$= \frac{4 - (-1)}{3 - (-3)}$$

$$= \frac{5}{6}$$

b.

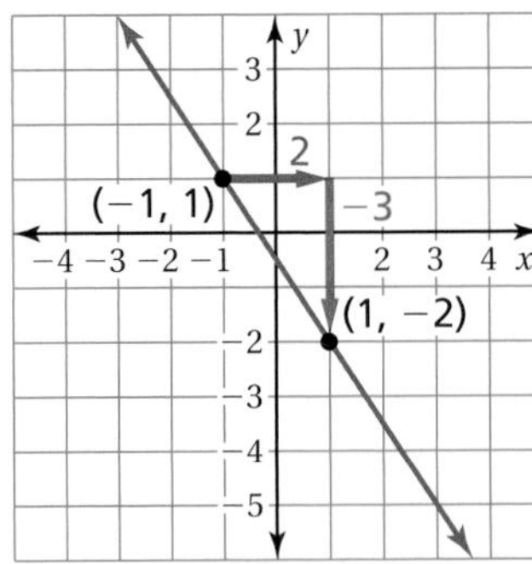

The line falls from left to right. So, the slope is negative. Let $(x_1, y_1) = (-1, 1)$ and $(x_2, y_2) = (1, -2)$.

$$m = \frac{y_2 - y_1}{x_2 - x_1}$$

$$= \frac{-2 - 1}{1 - (-1)}$$

$$= \frac{-3}{2}, \text{ or } -\frac{3}{2}$$

Study Tip

When finding slope, you can label either point as (x_1, y_1) and the other point as (x_2, y_2).

Multi-Language Glossary at BigIdeasMath.com

Now You're Ready Exercises 7–9

Find the slope of the line.

1.

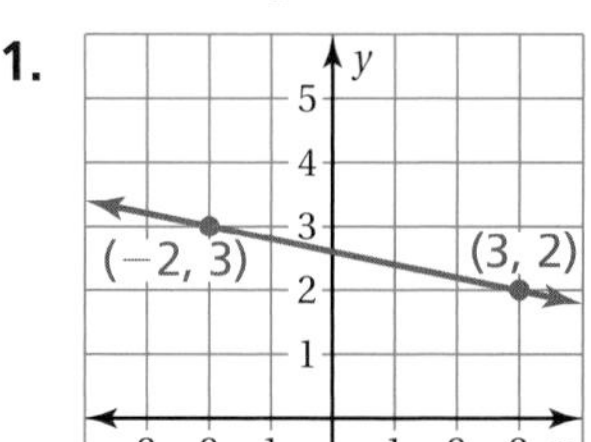

2.

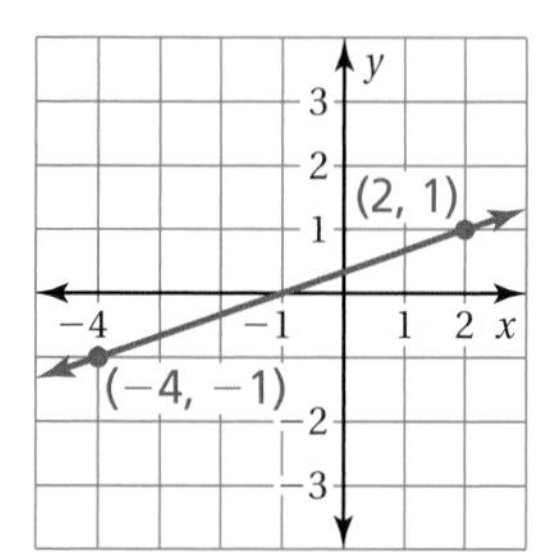

3.

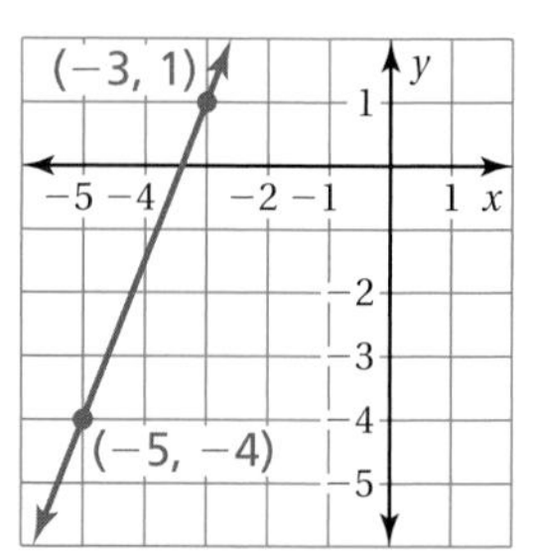

EXAMPLE 2 Finding the Slope of a Horizontal Line

Find the slope of the line.

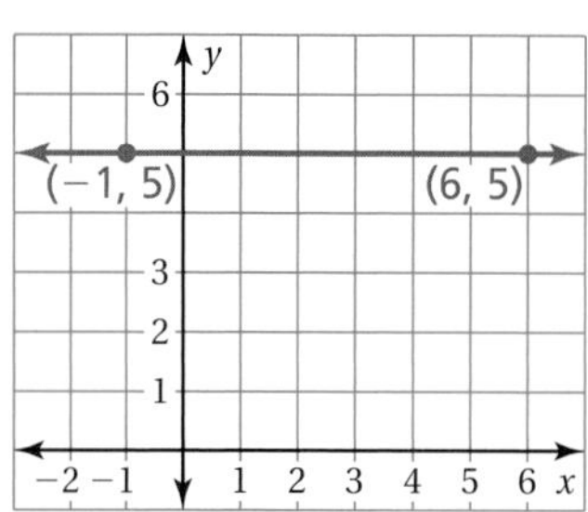

$$m = \frac{y_2 - y_1}{x_2 - x_1}$$

$$= \frac{5 - 5}{6 - (-1)}$$

$$= \frac{0}{7}, \text{ or } 0$$

The slope is 0.

EXAMPLE 3 Finding the Slope of a Vertical Line

Find the slope of the line.

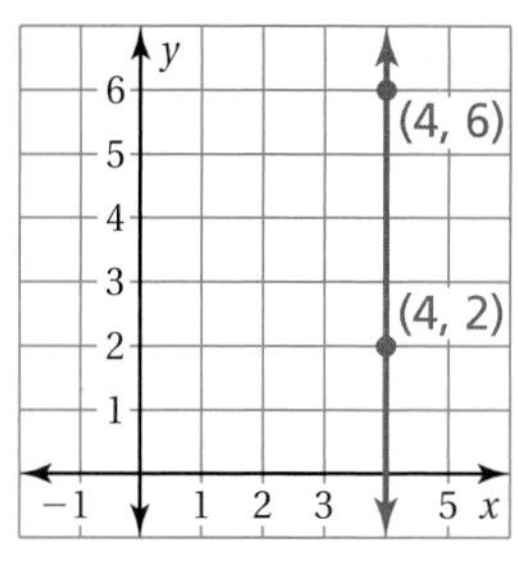

$$m = \frac{y_2 - y_1}{x_2 - x_1}$$

$$= \frac{6 - 2}{4 - 4}$$

$$= \frac{4}{0} \quad ✗$$

Because division by zero is undefined, the slope of the line is undefined.

Study Tip

The slope of every horizontal line is 0. The slope of every vertical line is undefined.

On Your Own

Now You're Ready Exercises 13–15

Find the slope of the line through the given points.

4. $(1, -2), (7, -2)$

5. $(-2, 4), (3, 4)$

6. $(-3, -3), (-3, -5)$

7. $(0, 8), (0, 0)$

8. How do you know that the slope of every horizontal line is 0? How do you know that the slope of every vertical line is undefined?

EXAMPLE 4 Finding Slope from a Table

The points in the table lie on a line. How can you find the slope of the line from the table? What is the slope?

x	1	4	7	10
y	8	6	4	2

Choose any two points from the table and use the slope formula.

Use the points $(x_1, y_1) = (1, 8)$ and $(x_2, y_2) = (4, 6)$.

$$m = \frac{y_2 - y_1}{x_2 - x_1}$$

$$= \frac{6 - 8}{4 - 1}$$

$$= \frac{-2}{3}, \text{ or } -\frac{2}{3}$$

The slope is $-\frac{2}{3}$.

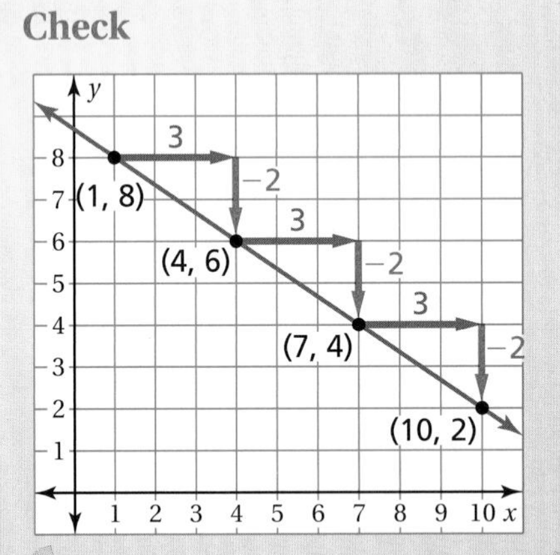

On Your Own

Now You're Ready
Exercises 21–24

The points in the table lie on a line. Find the slope of the line.

9.

x	1	3	5	7
y	2	5	8	11

10.

x	−3	−2	−1	0
y	6	4	2	0

Summary

Slope

Positive Slope

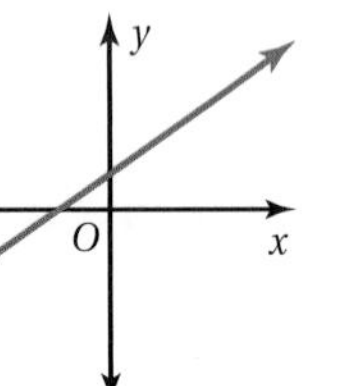

The line rises from left to right.

Negative Slope

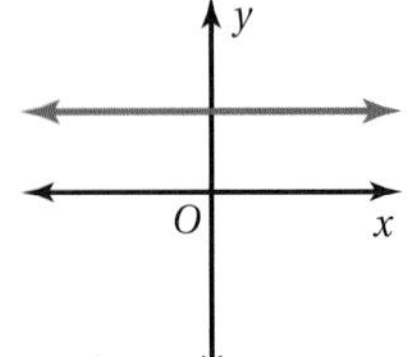

The line falls from left to right.

Slope of 0

The line is horizontal.

Undefined Slope

The line is vertical.

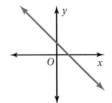

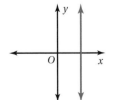

4.2 Exercises

Vocabulary and Concept Check

1. **CRITICAL THINKING** Refer to the graph.
   a. Which lines have positive slopes?
   b. Which line has the steepest slope?
   c. Do any lines have an undefined slope? Explain.

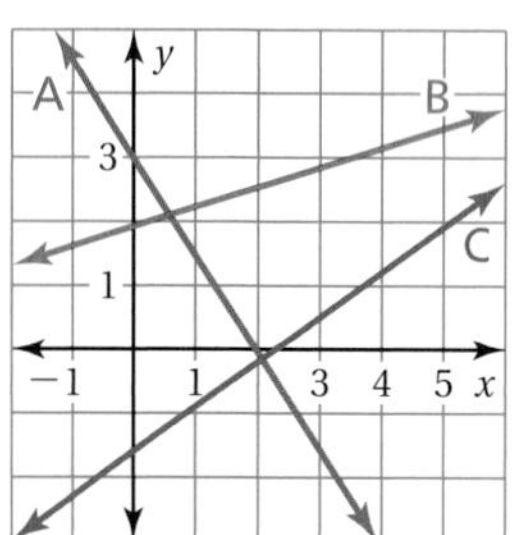

2. **OPEN-ENDED** Describe a real-life situation in which you need to know the slope.
3. **REASONING** The slope of a line is 0. What do you know about the line?

Practice and Problem Solving

Draw a line through each point using the given slope. What do you notice about the two lines?

4. slope = 1

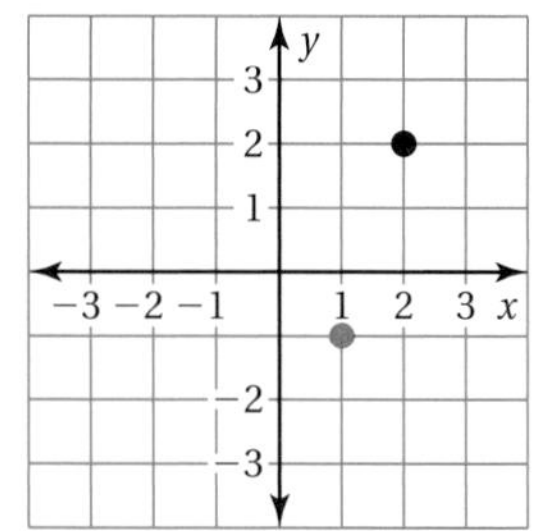

5. slope = −3

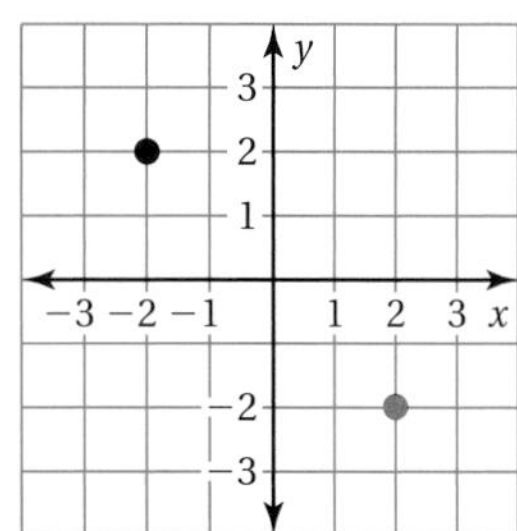

6. slope = $\frac{1}{4}$

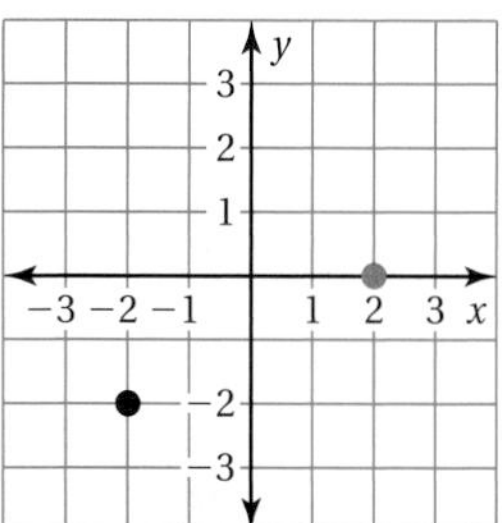

Find the slope of the line.

1 7.

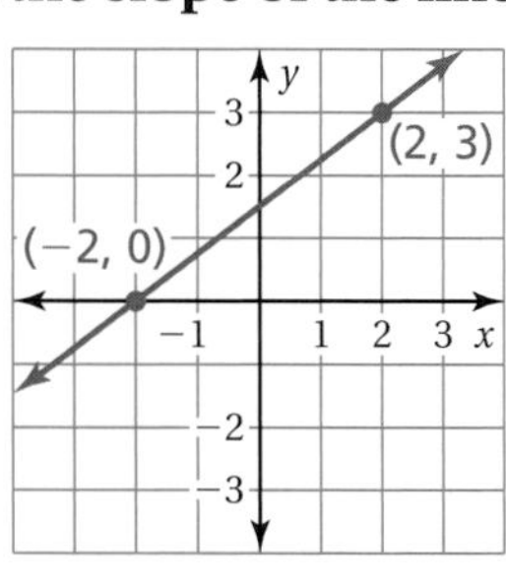

8.

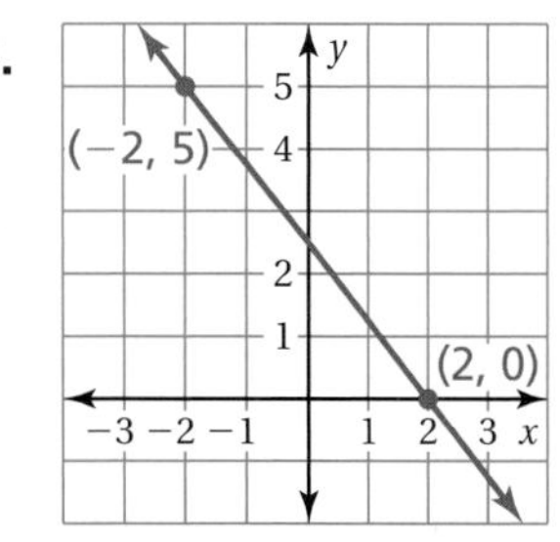

9.

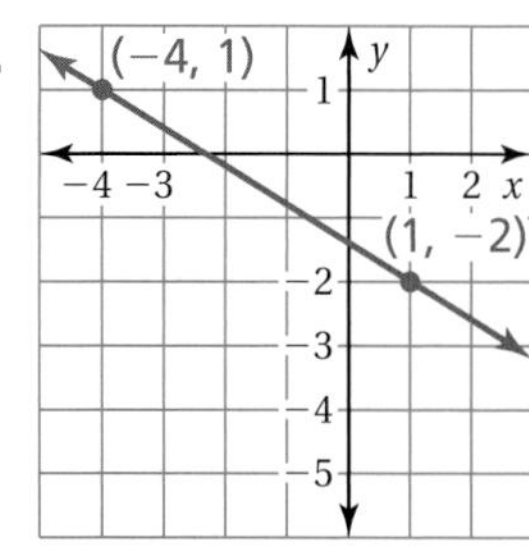

10.

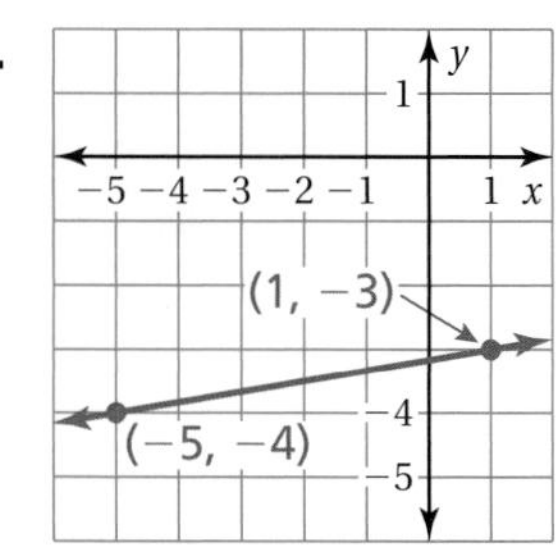

11.

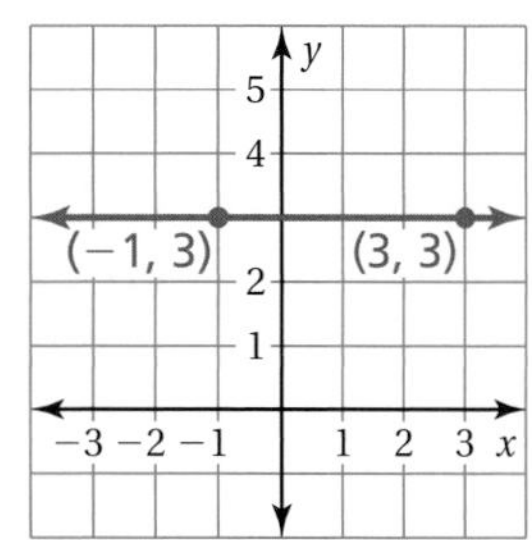

12.

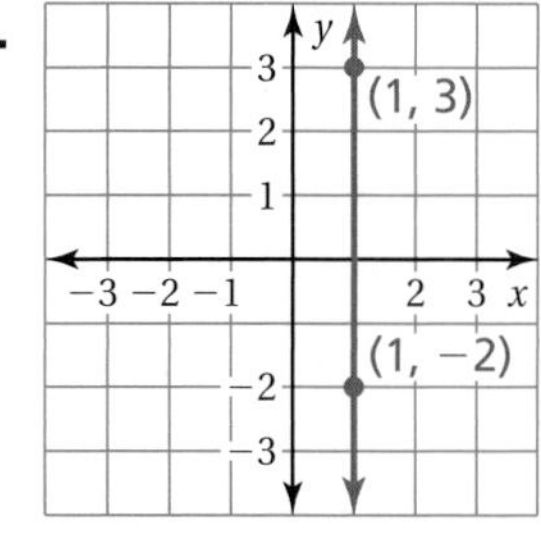

Find the slope of the line through the given points.

②③ **13.** $(4, -1), (-2, -1)$ **14.** $(5, -3), (5, 8)$ **15.** $(-7, 0), (-7, -6)$

16. $(-3, 1), (-1, 5)$ **17.** $(10, 4), (4, 15)$ **18.** $(-3, 6), (2, 6)$

19. ERROR ANALYSIS Describe and correct the error in finding the slope of the line.

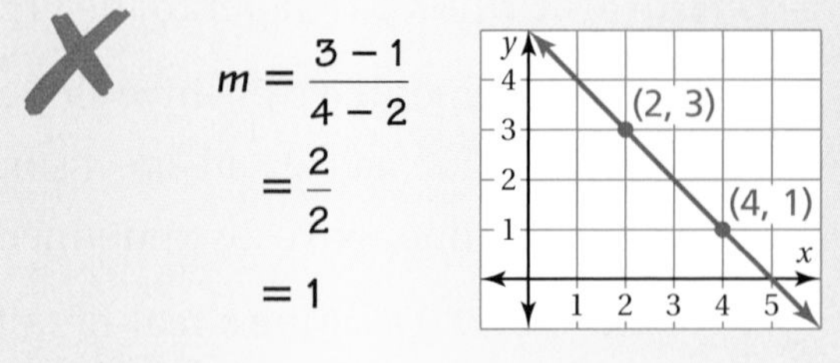

20. CRITICAL THINKING Is it more difficult to walk up the ramp or the hill? Explain.

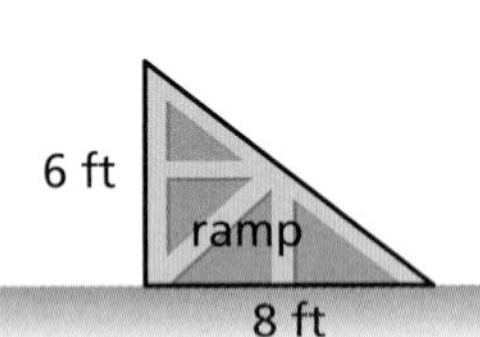

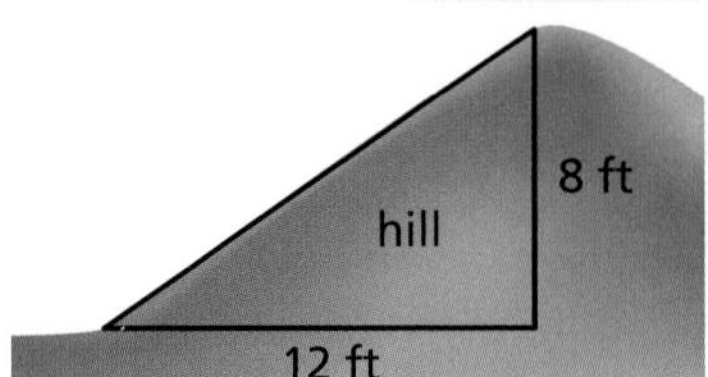

The points in the table lie on a line. Find the slope of the line.

④ **21.**

x	1	3	5	7
y	2	10	18	26

22.

x	−3	2	7	12
y	0	2	4	6

23.

x	−6	−2	2	6
y	8	5	2	−1

24.

x	−8	−2	4	10
y	8	1	−6	−13

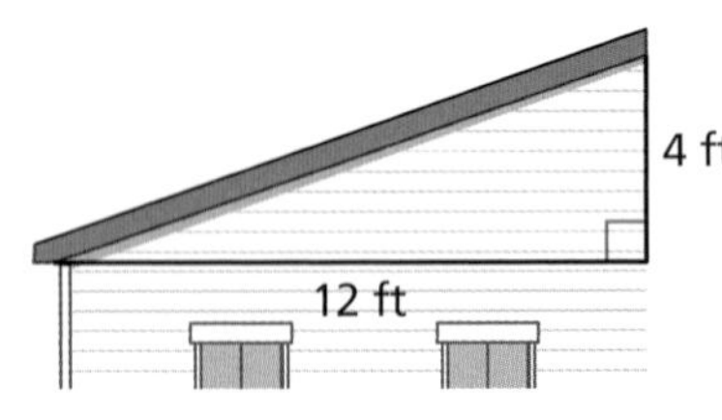

25. PITCH Carpenters refer to the slope of a roof as the *pitch* of the roof. Find the pitch of the roof.

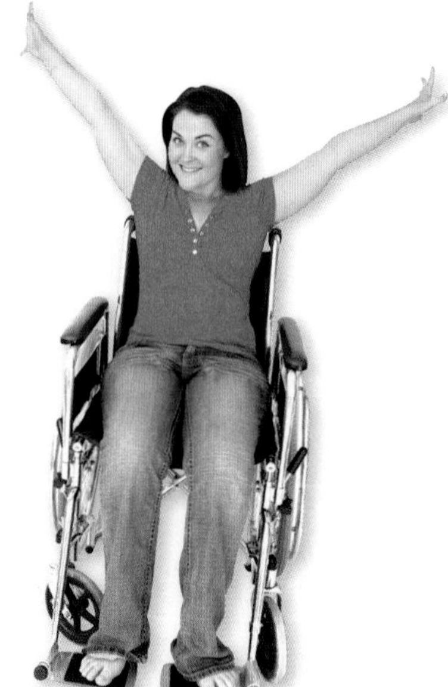

26. PROJECT The guidelines for a wheelchair ramp suggest that the ratio of the rise to the run be no greater than 1 : 12.

a. CHOOSE TOOLS Find a wheelchair ramp in your school or neighborhood. Measure its slope. Does the ramp follow the guidelines?

b. Design a wheelchair ramp that provides access to a building with a front door that is 2.5 feet above the sidewalk. Illustrate your design.

Use an equation to find the value of *k* so that the line that passes through the given points has the given slope.

27. $(1, 3), (5, k); m = 2$ **28.** $(-2, k), (2, 0); m = -1$

29. $(-4, k), (6, -7); m = -\frac{1}{5}$ **30.** $(4, -4), (k, -1); m = \frac{3}{4}$

31. TURNPIKE TRAVEL The graph shows the cost of traveling by car on a turnpike.

a. Find the slope of the line.

b. Explain the meaning of the slope as a rate of change.

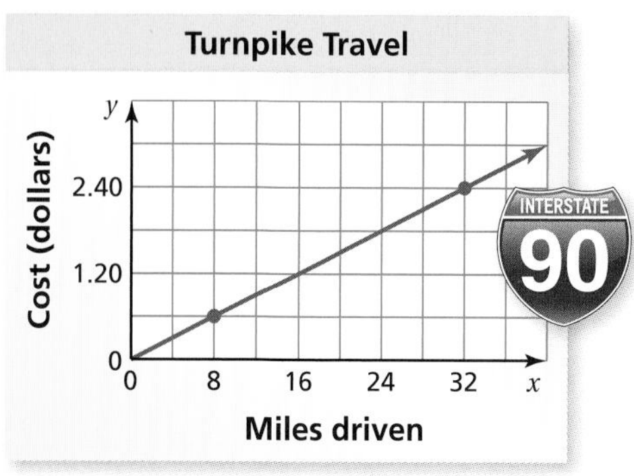

32. BOAT RAMP Which is steeper: the boat ramp or a road with a 12% grade? Explain. (*Note:* Road grade is the vertical increase divided by the horizontal distance.)

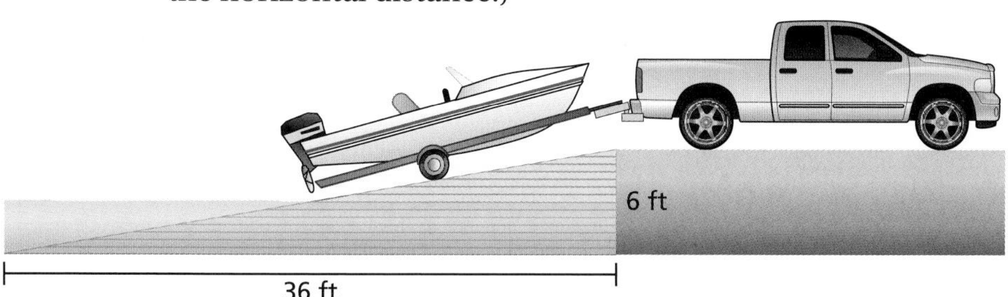

33. REASONING Do the points $A(-2, -1)$, $B(1, 5)$, and $C(4, 11)$ lie on the same line? Without using a graph, how do you know?

34. BUSINESS A small business earns a profit of \$6500 in January and \$17,500 in May. What is the rate of change in profit for this time period?

35. STRUCTURE Choose two points in the coordinate plane. Use the slope formula to find the slope of the line that passes through the two points. Then find the slope using the formula $\frac{y_1 - y_2}{x_1 - x_2}$. Explain why your results are the same.

36. Critical Thinking The top and the bottom of the slide are level with the ground, which has a slope of 0.

a. What is the slope of the main portion of the slide?

b. How does the slope change when the bottom of the slide is only 12 inches above the ground? Is the slide steeper? Explain.

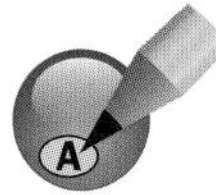

Fair Game Review What you learned in previous grades & lessons

Solve the proportion. *(Skills Review Handbook)*

37. $\frac{b}{30} = \frac{5}{6}$ **38.** $\frac{7}{4} = \frac{n}{32}$ **39.** $\frac{3}{8} = \frac{x}{20}$

40. MULTIPLE CHOICE What is the prime factorization of 84? *(Skills Review Handbook)*

Ⓐ $2 \times 3 \times 7$ Ⓑ $2^2 \times 3 \times 7$ Ⓒ $2 \times 3^2 \times 7$ Ⓓ $2^2 \times 21$

Slopes of Parallel and Perpendicular Lines

Graphing Equations

In this extension, you will

- identify parallel and perpendicular lines.

Parallel Lines and Slopes

Lines in the same plane that do not intersect are parallel lines. Nonvertical parallel lines have the same slope.

All vertical lines are parallel.

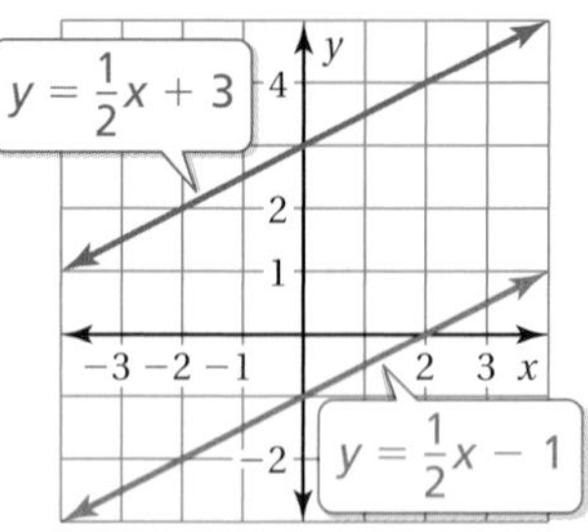

EXAMPLE 1 Identifying Parallel Lines

Which two lines are parallel? How do you know?

Find the slope of each line.

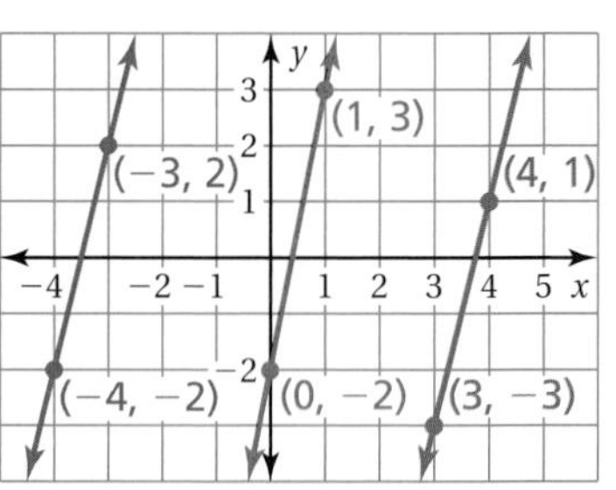

Blue Line

$$m = \frac{y_2 - y_1}{x_2 - x_1} = \frac{-2 - 2}{-4 - (-3)} = \frac{-4}{-1}, \text{ or } 4$$

Red Line

$$m = \frac{y_2 - y_1}{x_2 - x_1} = \frac{-2 - 3}{0 - 1} = \frac{-5}{-1}, \text{ or } 5$$

Green Line

$$m = \frac{y_2 - y_1}{x_2 - x_1} = \frac{-3 - 1}{3 - 4} = \frac{-4}{-1}, \text{ or } 4$$

The slopes of the blue and green lines are 4. The slope of the red line is 5.

∴ The blue and green lines have the same slope, so they are parallel.

Practice

Which lines are parallel? How do you know?

1. 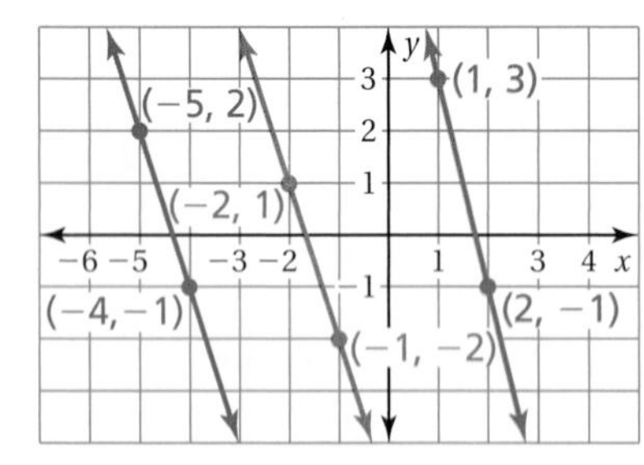

2. 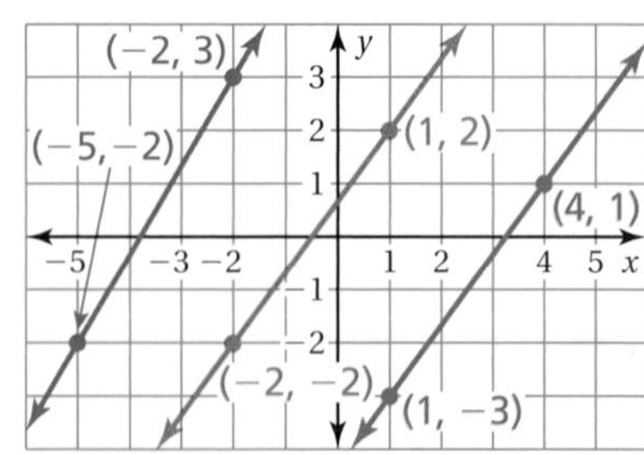

Are the given lines parallel? Explain your reasoning.

3. $y = -5, y = 3$
4. $y = 0, x = 0$
5. $x = -4, x = 1$

6. **GEOMETRY** The vertices of a quadrilateral are $A(-5, 3)$, $B(2, 2)$, $C(4, -3)$, and $D(-2, -2)$. How can you use slope to determine whether the quadrilateral is a parallelogram? Is it a parallelogram? Justify your answer.

Perpendicular Lines and Slope

Lines in the same plane that intersect at right angles are perpendicular lines. Two nonvertical lines are perpendicular when the product of their slopes is -1.

Vertical lines are perpendicular to horizontal lines.

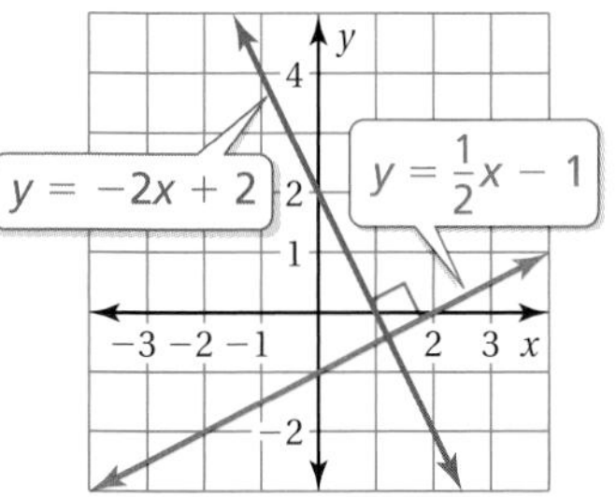

EXAMPLE 2 Identifying Perpendicular Lines

Which two lines are perpendicular? How do you know?

Find the slope of each line.

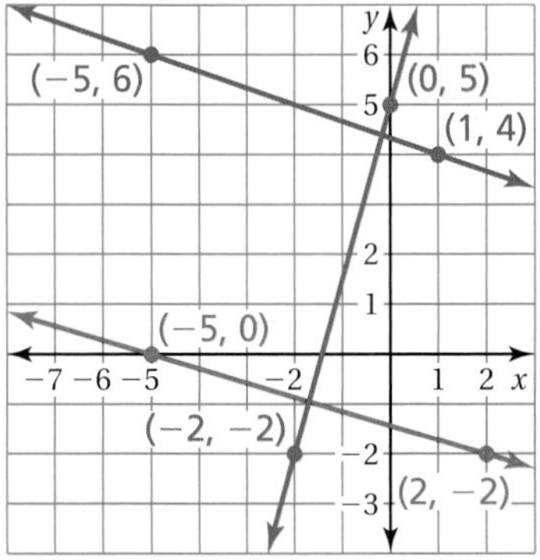

Blue Line	*Red Line*	*Green Line*
$m = \dfrac{y_2 - y_1}{x_2 - x_1}$	$m = \dfrac{y_2 - y_1}{x_2 - x_1}$	$m = \dfrac{y_2 - y_1}{x_2 - x_1}$
$= \dfrac{4 - 6}{1 - (-5)}$	$= \dfrac{-2 - 0}{2 - (-5)}$	$= \dfrac{5 - (-2)}{0 - (-2)}$
$= \dfrac{-2}{6}$, or $-\dfrac{1}{3}$	$= -\dfrac{2}{7}$	$= \dfrac{7}{2}$

The slope of the red line is $-\frac{2}{7}$. The slope of the green line is $\frac{7}{2}$.

∴ Because $-\frac{2}{7} \cdot \frac{7}{2} = -1$, the red and green lines are perpendicular.

Practice

Which lines are perpendicular? How do you know?

7.

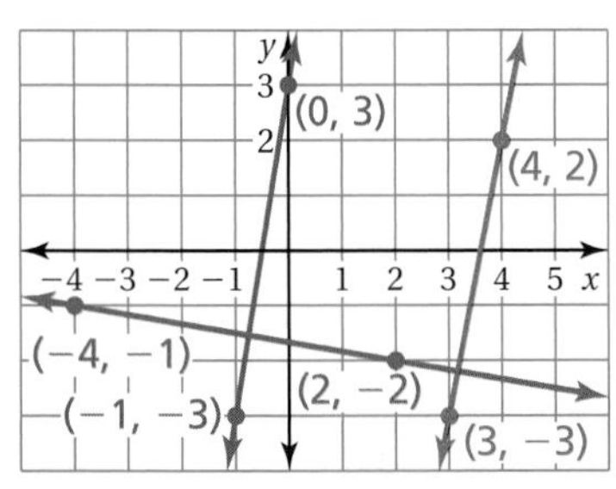

8.

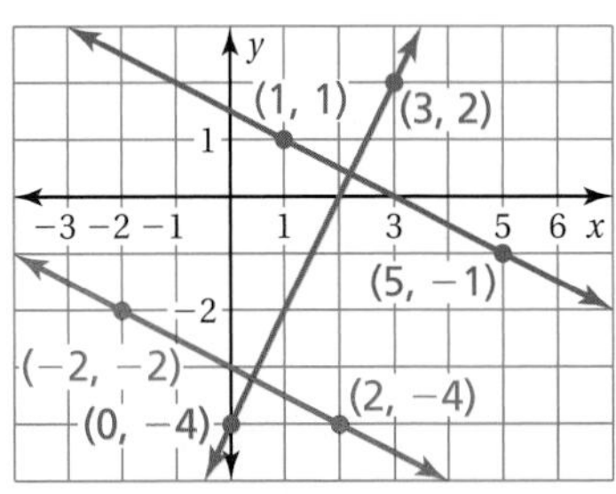

Are the given lines perpendicular? Explain your reasoning.

9. $x = -2, y = 8$ **10.** $x = -8, x = 7$ **11.** $y = 0, x = 0$

12. **GEOMETRY** The vertices of a parallelogram are $J(-5, 0)$, $K(1, 4)$, $L(3, 1)$, and $M(-3, -3)$. How can you use slope to determine whether the parallelogram is a rectangle? Is it a rectangle? Justify your answer.

4.3 Graphing Proportional Relationships

Essential Question How can you describe the graph of the equation $y = mx$?

1 ACTIVITY: Identifying Proportional Relationships

Work with a partner. Tell whether x and y are in a proportional relationship. Explain your reasoning.

a.

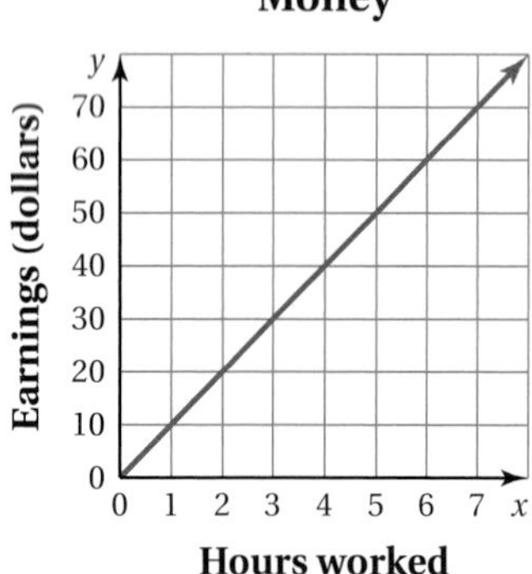

b.

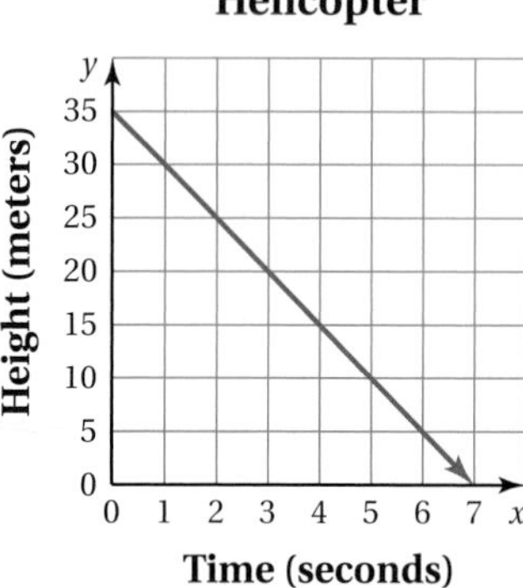

c.

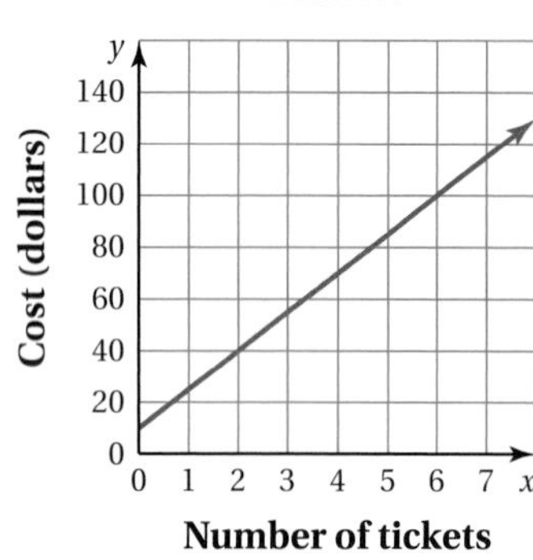

d.

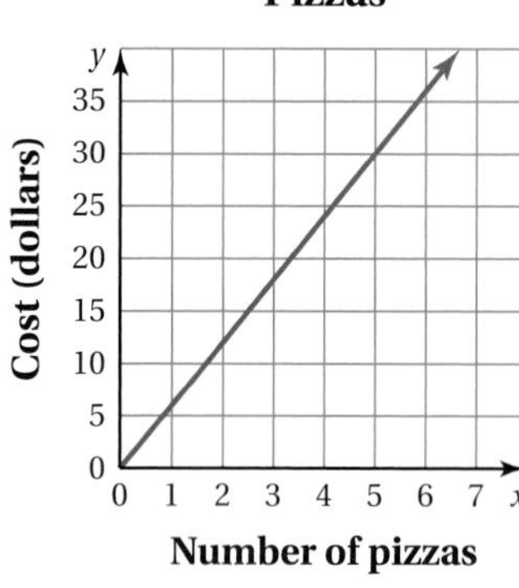

e.

Laps, x	1	2	3	4
Time (seconds), y	90	200	325	480

f.

Cups of Sugar, x	$\frac{1}{2}$	1	$1\frac{1}{2}$	2
Cups of Flour, y	1	2	3	4

Graphing Equations

In this lesson, you will

- write and graph proportional relationships.

2 ACTIVITY: Analyzing Proportional Relationships

Work with a partner. Use only the proportional relationships in Activity 1 to do the following.

- **Find the slope of the line.**
- **Find the value of y for the ordered pair $(1, y)$.**

What do you notice? What does the value of y represent?

3 ACTIVITY: Deriving an Equation

Work with a partner. Let (x, y) represent any point on the graph of a proportional relationship.

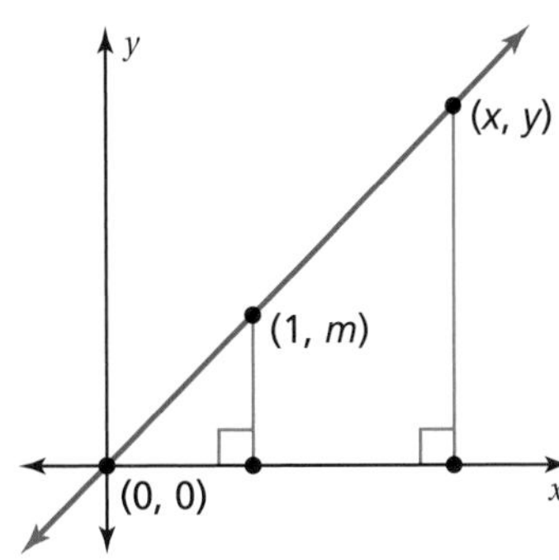

a. Explain why the two triangles are similar.

b. Because the triangles are similar, the corresponding side lengths are proportional. Use the vertical and horizontal side lengths to complete the steps below.

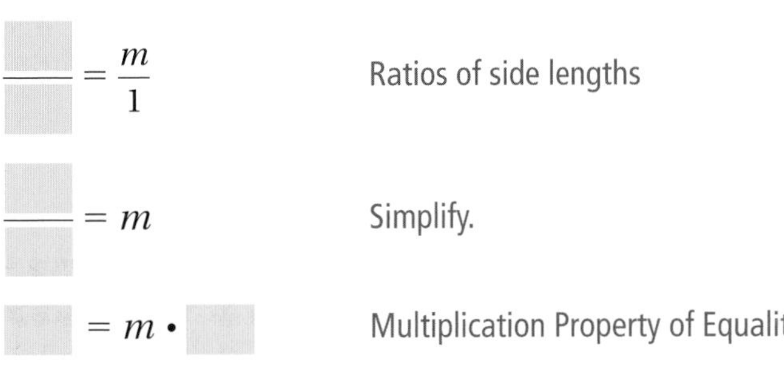

$\frac{\square}{\square} = \frac{m}{1}$ Ratios of side lengths
$\frac{\square}{\square} = m$ Simplify.
$\square = m \cdot \square$ Multiplication Property of Equality

Math Practice

View as Components

What part of the graph can you use to find the side lengths?

What does the final equation represent?

c. Use your result in part (b) to write an equation that represents each proportional relationship in Activity 1.

What Is Your Answer?

4. IN YOUR OWN WORDS How can you describe the graph of the equation $y = mx$? How does the value of m affect the graph of the equation?

5. Give a real-life example of two quantities that are in a proportional relationship. Write an equation that represents the relationship and sketch its graph.

Practice Use what you learned about proportional relationships to complete Exercises 3–6 on page 162.

4.3 Lesson

Study Tip

In the direct variation equation $y = mx$, m represents the constant of proportionality, the slope, and the unit rate.

Direct Variation

Words When two quantities x and y are proportional, the relationship can be represented by the direct variation equation $y = mx$, where m is the constant of proportionality.

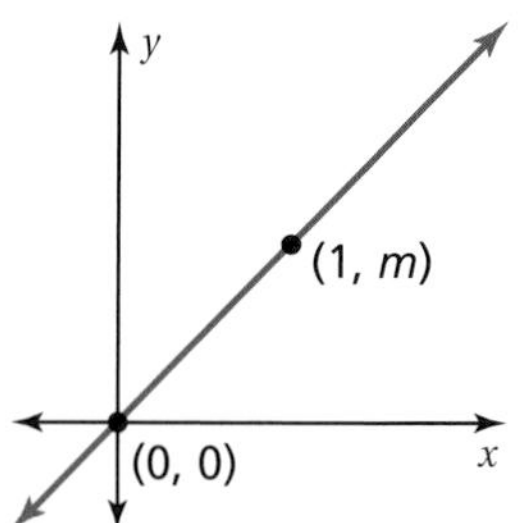

Graph The graph of $y = mx$ is a line with a slope of m that passes through the origin.

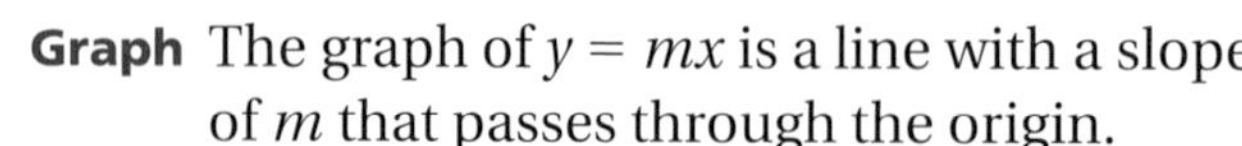

EXAMPLE 1 Graphing a Proportional Relationship

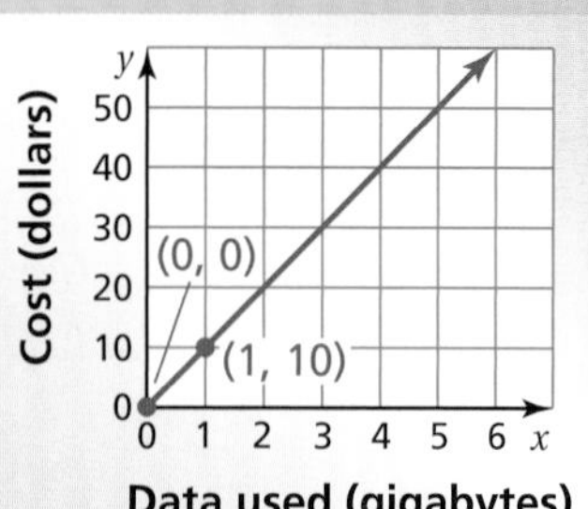

The cost y (in dollars) for x gigabytes of data on an Internet plan is represented by $y = 10x$. Graph the equation and interpret the slope.

The equation shows that the slope m is 10. So, the graph passes through (0, 0) and (1, 10).

Plot the points and draw a line through the points. Because negative values of x do not make sense in this context, graph in the first quadrant only.

∴ The slope indicates that the unit cost is \$10 per gigabyte.

EXAMPLE 2 Writing and Using a Direct Variation Equation

The weight y of an object on Titan, one of Saturn's moons, is proportional to the weight x of the object on Earth. An object that weighs 105 pounds on Earth would weigh 15 pounds on Titan.

a. Write an equation that represents the situation.

Use the point (105, 15) to find the slope of the line.

$y = mx$ Direct variation equation

$15 = m(105)$ Substitute 15 for y and 105 for x.

$\frac{1}{7} = m$ Simplify.

∴ So, an equation that represents the situation is $y = \frac{1}{7}x$.

b. How much would a chunk of ice that weighs 3.5 pounds on Titan weigh on Earth?

$3.5 = \frac{1}{7}x$ Substitute 3.5 for y.

$24.5 = x$ Multiply each side by 7.

∴ So, the chunk of ice would weigh 24.5 pounds on Earth.

Study Tip

In Example 2, the slope indicates that the weight of an object on Titan is one-seventh its weight on Earth.

On Your Own

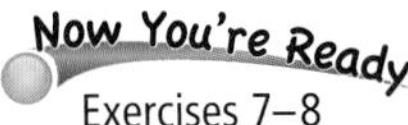

1. **WHAT IF?** In Example 1, the cost is represented by $y = 12x$. Graph the equation and interpret the slope.
2. In Example 2, how much would a spacecraft that weighs 3500 kilograms on Earth weigh on Titan?

EXAMPLE 3 Comparing Proportional Relationships

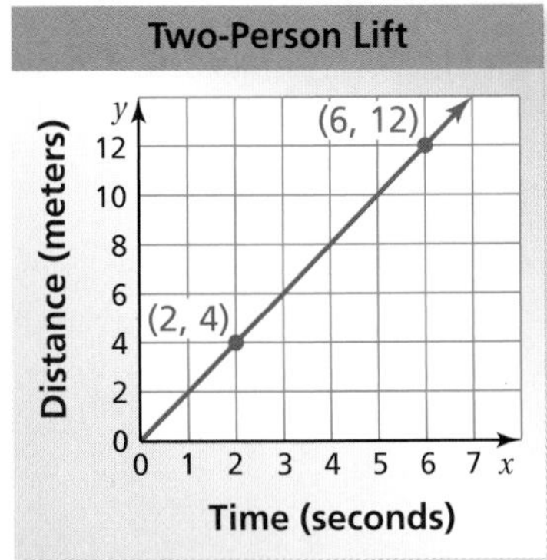

The distance y (in meters) that a four-person ski lift travels in x seconds is represented by the equation $y = 2.5x$. The graph shows the distance that a two-person ski lift travels.

a. Which ski lift is faster?

Interpret each slope as a unit rate.

Four-Person Lift	*Two-Person Lift*
$y = 2.5x$ (The slope is 2.5.)	$\text{slope} = \dfrac{\text{change in } y}{\text{change in } x} = \dfrac{8}{4} = 2$
The four-person lift travels 2.5 meters per second.	The two-person lift travels 2 meters per second.

So, the four-person lift is faster than the two-person lift.

b. Graph the equation that represents the four-person lift in the same coordinate plane as the two-person lift. Compare the steepness of the graphs. What does this mean in the context of the problem?

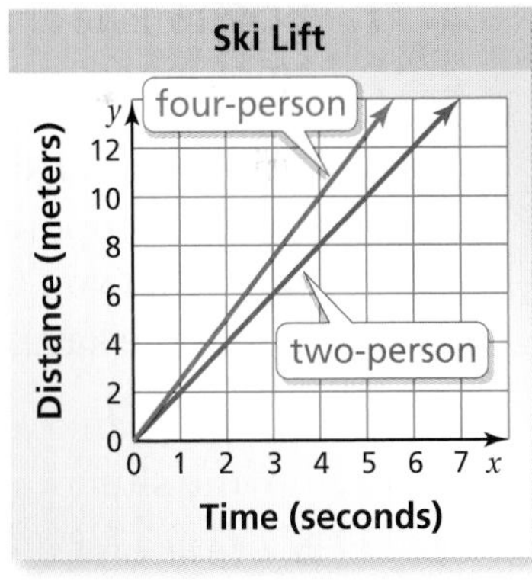

The graph that represents the four-person lift is steeper than the graph that represents the two-person lift. So, the four-person lift is faster.

On Your Own

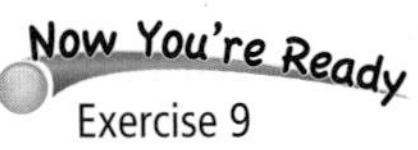

3. The table shows the distance y (in meters) that a T-bar ski lift travels in x seconds. Compare its speed to the ski lifts in Example 3.

x (seconds)	1	2	3	4
y (meters)	$2\frac{1}{4}$	$4\frac{1}{2}$	$6\frac{3}{4}$	9

4.3 Exercises

Vocabulary and Concept Check

1. **VOCABULARY** What point is on the graph of every direct variation equation?
2. **REASONING** Does the equation $y = 2x + 3$ represent a proportional relationship? Explain.

Practice and Problem Solving

Tell whether x and y are in a proportional relationship. Explain your reasoning. If so, write an equation that represents the relationship.

3. 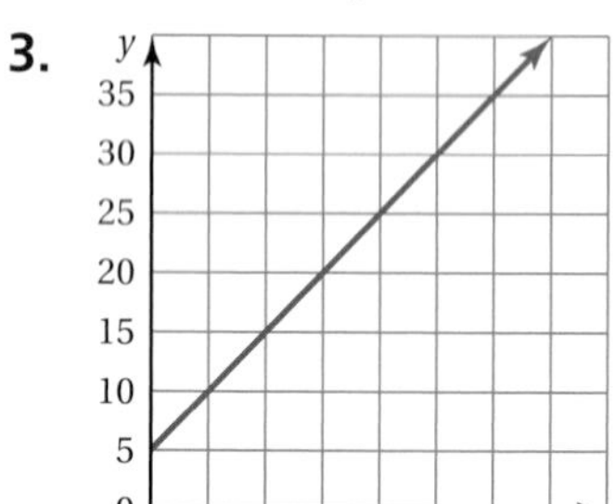

4. 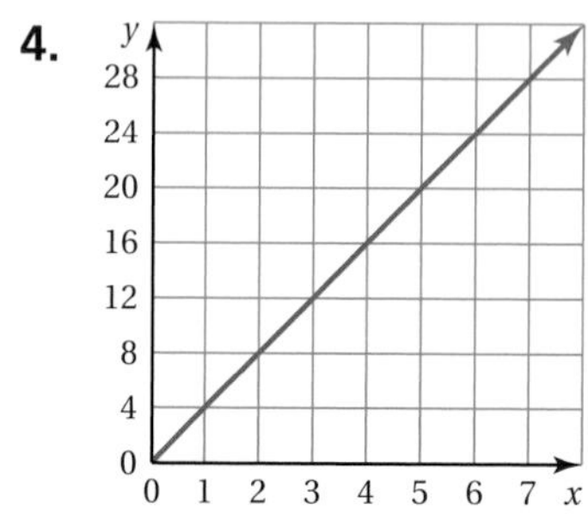

5.

x	3	6	9	12
y	1	2	3	4

6. 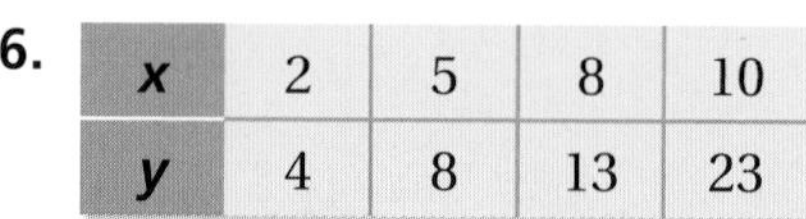

x	2	5	8	10
y	4	8	13	23

(1) 7. **TICKETS** The amount y (in dollars) that you raise by selling x fundraiser tickets is represented by the equation $y = 5x$. Graph the equation and interpret the slope.

(2) 8. **KAYAK** The cost y (in dollars) to rent a kayak is proportional to the number x of hours that you rent the kayak. It costs \$27 to rent the kayak for 3 hours.

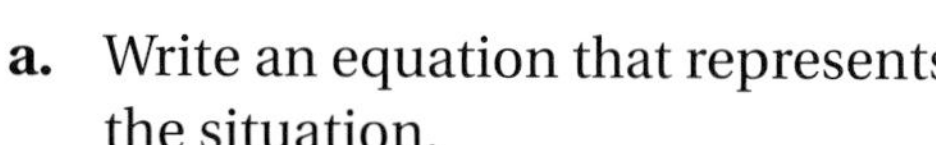

a. Write an equation that represents the situation.

b. Interpret the slope.

c. How much does it cost to rent the kayak for 5 hours?

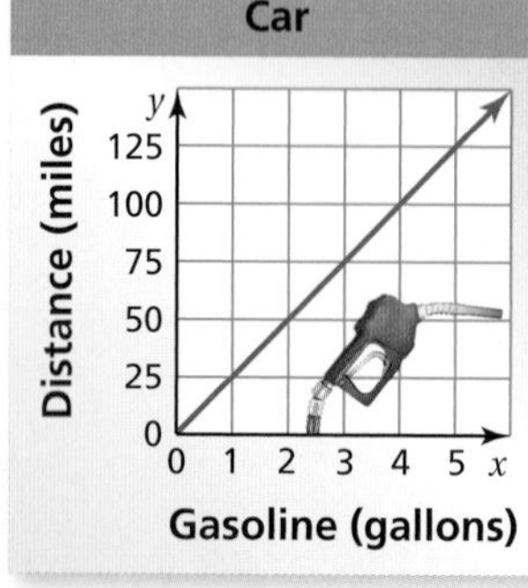

(3) 9. **MILEAGE** The distance y (in miles) that a truck travels on x gallons of gasoline is represented by the equation $y = 18x$. The graph shows the distance that a car travels.

a. Which vehicle gets better gas mileage? Explain how you found your answer.

b. How much farther can the vehicle you chose in part (a) travel than the other vehicle on 8 gallons of gasoline?

10. **BIOLOGY** Toenails grow about 13 millimeters per year. The table shows fingernail growth.

Weeks	1	2	3	4
Fingernail Growth (millimeters)	0.7	1.4	2.1	2.8

a. Do fingernails or toenails grow faster? Explain.

b. In the same coordinate plane, graph equations that represent the growth rates of toenails and fingernails. Compare the steepness of the graphs. What does this mean in the context of the problem?

11. **REASONING** The quantities x and y are in a proportional relationship. What do you know about the ratio of y to x for any point (x, y) on the line?

12. **PROBLEM SOLVING** The graph relates the temperature change y (in degrees Fahrenheit) to the altitude change x (in thousands of feet).

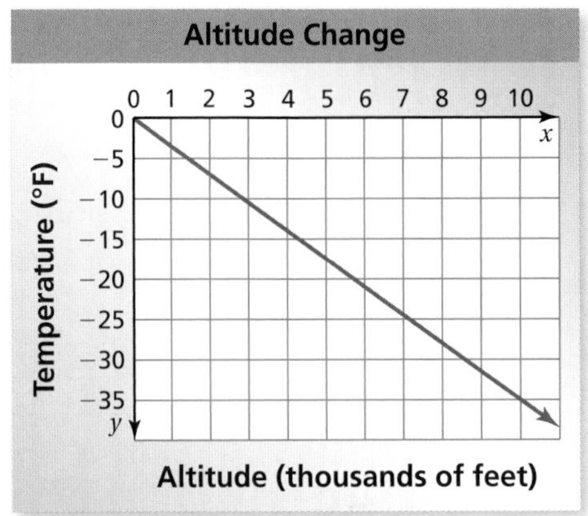

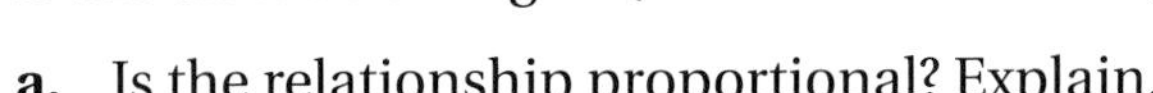

a. Is the relationship proportional? Explain.

b. Write an equation of the line. Interpret the slope.

c. You are at the bottom of a mountain where the temperature is 74°F. The top of the mountain is 5500 feet above you. What is the temperature at the top of the mountain?

13. **Critical Thinking** Consider the distance equation $d = rt$, where d is the distance (in feet), r is the rate (in feet per second), and t is the time (in seconds).

a. You run 6 feet per second. Are distance and time proportional? Explain. Graph the equation.

b. You run for 50 seconds. Are distance and rate proportional? Explain. Graph the equation.

c. You run 300 feet. Are rate and time proportional? Explain. Graph the equation.

d. One of these situations represents *inverse variation*. Which one is it? Why do you think it is called inverse variation?

Fair Game Review What you learned in previous grades & lessons

Graph the linear equation. *(Section 4.1)*

14. $y = -\frac{1}{2}x$

15. $y = 3x - \frac{3}{4}$

16. $y = -\frac{x}{3} - \frac{3}{2}$

17. **MULTIPLE CHOICE** What is the value of x? *(Section 3.3)*

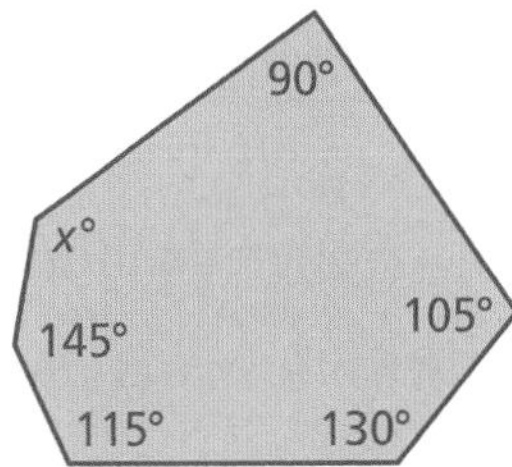

Ⓐ 110 Ⓑ 135

Ⓒ 315 Ⓓ 522

4 Study Help

You can use a **process diagram** to show the steps involved in a procedure. Here is an example of a process diagram for graphing a linear equation.

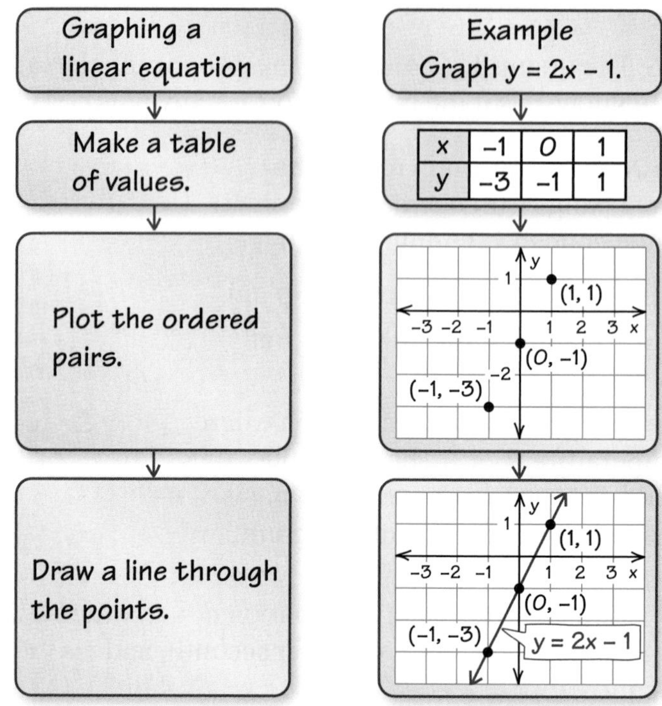

On Your Own

Make process diagrams with examples to help you study these topics.

1. finding the slope of a line
2. graphing a proportional relationship

After you complete this chapter, make process diagrams for the following topics.

3. graphing a linear equation using
   a. slope and y-intercept
   b. x- and y-intercepts
4. writing equations in slope-intercept form
5. writing equations in point-slope form

"Here is a process diagram with suggestions for what to do if a hyena knocks on your door."

4.1–4.3 Quiz

Graph the linear equation. *(Section 4.1)*

1. $y = -x + 8$ **2.** $y = \frac{x}{3} - 4$ **3.** $x = -1$ **4.** $y = 3.5$

Find the slope of the line. *(Section 4.2)*

5.

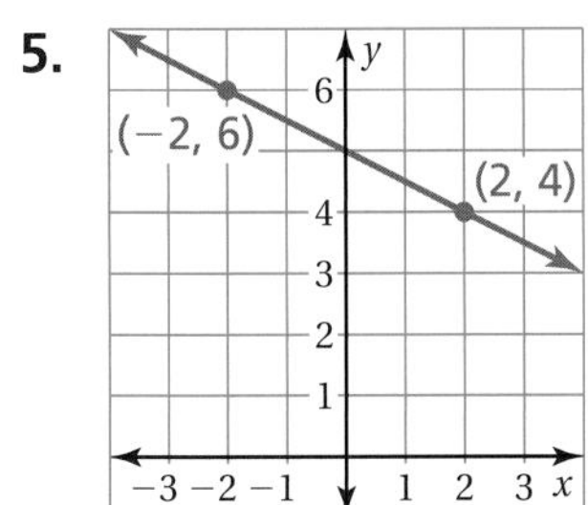

6.

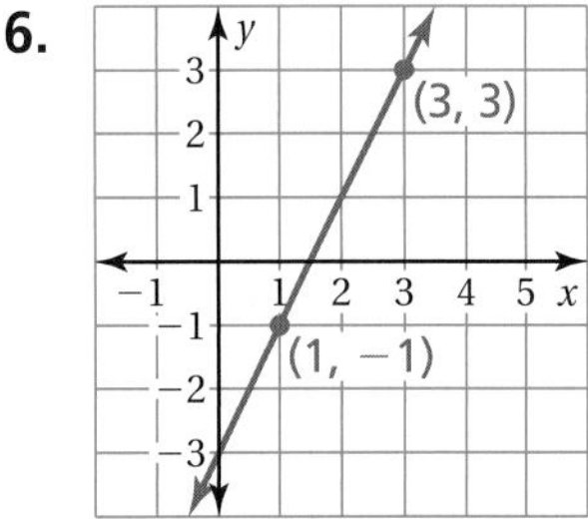

7.

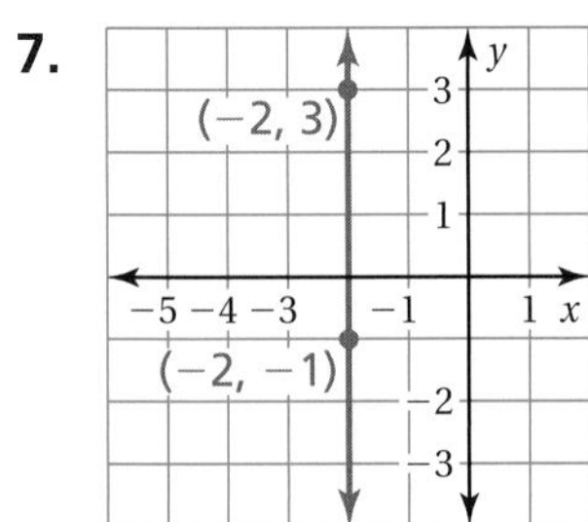

8.

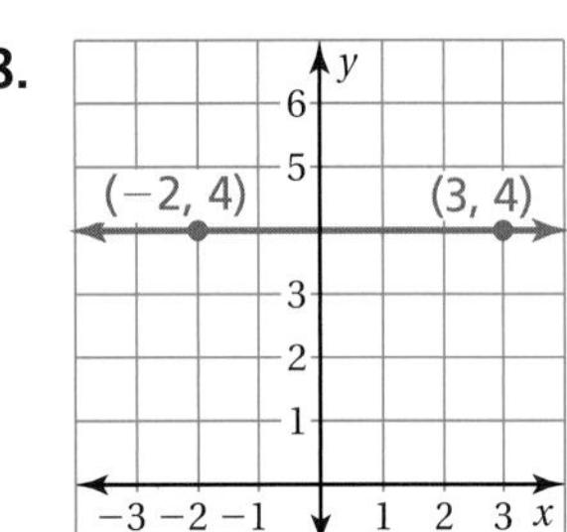

9. What is the slope of a line that is parallel to the line in Exercise 5? What is the slope of a line that is perpendicular to the line in Exercise 5? *(Section 4.2)*

10. Are the lines $y = -1$ and $x = 1$ parallel? Are they perpendicular? Justify your answer. *(Section 4.2)*

11. BANKING A bank charges \$3 each time you use an out-of-network ATM. At the beginning of the month, you have \$1500 in your bank account. You withdraw \$60 from your bank account each time you use an out-of-network ATM. Graph a linear equation that represents the balance in your account after you use an out-of-network ATM x times. *(Section 4.1)*

12. MUSIC The number y of hours of cello lessons that you take after x weeks is represented by the equation $y = 3x$. Graph the equation and interpret the slope. *(Section 4.3)*

13. DINNER PARTY The cost y (in dollars) to provide food for guests at a dinner party is proportional to the number x of guests attending the party. It costs \$30 to provide food for 4 guests. *(Section 4.3)*

a. Write an equation that represents the situation.

b. Interpret the slope.

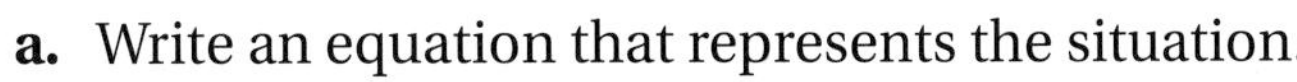

c. How much does it cost to provide food for 10 guests?

4.4 Graphing Linear Equations in Slope-Intercept Form

Essential Question How can you describe the graph of the equation $y = mx + b$?

1 ACTIVITY: Analyzing Graphs of Lines

Work with a partner.

- **Graph each equation.**
- **Find the slope of each line.**
- **Find the point where each line crosses the y-axis.**
- **Complete the table.**

Equation	Slope of Graph	Point of Intersection with y-axis
a. $y = -\frac{1}{2}x + 1$		
b. $y = -x + 2$		
c. $y = -x - 2$		
d. $y = \frac{1}{2}x + 1$		
e. $y = x + 2$		
f. $y = x - 2$		
g. $y = \frac{1}{2}x - 1$		
h. $y = -\frac{1}{2}x - 1$		
i. $y = 3x + 2$		
j. $y = 3x - 2$		

k. Do you notice any relationship between the slope of the graph and its equation? between the point of intersection with the y-axis and its equation? Compare the results with those of other students in your class.

Graphing Equations

In this lesson, you will

- find slopes and y-intercepts of graphs of linear equations.
- graph linear equations written in slope-intercept form.

2 ACTIVITY: Deriving an Equation

Work with a partner.

a. Look at the graph of each equation in Activity 1. Do any of the graphs represent a proportional relationship? Explain.

b. For a nonproportional linear relationship, the graph crosses the y-axis at some point $(0, b)$, where b does not equal 0. Let (x, y) represent any other point on the graph. You can use the formula for slope to write the equation for a nonproportional linear relationship.

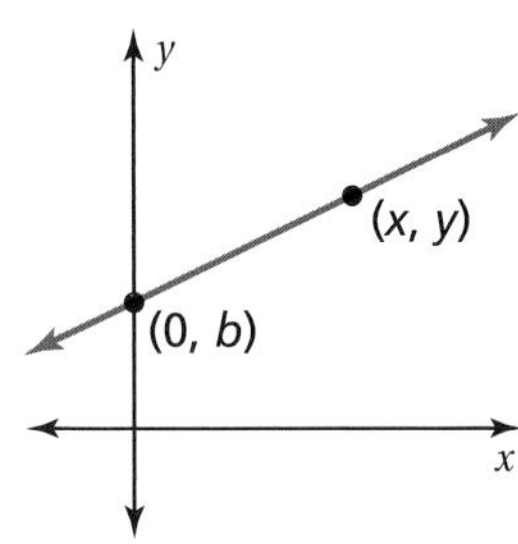

Use the graph to complete the steps.

$\dfrac{y_2 - y_1}{x_2 - x_1} = m$ Slope formula

$\dfrac{y - \square}{x - \square} = m$ Substitute values.

$\dfrac{\square}{\square} = m$ Simplify.

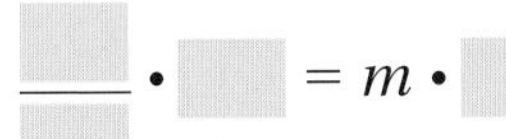

$\dfrac{\square}{\square} \cdot \square = m \cdot \square$ Multiplication Property of Equality

$y - \square = m \cdot \square$ Simplify.

$y = m\square + \square$ Addition Property of Equality

Math Practice

Use Prior Results

How can you use the results of Activity 1 to help support your answer?

c. What do m and b represent in the equation?

What Is Your Answer?

3. **IN YOUR OWN WORDS** How can you describe the graph of the equation $y = mx + b$?

   a. How does the value of m affect the graph of the equation?

   b. How does the value of b affect the graph of the equation?

   c. Check your answers to parts (a) and (b) with three equations that are not in Activity 1.

4. **LOGIC** Why do you think $y = mx + b$ is called the *slope-intercept form* of the equation of a line? Use drawings or diagrams to support your answer.

Practice Use what you learned about graphing linear equations in slope-intercept form to complete Exercises 4–6 on page 170.

4.4 Lesson

Key Vocabulary
x-intercept, *p. 168*
y-intercept, *p. 168*
slope-intercept form, *p. 168*

Key Ideas

Intercepts

The ***x*-intercept** of a line is the x-coordinate of the point where the line crosses the x-axis. It occurs when $y = 0$.

The ***y*-intercept** of a line is the y-coordinate of the point where the line crosses the y-axis. It occurs when $x = 0$.

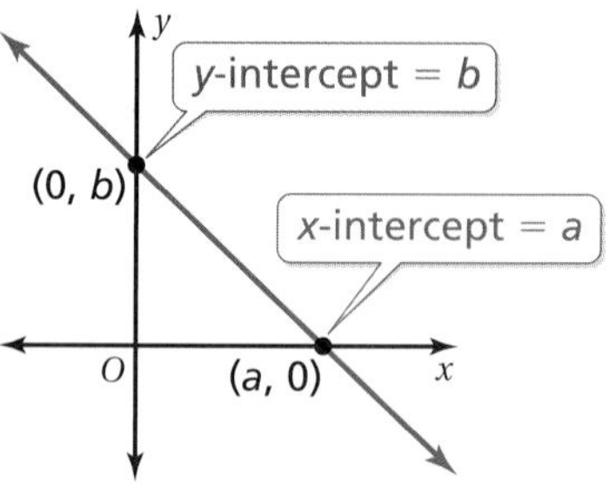

Study Tip

Linear equations can, but do not always, pass through the origin. So, proportional relationships are a special type of linear equation in which $b = 0$.

Slope-Intercept Form

Words A linear equation written in the form $y = mx + b$ is in **slope-intercept form**. The slope of the line is m, and the y-intercept of the line is b.

Algebra $y = mx + b$ (slope: m; y-intercept: b)

EXAMPLE 1 Identifying Slopes and y-Intercepts

Find the slope and the y-intercept of the graph of each linear equation.

a. $y = -4x - 2$

$y = -4x + (-2)$ Write in slope-intercept form.

The slope is -4, and the y-intercept is -2.

b. $y - 5 = \frac{3}{2}x$

$y = \frac{3}{2}x + 5$ Add 5 to each side.

The slope is $\frac{3}{2}$, and the y-intercept is 5.

On Your Own

Now You're Ready Exercises 7–15

Find the slope and the y-intercept of the graph of the linear equation.

1. $y = 3x - 7$

2. $y - 1 = -\frac{2}{3}x$

EXAMPLE 2 Graphing a Linear Equation in Slope-Intercept Form

Graph $y = -3x + 3$. Identify the x-intercept.

Step 1: Find the slope and the y-intercept.

$$y = -3x + 3$$

slope: -3 y-intercept: 3

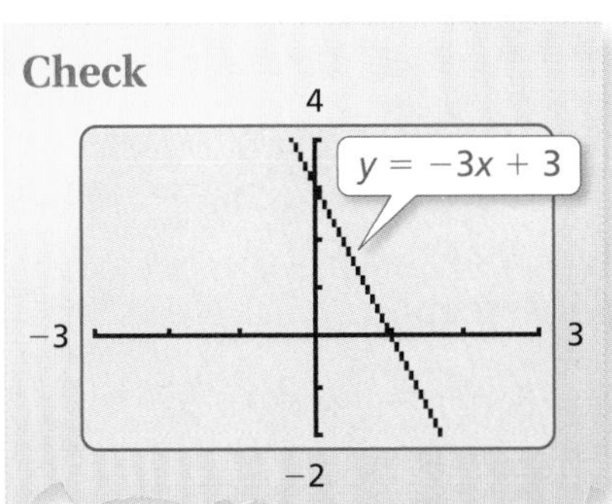

Step 2: The y-intercept is 3. So, plot (0, 3).

Step 3: Use the slope to find another point and draw the line.

$$m = \frac{\text{rise}}{\text{run}} = \frac{-3}{1}$$

Plot the point that is 1 unit right and 3 units down from (0, 3). Draw a line through the two points.

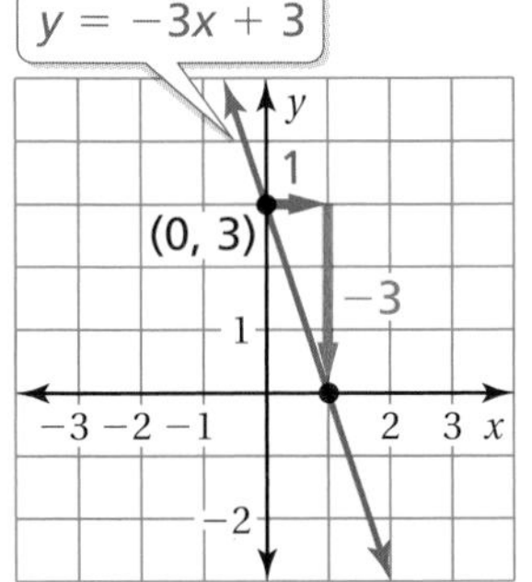

∴ The line crosses the x-axis at (1, 0). So, the x-intercept is 1.

EXAMPLE 3 Real-Life Application

The cost y (in dollars) of taking a taxi x miles is $y = 2.5x + 2$. (a) Graph the equation. (b) Interpret the y-intercept and the slope.

a. The slope of the line is $2.5 = \frac{5}{2}$. Use the slope and the y-intercept to graph the equation.

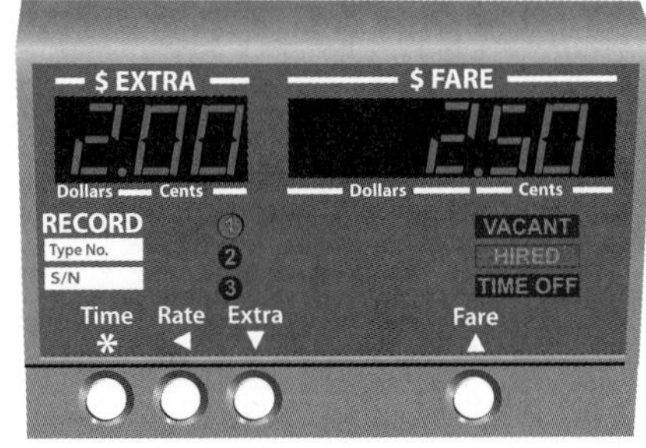

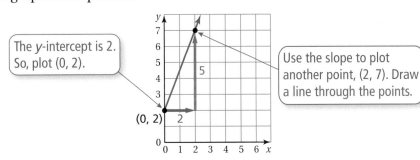

The y-intercept is 2. So, plot (0, 2).

Use the slope to plot another point, (2, 7). Draw a line through the points.

b. The slope is 2.5. So, the cost per mile is \$2.50. The y-intercept is 2. So, there is an initial fee of \$2 to take the taxi.

On Your Own

Now You're Ready
Exercises 18–23

Graph the linear equation. Identify the x-intercept. Use a graphing calculator to check your answer.

3. $y = x - 4$

4. $y = -\frac{1}{2}x + 1$

5. In Example 3, the cost y (in dollars) of taking a different taxi x miles is $y = 2x + 1.5$. Interpret the y-intercept and the slope.

4.4 Exercises

Vocabulary and Concept Check

1. **VOCABULARY** How can you find the x-intercept of the graph of $2x + 3y = 6$?
2. **CRITICAL THINKING** Is the equation $y = 3x$ in slope-intercept form? Explain.
3. **OPEN-ENDED** Describe a real-life situation that you can model with a linear equation. Write the equation. Interpret the y-intercept and the slope.

Practice and Problem Solving

Match the equation with its graph. Identify the slope and the y-intercept.

4. $y = 2x + 1$
5. $y = \frac{1}{3}x - 2$
6. $y = -\frac{2}{3}x + 1$

A. 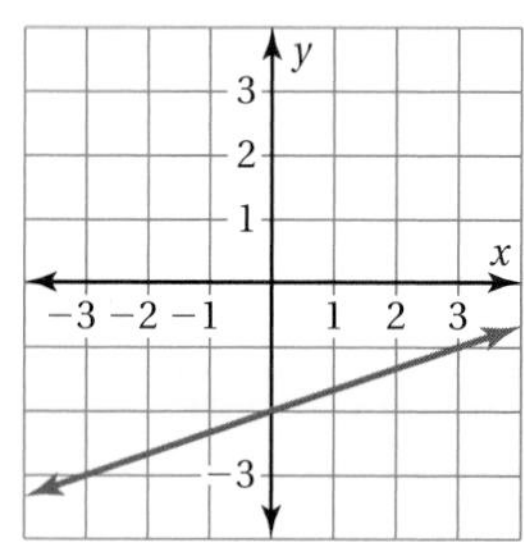

B. 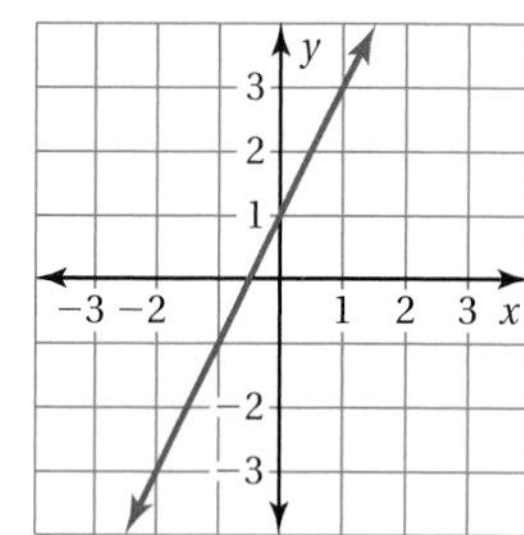

C. 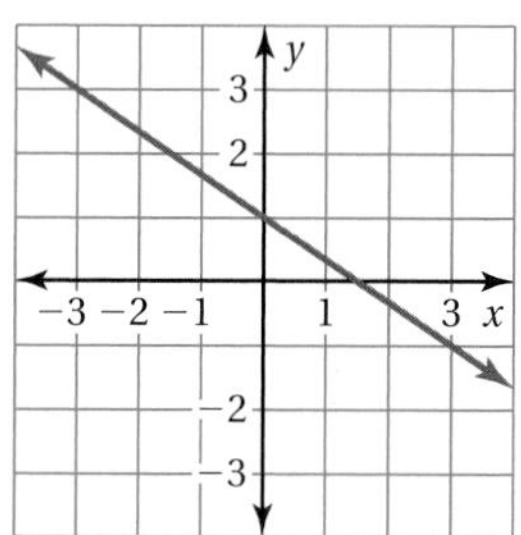

Find the slope and the y-intercept of the graph of the linear equation.

1 7. $y = 4x - 5$
8. $y = -7x + 12$
9. $y = -\frac{4}{5}x - 2$
10. $y = 2.25x + 3$
11. $y + 1 = \frac{4}{3}x$
12. $y - 6 = \frac{3}{8}x$
13. $y - 3.5 = -2x$
14. $y = -5 - \frac{1}{2}x$
15. $y = 11 + 1.5x$

16. **ERROR ANALYSIS** Describe and correct the error in finding the slope and the y-intercept of the graph of the linear equation.

$y = 4x - 3$

The slope is 4, and the y-intercept is 3.

17. **SKYDIVING** A skydiver parachutes to the ground. The height y (in feet) of the skydiver after x seconds is $y = -10x + 3000$.

    a. Graph the equation.

    b. Interpret the x-intercept and the slope.

Graph the linear equation. Identify the x-intercept. Use a graphing calculator to check your answer.

18. $y = \frac{1}{5}x + 3$ **19.** $y = 6x - 7$ **20.** $y = -\frac{8}{3}x + 9$

21. $y = -1.4x - 1$ **22.** $y + 9 = -3x$ **23.** $y = 4 - \frac{3}{5}x$

24. APPLES You go to a harvest festival and pick apples.

a. Which equation represents the cost (in dollars) of going to the festival and picking x pounds of apples? Explain.

$y = 5x + 0.75$ $y = 0.75x + 5$

b. Graph the equation you chose in part (a).

Admission: $5.00
Apples: $0.75 per lb

25. REASONING Without graphing, identify the equations of the lines that are (a) parallel and (b) perpendicular. Explain your reasoning.

$y = 2x + 4$	$y = -\frac{1}{3}x - 1$	$y = -3x - 2$	$y = \frac{1}{2}x + 1$
$y = 3x + 3$	$y = -\frac{1}{2}x + 2$	$y = -3x + 5$	$y = 2x - 3$

26. Critical Thinking Six friends create a website. The website earns money by selling banner ads. The site has 5 banner ads. It costs $120 a month to operate the website.

a. A banner ad earns $0.005 per click. Write a linear equation that represents the monthly income y (in dollars) for x clicks.

b. Graph the equation in part (a). On the graph, label the number of clicks needed for the friends to start making a profit.

Fair Game Review What you learned in previous grades & lessons

Solve the equation for y. *(Section 1.4)*

27. $y - 2x = 3$ **28.** $4x + 5y = 13$ **29.** $2x - 3y = 6$ **30.** $7x + 4y = 8$

31. MULTIPLE CHOICE Which point is a solution of the equation $3x - 8y = 11$? *(Section 4.1)*

Ⓐ (1, 1) Ⓑ (1, −1) Ⓒ (−1, 1) Ⓓ (−1, −1)

4.5 Graphing Linear Equations in Standard Form

Essential Question How can you describe the graph of the equation $ax + by = c$?

1 ACTIVITY: Using a Table to Plot Points

Work with a partner. You sold a total of $16 worth of tickets to a school concert. You lost track of how many of each type of ticket you sold.

 adult · Number of adult tickets +  student · Number of student tickets = 

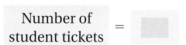

a. Let x represent the number of adult tickets.
Let y represent the number of student tickets.
Write an equation that relates x and y.

b. Copy and complete the table showing the different combinations of tickets you might have sold.

Number of Adult Tickets, x					
Number of Student Tickets, y					

c. Plot the points from the table. Describe the pattern formed by the points.

d. If you remember how many adult tickets you sold, can you determine how many student tickets you sold? Explain your reasoning.

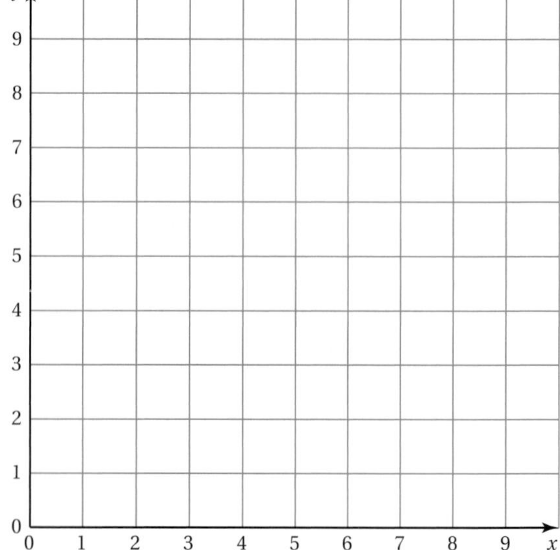

Graphing Equations
In this lesson, you will
- graph linear equations written in standard form.

2 ACTIVITY: Rewriting an Equation

Work with a partner. You sold a total of $16 worth of cheese. You forgot how many pounds of each type of cheese you sold.

CHEESE FOR SALE
Swiss: $4/lb Cheddar: $2/lb

$\dfrac{\square}{\text{pound}} \cdot$ Pounds of swiss $+ \dfrac{\square}{\text{pound}} \cdot$ Pounds of cheddar $= \square$

Math Practice

Understand Quantities

What do the equation and the graph represent? How can you use this information to solve the problem?

a. Let x represent the number of pounds of swiss cheese.

Let y represent the number of pounds of cheddar cheese.

Write an equation that relates x and y.

b. Rewrite the equation in slope-intercept form. Then graph the equation.

c. You sold 2 pounds of cheddar cheese. How many pounds of swiss cheese did you sell?

d. Does the value $x = 2.5$ make sense in the context of the problem? Explain.

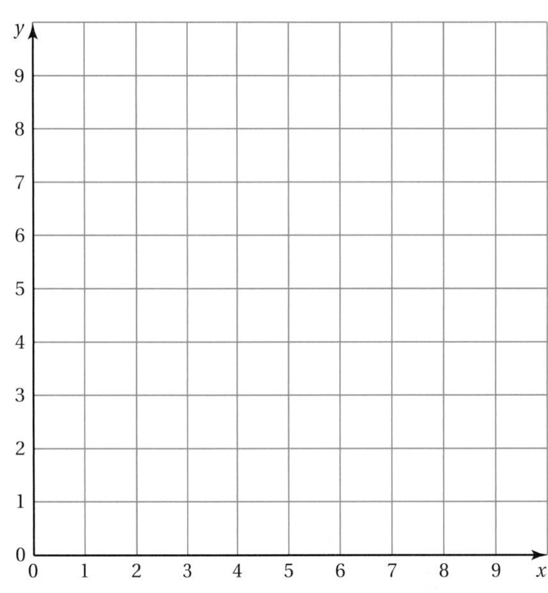

What Is Your Answer?

3. **IN YOUR OWN WORDS** How can you describe the graph of the equation $ax + by = c$?

4. Activities 1 and 2 show two different methods for graphing $ax + by = c$. Describe the two methods. Which method do you prefer? Explain.

5. Write a real-life problem that is similar to those shown in Activities 1 and 2.

6. Why do you think it might be easier to graph $x + y = 10$ without rewriting it in slope-intercept form and then graphing?

Practice Use what you learned about graphing linear equations in standard form to complete Exercises 3 and 4 on page 176.

4.5 Lesson

Key Vocabulary
standard form, *p. 174*

Study Tip
Any linear equation can be written in standard form.

Key Idea

Standard Form of a Linear Equation

The **standard form** of a linear equation is

$$ax + by = c$$

where a and b are not both zero.

EXAMPLE 1 Graphing a Linear Equation in Standard Form

Graph $-2x + 3y = -6$.

Step 1: Write the equation in slope-intercept form.

$-2x + 3y = -6$ Write the equation.

$3y = 2x - 6$ Add $2x$ to each side.

$y = \frac{2}{3}x - 2$ Divide each side by 3.

Step 2: Use the slope and the y-intercept to graph the equation.

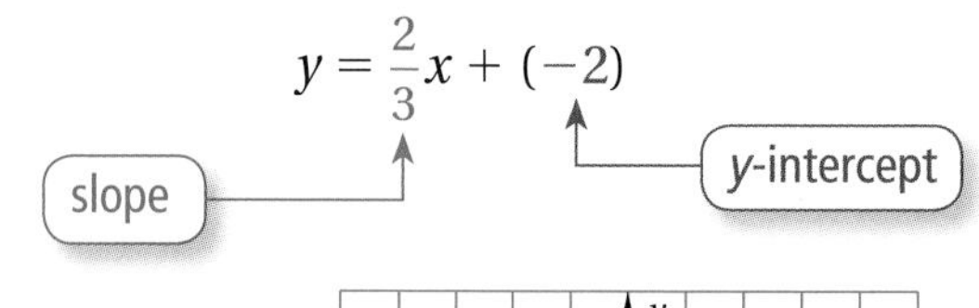

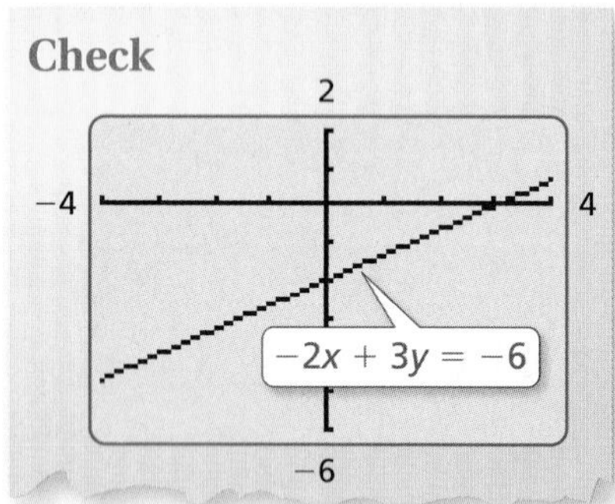

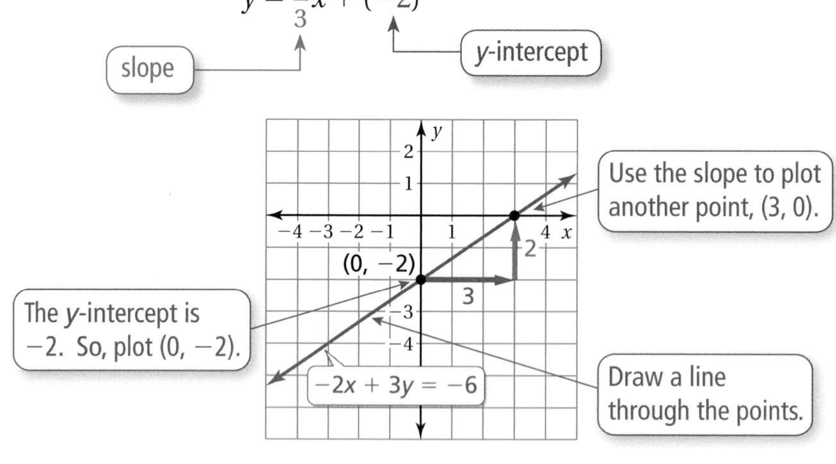

On Your Own

Now You're Ready
Exercises 5–10

Graph the linear equation. Use a graphing calculator to check your graph.

1. $x + y = -2$
2. $-\frac{1}{2}x + 2y = 6$
3. $-\frac{2}{3}x + y = 0$
4. $2x + y = 5$

EXAMPLE 2 Graphing a Linear Equation in Standard Form

Graph $x + 3y = -3$ using intercepts.

Step 1: To find the x-intercept, substitute 0 for y.

$$x + 3y = -3$$
$$x + 3(0) = -3$$
$$x = -3$$

To find the y-intercept, substitute 0 for x.

$$x + 3y = -3$$
$$0 + 3y = -3$$
$$y = -1$$

Step 2: Graph the equation.

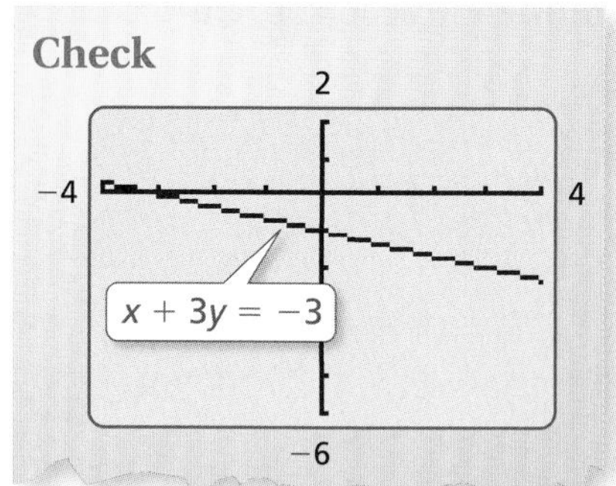

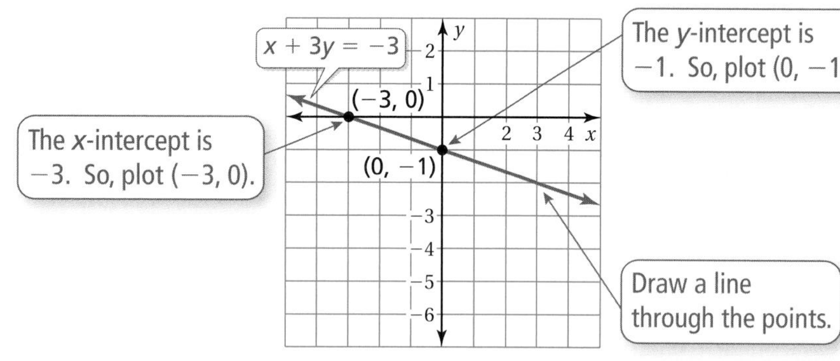

EXAMPLE 3 Real-Life Application

You have \$6 to spend on apples and bananas. (a) Graph the equation $1.5x + 0.6y = 6$, where x is the number of pounds of apples and y is the number of pounds of bananas. (b) Interpret the intercepts.

a. Find the intercepts and graph the equation.

***x*-intercept**

$$1.5x + 0.6y = 6$$
$$1.5x + 0.6(0) = 6$$
$$x = 4$$

***y*-intercept**

$$1.5x + 0.6y = 6$$
$$1.5(0) + 0.6y = 6$$
$$y = 10$$

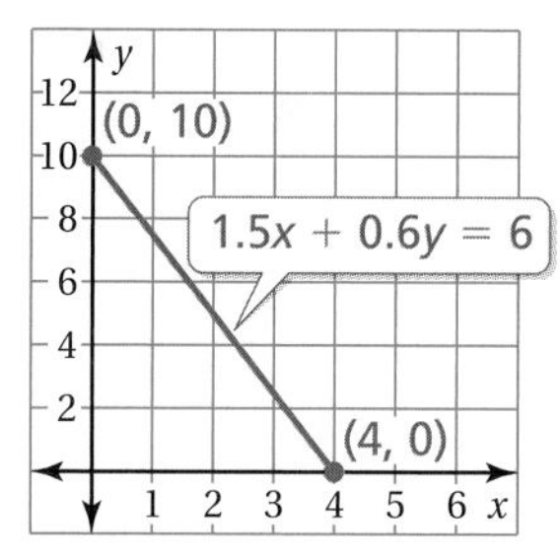

b. The x-intercept shows that you can buy 4 pounds of apples when you do not buy any bananas. The y-intercept shows that you can buy 10 pounds of bananas when you do not buy any apples.

On Your Own

Now You're Ready
Exercises 16–18

Graph the linear equation using intercepts. Use a graphing calculator to check your graph.

5. $2x - y = 8$

6. $x + 3y = 6$

7. **WHAT IF?** In Example 3, you buy y pounds of oranges instead of bananas. Oranges cost \$1.20 per pound. Graph the equation $1.5x + 1.2y = 6$. Interpret the intercepts.

4.5 Exercises

Vocabulary and Concept Check

1. **VOCABULARY** Is the equation $y = -2x + 5$ in standard form? Explain.
2. **WRITING** Describe two ways to graph the equation $4x + 2y = 6$.

Practice and Problem Solving

Define two variables for the verbal model. Write an equation in slope-intercept form that relates the variables. Graph the equation.

3. $\frac{\$2.00}{\text{pound}} \bullet \text{Pounds of peaches} + \frac{\$1.50}{\text{pound}} \bullet \text{Pounds of apples} = \$15$

4. $\frac{16 \text{ miles}}{\text{hour}} \bullet \text{Hours biked} + \frac{2 \text{ miles}}{\text{hour}} \bullet \text{Hours walked} = 32 \text{ miles}$

Write the linear equation in slope-intercept form.

1 5. $2x + y = 17$ 6. $5x - y = \frac{1}{4}$ 7. $-\frac{1}{2}x + y = 10$

Graph the linear equation. Use a graphing calculator to check your graph.

8. $-18x + 9y = 72$ 9. $16x - 4y = 2$ 10. $\frac{1}{4}x + \frac{3}{4}y = 1$

Match the equation with its graph.

11. $15x - 12y = 60$ 12. $5x + 4y = 20$ 13. $10x + 8y = -40$

A.
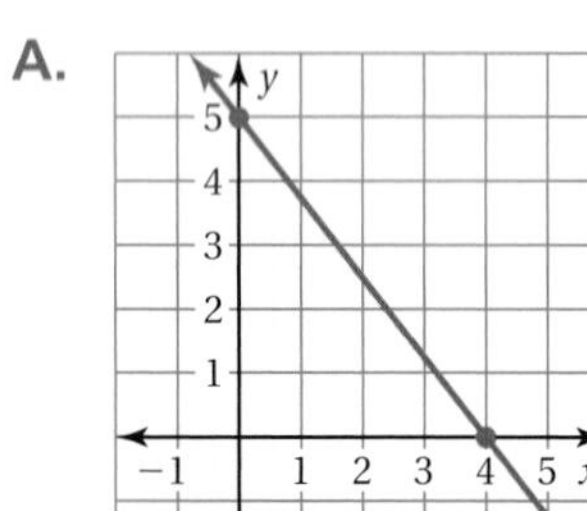

B.
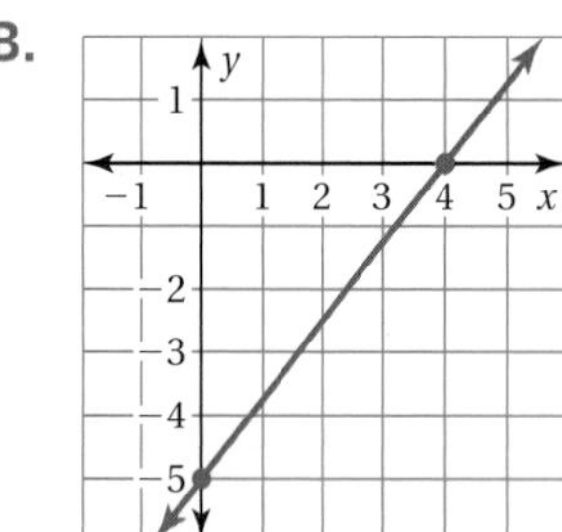

C.
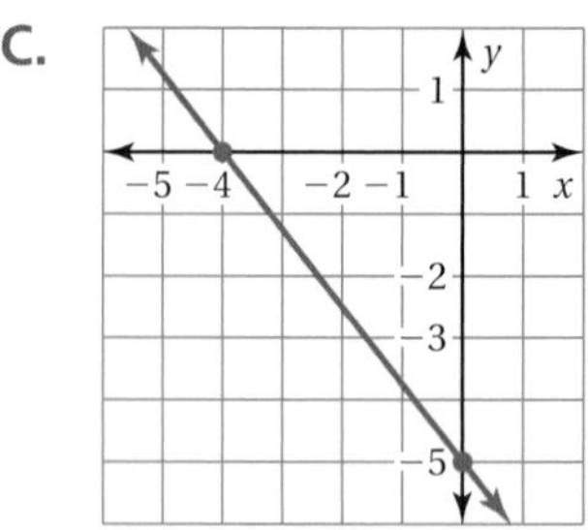

14. **ERROR ANALYSIS** Describe and correct the error in finding the x-intercept.

15. **BRACELET** A charm bracelet costs \$65, plus \$25 for each charm. The equation $-25x + y = 65$ represents the cost y of the bracelet, where x is the number of charms.
    a. Graph the equation.
    b. How much does the bracelet shown cost?

Graph the linear equation using intercepts. Use a graphing calculator to check your graph.

2 **16.** $3x - 4y = -12$ **17.** $2x + y = 8$ **18.** $\frac{1}{3}x - \frac{1}{6}y = -\frac{2}{3}$

19. SHOPPING The amount of money you spend on x CDs and y DVDs is given by the equation $14x + 18y = 126$. Find the intercepts and graph the equation.

20. SCUBA Five friends go scuba diving. They rent a boat for x days and scuba gear for y days. The total spent is $1000.

a. Write an equation in standard form that represents the situation.

b. Graph the equation and interpret the intercepts.

21. MODELING You work at a restaurant as a host and a server. You earn $9.45 for each hour you work as a host and $7.65 for each hour you work as a server.

a. Write an equation in standard form that models your earnings.

b. Graph the equation.

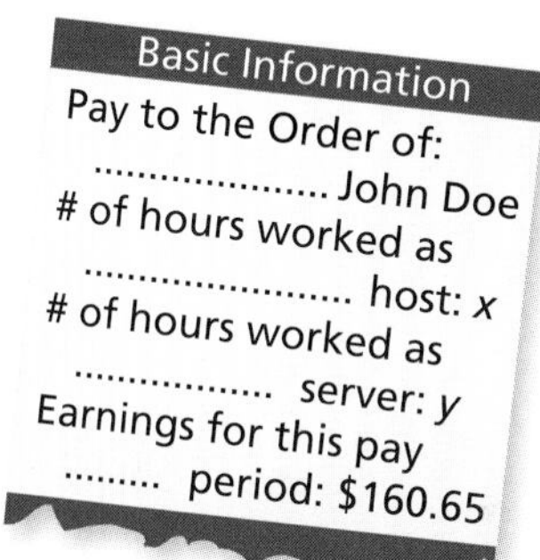
Basic Information
Pay to the Order of: .... John Doe
# of hours worked as .... host: x
# of hours worked as .... server: y
Earnings for this pay .... period: $160.65

22. LOGIC Does the graph of every linear equation have an x-intercept? Explain your reasoning. Include an example.

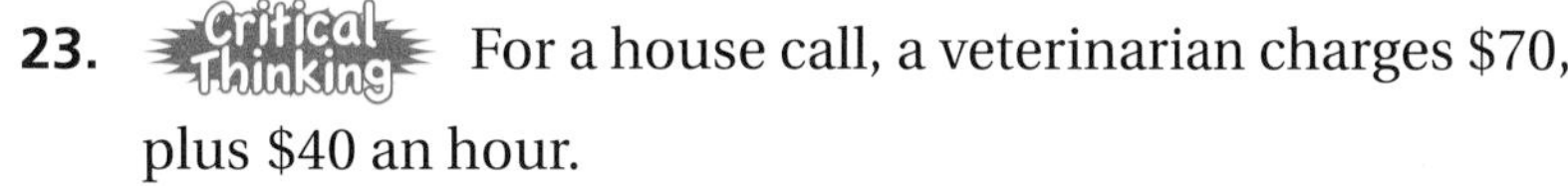

23. Critical Thinking For a house call, a veterinarian charges $70, plus $40 an hour.

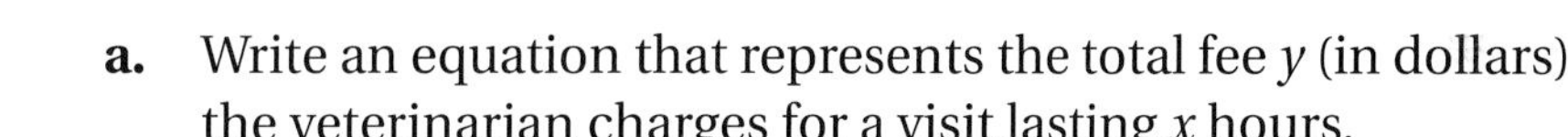

a. Write an equation that represents the total fee y (in dollars) the veterinarian charges for a visit lasting x hours.

b. Find the x-intercept. Does this value make sense in this context? Explain your reasoning.

c. Graph the equation.

Fair Game Review What you learned in previous grades & lessons

The points in the table lie on a line. Find the slope of the line. *(Section 4.2)*

24.

x	−2	−1	0	1
y	−10	−6	−2	2

25.

x	2	4	6	8
y	2	3	4	5

26. MULTIPLE CHOICE Which value of x makes the equation $4x - 12 = 3x - 9$ true? *(Section 1.3)*

(A) −1 (B) 0 (C) 1 (D) 3

4.6 Writing Equations in Slope-Intercept Form

Essential Question How can you write an equation of a line when you are given the slope and the *y*-intercept of the line?

1 ACTIVITY: Writing Equations of Lines

Work with a partner.

- **Find the slope of each line.**
- **Find the *y*-intercept of each line.**
- **Write an equation for each line.**
- **What do the three lines have in common?**

a.

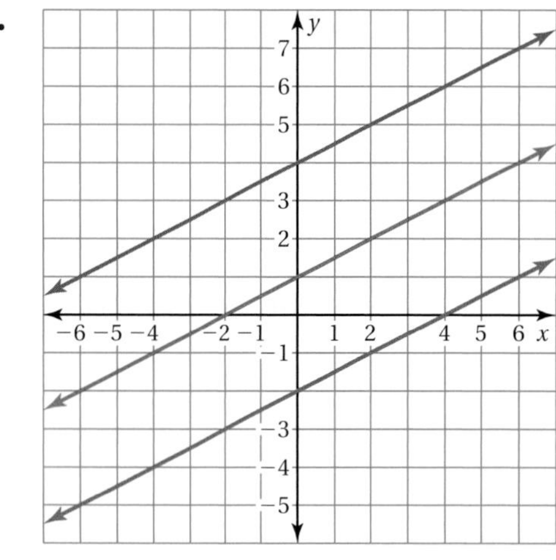

b.

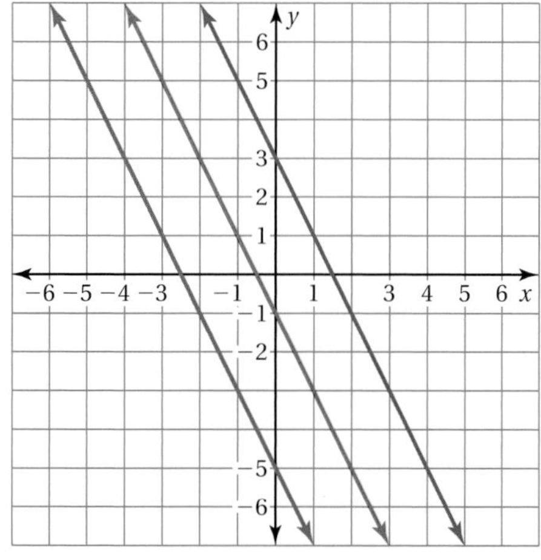

c.

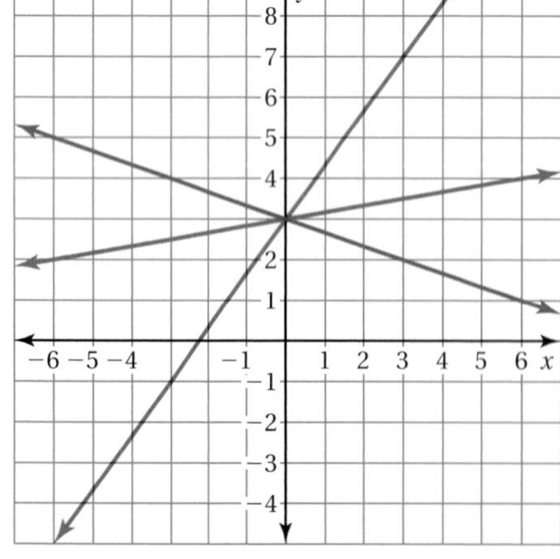

d.

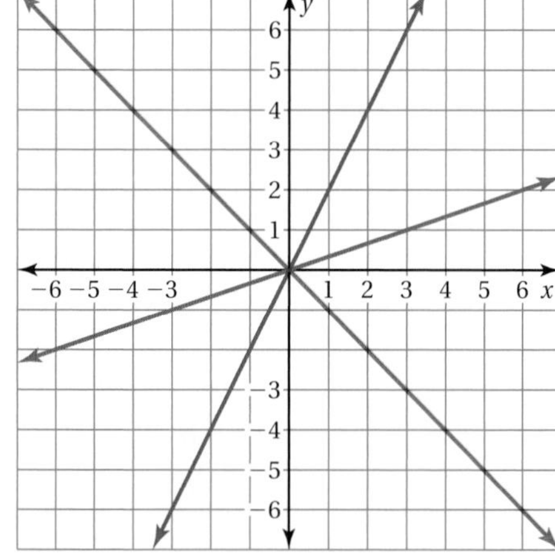

Writing Equations

In this lesson, you will

- write equations of lines in slope-intercept form.

2 ACTIVITY: Describing a Parallelogram

Math Practice

Analyze Givens

What do you need to know to write an equation?

Work with a partner.

- **Find the area of each parallelogram.**
- **Write an equation that represents each side of each parallelogram.**

a.

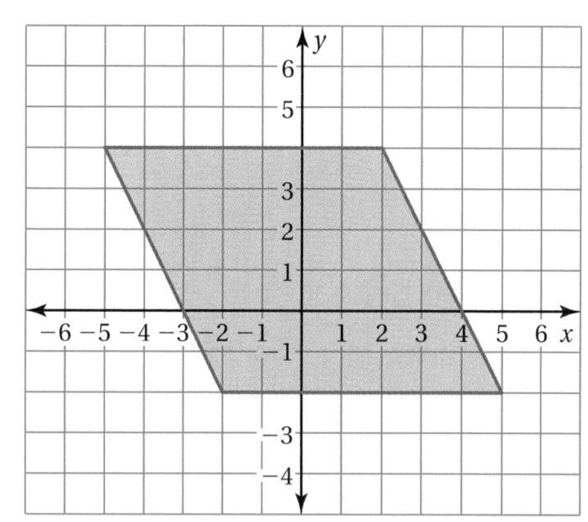

b.

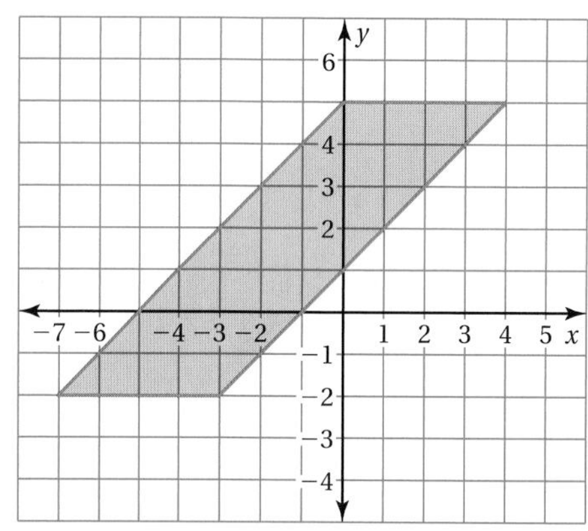

3 ACTIVITY: Interpreting the Slope and the *y*-Intercept

Work with a partner. The graph shows a trip taken by a car, where *t* is the time (in hours) and *y* is the distance (in miles) from Phoenix.

a. Find the *y*-intercept of the graph. What does it represent?

b. Find the slope of the graph. What does it represent?

c. How long did the trip last?

d. How far from Phoenix was the car at the end of the trip?

e. Write an equation that represents the graph.

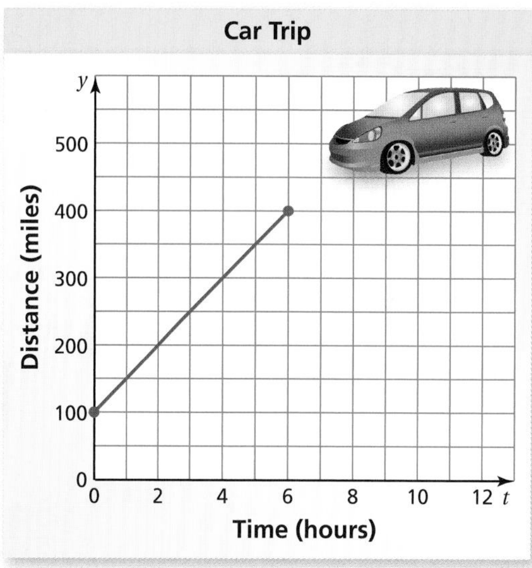

What Is Your Answer?

4. IN YOUR OWN WORDS How can you write an equation of a line when you are given the slope and the *y*-intercept of the line? Give an example that is different from those in Activities 1, 2, and 3.

5. Two sides of a parallelogram are represented by the equations $y = 2x + 1$ and $y = -x + 3$. Give two equations that can represent the other two sides.

Practice Use what you learned about writing equations in slope-intercept form to complete Exercises 3 and 4 on page 182.

4.6 Lesson

EXAMPLE 1 Writing Equations in Slope-Intercept Form

Write an equation of the line in slope-intercept form.

a.

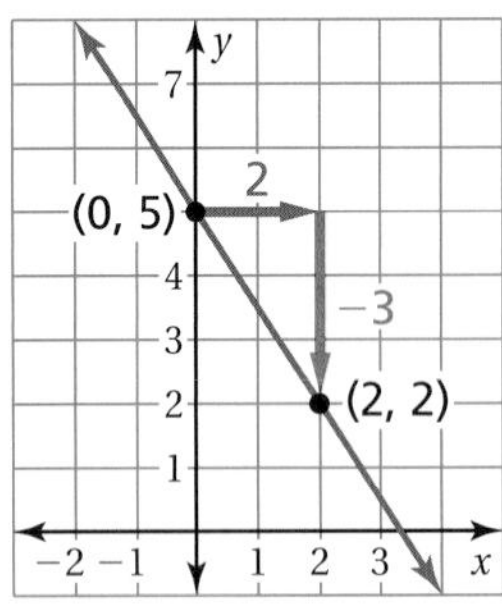

Find the slope and the y-intercept.

$$m = \frac{y_2 - y_1}{x_2 - x_1}$$

$$= \frac{2 - 5}{2 - 0}$$

$$= \frac{-3}{2}, \text{ or } -\frac{3}{2}$$

Because the line crosses the y-axis at (0, 5), the y-intercept is 5.

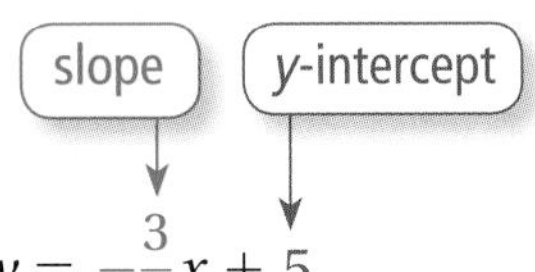

So, the equation is $y = -\frac{3}{2}x + 5$.

Study Tip

After writing an equation, check that the given points are solutions of the equation.

b.

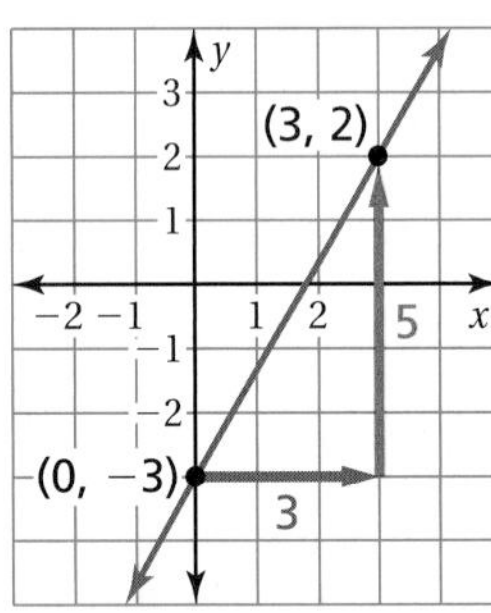

Find the slope and the y-intercept.

$$m = \frac{y_2 - y_1}{x_2 - x_1}$$

$$= \frac{-3 - 2}{0 - 3}$$

$$= \frac{-5}{-3}, \text{ or } \frac{5}{3}$$

Because the line crosses the y-axis at (0, −3), the y-intercept is −3.

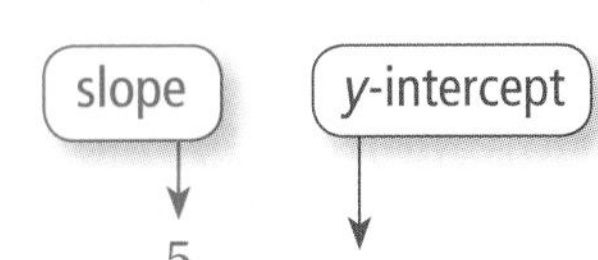

So, the equation is $y = \frac{5}{3}x + (-3)$, or $y = \frac{5}{3}x - 3$.

On Your Own

Now You're Ready
Exercises 5–10

Write an equation of the line in slope-intercept form.

1.

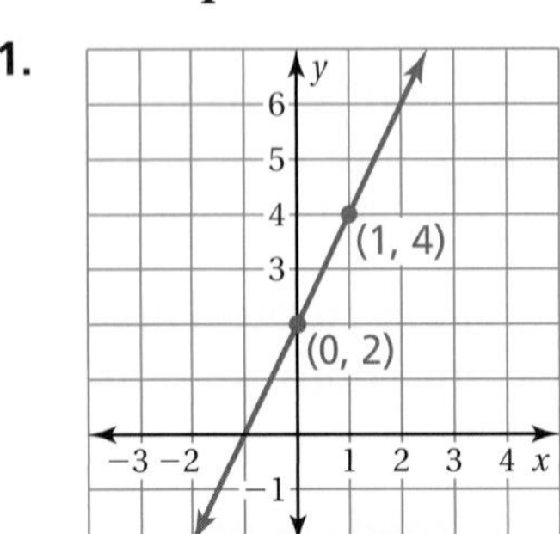

2.

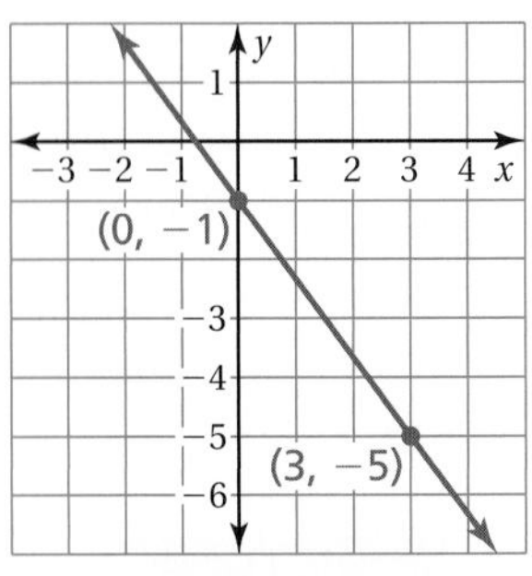

EXAMPLE 2 Writing an Equation

Which equation is shown in the graph?

Ⓐ $y = -4$ Ⓑ $y = -3$

Ⓒ $y = 0$ Ⓓ $y = -3x$

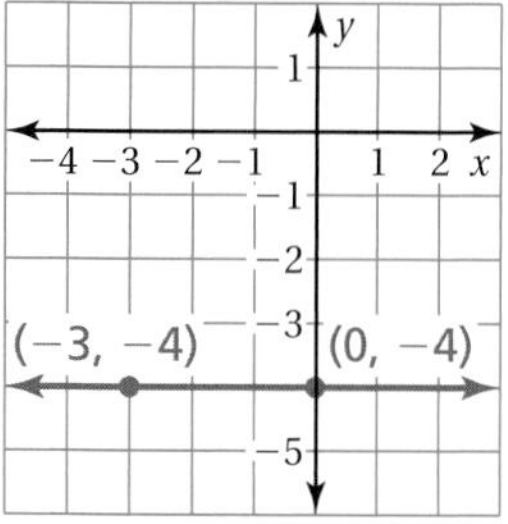

Remember

The graph of $y = a$ is a horizontal line that passes through $(0, a)$.

Find the slope and the y-intercept.

The line is horizontal, so the change in y is 0.

$$m = \frac{\text{change in } y}{\text{change in } x} = \frac{0}{3} = 0$$

Because the line crosses the y-axis at $(0, -4)$, the y-intercept is -4.

So, the equation is $y = 0x + (-4)$, or $y = -4$. The correct answer is Ⓐ.

EXAMPLE 3 Real-Life Application

The graph shows the distance remaining to complete a tunnel. (a) Write an equation that represents the distance y (in feet) remaining after x months. (b) How much time does it take to complete the tunnel?

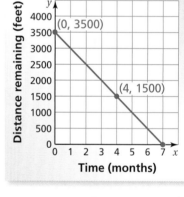

Engineers used tunnel boring machines like the ones shown above to dig an extension of the Metro Gold Line in Los Angeles. The new tunnels are 1.7 miles long and 21 feet wide.

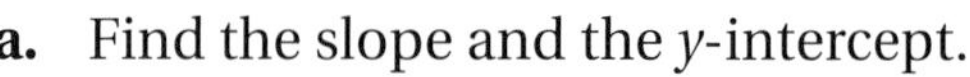

a. Find the slope and the y-intercept.

$$m = \frac{\text{change in } y}{\text{change in } x} = \frac{-2000}{4} = -500$$

Because the line crosses the y-axis at $(0, 3500)$, the y-intercept is 3500.

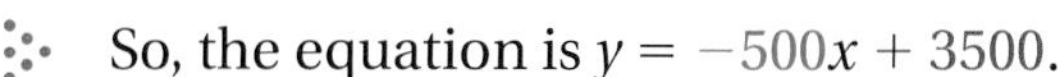

So, the equation is $y = -500x + 3500$.

b. The tunnel is complete when the distance remaining is 0 feet. So, find the value of x when $y = 0$.

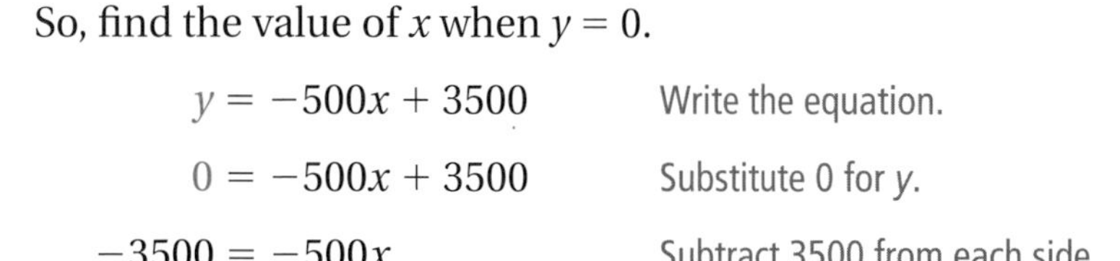

$y = -500x + 3500$	Write the equation.
$0 = -500x + 3500$	Substitute 0 for y.
$-3500 = -500x$	Subtract 3500 from each side.
$7 = x$	Divide each side by -500.

It takes 7 months to complete the tunnel.

On Your Own

Now You're Ready
Exercises 13–15

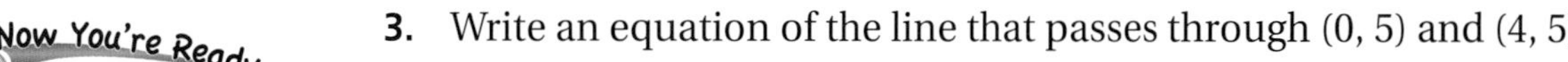

3. Write an equation of the line that passes through $(0, 5)$ and $(4, 5)$.

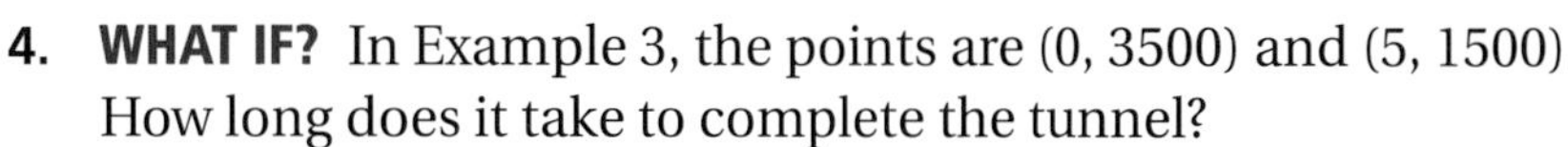

4. **WHAT IF?** In Example 3, the points are $(0, 3500)$ and $(5, 1500)$. How long does it take to complete the tunnel?

4.6 Exercises

Vocabulary and Concept Check

1. **PRECISION** Explain how to find the slope of a line given the intercepts of the line.
2. **WRITING** Explain how to write an equation of a line using its graph.

Practice and Problem Solving

Write an equation that represents each side of the figure.

3. 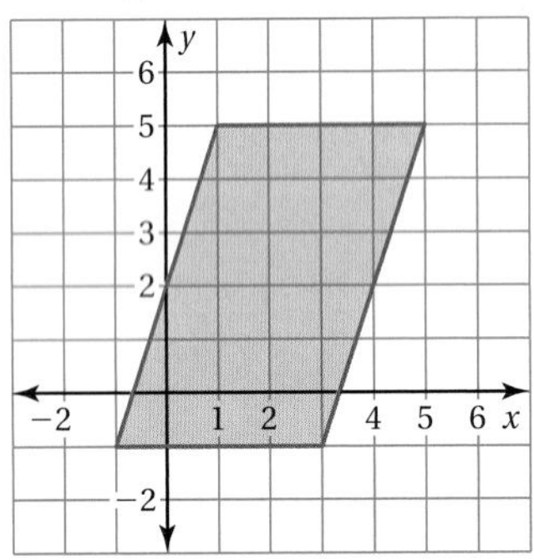

4. 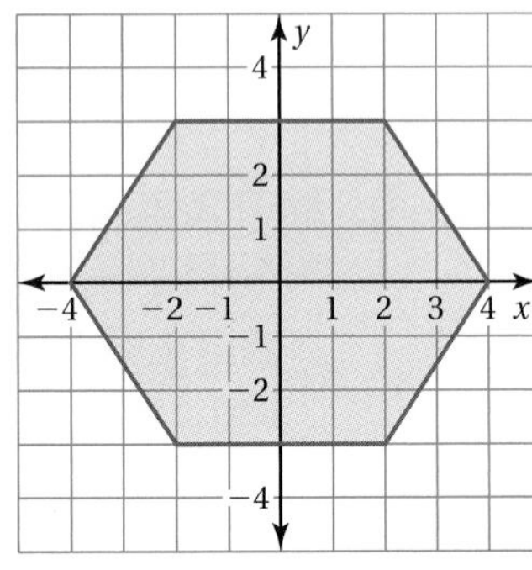

Write an equation of the line in slope-intercept form.

1 5. 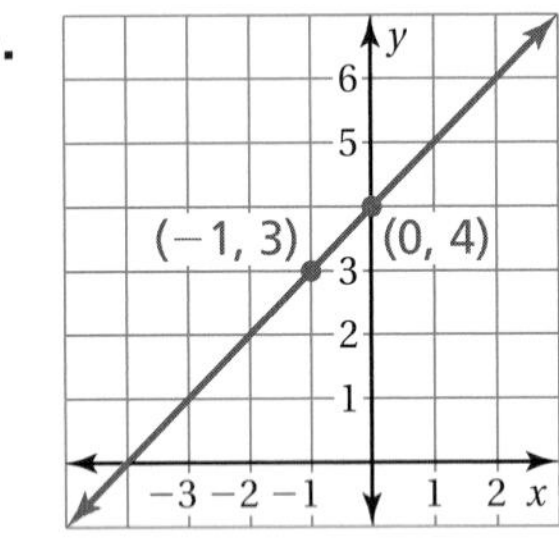

6. 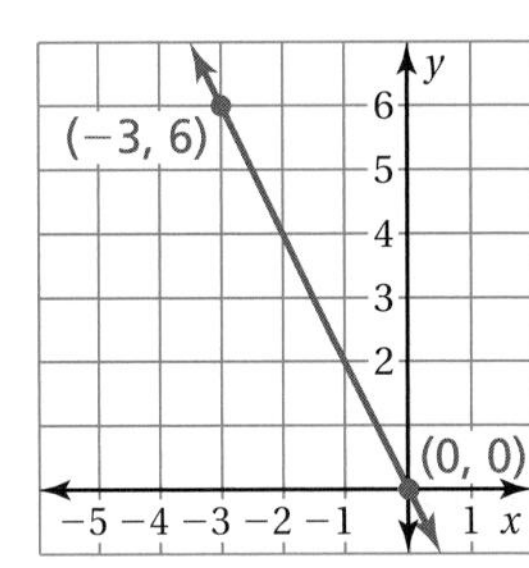

7. 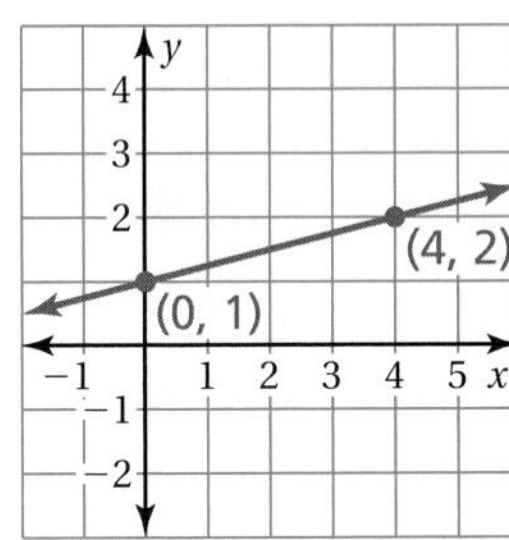

8. 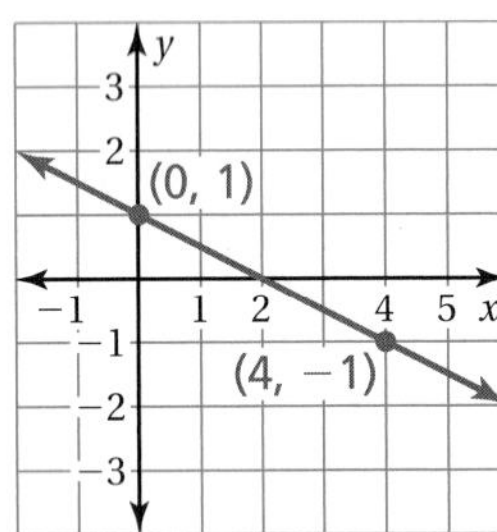

9. 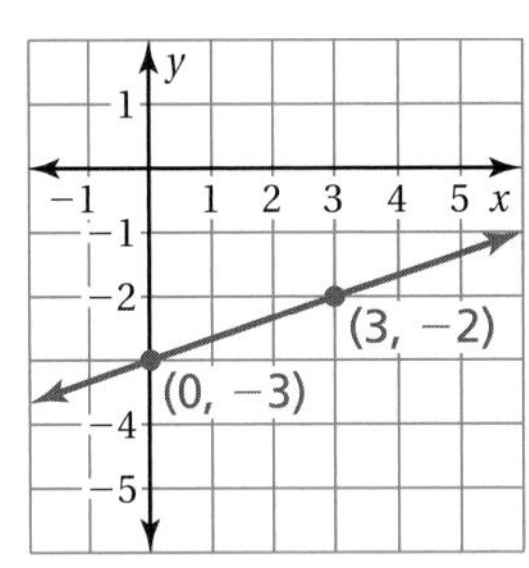

10. 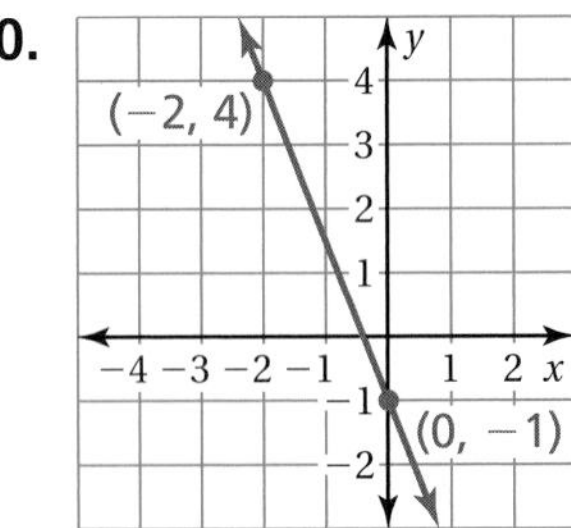

11. **ERROR ANALYSIS** Describe and correct the error in writing an equation of the line.

$$y = \frac{1}{2}x + 4$$

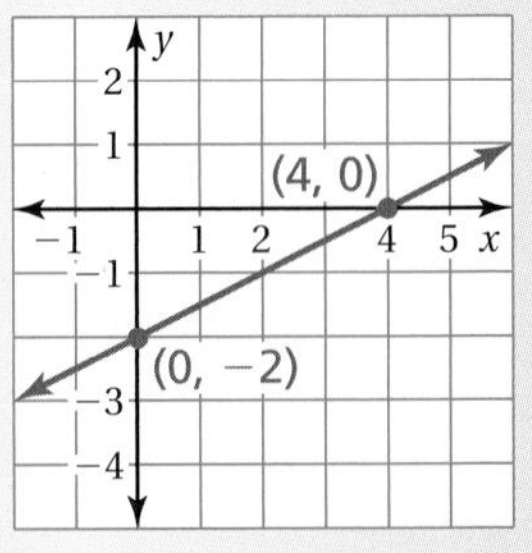

12. **BOA** A boa constrictor is 18 inches long at birth and grows 8 inches per year. Write an equation that represents the length y (in feet) of a boa constrictor that is x years old.

Write an equation of the line that passes through the points.

2 **13.** (2, 5), (0, 5) **14.** (−3, 0), (0, 0) **15.** (0, −2), (4, −2)

16. WALKATHON One of your friends gives you \$10 for a charity walkathon. Another friend gives you an amount per mile. After 5 miles, you have raised \$13.50 total. Write an equation that represents the amount y of money you have raised after x miles.

17. BRAKING TIME During each second of braking, an automobile slows by about 10 miles per hour.

a. Plot the points (0, 60) and (6, 0). What do the points represent?

b. Draw a line through the points. What does the line represent?

c. Write an equation of the line.

18. PAPER You have 500 sheets of notebook paper. After 1 week, you have 72% of the sheets left. You use the same number of sheets each week. Write an equation that represents the number y of pages remaining after x weeks.

19. Critical Thinking The palm tree on the left is 10 years old. The palm tree on the right is 8 years old. The trees grow at the same rate.

a. Estimate the height y (in feet) of each tree.

b. Plot the two points (x, y), where x is the age of each tree and y is the height of each tree.

c. What is the rate of growth of the trees?

d. Write an equation that represents the height of a palm tree in terms of its age.

Fair Game Review What you learned in previous grades & lessons

Plot the ordered pair in a coordinate plane. *(Skills Review Handbook)*

20. (1, 4) **21.** (−1, −2) **22.** (0, 1) **23.** (2, 7)

24. MULTIPLE CHOICE Which of the following statements is true? *(Section 4.4)*

Ⓐ The x-intercept is 5.

Ⓑ The x-intercept is −2.

Ⓒ The y-intercept is 5.

Ⓓ The y-intercept is −2.

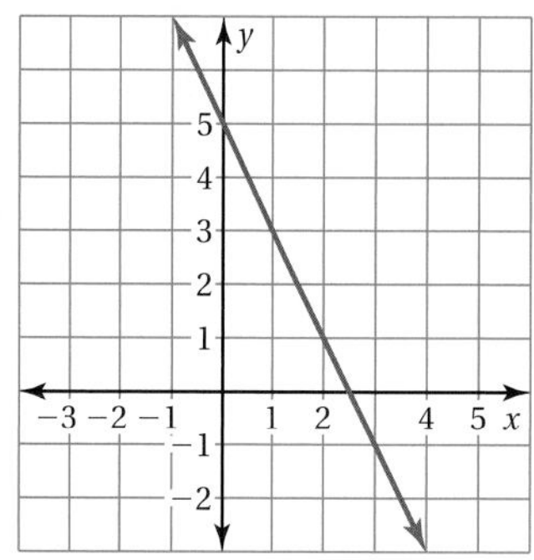

4.7 Writing Equations in Point-Slope Form

Essential Question How can you write an equation of a line when you are given the slope and a point on the line?

1 ACTIVITY: Writing Equations of Lines

Work with a partner.

- **Sketch the line that has the given slope and passes through the given point.**
- **Find the y-intercept of the line.**
- **Write an equation of the line.**

a. $m = -2$

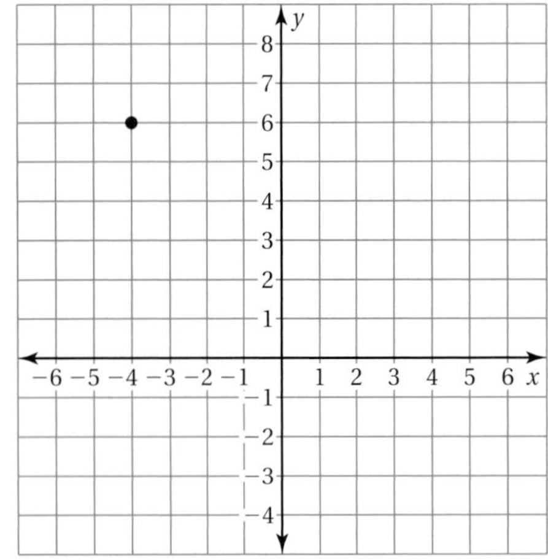

b. $m = \frac{1}{3}$

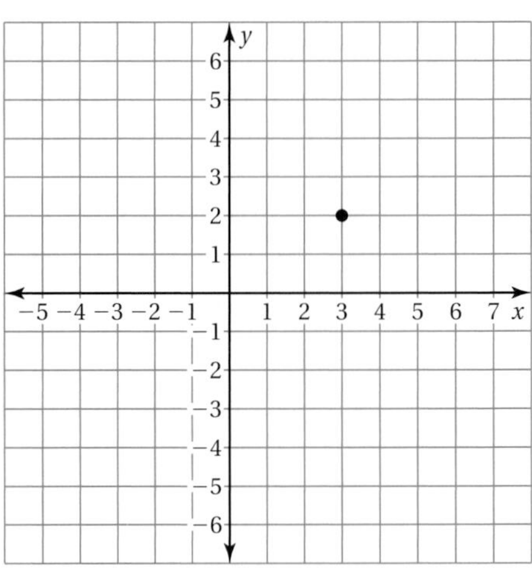

c. $m = -\frac{2}{3}$

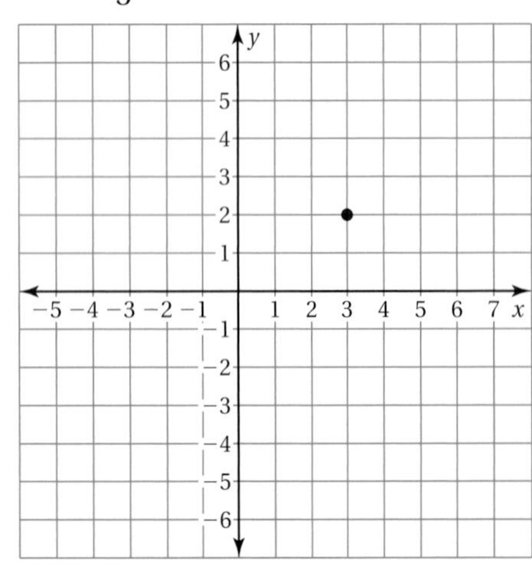

d. $m = \frac{5}{2}$

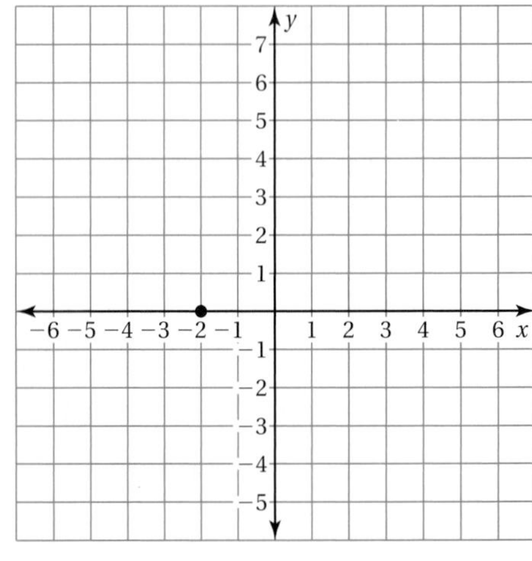

Writing Equations

In this lesson, you will

- write equations of lines using a slope and a point.
- write equations of lines using two points.

2 ACTIVITY: Deriving an Equation

Work with a partner.

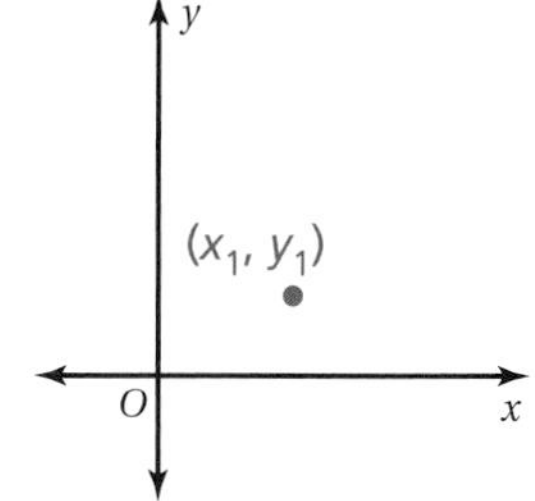

a. Draw a nonvertical line that passes through the point (x_1, y_1).

b. Plot another point on your line. Label this point as (x, y). This point represents any other point on the line.

c. Label the rise and the run of the line through the points (x_1, y_1) and (x, y).

d. The rise can be written as $y - y_1$. The run can be written as $x - x_1$. Explain why this is true.

e. Write an equation for the slope m of the line using the expressions from part (d).

f. Multiply each side of the equation by the expression in the denominator. Write your result. What does this result represent?

Math Practice

Construct Arguments

How does a graph help you derive an equation?

3 ACTIVITY: Writing an Equation

Work with a partner.

For 4 months, you saved $25 a month. You now have $175 in your savings account.

- Draw a graph that shows the balance in your account after t months.
- Use your result from Activity 2 to write an equation that represents the balance A after t months.

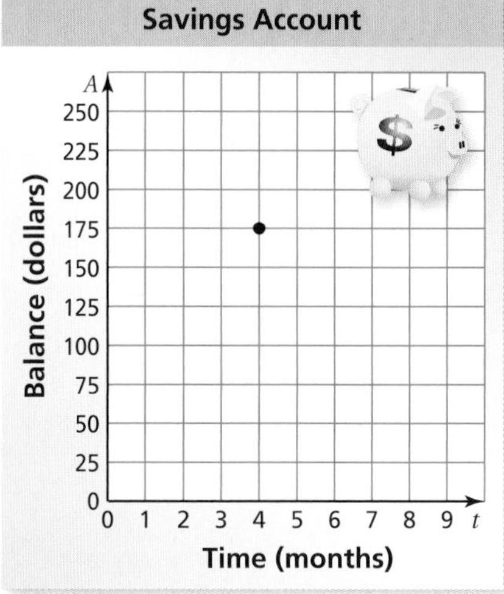

What Is Your Answer?

4. Redo Activity 1 using the equation you found in Activity 2. Compare the results. What do you notice?

5. Why do you think $y - y_1 = m(x - x_1)$ is called the *point-slope form* of the equation of a line? Why do you think it is important?

6. **IN YOUR OWN WORDS** How can you write an equation of a line when you are given the slope and a point on the line? Give an example that is different from those in Activity 1.

Practice Use what you learned about writing equations using a slope and a point to complete Exercises 3–5 on page 188.

4.7 Lesson

Key Vocabulary
point-slope form, *p. 186*

Key Idea

Point-Slope Form

Words A linear equation written in the form $y - y_1 = m(x - x_1)$ is in **point-slope form**. The line passes through the point (x_1, y_1), and the slope of the line is m.

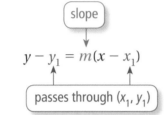

Algebra

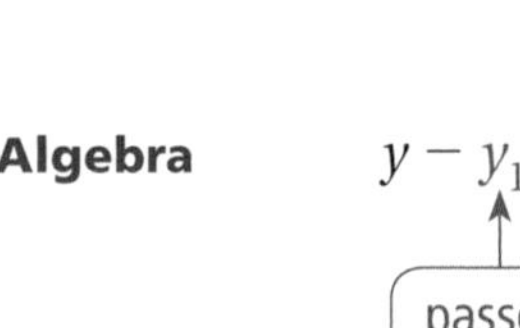

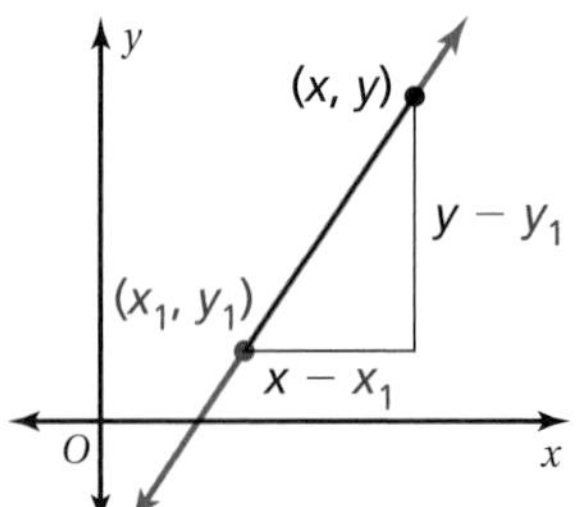

EXAMPLE 1 Writing an Equation Using a Slope and a Point

Write in point-slope form an equation of the line that passes through the point (−6, 1) with slope $\frac{2}{3}$.

$y - y_1 = m(x - x_1)$ Write the point-slope form.

$y - 1 = \frac{2}{3}[x - (-6)]$ Substitute $\frac{2}{3}$ for m, −6 for x_1, and 1 for y_1.

$y - 1 = \frac{2}{3}(x + 6)$ Simplify.

So, the equation is $y - 1 = \frac{2}{3}(x + 6)$.

Check Check that (−6, 1) is a solution of the equation.

$y - 1 = \frac{2}{3}(x + 6)$ Write the equation.

$1 - 1 \stackrel{?}{=} \frac{2}{3}(-6 + 6)$ Substitute.

$0 = 0$ ✓ Simplify.

On Your Own

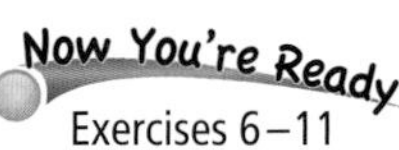

Write in point-slope form an equation of the line that passes through the given point and has the given slope.

1. $(1, 2)$; $m = -4$
2. $(7, 0)$; $m = 1$
3. $(-8, -5)$; $m = -\frac{3}{4}$

EXAMPLE 2 Writing an Equation Using Two Points

Write in slope-intercept form an equation of the line that passes through the points (2, 4) and (5, −2).

Find the slope: $m = \dfrac{y_2 - y_1}{x_2 - x_1} = \dfrac{-2 - 4}{5 - 2} = \dfrac{-6}{3} = -2$

Then use the slope $m = -2$ and the point (2, 4) to write an equation of the line.

$y - y_1 = m(x - x_1)$	Write the point-slope form.
$y - 4 = -2(x - 2)$	Substitute −2 for m, 2 for x_1, and 4 for y_1.
$y - 4 = -2x + 4$	Distributive Property
$y = -2x + 8$	Write in slope-intercept form.

Study Tip

You can use either of the given points to write the equation of the line.

Use $m = -2$ and $(5, -2)$.

$y - (-2) = -2(x - 5)$

$y + 2 = -2x + 10$

$y = -2x + 8$ ✓

EXAMPLE 3 Real-Life Application

You finish parasailing and are being pulled back to the boat. After 2 seconds, you are 25 feet above the boat. (a) Write and graph an equation that represents your height *y* (in feet) above the boat after *x* seconds. (b) At what height were you parasailing?

a. You are being pulled down at the rate of 10 feet per second. So, the slope is −10. You are 25 feet above the boat after 2 seconds. So, the line passes through (2, 25). Use the point-slope form.

$y - 25 = -10(x - 2)$	Substitute for m, x_1, and y_1.
$y - 25 = -10x + 20$	Distributive Property
$y = -10x + 45$	Write in slope-intercept form.

∴ So, the equation is $y = -10x + 45$.

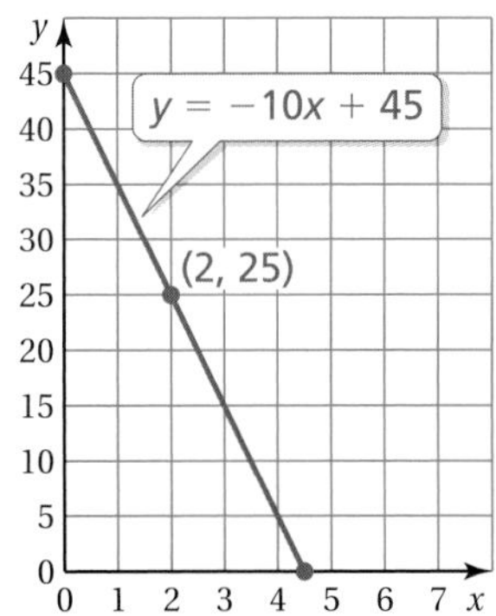

b. You start descending when $x = 0$. The y-intercept is 45. So, you were parasailing at a height of 45 feet.

On Your Own

Now You're Ready
Exercises 12–17

Write in slope-intercept form an equation of the line that passes through the given points.

4. (−2, 1), (3, −4) **5.** (−5, −5), (−3, 3) **6.** (−8, 6), (−2, 9)

7. WHAT IF? In Example 3, you are 35 feet above the boat after 2 seconds. Write and graph an equation that represents your height *y* (in feet) above the boat after *x* seconds.

4.7 Exercises

Vocabulary and Concept Check

1. **VOCABULARY** From the equation $y - 3 = -2(x + 1)$, identify the slope and a point on the line.

2. **WRITING** Describe how to write an equation of a line using (a) its slope and a point on the line and (b) two points on the line.

Practice and Problem Solving

Use the point-slope form to write an equation of the line with the given slope that passes through the given point.

3. $m = \frac{1}{2}$

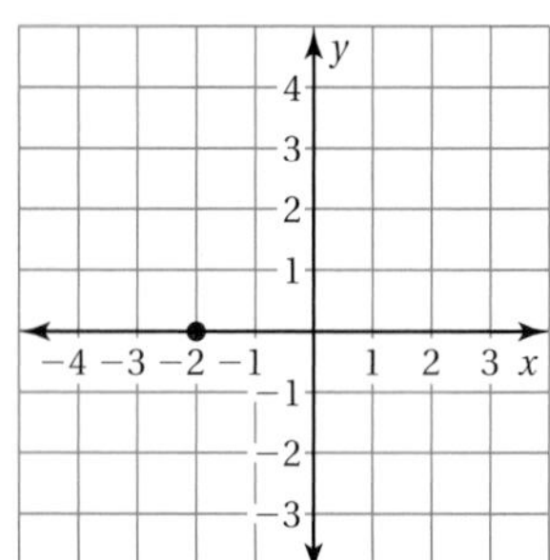

4. $m = -\frac{3}{4}$

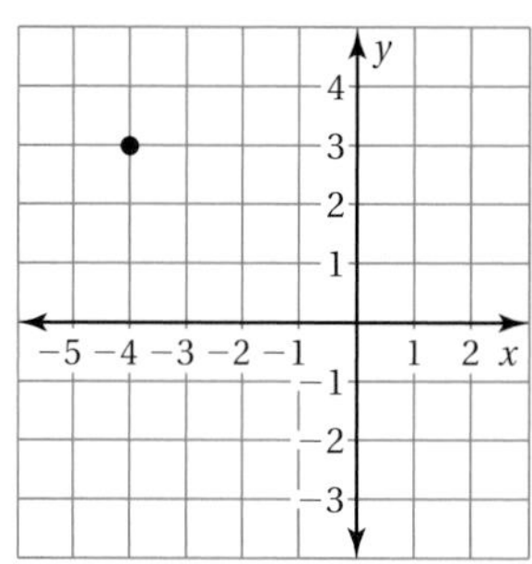

5. $m = -3$

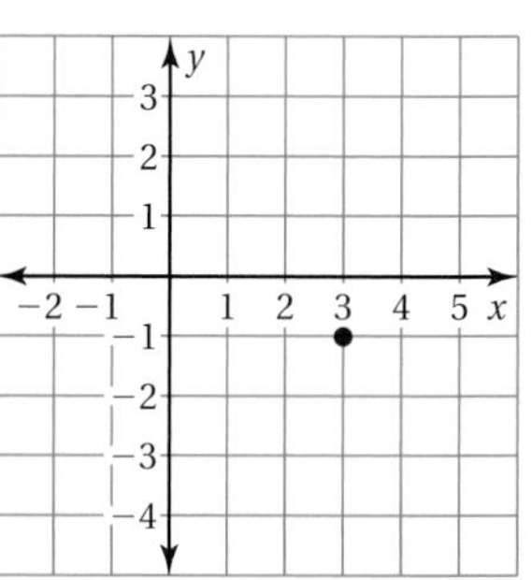

Write in point-slope form an equation of the line that passes through the given point and has the given slope.

1 6. $(3, 0);\ m = -\frac{2}{3}$

7. $(4, 8);\ m = \frac{3}{4}$

8. $(1, -3);\ m = 4$

9. $(7, -5);\ m = -\frac{1}{7}$

10. $(3, 3);\ m = \frac{5}{3}$

11. $(-1, -4);\ m = -2$

Write in slope-intercept form an equation of the line that passes through the given points.

2 12. $(-1, -1), (1, 5)$

13. $(2, 4), (3, 6)$

14. $(-2, 3), (2, 7)$

15. $(4, 1), (8, 2)$

16. $(-9, 5), (-3, 3)$

17. $(1, 2), (-2, -1)$

18. **CHEMISTRY** At 0 °C, the volume of a gas is 22 liters. For each degree the temperature T (in degrees Celsius) increases, the volume V (in liters) of the gas increases by $\frac{2}{25}$. Write an equation that represents the volume of the gas in terms of the temperature.

19. **CARS** After it is purchased, the value of a new car decreases \$4000 each year. After 3 years, the car is worth \$18,000.

a. Write an equation that represents the value V (in dollars) of the car x years after it is purchased.

b. What was the original value of the car?

20. **REASONING** Write an equation of a line that passes through the point (8, 2) that is (a) parallel and (b) perpendicular to the graph of the equation $y = 4x - 3$.

21. **CRICKETS** According to Dolbear's law, you can predict the temperature T (in degrees Fahrenheit) by counting the number x of chirps made by a snowy tree cricket in 1 minute. For each rise in temperature of 0.25°F, the cricket makes an additional chirp each minute.

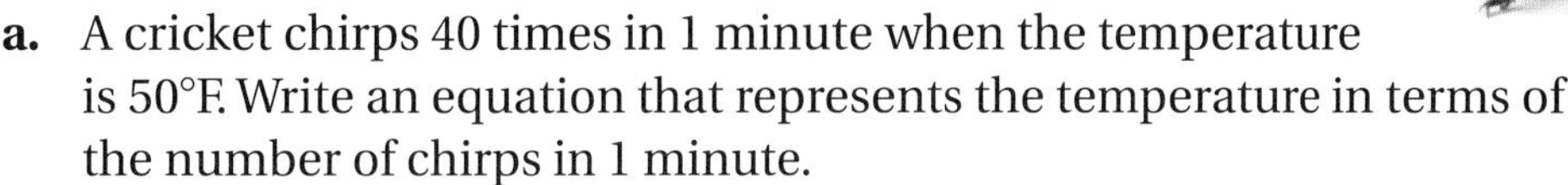

a. A cricket chirps 40 times in 1 minute when the temperature is 50°F. Write an equation that represents the temperature in terms of the number of chirps in 1 minute.

b. You count 100 chirps in 1 minute. What is the temperature?

c. The temperature is 96°F. How many chirps would you expect the cricket to make?

Leaning Tower of Pisa

22. **WATERING CAN** You water the plants in your classroom at a constant rate. After 5 seconds, your watering can contains 58 ounces of water. Fifteen seconds later, the can contains 28 ounces of water.

a. Write an equation that represents the amount y (in ounces) of water in the can after x seconds.

b. How much water was in the can when you started watering the plants?

c. When is the watering can empty?

23. Problem Solving The Leaning Tower of Pisa in Italy was built between 1173 and 1350.

a. Write an equation for the yellow line.

b. The tower is 56 meters tall. How far off center is the top of the tower?

Fair Game Review What you learned in previous grades & lessons

Graph the linear equation. *(Section 4.4)*

24. $y = 4x$

25. $y = -2x + 1$

26. $y = 3x - 5$

27. **MULTIPLE CHOICE** What is the x-intercept of the equation $3x + 5y = 30$? *(Section 4.5)*

Ⓐ -10 Ⓑ -6 Ⓒ 6 Ⓓ 10

4.4–4.7 Quiz

Find the slope and the y-intercept of the graph of the linear equation. *(Section 4.4)*

1. $y = \frac{1}{4}x - 8$

2. $y = -x + 3$

Find the x- and y-intercepts of the graph of the equation. *(Section 4.5)*

3. $3x - 2y = 12$

4. $x + 5y = 15$

Write an equation of the line in slope-intercept form. *(Section 4.6)*

5.

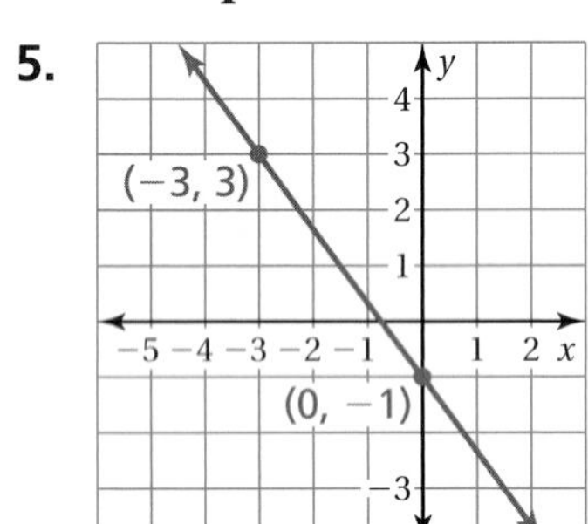

6.

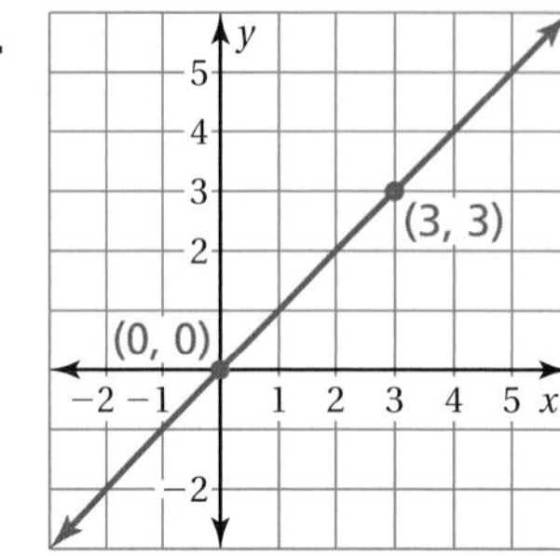

7.

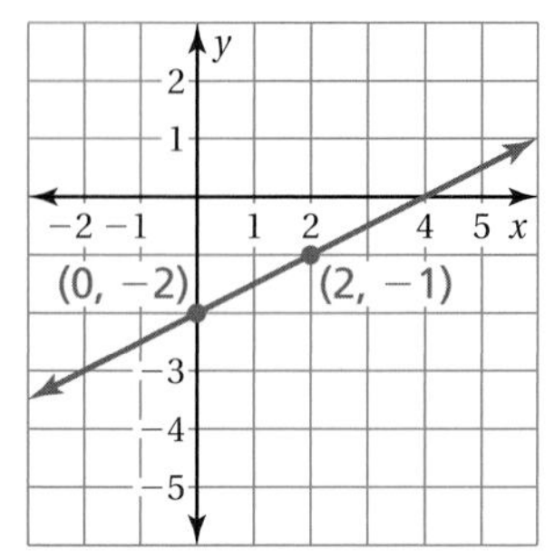

Write in point-slope form an equation of the line that passes through the given point and has the given slope. *(Section 4.7)*

8. $(1, 3);\ m = 2$

9. $(-3, -2);\ m = \frac{1}{3}$

10. $(-1, 4);\ m = -1$

11. $(8, -5);\ m = -\frac{1}{8}$

Write in slope-intercept form an equation of the line that passes through the given points. *(Section 4.7)*

12. $\left(0, -\frac{2}{3}\right) \left(-3, -\frac{2}{3}\right)$

13. $(4, 0), (0, 4)$

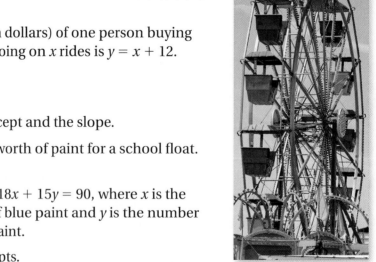

14. STATE FAIR The cost y (in dollars) of one person buying admission to a fair and going on x rides is $y = x + 12$. *(Section 4.4)*

a. Graph the equation.

b. Interpret the y-intercept and the slope.

15. PAINTING You used \$90 worth of paint for a school float. *(Section 4.5)*

a. Graph the equation $18x + 15y = 90$, where x is the number of gallons of blue paint and y is the number of gallons of white paint.

b. Interpret the intercepts.

16. CONSTRUCTION A construction crew is extending a highway sound barrier that is 13 miles long. The crew builds $\frac{1}{2}$ of a mile per week. Write an equation that represents the length y (in miles) of the barrier after x weeks. *(Section 4.6)*

4 Chapter Review

Check It Out
Vocabulary Help
BigIdeasMath.com

Review Key Vocabulary

linear equation *p. 144*
solution of a linear equation, *p. 144*
slope, *p. 150*
rise, *p. 150*
run, *p. 150*

x-intercept, *p. 168*
y-intercept, *p. 168*
slope-intercept form, *p. 168*
standard form, *p. 174*
point-slope form, *p. 186*

Review Examples and Exercises

4.1 Graphing Linear Equations *(pp. 142–147)*

Graph $y = 3x - 1$.

Step 1: Make a table of values.

x	$y = 3x - 1$	y	(x, y)
-2	$y = 3(-2) - 1$	-7	$(-2, -7)$
-1	$y = 3(-1) - 1$	-4	$(-1, -4)$
0	$y = 3(0) - 1$	-1	$(0, -1)$
1	$y = 3(1) - 1$	2	$(1, 2)$

Step 2: Plot the ordered pairs.

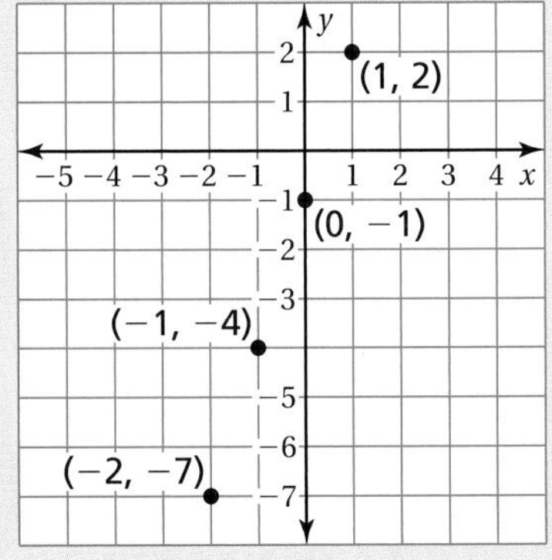

Step 3: Draw a line through the points.

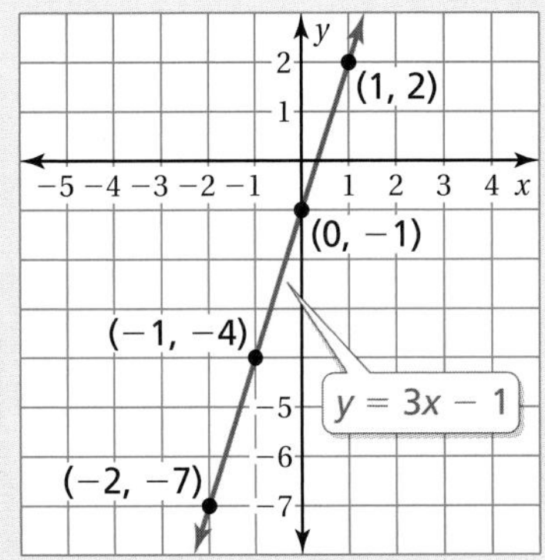

Exercises

Graph the linear equation.

1. $y = \frac{3}{5}x$

2. $y = -2$

3. $y = 9 - x$

4. $y = 1$

5. $y = \frac{2}{3}x + 2$

6. $x = -5$

4.2 Slope of a Line (pp. 148–157)

Find the slope of each line in the graph.

Red Line: $m = \dfrac{y_2 - y_1}{x_2 - x_1} = \dfrac{5 - (-3)}{2 - 2} = \dfrac{8}{0}$

∴ The slope of the red line is undefined.

Blue Line: $m = \dfrac{y_2 - y_1}{x_2 - x_1} = \dfrac{-1 - 2}{4 - (-3)} = \dfrac{-3}{7}$, or $-\dfrac{3}{7}$

Green Line: $m = \dfrac{y_2 - y_1}{x_2 - x_1} = \dfrac{4 - 4}{5 - 0} = \dfrac{0}{5}$, or 0

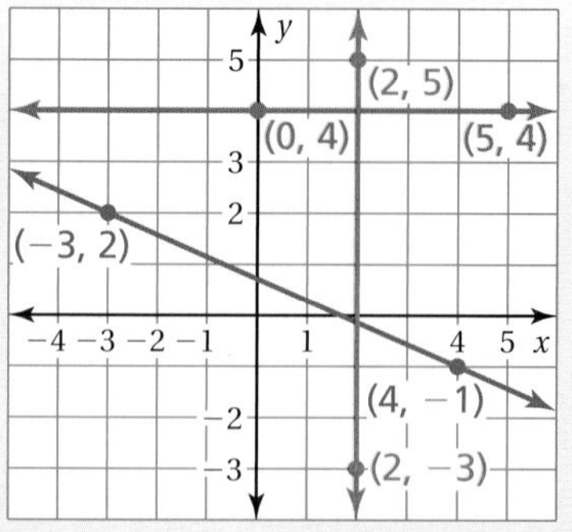

Exercises

The points in the table lie on a line. Find the slope of the line.

7.

x	0	1	2	3
y	−1	0	1	2

8.

x	−2	0	2	4
y	3	4	5	6

9. Are the lines $x = 2$ and $y = 4$ parallel? Are they perpendicular? Explain.

4.3 Graphing Proportional Relationships (pp. 158–163)

The cost y (in dollars) for x tickets to a movie is represented by the equation $y = 7x$. Graph the equation and interpret the slope.

The equation shows that the slope m is 7. So, the graph passes through (0, 0) and (1, 7).

Plot the points and draw a line through the points. Because negative values of x do not make sense in this context, graph in the first quadrant only.

∴ The slope indicates that the unit cost is $7 per ticket.

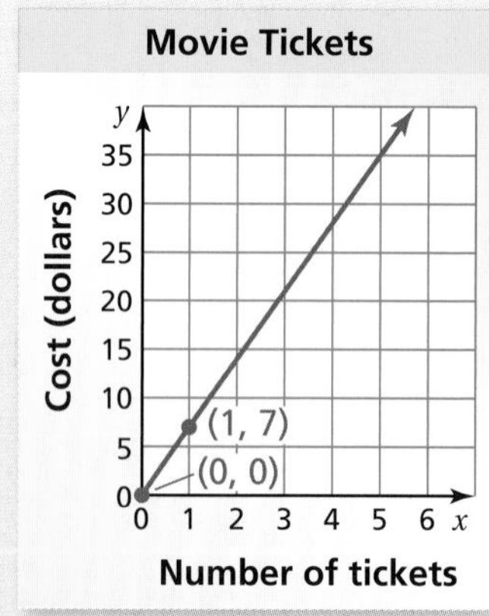

Exercises

10. RUNNING The number y of miles you run after x weeks is represented by the equation $y = 8x$. Graph the equation and interpret the slope.

11. STUDYING The number y of hours that you study after x days is represented by the equation $y = 1.5x$. Graph the equation and interpret the slope.

4.4 Graphing Linear Equations in Slope-Intercept Form (pp. 166–171)

Graph $y = 0.5x - 3$. Identify the x-intercept.

Step 1: Find the slope and the y-intercept.

$$y = 0.5x + (-3)$$

slope → 0.5; y-intercept → −3

Step 2: The y-intercept is −3. So, plot (0, −3).

Step 3: Use the slope to find another point and draw the line.

$$m = \frac{\text{rise}}{\text{run}} = \frac{1}{2}$$

Plot the point that is 2 units right and 1 unit up from (0, –3). Draw a line through the two points.

∴ The line crosses the x-axis at (6, 0). So, the x-intercept is 6.

Exercises

Graph the linear equation. Identify the x-intercept. Use a graphing calculator to check your answer.

12. $y = 2x - 6$ **13.** $y = -4x + 8$ **14.** $y = -x - 8$

4.5 Graphing Linear Equations in Standard Form (pp. 172–177)

Graph $8x + 4y = 16$.

Step 1: Write the equation in slope-intercept form.

$8x + 4y = 16$	Write the equation.
$4y = -8x + 16$	Subtract $8x$ from each side.
$y = -2x + 4$	Divide each side by 4.

Step 2: Use the slope and the y-intercept to graph the equation.

$$y = -2x + 4$$

slope → −2; y-intercept → 4

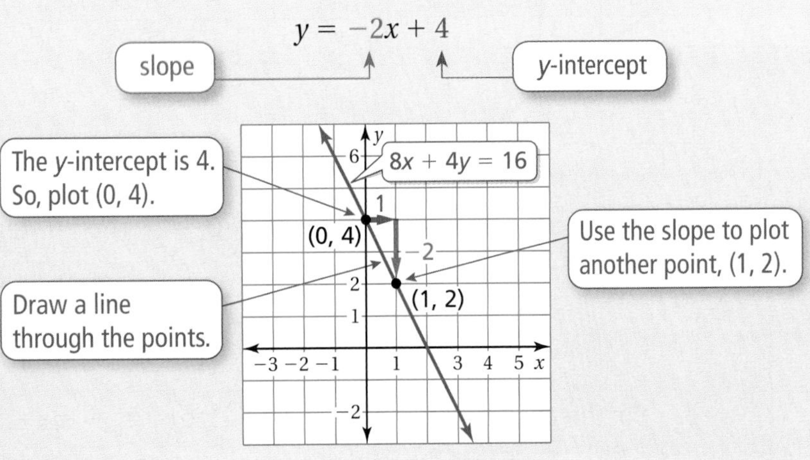
The y-intercept is 4. So, plot (0, 4).
Draw a line through the points.
8x + 4y = 16
(0, 4)
(1, 2)
Use the slope to plot another point, (1, 2).

Exercises

Graph the linear equation.

15. $\frac{1}{4}x + y = 3$

16. $-4x + 2y = 8$

17. $x + 5y = 10$

18. $-\frac{1}{2}x + \frac{1}{8}y = \frac{3}{4}$

19. A dog kennel charges \$30 per night to board your dog and \$6 for each hour of playtime. The amount of money you spend is given by $30x + 6y = 180$, where x is the number of nights and y is the number of hours of playtime. Graph the equation and interpret the intercepts.

4.6 Writing Equations in Slope-Intercept Form *(pp. 178–183)*

Write an equation of the line in slope-intercept form.

a.

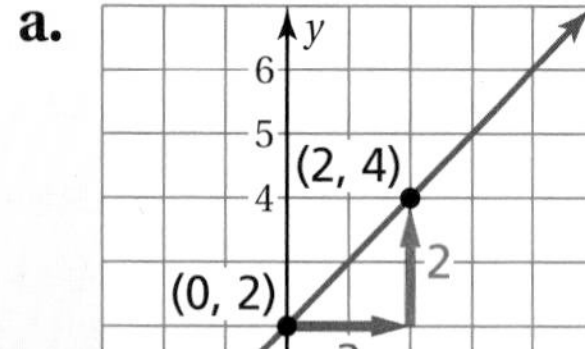

Find the slope and the y-intercept.

$$m = \frac{y_2 - y_1}{x_2 - x_1} = \frac{4 - 2}{2 - 0} = \frac{2}{2}, \text{ or } 1$$

Because the line crosses the y-axis at (0, 2), the y-intercept is 2.

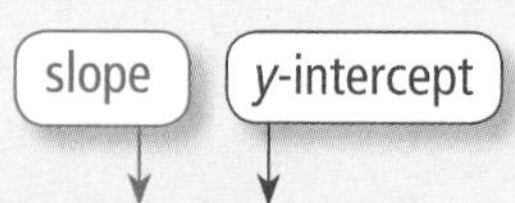

So, the equation is $y = 1x + 2$, or $y = x + 2$.

b.

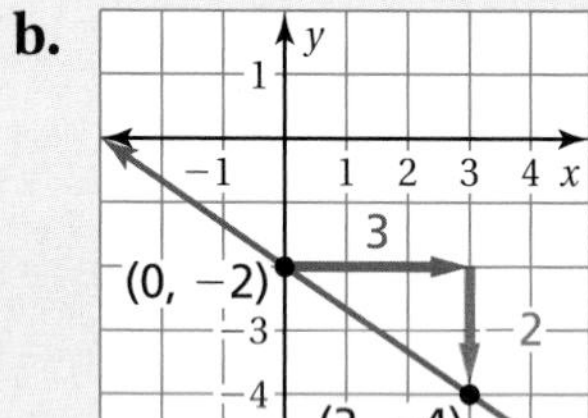

Find the slope and the y-intercept.

$$m = \frac{y_2 - y_1}{x_2 - x_1} = \frac{-4 - (-2)}{3 - 0} = \frac{-2}{3}, \text{ or } -\frac{2}{3}$$

Because the line crosses the y-axis at (0, −2), the y-intercept is −2.

slope
y-intercept

So, the equation is $y = -\frac{2}{3}x + (-2)$, or $y = -\frac{2}{3}x - 2$.

Exercises

Write an equation of the line in slope-intercept form.

20. 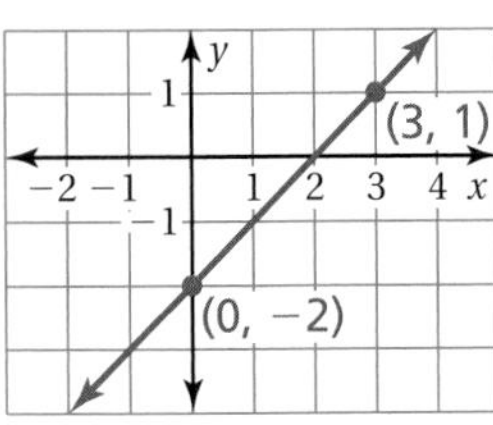

21. 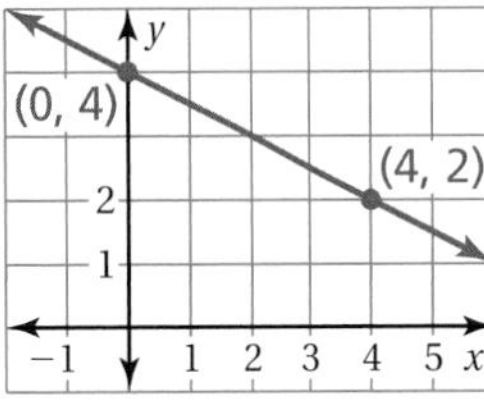

22. 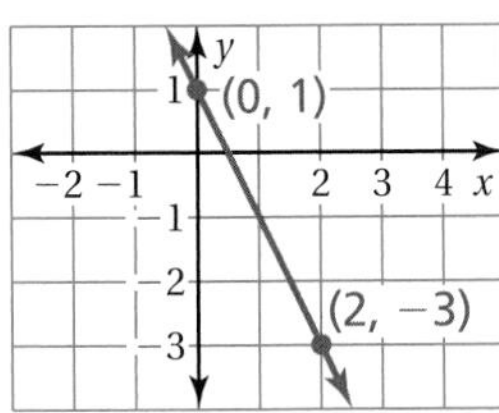

23. 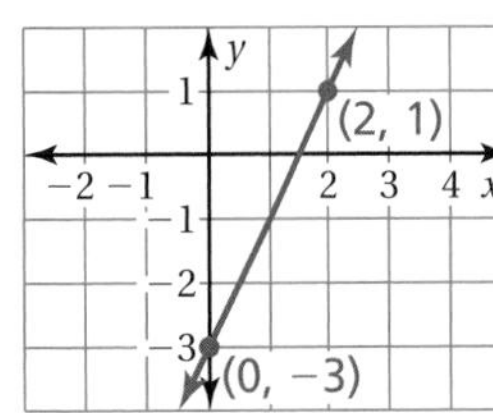

24. Write an equation of the line that passes through (0, 8) and (6, 8).

25. Write an equation of the line that passes through (0, −5) and (−5, −5).

4.7 Writing Equations in Point-Slope Form *(pp. 184–189)*

Write in slope-intercept form an equation of the line that passes through the points (2, 1) and (3, 5).

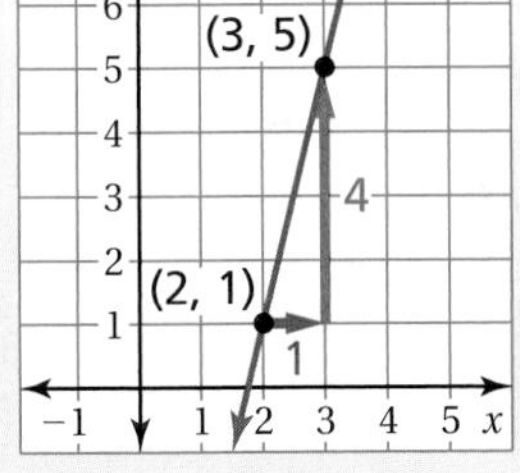

Find the slope.

$$m = \frac{y_2 - y_1}{x_2 - x_1} = \frac{5 - 1}{3 - 2} = \frac{4}{1}, \text{ or } 4$$

Then use the slope and one of the given points to write an equation of the line.

Use $m = 4$ and (2, 1).

$y - y_1 = m(x - x_1)$	Write the point-slope form.
$y - 1 = 4(x - 2)$	Substitute 4 for m, 2 for x_1, and 1 for y_1.
$y - 1 = 4x - 8$	Distributive Property
$y = 4x - 7$	Write in slope-intercept form.

So, the equation is $y = 4x - 7$.

Exercises

26. Write in point-slope form an equation of the line that passes through the point (4, 4) with slope 3.

27. Write in slope-intercept form an equation of the line that passes through the points (−4, 2) and (6, −3).

4 Chapter Test

Find the slope and the *y*-intercept of the graph of the linear equation.

1. $y = 6x - 5$
2. $y = 20x + 15$
3. $y = -5x - 16$
4. $y - 1 = 3x + 8.4$
5. $y + 4.3 = 0.1x$
6. $-\frac{1}{2}x + 2y = 7$

Graph the linear equation.

7. $y = 2x + 4$
8. $y = -\frac{1}{2}x - 5$
9. $-3x + 6y = 12$

10. Which lines are parallel? Which lines are perpendicular? Explain.

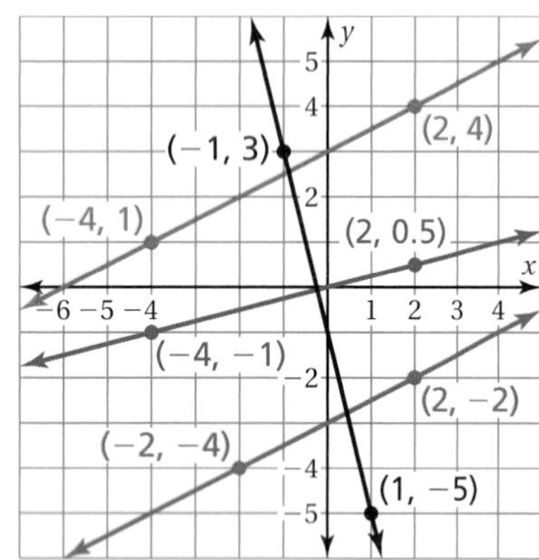

11. The points in the table lie on a line. Find the slope of the line.

x	y
−1	−4
0	−1
1	2
2	5

Write an equation of the line in slope-intercept form.

12.

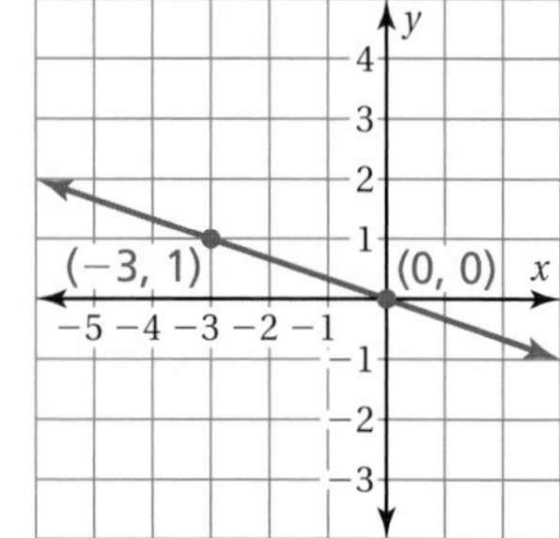

13.

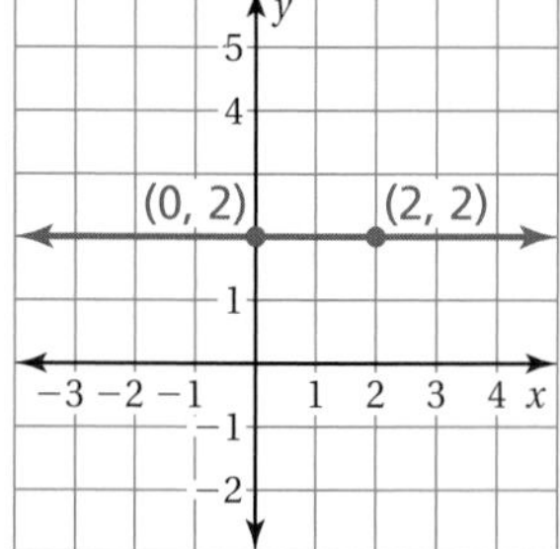

Write in slope-intercept form an equation of the line that passes through the given points.

14. $(-1, 5), (3, -3)$
15. $(-4, 1), (4, 3)$
16. $(-2, 5), (-1, 1)$

17. **VOCABULARY** The number y of new vocabulary words that you learn after x weeks is represented by the equation $y = 15x$.
    a. Graph the equation and interpret the slope.
    b. How many new vocabulary words do you learn after 5 weeks?
    c. How many more vocabulary words do you learn after 6 weeks than after 4 weeks?

4 Cumulative Assessment

1. Which equation matches the line shown in the graph?

A. $y = 2x - 2$

B. $y = 2x + 1$

C. $y = x - 2$

D. $y = x + 1$

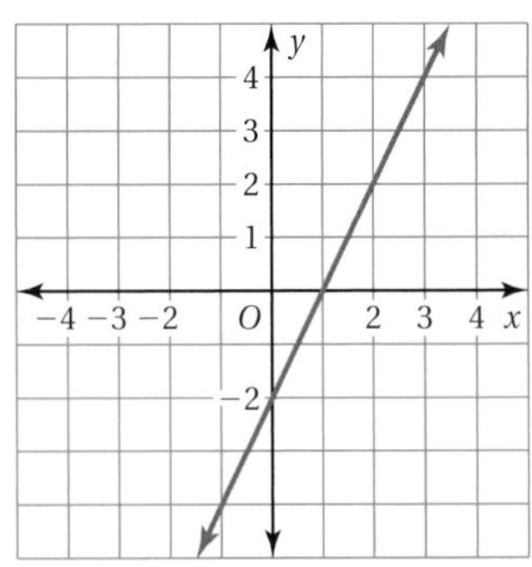

2. The equation $6x - 5y = 14$ is written in standard form. Which point lies on the graph of this equation?

F. $(-4, -1)$

G. $(-2, 4)$

H. $(-1, -4)$

I. $(4, -2)$

3. Which line has a slope of 0?

A.

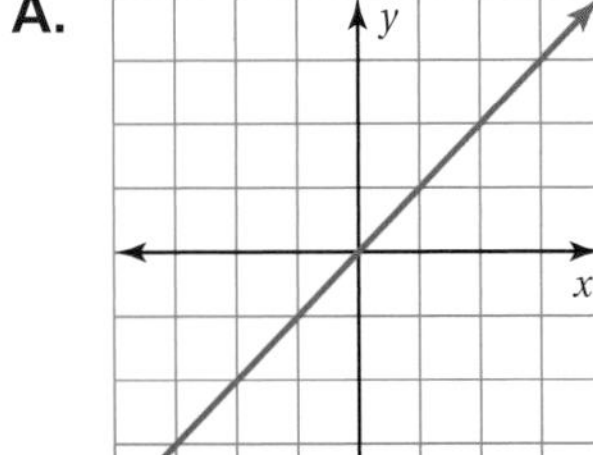

B.

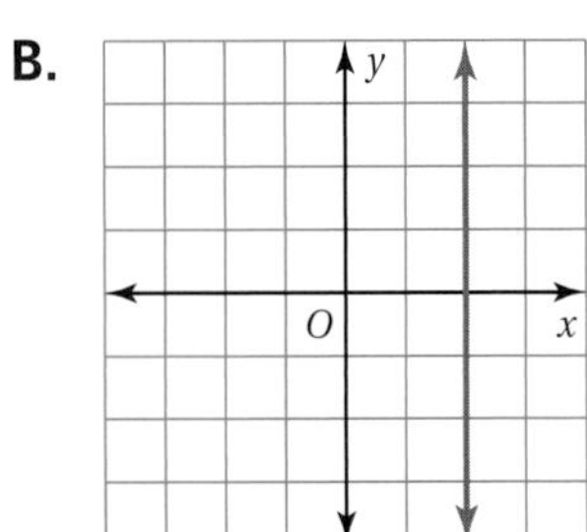

C.

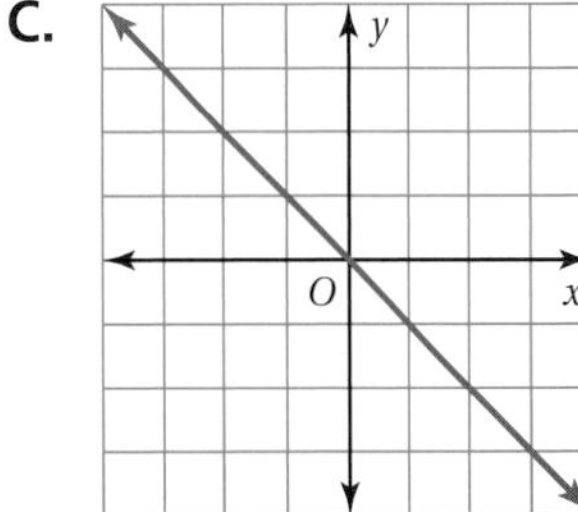

D.

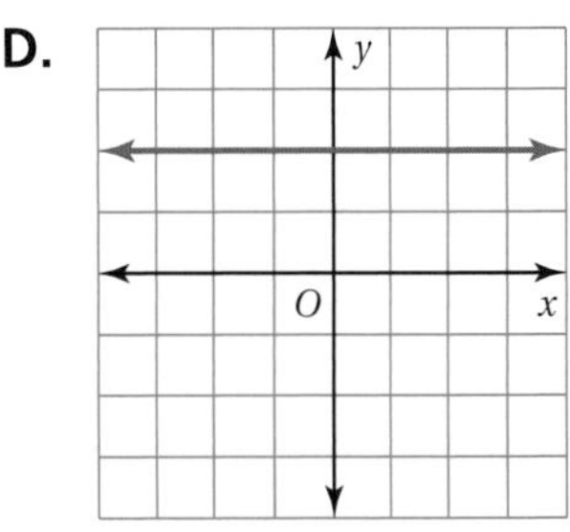

4. Which of the following is the equation of a line perpendicular to the line shown in the graph?

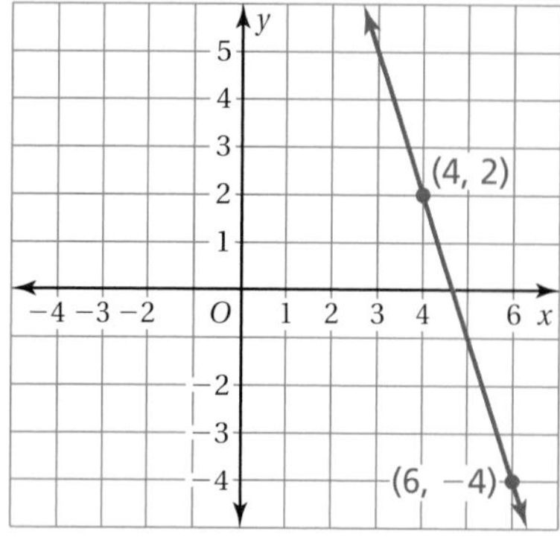

F. $y = 3x - 10$

H. $y = -3x + 5$

G. $y = \frac{1}{3}x + 12$

I. $y = -\frac{1}{3}x - 18$

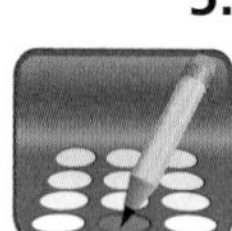

5. What is the slope of the line that passes through the points $(2, -2)$ and $(8, 1)$?

6. A cell phone plan costs \$10 per month plus \$0.10 for each minute used. Last month, you spent \$18.50 using this plan. This can be modeled by the equation below, where m represents the number of minutes used.

$$0.1m + 10 = 18.5$$

How many minutes did you use last month?

A. 8.4 min

C. 185 min

B. 85 min

D. 285 min

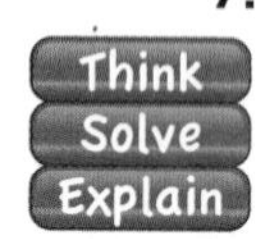

7. It costs \$40 to rent a car for one day. In addition, the rental agency charges you for each mile driven, as shown in the graph.

Part A Determine the slope of the line joining the points on the graph.

Part B Explain what the slope represents.

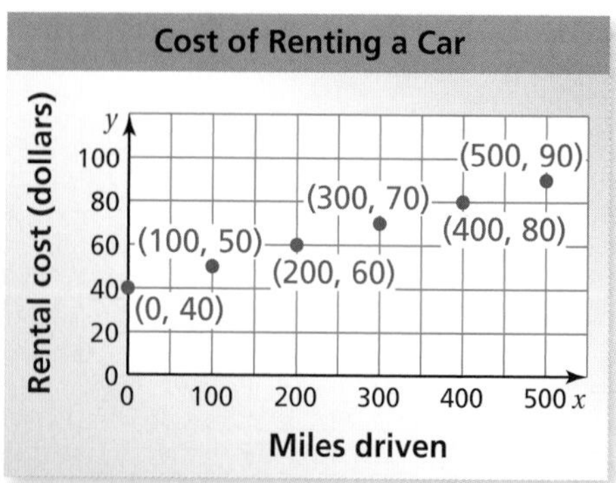

8. What value of x makes the equation below true?

$$7 + 2x = 4x - 5$$

9. Trapezoid *KLMN* is graphed in the coordinate plane shown.

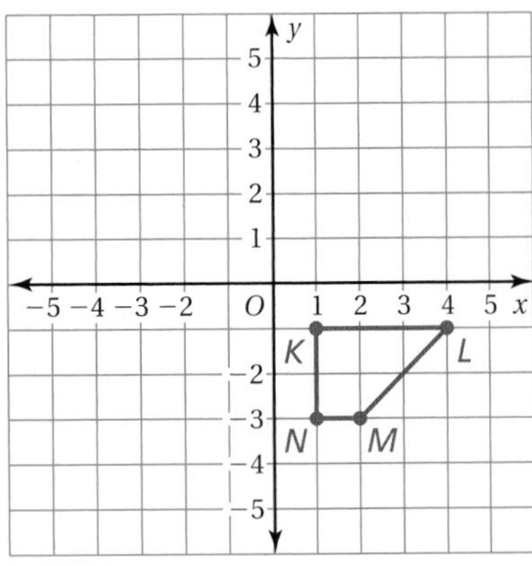

Rotate Trapezoid *KLMN* 90° clockwise about the origin. What are the coordinates of point M', the image of point M after the rotation?

F. $(-3, -2)$

G. $(-2, -3)$

H. $(-2, 3)$

I. $(3, 2)$

10. Solve the formula $K = 3M - 7$ for M.

A. $M = K + 7$

B. $M = \dfrac{K + 7}{3}$

C. $M = \dfrac{K}{3} + 7$

D. $M = \dfrac{K - 7}{3}$

11. What is the distance d across the canyon?

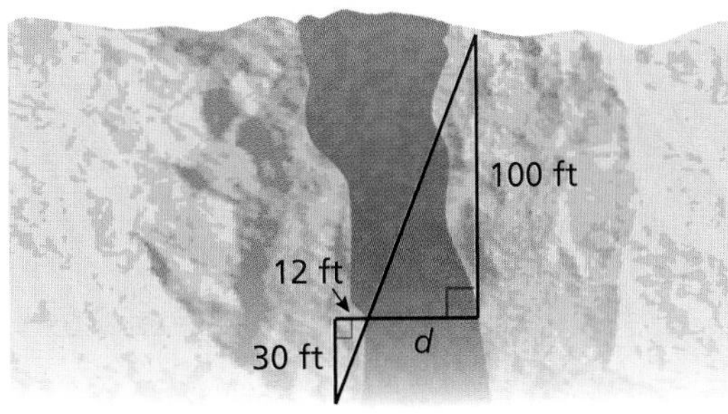

F. 3.6 ft

G. 12 ft

H. 40 ft

I. 250 ft

5 Systems of Linear Equations

5.1 Solving Systems of Linear Equations by Graphing

5.2 Solving Systems of Linear Equations by Substitution

5.3 Solving Systems of Linear Equations by Elimination

5.4 Solving Special Systems of Linear Equations

"Can you graph a system of linear equations that shows the number of biscuits and treats that I am going to share with you?"

"Hey, look over here. Can you estimate the solution of the system of linear equations that I made with these cattails?"

What You Learned Before

"Hold your tail a bit lower."

Combining Like Terms

Example 1 **Simplify each expression.**

a. $4x + 7 + 5x - 2$

$4x + 7 + 5x - 2 = 4x + 5x + 7 - 2$ — Commutative Property of Addition

$= (4 + 5)x + 7 - 2$ — Distributive Property

$= 9x + 5$ — Simplify.

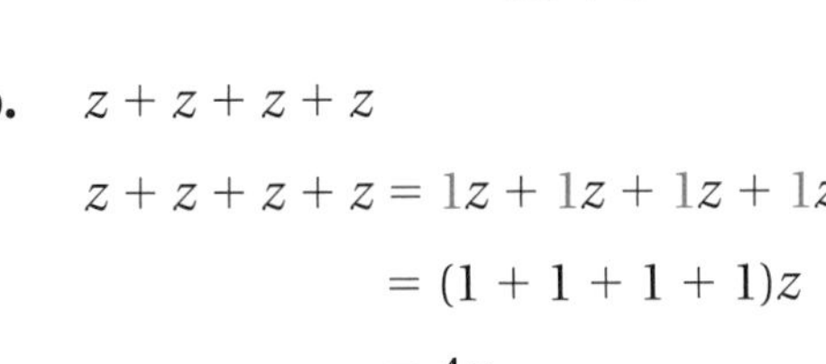

b. $z + z + z + z$

$z + z + z + z = 1z + 1z + 1z + 1z$ — Multiplication Property of One

$= (1 + 1 + 1 + 1)z$ — Distributive Property

$= 4z$ — Add coefficients.

Try It Yourself

Simplify the expression.

1. $5 + 4z - 2z$

2. $5(c + 8) + c + 3$

Solving Multi-Step Equations

Example 2 **Solve $4x - 2(3x + 1) = 16$.**

$4x - 2(3x + 1) = 16$ — Write the equation.

$4x - 6x - 2 = 16$ — Distributive Property

$-2x - 2 = 16$ — Combine like terms.

$-2x = 18$ — Add 2 to each side.

$x = -9$ — Divide each side by -2.

The solution is $x = -9$.

Try It Yourself

Solve the equation. Check your solution.

3. $-5x + 8 = -7$

4. $7w + w - 15 = 17$

5. $-3(z - 8) + 10 = -5$

6. $2 = 10c - 4(2c - 9)$

5.1 Solving Systems of Linear Equations by Graphing

Essential Question How can you solve a system of linear equations?

1 ACTIVITY: Writing a System of Linear Equations

Work with a partner.

Your family starts a bed-and-breakfast. It spends \$500 fixing up a bedroom to rent. The cost for food and utilities is \$10 per night. Your family charges \$60 per night to rent the bedroom.

a. Write an equation that represents the costs.

Cost, C (in dollars)	=	\$10 per night	•	Number of nights, x	+	\$500

b. Write an equation that represents the revenue (income).

Revenue, R (in dollars)	=	\$60 per night	•	Number of nights, x

c. A set of two (or more) linear equations is called a **system of linear equations**. Write the system of linear equations for this problem.

2 ACTIVITY: Using a Table to Solve a System

Work with a partner. Use the cost and revenue equations from Activity 1 to find how many nights your family needs to rent the bedroom before recovering the cost of fixing up the bedroom. This is the *break-even point*.

a. Copy and complete the table.

x	0	1	2	3	4	5	6	7	8	9	10	11
C												
R												

b. How many nights does your family need to rent the bedroom before breaking even?

Systems of Equations
In this lesson, you will
- write and solve systems of linear equations by graphing.
- solve real-life problems.

3 ACTIVITY: Using a Graph to Solve a System

Work with a partner.

a. Graph the cost equation from Activity 1.

b. In the same coordinate plane, graph the revenue equation from Activity 1.

c. Find the point of intersection of the two graphs. What does this point represent? How does this compare to the break-even point in Activity 2? Explain.

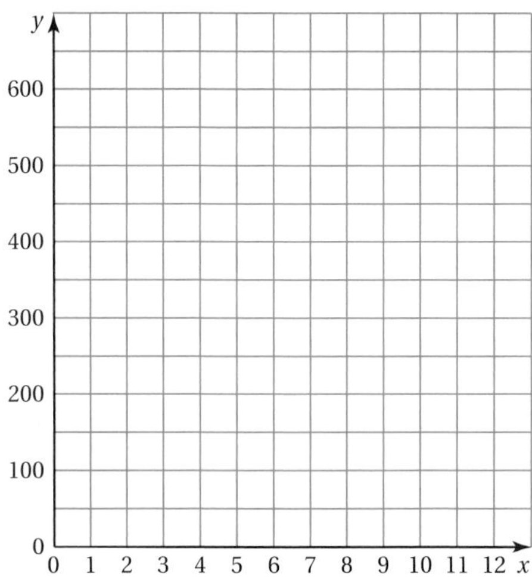

4 ACTIVITY: Using a Graphing Calculator

Work with a partner. Use a graphing calculator to solve the system.

$y = 10x + 500$ Equation 1

$y = 60x$ Equation 2

a. Enter the equations into your calculator. Then graph the equations. What is an appropriate window?

b. On your graph, how can you determine which line is the graph of which equation? Label the equations on the graph shown.

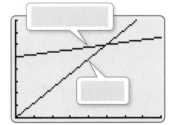

c. Visually estimate the point of intersection of the graphs.

d. To find the solution, use the *intersect* feature to find the point of intersection. The solution is (▢ , ▢).

Math Practice

Use Technology to Explore

How do you decide the values for the viewing window of your calculator? What other viewing windows could you use?

What Is Your Answer?

5. **IN YOUR OWN WORDS** How can you solve a system of linear equations? How can you check your solution?

6. **CHOOSE TOOLS** Solve one of the systems by using a table, another system by sketching a graph, and the remaining system by using a graphing calculator. Explain why you chose each method.

a. $y = 4.3x + 1.2$
$y = -1.7x - 2.4$

b. $y = x$
$y = -2x + 9$

c. $y = -x - 5$
$y = 3x + 1$

Practice Use what you learned about systems of linear equations to complete Exercises 4–6 on page 206.

5.1 Lesson

Key Vocabulary

system of linear equations, *p. 204*

solution of a system of linear equations, *p. 204*

A **system of linear equations** is a set of two or more linear equations in the same variables. An example is shown below.

$y = x + 1$ Equation 1

$y = 2x - 7$ Equation 2

A **solution of a system of linear equations** in two variables is an ordered pair that is a solution of each equation in the system. The solution of a system of linear equations is the point of intersection of the graphs of the equations.

Key Idea

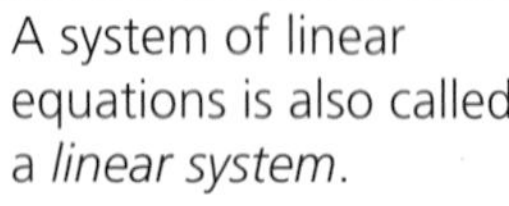

Reading

A system of linear equations is also called a *linear system*.

Solving a System of Linear Equations by Graphing

Step 1: Graph each equation in the same coordinate plane.

Step 2: Estimate the point of intersection.

Step 3: Check the point from Step 2 by substituting for x and y in each equation of the original system.

EXAMPLE 1 Solving a System of Linear Equations by Graphing

Solve the system by graphing. $\mathbf{y = 2x + 5}$ Equation 1

$\mathbf{y = -4x - 1}$ Equation 2

Step 1: Graph each equation.

Step 2: Estimate the point of intersection. The graphs appear to intersect at $(-1, 3)$.

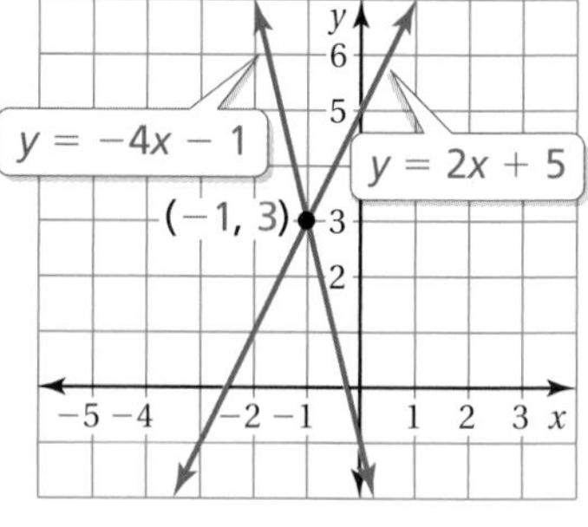

Step 3: Check the point from Step 2.

Equation 1	Equation 2
$y = 2x + 5$	$y = -4x - 1$
$3 \stackrel{?}{=} 2(-1) + 5$	$3 \stackrel{?}{=} -4(-1) - 1$
$3 = 3$ ✓	$3 = 3$ ✓

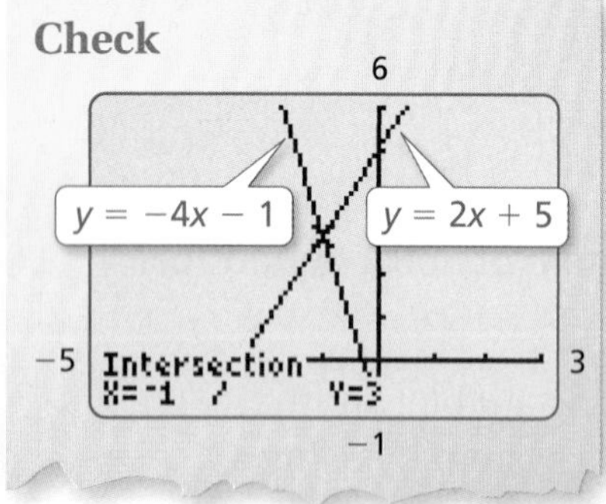

The solution is $(-1, 3)$.

On Your Own

Now You're Ready
Exercises 10–12

Solve the system of linear equations by graphing.

1. $y = x - 1$
   $y = -x + 3$

2. $y = -5x + 14$
   $y = x - 10$

3. $y = x$
   $y = 2x + 1$

EXAMPLE 2 Real-Life Application

A kicker on a football team scores 1 point for making an extra point and 3 points for making a field goal. The kicker makes a total of 8 extra points and field goals in a game and scores 12 points. Write and solve a system of linear equations to find the number x of extra points and the number y of field goals.

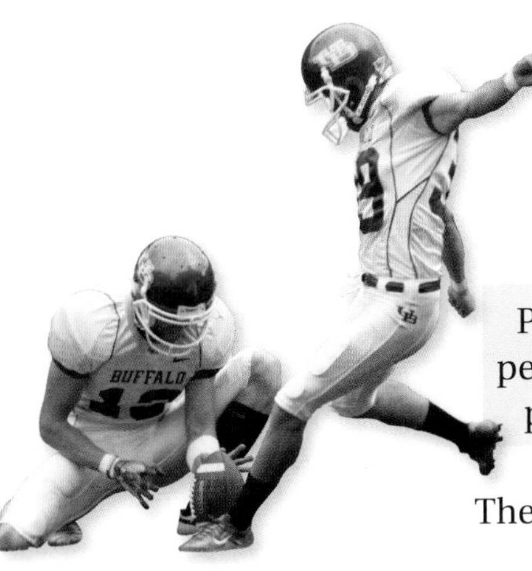

Use a verbal model to write a system of linear equations.

Number of extra points, x	+	Number of field goals, y	=	Total number of kicks

Points per extra point	•	Number of extra points, x	+	Points per field goal	•	Number of field goals, y	=	Total number of points

The system is: $x + y = 8$ Equation 1

$x + 3y = 12$ Equation 2

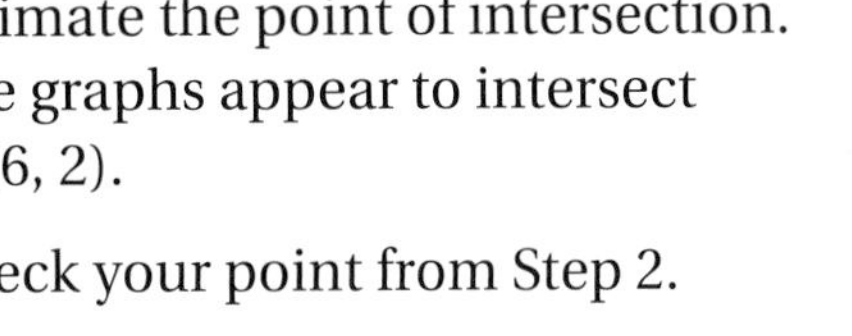

Step 1: Graph each equation.

Step 2: Estimate the point of intersection. The graphs appear to intersect at (6, 2).

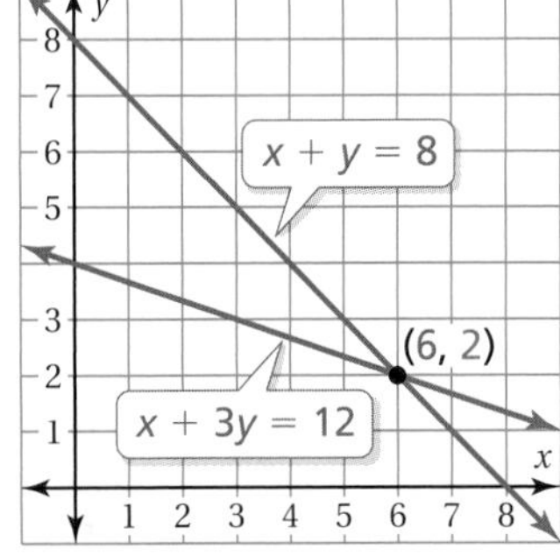

Step 3: Check your point from Step 2.

Equation 1	Equation 2
$x + y = 8$	$x + 3y = 12$
$6 + 2 \stackrel{?}{=} 8$	$6 + 3(2) \stackrel{?}{=} 12$
$8 = 8$ ✓	$12 = 12$ ✓

∴ The solution is (6, 2). So, the kicker made 6 extra points and 2 field goals.

Check

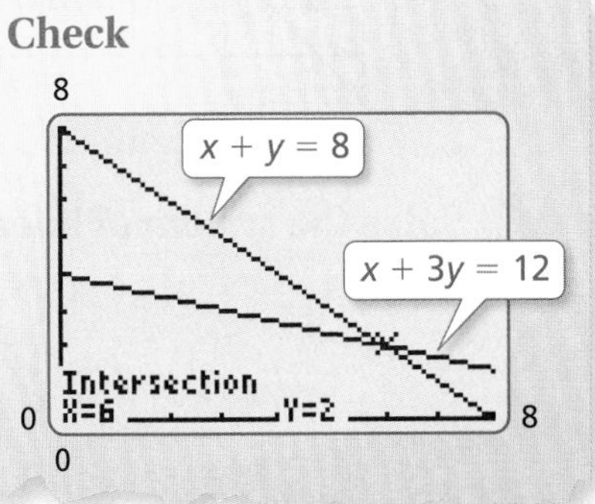

Study Tip

It may be easier to graph the equations in a system by rewriting the equations in slope-intercept form.

On Your Own

Now You're Ready
Exercises 13–15

Solve the system of linear equations by graphing.

4. $y = -4x - 7$
   $x + y = 2$

5. $x - y = 5$
   $-3x + y = -1$

6. $\frac{1}{2}x + y = -6$
   $6x + 2y = 8$

7. **WHAT IF?** The kicker makes a total of 7 extra points and field goals and scores 17 points. Write and solve a system of linear equations to find the numbers of extra points and field goals.

5.1 Exercises

Vocabulary and Concept Check

1. **VOCABULARY** Do the equations $4x - 3y = 5$ and $7y + 2x = -8$ form a system of linear equations? Explain.
2. **WRITING** What does it mean to solve a system of equations?
3. **WRITING** You graph a system of linear equations, and the solution appears to be (3, 4). How can you verify that the solution is (3, 4)?

Practice and Problem Solving

Use a table to find the break-even point. Check your solution.

4. $C = 15x + 150$
 $R = 45x$
5. $C = 24x + 80$
 $R = 44x$
6. $C = 36x + 200$
 $R = 76x$

Match the system of linear equations with the corresponding graph. Use the graph to estimate the solution. Check your solution.

7. $y = 1.5x - 2$
 $y = -x + 13$
8. $y = x + 4$
 $y = 3x - 1$
9. $y = \frac{2}{3}x - 3$
 $y = -2x + 5$

A. 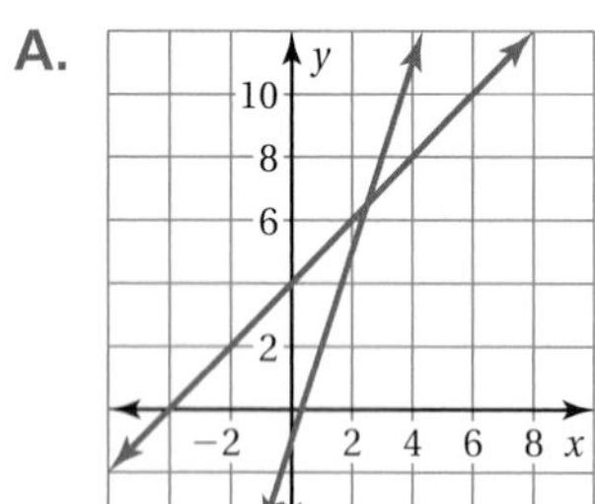

B. 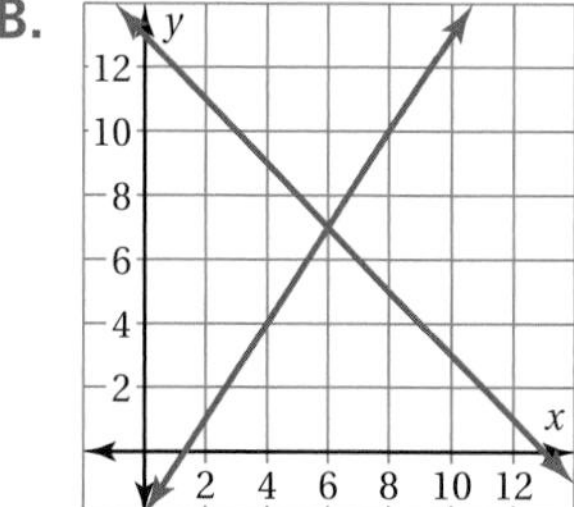

C. 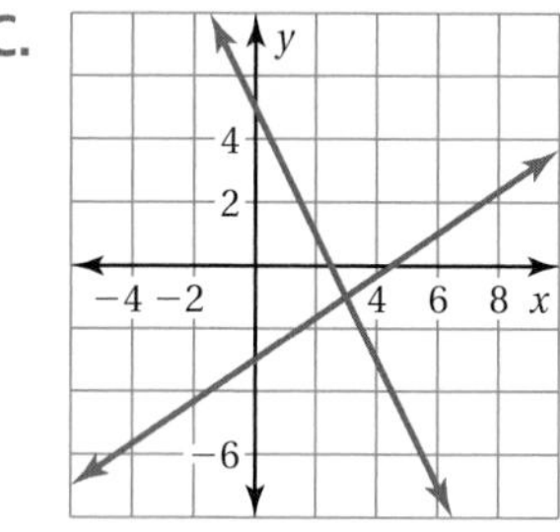

Solve the system of linear equations by graphing.

1 10. $y = 2x + 9$
 $y = 6 - x$

11. $y = -x - 4$
 $y = \frac{3}{5}x + 4$

12. $y = 2x + 5$
 $y = \frac{1}{2}x - 1$

2 13. $x + y = 27$
 $y = x + 3$

14. $y - x = 17$
 $y = 4x + 2$

15. $x - y = 7$
 $0.5x + y = 5$

16. **CARRIAGE RIDES** The cost C (in dollars) for the care and maintenance of a horse and carriage is $C = 15x + 2000$, where x is the number of rides.
    a. Write an equation for the revenue R in terms of the number of rides.
    b. How many rides are needed to break even?

Use a graphing calculator to solve the system of linear equations.

17. $2.2x + y = 12.5$
$1.4x - 4y = 1$

18. $2.1x + 4.2y = 14.7$
$-5.7x - 1.9y = -11.4$

19. $-1.1x - 5.5y = -4.4$
$0.8x - 3.2y = -11.2$

20. ERROR ANALYSIS Describe and correct the error in solving the system of linear equations.

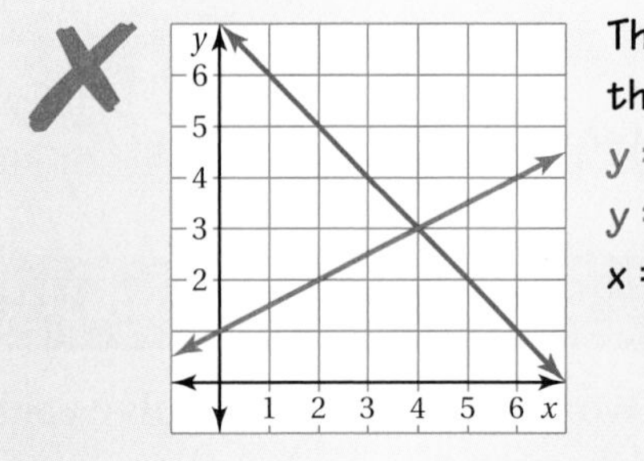

The solution of the linear system y = 0.5x + 1 and y = −x + 7 is x = 4.

21. REASONING Is it possible for a system of two linear equations to have exactly two solutions? Explain your reasoning.

22. MODELING You have a total of 42 math and science problems for homework. You have 10 more math problems than science problems. How many problems do you have in each subject? Use a system of linear equations to justify your answer.

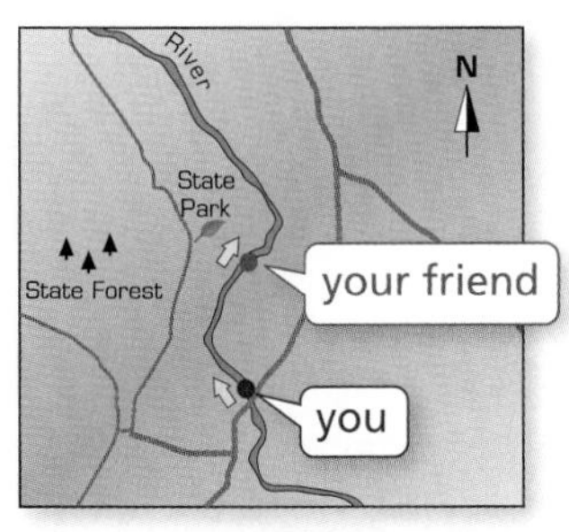

23. CANOE RACE You and your friend are in a canoe race. Your friend is a half mile in front of you and paddling 3 miles per hour. You are paddling 3.4 miles per hour.

a. You are 8.5 miles from the finish line. How long will it take you to catch up to your friend?

b. You both maintain your paddling rates for the remainder of the race. How far ahead of your friend will you be when you cross the finish line?

24. 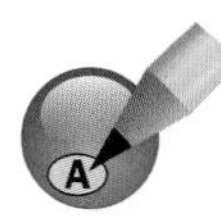

Your friend is trying to grow her hair as long as her cousin's hair. The table shows their hair lengths (in inches) in different months.

Month	Friend's Hair (in.)	Cousin's Hair (in.)
3	4	7
8	6.5	9

a. Write a system of linear equations that represents this situation.

b. Will your friend's hair ever be as long as her cousin's hair? If so, in what month?

Fair Game Review What you learned in previous grades & lessons

Solve the equation. Check your solution. *(Section 1.2)*

25. $\frac{3}{4}c - \frac{1}{4}c + 3 = 7$

26. $5(2 - y) + y = -6$

27. $6x - 3(x + 8) = 9$

28. MULTIPLE CHOICE What is the slope of the line that passes through $(-2, -2)$ and $(3, -1)$? *(Section 4.2)*

Ⓐ -5 Ⓑ $-\frac{1}{5}$ Ⓒ $\frac{1}{5}$ Ⓓ 5

5.2 Solving Systems of Linear Equations by Substitution

Essential Question How can you use substitution to solve a system of linear equations?

1 ACTIVITY: Using Substitution to Solve a System

Work with a partner. Solve each system of linear equations by using two methods.

Method 1: Solve for x first.

Solve for x in one of the equations. Use the expression for x to find the solution of the system. Explain how you did it.

Method 2: Solve for y first.

Solve for y in one of the equations. Use the expression for y to find the solution of the system. Explain how you did it.

$y = 6x - 11$

Is the solution the same using both methods?

a. $6x - y = 11$
$2x + 3y = 7$

b. $2x - 3y = -1$
$x - y = 1$

c. $3x + y = 5$
$5x - 4y = -3$

d. $5x - y = 2$
$3x - 6y = 12$

e. $x + y = -1$
$5x + y = -13$

f. $2x - 6y = -6$
$7x - 8y = 5$

2 ACTIVITY: Writing and Solving a System of Equations

Systems of Equations
In this lesson, you will
- write and solve systems of linear equations by substitution.
- solve real-life problems.

Work with a partner.

a. Roll a pair of number cubes that have different colors. Then write the ordered pair shown by the number cubes. The ordered pair at the right is (3, 4).

x-value

y-value

b. Write a system of linear equations that has this ordered pair as its solution.

c. Exchange systems with your partner. Use one of the methods from Activity 1 to solve the system.

3 ACTIVITY: Solving a Secret Code

Math Practice

Check Progress

As you complete each system of equations, how do you know your answer is correct?

Work with a partner. Decode the quote by Archimedes.

$\underline{\ \ }$ $\underline{\ \ }$ $\underline{\ \ }$ $\underline{\ \ }$ $\underline{\ \ }$ $\underline{\ \ }$ $\underline{\ \ }$ $\underline{\ \ }$ $\underline{\ \ }$ $\underline{\ \ }$ $\underline{\ \ }$ $\underline{\ \ }$ $\underline{\ \ }$ $\underline{\ \ }$ $\underline{\ \ }$ $\underline{\ \ }$ $\underline{\ \ }$ $\underline{\ \ }$ $\underline{\ \ }$,

−8 −7 7 −5   −4 −5   −3   −2 −1 −3 0 −5   1 2   3 1 −3 4 5,

−3 4 5   −7   6 −7 −1 −1   −4 2 7 −5   1 8 −5   −5 −3 9 1 8.

(A, C)	$x + y = -3$ $x - y = -3$	**(D, E)**	$x + y = 0$ $x - y = 10$	**(G, H)**	$x + y = 0$ $x - y = -16$
(I, L)	$x + 2y = -9$ $2x - y = -13$	**(M, N)**	$x + 2y = 4$ $2x - y = -12$	**(O, P)**	$x + 2y = -2$ $2x - y = 6$
(R, S)	$2x + y = 21$ $x - y = 6$	**(T, U)**	$2x + y = -7$ $x - y = 10$	**(V, W)**	$2x + y = 20$ $x - y = 1$

What Is Your Answer?

4. **IN YOUR OWN WORDS** How can you use substitution to solve a system of linear equations?

Practice Use what you learned about systems of linear equations to complete Exercises 4–6 on page 212.

5.2 Lesson

Another way to solve systems of linear equations is to use substitution.

Key Idea

Solving a System of Linear Equations by Substitution

Step 1: Solve one of the equations for one of the variables.

Step 2: Substitute the expression from Step 1 into the other equation and solve for the other variable.

Step 3: Substitute the value from Step 2 into one of the original equations and solve.

EXAMPLE 1 Solving a System of Linear Equations by Substitution

Solve the system by substitution. $\mathbf{y = 2x - 4}$ Equation 1

$\mathbf{7x - 2y = 5}$ Equation 2

Step 1: Equation 1 is already solved for y.

Step 2: Substitute $2x - 4$ for y in Equation 2.

$7x - 2y = 5$	Equation 2
$7x - 2(2x - 4) = 5$	Substitute $2x - 4$ for y.
$7x - 4x + 8 = 5$	Distributive Property
$3x + 8 = 5$	Combine like terms.
$3x = -3$	Subtract 8 from each side.
$x = -1$	Divide each side by 3.

Step 3: Substitute -1 for x in Equation 1 and solve for y.

$y = 2x - 4$	Equation 1
$= 2(-1) - 4$	Substitute -1 for x.
$= -2 - 4$	Multiply.
$= -6$	Subtract.

The solution is $(-1, -6)$.

Check

Equation 1

$y = 2x - 4$

$-6 \stackrel{?}{=} 2(-1) - 4$

$-6 = -6$ ✓

Equation 2

$7x - 2y = 5$

$7(-1) - 2(-6) \stackrel{?}{=} 5$

$5 = 5$ ✓

On Your Own

Now You're Ready
Exercises 10–15

Solve the system of linear equations by substitution. Check your solution.

1. $y = 2x + 3$
$y = 5x$

2. $4x + 2y = 0$
$y = \frac{1}{2}x - 5$

3. $x = 5y + 3$
$2x + 4y = -1$

EXAMPLE 2 Real-Life Application

You buy a total of 50 turkey burgers and veggie burgers for \$90. You pay \$2 per turkey burger and \$1.50 per veggie burger. Write and solve a system of linear equations to find the number x of turkey burgers and the number y of veggie burgers you buy.

Use a verbal model to write a system of linear equations.

Number of turkey burgers, x + Number of veggie burgers, y = Total number of burgers

Cost per turkey burger • Number of turkey burgers, x + Cost per veggie burger • Number of veggie burgers, y = Total cost

The system is: $x + y = 50$ Equation 1

$2x + 1.5y = 90$ Equation 2

Study Tip

It is easiest to solve for a variable that has a coefficient of 1 or −1.

Step 1: Solve Equation 1 for x.

$x + y = 50$ Equation 1

$x = 50 - y$ Subtract y from each side.

Step 2: Substitute $50 - y$ for x in Equation 2.

$2x + 1.5y = 90$ Equation 2

$2(50 - y) + 1.5y = 90$ Substitute $50 - y$ for x.

$100 - 2y + 1.5y = 90$ Distributive Property

$-0.5y = -10$ Simplify.

$y = 20$ Divide each side by −0.5.

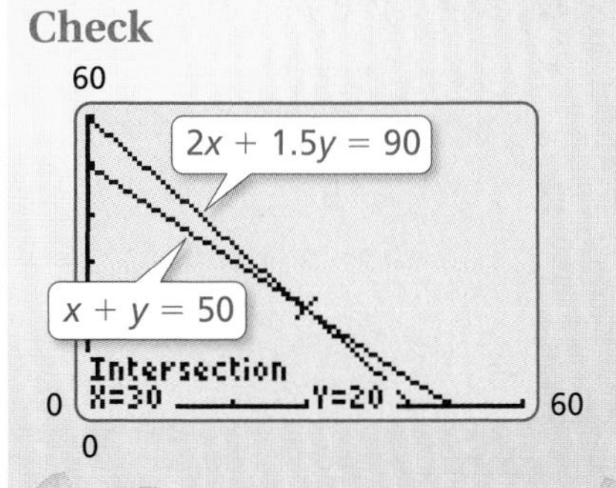

Step 3: Substitute 20 for y in Equation 1 and solve for x.

$x + y = 50$ Equation 1

$x + 20 = 50$ Substitute 20 for y.

$x = 30$ Subtract 20 from each side.

∴ You buy 30 turkey burgers and 20 veggie burgers.

On Your Own

Exercises 18–20

4. You sell lemonade for \$2 per cup and orange juice for \$3 per cup. You sell a total of 100 cups for \$240. Write and solve a system of linear equations to find the number of cups of lemonade and the number of cups of orange juice you sold.

5.2 Exercises

Vocabulary and Concept Check

1. **WRITING** Describe how to solve a system of linear equations by substitution.
2. **NUMBER SENSE** When solving a system of linear equations by substitution, how do you decide which variable to solve for in Step 1?
3. **REASONING** Does solving a system of linear equations by graphing give the same solution as solving by substitution? Explain your reasoning.

Practice and Problem Solving

Write a system of linear equations that has the ordered pair as its solution. Use a method from Activity 1 to solve the system.

4. 

5. 

6. 

Tell which equation you would choose to solve for one of the variables when solving the system by substitution. Explain your reasoning.

7. $2x + 3y = 5$
   $4x - y = 3$

8. $\frac{2}{3}x + 5y = -1$
   $x + 6y = 0$

9. $2x + 10y = 14$
   $5x - 9y = 1$

Solve the system of linear equations by substitution. Check your solution.

1 10. $y = x - 4$
   $y = 4x - 10$

11. $y = 2x + 5$
   $y = 3x - 1$

12. $x = 2y + 7$
   $3x - 2y = 3$

13. $4x - 2y = 14$
   $y = \frac{1}{2}x - 1$

14. $2x = y - 10$
   $x + 7 = y$

15. $8x - \frac{1}{3}y = 0$
   $12x + 3 = y$

16. **SCHOOL CLUBS** There are a total of 64 students in a drama club and a yearbook club. The drama club has 10 more students than the yearbook club.
   a. Write a system of linear equations that represents this situation.
   b. How many students are in the drama club? the yearbook club?

17. **THEATER** A drama club earns \$1040 from a production. It sells a total of 64 adult tickets and 132 student tickets. An adult ticket costs twice as much as a student ticket.
   a. Write a system of linear equations that represents this situation.
   b. What is the cost of each ticket?

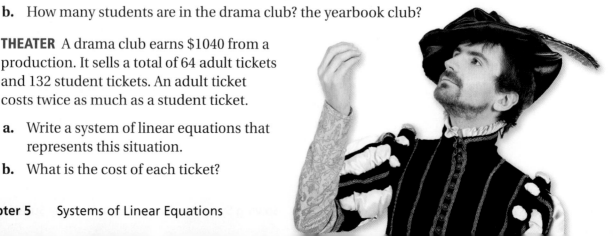

Solve the system of linear equations by substitution. Check your solution.

2 **18.** $y - x = 0$
$2x - 5y = 9$

19. $x + 4y = 14$
$3x + 7y = 22$

20. $-2x - 5y = 3$
$3x + 8y = -6$

21. ERROR ANALYSIS Describe and correct the error in solving the system of linear equations.

✗		Step 1:	Step 2:
$2x + y = 5$	Equation 1	$2x + y = 5$	$2x + (-2x + 5) = 5$
$3x - 2y = 4$	Equation 2	$y = -2x + 5$	$2x - 2x + 5 = 5$
			$5 = 5$

22. STRUCTURE The measure of the obtuse angle in the isosceles triangle is two and a half times the measure of one base angle. Write and solve a system of linear equations to find the measures of all the angles.

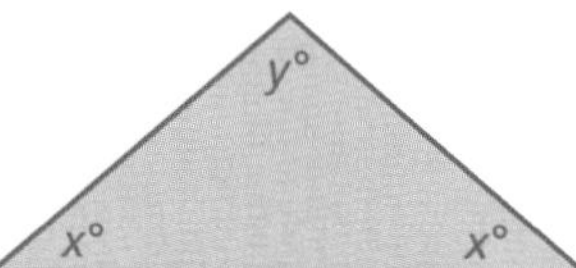

23. ANIMAL SHELTER An animal shelter has a total of 65 abandoned cats and dogs. The ratio of cats to dogs is 6 : 7. How many cats are in the shelter? How many dogs are in the shelter? Justify your answers.

24. NUMBER SENSE The sum of the digits of a two-digit number is 8. When the digits are reversed, the number increases by 36. Find the original number.

25. Repeated Reasoning A DJ has a total of 1075 dance, rock, and country songs on her system. The dance selection is three times the size of the rock selection. The country selection has 105 more songs than the rock selection. How many songs on the system are dance? rock? country?

Fair Game Review What you learned in previous grades & lessons

Write the equation in standard form. *(Section 4.5)*

26. $3x - 9 = 7y$

27. $8 - 5y = -2x$

28. $6x = y + 3$

29. MULTIPLE CHOICE Use the figure to find the measure of $\angle 2$. *(Section 3.1)*

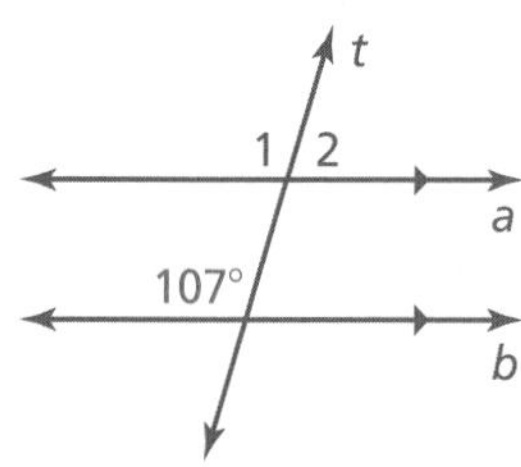

Ⓐ 17° Ⓑ 73°

Ⓒ 83° Ⓓ 107°

5 Study Help

You can use a **notetaking organizer** to write notes, vocabulary, and questions about a topic. Here is an example of a notetaking organizer for solving systems of linear equations by graphing.

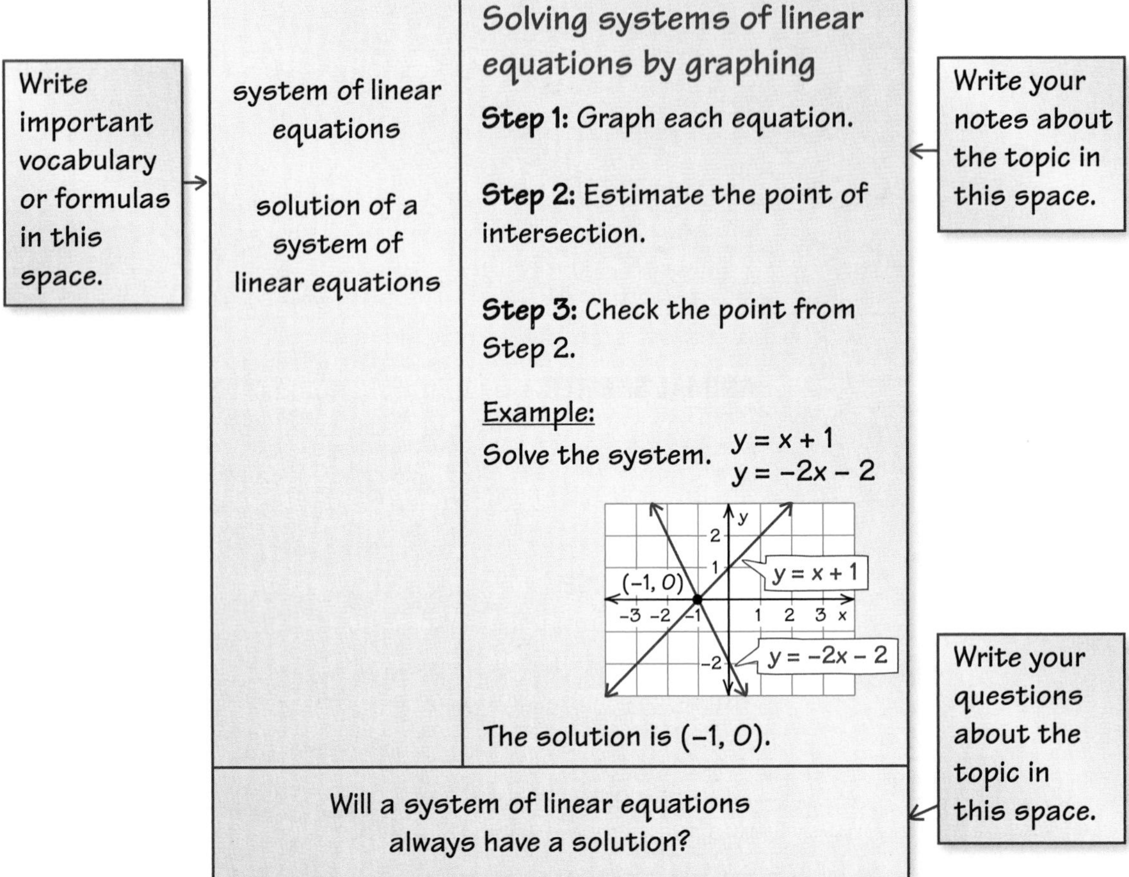

Write important vocabulary or formulas in this space.
system of linear equations
solution of a system of linear equations
Solving systems of linear equations by graphing
Step 1: Graph each equation.
Step 2: Estimate the point of intersection.
Step 3: Check the point from Step 2.
Example:
Solve the system. y = x + 1
y = −2x − 2
(−1, 0)
y = x + 1
y = −2x − 2
The solution is (−1, 0).
Write your notes about the topic in this space.
Will a system of linear equations always have a solution?
Write your questions about the topic in this space.

On Your Own

Make a notetaking organizer to help you study this topic.

1. solving systems of linear equations by substitution

After you complete this chapter, make notetaking organizers for the following topics.

2. solving systems of linear equations by elimination
3. solving systems of linear equations with no solution or infinitely many solutions
4. solving linear equations by graphing

"My notetaking organizer has me thinking about retirement when I won't have to fetch sticks anymore."

5.1–5.2 Quiz

Match the system of linear equations with the corresponding graph. Use the graph to estimate the solution. Check your solution. *(Section 5.1)*

1. $y = x - 2$

$y = -2x + 1$

2. $y = x - 3$

$y = -\frac{1}{3}x + 1$

3. $y = \frac{1}{2}x - 2$

$y = 4x + 5$

A. 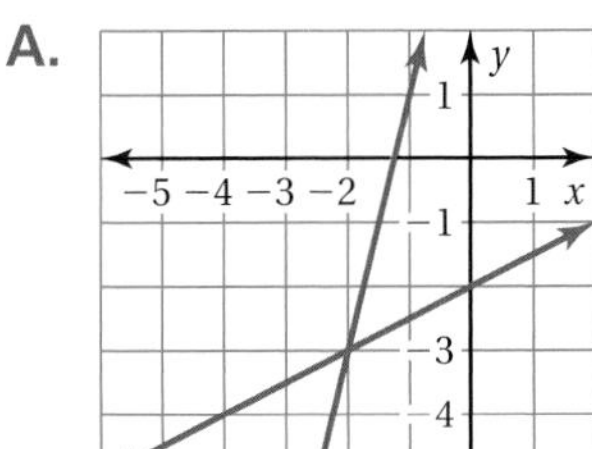

B. 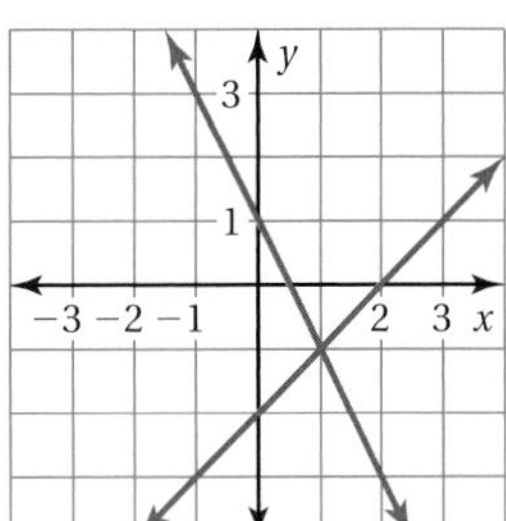

C. 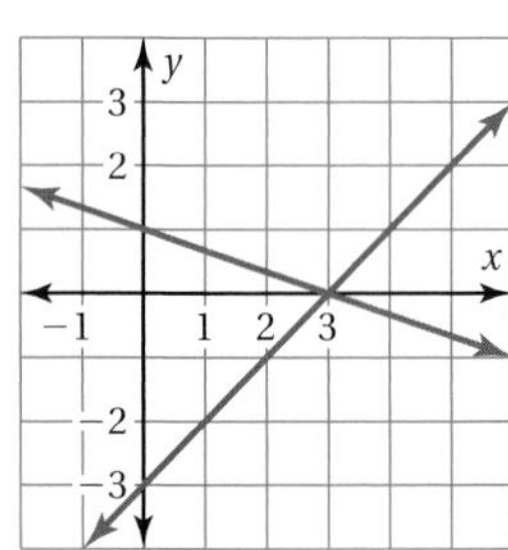

Solve the system of linear equations by graphing. *(Section 5.1)*

4. $y = 2x - 3$

$y = -x + 9$

5. $6x + y = -2$

$y = -3x + 1$

6. $4x + 2y = 2$

$3x = 4 - y$

Solve the system of linear equations by substitution. Check your solution. *(Section 5.2)*

7. $y = x - 8$

$y = 2x - 14$

8. $x = 2y + 2$

$2x - 5y = 1$

9. $x - 5y = 1$

$-2x + 9y = -1$

10. MOVIE CLUB Members of a movie rental club pay a \$15 annual membership fee and \$2 for new release movies. Nonmembers pay \$3 for new release movies. *(Section 5.1)*

a. Write a system of linear equations that represents this situation.

b. When is it beneficial to have a membership?

11. NUMBER SENSE The sum of two numbers is 38. The greater number is 8 more than the other number. Find each number. Use a system of linear equations to justify your answer. *(Section 5.1)*

12. VOLLEYBALL The length of a sand volleyball court is twice its width. The perimeter is 180 feet. Find the length and width of the sand volleyball court. *(Section 5.2)*

13. MEDICAL STAFF A hospital employs a total of 77 nurses and doctors. The ratio of nurses to doctors is 9 : 2. How many nurses are employed at the hospital? How many doctors are employed at the hospital? *(Section 5.2)*

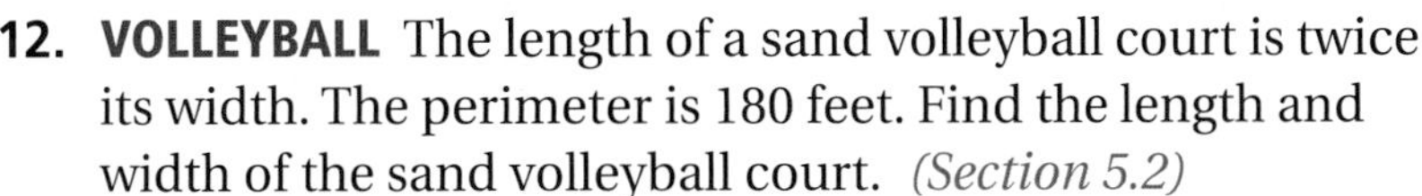

5.3 Solving Systems of Linear Equations by Elimination

Essential Question How can you use elimination to solve a system of linear equations?

1 ACTIVITY: Using Elimination to Solve a System

Work with a partner. Solve each system of linear equations by using two methods.

Method 1: Subtract.

Subtract Equation 2 from Equation 1. What is the result? Explain how you can use the result to solve the system of equations.

Method 2: Add.

Add the two equations. What is the result? Explain how you can use the result to solve the system of equations.

Is the solution the same using both methods?

a. $2x + y = 4$
$2x - y = 0$

b. $3x - y = 4$
$3x + y = 2$

c. $x + 2y = 7$
$x - 2y = -5$

2 ACTIVITY: Using Elimination to Solve a System

Work with a partner.

$2x + y = 2$ Equation 1

$x + 5y = 1$ Equation 2

a. Can you add or subtract the equations to solve the system of linear equations? Explain.

b. Explain what property you can apply to Equation 1 in the system so that the y-coefficients are the same.

c. Explain what property you can apply to Equation 2 in the system so that the x-coefficients are the same.

d. You solve the system in part (b). Your partner solves the system in part (c). Compare your solutions.

e. Use a graphing calculator to check your solution.

Systems of Equations
In this lesson, you will
- write and solve systems of linear equations by elimination.
- solve real-life problems.

3 ACTIVITY: Solving a Secret Code

Work with a partner. Solve the puzzle to find the name of a famous mathematician who lived in Egypt around 350 A.D.

Math Practice

Find Entry Points

What is the first thing you do to solve a system of linear equations? Why?

4	B	W	R	M	F	Y	K	N
3	O	J	A	S	I	D	X	Z
2	Q	P	C	E	G	B	T	J
1	M	R	C	Z	N	O	U	W
0	K	X	U	H	L	Y	S	Q
−1	F	E	A	S	W	K	R	M
−2	G	J	Z	N	H	V	D	G
−3	E	L	X	L	F	Q	O	B
	−3	−2	−1	0	1	2	3	4

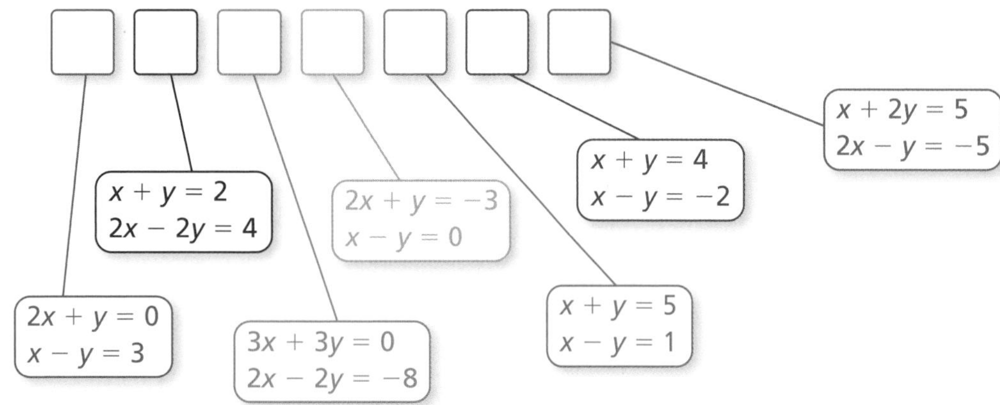

What Is Your Answer?

4. **IN YOUR OWN WORDS** How can you use elimination to solve a system of linear equations?

5. **STRUCTURE** When can you add or subtract equations in a system to solve the system? When do you have to multiply first? Justify your answers with examples.

6. **LOGIC** In Activity 2, why can you multiply equations in the system by a constant and not change the solution of the system? Explain your reasoning.

Practice Use what you learned about systems of linear equations to complete Exercises 4–6 on page 221.

5.3 Lesson

Key Idea

Solving a System of Linear Equations by Elimination

Step 1: Multiply, if necessary, one or both equations by a constant so at least 1 pair of like terms has the same or opposite coefficients.

Step 2: Add or subtract the equations to eliminate one of the variables.

Step 3: Solve the resulting equation for the remaining variable.

Step 4: Substitute the value from Step 3 into one of the original equations and solve.

EXAMPLE 1 Solving a System of Linear Equations by Elimination

Study Tip

Because the coefficients of x are the same, you can also solve the system by subtracting in Step 2.

$$\begin{aligned} x + 3y &= -2 \\ x - 3y &= 16 \\ \hline 6y &= -18 \end{aligned}$$

So, $y = -3$.

Solve the system by elimination. $\quad$ $x + 3y = -2$ $\quad$ Equation 1

$x - 3y = 16$ $\quad$ Equation 2

Step 1: The coefficients of the y-terms are already opposites.

Step 2: Add the equations.

$x + 3y = -2$ $\quad$ Equation 1

$x - 3y = 16$ $\quad$ Equation 2

$2x = 14$ $\quad$ Add the equations.

Step 3: Solve for x.

$2x = 14$ $\quad$ Equation from Step 2

$x = 7$ $\quad$ Divide each side by 2.

Step 4: Substitute 7 for x in one of the original equations and solve for y.

$x + 3y = -2$ $\quad$ Equation 1

$7 + 3y = -2$ $\quad$ Substitute 7 for x.

$3y = -9$ $\quad$ Subtract 7 from each side.

$y = -3$ $\quad$ Divide each side by 3.

The solution is $(7, -3)$.

Check

Equation 1

$x + 3y = -2$

$7 + 3(-3) \stackrel{?}{=} -2$

$-2 = -2$ ✓

Equation 2

$x - 3y = 16$

$7 - 3(-3) \stackrel{?}{=} 16$

$16 = 16$ ✓

On Your Own

Now You're Ready
Exercises 7–12

Solve the system of linear equations by elimination. Check your solution.

1. $2x - y = 9$
   $4x + y = 21$

2. $-5x + 2y = 13$
   $5x + y = -1$

3. $3x + 4y = -6$
   $7x + 4y = -14$

EXAMPLE 2 Solving a System of Linear Equations by Elimination

Solve the system by elimination.

$-6x + 5y = 25$ Equation 1

$-2x - 4y = 14$ Equation 2

Step 1: Multiply Equation 2 by 3.

$-6x + 5y = 25$ → $-6x + 5y = 25$ Equation 1

$-2x - 4y = 14$ **Multiply by 3.** → $-6x - 12y = 42$ Revised Equation 2

Step 2: Subtract the equations.

$-6x + 5y = 25$ Equation 1

$-6x - 12y = 42$ Revised Equation 2

$17y = -17$ Subtract the equations.

Step 3: Solve for y.

$17y = -17$ Equation from Step 2

$y = -1$ Divide each side by 17.

Step 4: Substitute -1 for y in one of the original equations and solve for x.

$-2x - 4y = 14$ Equation 2

$-2x - 4(-1) = 14$ Substitute -1 for y.

$-2x + 4 = 14$ Multiply.

$-2x = 10$ Subtract 4 from each side.

$x = -5$ Divide each side by -2.

The solution is $(-5, -1)$.

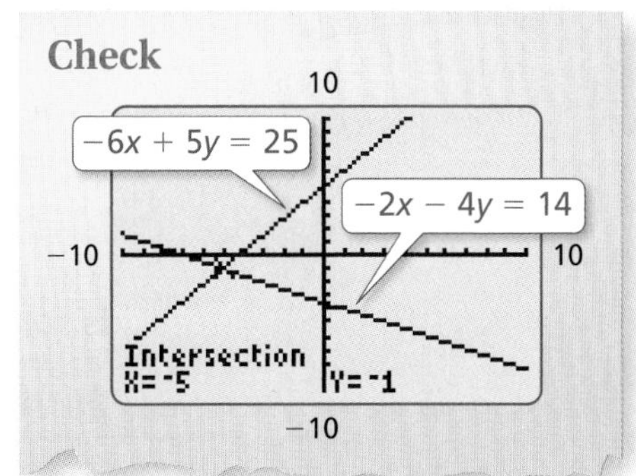

Study Tip

In Example 2, notice that you can also multiply Equation 2 by -3 and then add the equations.

On Your Own

Solve the system of linear equations by elimination. Check your solution.

4. $3x + y = 11$
   $6x + 3y = 24$

5. $4x - 5y = -19$
   $-x - 2y = 8$

6. $5y = 15 - 5x$
   $y = -2x + 3$

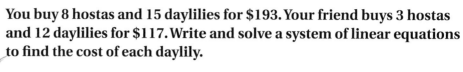

EXAMPLE 3 Real-Life Application

You buy 8 hostas and 15 daylilies for \$193. Your friend buys 3 hostas and 12 daylilies for \$117. Write and solve a system of linear equations to find the cost of each daylily.

Use a verbal model to write a system of linear equations.

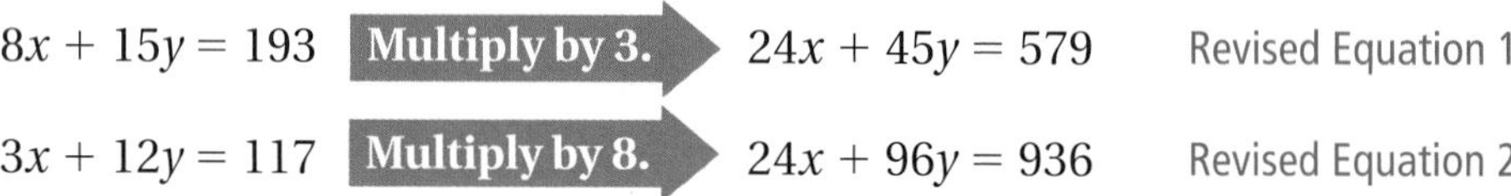

The system is: $8x + 15y = 193$ Equation 1 (You)

$3x + 12y = 117$ Equation 2 (Your friend)

Step 1: To find the cost y of each daylily, eliminate the x-terms. Multiply Equation 1 by 3. Multiply Equation 2 by 8.

$8x + 15y = 193$ **Multiply by 3.** → $24x + 45y = 579$ Revised Equation 1

$3x + 12y = 117$ **Multiply by 8.** → $24x + 96y = 936$ Revised Equation 2

Step 2: Subtract the revised equations.

$24x + 45y = 579$ Revised Equation 1

$24x + 96y = 936$ Revised Equation 2

$-51y = -357$ Subtract the equations.

Step 3: Solving the equation $-51y = -357$ gives $y = 7$.

So, each daylily costs \$7.

On Your Own

Exercises 16–21

7. A landscaper buys 4 peonies and 9 geraniums for \$190. Another landscaper buys 5 peonies and 6 geraniums for \$185. Write and solve a system of linear equations to find the cost of each peony.

Methods for Solving Systems of Linear Equations

Method	When to Use
Graphing *(Lesson 5.1)*	To estimate solutions
Substitution *(Lesson 5.2)*	When one of the variables in one of the equations has a coefficient of 1 or -1
Elimination *(Lesson 5.3)*	When at least 1 pair of like terms has the same or opposite coefficients
Elimination (Multiply First) *(Lesson 5.3)*	When one of the variables cannot be eliminated by adding or subtracting the equations

5.3 Exercises

Vocabulary and Concept Check

1. **WRITING** Describe how to solve a system of linear equations by elimination.

2. **NUMBER SENSE** When should you use multiplication to solve a system of linear equations by elimination?

3. **WHICH ONE DOESN'T BELONG?** Which system of equations does *not* belong with the other three? Explain your reasoning.

$3x + 3y = 3$ $2x - 3y = 7$	$-2x + y = 6$ $2x - 3y = -10$	$2x + 3y = 11$ $3x - 2y = 10$	$x + y = 5$ $3x - y = 3$

Practice and Problem Solving

Use a method from Activity 1 to solve the system.

4. $x + y = 3$
   $x - y = 1$

5. $-x + 3y = 0$
   $x + 3y = 12$

6. $3x + 2y = 3$
   $3x - 2y = -9$

Solve the system of linear equations by elimination. Check your solution.

7. $x + 3y = 5$
   $-x - y = -3$

8. $x - 2y = -7$
   $3x + 2y = 3$

9. $4x + 3y = -5$
   $-x + 3y = -10$

10. $2x + 7y = 1$
    $2x - 4y = 12$

11. $2x + 5y = 16$
    $3x - 5y = -1$

12. $3x - 2y = 4$
    $6x - 2y = -2$

13. **ERROR ANALYSIS** Describe and correct the error in solving the system of linear equations.

> ✗
> $5x + 2y = 9$ Equation 1
> $3x - 2y = -1$ Equation 2
> $2x \quad = 10$
> $x = 5$
> The solution is (5, −8).

14. **RAFFLE TICKETS** You and your friend are selling raffle tickets for a new laptop. You sell 14 more tickets than your friend sells. Together, you and your friend sell 58 tickets.

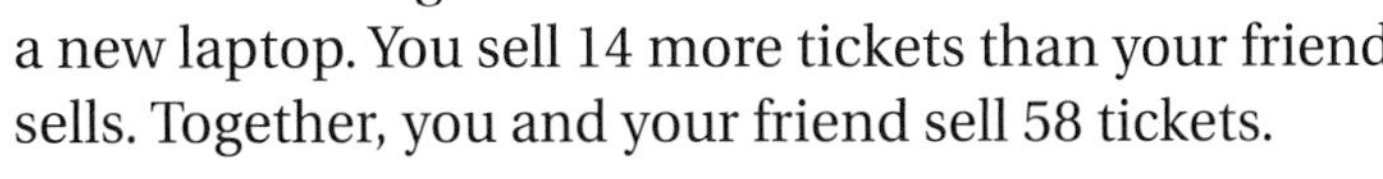

   a. Write a system of linear equations that represents this situation.
   b. How many tickets does each of you sell?

15. **JOGGING** You can jog around your block twice and the park once in 10 minutes. You can jog around your block twice and the park 3 times in 22 minutes.

   a. Write a system of linear equations that represents this situation.
   b. How long does it take you to jog around the park?

Solve the system of linear equations by elimination. Check your solution.

②③ **16.** $2x - y = 0$
$3x - 2y = -3$

17. $x + 4y = 1$
$3x + 5y = 10$

18. $-2x + 3y = 7$
$5x + 8y = -2$

19. $3x + 3 = 3y$
$2x - 6y = 2$

20. $2x - 6 = 4y$
$7y = -3x + 9$

21. $5x = 4y + 8$
$3y = 3x - 3$

22. ERROR ANALYSIS Describe and correct the error in solving the system of linear equations.

✗
$x + y = 1$ Equation 1 **Multiply by −5.** → $-5x + 5y = -5$
$5x + 3y = -3$ Equation 2 → $5x + 3y = -3$

$8y = -8$
$y = -1$

The solution is (2, −1).

23. REASONING For what values of *a* and *b* should you solve the system by elimination?

a. $4x - y = 3$
$ax + 10y = 6$

b. $x - 7y = 6$
$-6x + by = 9$

Determine whether the line through the first pair of points intersects the line through the second pair of points. Explain.

24. Line 1: $(-2, 1), (2, 7)$
Line 2: $(-4, -1), (0, 5)$

25. Line 1: $(3, -2), (7, -1)$
Line 2: $(5, 2), (6, -2)$

26. AIRPLANES Two airplanes are flying to the same airport. Their positions are shown in the graph. Write a system of linear equations that represents this situation. Solve the system by elimination to justify your answer.

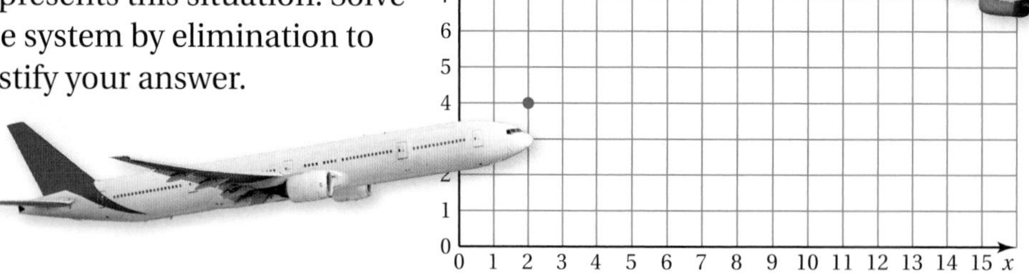

27. TEST PRACTICE The table shows the number of correct answers on a practice standardized test. You score 86 points on the test, and your friend scores 76 points.

	You	Your Friend
Multiple Choice	23	28
Short Response	10	5

a. Write a system of linear equations that represents this situation.

b. How many points is each type of question worth?

28. LOGIC You solve a system of equations in which x represents the number of adult tickets sold and y represents the number of student tickets sold. Can $(-6, 24)$ be the solution of the system? Explain your reasoning.

29. VACATION The table shows the activities of two tourists at a vacation resort. You want to go parasailing for 1 hour and horseback riding for 2 hours. How much do you expect to pay?

	Parasailing	Horseback Riding	Total Cost
Tourist 1	2 hours	5 hours	\$205
Tourist 2	3 hours	3 hours	\$240

30. REASONING The solution of a system of linear equations is $(2, -4)$. One equation in the system is $2x + y = 0$. Explain how you could find a second equation for the system. Then find a second equation. Solve the system by elimination to justify your answer.

31. JEWELER A metal alloy is a mixture of two or more metals. A jeweler wants to make 8 grams of 18-carat gold, which is 75% gold. The jeweler has an alloy that is 90% gold and an alloy that is 50% gold. How much of each alloy should the jeweler use?

32. PROBLEM SOLVING A powerboat takes 30 minutes to travel 10 miles downstream. The return trip takes 50 minutes. What is the speed of the current?

33. Critical Thinking Solve the system of equations by elimination.

$$2x - y + 3z = -1$$
$$x + 2y - 4z = -1$$
$$y - 2z = 0$$

Fair Game Review What you learned in previous grades & lessons

Decide whether the two equations are equivalent. *(Section 1.2 and Section 1.3)*

34. $4n + 1 = n - 8$

$3n = -9$

35. $2a + 6 = 12$

$a + 3 = 6$

36. $7v - \frac{3}{2} = 5$

$14v - 3 = 15$

37. MULTIPLE CHOICE Which line has the same slope as $y = \frac{1}{2}x - 3$? *(Section 4.4)*

Ⓐ $y = -2x + 4$ Ⓑ $y = 2x + 3$ Ⓒ $y - 2x = 5$ Ⓓ $2y - x = 7$

5.4 Solving Special Systems of Linear Equations

Essential Question Can a system of linear equations have no solution? Can a system of linear equations have many solutions?

1 ACTIVITY: Writing a System of Linear Equations

Work with a partner. Your cousin is 3 years older than you. You can represent your ages by two linear equations.

$y = t$ Your age

$y = t + 3$ Your cousin's age

a. Graph both equations in the same coordinate plane.

b. What is the vertical distance between the two graphs? What does this distance represent?

c. Do the two graphs intersect? Explain what this means in terms of your age and your cousin's age.

2 ACTIVITY: Using a Table to Solve a System

Work with a partner. You invest $500 for equipment to make dog backpacks. Each backpack costs you $15 for materials. You sell each backpack for $15.

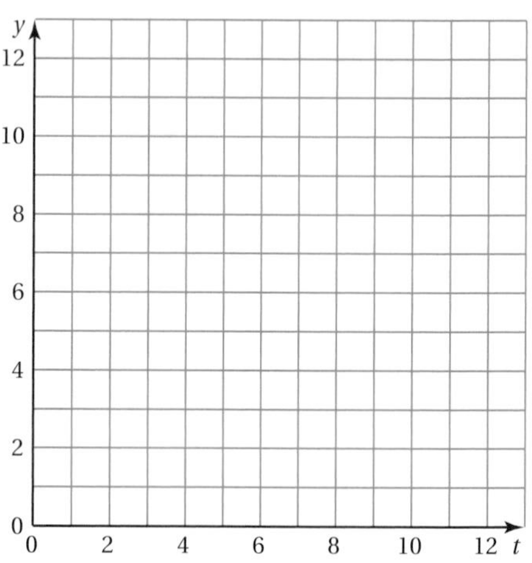

a. Copy and complete the table for your cost C and your revenue R.

x	0	1	2	3	4	5	6	7	8	9	10
C											
R											

b. When will you break even? What is wrong?

Systems of Equations
In this lesson, you will
- solve systems of linear equations with no solution or infinitely many solutions.

3 ACTIVITY: Using a Graph to Solve a Puzzle

Math Practice

Analyze Relationships

What do you know about the graphs of the two equations? How does this relate to the number of solutions?

Work with a partner. Let x and y be two numbers. Here are two clues about the values of x and y.

	Words	Equation
Clue 1:	y is 4 more than twice the value of x.	$y = 2x + 4$
Clue 2:	The difference of $3y$ and $6x$ is 12.	$3y - 6x = 12$

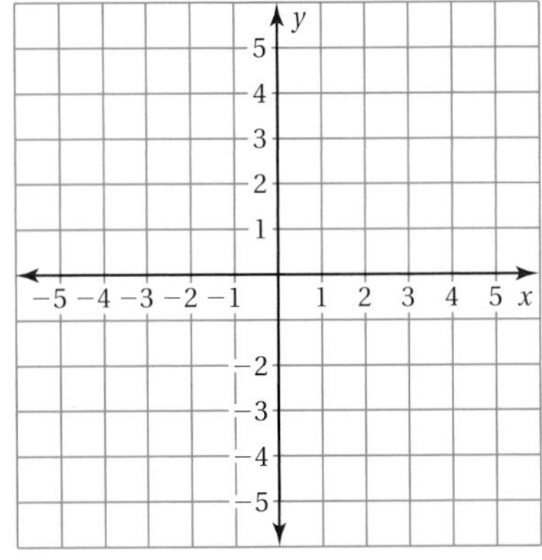

a. Graph both equations in the same coordinate plane.

b. Do the two lines intersect? Explain.

c. What is the solution of the puzzle?

d. Use the equation $y = 2x + 4$ to complete the table.

x	0	1	2	3	4	5	6	7	8	9	10
y											

e. Does each solution in the table satisfy *both* clues?

f. What can you conclude? How many solutions does the puzzle have? How can you describe them?

What Is Your Answer?

4. IN YOUR OWN WORDS Can a system of linear equations have no solution? Can a system of linear equations have many solutions? Give examples to support your answers.

Practice Use what you learned about special systems of linear equations to complete Exercises 3 and 4 on page 228.

5.4 Lesson

Solutions of Systems of Linear Equations

A system of linear equations can have *one solution, no solution,* or *infinitely many solutions.*

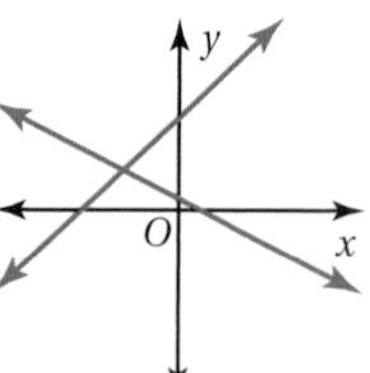

One solution

The lines intersect.

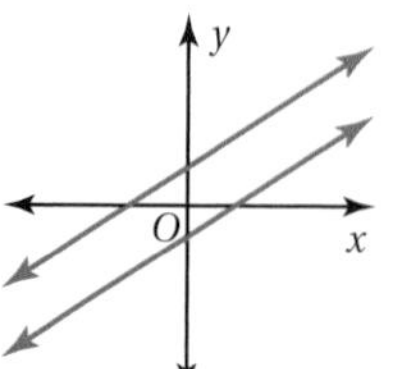

No solution

The lines are parallel.

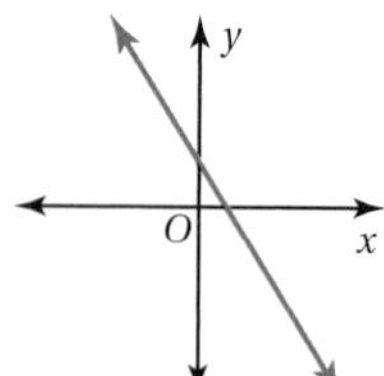

Infinitely many solutions

The lines are the same.

EXAMPLE 1 Solving a System: No Solution

Solve the system. $\quad \mathbf{y = 3x + 1}$ Equation 1

$\mathbf{y = 3x - 3}$ Equation 2

Method 1: Solve by graphing.

Graph each equation. The lines have the same slope and different y-intercepts. So, the lines are parallel.

Because parallel lines do not intersect, there is no point that is a solution of both equations.

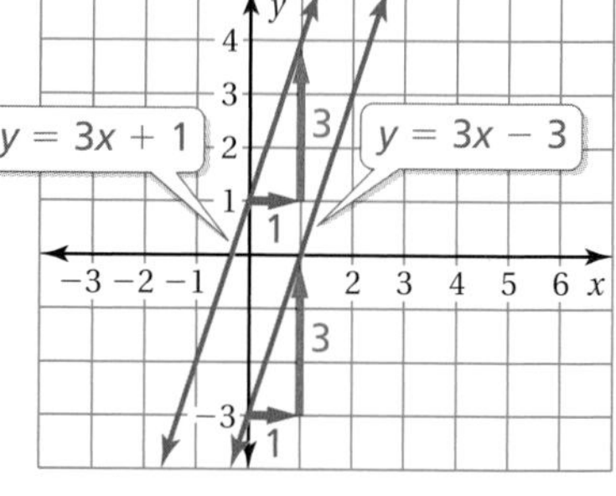

∴ So, the system of linear equations has no solution.

Method 2: Solve by substitution.

Substitute $3x - 3$ for y in Equation 1.

$y = 3x + 1$	Equation 1
$3x - 3 = 3x + 1$	Substitute $3x - 3$ for y.
$-3 = 1$ ✗	Subtract $3x$ from each side.

∴ The equation $-3 = 1$ is never true. So, the system of linear equations has no solution.

Study Tip

You can solve some linear systems by inspection. In Example 1, notice you can rewrite the system as

$-3x + y = 1$

$-3x + y = -3.$

This system has no solution because $-3x + y$ cannot be equal to 1 and -3 at the same time.

On Your Own

Now You're Ready
Exercises 8–10

Solve the system of linear equations. Check your solution.

1. $y = -x + 3$
   $y = -x + 5$
2. $y = -5x - 2$
   $5x + y = 0$
3. $x = 2y + 10$
   $2x + 3y = -1$

EXAMPLE 2 Solving a System: Infinitely Many Solutions

The perimeter of Rectangle A is 36 units. The perimeter of Rectangle B is 108 units. Write and solve a system of linear equations to find the values of *x* and *y*.

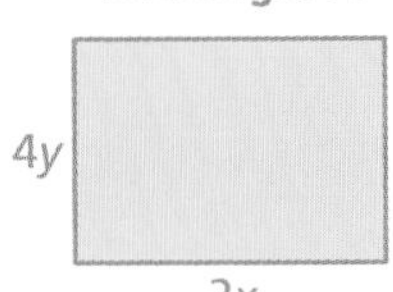

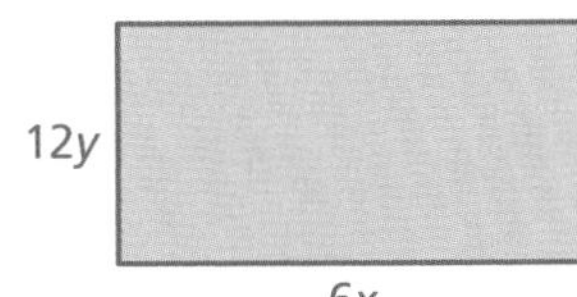

Perimeter of Rectangle A

$2(2x) + 2(4y) = 36$

$4x + 8y = 36$ Equation 1

Perimeter of Rectangle B

$2(6x) + 2(12y) = 108$

$12x + 24y = 108$ Equation 2

The system is: $4x + 8y = 36$ Equation 1

$12x + 24y = 108$ Equation 2

Method 1: Solve by graphing.

Graph each equation.

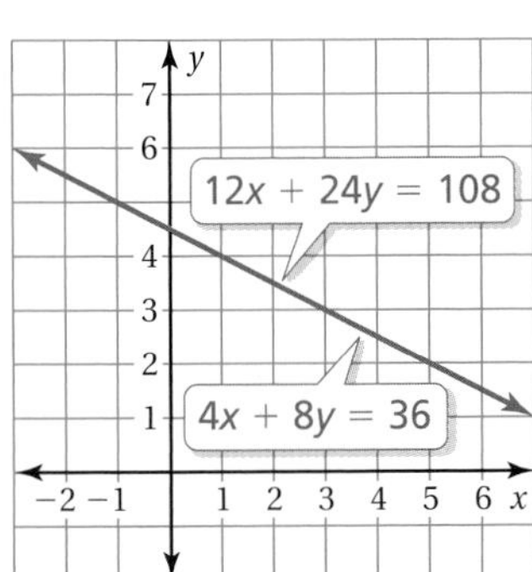

The lines have the same slope and the same y-intercept. So, the lines are the same.

In this context, x and y must be positive. Because the lines are the same, all the points on the line in Quadrant I are solutions of both equations. So, the system of linear equations has infinitely many solutions.

Method 2: Solve by elimination.

Multiply Equation 1 by 3 and subtract the equations.

$4x + 8y = 36$ **Multiply by 3.** → $12x + 24y = 108$ Revised Equation 1

$12x + 24y = 108$ → $12x + 24y = 108$ Equation 2

$0 = 0$ Subtract.

The equation $0 = 0$ is always true. In this context, x and y must be positive. So, the solutions are all the points on the line $4x + 8y = 36$ in Quadrant I. The system of linear equations has infinitely many solutions.

On Your Own

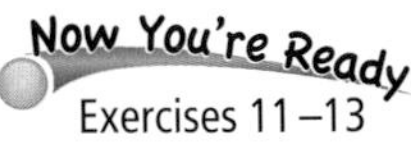

Solve the system of linear equations. Check your solution.

4. $x + y = 3$
   $x - y = -3$
5. $2x + y = 5$
   $4x + 2y = 0$
6. $2x - 4y = 10$
   $-12x + 24y = -60$
7. **WHAT IF?** What happens to the solution in Example 2 if the perimeter of Rectangle A is 54 units? Explain.

5.4 Exercises

Vocabulary and Concept Check

1. **WRITING** Describe the difference between the graph of a system of linear equations that has *no solution* and the graph of a system of linear equations that has *infinitely many solutions.*

2. **REASONING** When solving a system of linear equations algebraically, how do you know when the system has *no solution*? *infinitely many solutions*?

Practice and Problem Solving

Let x and y be two numbers. Find the solution of the puzzle.

3. y is $\frac{1}{3}$ more than 4 times the value of x.

   The difference of $3y$ and $12x$ is 1.

4. $\frac{1}{2}$ of x plus 3 is equal to y.

   x is 6 more than twice the value of y.

Without graphing, determine whether the system of linear equations has *one solution, infinitely many solutions,* or *no solution.* Explain your reasoning.

5. $y = 5x - 9$
   $y = 5x + 9$

6. $y = 6x + 2$
   $y = 3x + 1$

7. $y = 8x - 2$
   $y - 8x = -2$

Solve the system of linear equations. Check your solution.

1 8. $y = 2x - 2$
   $y = 2x + 9$

9. $y = 3x + 1$
   $-x + 2y = -3$

10. $y = \frac{\pi}{3}x + \pi$
    $-\pi x + 3y = -6\pi$

2 11. $y = -\frac{1}{6}x + 5$
    $x + 6y = 30$

12. $\frac{1}{3}x + y = 1$
    $2x + 6y = 6$

13. $-2x + y = 1.3$
    $2(0.5x - y) = 4.6$

14. **ERROR ANALYSIS** Describe and correct the error in solving the system of linear equations.

$y = -2x + 4$
$y = -2x + 6$

The lines have the same slope, so, there are infinitely many solutions.

15. **PIG RACE** In a pig race, your pig gets a head start of 3 feet and is running at a rate of 2 feet per second. Your friend's pig is also running at a rate of 2 feet per second. A system of linear equations that represents this situation is $y = 2x + 3$ and $y = 2x$. Will your friend's pig catch up to your pig? Explain.

16. **REASONING** One equation in a system of linear equations has a slope of -3. The other equation has a slope of 4. How many solutions does the system have? Explain.

17. **LOGIC** How can you use the slopes and the y-intercepts of equations in a system of linear equations to determine whether the system has *one solution, infinitely many solutions,* or *no solution*? Explain your reasoning.

$$4x + 8y = 64$$
$$8x + 16y = 128$$

18. **MONEY** You and a friend both work two different jobs. The system of linear equations represents the total earnings for x hours worked at the first job and y hours worked at the second job. Your friend earns twice as much as you.

 a. One week, both of you work 4 hours at the first job. How many hours do you and your friend work at the second job?

 b. Both of you work the same number of hours at the second job. Compare the number of hours each of you works at the first job.

19. **DOWNLOADS** You download a digital album for \$10. Then you and your friend download the same number of individual songs for \$0.99 each. Write a system of linear equations that represents this situation. Will you and your friend spend the same amount of money? Explain.

20. **REASONING** Does the system shown *always, sometimes,* or *never* have no solution when $a = b$? $a \geq b$? $a < b$? Explain your reasoning.

$$y = ax + 1$$
$$y = bx + 4$$

21. **SKIING** The table shows the number of lift tickets and ski rentals sold to two different groups. Is it possible to determine how much each lift ticket costs? Justify your answer.

Group	1	2
Number of Lift Tickets	36	24
Number of Ski Rentals	18	12
Total Cost (dollars)	684	456

22. Precision Find the values of a and b so the system shown has the solution (2, 3). Does the system have any other solutions? Explain.

$$12x - 2by = 12$$
$$3ax - by = 6$$

Fair Game Review What you learned in previous grades & lessons

Write an equation of the line that passes through the given points. *(Section 4.7)*

23. (0, 0), (2, 6)

24. (0, -3), (3, 3)

25. (-6, 5), (0, 2)

26. **MULTIPLE CHOICE** What is the solution of $-2(y + 5) \leq 16$? *(Skills Review Handbook)*

 Ⓐ $y \leq -13$ Ⓑ $y \geq -13$ Ⓒ $y \leq -3$ Ⓓ $y \geq -3$

Extension 5.4 Solving Linear Equations by Graphing

Systems of Equations

In this extension, you will

- solve linear equations by graphing a system of linear equations.
- solve real-life problems.

Key Idea

Solving Equations Using Graphs

Step 1: To solve the equation $ax + b = cx + d$, write two linear equations.

$$ax + b = cx + d$$

$y = ax + b$ and $y = cx + d$

Step 2: Graph the system of linear equations. The x-value of the solution of the system of linear equations is the solution of the equation $ax + b = cx + d$.

EXAMPLE 1 Solving an Equation Using a Graph

Solve $x - 2 = -\frac{1}{2}x + 1$ using a graph. Check your solution.

Step 1: Write a system of linear equations using each side of the equation.

$$x - 2 = -\frac{1}{2}x + 1$$

$y = x - 2$ and $y = -\frac{1}{2}x + 1$

Step 2: Graph the system.

$$y = x - 2$$

$$y = -\frac{1}{2}x + 1$$

The graphs intersect at (2, 0).

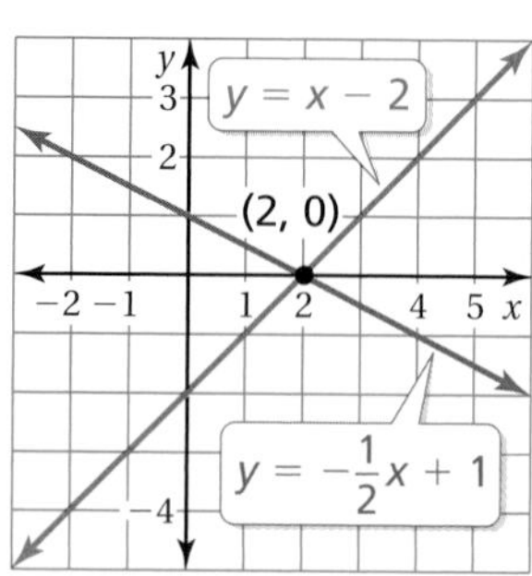

So, the solution of the equation is $x = 2$.

Check

$$x - 2 = -\frac{1}{2}x + 1$$

$$2 - 2 \stackrel{?}{=} -\frac{1}{2}(2) + 1$$

$$0 = 0 \checkmark$$

Practice

Use a graph to solve the equation. Check your solution.

1. $2x + 3 = 4$
2. $2x = x - 3$
3. $3x + 1 = 3x + 2$
4. $\frac{1}{3}x = x + 8$
5. $1.5x + 2 = 11 - 3x$
6. $3 - 2x = -2x + 3$
7. **STRUCTURE** Write an equation with variables on both sides that has no solution. How can you change the equation so that it has infinitely many solutions?

EXAMPLE 2 Real-Life Application

Plant A | Plant B

12 in. | 9 in.

Plant A grows 0.6 inch per month. Plant B grows twice as fast.

a. Use the model to write an equation.

b. After how many months x are the plants the same height?

Growth rate	$\cdot$	Months, x	$+$	Original height	$=$	Growth rate	$\cdot$	Months, x	$+$	Original height

a. The equation is $0.6x + 12 = 1.2x + 9$.

b. Write a system of linear equations using each side of the equation. Then use a graphing calculator to graph the system.

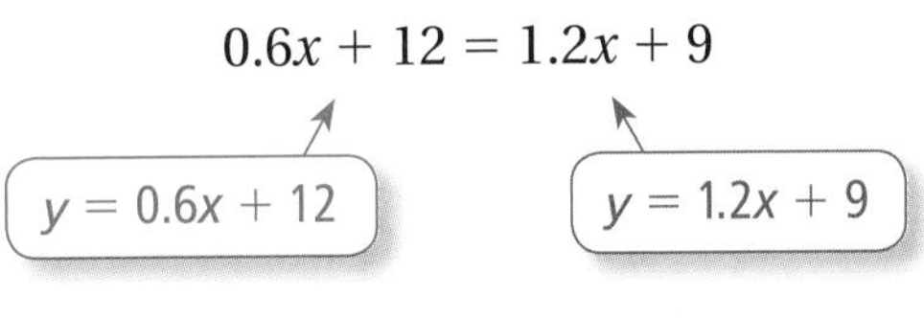

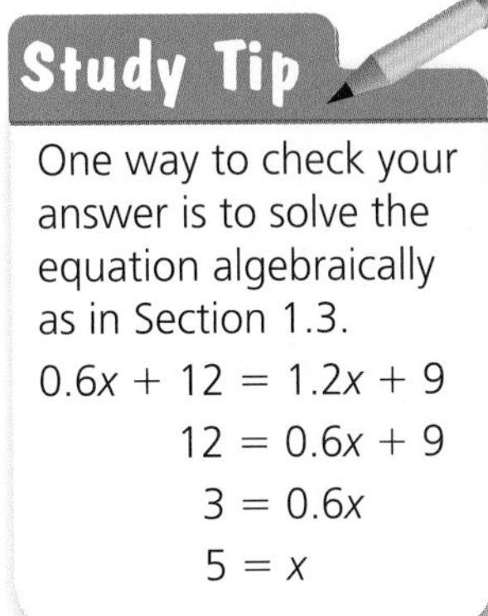

Study Tip

One way to check your answer is to solve the equation algebraically as in Section 1.3.

$0.6x + 12 = 1.2x + 9$

$12 = 0.6x + 9$

$3 = 0.6x$

$5 = x$

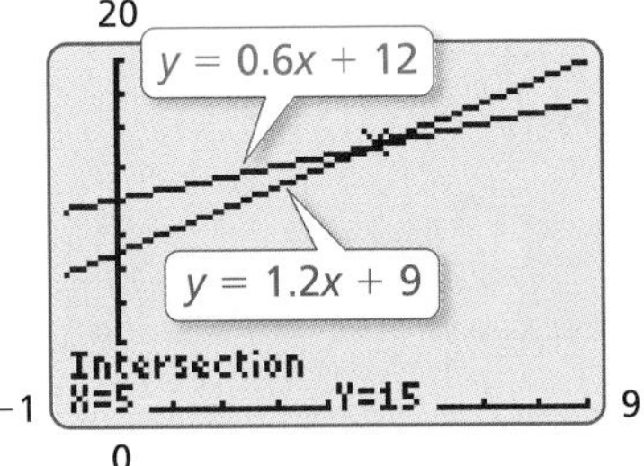

The solution of the system is (5, 15).

So, the plants are both 15 inches tall after 5 months.

Practice

Use a graph to solve the equation. Check your solution.

8. $6x - 2 = x + 11$

9. $\frac{4}{3}x - 1 = \frac{2}{3}x + 6$

10. $1.75x = 2.25x + 10.25$

11. WHAT IF? In Example 2, the growth rate of Plant A is 0.5 inch per month. After how many months x are the plants the same height?

5.3–5.4 Quiz

Solve the system of linear equations by elimination. Check your solution. *(Section 5.3)*

1. $x + 2y = 4$
$-x - y = 2$

2. $2x - y = 1$
$x + 3y - 4 = 0$

3. $3x = -4y + 10$
$4x + 3y = 11$

4. $2x + 5y = 60$
$2x - 5y = -20$

Solve the system of linear equations. Check your solution. *(Section 5.4)*

5. $3x - 2y = 16$
$6x - 4y = 32$

6. $4y = x - 8$
$-\frac{1}{4}x + y = -1$

7. $-2x + y = -2$
$3x + y = 3$

8. $3x = \frac{1}{3}y + 2$
$9x - y = -6$

Use a graph to solve the equation. Check your solution. *(Section 5.4)*

9. $4x - 1 = 2x$

10. $-\frac{1}{2}x + 1 = -x + 1$

11. $1 - 3x = -3x + 2$

12. $1 - 5x = 3 - 7x$

13. TOURS The table shows the activities of two visitors at a park. You want to take the boat tour for 2 hours and the walking tour for 3 hours. Can you determine how much you will pay? Explain. *(Section 5.4)*

	Boat Tour	Walking Tour	Total Cost
Visitor 1	1 hour	2 hours	\$19
Visitor 2	1.5 hours	3 hours	\$28.50

14. RENTALS A business rents bicycles and in-line skates. Bicycle rentals cost \$25 per day, and in-line skate rentals cost \$20 per day. The business has 20 rentals today and makes \$455. *(Section 5.3)*

a. Write a system of linear equations that represents this situation.

b. How many bicycle rentals and in-line skate rentals did the business have today?

5 Chapter Review

Check It Out
Vocabulary Help
BigIdeasMath.com

Review Key Vocabulary

system of linear equations, *p. 204*

solution of a system of linear equations, *p. 204*

Review Examples and Exercises

5.1 Solving Systems of Linear Equations by Graphing *(pp. 202–207)*

Solve the system by graphing. $y = -2x$ Equation 1

$y = 3x + 5$ Equation 2

Step 1: Graph each equation.

Step 2: Estimate the point of intersection. The graphs appear to intersect at $(-1, 2)$.

Step 3: Check the point from Step 2.

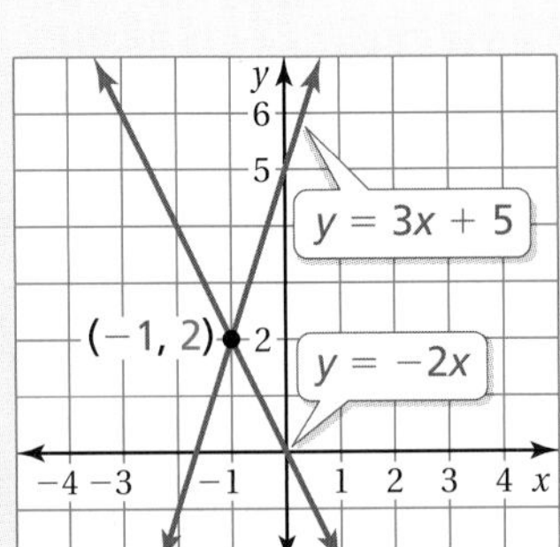

Equation 1	Equation 2
$y = -2x$	$y = 3x + 5$
$2 \stackrel{?}{=} -2(-1)$	$2 \stackrel{?}{=} 3(-1) + 5$
$2 = 2$ ✓	$2 = 2$ ✓

∴ The solution is $(-1, 2)$.

Exercises

Solve the system of linear equations by graphing.

1. $y = 2x - 3$
$y = x + 2$

2. $y = -x + 4$
$x + 3y = 0$

3. $x - y = -2$
$2x - 3y = -2$

5.2 Solving Systems of Linear Equations by Substitution *(pp. 208–213)*

Solve the system by substitution. $x = 1 + y$ Equation 1

$x + 3y = 13$ Equation 2

Step 1: Equation 1 is already solved for x.

Step 2: Substitute $1 + y$ for x in Equation 2.

$1 + y + 3y = 13$ Substitute $1 + y$ for x.

$y = 3$ Solve for y.

Step 3: Substituting 3 for y in Equation 1 gives $x = 4$.

∴ The solution is $(4, 3)$.

Exercises

Solve the system of linear equations by substitution. Check your solution.

4. $y = -3x - 7$
$y = x + 9$

5. $\frac{1}{2}x + y = -4$
$y = 2x + 16$

6. $-x + 5y = 28$
$x + 3y = 20$

5.3 Solving Systems of Linear Equations by Elimination *(pp. 216–223)*

You have a total of 5 quarters and dimes in your pocket. The value of the coins is \$0.80. Write and solve a system of linear equations to find the number x of dimes and the number y of quarters in your pocket.

Use a verbal model to write a system of linear equations.

Number of dimes, x + Number of quarters, y = Number of coins

Value of a dime • Number of dimes, x + Value of a quarter • Number of quarters, y = Total value

The system is $x + y = 5$ and $0.1x + 0.25y = 0.8$.

Step 1: Multiply Equation 2 by 10.

$x + y = 5$		$x + y = 5$	Equation 1
$0.1x + 0.25y = 0.8$	Multiply by 10.	$x + 2.5y = 8$	Revised Equation 2

Step 2: Subtract the equations.

$x + y = 5$ — Equation 1
$x + 2.5y = 8$ — Revised Equation 2
$-1.5y = -3$ — Subtract the equations.

Step 3: Solving the equation $-1.5y = -3$ gives $y = 2$.

Step 4: Substitute 2 for y in one of the original equations and solve for x.

$x + y = 5$ — Equation 1
$x + 2 = 5$ — Substitute 2 for y.
$x = 3$ — Subtract 2 from each side.

So, you have 3 dimes and 2 quarters in your pocket.

Exercises

7. GIFT BASKET A gift basket that contains jars of jam and packages of bread mix costs \$45. There are 8 items in the basket. Jars of jam cost \$6 each, and packages of bread mix cost \$5 each. Write and solve a system of linear equations to find the number of jars of jam and the number of packages of bread mix in the gift basket.

5.4 Solving Special Systems of Linear Equations *(pp. 224–231)*

a. Solve the system. $y = -5x - 8$ Equation 1

$y = -5x + 4$ Equation 2

Solve by substitution. Substitute $-5x + 4$ for y in Equation 1.

$y = -5x - 8$ Equation 1

$-5x + 4 = -5x - 8$ Substitute $-5x + 4$ for y.

$4 = -8$ ✗ Add $5x$ to each side.

∴ The equation $4 = -8$ is never true. So, the system of linear equations has no solution.

b. Solve $x + 1 = \frac{1}{3}x + 3$ using a graph. Check your solution.

Step 1: Write a system of linear equations using each side of the equation.

$$x + 1 = \frac{1}{3}x + 3$$

$y = x + 1$ $y = \frac{1}{3}x + 3$

Step 2: Graph the system.

$y = x + 1$

$y = \frac{1}{3}x + 3$

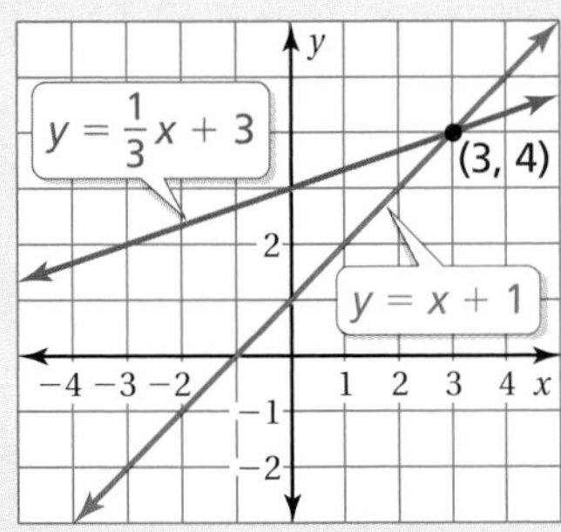

Check

$x + 1 = \frac{1}{3}x + 3$

$3 + 1 \stackrel{?}{=} \frac{1}{3}(3) + 3$

$4 = 4$ ✓

The graphs intersect at (3, 4).

∴ So, the solution is $x = 3$.

Exercises

Solve the system of linear equations. Check your solution.

8. $x + 2y = -5$
$x - 2y = -5$

9. $3x - 2y = 1$
$9x - 6y = 3$

10. $8x - 2y = 16$
$-4x + y = 8$

11. Use a graph to solve $2x - 9 = 7x + 11$. Check your solution.

5 Chapter Test

Solve the system of linear equations by graphing.

1. $y = 4 - x$
   $y = x - 4$

2. $y = \frac{1}{2}x + 10$
   $y = 4x - 4$

3. $y + x = 0$
   $3y + 6x = -9$

Solve the system of linear equations by substitution. Check your solution.

4. $-3x + y = 2$
   $-x + y - 4 = 0$

5. $x + y = 20$
   $y = 2x - 1$

6. $x - y = 3$
   $x + 2y = -6$

Solve the system of linear equations by elimination. Check your solution.

7. $2x + y = 3$
   $x - y = 3$

8. $x + y = 12$
   $3x = 2y + 6$

9. $-2x + y + 3 = 0$
   $3x + 4y = -1$

Without graphing, determine whether the system of linear equations has *one solution, infinitely many solutions,* or *no solution.* Explain your reasoning.

10. $y = 4x + 8$
    $y = 5x + 1$

11. $2y = 16x - 2$
    $y = 8x - 1$

12. $y = -3x + 2$
    $6x + 2y = 10$

Use a graph to solve the equation. Check your solution.

13. $\frac{1}{4}x - 4 = \frac{3}{4}x + 2$

14. $8x - 14 = -2x - 4$

15. **FRUIT** The price of 2 pears and 6 apples is \$14. The price of 3 pears and 9 apples is \$21. Can you determine the unit prices for pears and apples? Explain.

16. **BOUQUET** A bouquet of lilies and tulips has 12 flowers. Lilies cost \$3 each, and tulips cost \$2 each. The bouquet costs \$32. Write and solve a system of linear equations to find the number of lilies and tulips in the bouquet.

17. **DINNER** How much does it cost for 2 specials and 2 glasses of milk?

5 Cumulative Assessment

1. What is the solution of the system of equations shown below?

$$y = -\frac{2}{3}x - 1$$

$$4x + 6y = -6$$

A. $\left(-\frac{3}{2}, 0\right)$ **C.** no solution

B. $(0, -1)$ **D.** infinitely many solutions

2. What is the slope of a line that is perpendicular to the line $y = -0.25x + 3$?

3. On the grid below, Rectangle $EFGH$ is plotted and its vertices are labeled.

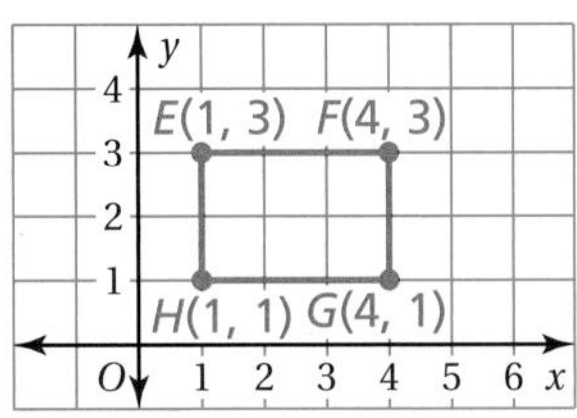

Which of the following shows Rectangle $E'F'G'H'$, the image of Rectangle $EFGH$ after it is reflected in the x-axis?

F.

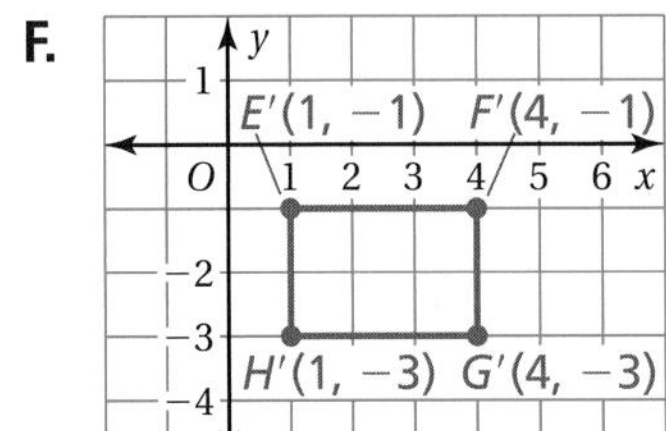

H.

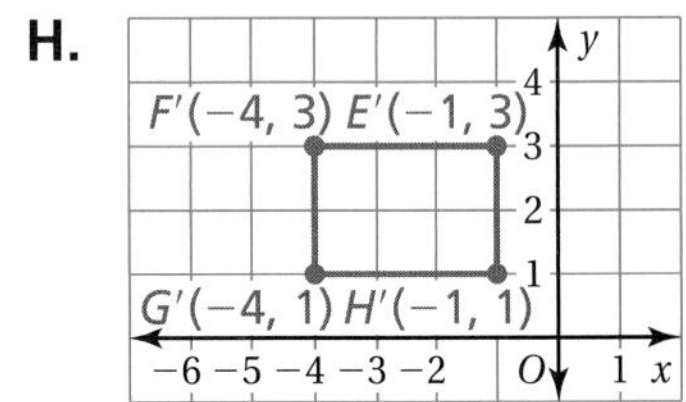

G.

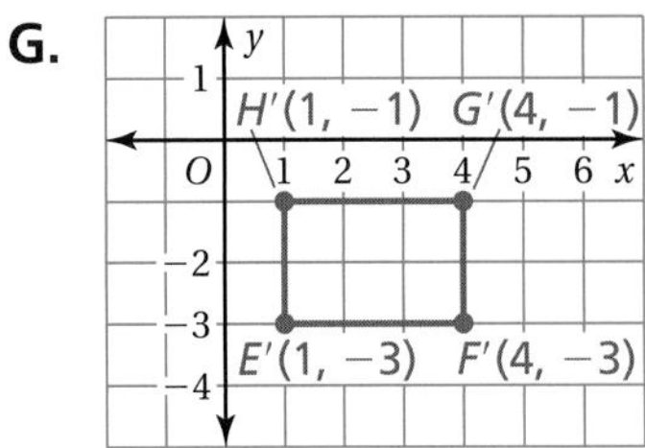

I.

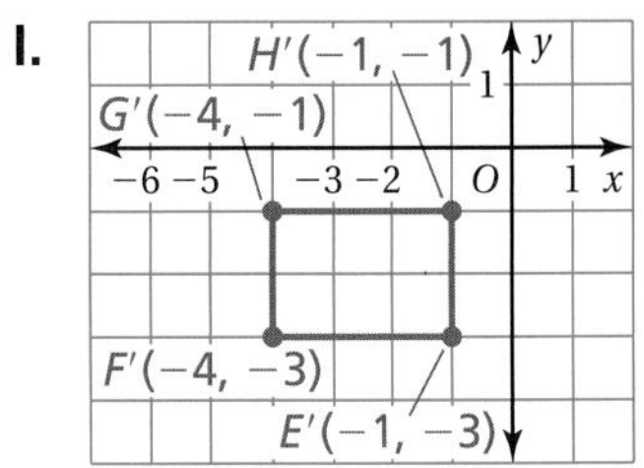

4. Which point is a solution of the system of equations shown below?

$$x + 3y = 10$$

$$x = 2y - 5$$

A. (1, 3) **C.** (55, −15)

B. (3, 1) **D.** (−35, −15)

5. The graph of a system of two linear equations is shown. How many solutions does the system have?

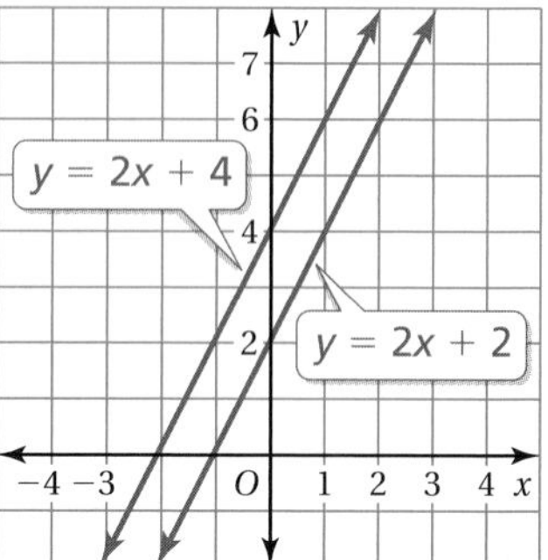

F. none

G. exactly one

H. exactly two

I. infinitely many

6. A scenic train ride has one price for adults and one price for children. One family of two adults and two children pays \$62 for the train ride. Another family of one adult and four children pays \$70. Which system of linear equations can you use to find the price x for an adult and the price y for a child?

A. $2x + 2y = 70$
$x + 4y = 62$

B. $x + y = 62$
$x + y = 70$

C. $2x + 2y = 62$
$4x + y = 70$

D. $2x + 2y = 62$
$x + 4y = 70$

7. Which of the following is true about the graph of the linear equation $y = -7x + 5$?

F. The slope is 5, and the y-intercept is −7.

G. The slope is −5, and the y-intercept is −7.

H. The slope is −7, and the y-intercept is −5.

I. The slope is −7, and the y-intercept is 5.

8. What value of w makes the equation below true?

$$7w - 3w = 2(3w + 11)$$

9. The graph of which equation is parallel to the line that passes through the points $(-1, 5)$ and $(4, 7)$?

A. $y = \frac{2}{3}x + 6$

B. $y = -\frac{5}{2}x + 4$

C. $y = \frac{2}{5}x + 1$

D. $y = \frac{5}{2}x - 1$

10. You buy 3 T-shirts and 2 pairs of shorts for \$42.50. Your friend buys 5 T-shirts and 3 pairs of shorts for \$67.50. Use a system of linear equations to find the cost of each T-shirt. Show your work and explain your reasoning.

Think Solve Explain

11. The two figures have the same area. What is the value of y?

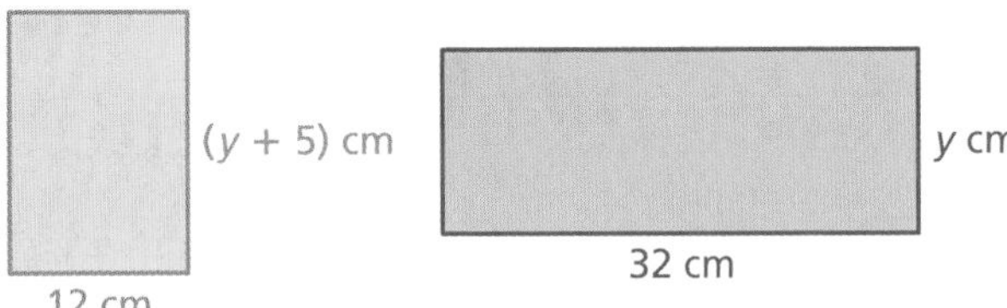

F. $\frac{1}{4}$

G. $\frac{15}{8}$

H. 3

I. 8

12. A system of two linear equations has infinitely many solutions. What can you conclude about the graphs of the two equations?

A. The lines have the same slope and the same y-intercept.

B. The lines have the same slope and different y-intercepts.

C. The lines have different slopes and the same y-intercept.

D. The lines have different slopes and different y-intercepts.

13. The sum of one-third of a number and 10 is equal to 13. What is the number?

F. $\frac{8}{3}$

G. 9

H. 29

I. 69

14. Solve the equation $4x + 7y = 16$ for x.

A. $x = 4 + \frac{7}{4}y$

B. $x = 4 - \frac{7}{4}y$

C. $x = 4 + \frac{4}{7}y$

D. $x = 16 - 7y$

6 Functions

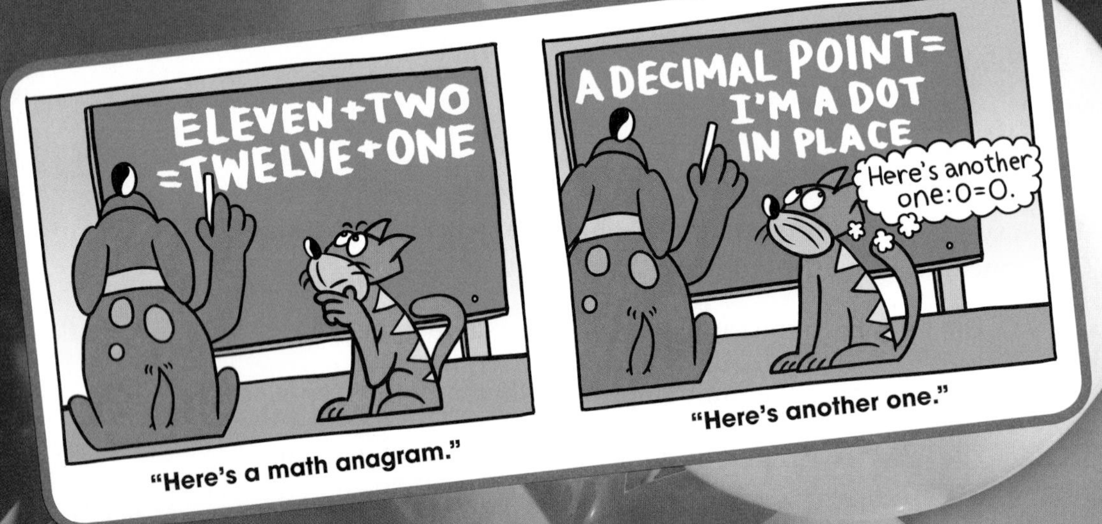

"Here's a math anagram."

"Here's another one."

"It is my treat-converter function machine. However many cat treats I input, the machine outputs TWICE that many dog biscuits. Isn't that cool?"

What You Learned Before

"Do you think the stripes in this shirt make me look too linear?"

Identifying Patterns

Example 1 **Find the missing value in the table.**

x	y
30	0
40	10
50	20
60	

Each y-value is 30 less than the x-value.

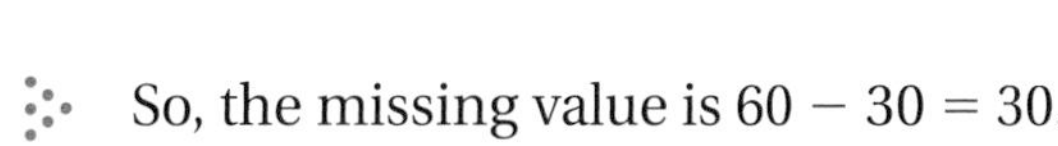

So, the missing value is $60 - 30 = 30$.

Try It Yourself

Find the missing value in the table.

1.

x	y
5	10
7	14
10	20
40	

2.

x	y
0.5	1
1.5	2
3	3.5
9.5	

3.

x	y
15	5
30	10
45	15
60	

Evaluating Algebraic Expressions

Example 2 **Evaluate $2x - 12$ when $x = 5$.**

$$2x - 12 = 2(5) - 12$$ Substitute 5 for x.

$$= 10 - 12$$ Using order of operations, multiply 2 and 5.

$$= 10 + (-12)$$ Add the opposite of 12.

$$= -2$$ Add.

Try It Yourself

Evaluate the expression when $y = 4$.

4. $-4y + 2$

5. $\frac{y}{2} - 8$

6. $-10 - 6y$

6.1 Relations and Functions

Essential Question How can you use a mapping diagram to show the relationship between two data sets?

1 ACTIVITY: Constructing Mapping Diagrams

Work with a partner. Copy and complete the mapping diagram.

a. Area A

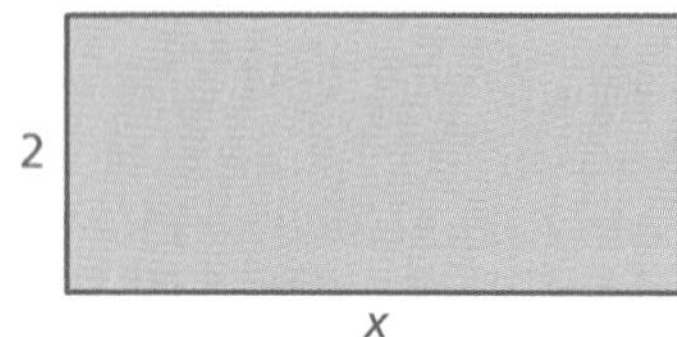

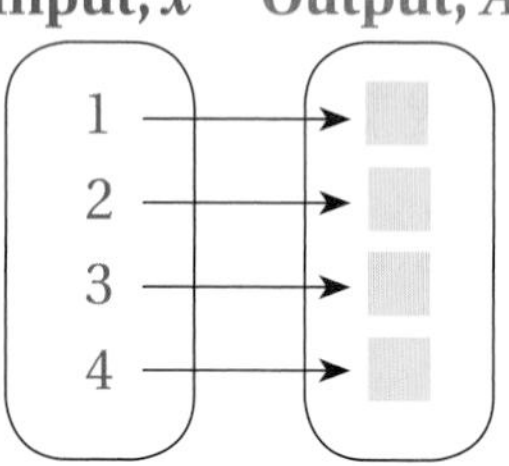

b. Perimeter P

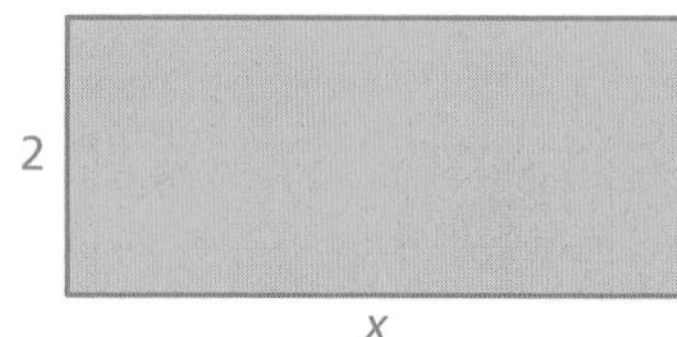

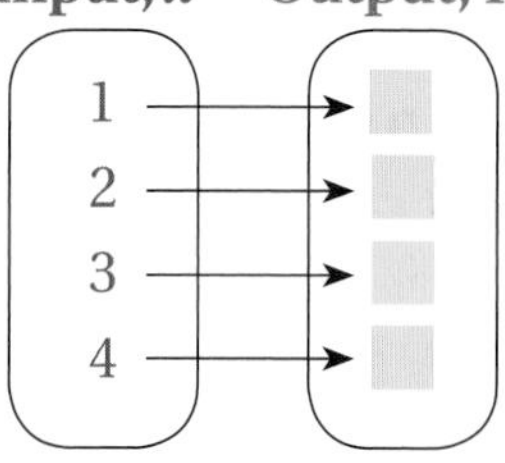

c. Circumference C

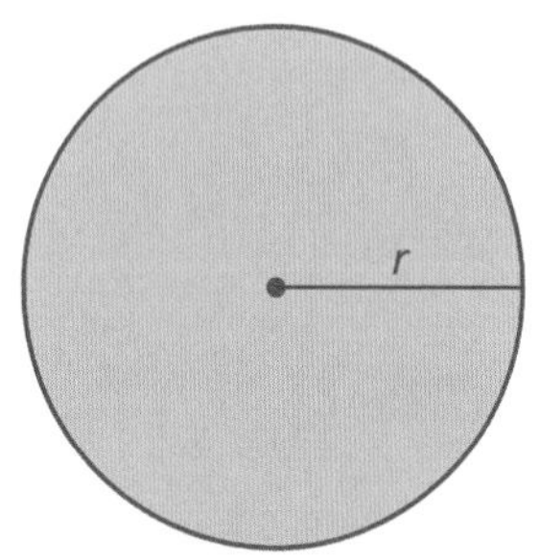

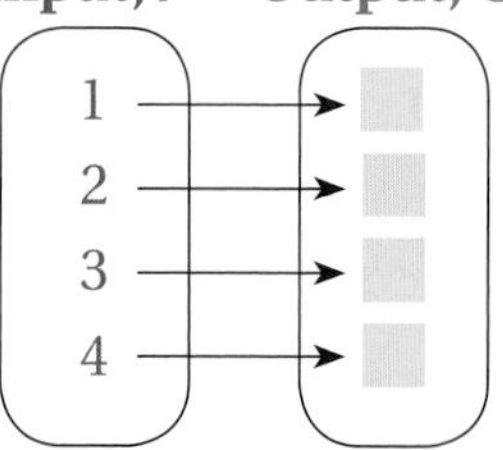

Functions

In this lesson, you will

- define relations and functions.
- determine whether relations are functions.
- describe patterns in mapping diagrams.

d. Volume V

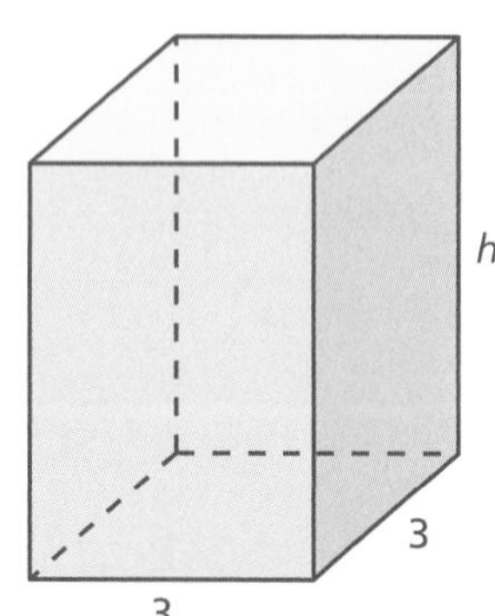

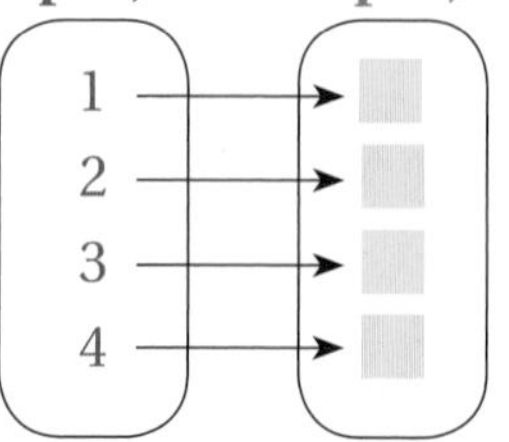

2 ACTIVITY: Describing Situations

Math Practice

View as Components

What are the input values? Do any of the input values point to more than one output value? How does this help you describe a possible situation?

Work with a partner. How many outputs are assigned to each input? Describe a possible situation for each mapping diagram.

a. Input, x Output, y

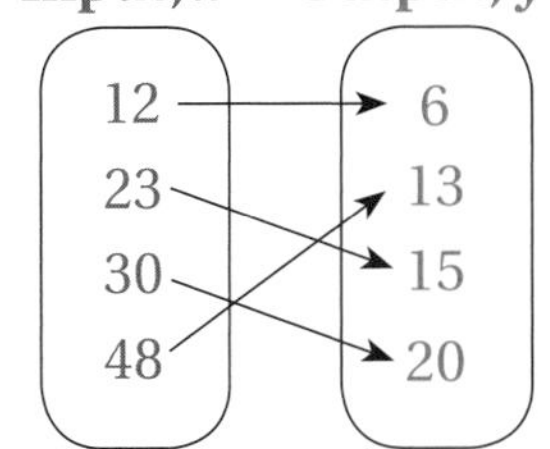

b. Input, x Output, y

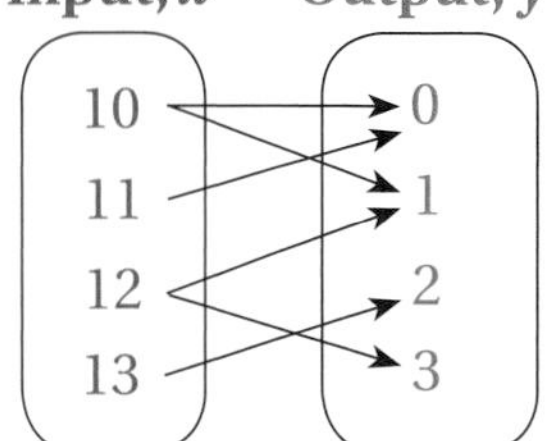

3 ACTIVITY: Interpreting Mapping Diagrams

Work with a partner. Describe the pattern in the mapping diagram. Copy and complete the diagram.

a. Input, t Output, M

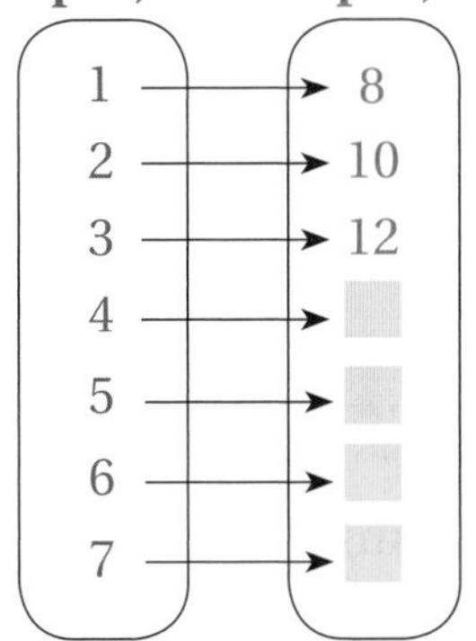

b. Input, x Output, A

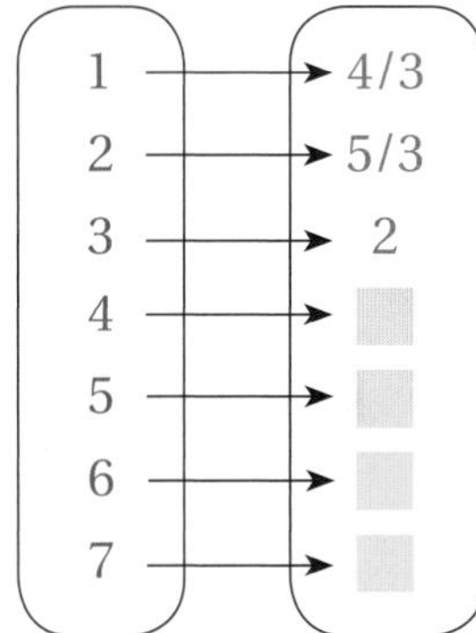

What Is Your Answer?

4. **IN YOUR OWN WORDS** How can you use a mapping diagram to show the relationship between two data sets?

"I made a mapping diagram."

"It shows how I feel about my skateboard with each passing day."

Practice Use what you learned about mapping diagrams to complete Exercises 3–5 on page 246.

6.1 Lesson

Key Vocabulary
input, *p. 244*
output, *p. 244*
relation, *p. 244*
mapping diagram, *p. 244*
function, *p. 245*

Ordered pairs can be used to show **inputs** and **outputs**.

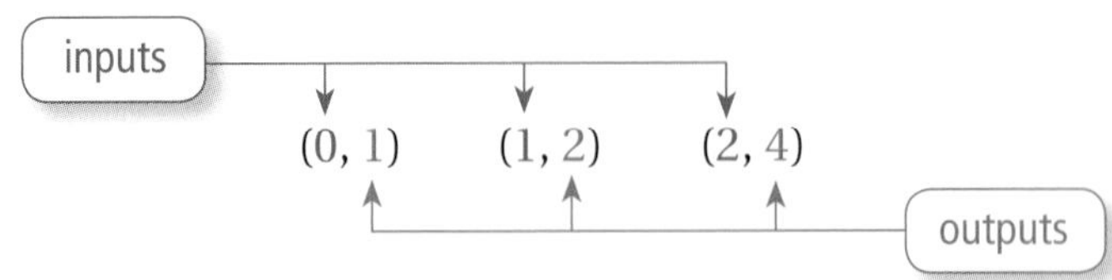

Key Idea

Relations and Mapping Diagrams

A **relation** pairs inputs with outputs. A relation can be represented by ordered pairs or a **mapping diagram**.

Ordered Pairs

(0, 1)

(1, 2)

(2, 4)

Mapping Diagram

Input	Output
0	1
1	2
2	4

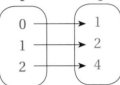

EXAMPLE 1 Listing Ordered Pairs of a Relation

List the ordered pairs shown in the mapping diagram.

a.

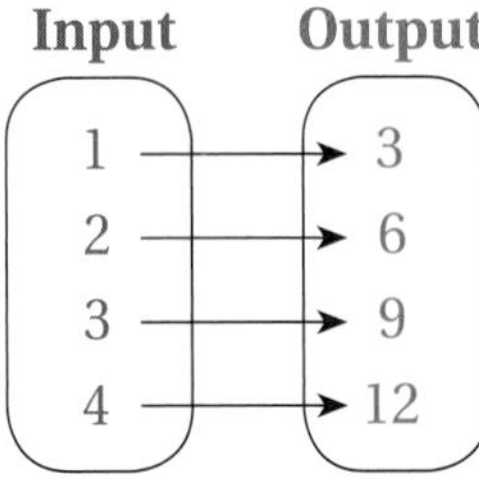

b.

Input	Output
0	1
2	0
4	−2
	−3

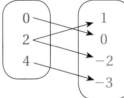

a. The ordered pairs are (1, 3), (2, 6), (3, 9), and (4, 12).

b. The ordered pairs are (0, 0), (2, 1), (2, −2), and (4, −3).

On Your Own

Now You're Ready
Exercises 6–8

List the ordered pairs shown in the mapping diagram.

1.

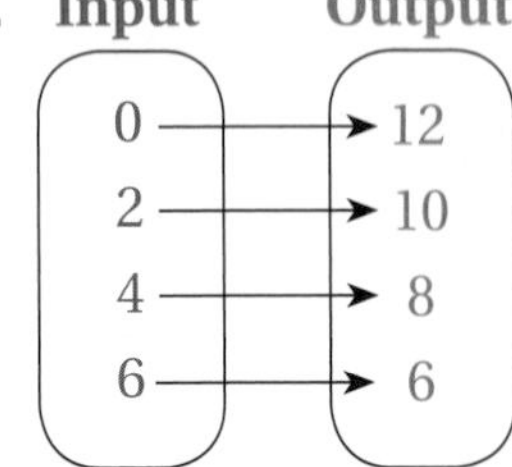

2.

Input	Output
1	−1
2	−2
	−3
	−4

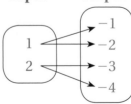

Multi-Language Glossary at BigIdeasMath.com

A relation that pairs each input with *exactly one* output is a **function**.

EXAMPLE 2 Determining Whether Relations Are Functions

Determine whether each relation is a function.

a.

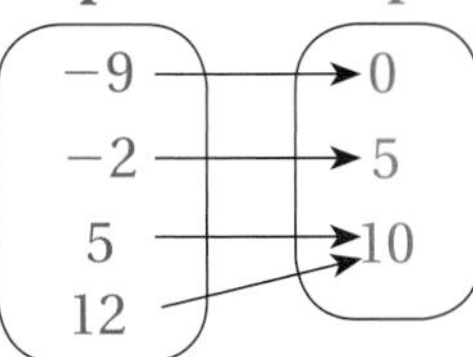

Each input has exactly one output. So, the relation is a function.

b.

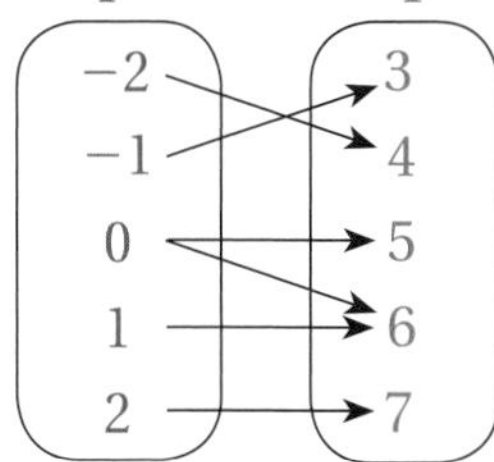

The input 0 has two outputs, 5 and 6. So, the relation is *not* a function.

EXAMPLE 3 Describing a Mapping Diagram

Input	Output
1	15
2	30
3	45
4	60

Consider the mapping diagram at the left.

a. Determine whether the relation is a function.

Each input has exactly one output.

So, the relation is a function.

b. Describe the pattern of inputs and outputs in the mapping diagram.

Look at the relationship between the inputs and the outputs.

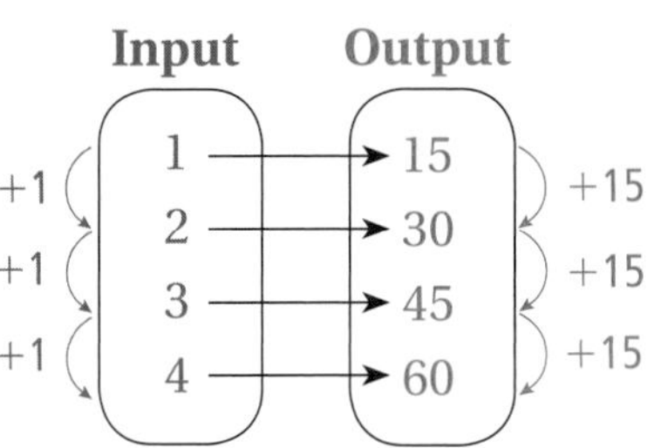

As each input increases by 1, the output increases by 15.

On Your Own

Now You're Ready
Exercises 9–11 and 13–15

Determine whether the relation is a function.

3.

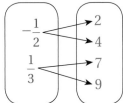

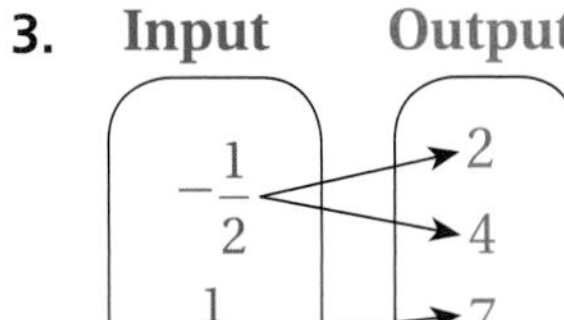

4.

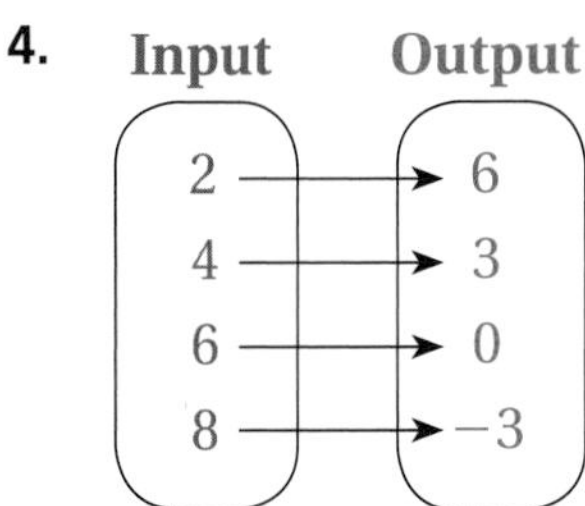

5. Describe the pattern of inputs and outputs in the mapping diagram in On Your Own 4.

6.1 Exercises

Vocabulary and Concept Check

1. **VOCABULARY** In an ordered pair, which number represents the input? the output?
2. **PRECISION** Describe how relations and functions are different.

Practice and Problem Solving

Describe the pattern in the mapping diagram. Copy and complete the diagram.

3. 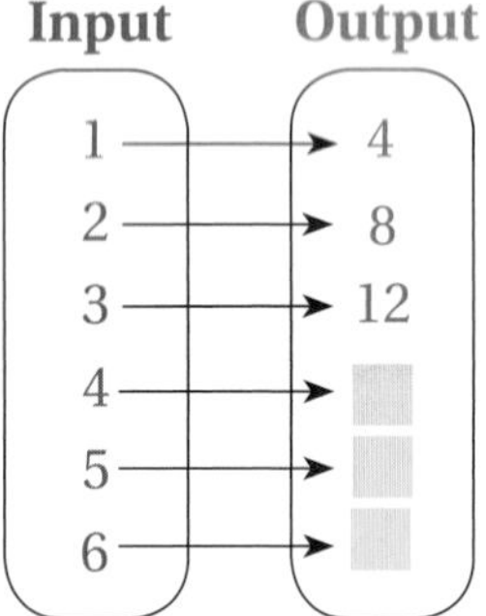

4. 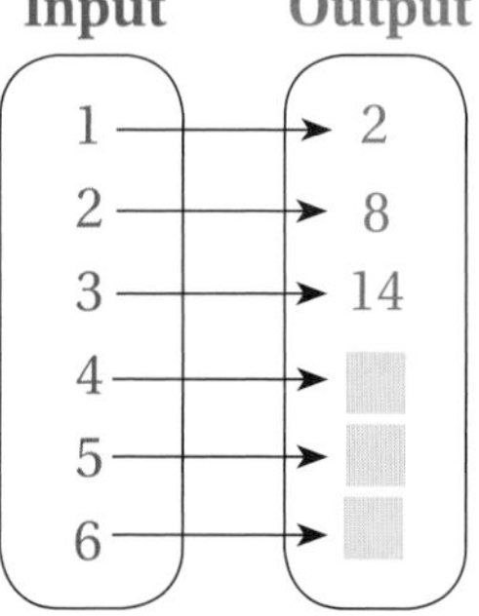

5. 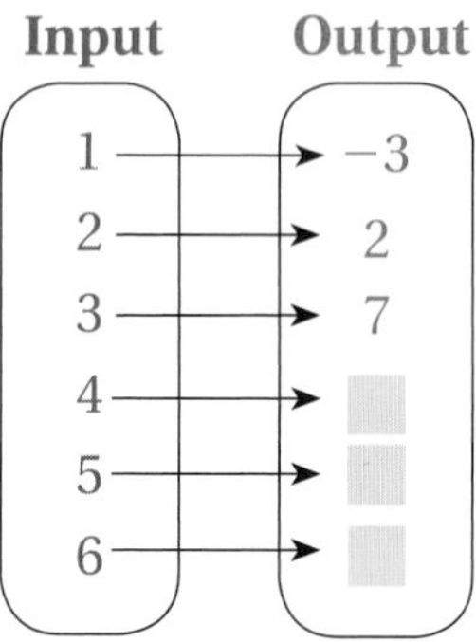

List the ordered pairs shown in the mapping diagram.

1 6. 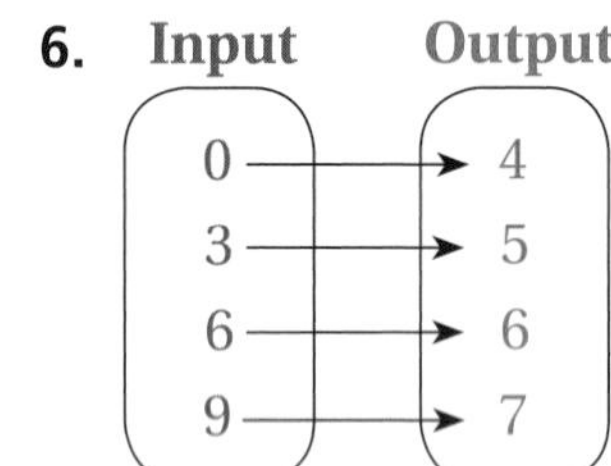

7. 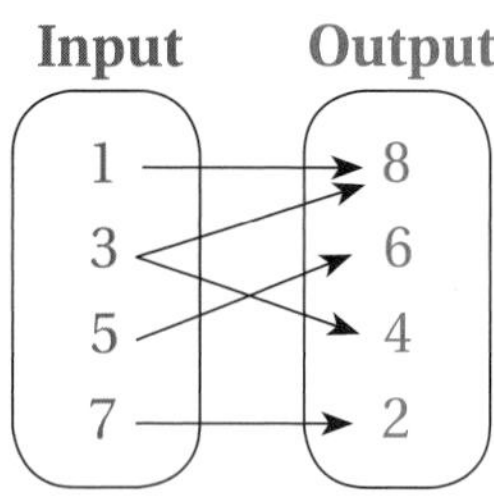

8. 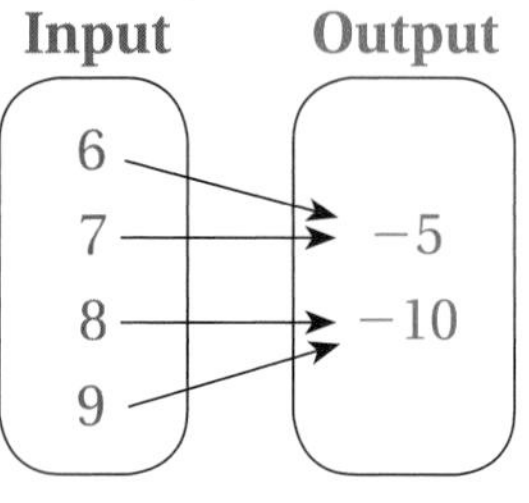

Determine whether the relation is a function.

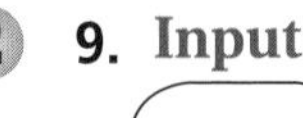

2 9. 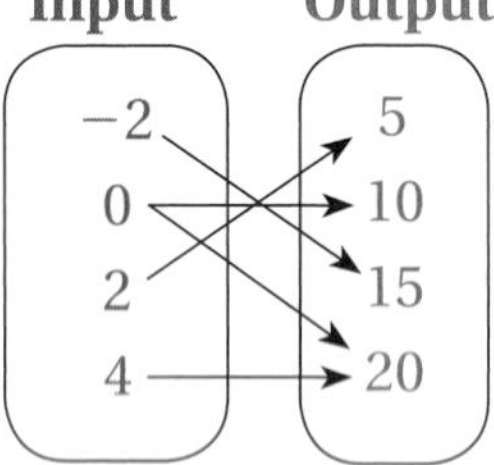

10. 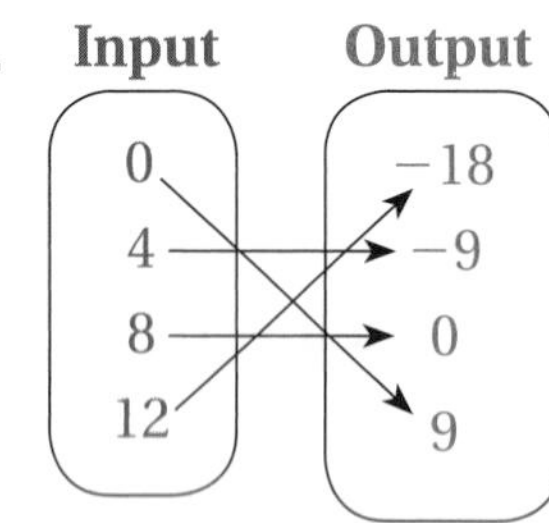

11. 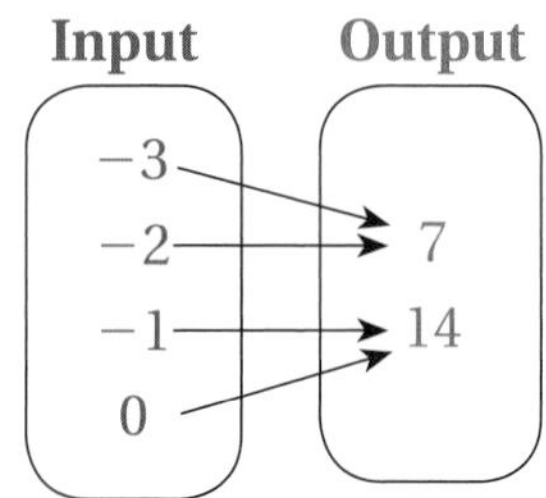

12. **ERROR ANALYSIS** Describe and correct the error in determining whether the relation is a function.

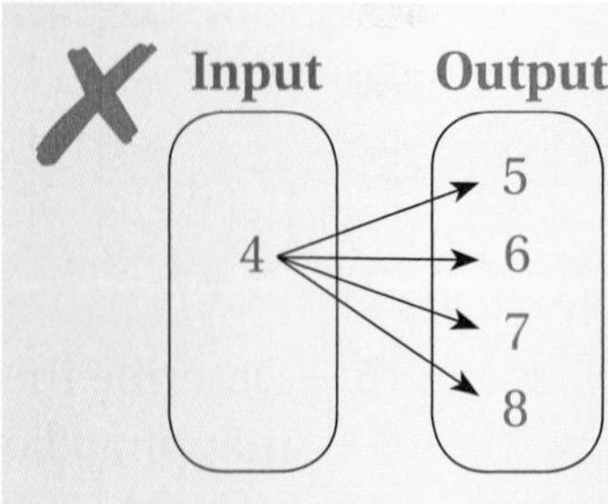

Each output is paired with exactly one input. So, the relation is a function.

Draw a mapping diagram for the graph. Then describe the pattern of inputs and outputs.

3 **13.**

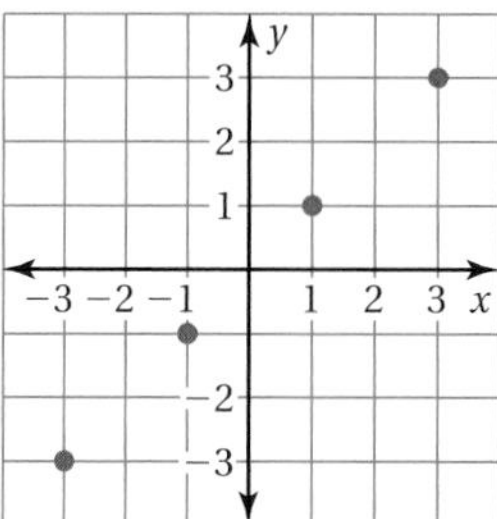

14.

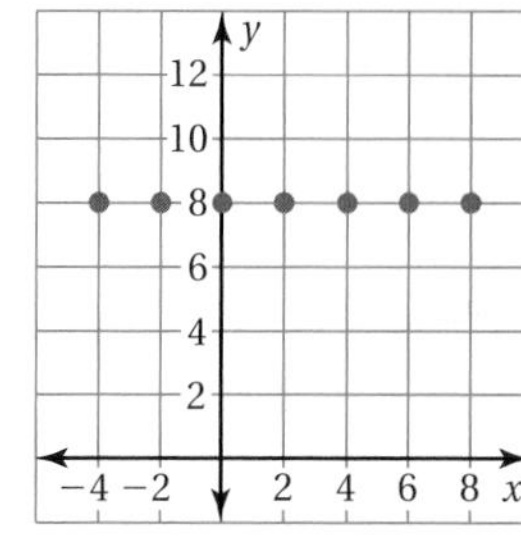

15.

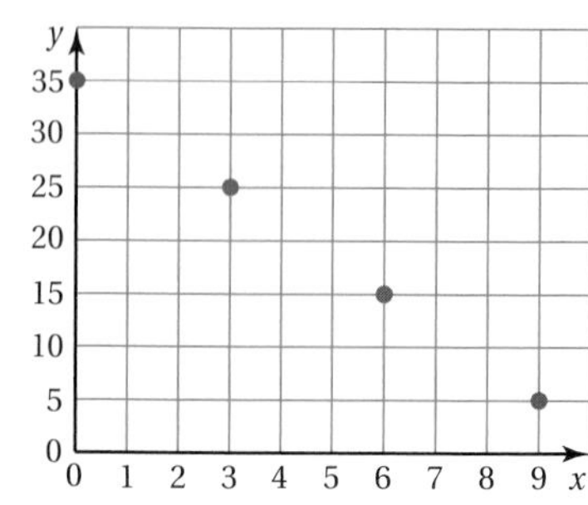

16. SCUBA DIVING The normal pressure at sea level is one atmosphere of pressure (1 ATM). As you dive below sea level, the pressure increases by 1 ATM for each 10 meters of depth.

a. Complete the mapping diagram.

b. Is the relation a function? Explain.

c. List the ordered pairs. Then plot the ordered pairs in a coordinate plane.

d. Compare the mapping diagram and graph. Which do you prefer? Why?

e. **RESEARCH** What are common depths for people who are just learning to scuba dive? What are common depths for experienced scuba divers?

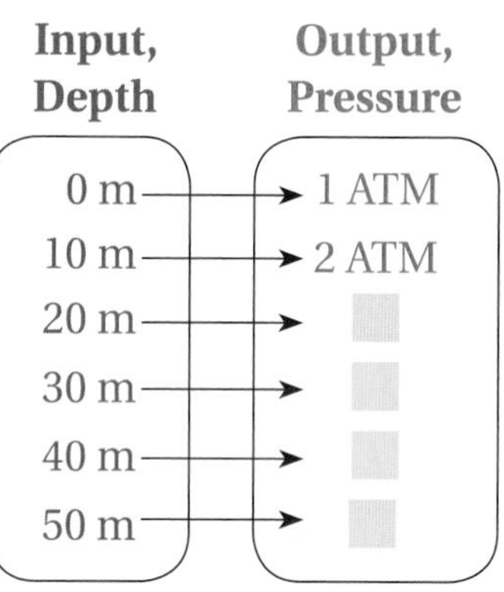

17. MOVIES A store sells previously viewed movies. The table shows the cost of buying 1, 2, 3, or 4 movies.

a. Use the table to draw a mapping diagram.

b. Is the relation a function? Explain.

c. Describe the pattern. How does the cost per movie change as you buy more movies?

Movies	Cost
1	$10
2	$18
3	$24
4	$28

18. Repeated Reasoning The table shows the outputs for several inputs. Use two methods to find the output for an input of 200.

Input, x	0	1	2	3	4
Output, y	25	30	35	40	45

Fair Game Review What you learned in previous grades & lessons

The coordinates of a point and its image are given. Is the reflection in the *x*-axis or *y*-axis? *(Section 2.3)*

19. $(3, -3) \rightarrow (-3, -3)$

20. $(-5, 1) \rightarrow (-5, -1)$

21. $(-2, -4) \rightarrow (-2, 4)$

22. MULTIPLE CHOICE Which word best describes two figures that have the same size and the same shape? *(Section 2.1)*

Ⓐ congruent Ⓑ dilation Ⓒ parallel Ⓓ similar

6.2 Representations of Functions

Essential Question How can you represent a function in different ways?

1 ACTIVITY: Describing a Function

Work with a partner. Copy and complete the mapping diagram for the area of the figure. Then write an equation that describes the function.

a.

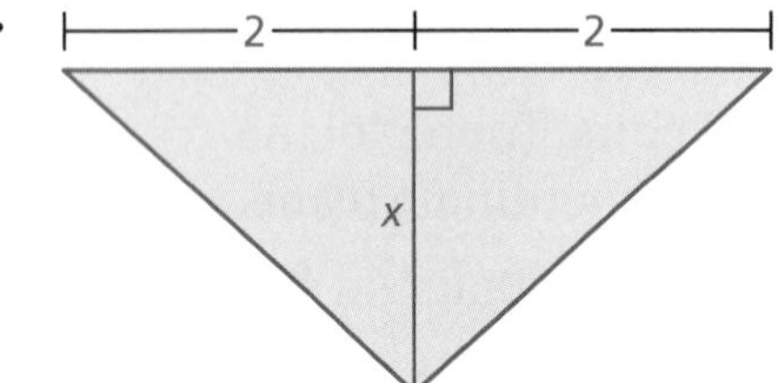

b.

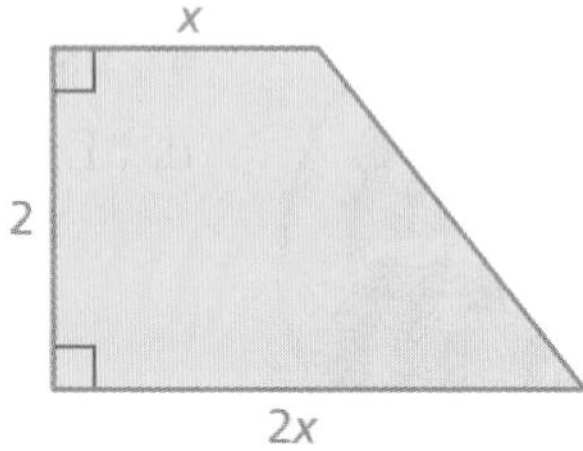

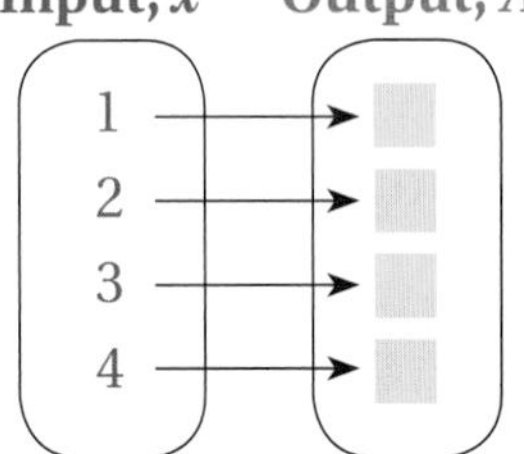

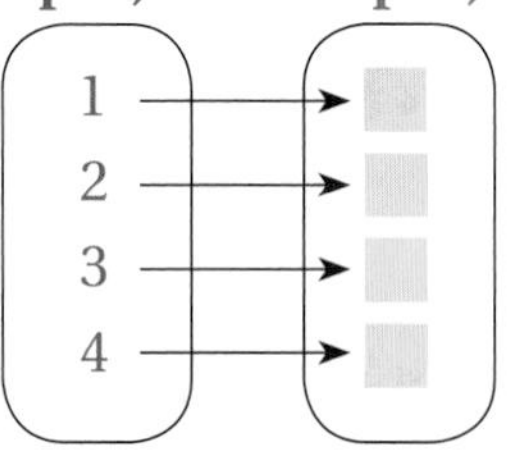

2 ACTIVITY: Using a Table

Work with a partner. Make a table that shows the pattern for the area, where the input is the figure number x and the output is the area A. Write an equation that describes the function. Then use your equation to find which figure has an area of 81 when the pattern continues.

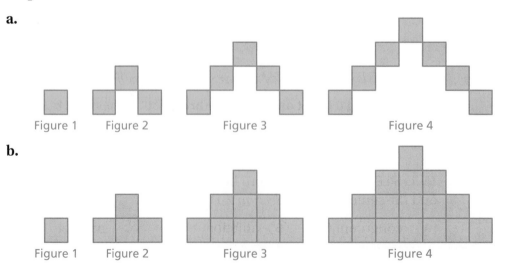

1 square unit

a.

Figure 1 Figure 2 Figure 3 Figure 4

b.

Figure 1 Figure 2 Figure 3 Figure 4

Functions

In this lesson, you will

- write function rules.
- use input-output tables to represent functions.
- use graphs to represent functions.

3 ACTIVITY: Using a Graph

Math Practice

Construct Arguments

How does the graph help you determine whether the statement is true?

Work with a partner. Graph the data. Use the graph to test the truth of each statement. If the statement is true, write an equation that shows how to obtain one measurement from the other measurement.

a. "You can find the horsepower of a race car engine if you know its volume in cubic inches."

Volume (cubic inches), x	200	350	350	500
Horsepower, y	375	650	250	600

b. "You can find the volume of a race car engine in cubic centimeters if you know its volume in cubic inches."

Volume (cubic inches), x	100	200	300
Volume (cubic centimeters), y	1640	3280	4920

4 ACTIVITY: Interpreting a Graph

Work with a partner. The table shows the average speeds of the winners of the Daytona 500. Graph the data. Can you use the graph to predict future winning speeds? Explain why or why not.

Year, x	2004	2005	2006	2007	2008	2009	2010	2011	2012
Speed (mi/h), y	156	135	143	149	153	133	137	130	140

What Is Your Answer?

5. IN YOUR OWN WORDS How can you represent a function in different ways?

"I graphed our profits."

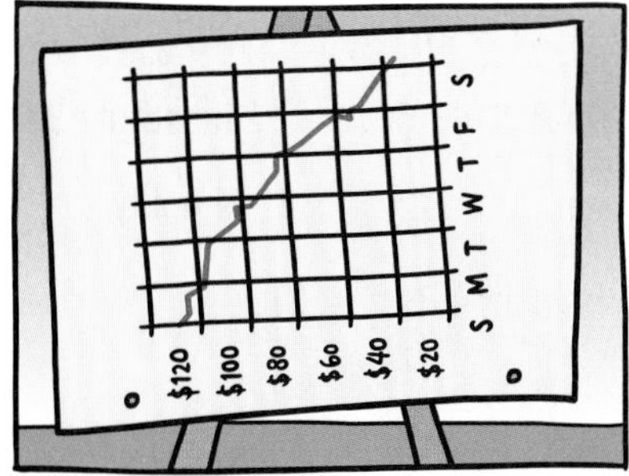

"And I am happy to say that they are going up every day!"

Practice Use what you learned about representing functions to complete Exercises 4–6 on page 253.

6.2 Lesson

Key Vocabulary
function rule, *p. 250*

Remember

An independent variable represents a quantity that can change freely. A dependent variable *depends* on the independent variable.

Functions as Equations

A **function rule** is an equation that describes the relationship between inputs (independent variable) and outputs (dependent variable).

EXAMPLE 1 Writing Function Rules

a. Write a function rule for "The output is five less than the input."

Words	The output	is	five less than	the input.
Equation	y	$=$	$-\ 5$	x

A function rule is $y = x - 5$.

b. Write a function rule for "The output is the square of the input."

Words	The output	is	the square of the input.
Equation	y	$=$	x^2

A function rule is $y = x^2$.

EXAMPLE 2 Evaluating a Function

What is the value of $y = 2x + 5$ when $x = 3$?

$y = 2x + 5$ Write the equation.

$= 2(3) + 5$ Substitute 3 for x.

$= 11$ Simplify.

When $x = 3$, $y = 11$.

On Your Own

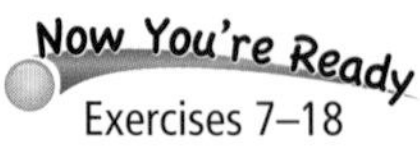

1. Write a function rule for "The output is one-fourth of the input."

Find the value of y when $x = 5$.

2. $y = 4x - 1$
3. $y = 10x$
4. $y = 7 - 3x$

Key Idea

Functions as Tables and Graphs

A function can be represented by an input-output table and by a graph. The table and graph below represent the function $y = x + 2$.

Input, x	Output, y	Ordered Pair, (x, y)
1	3	(1, 3)
2	4	(2, 4)
3	5	(3, 5)

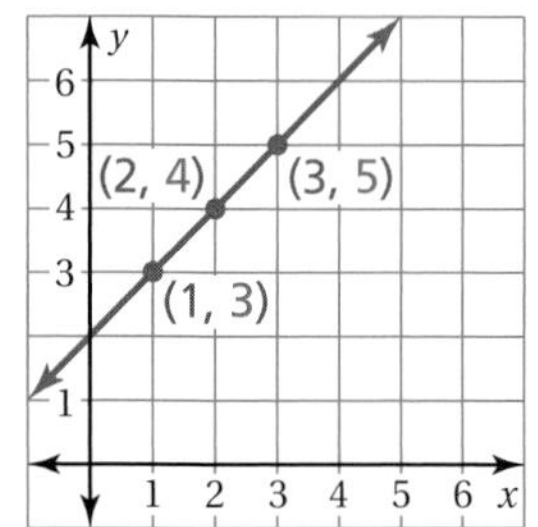

By drawing a line through the points, you graph *all* of the solutions of the function $y = x + 2$.

EXAMPLE 3 Graphing a Function

Graph the function $y = -2x + 1$ using inputs of -1, 0, 1, and 2.

Make an input-output table.

Input, x	$-2x + 1$	Output, y	Ordered Pair, (x, y)
-1	$-2(-1) + 1$	3	$(-1, 3)$
0	$-2(0) + 1$	1	(0, 1)
1	$-2(1) + 1$	-1	$(1, -1)$
2	$-2(2) + 1$	-3	$(2, -3)$

Plot the ordered pairs and draw a line through the points.

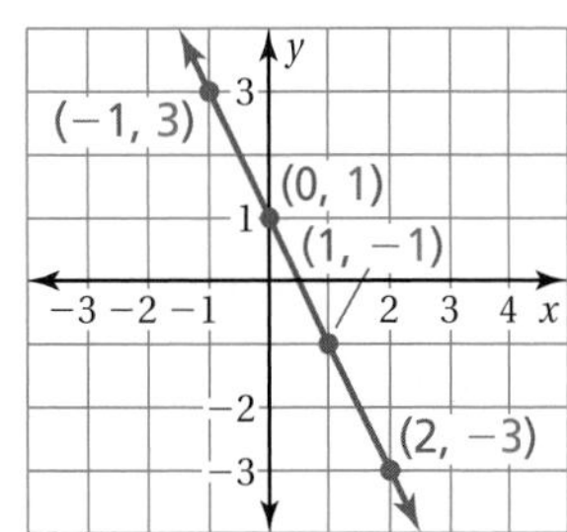

On Your Own

Now You're Ready
Exercises 19–24

Graph the function.

5. $y = x + 1$ **6.** $y = -3x$ **7.** $y = 3x + 2$

EXAMPLE  Real-Life Application

The number of pounds p of carbon dioxide produced by a car is 20 times the number of gallons g of gasoline used by the car. Write and graph a function that describes the relationship between g and p.

Write a function rule using the variables g and p.

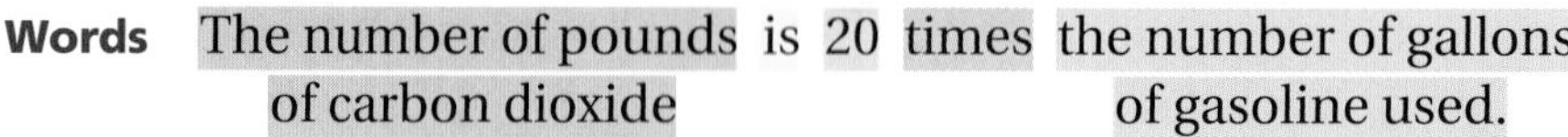

Words The number of pounds of carbon dioxide is 20 times the number of gallons of gasoline used.

Equation $p = 20 \cdot g$

Make an input-output table that represents the function $p = 20g$.

Input, g	$20g$	Output, p	Ordered Pair, (g, p)
1	20(1)	20	(1, 20)
2	20(2)	40	(2, 40)
3	20(3)	60	(3, 60)

Plot the ordered pairs and draw a line through the points.

Because you cannot have a negative number of gallons, use only positive values of g.

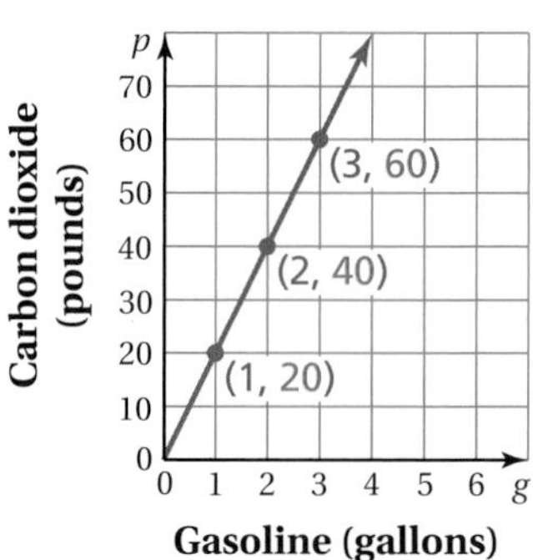

 On Your Own

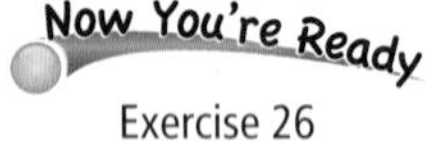
Now You're Ready
Exercise 26

8. **WHAT IF?** For a truck, p is 25 times g. Write and graph a function that describes the relationship between g and p.

Summary

Representations of Functions

Words An output is 2 more than the input.

Equation $y = x + 2$

Input-Output Table

Input, x	Output, y
−1	1
0	2
1	3
2	4

Mapping Diagram

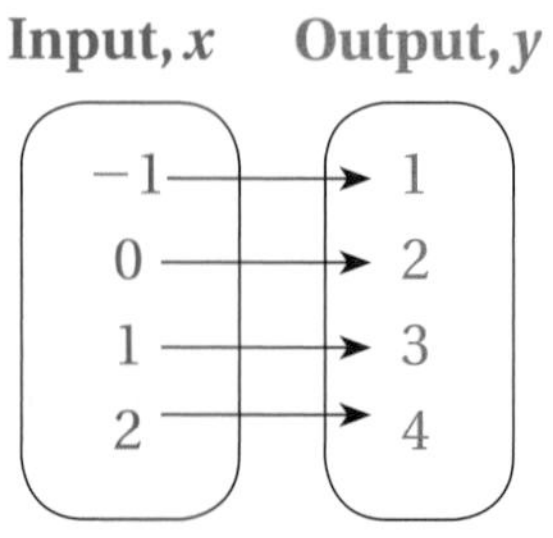

Graph

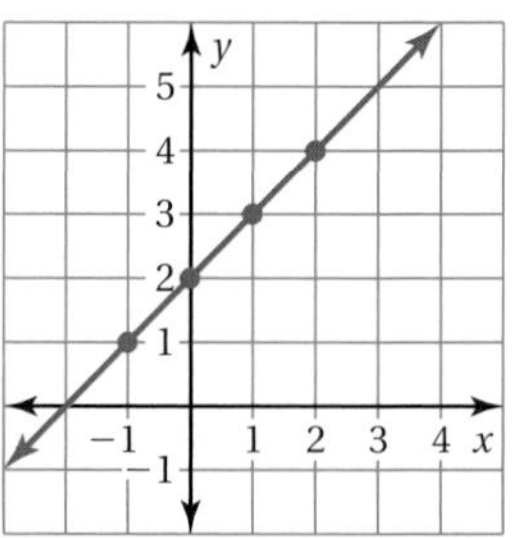

6.2 Exercises

Vocabulary and Concept Check

1. **VOCABULARY** Identify the input variable and the output variable for the function rule $y = 2x + 5$.
2. **WRITING** Describe five ways to represent a function.
3. **DIFFERENT WORDS, SAME QUESTION** Which is different? Find "both" answers.

What output is 4 more than twice the input 3?	What output is twice the sum of the input 3 and 4?
What output is the sum of 2 times the input 3 and 4?	What output is 4 increased by twice the input 3?

Practice and Problem Solving

Write an equation that describes the function.

4\.

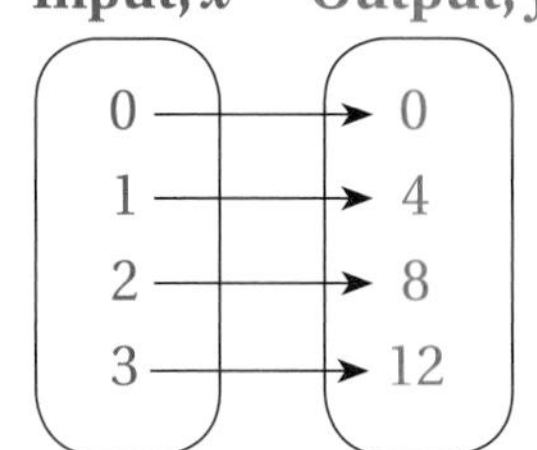

5\.

Input, x	Output, y
1	8
2	9
3	10
4	11

6\.

Input, x	Output, y
1	0
3	-2
5	-4
7	-6

Write a function rule for the statement.

1 7. The output is half of the input.
8. The output is eleven more than the input.
9. The output is three less than the input.
10. The output is the cube of the input.
11. The output is six times the input.
12. The output is one more than twice the input.

Find the value of y for the given value of x.

2 13. $y = x + 5;\ x = 3$
14. $y = 7x;\ x = -5$
15. $y = 1 - 2x;\ x = 9$
16. $y = 3x + 2;\ x = 0.5$
17. $y = 2x^3;\ x = 3$
18. $y = \frac{x}{2} + 9;\ x = -12$

Graph the function.

3 19. $y = x + 4$
20. $y = 2x$
21. $y = -5x + 3$
22. $y = \frac{x}{4}$
23. $y = \frac{3}{2}x + 1$
24. $y = 1 + 0.5x$

25. **ERROR ANALYSIS** Describe and correct the error in graphing the function represented by the input-output table.

Input, x	-4	-2	0	2
Output, y	-1	1	3	5

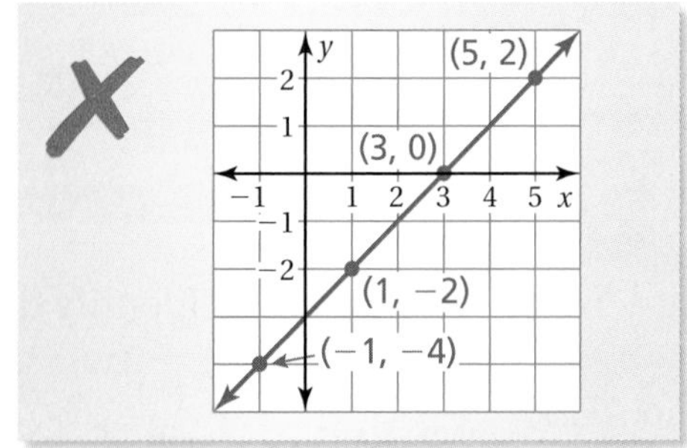

④ 26. **DOLPHIN** A dolphin eats 30 pounds of fish per day.

a. Write and graph a function that relates the number of pounds p of fish that a dolphin eats in d days.

b. How many pounds of fish does a dolphin eat in 30 days?

Match the graph with the function it represents.

27. 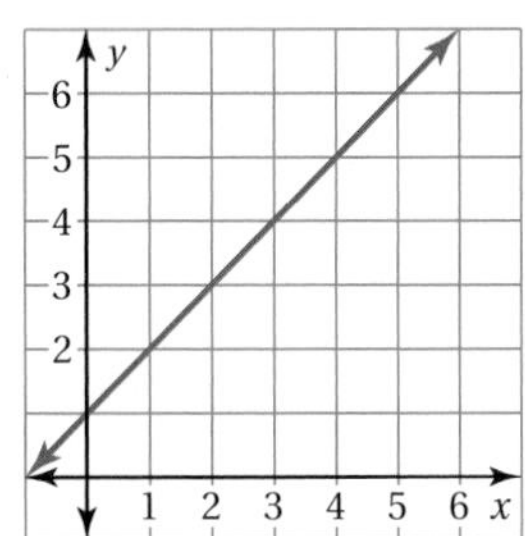

28. 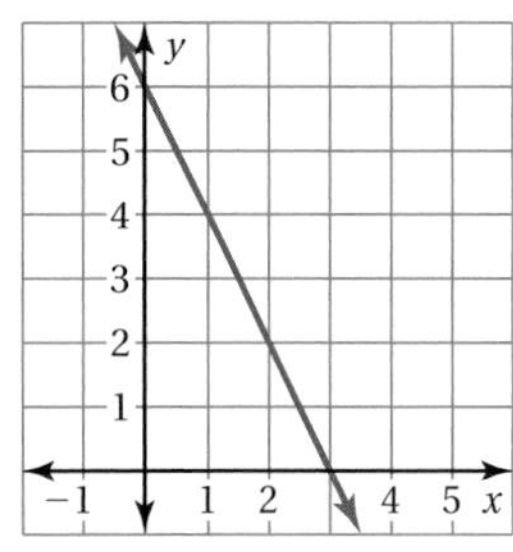

29. 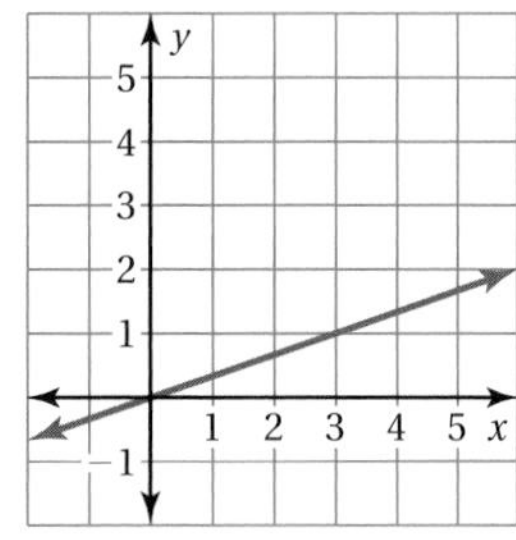

A. $y = \frac{x}{3}$

B. $y = x + 1$

C. $y = -2x + 6$

Find the value of x for the given value of y.

30. $y = 5x - 7;\ y = -22$

31. $y = 9 - 7x;\ y = 37$

32. $y = \frac{x}{4} - 7;\ y = 2$

33. **BRACELETS** You decide to make and sell bracelets. The cost of your materials is \$84. You charge \$3.50 for each bracelet.

a. Write a function that represents the profit P for selling b bracelets.

b. Which variable is independent? dependent? Explain.

c. You will *break even* when the cost of your materials equals your income. How many bracelets must you sell to break even?

34. **SALE** A furniture store is having a sale where everything is 40% off.

a. Write a function that represents the amount of discount d on an item with a regular price p.

b. Graph the function using the inputs 100, 200, 300, 400, and 500 for p.

c. You buy a bookshelf that has a regular price of \$85. What is the sale price of the bookshelf?

35. AIRBOAT TOURS You want to take a two-hour airboat tour.

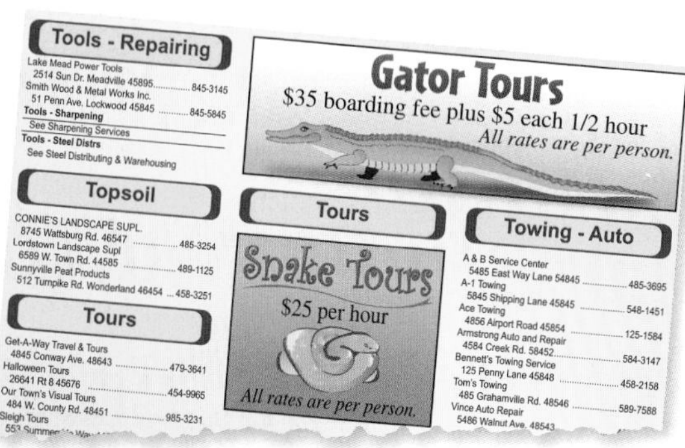

a. Write a function that represents the cost G of a tour at Gator Tours.

b. Write a function that represents the cost S of a tour at Snake Tours.

c. Which is a better deal? Explain.

36. REASONING The graph of a function is a line that goes through the points (3, 2), (5, 8), and (8, y). What is the value of y?

37. CRITICAL THINKING Make a table where the independent variable is the side length of a square and the dependent variable is the *perimeter*. Make a second table where the independent variable is the side length of a square and the dependent variable is the *area*. Graph both functions in the same coordinate plane. Compare the functions and graphs.

38. Puzzle The blocks that form the diagonals of each square are shaded. Each block is one square unit. Find the "green area" of Square 20. Find the "green area" of Square 21. Explain your reasoning.

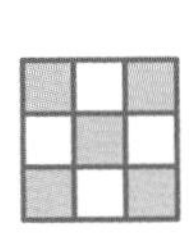

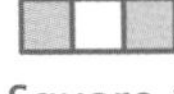

Square 1

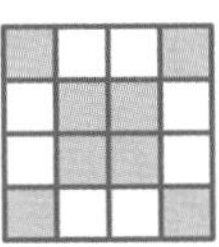

Square 2

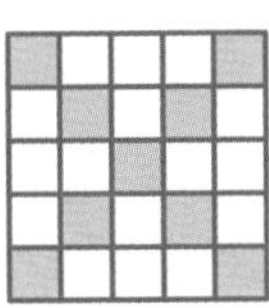

Square 3

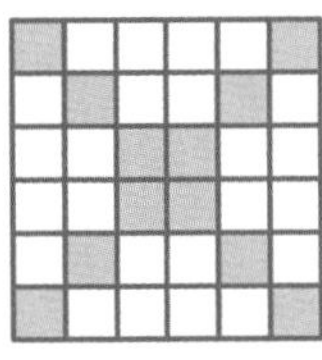

Square 4

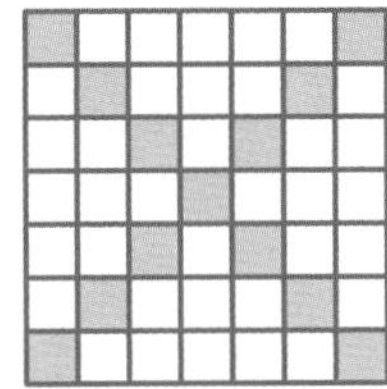

Square 5

Fair Game Review What you learned in previous grades & lessons

Find the slope of the line. *(Section 4.2)*

39.

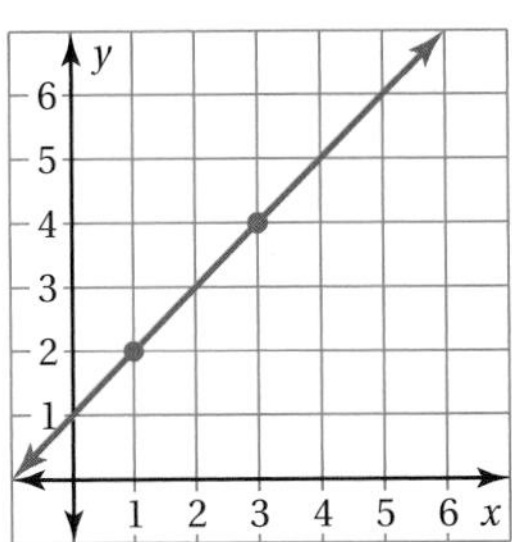

40.

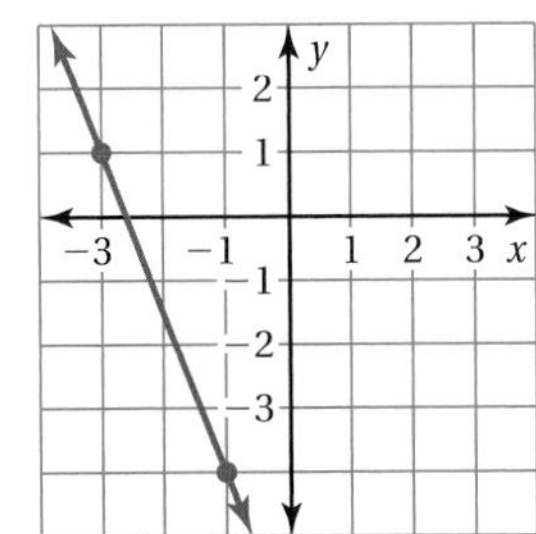

41.

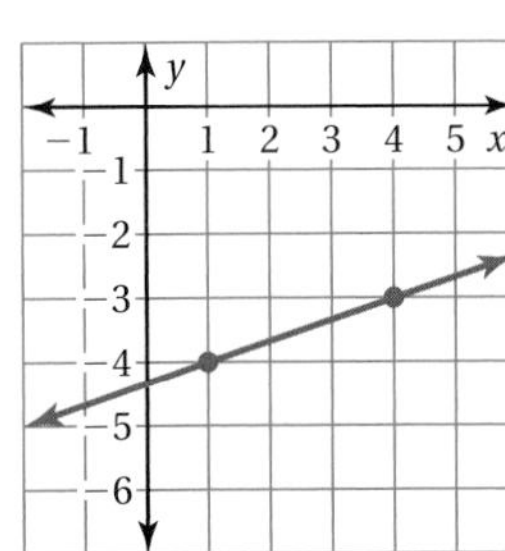

42. MULTIPLE CHOICE You want to volunteer for at most 20 hours each month. So far, you have volunteered for 7 hours this month. Which inequality represents the number of hours h you can volunteer for the rest of this month? *(Skills Review Handbook)*

(A) $h \geq 13$ (B) $h \geq 27$ (C) $h \leq 13$ (D) $h < 27$

6.3 Linear Functions

Essential Question How can you use a function to describe a linear pattern?

1 ACTIVITY: Finding Linear Patterns

Work with a partner.

- **Plot the points from the table in a coordinate plane.**
- **Write a linear equation for the function represented by the graph.**

a.

x	0	2	4	6	8
y	150	125	100	75	50

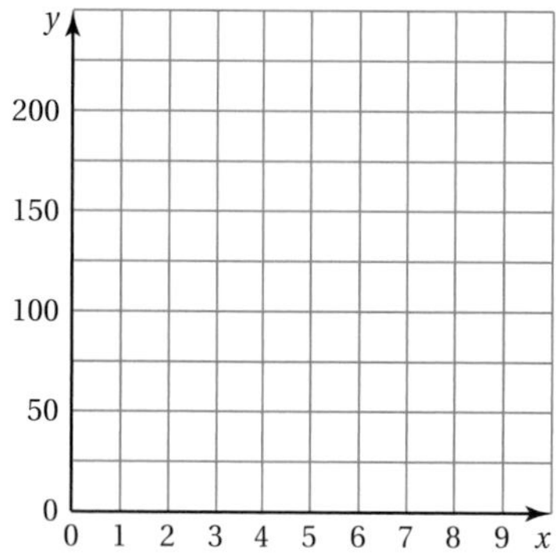

b.

x	4	6	8	10	12
y	15	20	25	30	35

c.

x	−4	−2	0	2	4
y	4	6	8	10	12

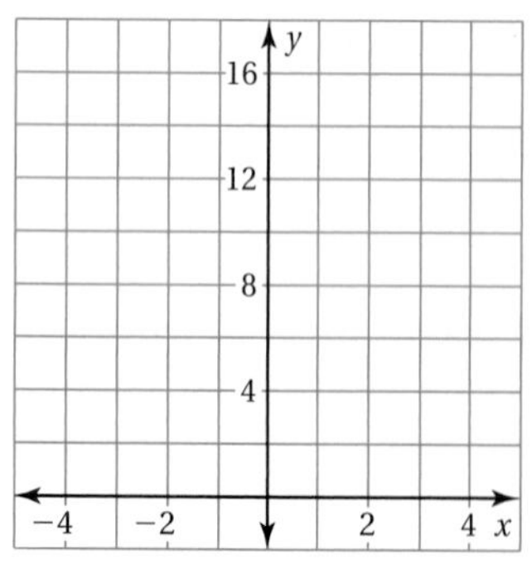

d.

x	−4	−2	0	2	4
y	1	0	−1	−2	−3

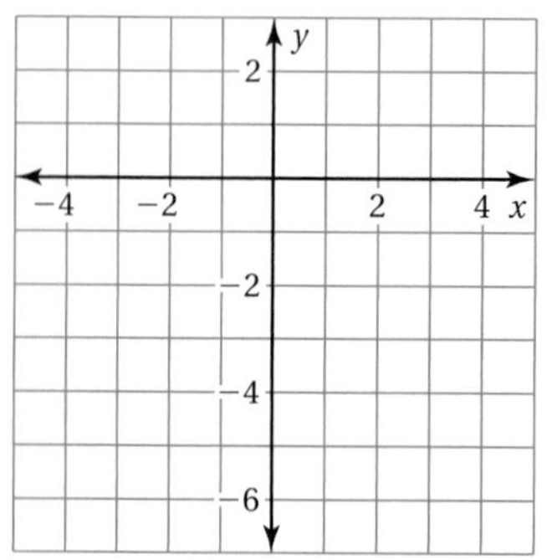

Functions

In this lesson, you will

- understand that the equation $y = mx + b$ defines a linear function.
- write linear functions using graphs or tables.
- compare linear functions.

2 ACTIVITY: Finding Linear Patterns

Math Practice

Label Axes

How do you know what to label the axes? How does this help you accurately graph the data?

Work with a partner. The table shows a familiar linear pattern from geometry.

- **Write a function that relates *y* to *x*.**
- **What do the variables *x* and *y* represent?**
- **Graph the function.**

a.

x	1	2	3	4	5
y	2π	4π	6π	8π	10π

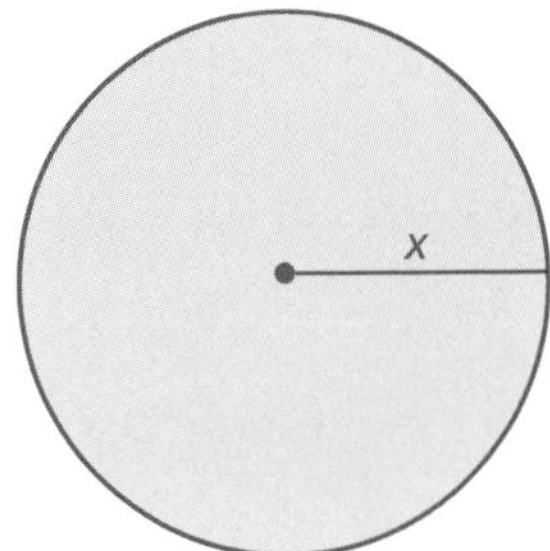

b.

x	1	2	3	4	5
y	10	12	14	16	18

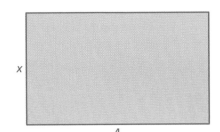

c.

x	1	2	3	4	5
y	5	6	7	8	9

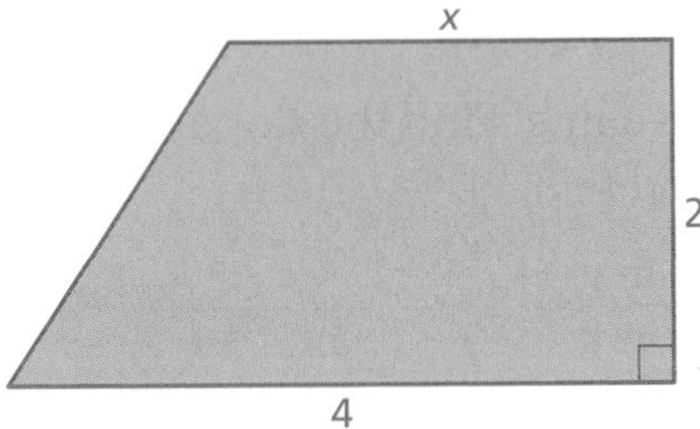

d.

x	1	2	3	4	5
y	28	40	52	64	76

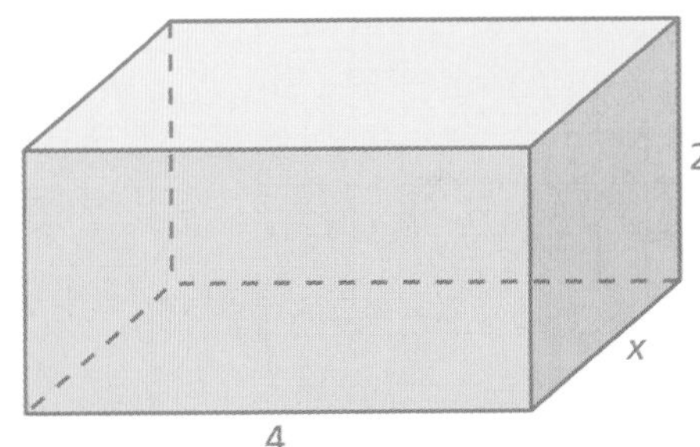

What Is Your Answer?

3. IN YOUR OWN WORDS How can you use a function to describe a linear pattern?

4. Describe the strategy you used to find the functions in Activities 1 and 2.

Practice Use what you learned about linear patterns to complete Exercises 3 and 4 on page 261.

6.3 Lesson

Key Vocabulary
linear function, p. 258

A **linear function** is a function whose graph is a nonvertical line. A linear function can be written in the form $y = mx + b$, where m is the slope and b is the y-intercept.

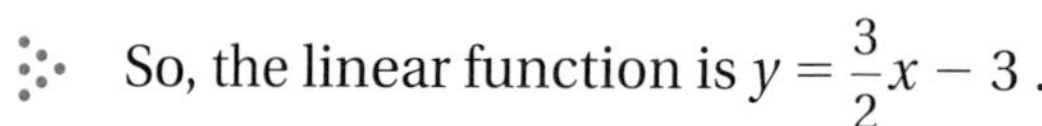

EXAMPLE 1 Writing a Linear Function Using a Graph

Use the graph to write a linear function that relates y to x.

The points lie on a line. Find the slope by using the points (2, 0) and (4, 3).

$$m = \frac{\text{change in } y}{\text{change in } x} = \frac{3 - 0}{4 - 2} = \frac{3}{2}$$

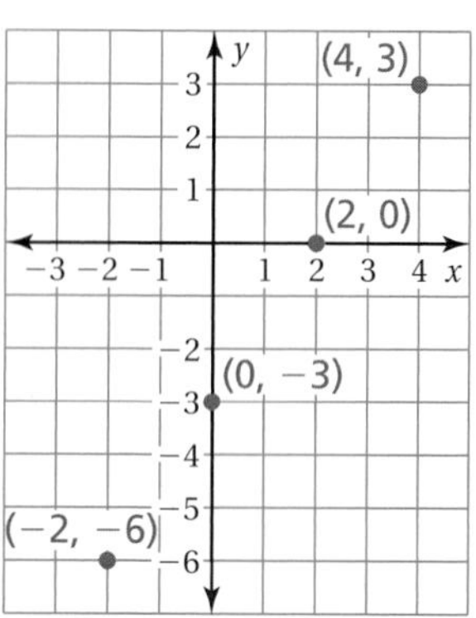

Because the line crosses the y-axis at $(0, -3)$, the y-intercept is -3.

So, the linear function is $y = \frac{3}{2}x - 3$.

EXAMPLE 2 Writing a Linear Function Using a Table

Use the table to write a linear function that relates y to x.

x	−3	−2	−1	0
y	9	7	5	3

Plot the points in the table.

The points lie on a line. Find the slope by using the points $(-2, 7)$ and $(-3, 9)$.

$$m = \frac{\text{change in } y}{\text{change in } x} = \frac{9 - 7}{-3 - (-2)} = \frac{2}{-1} = -2$$

Because the line crosses the y-axis at (0, 3), the y-intercept is 3.

So, the linear function is $y = -2x + 3$.

On Your Own

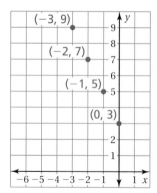

Now You're Ready Exercises 5–10

Use the graph or table to write a linear function that relates y to x.

1.

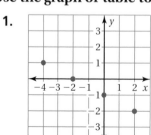

2.

x	−2	−1	0	1
y	2	2	2	2

Multi-Language Glossary at BigIdeasMath.com

EXAMPLE 3 Real-Life Application

Minutes, x	Height (thousands of feet), y
0	65
10	60
20	55
30	50

You are controlling an unmanned aerial vehicle (UAV) for surveillance. The table shows the height y (in thousands of feet) of the UAV x minutes after you start its descent from cruising altitude.

a. Write a linear function that relates y to x. Interpret the slope and the y-intercept.

You can write a linear function that relates the dependent variable y to the independent variable x because the table shows a constant rate of change. Find the slope by using the points (0, 65) and (10, 60).

$$m = \frac{\text{change in } y}{\text{change in } x} = \frac{60 - 65}{10 - 0} = \frac{-5}{10} = -0.5$$

Because the line crosses the y-axis at (0, 65), the y-intercept is 65.

So, the linear function is $y = -0.5x + 65$. The slope indicates that the height decreases 500 feet per minute. The y-intercept indicates that the descent begins at a cruising altitude of 65,000 feet.

Common Error

Make sure you consider the units when interpreting the slope and the y-intercept.

b. Graph the linear function.

Plot the points in the table and draw a line through the points.

Because time cannot be negative in this context, use only positive values of x.

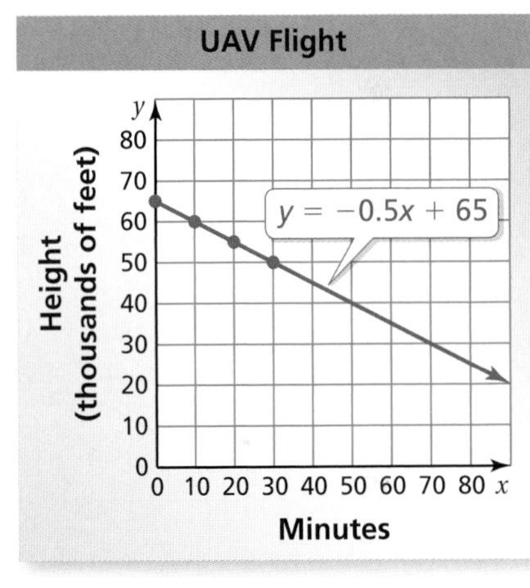

c. Find the height of the UAV when you stop the descent after 1 hour.

Because 1 hour = 60 minutes, find the value of y when $x = 60$.

$y = -0.5x + 65$ — Write the equation.

$= -0.5(60) + 65$ — Substitute 60 for x.

$= 35$ — Simplify.

So, the descent of the UAV stops at a height of 35,000 feet.

On Your Own

Now You're Ready
Exercises 11–13

3. **WHAT IF?** You double the rate of descent. Repeat parts (a)–(c).

EXAMPLE 4 Comparing Linear Functions

The earnings y (in dollars) of a nighttime employee working x hours are represented by the linear function $y = 7.5x + 30$. The table shows the earnings of a daytime employee.

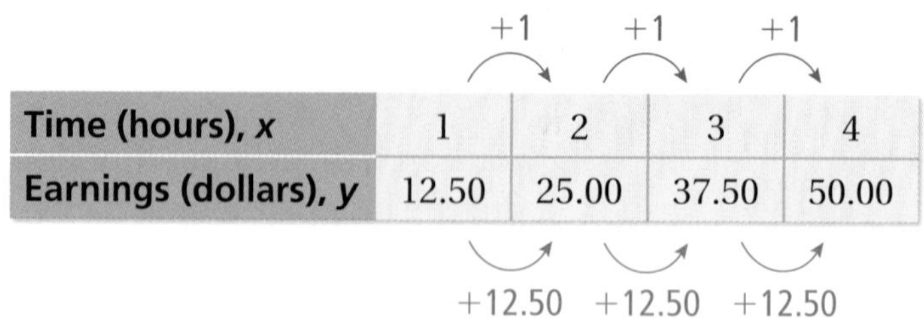

Time (hours), x	1	2	3	4
Earnings (dollars), y	12.50	25.00	37.50	50.00

a. Which employee has a higher hourly wage?

Nighttime Employee

$y = 7.5x + 30$

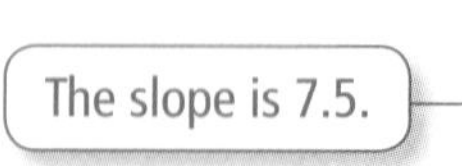

The nighttime employee earns \$7.50 per hour.

Daytime Employee

$$\frac{\text{change in earnings}}{\text{change in time}} = \frac{\$12.50}{1 \text{ hour}}$$

The daytime employee earns \$12.50 per hour.

∴ So, the daytime employee has a higher hourly wage.

b. Write a linear function that relates the daytime employee's earnings to the number of hours worked. In the same coordinate plane, graph the linear functions that represent the earnings of the two employees. Interpret the graphs.

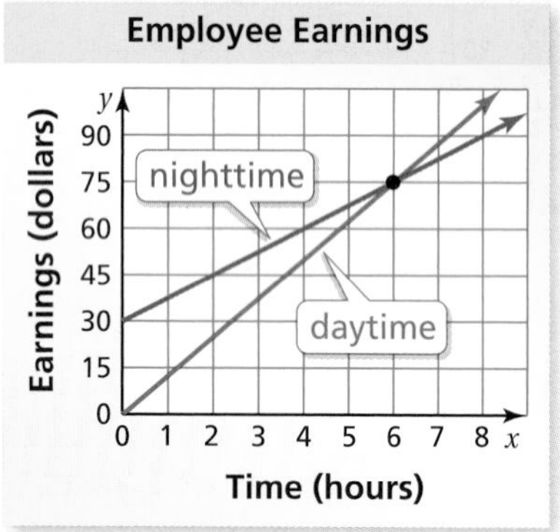

Use a verbal model to write a linear function that represents the earnings of the daytime employee.

$$\text{Earnings} = \text{Hourly wage} \cdot \text{Hours worked}$$

$$y = 12.5x$$

∴ The graph shows that the daytime employee has a higher hourly wage but does not earn more money than the nighttime employee until each person has worked more than 6 hours.

On Your Own

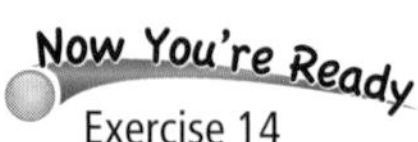

4. Manager A earns \$15 per hour and receives a \$50 bonus. The graph shows the earnings of Manager B.

   a. Which manager has a higher hourly wage?

   b. After how many hours does Manager B earn more money than Manager A?

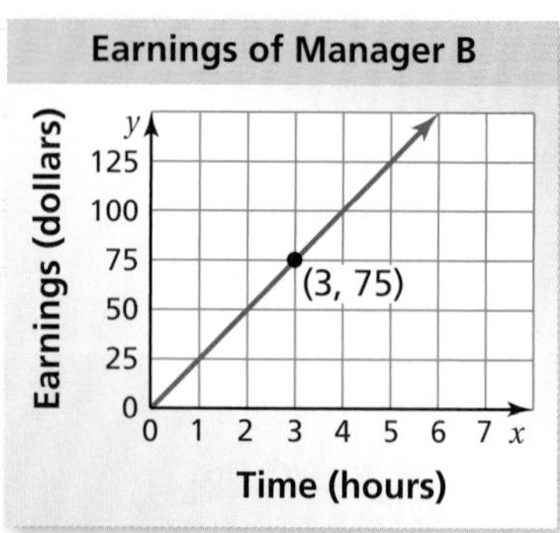

6.3 Exercises

Vocabulary and Concept Check

1. **STRUCTURE** Is $y = mx + b$ a linear function when $b = 0$? Explain.
2. **WRITING** Explain why the vertical line does not represent a linear function.

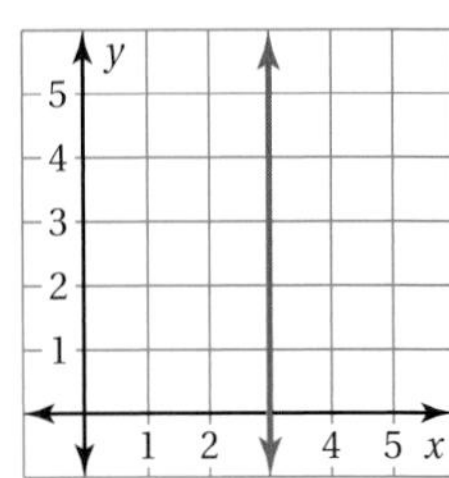

Practice and Problem Solving

The table shows a familiar linear pattern from geometry. Write a function that relates *y* to *x*. What do the variables *x* and *y* represent? Graph the function.

3.

x	1	2	3	4	5
y	π	2π	3π	4π	5π

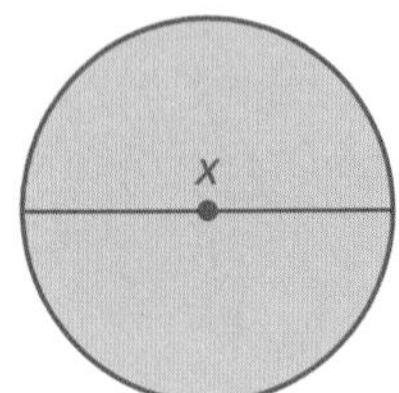

4.

x	1	2	3	4	5
y	2	4	6	8	10

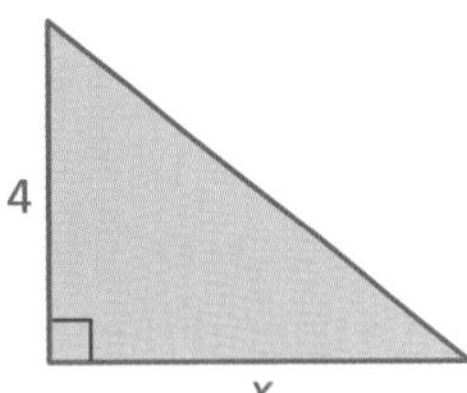

Use the graph or table to write a linear function that relates *y* to *x*.

1 2 5.

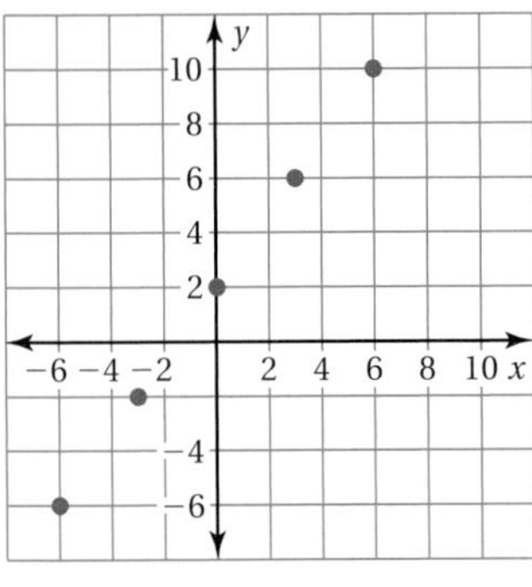

6.

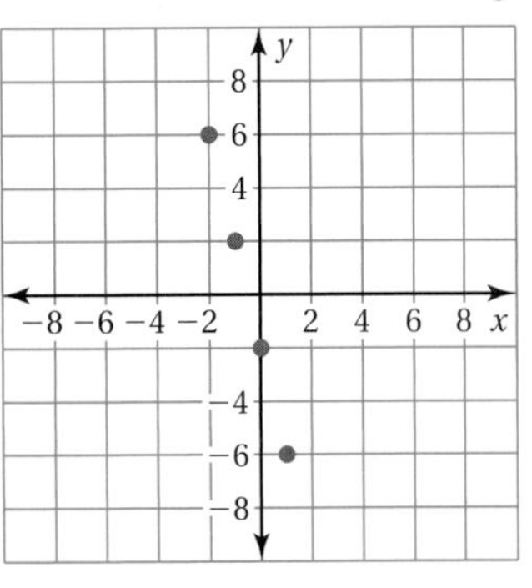

7.

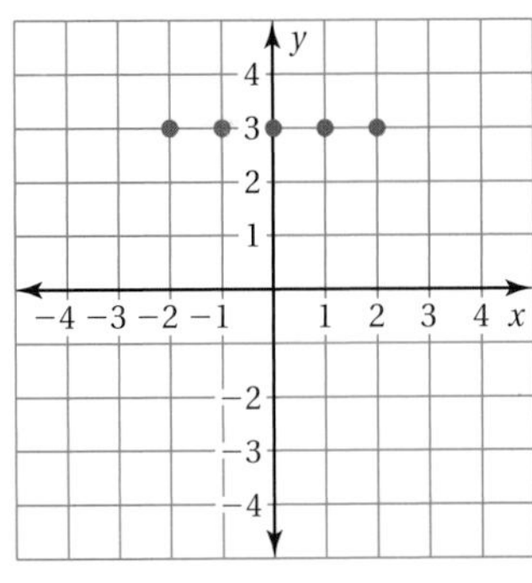

8.

x	−2	−1	0	1
y	−4	−2	0	2

9.

x	−8	−4	0	4
y	2	1	0	−1

10.

x	−3	0	3	6
y	3	5	7	9

3 11. **MOVIES** The table shows the cost *y* (in dollars) of renting *x* movies.

a. Which variable is independent? dependent?

b. Write a linear function that relates *y* to *x*. Interpret the slope.

c. Graph the linear function.

d. How much does it cost to rent three movies?

Number of Movies, *x*	0	1	2	4
Cost, *y*	0	3	6	12

12. **BIKE JUMPS** A *bunny hop* is a bike trick in which the rider brings both tires off the ground without using a ramp. The table shows the height y (in inches) of a bunny hop on a bike that weighs x pounds.

Weight (pounds), x	19	21	23
Height (inches), y	10.2	9.8	9.4

a. Write a linear function that relates the height of a bunny hop to the weight of the bike.

b. Graph the linear function.

c. What is the height of a bunny hop on a bike that weighs 21.5 pounds?

13. **BATTERY** The graph shows the percent y (in decimal form) of battery power remaining x hours after you turn on a laptop computer.

a. Write a linear function that relates y to x.

b. Interpret the slope, the x-intercept, and the y-intercept.

c. After how many hours is the battery power at 75%?

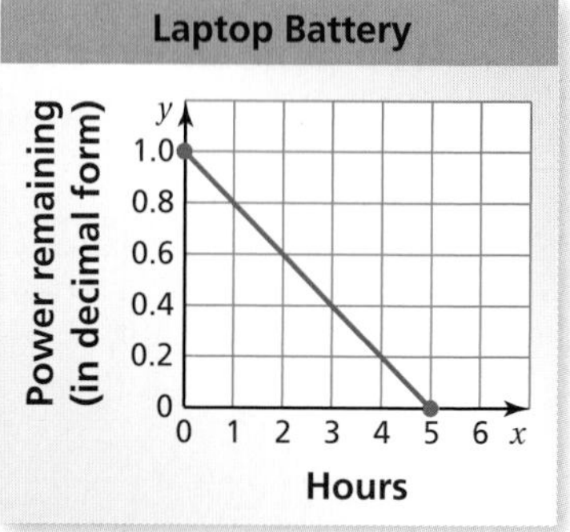

4 14. **RACE** You and a friend race each other. You give your friend a 50-foot head start. The distance y (in feet) your friend runs after x seconds is represented by the linear function $y = 14x + 50$. The table shows the distances you run.

Time (seconds), x	2	4	6	8
Distance (feet), y	38	76	114	152

a. Who runs at a faster rate? What is that rate?

b. Write a linear function that relates your distance to the number of seconds. In the same coordinate plane, graph the linear functions that represent the distances of you and your friend.

c. For what distances will you win the race? Explain.

15. **CALORIES** The number of calories burned y after x minutes of kayaking is represented by the linear function $y = 4.5x$. The graph shows the calories burned by hiking.

a. Which activity burns more calories per minute?

b. How many more calories are burned by doing the activity in part (a) than the other activity for 45 minutes?

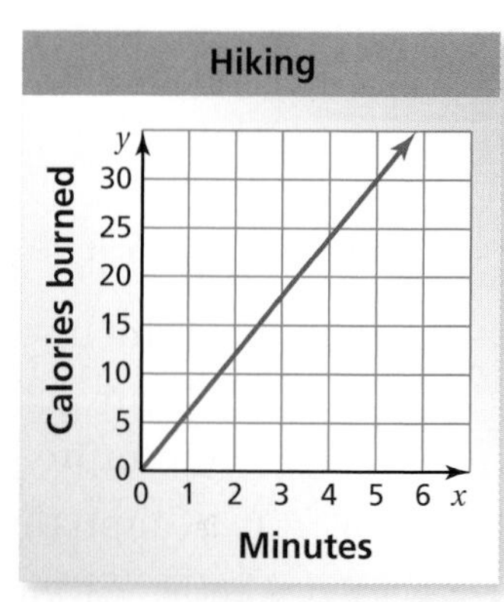

16. **SAVINGS** You and your friend are saving money to buy bicycles that cost \$175 each. The amount y (in dollars) you save after x weeks is represented by the equation $y = 5x + 45$. The graph shows your friend's savings.

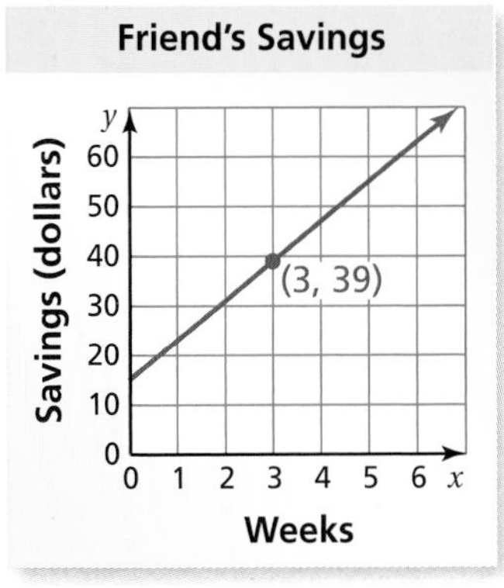

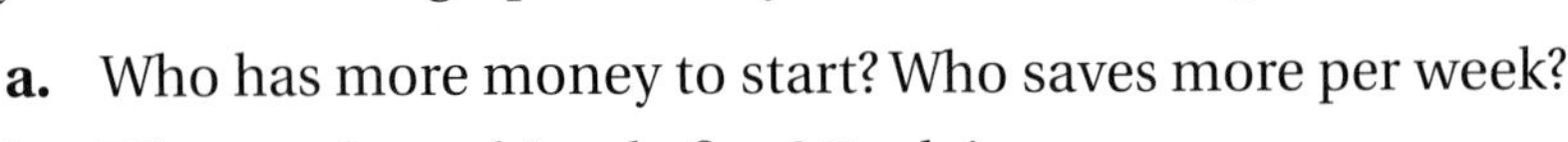

a. Who has more money to start? Who saves more per week?

b. Who can buy a bicycle first? Explain.

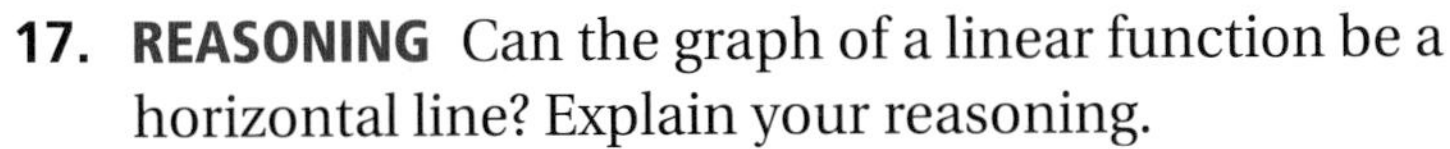

17. **REASONING** Can the graph of a linear function be a horizontal line? Explain your reasoning.

Years of Education, x	Annual Salary, y
0	28
2	40
4	52
6	64
10	88

18. **SALARY** The table shows a person's annual salary y (in thousands of dollars) after x years of education beyond high school.

a. Graph the data. Then describe the pattern.

b. What is the annual salary of the person after 8 years of education beyond high school?

c. Find the annual salary of a person with 30 years of education beyond high school. Do you think this situation makes sense? Explain.

19. Problem Solving The Heat Index is calculated using the relative humidity and the temperature. For every 1 degree increase in the temperature from 94° F to 98° F at 75% relative humidity, the Heat Index rises 4° F.

a. On a summer day, the relative humidity is 75%, the temperature is 94° F, and the Heat Index is 122° F. Construct a table that relates the temperature t to the Heat Index H. Start the table at 94° F and end it at 98° F.

b. Identify the independent and dependent variables.

c. Write a linear function that represents this situation.

d. Estimate the Heat Index when the temperature is 100° F.

Fair Game Review What you learned in previous grades & lessons

Solve the equation. *(Section 1.1)*

20. $b - 1.6 \div 4 = -3$

21. $w + |-2.8| = 4.3$

22. $\frac{3}{4} = y - \frac{1}{5}(8)$

23. **MULTIPLE CHOICE** Which of the following describes the translation from the red figure to the blue figure? *(Section 2.2)*

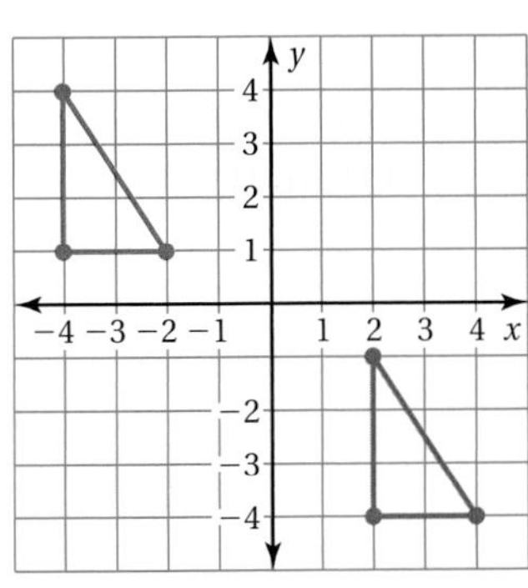

Ⓐ $(x - 6, y + 5)$ Ⓑ $(x - 5, y + 6)$

Ⓒ $(x + 6, y - 5)$ Ⓓ $(x + 5, y - 6)$

6 Study Help

You can use a **comparison chart** to compare two topics. Here is an example of a comparison chart for relations and functions.

	Relations	Functions
Definition	A relation pairs inputs with outputs.	A relation that pairs each input with *exactly one* output is a function.
Ordered pairs	(1, 0) (3, −1) (3, 6) (7, 14)	(1, 0) (2, −1) (5, 7) (6, 20)
Mapping diagram	Input: 1, 3, 7 Output: −1, 0, 6, 14 1 → 0; 3 → −1; 3 → 6; 7 → 14	Input: 1, 2, 5, 6 Output: −1, 0, 7, 20 1 → 0; 2 → −1; 5 → 7; 6 → 20

On Your Own

Make comparison charts to help you study and compare these topics.

1. functions as tables and functions as graphs
2. linear functions with positive slopes and linear functions with negative slopes

After you complete this chapter, make comparison charts for the following topics.

3. linear functions and nonlinear functions
4. graphs with numerical values on the axes and graphs without numerical values on the axes

Comparison of Beagles & Calicos
	Beagles	Calicos
Example	Newton	Descartes
Treats per day	Lots	1 or 2
Either that… or you just wanted more doggy treats.

"Creating a comparison chart causes canines to crystalize concepts."

6.1–6.3 Quiz

Check It Out
Progress Check
BigIdeasMath.com

List the ordered pairs shown in the mapping diagram. Then determine whether the relation is a function. *(Section 6.1)*

1. 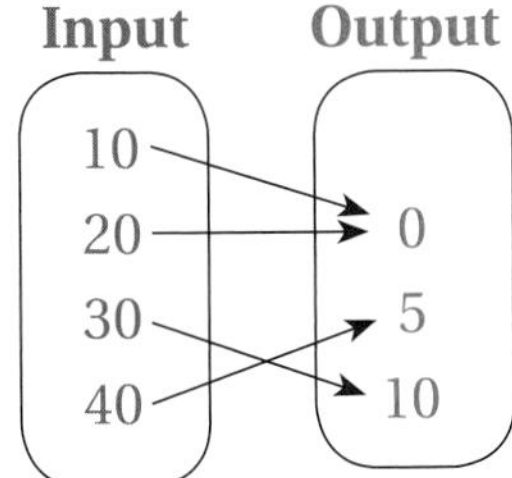

2. 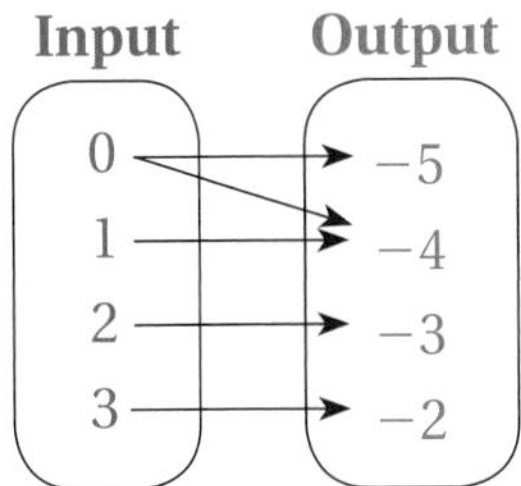

Find the value of y for the given value of x. *(Section 6.2)*

3. $y = 10x;\ x = -3$

4. $y = 6 - 2x;\ x = 11$

5. $y = 4x + 5;\ x = \frac{1}{2}$

Graph the function. *(Section 6.2)*

6. $y = x - 10$

7. $y = 2x + 3$

8. $y = \frac{x}{2}$

Use the graph or table to write a linear function that relates y to x. *(Section 6.3)*

9. 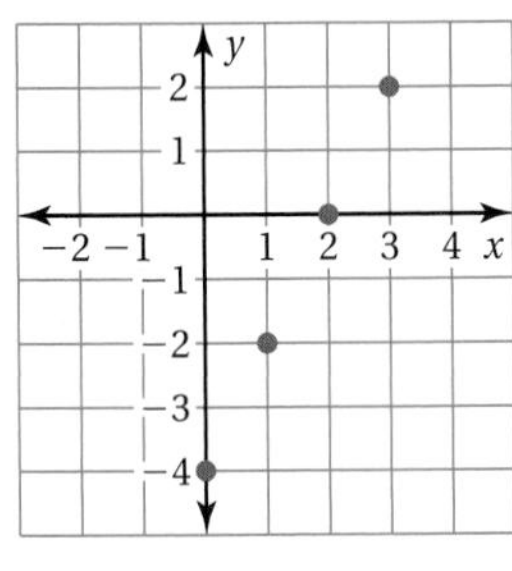

10.

x	y
−3	−3
0	−1
3	1
6	3

11. PUPPIES The table shows the ages of four puppies and their weights. Use the table to draw a mapping diagram. *(Section 6.1)*

Age (weeks)	Weight (oz)
3	11
4	85
6	85
10	480

12. MUSIC An online music store sells songs for \$0.90 each. *(Section 6.2)*

a. Write a function that you can use to find the cost C of buying s songs.

b. What is the cost of buying 5 songs?

13. ADVERTISING The table shows the revenue R (in millions of dollars) of a company when it spends A (in millions of dollars) on advertising. *(Section 6.3)*

Advertising, A	Revenue, R
0	2
2	6
4	10
6	14
8	18

a. Write and graph a linear function that relates the revenue to the advertising cost.

b. What is the revenue of the company when it spends \$15 million on advertising?

6.4 Comparing Linear and Nonlinear Functions

Essential Question How can you recognize when a pattern in real life is linear or nonlinear?

1 ACTIVITY: Finding Patterns for Similar Figures

Work with a partner. Copy and complete each table for the sequence of similar rectangles. Graph the data in each table. Decide whether each pattern is linear or nonlinear.

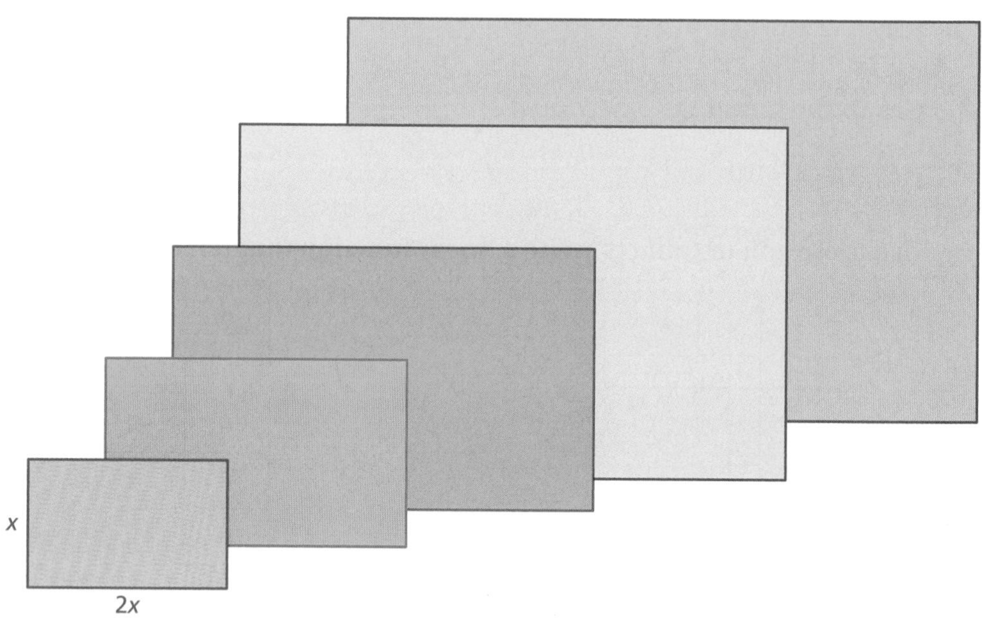

a. Perimeters of similar rectangles

x	1	2	3	4	5
P					

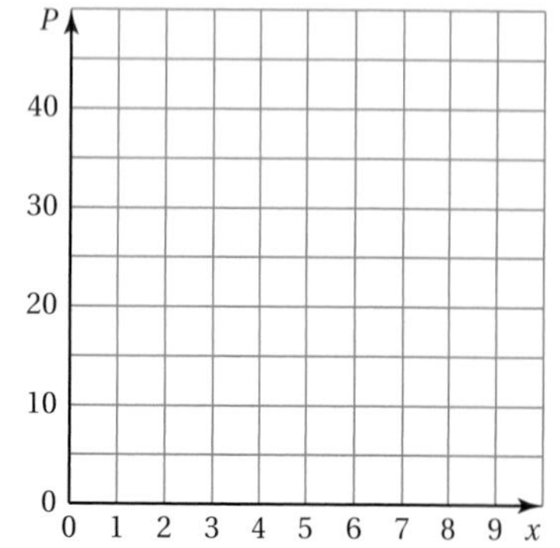

b. Areas of similar rectangles

x	1	2	3	4	5
A					

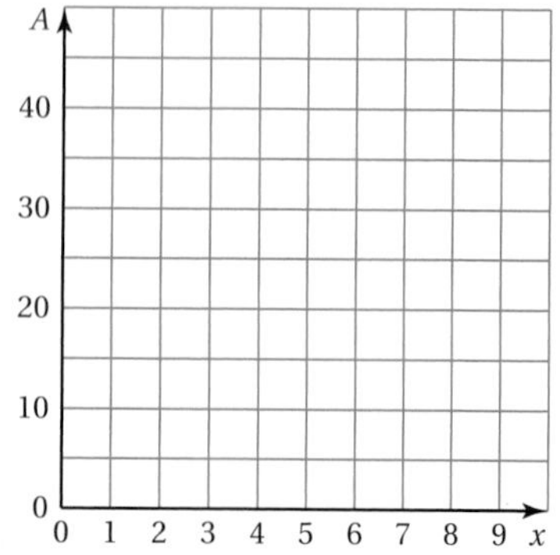

Functions

In this lesson, you will

- identify linear and nonlinear functions from tables or graphs.
- compare linear and nonlinear functions.

2 ACTIVITY: Comparing Linear and Nonlinear Functions

Work with a partner. Each table shows the height *h* (in feet) of a falling object at *t* seconds.

- **Graph the data in each table.**
- **Decide whether each graph is linear or nonlinear.**
- **Compare the two falling objects. Which one has an increasing speed?**

a. Falling parachute jumper

t	0	1	2	3	4
h	300	285	270	255	240

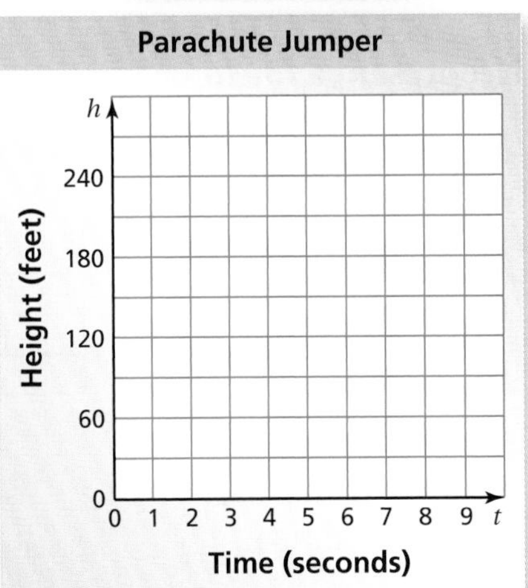

b. Falling bowling ball

t	0	1	2	3	4
h	300	284	236	156	44

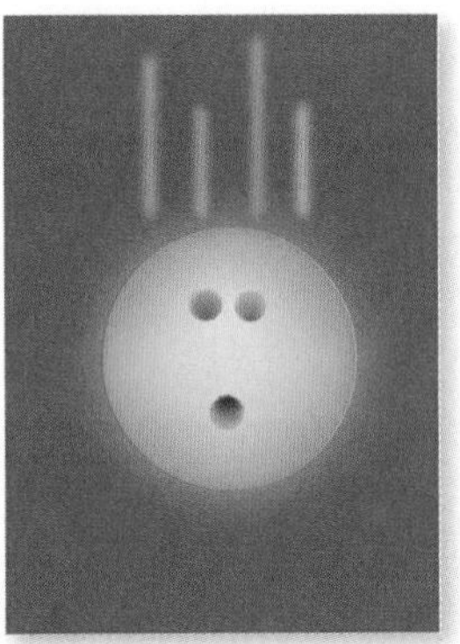

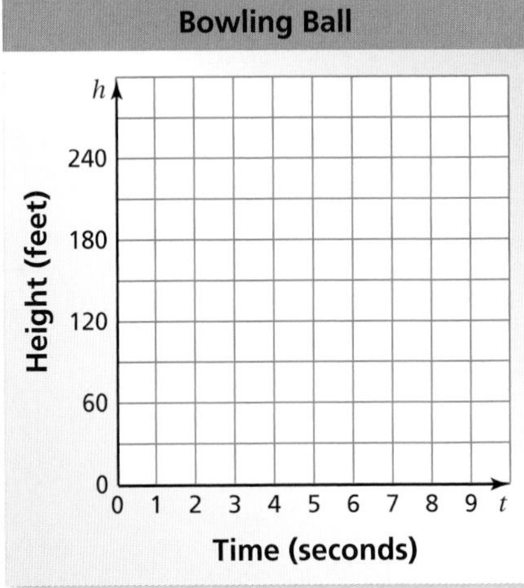

Math Practice

Apply Mathematics

What will the graph look like for an object that has a constant speed? an increasing speed? Explain.

What Is Your Answer?

3. **IN YOUR OWN WORDS** How can you recognize when a pattern in real life is linear or nonlinear? Describe two real-life patterns: one that is linear and one that is nonlinear. Use patterns that are different from those described in Activities 1 and 2.

Use what you learned about comparing linear and nonlinear functions to complete Exercises 3–6 on page 270.

6.4 Lesson

Key Vocabulary
nonlinear function, *p. 268*

The graph of a linear function shows a constant rate of change. A **nonlinear function** does not have a constant rate of change. So, its graph is *not* a line.

EXAMPLE 1 Identifying Functions from Tables

Study Tip
A constant rate of change describes a quantity that changes by equal amounts over equal intervals.

Does the table represent a *linear* or *nonlinear* function? Explain.

a.

x increases by +3, +3, +3

x	3	6	9	12
y	40	32	24	16

y changes by −8, −8, −8

As x increases by 3, y decreases by 8. The rate of change is constant. So, the function is linear.

b.

x increases by +2, +2, +2

x	1	3	5	7
y	2	11	33	88

y changes by +9, +22, +55

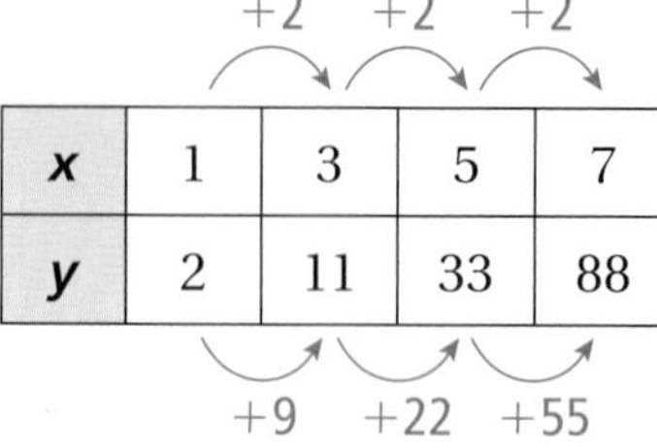

As x increases by 2, y increases by different amounts. The rate of change is *not* constant. So, the function is nonlinear.

EXAMPLE 2 Identifying Functions from Graphs

Does the graph represent a *linear* or *nonlinear* function? Explain.

a.

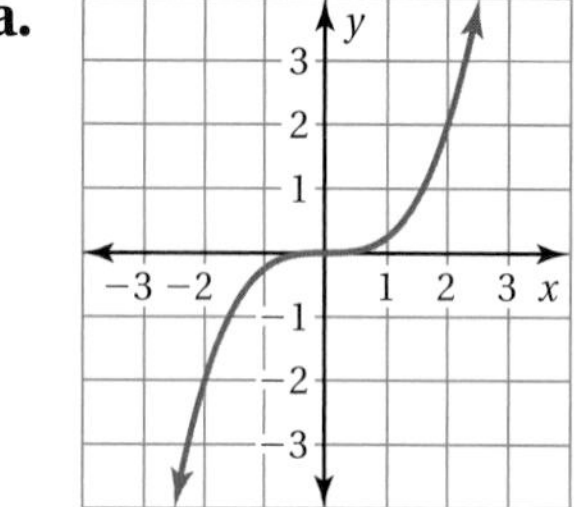

The graph is *not* a line. So, the function is nonlinear.

b.

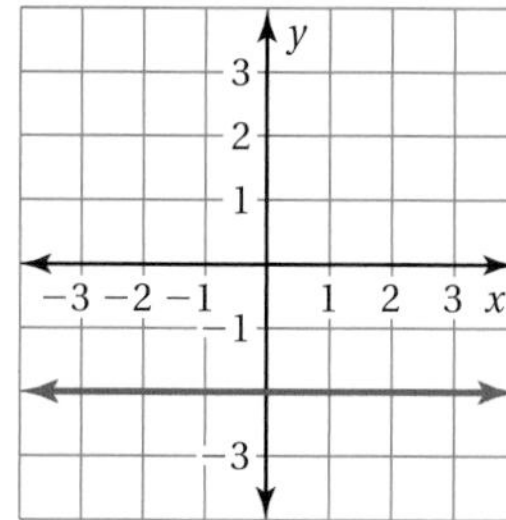

The graph is a line. So, the function is linear.

On Your Own

Now You're Ready
Exercises 7–10

Does the table or graph represent a *linear* or *nonlinear* function? Explain.

1.

x	y
0	25
7	20
14	15
21	10

2.

x	y
2	8
4	4
6	0
8	−4

3.

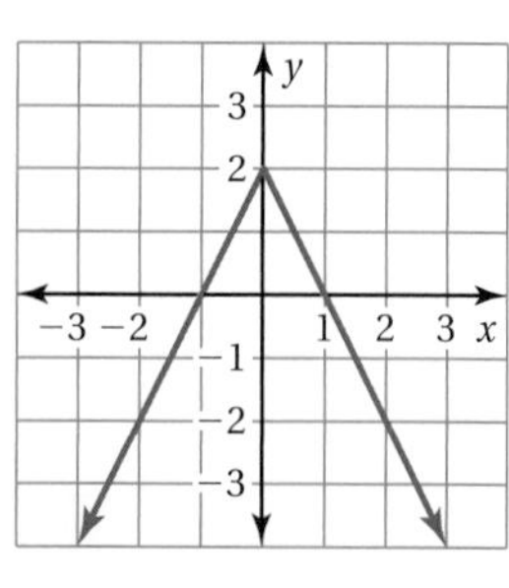

EXAMPLE 3 Identifying a Nonlinear Function

Which equation represents a *nonlinear* function?

Ⓐ $y = 4.7$ Ⓑ $y = \pi x$

Ⓒ $y = \frac{4}{x}$ Ⓓ $y = 4(x - 1)$

You can rewrite the equations $y = 4.7$, $y = \pi x$, and $y = 4(x - 1)$ in slope-intercept form. So, they are linear functions.

You cannot rewrite the equation $y = \frac{4}{x}$ in slope-intercept form. So, it is a nonlinear function.

∴ The correct answer is Ⓒ.

EXAMPLE 4 Real-Life Application

Account A earns simple interest. Account B earns compound interest. The table shows the balances for 5 years. Graph the data and compare the graphs.

Year, t	Account A Balance	Account B Balance
0	\$100	\$100
1	\$110	\$110
2	\$120	\$121
3	\$130	\$133.10
4	\$140	\$146.41
5	\$150	\$161.05

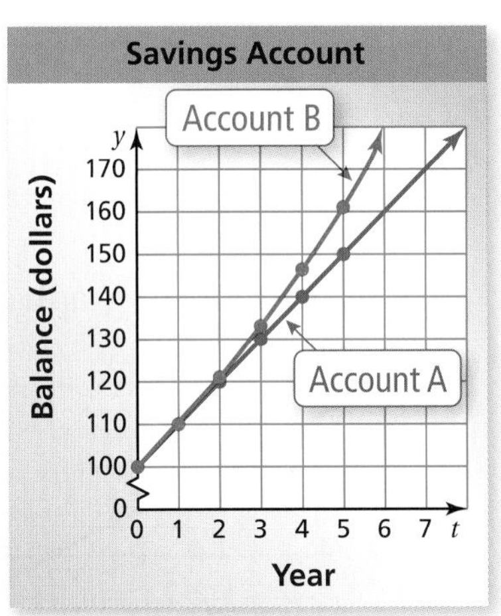

Both graphs show that the balances are positive and increasing.

The balance of Account A has a constant rate of change of \$10. So, the function representing the balance of Account A is linear.

The balance of Account B increases by different amounts each year. Because the rate of change is not constant, the function representing the balance of Account B is nonlinear.

On Your Own

Now You're Ready
Exercises 12–14

Does the equation represent a *linear* or *nonlinear* function? Explain.

4. $y = x + 5$
5. $y = \frac{4x}{3}$
6. $y = 1 - x^2$

6.4 Exercises

Vocabulary and Concept Check

1. **VOCABULARY** Describe how linear functions and nonlinear functions are different.

2. **WHICH ONE DOESN'T BELONG?** Which equation does *not* belong with the other three? Explain your reasoning.

$5y = 2x$ $\qquad$ $y = \frac{2}{5}x$ $\qquad$ $10y = 4x$ $\qquad$ $5xy = 2$

Practice and Problem Solving

Graph the data in the table. Decide whether the graph is *linear* or *nonlinear*.

3.

x	0	1	2	3
y	4	8	12	16

4.

x	1	2	3	4
y	1	2	6	24

5.

x	6	5	4	3
y	21	15	10	6

6.

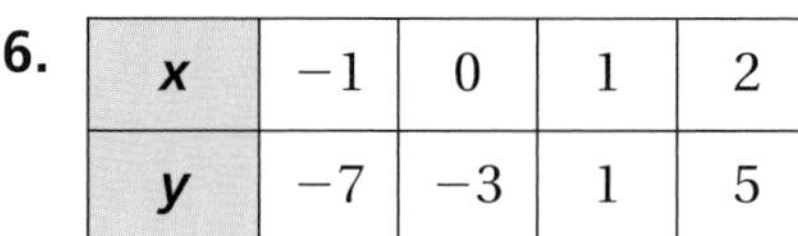

x	−1	0	1	2
y	−7	−3	1	5

Does the table or graph represent a *linear* or *nonlinear* function? Explain.

7.

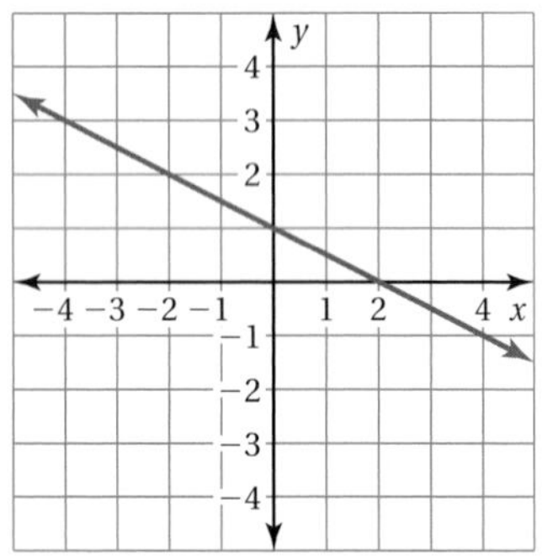

8.

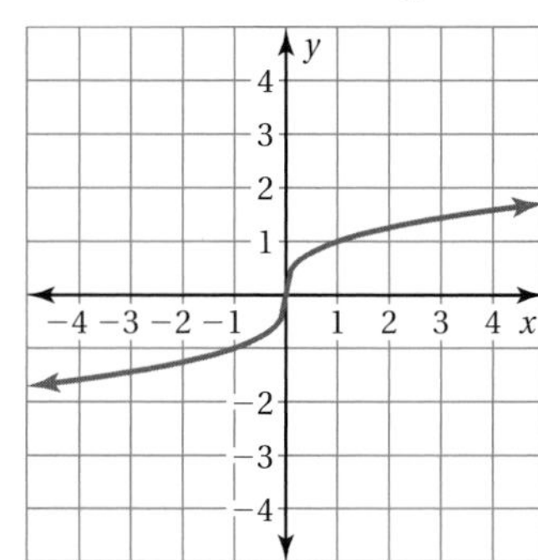

9.

x	5	11	17	23
y	7	11	15	19

10.

x	−3	−1	1	3
y	9	1	1	9

11. **VOLUME** The table shows the volume V (in cubic feet) of a cube with an edge length of x feet. Does the table represent a linear or nonlinear function? Explain.

Edge Length, x	1	2	3	4	5	6	7	8
Volume, V	1	8	27	64	125	216	343	512

Does the equation represent a *linear* or *nonlinear* function? Explain.

3 **12.** $2x + 3y = 7$ **13.** $y + x = 4x + 5$ **14.** $y = \frac{8}{x^2}$

15. LIGHT The frequency y (in terahertz) of a light wave is a function of its wavelength x (in nanometers). Does the table represent a linear or nonlinear function? Explain.

Color	Red	Yellow	Green	Blue	Violet
Wavelength, x	660	595	530	465	400
Frequency, y	454	504	566	645	749

16. MODELING The table shows the cost y (in dollars) of x pounds of sunflower seeds.

Pounds, x	Cost, y
2	2.80
3	?
4	5.60

a. What is the missing y-value that makes the table represent a linear function?

b. Write a linear function that represents the cost y of x pounds of seeds. Interpret the slope.

c. Does the function have a maximum value? Explain your reasoning.

17. TREES Tree A is 5 feet tall and grows at a rate of 1.5 feet per year. The table shows the height h (in feet) of Tree B after x years.

Years, x	Height, h
0	5
1	11
4	17
9	23

a. Does the table represent a linear or nonlinear function? Explain.

b. Which tree is taller after 10 years? Explain.

18. 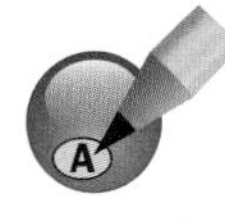

The ordered pairs represent a function.

(0, −1), (1, 0), (2, 3), (3, 8), and (4, 15)

a. Graph the ordered pairs and describe the pattern. Is the function linear or nonlinear?

b. Write an equation that represents the function.

Fair Game Review What you learned in previous grades & lessons

The vertices of a figure are given. Draw the figure and its image after a dilation with the given scale factor *k*. Identify the type of dilation. *(Section 2.7)*

19. $A(-3, 1), B(-1, 3), C(-1, 1); k = 3$

20. $J(-8, -4), K(2, -4), L(6, -10), M(-8, -10); k = \frac{1}{4}$

21. MULTIPLE CHOICE What is the value of x? *(Section 3.3)*

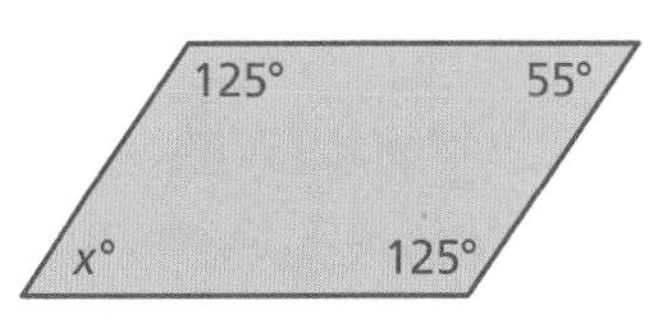

Ⓐ 25 Ⓑ 35

Ⓒ 55 Ⓓ 125

6.5 Analyzing and Sketching Graphs

Essential Question How can you use a graph to represent relationships between quantities without using numbers?

1 ACTIVITY: Interpreting a Graph

Work with a partner. Use the graph shown.

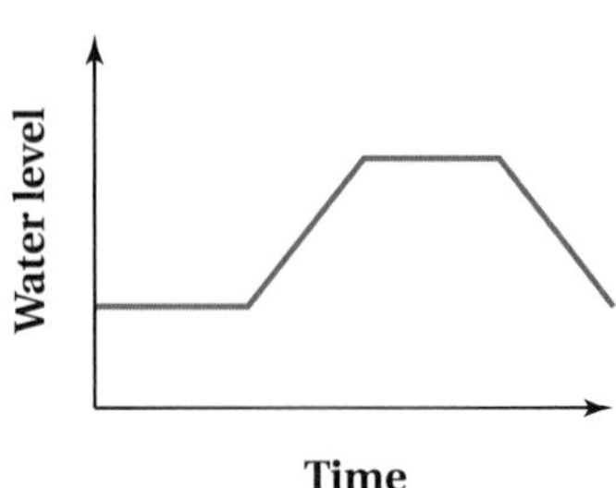

a. How is this graph different from the other graphs you have studied?

b. Write a short paragraph that describes how the water level changes over time.

c. What situation can this graph represent?

2 ACTIVITY: Matching Situations to Graphs

Work with a partner. You are riding your bike. Match each situation with the appropriate graph. Explain your reasoning.

A.

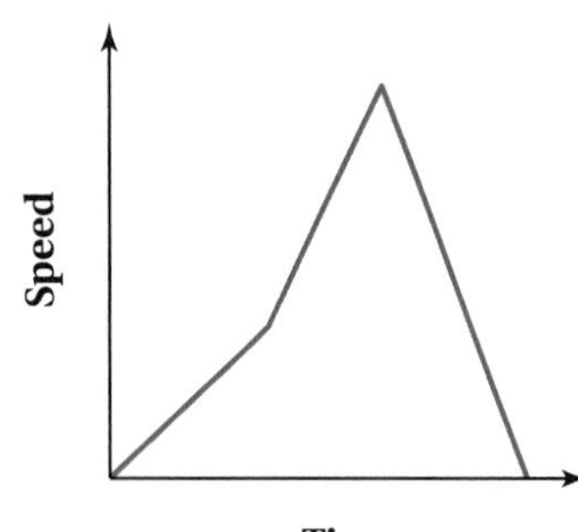

B.

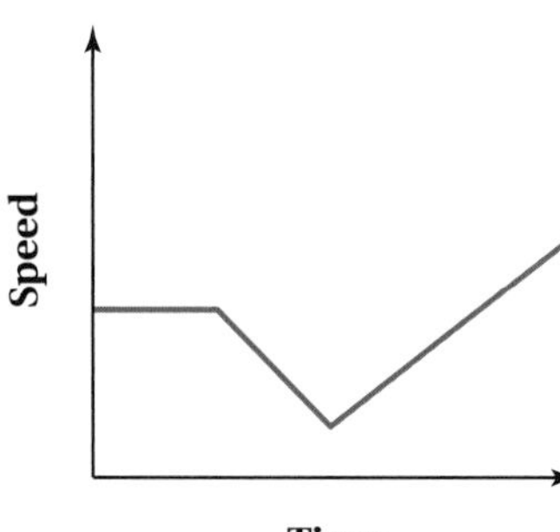

C.

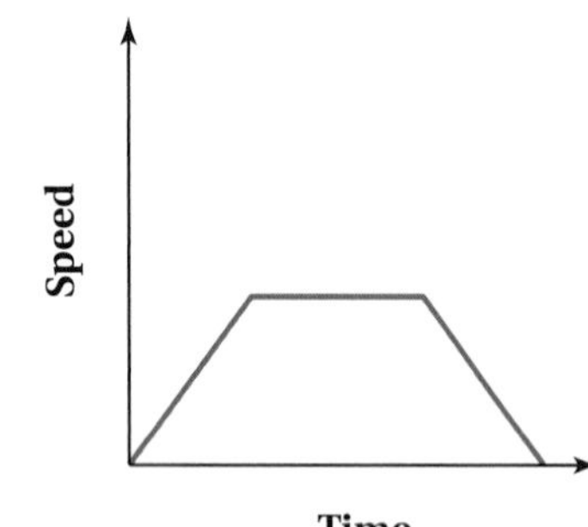

D.

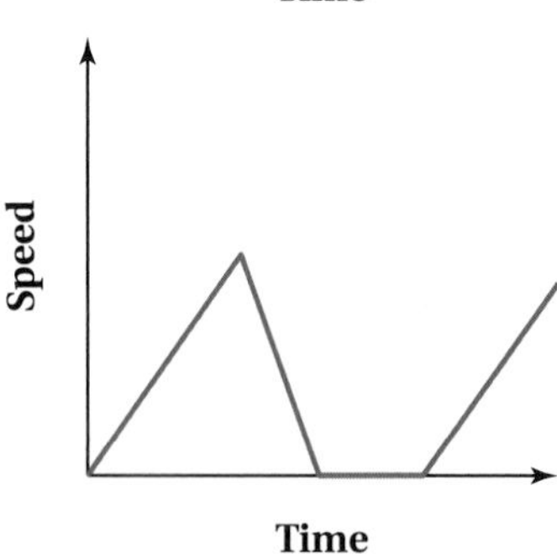

a. You gradually increase your speed, then ride at a constant speed along a bike path. You then slow down until you reach your friend's house.

b. You gradually increase your speed, then go down a hill. You then quickly come to a stop at an intersection.

c. You gradually increase your speed, then stop at a store for a couple of minutes. You then continue to ride, gradually increasing your speed.

d. You ride at a constant speed, then go up a hill. Once on top of the hill, you gradually increase your speed.

Functions

In this lesson, you will

- analyze the relationship between two quantities using graphs.
- sketch graphs to represent the relationship between two quantities.

3 ACTIVITY: Comparing Graphs

Work with a partner. The graphs represent the heights of a rocket and a weather balloon after they are launched.

a. How are the graphs similar? How are they different? Explain.

b. Compare the steepness of each graph.

c. Which graph do you think represents the height of the rocket? Explain.

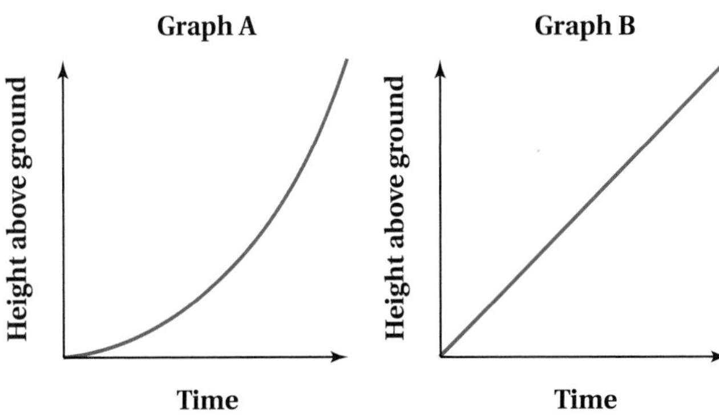

4 ACTIVITY: Comparing Graphs

Math Practice

Consider Similar Problems

How is this activity similar to the previous activity?

Work with a partner. The graphs represent the speeds of two cars. One car is approaching a stop sign. The other car is approaching a yield sign.

a. How are the graphs similar? How are they different? Explain.

b. Compare the steepness of each graph.

c. Which graph do you think represents the car approaching a stop sign? Explain.

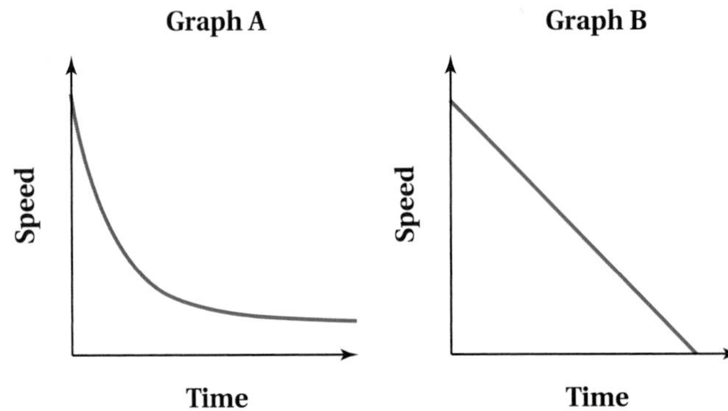

What Is Your Answer?

5. **IN YOUR OWN WORDS** How can you use a graph to represent relationships between quantities without using numbers?

6. Describe a possible situation represented by the graph shown.

7. Sketch a graph similar to the graphs in Activities 1 and 2. Exchange graphs with a classmate and describe a possible situation represented by the graph. Discuss the results.

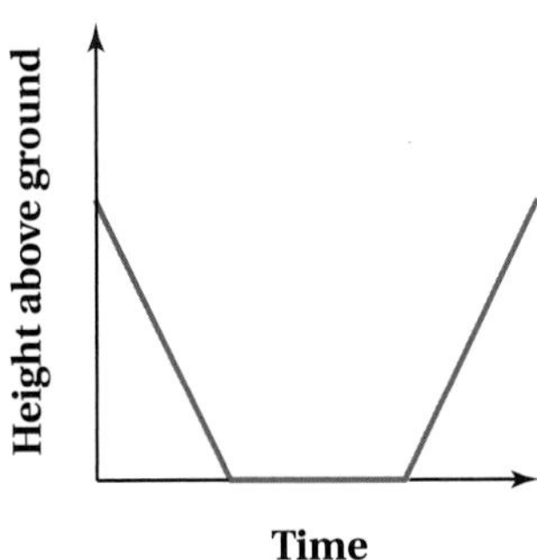

Practice Use what you learned about analyzing and sketching graphs to complete Exercises 7–9 on page 276.

6.5 Lesson

Graphs can show the relationship between quantities without using specific numbers on the axes.

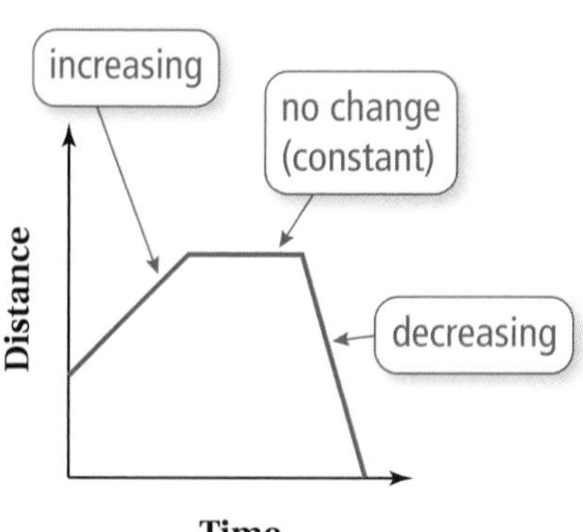

EXAMPLE 1 Analyzing Graphs

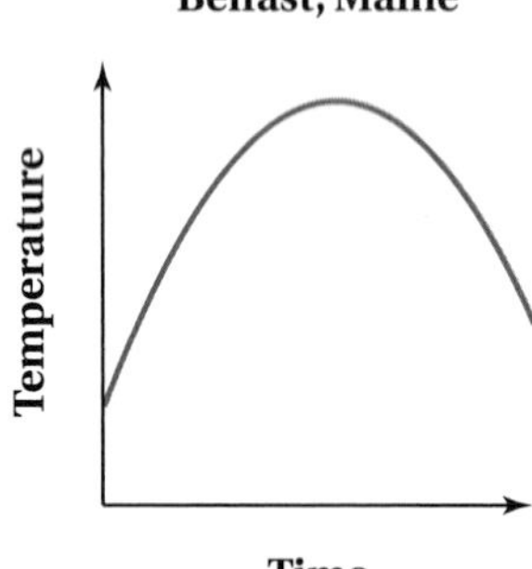

The graphs show the temperatures throughout the day in two cities.

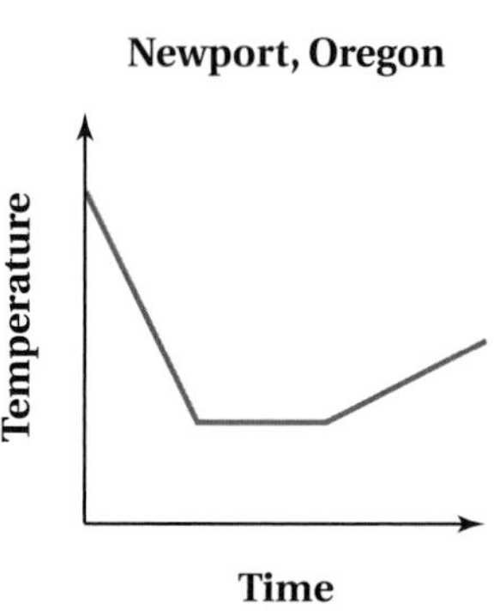

a. Describe the change in temperature in each city.

Belfast: The temperature increases at the beginning of the day. Then the temperature begins to decrease at a faster and faster rate for the rest of the day.

Newport: The temperature decreases at a constant rate at the beginning of the day. Then the temperature stays the same for a while before increasing at a constant rate for the rest of the day.

b. Make three comparisons from the graphs.

Three possible comparisons follow:

- Both graphs show increasing and decreasing temperatures.
- Both graphs are nonlinear, but the graph of the temperatures in Newport consists of three linear sections.
- In Belfast, it was warmer at the end of the day than at the beginning. In Newport, it was colder at the end of the day than at the beginning.

Study Tip

The comparisons given in Example 1(b) are sample answers. You can make many other correct comparisons.

On Your Own

Now You're Ready
Exercises 7–12

1. The graphs show the paths of two birds diving to catch fish.
   a. Describe the path of each bird.
   b. Make three comparisons from the graphs.

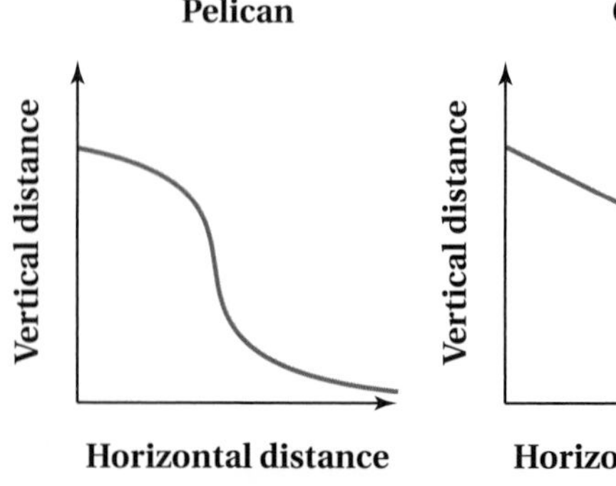

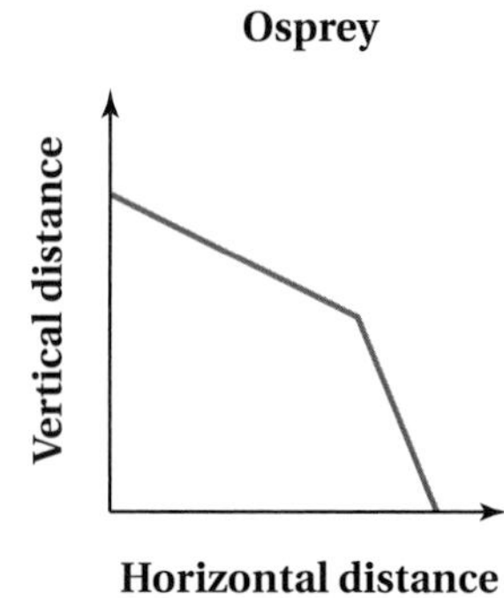

You can sketch graphs showing relationships between quantities that are described verbally.

EXAMPLE 2 Sketching Graphs

Sketch a graph that represents each situation.

a. A stopped subway train gains speed at a constant rate until it reaches its maximum speed. It travels at this speed for a while, and then slows down at a constant rate until coming to a stop at the next station.

Step 1: Draw the axes. Label the vertical axis "Speed" and the horizontal axis "Time."

Step 2: Sketch the graph.

Speed

Time

Words	Graph
A stopped subway train gains speed at a constant rate . . .	increasing line segment starting at the origin
until it reaches its maximum speed. It travels at this speed for a while, . . .	horizontal line segment
and then slows down at a constant rate until coming to a stop at the next station.	decreasing line segment ending at the horizontal axis

b. As television size increases, the price increases at an increasing rate.

Step 1: Draw the axes. Label the vertical axis "Price" and the horizontal axis "TV size."

Step 2: Sketch the graph.

The price *increases at an increasing rate*. So, the graph is nonlinear and becomes steeper and steeper as the TV size increases.

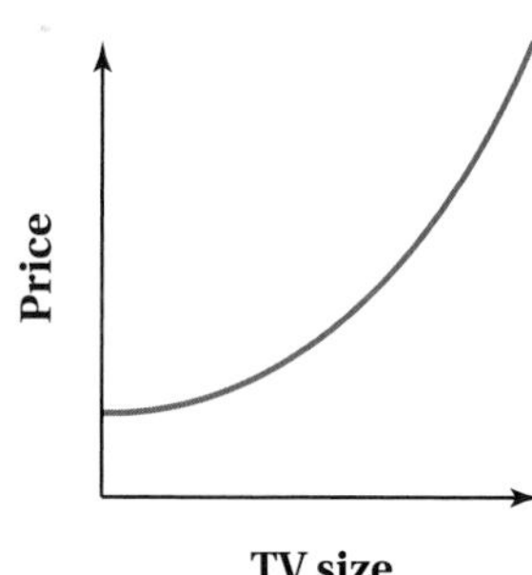

On Your Own

Sketch a graph that represents the situation.

2. A fully charged battery loses its charge at a constant rate until it has no charge left. You plug it in and recharge it fully. Then it loses its charge at a constant rate until it has no charge left.
3. As the available quantity of a product increases, the price decreases at a decreasing rate.

6.5 Exercises

Vocabulary and Concept Check

MATCHING **Match the verbal description with the part of the graph it describes.**

1. stays the same
2. slowly decreases at a constant rate
3. slowly increases at a constant rate
4. increases at an increasing rate
5. quickly decreases at a constant rate
6. quickly increases at a constant rate

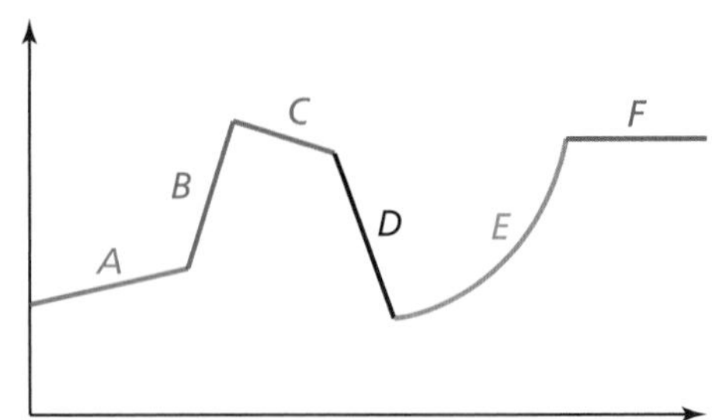

Practice and Problem Solving

Describe the relationship between the two quantities.

1 7. **Balloon**

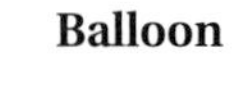

Volume

Time

8. **Sales**

Dollars

Time

9. 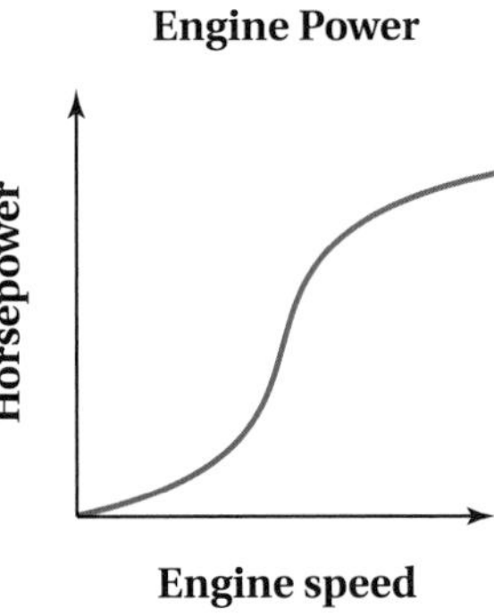

10. **Decay**

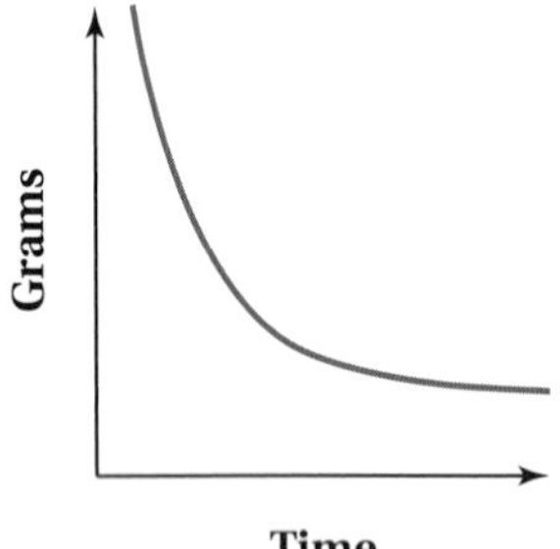

11. **Hair**

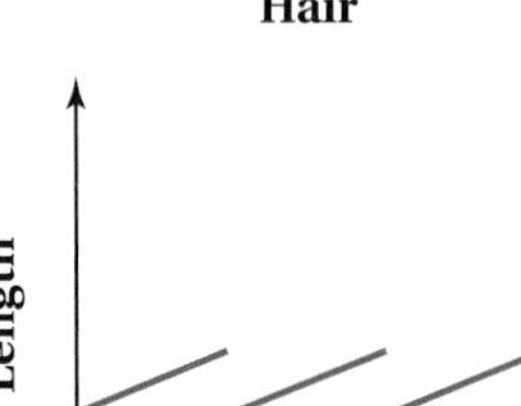

Length

Time

12. 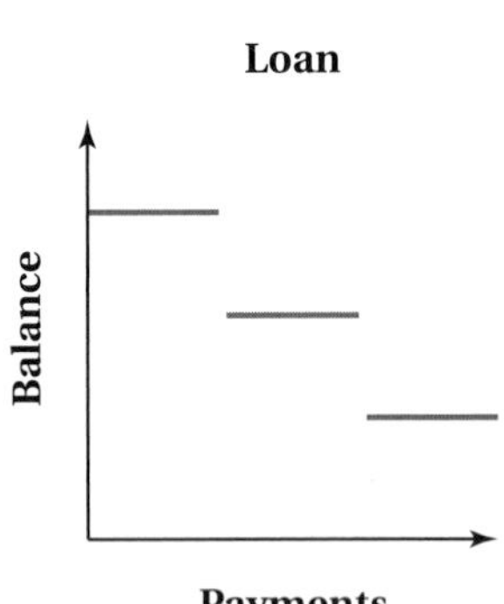

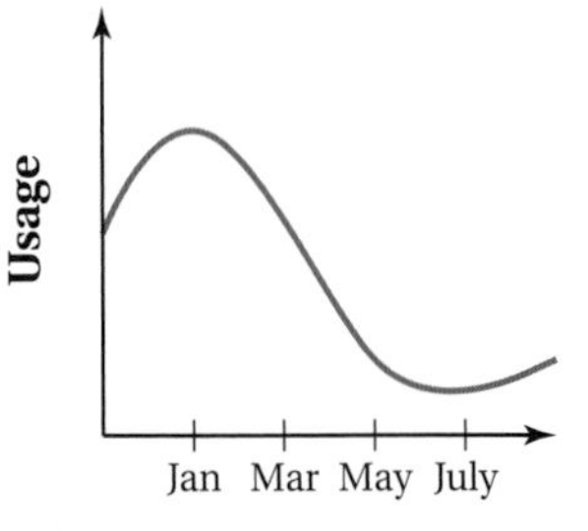

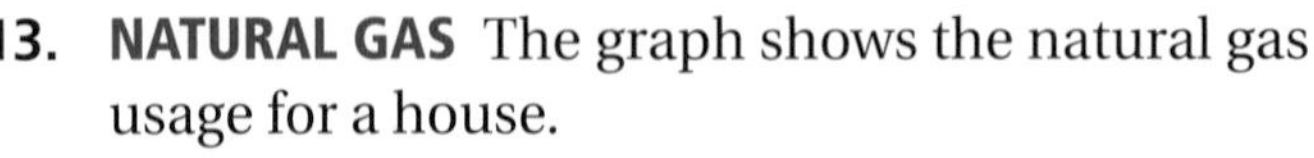

13. **NATURAL GAS** The graph shows the natural gas usage for a house.

   a. Describe the change in usage from January to March.

   b. Describe the change in usage from March to May.

14. **REASONING** The graph shows two bowlers' averages during a bowling season.

    a. Describe each bowler's performance.

    b. Who had a greater average most of the season? Who had a greater average at the end of the season?

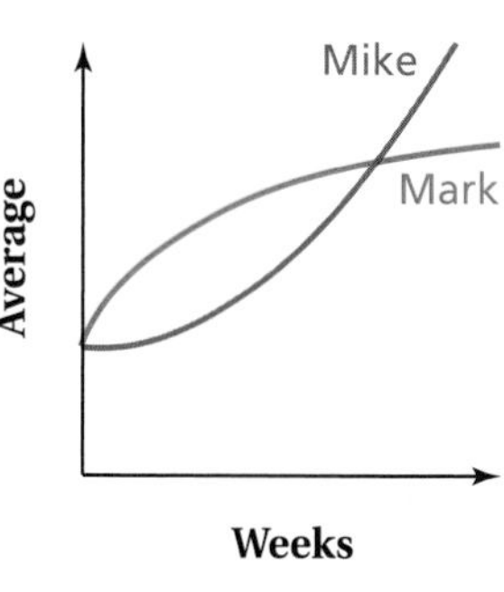

Sketch a graph that represents the situation.

2 15. The value of a car depreciates. The value decreases quickly at first and then more slowly.

16. The distance from the ground changes as your friend swings on a swing.

17. The value of a rare coin increases at an increasing rate.

18. You are typing at a constant rate. You pause to think about your next paragraph, and then you resume typing at the same constant rate.

19. **Economics** You can use a *supply and demand model* to understand how the price of a product changes in a market. The *supply curve* of a particular product represents the quantity suppliers will produce at various prices. The *demand curve* for the product represents the quantity consumers are willing to buy at various prices.

    a. Describe and interpret each curve.

    b. Which part of the graph represents a surplus? a shortage? Explain your reasoning.

    c. The curves intersect at the *equilibrium point*, which is where the quantity produced equals the quantity demanded. Suppose that demand for a product suddenly increases, causing the entire demand curve to shift to the right. What happens to the equilibrium point?

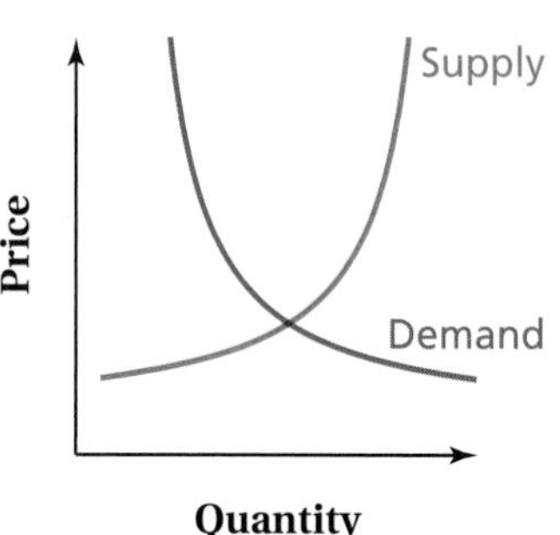

Fair Game Review What you learned in previous grades & lessons

Solve the system of linear equations by graphing. *(Section 5.1)*

20. $y = x + 2$
    $y = -x - 4$

21. $x - y = 3$
    $-2x + y = -5$

22. $3x + 2y = 2$
    $5x - 3y = -22$

23. **MULTIPLE CHOICE** Which triangle is a rotation of Triangle D? *(Section 2.4)*

    (A) Triangle A
    (B) Triangle B
    (C) Triangle C
    (D) none

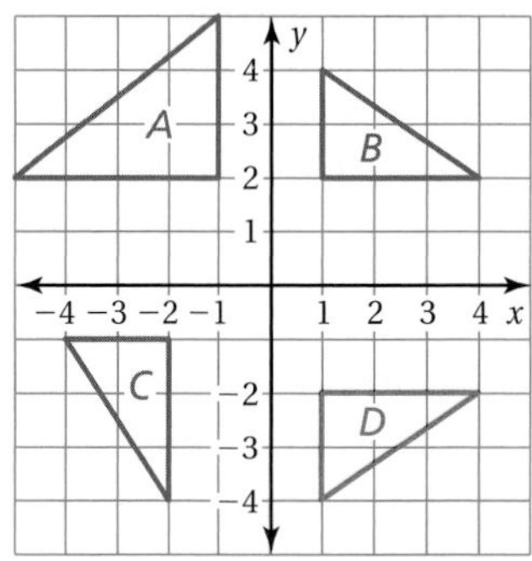

6.4–6.5 Quiz

Check It Out
Progress Check
BigIdeasMath.com

Does the table or graph represent a *linear* or *nonlinear* function? Explain. *(Section 6.4)*

1.

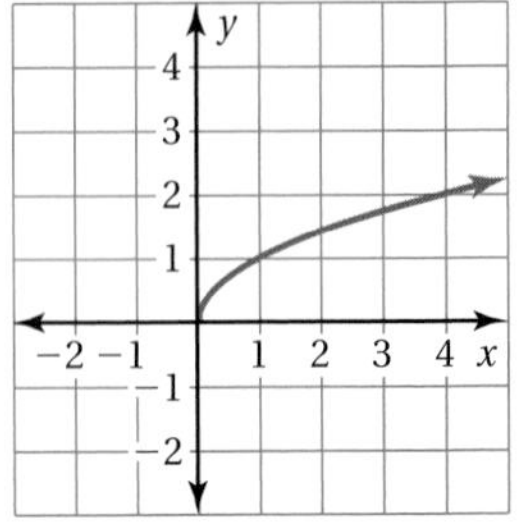

2.

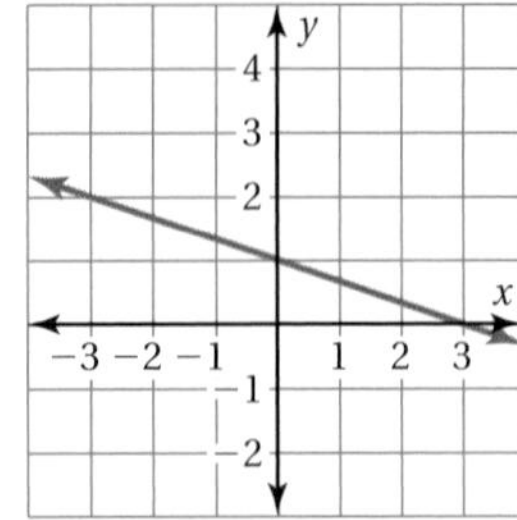

3.

x	y
0	3
3	0
6	3
9	6

4.

x	y
−1	3
1	7
3	11
5	15

5. CHICKEN SALAD The equation $y = 7.9x$ represents the cost y (in dollars) of buying x pounds of chicken salad. Does this equation represent a linear or nonlinear function? Explain. *(Section 6.4)*

6. HEIGHTS The graphs show the heights of two people over time. *(Section 6.5)*

a. Describe the change in height of each person.

b. Make three comparisons from the graphs.

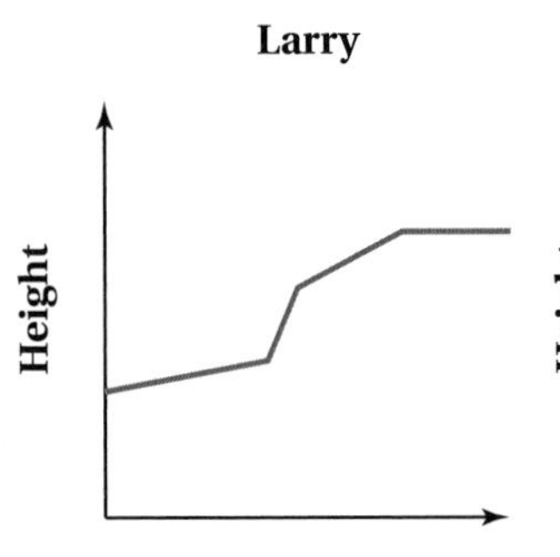

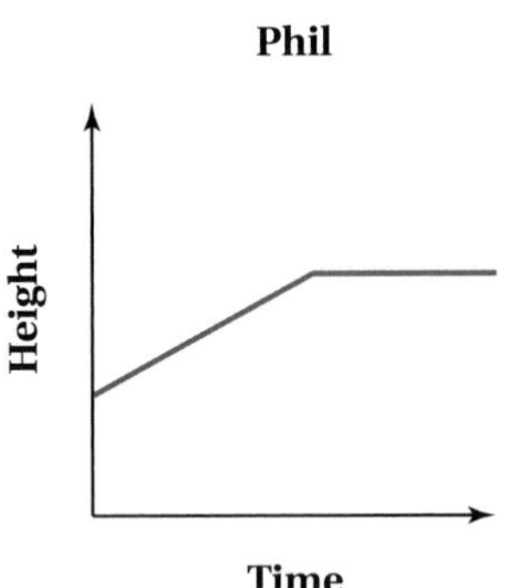

You are snowboarding down a hill. Sketch a graph that represents the situation. *(Section 6.5)*

7. You gradually increase your speed at a constant rate over time but fall about halfway down the hill. You take a short break, then get up, and gradually increase your speed again.

8. You gradually increase your speed at a constant rate over time. You come to a steep section of the hill and rapidly increase your speed at a constant rate. You then decrease your speed at a constant rate until you come to a stop.

6 Chapter Review

Check It Out
Vocabulary Help
BigIdeasMath.com

Review Key Vocabulary

input, *p. 244*
output, *p. 244*
relation, *p. 244*
mapping diagram, *p. 244*
function, *p. 245*
function rule, *p. 250*
linear function, *p. 258*
nonlinear function, *p. 268*

Review Examples and Exercises

6.1 Relations and Functions *(pp. 242–247)*

Determine whether the relation is a function.

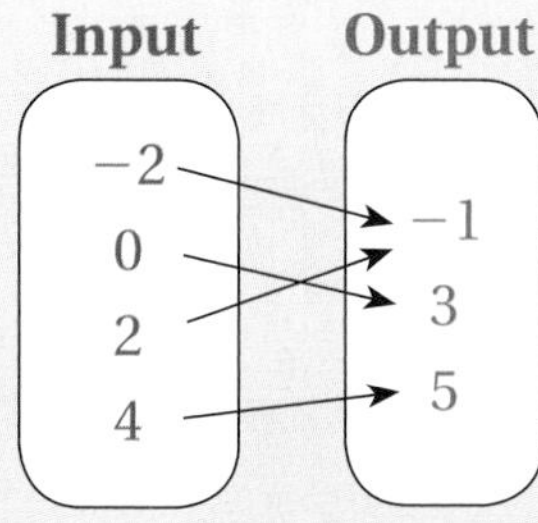

Each input has exactly one output.

∴ So, the relation is a function.

Exercises

Determine whether the relation is a function.

1.

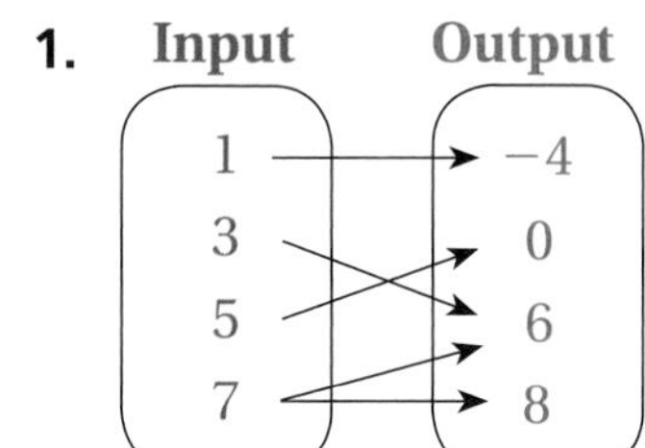

2.

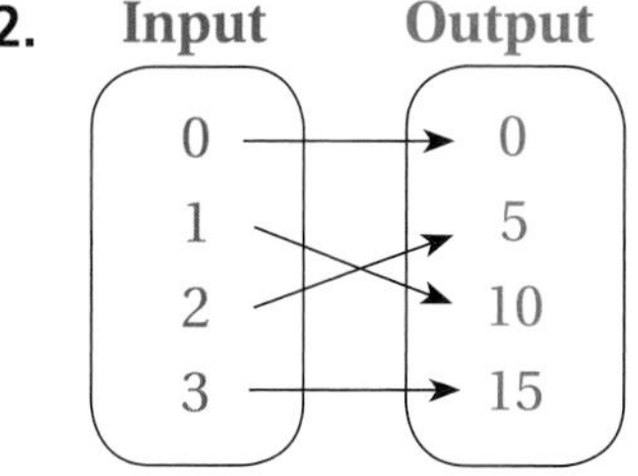

6.2 Representations of Functions *(pp. 248–255)*

Graph the function $y = x - 1$ using inputs of $-1, 0, 1$, and 2.

Make an input-output table.

Input, x	$x - 1$	Output, y	Ordered Pair, (x, y)
−1	−1 − 1	−2	(−1, −2)
0	0 − 1	−1	(0, −1)
1	1 − 1	0	(1, 0)
2	2 − 1	1	(2, 1)

Plot the ordered pairs and draw a line through the points.

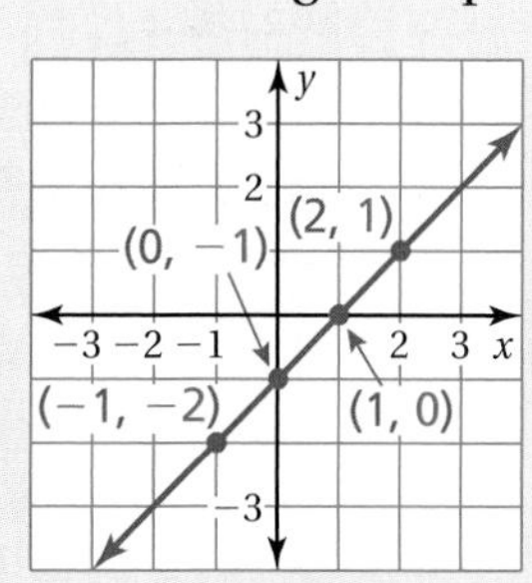

Exercises

Find the value of y for the given value of x.

3. $y = 2x - 3;\ x = -4$

4. $y = 2 - 9x;\ x = \frac{2}{3}$

5. $y = \frac{x}{3} + 5;\ x = 6$

Graph the function.

6. $y = x + 3$

7. $y = -5x$

8. $y = 3 - 3x$

6.3 Linear Functions *(pp. 256–263)*

Use the graph to write a linear function that relates y to x.

The points lie on a line. Find the slope by using the points (1, 1) and (2, 3).

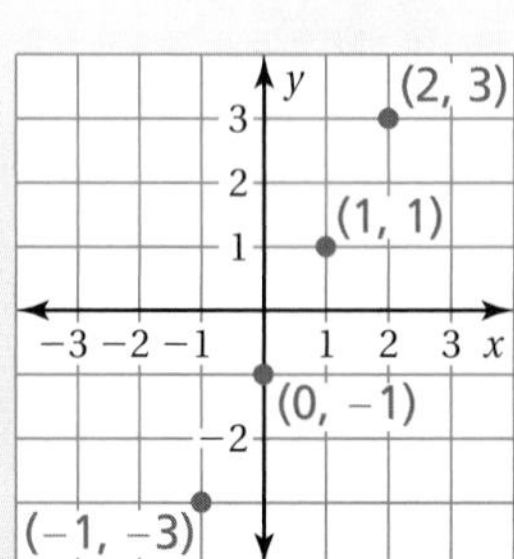

$$m = \frac{\text{change in } y}{\text{change in } x} = \frac{3-1}{2-1} = \frac{2}{1} = 2$$

Because the line crosses the y-axis at $(0, -1)$, the y-intercept is -1.

∴ So, the linear function is $y = 2x - 1$.

Exercises

Use the graph or table to write a linear function that relates y to x.

9.

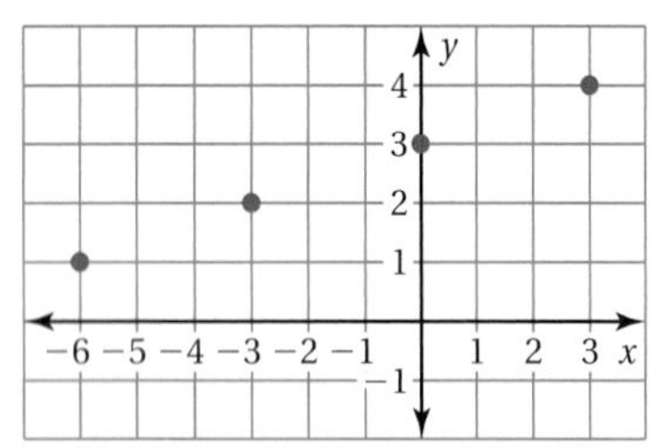

10.

x	−2	0	2	4
y	−7	−7	−7	−7

6.4 Comparing Linear and Nonlinear Functions *(pp. 266–271)*

Does the table represent a *linear* or *nonlinear* function? Explain.

a.

+2 +2 +2

x	0	2	4	6
y	0	1	4	9

+1 +3 +5

∴ As x increases by 2, y increases by different amounts. The rate of change is *not* constant. So, the function is nonlinear.

b.

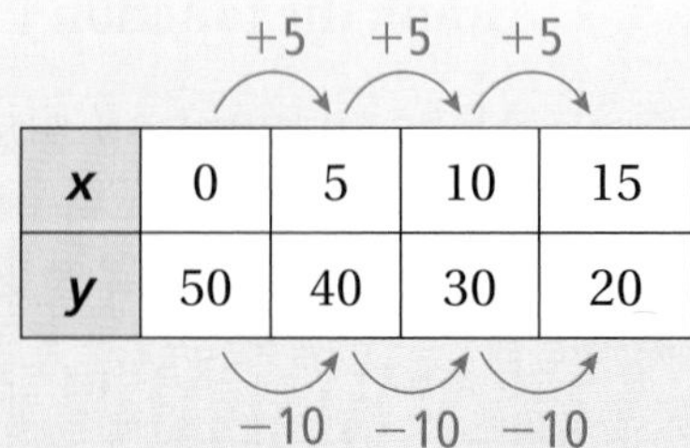

+5 +5 +5

x	0	5	10	15
y	50	40	30	20

−10 −10 −10

∴ As x increases by 5, y decreases by 10. The rate of change is constant. So, the function is linear.

Exercises

Does the table represent a *linear* or *nonlinear* function? Explain.

11.

x	3	6	9	12
y	1	10	19	28

12.

x	1	3	5	7
y	3	1	1	3

6.5 Analyzing and Sketching Graphs (pp. 272–277)

The graphs show the populations of two cities over several years.

a. Describe the change in population in each city.

Lake City: The population gradually decreases at a constant rate, then gradually increases at a constant rate. Then the population rapidly increases at a constant rate.

Gold Point: The population rapidly increases at a constant rate. Then the population stays the same for a short period of time before gradually decreasing at a constant rate.

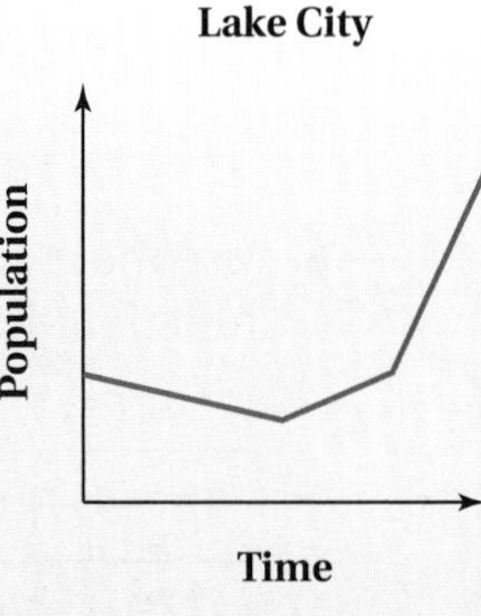

b. Make three comparisons from the graphs.

- Both graphs show increasing and decreasing populations.
- Both graphs are nonlinear, but both graphs consist of three linear sections.
- Both populations at the end of the time period are greater than the populations at the beginning of the time period.

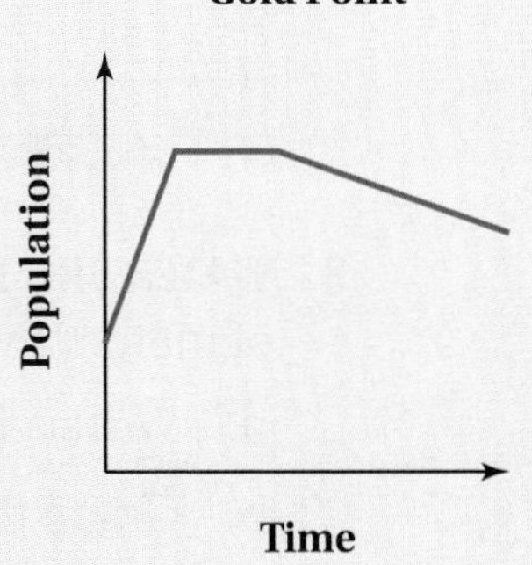

Exercises

13. **SALES** The graphs show the sales of two companies.

  a. Describe the sales of each company.

  b. Make three comparisons from the graphs.

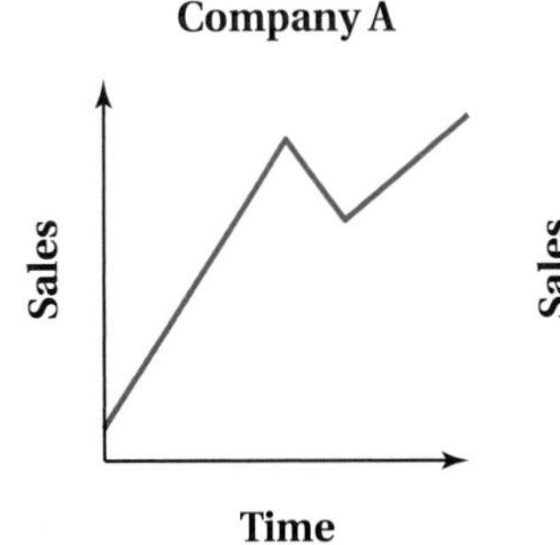

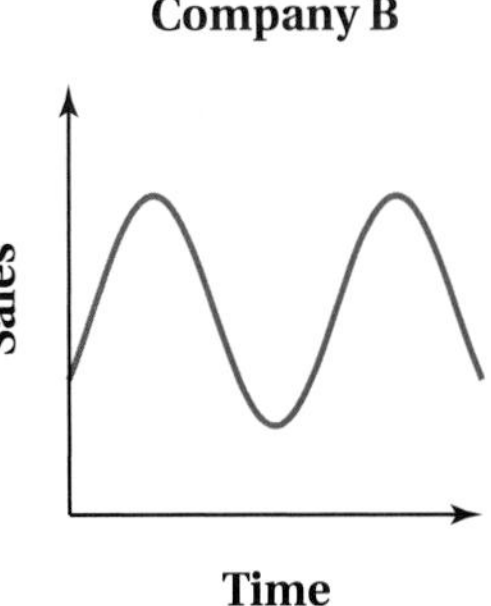

Sketch a graph that represents the situation.

14. You climb up a climbing wall. You gradually climb halfway up the wall at a constant rate, then stop and take a break. You then climb to the top of the wall at a constant rate.

15. The price of a stock steadily increases at a constant rate for several months before the stock market crashes. The price then quickly decreases at a constant rate.

6 Chapter Test

Determine whether the relation is a function.

1.

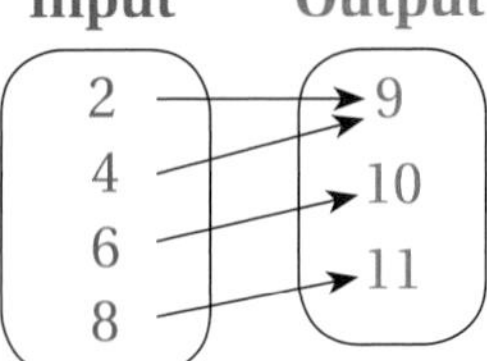

2.

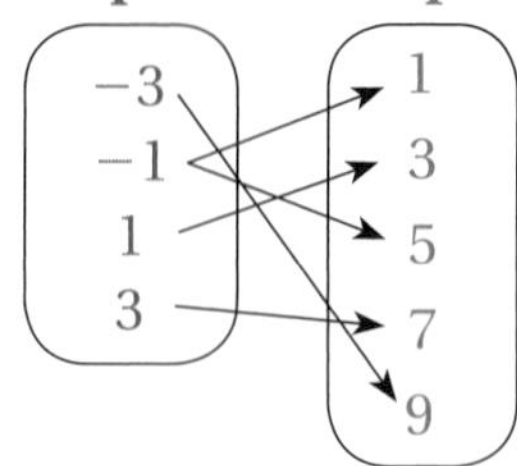

Graph the function.

3. $y = x + 8$

4. $y = 1 - 3x$

5. $y = x - 4$

6. Use the graph to write a linear function that relates y to x.

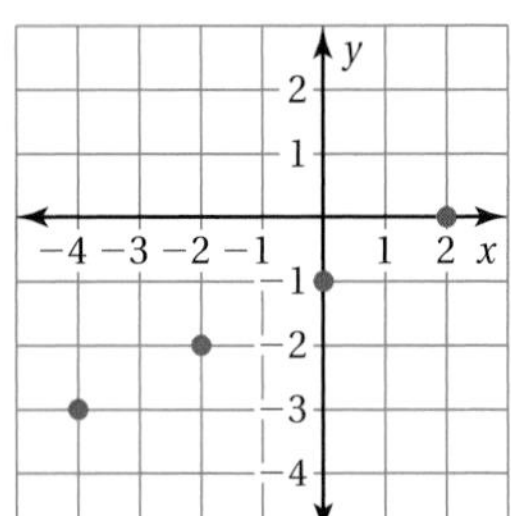

7. Does the table represent a *linear* or *nonlinear* function? Explain.

x	0	2	4	6
y	8	0	−8	−16

8. WATER SKI The table shows the number of meters a water skier travels in x minutes.

Minutes, x	1	2	3	4	5
Meters, y	600	1200	1800	2400	3000

a. Write a function that relates x to y.

b. Graph the linear function.

c. At this rate, how many *kilometers* would the water skier travel in 12 minutes?

9. STOCKS The graphs show the prices of two stocks during one day.

a. Describe the prices of each stock.

b. Make three comparisons from the graphs.

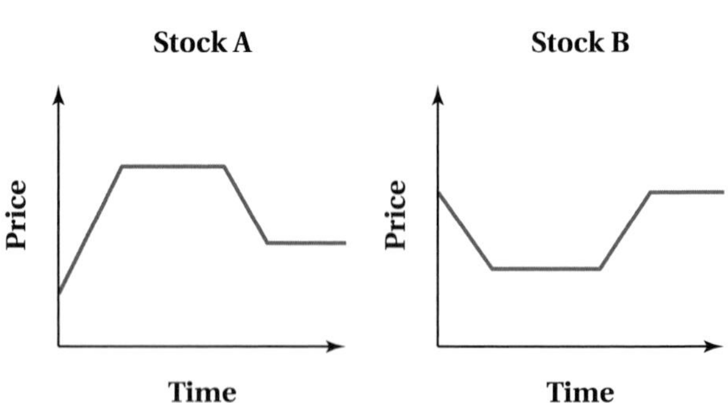

10. RACE You are competing in a race. You begin the race by increasing your speed at a constant rate. You then run at a constant speed until you get a cramp and have to stop. You wait until your cramp goes away before you start gradually increasing your speed again at a constant rate. Sketch a graph that represents the situation.

6 Cumulative Assessment

1. What is the slope of the line shown in the graph below?

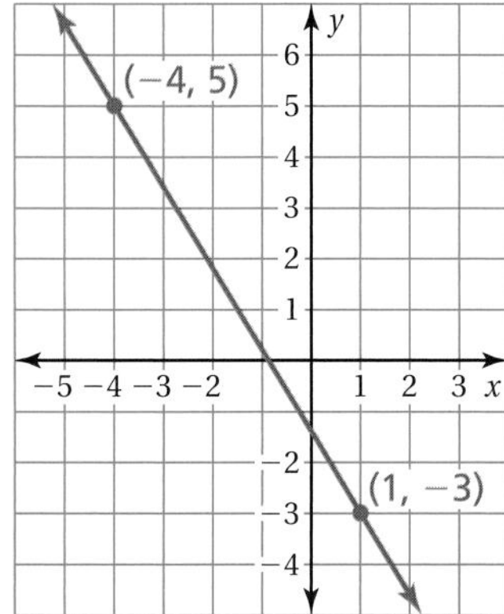

A. $-\frac{8}{3}$

B. $-\frac{8}{5}$

C. $-\frac{2}{3}$

D. $-\frac{2}{5}$

Test-Taking Strategy
Work Backwards
For x cats, a litter box is changed y=3x times per month. How many cats are there when y=12?
Ⓐ1 Ⓑ2 Ⓒ3 Ⓓ4
Share a litter box? Please!
KEEP OFF!
"Work backwards by trying 1, 2, 3, and 4. You will see that 3(4) = 12. So, D is correct."

2. Which value of a makes the equation below true?

$$24 = \frac{a}{3} - 9$$

F. 5

G. 11

H. 45

I. 99

3. A mapping diagram is shown.

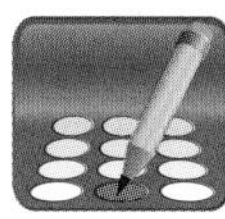

What number belongs in the box below so that the equation will correctly describe the function represented by the mapping diagram?

$$y = \square x + 5$$

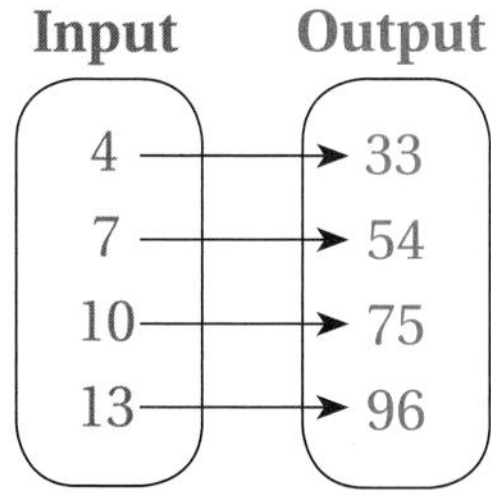

4. What is the solution of the system of linear equations shown below?

$$y = 2x - 1$$

$$y = 3x + 5$$

A. $(-13, -6)$

B. $(-6, -13)$

C. $(-13, 6)$

D. $(-6, 13)$

5. A system of two linear equations has no solution. What can you conclude about the graphs of the two equations?

F. The lines have the same slope and the same y-intercept.

G. The lines have the same slope and different y-intercepts.

H. The lines have different slopes and the same y-intercept.

I. The lines have different slopes and different y-intercepts.

6. Which graph shows a nonlinear function?

A. 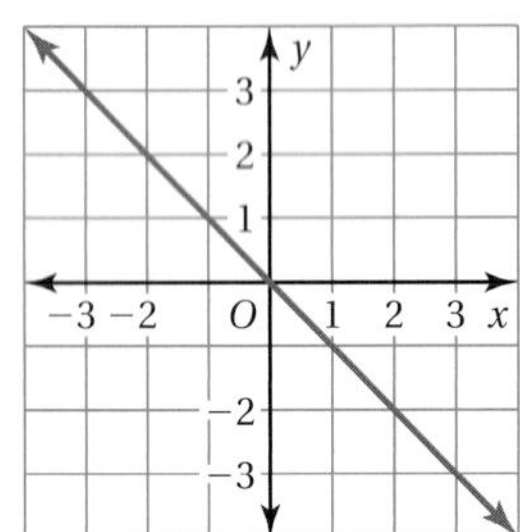

C. 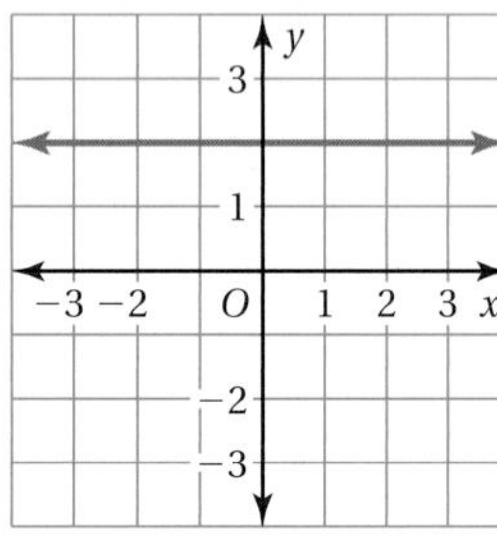

B. 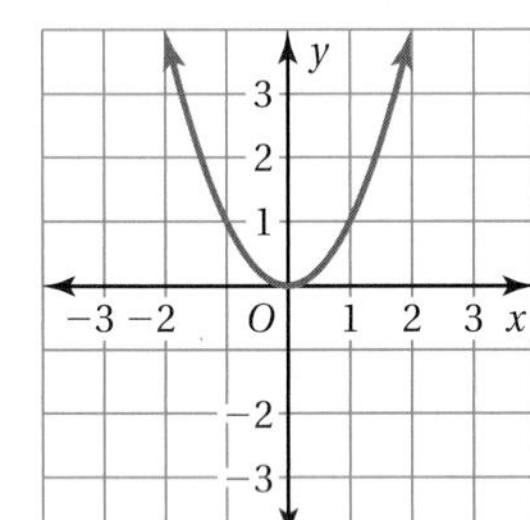

D. 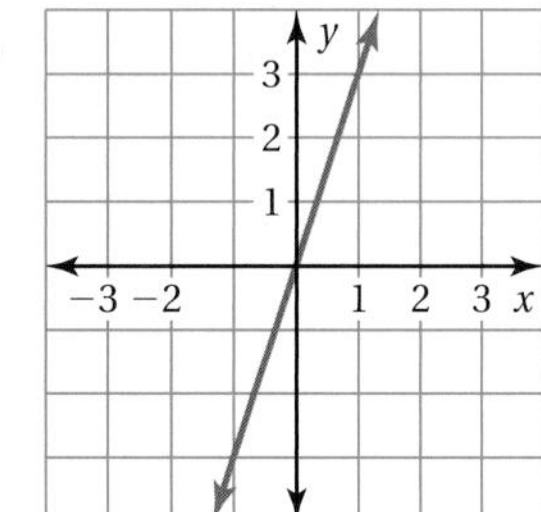

7. What is the value of x?

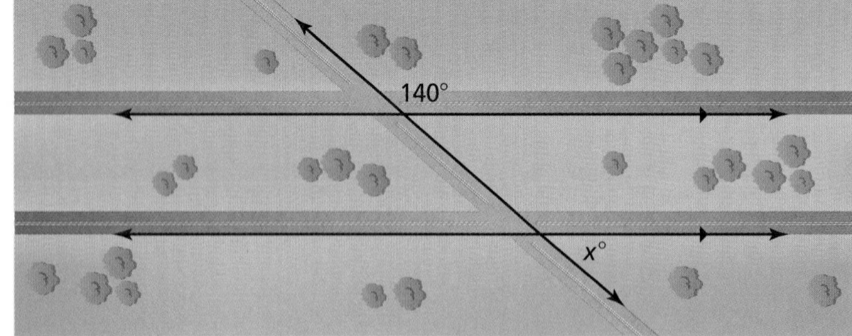

F. 40

H. 140

G. 50

I. 220

8. The tables show the sales (in millions of dollars) for two companies over a 5-year period. Examine the data in the tables.

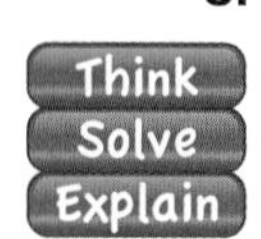

Year	1	2	3	4	5
Sales	2	4	6	8	10

Year	1	2	3	4	5
Sales	1	1	2	3	5

Part A Does the first table show a linear function? Explain your reasoning.

Part B Does the second table show a linear function? Explain your reasoning.

9. The equations $y = -x + 4$ and $y = \frac{1}{2}x - 8$ form a system of linear equations. The table below shows the y-value for each equation at six different values of x.

x	0	2	4	6	8	10
$y = -x + 4$	4	2	0	−2	−4	−6
$y = \frac{1}{2}x - 8$	−8	−7	−6	−5	−4	−3

What can you conclude from the table?

A. The system has one solution, when $x = 0$.

B. The system has one solution, when $x = 4$.

C. The system has one solution, when $x = 8$.

D. The system has no solution.

10. In the diagram below, Triangle *ABC* is a dilation of Triangle *DEF*. What is the value of x?

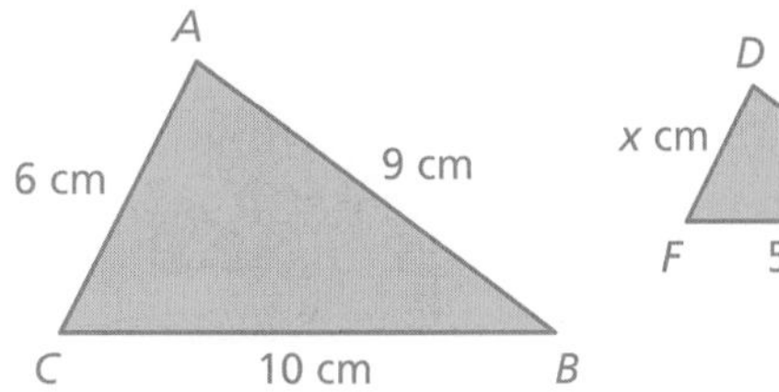

7 Real Numbers and the Pythagorean Theorem

7.1 Finding Square Roots
7.2 Finding Cube Roots
7.3 The Pythagorean Theorem
7.4 Approximating Square Roots
7.5 Using the Pythagorean Theorem

"I'm pretty sure that Pythagoras was a Greek."

"I said 'Greek,' not 'Geek.' "

"Here's how I remember the square root of 2."

"February is the 2nd month. It has 28 days. Split 28 into 14 and 14. Move the decimal to get 1.414."

What You Learned Before

Pythagorean Theorem

$(\text{shorter})^2 + (\text{shorter})^2 = (\text{longest})^2$

Speaking of being a square...

"I just remember that the sum of the squares of the shorter sides is equal to the square of the longest side."

Comparing Decimals

Complete the number sentence with <, >, or =.

Example 1 **1.1 $\square$ 1.01**

Because $\frac{110}{100}$ is greater than $\frac{101}{100}$, 1.1 is greater than 1.01.

∴ So, $1.1 > 1.01$.

Example 2 **-0.3 $\square$ -0.003**

Because $-\frac{300}{1000}$ is less than $-\frac{3}{1000}$, -0.3 is less than -0.003.

∴ So, $-0.3 < -0.003$.

Example 3 **Find three decimals that make the number sentence $-5.12 > \square$ true.**

Any decimal less than -5.12 will make the sentence true.

∴ *Sample answer:* $-10.1, -9.05, -8.25$

Try It Yourself

Complete the number sentence with <, >, or =.

1. $2.10 \ \square \ 2.1$ **2.** $-4.5 \ \square \ -4.25$ **3.** $\pi \ \square \ 3.2$

Find three decimals that make the number sentence true.

4. $-0.01 \leq \square$ **5.** $1.75 > \square$ **6.** $0.75 \geq \square$

Using Order of Operations

Example 4 **Evaluate $8^2 \div (32 \div 2) - 2(3 - 5)$.**

First:	Parentheses	$8^2 \div (32 \div 2) - 2(3 - 5) = 8^2 \div 16 - 2(-2)$
Second:	Exponents	$= 64 \div 16 - 2(-2)$
Third:	Multiplication and Division (from left to right)	$= 4 + 4$
Fourth:	Addition and Subtraction (from left to right)	$= 8$

Try It Yourself

Evaluate the expression.

7. $15\left(\frac{12}{3}\right) - 7^2 - 2 \cdot 7$ **8.** $3^2 \cdot 4 \div 18 + 30 \cdot 6 - 1$ **9.** $-1 + \left(\frac{4}{2}(6 - 1)\right)^2$

7.1 Finding Square Roots

Essential Question How can you find the dimensions of a square or a circle when you are given its area?

When you multiply a number by itself, you square the number.

Symbol for squaring is the exponent 2. → $4^2 = 4 \cdot 4$

$= 16$ 4 squared is 16.

To "undo" this, take the *square root* of the number.

Symbol for square root is a *radical sign*, $\sqrt{\ }$. → $\sqrt{16} = \sqrt{4^2} = 4$ The square root of 16 is 4.

1 ACTIVITY: Finding Square Roots

Work with a partner. Use a square root symbol to write the side length of the square. Then find the square root. Check your answer by multiplying.

a. Sample: $s = \sqrt{121} = 11$ ft

Area = 121 ft²

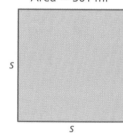

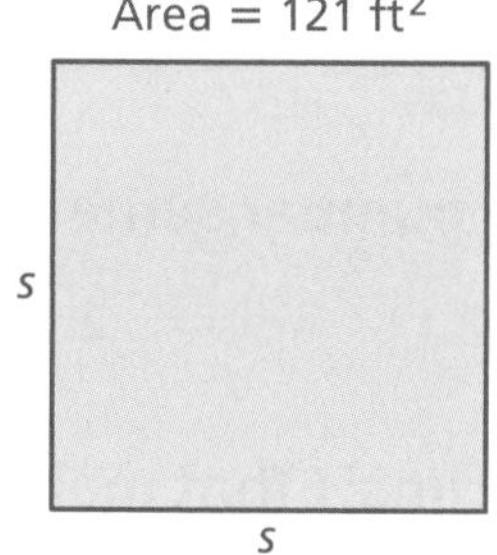

Check

$$\begin{array}{r} 11 \\ \times\, 11 \\ \hline 11 \\ 110 \\ \hline 121 \end{array}$$ ✓

The side length of the square is 11 feet.

b. Area = 81 yd²

s, s

c. Area = 324 cm²

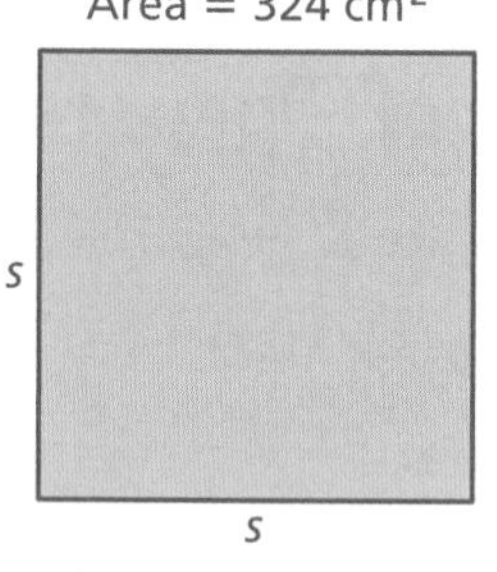

d. Area = 361 mi²

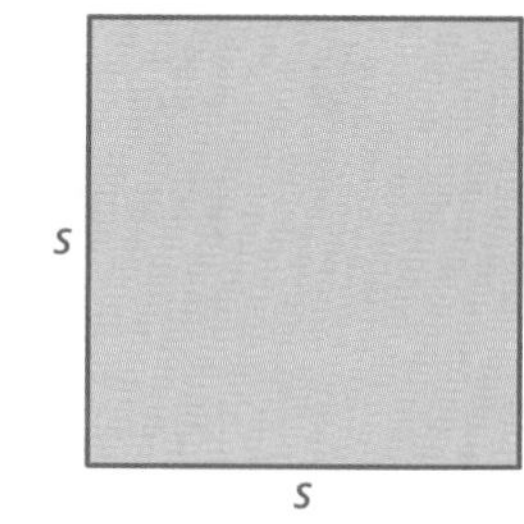

e. Area = 225 mi²

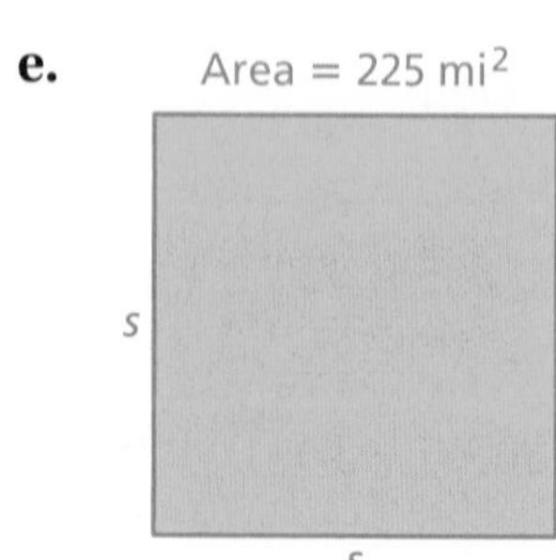

f. Area = 2.89 in.²

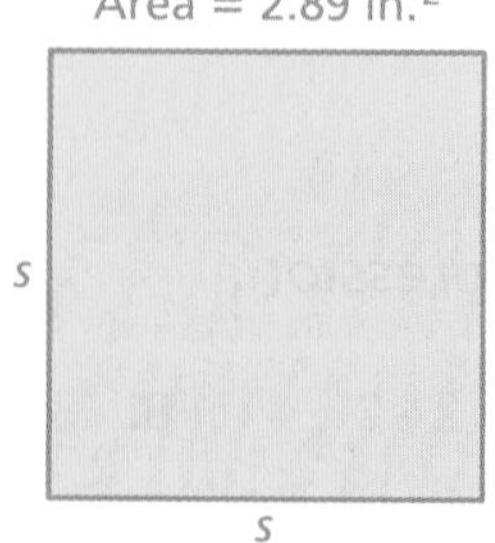

g. Area = $\frac{4}{9}$ ft²

s, s

Square Roots

In this lesson, you will

- find square roots of perfect squares.
- evaluate expressions involving square roots.
- use square roots to solve equations.

2 ACTIVITY: Using Square Roots

Work with a partner. Find the radius of each circle.

a.

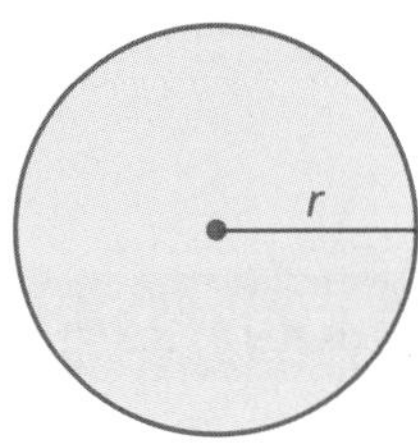

Area = 36π in.^{2}

b.

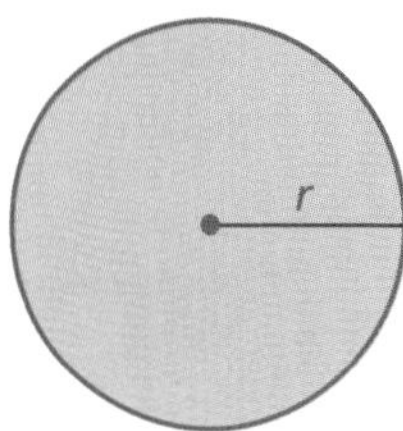

Area = π yd^2

c.

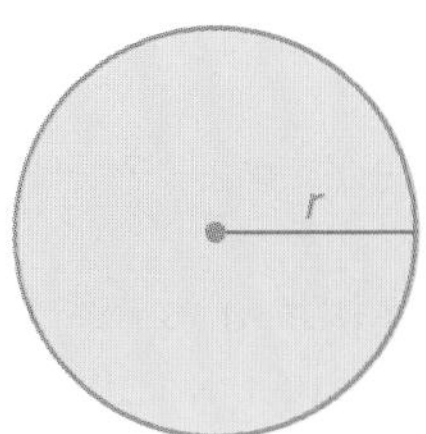

Area = 0.25π ft^2

d.

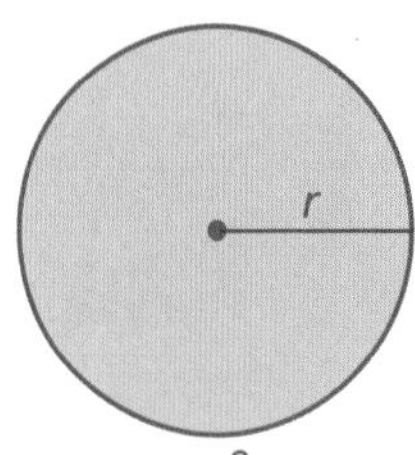

Area = $\frac{9}{16}\pi$ m^2

3 ACTIVITY: The Period of a Pendulum

Math Practice

Calculate Accurately

How can you use the graph to help you determine whether you calculated the values of T correctly?

Work with a partner.

The period of a pendulum is the time (in seconds) it takes the pendulum to swing back *and* forth.

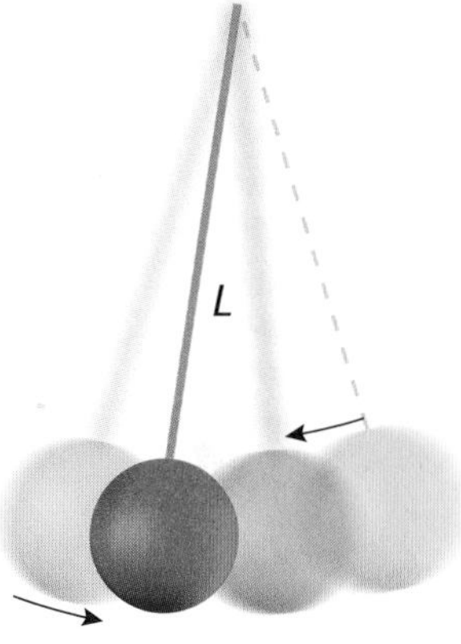

The period T is represented by $T = 1.1\sqrt{L}$, where L is the length of the pendulum (in feet).

Copy and complete the table. Then graph the function. Is the function linear?

L	1.00	1.96	3.24	4.00	4.84	6.25	7.29	7.84	9.00
T									

What Is Your Answer?

4. **IN YOUR OWN WORDS** How can you find the dimensions of a square or a circle when you are given its area? Give an example of each. How can you check your answers?

Practice Use what you learned about finding square roots to complete Exercises 4–6 on page 292.

7.1 Lesson

square root, *p. 290*
perfect square, *p. 290*
radical sign, *p. 290*
radicand, *p. 290*

A **square root** of a number is a number that, when multiplied by itself, equals the given number. Every positive number has a positive *and* a negative square root. A **perfect square** is a number with integers as its square roots.

EXAMPLE 1 Finding Square Roots of a Perfect Square

Study Tip

Zero has one square root, which is 0.

Find the two square roots of 49.

$7 \cdot 7 = 49$ and $(-7) \cdot (-7) = 49$

So, the square roots of 49 are 7 and −7.

The symbol $\sqrt{}$ is called a **radical sign**. It is used to represent a square root. The number under the radical sign is called the **radicand**.

Positive Square Root, $\sqrt{}$	Negative Square Root, $-\sqrt{}$	Both Square Roots, $\pm\sqrt{}$
$\sqrt{16} = 4$	$-\sqrt{16} = -4$	$\pm\sqrt{16} = \pm 4$

EXAMPLE 2 Finding Square Roots

Find the square root(s).

a. $\sqrt{25}$

$\sqrt{25}$ represents the *positive* square root.

Because $5^2 = 25$, $\sqrt{25} = \sqrt{5^2} = 5$.

b. $-\sqrt{\frac{9}{16}}$

$-\sqrt{\frac{9}{16}}$ represents the *negative* square root.

Because $\left(\frac{3}{4}\right)^2 = \frac{9}{16}$, $-\sqrt{\frac{9}{16}} = -\sqrt{\left(\frac{3}{4}\right)^2} = -\frac{3}{4}$.

c. $\pm\sqrt{2.25}$

$\pm\sqrt{2.25}$ represents both the *positive* and the *negative* square roots.

Because $1.5^2 = 2.25$, $\pm\sqrt{2.25} = \pm\sqrt{1.5^2} = 1.5$ and -1.5.

On Your Own

Exercises 7–18

Find the two square roots of the number.

1. 36 **2.** 100 **3.** 121

Find the square root(s).

4. $-\sqrt{1}$ **5.** $\pm\sqrt{\frac{4}{25}}$ **6.** $\sqrt{12.25}$

Squaring a positive number and finding a square root are inverse operations. You can use this relationship to evaluate expressions and solve equations involving squares.

EXAMPLE 3 Evaluating Expressions Involving Square Roots

Evaluate each expression.

a. $5\sqrt{36} + 7 = 5(6) + 7$ — Evaluate the square root.

$= 30 + 7$ — Multiply.

$= 37$ — Add.

b. $\frac{1}{4} + \sqrt{\frac{18}{2}} = \frac{1}{4} + \sqrt{9}$ — Simplify.

$= \frac{1}{4} + 3$ — Evaluate the square root.

$= 3\frac{1}{4}$ — Add.

c. $(\sqrt{81})^2 - 5 = 81 - 5$ — Evaluate the power using inverse operations.

$= 76$ — Subtract.

EXAMPLE 4 Real-Life Application

The area of a crop circle is 45,216 square feet. What is the radius of the crop circle? Use 3.14 for π.

$A = \pi r^2$ — Write the formula for the area of a circle.

$45{,}216 \approx 3.14r^2$ — Substitute 45,216 for A and 3.14 for π.

$14{,}400 = r^2$ — Divide each side by 3.14.

$\sqrt{14{,}400} = \sqrt{r^2}$ — Take positive square root of each side.

$120 = r$ — Simplify.

The radius of the crop circle is about 120 feet.

On Your Own

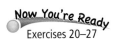

Now You're Ready
Exercises 20–27

Evaluate the expression.

7. $12 - 3\sqrt{25}$

8. $\sqrt{\frac{28}{7}} + 2.4$

9. $15 - (\sqrt{4})^2$

10. The area of a circle is 2826 square feet. Write and solve an equation to find the radius of the circle. Use 3.14 for π.

7.1 Exercises

Vocabulary and Concept Check

1. **VOCABULARY** Is 26 a perfect square? Explain.
2. **REASONING** Can the square of an integer be a negative number? Explain.
3. **NUMBER SENSE** Does $\sqrt{256}$ represent the positive square root of 256, the negative square root of 256, or both? Explain.

Practice and Problem Solving

Find the dimensions of the square or circle. Check your answer.

4. Area = 441 cm²

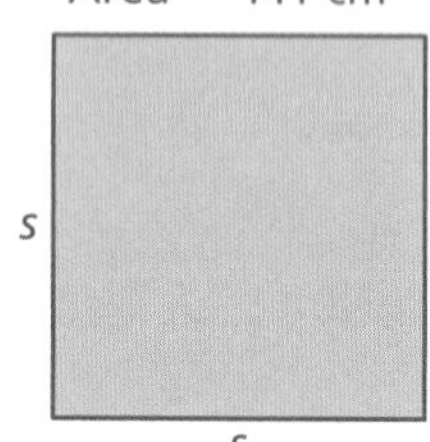

5. Area = 1.69 km²

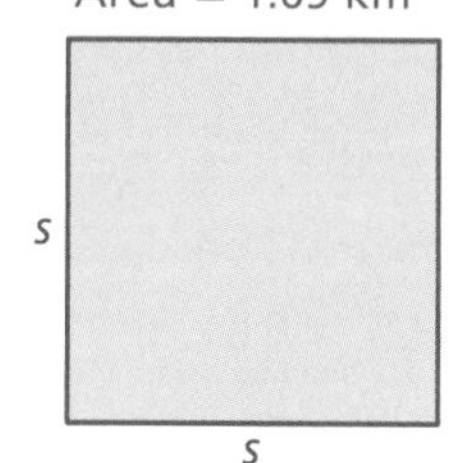

6. Area = 64π in.²

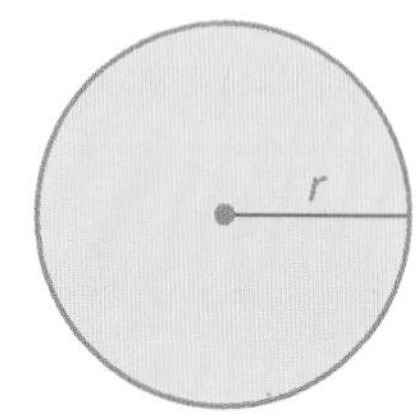

Find the two square roots of the number.

1 7. 9   8. 64   9. 4   10. 144

Find the square root(s).

2 11. $\sqrt{625}$ 12. $\pm\sqrt{196}$ 13. $\pm\sqrt{\frac{1}{961}}$ 14. $-\sqrt{\frac{9}{100}}$

15. $\pm\sqrt{4.84}$ 16. $\sqrt{7.29}$ 17. $-\sqrt{361}$ 18. $-\sqrt{2.25}$

19. **ERROR ANALYSIS** Describe and correct the error in finding the square roots.

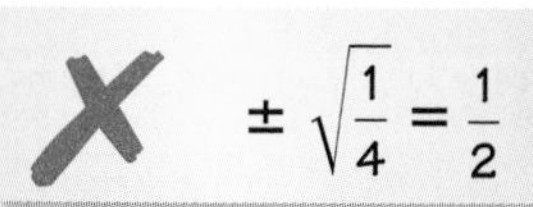

Evaluate the expression.

3 20. $(\sqrt{9})^2 + 5$ 21. $28 - (\sqrt{144})^2$ 22. $3\sqrt{16} - 5$ 23. $10 - 4\sqrt{\frac{1}{16}}$

24. $\sqrt{6.76} + 5.4$ 25. $8\sqrt{8.41} + 1.8$ 26. $2\left(\sqrt{\frac{80}{5}} - 5\right)$ 27. $4\left(\sqrt{\frac{147}{3}} + 3\right)$

28. **NOTEPAD** The area of the base of a square notepad is 2.25 square inches. What is the length of one side of the base of the notepad?

29. **CRITICAL THINKING** There are two square roots of 25. Why is there only one answer for the radius of the button?

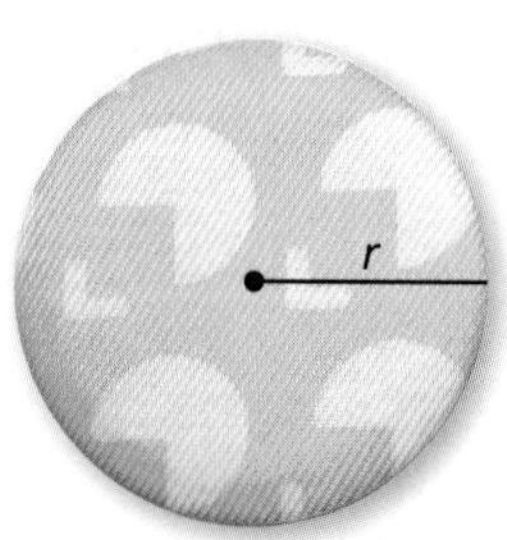

$A = 25\pi$ mm²

Copy and complete the statement with <, >, or =.

30. $\sqrt{81}$ ☐ 8 **31.** 0.5 ☐ $\sqrt{0.25}$ **32.** $\frac{3}{2}$ ☐ $\sqrt{\frac{25}{4}}$

33. SAILBOAT The area of a sail is $40\frac{1}{2}$ square feet. The base and the height of the sail are equal. What is the height of the sail (in feet)?

34. REASONING Is the product of two perfect squares always a perfect square? Explain your reasoning.

35. ENERGY The kinetic energy K (in joules) of a falling apple is represented by $K = \frac{v^2}{2}$, where v is the speed of the apple (in meters per second). How fast is the apple traveling when the kinetic energy is 32 joules?

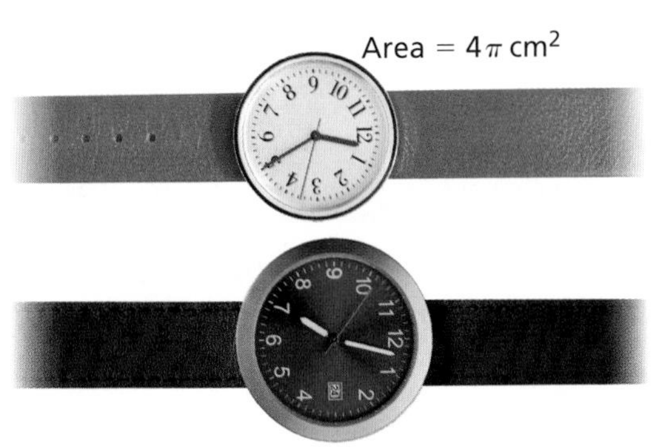

36. PRECISION The areas of the two watch faces have a ratio of 16 : 25.

a. What is the ratio of the radius of the smaller watch face to the radius of the larger watch face?

b. What is the radius of the larger watch face?

37. WINDOW The cost C (in dollars) of making a square window with a side length of n inches is represented by $C = \frac{n^2}{5} + 175$. A window costs \$355. What is the length (in feet) of the window?

38. Geometry The area of the triangle is represented by the formula $A = \sqrt{s(s-21)(s-17)(s-10)}$, where s is equal to half the perimeter. What is the height of the triangle?

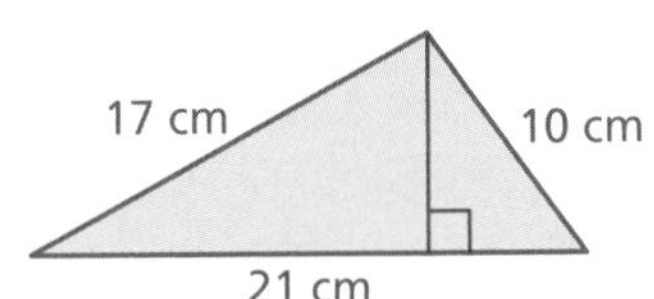

Fair Game Review What you learned in previous grades & lessons

Write in slope-intercept form an equation of the line that passes through the given points. *(Section 4.7)*

39. (2, 4), (5, 13) **40.** (−1, 7), (3, −1) **41.** (−5, −2), (5, 4)

42. MULTIPLE CHOICE What is the value of x? *(Section 3.2)*

Ⓐ 41 Ⓑ 44

Ⓒ 88 Ⓓ 134

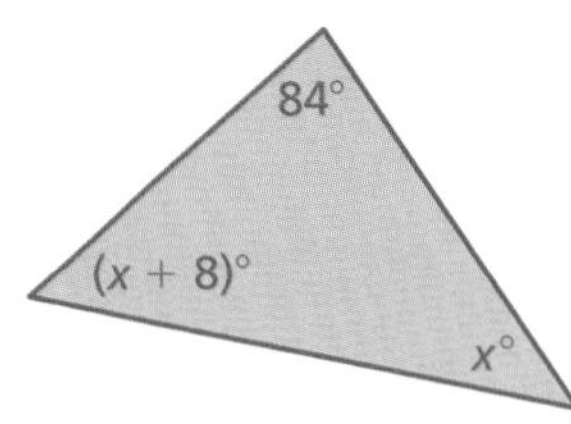

7.2 Finding Cube Roots

Essential Question How is the cube root of a number different from the square root of a number?

When you multiply a number by itself twice, you cube the number.

Symbol for cubing is the exponent 3.

$$4^3 = 4 \cdot 4 \cdot 4$$
$$= 64$$ 4 cubed is 64.

To "undo" this, take the *cube root* of the number.

Symbol for cube root is $\sqrt[3]{}$.

$$\sqrt[3]{64} = \sqrt[3]{4^3} = 4$$ The cube root of 64 is 4.

1 ACTIVITY: Finding Cube Roots

Work with a partner. Use a cube root symbol to write the edge length of the cube. Then find the cube root. Check your answer by multiplying.

a. Sample:

$s = \sqrt[3]{343} = \sqrt[3]{7^3} = 7$ inches

Volume = 343 in.^{3}

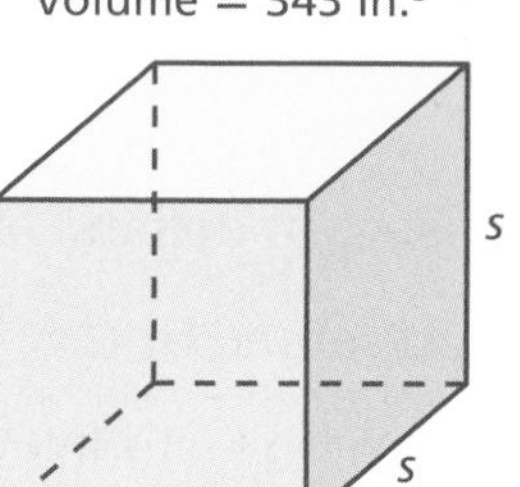

Check

$7 \cdot 7 \cdot 7 = 49 \cdot 7$

$= 343$ ✓

The edge length of the cube is 7 inches.

b. Volume = 27 ft^3

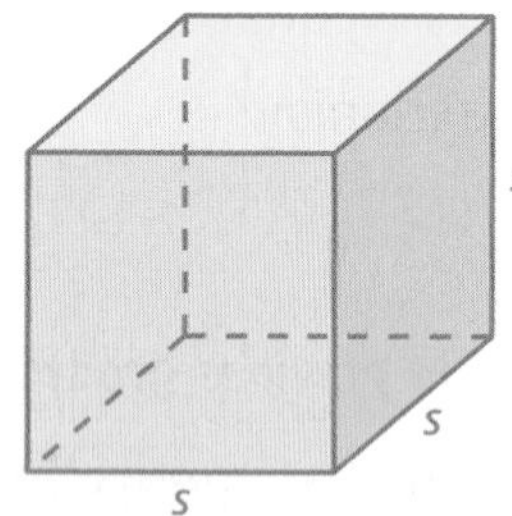

c. Volume = 125 m^3

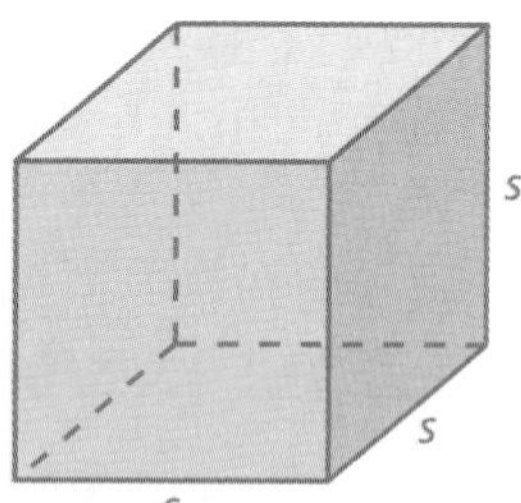

d. Volume = 0.001 cm^3

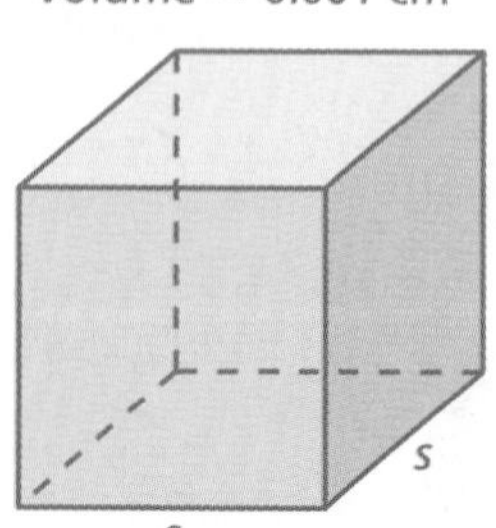

e. Volume = $\frac{1}{8}$ yd^3

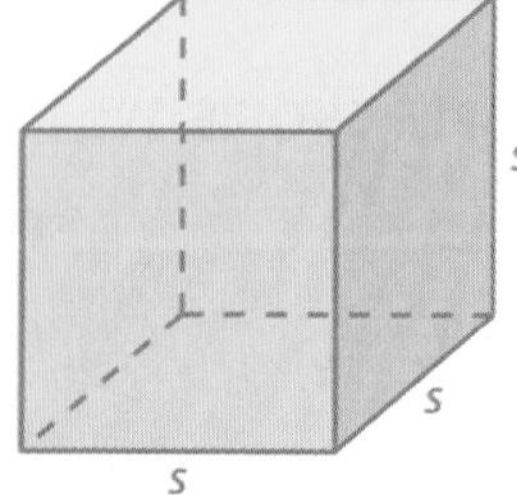

Cube Roots

In this lesson, you will

- find cube roots of perfect cubes.
- evaluate expressions involving cube roots.
- use cube roots to solve equations.

2 ACTIVITY: Using Prime Factorizations to Find Cube Roots

Math Practice

View as Components

When writing the prime factorizations in Activity 2, how many times do you expect to see each factor? Why?

Work with a partner. Write the prime factorization of each number. Then use the prime factorization to find the cube root of the number.

a. 216

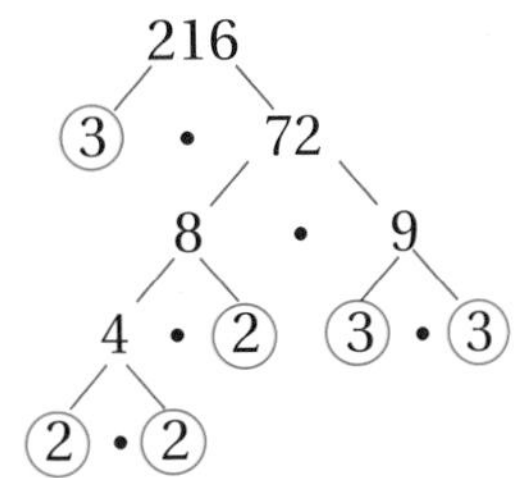

$216 = 3 \cdot 2 \cdot 3 \cdot 3 \cdot 2 \cdot 2$ Prime factorization

$= (3 \cdot \square) \cdot (3 \cdot \square) \cdot (3 \cdot \square)$ Commutative Property of Multiplication

$= \square \cdot \square \cdot \square$ Simplify.

∴ The cube root of 216 is $\square$.

b. 1000

c. 3375

d. **STRUCTURE** Does this procedure work for every number? Explain why or why not.

What Is Your Answer?

3. Complete each statement using *positive* or *negative*.

a. A positive number times a positive number is a __________ number.

b. A negative number times a negative number is a __________ number.

c. A positive number multiplied by itself twice is a __________ number.

d. A negative number multiplied by itself twice is a __________ number.

4. **REASONING** Can a negative number have a cube root? Give an example to support your explanation.

5. **IN YOUR OWN WORDS** How is the cube root of a number different from the square root of a number?

6. Give an example of a number whose square root and cube root are equal.

7. A cube has a volume of 13,824 cubic meters. Use a calculator to find the edge length.

Practice Use what you learned about cube roots to complete Exercises 3–5 on page 298.

7.2 Lesson

Key Vocabulary
cube root, *p. 296*
perfect cube, *p. 296*

A **cube root** of a number is a number that, when multiplied by itself, and then multiplied by itself again, equals the given number. A **perfect cube** is a number that can be written as the cube of an integer. The symbol $\sqrt[3]{\ }$ is used to represent a cube root.

EXAMPLE 1 Finding Cube Roots

Find each cube root.

a. $\sqrt[3]{8}$

Because $2^3 = 8$, $\sqrt[3]{8} = \sqrt[3]{2^3} = 2$.

b. $\sqrt[3]{-27}$

Because $(-3)^3 = -27$, $\sqrt[3]{-27} = \sqrt[3]{(-3)^3} = -3$.

c. $\sqrt[3]{\frac{1}{64}}$

Because $\left(\frac{1}{4}\right)^3 = \frac{1}{64}$, $\sqrt[3]{\frac{1}{64}} = \sqrt[3]{\left(\frac{1}{4}\right)^3} = \frac{1}{4}$.

Cubing a number and finding a cube root are inverse operations. You can use this relationship to evaluate expressions and solve equations involving cubes.

EXAMPLE 2 Evaluating Expressions Involving Cube Roots

Evaluate each expression.

a. $2\sqrt[3]{-216} - 3 = 2(-6) - 3$ — Evaluate the cube root.

$= -12 - 3$ — Multiply.

$= -15$ — Subtract.

b. $\left(\sqrt[3]{125}\right)^3 + 21 = 125 + 21$ — Evaluate the power using inverse operations.

$= 146$ — Add.

On Your Own

Now You're Ready
Exercises 6–17

Find the cube root.

1. $\sqrt[3]{1}$
2. $\sqrt[3]{-343}$
3. $\sqrt[3]{-\frac{27}{1000}}$

Evaluate the expression.

4. $18 - 4\sqrt[3]{8}$
5. $\left(\sqrt[3]{-64}\right)^3 + 43$
6. $5\sqrt[3]{512} - 19$

Multi-Language Glossary at BigIdeasMath.com

EXAMPLE 3 Evaluating an Algebraic Expression

Evaluate $\frac{x}{4} + \sqrt[3]{\frac{x}{3}}$ when $x = 192$.

$\frac{x}{4} + \sqrt[3]{\frac{x}{3}} = \frac{192}{4} + \sqrt[3]{\frac{192}{3}}$ Substitute 192 for x.

$= 48 + \sqrt[3]{64}$ Simplify.

$= 48 + 4$ Evaluate the cube root.

$= 52$ Add.

On Your Own

Now You're Ready
Exercises 18–20

Evaluate the expression for the given value of the variable.

7. $\sqrt[3]{8y} + y, y = 64$

8. $2b - \sqrt[3]{9b}, b = -3$

EXAMPLE 4 Real-Life Application

Find the surface area of the baseball display case.

The baseball display case is in the shape of a cube. Use the formula for the volume of a cube to find the edge length s.

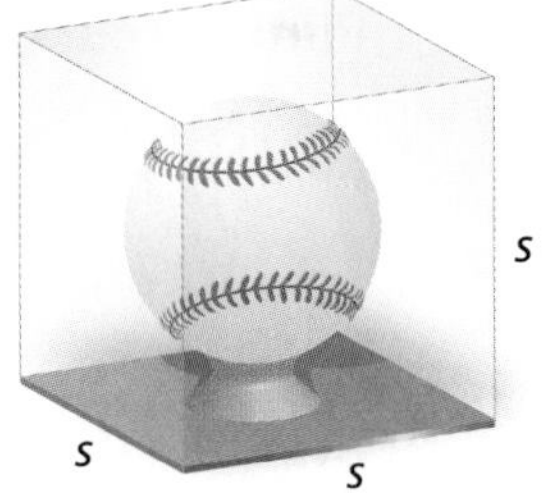

Volume = 125 in.^{3}

$V = s^3$ Write formula for volume.

$125 = s^3$ Substitute 125 for V.

$\sqrt[3]{125} = \sqrt[3]{s^3}$ Take the cube root of each side.

$5 = s$ Simplify.

Remember

The volume V of a cube with edge length s is given by $V = s^3$. The surface area S is given by $S = 6s^2$.

The edge length is 5 inches. Use a formula to find the surface area of the cube.

$S = 6s^2$ Write formula for surface area.

$= 6(5)^2$ Substitute 5 for s.

$= 150$ Simplify.

So, the surface area of the baseball display case is 150 square inches.

On Your Own

9. The volume of a music box that is shaped like a cube is 512 cubic centimeters. Find the surface area of the music box.

7.2 Exercises

Vocabulary and Concept Check

1. **VOCABULARY** Is 25 a perfect cube? Explain.
2. **REASONING** Can the cube of an integer be a negative number? Explain.

Practice and Problem Solving

Find the edge length of the cube.

3. Volume = 125,000 in.^{3}

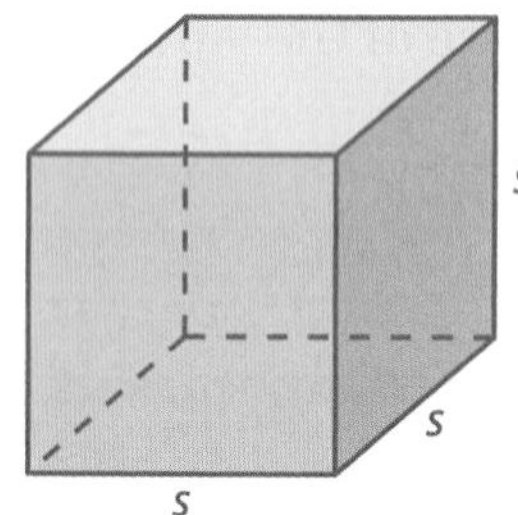

4. Volume = $\frac{1}{27}$ ft^3

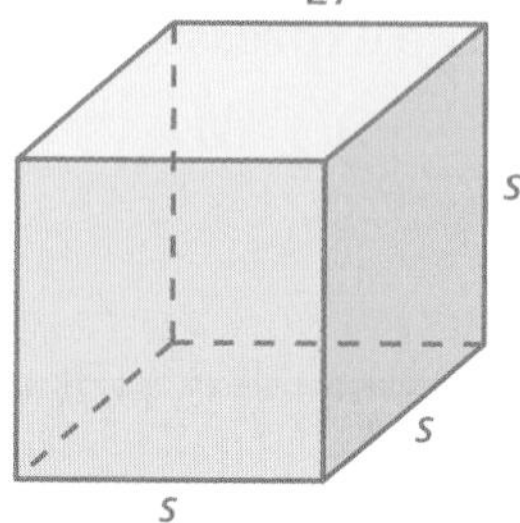

5. Volume = 0.064 m^3

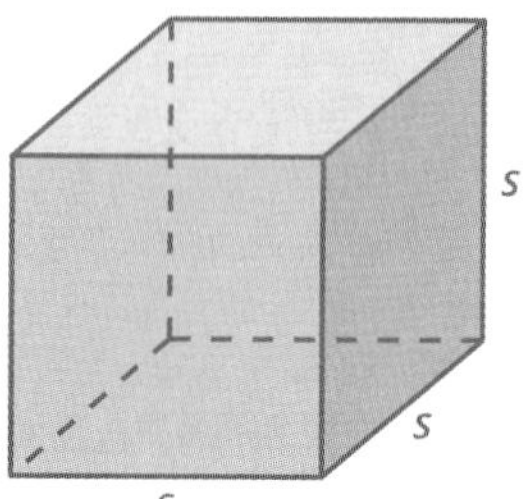

Find the cube root.

1 6. $\sqrt[3]{729}$
7. $\sqrt[3]{-125}$
8. $\sqrt[3]{-1000}$
9. $\sqrt[3]{1728}$
10. $\sqrt[3]{-\frac{1}{512}}$
11. $\sqrt[3]{\frac{343}{64}}$

Evaluate the expression.

2 12. $18 - \left(\sqrt[3]{27}\right)^3$
13. $\left(\sqrt[3]{-\frac{1}{8}}\right)^3 + 3\frac{3}{4}$
14. $5\sqrt[3]{729} - 24$
15. $\frac{1}{4} - 2\sqrt[3]{-\frac{1}{216}}$
16. $54 + \sqrt[3]{-4096}$
17. $4\sqrt[3]{8000} - 6$

Evaluate the expression for the given value of the variable.

3 18. $\sqrt[3]{\frac{n}{4}} + \frac{n}{10}$, $n = 500$
19. $\sqrt[3]{6w} - w$, $w = 288$
20. $2d + \sqrt[3]{-45d}$, $d = 75$

21. **STORAGE CUBE** The volume of a plastic storage cube is 27,000 cubic centimeters. What is the edge length of the storage cube?

22. **ICE SCULPTURE** The volume of a cube of ice for an ice sculpture is 64,000 cubic inches.
    a. What is the edge length of the cube of ice?
    b. What is the surface area of the cube of ice?

Copy and complete the statement with <, >, or =.

23. $-\frac{1}{4}$ ☐ $\sqrt[3]{-\frac{8}{125}}$

24. $\sqrt[3]{0.001}$ ☐ 0.01

25. $\sqrt[3]{64}$ ☐ $\sqrt{64}$

26. DRAG RACE The estimated velocity v (in miles per hour) of a car at the end of a drag race is $v = 234\sqrt[3]{\frac{p}{w}}$, where p is the horsepower of the car and w is the weight (in pounds) of the car. A car has a horsepower of 1311 and weighs 2744 pounds. Find the velocity of the car at the end of a drag race. Round your answer to the nearest whole number.

27. NUMBER SENSE There are three numbers that are their own cube roots. What are the numbers?

28. LOGIC Each statement below is true for square roots. Determine whether the statement is also true for cube roots. Explain your reasoning and give an example to support your explanation.

a. You cannot find the square root of a negative number.

b. Every positive number has a positive square root and a negative square root.

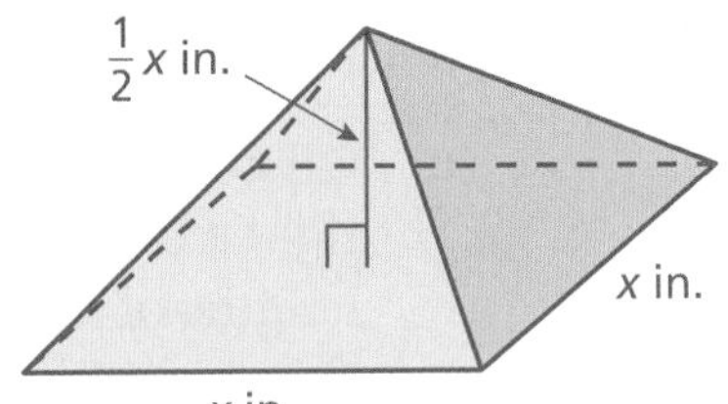

29. GEOMETRY The pyramid has a volume of 972 cubic inches. What are the dimensions of the pyramid?

30. RATIOS The ratio $125 : x$ is equivalent to the ratio $x^2 : 125$. What is the value of x?

Solve the equation.

31. $(3x + 4)^3 = 2197$

32. $(8x^3 - 9)^3 = 5832$

33. $\left((5x - 16)^3 - 4\right)^3 = 216{,}000$

Fair Game Review What you learned in previous grades & lessons

Evaluate the expression. *(Skills Review Handbook)*

34. $3^2 + 4^2$

35. $8^2 + 15^2$

36. $13^2 - 5^2$

37. $25^2 - 24^2$

38. MULTIPLE CHOICE Which linear function is shown by the table? *(Section 6.3)*

x	1	2	3	4
y	4	7	10	13

Ⓐ $y = \frac{1}{3}x + 1$

Ⓑ $y = 4x$

Ⓒ $y = 3x + 1$

Ⓓ $y = \frac{1}{4}x$

7.3 The Pythagorean Theorem

Essential Question How are the lengths of the sides of a right triangle related?

Pythagoras was a Greek mathematician and philosopher who discovered one of the most famous rules in mathematics. In mathematics, a rule is called a **theorem**. So, the rule that Pythagoras discovered is called the Pythagorean Theorem.

Pythagoras
(c. 570–c. 490 B.C.)

1 ACTIVITY: Discovering the Pythagorean Theorem

Work with a partner.

a. On grid paper, draw any right triangle. Label the lengths of the two shorter sides a and b.

b. Label the length of the longest side c.

c. Draw squares along each of the three sides. Label the areas of the three squares a^2, b^2, and c^2.

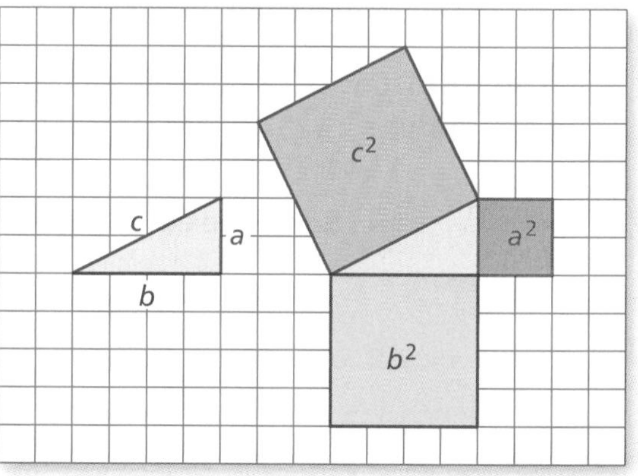

d. Cut out the three squares. Make eight copies of the right triangle and cut them out. Arrange the figures to form two identical larger squares.

e. MODELING The Pythagorean Theorem describes the relationship among a^2, b^2, and c^2. Use your result from part (d) to write an equation that describes this relationship.

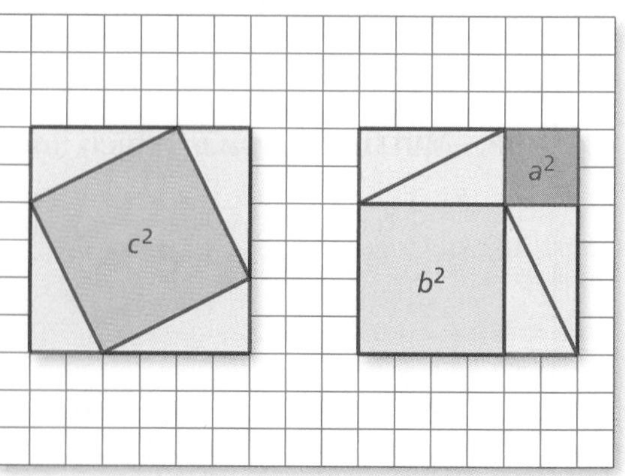

Pythagorean Theorem

In this lesson, you will

- provide geometric proof of the Pythagorean Theorem.
- use the Pythagorean Theorem to find missing side lengths of right triangles.
- solve real-life problems.

2 ACTIVITY: Using the Pythagorean Theorem in Two Dimensions

Work with a partner. Use a ruler to measure the longest side of each right triangle. Verify the result of Activity 1 for each right triangle.

a.

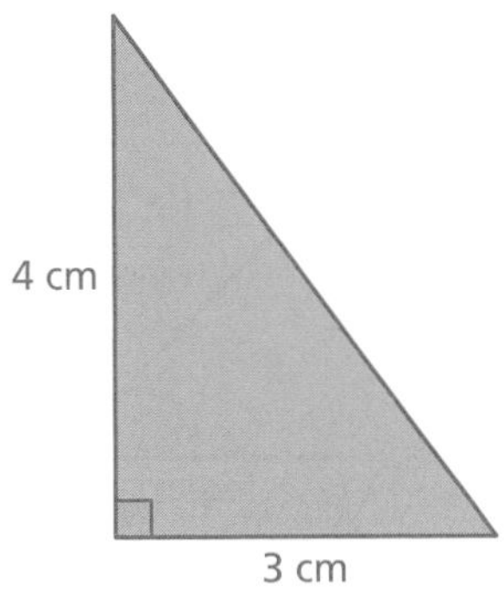

b.

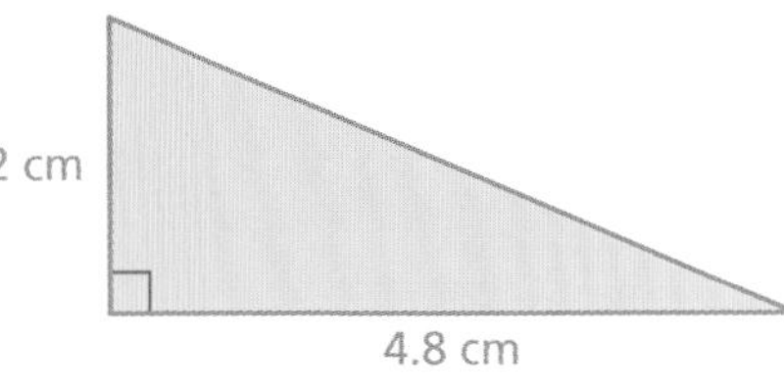

c.

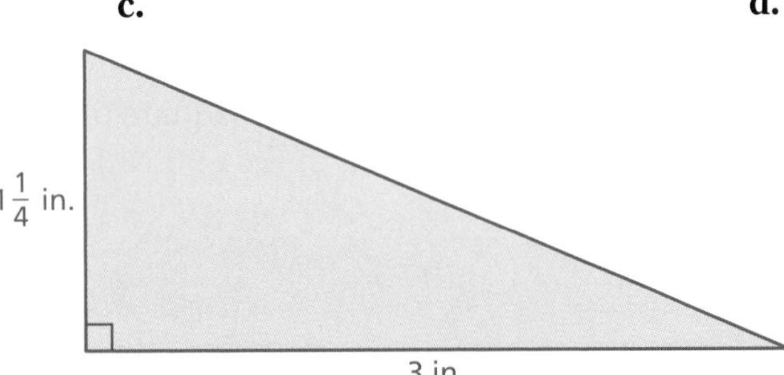

d.

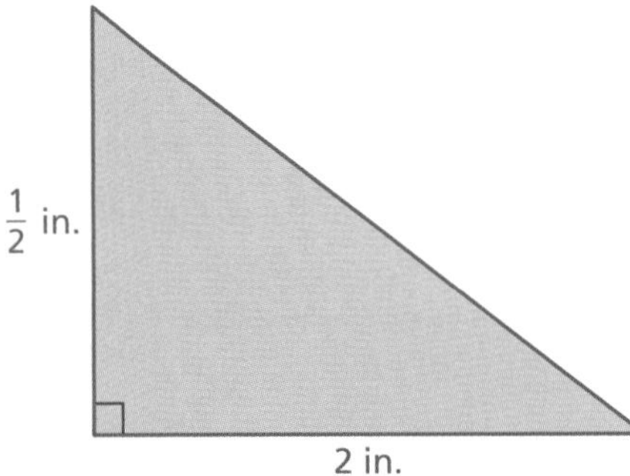

3 ACTIVITY: Using the Pythagorean Theorem in Three Dimensions

Math Practice

Use Definitions

How can you use what you know about the Pythagorean Theorem to describe the procedure for finding the length of the guy wire?

Work with a partner. A guy wire attached 24 feet above ground level on a telephone pole provides support for the pole.

a. **PROBLEM SOLVING** Describe a procedure that you could use to find the length of the guy wire without directly measuring the wire.

b. Find the length of the wire when it meets the ground 10 feet from the base of the pole.

What Is Your Answer?

4. **IN YOUR OWN WORDS** How are the lengths of the sides of a right triangle related? Give an example using whole numbers.

Practice Use what you learned about the Pythagorean Theorem to complete Exercises 3 and 4 on page 304.

Key Vocabulary
theorem, *p. 300*
legs, *p. 302*
hypotenuse, *p. 302*
Pythagorean Theorem, *p. 302*

Sides of a Right Triangle

The sides of a right triangle have special names.

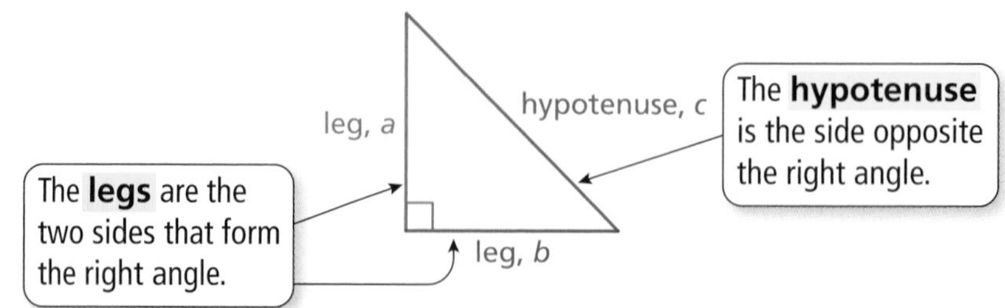

Study Tip

In a right triangle, the legs are the shorter sides and the hypotenuse is always the longest side.

The Pythagorean Theorem

Words In any right triangle, the sum of the squares of the lengths of the legs is equal to the square of the length of the hypotenuse.

Algebra $a^2 + b^2 = c^2$

EXAMPLE 1 Finding the Length of a Hypotenuse

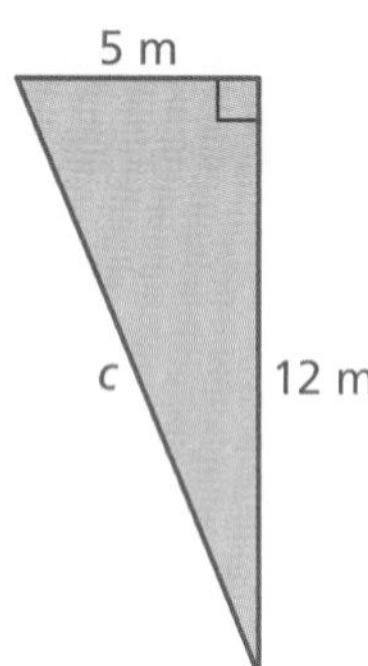

Find the length of the hypotenuse of the triangle.

$a^2 + b^2 = c^2$ Write the Pythagorean Theorem.

$5^2 + 12^2 = c^2$ Substitute 5 for *a* and 12 for *b*.

$25 + 144 = c^2$ Evaluate powers.

$169 = c^2$ Add.

$\sqrt{169} = \sqrt{c^2}$ Take positive square root of each side.

$13 = c$ Simplify.

The length of the hypotenuse is 13 meters.

On Your Own

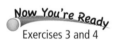

Now You're Ready
Exercises 3 and 4

Find the length of the hypotenuse of the triangle.

1.

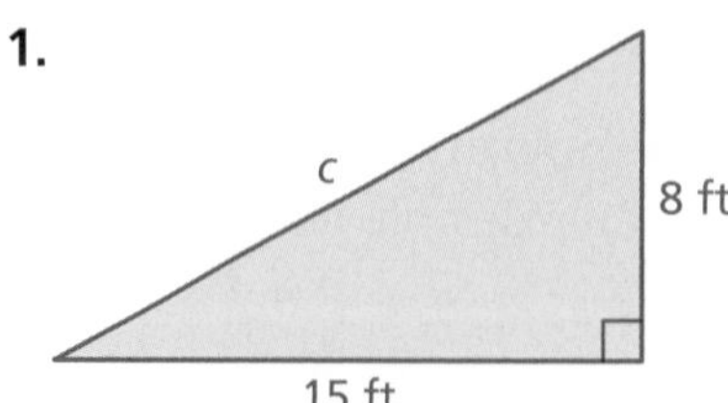

2.

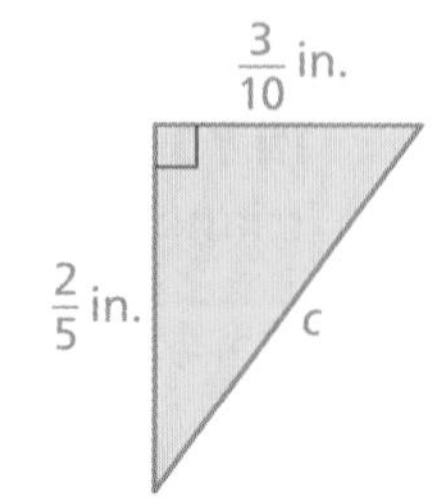

EXAMPLE 2 Finding the Length of a Leg

Find the missing length of the triangle.

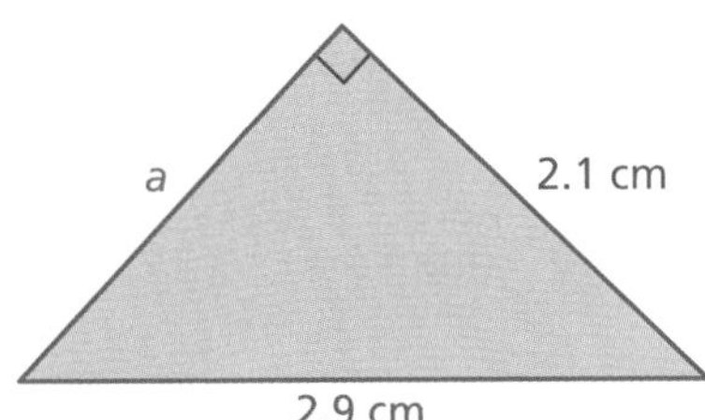

$a^2 + b^2 = c^2$	Write the Pythagorean Theorem.
$a^2 + 2.1^2 = 2.9^2$	Substitute 2.1 for *b* and 2.9 for *c*.
$a^2 + 4.41 = 8.41$	Evaluate powers.
$a^2 = 4$	Subtract 4.41 from each side.
$a = 2$	Take positive square root of each side.

∴ The missing length is 2 centimeters.

EXAMPLE 3 Real-Life Application

You are playing capture the flag. You are 50 yards north and 20 yards east of your team's base. The other team's base is 80 yards north and 60 yards east of your base. How far are you from the other team's base?

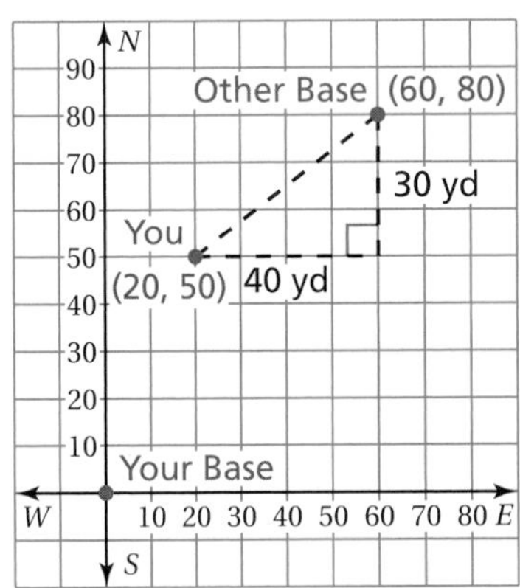

Step 1: Draw the situation in a coordinate plane. Let the origin represent your team's base. From the descriptions, you are at (20, 50) and the other team's base is at (60, 80).

Step 2: Draw a right triangle with a hypotenuse that represents the distance between you and the other team's base. The lengths of the legs are 30 yards and 40 yards.

Step 3: Use the Pythagorean Theorem to find the length of the hypotenuse.

$a^2 + b^2 = c^2$	Write the Pythagorean Theorem.
$30^2 + 40^2 = c^2$	Substitute 30 for *a* and 40 for *b*.
$900 + 1600 = c^2$	Evaluate powers.
$2500 = c^2$	Add.
$50 = c$	Take positive square root of each side.

∴ So, you are 50 yards from the other team's base.

On Your Own

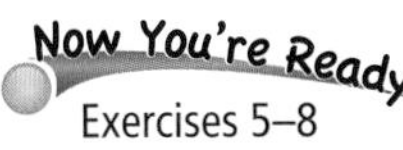

Find the missing length of the triangle.

3. 34 yd, 16 yd, *b*

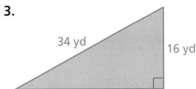

4.

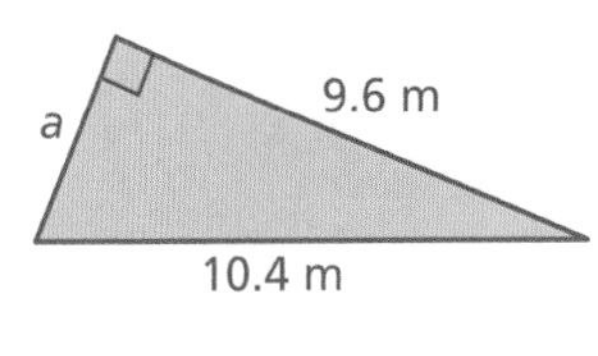

5. In Example 3, what is the distance between the bases?

7.3 Exercises

Vocabulary and Concept Check

1. **VOCABULARY** In a right triangle, how can you tell which sides are the legs and which side is the hypotenuse?

2. **DIFFERENT WORDS, SAME QUESTION** Which is different? Find "both" answers.

Which side is the hypotenuse?

Which side is the longest?

Which side is a leg?

Which side is opposite the right angle?

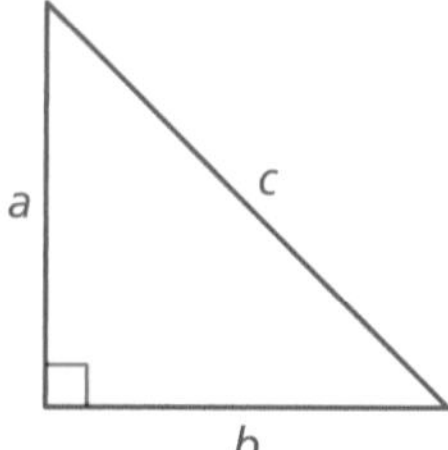

Practice and Problem Solving

Find the missing length of the triangle.

1 2 3.

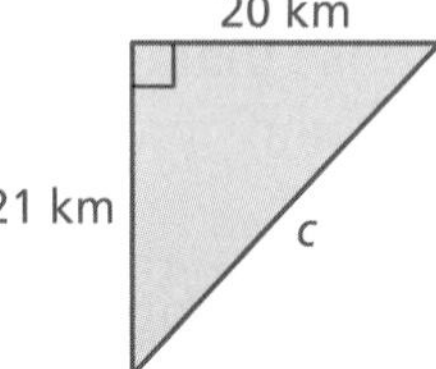

4.

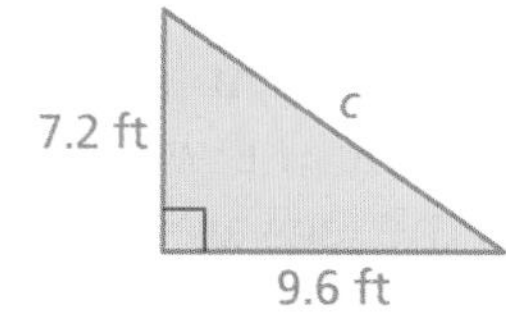

5.

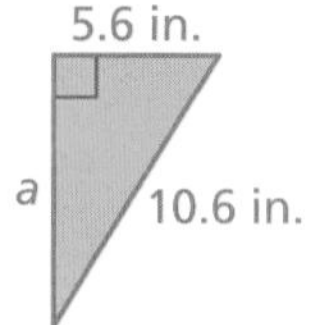

6.

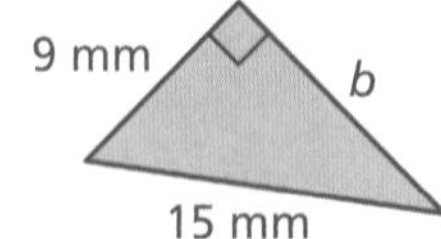

7.

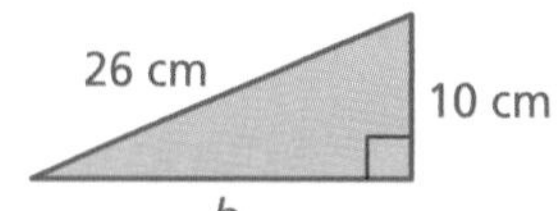

8.

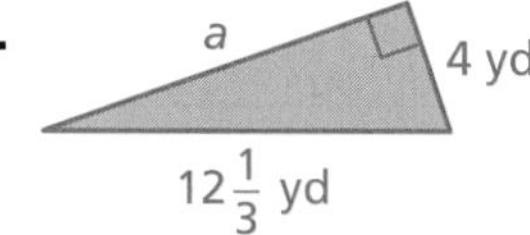

9. **ERROR ANALYSIS** Describe and correct the error in finding the missing length of the triangle.

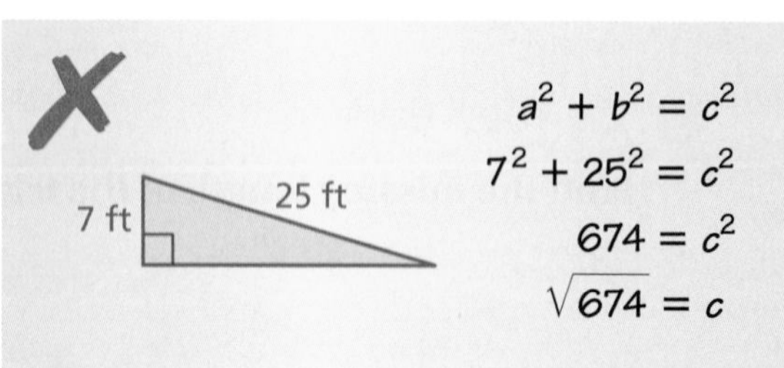

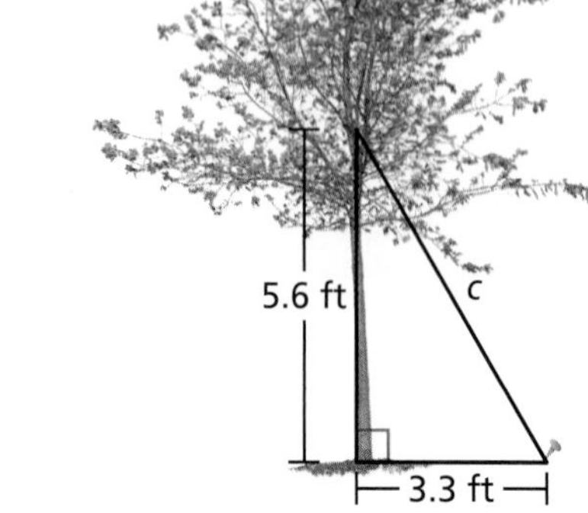

10. **TREE SUPPORT** How long is the wire that supports the tree?

Find the missing length of the figure.

11. 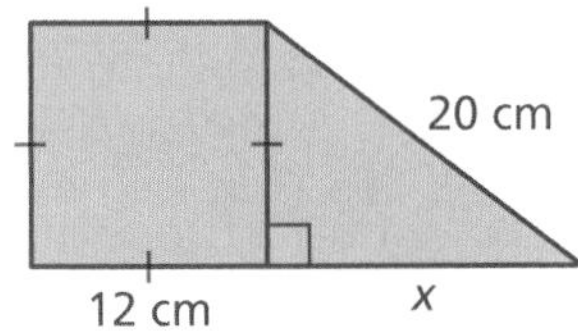

12. 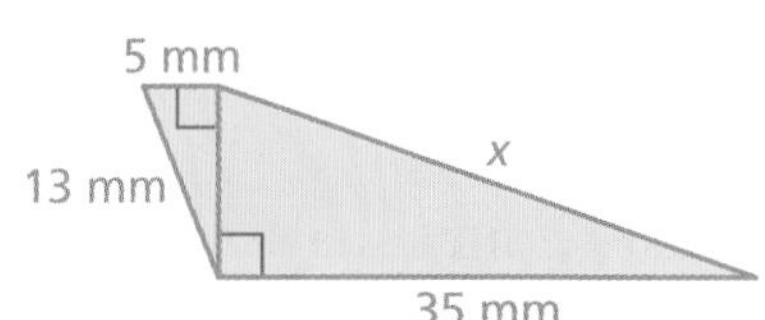

13. GOLF The figure shows the location of a golf ball after a tee shot. How many feet from the hole is the ball?

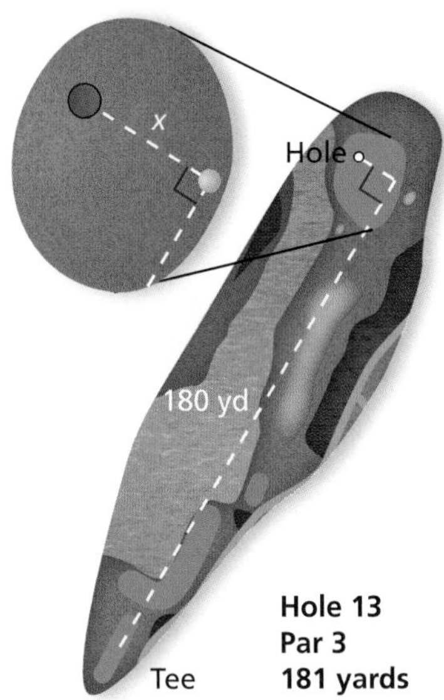

14. TENNIS A tennis player asks the referee a question. The sound of the player's voice travels only 30 feet. Can the referee hear the question? Explain.

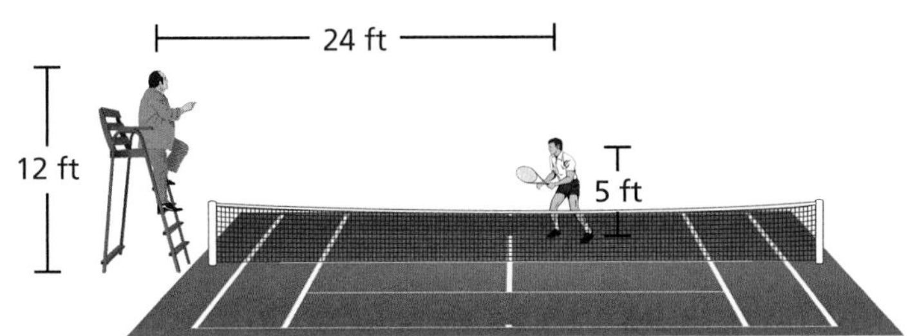

15. PROJECT Measure the length, width, and height of a rectangular room. Use the Pythagorean Theorem to find length BC and length AB.

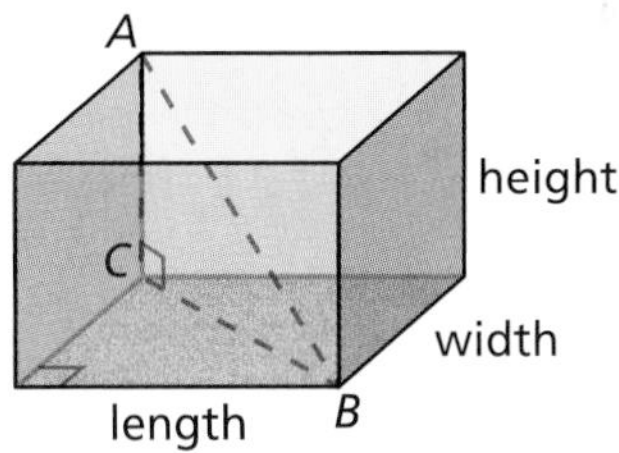

16. ALGEBRA The legs of a right triangle have lengths of 28 meters and 21 meters. The hypotenuse has a length of $5x$ meters. What is the value of x?

17. SNOWBALLS You and a friend stand back-to-back. You run 20 feet forward, then 15 feet to your right. At the same time, your friend runs 16 feet forward, then 12 feet to her right. She stops and hits you with a snowball.

a. Draw the situation in a coordinate plane.

b. How far does your friend throw the snowball?

18. Precision A box has a length of 6 inches, a width of 8 inches, and a height of 24 inches. Can a cylindrical rod with a length of 63.5 centimeters fit in the box? Explain your reasoning.

Fair Game Review What you learned in previous grades & lessons

Find the square root(s). *(Section 7.1)*

19. $\pm\sqrt{36}$ **20.** $-\sqrt{121}$ **21.** $\sqrt{169}$ **22.** $-\sqrt{225}$

23. MULTIPLE CHOICE What is the solution of the system of linear equations $y = 4x + 1$ and $2x + y = 13$? *(Section 5.2)*

Ⓐ $x = 1, y = 5$ Ⓑ $x = 5, y = 3$ Ⓒ $x = 2, y = 9$ Ⓓ $x = 9, y = 2$

7 Study Help

You can use a **four square** to organize information about a topic. Each of the four squares can be a category, such as *definition, vocabulary, example, non-example, words, algebra, table, numbers, visual, graph,* or *equation*. Here is an example of a four square for the Pythagorean Theorem.

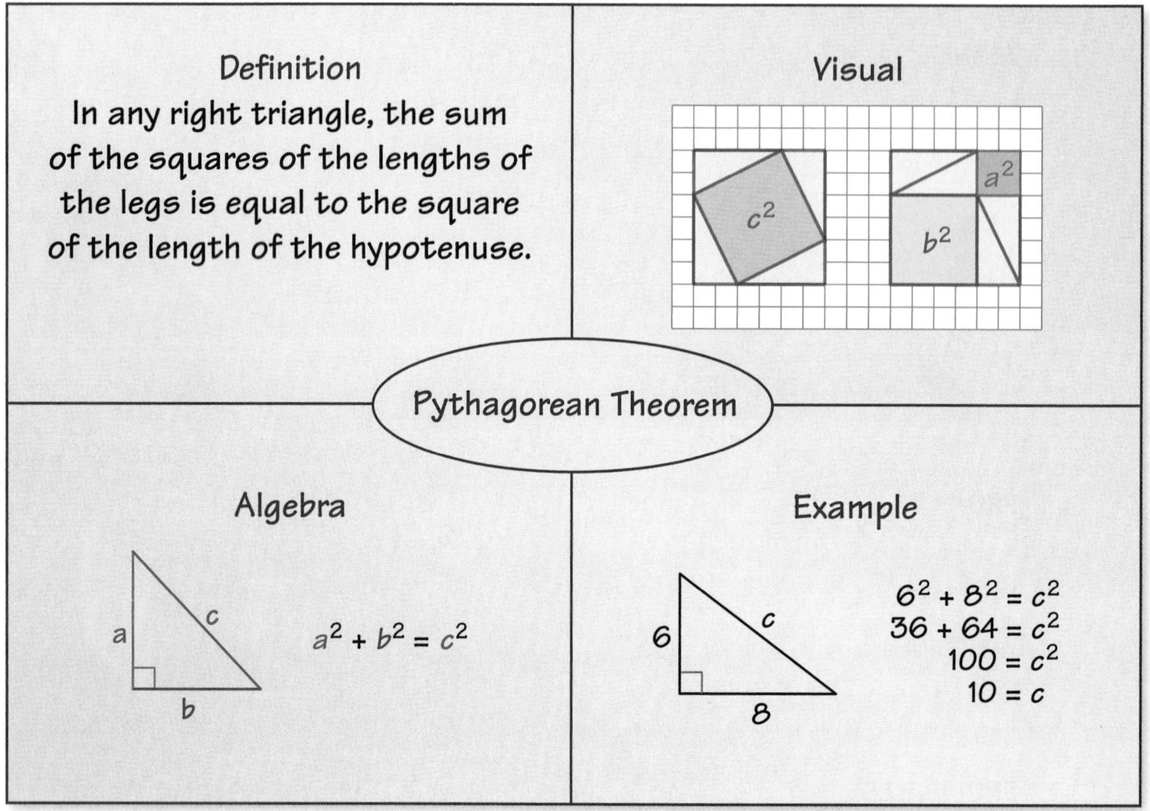

On Your Own

Make four squares to help you study these topics.

1. square roots
2. cube roots

After you complete this chapter, make four squares for the following topics.

3. irrational numbers
4. real numbers
5. converse of the Pythagorean Theorem
6. distance formula

"I'm taking a survey for my four square. How many fleas do you have?"

7.1–7.3 Quiz

Check It Out
Progress Check
BigIdeasMath.com

Find the square root(s). *(Section 7.1)*

1. $-\sqrt{4}$
2. $\sqrt{\frac{16}{25}}$
3. $\pm\sqrt{6.25}$

Find the cube root. *(Section 7.2)*

4. $\sqrt[3]{64}$
5. $\sqrt[3]{-216}$
6. $\sqrt[3]{-\frac{343}{1000}}$

Evaluate the expression. *(Section 7.1 and Section 7.2)*

7. $3\sqrt{49} + 5$
8. $10 - 4\sqrt{16}$
9. $\frac{1}{4} + \sqrt{\frac{100}{4}}$
10. $\left(\sqrt[3]{-27}\right)^3 + 61$
11. $15 + 3\sqrt[3]{125}$
12. $2\sqrt[3]{-729} - 5$

Find the missing length of the triangle. *(Section 7.3)*

13.
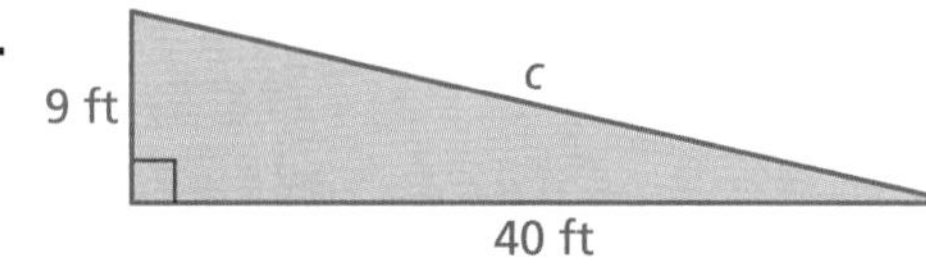

14.
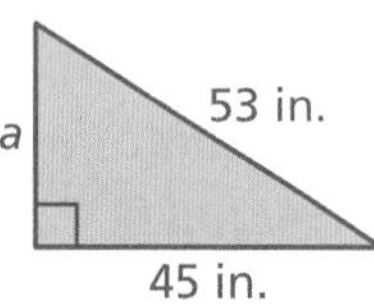

15.
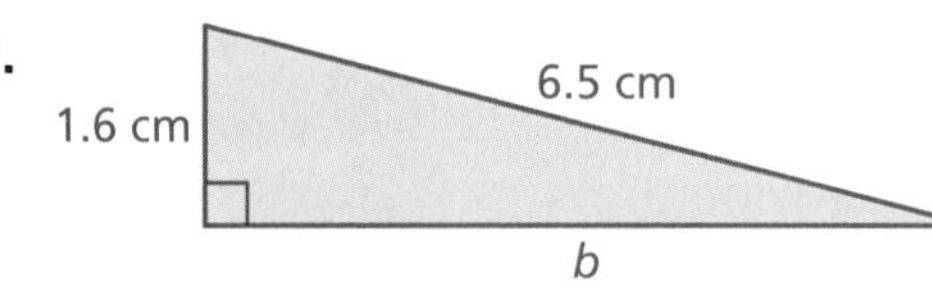

16.
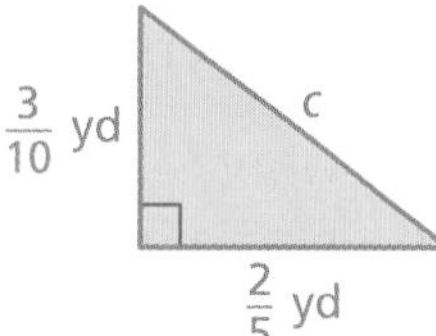

17. **POOL** The area of a circular pool cover is 314 square feet. Write and solve an equation to find the diameter of the pool cover. Use 3.14 for π. *(Section 7.1)*

18. **PACKAGE** A cube-shaped package has a volume of 5832 cubic inches. What is the edge length of the package? *(Section 7.2)*

19. **FABRIC** You are cutting a rectangular piece of fabric in half along the diagonal. The fabric measures 28 inches wide and $1\frac{1}{4}$ yards long. What is the length (in inches) of the diagonal? *(Section 7.3)*

7.4 Approximating Square Roots

Essential Question How can you find decimal approximations of square roots that are not rational?

1 ACTIVITY: Approximating Square Roots

Work with a partner. Archimedes was a Greek mathematician, physicist, engineer, inventor, and astronomer. He tried to find a rational number whose square is 3. Two that he tried were $\frac{265}{153}$ and $\frac{1351}{780}$.

a. Are either of these numbers equal to $\sqrt{3}$? Explain.

b. Use a calculator to approximate $\sqrt{3}$. Write the number on a piece of paper. Enter it into the calculator and square it. Then subtract 3. Do you get 0? What does this mean?

c. The value of $\sqrt{3}$ is between which two integers?

d. Tell whether the value of $\sqrt{3}$ is between the given numbers. Explain your reasoning.

1.7 and 1.8	1.72 and 1.73	1.731 and 1.732

2 ACTIVITY: Approximating Square Roots Geometrically

Work with a partner. Refer to the square on the number line below.

Square Roots

In this lesson, you will

- define irrational numbers.
- approximate square roots.
- approximate values of expressions involving irrational numbers.

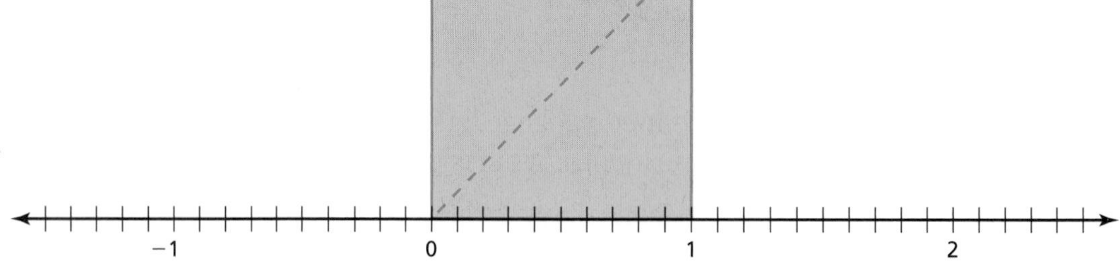

a. What is the length of the diagonal of the square?

b. Copy the square and its diagonal onto a piece of transparent paper. Rotate it about zero on the number line so that the diagonal aligns with the number line. Use the number line to estimate the length of the diagonal.

c. **STRUCTURE** How do you think your answers in parts (a) and (b) are related?

3 ACTIVITY: Approximating Square Roots Geometrically

Math Practice

Recognize Usefulness of Tools

Why is the Pythagorean Theorem a useful tool when approximating a square root?

Work with a partner.

a. Use grid paper and the given scale to draw a horizontal line segment 1 unit in length. Label this segment *AC*.

b. Draw a vertical line segment 2 units in length. Label this segment *DC*.

c. Set the point of a compass on *A*. Set the compass to 2 units. Swing the compass to intersect segment *DC*. Label this intersection as *B*.

d. Use the Pythagorean Theorem to find the length of segment *BC*.

e. Use the grid paper to approximate $\sqrt{3}$ to the nearest tenth.

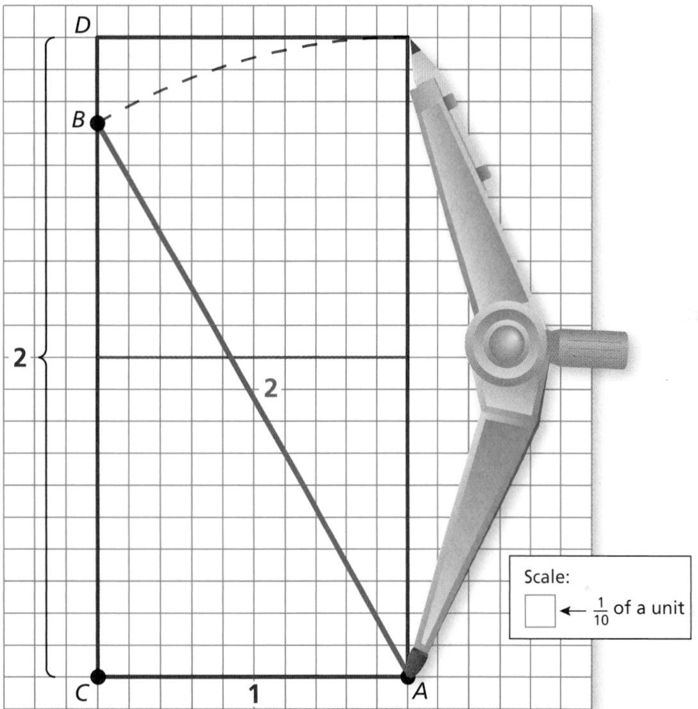

What Is Your Answer?

4. Compare your approximation in Activity 3 with your results from Activity 1.

5. Repeat Activity 3 for a triangle in which segment *AC* is 2 units and segment *BA* is 3 units. Use the Pythagorean Theorem to find the length of segment *BC*. Use the grid paper to approximate $\sqrt{5}$ to the nearest tenth.

6. **IN YOUR OWN WORDS** How can you find decimal approximations of square roots that are not rational?

Practice Use what you learned about approximating square roots to complete Exercises 5–8 on page 313.

7.4 Lesson

Key Vocabulary
irrational number, *p. 310*
real numbers, *p. 310*

A rational number is a number that can be written as the ratio of two integers. An **irrational number** cannot be written as the ratio of two integers.

- The square root of any whole number that is not a perfect square is irrational. The cube root of any integer that is not a perfect cube is irrational.
- The decimal form of an irrational number neither terminates nor repeats.

Real Numbers

Rational numbers and irrational numbers together form the set of **real numbers**.

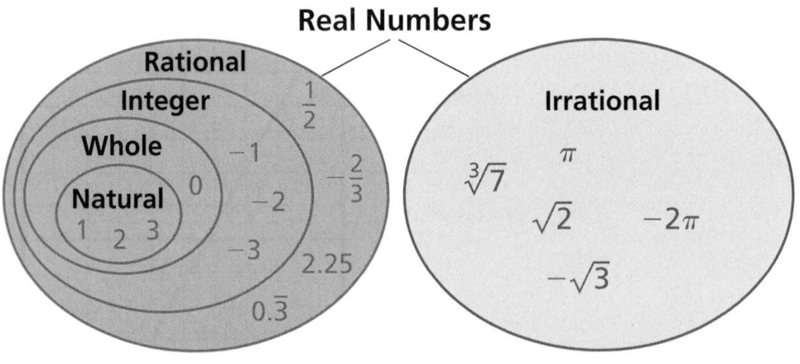

Remember

The decimal form of a rational number either terminates or repeats.

EXAMPLE 1 Classifying Real Numbers

Classify each real number.

	Number	Subset(s)	Reasoning
a.	$\sqrt{12}$	Irrational	12 is not a perfect square.
b.	$-0.\overline{25}$	Rational	$-0.\overline{25}$ is a repeating decimal.
c.	$-\sqrt{9}$	Integer, Rational	$-\sqrt{9}$ is equal to -3.
d.	$\frac{72}{4}$	Natural, Whole, Integer, Rational	$\frac{72}{4}$ is equal to 18.
e.	π	Irrational	The decimal form of π neither terminates nor repeats.

Study Tip

When classifying a real number, list all the subsets in which the number belongs.

On Your Own

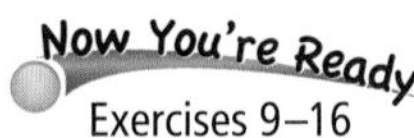

Classify the real number.

1. $0.121221222\ldots$ **2.** $-\sqrt{196}$ **3.** $\sqrt[3]{2}$

Multi-Language Glossary at BigIdeasMath.com

EXAMPLE 2 Approximating a Square Root

Estimate $\sqrt{71}$ to the nearest (a) integer and (b) tenth.

a. Make a table of numbers whose squares are close to 71.

Number	7	8	9	10
Square of Number	49	64	81	100

The table shows that 71 is between the perfect squares 64 and 81. Because 71 is closer to 64 than to 81, $\sqrt{71}$ is closer to 8 than to 9.

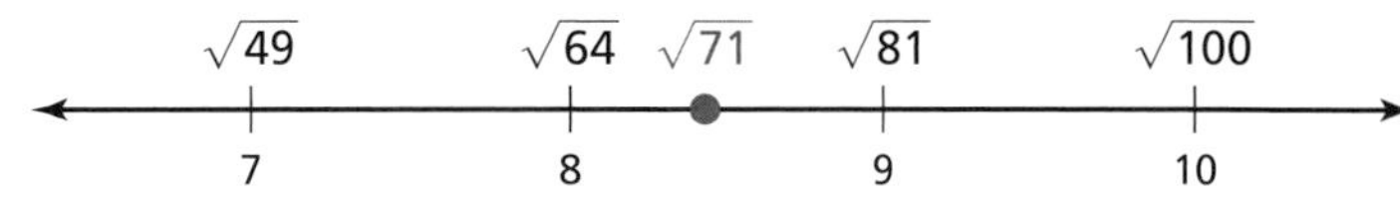

So, $\sqrt{71} \approx 8$.

b. Make a table of numbers between 8 and 9 whose squares are close to 71.

Number	8.3	8.4	8.5	8.6
Square of Number	68.89	70.56	72.25	73.96

Study Tip

You can continue the process shown in Example 2 to approximate square roots using more decimal places.

Because 71 is closer to 70.56 than to 72.25, $\sqrt{71}$ is closer to 8.4 than to 8.5.

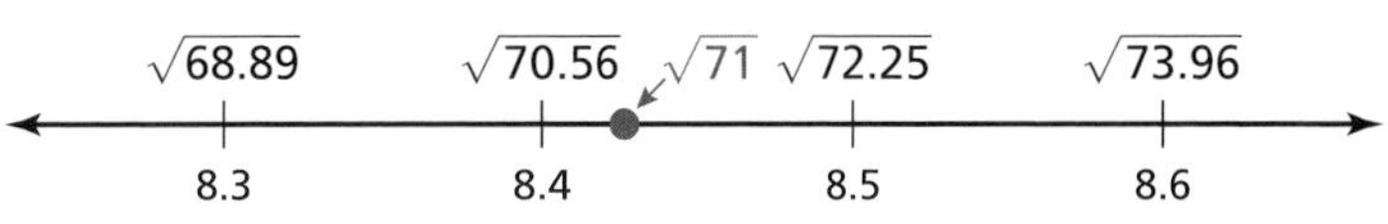

So, $\sqrt{71} \approx 8.4$.

On Your Own

Now You're Ready
Exercises 20–25

Estimate the square root to the nearest (a) integer and (b) tenth.

4. $\sqrt{8}$ **5.** $-\sqrt{13}$ **6.** $-\sqrt{24}$ **7.** $\sqrt{110}$

EXAMPLE 3 Comparing Real Numbers

Which is greater, $\sqrt{5}$ or $2\frac{2}{3}$?

Estimate $\sqrt{5}$ to the nearest integer. Then graph the numbers on a number line.

$\sqrt{5} \approx 2$ $2\frac{2}{3} = 2.\overline{6}$

$\sqrt{4} = 2$ $\sqrt{9} = 3$

$2\frac{2}{3}$ is to the right of $\sqrt{5}$. So, $2\frac{2}{3}$ is greater.

EXAMPLE 4 Approximating the Value of an Expression

The radius of a circle with area A is approximately $\sqrt{\dfrac{A}{3}}$. The area of a circular mouse pad is 51 square inches. Estimate its radius to the nearest integer.

$$\sqrt{\frac{A}{3}} = \sqrt{\frac{51}{3}} \qquad \text{Substitute 51 for } A.$$

$$= \sqrt{17} \qquad \text{Divide.}$$

The nearest perfect square less than 17 is 16. The nearest perfect square greater than 17 is 25.

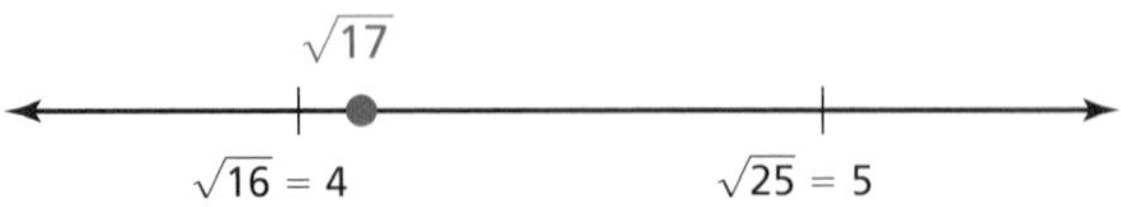

Because 17 is closer to 16 than to 25, $\sqrt{17}$ is closer to 4 than to 5.

∴ So, the radius is about 4 inches.

EXAMPLE 5 Real-Life Application

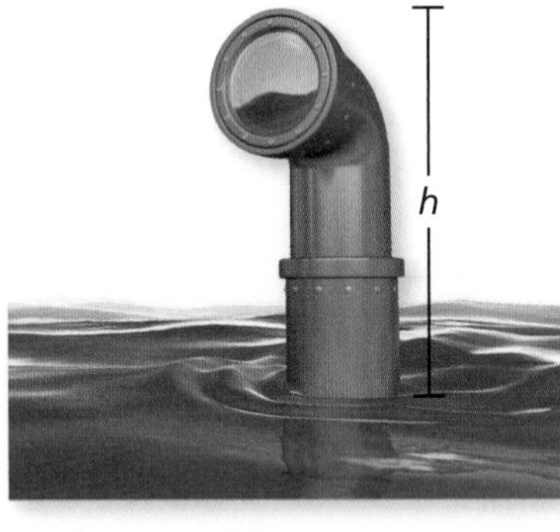

The distance (in nautical miles) you can see with a periscope is $1.17\sqrt{h}$, where h is the height of the periscope above the water. Can you see twice as far with a periscope that is 6 feet above the water than with a periscope that is 3 feet above the water? Explain.

Use a calculator to find the distances.

3 Feet Above Water		*6 Feet Above Water*
$1.17\sqrt{h} = 1.17\sqrt{3}$	Substitute for h.	$1.17\sqrt{h} = 1.17\sqrt{6}$
≈ 2.03	Use a calculator.	≈ 2.87

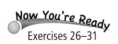

```
1.17√(3)
     2.026499445
1.17√(6)
     2.865902999
```

You can see $\dfrac{2.87}{2.03} \approx 1.41$ times farther with the periscope that is 6 feet above the water than with the periscope that is 3 feet above the water.

∴ No, you cannot see twice as far with the periscope that is 6 feet above the water.

On Your Own

Now You're Ready Exercises 26–31

Which number is greater? Explain.

8. $4\frac{1}{5}$, $\sqrt{23}$ **9.** $\sqrt{10}$, $-\sqrt{5}$ **10.** $-\sqrt{2}$, -2

11. The area of a circular mouse pad is 64 square inches. Estimate its radius to the nearest integer.

12. In Example 5, you use a periscope that is 10 feet above the water. Can you see farther than 4 nautical miles? Explain.

7.4 Exercises

Vocabulary and Concept Check

1. **VOCABULARY** How are rational numbers and irrational numbers different?
2. **WRITING** Describe a method of approximating $\sqrt{32}$.
3. **VOCABULARY** What are real numbers? Give three examples.
4. **WHICH ONE DOESN'T BELONG?** Which number does *not* belong with the other three? Explain your reasoning.

$-\frac{11}{12}$ | 25.075 | $\sqrt{8}$ | $-3.\overline{3}$

Practice and Problem Solving

Tell whether the rational number is a reasonable approximation of the square root.

5. $\frac{559}{250}, \sqrt{5}$
6. $\frac{3021}{250}, \sqrt{11}$
7. $\frac{678}{250}, \sqrt{28}$
8. $\frac{1677}{250}, \sqrt{45}$

Classify the real number.

1 9. 0
10. $\sqrt[3]{343}$
11. $\frac{\pi}{6}$
12. $-\sqrt{81}$
13. -1.125
14. $\frac{52}{13}$
15. $\sqrt[3]{-49}$
16. $\sqrt{15}$

17. **ERROR ANALYSIS** Describe and correct the error in classifying the number.

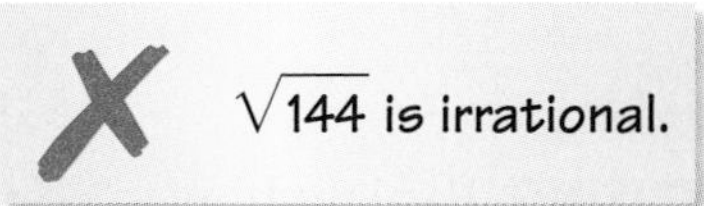

18. **SCRAPBOOKING** You cut a picture into a right triangle for your scrapbook. The lengths of the legs of the triangle are 4 inches and 6 inches. Is the length of the hypotenuse a rational number? Explain.

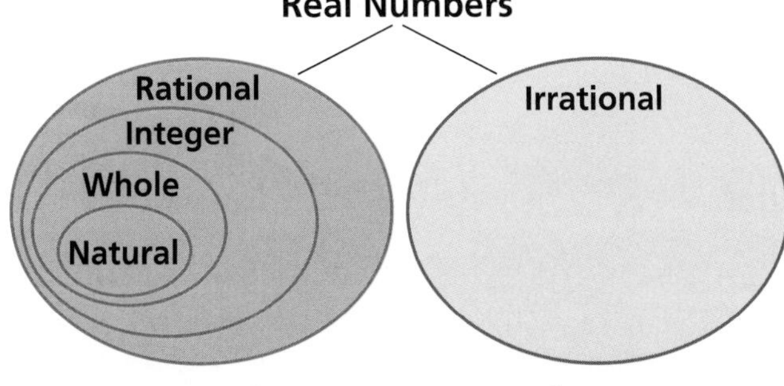

19. **VENN DIAGRAM** Place each number in the correct area of the Venn Diagram.
    a. the last digit of your phone number
    b. the square root of any prime number
    c. the ratio of the circumference of a circle to its diameter

Estimate the square root to the nearest (a) integer and (b) tenth.

2 20. $\sqrt{46}$
21. $\sqrt{685}$
22. $-\sqrt{61}$
23. $-\sqrt{105}$
24. $\sqrt{\frac{27}{4}}$
25. $-\sqrt{\frac{335}{2}}$

Which number is greater? Explain.

3 **26.** $\sqrt{20}$, 10

27. $\sqrt{15}$, -3.5

28. $\sqrt{133}$, $10\frac{3}{4}$

29. $\frac{2}{3}$, $\sqrt{\frac{16}{81}}$

30. $-\sqrt{0.25}$, -0.25

31. $-\sqrt{182}$, $-\sqrt{192}$

Use the graphing calculator screen to determine whether the statement is *true* or *false*.

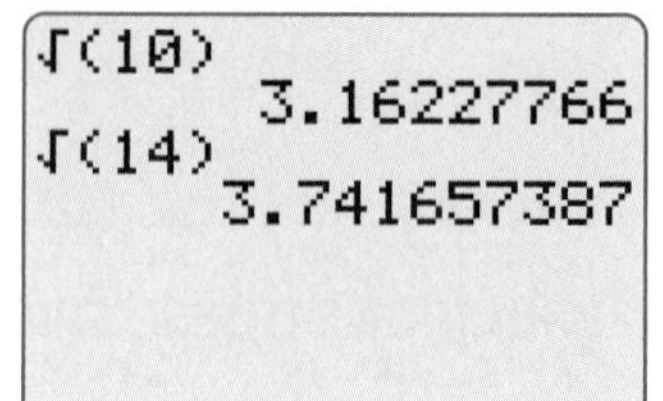

32. To the nearest tenth, $\sqrt{10} = 3.1$.

33. The value of $\sqrt{14}$ is between 3.74 and 3.75.

34. $\sqrt{10}$ lies between 3.1 and 3.16 on a number line.

35. FOUR SQUARE The area of a four square court is 66 square feet. Estimate the side length s to the nearest tenth of a foot.

36. CHECKERS A checkers board is 8 squares long and 8 squares wide. The area of each square is 14 square centimeters. Estimate the perimeter of the checkers board to the nearest tenth of a centimeter.

Approximate the length of the diagonal of the square or rectangle to the nearest tenth.

37.

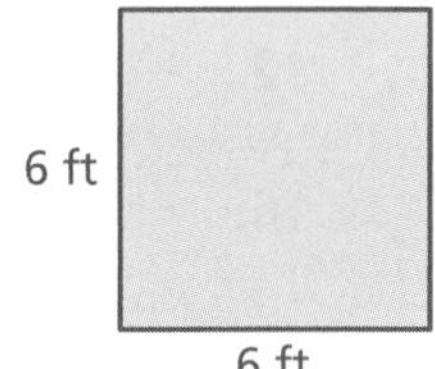

38.

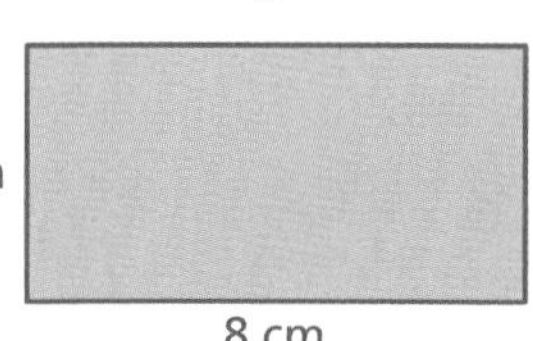

39.

40. WRITING Explain how to continue the method in Example 2 to estimate $\sqrt{71}$ to the nearest hundredth.

41. REPEATED REASONING Describe a method that you can use to estimate a cube root to the nearest tenth. Use your method to estimate $\sqrt[3]{14}$ to the nearest tenth.

42. RADIO SIGNAL The maximum distance (in nautical miles) that a radio transmitter signal can be sent is represented by the expression $1.23\sqrt{h}$, where h is the height (in feet) above the transmitter.

Estimate the maximum distance x (in nautical miles) between the plane that is receiving the signal and the transmitter. Round your answer to the nearest tenth.

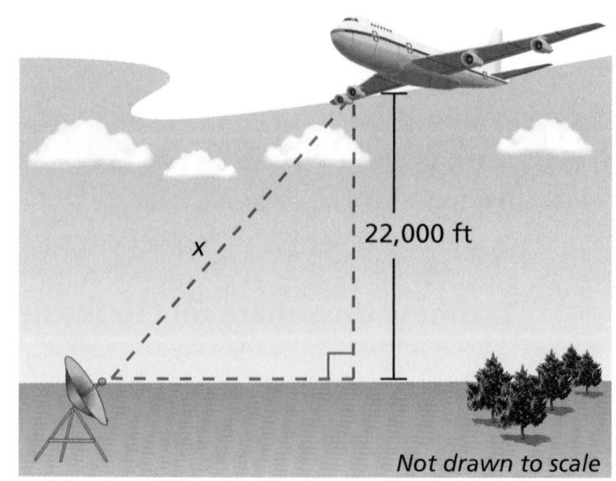

43. **OPEN-ENDED** Find two numbers a and b that satisfy the diagram.

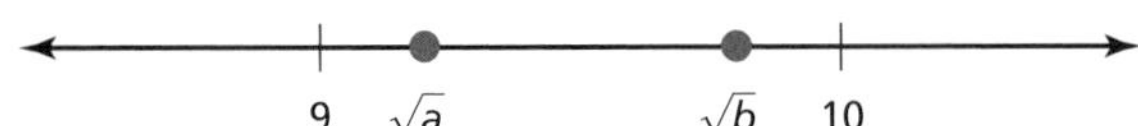

Estimate the square root to the nearest tenth.

44. $\sqrt{0.39}$ 45. $\sqrt{1.19}$ 46. $\sqrt{1.52}$

47. **ROLLER COASTER** The speed s (in meters per second) of a roller-coaster car is approximated by the equation $s = 3\sqrt{6r}$, where r is the radius of the loop. Estimate the speed of a car going around the loop. Round your answer to the nearest tenth.

48. **STRUCTURE** Is $\sqrt{\frac{1}{4}}$ a rational number? Is $\sqrt{\frac{3}{16}}$ a rational number? Explain.

49. **WATER BALLOON** The time t (in seconds) it takes a water balloon to fall d meters is represented by the equation $t = \sqrt{\frac{d}{4.9}}$. Estimate the time it takes the balloon to fall to the ground from a window that is 14 meters above the ground. Round your answer to the nearest tenth.

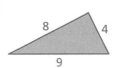

50. **Number Sense** Determine if the statement is *sometimes*, *always*, or *never* true. Explain your reasoning and give an example of each.

a. A rational number multiplied by a rational number is rational.
b. A rational number multiplied by an irrational number is rational.
c. An irrational number multiplied by an irrational number is rational.

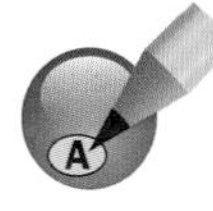

Fair Game Review What you learned in previous grades & lessons

Find the missing length of the triangle. *(Section 7.3)*

51.
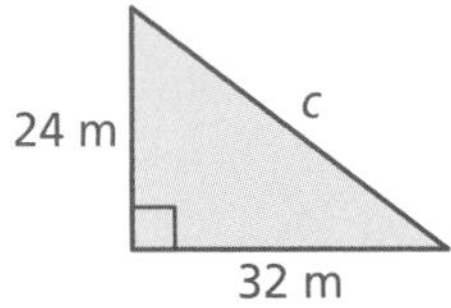

52.
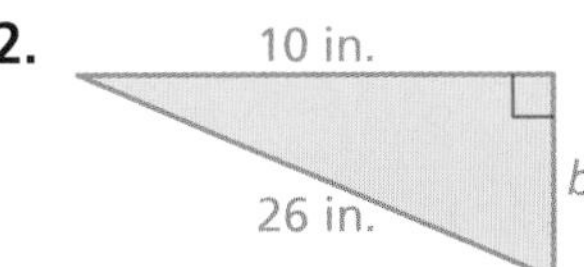

53.
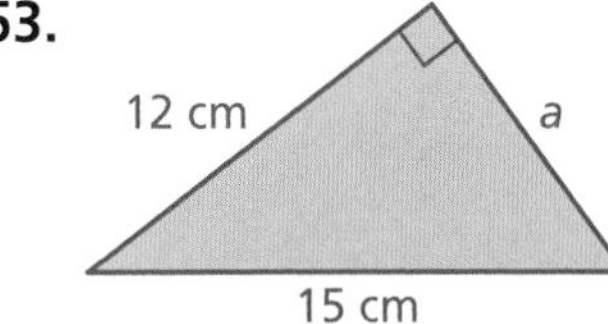

54. **MULTIPLE CHOICE** What is the ratio (red to blue) of the corresponding side lengths of the similar triangles? *(Section 2.5)*

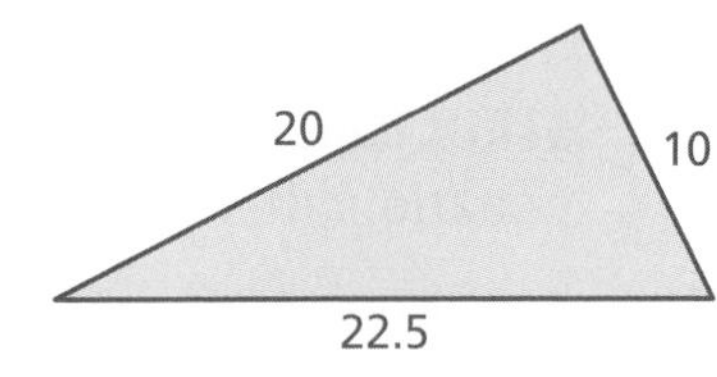

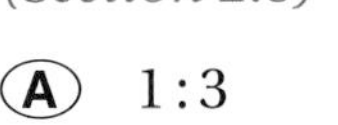

Ⓐ 1 : 3 Ⓑ 5 : 2

Ⓒ 3 : 4 Ⓓ 2 : 5

Extension 7.4 Repeating Decimals

You have written terminating decimals as fractions. Because repeating decimals are rational numbers, you can also write repeating decimals as fractions.

Rational Numbers
In this extension, you will
- write a repeating decimal as a fraction.

Key Idea

Writing a Repeating Decimal as a Fraction

Let a variable x equal the repeating decimal d.

Step 1: Write the equation $x = d$.

Step 2: Multiply each side of the equation by 10^n to form a new equation, where n is the number of repeating digits.

Step 3: Subtract the original equation from the new equation.

Step 4: Solve for x.

EXAMPLE 1 Writing a Repeating Decimal as a Fraction (1 Digit Repeats)

Write $0.\overline{4}$ as a fraction in simplest form.

Let $x = 0.\overline{4}$.

$x = 0.\overline{4}$ — Step 1: Write the equation.

$10 \cdot x = 10 \cdot 0.\overline{4}$ — Step 2: There is 1 repeating digit, so multiply each side by $10^1 = 10$.

$10x = 4.\overline{4}$ — Simplify.

$-(x = 0.\overline{4})$ — Step 3: Subtract the original equation.

$9x = 4$ — Simplify.

$x = \frac{4}{9}$ — Step 4: Solve for x.

So, $0.\overline{4} = \frac{4}{9}$.

Check

$$\begin{array}{r} 0.44\ldots \\ 9\overline{)4.00} \\ \underline{36} \\ 40 \\ \underline{36} \\ 40 \end{array}$$ ✓

Practice

Write the decimal as a fraction or a mixed number.

1. $0.\overline{1}$
2. $-0.\overline{5}$
3. $-1.\overline{2}$
4. $5.\overline{8}$
5. **STRUCTURE** In Example 1, why can you subtract the original equation from the new equation after multiplying by 10? Explain why these two steps are performed.
6. **REPEATED REASONING** Compare the repeating decimals and their equivalent fractions in Exercises 1–4. Describe the pattern. Use the pattern to explain how to write a repeating decimal as a fraction when only the tenths digit repeats.

EXAMPLE 2 Writing a Repeating Decimal as a Fraction (1 Digit Repeats)

Write $-0.2\overline{3}$ as a fraction in simplest form.

Let $x = -0.2\overline{3}$.

Check

```
-7/30
     -.2333333333
```

$x = -0.2\overline{3}$	Step 1: Write the equation.
$10 \cdot x = 10 \cdot (-0.2\overline{3})$	Step 2: There is 1 repeating digit, so multiply each side by $10^1 = 10$.
$10x = -2.\overline{3}$	Simplify.
$-(x = -0.2\overline{3})$	Step 3: Subtract the original equation.
$9x = -2.1$	Simplify.
$x = \frac{-2.1}{9}$	Step 4: Solve for x.

So, $-0.2\overline{3} = \frac{-2.1}{9} = -\frac{21}{90} = -\frac{7}{30}$.

EXAMPLE 3 Writing a Repeating Decimal as a Fraction (2 Digits Repeat)

Write $1.\overline{25}$ as a mixed number.

Let $x = 1.\overline{25}$.

Check

```
124/99
      1.252525253
```

$x = 1.\overline{25}$	Step 1: Write the equation.
$100 \cdot x = 100 \cdot 1.\overline{25}$	Step 2: There are 2 repeating digits, so multiply each side by $10^2 = 100$.
$100x = 125.\overline{25}$	Simplify.
$-(x = 1.\overline{25})$	Step 3: Subtract the original equation.
$99x = 124$	Simplify.
$x = \frac{124}{99}$	Step 4: Solve for x.

So, $1.\overline{25} = \frac{124}{99} = 1\frac{25}{99}$.

Practice

Write the decimal as a fraction or a mixed number.

7. $-0.4\overline{3}$ **8.** $2.0\overline{6}$ **9.** $0.\overline{27}$ **10.** $-4.\overline{50}$

11. REPEATED REASONING Find a pattern in the fractional representations of repeating decimals in which only the tenths and hundredths digits repeat. Use the pattern to explain how to write $9.\overline{04}$ as a mixed number.

7.5 Using the Pythagorean Theorem

Essential Question In what other ways can you use the Pythagorean Theorem?

The *converse* of a statement switches the hypothesis and the conclusion.

Statement:	Converse of the statement:
If p, then q.	If q, then p.

1 ACTIVITY: Analyzing Converses of Statements

Work with a partner. Write the converse of the true statement. Determine whether the converse is *true* or *false*. If it is true, justify your reasoning. If it is false, give a counterexample.

a. If $a = b$, then $a^2 = b^2$.

b. If $a = b$, then $a^3 = b^3$.

c. If one figure is a translation of another figure, then the figures are congruent.

d. If two triangles are similar, then the triangles have the same angle measures.

Is the converse of a true statement always true? always false? Explain.

2 ACTIVITY: The Converse of the Pythagorean Theorem

Work with a partner. The converse of the Pythagorean Theorem states: "If the equation $a^2 + b^2 = c^2$ is true for the side lengths of a triangle, then the triangle is a right triangle."

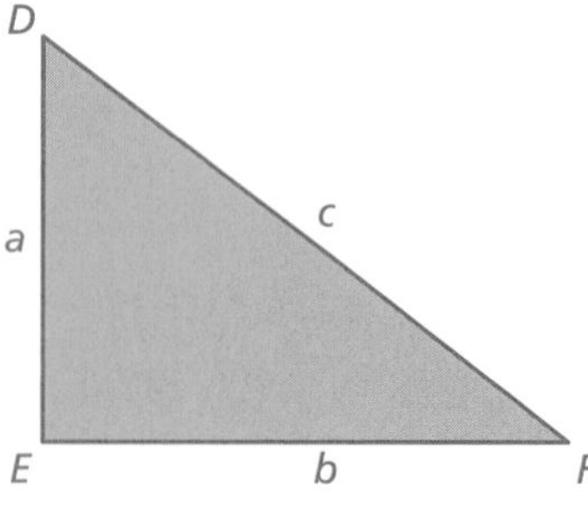

a. Do you think the converse of the Pythagorean Theorem is *true* or *false*? How could you use deductive reasoning to support your answer?

b. Consider $\triangle DEF$ with side lengths a, b, and c, such that $a^2 + b^2 = c^2$. Also consider $\triangle JKL$ with leg lengths a and b, where $\angle K = 90°$.

- What does the Pythagorean Theorem tell you about $\triangle JKL$?
- What does this tell you about c and x?
- What does this tell you about $\triangle DEF$ and $\triangle JKL$?
- What does this tell you about $\angle E$?
- What can you conclude?

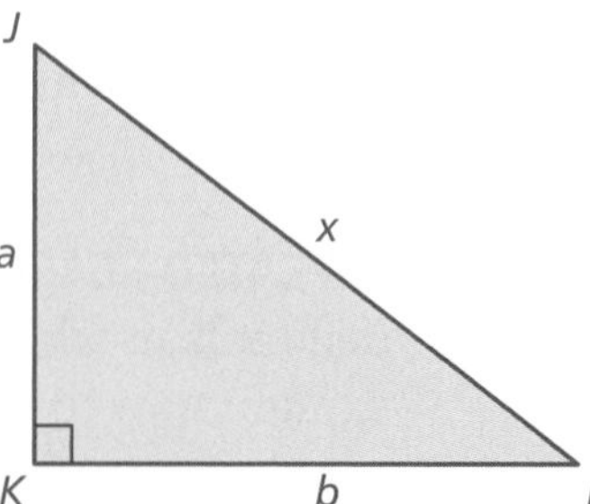

Pythagorean Theorem

In this lesson, you will
- use the converse of the Pythagorean Theorem to identify right triangles.
- use the Pythagorean Theorem to find distances in a coordinate plane.
- solve real-life problems.

3 ACTIVITY: Developing the Distance Formula

Work with a partner. Follow the steps below to write a formula that you can use to find the distance between any two points in a coordinate plane.

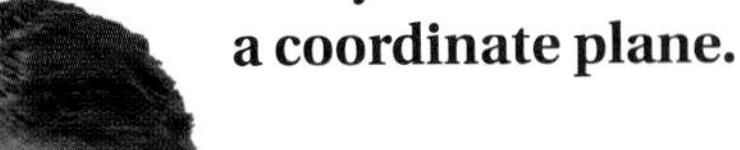

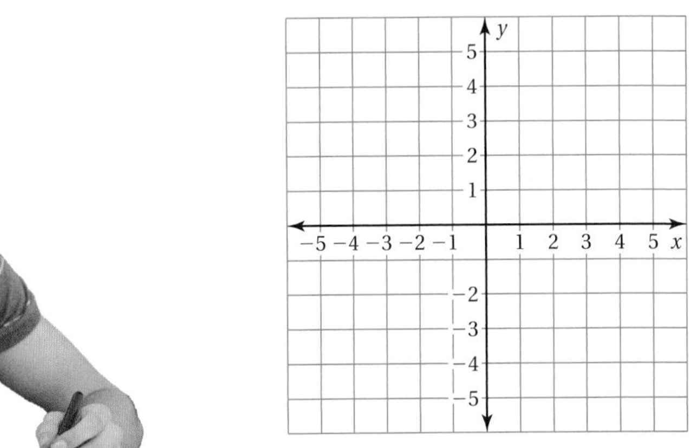

Step 1: Choose two points in the coordinate plane that do not lie on the same horizontal or vertical line. Label the points (x_1, y_1) and (x_2, y_2).

Step 2: Draw a line segment connecting the points. This will be the hypotenuse of a right triangle.

Step 3: Draw horizontal and vertical line segments from the points to form the legs of the right triangle.

Step 4: Use the x-coordinates to write an expression for the length of the horizontal leg.

Step 5: Use the y-coordinates to write an expression for the length of the vertical leg.

Step 6: Substitute the expressions for the lengths of the legs into the Pythagorean Theorem.

Step 7: Solve the equation in Step 6 for the hypotenuse c.

What does the length of the hypotenuse tell you about the two points?

Math Practice

Communicate Precisely

What steps can you take to make sure that you have written the distance formula accurately?

What Is Your Answer?

4. **IN YOUR OWN WORDS** In what other ways can you use the Pythagorean Theorem?

5. What kind of real-life problems do you think the converse of the Pythagorean Theorem can help you solve?

Practice Use what you learned about the converse of a true statement to complete Exercises 3 and 4 on page 322.

7.5 Lesson

Key Vocabulary
distance formula, p. 320

Converse of the Pythagorean Theorem

If the equation $a^2 + b^2 = c^2$ is true for the side lengths of a triangle, then the triangle is a right triangle.

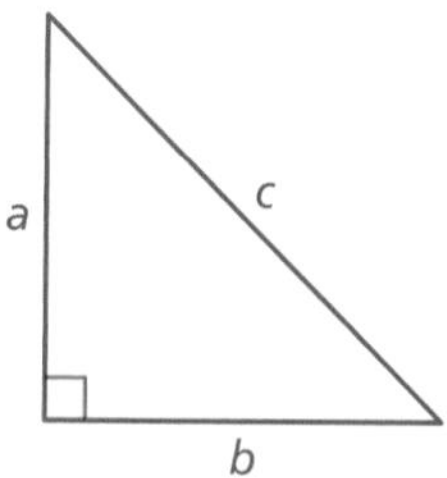

EXAMPLE 1 Identifying a Right Triangle

Tell whether each triangle is a right triangle.

a.

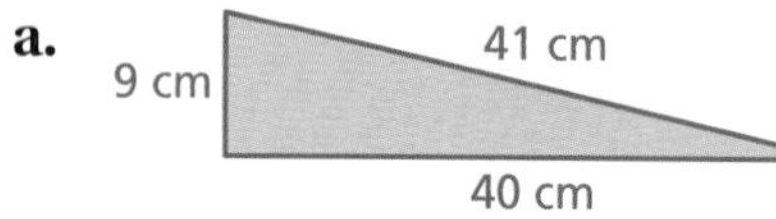

$$a^2 + b^2 = c^2$$
$$9^2 + 40^2 \stackrel{?}{=} 41^2$$
$$81 + 1600 \stackrel{?}{=} 1681$$
$$1681 = 1681 \checkmark$$

It *is* a right triangle.

b.

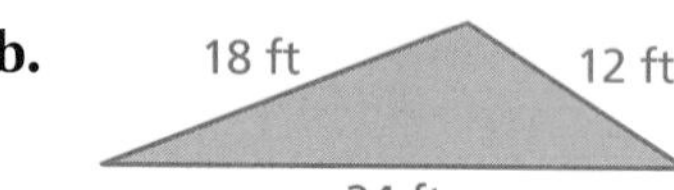

$$a^2 + b^2 = c^2$$
$$12^2 + 18^2 \stackrel{?}{=} 24^2$$
$$144 + 324 \stackrel{?}{=} 576$$
$$468 \neq 576 \times$$

It is *not* a right triangle.

Study Tip

A *Pythagorean triple* is a set of three positive integers *a*, *b*, and *c*, where $a^2 + b^2 = c^2$.

Common Error

When using the converse of the Pythagorean Theorem, always substitute the length of the longest side for *c*.

On Your Own

Tell whether the triangle with the given side lengths is a right triangle.

1. 28 in., 21 in., 20 in.
2. 1.25 mm, 1 mm, 0.75 mm

Now You're Ready
Exercises 5–10

On page 319, you used the Pythagorean Theorem to develop the *distance formula.* You can use the **distance formula** to find the distance between any two points in a coordinate plane.

Distance Formula

The distance *d* between any two points (x_1, y_1) and (x_2, y_2) is given by the formula

$$d = \sqrt{(x_2 - x_1)^2 + (y_2 - y_1)^2}.$$

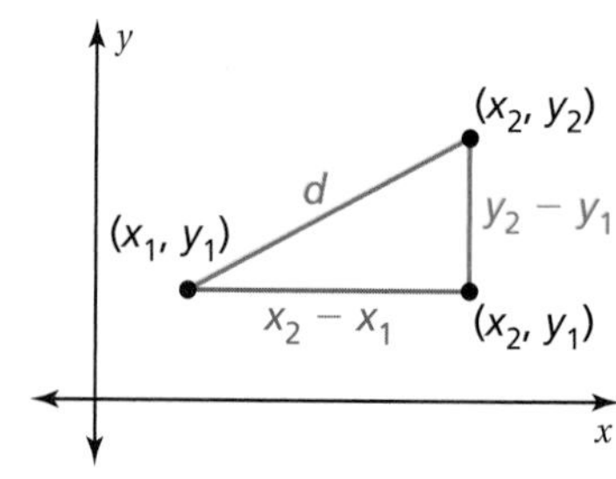

EXAMPLE 2 Finding the Distance Between Two Points

Find the distance between (1, 5) and (−4, −2).

Let $(x_1, y_1) = (1, 5)$ and $(x_2, y_2) = (-4, -2)$.

$$d = \sqrt{(x_2 - x_1)^2 + (y_2 - y_1)^2}$$ Write the distance formula.

$$= \sqrt{(-4 - 1)^2 + (-2 - 5)^2}$$ Substitute.

$$= \sqrt{(-5)^2 + (-7)^2}$$ Simplify.

$$= \sqrt{25 + 49}$$ Evaluate powers.

$$= \sqrt{74}$$ Add.

EXAMPLE 3 Real-Life Application

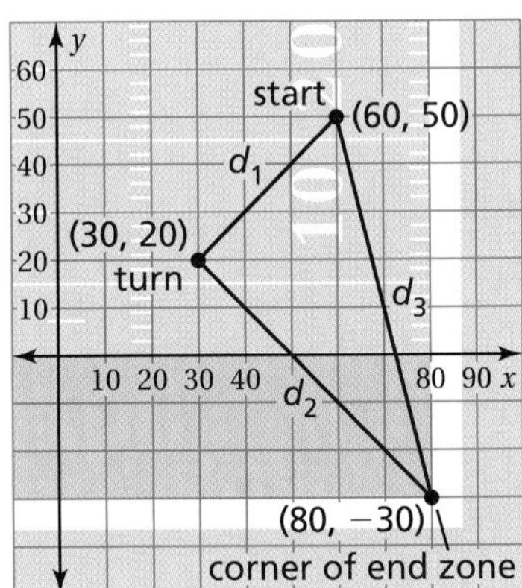

You design a football play in which a player runs down the field, makes a 90° turn, and runs to the corner of the end zone. Your friend runs the play as shown. Did your friend make a 90° turn? Each unit of the grid represents 10 feet.

Use the distance formula to find the lengths of the three sides.

$$d_1 = \sqrt{(60 - 30)^2 + (50 - 20)^2} = \sqrt{30^2 + 30^2} = \sqrt{1800} \text{ feet}$$

$$d_2 = \sqrt{(80 - 30)^2 + (-30 - 20)^2} = \sqrt{50^2 + (-50)^2} = \sqrt{5000} \text{ feet}$$

$$d_3 = \sqrt{(80 - 60)^2 + (-30 - 50)^2} = \sqrt{20^2 + (-80)^2} = \sqrt{6800} \text{ feet}$$

Use the converse of the Pythagorean Theorem to determine if the side lengths form a right triangle.

$$(\sqrt{1800})^2 + (\sqrt{5000})^2 \stackrel{?}{=} (\sqrt{6800})^2$$

$$1800 + 5000 \stackrel{?}{=} 6800$$

$$6800 = 6800 \checkmark$$

The sides form a right triangle.

So, your friend made a 90° turn.

On Your Own

Now You're Ready
Exercises 11–16

Find the distance between the two points.

3. (0, 0), (4, 5) **4.** (7, −3), (9, 6) **5.** (−2, −3), (−5, 1)

6. **WHAT IF?** In Example 3, your friend made the turn at (20, 10). Did your friend make a 90° turn?

7.5 Exercises

Vocabulary and Concept Check

1. **WRITING** Describe two ways to find the distance between two points in a coordinate plane.
2. **WHICH ONE DOESN'T BELONG?** Which set of numbers does *not* belong with the other three? Explain your reasoning.

3, 6, 8	6, 8, 10	5, 12, 13	7, 24, 25

Practice and Problem Solving

Write the converse of the true statement. Determine whether the converse is *true* or *false*. If it is true, justify your reasoning. If it is false, give a counterexample.

3. If a is an odd number, then a^2 is odd.
4. If $ABCD$ is a square, then $ABCD$ is a parallelogram.

Tell whether the triangle with the given side lengths is a right triangle.

1 5.

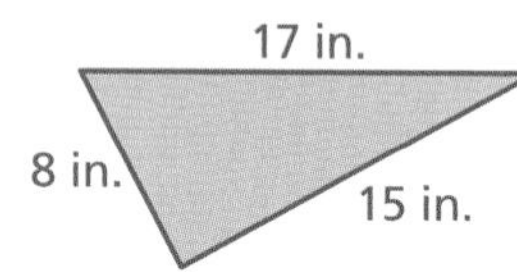

6.

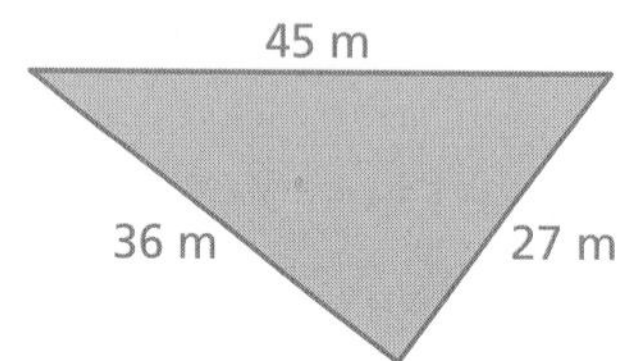

7.

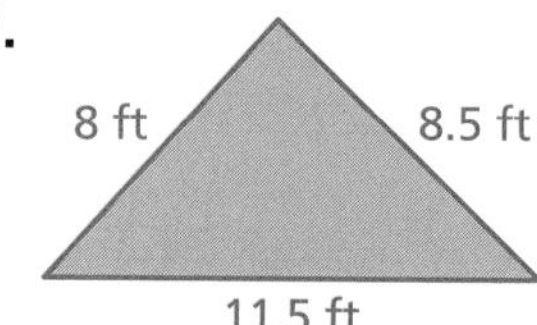

8. 14 mm, 19 mm, 23 mm
9. $\frac{9}{10}$ mi, $1\frac{1}{5}$ mi, $1\frac{1}{2}$ mi
10. 1.4 m, 4.8 m, 5 m

Find the distance between the two points.

2 11. (1, 2), (7, 6)
12. (4, −5), (−1, 7)
13. (2, 4), (7, 2)
14. (−1, −3), (1, 3)
15. (−6, −7), (0, 0)
16. (12, 5), (−12, −2)

17. **ERROR ANALYSIS** Describe and correct the error in finding the distance between the points (−3, −2) and (7, 4).

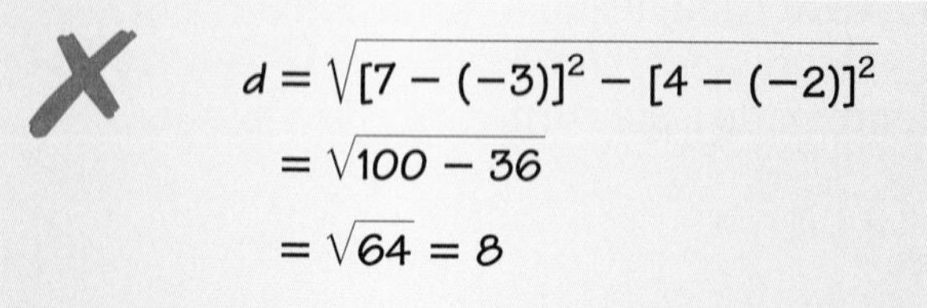

18. **CONSTRUCTION** A post and beam frame for a shed is shown in the diagram. Does the brace form a right triangle with the post and beam? Explain.

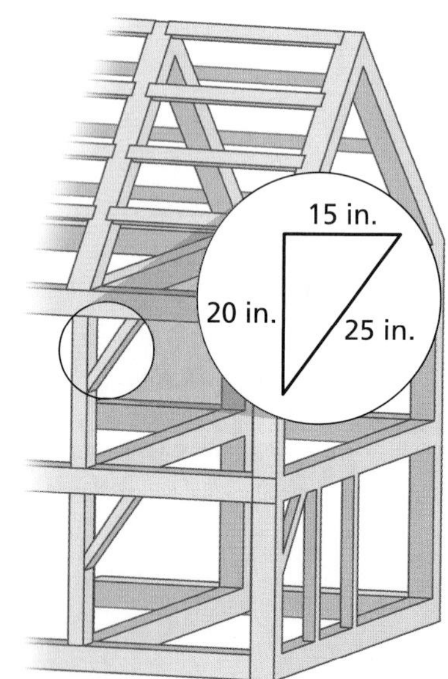

Tell whether a triangle with the given side lengths is a right triangle.

19. $\sqrt{63}$, 9, 12 **20.** 4, $\sqrt{15}$, 6 **21.** $\sqrt{18}$, $\sqrt{24}$, $\sqrt{42}$

22. REASONING Plot the points (−1, 3), (4, −2), and (1, −5) in a coordinate plane. Are the points the vertices of a right triangle? Explain.

23. GEOCACHING You spend the day looking for hidden containers in a wooded area using a Global Positioning System (GPS). You park your car on the side of the road, and then locate Container 1 and Container 2 before going back to the car. Does your path form a right triangle? Explain. Each unit of the grid represents 10 yards.

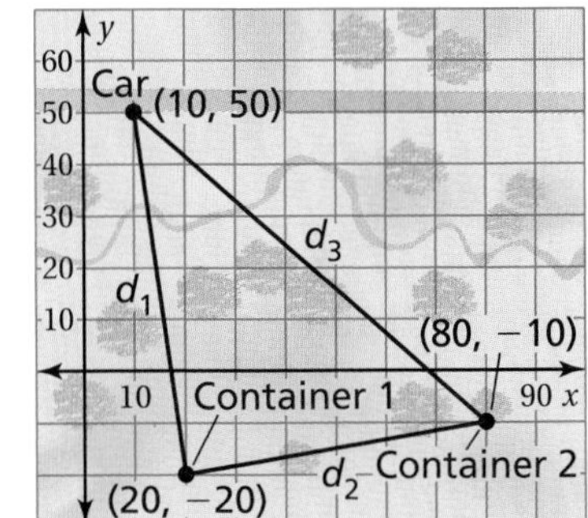

24. REASONING Your teacher wants the class to find the distance between the two points (2, 4) and (9, 7). You use (2, 4) for (x_1, y_1), and your friend uses (9, 7) for (x_1, y_1). Do you and your friend obtain the same result? Justify your answer.

25. AIRPORT Which plane is closer to the base of the airport tower? Explain.

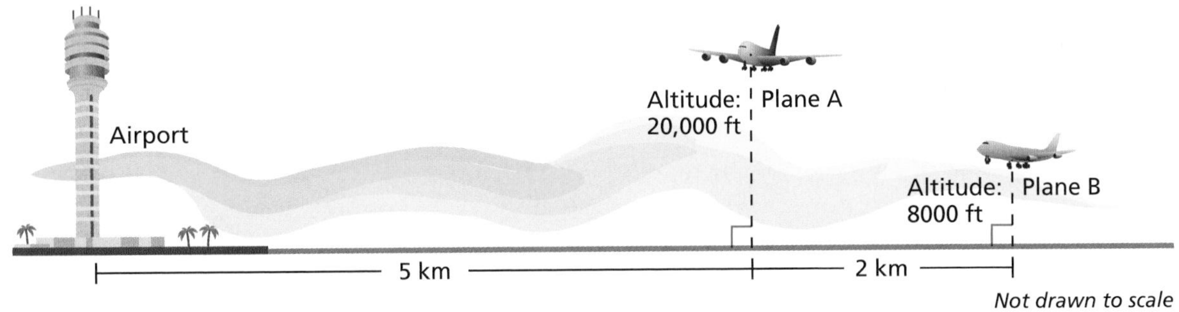

26. **Structure** Consider the two points (x_1, y_1) and (x_2, y_2) in the coordinate plane. How can you find the point (x_m, y_m) located in the middle of the two given points? Justify your answer using the distance formula.

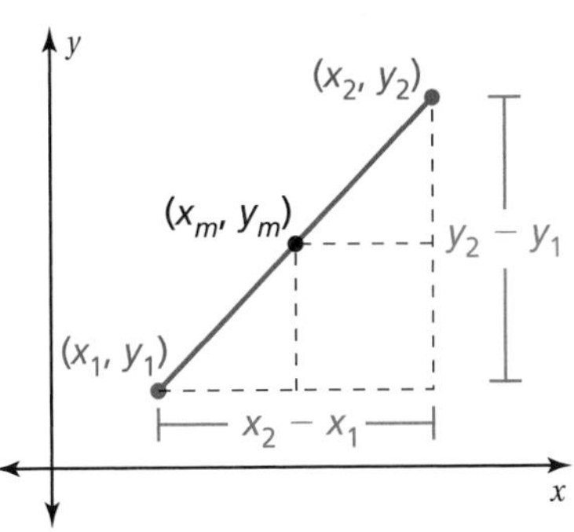

Fair Game Review What you learned in previous grades & lessons

Find the mean, median, and mode of the data. *(Skills Review Handbook)*

27. 12, 9, 17, 15, 12, 13 **28.** 21, 32, 16, 27, 22, 19, 10 **29.** 67, 59, 34, 71, 59

30. MULTIPLE CHOICE What is the sum of the interior angle measures of an octagon? *(Section 3.3)*

Ⓐ 720° Ⓑ 1080° Ⓒ 1440° Ⓓ 1800°

7.4–7.5 Quiz

Classify the real number. *(Section 7.4)*

1. $-\sqrt{225}$
2. $-1\frac{1}{9}$
3. $\sqrt{41}$
4. $\sqrt{17}$

Estimate the square root to the nearest (a) integer and (b) tenth. *(Section 7.4)*

5. $\sqrt{38}$
6. $-\sqrt{99}$
7. $\sqrt{172}$
8. $\sqrt{115}$

Which number is greater? Explain. *(Section 7.4)*

9. $\sqrt{11}, 3\frac{3}{5}$
10. $\sqrt{1.44}, 1.1\overline{8}$

Write the decimal as a fraction or a mixed number. *(Section 7.4)*

11. $0.\overline{7}$
12. $-1.\overline{63}$

Tell whether the triangle with the given side lengths is a right triangle. *(Section 7.5)*

13.

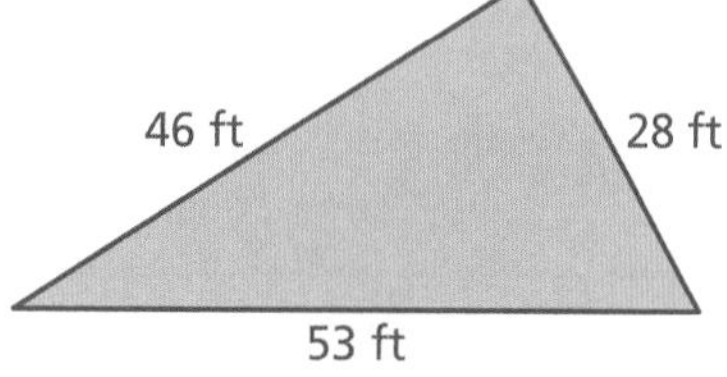

14.

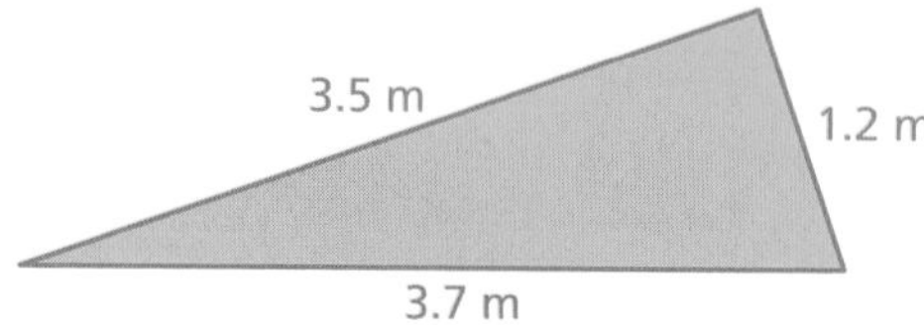

Find the distance between the two points. *(Section 7.5)*

15. $(-3, -1), (-1, -5)$
16. $(-4, 2), (5, 1)$
17. $(1, -2), (4, -5)$
18. $(-1, 1), (7, 4)$
19. $(-6, 5), (-4, -6)$
20. $(-1, 4), (1, 3)$

Use the figure to answer Exercises 21–24. Round your answer to the nearest tenth. *(Section 7.5)*

21. How far is the cabin from the peak?
22. How far is the fire tower from the lake?
23. How far is the lake from the peak?
24. You are standing at $(-5, -6)$. How far are you from the lake?

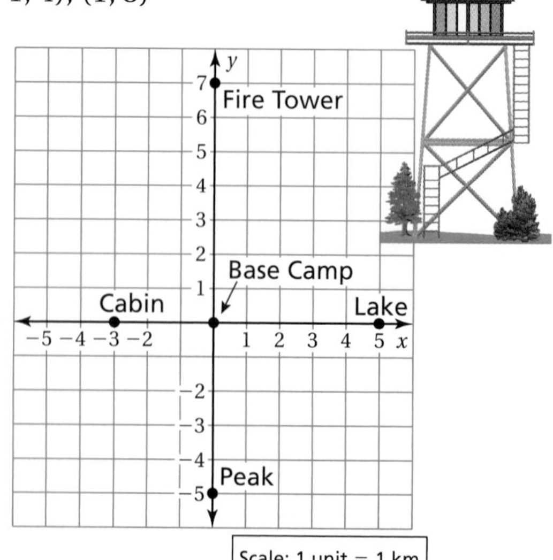

7 Chapter Review

Check It Out
Vocabulary Help
BigIdeasMath.com

Review Key Vocabulary

square root, *p. 290*
perfect square, *p. 290*
radical sign, *p. 290*
radicand, *p. 290*
cube root, *p. 296*
perfect cube, *p. 296*
theorem, *p. 300*
legs, *p. 302*
hypotenuse, *p. 302*
Pythagorean Theorem, *p. 302*
irrational number, *p. 310*
real numbers, *p. 310*
distance formula, *p. 320*

Review Examples and Exercises

7.1 Finding Square Roots *(pp. 288–293)*

Find $-\sqrt{36}$.

$-\sqrt{36}$ represents the *negative* square root.

Because $6^2 = 36$, $-\sqrt{36} = -\sqrt{6^2} = -6$.

Exercises

Find the square root(s).

1. $\sqrt{1}$

2. $-\sqrt{\frac{9}{25}}$

3. $\pm\sqrt{1.69}$

Evaluate the expression.

4. $15 - 4\sqrt{36}$

5. $\sqrt{\frac{54}{6}} + \frac{2}{3}$

6. $10(\sqrt{81} - 12)$

7.2 Finding Cube Roots *(pp. 294–299)*

Find $\sqrt[3]{\frac{125}{216}}$.

Because $\left(\frac{5}{6}\right)^3 = \frac{125}{216}$, $\sqrt[3]{\frac{125}{216}} = \sqrt[3]{\left(\frac{5}{6}\right)^3} = \frac{5}{6}$.

Exercises

Find the cube root.

7. $\sqrt[3]{729}$

8. $\sqrt[3]{\frac{64}{343}}$

9. $\sqrt[3]{-\frac{8}{27}}$

Evaluate the expression.

10. $\sqrt[3]{27} - 16$

11. $25 + 2\sqrt[3]{-64}$

12. $3\sqrt[3]{-125} - 27$

7.3 The Pythagorean Theorem (pp. 300–305)

Find the length of the hypotenuse of the triangle.

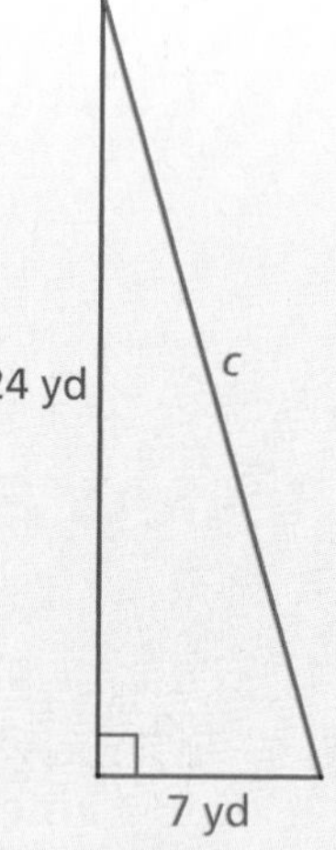

$a^2 + b^2 = c^2$	Write the Pythagorean Theorem.
$7^2 + 24^2 = c^2$	Substitute.
$49 + 576 = c^2$	Evaluate powers.
$625 = c^2$	Add.
$\sqrt{625} = \sqrt{c^2}$	Take positive square root of each side.
$25 = c$	Simplify.

∴ The length of the hypotenuse is 25 yards.

Exercises

Find the missing length of the triangle.

13.

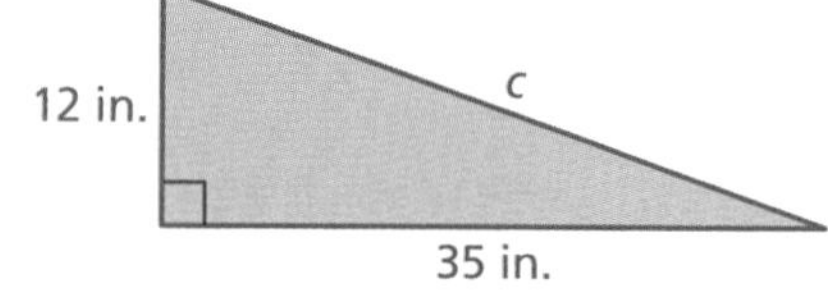

14.

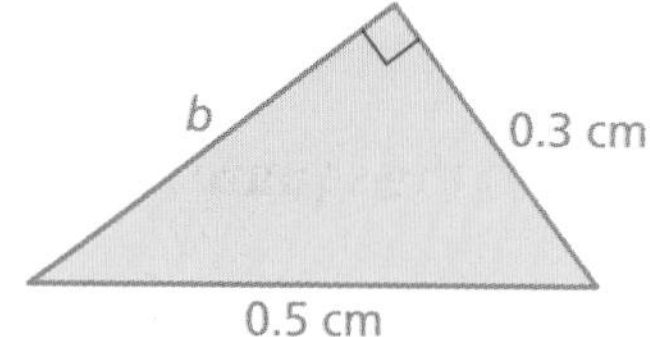

7.4 Approximating Square Roots (pp. 308–317)

a. Classify $\sqrt{19}$.

∴ The number $\sqrt{19}$ is irrational because 19 is not a perfect square.

b. Estimate $\sqrt{34}$ to the nearest integer.

Make a table of numbers whose squares are close to the radicand, 34.

Number	4	5	6	7
Square of Number	16	25	36	49

The table shows that 34 is between the perfect squares 25 and 36. Because 34 is closer to 36 than to 25, $\sqrt{34}$ is closer to 6 than to 5.

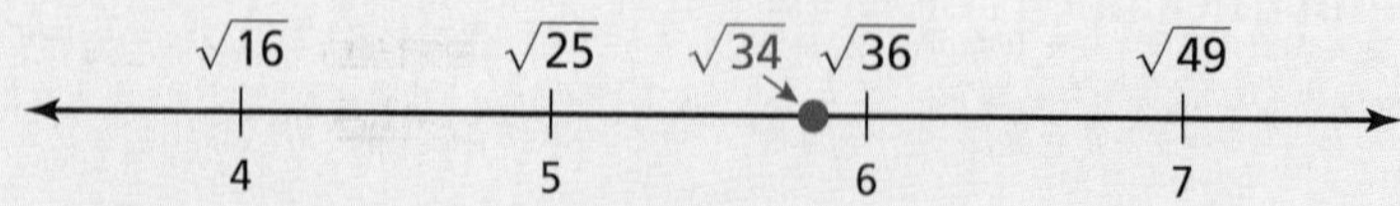

∴ So, $\sqrt{34} \approx 6$.

Exercises

Classify the real number.

15. $0.81\overline{5}$ **16.** $\sqrt{101}$ **17.** $\sqrt{4}$

Estimate the square root to the nearest (a) integer and (b) tenth.

18. $\sqrt{14}$ **19.** $\sqrt{90}$ **20.** $\sqrt{175}$

Write the decimal as a fraction.

21. $0.\overline{8}$ **22.** $0.\overline{36}$ **23.** $-1.\overline{6}$

7.5 Using the Pythagorean Theorem *(pp. 318–323)*

a. Is the triangle formed by the rope and the tent a right triangle?

$$a^2 + b^2 = c^2$$

$$64^2 + 48^2 \stackrel{?}{=} 80^2$$

$$4096 + 2304 \stackrel{?}{=} 6400$$

$$6400 = 6400 \checkmark$$

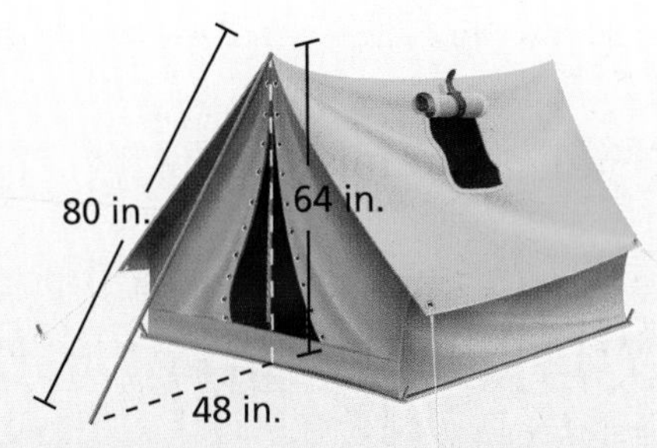

∴ It *is* a right triangle.

b. Find the distance between (−3, 1) and (4, 7).

Let $(x_1, y_1) = (-3, 1)$ and $(x_2, y_2) = (4, 7)$.

$d = \sqrt{(x_2 - x_1)^2 + (y_2 - y_1)^2}$ Write the distance formula.

$= \sqrt{[4 - (-3)]^2 + (7 - 1)^2}$ Substitute.

$= \sqrt{7^2 + 6^2}$ Simplify.

$= \sqrt{49 + 36}$ Evaluate powers.

$= \sqrt{85}$ Add.

Exercises

Tell whether the triangle is a right triangle.

24.

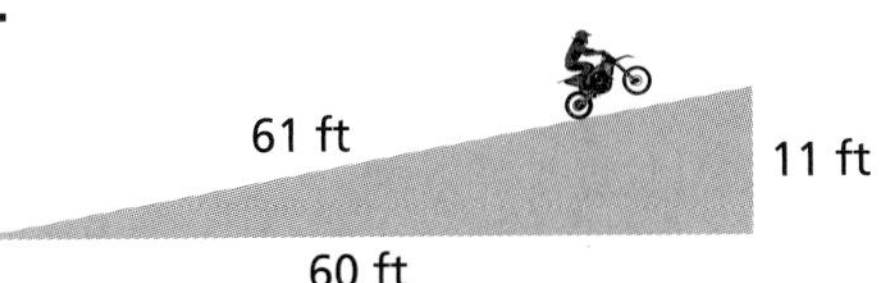

25.

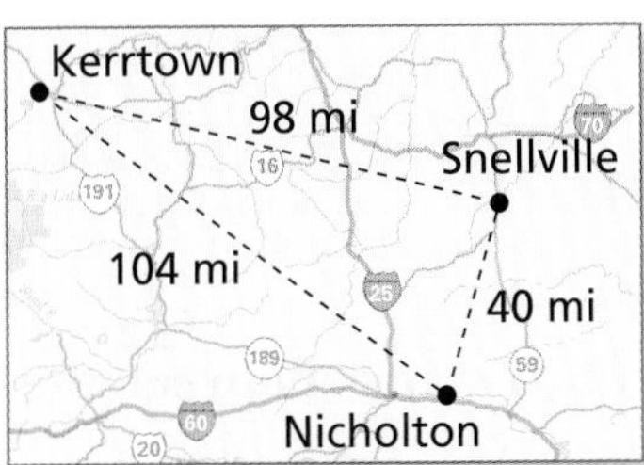

Find the distance between the two points.

26. $(-2, -5), (3, 5)$ **27.** $(-4, 7), (4, 0)$

7 Chapter Test

Find the square root(s).

1. $-\sqrt{1600}$

2. $\sqrt{\frac{25}{49}}$

3. $\pm\sqrt{\frac{100}{9}}$

Find the cube root.

4. $\sqrt[3]{-27}$

5. $\sqrt[3]{\frac{8}{125}}$

6. $\sqrt[3]{-\frac{729}{64}}$

Evaluate the expression.

7. $12 + 8\sqrt{16}$

8. $\frac{1}{2} + \sqrt{\frac{72}{2}}$

9. $\left(\sqrt[3]{-125}\right)^3 + 75$

10. $50\sqrt[3]{\frac{512}{1000}} + 14$

11. Find the missing length of the triangle.

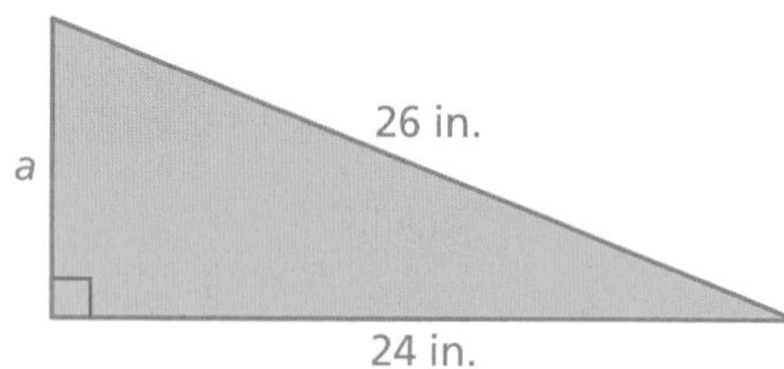

Classify the real number.

12. 16π

13. $-\sqrt{49}$

Estimate the square root to the nearest (a) integer and (b) tenth.

14. $\sqrt{58}$

15. $\sqrt{83}$

Write the decimal as a fraction or a mixed number.

16. $-0.\overline{3}$

17. $1.\overline{24}$

18. Tell whether the triangle is a right triangle.

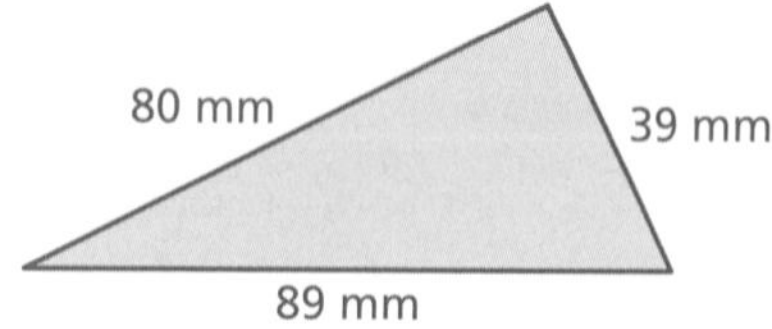

Find the distance between the two points.

19. $(-2, 3)$, $(6, 9)$

20. $(0, -5)$, $(4, 1)$

21. SUPERHERO Find the altitude of the superhero balloon.

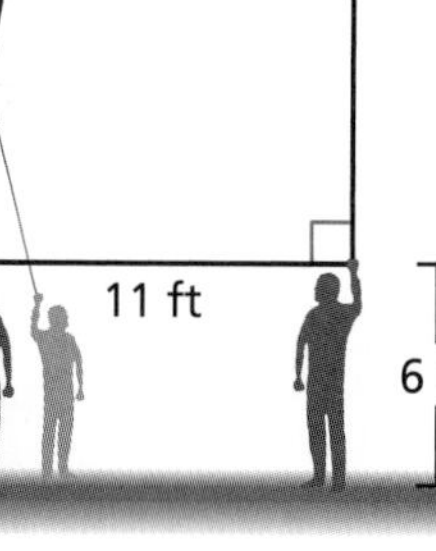

7 Cumulative Assessment

1. The period T of a pendulum is the time, in seconds, it takes the pendulum to swing back and forth. The period can be found using the formula $T = 1.1\sqrt{L}$, where L is the length, in feet, of the pendulum. A pendulum has a length of 4 feet. Find its period.

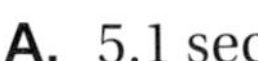

A. 5.1 sec　　**C.** 3.1 sec

B. 4.4 sec　　**D.** 2.2 sec

2. Which parallelogram is a dilation of parallelogram $JKLM$? (Figures not drawn to scale.)

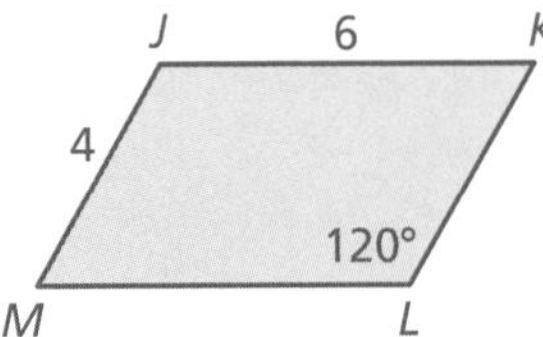

F.

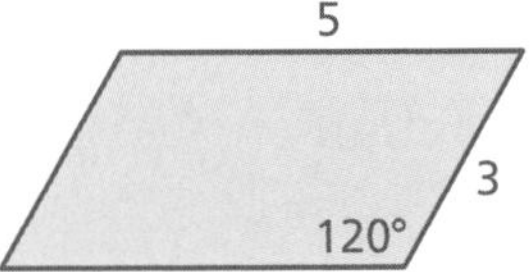

H.

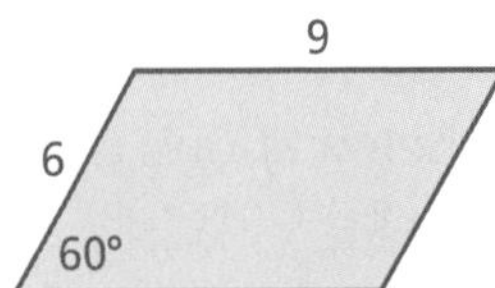

G.

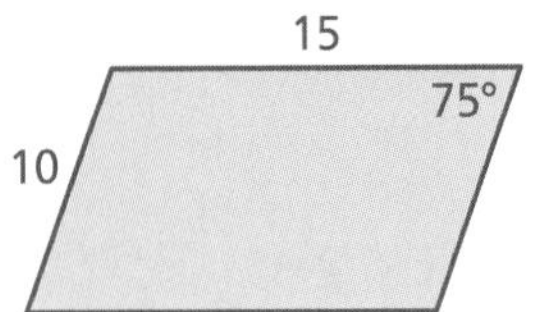

I.

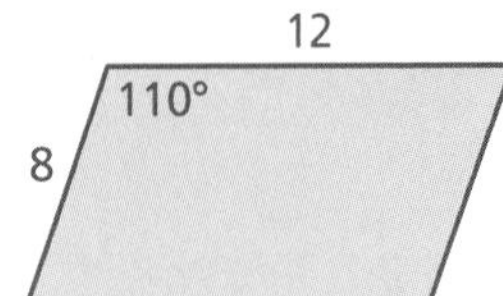

3. Which equation represents a linear function?

A. $y = x^2$　　**C.** $xy = 1$

B. $y = \dfrac{2}{x}$　　**D.** $x + y = 1$

4. Which linear function matches the line shown in the graph?

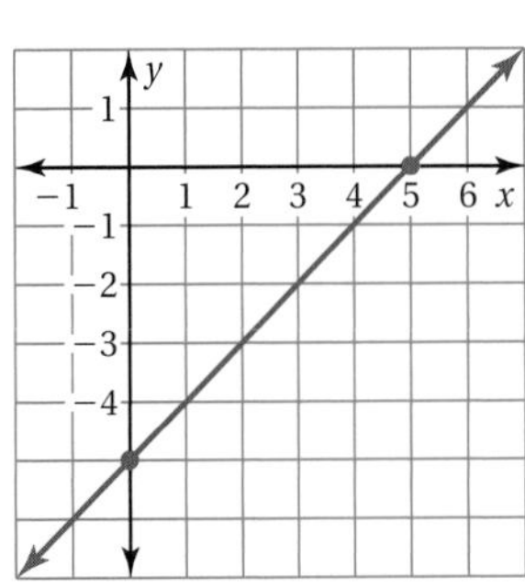

F. $y = x - 5$　　**H.** $y = -x - 5$

G. $y = x + 5$　　**I.** $y = -x + 5$

5. A football field is 40 yards wide and 120 yards long. Find the distance between opposite corners of the football field. Show your work and explain your reasoning.

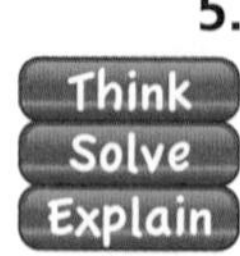

6. A computer consultant charges $50 plus $40 for each hour she works. The consultant charged $650 for one job. This can be represented by the equation below, where h represents the number of hours worked.

$$40h + 50 = 650$$

How many hours did the consultant work?

7. You can use the formula below to find the sum S of the interior angle measures of a polygon with n sides. Solve the formula for n.

$$S = 180(n - 2)$$

A. $n = 180(S - 2)$

B. $n = \frac{S}{180} + 2$

C. $n = \frac{S}{180} - 2$

D. $n = \frac{S}{180} + \frac{1}{90}$

8. The table below shows a linear pattern. Which linear function relates y to x?

x	1	2	3	4	5
y	4	2	0	−2	−4

F. $y = 2x + 2$

G. $y = 4x$

H. $y = -2x + 2$

I. $y = -2x + 6$

9. An airplane flies from City 1 at (0, 0) to City 2 at (33, 56) and then to City 3 at (23, 32). What is the total number of miles it flies? Each unit of the coordinate grid represents 1 mile.

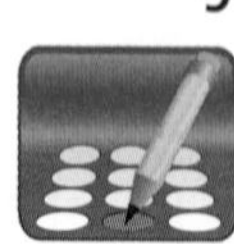

10. What is the missing length of the right triangle shown?

x
7 cm
25 cm

A. 16 cm

B. 18 cm

C. 24 cm

D. $\sqrt{674}$ cm

11. A system of linear equations is shown in the coordinate plane below. What is the solution for this system?

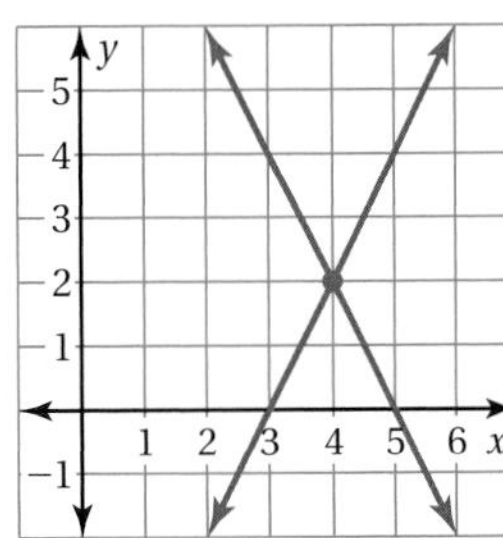

F. (0, 10)

G. (3, 0)

H. (4, 2)

I. (5, 0)

12. In the diagram, lines ℓ and m are parallel. Which angle has the same measure as $\angle 1$?

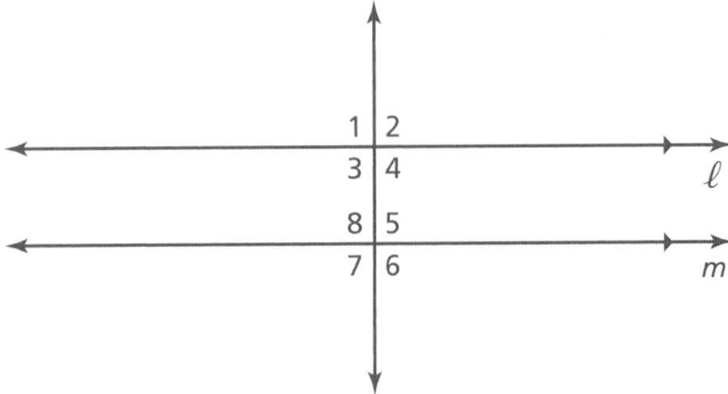

A. $\angle 2$

B. $\angle 5$

C. $\angle 7$

D. $\angle 8$

13. Which graph represents the linear equation $y = -2x - 2$?

F.

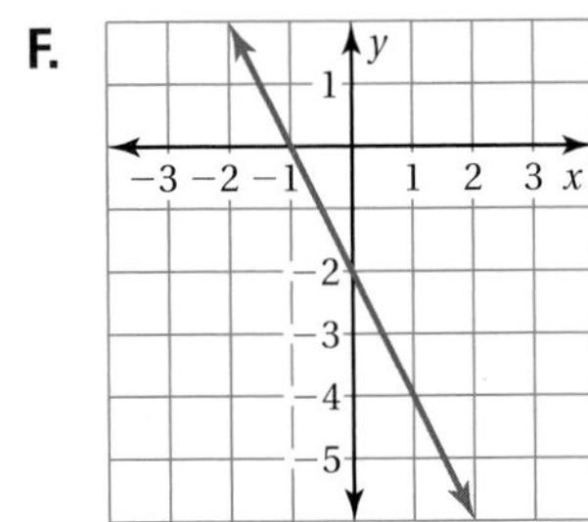

H.

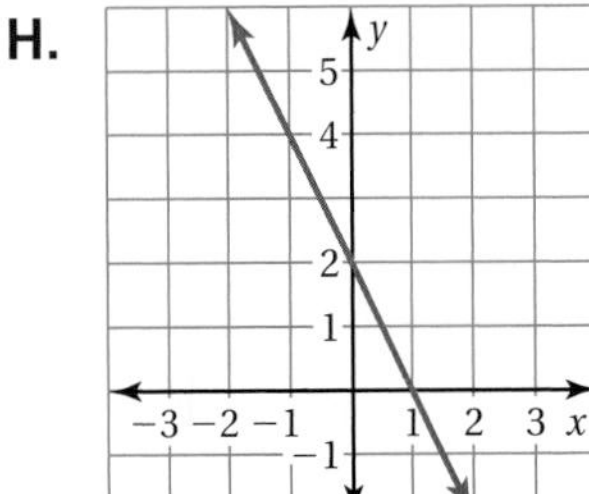

G.

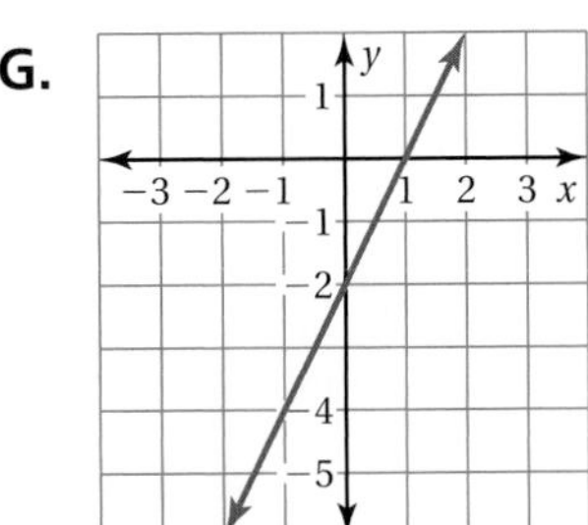

I.

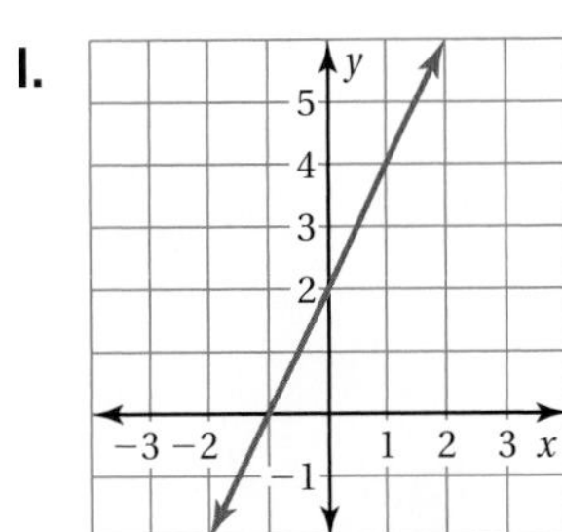

8 Volume and Similar Solids

8.1 Volumes of Cylinders

8.2 Volumes of Cones

8.3 Volumes of Spheres

8.4 Surface Areas and Volumes of Similar Solids

"Dear Sir: Why do you sell dog food in tall cans and sell cat food in short cans?"

"Neither of these shapes is the optimal use of surface area when compared to volume."

"Do you know why the volume of a cone is one-third the volume of a cylinder with the same height and base?"

What You Learned Before

Number one on America's list of 10 worst ideas.

"I just figured out how to find your volume. We'll immerse you in a barrel of water and measure the water that overflows."

Finding the Area of a Composite Figure

Example 1 Find the area of the figure.

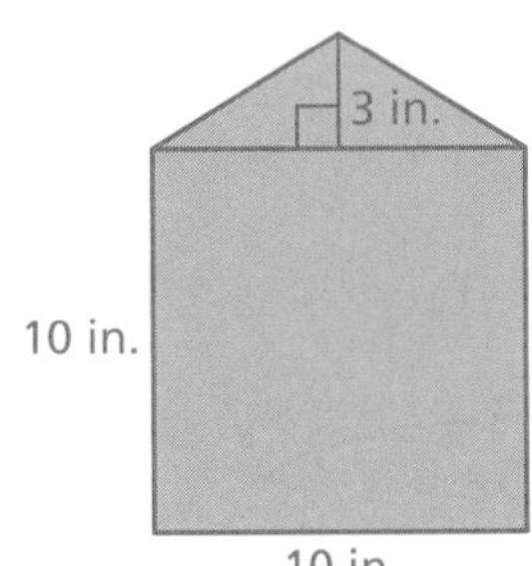

Area = Area of square + Area of triangle

$$A = s^2 + \frac{1}{2}bh$$

$$= 10^2 + \left(\frac{1}{2} \cdot 10 \cdot 3\right)$$

$$= 100 + 15$$

$$= 115 \text{ in.}^2$$

Try It Yourself

Find the area of the figure.

1. 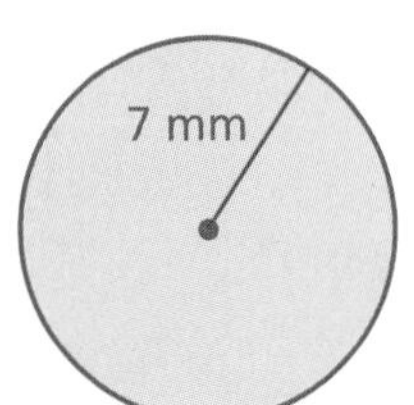

2. 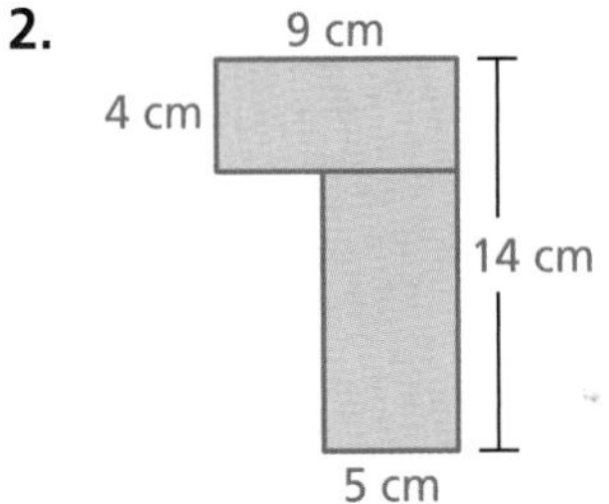

Finding the Areas of Circles

Example 2 Find the area of the circle.

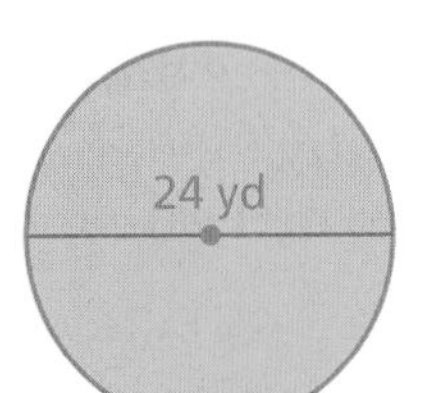

$$A = \pi r^2$$

$$\approx \frac{22}{7} \cdot 7^2$$

$$= \frac{22}{7} \cdot 49$$

$$= 154 \text{ mm}^2$$

Example 3 Find the area of the circle.

$$A = \pi r^2$$

$$\approx 3.14 \cdot 12^2$$

$$= 3.14 \cdot 144$$

$$= 452.16 \text{ yd}^2$$

Try It Yourself

Find the area of the circle.

3. 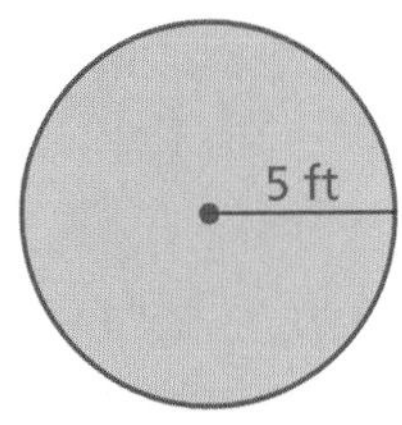

4. 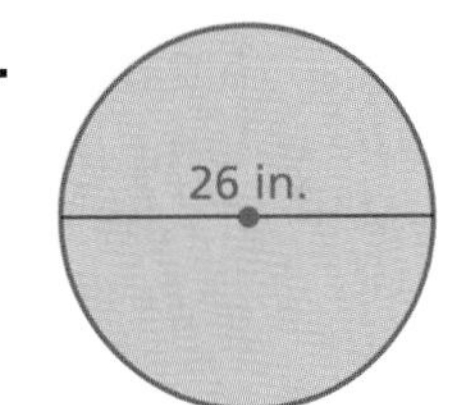

5. 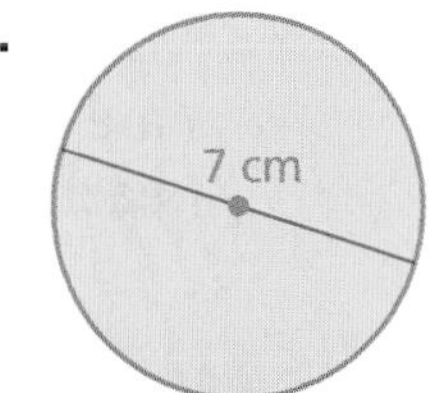

8.1 Volumes of Cylinders

Essential Question How can you find the volume of a cylinder?

1 ACTIVITY: Finding a Formula Experimentally

Work with a partner.

a. Find the area of the face of a coin.

b. Find the volume of a stack of a dozen coins.

c. Write a formula for the volume of a cylinder.

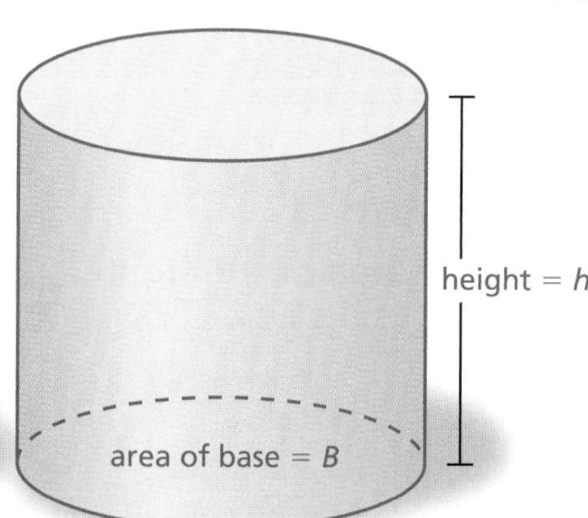

2 ACTIVITY: Making a Business Plan

Work with a partner. You are planning to make and sell three different sizes of cylindrical candles. You buy 1 cubic foot of candle wax for \$20 to make 8 candles of each size.

a. Design the candles. What are the dimensions of each size of candle?

b. You want to make a profit of \$100. Decide on a price for each size of candle.

c. Did you set the prices so that they are proportional to the volume of each size of candle? Why or why not?

Geometry

In this lesson, you will

- find the volumes of cylinders.
- find the heights of cylinders given the volumes.
- solve real-life problems.

3 ACTIVITY: Science Experiment

Work with a partner. Use the diagram to describe how you can find the volume of a small object.

4 ACTIVITY: Comparing Cylinders

Math Practice

Consider Similar Problems

How can you use the results of Activity 1 to find the volumes of the cylinders?

Work with a partner.

a. Just by looking at the two cylinders, which one do you think has the greater volume? Explain your reasoning.

b. Find the volume of each cylinder. Was your prediction in part (a) correct? Explain your reasoning.

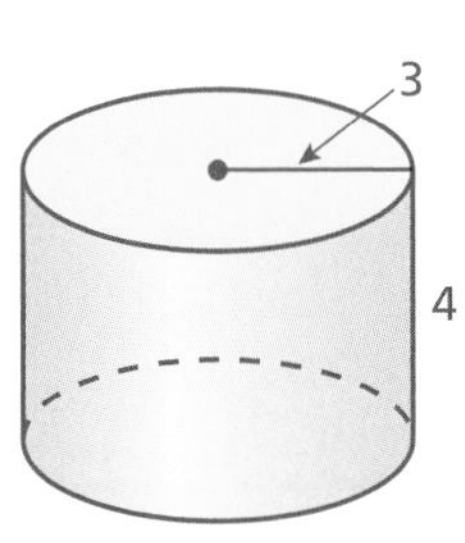

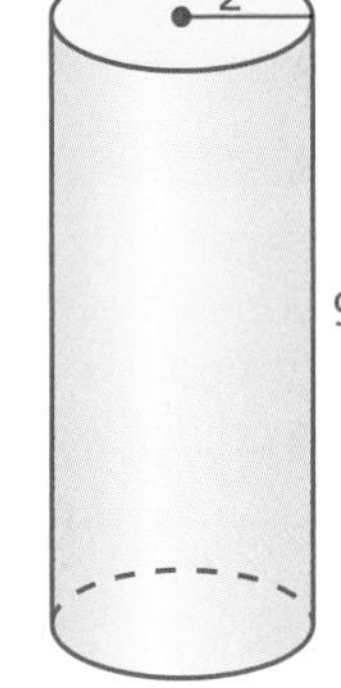

What Is Your Answer?

5. **IN YOUR OWN WORDS** How can you find the volume of a cylinder?

6. Compare your formula for the volume of a cylinder with the formula for the volume of a prism. How are they the same?

"Here's how I remember how to find the volume of any prism or cylinder."

"Base times tall, will fill 'em all."

Practice Use what you learned about the volumes of cylinders to complete Exercises 3–5 on page 338.

8.1 Lesson

Volume of a Cylinder

Words The volume V of a cylinder is the product of the area of the base and the height of the cylinder.

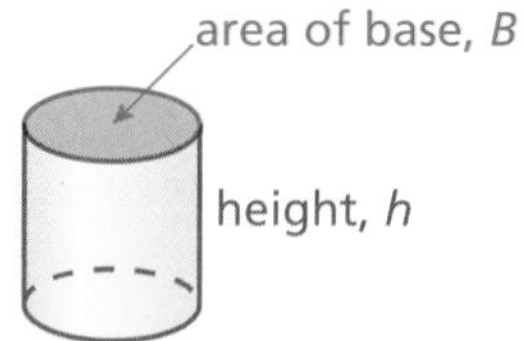

Algebra $V = Bh$

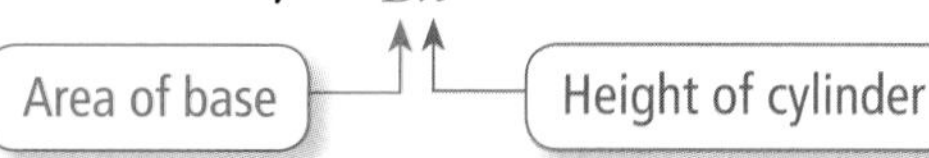

EXAMPLE 1 Finding the Volume of a Cylinder

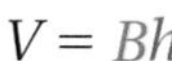

Because $B = \pi r^2$, you can use $V = \pi r^2 h$ to find the volume of a cylinder.

Find the volume of the cylinder. Round your answer to the nearest tenth.

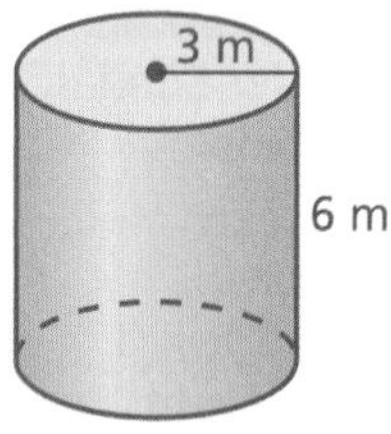

$V = Bh$	Write formula for volume.
$= \pi(3)^2(6)$	Substitute.
$= 54\pi \approx 169.6$	Use a calculator.

The volume is about 169.6 cubic meters.

EXAMPLE 2 Finding the Height of a Cylinder

Find the height of the cylinder. Round your answer to the nearest whole number.

The diameter is 10 inches. So, the radius is 5 inches.

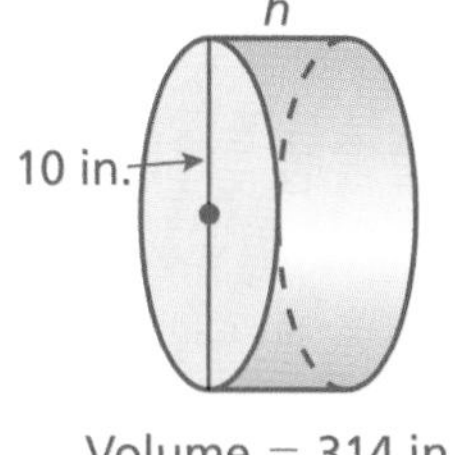

Volume = 314 in.^{3}

$V = Bh$	Write formula for volume.
$314 = \pi(5)^2(h)$	Substitute.
$314 = 25\pi h$	Simplify.
$4 \approx h$	Divide each side by 25π.

The height is about 4 inches.

On Your Own

Now You're Ready
Exercises 3–11 and 13–15

Find the volume V or height h of the cylinder. Round your answer to the nearest tenth.

1. 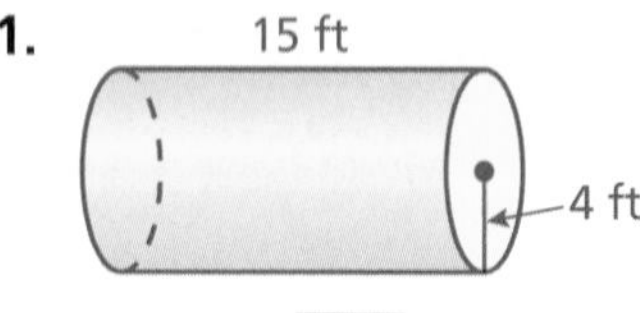

2. 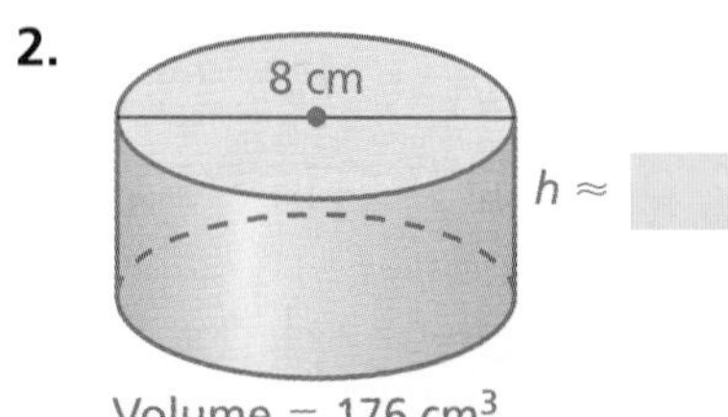

Volume = 176 cm^3

EXAMPLE 3 Real-Life Application

How much salsa is missing from the jar?

The empty space in the jar is a cylinder with a height of $10 - 4 = 6$ centimeters and a radius of 5 centimeters.

$V = Bh$	Write formula for volume.
$= \pi(5)^2(6)$	Substitute.
$= 150\pi \approx 471$	Use a calculator.

So, about 471 cubic centimeters of salsa are missing from the jar.

EXAMPLE 4 Real-Life Application

About how many gallons of water does the watercooler bottle contain? ($1 \text{ ft}^3 \approx 7.5$ gal)

Ⓐ 5.3 gallons Ⓑ 10 gallons Ⓒ 17 gallons Ⓓ 40 gallons

Find the volume of the cylinder. The diameter is 1 foot. So, the radius is 0.5 foot.

$V = Bh$	Write formula for volume.
$= \pi(0.5)^2(1.7)$	Substitute.
$= 0.425\pi \approx 1.3352$	Use a calculator.

So, the bottle contains about 1.3352 cubic feet of water. To find the number of gallons it contains, multiply by the conversion factor $\frac{7.5 \text{ gal}}{1 \text{ ft}^3}$.

$$1.3352 \cancel{\text{ft}^3} \times \frac{7.5 \text{ gal}}{1 \cancel{\text{ft}^3}} \approx 10 \text{ gal}$$

The watercooler bottle contains about 10 gallons of water. So, the correct answer is Ⓑ.

On Your Own

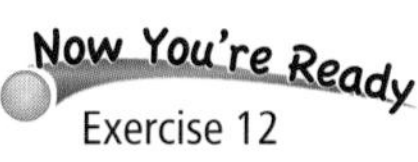

3. **WHAT IF?** In Example 3, the height of the salsa in the jar is 5 centimeters. How much salsa is missing from the jar?

4. A cylindrical water tower has a diameter of 15 meters and a height of 5 meters. About how many gallons of water can the tower contain? ($1 \text{ m}^3 \approx 264$ gal)

8.1 Exercises

Vocabulary and Concept Check

1. **DIFFERENT WORDS, SAME QUESTION** Which is different? Find "both" answers.

How much does it take to fill the cylinder?

What is the capacity of the cylinder?

How much does it take to cover the cylinder?

How much does the cylinder contain?

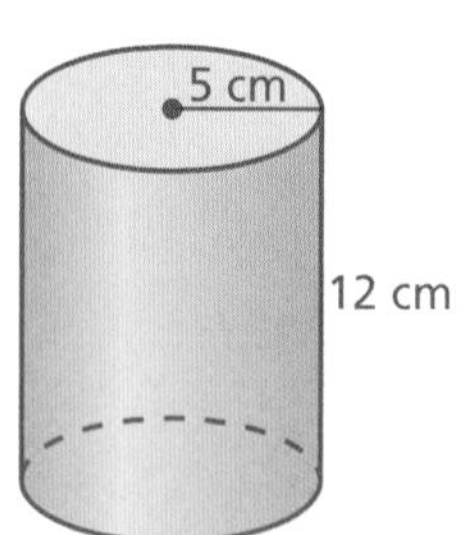

2. **REASONING** Without calculating, which of the solids has the greater volume? Explain.

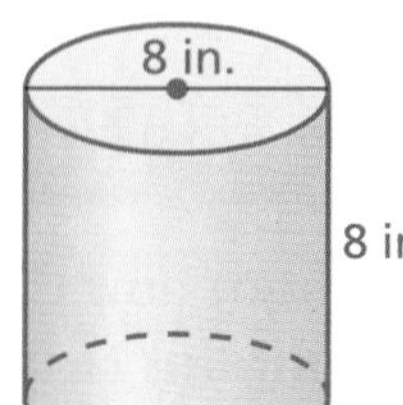

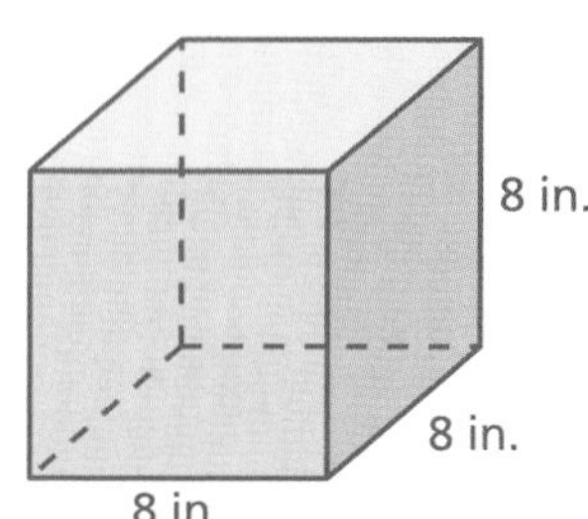

Practice and Problem Solving

Find the volume of the cylinder. Round your answer to the nearest tenth.

① 3.

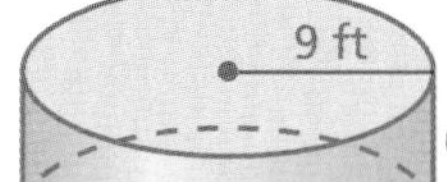

4.

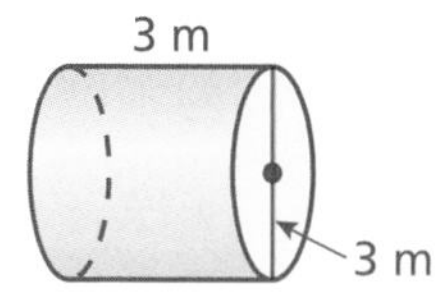

5.

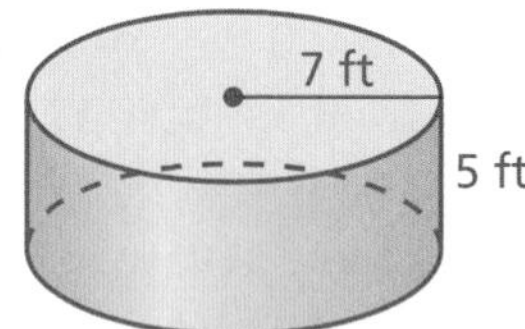

6.

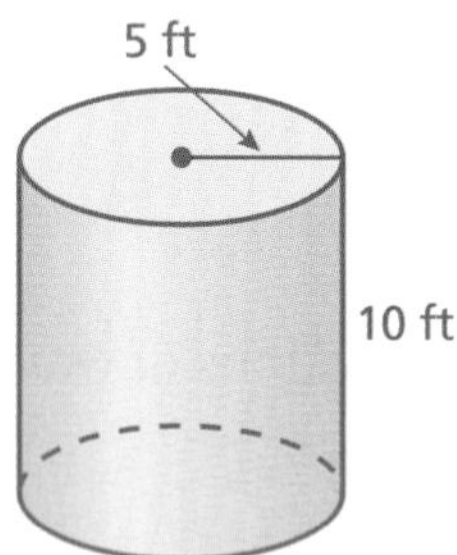

7.

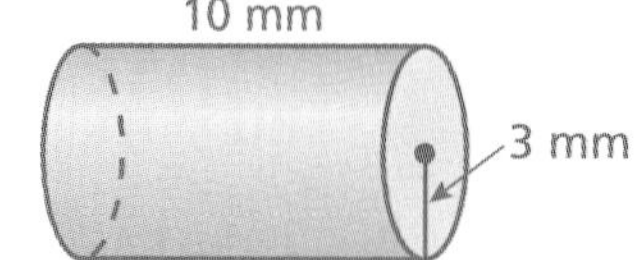

8.

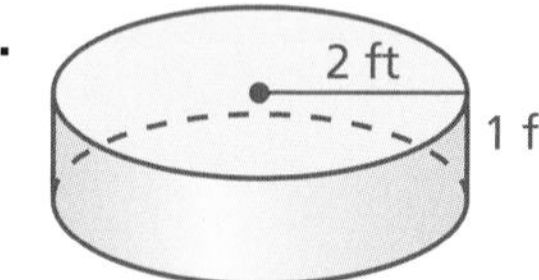

9.

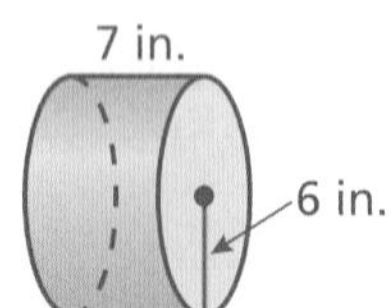

10.

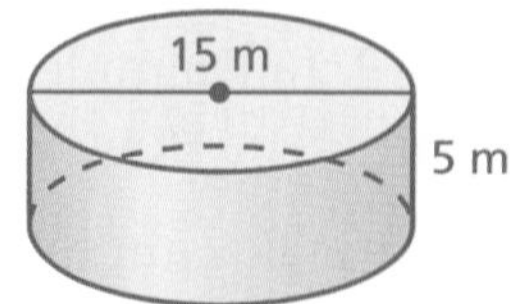

11.

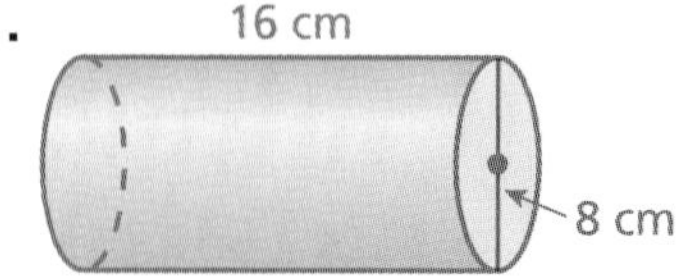

④ 12. **SWIMMING POOL** A cylindrical swimming pool has a diameter of 16 feet and a height of 4 feet. About how many gallons of water can the pool contain? Round your answer to the nearest whole number. ($1 \text{ ft}^3 \approx 7.5$ gal)

Find the missing dimension of the cylinder. Round your answer to the nearest whole number.

2 **13.** Volume = 250 ft^3

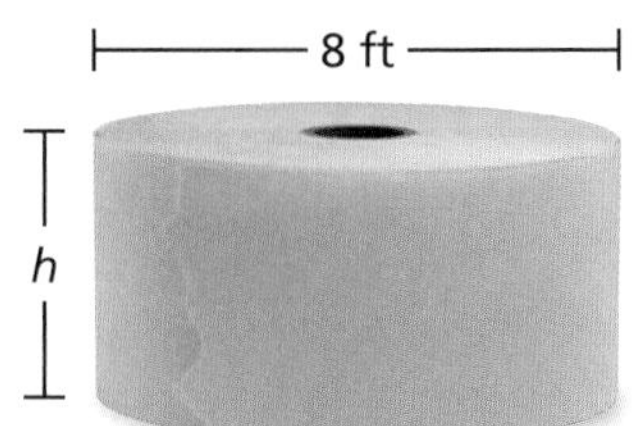

14. Volume = $10{,}000\pi$ $in.^3$

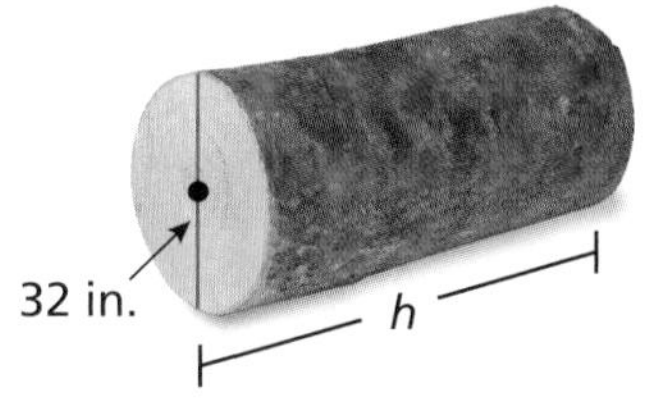

15. Volume = 600,000 cm^3

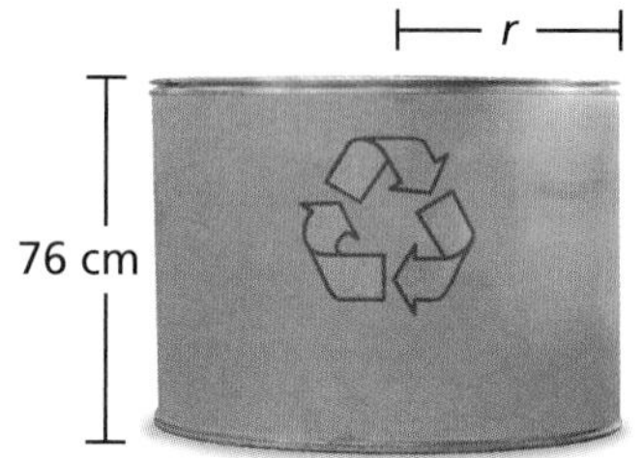

16. CRITICAL THINKING How does the volume of a cylinder change when its diameter is halved? Explain.

Round hay bale

17. MODELING A traditional "square" bale of hay is actually in the shape of a rectangular prism. Its dimensions are 2 feet by 2 feet by 4 feet. How many square bales contain the same amount of hay as one large "round" bale?

18. ROAD ROLLER A tank on a road roller is filled with water to make the roller heavy. The tank is a cylinder that has a height of 6 feet and a radius of 2 feet. One cubic foot of water weighs 62.5 pounds. Find the weight of the water in the tank.

19. VOLUME A cylinder has a surface area of 1850 square meters and a radius of 9 meters. Estimate the volume of the cylinder to the nearest whole number.

20. Problem Solving Water flows at 2 feet per second through a pipe with a diameter of 8 inches. A cylindrical tank with a diameter of 15 feet and a height of 6 feet collects the water.

a. What is the volume, in cubic inches, of water flowing out of the pipe every second?

b. What is the height, in inches, of the water in the tank after 5 minutes?

c. How many minutes will it take to fill 75% of the tank?

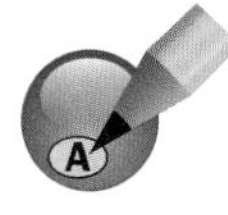

Fair Game Review What you learned in previous grades & lessons

Tell whether the triangle with the given side lengths is a right triangle. *(Section 7.5)*

21. 20 m, 21 m, 29 m

22. 1 in., 2.4 in., 2.6 in.

23. 5.6 ft, 8 ft, 10.6 ft

24. MULTIPLE CHOICE Which ordered pair is the solution of the linear system $3x + 4y = -10$ and $2x - 4y = 0$? *(Section 5.3)*

Ⓐ $(-6, 2)$ Ⓑ $(2, -6)$ Ⓒ $(-2, -1)$ Ⓓ $(-1, -2)$

8.2 Volumes of Cones

Essential Question How can you find the volume of a cone?

You already know how the volume of a pyramid relates to the volume of a prism. In this activity, you will discover how the volume of a cone relates to the volume of a cylinder.

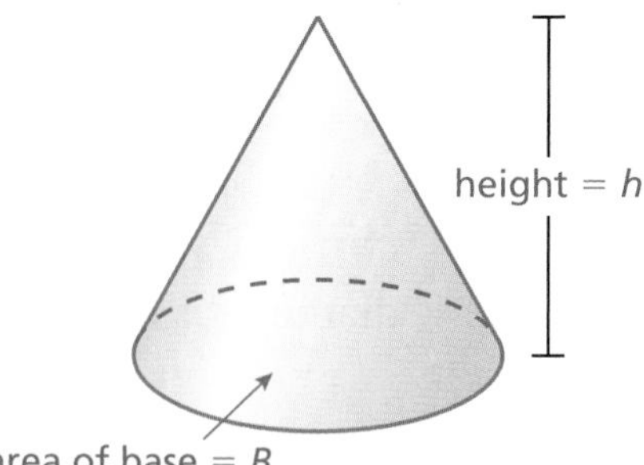

1 ACTIVITY: Finding a Formula Experimentally

Work with a partner. Use a paper cup that is shaped like a cone.

- Estimate the height of the cup.
- Trace the top of the cup on a piece of paper. Find the diameter of the circle.
- Use these measurements to draw a net for a cylinder with the same base and height as the paper cup.
- Cut out the net. Then fold and tape it to form an open cylinder.
- Fill the paper cup with rice. Then pour the rice into the cylinder. Repeat this until the cylinder is full. How many cones does it take to fill the cylinder?
- Use your result to write a formula for the volume of a cone.

2 ACTIVITY: Summarizing Volume Formulas

Geometry

In this lesson, you will

- find the volumes of cones.
- find the heights of cones given the volumes.
- solve real-life problems.

Work with a partner. You can remember the volume formulas for prisms, cylinders, pyramids, and cones with just two concepts.

Volumes of Prisms and Cylinders

Volume = Area of base × ______

Volumes of Pyramids and Cones

Volume = ______ Volume of prism or cylinder with same base and height

Make a list of all the formulas you need to remember to find the area of a base. Talk about strategies for remembering these formulas.

3 ACTIVITY: Volumes of Oblique Solids

Work with a partner. Think of a stack of paper. When you adjust the stack so that the sides are oblique (slanted), do you change the volume of the stack? If the volume of the stack does not change, then the formulas for volumes of right solids also apply to oblique solids.

Math Practice

Use Equations

What equation would you use to find the volume of the oblique solid? Explain.

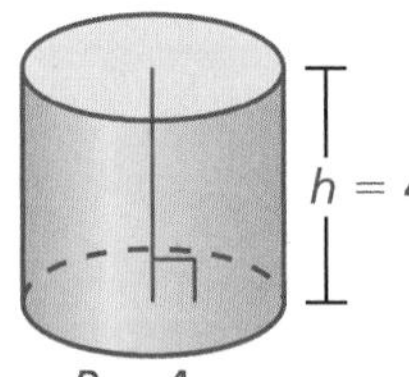

Right cylinder

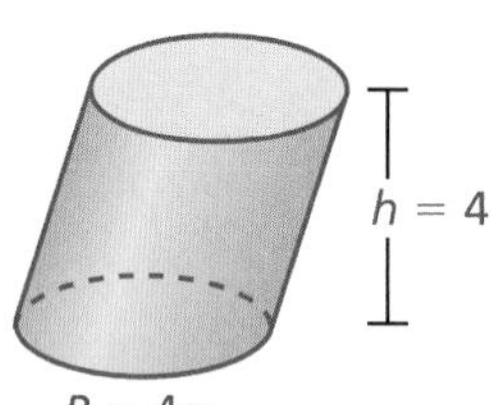

Oblique cylinder

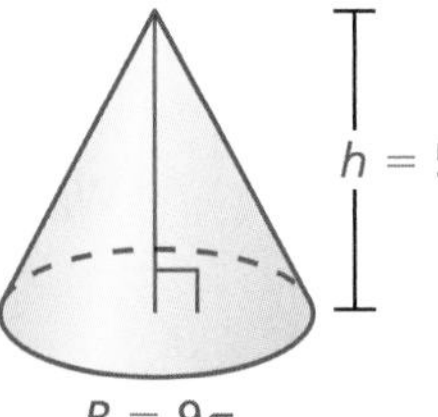

Right cone

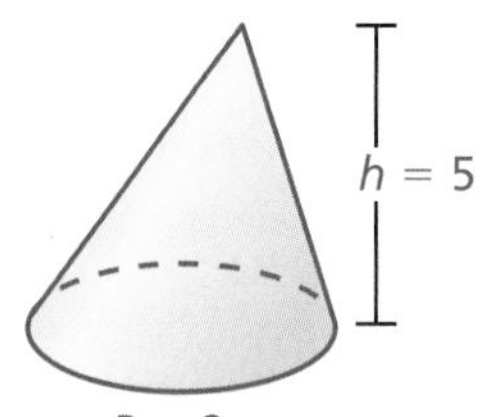

Oblique cone

What Is Your Answer?

4. **IN YOUR OWN WORDS** How can you find the volume of a cone?

5. Describe the intersection of the plane and the cone. Then explain how to find the volume of each section of the solid.

a.

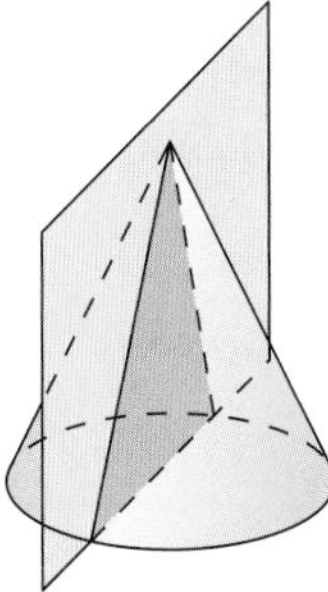

b.

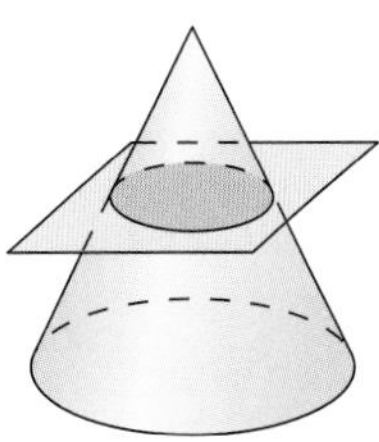

Practice Use what you learned about the volumes of cones to complete Exercises 4–6 on page 344.

8.2 Lesson

Study Tip

The *height* of a cone is the perpendicular distance from the base to the vertex.

Volume of a Cone

Words The volume V of a cone is one-third the product of the area of the base and the height of the cone.

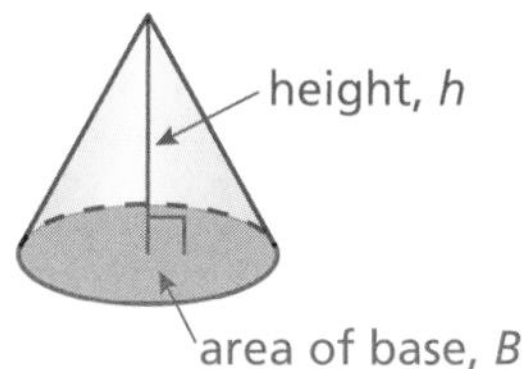

Algebra $V = \frac{1}{3}Bh$

Area of base (B)

Height of cone (h)

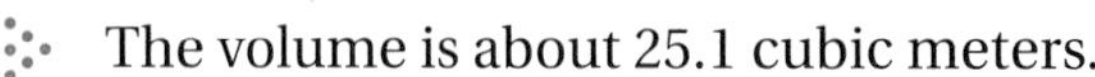

EXAMPLE 1 Finding the Volume of a Cone

Study Tip

Because $B = \pi r^2$, you can use $V = \frac{1}{3}\pi r^2 h$ to find the volume of a cone.

Find the volume of the cone. Round your answer to the nearest tenth.

The diameter is 4 meters. So, the radius is 2 meters.

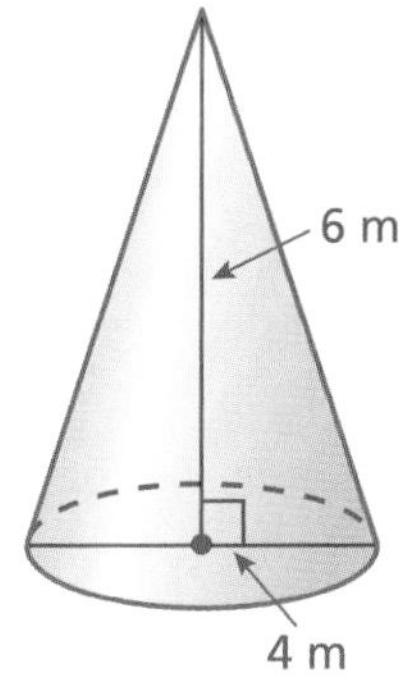

$V = \frac{1}{3}Bh$	Write formula for volume.
$= \frac{1}{3}\pi(2)^2(6)$	Substitute.
$= 8\pi \approx 25.1$	Use a calculator.

∴ The volume is about 25.1 cubic meters.

EXAMPLE 2 Finding the Height of a Cone

Find the height of the cone. Round your answer to the nearest tenth.

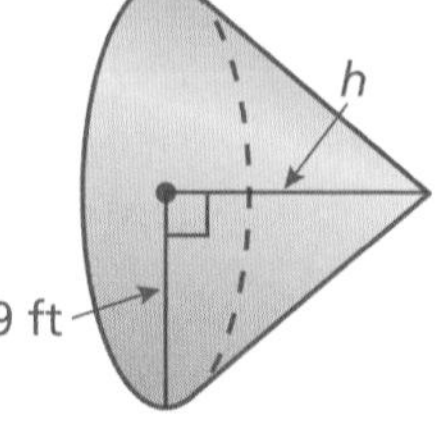

Volume = 956 ft^3

$V = \frac{1}{3}Bh$	Write formula for volume.
$956 = \frac{1}{3}\pi(9)^2(h)$	Substitute.
$956 = 27\pi h$	Simplify.
$11.3 \approx h$	Divide each side by 27π.

∴ The height is about 11.3 feet.

On Your Own

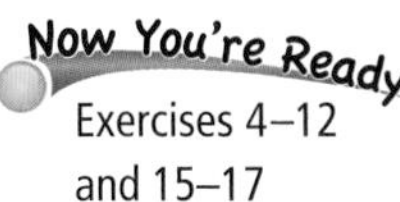

Exercises 4–12 and 15–17

Find the volume V or height h of the cone. Round your answer to the nearest tenth.

1. 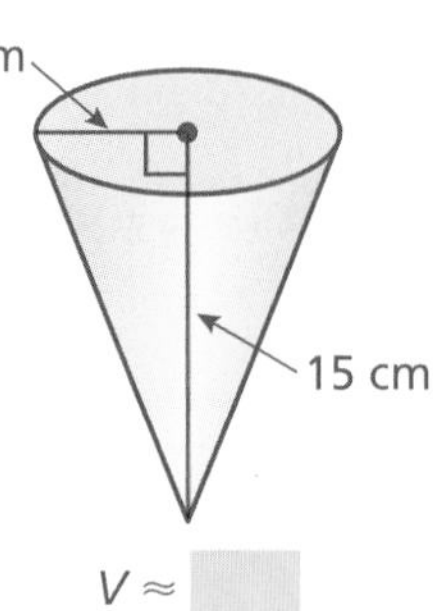

2. 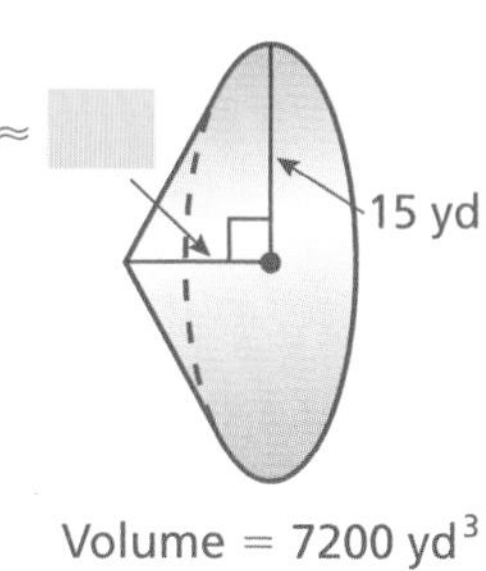

EXAMPLE  3 **Real-Life Application**

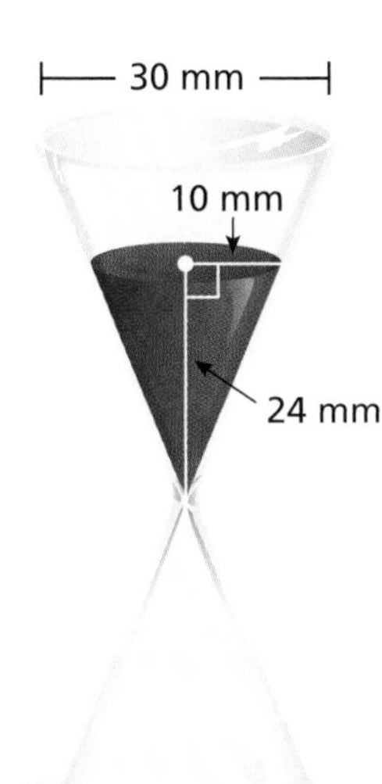

You must answer a trivia question before the sand in the timer falls to the bottom. The sand falls at a rate of 50 cubic millimeters per second. How much time do you have to answer the question?

Use the formula for the volume of a cone to find the volume of the sand in the timer.

$$V = \frac{1}{3}Bh \qquad \text{Write formula for volume.}$$

$$= \frac{1}{3}\pi(10)^2(24) \qquad \text{Substitute.}$$

$$= 800\pi \approx 2513 \qquad \text{Use a calculator.}$$

The volume of the sand is about 2513 cubic millimeters. To find the amount of time you have to answer the question, multiply the volume by the rate at which the sand falls.

$$2513 \cancel{\text{mm}^3} \times \frac{1 \text{ sec}}{50 \cancel{\text{mm}^3}} = 50.26 \text{ sec}$$

∴ So, you have about 50 seconds to answer the question.

On Your Own

3. **WHAT IF?** The sand falls at a rate of 60 cubic millimeters per second. How much time do you have to answer the question?

4. **WHAT IF?** The height of the sand in the timer is 12 millimeters, and the radius is 5 millimeters. How much time do you have to answer the question?

8.2 Exercises

Vocabulary and Concept Check

1. **VOCABULARY** Describe the height of a cone.
2. **WRITING** Compare and contrast the formulas for the volume of a pyramid and the volume of a cone.
3. **REASONING** You know the volume of a cylinder. How can you find the volume of a cone with the same base and height?

Practice and Problem Solving

Find the volume of the cone. Round your answer to the nearest tenth.

1 **4.**

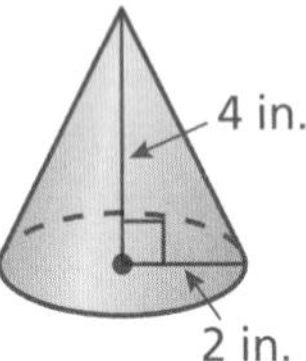

5.

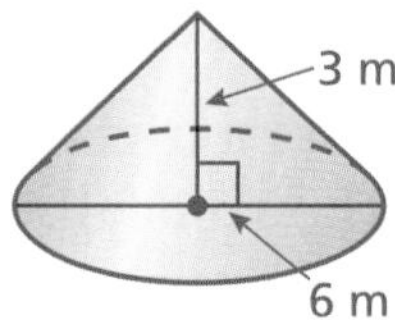

6.

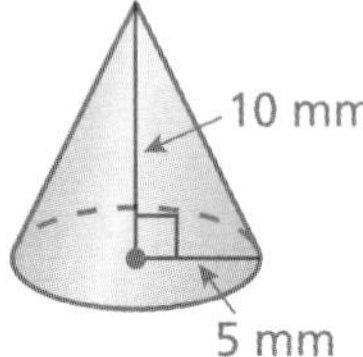

7.

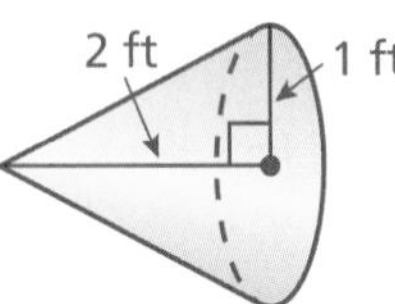

8.

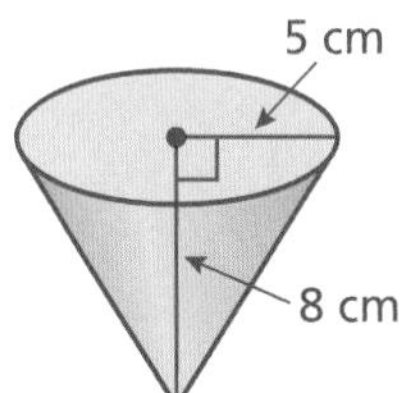

9.

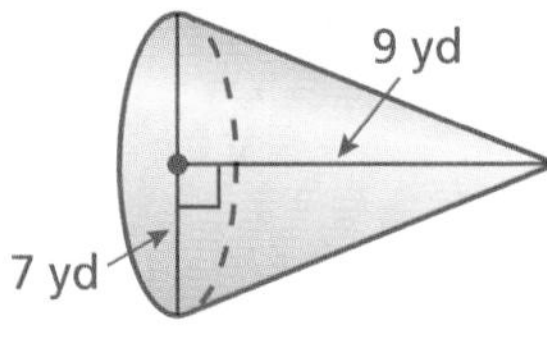

10.

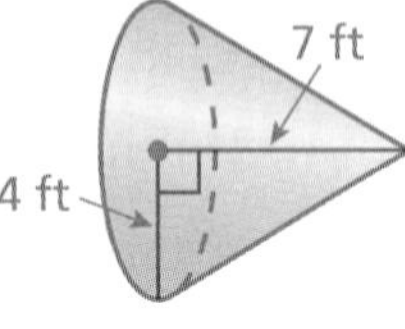

11.

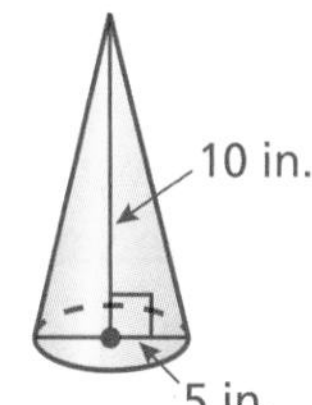

12.

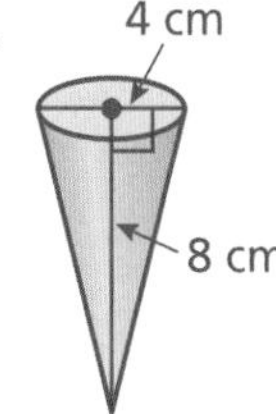

13. **ERROR ANALYSIS** Describe and correct the error in finding the volume of the cone.

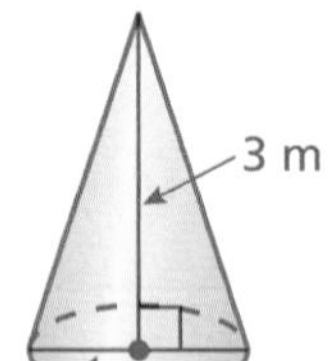

$$V = \frac{1}{3}Bh$$
$$= \frac{1}{3}(\pi)(2)^2(3)$$
$$= 4\pi \text{ m}^3$$

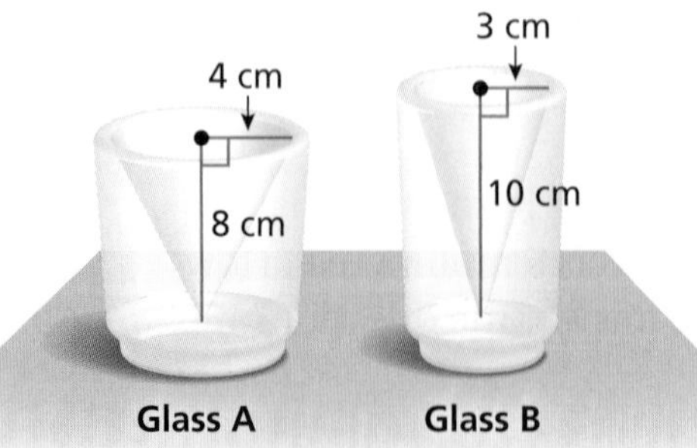

14. **GLASS** The inside of each glass is shaped like a cone. Which glass can hold more liquid? How much more?

Find the missing dimension of the cone. Round your answer to the nearest tenth.

② **15.** Volume $= \frac{1}{18}\pi \text{ ft}^3$

16. Volume $= 225 \text{ cm}^3$

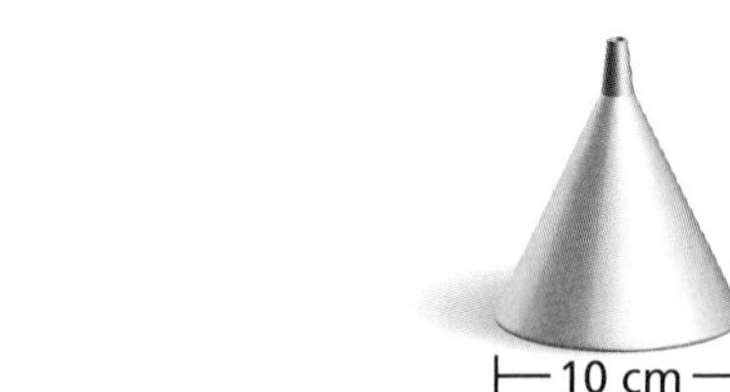

17. Volume $= 3.6 \text{ in.}^3$

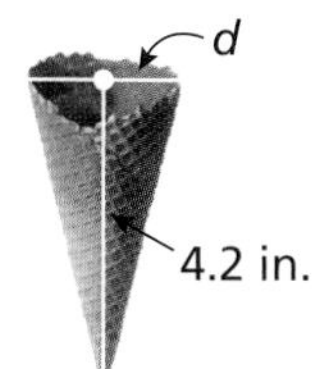

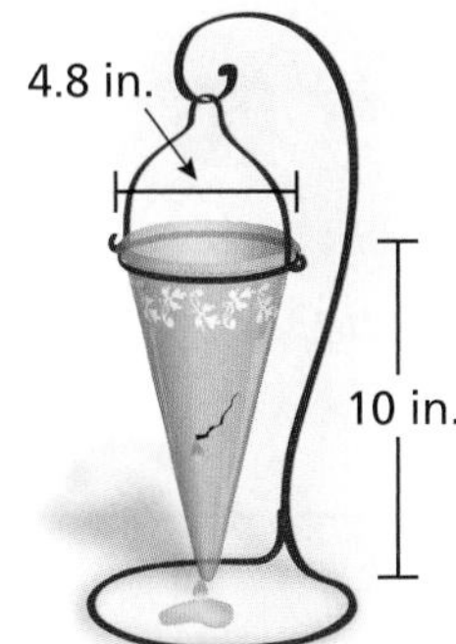

18. REASONING The volume of a cone is 20π cubic meters. What is the volume of a cylinder with the same base and height?

19. VASE Water leaks from a crack in a vase at a rate of 0.5 cubic inch per minute. How long does it take for 20% of the water to leak from a full vase?

20. LEMONADE STAND You have 10 gallons of lemonade to sell. (1 gal $\approx$ 3785 cm^3)

a. Each customer uses one paper cup. How many paper cups will you need?

b. The cups are sold in packages of 50. How many packages should you buy?

c. How many cups will be left over if you sell 80% of the lemonade?

21. STRUCTURE The cylinder and the cone have the same volume. What is the height of the cone?

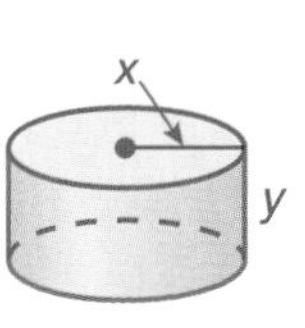

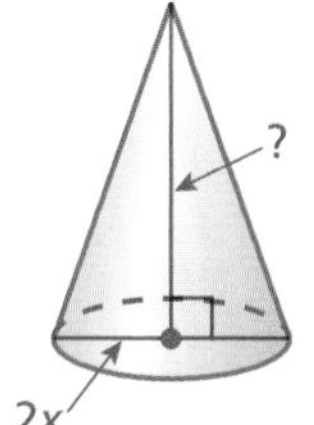

22. 

In Example 3, you use a different timer with the same dimensions. The sand in this timer has a height of 30 millimeters. How much time do you have to answer the question?

Fair Game Review What you learned in previous grades & lessons

The vertices of a figure are given. Rotate the figure as described. Find the coordinates of the image. *(Section 2.4)*

23. $A(-1, 1)$, $B(2, 3)$, $C(2, 1)$
90° counterclockwise about vertex A

24. $E(-4, 1)$, $F(-3, 3)$, $G(-2, 3)$, $H(-1, 1)$
180° about the origin

25. MULTIPLE CHOICE $\triangle ABC$ is similar to $\triangle XYZ$. How many times greater is the area of $\triangle XYZ$ than the area of $\triangle ABC$? *(Section 2.6)*

Ⓐ $\frac{1}{9}$ Ⓑ $\frac{1}{3}$

Ⓒ 3 Ⓓ 9

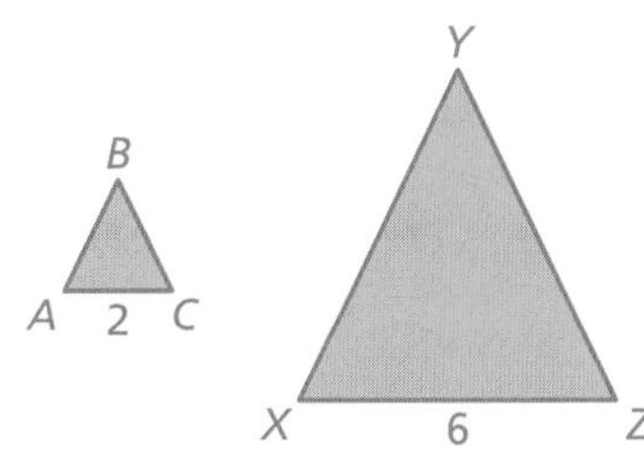

8 Study Help

Check It Out
Graphic Organizer
BigIdeasMath.com

You can use a **formula triangle** to arrange variables and operations of a formula. Here is an example of a formula triangle for the volume of a cylinder.

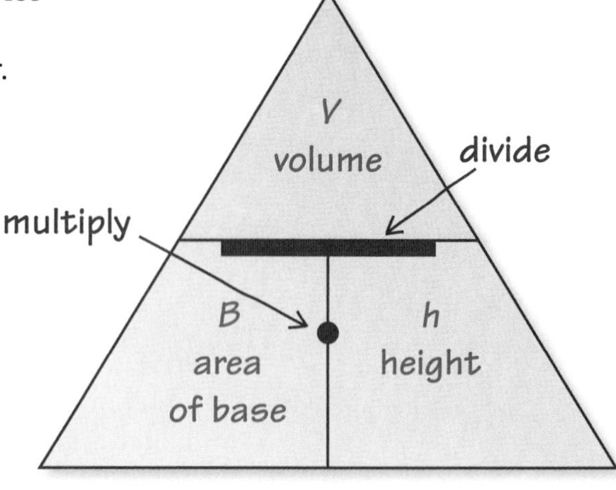

To find an unknown variable, use the other variables and the operation between them. For example, to find the area B of the base, cover up the B. Then you can see that you divide the volume V by the height h.

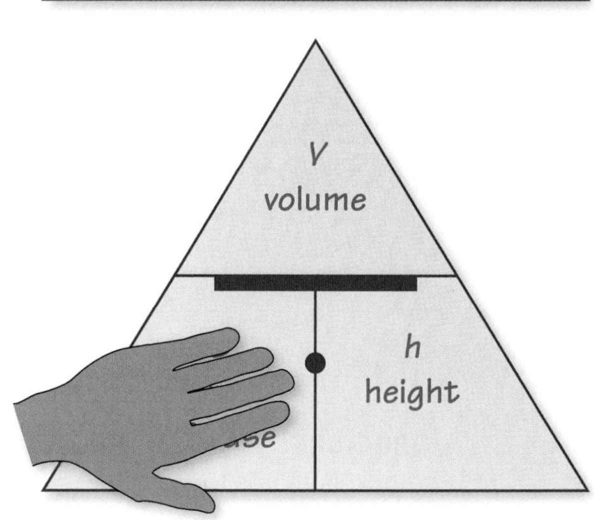

On Your Own

Make a formula triangle to help you study this topic. (*Hint:* Your formula triangle may have a different form than what is shown in the example.)

1. volume of a cone

After you complete this chapter, make formula triangles for the following topics.

2. volume of a sphere
3. volume of a composite solid
4. surface areas of similar solids
5. volumes of similar solids

"See how a formula triangle works? Cover any variable and you get its formula."

8.1–8.2 Quiz

Find the volume of the solid. Round your answer to the nearest tenth. *(Section 8.1 and Section 8.2)*

1. 4 yd; 3.5 yd

2. 3 ft; 4 ft

3. 5 cm; 6 cm

4. 11 in.; 12 in.

Find the missing dimension of the solid. Round your answer to the nearest tenth. *(Section 8.1 and Section 8.2)*

5. 3 ft; h

Volume = 340 ft^3

6. 4.7 cm; r

Volume = 938 cm^3

7. **PAPER CONE** The paper cone can hold 84.78 cubic centimeters of water. What is the height of the cone? *(Section 8.2)*

6 cm; h

8. **GEOMETRY** Triple both dimensions of the cylinder. How many times greater is the volume of the new cylinder than the volume of the original cylinder? *(Section 8.1)*

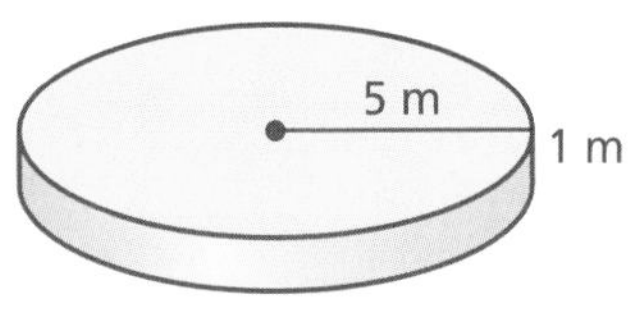

1.5 in.; 16 in.

9. **SAND ART** There are 42.39 cubic inches of blue sand and 28.26 cubic inches of red sand in the cylindrical container. How many cubic inches of white sand are in the container? *(Section 8.1)*

10. **JUICE CAN** You are buying two cylindrical cans of juice. Each can holds the same amount of juice. What is the height of Can B? *(Section 8.1)*

8.3 Volumes of Spheres

Essential Question How can you find the volume of a sphere?

A **sphere** is the set of all points in space that are the same distance from a point called the *center*. The *radius r* is the distance from the center to any point on the sphere.

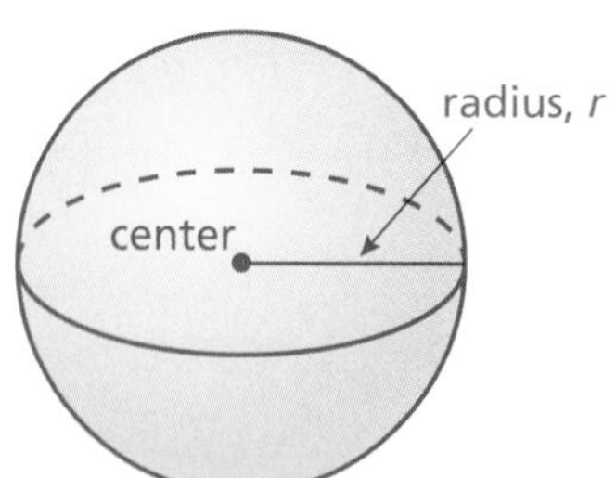

A sphere is different from the other solids you have studied so far because it does not have a base. To discover the volume of a sphere, you can use an activity similar to the one in the previous section.

1 ACTIVITY: Exploring the Volume of a Sphere

Work with a partner. Use a plastic ball similar to the one shown.

- Estimate the diameter and the radius of the ball.

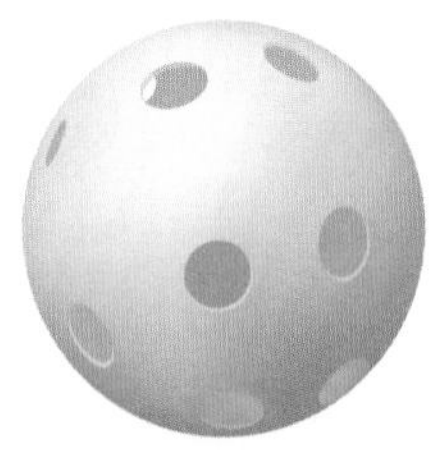

- Use these measurements to draw a net for a cylinder with a diameter and a height equal to the diameter of the ball. How is the height h of the cylinder related to the radius r of the ball? Explain.

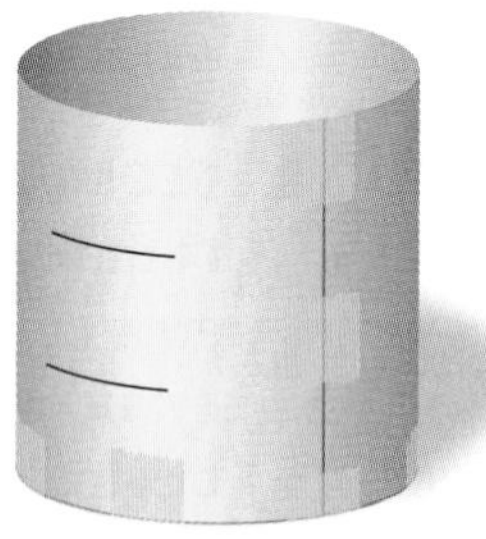

- Cut out the net. Then fold and tape it to form an open cylinder. Make two marks on the cylinder that divide it into thirds, as shown.

- Cover the ball with aluminum foil or tape. Leave one hole open. Fill the ball with rice. Then pour the rice into the cylinder. What fraction of the cylinder is filled with rice?

Geometry

In this lesson, you will

- find the volumes of spheres.
- find the radii of spheres given the volumes.
- solve real-life problems.

2 ACTIVITY: Deriving the Formula for the Volume of a Sphere

Work with a partner. Use the results from Activity 1 and the formula for the volume of a cylinder to complete the steps.

Math Practice

Analyze Relationships

What is the relationship between the volume of a sphere and the volume of a cylinder? How does this help you derive a formula for the volume of a sphere?

$V = \pi r^2 h$	Write formula for volume of a cylinder.
$= \frac{\square}{\square} \pi r^2 h$	Multiply by $\frac{\square}{\square}$ because the volume of a sphere is $\frac{\square}{\square}$ of the volume of the cylinder.
$= \frac{\square}{\square} \pi r^2 \square$	Substitute $\square$ for h.
$= \frac{\square}{\square} \pi \square$	Simplify.

3 ACTIVITY: Deriving the Formula for the Volume of a Sphere

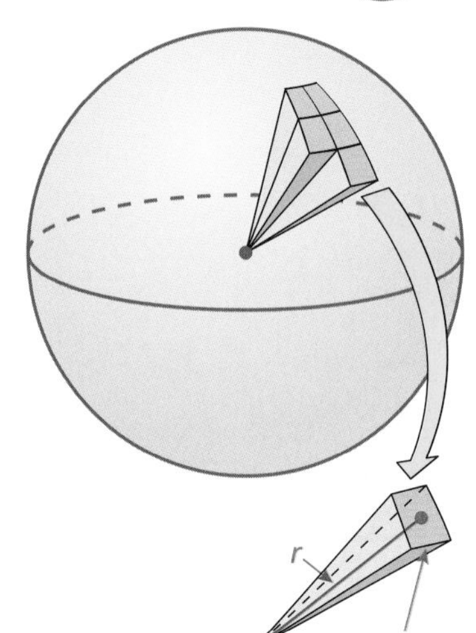

Work with a partner. Imagine filling the inside of a sphere with n small pyramids. The vertex of each pyramid is at the center of the sphere. The height of each pyramid is approximately equal to r, as shown. Complete the steps. (The surface area of a sphere is equal to $4\pi r^2$.)

$V = \frac{1}{3}Bh$	Write formula for volume of a pyramid.
$= n \frac{1}{3} B \square$	Multiply by the number of small pyramids n and substitute $\square$ for h.
$= \frac{1}{3}(4\pi r^2) \square$	$4\pi r^2 \approx n \cdot \square$

Show how this result is equal to the result in Activity 2.

What Is Your Answer?

4. **IN YOUR OWN WORDS** How can you find the volume of a sphere?

5. Describe the intersection of the plane and the sphere. Then explain how to find the volume of each section of the solid.

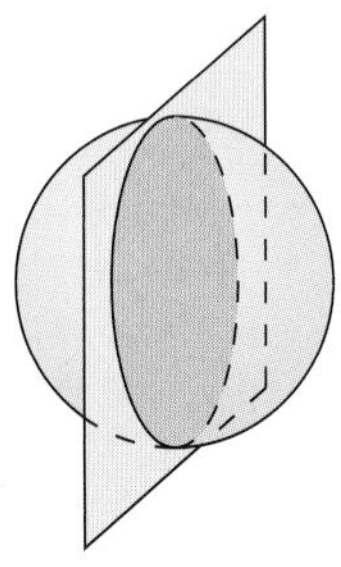

Practice Use what you learned about the volumes of spheres to complete Exercises 3–5 on page 352.

8.3 Lesson

Key Vocabulary
sphere, *p. 348*
hemisphere, *p. 351*

Volume of a Sphere

Words The volume V of a sphere is the product of $\frac{4}{3}\pi$ and the cube of the radius of the sphere.

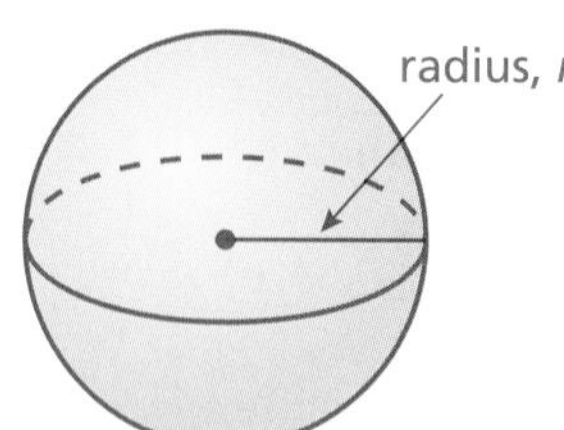

Algebra $V = \frac{4}{3}\pi r^3$ — Cube of radius of sphere

EXAMPLE 1 Finding the Volume of a Sphere

Find the volume of the sphere. Round your answer to the nearest tenth.

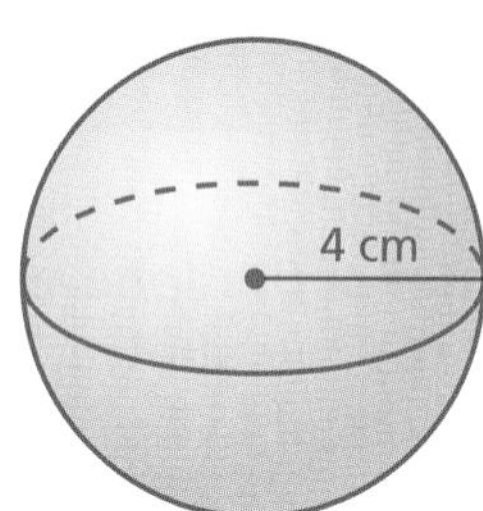

$V = \frac{4}{3}\pi r^3$	Write formula for volume.
$= \frac{4}{3}\pi(4)^3$	Substitute 4 for r.
$= \frac{256}{3}\pi$	Simplify.
≈ 268.1	Use a calculator.

The volume is about 268.1 cubic centimeters.

EXAMPLE 2 Finding the Radius of a Sphere

Find the radius of the sphere.

Volume = 288π in.^{3}

$V = \frac{4}{3}\pi r^3$	Write formula.
$288\pi = \frac{4}{3}\pi r^3$	Substitute.
$288\pi = \frac{4\pi}{3}r^3$	Multiply.
$\frac{3}{4\pi} \cdot 288\pi = \frac{3}{4\pi} \cdot \frac{4\pi}{3}r^3$	Multiplication Property of Equality
$216 = r^3$	Simplify.
$6 = r$	Take the cube root of each side.

The radius is 6 inches.

On Your Own

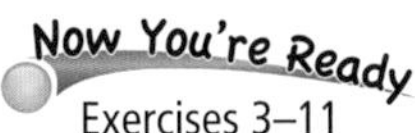

Exercises 3–11

Find the volume V or radius r of the sphere. Round your answer to the nearest tenth, if necessary.

1.

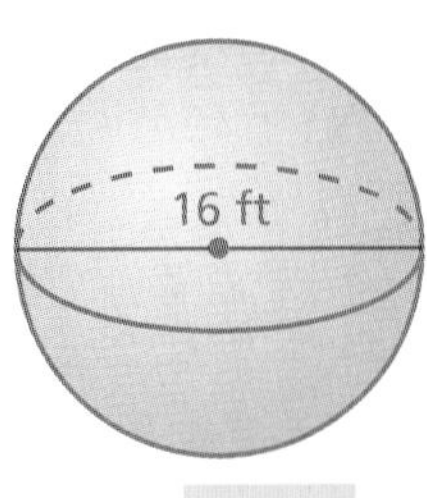

2.

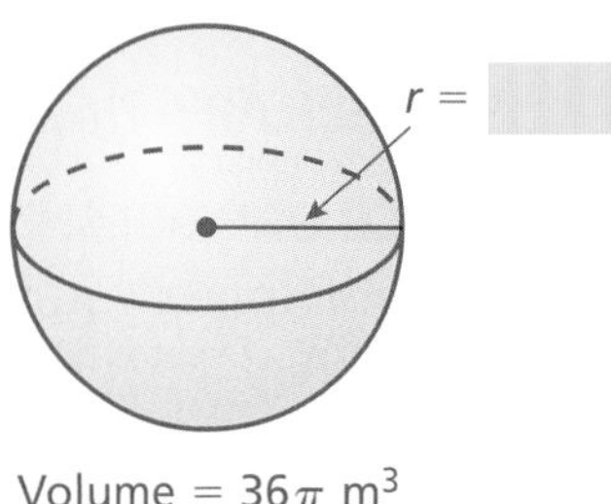

EXAMPLE  **3**

Finding the Volume of a Composite Solid

A hemisphere is one-half of a sphere. The top of the silo is a hemisphere with a radius of 12 feet. What is the volume of the silo? Round your answer to the nearest thousand.

The silo is made up of a cylinder and a hemisphere. Find the volume of each solid.

Cylinder

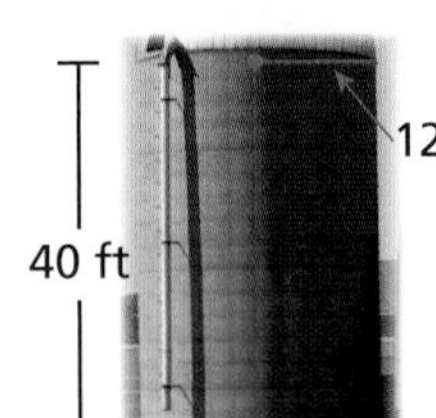
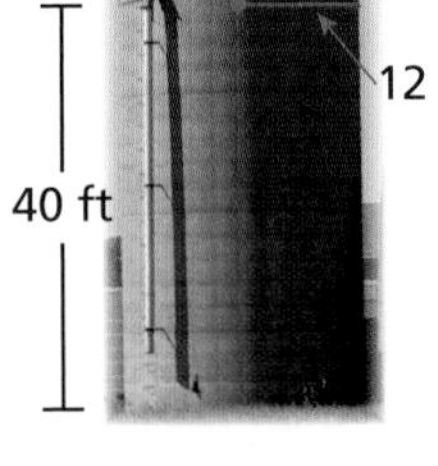

$V = Bh$

$= \pi(12)^2(40)$

$= 5760\pi$

Hemisphere

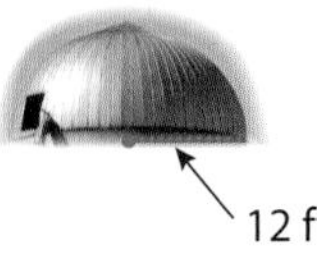

$V = \frac{1}{2} \cdot \frac{4}{3}\pi r^3$

$= \frac{1}{2} \cdot \frac{4}{3}\pi(12)^3$

$= 1152\pi$

So, the volume is $5760\pi + 1152\pi = 6912\pi \approx 22{,}000$ cubic feet.

Study Tip

In Example 3, the height of the cylindrical part of the silo is the difference of the silo height and the radius of the hemisphere.

$52 - 12 = 40$ ft

On Your Own

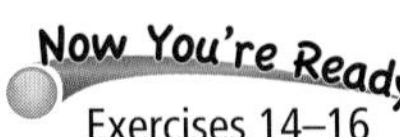

Exercises 14–16

Find the volume of the composite solid. Round your answer to the nearest tenth.

3.

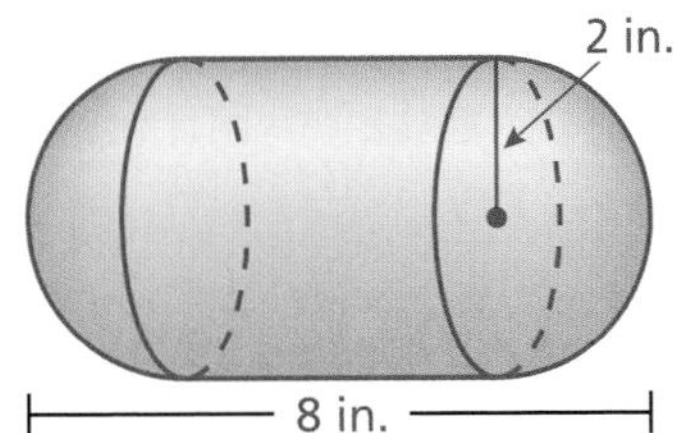

4.

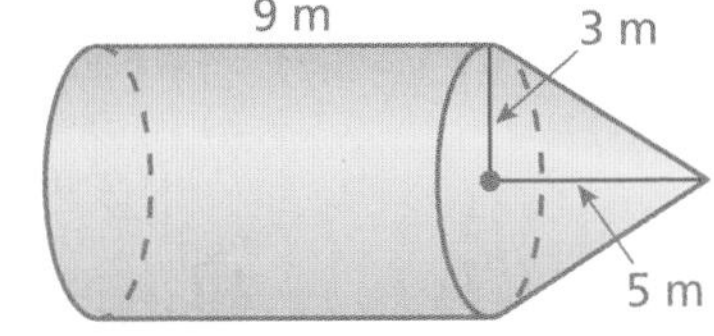

8.3 Exercises

Vocabulary and Concept Check

1. **VOCABULARY** How is a sphere different from a hemisphere?
2. **WHICH ONE DOESN'T BELONG?** Which figure does *not* belong with the other three? Explain your reasoning.

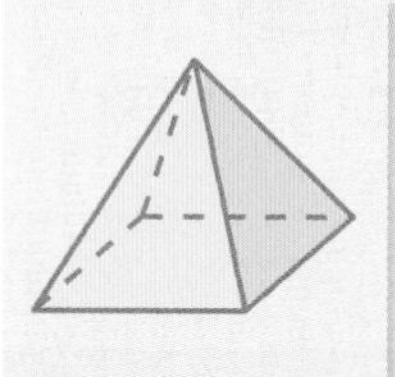 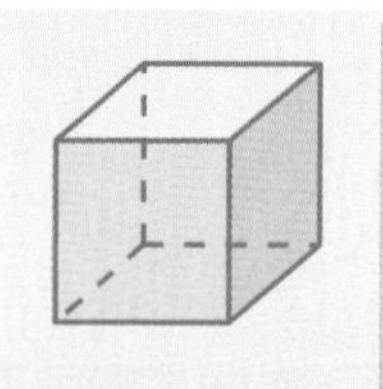 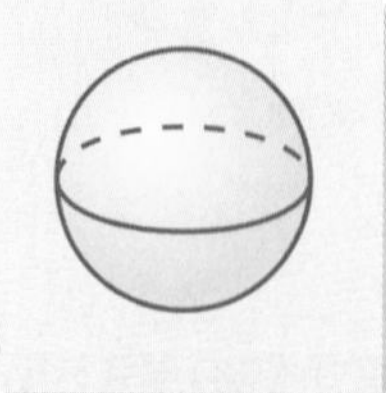 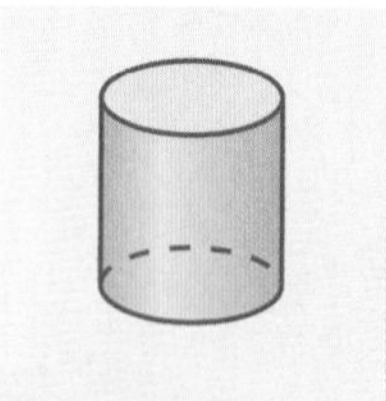

Practice and Problem Solving

Find the volume of the sphere. Round your answer to the nearest tenth.

① 3.

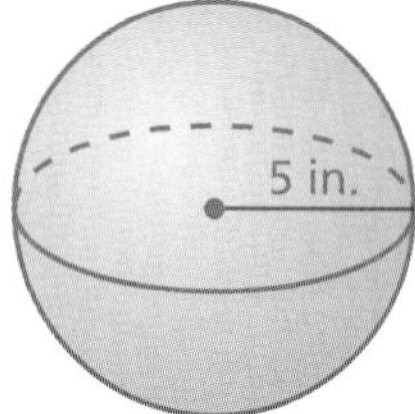

4.

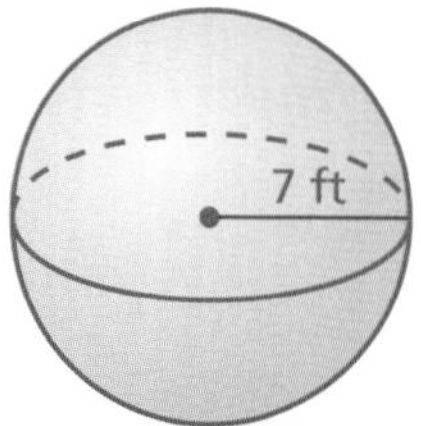

5.

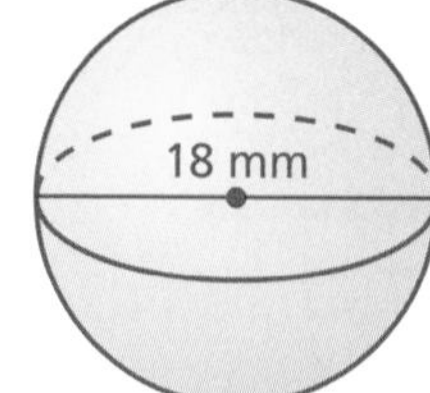

6.

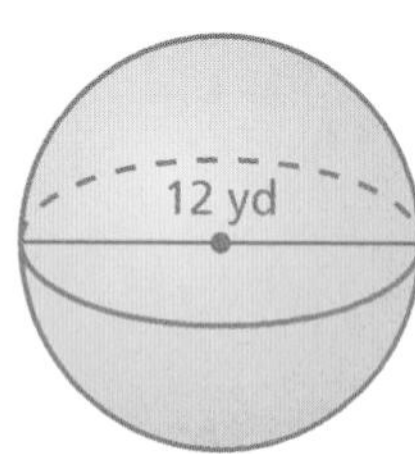

7.

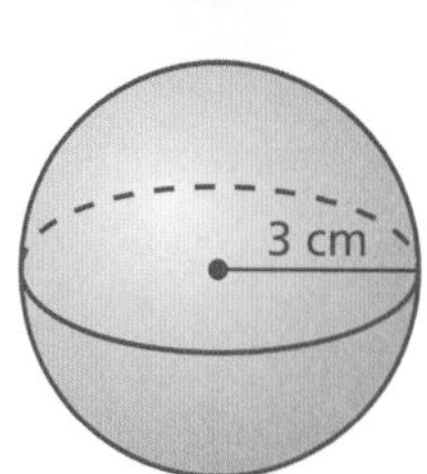

8.

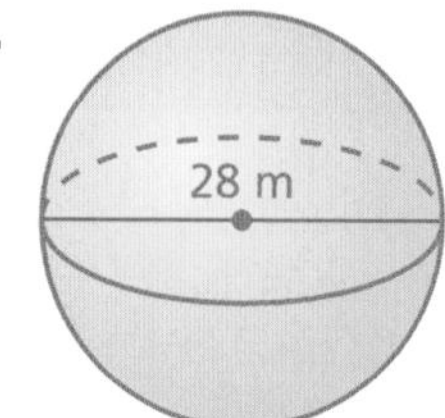

Find the radius of the sphere with the given volume.

② 9. Volume $= 972\pi$ mm^3

10. Volume $= 4.5\pi$ cm^3

11. Volume $= 121.5\pi$ ft^3

12. **GLOBE** The globe of the Moon has a radius of 10 inches. Find the volume of the globe. Round your answer to the nearest whole number.

13. **SOFTBALL** A softball has a volume of $\frac{125}{6}\pi$ cubic inches. Find the radius of the softball.

Find the volume of the composite solid. Round your answer to the nearest tenth.

3 **14.**

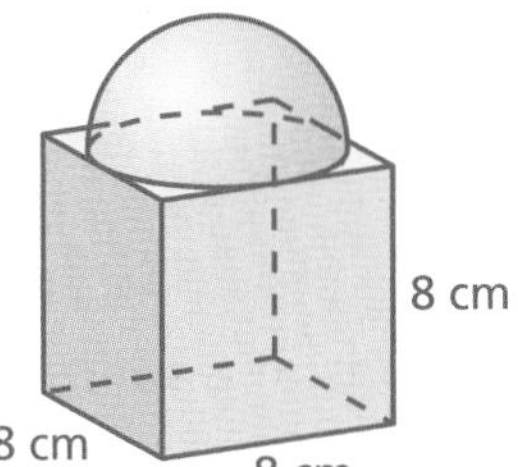

15.

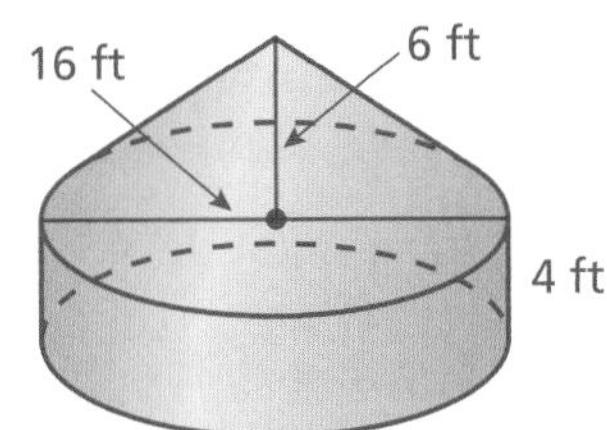

16.

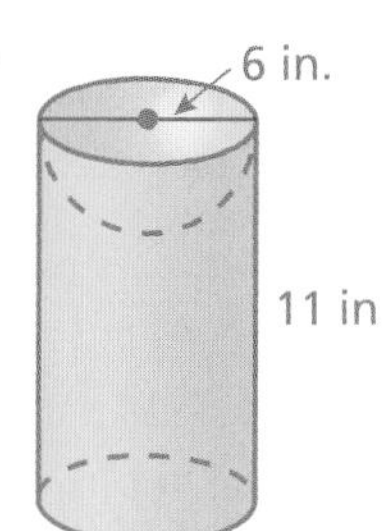

17. **REASONING** A sphere and a right cylinder have the same radius and volume. Find the radius r in terms of the height h of the cylinder.

18. **PACKAGING** A cylindrical container of three rubber balls has a height of 18 centimeters and a diameter of 6 centimeters. Each ball in the container has a radius of 3 centimeters. Find the amount of space in the container that is not occupied by rubber balls. Round your answer to the nearest whole number.

Volume = 4500π in.^{3}

19. **BASKETBALL** The basketball shown is packaged in a box that is in the shape of a cube. The edge length of the box is equal to the diameter of the basketball. What is the surface area and the volume of the box?

20. **Logic** Your friend says that the volume of a sphere with radius r is four times the volume of a cone with radius r. When is this true? Justify your answer.

Fair Game Review What you learned in previous grades & lessons

The blue figure is a dilation of the red figure. Identify the type of dilation and find the scale factor. *(Section 2.7)*

21.

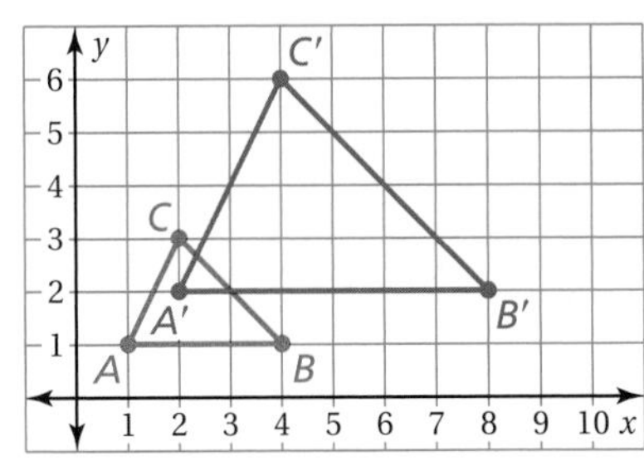

22.

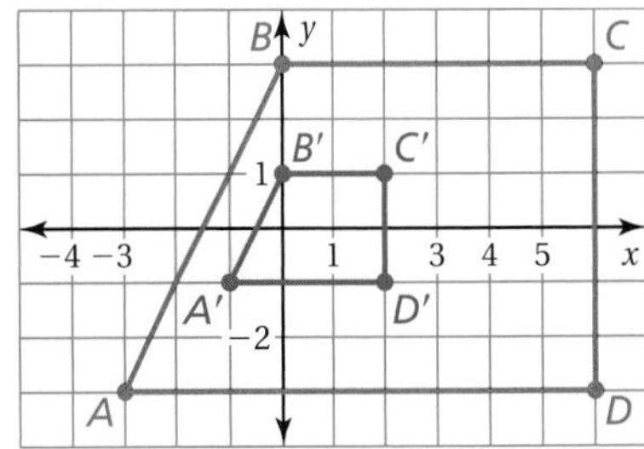

23. **MULTIPLE CHOICE** A person who is 5 feet tall casts a 6-foot-long shadow. A nearby flagpole casts a 30-foot-long shadow. What is the height of the flagpole? *(Section 3.4)*

Ⓐ 25 ft Ⓑ 29 ft Ⓒ 36 ft Ⓓ 40 ft

8.4 Surface Areas and Volumes of Similar Solids

Essential Question When the dimensions of a solid increase by a factor of k, how does the surface area change? How does the volume change?

1 ACTIVITY: Comparing Surface Areas and Volumes

Work with a partner. Copy and complete the table. Describe the pattern. Are the dimensions proportional? Explain your reasoning.

a.

Radius	1	1	1	1	1
Height	1	2	3	4	5
Surface Area					
Volume					

b.

Radius	1	2	3	4	5
Height	1	2	3	4	5
Surface Area					
Volume					

Geometry

In this lesson, you will

- identify similar solids.
- use properties of similar solids to find missing measures.
- understand the relationship between surface areas of similar solids.
- understand the relationship between volumes of similar solids.
- solve real-life problems.

2 ACTIVITY: Comparing Surface Areas and Volumes

Work with a partner. Copy and complete the table. Describe the pattern. Are the dimensions proportional? Explain.

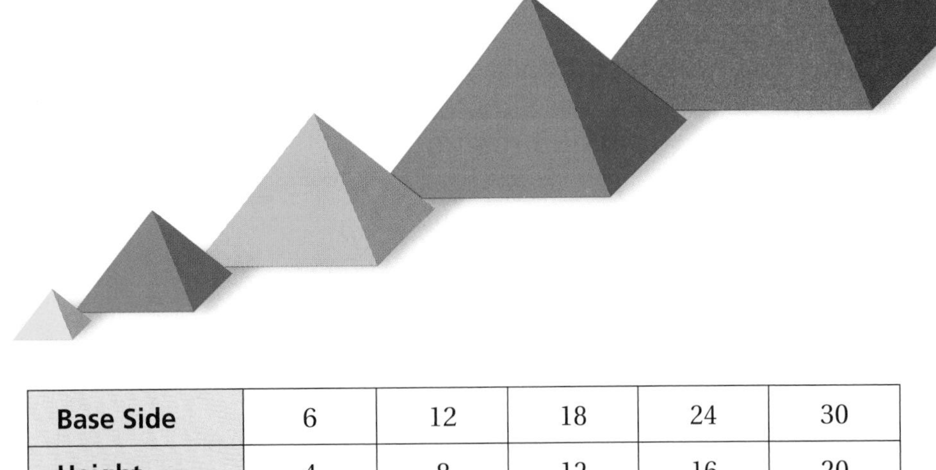

Math Practice

Repeat Calculations

Which calculations are repeated? How does this help you describe the pattern?

Base Side	6	12	18	24	30
Height	4	8	12	16	20
Slant Height	5	10	15	20	25
Surface Area					
Volume					

What Is Your Answer?

3. **IN YOUR OWN WORDS** When the dimensions of a solid increase by a factor of k, how does the surface area change?

4. **IN YOUR OWN WORDS** When the dimensions of a solid increase by a factor of k, how does the volume change?

5. **REPEATED REASONING** All the dimensions of a prism increase by a factor of 5.

   a. How many times greater is the surface area? Explain.

   5   10   25   125

   b. How many times greater is the volume? Explain.

   5   10   25   125

Practice Use what you learned about surface areas and volumes of similar solids to complete Exercise 3 on page 359.

8.4 Lesson

Key Vocabulary
similar solids, *p. 356*

Similar solids are solids that have the same shape and proportional corresponding dimensions.

EXAMPLE 1 Identifying Similar Solids

Cylinder B

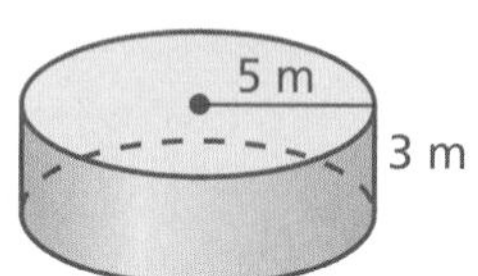

Cylinder C

7.5 m

5 m

Cylinder A

6 m

4 m

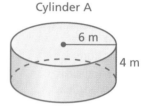

Which cylinder is similar to Cylinder A?

Check to see if corresponding dimensions are proportional.

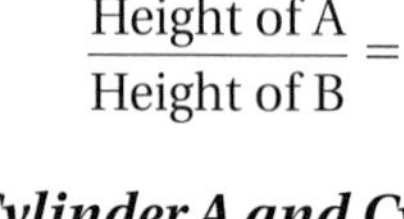

Cylinder A and Cylinder B

$\frac{\text{Height of A}}{\text{Height of B}} = \frac{4}{3}$ $\qquad$ $\frac{\text{Radius of A}}{\text{Radius of B}} = \frac{6}{5}$ $\qquad$ Not proportional

Cylinder A and Cylinder C

$\frac{\text{Height of A}}{\text{Height of C}} = \frac{4}{5}$ $\qquad$ $\frac{\text{Radius of A}}{\text{Radius of C}} = \frac{6}{7.5} = \frac{4}{5}$ $\qquad$ Proportional

So, Cylinder C is similar to Cylinder A.

EXAMPLE 2 Finding Missing Measures in Similar Solids

Cone X

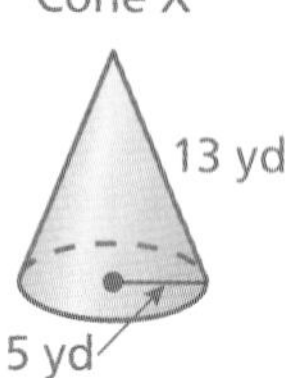

Cone Y

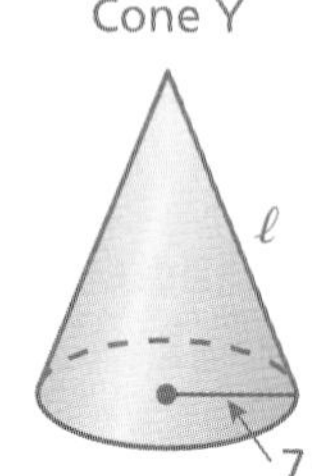

The cones are similar. Find the missing slant height ℓ.

$$\frac{\text{Radius of X}}{\text{Radius of Y}} = \frac{\text{Slant height of X}}{\text{Slant height of Y}}$$

$\frac{5}{7} = \frac{13}{\ell}$ $\qquad$ Substitute.

$5\ell = 91$ $\qquad$ Cross Products Property

$\ell = 18.2$ $\qquad$ Divide each side by 5.

The slant height is 18.2 yards.

On Your Own

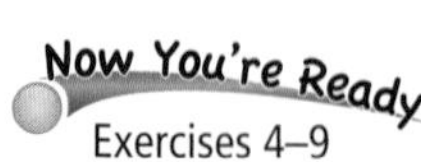

1. Cylinder D has a radius of 7.5 meters and a height of 4.5 meters. Which cylinder in Example 1 is similar to Cylinder D?
2. The prisms at the right are similar. Find the missing width and length.

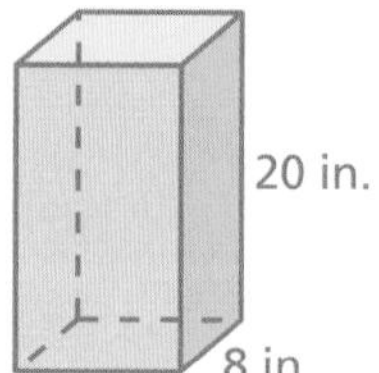

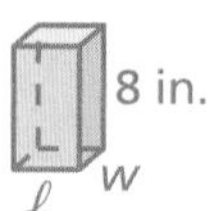

Multi-Language Glossary at BigIdeasMath.com

Key Ideas

Linear Measures

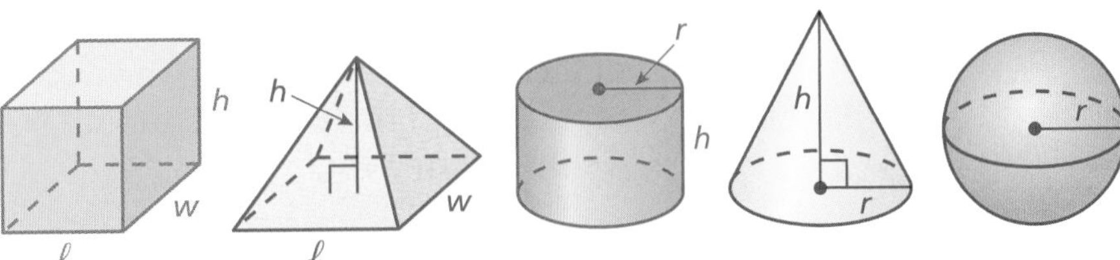

Surface Areas of Similar Solids

When two solids are similar, the ratio of their surface areas is equal to the square of the ratio of their corresponding linear measures.

$$\frac{\text{Surface Area of A}}{\text{Surface Area of B}} = \left(\frac{a}{b}\right)^2$$

Solid A

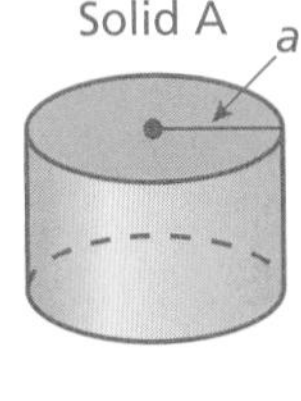

Solid B

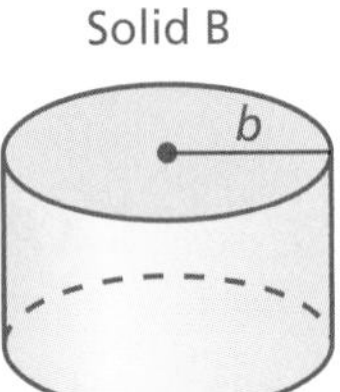

EXAMPLE 3 Finding Surface Area

Pyramid A

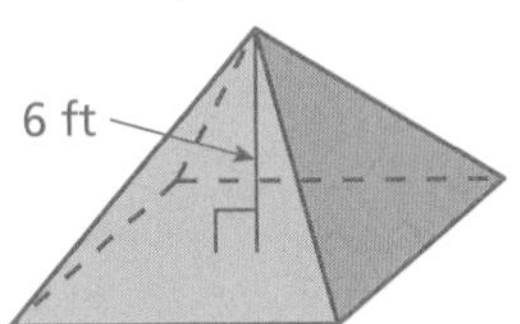

Pyramid B

10 ft

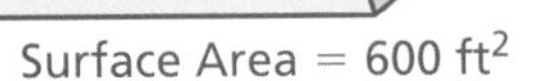

Surface Area = 600 ft²

The pyramids are similar. What is the surface area of Pyramid A?

$$\frac{\text{Surface Area of A}}{\text{Surface Area of B}} = \left(\frac{\text{Height of A}}{\text{Height of B}}\right)^2$$

$$\frac{S}{600} = \left(\frac{6}{10}\right)^2 \qquad \text{Substitute.}$$

$$\frac{S}{600} = \frac{36}{100} \qquad \text{Evaluate.}$$

$$\frac{S}{600} \cdot 600 = \frac{36}{100} \cdot 600 \qquad \text{Multiplication Property of Equality}$$

$$S = 216 \qquad \text{Simplify.}$$

The surface area of Pyramid A is 216 square feet.

On Your Own

The solids are similar. Find the surface area of the red solid. Round your answer to the nearest tenth.

3.

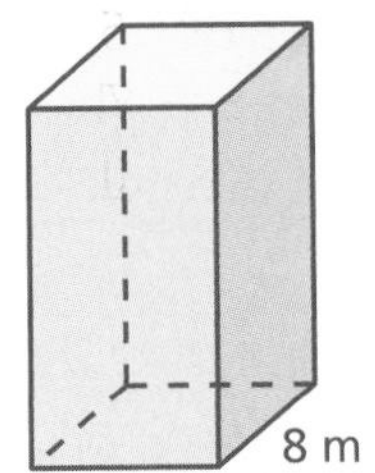

5 m

Surface Area = 608 m²

4.

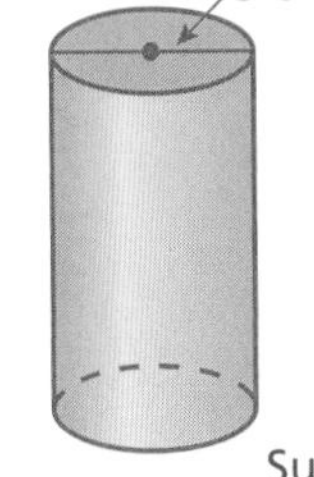

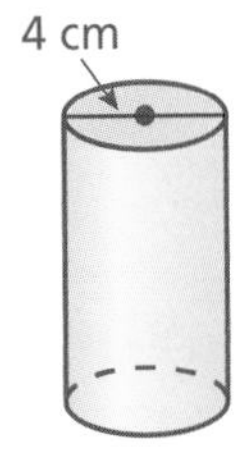

Surface Area = 110 cm²

Key Idea

Volumes of Similar Solids

When two solids are similar, the ratio of their volumes is equal to the cube of the ratio of their corresponding linear measures.

$$\frac{\text{Volume of A}}{\text{Volume of B}} = \left(\frac{a}{b}\right)^3$$

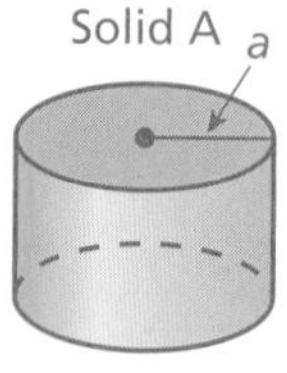

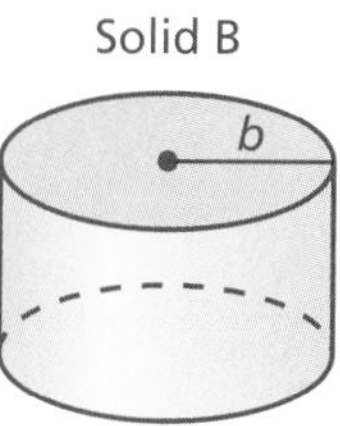

EXAMPLE 4 Finding Volume

Original Tank

Volume = 2000 ft^3

The dimensions of the touch tank at an aquarium are doubled. What is the volume of the new touch tank?

Ⓐ 150 ft^3　　Ⓑ 4000 ft^3

Ⓒ 8000 ft^3　　Ⓓ 16,000 ft^3

The dimensions are doubled, so the ratio of the dimensions of the original tank to the dimensions of the new tank is 1 : 2.

$$\frac{\text{Original volume}}{\text{New volume}} = \left(\frac{\text{Original dimension}}{\text{New dimension}}\right)^3$$

$$\frac{2000}{V} = \left(\frac{1}{2}\right)^3 \qquad \text{Substitute.}$$

$$\frac{2000}{V} = \frac{1}{8} \qquad \text{Evaluate.}$$

$$16{,}000 = V \qquad \text{Cross Products Property}$$

The volume of the new tank is 16,000 cubic feet. So, the correct answer is Ⓓ.

Study Tip

When the dimensions of a solid are multiplied by k, the surface area is multiplied by k^2 and the volume is multiplied by k^3.

On Your Own

Now You're Ready
Exercises 10–13

The solids are similar. Find the volume of the red solid. Round your answer to the nearest tenth.

5.

6.

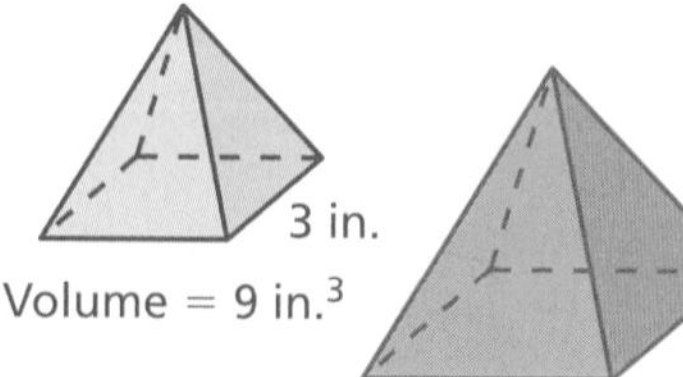

8.4 Exercises

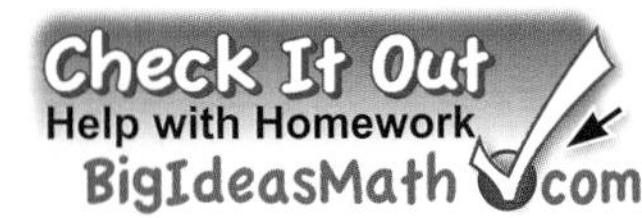

Vocabulary and Concept Check

1. **VOCABULARY** What are similar solids?
2. **OPEN-ENDED** Draw two similar solids and label their corresponding linear measures.

Practice and Problem Solving

3. **NUMBER SENSE** All the dimensions of a cube increase by a factor of $\frac{3}{2}$.
   a. How many times greater is the surface area? Explain.
   b. How many times greater is the volume? Explain.

Determine whether the solids are similar.

1 4.

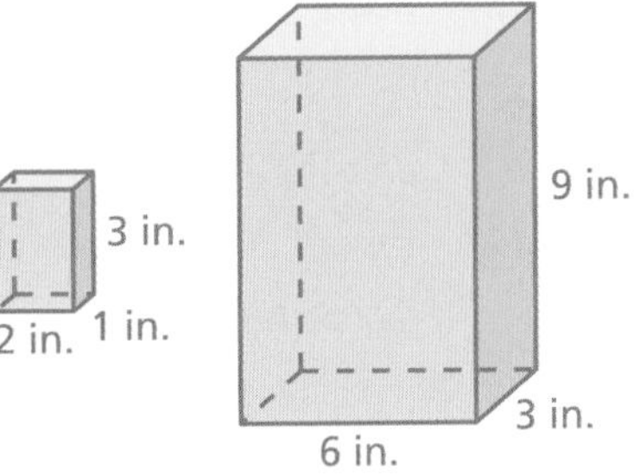

5.

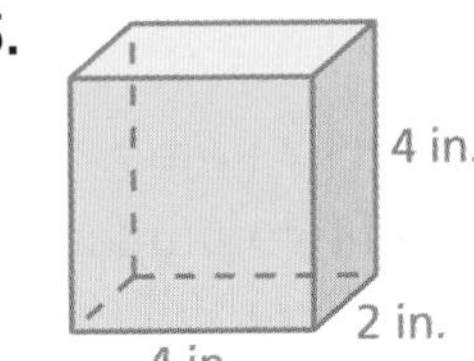

6.

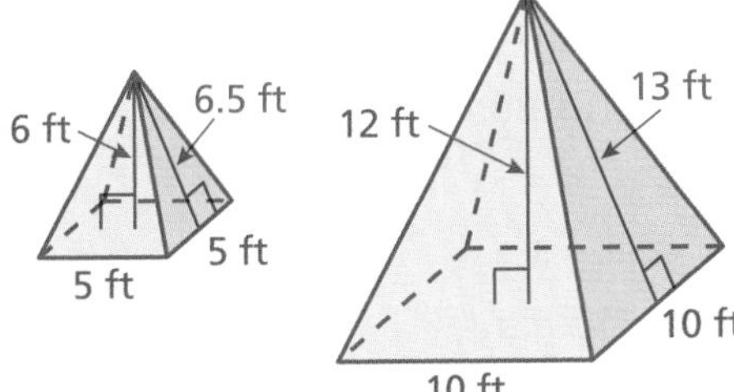

7.

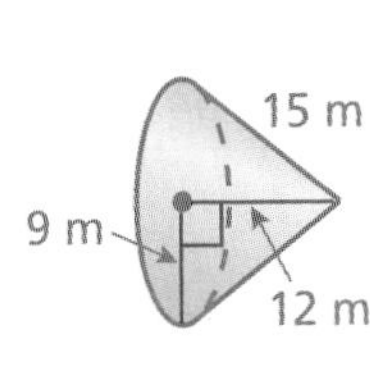

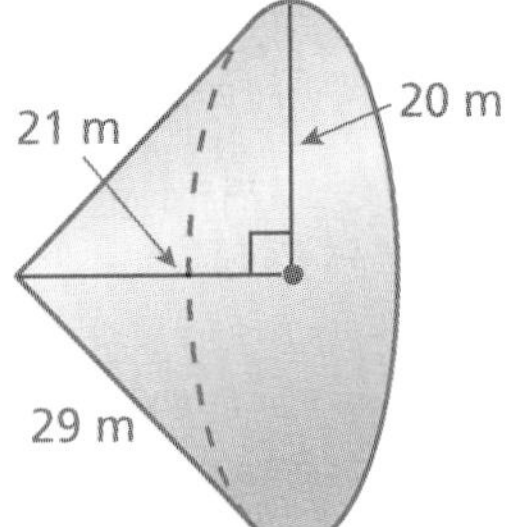

The solids are similar. Find the missing dimension(s).

2 8.

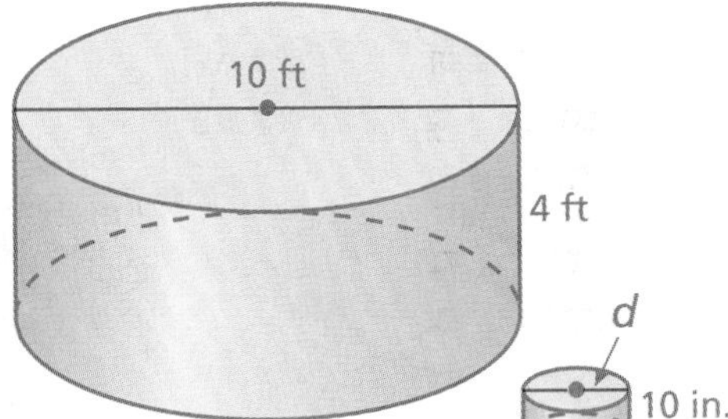

9.

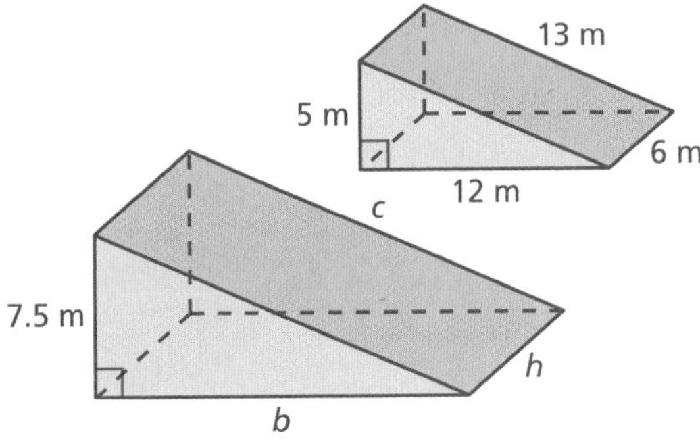

The solids are similar. Find the surface area S or volume V of the red solid. Round your answer to the nearest tenth.

③ ④ **10.**

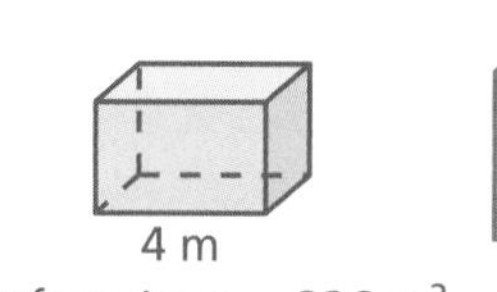

Surface Area = 336 m^2

11.

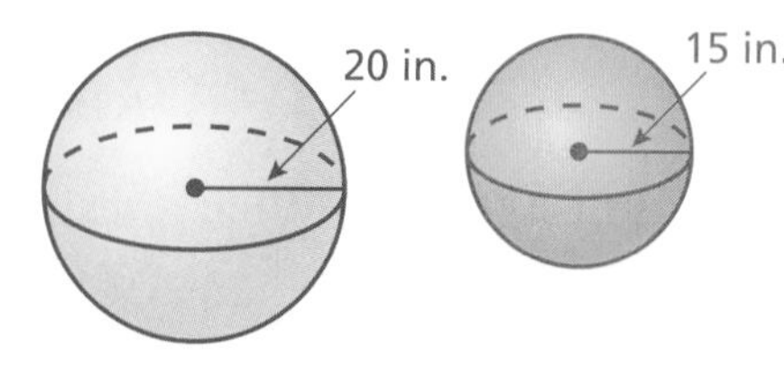

Surface Area = 1800 in.^{2}

12.

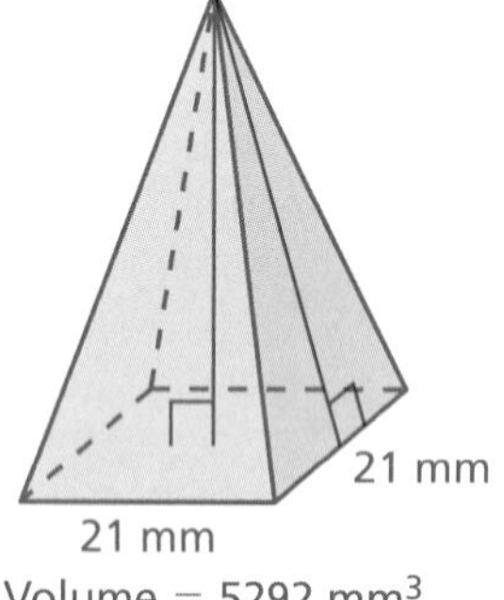

7 mm 7 mm

Volume = 5292 mm^3

13.

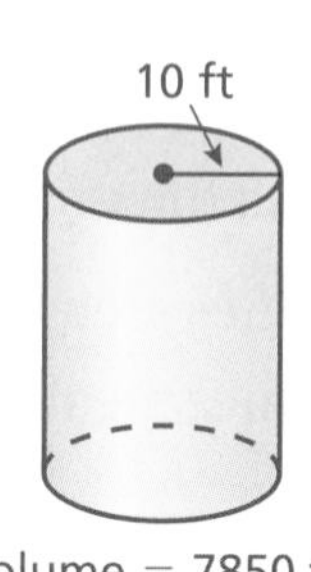

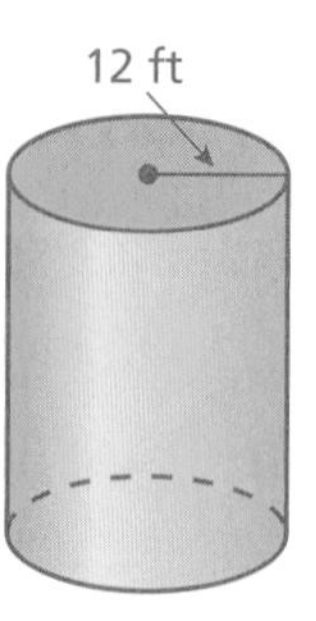

Volume = 7850 ft^3

14. ERROR ANALYSIS The ratio of the corresponding linear measures of two similar solids is 3 : 5. The volume of the smaller solid is 108 cubic inches. Describe and correct the error in finding the volume of the larger solid.

✗

$$\frac{108}{V} = \left(\frac{3}{5}\right)^2$$

$$\frac{108}{V} = \frac{9}{25}$$

$$300 = V$$

The volume of the larger solid is 300 cubic inches.

15. MIXED FRUIT The ratio of the corresponding linear measures of two similar cans of fruit is 4 to 7. The smaller can has a surface area of 220 square centimeters. Find the surface area of the larger can.

16. CLASSIC MUSTANG The volume of a 1968 Ford Mustang GT engine is 390 cubic inches. Which scale model of the Mustang has the greater engine volume, a 1 : 18 scale model or a 1 : 24 scale model? How much greater is it?

17. **MARBLE STATUE** You have a small marble statue of Wolfgang Mozart. It is 10 inches tall and weighs 16 pounds. The original statue is 7 feet tall.

a. Estimate the weight of the original statue. Explain your reasoning.

b. If the original statue were 20 feet tall, how much would it weigh?

Wolfgang Mozart

18. **REPEATED REASONING** The largest doll is 7 inches tall. Each of the other dolls is 1 inch shorter than the next larger doll. Make a table that compares the surface areas and the volumes of the seven dolls.

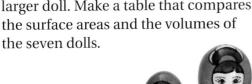

19. Precision You and a friend make paper cones to collect beach glass. You cut out the largest possible three-fourths circle from each piece of paper.

a. Are the cones similar? Explain your reasoning.

b. Your friend says that because your sheet of paper is twice as large, your cone will hold exactly twice the volume of beach glass. Is this true? Explain your reasoning.

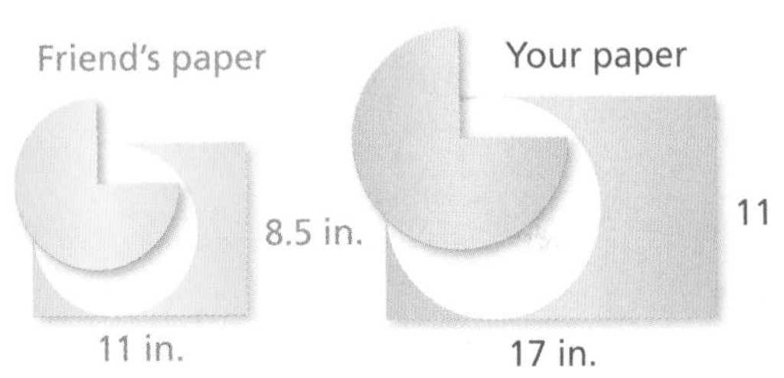

Fair Game Review What you learned in previous grades & lessons

Draw the figure and its reflection in the x-axis. Identify the coordinates of the image. *(Section 2.3)*

20. $A(1, 1)$, $B(3, 4)$, $C(4, 2)$

21. $J(-3, 0)$, $K(-4, 3)$, $L(-1, 4)$

22. **MULTIPLE CHOICE** Which system of linear equations has no solution? *(Section 5.4)*

Ⓐ $y = 4x + 1$; $y = -4x + 1$

Ⓑ $y = 2x - 7$; $y = 2x + 7$

Ⓒ $3x + y = 1$; $6x + 2y = 2$

Ⓓ $5x + y = 3$; $x + 5y = 15$

8.3–8.4 Quiz

Find the volume of the sphere. Round your answer to the nearest tenth. *(Section 8.3)*

1. 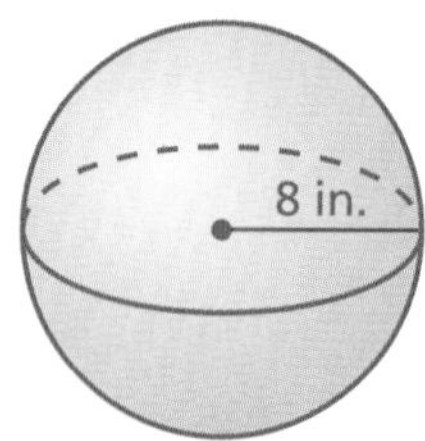

2. 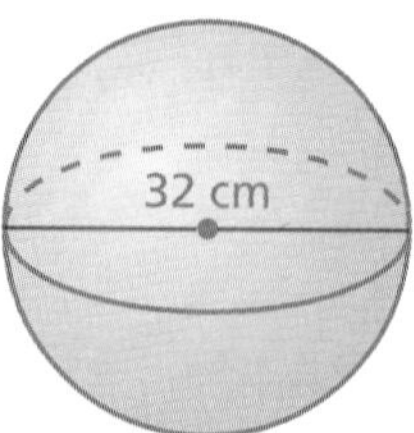

Find the radius of the sphere with the given volume. *(Section 8.3)*

3. Volume $= 4500\pi$ yd^3

4. Volume $= \frac{32}{3}\pi$ ft^3

5. Find the volume of the composite solid. Round your answer to the nearest tenth. *(Section 8.3)*

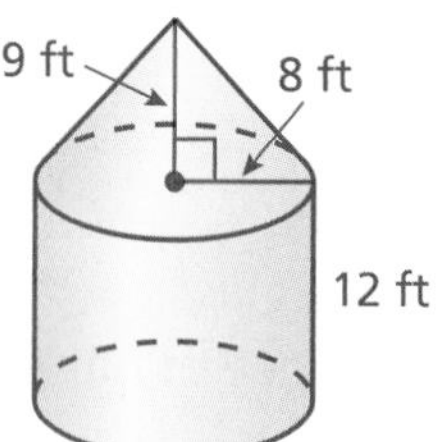

6. Determine whether the solids are similar. *(Section 8.4)*

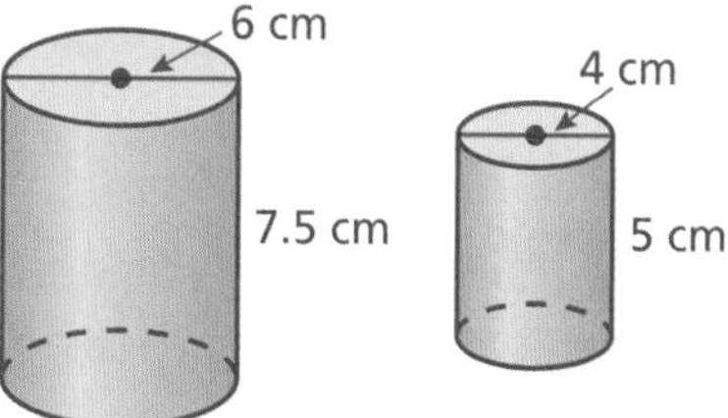

7. The prisms are similar. Find the missing width and height. *(Section 8.4)*

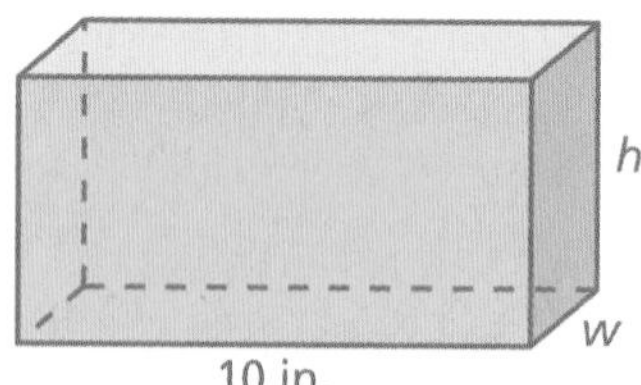

8. The solids are similar. Find the surface area of the red solid. *(Section 8.4)*

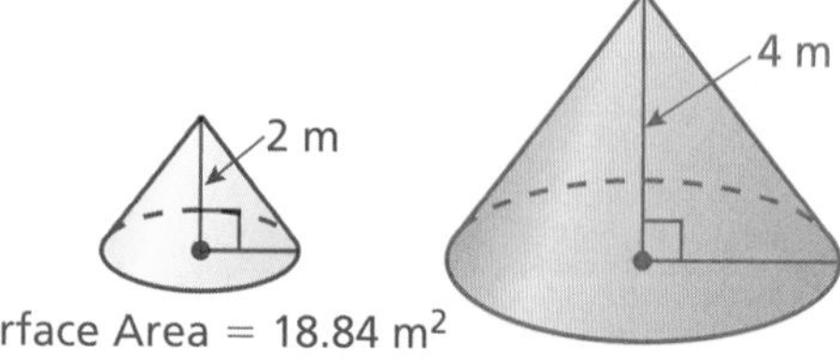

9. **HAMSTER** A hamster toy is in the shape of a sphere. What is the volume of the toy? Round your answer to the nearest whole number. *(Section 8.3)*

10. **JEWELRY BOXES** The ratio of the corresponding linear measures of two similar jewelry boxes is 2 to 3. The larger box has a volume of 162 cubic inches. Find the volume of the smaller jewelry box. *(Section 8.4)*

11. **ARCADE** You win a token after playing an arcade game. What is the volume of the gold ring? Round your answer to the nearest tenth. *(Section 8.3)*

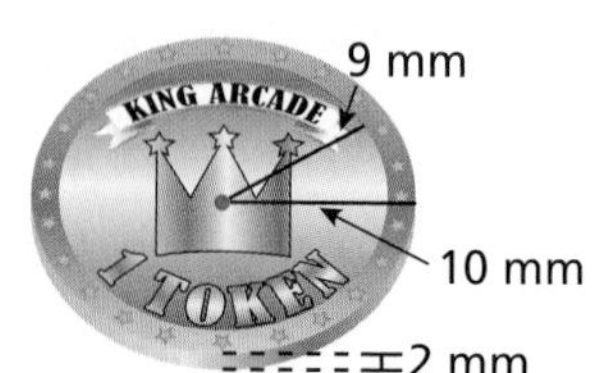

8 Chapter Review

Check It Out
Vocabulary Help
BigIdeasMath.com

Review Key Vocabulary

sphere, *p. 348* hemisphere, *p. 351* similar solids, *p. 356*

Review Examples and Exercises

8.1 Volumes of Cylinders *(pp. 334–339)*

Find the volume of the cylinder. Round your answer to the nearest tenth.

$V = Bh$	Write formula for volume.
$= \pi(2)^2(8)$	Substitute.
$= 32\pi \approx 100.5$	Use a calculator.

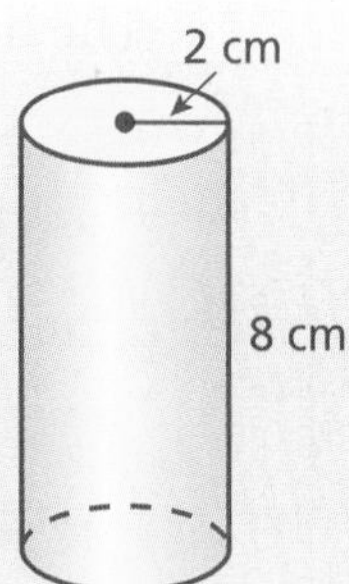

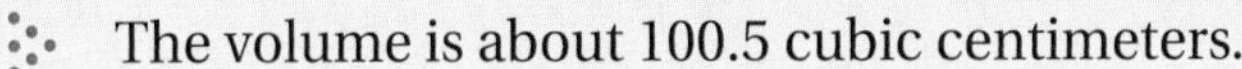

∴ The volume is about 100.5 cubic centimeters.

Exercises

Find the volume of the cylinder. Round your answer to the nearest tenth.

1. 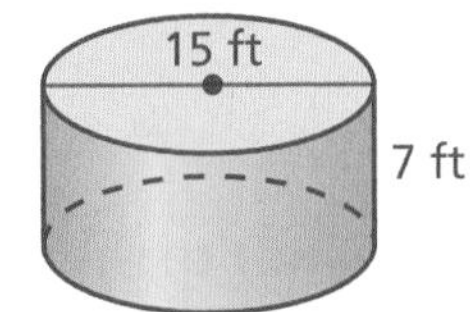

2. 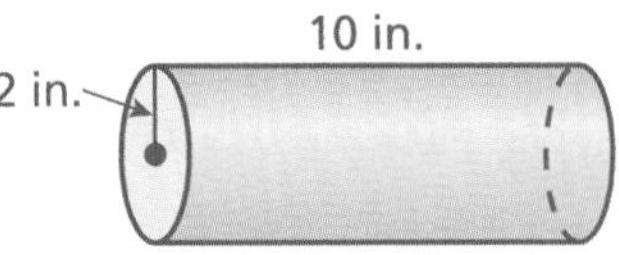

3. 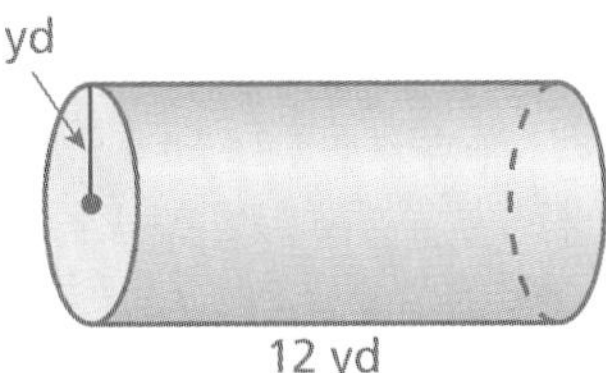

4. 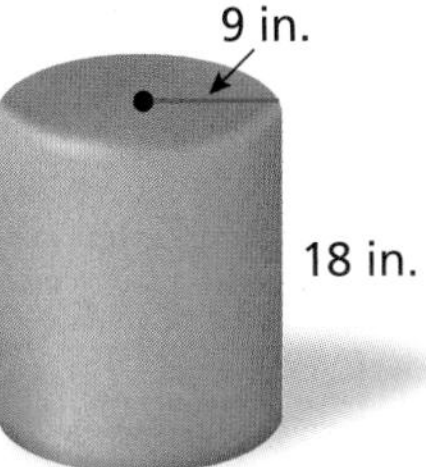

Find the missing dimension of the cylinder. Round your answer to the nearest whole number.

5. Volume = 25 in.^{3}

6. Volume = 7599 m^3

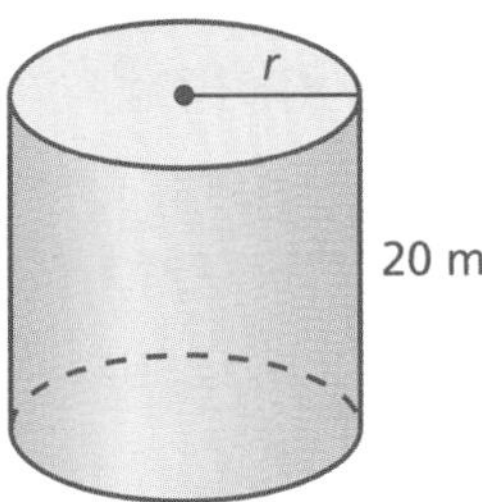

8.2 Volumes of Cones (pp. 340–345)

Find the height of the cone. Round your answer to the nearest tenth.

$V = \frac{1}{3}Bh$ — Write formula for volume.

$900 = \frac{1}{3}\pi(6)^2(h)$ — Substitute.

$900 = 12\pi h$ — Simplify.

$23.9 \approx h$ — Divide each side by 12π.

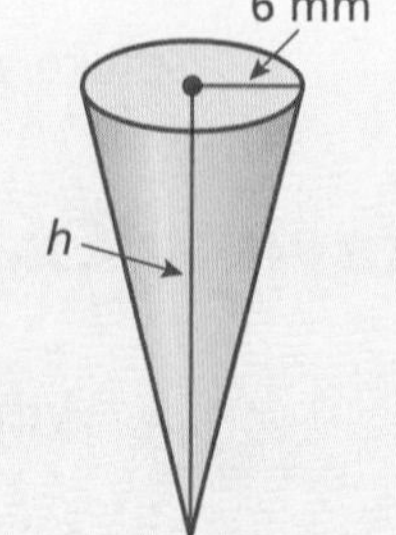

The height is about 23.9 millimeters.

Exercises

Find the volume V or height h of the cone. Round your answer to the nearest tenth.

7\.

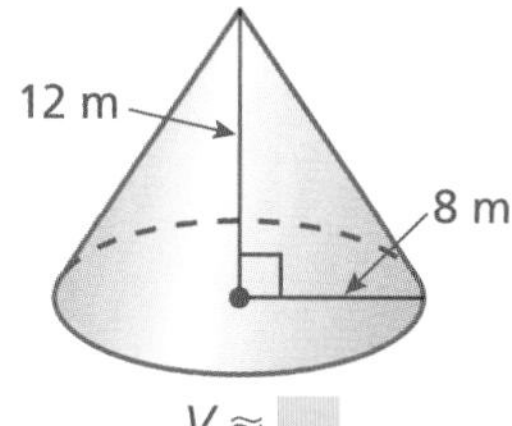

8\.

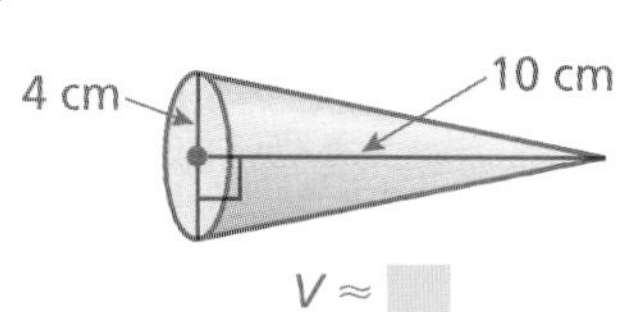

9\.

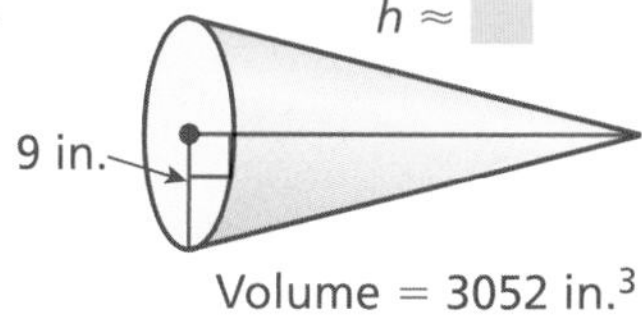

8.3 Volumes of Spheres (pp. 348–353)

a. Find the volume of the sphere. Round your answer to the nearest tenth.

$V = \frac{4}{3}\pi r^3$ — Write formula for volume.

$= \frac{4}{3}\pi(11)^3$ — Substitute 11 for r.

$= \frac{5324}{3}\pi$ — Simplify.

≈ 5575.3 — Use a calculator.

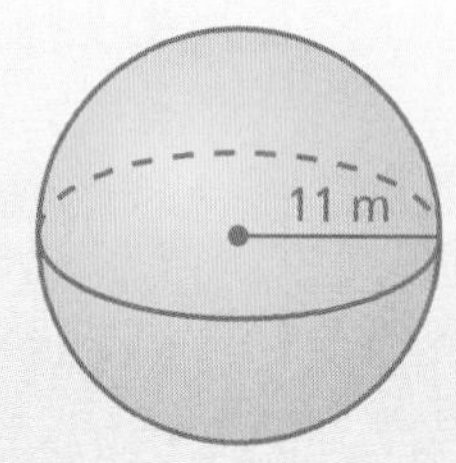

The volume is about 5575.3 cubic meters.

b. Find the volume of the composite solid. Round your answer to the nearest tenth.

Square Prism	***Cylinder***
$V = Bh$	$V = Bh$
$= (12)(12)(9)$	$= \pi(5)^2(9)$
$= 1296$	$= 225\pi \approx 706.9$

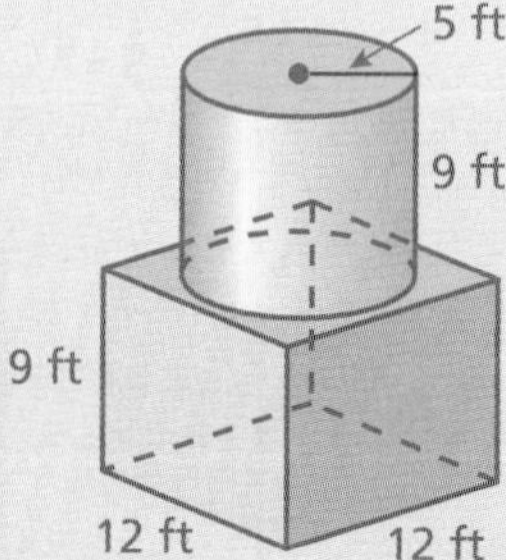

So, the volume is about $1296 + 706.9 = 2002.9$ cubic feet.

Exercises

Find the volume *V* or radius *r* of the sphere. Round your answer to the nearest tenth, if necessary.

10.

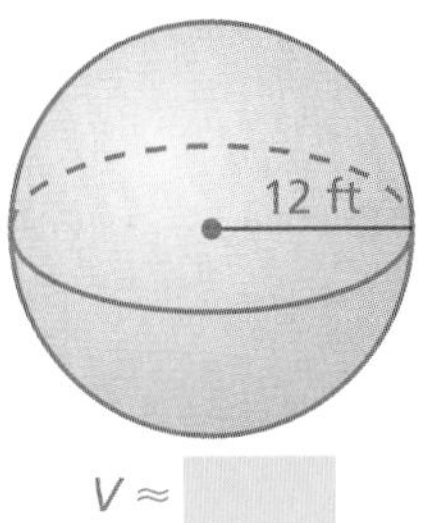

11.

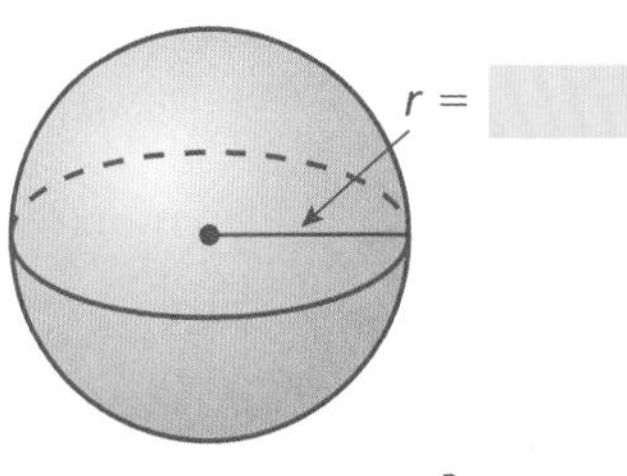

Find the volume of the composite solid. Round your answer to the nearest tenth.

12.

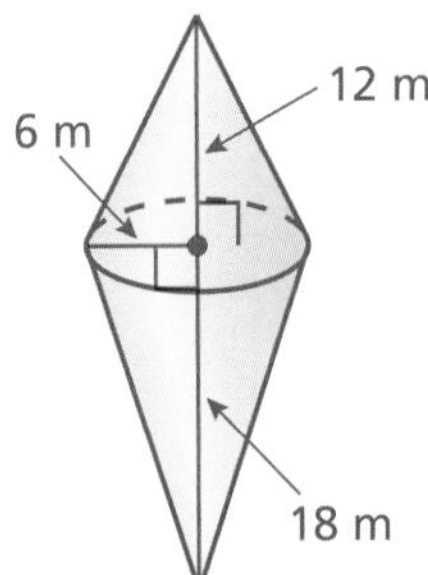

13.

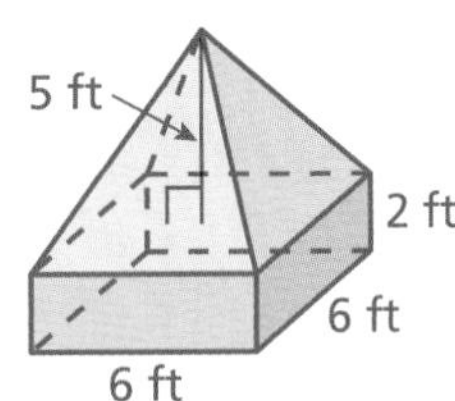

14.

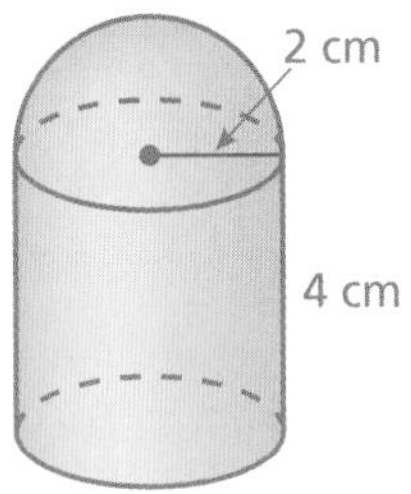

8.4 Surface Areas and Volumes of Similar Solids *(pp. 354–361)*

The cones are similar. What is the volume of the red cone? Round your answer to the nearest tenth.

$$\frac{\text{Volume of A}}{\text{Volume of B}} = \left(\frac{\text{Height of A}}{\text{Height of B}}\right)^3$$

$\frac{V}{157} = \left(\frac{4}{6}\right)^3$ Substitute.

$\frac{V}{157} = \frac{64}{216}$ Evaluate.

$V \approx 46.5$ Solve for V.

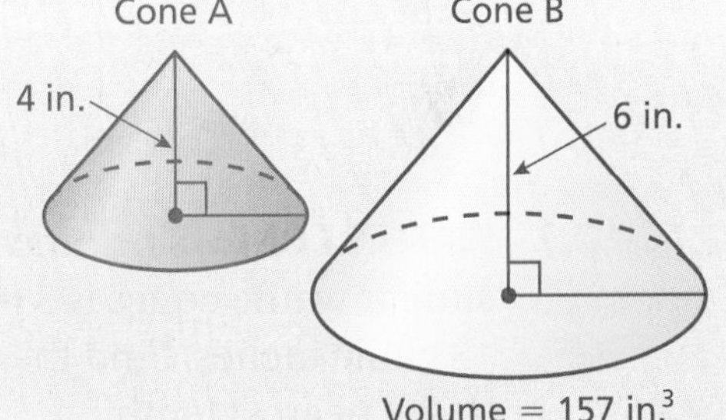

∴ The volume is about 46.5 cubic inches.

Exercises

The solids are similar. Find the surface area *S* or volume *V* of the red solid. Round your answer to the nearest tenth.

15.

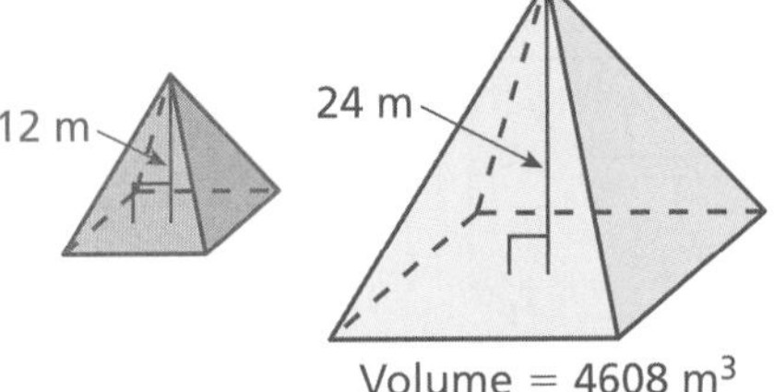

16.

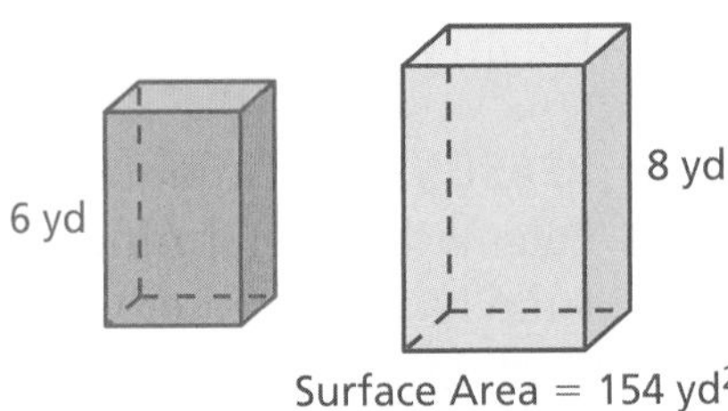

8 Chapter Test

Find the volume of the solid. Round your answer to the nearest tenth.

1.

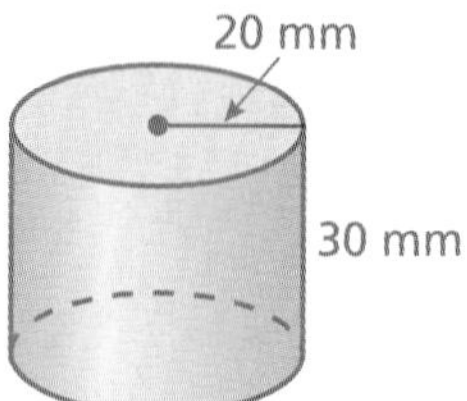

2.

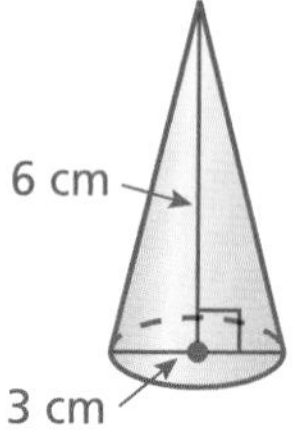

3.

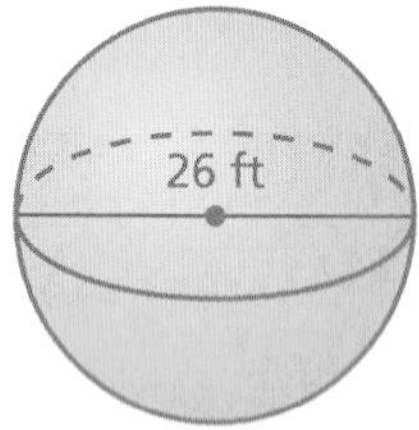

4.

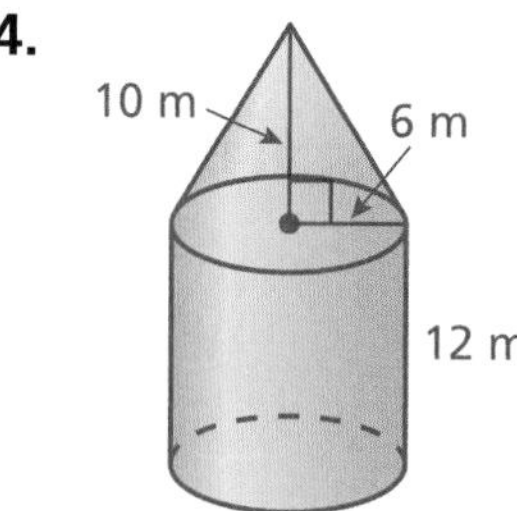

5. The pyramids are similar.

a. Find the missing dimension.

b. Find the surface area of the red pyramid.

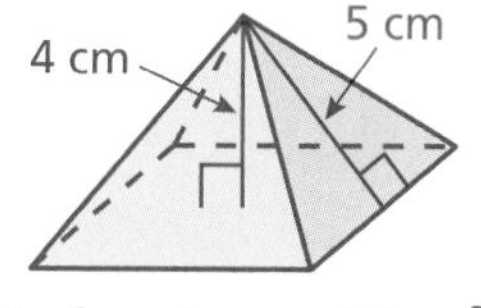

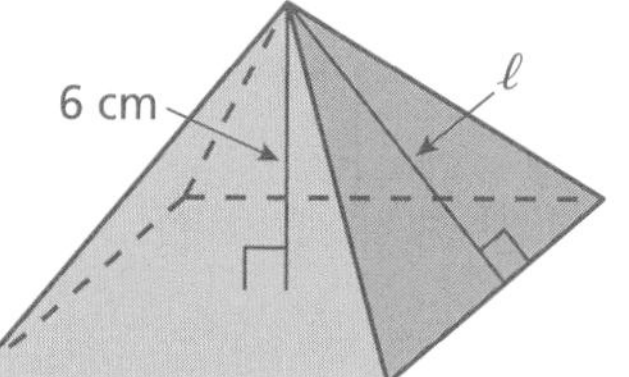

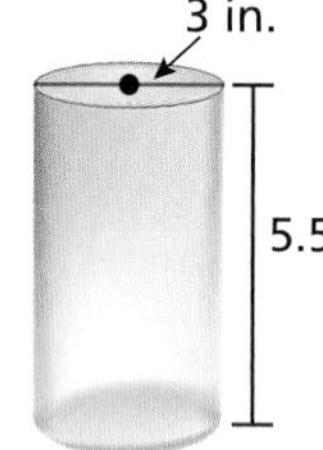

6. SMOOTHIES You are making smoothies. You will use either the cone-shaped glass or the cylindrical glass. Which glass holds more? About how much more?

7. WAFFLE CONES The ratio of the corresponding linear measures of two similar waffle cones is 3 to 4. The smaller cone has a volume of about 18 cubic inches. Find the volume of the larger cone. Round your answer to the nearest tenth.

8. OPEN-ENDED Draw two different composite solids that have the same volume but different surface areas. Explain your reasoning.

9. MILK Glass A has a diameter of 3.5 inches and a height of 4 inches. Glass B has a radius of 1.5 inches and a height of 5 inches. Which glass can hold more milk?

10. REASONING Without calculating, determine which solid has the greater volume. Explain your reasoning.

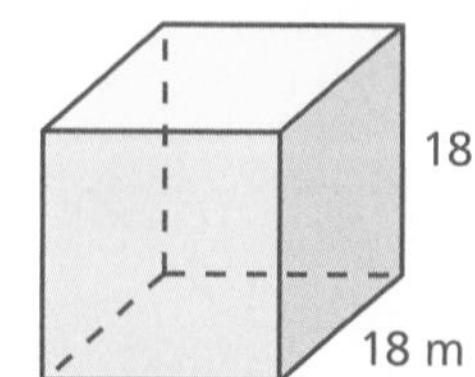

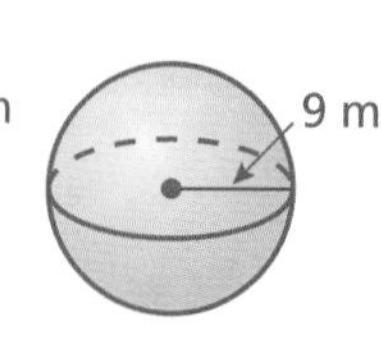

8 Cumulative Assessment

Test-Taking Strategy
After Answering Easy Questions, Relax

How much catnip fits in a cylinder whose radius is 1 inch and height is 2 inches?
Ⓐ 2π in.³ Ⓑ 4π in.³ Ⓒ 8π in.³ Ⓓ 2 in.³
Catnip pie, yummy for me!

"After answering the easy questions, relax and try the harder ones. For this, $\pi r^2 h = 2\pi$. So, it's A."

1. What value of w makes the equation below true?

$$\frac{w}{3} = 3(w - 1) - 1$$

A. $\frac{1}{2}$

B. $\frac{3}{4}$

C. $\frac{5}{4}$

D. $\frac{3}{2}$

2. A right circular cone and its dimensions are shown below.

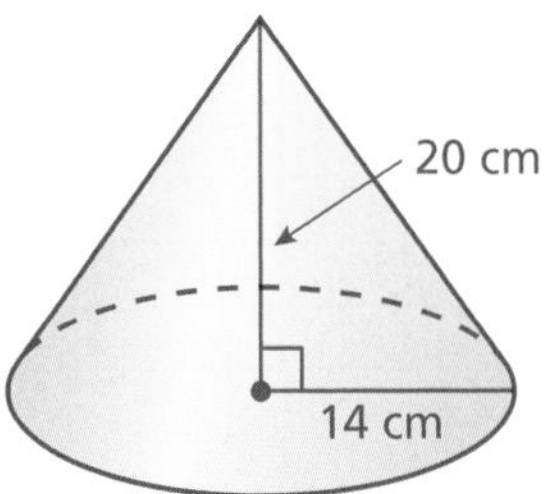

What is the volume of the right circular cone? $\left(\text{Use } \frac{22}{7} \text{ for } \pi.\right)$

F. $1{,}026\frac{2}{3}$ cm^3

G. $3{,}080$ cm^3

H. $4{,}106\frac{2}{3}$ cm^3

I. $12{,}320$ cm^3

3. Patricia solved the equation in the box shown.

What should Patricia do to correct the error that she made?

$$-\frac{3}{2}(8x - 10) = -20$$
$$8x - 10 = -20\left(-\frac{3}{2}\right)$$
$$8x - 10 = 30$$
$$8x - 10 + 10 = 30 + 10$$
$$8x = 40$$
$$\frac{8x}{8} = \frac{40}{8}$$
$$x = 5$$

A. Add 10 to -20.

B. Distribute $-\frac{3}{2}$ to get $-12x - 15$.

C. Multiply both sides by $-\frac{2}{3}$ instead of $-\frac{3}{2}$.

D. Multiply both sides by $\frac{3}{2}$ instead of $-\frac{3}{2}$.

4. On the grid below, Rectangle *EFGH* is plotted and its vertices are labeled.

E (−1, 2)
F (3, 2)
H (−1, −3)
G (3, −3)

Which of the following shows Rectangle *E′F′G′H′*, the image of Rectangle *EFGH* after it is reflected in the *x*-axis?

F.

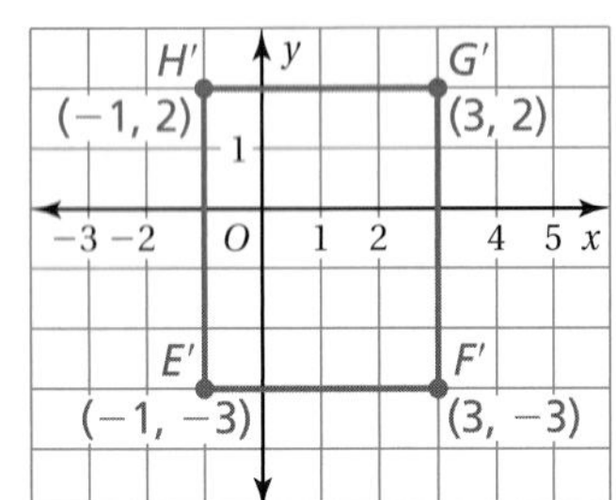

H.

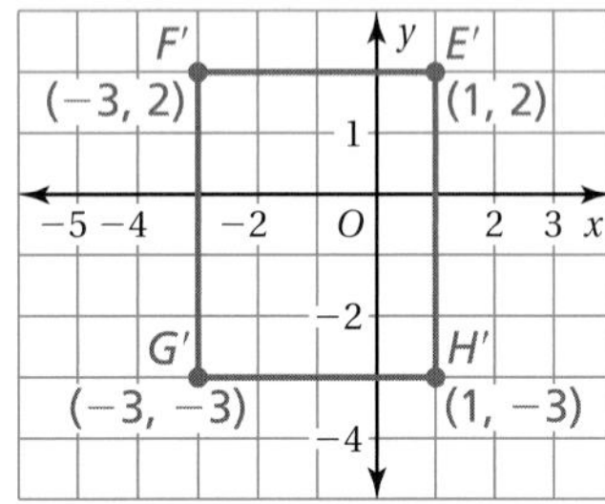

G.

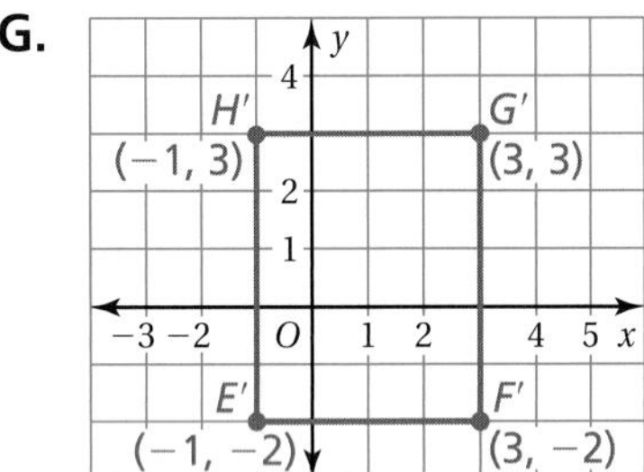

I.

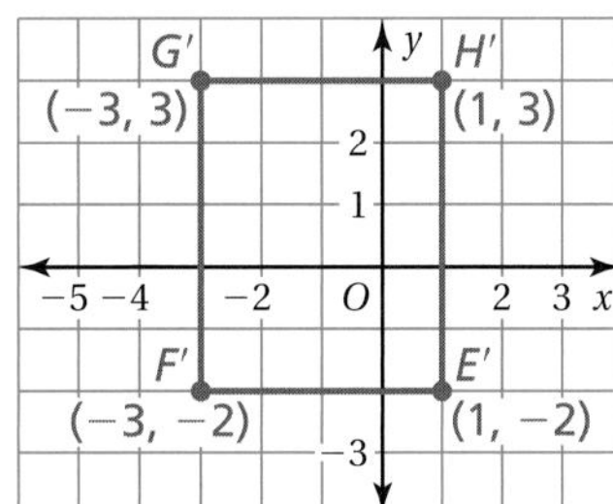

5. List the ordered pairs shown in the mapping diagram below.

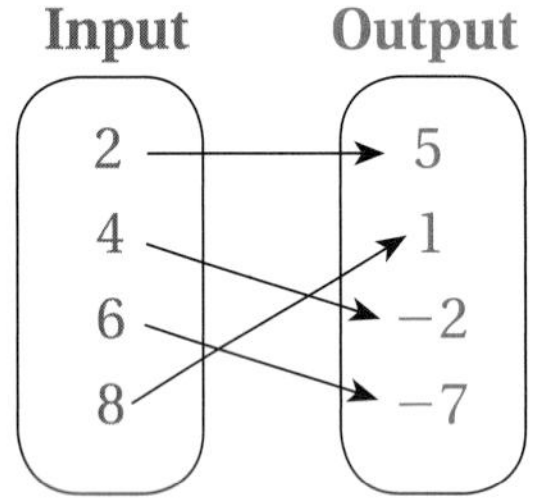

A. (2, 5), (4, −2), (6, −7), (8, 1)

B. (2, −7), (4, −2), (6, 1), (8, 5)

C. (2, 5), (4, 1), (6, −2), (8, −7)

D. (5, 2), (−2, 4), (−7, 6), (1, 8)

6. The temperature fell from 54 degrees Fahrenheit to 36 degrees Fahrenheit over a 6-hour period. The temperature fell by the same number of degrees each hour. How many degrees Fahrenheit did the temperature fall each hour?

7. Solve the formula below for I.

$$A = P + PI$$

F. $I = A - 2P$

G. $I = \dfrac{A}{P} - P$

H. $I = A - \dfrac{P}{P}$

I. $I = \dfrac{A - P}{P}$

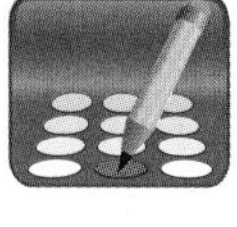

8. A right circular cylinder has a volume of 1296 cubic inches. If you divide the radius of the cylinder by 12, what would be the volume, in cubic inches, of the smaller cylinder?

9. Which graph represents a linear function?

A.

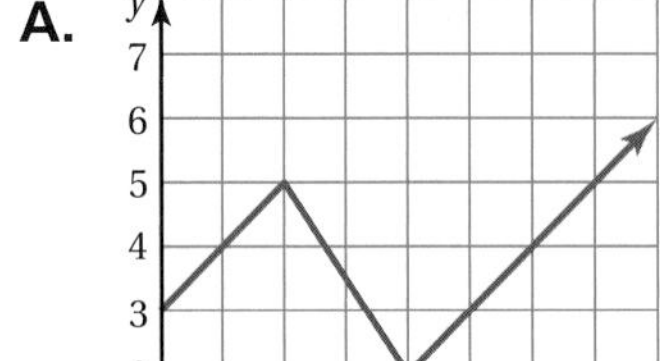
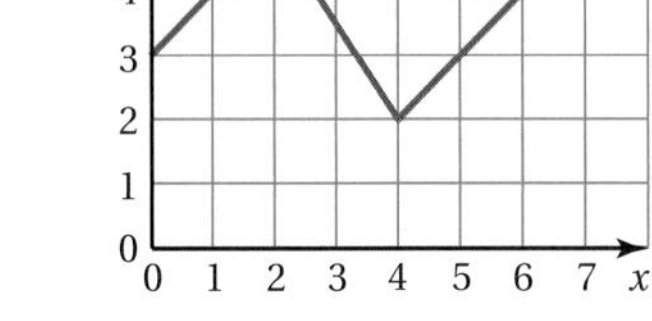

B.

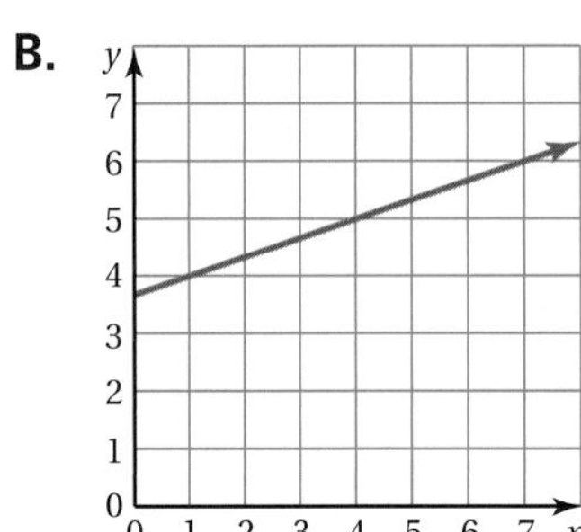

C.

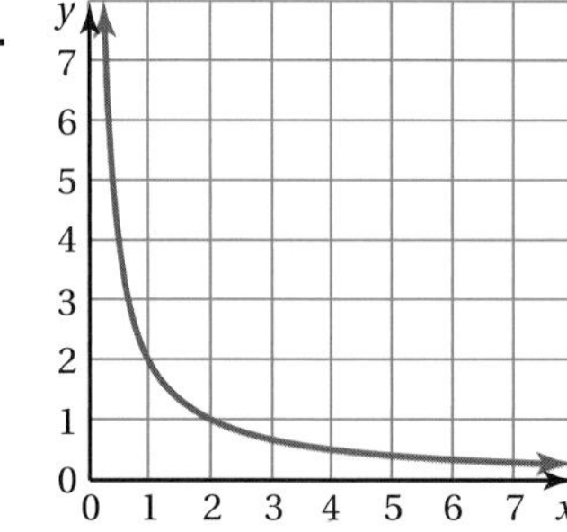

D.

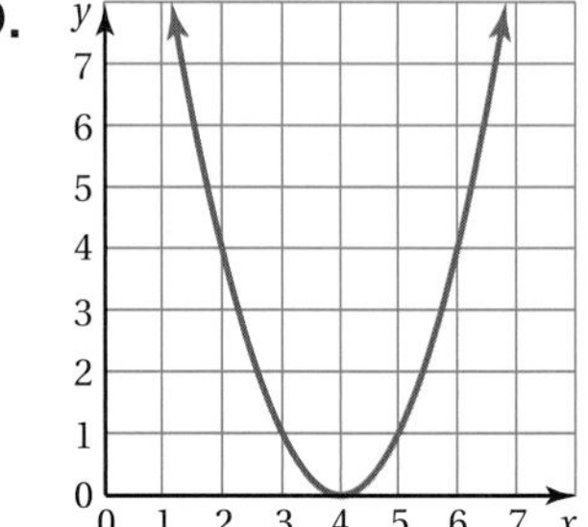

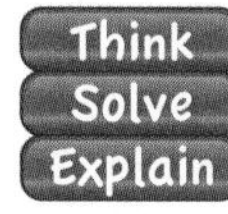

10. The figure below is a diagram for making a tin lantern.

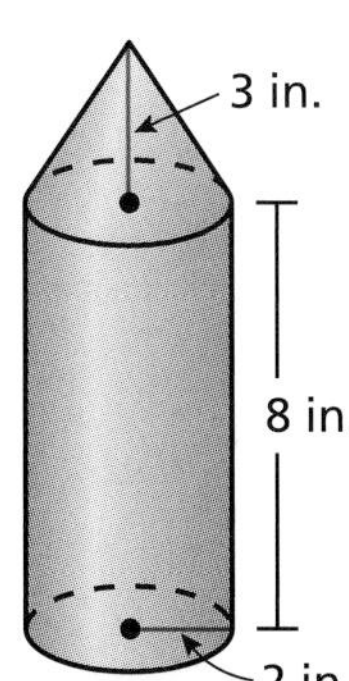

The figure consists of a right circular cylinder without its top base and a right circular cone without its base. What is the volume, in cubic inches, of the entire lantern? Show your work and explain your reasoning. (Use 3.14 for π.)

9 Data Analysis and Displays

"Wow. The number of minutes I can dog paddle is growing like crazy!"

"But I have a really good feeling about November."

What You Learned Before

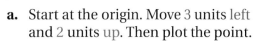

"Here's an interesting survey about favorite dog toys."

Plotting Points

Example 1 **Plot (a) (−3, 2) and (b) (4, −2.5) in a coordinate plane. Describe the location of each point.**

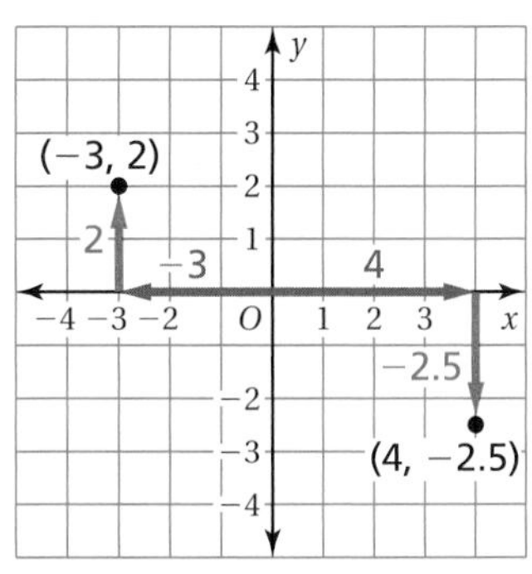

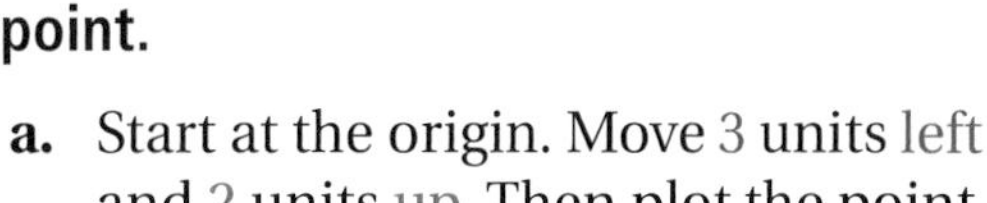

a. Start at the origin. Move 3 units left and 2 units up. Then plot the point.

∴ The point is in Quadrant II.

b. Start at the origin. Move 4 units right and −2.5 units down. Then plot the point.

∴ The point is in Quadrant IV.

Try It Yourself

Plot the ordered pair in a coordinate plane. Describe the location of the point.

1. (1, 3) **2.** (−2, 4) **3.** (1, −3.5) **4.** $\left(-1\frac{3}{4}, -2\frac{1}{4}\right)$

Writing an Equation Using Two Points

Example 2 **Write in slope-intercept form an equation of the line that passes through the points (4, 2) and (−1, −8).**

Find the slope:

$$m = \frac{y_2 - y_1}{x_2 - x_1} = \frac{-8 - 2}{-1 - 4} = \frac{-10}{-5} = 2$$

Then use the slope $m = 2$ and the point (4, 2) to write an equation of the line.

$y - y_1 = m(x - x_1)$	Write the point-slope form.
$y - 2 = 2(x - 4)$	Substitute 2 for m, 4 for x_1, and 2 for y_1.
$y - 2 = 2x - 8$	Distributive Property
$y = 2x - 6$	Write in slope-intercept form.

Try It Yourself

Write in slope-intercept form an equation of the line that passes through the given points.

5. (−1, 4), (3, 8) **6.** (0, −1), (−8, −2) **7.** (6, 8), (3, −9)

9.1 Scatter Plots

Essential Question How can you construct and interpret a scatter plot?

1 ACTIVITY: Constructing a Scatter Plot

Work with a partner. The weights x (in ounces) and circumferences C (in inches) of several sports balls are shown.

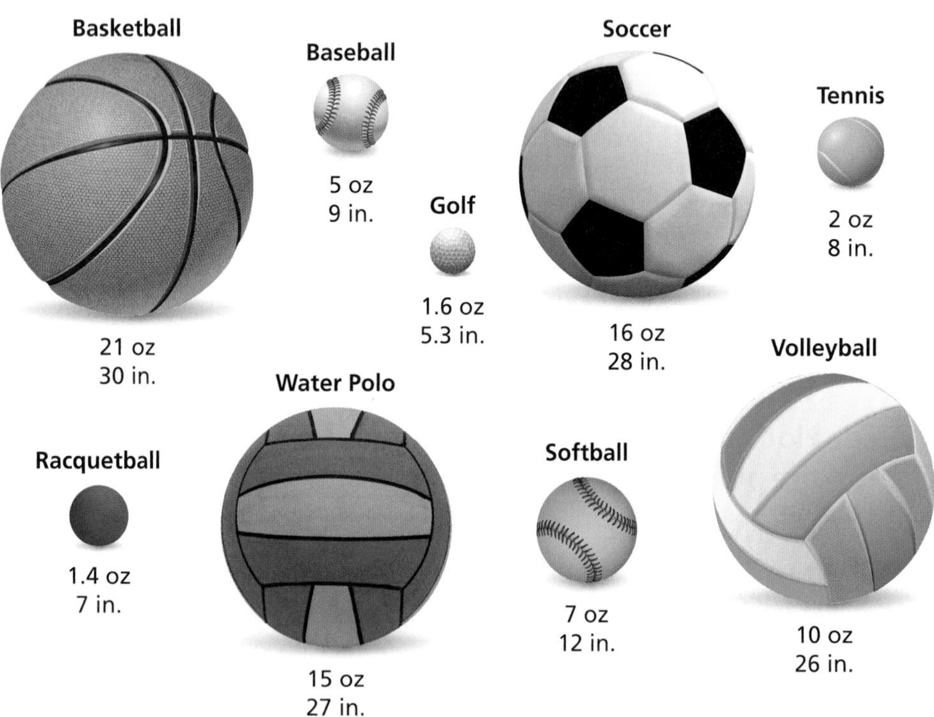

a. Choose a scale for the horizontal axis and the vertical axis of the coordinate plane shown.

b. Write the weight x and circumference C of each ball as an ordered pair. Then plot the ordered pairs in the coordinate plane.

c. Describe the relationship between weight and circumference. Are any of the points close together?

d. In general, do you think you can describe this relationship as *positive* or *negative*? *linear* or *nonlinear*? Explain.

Circumference (inches)

Weight (ounces)

e. A bowling ball has a weight of 225 ounces and a circumference of 27 inches. Describe the location of the ordered pair that represents this data point in the coordinate plane. How does this point compare to the others? Explain your reasoning.

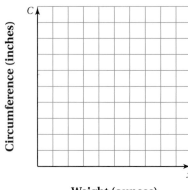

Data Analysis

In this lesson, you will
- construct and interpret scatter plots.
- describe patterns in scatter plots.

2 ACTIVITY: Constructing a Scatter Plot

Math Practice

Recognize Usefulness of Tools

How do you know when a scatter plot is a useful tool for making a prediction?

Work with a partner. The table shows the number of absences and the final grade for each student in a sample.

Absences	Final Grade
0	95
3	88
2	90
5	83
7	79
9	70
4	85
1	94
10	65
8	75

a. Write the ordered pairs from the table. Then plot them in a coordinate plane.

b. Describe the relationship between absences and final grade. How is this relationship similar to the relationship between weight and circumference in Activity 1? How is it different?

c. **MODELING** A student has been absent 6 days. Use the data to predict the student's final grade. Explain how you found your answer.

3 ACTIVITY: Identifying Scatter Plots

Work with a partner. Match the data sets with the most appropriate scatter plot. Explain your reasoning.

a. month of birth and birth weight for infants at a day care

b. quiz score and test score of each student in a class

c. age and value of laptop computers

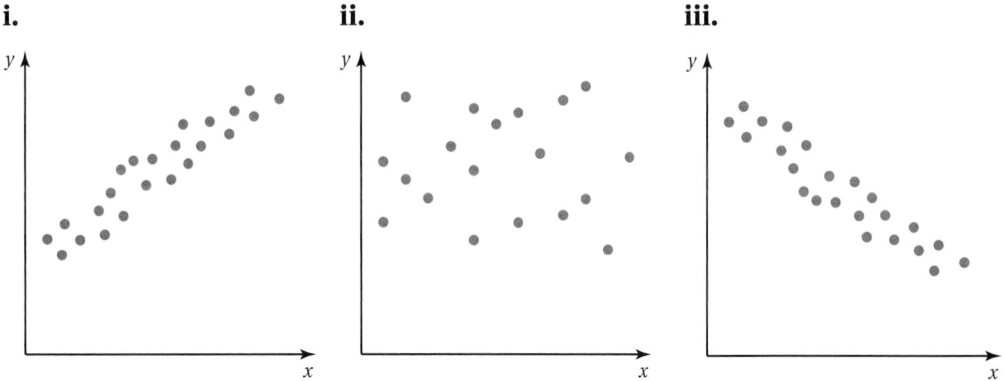

What Is Your Answer?

4. How would you define the term *scatter plot*?

5. **IN YOUR OWN WORDS** How can you construct and interpret a scatter plot?

Practice Use what you learned about scatter plots to complete Exercise 7 on page 376.

9.1 Lesson

Key Vocabulary
scatter plot, *p. 374*

Scatter Plot

A **scatter plot** is a graph that shows the relationship between two data sets. The two sets of data are graphed as ordered pairs in a coordinate plane.

EXAMPLE  Interpreting a Scatter Plot

Restaurant Sandwiches

Calories (y: 0, 300, 350, 400, 450, 500, 550, 600, 650, 700, 750, 800)

Fat (grams) (x: 0, 5, 10, 15, 20, 25, 30, 35, 40, 45)

The scatter plot at the left shows the amounts of fat (in grams) and the numbers of calories in 12 restaurant sandwiches.

a. How many calories are in the sandwich that contains 17 grams of fat?

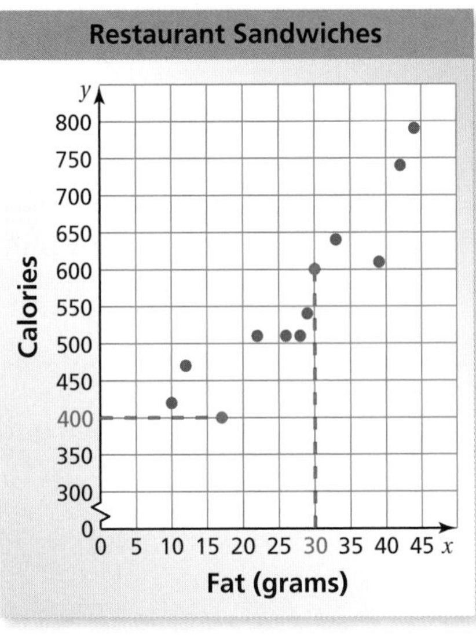

Draw a horizontal line from the point that has an x-value of 17. It crosses the y-axis at 400.

So, the sandwich has 400 calories.

b. How many grams of fat are in the sandwich that contains 600 calories?

Draw a vertical line from the point that has a y-value of 600. It crosses the x-axis at 30.

So, the sandwich has 30 grams of fat.

c. What tends to happen to the number of calories as the number of grams of fat increases?

Looking at the graph, the plotted points go up from left to right.

So, as the number of grams of fat increases, the number of calories increases.

On Your Own

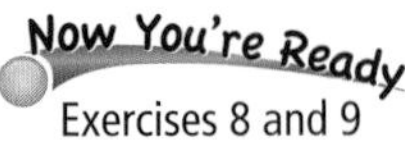

1. **WHAT IF?** A sandwich has 650 calories. Based on the scatter plot in Example 1, how many grams of fat would you expect the sandwich to have? Explain your reasoning.

Multi-Language Glossary at BigIdeasMath.com

A scatter plot can show that a relationship exists between two data sets.

Positive Linear Relationship

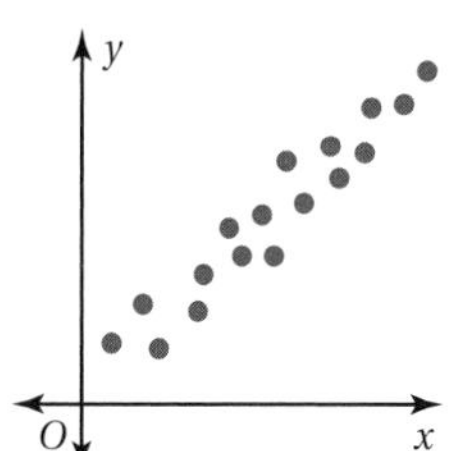

The points lie close to a line. As x increases, y increases.

Negative Linear Relationship

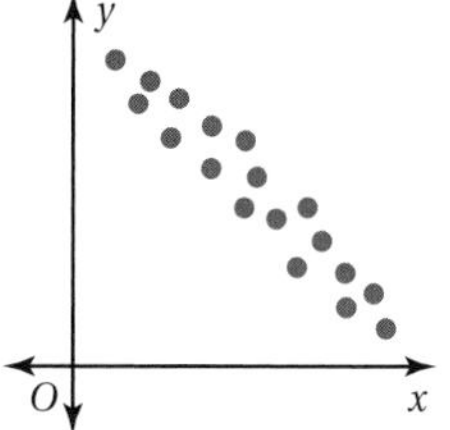

The points lie close to a line. As x increases, y decreases.

Nonlinear Relationship

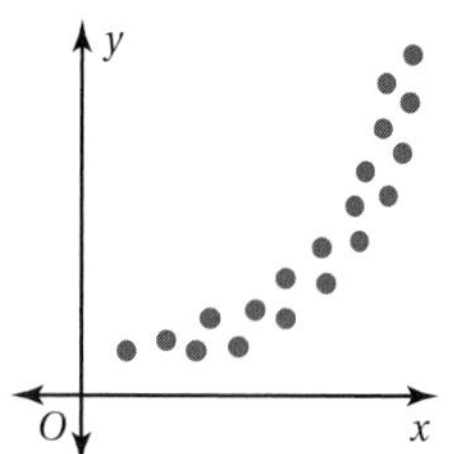

The points lie in the shape of a curve.

No Relationship

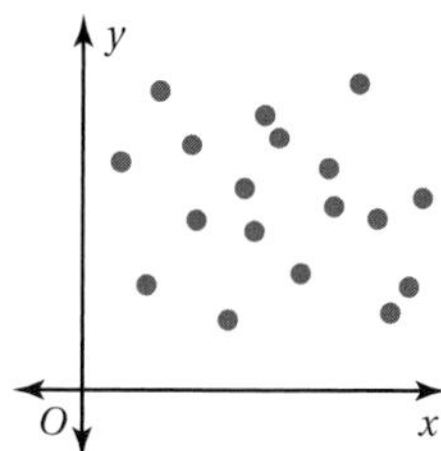

The points show no pattern.

EXAMPLE 2 Identifying Relationships

Describe the relationship between the data. Identify any outliers, gaps, or clusters.

a. television size and price

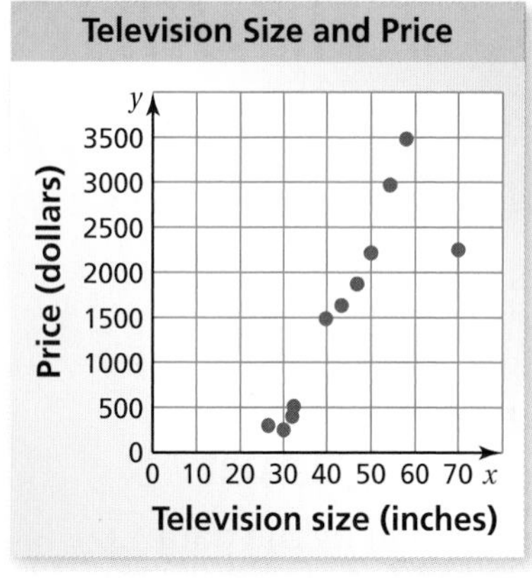

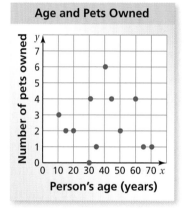

The points appear to lie close to a line. As x increases, y increases.

So, the scatter plot shows a positive linear relationship. There is an outlier at (70, 2250), a cluster of data under $500, and a gap in the data from $500 to $1500.

b. age and number of pets owned

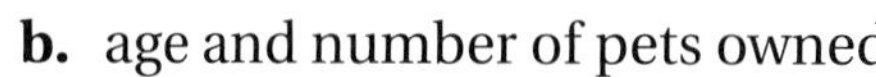

Age and Pets Owned

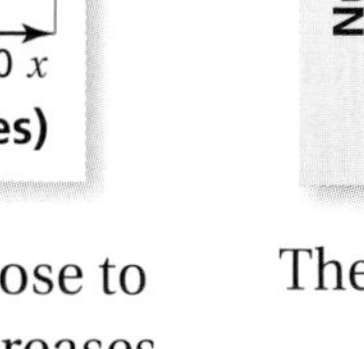

The points show no pattern.

So, the scatter plot shows no relationship. There are no obvious outliers, gaps, or clusters in the data.

On Your Own

2. Make a scatter plot of the data and describe the relationship between the data. Identify any outliers, gaps, or clusters.

Study Time (min), x	30	20	60	90	45	10	30	75	120	80
Test Score, y	80	74	92	97	85	62	83	90	70	91

9.1 Exercises

Vocabulary and Concept Check

1. **VOCABULARY** What type of data do you need to make a scatter plot? Explain.
2. **REASONING** How can you identify an outlier in a scatter plot?

LOGIC Describe the relationship you would expect between the data. Explain.

3. shoe size of a student and the student's IQ
4. time since a train's departure and the distance to its destination
5. height of a bouncing ball and the time since it was dropped
6. number of toppings on a pizza and the price of the pizza

Practice and Problem Solving

7. **JEANS** The table shows the average price (in dollars) of jeans sold at different stores and the number of pairs of jeans sold at each store in one month.

Average Price	22	40	28	35	46
Number Sold	152	94	134	110	81

   a. Write the ordered pairs from the table and plot them in a coordinate plane.
   b. Describe the relationship between the two data sets.

1 8. **SUVS** The scatter plot shows the numbers of sport utility vehicles sold in a city from 2009 to 2014.

   a. In what year were 1000 SUVs sold?
   b. About how many SUVs were sold in 2013?
   c. Describe the relationship shown by the data.

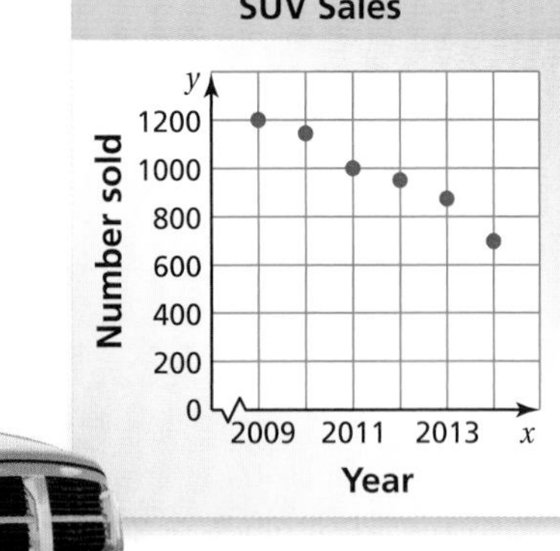

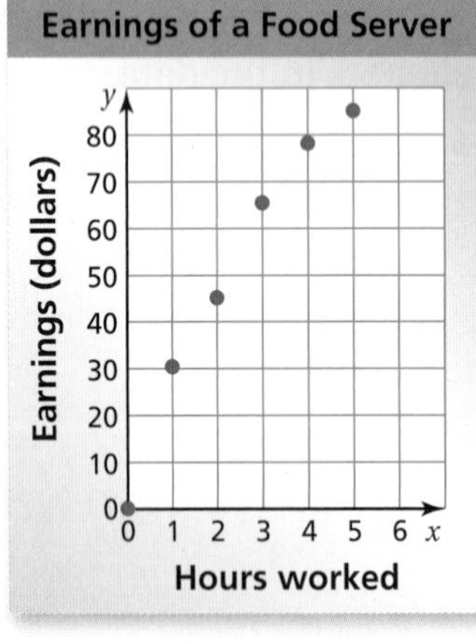

9. **EARNINGS** The scatter plot shows the total earnings (wages and tips) of a food server during one day.

   a. About how many hours must the server work to earn $70?
   b. About how much did the server earn for 5 hours of work?
   c. Describe the relationship shown by the data.

Describe the relationship between the data. Identify any outliers, gaps, or clusters.

2 **10.**

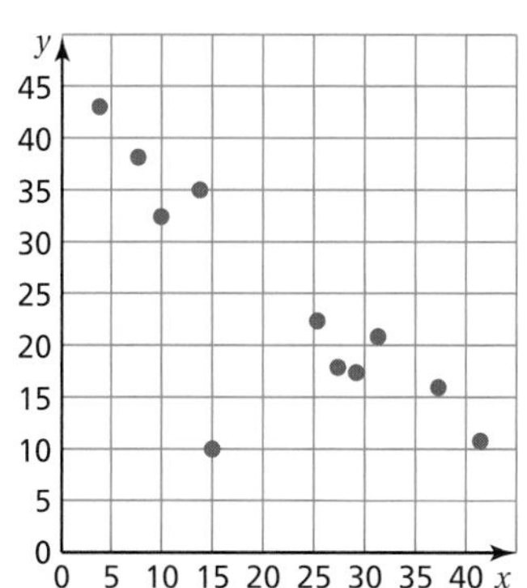

11.

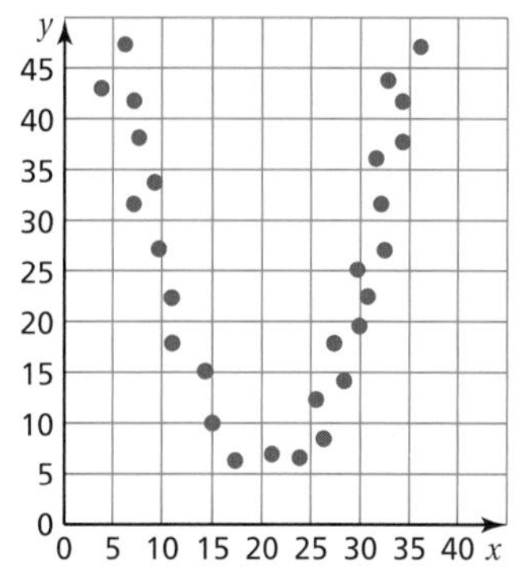

12.

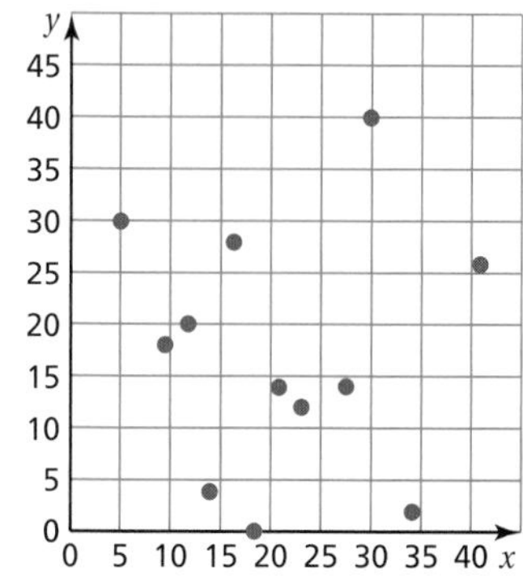

13. HONEY The table shows the average price per pound for honey in the United States from 2009 to 2012. What type of relationship do the data show?

Year, x	2009	2010	2011	2012
Average Price per Pound, y	\$4.65	\$4.85	\$5.15	\$5.53

14. TEST SCORES The scatter plot shows the numbers of minutes spent studying and the test scores for a science class. (a) What type of relationship do the data show? (b) Interpret the relationship.

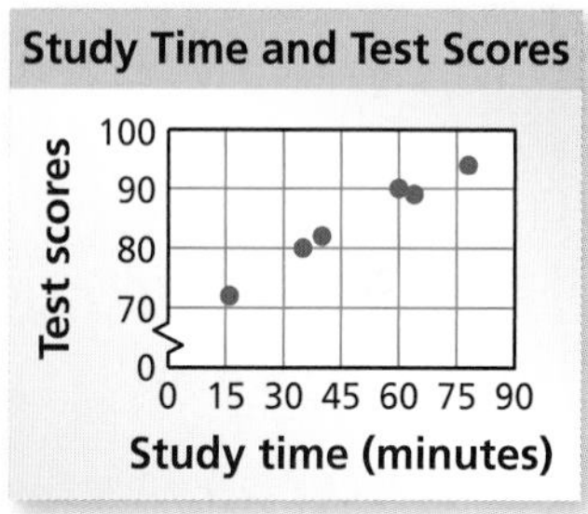

15. OPEN-ENDED Describe a set of real-life data that has a negative linear relationship.

16. PROBLEM SOLVING The table shows the memory capacities (in gigabytes) and prices (in dollars) of 7-inch tablet computers at a store. (a) Make a scatter plot of the data. Then describe the relationship between the data. (b) Identify any outliers, gaps, or clusters. Explain why you think they exist.

Memory (GB), x	8	16	4	32	4	16	4	8	16	8	16	8
Price (dollars), y	200	230	120	250	100	200	90	160	150	180	220	150

17. Reasoning Sales of sunglasses and beach towels at a store show a positive linear relationship in the summer. Does this mean that the sales of one item *cause* the sales of the other item to increase? Explain.

Fair Game Review What you learned in previous grades & lessons

Use a graph to solve the equation. Check your solution. *(Section 5.4)*

18. $5x = 2x + 6$

19. $7x + 3 = 9x - 13$

20. $\frac{2}{3}x = -\frac{1}{3}x - 4$

21. MULTIPLE CHOICE When graphing a proportional relationship represented by $y = mx$, which point is not on the graph? *(Section 4.3)*

(A) $(0, 0)$ (B) $(0, m)$ (C) $(1, m)$ (D) $(2, 2m)$

9.2 Lines of Fit

Essential Question How can you use data to predict an event?

1 ACTIVITY: Representing Data by a Linear Equation

Work with a partner. You have been working on a science project for 8 months. Each month, you measured the length of a baby alligator.

The table shows your measurements.

September → 0; April → 7

Month, x	0	1	2	3	4	5	6	7
Length (in.), y	22.0	22.5	23.5	25.0	26.0	27.5	28.5	29.5

Data Analysis

In this lesson, you will

- find lines of fit.
- use lines of fit to solve problems.

Use the following steps to predict the baby alligator's length next September.

a. Graph the data in the table.

b. Draw a line that you think best approximates the points.

c. Write an equation for your line.

d. **MODELING** Use the equation to predict the baby alligator's length next September.

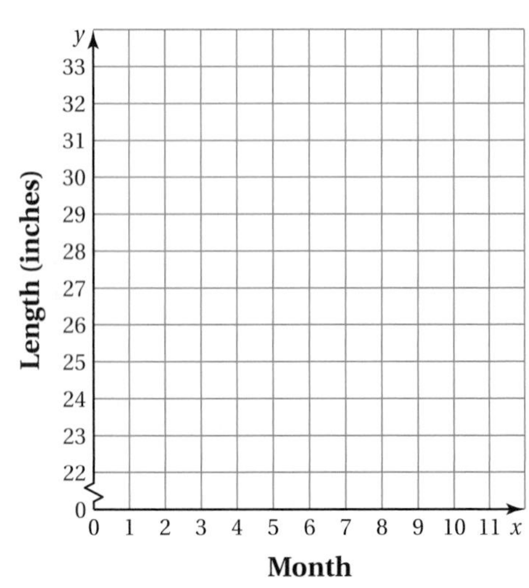

2 ACTIVITY: Representing Data by a Linear Equation

Work with a partner. You are a biologist and study bat populations.

You are asked to predict the number of bats that will be living in an abandoned mine after 3 years.

To start, you find the number of bats that have been living in the mine during the past 8 years.

The table shows the results of your research.

Math Practice

Use a Graph

How can you draw a line that "fits" the collection of points? How should the points be positioned around the line?

7 years ago → Year 0; this year → Year 7

Year, x	0	1	2	3	4	5	6	7
Bats (thousands), y	327	306	299	270	254	232	215	197

Use the following steps to predict the number of bats that will be living in the mine after 3 years.

a. Graph the data in the table.

b. Draw a line that you think best approximates the points.

c. Write an equation for your line.

d. **MODELING** Use the equation to predict the number of bats in 3 years.

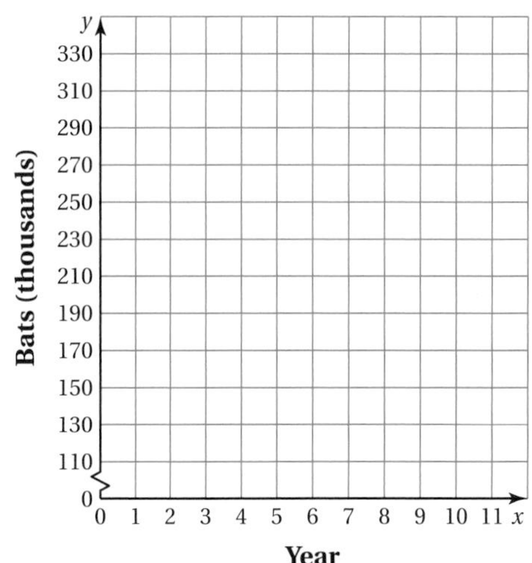

What Is Your Answer?

3. **IN YOUR OWN WORDS** How can you use data to predict an event?

4. **MODELING** Use the Internet or some other reference to find data that appear to have a linear pattern. List the data in a table, and then graph the data. Use an equation that is based on the data to predict a future event.

Practice Use what you learned about lines of fit to complete Exercise 4 on page 382.

9.2 Lesson

Key Vocabulary
line of fit, *p. 380*
line of best fit, *p. 381*

A **line of fit** is a line drawn on a scatter plot close to most of the data points. It can be used to estimate data on a graph.

EXAMPLE 1 Finding a Line of Fit

Month, x	Depth (feet), y
0	20
1	19
2	15
3	13
4	11
5	10
6	8
7	7
8	5

The table shows the depth of a river x months after a monsoon season ends. (a) Make a scatter plot of the data and draw a line of fit. (b) Write an equation of the line of fit. (c) Interpret the slope and the y-intercept of the line of fit. (d) Predict the depth in month 9.

a. Plot the points in a coordinate plane. The scatter plot shows a negative linear relationship. Draw a line that is close to the data points. Try to have as many points above the line as below it.

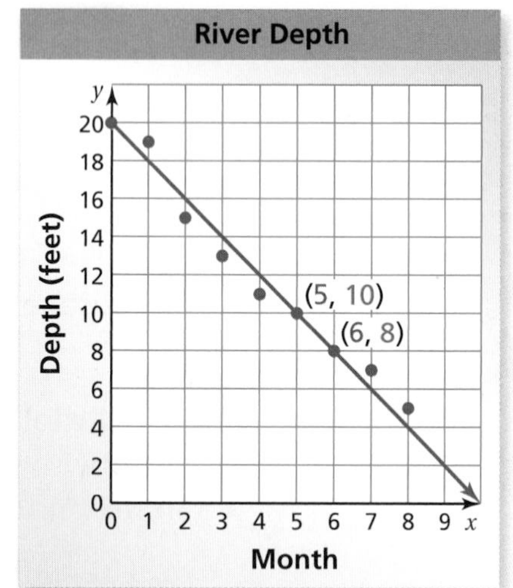

b. The line passes through (5, 10) and (6, 8).

$$\text{slope} = \frac{\text{rise}}{\text{run}} = \frac{-2}{1} = -2$$

Because the line crosses the y-axis at (0, 20), the y-intercept is 20.

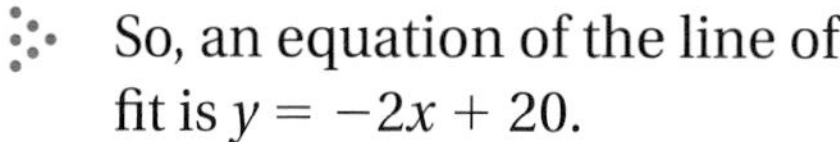

So, an equation of the line of fit is $y = -2x + 20$.

c. The slope is -2, and the y-intercept is 20. So, the depth of the river is 20 feet at the end of the monsoon season and decreases by about 2 feet per month.

Study Tip
A line of fit does not need to pass through any of the data points.

d. To predict the depth in month 9, substitute 9 for x in the equation of the line of fit.

$$y = -2x + 20 = -2(9) + 20 = 2$$

The depth in month 9 should be about 2 feet.

On Your Own

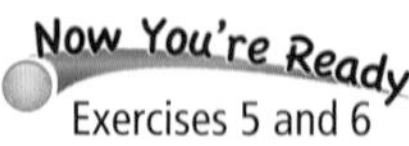

Exercises 5 and 6

1. The table shows the numbers of people who have attended a festival over an 8-year period. (a) Make a scatter plot of the data and draw a line of fit. (b) Write an equation of the line of fit. (c) Interpret the slope and the y-intercept of the line of fit. (d) Predict the number of people who will attend the festival in year 10.

Year, x	1	2	3	4	5	6	7	8
Attendance, y	420	500	650	900	1100	1500	1750	2400

Study Tip

You know how to use two points to find an equation of a line of fit. When finding an equation of the line of best fit, every point in the data set is used.

Graphing calculators use a method called *linear regression* to find a precise line of fit called a **line of best fit**. This line best models a set of data. A calculator often gives a value r called the *correlation coefficient*. This value tells whether the correlation is positive or negative, and how closely the equation models the data. Values of r range from -1 to 1. When r is close to 1 or -1, there is a strong correlation between the variables. As r gets closer to 0, the correlation becomes weaker.

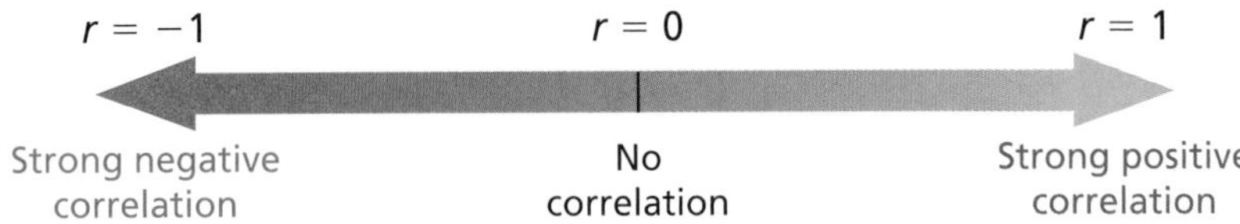

Strong negative correlation — No correlation — Strong positive correlation

EXAMPLE 2 Finding a Line of Best Fit Using Technology

The table shows the worldwide movie ticket sales y (in billions of dollars) from 2000 to 2011, where $x = 0$ represents the year 2000. Use a graphing calculator to find an equation of the line of best fit. Identify and interpret the correlation coefficient.

Year, x	0	1	2	3	4	5	6	7	8	9	10	11
Ticket Sales, y	16	17	20	20	25	23	26	26	28	29	32	33

Step 1: Enter the data from the table into your calculator.

L1	L2	L3 1
0	16	------
1	17	
2	20	
3	20	
4	25	
5	23	
6	26	

L1(1)=0

Step 2: Use the *linear regression* feature.

∴ An equation of the line of best fit is $y = 1.5x + 16$. The correlation coefficient is about 0.982. This means that the relationship between years and ticket sales is a strong positive correlation and that the equation closely models the data.

Study Tip

The slope of 1.5 indicates that sales are increasing by about \$1.5 billion each year. The y-intercept of 16 represents the ticket sales of \$16 billion for 2000.

Check Use a graphing calculator to make a scatter plot and graph the line of best fit.

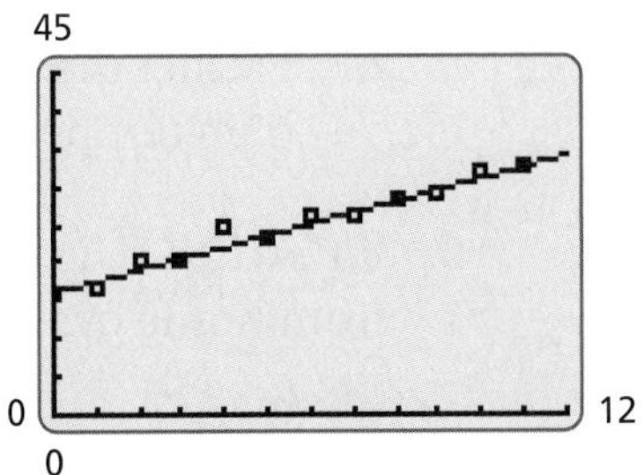

On Your Own

Now You're Ready Exercises 8–10

2. Use a graphing calculator to find an equation of the line of best fit for the data in Example 1. Identify and interpret the correlation coefficient.

9.2 Exercises

Vocabulary and Concept Check

1. **WRITING** Explain why a line of fit is helpful when analyzing data.
2. **REASONING** Tell whether the line drawn on the graph is a good fit for the data. Explain your reasoning.
3. **NUMBER SENSE** Which correlation coefficient indicates a stronger relationship: -0.98 or 0.91? Explain.

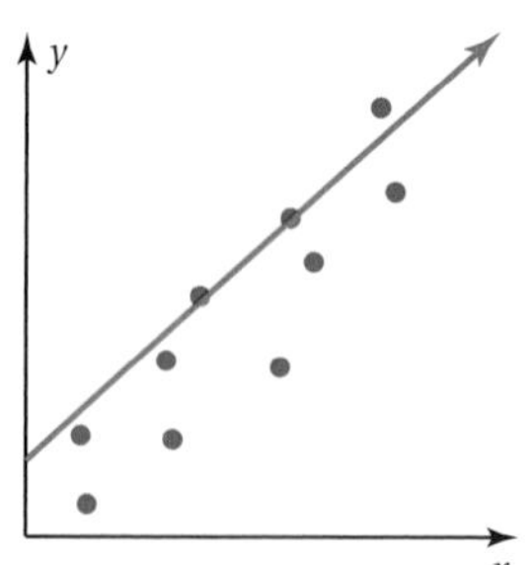

Practice and Problem Solving

4. **BLUEBERRIES** The table shows the weights y of x pints of blueberries.

Number of Pints, x	0	1	2	3	4	5
Weight (pounds), y	0	0.8	1.50	2.20	3.0	3.75

   a. Graph the data in the table.
   b. Draw a line that you think best approximates the points.
   c. Write an equation for your line.
   d. Use the equation to predict the weight of 10 pints of blueberries.
   e. Blueberries cost $2.25 per pound. How much do 10 pints of blueberries cost?

1 5. **HOT CHOCOLATE** The table shows the daily high temperature (°F) and the number of hot chocolates sold at a coffee shop for eight randomly selected days.

Temperature (°F), x	30	36	44	51	60	68	75	82
Hot Chocolates, y	45	43	36	35	30	27	23	17

   a. Make a scatter plot of the data and draw a line of fit.
   b. Write an equation of the line of fit.
   c. Interpret the slope and the y-intercept of the line of fit.
   d. Predict the number of hot chocolates sold when the high temperature is 20°F.

6. **VACATION** The table shows the distance you are away from home over a 6-hour period of your vacation.

Hours, x	Distance (miles), y
1	62
2	123
3	188
4	228
5	280
6	344

   a. Make a scatter plot of the data and draw a line of fit.
   b. Write an equation of the line of fit.
   c. About how many miles per hour do you travel?
   d. About how far were you from home when you started?
   e. Predict the distance from home in 7 hours.

7. **REASONING** A data set has no relationship. Is it possible to find a line of fit for the data? Explain.

2 8. **AMUSEMENT PARK** The table shows the attendance y (in thousands) at an amusement park from 2004 to 2013, where $x = 4$ represents the year 2004. Use a graphing calculator to find an equation of the line of best fit. Identify and interpret the correlation coefficient.

Year, x	4	5	6	7	8	9	10	11	12	13
Attendance (thousands), y	850	845	828	798	800	792	785	781	775	760

9. **SNOWSTORM** The table shows the total snow depth y (in inches) on the ground during a snowstorm x hours after it began. Use a graphing calculator to find an equation of the line of best fit. Identify and interpret the correlation coefficient. Use your equation to estimate how much snow was on the ground before the snowstorm began.

Hours, x	1	2	3	4	5	6	7	8
Snow Depth (inches), y	5	6	6.75	7.75	8.5	9.5	10.5	11.5

10. **TEXTING** The table shows the numbers y (in billions) of text messages sent from 2006 to 2011, where $x = 6$ represents the year 2006.

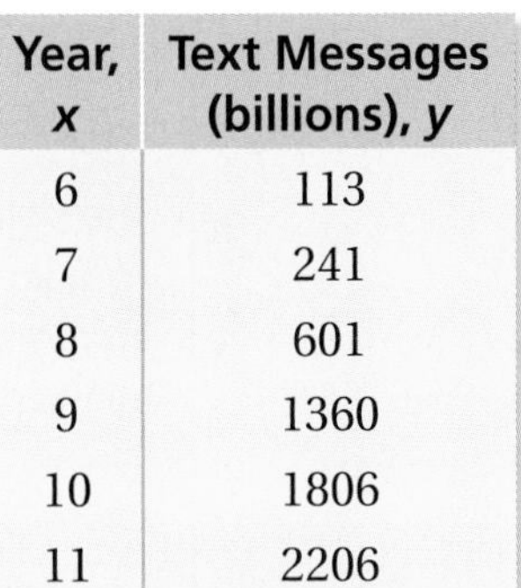

Year, x	Text Messages (billions), y
6	113
7	241
8	601
9	1360
10	1806
11	2206

a. Use a graphing calculator to find an equation of the line of best fit. Identify and interpret the correlation coefficient.

b. Interpret the slope of the line of best fit. Does the y-intercept make sense for this problem? Explain.

c. Predict the number of text messages sent in 2015.

11. **Modeling** The table shows the height y (in feet) of a baseball x seconds after it was hit.

Seconds, x	Height (feet), y
0	3
0.5	39
1	67
1.5	87
2	99

a. Use a graphing calculator to find an equation of the line of best fit. Identify and interpret the correlation coefficient.

b. Predict the height after 5 seconds.

c. The actual height after 5 seconds is about 3 feet. Why do you think this is different from your prediction?

A Fair Game Review What you learned in previous grades & lessons

Write the decimal as a fraction or a mixed number. *(Section 7.4)*

12. $0.\overline{2}$ 13. $-2.\overline{7}$ 14. $-1.4\overline{6}$ 15. $0.\overline{81}$

16. **MULTIPLE CHOICE** Which expression represents the volume of a sphere with radius r? *(Section 8.3)*

(A) $\frac{1}{3}\pi r^2 h$ (B) $\pi r^2 h$ (C) $4\pi r^2$ (D) $\frac{4}{3}\pi r^3$

9 Study Help

You can use an **information frame** to help you organize and remember concepts. Here is an example of an information frame for scatter plots.

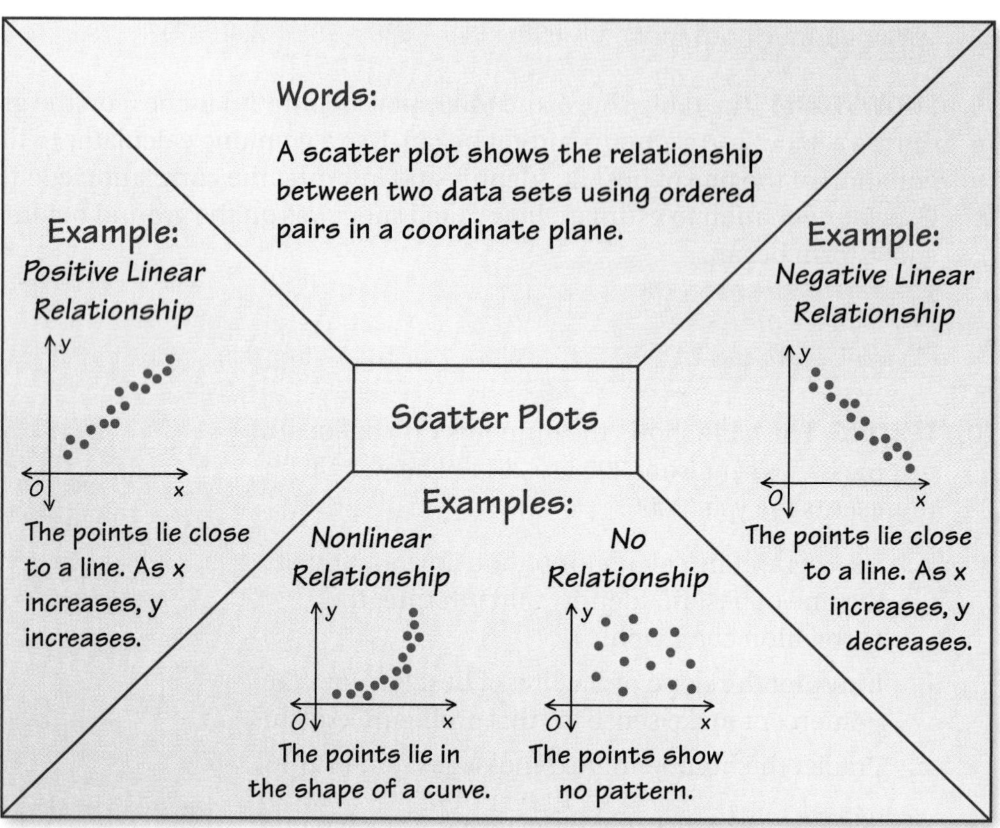

Words:
A scatter plot shows the relationship between two data sets using ordered pairs in a coordinate plane.
Example:
Positive Linear Relationship
The points lie close to a line. As x increases, y increases.
Example:
Negative Linear Relationship
The points lie close to a line. As x increases, y decreases.
Scatter Plots
Examples:
Nonlinear Relationship
The points lie in the shape of a curve.
No Relationship
The points show no pattern.

On Your Own

Make an information frame to help you study this topic.

1. lines of fit

After you complete this chapter, make information frames for the following topics.

2. two-way tables
3. data displays

"Dear Teacher, I am emailing my information frame showing the characteristics of circles."

9.1–9.2 Quiz

Check It Out
Progress Check
BigIdeasMath.com

1. **CHARITY** The scatter plot shows the amount of money donated to a charity from 2007 to 2012. *(Section 9.1)*

   a. In what year did the charity receive $150,000?

   b. How much did the charity receive in 2010?

   c. Describe the relationship shown by the data.

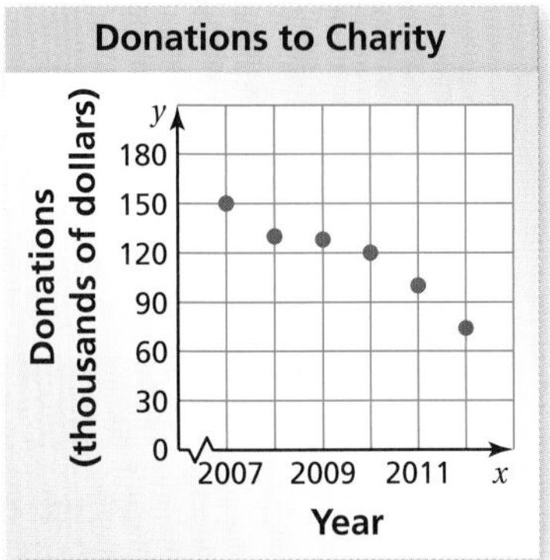

Describe the relationship between the data. Identify any outliers, gaps, or clusters. *(Section 9.1)*

2. 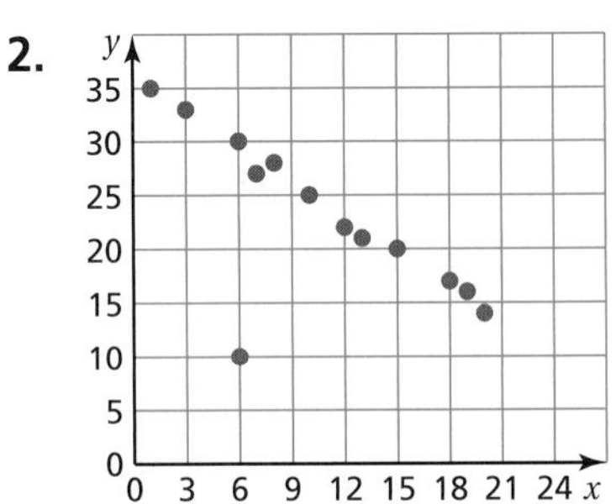

3. 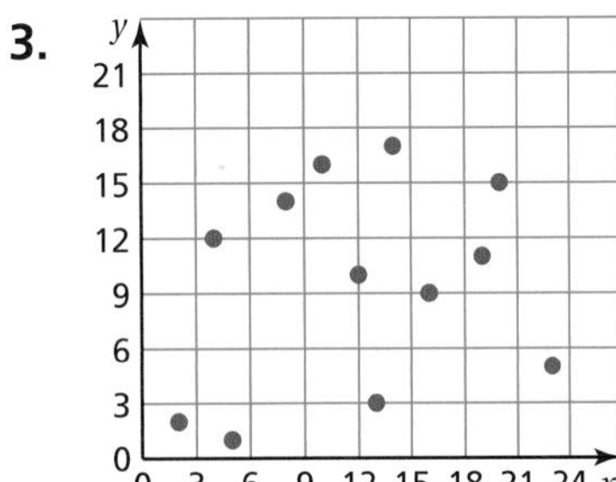

4. 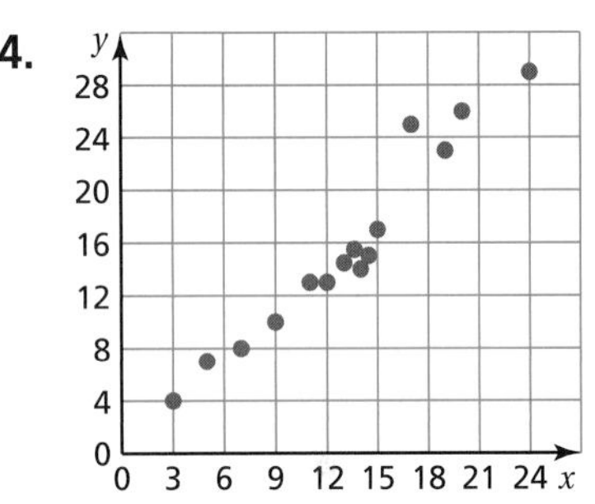

Year, x	Millions of Customers, y
0	12
1	18
2	21
3	28
4	33
5	38
6	42
7	48

5. **CUSTOMERS SERVED** The table shows the numbers of customers (in millions) served by a restaurant chain over an 8-year period. *(Section 9.1 and Section 9.2)*

   a. Make a scatter plot of the data. Then describe the relationship between the data. Identify any outliers, gaps, or clusters.

   b. Use a graphing calculator to find the equation of the line of best fit for the data. Identify and interpret the correlation coefficient.

6. **CATS** An animal shelter opens in December. The table shows the number of cats adopted from the shelter each month from January to September. *(Section 9.2)*

Month	1	2	3	4	5	6	7	8	9
Cats	3	6	7	11	13	14	15	18	19

   a. Make a scatter plot of the data and draw a line of fit.

   b. Write an equation of the line of fit.

   c. Interpret the slope and the y-intercept of the line of fit.

   d. Predict how many cats will be adopted in October.

9.3 Two-Way Tables

Essential Question How can you read and make a two-way table?

Two categories of data can be displayed in a *two-way table.*

1 ACTIVITY: Reading a Two-Way Table

Work with a partner. You are the manager of a sports shop. The two-way table shows the numbers of soccer T-shirts that your shop has left in stock at the end of the season.

		T-Shirt Size					
		S	M	L	XL	XXL	Total
Color	Blue/White	5	4	1	0	2	
	Blue/Gold	3	6	5	2	0	
	Red/White	4	2	4	1	3	
	Black/White	3	4	1	2	1	
	Black/Gold	5	2	3	0	2	
	Total						65

a. Complete the totals for the rows and columns.

b. Are there any black-and-gold XL T-shirts in stock? Justify your answer.

c. The numbers of T-shirts you ordered at the beginning of the season are shown below. Complete the two-way table.

		T-Shirt Size					
		S	M	L	XL	XXL	Total
Color	Blue/White	5	6	7	6	5	
	Blue/Gold	5	6	7	6	5	
	Red/White	5	6	7	6	5	
	Black/White	5	6	7	6	5	
	Black/Gold	5	6	7	6	5	
	Total						

d. REASONING How would you alter the numbers of T-shirts you order for next season? Explain your reasoning.

Data Analysis

In this lesson, you will

- read two-way tables.
- make and interpret two-way tables.

2 ACTIVITY: Analyzing Data

Math Practice

Construct Arguments

What are the advantages of using a table instead of a graph to analyze data?

Work with a partner. The three-dimensional two-way table shows information about the numbers of hours students at a high school work at part-time jobs during the school year.

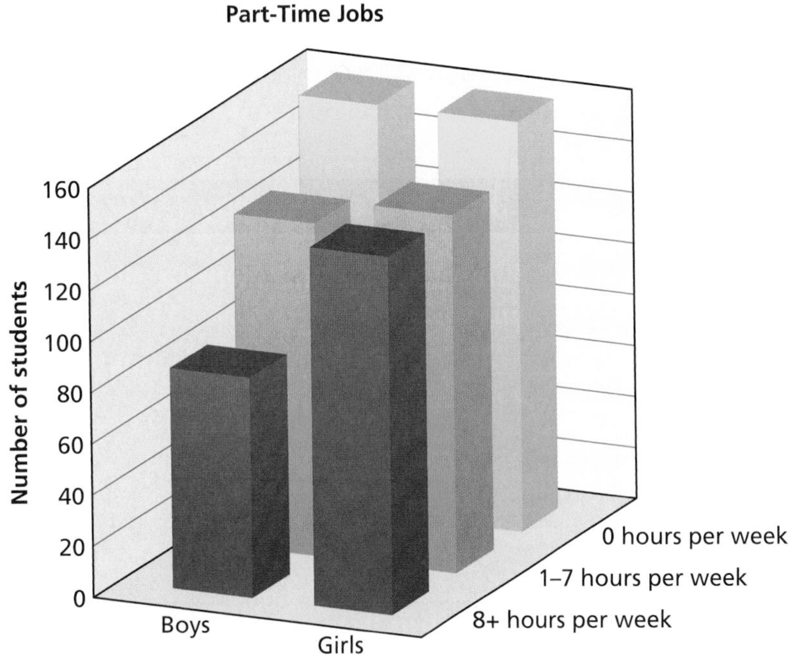

a. Make a two-way table showing the data. Use estimation to find the entries in your table.

b. Write two observations you can make that summarize the data in your table.

c. **REASONING** A newspaper article claims that more boys than girls drop out of high school to work full-time. Do the data support this claim? Explain your reasoning.

What Is Your Answer?

3. **IN YOUR OWN WORDS** How can you read and make a two-way table?

4. Find a real-life data set that you can represent by a two-way table. Then make a two-way table for the data set.

Use what you learned about two-way tables to complete Exercises 3–6 on page 390.

9.3 Lesson

Key Vocabulary

two-way table, *p. 388*
joint frequency, *p. 388*
marginal frequency, *p. 388*

A **two-way table** displays two categories of data collected from the same source.

You randomly survey students in your school about their grades on the last test and whether they studied for the test. The two-way table shows your results. Each entry in the table is called a **joint frequency**.

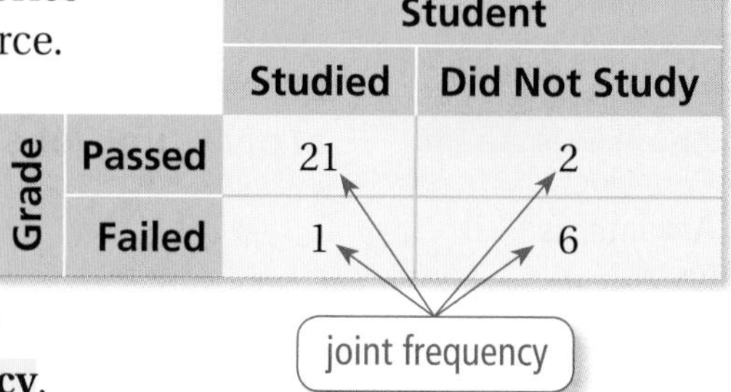

		Student	
		Studied	Did Not Study
Grade	Passed	21	2
	Failed	1	6

joint frequency

EXAMPLE 1 Reading a Two-Way Table

How many of the students in the survey above studied for the test and passed?

The entry in the "Studied" column and "Passed" row is 21.

So, 21 of the students in the survey studied for the test and passed.

The sums of the rows and columns in a two-way table are called **marginal frequencies**.

EXAMPLE 2 Finding Marginal Frequencies

Find and interpret the marginal frequencies for the survey above.

Create a new column and a new row for the sums. Then add the entries.

		Student		
		Studied	Did Not Study	Total
Grade	Passed	21	2	23
	Failed	1	6	7
	Total	22	8	30

23 students passed.
7 students failed.
30 students were surveyed.
22 students studied.
8 students did not study.

On Your Own

Now You're Ready
Exercises 5–8

1. You randomly survey students in a cafeteria about their plans for a football game and a school dance. The two-way table shows your results.

   a. How many students will attend the dance but not the football game?

   b. Find and interpret the marginal frequencies for the survey.

		Football Game	
		Attend	Not Attend
Dance	Attend	35	5
	Not Attend	16	20

Multi-Language Glossary at BigIdeasMath.com

EXAMPLE 3 Making a Two-Way Table

Rides Bus

Age	Tally
12-13	卌 卌 卌 卌 ||||
14-15	卌 卌 ||
16-17	卌 卌 ||||

Does Not Ride Bus

Age	Tally
12-13	卌 卌 卌 |
14-15	卌 卌 |||
16-17	卌 卌 卌 卌 |

You randomly survey students between the ages of 12 and 17 about whether they ride the bus to school. The results are shown in the tally sheets. Make a two-way table that includes the marginal frequencies.

The two categories for the table are the ages and whether or not they ride the bus. Use the tally sheets to calculate each joint frequency. Then add to find each marginal frequency.

		Age			
		12–13	14–15	16–17	Total
Student	Rides Bus	24	12	14	50
	Does Not Ride Bus	16	13	21	50
	Total	40	25	35	100

EXAMPLE 4 Finding a Relationship in a Two-Way Table

Use the two-way table in Example 3.

a. For each age group, what percent of the students in the survey ride the bus to school? do not ride the bus to school? Organize the results in a two-way table. Explain what one of the entries represents.

		Age		
		12–13	14–15	16–17
Student	Rides Bus	60%	48%	40%
	Does Not Ride Bus	40%	52%	60%

$\frac{14}{35} = 0.4$

So, 40% of the 16- and 17-year-old students in the survey ride the bus to school.

b. Does the table in part (a) show a relationship between age and whether students ride the bus to school? Explain.

Yes, the table shows that as age increases, students are less likely to ride the bus to school.

On Your Own

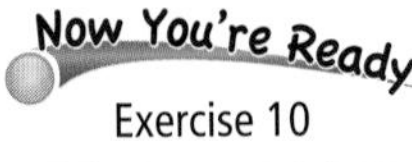

Exercise 10

Grade 6 Students
11 pack lunch, 9 buy school lunch

Grade 7 Students
23 pack lunch, 27 buy school lunch

Grade 8 Students
16 pack lunch, 14 buy school lunch

2. You randomly survey students in a school about whether they buy a school lunch or pack a lunch. Your results are shown.

   a. Make a two-way table that includes the marginal frequencies.

   b. For each grade level, what percent of the students in the survey pack a lunch? buy a school lunch? Organize the results in a two-way table. Explain what one of the entries represents.

   c. Does the table in part (b) show a relationship between grade level and lunch choice? Explain.

9.3 Exercises

Vocabulary and Concept Check

1. **VOCABULARY** Explain the relationship between joint frequencies and marginal frequencies.

2. **OPEN-ENDED** Describe how you can use a two-way table to organize data you collect from a survey.

Practice and Problem Solving

You randomly survey students about participating in their class's yearly fundraiser. You display the two categories of data in the two-way table.

3. Find the total of each row.

4. Find the total of each column.

① 5. How many female students will be participating in the fundraiser?

6. How many male students will *not* be participating in the fundraiser?

		Fundraiser	
		No	Yes
Gender	Female	22	51
	Male	30	29

Find and interpret the marginal frequencies.

② 7.

		School Play	
		Attend	Not Attend
Class	Junior	41	30
	Senior	52	23

8.

		Cell Phone Minutes	
		Limited	Unlimited
Text Plan	Limited	78	0
	Unlimited	175	15

9. **GOALS** You randomly survey students in your school. You ask what is most important to them: grades, popularity, or sports. You display your results in the two-way table.

   a. How many 7th graders chose sports? How many 8th graders chose grades?

   b. Find and interpret the marginal frequencies for the survey.

   c. What percent of students in the survey are 6th graders who chose popularity?

		Goal		
		Grades	Popularity	Sports
Grade	6th	31	18	23
	7th	39	16	19
	8th	42	6	17

③ ④ **10. SAVINGS** You randomly survey people in your neighborhood about whether they have at least \$1000 in savings. The results are shown in the tally sheets.

Have at Least \$1000 in Savings

Age	Tally
20-29	𝍸 𝍸 ||||
30-39	𝍸 𝍸 𝍸 𝍸 𝍸 ||
40-49	𝍸 𝍸 𝍸 𝍸 𝍸

Don't Have at Least \$1000 in Savings

Age	Tally
20-29	𝍸 𝍸 𝍸 𝍸 𝍸 𝍸 𝍸 |
30-39	𝍸 𝍸 𝍸 𝍸 𝍸 𝍸 |||
40-49	𝍸 𝍸 𝍸

a. Make a two-way table that includes the marginal frequencies.

b. For each age group, what percent of the people have at least \$1000 in savings? do not have at least \$1000 in savings? Organize the results in a two-way table.

c. Does the table in part (b) show a relationship between age and whether people have at least \$1000 in savings? Explain.

11. EYE COLOR You randomly survey students in your school about the color of their eyes. The results are shown in the tables.

Eye Color of Males Surveyed		
Green	Blue	Brown
5	16	27

Eye Color of Females Surveyed		
Green	Blue	Brown
3	19	18

a. Make a two-way table.

b. Find and interpret the marginal frequencies for the survey.

c. For each eye color, what percent of the students in the survey are male? female? Organize the results in a two-way table. Explain what two of the entries represent.

12. REASONING Use the information from Exercise 11. For each gender, what percent of the students in the survey have green eyes? blue eyes? brown eyes? Organize the results in a two-way table. Explain what two of the entries represent.

13. Precision What percent of students in the survey in Exercise 11 are either female or have green eyes? What percent of students in the survey are males who do not have green eyes? Find and explain the sum of these two percents.

Fair Game Review What you learned in previous grades & lessons

Write an equation of the line that passes through the points. *(Section 4.6)*

14. (0, 1), (−2, −5) **15.** (0, −2), (3, 13) **16.** (−4, 1), (0, 3)

17. MULTIPLE CHOICE Which equation does not represent a linear function? *(Section 6.4)*

Ⓐ $y = 4x$ Ⓑ $xy = 8$ Ⓒ $y = -3$ Ⓓ $6x + 5y = -2$

9.4 Choosing a Data Display

Essential Question How can you display data in a way that helps you make decisions?

1 ACTIVITY: Displaying Data

Work with a partner. Analyze and display each data set in a way that best describes the data. Explain your choice of display.

a. ROADKILL A group of schools in New England participated in a 2-month study. They reported 3962 dead animals.

Birds: 307 Mammals: 2746
Amphibians: 145 Reptiles: 75
Unknown: 689

b. BLACK BEAR ROADKILL The data below show the numbers of black bears killed on a state's roads from 1993 to 2012.

1993:	30	2000:	47	2007:	99
1994:	37	2001:	49	2008:	129
1995:	46	2002:	61	2009:	111
1996:	33	2003:	74	2010:	127
1997:	43	2004:	88	2011:	141
1998:	35	2005:	82	2012:	135
1999:	43	2006:	109		

c. RACCOON ROADKILL A 1-week study along a 4-mile section of road found the following weights (in pounds) of raccoons that had been killed by vehicles.

13.4	14.8	17.0	12.9
21.3	21.5	16.8	14.8
15.2	18.7	18.6	17.2
18.5	9.4	19.4	15.7
14.5	9.5	25.4	21.5
17.3	19.1	11.0	12.4
20.4	13.6	17.5	18.5
21.5	14.0	13.9	19.0

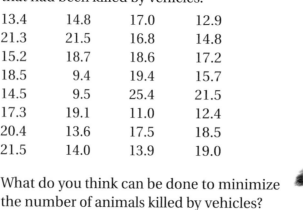

d. What do you think can be done to minimize the number of animals killed by vehicles?

Data Analysis
In this lesson, you will
- choose appropriate data displays.
- identify and analyze misleading data displays.

2 ACTIVITY: Statistics Project

Math Practice

Use a Graph

How can you use a graph to represent the data you have gathered for your report? What does the graph tell you about the data?

ENDANGERED SPECIES PROJECT Use the Internet or some other reference to write a report about an animal species that is (or has been) endangered. Include graphical displays of the data you have gathered.

Sample: Florida Key Deer

In 1939, Florida banned the hunting of Key deer. The numbers of Key deer fell to about 100 in the 1940s.

In 1947, public sentiment was stirred by 11-year-old Glenn Allen from Miami. Allen organized Boy Scouts and others in a letter-writing campaign that led to the establishment of the National Key Deer Refuge in 1957. The approximately 8600-acre refuge includes 2280 acres of designated wilderness.

The Key Deer Refuge has increased the population of Key deer. A recent study estimated the total Key deer population to be approximately 800.

About half of Key deer deaths are due to vehicles.

One of two Key deer wildlife underpasses on Big Pine Key

What Is Your Answer?

3. **IN YOUR OWN WORDS** How can you display data in a way that helps you make decisions? Use the Internet or some other reference to find examples of the following types of data displays.

- Bar graph
- Circle graph
- Scatter plot
- Stem-and-leaf plot
- Box-and-whisker plot

Use what you learned about choosing data displays to complete Exercise 3 on page 397.

Data Display	What does it do?
Pictograph	shows data using pictures
Bar Graph	shows data in specific categories
Circle Graph	shows data as parts of a whole
Line Graph	shows how data change over time
Histogram	shows frequencies of data values in intervals of the same size
Stem-and-Leaf Plot	orders numerical data and shows how they are distributed
Box-and-Whisker Plot	shows the variability of a data set by using quartiles
Dot Plot	shows the number of times each value occurs in a data set
Scatter Plot	shows the relationship between two data sets by using ordered pairs in a coordinate plane

EXAMPLE 1 Choosing an Appropriate Data Display

Choose an appropriate data display for the situation. Explain your reasoning.

a. the number of students in a marching band each year

A line graph shows change over time. So, a line graph is an appropriate data display.

b. a comparison of people's shoe sizes and their heights

You want to compare two different data sets. So, a scatter plot is an appropriate data display.

On Your Own

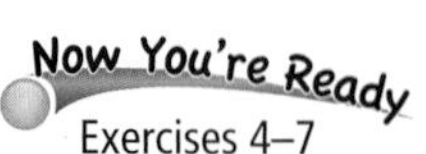

Choose an appropriate data display for the situation. Explain your reasoning.

1. the population of the United States divided into age groups
2. the percents of students in your school who play basketball, football, soccer, or lacrosse

EXAMPLE 2 Identifying an Appropriate Data Display

You record the number of hits for your school's new website for 5 months. Tell whether the data display is appropriate for representing how the number of hits changed during the 5 months. Explain your reasoning.

Month	Hits
August	250
September	320
October	485
November	650
December	925

a.

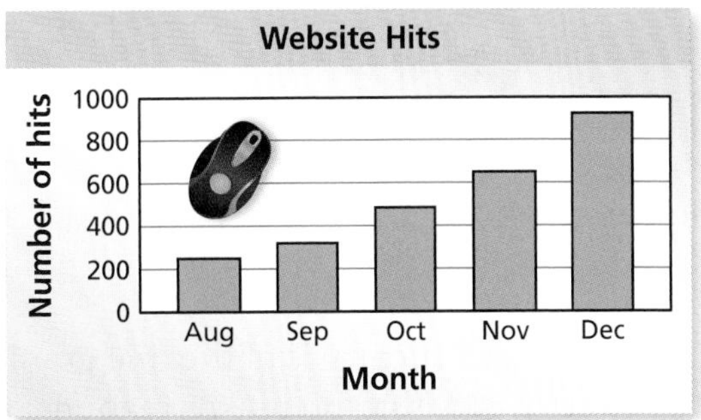

 The bar graph shows the number of hits for each month. So, it is an appropriate data display.

b.

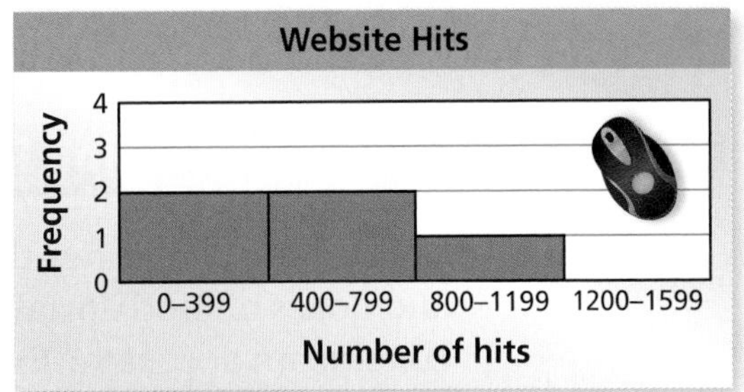

 The histogram does not show the number of hits for each month or how the number of hits changes over time. So, it is *not* an appropriate data display.

c.

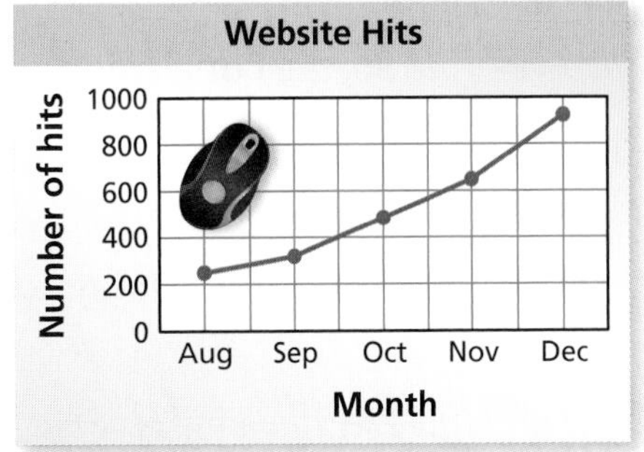

The line graph shows how the number of hits changes over time. So, it is an appropriate data display.

On Your Own

Now You're Ready
Exercises 8 and 9

Tell whether the data display is appropriate for representing the data in Example 2. Explain your reasoning.

3. dot plot **4.** circle graph **5.** stem-and-leaf plot

EXAMPLE 3 Identifying a Misleading Data Display

Which line graph is misleading? Explain.

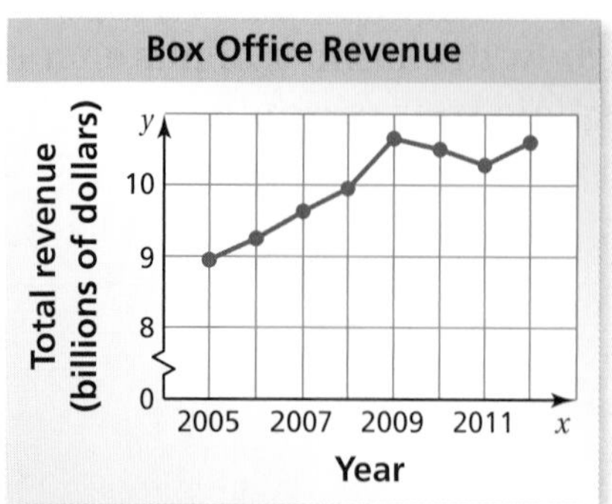

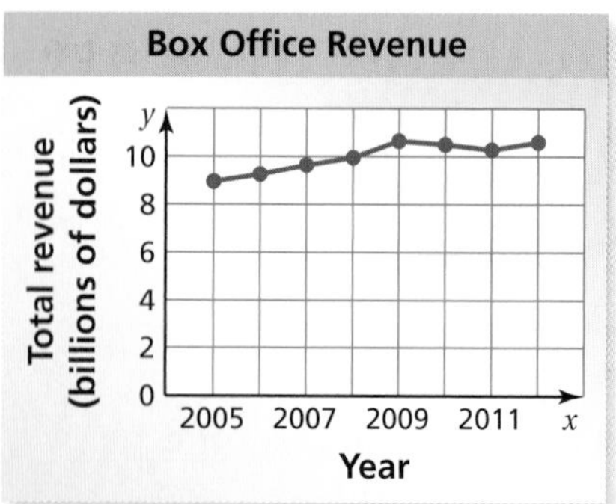

The vertical axis of the line graph on the left has a break (⌇) and begins at 8. This graph makes it appear that the total revenue increased rapidly from 2005 to 2009. The graph on the right has an unbroken axis. It is more honest and shows that the total revenue increased slowly.

∴ So, the graph on the left is misleading.

EXAMPLE 4 Analyzing a Misleading Data Display

Food Drive Donations Totals

Canned food

Boxed food

Juice

= 20 cans = 20 boxes = 20 bottles

A volunteer concludes that the numbers of cans of food and boxes of food donated were about the same. Is this conclusion accurate? Explain.

Each icon represents the same number of items. Because the box icon is larger than the can icon, it looks like the number of boxes is about the same as the number of cans. But the number of boxes is actually about half of the number of cans.

∴ So, the conclusion is not accurate.

On Your Own

Now You're Ready
Exercises 11–14

Explain why the data display is misleading.

6.

Concert Ticket Prices

Concert ticket price (dollars)

55, 50, 45, 40, 35, 0

A B C

Rock band

7.

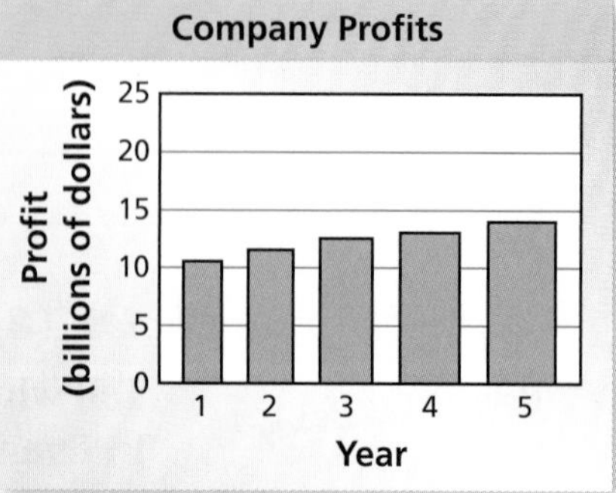

9.4 Exercises

Vocabulary and Concept Check

1. **REASONING** Can more than one display be appropriate for a data set? Explain.
2. **OPEN-ENDED** Describe how a histogram can be misleading.

Practice and Problem Solving

3. Analyze and display the data in a way that best describes the data. Explain your choice of display.

Notebooks Sold in One Week				
192 red	170 green	203 black	183 pink	230 blue
165 yellow	210 purple	250 orange	179 white	218 other

Choose an appropriate data display for the situation. Explain your reasoning.

1 4. a student's test scores and how the scores are spread out

5. the distance a person drives each month

6. the outcome of rolling a number cube

7. homework problems assigned each day

2 8. **LIFEGUARD** The table shows how many hours you worked as a lifeguard from May to August. Tell whether the data display is appropriate for representing how the number of hours worked changed during the 4 months. Explain your reasoning.

Lifeguard Schedule	
Month	**Hours Worked**
May	40
June	80
July	160
August	120

a.

May
June
July
August

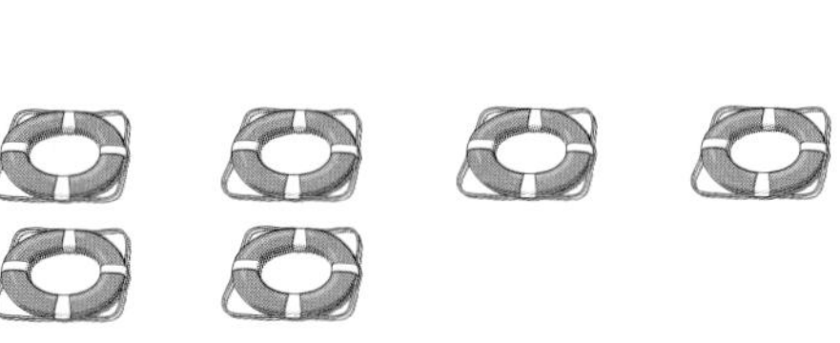

Key: = 20 hours

b.

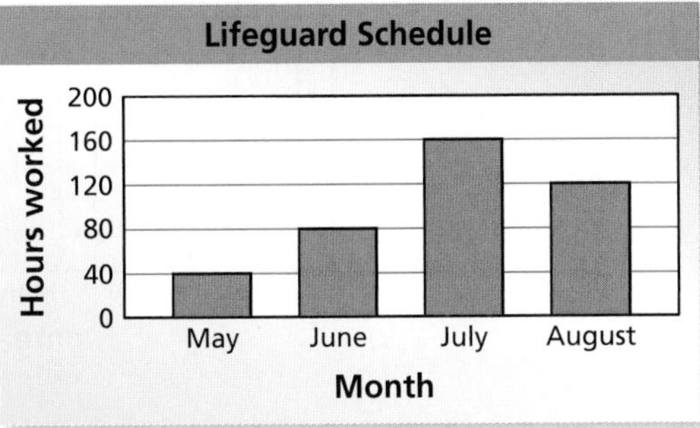

9. **FAVORITE SUBJECT** A survey asked 800 students to choose their favorite subject. The results are shown in the table. Tell whether the data display is appropriate for representing the portion of students who prefer math. Explain your reasoning.

Favorite School Subject	
Subject	**Number of Students**
Science	200
Math	160
Literature	240
Social Studies	120
Other	80

a.

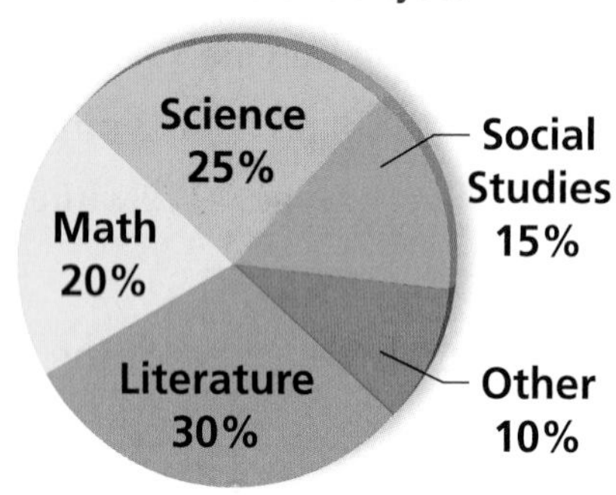

b.

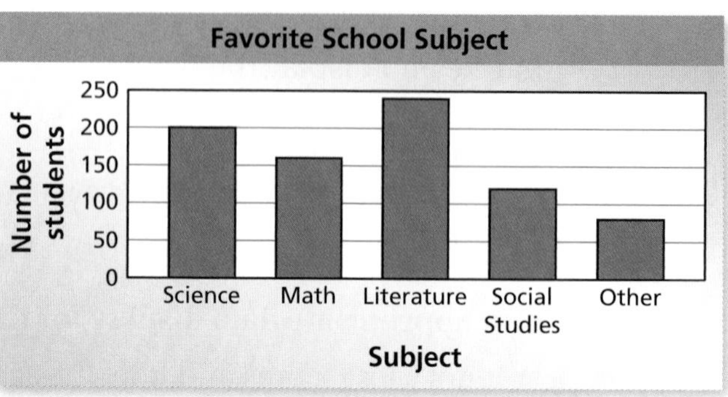

10. **WRITING** When should you use a histogram instead of a bar graph to display data? Use an example to support your answer.

Explain why the data display is misleading.

③ ④ 11. Miles Biked

Monday

Tuesday

Wednesday

= 1 mile

12.

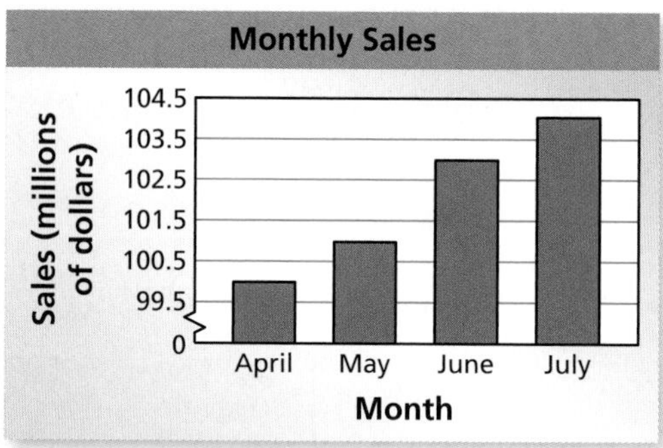

13.

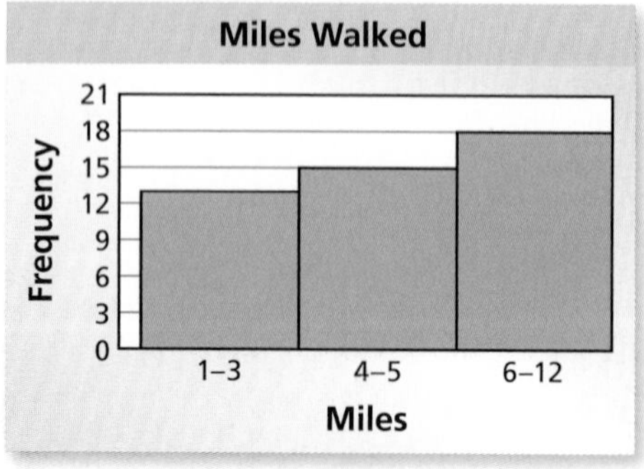

14.

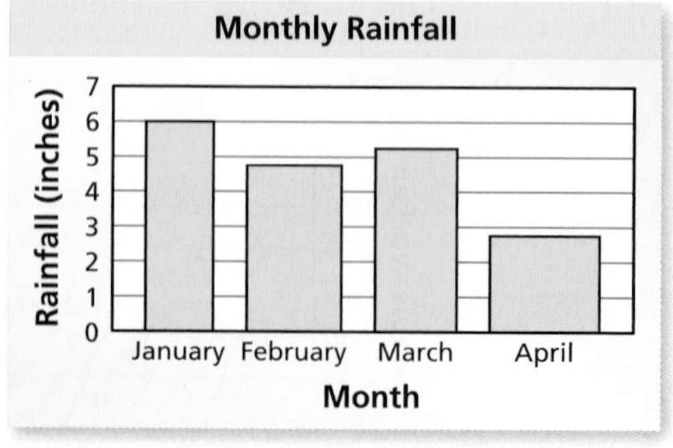

15. **VEGETABLES** A nutritionist wants to use a data display to show the favorite vegetables of the students at a school. Choose an appropriate data display for the situation. Explain your reasoning.

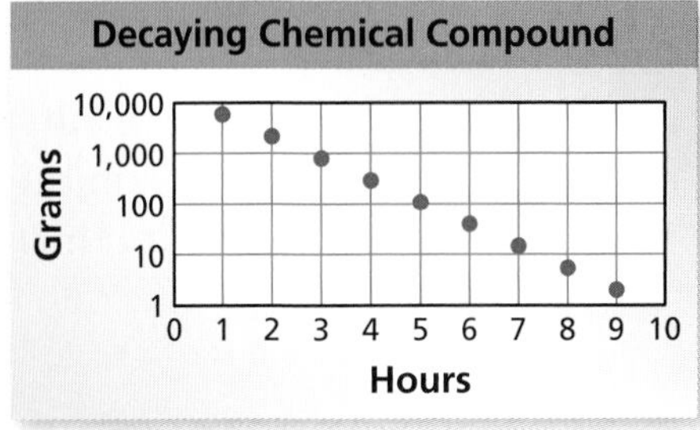

16. **CHEMICALS** A scientist gathers data about a decaying chemical compound. The results are shown in the scatter plot. Is the data display misleading? Explain.

17. **REASONING** What type of data display is appropriate for showing the mode of a data set?

18. **SPORTS** A survey asked 100 students to choose their favorite sports. The results are shown in the circle graph.

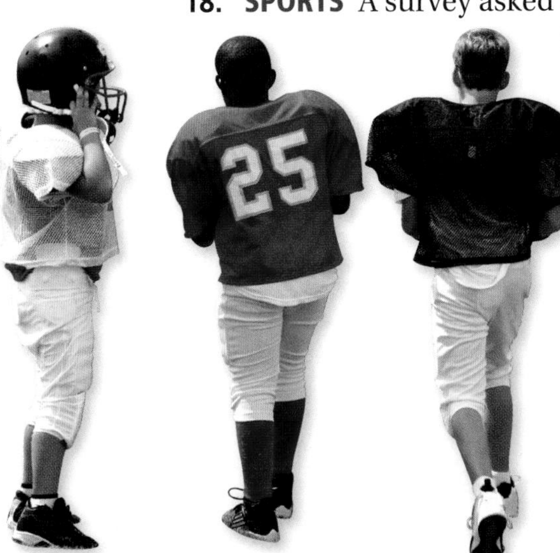

a. Explain why the graph is misleading.

b. What type of data display would be more appropriate for the data? Explain.

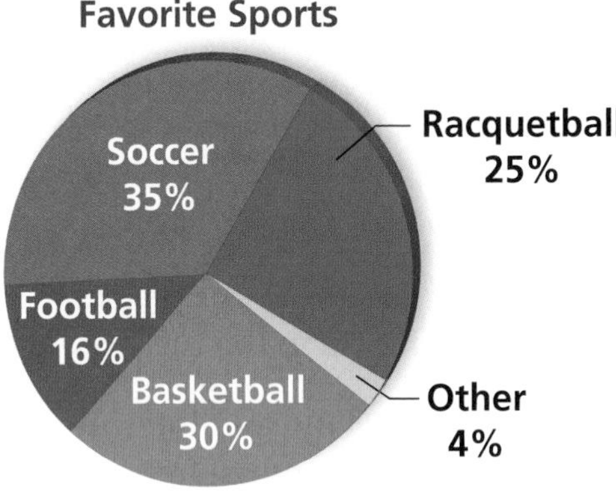

19. **Structure** With the help of computers, mathematicians have computed and analyzed billions of digits of the irrational number π. One of the things they analyze is the frequency of each of the numbers 0 through 9. The table shows the frequency of each number in the first 100,000 digits of π.

a. Display the data in a bar graph.

b. Display the data in a circle graph.

c. Which data display is more appropriate? Explain.

d. Describe the distribution.

Number	0	1	2	3	4	5	6	7	8	9
Frequency	9999	10,137	9908	10,025	9971	10,026	10,029	10,025	9978	9902

Fair Game Review What you learned in previous grades & lessons

Estimate the square root to the nearest (a) integer and (b) tenth. *(Section 7.4)*

20. $\sqrt{20}$ 21. $-\sqrt{74}$ 22. $\sqrt{140}$

23. **MULTIPLE CHOICE** What is 20% of 25% of 400? *(Skills Review Handbook)*

(A) 20 (B) 200 (C) 240 (D) 380

9.3–9.4 Quiz

1. **RECYCLING** The results of a recycling survey are shown in the two-way table. Find and interpret the marginal frequencies. *(Section 9.3)*

		Recycle	
		Yes	No
Gender	Female	28	9
	Male	24	14

2. **MUSIC** The results of a music survey are shown in the two-way table. Find and interpret the marginal frequencies. *(Section 9.3)*

		Jazz	
		Likes	Dislikes
Country	Likes	26	14
	Dislikes	17	8

3. **ELECTION** The results of a voting survey are shown in the two-way table. *(Section 9.3)*

   a. Find and interpret the marginal frequencies.

   b. For each age group, what percent of voters prefer Smith? prefer Jackson? Organize your results in a two-way table.

   c. Does your table in part (b) show a relationship between age and candidate preference? Explain.

		Voter's Age		
		18–34	35–64	65+
Candidate	Smith	36	25	6
	Jackson	12	32	24

Choose an appropriate data display for the situation. Explain your reasoning. *(Section 9.4)*

4. the percent of band students in each section of instruments

5. a company's profit for each week

6. **TURTLES** The tables show the weights (in pounds) of turtles caught in two ponds. Which type of data display would you use for this information? Explain. *(Section 9.4)*

Pond A			
12	13	15	6
7	8	12	7

Pond B			
9	12	5	8
12	15	16	19

Funds Raised for Class Trip

Amount raised (dollars): 0, 500, 600, 700, 800

Month: 0, 1, 2, 3, 4, 5, 6

7. **FUNDRAISER** The line graph shows the amount of money that the eighth-grade students at a school raised each month to pay for a class trip. Is the graph misleading? Explain. *(Section 9.4)*

9 Chapter Review

Review Key Vocabulary

scatter plot, *p. 374*
line of fit, *p. 380*
line of best fit, *p. 381*
two-way table, *p. 388*
joint frequency, *p. 388*
marginal frequency, *p. 388*

Review Examples and Exercises

9.1 Scatter Plots *(pp. 372–377)*

Your school is ordering custom T-shirts. The scatter plot shows the number of T-shirts ordered and the cost per shirt. What tends to happen to the cost per shirt as the number of T-shirts ordered increases?

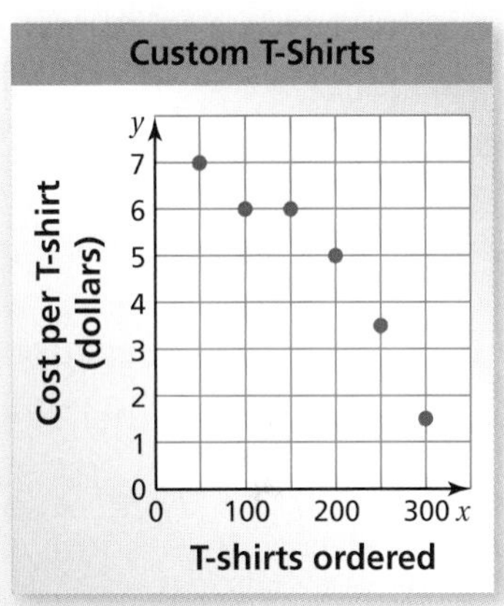

Looking at the graph, the plotted points go down from left to right.

So, as the number of T-shirts ordered increases, the cost per shirt decreases.

Exercises

1. **MIGRATION** The scatter plot shows the number of geese that migrated to a park each season.

   a. In what year did 270 geese migrate?

   b. How many geese migrated in 2010?

   c. Describe the relationship shown by the data.

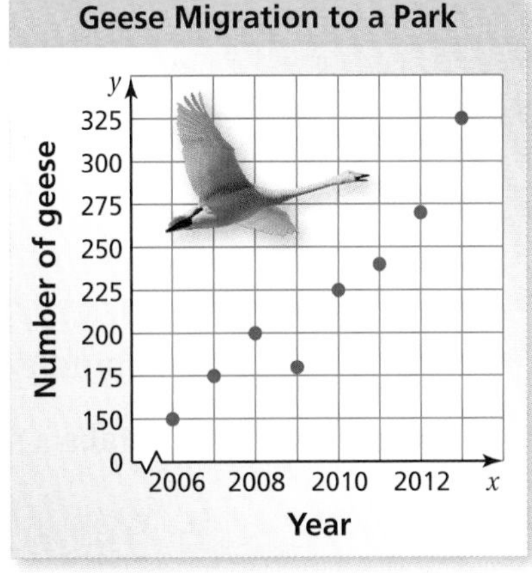

Describe the relationship between the data. Identify any outliers, gaps, or clusters.

2. 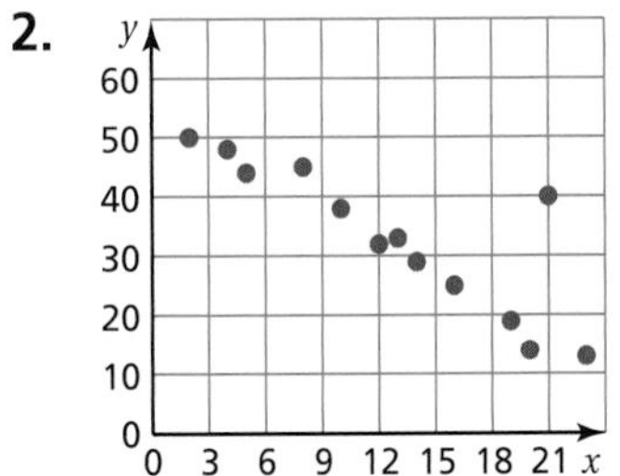

3. 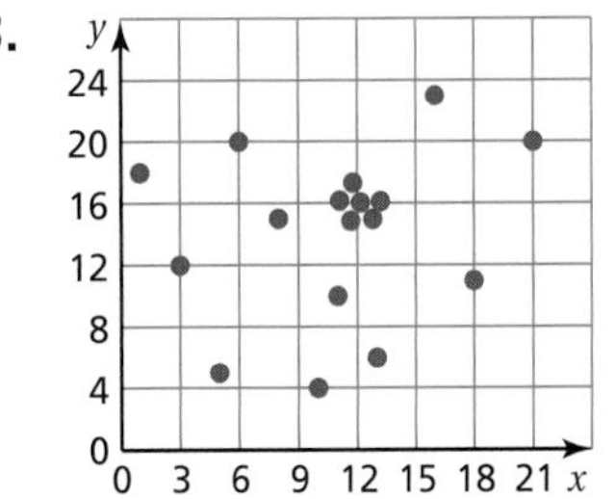

4. 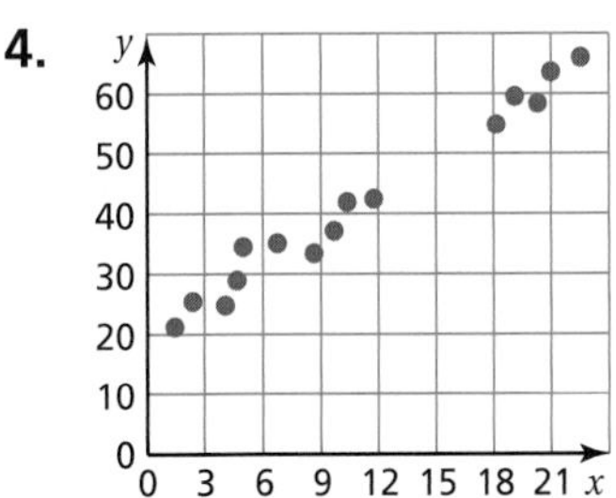

9.2 Lines of Fit (pp. 378–383)

The table shows the revenue (in millions of dollars) for a company over an 8-year period. (a) Make a scatter plot of the data and draw a line of fit. (b) Write an equation of the line of fit. (c) Interpret the slope and the y-intercept of the line of fit. (d) Predict what the revenue will be in year 9.

Year, x	1	2	3	4	5	6	7	8
Revenue (millions of dollars), y	20	35	46	56	68	82	92	108

a. Plot the points in a coordinate plane. The scatter plot shows a positive linear relationship. Draw a line that is close to the data points.

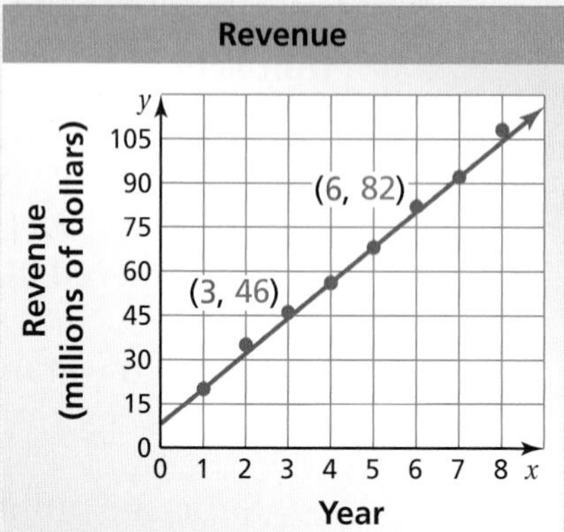

b. slope $= \dfrac{\text{rise}}{\text{run}} = \dfrac{36}{3} = 12$

Because the line crosses the y-axis at $(0, 8)$, the y-intercept is 8.

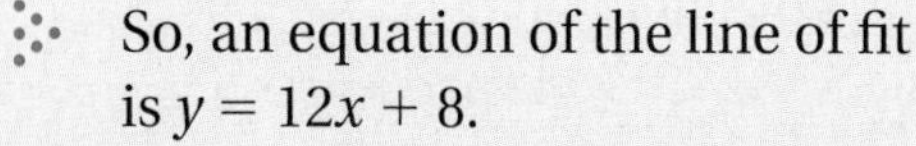

So, an equation of the line of fit is $y = 12x + 8$.

c. The slope is 12. So, the revenue increased by about \$12 million each year. The y-intercept is 8. So, you can estimate that the revenue was \$8 million in the year before this 8-year period.

d. $y = 12x + 8 = 12(9) + 8 = 116$

The revenue in year 9 will be about \$116 million.

Exercises

5. **STUDENTS** The table shows the number of students at a middle school over a 10-year period.

   a. Make a scatter plot of the data and draw a line of fit.

   b. Write an equation of the line of fit.

   c. Interpret the slope and the y-intercept of the line of fit.

   d. Predict the number of students in year 11.

Year, x	Number of Students, y
1	492
2	507
3	520
4	535
5	550
6	562
7	577
8	591
9	604
10	618

6. **LINE OF BEST FIT** Use a graphing calculator to find an equation of the line of best fit for the data in Exercise 5. Identify and interpret the correlation coefficient.

9.3 Two-Way Tables (pp. 386–391)

You randomly survey students in your school about whether they liked a recent school play. The results are shown. Make a two-way table that includes the marginal frequencies. What percent of the students surveyed liked the play?

Male Students
48 likes, 12 dislikes
Female Students
56 likes, 14 dislikes

Of the 130 students surveyed, 104 students liked the play.

		Student		
		Liked	Did Not Like	Total
Gender	Male	48	12	60
	Female	56	14	70
	Total	104	26	130

∴ Because $\frac{104}{130} = 0.8$, 80% of the students in the survey liked the play.

Exercises

You randomly survey people at a mall about whether they like the new food court. The results are shown.

Teenagers
96 likes, 4 dislikes
Adults
21 likes, 79 dislikes
Senior Citizens
18 likes, 82 dislikes

7. Make a two-way table that includes the marginal frequencies.
8. For each group, what percent of the people surveyed like the food court? dislike the food court? Organize your results in a two-way table.
9. Does your table in Exercise 8 show a relationship between age and whether people like the food court?

9.4 Choosing a Data Display (pp. 392–399)

Choose an appropriate data display for the situation. Explain your reasoning.

a. the percent of votes that each candidate received in an election

∴ A circle graph shows data as parts of a whole. So, a circle graph is an appropriate data display.

b. the distribution of the ages of U.S. presidents

∴ A stem-and-leaf plot orders numerical data and shows how they are distributed. So, a stem-and-leaf plot is an appropriate data display.

Exercises

Choose an appropriate data display for the situation. Explain your reasoning.

10. the number of pairs of shoes sold by a store each week
11. a comparison of the heights of brothers and sisters

9 Chapter Test

1. **POPULATION** The graph shows the population (in millions) of the United States from 1960 to 2010.

   a. In what year was the population of the United States about 180 million?

   b. What was the approximate population of the United States in 1990?

   c. Describe the trend shown by the data.

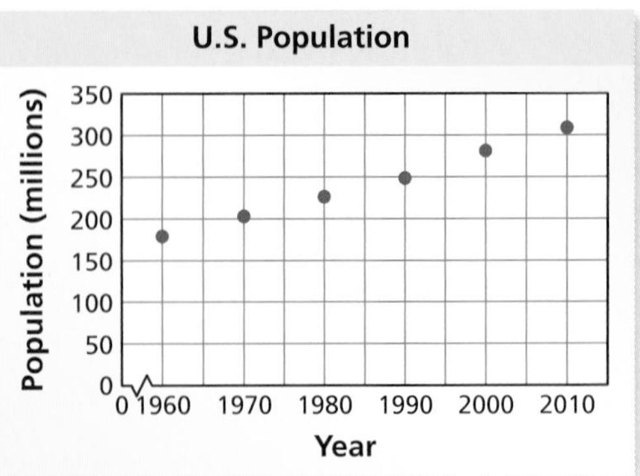

2. **WEIGHT** The table shows the weight of a baby over several months.

   a. Make a scatter plot of the data and draw a line of fit.

   b. Write an equation of the line of fit.

   c. Interpret the slope and the y-intercept of the line of fit.

   d. Predict how much the baby will weigh at 7 months.

Age (months)	Weight (pounds)
1	8
2	9.25
3	11.75
4	13
5	14.5
6	16

		Nonfiction	
		Likes	Dislikes
Fiction	Likes	26	20
	Dislikes	22	2

3. **READING** You randomly survey students at your school about what type of books they like to read. The two-way table shows your results. Find and interpret the marginal frequencies.

Choose an appropriate data display for the situation. Explain your reasoning.

4. magazine sales grouped by price

5. the distance a person hikes each week

6. **SAT** The table shows the numbers y of students (in thousands) who took the SAT from 2006 to 2010, where $x = 6$ represents the year 2006. Use a graphing calculator to find an equation of the line of best fit. Identify and interpret the correlation coefficient.

Year, x	6	7	8	9	10
Number of Students, y	1466	1495	1519	1530	1548

7. **RECYCLING** You randomly survey shoppers at a supermarket about whether they use reusable bags. Of 60 male shoppers, 15 use reusable bags. Of 110 female shoppers, 60 use reusable bags. Organize your results in a two-way table. Include the marginal frequencies.

9 Cumulative Assessment

Test-Taking Strategy
Read All Choices Before Answering
Which type of graph would best show the percent of cats who said tuna is their favorite food?
Did someone say "tuna"?
"Reading all choices before answering can sometimes point out the obvious answer!"

1. What is the volume of the trash bin?

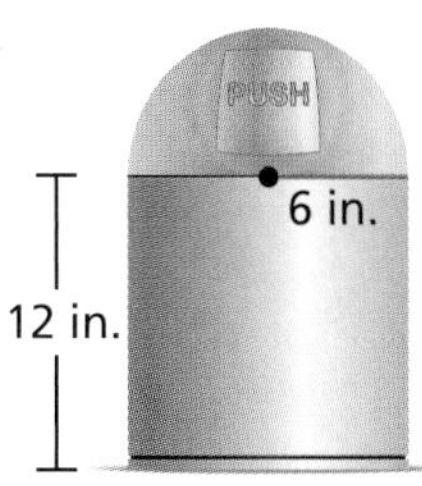

A. 288π in.^{3}

B. 576π in.^{3}

C. 648π in.^{3}

D. 720π in.^{3}

2. The diagram below shows parallel lines cut by a transversal. Which angle is the corresponding angle for $\angle 6$?

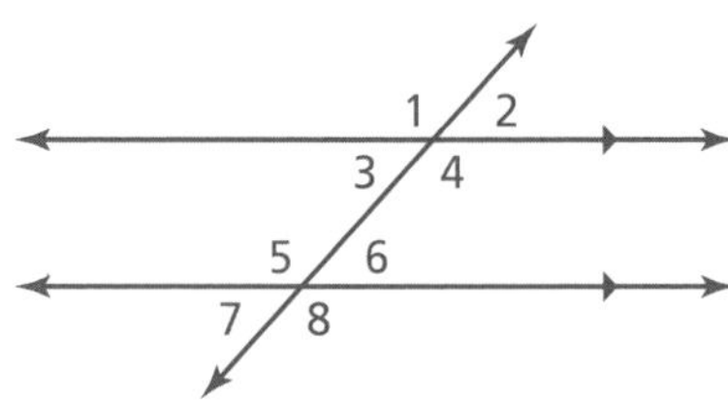

F. $\angle 2$

G. $\angle 3$

H. $\angle 4$

I. $\angle 8$

3. You randomly survey students in your school. You ask whether they have jobs. You display your results in the two-way table. How many male students do *not* have a job?

		Job	
		Yes	No
Gender	Male	27	12
	Female	31	17

4. Which scatter plot shows a negative relationship between x and y?

A. 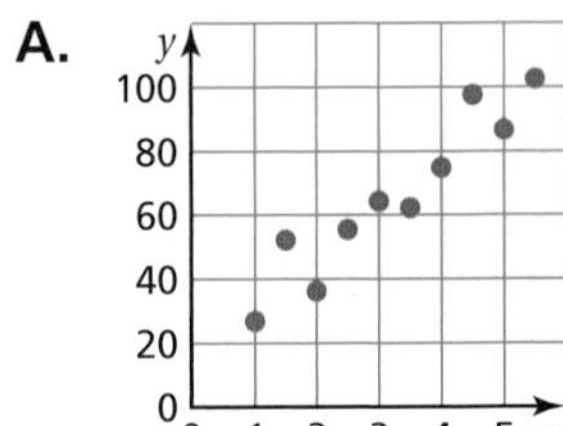

C. 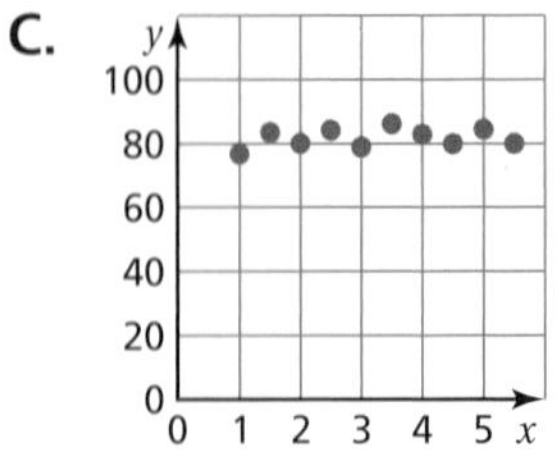

B. 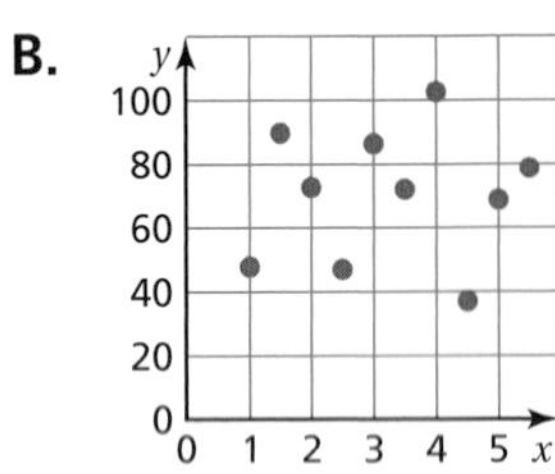

D. 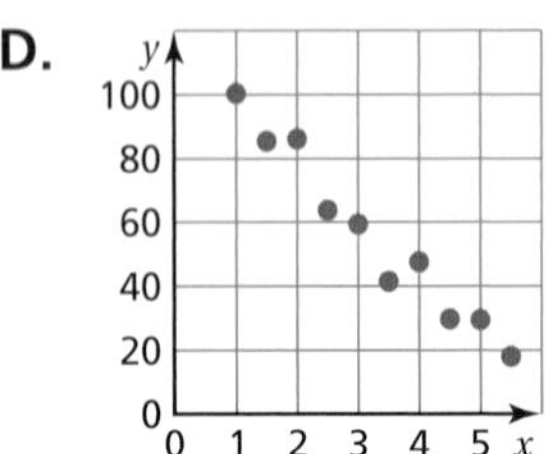

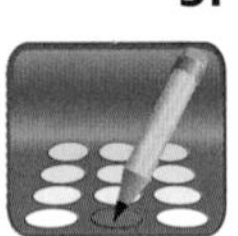

5. The legs of a right triangle have the lengths of 8 centimeters and 15 centimeters. What is the length of the hypotenuse, in centimeters?

6. What is the solution of the equation?

$$0.22(x + 6) = 0.2x + 1.8$$

F. $x = 2.4$

H. $x = 24$

G. $x = 15.6$

I. $x = 156$

7. Which triangle is *not* a right triangle?

A. 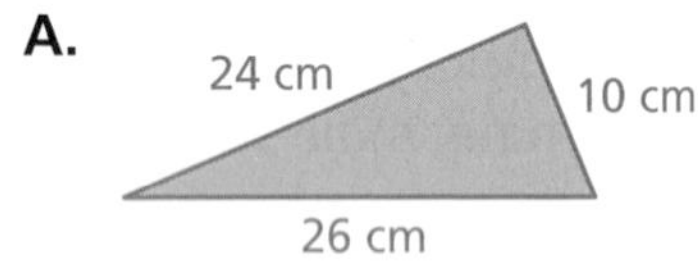

C. 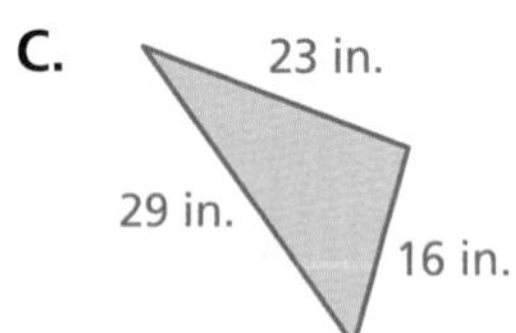

B. 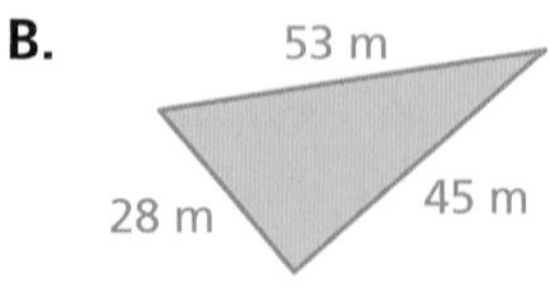

D. 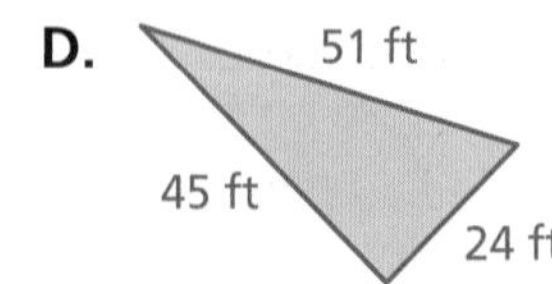

8. A store has recorded total dollar sales each month for the past three years. Which type of graph would best show how sales have increased over this time period?

F. circle graph

G. line graph

H. histogram

I. stem-and-leaf plot

9. Trapezoid *KLMN* is graphed in the coordinate plane shown.

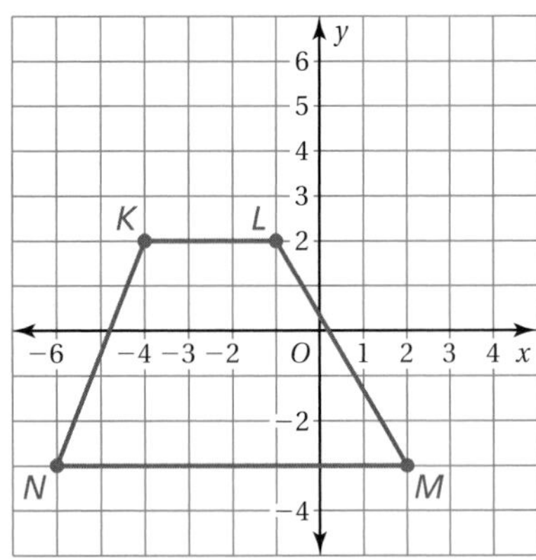

Rotate Trapezoid *KLMN* 90° clockwise about the origin. What are the coordinates of point M', the image of point M after the rotation?

A. $(-3, -2)$

B. $(-2, -3)$

C. $(-2, 3)$

D. $(3, 2)$

10. The table shows the numbers of hours students spent watching television from Monday through Friday for one week and their scores on a test that Friday.

Think
Solve
Explain

Hours of Television, x	5	2	10	15	3	4	8	2	12	9
Test Score, y	92	98	79	66	97	88	82	95	72	81

Part A Make a scatter plot of the data.

Part B Describe the relationship between hours of television watched and test score.

Part C Explain how to justify your answer in Part B using the linear regression feature of a graphing calculator.

10 Exponents and Scientific Notation

"Here's how it goes, Descartes."

"The friends of my friends are my friends. The friends of my enemies are my enemies."

"The enemies of my friends are my enemies. The enemies of my enemies are my friends."

"If one flea had 100 babies, and each baby grew up and had 100 babies, ..."

"... and each of those babies grew up and had 100 babies, you would have 1,010,101 fleas."

What You Learned Before

I love it!
"It's called the Power of Negative One, Descartes!"

Using Order of Operations

Example 1 **Evaluate $6^2 \div 4 - 2(9 - 5)$.**

First:	Parentheses	$6^2 \div 4 - 2(9 - 5) = 6^2 \div 4 - 2 \cdot 4$
Second:	Exponents	$= 36 \div 4 - 2 \cdot 4$
Third:	Multiplication and Division (from left to right)	$= 9 - 8$
Fourth:	Addition and Subtraction (from left to right)	$= 1$

Try It Yourself

Evaluate the expression.

1. $15\left(\frac{8}{4}\right) + 2^2 - 3 \cdot 7$
2. $5^2 \cdot 2 \div 10 + 3 \cdot 2 - 1$
3. $3^2 - 1 + 2(4(3 + 2))$

Multiplying and Dividing Decimals

Example 2 **Find $2.1 \cdot 0.35$.**

$$\begin{array}{r} 2.1 \\ \times\ 0.35 \\ \hline 105 \\ 63 \\ \hline 0.735 \end{array}$$

2.1 ← 1 decimal place
× 0.35 ← + 2 decimal places
0.735 ← 3 decimal places

Example 3 **Find $1.08 \div 0.9$.**

$0.9\overline{)1.08}$ Multiply each number by 10.

$$\begin{array}{r} 1.2 \\ 9\overline{)10.8} \\ -\ 9 \\ \hline 1\,8 \\ -\ 1\,8 \\ \hline 0 \end{array}$$

Place the decimal point above the decimal point in the dividend 10.8.

Try It Yourself

Find the product or quotient.

4. $1.75 \cdot 0.2$
5. $1.4 \cdot 0.6$
6. $\begin{array}{r} 7.03 \\ \times\ 4.3 \\ \hline \end{array}$
7. $\begin{array}{r} 0.894 \\ \times\ 0.2 \\ \hline \end{array}$
8. $5.40 \div 0.09$
9. $4.17 \div 0.3$
10. $0.15\overline{)3.6}$
11. $0.004\overline{)7.2}$

10.1 Exponents

Essential Question How can you use exponents to write numbers?

The expression 3^5 is called a *power*. The *base* is 3. The *exponent* is 5.

base → 3^5 ← exponent

1 ACTIVITY: Using Exponent Notation

Work with a partner.

a. Copy and complete the table.

Power	Repeated Multiplication Form	Value
$(-3)^1$	-3	-3
$(-3)^2$	$(-3) \cdot (-3)$	9
$(-3)^3$		
$(-3)^4$		
$(-3)^5$		
$(-3)^6$		
$(-3)^7$		

b. **REPEATED REASONING** Describe what is meant by the expression $(-3)^n$. How can you find the value of $(-3)^n$?

2 ACTIVITY: Using Exponent Notation

Work with a partner.

a. The cube at the right has \$3 in each of its small cubes. Write a power that represents the total amount of money in the large cube.

b. Evaluate the power to find the total amount of money in the large cube.

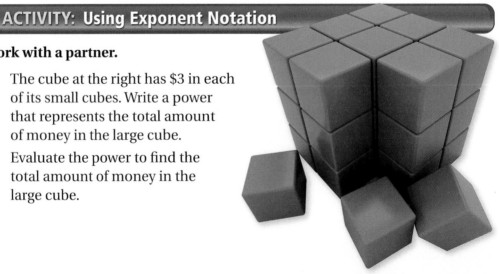

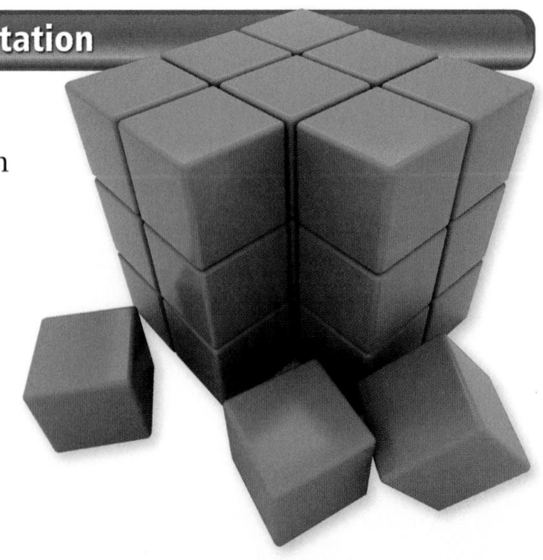

Exponents

In this lesson, you will

- write expressions using integer exponents.
- evaluate expressions involving integer exponents.

3 ACTIVITY: Writing Powers as Whole Numbers

Work with a partner. Write each distance as a whole number. Which numbers do you know how to write in words? For instance, in words, 10^3 is equal to *one thousand*.

a. 10^{26} meters: diameter of observable universe

b. 10^{21} meters: diameter of Milky Way galaxy

c. 10^{16} meters: diameter of solar system

d. 10^{7} meters: diameter of Earth

e. 10^{6} meters: length of Lake Erie shoreline

f. 10^{5} meters: width of Lake Erie

4 ACTIVITY: Writing a Power

Math Practice

Analyze Givens
What information is given in the poem? What are you trying to find?

Work with a partner. Write the numbers of kits, cats, sacks, and wives as powers.

As I was going to St. Ives
I met a man with seven wives
Each wife had seven sacks
Each sack had seven cats
Each cat had seven kits
Kits, cats, sacks, wives
How many were going to St. Ives?

Nursery Rhyme, 1730

What Is Your Answer?

5. **IN YOUR OWN WORDS** How can you use exponents to write numbers? Give some examples of how exponents are used in real life.

Practice Use what you learned about exponents to complete Exercises 3–5 on page 414.

10.1 Lesson

Check It Out
Lesson Tutorials
BigIdeasMath.com

Key Vocabulary
power, *p. 412*
base, *p. 412*
exponent, *p. 412*

A **power** is a product of repeated factors. The **base** of a power is the common factor. The **exponent** of a power indicates the number of times the base is used as a factor.

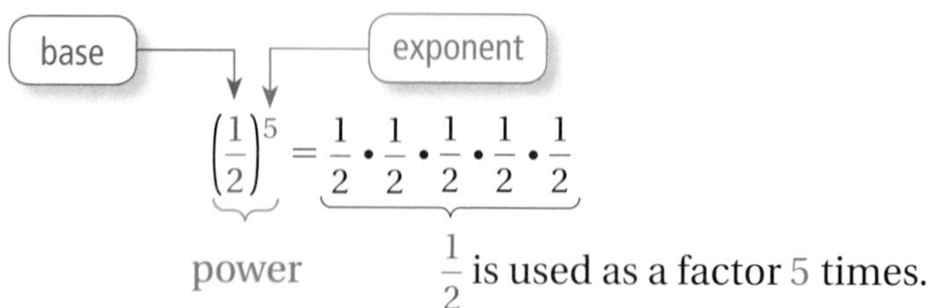

EXAMPLE 1 Writing Expressions Using Exponents

Write each product using exponents.

a. $(-7) \cdot (-7) \cdot (-7)$

Because -7 is used as a factor 3 times, its exponent is 3.

So, $(-7) \cdot (-7) \cdot (-7) = (-7)^3$.

Study Tip

Use parentheses to write powers with negative bases.

b. $\pi \cdot \pi \cdot r \cdot r \cdot r$

Because π is used as a factor 2 times, its exponent is 2. Because r is used as a factor 3 times, its exponent is 3.

So, $\pi \cdot \pi \cdot r \cdot r \cdot r = \pi^2 r^3$.

On Your Own

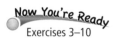

Now You're Ready
Exercises 3–10

Write the product using exponents.

1. $\frac{1}{4} \cdot \frac{1}{4} \cdot \frac{1}{4} \cdot \frac{1}{4} \cdot \frac{1}{4}$
2. $0.3 \cdot 0.3 \cdot 0.3 \cdot 0.3 \cdot x \cdot x$

EXAMPLE 2 Evaluating Expressions

Evaluate each expression.

a. $(-2)^4$

The base is -2.

$(-2)^4 = (-2) \cdot (-2) \cdot (-2) \cdot (-2)$ — Write as repeated multiplication.

$= 16$ — Simplify.

b. -2^4

The base is 2.

$-2^4 = -(2 \cdot 2 \cdot 2 \cdot 2)$ — Write as repeated multiplication.

$= -16$ — Simplify.

Multi-Language Glossary at BigIdeasMath.com

EXAMPLE 3 Using Order of Operations

Evaluate each expression.

a. $3 + 2 \cdot 3^4 = 3 + 2 \cdot 81$ Evaluate the power.

$= 3 + 162$ Multiply.

$= 165$ Add.

b. $3^3 - 8^2 \div 2 = 27 - 64 \div 2$ Evaluate the powers.

$= 27 - 32$ Divide.

$= -5$ Subtract.

On Your Own

Now You're Ready
Exercises 11–16 and 21–26

Evaluate the expression.

3. -5^4 **4.** $\left(-\frac{1}{6}\right)^3$ **5.** $\left|-3^3 \div 27\right|$ **6.** $9 - 2^5 \cdot 0.5$

EXAMPLE 4 Real-Life Application

In sphering, a person is secured inside a small, hollow sphere that is surrounded by a larger sphere. The space between the spheres is inflated with air. What is the volume of the inflated space?

You can find the radius of each sphere by dividing each diameter given in the diagram by 2.

Outer Sphere		***Inner Sphere***
$V = \frac{4}{3}\pi r^3$	Write formula.	$V = \frac{4}{3}\pi r^3$
$= \frac{4}{3}\pi\left(\frac{3}{2}\right)^3$	Substitute.	$= \frac{4}{3}\pi(1)^3$
$= \frac{4}{3}\pi\left(\frac{27}{8}\right)$	Evaluate the power.	$= \frac{4}{3}\pi(1)$
$= \frac{9}{2}\pi$	Multiply.	$= \frac{4}{3}\pi$

So, the volume of the inflated space is $\frac{9}{2}\pi - \frac{4}{3}\pi = \frac{19}{6}\pi$, or about 10 cubic meters.

On Your Own

7. WHAT IF? The diameter of the inner sphere is 1.8 meters. What is the volume of the inflated space?

10.1 Exercises

Vocabulary and Concept Check

1. **NUMBER SENSE** Describe the difference between -3^4 and $(-3)^4$.

2. **WHICH ONE DOESN'T BELONG?** Which one does *not* belong with the other three? Explain your reasoning.

5^3 The exponent is 3.	5^3 The power is 5.	5^3 The base is 5.	5^3 Five is used as a factor 3 times.

Practice and Problem Solving

Write the product using exponents.

3. $3 \cdot 3 \cdot 3 \cdot 3$

4. $(-6) \cdot (-6)$

5. $\left(-\frac{1}{2}\right) \cdot \left(-\frac{1}{2}\right) \cdot \left(-\frac{1}{2}\right)$

6. $\frac{1}{3} \cdot \frac{1}{3} \cdot \frac{1}{3}$

7. $\pi \cdot \pi \cdot \pi \cdot x \cdot x \cdot x \cdot x$

8. $(-4) \cdot (-4) \cdot (-4) \cdot y \cdot y$

9. $6.4 \cdot 6.4 \cdot 6.4 \cdot 6.4 \cdot b \cdot b \cdot b$

10. $(-t) \cdot (-t) \cdot (-t) \cdot (-t) \cdot (-t)$

Evaluate the expression.

2 11. 5^2

12. -11^3

13. $(-1)^6$

14. $\left(\frac{1}{2}\right)^6$

15. $\left(-\frac{1}{12}\right)^2$

16. $-\left(\frac{1}{9}\right)^3$

17. **ERROR ANALYSIS** Describe and correct the error in evaluating the expression.

✗ $-6^2 = (-6) \cdot (-6) = 36$

18. **PRIME FACTORIZATION** Write the prime factorization of 675 using exponents.

19. **STRUCTURE** Write $-\left(\frac{1}{4} \cdot \frac{1}{4} \cdot \frac{1}{4} \cdot \frac{1}{4}\right)$ using exponents.

20. **RUSSIAN DOLLS** The largest doll is 12 inches tall. The height of each of the other dolls is $\frac{7}{10}$ the height of the next larger doll. Write an expression involving a power for the height of the smallest doll. What is the height of the smallest doll?

Evaluate the expression.

3 **21.** $5 + 3 \cdot 2^3$ **22.** $2 + 7 \cdot (-3)^2$ **23.** $(13^2 - 12^2) \div 5$

24. $\frac{1}{2}(4^3 - 6 \cdot 3^2)$ **25.** $\left|\frac{1}{2}(7 + 5^3)\right|$ **26.** $\left|\left(-\frac{1}{2}\right)^3 \div \left(\frac{1}{4}\right)^2\right|$

27. MONEY You have a part-time job. One day your boss offers to pay you either $2^h - 1$ or 2^{h-1} dollars for each hour h you work that day. Copy and complete the table. Which option should you choose? Explain.

h	1	2	3	4	5
$2^h - 1$					
2^{h-1}					

28. CARBON-14 DATING Scientists use carbon-14 dating to determine the age of a sample of organic material.

a. The amount C (in grams) of a 100-gram sample of carbon-14 remaining after t years is represented by the equation $C = 100(0.99988)^t$. Use a calculator to find the amount of carbon-14 remaining after 4 years.

b. What percent of the carbon-14 remains after 4 years?

29. Critical Thinking The frequency (in vibrations per second) of a note on a piano is represented by the equation $F = 440(1.0595)^n$, where n is the number of notes above A-440. Each black or white key represents one note.

a. How many notes do you take to travel from A-440 to A?

b. What is the frequency of A?

c. Describe the relationship between the number of notes between A-440 and A and the increase in frequency.

Fair Game Review What you learned in previous grades & lessons

Tell which property is illustrated by the statement. *(Skills Review Handbook)*

30. $8 \cdot x = x \cdot 8$ **31.** $(2 \cdot 10)x = 2(10 \cdot x)$ **32.** $3(x \cdot 1) = 3x$

33. MULTIPLE CHOICE The polygons are similar. What is the value of x? *(Section 2.5)*

Ⓐ 15 Ⓑ 16

Ⓒ 17 Ⓓ 36

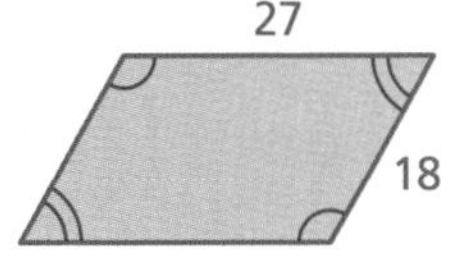

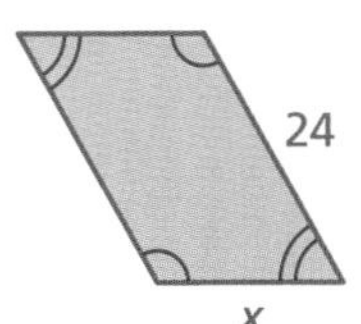

10.2 Product of Powers Property

Essential Question How can you use inductive reasoning to observe patterns and write general rules involving properties of exponents?

1 ACTIVITY: Finding Products of Powers

Work with a partner.

a. Copy and complete the table.

Product	Repeated Multiplication Form	Power
$2^2 \cdot 2^4$		
$(-3)^2 \cdot (-3)^4$		
$7^3 \cdot 7^2$		
$5.1^1 \cdot 5.1^6$		
$(-4)^2 \cdot (-4)^2$		
$10^3 \cdot 10^5$		
$\left(\frac{1}{2}\right)^5 \cdot \left(\frac{1}{2}\right)^5$		

b. **INDUCTIVE REASONING** Describe the pattern in the table. Then write a *general rule* for multiplying two powers that have the same base.

$$a^m \cdot a^n = a^{\square}$$

c. Use your rule to simplify the products in the first column of the table above. Does your rule give the results in the third column?

d. Most calculators have *exponent* keys that you can use to evaluate powers. Use a calculator with an exponent key to evaluate the products in part (a).

Exponents
In this lesson, you will
- multiply powers with the same base.
- find a power of a power.
- find a power of a product.

2 ACTIVITY: Writing a Rule for Powers of Powers

Work with a partner. Write the expression as a single power. Then write a *general rule* for finding a power of a power.

a. $(3^2)^3 = (3 \cdot 3)(3 \cdot 3)(3 \cdot 3) = \square^{\square}$

b. $(2^2)^4 =$

c. $(7^3)^2 =$

d. $(y^3)^3 =$

e. $(x^4)^2 =$

3 ACTIVITY: Writing a Rule for Powers of Products

Work with a partner. Write the expression as the product of two powers. Then write a *general rule* for finding a power of a product.

a. $(2 \cdot 3)^3 = (2 \cdot 3)(2 \cdot 3)(2 \cdot 3) = \underline{\quad}^{\underline{\quad}} \cdot \underline{\quad}^{\underline{\quad}}$

b. $(2 \cdot 5)^2 = \underline{\qquad}$

c. $(5 \cdot 4)^3 = \underline{\qquad}$

d. $(6a)^4 = \underline{\qquad}$

e. $(3x)^2 = \underline{\qquad}$

4 ACTIVITY: The Penny Puzzle

Work with a partner.

- The rows y and columns x of a chessboard are numbered as shown.
- Each position on the chessboard has a stack of pennies. (Only the first row is shown.)
- The number of pennies in each stack is $2^x \cdot 2^y$.

Math Practice

Look for Patterns

What patterns do you notice? How does this help you determine which stack is the tallest?

a. How many pennies are in the stack in location (3, 5)?

b. Which locations have 32 pennies in their stacks?

c. How much money (in dollars) is in the location with the tallest stack?

d. A penny is about 0.06 inch thick. About how tall (in inches) is the tallest stack?

What Is Your Answer?

5. IN YOUR OWN WORDS How can you use inductive reasoning to observe patterns and write general rules involving properties of exponents?

Practice Use what you learned about properties of exponents to complete Exercises 3–5 on page 420.

10.2 Lesson

Product of Powers Property

Words To multiply powers with the same base, add their exponents.

Numbers $4^2 \cdot 4^3 = 4^{2+3} = 4^5$ **Algebra** $a^m \cdot a^n = a^{m+n}$

Power of a Power Property

Words To find a power of a power, multiply the exponents.

Numbers $(4^6)^3 = 4^{6 \cdot 3} = 4^{18}$ **Algebra** $(a^m)^n = a^{mn}$

Power of a Product Property

Words To find a power of a product, find the power of each factor and multiply.

Numbers $(3 \cdot 2)^5 = 3^5 \cdot 2^5$ **Algebra** $(ab)^m = a^m b^m$

EXAMPLE 1 Multiplying Powers with the Same Base

a. $2^4 \cdot 2^5 = 2^{4+5}$ Product of Powers Property

$= 2^9$ Simplify.

Study Tip

When a number is written without an exponent, its exponent is 1.

b. $-5 \cdot (-5)^6 = (-5)^1 \cdot (-5)^6$ Rewrite -5 as $(-5)^1$.

$= (-5)^{1+6}$ Product of Powers Property

$= (-5)^7$ Simplify.

c. $x^3 \cdot x^7 = x^{3+7}$ Product of Powers Property

$= x^{10}$ Simplify.

EXAMPLE 2 Finding a Power of a Power

a. $(3^4)^3 = 3^{4 \cdot 3}$ Power of a Power Property

$= 3^{12}$ Simplify.

b. $(w^5)^4 = w^{5 \cdot 4}$ Power of a Power Property

$= w^{20}$ Simplify.

EXAMPLE 3 Finding a Power of a Product

a. $(2x)^3 = 2^3 \cdot x^3$ — Power of a Product Property

$= 8x^3$ — Simplify.

b. $(3xy)^2 = 3^2 \cdot x^2 \cdot y^2$ — Power of a Product Property

$= 9x^2y^2$ — Simplify.

On Your Own

Now You're Ready
Exercises 3–14
and 17–22

Simplify the expression.

1. $6^2 \cdot 6^4$ **2.** $\left(-\frac{1}{2}\right)^3 \cdot \left(-\frac{1}{2}\right)^6$ **3.** $z \cdot z^{12}$

4. $(4^4)^3$ **5.** $(y^2)^4$ **6.** $((-4)^3)^2$

7. $(5y)^4$ **8.** $(ab)^5$ **9.** $(0.5mn)^2$

EXAMPLE 4 Simplifying an Expression

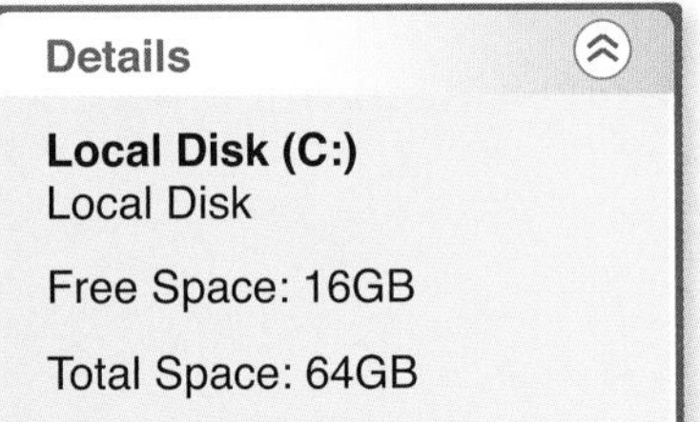

A gigabyte (GB) of computer storage space is 2^{30} bytes. The details of a computer are shown. How many bytes of total storage space does the computer have?

Ⓐ 2^{34} Ⓑ 2^{36} Ⓒ 2^{180} Ⓓ 128^{30}

The computer has 64 gigabytes of total storage space. Notice that you can write 64 as a power, 2^6. Use a model to solve the problem.

Total number of bytes = Number of bytes in a gigabyte • Number of gigabytes

$= 2^{30} \cdot 2^6$ — Substitute.

$= 2^{30+6}$ — Product of Powers Property

$= 2^{36}$ — Simplify.

∴ The computer has 2^{36} bytes of total storage space. The correct answer is Ⓑ.

On Your Own

10. How many bytes of free storage space does the computer have?

10.2 Exercises

Vocabulary and Concept Check

1. **REASONING** When should you use the Product of Powers Property?
2. **CRITICAL THINKING** Can you use the Product of Powers Property to multiply $5^2 \cdot 6^4$? Explain.

Practice and Problem Solving

Simplify the expression. Write your answer as a power.

1 2 3. $3^2 \cdot 3^2$

4. $8^{10} \cdot 8^4$

5. $(-4)^5 \cdot (-4)^7$

6. $a^3 \cdot a^3$

7. $h^6 \cdot h$

8. $\left(\frac{2}{3}\right)^2 \cdot \left(\frac{2}{3}\right)^6$

9. $\left(-\frac{5}{7}\right)^8 \cdot \left(-\frac{5}{7}\right)^9$

10. $(-2.9) \cdot (-2.9)^7$

11. $(5^4)^3$

12. $(b^{12})^3$

13. $(3.8^3)^4$

14. $\left(\left(-\frac{3}{4}\right)^5\right)^2$

ERROR ANALYSIS **Describe and correct the error in simplifying the expression.**

15. ✗ $5^2 \cdot 5^9 = (5 \cdot 5)^{2+9}$
$= 25^{11}$

16. ✗ $(r^6)^4 = r^{6+4}$
$= r^{10}$

Simplify the expression.

3 17. $(6g)^3$

18. $(-3v)^5$

19. $\left(\frac{1}{5}k\right)^2$

20. $(1.2m)^4$

21. $(rt)^{12}$

22. $\left(-\frac{3}{4}p\right)^3$

23. **PRECISION** Is $3^2 + 3^3$ equal to 3^5? Explain.

24. **ARTIFACT** A display case for the artifact is in the shape of a cube. Each side of the display case is three times longer than the width of the artifact.

    a. Write an expression for the volume of the case. Write your answer as a power.

    b. Simplify the expression.

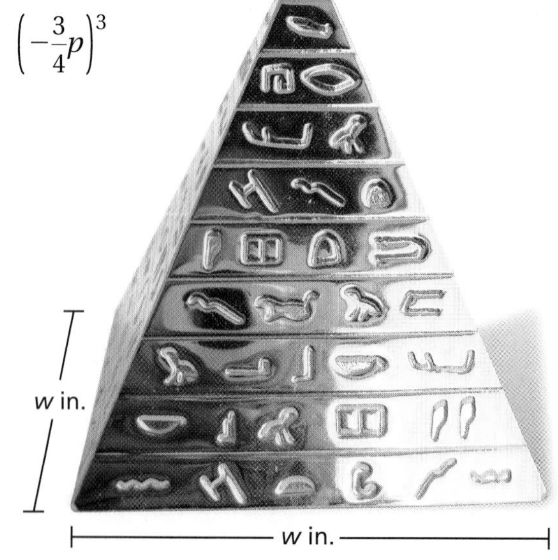

Simplify the expression.

25. $2^4 \cdot 2^5 - (2^2)^2$

26. $16\left(\frac{1}{2}x\right)^4$

27. $5^2(5^3 \cdot 5^2)$

28. **CLOUDS** The lowest altitude of an altocumulus cloud is about 3^8 feet. The highest altitude of an altocumulus cloud is about 3 times the lowest altitude. What is the highest altitude of an altocumulus cloud? Write your answer as a power.

29. **PYTHON EGG** The volume V of a python egg is given by the formula $V = \frac{4}{3}\pi abc$. For the python eggs shown, $a = 2$ inches, $b = 2$ inches, and $c = 3$ inches.

a. Find the volume of a python egg.

b. Square the dimensions of the python egg. Then evaluate the formula. How does this volume compare to your answer in part (a)?

30. **PYRAMID** A square pyramid has a height h and a base with side length b. The side lengths of the base increase by 50%. Write a formula for the volume of the new pyramid in terms of b and h.

31. **MAIL** The United States Postal Service delivers about $2^8 \cdot 5^2$ pieces of mail each second. There are $2^8 \cdot 3^4 \cdot 5^2$ seconds in 6 days. How many pieces of mail does the United States Postal Service deliver in 6 days? Write your answer as an expression involving powers.

32. **Critical Thinking** Find the value of x in the equation without evaluating the power.

a. $2^5 \cdot 2^x = 256$

b. $\left(\frac{1}{3}\right)^2 \cdot \left(\frac{1}{3}\right)^x = \frac{1}{729}$

Fair Game Review What you learned in previous grades & lessons

Simplify. *(Skills Review Handbook)*

33. $\frac{4 \cdot 4}{4}$

34. $\frac{5 \cdot 5 \cdot 5}{5}$

35. $\frac{2 \cdot 3}{2}$

36. $\frac{8 \cdot 6 \cdot 6}{6 \cdot 8}$

37. **MULTIPLE CHOICE** What is the measure of each interior angle of the regular polygon? *(Section 3.3)*

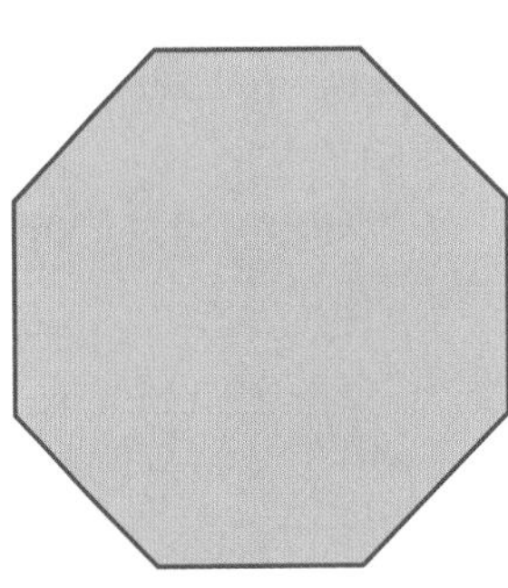

(A) 45°

(B) 135°

(C) 1080°

(D) 1440°

10.3 Quotient of Powers Property

Essential Question How can you divide two powers that have the same base?

1 ACTIVITY: Finding Quotients of Powers

Work with a partner.

a. Copy and complete the table.

Quotient	Repeated Multiplication Form	Power
$\frac{2^4}{2^2}$		
$\frac{(-4)^5}{(-4)^2}$		
$\frac{7^7}{7^3}$		
$\frac{8.5^9}{8.5^6}$		
$\frac{10^8}{10^5}$		
$\frac{3^{12}}{3^4}$		
$\frac{(-5)^7}{(-5)^5}$		
$\frac{11^4}{11^1}$		

Exponents

In this lesson, you will

- divide powers with the same base.
- simplify expressions involving the quotient of powers.

b. **INDUCTIVE REASONING** Describe the pattern in the table. Then write a rule for dividing two powers that have the same base.

$$\frac{a^m}{a^n} = a^{\square}$$

c. Use your rule to simplify the quotients in the first column of the table above. Does your rule give the results in the third column?

2 ACTIVITY: Comparing Volumes

Math Practice

Repeat Calculations

What calculations are repeated in the table?

Work with a partner.

How many of the smaller cubes will fit inside the larger cube? Record your results in the table. Describe the pattern in the table.

a.

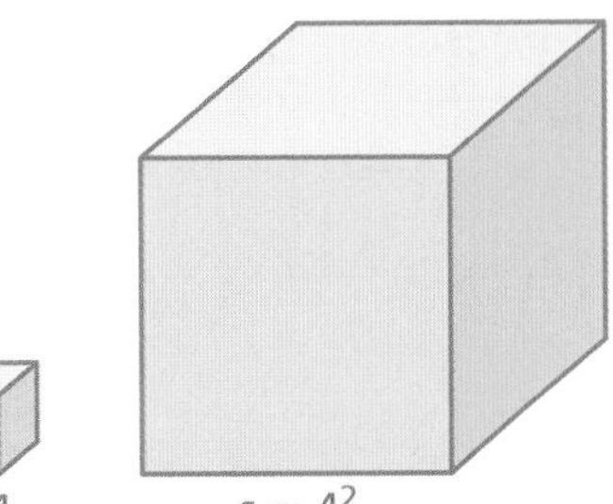

b.

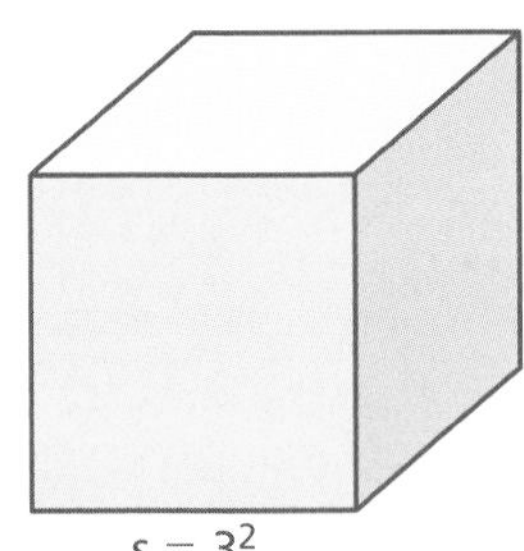

c.

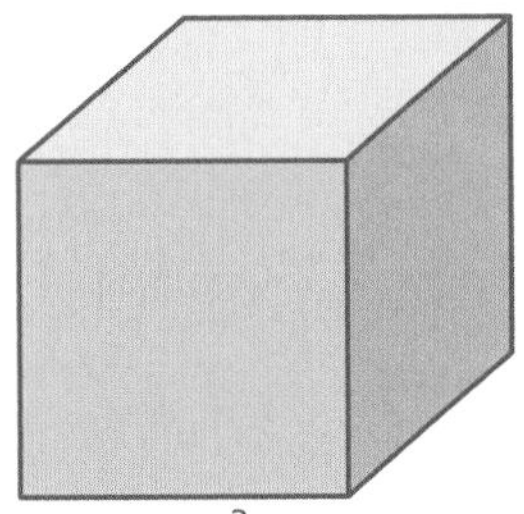

d.

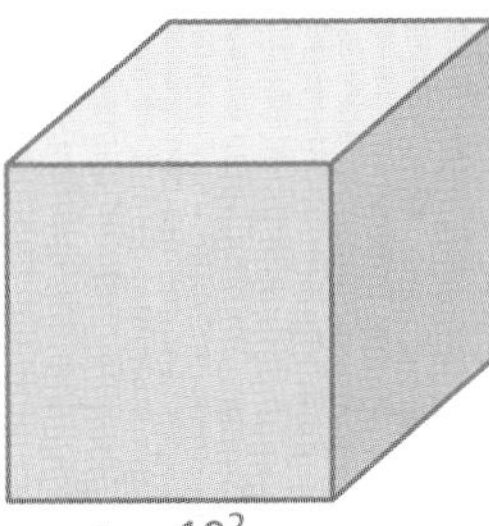

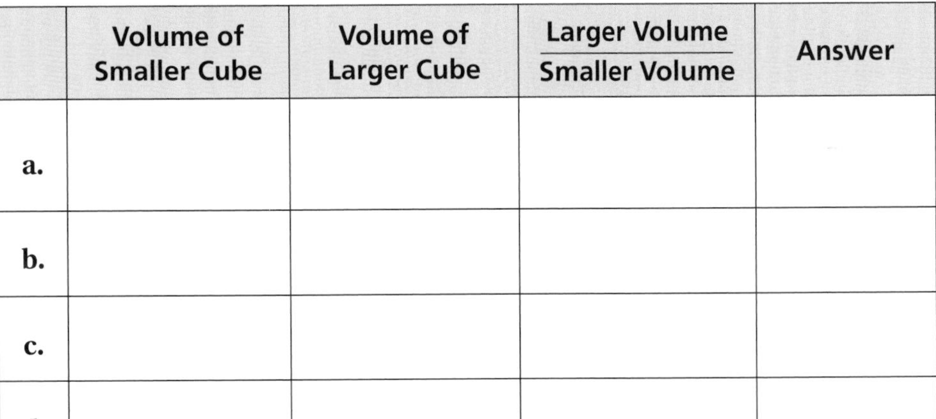

	Volume of Smaller Cube	Volume of Larger Cube	$\frac{\text{Larger Volume}}{\text{Smaller Volume}}$	Answer
a.				
b.				
c.				
d.				

What Is Your Answer?

3. **IN YOUR OWN WORDS** How can you divide two powers that have the same base? Give two examples of your rule.

Practice Use what you learned about dividing powers with the same base to complete Exercises 3–6 on page 426.

10.3 Lesson

Quotient of Powers Property

Words To divide powers with the same base, subtract their exponents.

Numbers $\frac{4^5}{4^2} = 4^{5-2} = 4^3$ **Algebra** $\frac{a^m}{a^n} = a^{m-n}$, where $a \neq 0$

EXAMPLE 1 Dividing Powers with the Same Base

a. $\frac{2^6}{2^4} = 2^{6-4}$ Quotient of Powers Property

$= 2^2$ Simplify.

Common Error

When dividing powers, do not divide the bases.

$\frac{2^6}{2^4} = 2^2$, not 1^2.

b. $\frac{(-7)^9}{(-7)^3} = (-7)^{9-3}$ Quotient of Powers Property

$= (-7)^6$ Simplify.

c. $\frac{h^7}{h^6} = h^{7-6}$ Quotient of Powers Property

$= h^1 = h$ Simplify.

On Your Own

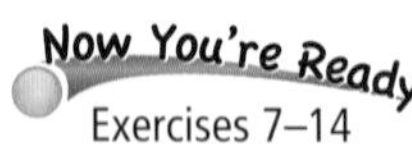

Exercises 7–14

Simplify the expression. Write your answer as a power.

1. $\frac{9^7}{9^4}$
2. $\frac{4.2^6}{4.2^5}$
3. $\frac{(-8)^8}{(-8)^4}$
4. $\frac{x^8}{x^3}$

EXAMPLE 2 Simplifying an Expression

Simplify $\frac{3^4 \cdot 3^2}{3^3}$. Write your answer as a power.

The numerator is a product of powers. Add the exponents in the numerator.

$\frac{3^4 \cdot 3^2}{3^3} = \frac{3^{4+2}}{3^3}$ Product of Powers Property

$= \frac{3^6}{3^3}$ Simplify.

$= 3^{6-3}$ Quotient of Powers Property

$= 3^3$ Simplify.

EXAMPLE 3 Simplifying an Expression

Simplify $\frac{a^{10}}{a^6} \cdot \frac{a^7}{a^4}$. Write your answer as a power.

$$\frac{a^{10}}{a^6} \cdot \frac{a^7}{a^4} = a^{10-6} \cdot a^{7-4} \qquad \text{Quotient of Powers Property}$$

$$= a^4 \cdot a^3 \qquad \text{Simplify.}$$

$$= a^{4+3} \qquad \text{Product of Powers Property}$$

$$= a^7 \qquad \text{Simplify.}$$

Study Tip

You can also simplify the expression in Example 3 as follows.

$$\frac{a^{10}}{a^6} \cdot \frac{a^7}{a^4} = \frac{a^{10} \cdot a^7}{a^6 \cdot a^4} = \frac{a^{17}}{a^{10}} = a^{17-10} = a^7$$

On Your Own

Now You're Ready Exercises 16–21

Simplify the expression. Write your answer as a power.

5. $\frac{2^{15}}{2^3 \cdot 2^5}$

6. $\frac{d^5}{d} \cdot \frac{d^9}{d^8}$

7. $\frac{5^9}{5^4} \cdot \frac{5^5}{5^2}$

EXAMPLE 4 Real-Life Application

The projected population of Tennessee in 2030 is about $5 \cdot 5.9^8$. Predict the average number of people per square mile in 2030.

Use a model to solve the problem.

$$\text{People per square mile} = \frac{\text{Population in 2030}}{\text{Land area}}$$

Land area: about 5.9^6 mi^2

$$= \frac{5 \cdot 5.9^8}{5.9^6} \qquad \text{Substitute.}$$

$$= 5 \cdot \frac{5.9^8}{5.9^6} \qquad \text{Rewrite.}$$

$$= 5 \cdot 5.9^2 \qquad \text{Quotient of Powers Property}$$

$$= 174.05 \qquad \text{Evaluate.}$$

So, there will be about 174 people per square mile in Tennessee in 2030.

On Your Own

Now You're Ready Exercises 23–28

8. The projected population of Alabama in 2030 is about $2.25 \cdot 2^{21}$. The land area of Alabama is about 2^{17} square kilometers. Predict the average number of people per square kilometer in 2030.

10.3 Exercises

Vocabulary and Concept Check

1. **WRITING** Describe in your own words how to divide powers.

2. **WHICH ONE DOESN'T BELONG?** Which quotient does *not* belong with the other three? Explain your reasoning.

$\frac{(-10)^7}{(-10)^2}$ $\frac{6^3}{6^2}$ $\frac{(-4)^8}{(-3)^4}$ $\frac{5^6}{5^3}$

Practice and Problem Solving

Simplify the expression. Write your answer as a power.

3. $\frac{6^{10}}{6^4}$
4. $\frac{8^9}{8^7}$
5. $\frac{(-3)^4}{(-3)^1}$
6. $\frac{4.5^5}{4.5^3}$

1 7. $\frac{5^9}{5^3}$
8. $\frac{64^4}{64^3}$
9. $\frac{(-17)^5}{(-17)^2}$
10. $\frac{(-7.9)^{10}}{(-7.9)^4}$
11. $\frac{(-6.4)^8}{(-6.4)^6}$
12. $\frac{\pi^{11}}{\pi^7}$
13. $\frac{b^{24}}{b^{11}}$
14. $\frac{n^{18}}{n^7}$

15. **ERROR ANALYSIS** Describe and correct the error in simplifying the quotient.

$$\frac{6^{15}}{6^5} = 6^{\frac{15}{5}} = 6^3$$

Simplify the expression. Write your answer as a power.

2 3 16. $\frac{7^5 \cdot 7^3}{7^2}$
17. $\frac{2^{19} \cdot 2^5}{2^{12} \cdot 2^3}$
18. $\frac{(-8.3)^8}{(-8.3)^7} \cdot \frac{(-8.3)^4}{(-8.3)^3}$
19. $\frac{\pi^{30}}{\pi^{18} \cdot \pi^4}$
20. $\frac{c^{22}}{c^8 \cdot c^9}$
21. $\frac{k^{13}}{k^5} \cdot \frac{k^{17}}{k^{11}}$

22. **SOUND INTENSITY** The sound intensity of a normal conversation is 10^6 times greater than the quietest noise a person can hear. The sound intensity of a jet at takeoff is 10^{14} times greater than the quietest noise a person can hear. How many times more intense is the sound of a jet at takeoff than the sound of a normal conversation?

Simplify the expression.

④ **23.** $\dfrac{x \cdot 4^8}{4^5}$ **24.** $\dfrac{6^3 \cdot w}{6^2}$ **25.** $\dfrac{a^3 \cdot b^4 \cdot 5^4}{b^2 \cdot 5}$

26. $\dfrac{5^{12} \cdot c^{10} \cdot d^2}{5^9 \cdot c^9}$ **27.** $\dfrac{x^{15}y^9}{x^8y^3}$ **28.** $\dfrac{m^{10}n^7}{m^1n^6}$

MP3 Player	Memory (GB)	Price
A	2^1	\$70
B	2^2	\$120
C	2^3	\$170
D	2^4	\$220
E	2^5	\$270

29. MEMORY The memory capacities and prices of five MP3 players are shown in the table.

a. How many times more memory does MP3 Player D have than MP3 Player B?

b. Do memory and price show a linear relationship? Explain.

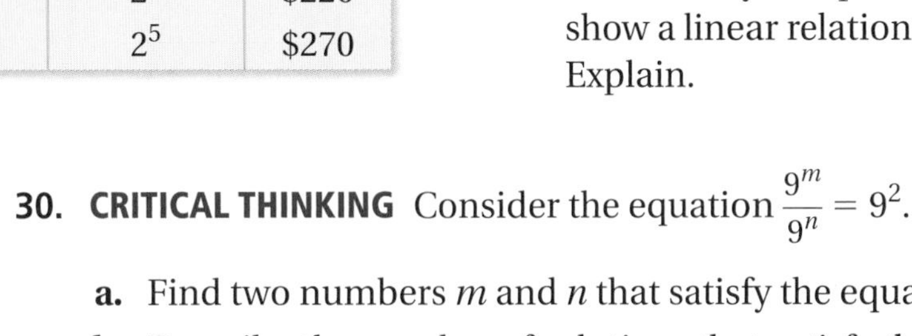

30. CRITICAL THINKING Consider the equation $\dfrac{9^m}{9^n} = 9^2$.

a. Find two numbers m and n that satisfy the equation.

b. Describe the number of solutions that satisfy the equation. Explain your reasoning.

Milky Way galaxy
$10 \cdot 10^{10}$ stars

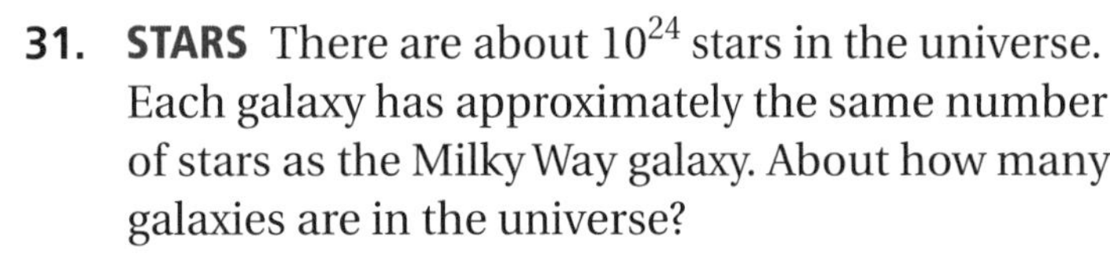

31. STARS There are about 10^{24} stars in the universe. Each galaxy has approximately the same number of stars as the Milky Way galaxy. About how many galaxies are in the universe?

32. Number Sense Find the value of x that makes $\dfrac{8^{3x}}{8^{2x+1}} = 8^9$ true. Explain how you found your answer.

Fair Game Review What you learned in previous grades & lessons

Subtract. *(Skills Review Handbook)*

33. $-4 - 5$ **34.** $-23 - (-15)$ **35.** $33 - (-28)$ **36.** $18 - 22$

37. MULTIPLE CHOICE What is the value of x?
(Skills Review Handbook)

Ⓐ 20 Ⓑ 30

Ⓒ 45 Ⓓ 60

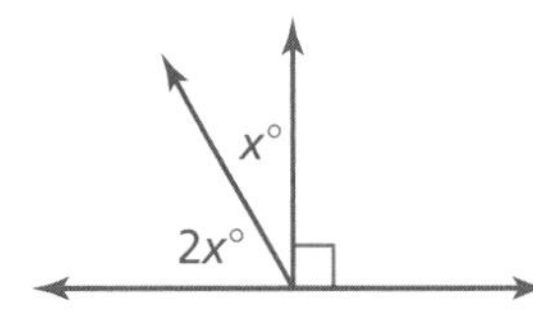

10.4 Zero and Negative Exponents

Essential Question How can you evaluate a nonzero number with an exponent of zero? How can you evaluate a nonzero number with a negative integer exponent?

1 ACTIVITY: Using the Quotient of Powers Property

Work with a partner.

a. Copy and complete the table.

Quotient	Quotient of Powers Property	Power
$\frac{5^3}{5^3}$		
$\frac{6^2}{6^2}$		
$\frac{(-3)^4}{(-3)^4}$		
$\frac{(-4)^5}{(-4)^5}$		

b. **REPEATED REASONING** Evaluate each expression in the first column of the table. What do you notice?

c. How can you use these results to define a^0 where $a \neq 0$?

2 ACTIVITY: Using the Product of Powers Property

Work with a partner.

a. Copy and complete the table.

Product	Product of Powers Property	Power
$3^0 \cdot 3^4$		
$8^2 \cdot 8^0$		
$(-2)^3 \cdot (-2)^0$		
$\left(-\frac{1}{3}\right)^0 \cdot \left(-\frac{1}{3}\right)^5$		

b. Do these results support your definition in Activity 1(c)?

Exponents
In this lesson, you will
- evaluate expressions involving numbers with zero as an exponent.
- evaluate expressions involving negative integer exponents.

3 ACTIVITY: Using the Product of Powers Property

Work with a partner.

a. Copy and complete the table.

Product	Product of Powers Property	Power
$5^{-3} \cdot 5^{3}$		
$6^{2} \cdot 6^{-2}$		
$(-3)^{4} \cdot (-3)^{-4}$		
$(-4)^{-5} \cdot (-4)^{5}$		

b. According to your results from Activities 1 and 2, the products in the first column are equal to what value?

c. **REASONING** How does the Multiplicative Inverse Property help you rewrite the numbers with negative exponents?

d. **STRUCTURE** Use these results to define a^{-n} where $a \neq 0$ and n is an integer.

4 ACTIVITY: Using a Place Value Chart

Math Practice

Use Operations

What operations are used when writing the expanded form?

Work with a partner. Use the place value chart that shows the number 3452.867.

Place Value Chart

thousands	hundreds	tens	ones	and	tenths	hundredths	thousandths
10^3	10^2	10^1	$10^{\square}$		$10^{\square}$	$10^{\square}$	$10^{\square}$
3	4	5	2	.	8	6	7

a. **REPEATED REASONING** What pattern do you see in the exponents? Continue the pattern to find the other exponents.

b. **STRUCTURE** Show how to write the expanded form of 3452.867.

What Is Your Answer?

5. **IN YOUR OWN WORDS** How can you evaluate a nonzero number with an exponent of zero? How can you evaluate a nonzero number with a negative integer exponent?

Practice Use what you learned about zero and negative exponents to complete Exercises 5–8 on page 432.

10.4 Lesson

Key Ideas

Zero Exponents

Words For any nonzero number a, $a^0 = 1$. The power 0^0 is *undefined*.

Numbers $4^0 = 1$

Algebra $a^0 = 1$, where $a \neq 0$

Negative Exponents

Words For any integer n and any nonzero number a, a^{-n} is the reciprocal of a^n.

Numbers $4^{-2} = \frac{1}{4^2}$

Algebra $a^{-n} = \frac{1}{a^n}$, where $a \neq 0$

EXAMPLE 1 Evaluating Expressions

a. $3^{-4} = \frac{1}{3^4}$ Definition of negative exponent

$= \frac{1}{81}$ Evaluate power.

b. $(-8.5)^{-4} \cdot (-8.5)^4 = (-8.5)^{-4+4}$ Product of Powers Property

$= (-8.5)^0$ Simplify.

$= 1$ Definition of zero exponent

c. $\frac{2^6}{2^8} = 2^{6-8}$ Quotient of Powers Property

$= 2^{-2}$ Simplify.

$= \frac{1}{2^2}$ Definition of negative exponent

$= \frac{1}{4}$ Evaluate power.

On Your Own

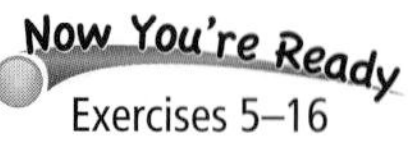

Exercises 5–16

Evaluate the expression.

1. 4^{-2}

2. $(-2)^{-5}$

3. $6^{-8} \cdot 6^8$

4. $\frac{(-3)^5}{(-3)^6}$

5. $\frac{1}{5^7} \cdot \frac{1}{5^{-4}}$

6. $\frac{4^5 \cdot 4^{-3}}{4^2}$

EXAMPLE 2 Simplifying Expressions

a. $-5x^0 = -5(1)$ Definition of zero exponent

$= -5$ Multiply.

b. $\dfrac{9y^{-3}}{y^5} = 9y^{-3-5}$ Quotient of Powers Property

$= 9y^{-8}$ Simplify.

$= \dfrac{9}{y^8}$ Definition of negative exponent

On Your Own

Simplify. Write the expression using only positive exponents.

7. $8x^{-2}$

8. $b^0 \cdot b^{-10}$

9. $\dfrac{z^6}{15z^9}$

EXAMPLE 3 Real-Life Application

A drop of water leaks from a faucet every second. How many liters of water leak from the faucet in 1 hour?

Drop of water: 50^{-2} liter

Convert 1 hour to seconds.

$$1\text{ h} \times \frac{60\text{ min}}{1\text{ h}} \times \frac{60\text{ sec}}{1\text{ min}} = 3600\text{ sec}$$

Water leaks from the faucet at a rate of 50^{-2} liter per second. Multiply the time by the rate.

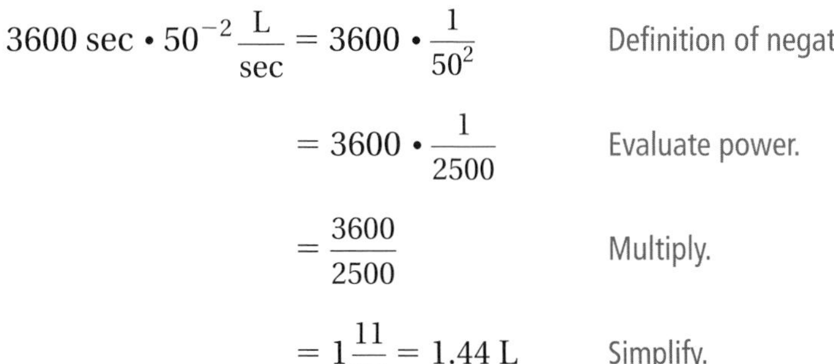

$3600\text{ sec} \cdot 50^{-2}\dfrac{\text{L}}{\text{sec}} = 3600 \cdot \dfrac{1}{50^2}$ Definition of negative exponent

$= 3600 \cdot \dfrac{1}{2500}$ Evaluate power.

$= \dfrac{3600}{2500}$ Multiply.

$= 1\dfrac{11}{25} = 1.44\text{ L}$ Simplify.

So, 1.44 liters of water leak from the faucet in 1 hour.

On Your Own

10. **WHAT IF?** The faucet leaks water at a rate of 5^{-5} liter per second. How many liters of water leak from the faucet in 1 hour?

10.4 Exercises

Vocabulary and Concept Check

1. **VOCABULARY** If a is a nonzero number, does the value of a^0 depend on the value of a? Explain.
2. **WRITING** Explain how to evaluate 10^{-3}.
3. **NUMBER SENSE** Without evaluating, order 5^0, 5^4, and 5^{-5} from least to greatest.
4. **DIFFERENT WORDS, SAME QUESTION** Which is different? Find "both" answers.

Rewrite $\frac{1}{3 \cdot 3 \cdot 3}$ using a negative exponent.	Write 3 to the negative third.
Write $\frac{1}{3}$ cubed as a power.	Write $(-3) \cdot (-3) \cdot (-3)$ as a power.

Practice and Problem Solving

Evaluate the expression.

1 5. $\frac{8^7}{8^7}$ 6. $5^0 \cdot 5^3$ 7. $(-2)^{-8} \cdot (-2)^8$ 8. $9^4 \cdot 9^{-4}$

9. 6^{-2} 10. 158^0 11. $\frac{4^3}{4^5}$ 12. $\frac{-3}{(-3)^2}$

13. $4 \cdot 2^{-4} + 5$ 14. $3^{-3} \cdot 3^{-2}$ 15. $\frac{1}{5^{-3}} \cdot \frac{1}{5^6}$ 16. $\frac{(1.5)^2}{(1.5)^{-2} \cdot (1.5)^4}$

17. **ERROR ANALYSIS** Describe and correct the error in evaluating the expression.

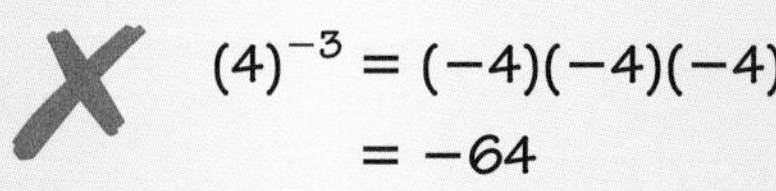

$$(4)^{-3} = (-4)(-4)(-4) = -64$$

18. **SAND** The mass of a grain of sand is about 10^{-3} gram. About how many grains of sand are in the bag of sand?

19. **CRITICAL THINKING** How can you write the number 1 as 2 to a power? 10 to a power?

Simplify. Write the expression using only positive exponents.

2 20. $6y^{-4}$ 21. $8^{-2} \cdot a^7$ 22. $\frac{9c^3}{c^{-4}}$ 23. $\frac{5b^{-2}}{b^{-3}}$

24. $\frac{8x^3}{2x^9}$ 25. $3d^{-4} \cdot 4d^4$ 26. $m^{-2} \cdot n^3$ 27. $\frac{3^{-2} \cdot k^0 \cdot w^0}{w^{-6}}$

28. OPEN-ENDED Write two different powers with negative exponents that have the same value.

METRIC UNITS In Exercises 29–32, use the table.

29. How many millimeters are in a decimeter?

30. How many micrometers are in a centimeter?

31. How many nanometers are in a millimeter?

32. How many micrometers are in a meter?

Unit of Length	Length (meter)
Decimeter	10^{-1}
Centimeter	10^{-2}
Millimeter	10^{-3}
Micrometer	10^{-6}
Nanometer	10^{-9}

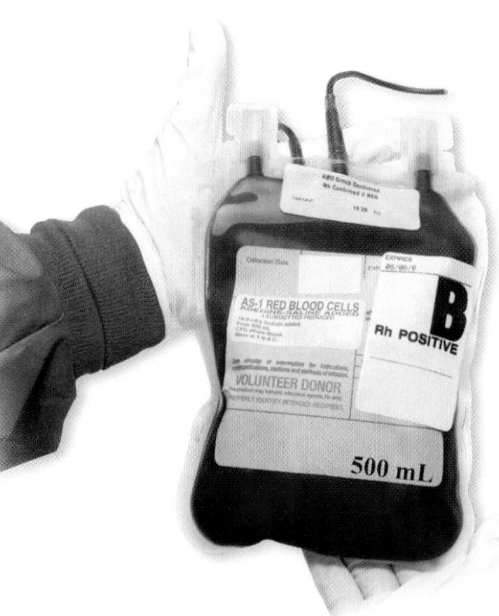

33. BACTERIA A species of bacteria is 10 micrometers long. A virus is 10,000 times smaller than the bacteria.

a. Using the table above, find the length of the virus in meters.

b. Is the answer to part (a) *less than, greater than,* or *equal to* one nanometer?

34. BLOOD DONATION Every 2 seconds, someone in the United States needs blood. A sample blood donation is shown. ($1 \text{ mm}^3 = 10^{-3} \text{ mL}$)

a. One cubic millimeter of blood contains about 10^4 white blood cells. How many white blood cells are in the donation? Write your answer in words.

b. One cubic millimeter of blood contains about 5×10^6 red blood cells. How many red blood cells are in the donation? Write your answer in words.

c. Compare your answers for parts (a) and (b).

35. PRECISION Describe how to rewrite a power with a positive exponent so that the exponent is in the denominator. Use the definition of negative exponents to justify your reasoning.

36.  The rule for negative exponents states that $a^{-n} = \frac{1}{a^n}$. Explain why this rule does not apply when $a = 0$.

Fair Game Review What you learned in previous grades & lessons

Simplify the expression. Write your answer as a power. *(Section 10.2 and Section 10.3)*

37. $10^3 \cdot 10^6$

38. $10^2 \cdot 10$

39. $\frac{10^8}{10^4}$

40. MULTIPLE CHOICE Which data display best orders numerical data and shows how they are distributed? *(Section 9.4)*

Ⓐ bar graph

Ⓑ line graph

Ⓒ scatter plot

Ⓓ stem-and-leaf plot

10 Study Help

You can use an **information wheel** to organize information about a topic. Here is an example of an information wheel for exponents.

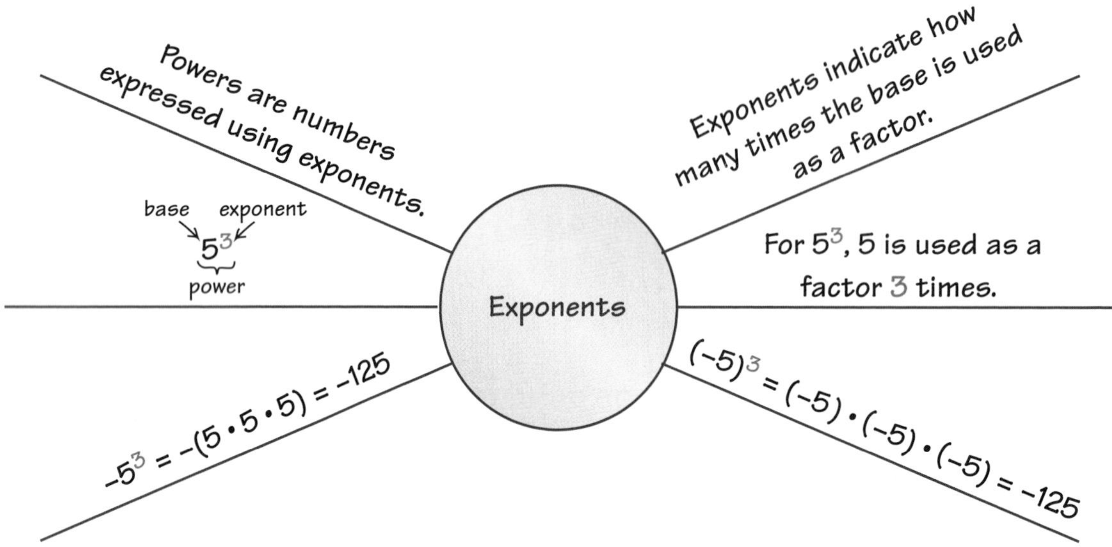

On Your Own

Make information wheels to help you study these topics.

1. Product of Powers Property
2. Quotient of Powers Property
3. zero and negative exponents

After you complete this chapter, make information wheels for the following topics.

4. writing numbers in scientific notation
5. writing numbers in standard form
6. adding and subtracting numbers in scientific notation
7. multiplying and dividing numbers in scientific notation
8. Choose three other topics you studied earlier in this course. Make an information wheel for each topic to summarize what you know about them.

"I decided to color code the different flavors in my information wheel."

10.1–10.4 Quiz

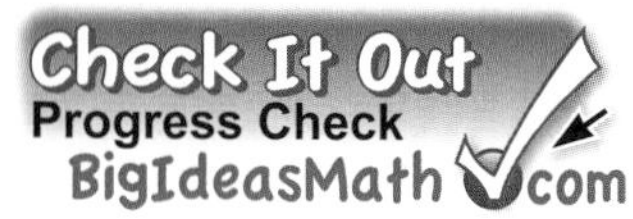

Write the product using exponents. *(Section 10.1)*

1. $(-5) \cdot (-5) \cdot (-5) \cdot (-5)$

2. $7 \cdot 7 \cdot m \cdot m \cdot m$

Evaluate the expression. *(Section 10.1 and Section 10.4)*

3. 5^4

4. $(-2)^6$

5. $(-4.8)^{-9} \cdot (-4.8)^9$

6. $\frac{5^4}{5^7}$

Simplify the expression. Write your answer as a power. *(Section 10.2)*

7. $3^8 \cdot 3$

8. $(a^5)^3$

Simplify the expression. *(Section 10.2)*

9. $(3c)^4$

10. $\left(-\frac{2}{7}p\right)^2$

Simplify the expression. Write your answer as a power. *(Section 10.3)*

11. $\frac{8^7}{8^4}$

12. $\frac{6^3 \cdot 6^7}{6^2}$

13. $\frac{\pi^{15}}{\pi^3 \cdot \pi^9}$

14. $\frac{t^{13}}{t^5} \cdot \frac{t^8}{t^6}$

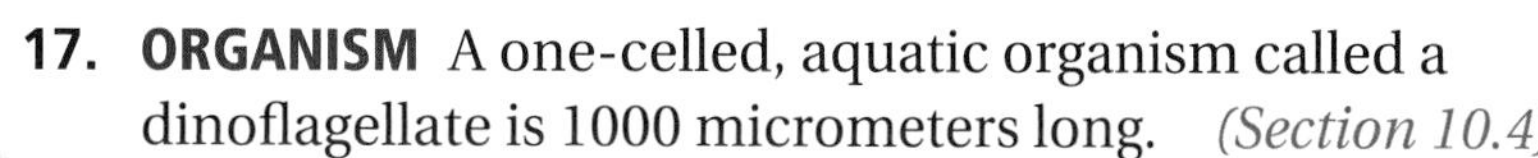

Simplify. Write the expression using only positive exponents. *(Section 10.4)*

15. $8d^{-6}$

16. $\frac{12x^5}{4x^7}$

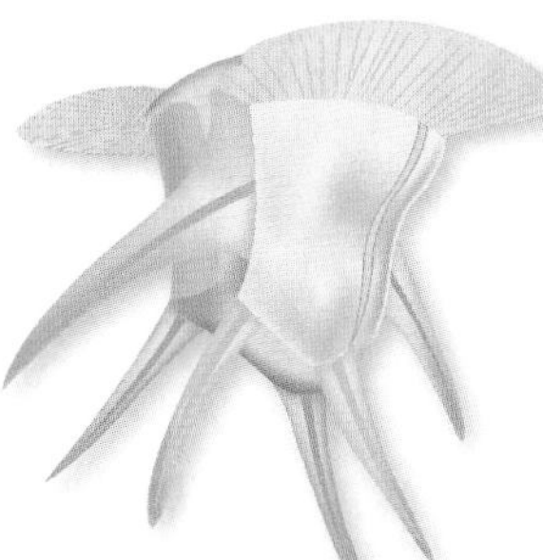

17. ORGANISM A one-celled, aquatic organism called a dinoflagellate is 1000 micrometers long. *(Section 10.4)*

a. One micrometer is 10^{-6} meter. What is the length of the dinoflagellate in meters?

b. Is the length of the dinoflagellate equal to 1 millimeter or 1 kilometer? Explain.

18. EARTHQUAKES An earthquake of magnitude 3.0 is 10^2 times stronger than an earthquake of magnitude 1.0. An earthquake of magnitude 8.0 is 10^7 times stronger than an earthquake of magnitude 1.0. How many times stronger is an earthquake of magnitude 8.0 than an earthquake of magnitude 3.0? *(Section 10.3)*

10.5 Reading Scientific Notation

Essential Question How can you read numbers that are written in scientific notation?

1 ACTIVITY: Very Large Numbers

Work with a partner.

- Use a calculator. Experiment with multiplying large numbers until your calculator displays an answer that is *not* in standard form.
- When the calculator at the right was used to multiply 2 billion by 3 billion, it listed the result as

  6.0E+18.

- Multiply 2 billion by 3 billion by hand. Use the result to explain what 6.0E+18 means.
- Check your explanation by calculating the products of other large numbers.
- Why didn't the calculator show the answer in standard form?
- Experiment to find the maximum number of digits your calculator displays. For instance, if you multiply 1000 by 1000 and your calculator shows 1,000,000, then it can display seven digits.

2 ACTIVITY: Very Small Numbers

Work with a partner.

- Use a calculator. Experiment with multiplying very small numbers until your calculator displays an answer that is *not* in standard form.
- When the calculator at the right was used to multiply 2 billionths by 3 billionths, it listed the result as

  6.0E–18.

- Multiply 2 billionths by 3 billionths by hand. Use the result to explain what 6.0E–18 means.
- Check your explanation by calculating the products of other very small numbers.

Scientific Notation

In this lesson, you will

- identify numbers written in scientific notation.
- write numbers in standard form.
- compare numbers in scientific notation.

3 ACTIVITY: Powers of 10 Matching Game

Math Practice

Analyze Relationships

How are the pictures related? How can you order the pictures to find the correct power of 10?

Work with a partner. Match each picture with its power of 10. Explain your reasoning.

10^5 m $\quad 10^2$ m $\quad 10^0$ m $\quad 10^{-1}$ m $\quad 10^{-2}$ m $\quad 10^{-5}$ m

A. 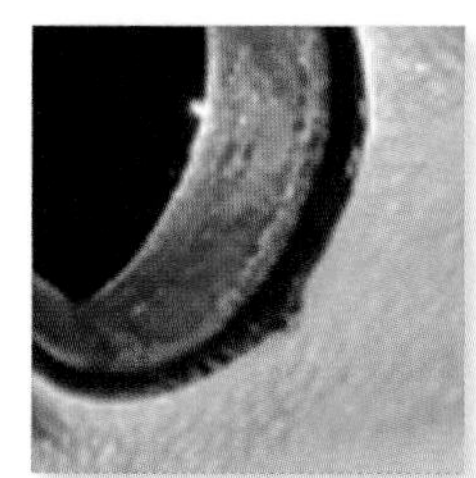

B. 

C. 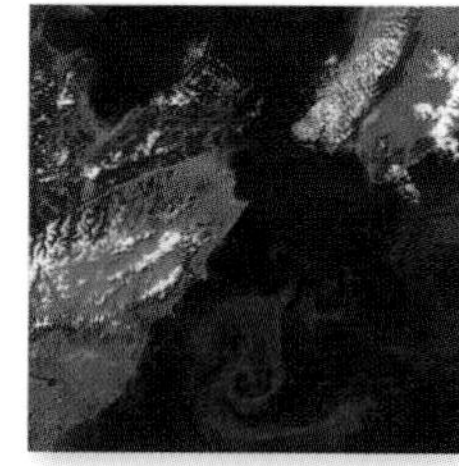

D. 

E. 

F. 

4 ACTIVITY: Choosing Appropriate Units

Work with a partner. Match each unit with its most appropriate measurement.

inches $\quad$ centimeters $\quad$ feet $\quad$ millimeters $\quad$ meters

A. Height of a door: 2×10^0

B. Height of a volcano: 1.6×10^4

C. Length of a pen: 1.4×10^2

D. Diameter of a steel ball bearing: 6.3×10^{-1}

E. Circumference of a beach ball: 7.5×10^1

What Is Your Answer?

5. **IN YOUR OWN WORDS** How can you read numbers that are written in scientific notation? Why do you think this type of notation is called *scientific notation*? Why is scientific notation important?

Practice Use what you learned about reading scientific notation to complete Exercises 3–5 on page 440.

10.5 Lesson

Key Vocabulary
scientific notation, *p. 438*

Scientific Notation

A number is written in **scientific notation** when it is represented as the product of a factor and a power of 10. The factor must be greater than or equal to 1 and less than 10.

The factor is greater than or equal to 1 and less than 10. → 8.3×10^{-7} ← The power of 10 has an integer exponent.

Study Tip

Scientific notation is used to write very small and very large numbers.

EXAMPLE 1 Identifying Numbers Written in Scientific Notation

Tell whether the number is written in scientific notation. Explain.

a. 5.9×10^{-6}

The factor is greater than or equal to 1 and less than 10. The power of 10 has an integer exponent. So, the number is written in scientific notation.

b. 0.9×10^{8}

The factor is less than 1. So, the number is not written in scientific notation.

Writing Numbers in Standard Form

The absolute value of the exponent indicates how many places to move the decimal point.

- If the exponent is negative, move the decimal point to the left.
- If the exponent is positive, move the decimal point to the right.

EXAMPLE 2 Writing Numbers in Standard Form

a. Write 3.22×10^{-4} in standard form.

$3.22 \times 10^{-4} = 0.000322$ (4) Move decimal point $|-4| = 4$ places to the left.

b. Write 7.9×10^{5} in standard form.

$7.9 \times 10^{5} = 790{,}000$ (5) Move decimal point $|5| = 5$ places to the right.

 Multi-Language Glossary at BigIdeasMath.com

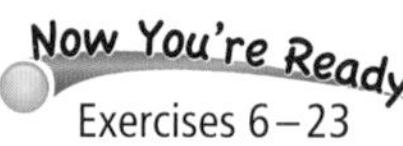

On Your Own

Now You're Ready
Exercises 6–23

1. Is 12×10^4 written in scientific notation? Explain.

Write the number in standard form.

2. 6×10^7
3. 9.9×10^{-5}
4. 1.285×10^4

EXAMPLE 3 Comparing Numbers in Scientific Notation

An object with a lesser density than water will float. An object with a greater density than water will sink. Use each given density (in kilograms per cubic meter) to explain what happens when you place a brick and an apple in water.

Water: 1.0×10^3 **Brick:** 1.84×10^3 **Apple:** 6.41×10^2

You can compare the densities by writing each in standard form.

Water	Brick	Apple
$1.0 \times 10^3 = 1000$	$1.84 \times 10^3 = 1840$	$6.41 \times 10^2 = 641$

The apple is less dense than water, so it will float. The brick is denser than water, so it will sink.

EXAMPLE 4 Real-Life Application

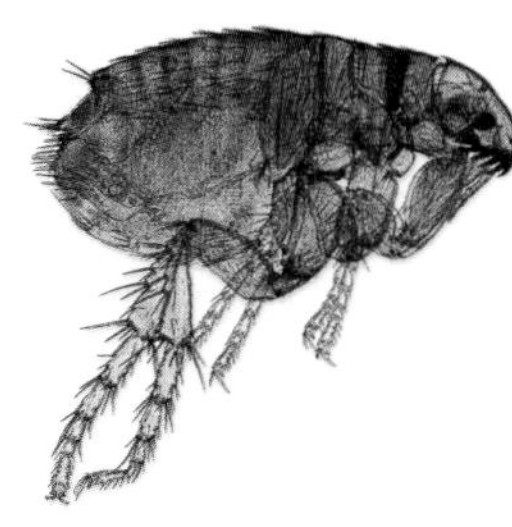

A female flea consumes about 1.4×10^{-5} liter of blood per day.

A dog has 100 female fleas. How much blood do the fleas consume per day?

$$1.4 \times 10^{-5} \cdot 100 = 0.000014 \cdot 100 \quad \text{Write in standard form.}$$
$$= 0.0014 \quad \text{Multiply.}$$

The fleas consume about 0.0014 liter, or 1.4 milliliters of blood per day.

On Your Own

Now You're Ready
Exercise 27

5. **WHAT IF?** In Example 3, the density of lead is 1.14×10^4 kilograms per cubic meter. What happens when you place lead in water?
6. **WHAT IF?** In Example 4, a dog has 75 female fleas. How much blood do the fleas consume per day?

10.5 Exercises

Vocabulary and Concept Check

1. **WRITING** Describe the difference between scientific notation and standard form.
2. **WHICH ONE DOESN'T BELONG?** Which number does *not* belong with the other three? Explain.

2.8×10^{15}	4.3×10^{-30}	1.05×10^{28}	10×9.2^{-13}

Practice and Problem Solving

Write the number shown on the calculator display in standard form.

3. 5.6E12
4. 2.1E-10
5. 8.73E16

Tell whether the number is written in scientific notation. Explain.

1 6. 1.8×10^9
7. 3.45×10^{14}
8. 0.26×10^{-25}
9. 10.5×10^{12}
10. 46×10^{-17}
11. 5×10^{-19}
12. 7.814×10^{-36}
13. 0.999×10^{42}
14. 6.022×10^{23}

Write the number in standard form.

2 15. 7×10^7
16. 8×10^{-3}
17. 5×10^2
18. 2.7×10^{-4}
19. 4.4×10^{-5}
20. 2.1×10^3
21. 1.66×10^9
22. 3.85×10^{-8}
23. 9.725×10^6

24. **ERROR ANALYSIS** Describe and correct the error in writing the number in standard form.

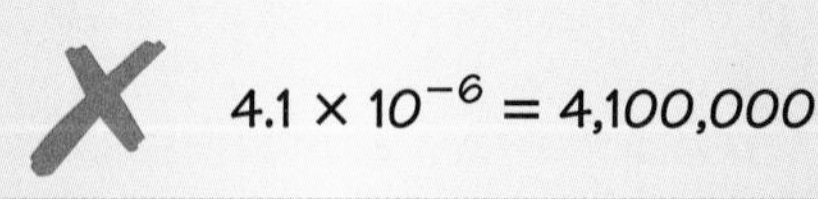

2.7×10^8 platelets per milliliter

25. **PLATELETS** Platelets are cell-like particles in the blood that help form blood clots.
    a. How many platelets are in 3 milliliters of blood? Write your answer in standard form.
    b. An adult human body contains about 5 liters of blood. How many platelets are in an adult human body?

26. **REASONING** A googol is 1.0×10^{100}. How many zeros are in a googol?

3 27. **STARS** The table shows the surface temperatures of five stars.

a. Which star has the highest surface temperature?

b. Which star has the lowest surface temperature?

Star	Betelgeuse	Bellatrix	Sun	Aldebaran	Rigel
Surface Temperature (°F)	6.2×10^3	3.8×10^4	1.1×10^4	7.2×10^3	2.2×10^4

28. **NUMBER SENSE** Describe how the value of a number written in scientific notation changes when you increase the exponent by 1.

29. **CORAL REEF** The area of the Florida Keys National Marine Sanctuary is about 9.6×10^3 square kilometers. The area of the Florida Reef Tract is about 16.2% of the area of the sanctuary. What is the area of the Florida Reef Tract in square kilometers?

30. **REASONING** A gigameter is 1.0×10^6 kilometers. How many square kilometers are in 5 square gigameters?

31. **WATER** There are about 1.4×10^9 cubic kilometers of water on Earth. About 2.5% of the water is fresh water. How much fresh water is on Earth?

32. 

The table shows the speed of light through five media.

a. In which medium does light travel the fastest?

b. In which medium does light travel the slowest?

Medium	Speed
Air	6.7×10^8 mi/h
Glass	6.6×10^8 ft/sec
Ice	2.3×10^5 km/sec
Vacuum	3.0×10^8 m/sec
Water	2.3×10^{10} cm/sec

Fair Game Review What you learned in previous grades & lessons

Write the product using exponents. *(Section 10.1)*

33. $4 \cdot 4 \cdot 4 \cdot 4 \cdot 4$

34. $3 \cdot 3 \cdot 3 \cdot y \cdot y \cdot y$

35. $(-2) \cdot (-2) \cdot (-2)$

36. **MULTIPLE CHOICE** What is the length of the hypotenuse of the right triangle? *(Section 7.3)*

Ⓐ $\sqrt{18}$ in. Ⓑ $\sqrt{41}$ in.

Ⓒ 18 in. Ⓓ 41 in.

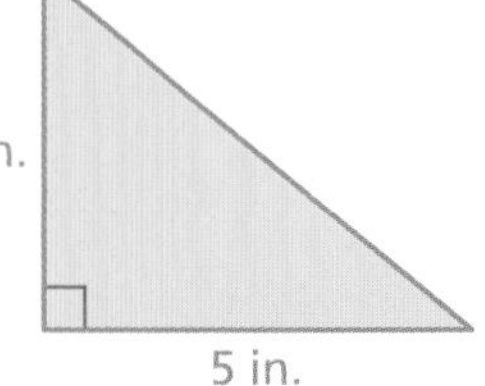

10.6 Writing Scientific Notation

Essential Question How can you write a number in scientific notation?

1 ACTIVITY: Finding pH Levels

Work with a partner. In chemistry, pH is a measure of the activity of dissolved hydrogen ions (H^+). Liquids with low pH values are called *acids*. Liquids with high pH values are called *bases*.

Find the pH of each liquid. Is the liquid a base, neutral, or an acid?

a. Lime juice:
$[H^+] = 0.01$

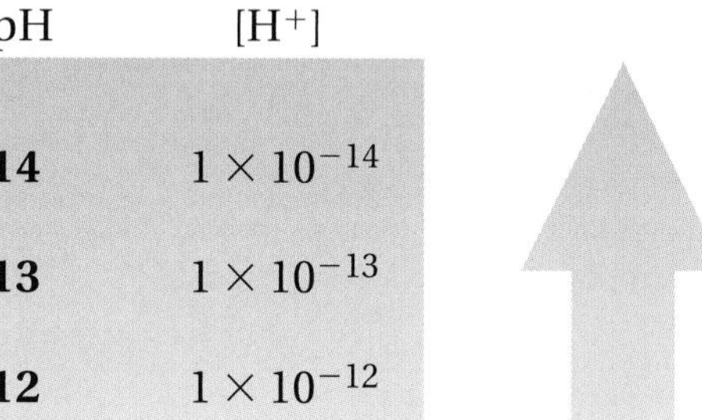

b. Egg:
$[H^+] = 0.00000001$

c. Distilled water:
$[H^+] = 0.0000001$

d. Ammonia water:
$[H^+] = 0.00000000001$

e. Tomato juice:
$[H^+] = 0.0001$

f. Hydrochloric acid:
$[H^+] = 1$

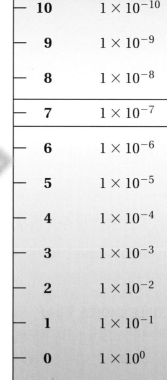

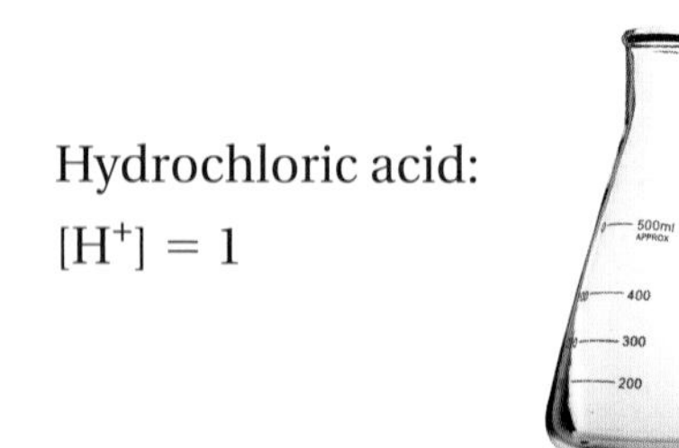

pH	$[H^+]$	
14	1×10^{-14}	Bases
13	1×10^{-13}	
12	1×10^{-12}	
11	1×10^{-11}	
10	1×10^{-10}	
9	1×10^{-9}	
8	1×10^{-8}	
7	1×10^{-7}	**Neutral**
6	1×10^{-6}	Acids
5	1×10^{-5}	
4	1×10^{-4}	
3	1×10^{-3}	
2	1×10^{-2}	
1	1×10^{-1}	
0	1×10^{0}	

Scientific Notation
In this lesson, you will
- write large and small numbers in scientific notation.
- perform operations with numbers written in scientific notation.

2 ACTIVITY: Writing Scientific Notation

Work with a partner. Match each planet with its distance from the Sun. Then write each distance in scientific notation. Do you think it is easier to match the distances when they are written in standard form or in scientific notation? Explain.

a. 1,800,000,000 miles

b. 67,000,000 miles

c. 890,000,000 miles

d. 93,000,000 miles

e. 140,000,000 miles

f. 2,800,000,000 miles

g. 480,000,000 miles

h. 36,000,000 miles

3 ACTIVITY: Making a Scale Drawing

Math Practice

Calculate Accurately

How can you verify that you have accurately written each distance in scientific notation?

Work with a partner. The illustration in Activity 2 is not drawn to scale. Use the instructions below to make a scale drawing of the distances in our solar system.

- **Cut a sheet of paper into three strips of equal width. Tape the strips together to make one long piece.**
- **Draw a long number line. Label the number line in hundreds of millions of miles.**
- **Locate each planet's position on the number line.**

What Is Your Answer?

4. **IN YOUR OWN WORDS** How can you write a number in scientific notation?

Practice Use what you learned about writing scientific notation to complete Exercises 3–5 on page 446.

10.6 Lesson

Writing Numbers in Scientific Notation

Step 1: Move the decimal point so it is located to the right of the leading nonzero digit.

Step 2: Count the number of places you moved the decimal point. This indicates the exponent of the power of 10, as shown below.

Number Greater Than or Equal to 10

Use a positive exponent when you move the decimal point to the left.

$8600 = 8.6 \times 10^3$ (3)

Number Between 0 and 1

Use a negative exponent when you move the decimal point to the right.

$0.0024 = 2.4 \times 10^{-3}$ (3)

Study Tip

When you write a number greater than or equal to 1 and less than 10 in scientific notation, use zero as the exponent.

$6 = 6 \times 10^0$

EXAMPLE 1 Writing Large Numbers in Scientific Notation

Google purchased YouTube for $1,650,000,000. Write this number in scientific notation.

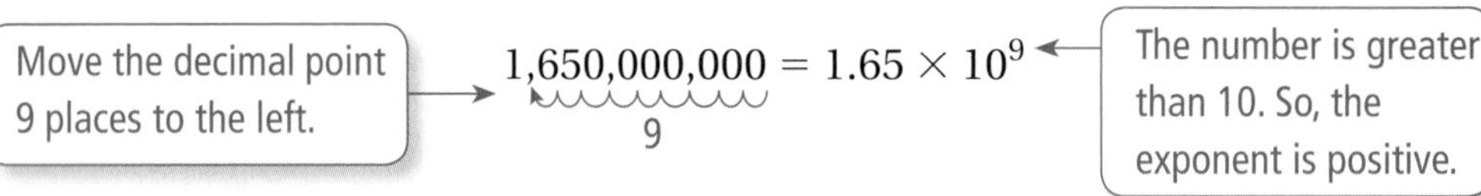

Move the decimal point 9 places to the left.

$1{,}650{,}000{,}000 = 1.65 \times 10^9$ (9)

The number is greater than 10. So, the exponent is positive.

EXAMPLE 2 Writing Small Numbers in Scientific Notation

The 2004 Indonesian earthquake slowed the rotation of Earth, making the length of a day 0.00000268 second shorter. Write this number in scientific notation.

Move the decimal point 6 places to the right.

$0.00000268 = 2.68 \times 10^{-6}$ (6)

The number is between 0 and 1. So, the exponent is negative.

On Your Own

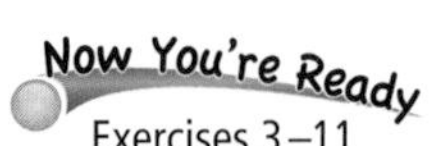

Write the number in scientific notation.

1. 50,000
2. 25,000,000
3. 683
4. 0.005
5. 0.00000033
6. 0.000506

EXAMPLE 3 Using Scientific Notation

An album receives an award when it sells 10,000,000 copies.

An album has sold 8,780,000 copies. How many more copies does it need to sell to receive the award?

Ⓐ 1.22×10^{-7} Ⓑ 1.22×10^{-6}

Ⓒ 1.22×10^{6} Ⓓ 1.22×10^{7}

Use a model to solve the problem.

Remaining sales needed for award = Sales required for award − Current sales total

$= 10{,}000{,}000 - 8{,}780{,}000$

$= 1{,}220{,}000$

$= 1.22 \times 10^{6}$

The album must sell 1.22×10^{6} more copies to receive the award. So, the correct answer is Ⓒ.

EXAMPLE  4 Real-Life Application

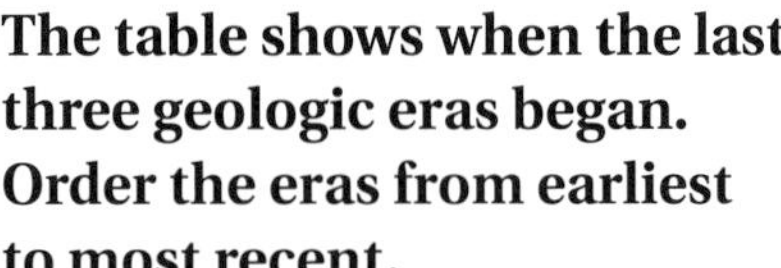

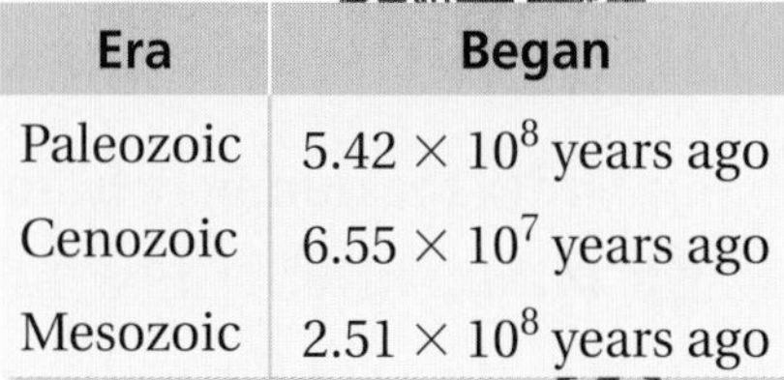

The table shows when the last three geologic eras began. Order the eras from earliest to most recent.

Era	Began
Paleozoic	5.42×10^{8} years ago
Cenozoic	6.55×10^{7} years ago
Mesozoic	2.51×10^{8} years ago

Step 1: Compare the powers of 10.

Because $10^7 < 10^8$,

$6.55 \times 10^7 < 5.42 \times 10^8$ and

$6.55 \times 10^7 < 2.51 \times 10^8$.

Step 2: Compare the factors when the powers of 10 are the same.

Because $2.51 < 5.42$,

$2.51 \times 10^8 < 5.42 \times 10^8$.

From greatest to least, the order is 5.42×10^8, 2.51×10^8, and 6.55×10^7.

So, the eras in order from earliest to most recent are the Paleozoic era, Mesozoic era, and Cenozoic era.

Common Error

To use the method in Example 4, the numbers must be written in scientific notation.

On Your Own

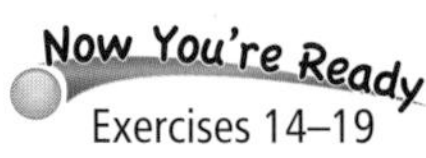

Now You're Ready
Exercises 14–19

7. **WHAT IF?** In Example 3, an album has sold 955,000 copies. How many more copies does it need to sell to receive the award? Write your answer in scientific notation.

8. The *Tyrannosaurus rex* lived 7.0×10^{7} years ago. Consider the eras given in Example 4. During which era did the *Tyrannosaurus rex* live?

10.6 Exercises

Vocabulary and Concept Check

1. **REASONING** How do you know whether a number written in standard form will have a positive or a negative exponent when written in scientific notation?

2. **WRITING** When is it appropriate to use scientific notation instead of standard form?

Practice and Problem Solving

Write the number in scientific notation.

(1) (2) **3.** 0.0021 **4.** 5,430,000 **5.** 321,000,000

6. 0.00000625 **7.** 0.00004 **8.** 10,700,000

9. 45,600,000,000 **10.** 0.000000000009256 **11.** 840,000

ERROR ANALYSIS Describe and correct the error in writing the number in scientific notation.

12.

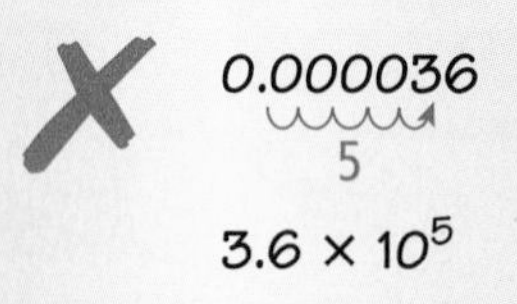

13.

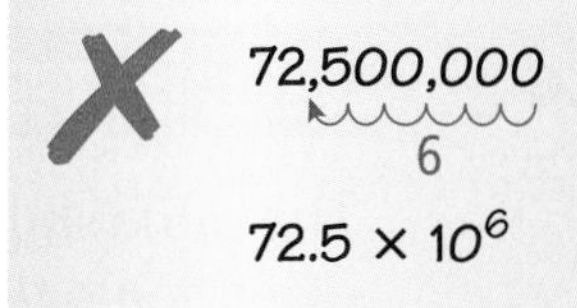

Order the numbers from least to greatest.

(4) **14.** 1.2×10^8, 1.19×10^8, 1.12×10^8

15. 6.8×10^{-5}, 6.09×10^{-5}, 6.78×10^{-5}

16. 5.76×10^{12}, 9.66×10^{11}, 5.7×10^{10}

17. 4.8×10^{-6}, 4.8×10^{-5}, 4.8×10^{-8}

18. 9.9×10^{-15}, 1.01×10^{-14}, 7.6×10^{-15}

19. 5.78×10^{23}, 6.88×10^{-23}, 5.82×10^{23}

20. **HAIR** What is the diameter of a human hair written in scientific notation?

Diameter: 0.000099 meter

21. **EARTH** What is the circumference of Earth written in scientific notation?

Circumference at the equator: about 40,100,000 meters

22. **CHOOSING UNITS** In Exercise 21, name a unit of measurement that would be more appropriate for the circumference. Explain.

Order the numbers from least to greatest.

23. $\frac{68{,}500}{10}$, 680, 6.8×10^3

24. $\frac{5}{241}$, 0.02, 2.1×10^{-2}

25. 6.3%, 6.25×10^{-3}, $6\frac{1}{4}$, 0.625

26. 3033.4, 305%, $\frac{10{,}000}{3}$, 3.3×10^2

27. SPACE SHUTTLE The total power of a space shuttle during launch is the sum of the power from its solid rocket boosters and the power from its main engines. The power from the solid rocket boosters is 9,750,000,000 watts. What is the power from the main engines?

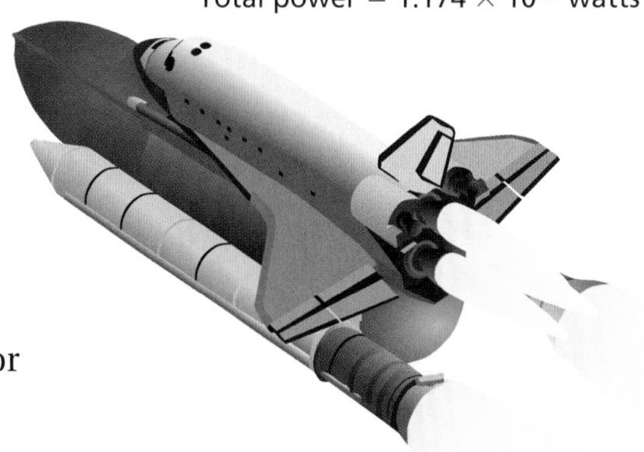

28. CHOOSE TOOLS Explain how to use a calculator to verify your answer to Exercise 27.

Equivalent to 1 Atomic Mass Unit
8.3×10^{-24} carat
1.66×10^{-21} milligram

29. ATOMIC MASS The mass of an atom or molecule is measured in atomic mass units. Which is greater, a *carat* or a *milligram*? Explain.

30. Reasoning In Example 4, the Paleozoic era ended when the Mesozoic era began. The Mesozoic era ended when the Cenozoic era began. The Cenozoic era is the current era.

a. Write the lengths of the three eras in scientific notation. Order the lengths from least to greatest.

b. Make a time line to show when the three eras occurred and how long each era lasted.

c. What do you notice about the lengths of the three eras? Use the Internet to determine whether your observation is true for *all* the geologic eras. Explain your results.

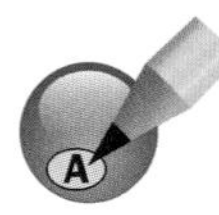

Fair Game Review What you learned in previous grades & lessons

Classify the real number. *(Section 7.4)*

31. 15

32. $\sqrt[3]{-8}$

33. $\sqrt{73}$

34. What is the surface area of the prism? *(Skills Review Handbook)*

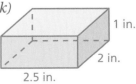

1 in.
2 in.
2.5 in.

Ⓐ 5 in.^{2}

Ⓑ 5.5 in.^{2}

Ⓒ 10 in.^{2}

Ⓓ 19 in.^{2}

10.7 Operations in Scientific Notation

Essential Question How can you perform operations with numbers written in scientific notation?

1 ACTIVITY: Adding Numbers in Scientific Notation

Work with a partner. Consider the numbers 2.4×10^3 and 7.1×10^3.

a. Explain how to use order of operations to find the sum of these numbers. Then find the sum.

$$2.4 \times 10^3 + 7.1 \times 10^3$$

b. The factor ____ is common to both numbers. How can you use the Distributive Property to rewrite the sum $(2.4 \times 10^3) + (7.1 \times 10^3)$?

$(2.4 \times 10^3) + (7.1 \times 10^3) =$ ____ Distributive Property

c. Use order of operations to evaluate the expression you wrote in part (b). Compare the result with your answer in part (a).

d. **STRUCTURE** Write a rule you can use to add numbers written in scientific notation where the powers of 10 are the same. Then test your rule using the sums below.

- $(4.9 \times 10^5) + (1.8 \times 10^5) =$ ____
- $(3.85 \times 10^4) + (5.72 \times 10^4) =$ ____

2 ACTIVITY: Adding Numbers in Scientific Notation

Work with a partner. Consider the numbers 2.4×10^3 and 7.1×10^4.

a. Explain how to use order of operations to find the sum of these numbers. Then find the sum.

$$2.4 \times 10^3 + 7.1 \times 10^4$$

b. How is this pair of numbers different from the pairs of numbers in Activity 1?

c. Explain why you cannot immediately use the rule you wrote in Activity 1(d) to find this sum.

d. **STRUCTURE** How can you rewrite one of the numbers so that you can use the rule you wrote in Activity 1(d)? Rewrite one of the numbers. Then find the sum using your rule and compare the result with your answer in part (a).

e. **REASONING** Do these procedures work when subtracting numbers written in scientific notation? Justify your answer by evaluating the differences below.

- $(8.2 \times 10^5) - (4.6 \times 10^5) =$ ____
- $(5.88 \times 10^5) - (1.5 \times 10^4) =$ ____

Scientific Notation

In this lesson, you will

- add, subtract, multiply, and divide numbers written in scientific notation.

3 ACTIVITY: Multiplying Numbers in Scientific Notation

Math Practice

Justify Conclusions

Which step of the procedure would be affected if the powers of 10 were different? Explain.

Work with a partner. Match each step with the correct description.

Step	Description
$(2.4 \times 10^3) \times (7.1 \times 10^3)$	Original expression
1. $= 2.4 \times 7.1 \times 10^3 \times 10^3$	**A.** Write in standard form.
2. $= (2.4 \times 7.1) \times (10^3 \times 10^3)$	**B.** Product of Powers Property
3. $= 17.04 \times 10^6$	**C.** Write in scientific notation.
4. $= 1.704 \times 10^1 \times 10^6$	**D.** Commutative Property of Multiplication
5. $= 1.704 \times 10^7$	**E.** Simplify.
6. $= 17{,}040{,}000$	**F.** Associative Property of Multiplication

Does this procedure work when the numbers have different powers of 10? Justify your answer by using this procedure to evaluate the products below.

- $(1.9 \times 10^2) \times (2.3 \times 10^5) =$ ______
- $(8.4 \times 10^6) \times (5.7 \times 10^{-4}) =$ ______

4 ACTIVITY: Using Scientific Notation to Estimate

Work with a partner. A person normally breathes about 6 liters of air per minute. The life expectancy of a person in the United States at birth is about 80 years. Use scientific notation to estimate the total amount of air a person born in the United States breathes over a lifetime.

What Is Your Answer?

5. **IN YOUR OWN WORDS** How can you perform operations with numbers written in scientific notation?

6. Use a calculator to evaluate the expression. Write your answer in scientific notation and in standard form.

   a. $(1.5 \times 10^4) + (6.3 \times 10^4)$ **b.** $(7.2 \times 10^5) - (2.2 \times 10^3)$

   c. $(4.1 \times 10^{-3}) \times (4.3 \times 10^{-3})$ **d.** $(4.75 \times 10^{-6}) \times (1.34 \times 10^7)$

Practice Use what you learned about evaluating expressions involving scientific notation to complete Exercises 3–6 on page 452.

10.7 Lesson

To add or subtract numbers written in scientific notation with the same power of 10, add or subtract the factors. When the numbers have different powers of 10, first rewrite the numbers so they have the same power of 10.

EXAMPLE 1 Adding and Subtracting Numbers in Scientific Notation

Find the sum or difference. Write your answer in scientific notation.

a. $(4.6 \times 10^3) + (8.72 \times 10^3)$

$= (4.6 + 8.72) \times 10^3$ — Distributive Property

$= 13.32 \times 10^3$ — Add.

$= (1.332 \times 10^1) \times 10^3$ — Write 13.32 in scientific notation.

$= 1.332 \times 10^4$ — Product of Powers Property

Study Tip

In Example 1(b), you will get the same answer when you start by rewriting 3.5×10^{-2} as 35×10^{-3}.

b. $(3.5 \times 10^{-2}) - (6.6 \times 10^{-3})$

Rewrite 6.6×10^{-3} so that it has the same power of 10 as 3.5×10^{-2}.

$6.6 \times 10^{-3} = 6.6 \times 10^{-1} \times 10^{-2}$ — Rewrite 10^{-3} as $10^{-1} \times 10^{-2}$.

$= 0.66 \times 10^{-2}$ — Rewrite 6.6×10^{-1} as 0.66.

Subtract the factors.

$(3.5 \times 10^{-2}) - (0.66 \times 10^{-2})$

$= (3.5 - 0.66) \times 10^{-2}$ — Distributive Property

$= 2.84 \times 10^{-2}$ — Subtract.

On Your Own

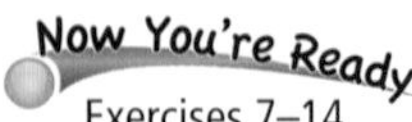

Exercises 7–14

Find the sum or difference. Write your answer in scientific notation.

1. $(8.2 \times 10^2) + (3.41 \times 10^{-1})$ **2.** $(7.8 \times 10^{-5}) - (4.5 \times 10^{-5})$

To multiply or divide numbers written in scientific notation, multiply or divide the factors and powers of 10 separately.

EXAMPLE 2 Multiplying Numbers in Scientific Notation

Study Tip

You can check your answer using standard form.

$(3 \times 10^{-5}) \times (5 \times 10^{-2})$
$= 0.00003 \times 0.05$
$= 0.0000015$
$= 1.5 \times 10^{-6}$

Find $(3 \times 10^{-5}) \times (5 \times 10^{-2})$. Write your answer in scientific notation.

$(3 \times 10^{-5}) \times (5 \times 10^{-2})$

$= 3 \times 5 \times 10^{-5} \times 10^{-2}$ — Commutative Property of Multiplication

$= (3 \times 5) \times (10^{-5} \times 10^{-2})$ — Associative Property of Multiplication

$= 15 \times 10^{-7}$ — Simplify.

$= 1.5 \times 10^1 \times 10^{-7}$ — Write 15 in scientific notation.

$= 1.5 \times 10^{-6}$ — Product of Powers Property

EXAMPLE 3 Dividing Numbers in Scientific Notation

Find $\frac{1.5 \times 10^{-8}}{6 \times 10^{7}}$. Write your answer in scientific notation.

$$\frac{1.5 \times 10^{-8}}{6 \times 10^{7}} = \frac{1.5}{6} \times \frac{10^{-8}}{10^{7}}$$ Rewrite as a product of fractions.

$$= 0.25 \times \frac{10^{-8}}{10^{7}}$$ Divide 1.5 by 6.

$$= 0.25 \times 10^{-15}$$ Quotient of Powers Property

$$= 2.5 \times 10^{-1} \times 10^{-15}$$ Write 0.25 in scientific notation.

$$= 2.5 \times 10^{-16}$$ Product of Powers Property

On Your Own

Find the product or quotient. Write your answer in scientific notation.

3. $6 \times (8 \times 10^{-5})$
4. $(7 \times 10^{2}) \times (3 \times 10^{5})$
5. $(9.2 \times 10^{12}) \div 4.6$
6. $(1.5 \times 10^{-3}) \div (7.5 \times 10^{2})$

EXAMPLE 4 Real-Life Application

Diameter = 1,400,000 km

How many times greater is the diameter of the Sun than the diameter of Earth?

Diameter = 1.28×10^4 km

Write the diameter of the Sun in scientific notation.

$$1,400,000 = 1.4 \times 10^{6}$$

6

Divide the diameter of the Sun by the diameter of Earth.

$$\frac{1.4 \times 10^{6}}{1.28 \times 10^{4}} = \frac{1.4}{1.28} \times \frac{10^{6}}{10^{4}}$$ Rewrite as a product of fractions.

$$= 1.09375 \times \frac{10^{6}}{10^{4}}$$ Divide 1.4 by 1.28.

$$= 1.09375 \times 10^{2}$$ Quotient of Powers Property

$$= 109.375$$ Write in standard form.

∴ The diameter of the Sun is about 109 times greater than the diameter of Earth.

On Your Own

7. How many more kilometers is the radius of the Sun than the radius of Earth? Write your answer in standard form.

10.7 Exercises

Vocabulary and Concept Check

1. **WRITING** Describe how to subtract two numbers written in scientific notation with the same power of 10.
2. **NUMBER SENSE** You are multiplying two numbers written in scientific notation with different powers of 10. Do you have to rewrite the numbers so they have the same power of 10 before multiplying? Explain.

Practice and Problem Solving

Evaluate the expression using two different methods. Write your answer in scientific notation.

3. $(2.74 \times 10^7) + (5.6 \times 10^7)$
4. $(8.3 \times 10^6) + (3.4 \times 10^5)$
5. $(5.1 \times 10^5) \times (9.7 \times 10^5)$
6. $(4.5 \times 10^4) \times (6.2 \times 10^3)$

Find the sum or difference. Write your answer in scientific notation.

(1) 7. $(2 \times 10^5) + (3.8 \times 10^5)$
8. $(6.33 \times 10^{-9}) - (4.5 \times 10^{-9})$
9. $(9.2 \times 10^8) - (4 \times 10^8)$
10. $(7.2 \times 10^{-6}) + (5.44 \times 10^{-6})$
11. $(7.8 \times 10^7) - (2.45 \times 10^6)$
12. $(5 \times 10^{-5}) + (2.46 \times 10^{-3})$
13. $(9.7 \times 10^6) + (6.7 \times 10^5)$
14. $(2.4 \times 10^{-1}) - (5.5 \times 10^{-2})$

15. **ERROR ANALYSIS** Describe and correct the error in finding the sum of the numbers.

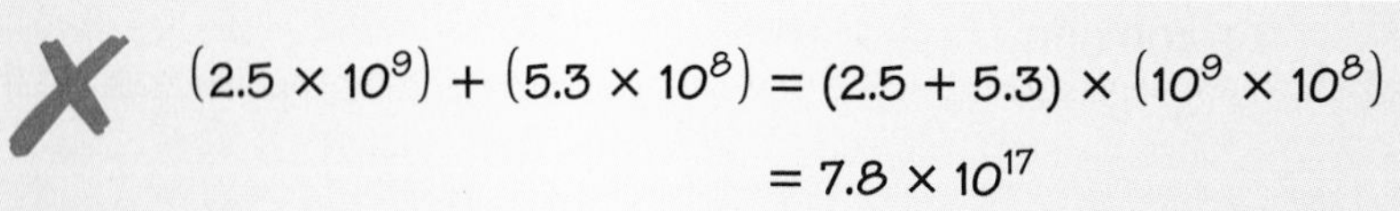

$(2.5 \times 10^9) + (5.3 \times 10^8) = (2.5 + 5.3) \times (10^9 \times 10^8)$
$= 7.8 \times 10^{17}$

Find the product or quotient. Write your answer in scientific notation.

(2) (3) 16. $5 \times (7 \times 10^7)$
17. $(5.8 \times 10^{-6}) \div (2 \times 10^{-3})$
18. $(1.2 \times 10^{-5}) \div 4$
19. $(5 \times 10^{-7}) \times (3 \times 10^6)$
20. $(3.6 \times 10^7) \div (7.2 \times 10^7)$
21. $(7.2 \times 10^{-1}) \times (4 \times 10^{-7})$
22. $(6.5 \times 10^8) \times (1.4 \times 10^{-5})$
23. $(2.8 \times 10^4) \div (2.5 \times 10^6)$

24. **MONEY** How many times greater is the thickness of a dime than the thickness of a dollar bill?

Thickness = 0.135 cm

Thickness = 1.0922×10^{-2} cm

Evaluate the expression. Write your answer in scientific notation.

25. $5{,}200{,}000 \times (8.3 \times 10^2) - (3.1 \times 10^8)$

26. $(9 \times 10^{-3}) + (2.4 \times 10^{-5}) \div 0.0012$

27. GEOMETRY Find the perimeter of the rectangle.

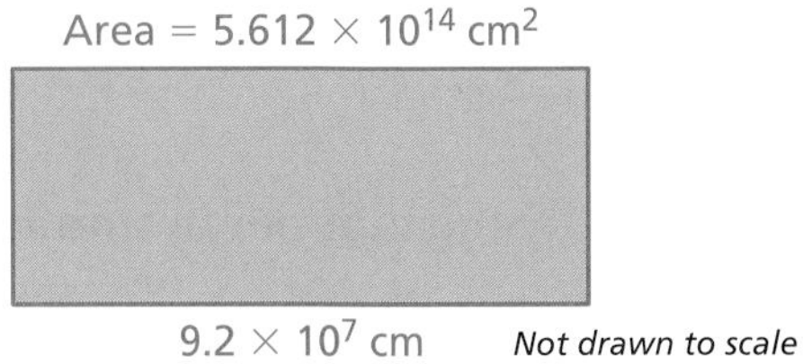

28. BLOOD SUPPLY A human heart pumps about 7×10^{-2} liter of blood per heartbeat. The average human heart beats about 72 times per minute. How many liters of blood does a heart pump in 1 year? in 70 years? Write your answers in scientific notation. Then use estimation to justify your answers.

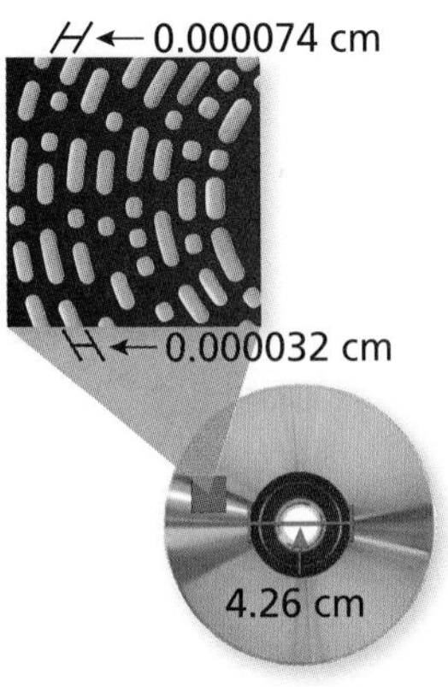

29. DVDS On a DVD, information is stored on bumps that spiral around the disk. There are 73,000 ridges (with bumps) and 73,000 valleys (without bumps) across the diameter of the DVD. What is the diameter of the DVD in centimeters?

30. PROJECT Use the Internet or some other reference to find the populations and areas (in square miles) of India, China, Argentina, the United States, and Egypt. Round each population to the nearest million and each area to the nearest thousand square miles.

a. Write each population and area in scientific notation.

b. Use your answers to part (a) to find and order the population densities (people per square mile) of each country from least to greatest.

31. Critical Thinking Albert Einstein's most famous equation is $E = mc^2$, where E is the energy of an object (in joules), m is the mass of an object (in kilograms), and c is the speed of light (in meters per second). A hydrogen atom has 15.066×10^{-11} joule of energy and a mass of 1.674×10^{-27} kilogram. What is the speed of light? Write your answer in scientific notation.

Fair Game Review What you learned in previous grades & lessons

Find the cube root. *(Section 7.2)*

32. $\sqrt[3]{-729}$

33. $\sqrt[3]{\frac{1}{512}}$

34. $\sqrt[3]{-\frac{125}{343}}$

35. MULTIPLE CHOICE What is the volume of the cone? *(Section 8.2)*

Ⓐ $16\pi \text{ cm}^3$

Ⓑ $108\pi \text{ cm}^3$

Ⓒ $48\pi \text{ cm}^3$

Ⓓ $144\pi \text{ cm}^3$

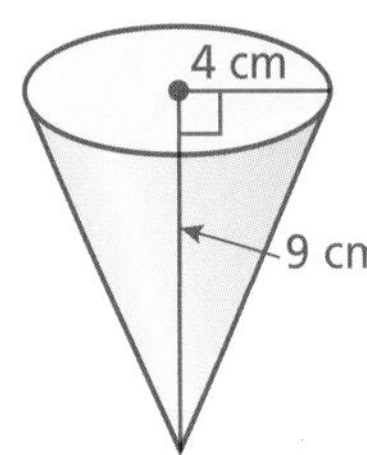

10.5–10.7 Quiz

Tell whether the number is written in scientific notation. Explain. *(Section 10.5)*

1. 23×10^9

2. 0.6×10^{-7}

Write the number in standard form. *(Section 10.5)*

3. 8×10^6

4. 1.6×10^{-2}

Write the number in scientific notation. *(Section 10.6)*

5. 0.00524

6. 892,000,000

Evaluate the expression. Write your answer in scientific notation. *(Section 10.7)*

7. $(7.26 \times 10^4) + (3.4 \times 10^4)$

8. $(2.8 \times 10^{-5}) - (1.6 \times 10^{-6})$

9. $(2.4 \times 10^4) \times (3.8 \times 10^{-6})$

10. $(5.2 \times 10^{-3}) \div (1.3 \times 10^{-12})$

11. PLANETS The table shows the equatorial radii of the eight planets in our solar system. *(Section 10.5)*

a. Which planet has the second-smallest equatorial radius?

b. Which planet has the second-largest equatorial radius?

Planet	Equatorial Radius (km)
Mercury	2.44×10^3
Venus	6.05×10^3
Earth	6.38×10^3
Mars	3.4×10^3
Jupiter	7.15×10^4
Saturn	6.03×10^4
Uranus	2.56×10^4
Neptune	2.48×10^4

12. OORT CLOUD The Oort cloud is a spherical cloud that surrounds our solar system. It is about 2×10^5 astronomical units from the Sun. An astronomical unit is about 1.5×10^8 kilometers. How far is the Oort cloud from the Sun in kilometers? *(Section 10.6)*

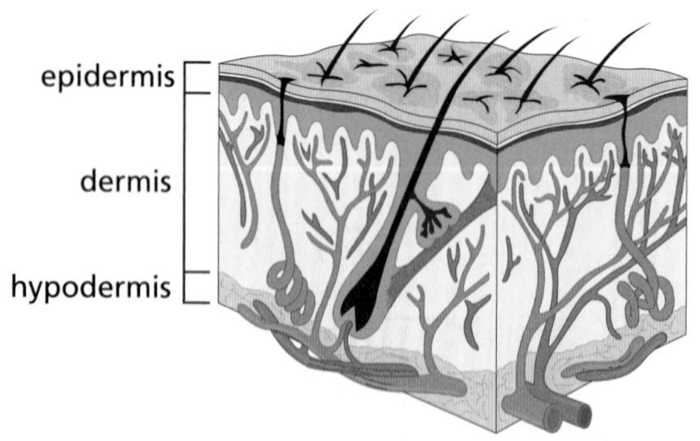

13. EPIDERMIS The outer layer of skin is called the *epidermis*. On the palm of your hand, the epidermis is 0.0015 meter thick. Write this number in scientific notation. *(Section 10.6)*

14. ORBITS It takes the Sun about 2.3×10^8 years to orbit the center of the Milky Way. It takes Pluto about 2.5×10^2 years to orbit the Sun. How many times does Pluto orbit the Sun while the Sun completes one orbit around the Milky Way? Write your answer in standard form. *(Section 10.7)*

10 Chapter Review

Review Key Vocabulary

power, *p. 412*
base, *p. 412*
exponent, *p. 412*
scientific notation, *p. 438*

Review Examples and Exercises

10.1 Exponents *(pp. 410–415)*

Write $(-4) \cdot (-4) \cdot (-4) \cdot y \cdot y$ using exponents.

Because -4 is used as a factor 3 times, its exponent is 3. Because y is used as a factor 2 times, its exponent is 2.

So, $(-4) \cdot (-4) \cdot (-4) \cdot y \cdot y = (-4)^3y^2$.

Exercises

Write the product using exponents.

1. $(-9) \cdot (-9) \cdot (-9) \cdot (-9) \cdot (-9)$
2. $2 \cdot 2 \cdot 2 \cdot n \cdot n$

Evaluate the expression.

3. 6^3
4. $-\left(\frac{1}{2}\right)^4$
5. $\left|\frac{1}{2}(16 - 6^3)\right|$

10.2 Product of Powers Property *(pp. 416–421)*

a. $\left(-\frac{1}{8}\right)^7 \cdot \left(-\frac{1}{8}\right)^4 = \left(-\frac{1}{8}\right)^{7+4}$ — Product of Powers Property

$= \left(-\frac{1}{8}\right)^{11}$ — Simplify.

b. $(2.5^7)^2 = 2.5^{7 \cdot 2}$ — Power of a Power Property

$= 2.5^{14}$ — Simplify.

c. $(3m)^2 = 3^2 \cdot m^2$ — Power of a Product Property

$= 9m^2$ — Simplify.

Exercises

Simplify the expression.

6. $p^5 \cdot p^2$
7. $(n^{11})^2$
8. $(5y)^3$
9. $(-2k)^4$

10.3 Quotient of Powers Property (pp. 422–427)

a. $\dfrac{(-4)^9}{(-4)^6} = (-4)^{9-6}$ Quotient of Powers Property

$= (-4)^3$ Simplify.

b. $\dfrac{x^4}{x^3} = x^{4-3}$ Quotient of Powers Property

$= x^1$

$= x$ Simplify.

Exercises

Simplify the expression. Write your answer as a power.

10. $\dfrac{8^8}{8^3}$

11. $\dfrac{5^2 \cdot 5^9}{5}$

12. $\dfrac{w^8}{w^7} \cdot \dfrac{w^5}{w^2}$

Simplify the expression.

13. $\dfrac{2^2 \cdot 2^5}{2^3}$

14. $\dfrac{(6c)^3}{c}$

15. $\dfrac{m^8}{m^6} \cdot \dfrac{m^{10}}{m^9}$

10.4 Zero and Negative Exponents (pp. 428–433)

a. $10^{-3} = \dfrac{1}{10^3}$ Definition of negative exponent

$= \dfrac{1}{1000}$ Evaluate power.

b. $(-0.5)^{-5} \cdot (-0.5)^5 = (-0.5)^{-5+5}$ Product of Powers Property

$= (-0.5)^0$ Simplify.

$= 1$ Definition of zero exponent

Exercises

Evaluate the expression.

16. 2^{-4}

17. 95^0

18. $\dfrac{8^2}{8^4}$

19. $(-12)^{-7} \cdot (-12)^7$

20. $\dfrac{1}{7^9} \cdot \dfrac{1}{7^{-6}}$

21. $\dfrac{9^4 \cdot 9^{-2}}{9^2}$

10.5 Reading Scientific Notation (pp. 436–441)

Write (a) 5.9×10^4 and (b) 7.31×10^{-6} in standard notation.

a. $5.9 \times 10^4 = 59{,}000$ (4)

Move decimal point $|4| = 4$ places to the right.

b. $7.31 \times 10^{-6} = 0.00000731$ (6)

Move decimal point $|-6| = 6$ places to the left.

Exercises

Write the number in standard form.

22. 2×10^7

23. 3.4×10^{-2}

24. 1.5×10^{-9}

25. 5.9×10^{10}

26. 4.8×10^{-3}

27. 6.25×10^5

10.6 Writing Scientific Notation (pp. 442–447)

Write (a) 309,000,000 and (b) 0.00056 in scientific notation.

a. $309{,}000{,}000 = 3.09 \times 10^8$ (8)

The number is greater than 10. So, the exponent is positive.

b. $0.00056 = 5.6 \times 10^{-4}$ (4)

The number is between 0 and 1. So, the exponent is negative.

Exercises

Write the number in scientific notation.

28. 0.00036

29. 800,000

30. 79,200,000

10.7 Operations in Scientific Notation (pp. 448–453)

Find $(2.6 \times 10^5) + (3.1 \times 10^5)$.

$(2.6 \times 10^5) + (3.1 \times 10^5) = (2.6 + 3.1) \times 10^5$ Distributive Property

$= 5.7 \times 10^5$ Add.

Exercises

Evaluate the expression. Write your answer in scientific notation.

31. $(4.2 \times 10^8) + (5.9 \times 10^9)$

32. $(5.9 \times 10^{-4}) - (1.8 \times 10^{-4})$

33. $(7.7 \times 10^8) \times (4.9 \times 10^{-5})$

34. $(3.6 \times 10^5) \div (1.8 \times 10^9)$

10 Chapter Test

Write the product using exponents.

1. $(-15) \cdot (-15) \cdot (-15)$
2. $\left(\frac{1}{12}\right) \cdot \left(\frac{1}{12}\right) \cdot \left(\frac{1}{12}\right) \cdot \left(\frac{1}{12}\right) \cdot \left(\frac{1}{12}\right)$

Evaluate the expression.

3. -2^3
4. $10 + 3^3 \div 9$

Simplify the expression. Write your answer as a power.

5. $9^{10} \cdot 9$
6. $(6^6)^5$
7. $\frac{10^4}{10^0}$
8. $\frac{(-3.5)^{13}}{(-3.5)^9}$

Evaluate the expression.

9. $5^{-2} \cdot 5^2$
10. $\frac{-8}{(-8)^3}$

Write the number in standard form.

11. 3×10^7
12. 9.05×10^{-3}

Evaluate the expression. Write your answer in scientific notation.

13. $(7.8 \times 10^7) + (9.9 \times 10^7)$
14. $(6.4 \times 10^5) - (5.4 \times 10^4)$
15. $(3.1 \times 10^6) \times (2.7 \times 10^{-2})$
16. $(9.6 \times 10^7) \div (1.2 \times 10^{-4})$
17. **CRITICAL THINKING** Is $(xy^2)^3$ the same as $(xy^3)^2$? Explain.
18. **RICE** A grain of rice weighs about 3^3 milligrams. About how many grains of rice are in one scoop?

One scoop of rice weighs about 3^9 milligrams.

19. **TASTE BUDS** There are about 10,000 taste buds on a human tongue. Write this number in scientific notation.
20. **LEAD** From 1978 to 2008, the amount of lead allowed in the air in the United States was 1.5×10^{-6} gram per cubic meter. In 2008, the amount allowed was reduced by 90%. What is the new amount of lead allowed in the air?

10 Cumulative Assessment

1. Mercury's distance from the Sun is approximately 5.79×10^7 kilometers. What is this distance in standard form?

   A. 5,790,000,000 km **C.** 57,900,000 km

   B. 579,000,000 km **D.** 5,790,000 km

2. The steps Jim took to answer the question are shown below. What should Jim change to correctly answer the question?

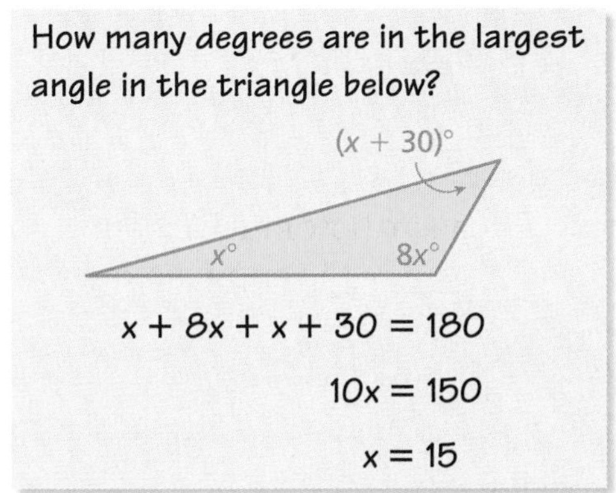

How many degrees are in the largest angle in the triangle below?
(x + 30)°
x°
8x°
x + 8x + x + 30 = 180
10x = 150
x = 15

   F. The left side of the equation should equal 360° instead of 180°.

   G. The sum of the acute angles should equal 90°.

   H. Evaluate the smallest angle when $x = 15$.

   I. Evaluate the largest angle when $x = 15$.

3. Which expression is equivalent to the expression below?

$$2^4 2^3$$

   A. 2^{12} **C.** 48

   B. 4^7 **D.** 128

4. In the figure below, $\triangle ABC$ is a dilation of $\triangle DEF$.

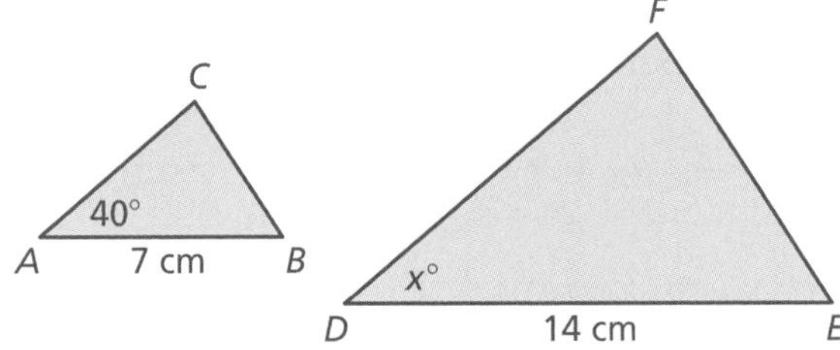

What is the value of x?

5. A bank account pays interest so that the amount in the account doubles every 10 years. The account started with \$5,000 in 1940. Which expression represents the amount (in dollars) in the account n decades later?

F. $2^n \cdot 5000$

G. $5000(n + 1)$

H. 5000^n

I. $2^n + 5000$

6. The formula for the volume V of a pyramid is $V = \frac{1}{3}Bh$. Solve the formula for the height h.

A. $h = \frac{1}{3}VB$

B. $h = \frac{3V}{B}$

C. $h = \frac{V}{3B}$

D. $h = V - \frac{1}{3}B$

7. The gross domestic product (GDP) is a way to measure how much a country produces economically in a year. The table below shows the approximate population and GDP for the United States.

United States 2012	
Population	312 million (312,000,000)
GDP	15.1 trillion dollars (\$15,100,000,000,000)

Part A Find the GDP per person for the United States. Show your work and explain your reasoning.

Part B Write the population and the GDP using scientific notation.

Part C Find the GDP per person for the United States using your answers from Part B. Write your answer in scientific notation. Show your work and explain your reasoning.

8. What is the equation of the line shown in the graph?

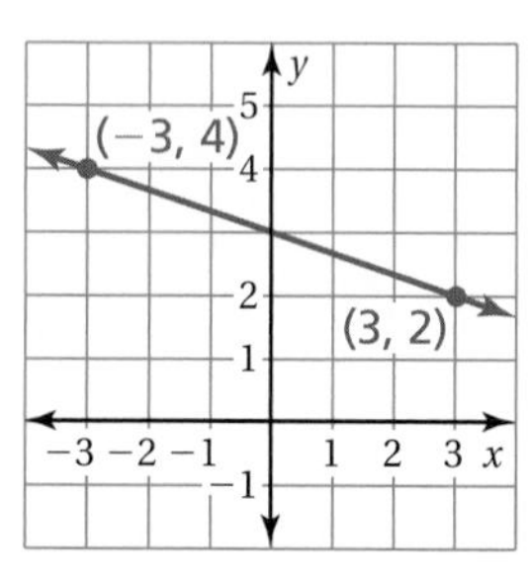

F. $y = -\frac{1}{3}x + 3$

G. $y = \frac{1}{3}x + 1$

H. $y = -3x + 3$

I. $y = 3x - \frac{1}{3}$

9. A cylinder and its dimensions are shown below.

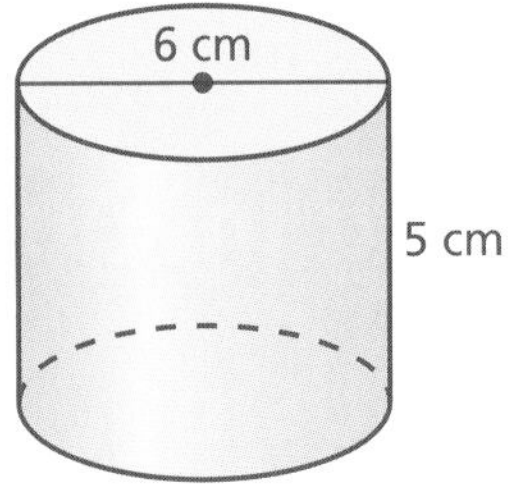

What is the volume of the cylinder? (Use 3.14 for π.)

A. 47.1 cm^3

B. 94.2 cm^3

C. 141.3 cm^3

D. 565.2 cm^3

10. Find $(-2.5)^{-2}$.

11. Two lines have the same y-intercept. The slope of one line is 1, and the slope of the other line is -1. What can you conclude?

F. The lines are parallel.

G. The lines meet at exactly one point.

H. The lines meet at more than one point.

I. The situation described is impossible.

12. The director of a research lab wants to present data to donors. The data show how the lab uses a great deal of donated money for research and only a small amount of money for other expenses. Which type of display is best suited for showing these data?

A. box-and-whisker plot

B. circle graph

C. line graph

D. scatter plot

11 Inequalities

11.1 Writing and Graphing Inequalities

11.2 Solving Inequalities Using Addition or Subtraction

11.3 Solving Inequalities Using Multiplication or Division

11.4 Solving Two-Step Inequalities

"If you reached into your water bowl and found more than \$20..."

"And then reached into your cat food bowl and found more than \$40..."

"What would you have?"

"Dear Precious Pet World: Your ad says 'Up to 75% off on selected items.'"

"I select Yummy Tummy Bacon-Flavored Dog Biscuits."

What You Learned Before

Why do I always have to be farther from the fulcrum?

"Move farther back. We still have an inequality."

Graphing Inequalities

Example 1 **Graph $x \geq 2$.**

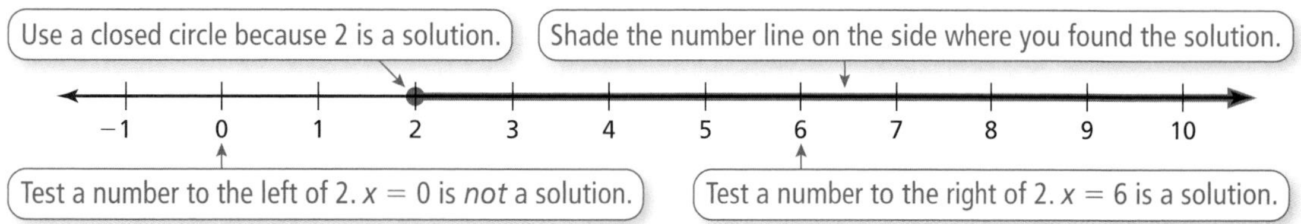

Use a closed circle because 2 is a solution.

−1 0 1 2 3 4 5 6 7 8 9 10
Test a number to the left of 2. x = 0 is not a solution.
Test a number to the right of 2. x = 6 is a solution.

Try It Yourself

Graph the inequality.

1. $x \geq 1$ **2.** $x < 5$ **3.** $x \leq 20$ **4.** $x > 13$

Comparing Numbers

Example 2 **Compare $-\frac{1}{3}$ and $-\frac{5}{6}$.**

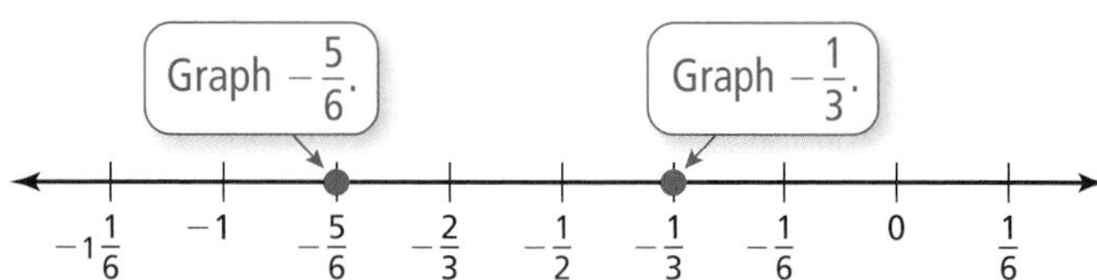

$-\frac{5}{6}$ is to the left of $-\frac{1}{3}$.

So, $-\frac{5}{6} < -\frac{1}{3}$.

Try It Yourself

Copy and complete the statement using < or >.

5. $-\frac{2}{3}$ ▢ $\frac{3}{8}$ **6.** $-\frac{1}{2}$ ▢ $-\frac{7}{8}$

7. $-\frac{1}{5}$ ▢ $\frac{1}{10}$ **8.** -1.4 ▢ 1.2

9. -2.2 ▢ -4.6 **10.** -1.9 ▢ -1.1

11.1 Writing and Graphing Inequalities

Essential Question How can you use a number line to represent solutions of an inequality?

1 ACTIVITY: Understanding Inequality Statements

Work with a partner. Read the statement. Circle each number that makes the statement true, and then answer the questions.

a. "You are in at least 5 of the photos."

$-3 \quad -2 \quad -1 \quad 0 \quad 1 \quad 2 \quad 3 \quad 4 \quad 5 \quad 6$

- What do you notice about the numbers that you circled?
- Is the number 5 included? Why or why not?
- Write four other numbers that make the statement true.

b. "The temperature is less than -4 degrees Fahrenheit."

$-7 \quad -6 \quad -5 \quad -4 \quad -3 \quad -2 \quad -1 \quad 0 \quad 1 \quad 2$

- What do you notice about the numbers that you circled?
- Can the temperature be exactly -4 degrees Fahrenheit? Explain.
- Write four other numbers that make the statement true.

c. "More than 3 students from our school are in the chess tournament."

$-3 \quad -2 \quad -1 \quad 0 \quad 1 \quad 2 \quad 3 \quad 4 \quad 5 \quad 6$

- What do you notice about the numbers that you circled?
- Is the number 3 included? Why or why not?
- Write four other numbers that make the statement true.

d. "The balance in a yearbook fund is no more than $-\$5$."

$-7 \quad -6 \quad -5 \quad -4 \quad -3 \quad -2 \quad -1 \quad 0 \quad 1 \quad 2$

- What do you notice about the numbers that you circled?
- Is the number -5 included? Why or why not?
- Write four other numbers that make the statement true.

Inequalities

In this lesson, you will

- write and graph inequalities.
- use substitution to check whether a number is a solution of an inequality.

2 ACTIVITY: Understanding Inequality Symbols

Work with a partner.

a. Consider the statement "x is a number such that $x > -1.5$."

- Can the number be exactly -1.5? Explain.
- Make a number line. Shade the part of the number line that shows the numbers that make the statement true.
- Write four other numbers that are not integers that make the statement true.

b. Consider the statement "x is a number such that $x \leq \frac{5}{2}$."

- Can the number be exactly $\frac{5}{2}$? Explain.
- Make a number line. Shade the part of the number line that shows the numbers that make the statement true.
- Write four other numbers that are not integers that make the statement true.

3 ACTIVITY: Writing and Graphing Inequalities

Math Practice

Check Progress

All the graphs are similar. So, what can you do to make sure that you have correctly written each inequality?

Work with a partner. Write an inequality for each graph. Then, in words, describe all the values of x that make the inequality true.

a.

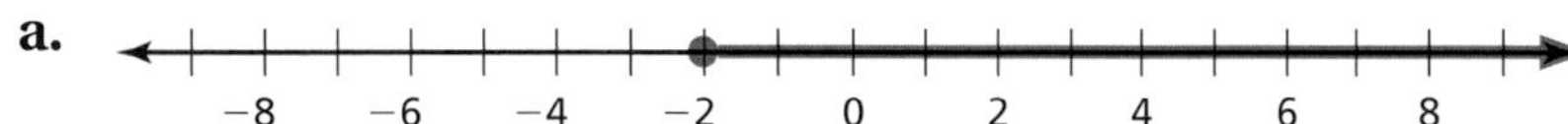

b.

−8 −6 −4 −2 0 2 4 6 8

c.

−8 −6 −4 −2 0 2 4 6 8

d.

−8 −6 −4 −2 0 2 4 6 8

What Is Your Answer?

4. **IN YOUR OWN WORDS** How can you use a number line to represent solutions of an inequality?
5. **STRUCTURE** Is $x \geq -1.4$ the same as $-1.4 \leq x$? Explain.

Practice

Use what you learned about writing and graphing inequalities to complete Exercises 4 and 5 on page 468.

11.1 Lesson

Key Vocabulary
inequality, *p. 466*
solution of an inequality, *p. 466*
solution set, *p. 466*
graph of an inequality, *p. 467*

An **inequality** is a mathematical sentence that compares expressions. It contains the symbols $<$, $>$, $\leq$, or $\geq$. To write an inequality, look for the following phrases to determine where to place the inequality symbol.

Inequality Symbols				
Symbol	$<$	$>$	$\leq$	$\geq$
Key Phrases	• is less than • is fewer than	• is greater than • is more than	• is less than or equal to • is at most • is no more than	• is greater than or equal to • is at least • is no less than

EXAMPLE 1 Writing an Inequality

A number q plus 5 is greater than or equal to -7.9. Write this word sentence as an inequality.

A number q plus 5 | is greater than or equal to | -7.9.

$q + 5$ | $\geq$ | -7.9

An inequality is $q + 5 \geq -7.9$.

On Your Own

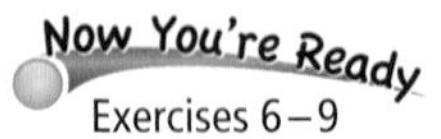

Exercises 6–9

Write the word sentence as an inequality.

1. A number x is at most -10.
2. Twice a number y is more than $-\frac{5}{2}$.

A **solution of an inequality** is a value that makes the inequality true. An inequality can have more than one solution. The set of all solutions of an inequality is called the **solution set**.

Value of x	$x + 2 \leq -1$	Is the inequality true?
-2	$-2 + 2 \overset{?}{\leq} -1$ $0 \nleq -1$ ✗	no
-3	$-3 + 2 \overset{?}{\leq} -1$ $-1 \leq -1$ ✓	yes
-4	$-4 + 2 \overset{?}{\leq} -1$ $-2 \leq -1$ ✓	yes

Reading

The symbol $\nleq$ means *is not less than or equal to.*

Multi-Language Glossary at BigIdeasMath.com

EXAMPLE 2 Checking Solutions

Tell whether −2 is a solution of each inequality.

a. $y - 5 \geq -6$

$y - 5 \geq -6$	Write the inequality.	$-5.5y < 14$
$-2 - 5 \overset{?}{\geq} -6$	Substitute −2 for *y*.	$-5.5(-2) \overset{?}{<} 14$
$-7 \ngeq -6$ ✗	Simplify.	$11 < 14$ ✓

−7 is *not* greater than or equal to −6.

So, −2 is *not* a solution of the inequality.

b. $-5.5y < 14$

11 is less than 14.

So, −2 is a solution of the inequality.

On Your Own

Now You're Ready
Exercises 11–16

Tell whether −5 is a solution of the inequality.

3. $x + 12 > 7$ **4.** $1 - 2p \leq -9$ **5.** $n \div 2.5 \geq -3$

The **graph of an inequality** shows all the solutions of the inequality on a number line. An open circle ○ is used when a number is *not* a solution. A closed circle ● is used when a number is a solution. An arrow to the left or right shows that the graph continues in that direction.

EXAMPLE 3 Graphing an Inequality

Graph $y > -8$.

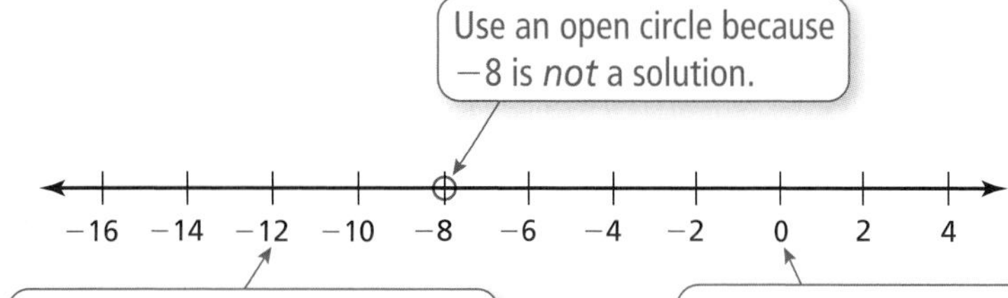
Use an open circle because −8 is *not* a solution.
−16 −14 −12 −10 −8 −6 −4 −2 0 2 4
Test a number to the left of −8. y = −12 is *not* a solution.
Test a number to the right of −8. y = 0 is a solution.

Shade the number line on the side where you found the solution.

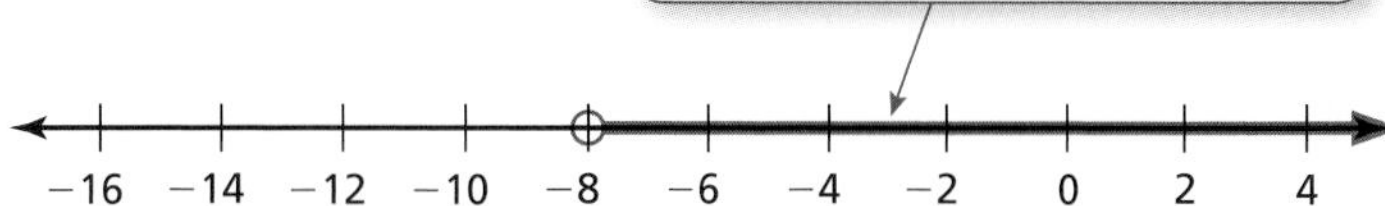
−16 −14 −12 −10 −8 −6 −4 −2 0 2 4

Study Tip

The graph in Example 3 shows that the inequality has *infinitely many* solutions.

On Your Own

Now You're Ready
Exercises 17–20

Graph the inequality on a number line.

6. $x < -1$ **7.** $z \geq 4$ **8.** $s \leq 1.4$ **9.** $-\frac{1}{2} < t$

11.1 Exercises

Vocabulary and Concept Check

1. **PRECISION** Should you use an open circle or a closed circle in the graph of the inequality $b \geq -42$? Explain.

2. **DIFFERENT WORDS, SAME QUESTION** Which is different? Write "both" inequalities.

k is less than or equal to -3.	k is no more than -3.
k is at most -3.	k is at least -3.

3. **REASONING** Do $x < 5$ and $5 < x$ represent the same inequality? Explain.

Practice and Problem Solving

Write an inequality for the graph. Then, in words, describe all the values of x that make the inequality true.

4. −4 0 4 8 12 16 20

5. 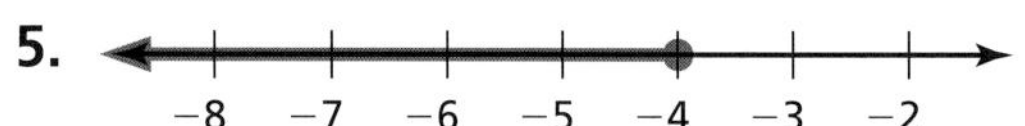

Write the word sentence as an inequality.

1 6. A number y is no more than -8.

7. A number w added to 2.3 is more than 18.

8. A number t multiplied by -4 is at least $-\frac{2}{5}$.

9. A number b minus 4.2 is less than -7.5.

10. **ERROR ANALYSIS** Describe and correct the error in writing the word sentence as an inequality.

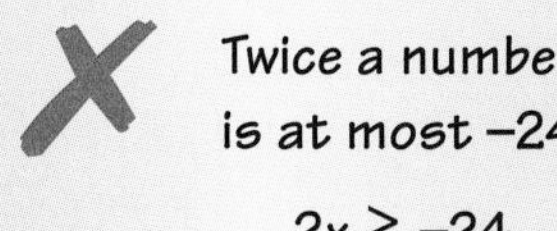

Tell whether the given value is a solution of the inequality.

2 11. $n + 8 \leq 13;\ n = 4$

12. $5h > -15;\ h = -5$

13. $p + 1.4 \leq 0.5;\ p = 0.1$

14. $\frac{a}{6} > -4;\ a = -18$

15. $-\frac{2}{3}s \geq 6;\ s = -9$

16. $\frac{7}{8} - 3k < -\frac{1}{2};\ k = \frac{1}{4}$

Graph the inequality on a number line.

3 17. $r \leq -9$

18. $g > 2.75$

19. $x \geq -3\frac{1}{2}$

20. $z < 1\frac{1}{4}$

21. **FOOD TRUCK** Each day at lunchtime, at least 53 people buy food from a food truck. Write an inequality that represents this situation.

Tell whether the given value is a solution of the inequality.

22. $4k < k + 8;\ k = 3$

23. $\frac{w}{3} \geq w - 12;\ w = 15$

24. $7 - 2y > 3y + 13;\ y = -1$

25. $\frac{3}{4}b - 2 \leq 2b + 8;\ b = -4$

26. MODELING A subway ride for a student costs \$1.25. A monthly pass costs \$35.

a. Write an inequality that represents the number of times you must ride the subway for the monthly pass to be a better deal.

b. You ride the subway about 45 times per month. Should you buy the monthly pass? Explain.

27. LOGIC Consider the inequality $b > -2$.

a. Describe the values of b that are solutions of the inequality.

b. Describe the values of b that are *not* solutions of the inequality. Write an inequality for these values.

c. What do all the values in parts (a) and (b) represent? Is this true for any inequality?

28. Critical Thinking A postal service says that a rectangular package can have a maximum combined length and *girth* of 108 inches. The girth of a package is the distance around the perimeter of a face that does not include the length.

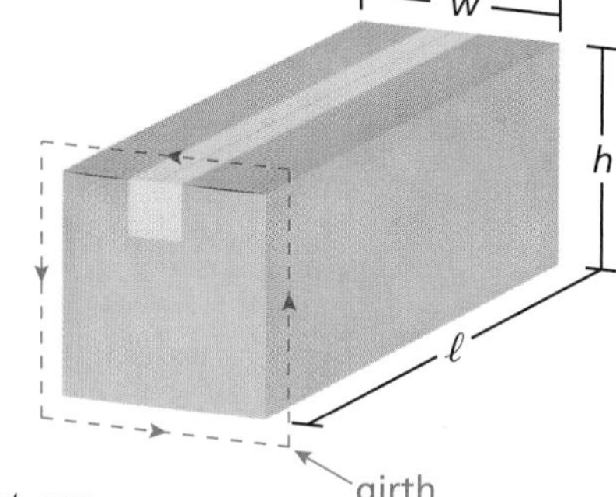

a. Write an inequality that represents the allowable dimensions for the package.

b. Find three different sets of allowable dimensions that are reasonable for the package. Find the volume of each package.

Fair Game Review What you learned in previous grades & lessons

Solve the equation. Check your solution. *(Section 1.1)*

29. $p - 8 = 3$ **30.** $8.7 + w = 5.1$ **31.** $x - 2 = -9$

32. MULTIPLE CHOICE Which expression has a value less than -5? *(Skills Review Handbook)*

Ⓐ $5 + 8$ Ⓑ $-9 + 5$ Ⓒ $1 + (-8)$ Ⓓ $7 + (-2)$

11.2 Solving Inequalities Using Addition or Subtraction

Essential Question How can you use addition or subtraction to solve an inequality?

1 ACTIVITY: Writing an Inequality

Work with a partner. Members of the Boy Scouts must be less than 18 years old. In 4 years, your friend will still be eligible to be a scout.

a. Which of the following represents your friend's situation? What does x represent? Explain your reasoning.

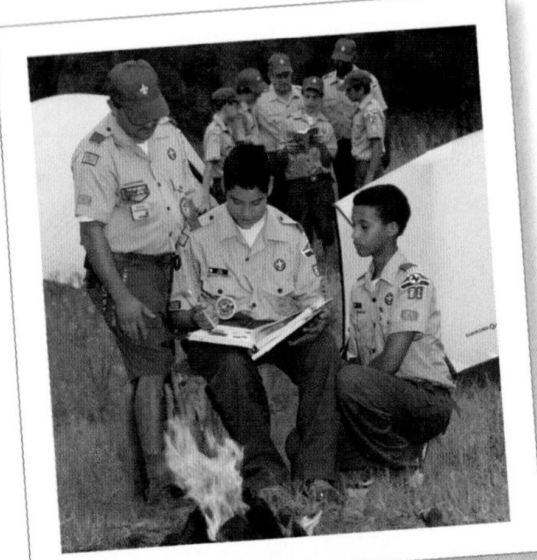

$x + 4 > 18$ $\quad$ $x + 4 < 18$

$x + 4 \geq 18$ $\quad$ $x + 4 \leq 18$

b. Graph the possible ages of your friend on a number line. Explain how you decided what to graph.

2 ACTIVITY: Writing an Inequality

Work with a partner. Supercooling is the process of lowering the temperature of a liquid or a gas below its freezing point without it becoming a solid. Water can be supercooled to 86°F below its normal freezing point (32°F) and still not freeze.

a. Let x represent the temperature of water. Which inequality represents the temperature at which water can be a liquid or a gas? Explain your reasoning.

$x - 32 > -86$ $\quad$ $x - 32 < -86$

$x - 32 \geq -86$ $\quad$ $x - 32 \leq -86$

b. On a number line, graph the possible temperatures at which water can be a liquid or a gas. Explain how you decided what to graph.

Inequalities

In this lesson, you will

- solve inequalities using addition or subtraction.
- solve real-life problems.

3 ACTIVITY: Solving Inequalities

Math Practice

Interpret Results

What does the solution of the inequality represent?

Work with a partner. Complete the following steps for Activity 1. Then repeat the steps for Activity 2.

- Use your inequality from part (a). Replace the inequality symbol with an equal sign.
- Solve the equation.
- Replace the equal sign with the original inequality symbol.
- Graph this new inequality.
- Compare the graph with your graph in part (b). What do you notice?

4 ACTIVITY: Temperatures of Continents

Work with a partner. The table shows the lowest recorded temperature on each continent. Write an inequality that represents each statement. Then solve and graph the inequality.

a. The temperature at a weather station in Asia is more than 150°F greater than the record low in Asia.

b. The temperature at a research station in Antarctica is at least 80°F greater than the record low in Antarctica.

Continent	Lowest Temperature
Africa	−11°F
Antarctica	−129°F
Asia	−90°F
Australia	−9.4°F
Europe	−67°F
North America	−81.4°F
South America	−27°F

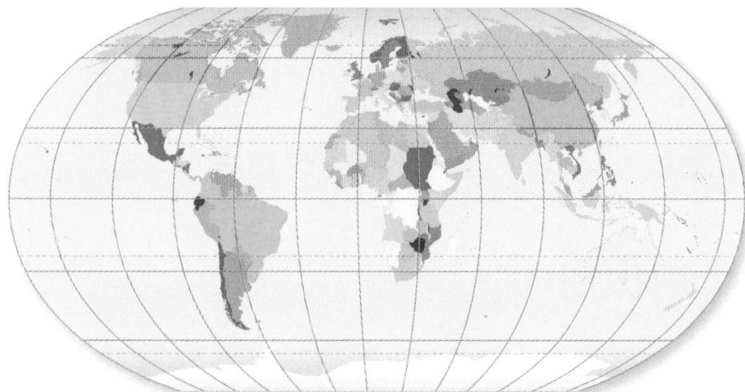

What Is Your Answer?

5. IN YOUR OWN WORDS How can you use addition or subtraction to solve an inequality?

6. Describe a real-life situation that you can represent with an inequality. Write the inequality. Graph the solution on a number line.

Practice Use what you learned about solving inequalities to complete Exercises 3–5 on page 474.

11.2 Lesson

Key Ideas

Study Tip

You can solve inequalities in the same way you solve equations. Use inverse operations to get the variable by itself.

Addition Property of Inequality

Words When you add the same number to each side of an inequality, the inequality remains true.

Numbers

$$\begin{array}{rcr} -4 & < & 3 \\ +2 & & +2 \\ \hline -2 & < & 5 \end{array}$$

Algebra If $a < b$, then $a + c < b + c$.

If $a > b$, then $a + c > b + c$.

Subtraction Property of Inequality

Words When you subtract the same number from each side of an inequality, the inequality remains true.

Numbers

$$\begin{array}{rcr} -2 & < & 2 \\ -3 & & -3 \\ \hline -5 & < & -1 \end{array}$$

Algebra If $a < b$, then $a - c < b - c$.

If $a > b$, then $a - c > b - c$.

These properties are also true for $\leq$ and $\geq$.

EXAMPLE 1 Solving an Inequality Using Addition

Solve $x - 5 < -3$. Graph the solution.

$x - 5 < -3$ Write the inequality.

Undo the subtraction. → $+5 \quad +5$ Addition Property of Inequality

$x < 2$ Simplify.

The solution is $x < 2$.

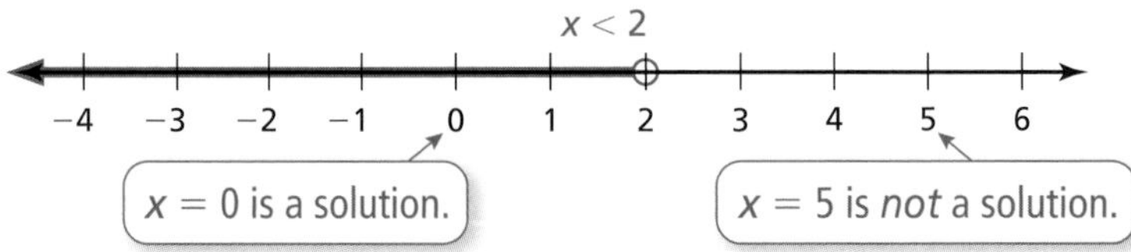

Check:

$x = 0$: $0 - 5 \stackrel{?}{<} -3$

$-5 < -3$ ✓

$x = 5$: $5 - 5 \stackrel{?}{<} -3$

$0 \not< -3$ ✗

On Your Own

Solve the inequality. Graph the solution.

1. $y - 6 > -7$
2. $b - 3.8 \leq 1.7$
3. $-\frac{1}{2} > z - \frac{1}{4}$

EXAMPLE 2 Solving an Inequality Using Subtraction

Solve $13 \le x + 14$. Graph the solution.

Undo the addition.

$$\begin{aligned} 13 &\le x + 14 && \text{Write the inequality.} \\ -14 & \quad\;\; -14 && \text{Subtraction Property of Inequality} \\ -1 &\le x && \text{Simplify.} \end{aligned}$$

The solution is $x \ge -1$.

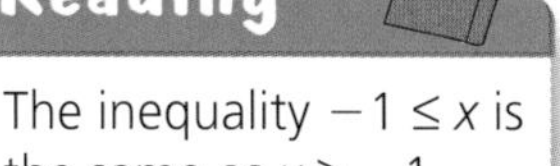

Reading

The inequality $-1 \le x$ is the same as $x \ge -1$.

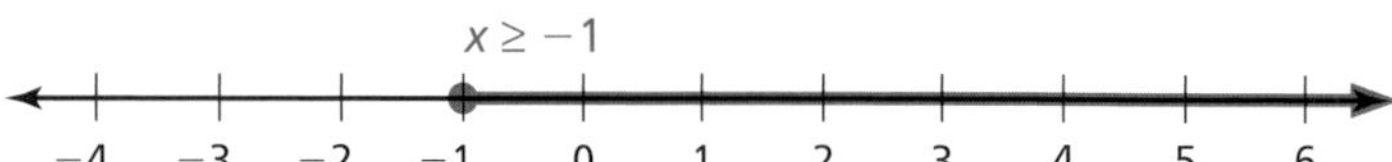

On Your Own

Now You're Ready
Exercises 3–17

Solve the inequality. Graph the solution.

4. $w + 7 \le 4$ **5.** $12.5 \ge d + 10$ **6.** $x + \frac{3}{4} > 1\frac{1}{2}$

EXAMPLE 3 Real-Life Application

A person can be no taller than 6.25 feet to become an astronaut pilot for NASA. Your friend is 5 feet 9 inches tall. Write and solve an inequality that represents how much your friend can grow and still meet the requirement.

Words	Current height	plus	amount your friend can grow	is no more than	the height limit.
Variable	Let h be the possible amounts your friend can grow.				
Inequality	5.75	+	h	$\le$	6.25

5 ft 9 in. = 60 + 9 = 69 in.

$69 \text{ in.} \times \frac{1 \text{ ft}}{12 \text{ in.}} = 5.75 \text{ ft}$

$$\begin{aligned} 5.75 + h &\le 6.25 && \text{Write the inequality.} \\ -5.75 \quad & \quad\; -5.75 && \text{Subtraction Property of Inequality} \\ h &\le 0.5 && \text{Simplify.} \end{aligned}$$

So, your friend can grow no more than 0.5 foot, or 6 inches.

On Your Own

7. Your cousin is 5 feet 3 inches tall. Write and solve an inequality that represents how much your cousin can grow and still meet the requirement.

11.2 Exercises

Vocabulary and Concept Check

1. **REASONING** Is the inequality $c + 3 > 5$ the same as $c > 5 - 3$? Explain.

2. **WHICH ONE DOESN'T BELONG?** Which inequality does *not* belong with the other three? Explain your reasoning.

   $w + \frac{7}{4} > \frac{3}{4}$ $w - \frac{3}{4} > -\frac{7}{4}$ $w + \frac{7}{4} < \frac{3}{4}$ $\frac{3}{4} < w + \frac{7}{4}$

Practice and Problem Solving

Solve the inequality. Graph the solution.

3. $x + 7 \geq 18$
4. $a - 2 > 4$
5. $3 \leq 7 + g$
6. $8 + k \leq -3$
7. $-12 < y - 6$
8. $n - 4 < 5$
9. $t - 5 \leq -7$
10. $p + \frac{1}{4} \geq 2$
11. $\frac{2}{7} > b + \frac{5}{7}$
12. $z - 4.7 \geq -1.6$
13. $-9.1 < d - 6.3$
14. $\frac{8}{5} > s + \frac{12}{5}$
15. $-\frac{7}{8} \geq m - \frac{13}{8}$
16. $r + 0.2 < -0.7$
17. $h - 6 \leq -8.4$

ERROR ANALYSIS Describe and correct the error in solving the inequality or graphing the solution of the inequality.

18.

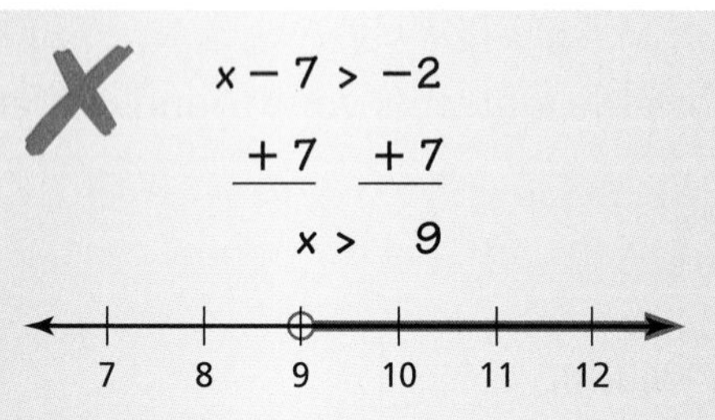

19.

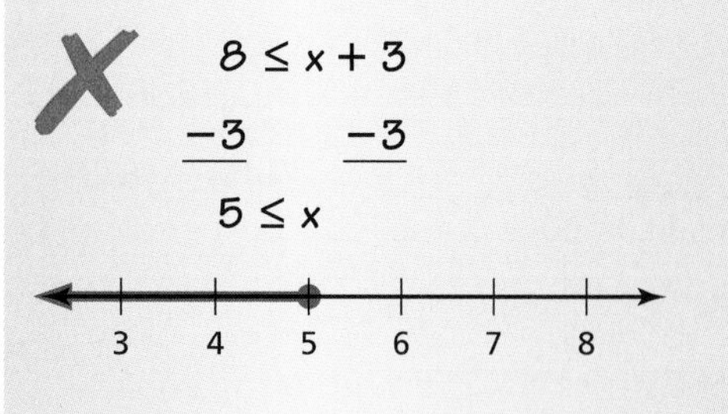

20. **AIRPLANE** A small airplane can hold 44 passengers. Fifteen passengers board the plane.

    a. Write and solve an inequality that represents the additional number of passengers that can board the plane.

    b. Can 30 more passengers board the plane? Explain.

Write and solve an inequality that represents x.

21. The perimeter is less than 28 feet.

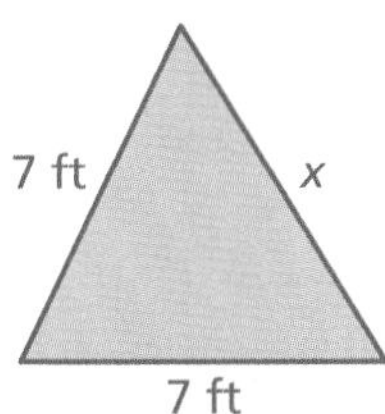

22. The base is greater than the height.

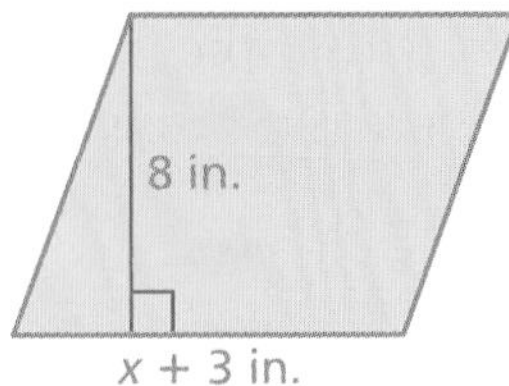

23. The perimeter is less than or equal to 51 meters.

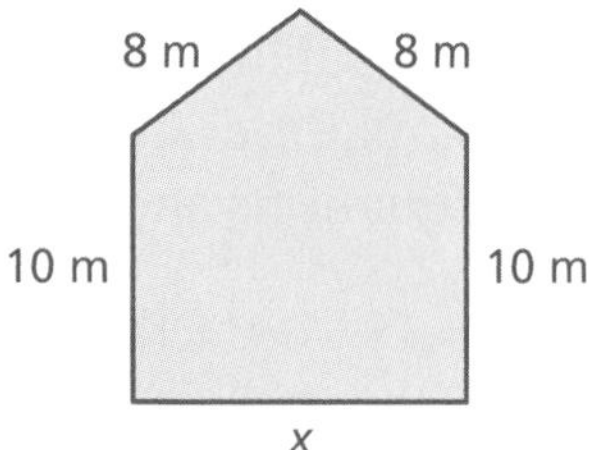

24. REASONING The solution of $d + s > -3$ is $d > -7$. What is the value of s?

25. BIRDFEEDER The hole for a birdfeeder post is 3 feet deep. The top of the post needs to be at least 5 feet above the ground. Write and solve an inequality that represents the required length of the post.

26. SHOPPING You can spend up to \$35 on a shopping trip.

a. You want to buy a shirt that costs \$14. Write and solve an inequality that represents the amount of money you will have left if you buy the shirt.

b. You notice that the shirt is on sale for 30% off. How does this change the inequality?

c. Do you have enough money to buy the shirt that is on sale and a pair of pants that costs \$23? Explain.

27. POWER A circuit overloads at 2400 watts of electricity. A portable heater that uses 1050 watts of electricity is plugged into the circuit.

a. Write and solve an inequality that represents the additional number of watts you can plug in without overloading the circuit.

b. In addition to the portable heater, what two other items in the table can you plug in at the same time without overloading the circuit? Is there more than one possibility? Explain.

Item	Watts
Aquarium	200
Hair dryer	1200
Television	150
Vacuum cleaner	1100

28. 

The possible values of x are given by $x + 8 \le 6$. What is the greatest possible value of $7x$?

Fair Game Review What you learned in previous grades & lessons

Solve the equation. Check your solution. *(Section 1.1)*

29. $4x = 36$

30. $\frac{w}{3} = -9$

31. $-2b = 44$

32. $60 = \frac{3}{4}h$

33. MULTIPLE CHOICE Which fraction is equivalent to –2.4? *(Skills Review Handbook)*

Ⓐ $-\frac{12}{5}$ Ⓑ $-\frac{51}{25}$ Ⓒ $-\frac{8}{5}$ Ⓓ $-\frac{6}{25}$

11 Study Help

You can use a **Y chart** to compare two topics. List differences in the branches and similarities in the base of the Y. Here is an example of a Y chart that compares solving equations and solving inequalities.

Solving Equations

- The sign between two expressions is an equal sign, $=$.
- One number is the solution.

Solving Inequalities

- The sign between two expressions is an inequality symbol: $<$, $>$, $\leq$, or $\geq$.
- More than one number can be a solution.

Both

- Use inverse operations to group numbers on one side.
- Use inverse operations to group variables on one side.
- Solve for the variable.

On Your Own

Make Y charts to help you study and compare these topics.

1. writing equations and writing inequalities
2. graphing the solution of an equation and graphing the solution of an inequality
3. graphing inequalities that use $>$ and graphing inequalities that use $<$
4. graphing inequalities that use $>$ or $<$ and graphing inequalities that use $\geq$ or $\leq$
5. solving inequalities using addition and solving inequalities using subtraction

"Hey Descartes, do you have any suggestions for the Y chart I am making?"

After you complete this chapter, make Y charts for the following topics.

6. solving inequalities using multiplication and solving inequalities using division
7. solving two-step equations and solving two-step inequalities
8. Pick two other topics that you studied earlier in this course and make a Y chart to compare them.

11.1–11.2 Quiz

Write the word sentence as an inequality. *(Section 11.1)*

1. A number y plus 2 is greater than -5.

2. A number s minus 2.4 is at least 8.

Tell whether the given value is a solution of the inequality. *(Section 11.1)*

3. $8p < -3$; $p = -2$

4. $z + 2 > -4$; $z = -8$

Graph the inequality on a number line. *(Section 11.1)*

5. $x < -12$

6. $v > \frac{5}{4}$

7. $b \geq -\frac{1}{3}$

8. $q \leq 4.2$

Solve the inequality. Graph the solution. *(Section 11.2)*

9. $n + 2 \leq -6$

10. $t - \frac{3}{7} > \frac{6}{7}$

11. $-\frac{3}{4} \geq w + 1$

12. $y - 2.6 < -3.4$

13. STUDYING You plan to study at least 1.5 hours for a geography test. Write an inequality that represents this situation. *(Section 11.1)*

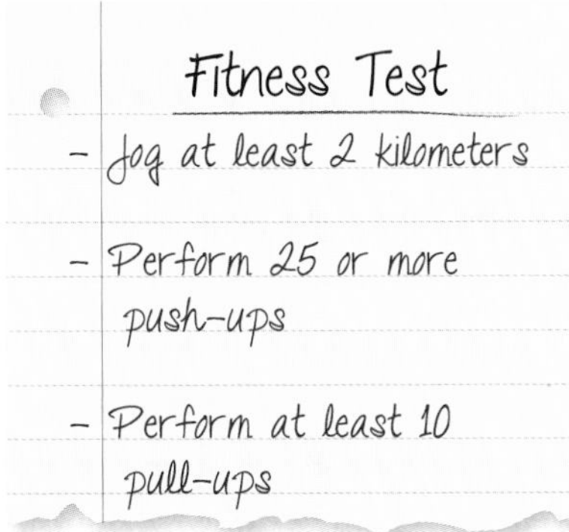

14. FITNESS TEST The three requirements to pass a fitness test are shown. *(Section 11.1)*

a. Write and graph three inequalities that represent the requirements.

b. You can jog 2500 meters, perform 30 push-ups, and perform 20 pull-ups. Do you satisfy the requirements of the test? Explain.

15. NUMBER LINE Use tape on the floor to make the number line shown. All units are in feet. You are standing at $-\frac{7}{2}$. You want to move to a number greater than $-\frac{3}{2}$. Write and solve an inequality that represents the distance you must move. *(Section 11.2)*

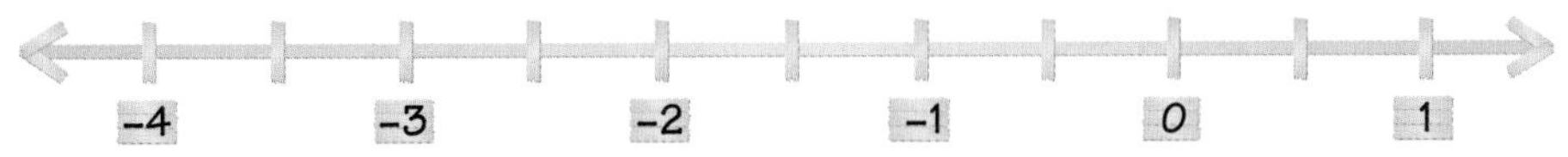

11.3 Solving Inequalities Using Multiplication or Division

Essential Question How can you use multiplication or division to solve an inequality?

1 ACTIVITY: Using a Table to Solve an Inequality

Work with a partner.

- **Copy and complete the table.**
- **Decide which graph represents the solution of the inequality.**
- **Write the solution of the inequality.**

a. $4x > 12$

x	−1	0	1	2	3	4	5
$4x$							
$4x \overset{?}{>} 12$							

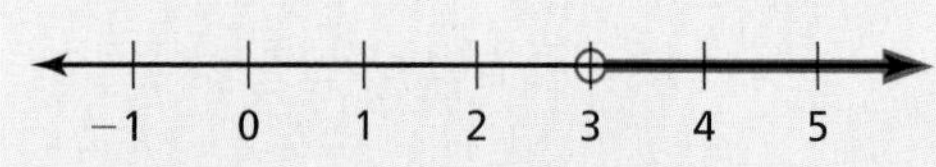

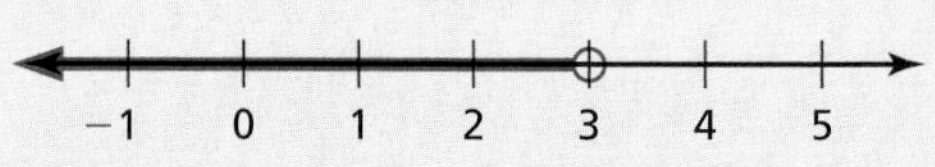

b. $-3x \le 9$

x	−5	−4	−3	−2	−1	0	1
$-3x$							
$-3x \overset{?}{\le} 9$							

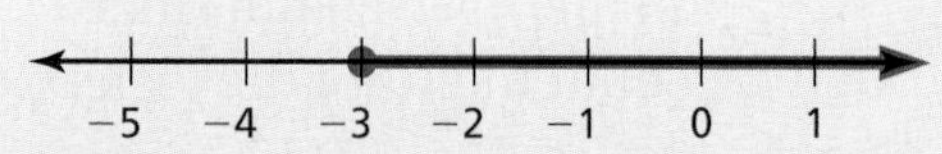

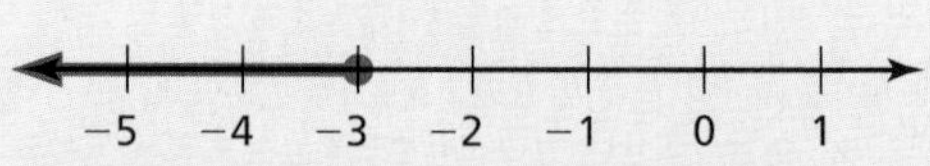

Inequalities

In this lesson, you will

- solve inequalities using multiplication or division.
- solve real-life problems.

2 ACTIVITY: Solving an Inequality

Work with a partner.

a. Solve $-3x \le 9$ by adding $3x$ to each side of the inequality first. Then solve the resulting inequality.

b. Compare the solution in part (a) with the solution in Activity 1(b).

3 ACTIVITY: Using a Table to Solve an Inequality

Work with a partner.

- **Copy and complete the table.**
- **Decide which graph represents the solution of the inequality.**
- **Write the solution of the inequality.**

a. $\frac{x}{3} < 1$

x	−1	0	1	2	3	4	5
$\frac{x}{3}$							
$\frac{x}{3} \overset{?}{<} 1$							

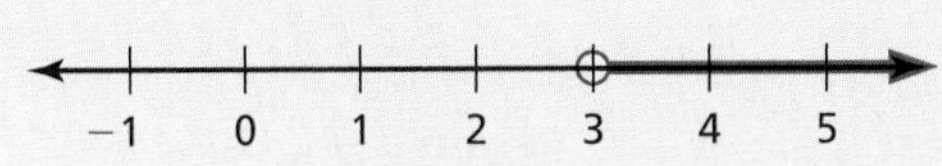

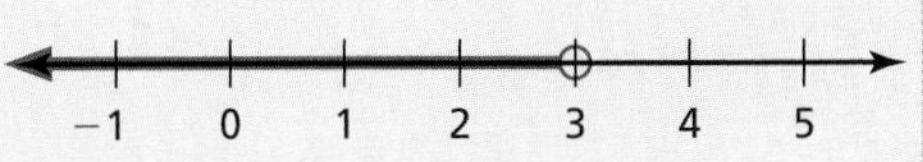

b. $\frac{x}{-4} \geq \frac{3}{4}$

x	−5	−4	−3	−2	−1	0	1
$\frac{x}{-4}$							
$\frac{x}{-4} \overset{?}{\geq} \frac{3}{4}$							

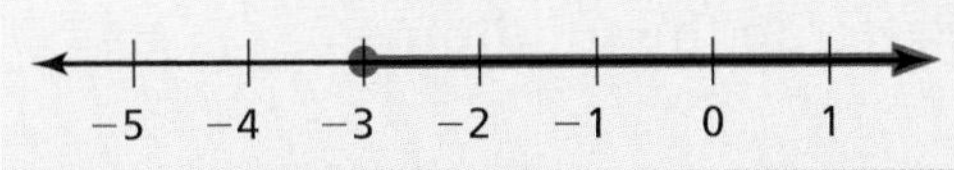

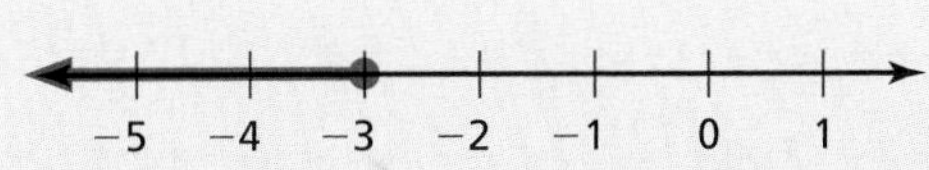

4 ACTIVITY: Writing Rules

Work with a partner. Use a table to solve each inequality.

a. $-2x \leq 10$ **b.** $-6x > 0$ **c.** $\frac{x}{-4} < 1$ **d.** $\frac{x}{-8} \geq \frac{1}{8}$

Write a set of rules that describes how to solve inequalities like those in Activities 1 and 3. Then use your rules to solve each of the four inequalities above.

Math Practice

Analyze Conjectures

When you apply your rules to parts (a)–(d), do you get the same solutions? Explain.

What Is Your Answer?

5. **IN YOUR OWN WORDS** How can you use multiplication or division to solve an inequality?

Practice Use what you learned about solving inequalities using multiplication or division to complete Exercises 4–9 on page 483.

11.3 Lesson

Remember

Multiplication and division are inverse operations.

Multiplication and Division Properties of Inequality (Case 1)

Words When you multiply or divide each side of an inequality by the same *positive* number, the inequality remains true.

Numbers

$-4 < 6$ $\qquad$ $4 > -6$

$2 \cdot (-4) < 2 \cdot 6$ $\qquad$ $\frac{4}{2} > \frac{-6}{2}$

$-8 < 12$ $\qquad$ $2 > -3$

Algebra If $a < b$ and c is positive, then

$$a \cdot c < b \cdot c \quad \text{and} \quad \frac{a}{c} < \frac{b}{c}.$$

If $a > b$ and c is positive, then

$$a \cdot c > b \cdot c \quad \text{and} \quad \frac{a}{c} > \frac{b}{c}.$$

These properties are also true for $\leq$ and $\geq$.

EXAMPLE 1 Solving an Inequality Using Multiplication

Solve $\frac{x}{5} \leq -3$. Graph the solution.

$\frac{x}{5} \leq -3$ — Write the inequality.

Undo the division. → $5 \cdot \frac{x}{5} \leq 5 \cdot (-3)$ — Multiplication Property of Inequality

$x \leq -15$ — Simplify.

∴ The solution is $x \leq -15$.

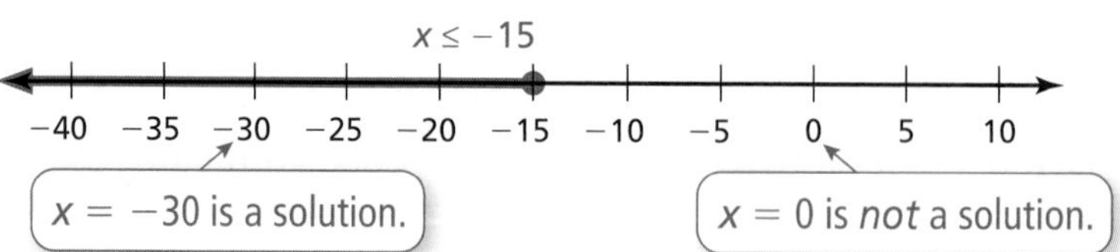

On Your Own

Solve the inequality. Graph the solution.

1. $n \div 3 < 1$
2. $-0.5 \leq \frac{m}{10}$
3. $-3 > \frac{2}{3}p$

EXAMPLE 2 Solving an Inequality Using Division

Solve $6x > -18$. Graph the solution.

	$6x > -18$	Write the inequality.
Undo the multiplication. →	$\frac{6x}{6} > \frac{-18}{6}$	Division Property of Inequality
	$x > -3$	Simplify.

The solution is $x > -3$.

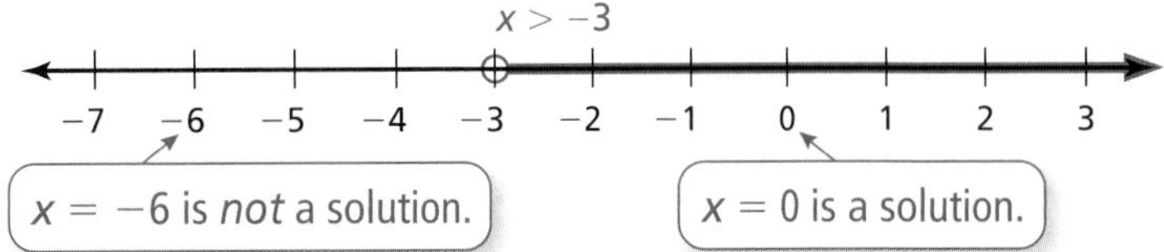

On Your Own

Solve the inequality. Graph the solution.

4. $4b \geq 2$
5. $12k \leq -24$
6. $-15 < 2.5q$

Key Idea

Multiplication and Division Properties of Inequality (Case 2)

Words When you multiply or divide each side of an inequality by the same *negative* number, the direction of the inequality symbol must be reversed for the inequality to remain true.

Numbers

$-4 < 6$	$4 > -6$
$-2 \cdot (-4) > -2 \cdot 6$	$\frac{4}{-2} < \frac{-6}{-2}$
$8 > -12$	$-2 < 3$

Algebra If $a < b$ and c is negative, then

$$a \cdot c > b \cdot c \quad \text{and} \quad \frac{a}{c} > \frac{b}{c}.$$

If $a > b$ and c is negative, then

$$a \cdot c < b \cdot c \quad \text{and} \quad \frac{a}{c} < \frac{b}{c}.$$

These properties are also true for ≤ and ≥.

Common Error

A negative sign in an inequality does not necessarily mean you must reverse the inequality symbol.

Only reverse the inequality symbol when you multiply or divide both sides by a negative number.

EXAMPLE 3 Solving an Inequality Using Multiplication

Solve $-\frac{3}{2}n \le 6$. Graph the solution.

$-\frac{3}{2}n \le 6$	Write the inequality.
$-\frac{2}{3} \cdot \left(-\frac{3}{2}n\right) \ge -\frac{2}{3} \cdot 6$	Use the Multiplication Property of Inequality. Reverse the inequality symbol.
$n \ge -4$	Simplify.

The solution is $n \ge -4$.

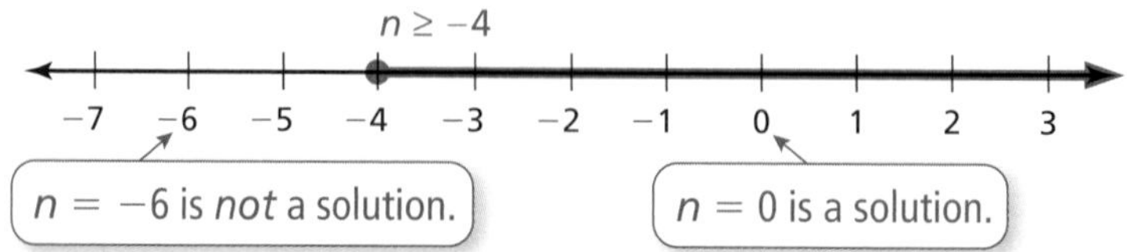

On Your Own

Solve the inequality. Graph the solution.

7. $\frac{x}{-3} > -4$

8. $0.5 \le -\frac{y}{2}$

9. $-12 \ge \frac{6}{5}m$

10. $-\frac{2}{5}h \le -8$

EXAMPLE 4 Solving an Inequality Using Division

Solve $-3z > -4.5$. Graph the solution.

$-3z > -4.5$	Write the inequality.
Undo the multiplication. → $\frac{-3z}{-3} < \frac{-4.5}{-3}$	Use the Division Property of Inequality. Reverse the inequality symbol.
$z < 1.5$	Simplify.

The solution is $z < 1.5$.

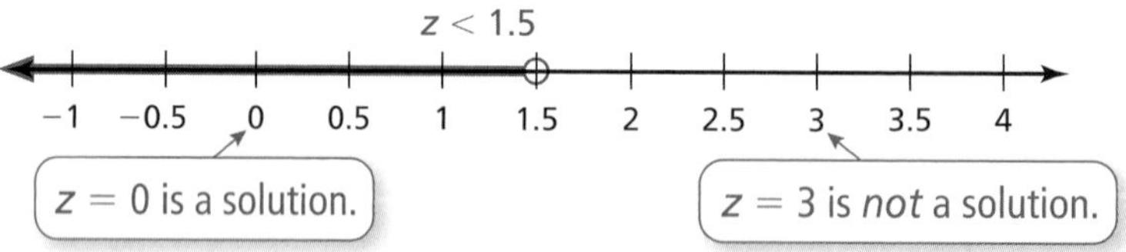

On Your Own

Solve the inequality. Graph the solution.

11. $-5z < 35$

12. $-2a > -9$

13. $-1.5 < 3n$

14. $-4.2 \ge -0.7w$

11.3 Exercises

Vocabulary and Concept Check

1. **WRITING** Explain how to solve $\frac{x}{3} < -2$.
2. **PRECISION** Explain how solving $4x < -16$ is different from solving $-4x < 16$.
3. **OPEN-ENDED** Write an inequality that you can solve using the Division Property of Inequality where the direction of the inequality symbol must be reversed.

Practice and Problem Solving

Use a table to solve the inequality.

4. $2x < 2$
5. $-3x \leq 3$
6. $-6x > 18$
7. $\frac{x}{-5} \geq 7$
8. $\frac{x}{-1} > \frac{2}{5}$
9. $\frac{x}{3} \leq \frac{1}{2}$

Solve the inequality. Graph the solution.

1 2 10. $2n > 20$
11. $\frac{c}{9} \leq -4$
12. $2.2m < 11$
13. $-16 > x \div 2$
14. $\frac{1}{6}w \geq 2.5$
15. $7 < 3.5k$
16. $3x \leq -\frac{5}{4}$
17. $4.2y \leq -12.6$
18. $11.3 > \frac{b}{4.3}$

19. **ERROR ANALYSIS** Describe and correct the error in solving the inequality.

✗
$$\frac{x}{3} < -9$$
$$3 \cdot \frac{x}{3} > 3 \cdot (-9)$$
$$x > -27$$

Write the word sentence as an inequality. Then solve the inequality.

20. The quotient of a number and 4 is at most 5.
21. A number divided by 7 is less than -3.
22. Six times a number is at least -24.
23. The product of -2 and a number is greater than 30.
24. **SMART PHONE** You earn \$9.20 per hour at your summer job. Write and solve an inequality that represents the number of hours you need to work in order to buy a smart phone that costs \$299.

25. **AVOCADOS** You have \$9.60 to buy avocados for a guacamole recipe. Avocados cost \$2.40 each.

a. Write and solve an inequality that represents the number of avocados you can buy.

b. Are there infinitely many solutions in this context? Explain.

26. **SCIENCE PROJECT** Students in a science class are divided into 6 equal groups with at least 4 students in each group for a project. Write and solve an inequality that represents the number of students in the class.

Solve the inequality. Graph the solution.

③ ④ 27. $-5n \leq 15$

28. $-7w > 49$

29. $-\frac{1}{3}h \geq 8$

30. $-9 < -\frac{1}{5}x$

31. $-3y < -14$

32. $-2d \geq 26$

33. $4.5 > -\frac{m}{6}$

34. $\frac{k}{-0.25} \leq 36$

35. $-2.4 > \frac{b}{-2.5}$

36. **ERROR ANALYSIS** Describe and correct the error in solving the inequality.

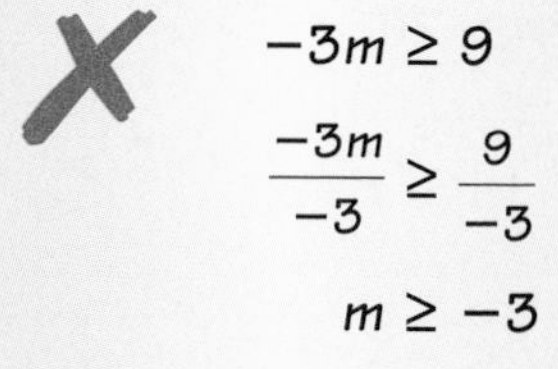

$-3m \geq 9$

$\frac{-3m}{-3} \geq \frac{9}{-3}$

$m \geq -3$

37. **TEMPERATURE** It is currently 0°C outside. The temperature is dropping 2.5°C every hour. Write and solve an inequality that represents the number of hours that must pass for the temperature to drop below −20°C.

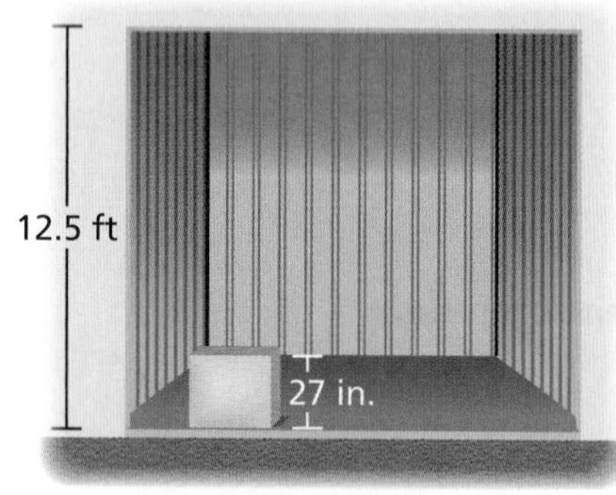

38. **STORAGE** You are moving some of your belongings into a storage facility.

a. Write and solve an inequality that represents the number of boxes that you can stack vertically in the storage unit.

b. Can you stack 6 boxes vertically in the storage unit? Explain.

Write and solve an inequality that represents *x*.

39. Area $\geq$ 120 cm^2

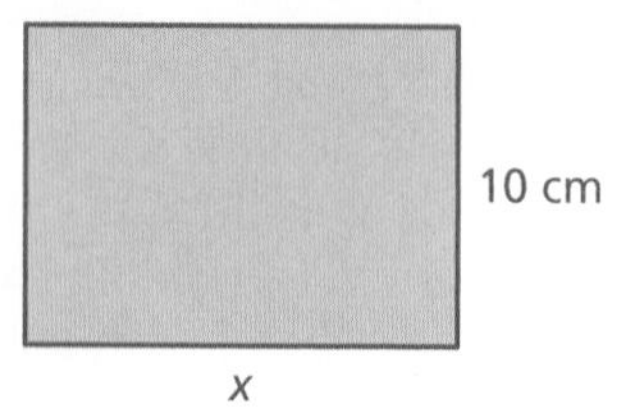

40. Area < 20 ft^2

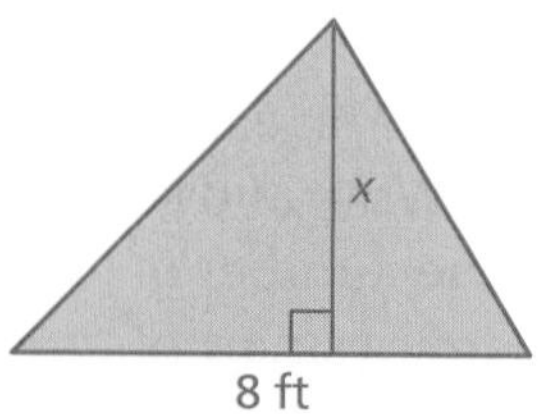

41. **AMUSEMENT PARK** You and four friends are planning a visit to an amusement park. You want to keep the cost below $100 per person. Write and solve an inequality that represents the total cost of visiting the amusement park.

42. **LOGIC** When you multiply or divide each side of an inequality by the same negative number, you must reverse the direction of the inequality symbol. Explain why.

43. **PROJECT** Choose two novels to research.

a. Use the Internet or a magazine to complete the table.

b. Use the table to find and compare the average number of copies sold per month for each novel. Which novel do you consider to be the most successful? Explain.

c. Assume each novel continues to sell at the average rate. Write and solve an inequality that represents the number of months it will take for the total number of copies sold to exceed twice the current number sold.

Author	Name of Novel	Release Date	Current Number of Copies Sold
1.			
2.			

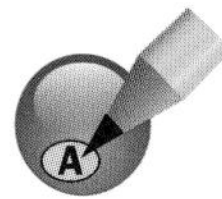

Describe all numbers that satisfy *both* inequalities. Include a graph with your description.

44. $4m > -4$ and $3m < 15$

45. $\frac{n}{3} \geq -4$ and $\frac{n}{-5} \geq 1$

46. $2x \geq -6$ and $2x \geq 6$

47. $-\frac{1}{2}s > -7$ and $\frac{1}{3}s < 12$

Fair Game Review What you learned in previous grades & lessons

Solve the equation. Check your solution. *(Section 1.2)*

48. $-2w + 4 = -12$

49. $\frac{v}{5} - 6 = 3$

50. $3(x - 1) = 18$

51. $\frac{m + 200}{4} = 51$

52. **MULTIPLE CHOICE** What is the value of $\frac{2}{3} + \left(-\frac{5}{7}\right)$? *(Skills Review Handbook)*

Ⓐ $-\frac{3}{4}$ Ⓑ $-\frac{1}{21}$ Ⓒ $\frac{7}{10}$ Ⓓ $1\frac{8}{21}$

11.4 Solving Two-Step Inequalities

Essential Question How can you use an inequality to describe the dimensions of a figure?

1 ACTIVITY: Areas and Perimeters of Figures

Work with a partner.

- **Use the given condition to choose the inequality that you can use to find the possible values of the variable. Justify your answer.**
- **Write four values of the variable that satisfy the inequality you chose.**

a. You want to find the values of x so that the area of the rectangle is more than 22 square units.

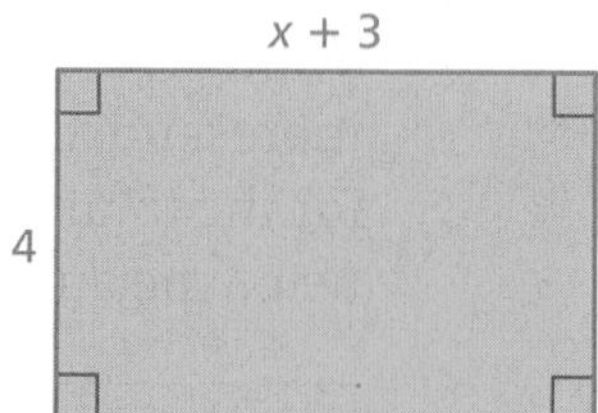

$4x + 12 > 22$ $\quad$ $4x + 3 > 22$

$4x + 12 \geq 22$ $\quad$ $2x + 14 > 22$

b. You want to find the values of x so that the perimeter of the rectangle is greater than or equal to 28 units.

$x + 7 \geq 28$ $\quad$ $4x + 12 \geq 28$ $\quad$ $2x + 14 \geq 28$ $\quad$ $2x + 14 \leq 28$

c. You want to find the values of y so that the area of the parallelogram is fewer than 41 square units.

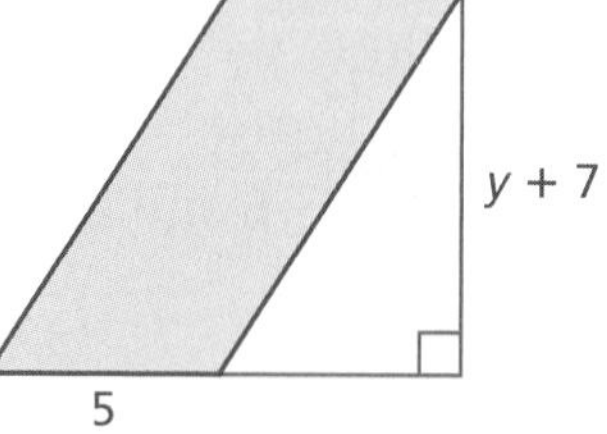

$5y + 7 < 41$ $\quad$ $5y + 35 < 41$

$5y + 7 \leq 41$ $\quad$ $5y + 35 \leq 41$

Inequalities

In this lesson, you will

- solve multi-step inequalities.
- solve real-life problems.

d. You want to find the values of z so that the area of the trapezoid is at most 100 square units.

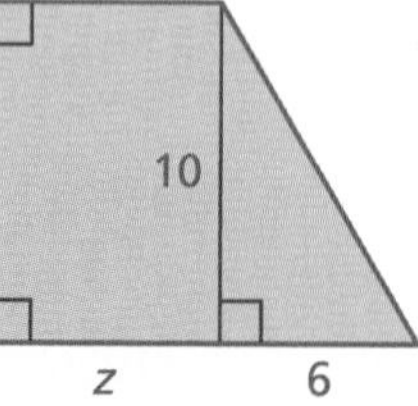

$5z + 30 \leq 100$ $\quad$ $10z + 30 \leq 100$

$5z + 30 < 100$ $\quad$ $10z + 30 < 100$

2 ACTIVITY: Volumes of Rectangular Prisms

Work with a partner.

- **Use the given condition to choose the inequality that you can use to find the possible values of the variable. Justify your answer.**
- **Write four values of the variable that satisfy the inequality you chose.**

Math Practice

State the Meaning of Symbols

What inequality symbols do the phrases *at least* and *no more than* represent? Explain.

a. You want to find the values of x so that the volume of the rectangular prism is at least 50 cubic units.

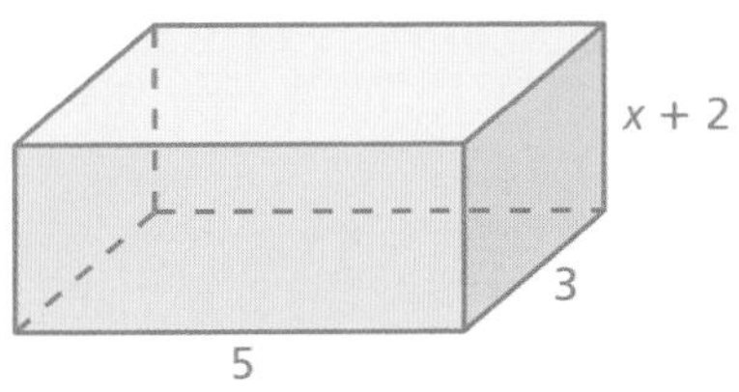

$15x + 30 > 50$	$x + 10 \geq 50$	$15x + 30 \geq 50$	$15x + 2 \geq 50$

b. You want to find the values of x so that the volume of the rectangular prism is no more than 36 cubic units.

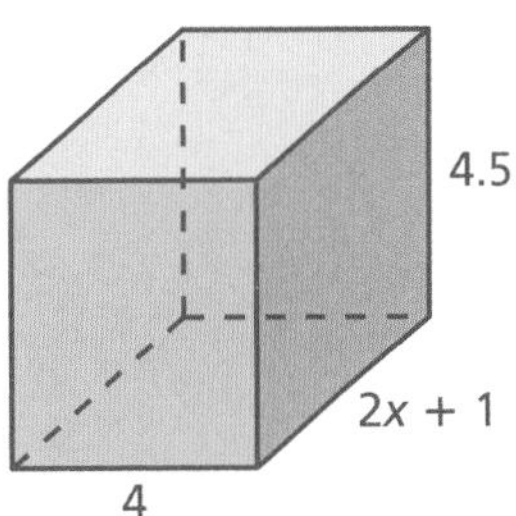

$8x + 4 < 36$	$36x + 18 < 36$	$2x + 9.5 \leq 36$	$36x + 18 \leq 36$

What Is Your Answer?

3. **IN YOUR OWN WORDS** How can you use an inequality to describe the dimensions of a figure?

4. Use what you know about solving equations and inequalities to describe how you can solve a two-step inequality. Give an example to support your explanation.

Practice Use what you learned about solving two-step inequalities to complete Exercises 3 and 4 on page 490.

11.4 Lesson

You can solve two-step inequalities in the same way you solve two-step equations.

EXAMPLE 1 Solving Two-Step Inequalities

a. Solve $5x - 4 \ge 11$. Graph the solution.

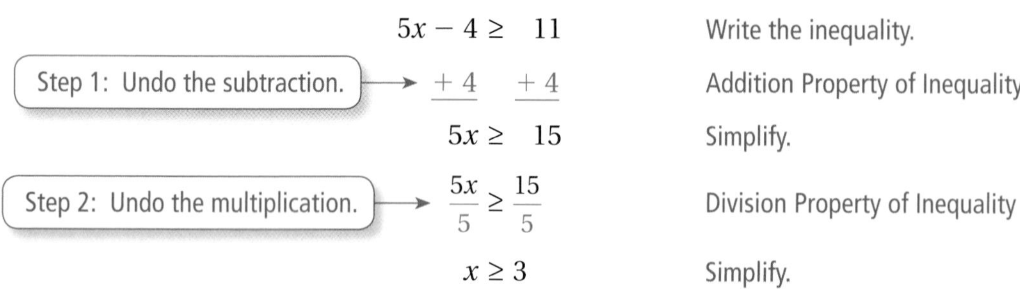

∴ The solution is $x \ge 3$.

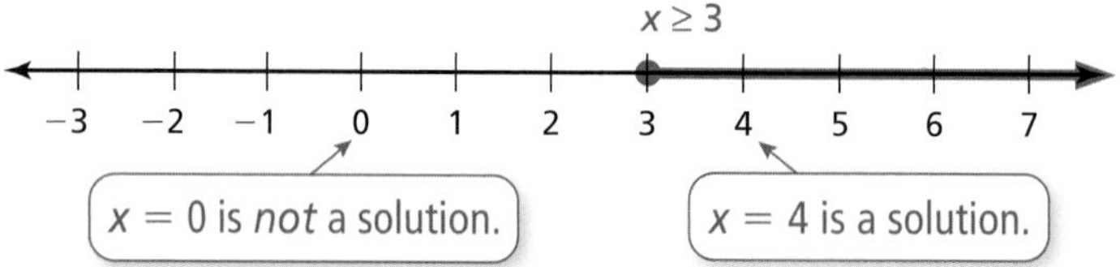

−3 −2 −1 0 1 2 3 4 5 6 7
x = 0 is not a solution.
x = 4 is a solution.

b. Solve $\dfrac{b}{-3} + 4 < 13$. Graph the solution.

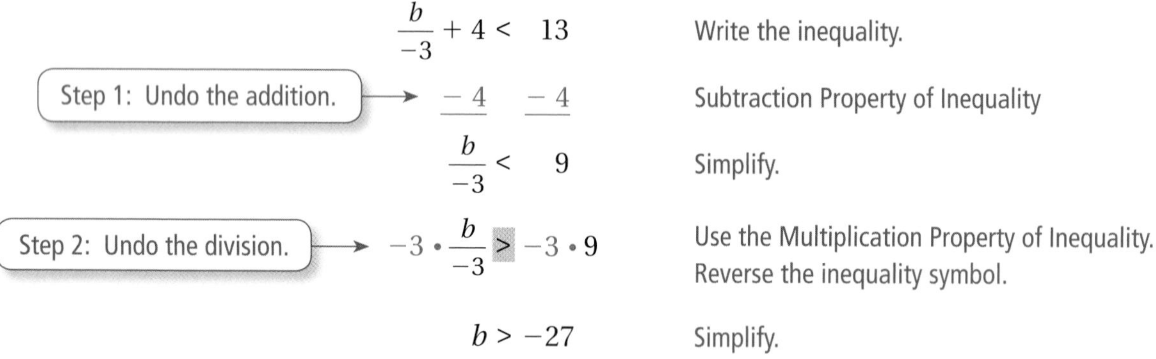

b/−3 + 4 < 13 Write the inequality.
Step 1: Undo the addition.
− 4 − 4 Subtraction Property of Inequality
b/−3 < 9 Simplify.
Step 2: Undo the division.
−3 • b/−3 > −3 • 9 Use the Multiplication Property of Inequality. Reverse the inequality symbol.
b > −27 Simplify.

∴ The solution is $b > -27$.

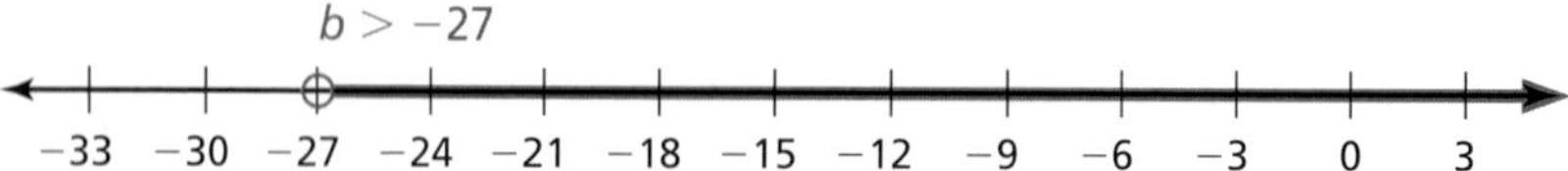

On Your Own

Now You're Ready
Exercises 5–10

Solve the inequality. Graph the solution.

1. $6y - 7 > 5$
2. $4 - 3d \ge 19$
3. $\dfrac{w}{-4} + 8 > 9$

EXAMPLE 2 Graphing an Inequality

Which graph represents the solution of $-7(x + 3) \leq 28$?

Ⓐ 

Ⓑ 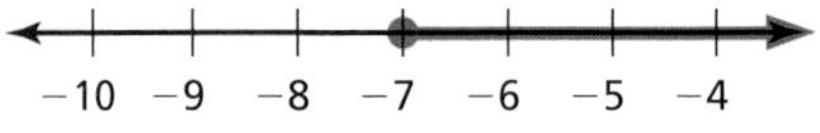

Ⓒ 

Ⓓ 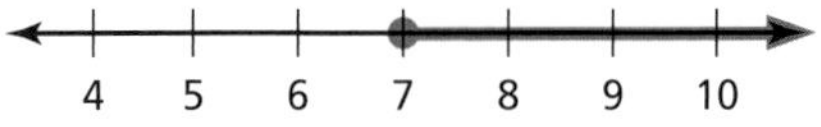

$-7(x + 3) \leq 28$	Write the inequality.
$-7x - 21 \leq 28$	Distributive Property
$+21 \quad +21$	Addition Property of Inequality
$-7x \leq 49$	Simplify.
$\dfrac{-7x}{-7} \geq \dfrac{49}{-7}$	Use the Division Property of Inequality. Reverse the inequality symbol.
$x \geq -7$	Simplify.

Step 1: Undo the subtraction. → $+21 \quad +21$

Step 2: Undo the multiplication. → $\dfrac{-7x}{-7} \geq \dfrac{49}{-7}$

∴ The correct answer is Ⓑ.

EXAMPLE 3 Real-Life Application

Progress Report	
Month	**Pounds Lost**
1	12
2	9
3	5
4	8

A contestant in a weight-loss competition wants to lose an average of at least 8 pounds per month during a 5-month period. How many pounds must the contestant lose in the fifth month to meet the goal?

Write and solve an inequality. Let x be the number of pounds lost in the fifth month.

$$\frac{12 + 9 + 5 + 8 + x}{5} \geq 8$$

The phrase *at least* means *greater than or equal to.*

$\dfrac{34 + x}{5} \geq 8$	Simplify.
$5 \cdot \dfrac{34 + x}{5} \geq 5 \cdot 8$	Multiplication Property of Inequality
$34 + x \geq 40$	Simplify.
$x \geq 6$	Subtract 34 from each side.

∴ So, the contestant must lose at least 6 pounds to meet the goal.

Remember

In Example 3, the average is equal to the sum of the pounds lost divided by the number of months.

On Your Own

Now You're Ready
Exercises 12–17

Solve the inequality. Graph the solution.

4. $2(k - 5) < 6$ **5.** $-4(n - 10) < 32$ **6.** $-3 \leq 0.5(8 + y)$

7. WHAT IF? In Example 3, the contestant wants to lose an average of at least 9 pounds per month. How many pounds must the contestant lose in the fifth month to meet the goal?

11.4 Exercises

Vocabulary and Concept Check

1. **WRITING** Compare and contrast solving two-step inequalities and solving two-step equations.

2. **OPEN-ENDED** Describe how to solve the inequality $3(a + 5) < 9$.

Practice and Problem Solving

Match the inequality with its graph.

3. $\dfrac{t}{3} - 1 \geq -3$

   A. −9 −8 −7 −6 −5 −4 −3 −2

   B. −9 −8 −7 −6 −5 −4 −3 −2

   C. −9 −8 −7 −6 −5 −4 −3 −2

4. $5x + 7 \leq 32$

   A. 2 3 4 5 6 7 8 9

   B. 2 3 4 5 6 7 8 9

   C. 2 3 4 5 6 7 8 9

Solve the inequality. Graph the solution.

1 5. $8y - 5 < 3$

6. $3p + 2 \geq -10$

7. $2 > 8 - \dfrac{4}{3}h$

8. $-2 > \dfrac{m}{6} - 7$

9. $-1.2b - 5.3 \geq 1.9$

10. $-1.3 \geq 2.9 - 0.6r$

11. **ERROR ANALYSIS** Describe and correct the error in solving the inequality.

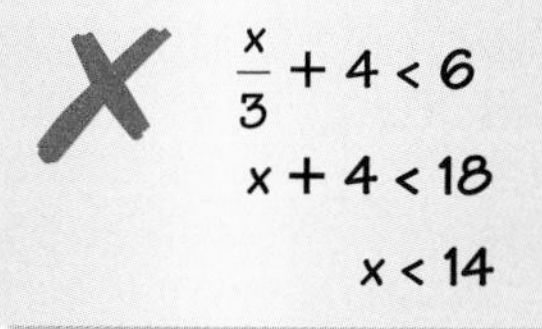

Solve the inequality. Graph the solution.

2 12. $5(g + 4) > 15$

13. $4(w - 6) \leq -12$

14. $-8 \leq \dfrac{2}{5}(k - 2)$

15. $-\dfrac{1}{4}(d + 1) < 2$

16. $7.2 > 0.9(n + 8.6)$

17. $20 \geq -3.2(c - 4.3)$

18. **UNICYCLE** The first jump in a unicycle high-jump contest is shown. The bar is raised 2 centimeters after each jump. Solve the inequality $2n + 10 \geq 26$ to find the number of additional jumps needed to meet or exceed the goal of clearing a height of 26 centimeters.

Solve the inequality. Graph the solution.

19. $9x - 4x + 4 \geq 36 - 12$

20. $3d - 7d + 2.8 < 5.8 - 27$

21. SCUBA DIVER A scuba diver is at an elevation of −38 feet. The diver starts moving at a rate of −12 feet per minute. Write and solve an inequality that represents how long it will take the diver to reach an elevation deeper than −200 feet.

22. KILLER WHALES A killer whale has eaten 75 pounds of fish today. It needs to eat at least 140 pounds of fish each day.

a. A bucket holds 15 pounds of fish. Write and solve an inequality that represents how many more buckets of fish the whale needs to eat.

b. Should the whale eat *four* or *five* more buckets of fish? Explain.

23. REASONING A student theater charges $9.50 per ticket.

a. The theater has already sold 70 tickets. Write and solve an inequality that represents how many more tickets the theater needs to sell to earn at least $1000.

b. The theater increases the ticket price by $1. Without solving an inequality, describe how this affects the total number of tickets needed to earn at least $1000.

24. Problem Solving For what values of r will the area of the shaded region be greater than or equal to 12 square units?

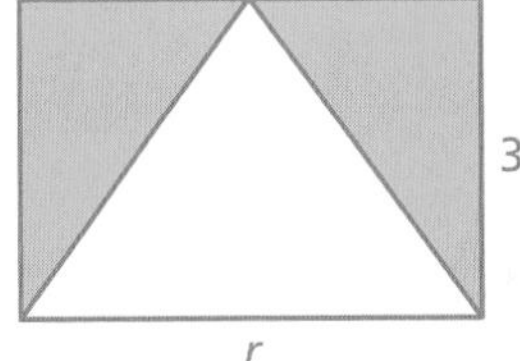

Fair Game Review What you learned in previous grades & lessons

Find the missing values in the ratio table. Then write the equivalent ratios. *(Skills Review Handbook)*

25.

Flutes	7		28
Clarinets	4	12	

26.

Boys	6	3	
Girls	10		50

27. MULTIPLE CHOICE What is the volume of the cube? *(Skills Review Handbook)*

Ⓐ 8 ft^3 Ⓑ 16 ft^3

Ⓒ 24 ft^3 Ⓓ 32 ft^3

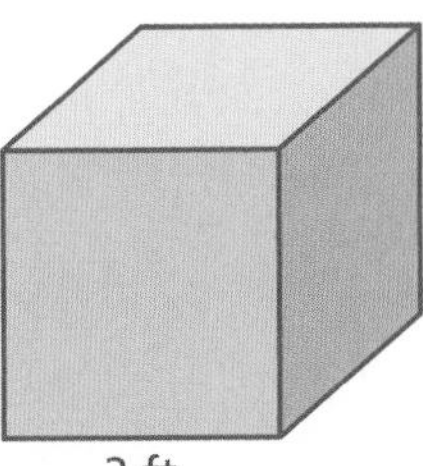

11.3–11.4 Quiz

Check It Out
Progress Check
BigIdeasMath.com

Solve the inequality. Graph the solution. *(Section 11.3 and Section 11.4)*

1. $3p \le 18$

2. $2x > -\frac{3}{5}$

3. $\frac{r}{3} \ge -5$

4. $-\frac{z}{8} < 1.5$

5. $3n + 2 \le 11$

6. $-2 < 5 - \frac{k}{2}$

7. $1.3m - 3.8 < -1.2$

8. $4.8 \ge 0.3(12 - y)$

Write the word sentence as an inequality. Then solve the inequality. *(Section 11.3)*

9. The quotient of a number and 5 is less than 4.

10. Six times a number is at least −14.

11. PEPPERS You have \$18 to buy peppers. Peppers cost \$1.50 each. Write and solve an inequality that represents the number of peppers you can buy. *(Section 11.3)*

12. MOVIES You have a gift card worth \$90. You want to buy several movies that cost \$12 each. Write and solve an inequality that represents the number of movies you can buy and still have at least \$30 on the gift card. *(Section 11.4)*

13. CHOCOLATES Your class sells boxes of chocolates to raise \$500 for a field trip. You earn \$6.25 for each box of chocolates sold. Write and solve an inequality that represents the number of boxes your class must sell to meet or exceed the fundraising goal. *(Section 11.3)*

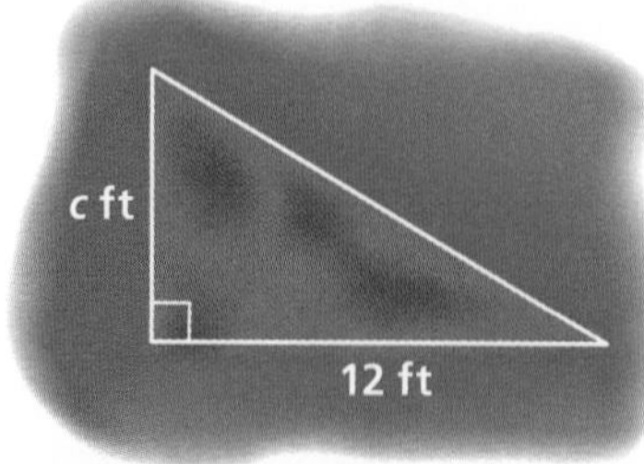

14. FENCE You want to put up a fence that encloses a triangular region with an area greater than or equal to 60 square feet. What is the least possible value of c? Explain. *(Section 11.3)*

11 Chapter Review

Check It Out
Vocabulary Help
BigIdeasMath .com

Review Key Vocabulary

inequality, *p. 466*
solution of an inequality, *p. 466*
solution set, *p. 466*
graph of an inequality, *p. 467*

Review Examples and Exercises

11.1 Writing and Graphing Inequalities *(pp. 464–469)*

a. Six plus a number x is at most $-\frac{1}{4}$. Write this word sentence as an inequality.

Six plus a number x is at most $-\frac{1}{4}$.

$6 + x \quad \leq \quad -\frac{1}{4}$

An inequality is $6 + x \leq -\frac{1}{4}$.

b. Graph $m > 3$.

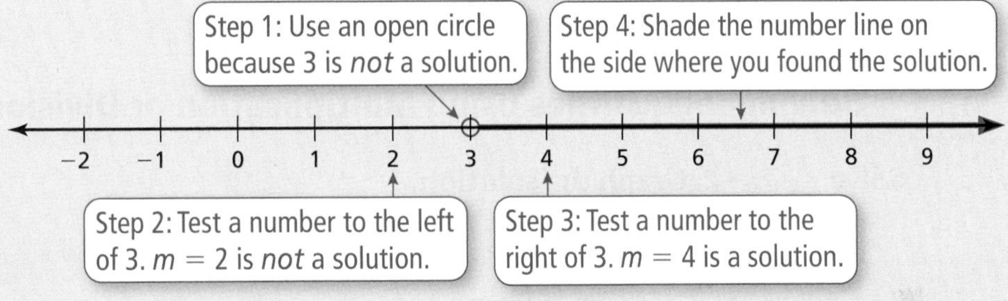

Step 1: Use an open circle because 3 is *not* a solution.
Step 4: Shade the number line on the side where you found the solution.
−2 −1 0 1 2 3 4 5 6 7 8 9
Step 2: Test a number to the left of 3. $m = 2$ is *not* a solution.
Step 3: Test a number to the right of 3. $m = 4$ is a solution.

Exercises

Write the word sentence as an inequality.

1. A number w is greater than -3.
2. A number y minus $\frac{1}{2}$ is no more than $-\frac{3}{2}$.

Tell whether the given value is a solution of the inequality.

3. $5 + j > 8; j = 7$
4. $6 \div n \leq -5; n = -3$

Graph the inequality on a number line.

5. $q > -1.3$
6. $s < 1\frac{3}{4}$
7. **BUMPER CARS** You must be at least 42 inches tall to ride the bumper cars at an amusement park. Write an inequality that represents this situation.

11.2 Solving Inequalities Using Addition or Subtraction (pp. 470–475)

Solve $-5 < m - 3$. Graph the solution.

Undo the subtraction.	$-5 < m - 3$	Write the inequality.
	$\underline{+3} \qquad \underline{+3}$	Addition Property of Inequality
	$-2 < m$	Simplify.

∴ The solution is $m > -2$.

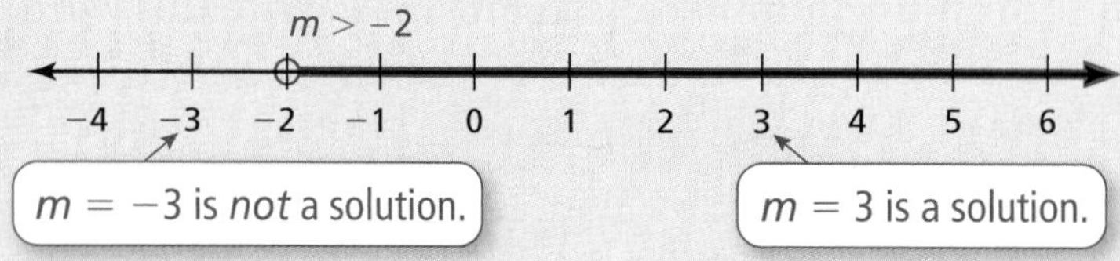

Exercises

Solve the inequality. Graph the solution.

8. $d + 12 < 19$
9. $t - 4 \leq -14$
10. $-8 \leq z + 6.4$

11.3 Solving Inequalities Using Multiplication or Division (pp. 478–485)

Solve $\frac{c}{-3} \geq -2$. Graph the solution.

	$\frac{c}{-3} \geq -2$	Write the inequality.
Undo the division.	$-3 \cdot \frac{c}{-3} \leq -3 \cdot (-2)$	Use the Multiplication Property of Inequality. Reverse the inequality symbol.
	$c \leq 6$	Simplify.

∴ The solution is $c \leq 6$.

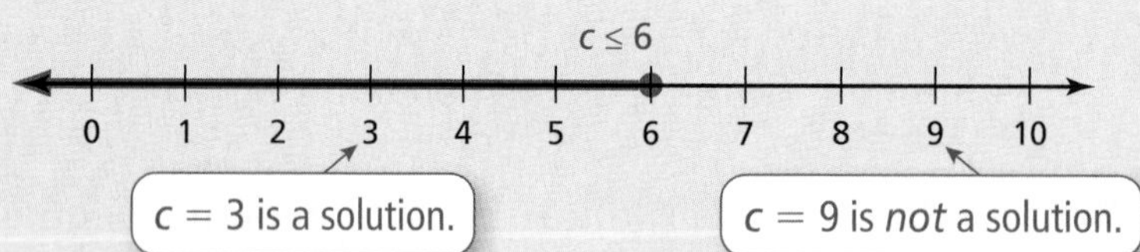

Exercises

Solve the inequality. Graph the solution.

11. $6q < -18$
12. $-\frac{r}{3} \leq 6$
13. $-4 > -\frac{4}{3}s$

11.4 Solving Two-Step Inequalities (pp. 486–491)

a. Solve $6x - 8 \leq 10$. Graph the solution.

$$6x - 8 \leq 10$$ Write the inequality.

$$\underline{+8 \quad +8}$$ Addition Property of Inequality

$$6x \leq 18$$ Simplify.

Step 2: Undo the multiplication.

$$\frac{6x}{6} \leq \frac{18}{6}$$ Division Property of Inequality

$$x \leq 3$$ Simplify.

∴ The solution is $x \leq 3$.

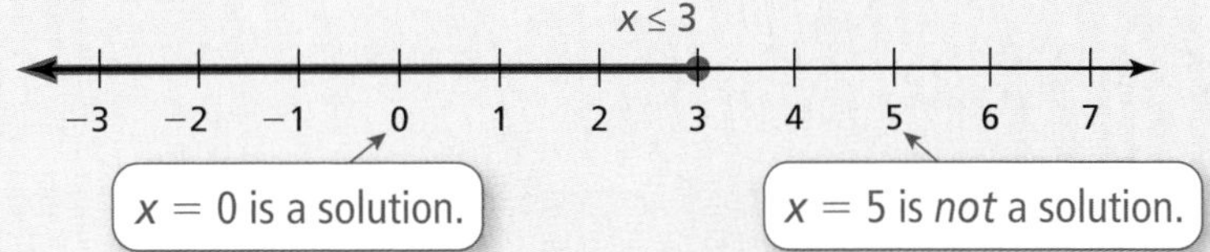

x = 0 is a solution.

x = 5 is *not* a solution.

b. Solve $\frac{q}{-4} + 7 < 11$. Graph the solution.

$$\frac{q}{-4} + 7 < 11$$ Write the inequality.

$$\underline{-7 \quad -7}$$ Subtraction Property of Inequality

$$\frac{q}{-4} < 4$$ Simplify.

Step 2: Undo the division.

$$-4 \cdot \frac{q}{-4} > -4 \cdot 4$$ Use the Multiplication Property of Inequality. Reverse the inequality symbol.

$$q > -16$$ Simplify.

∴ The solution is $q > -16$.

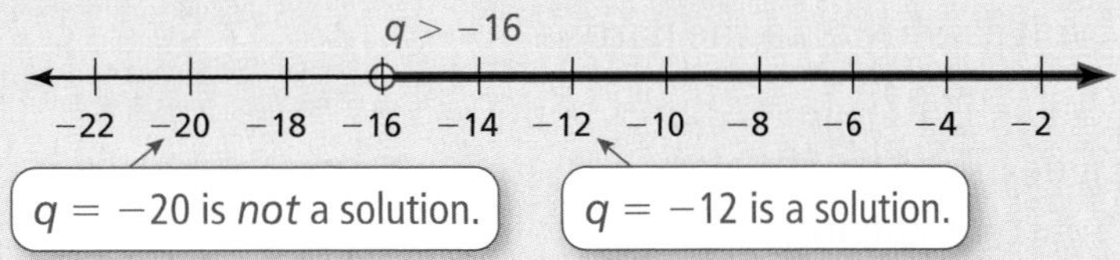

q = −20 is *not* a solution.

q = −12 is a solution.

Exercises

Solve the inequality. Graph the solution.

14. $3x + 4 > 16$

15. $\frac{z}{-2} - 6 \leq -2$

16. $-2t - 5 < 9$

17. $7(q + 2) < -77$

18. $-\frac{1}{3}(p + 9) \leq 4$

19. $1.2(j + 3.5) \geq 4.8$

11 Chapter Test

Write the word sentence as an inequality.

1. A number k plus 19.5 is less than or equal to 40.

2. A number q multiplied by $\frac{1}{4}$ is greater than -16.

Tell whether the given value is a solution of the inequality.

3. $n - 3 \le 4;\ n = 7$

4. $-\frac{3}{7}m < 1;\ m = -7$

5. $-4c \ge 7;\ c = -2$

6. $-2.4m > -6.8;\ m = -3$

Solve the inequality. Graph the solution.

7. $w + 4 \le 3$

8. $x - 4 > -6$

9. $-\frac{2}{9} + y \le \frac{5}{9}$

10. $-6z \ge 36$

11. $-5.2 \ge \frac{p}{4}$

12. $4k - 8 \ge 20$

13. $\frac{4}{7} - b \ge -\frac{1}{7}$

14. $-0.6 > -0.3(d + 6)$

15. **GUMBALLS** You have \$2.50. Each gumball in a gumball machine costs \$0.25. Write and solve an inequality that represents the number of gumballs you can buy.

16. **PARTY** You can spend no more than \$100 on a party you are hosting. The cost per guest is \$8.

 a. Write and solve an inequality that represents the number of guests you can invite to the party.

 b. What is the greatest number of guests that you can invite to the party? Explain your reasoning.

17. **BASEBALL CARDS** You have \$30 to buy baseball cards. Each pack of cards costs \$5. Write and solve an inequality that represents the number of packs of baseball cards you can buy and still have at least \$10 left.

11 Cumulative Assessment

1. What is the value of the expression below when $x = -5$, $y = 3$, and $z = -1$?

$$\frac{x^2 - 3y}{z}$$

A. -34 C. 16

B. -16 D. 34

2. What is the value of the expression below?

$$-\frac{3}{8} \cdot \frac{2}{5}$$

F. $-\frac{20}{3}$ H. $-\frac{15}{16}$

G. $-\frac{16}{15}$ I. $-\frac{3}{20}$

3. Which graph represents the inequality below?

$$\frac{x}{-4} - 8 \geq -9$$

A. −3 −2 −1 0 1 2 3 4 5 6

B. 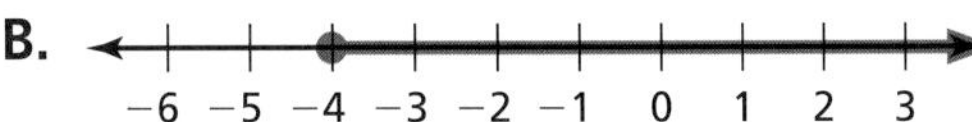

C. 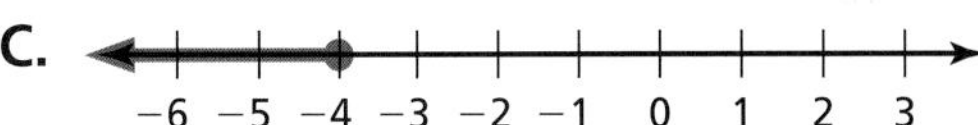

D. −3 −2 −1 0 1 2 3 4 5 6

4. Which value of p makes the equation below true?

$$5(p + 6) = 25$$

F. -1 H. 11

G. $3\frac{4}{5}$ I. 14

5. You set up the lemonade stand. Your profit is equal to your revenue from lemonade sales minus your cost to operate the stand. Your cost is \$8. How many cups of lemonade must you sell to earn a profit of \$30?

A. 4

B. 44

C. 60

D. 76

6. Which value is a solution of the inequality below?

$$3 - 2y < 7$$

F. -6

G. -3

H. -2

I. -1

7. What value of y makes the equation below true?

$$12 - 3y = -6$$

8. What is the mean distance of the four points from -3?

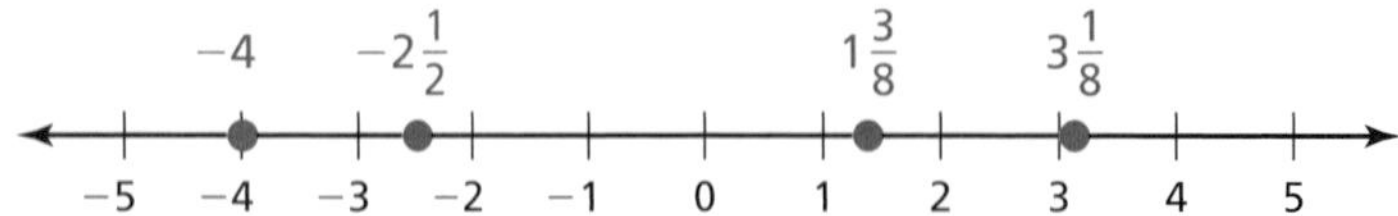

A. $-\frac{1}{2}$

B. $2\frac{1}{2}$

C. 3

D. $7\frac{1}{8}$

9. Martin graphed the solution of the inequality $-4x + 18 > 6$ in the box below.

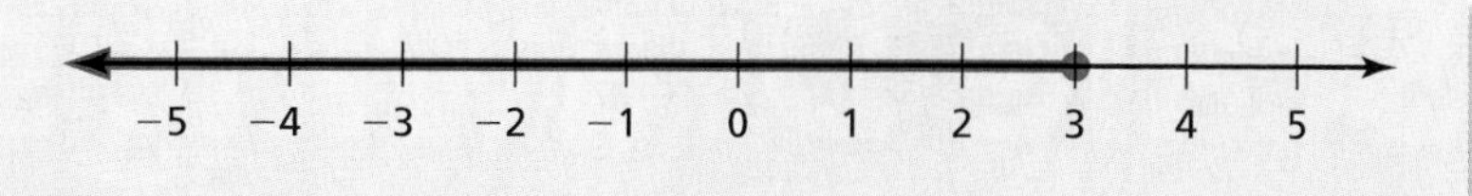

What should Martin do to correct the error that he made?

F. Use an open circle at 3 and shade to the left of 3.

G. Use an open circle at 3 and shade to the right of 3.

H. Use a closed circle and shade to the right of 3.

I. Use an open circle and shade to the left of -3.

10. What is the value of the expression below?

$$\frac{5}{12} - \frac{7}{8}$$

11. You are selling T-shirts to raise money for a charity. You sell the T-shirts for $10 each.

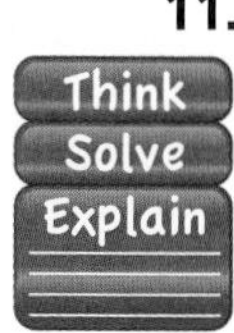

Part A You have already sold 2 T-shirts. How many more T-shirts must you sell to raise at least $500? Explain.

Part B Your friend is raising money for the same charity and has not sold any T-shirts previously. He sells the T-shirts for $8 each. What is the total number of T-shirts he must sell to raise at least $500? Explain.

Part C Who has to sell more T-shirts in total? How many more? Explain.

12. Which expression is equivalent to the expression below?

$$-\frac{2}{3} - \left(-\frac{4}{9}\right)$$

A. $-\frac{1}{3} + \frac{1}{9}$

B. $-\frac{2}{3} \times \left(-\frac{1}{3}\right)$

C. $-\frac{1}{3} - \frac{7}{9}$

D. $\frac{3}{2} \div \left(-\frac{1}{3}\right)$

12 Constructions and Scale Drawings

12.1 Adjacent and Vertical Angles

12.2 Complementary and Supplementary Angles

12.3 Triangles

12.4 Quadrilaterals

12.5 Scale Drawings

"Move 4 of the lines to make 3 equilateral triangles."

"Well done, Descartes!"

"I'm at 3rd base. You are running to 1st base, and Fluffy is running to 2nd base."

"Should I throw the ball to 2nd to get Fluffy out or throw it to 1st to get you out?"

What You Learned Before

Yes, it's very acute.
"Look at this baby crocodile! Isn't it cute?"

Measuring Angles

Example 1 **Use a protractor to find the measure of each angle. Then classify the angle as *acute*, *obtuse*, *right*, or *straight*.**

a.

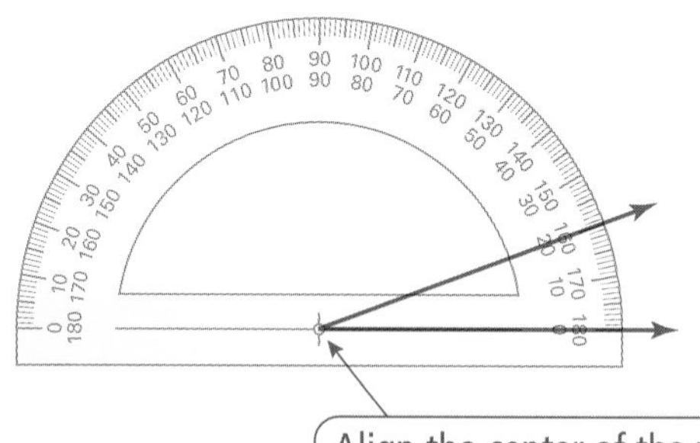

b.

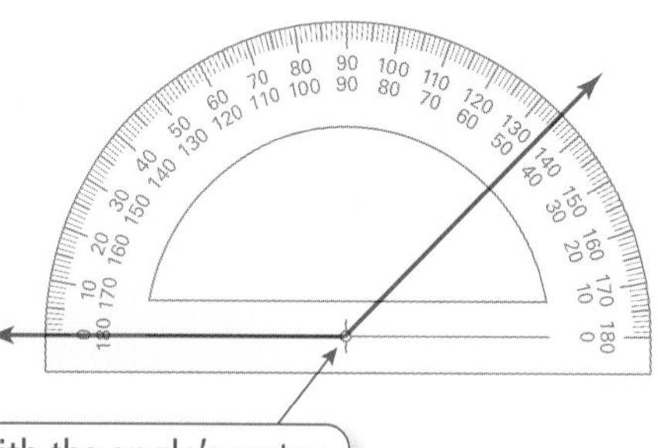

Align the center of the protractor with the angle's vertex.

The angle measure is 20°. So, the angle is acute.

The angle measure is 135°. So, the angle is obtuse.

Drawing Angles

Example 2 **Use a protractor to draw a 45° angle.**

Draw a ray. Place the center of the protractor on the endpoint of the ray and align the protractor so the ray passes through the 0° mark. Make a mark at 45°. Then draw a ray from the endpoint at the center of the protractor through the mark at 45°.

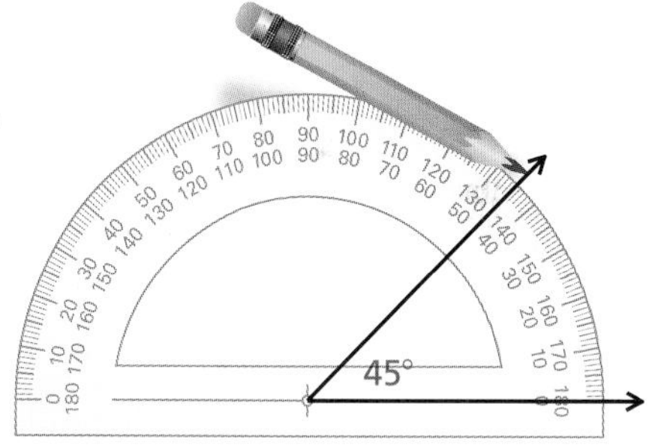

Try It Yourself

Use a protractor to find the measure of the angle. Then classify the angle as *acute*, *obtuse*, *right*, or *straight*.

1.

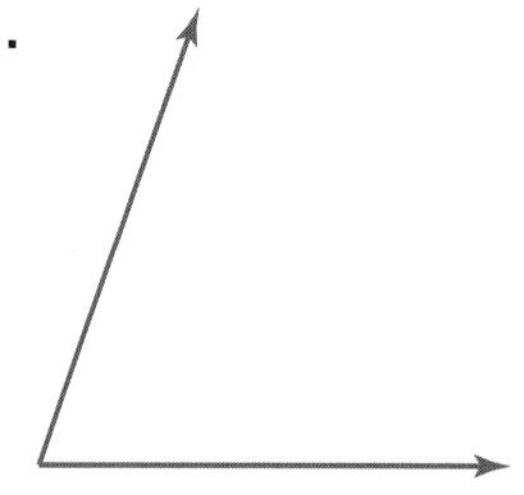

2.

3.

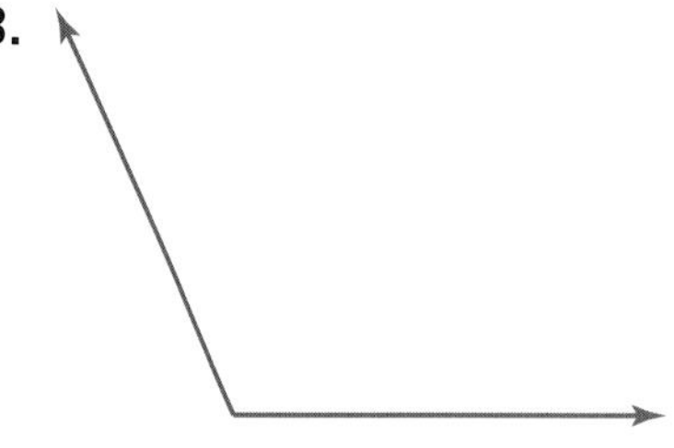

Use a protractor to draw an angle with the given measure.

4. 55° **5.** 160° **6.** 85° **7.** 180°

12.1 Adjacent and Vertical Angles

Essential Question What can you conclude about the angles formed by two intersecting lines?

Classification of Angles

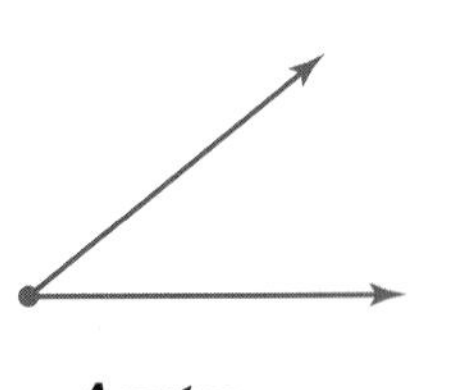

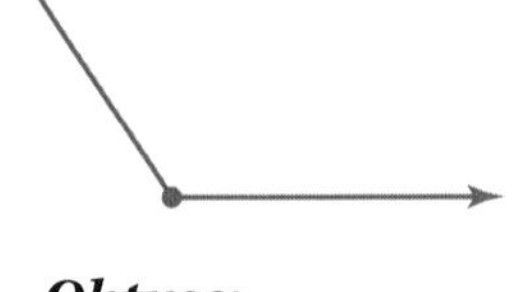

Acute:	***Right:***	***Obtuse:***	***Straight:***
Less than 90°	Equal to 90°	Greater than 90° and less than 180°	Equal to 180°

1 ACTIVITY: Drawing Angles

Work with a partner.

a. Draw the hands of the clock to represent the given type of angle.

Acute Straight Right Obtuse

b. What is the measure of the angle formed by the hands of the clock at the given time?

9:00 6:00 12:00

The Meaning of a Word  Adjacent

When two states are **adjacent**, they are next to each other and they share a common border.

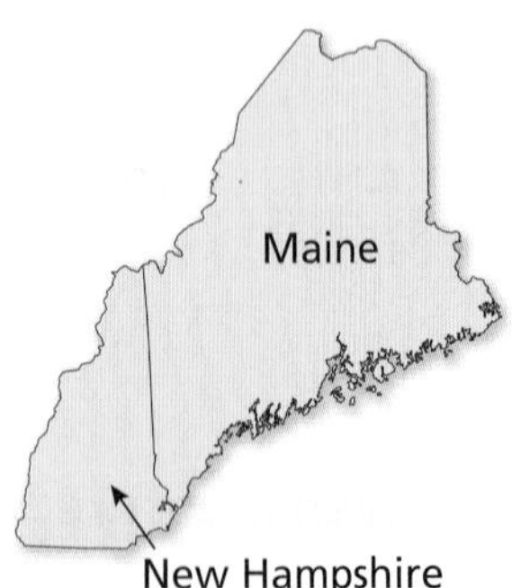

Geometry

In this lesson, you will
- identify adjacent and vertical angles.
- find angle measures using adjacent and vertical angles.

2 ACTIVITY: Naming Angles

Work with a partner. Some angles, such as $\angle A$, can be named by a single letter. When this does not clearly identify an angle, you should use three letters, as shown.

Math Practice

Justify Conclusions

When you name an angle, does the order in which you write the letters matter? Explain.

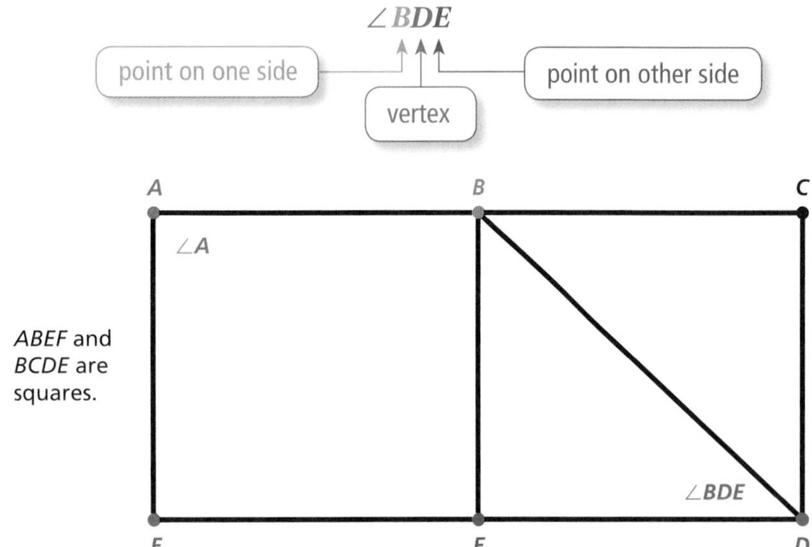

a. Name all the right angles, acute angles, and obtuse angles.

b. Which pairs of angles do you think are *adjacent?* Explain.

3 ACTIVITY: Measuring Angles

Work with a partner.

a. How many angles are formed by the intersecting roads? Number the angles.

b. **CHOOSE TOOLS** Measure each angle formed by the intersecting roads. What do you notice?

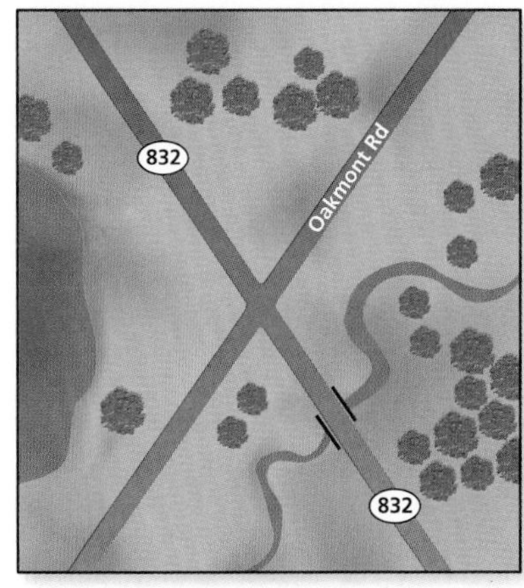

What Is Your Answer?

4. **IN YOUR OWN WORDS** What can you conclude about the angles formed by two intersecting lines?

5. Draw two acute angles that are adjacent.

Use what you learned about angles and intersecting lines to complete Exercises 3 and 4 on page 506.

12.1 Lesson

Key Vocabulary
adjacent angles, *p. 504*
vertical angles, *p. 504*
congruent angles, *p. 504*

Key Ideas

Adjacent Angles

Words Two angles are **adjacent angles** when they share a common side and have the same vertex.

Examples

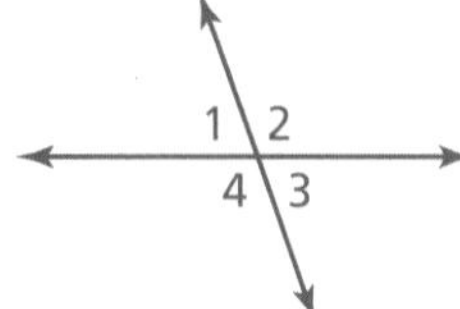

$\angle 1$ and $\angle 2$ are adjacent.

$\angle 2$ and $\angle 4$ are not adjacent.

Vertical Angles

Words Two angles are **vertical angles** when they are opposite angles formed by the intersection of two lines. Vertical angles are **congruent angles**, meaning they have the same measure.

Examples

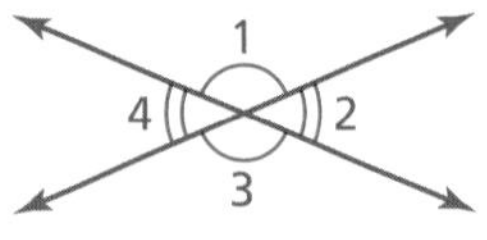

$\angle 1$ and $\angle 3$ are vertical angles.

$\angle 2$ and $\angle 4$ are vertical angles.

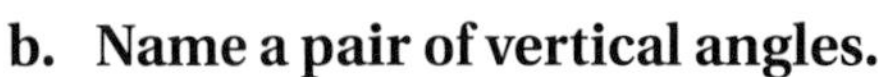

EXAMPLE 1 Naming Angles

Use the figure shown.

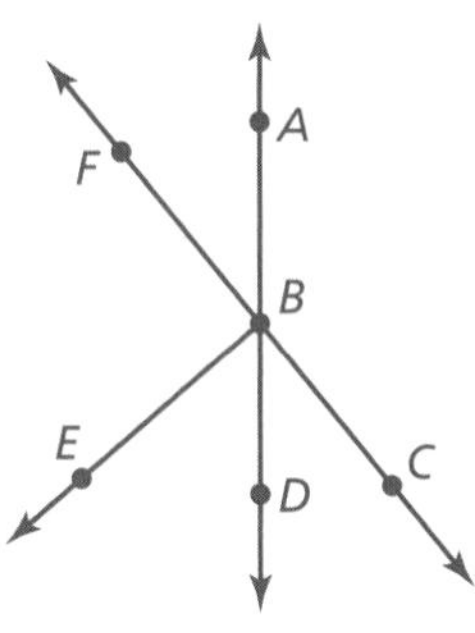

a. Name a pair of adjacent angles.

$\angle ABC$ and $\angle ABF$ share a common side and have the same vertex B.

So, $\angle ABC$ and $\angle ABF$ are adjacent angles.

b. Name a pair of vertical angles.

$\angle ABF$ and $\angle CBD$ are opposite angles formed by the intersection of two lines.

So, $\angle ABF$ and $\angle CBD$ are vertical angles.

On Your Own

Now You're Ready
Exercises 5 and 6

Name two pairs of adjacent angles and two pairs of vertical angles in the figure.

1.

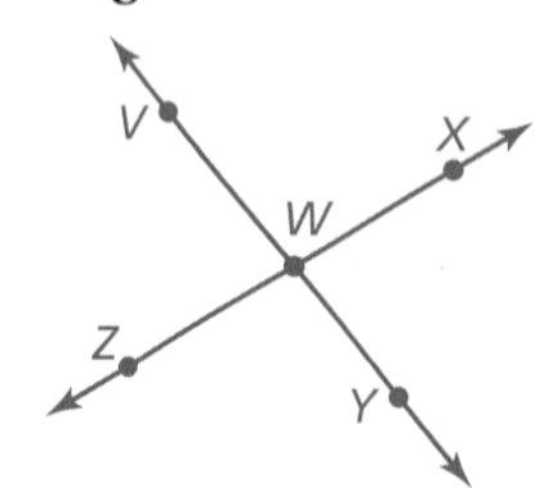

2.

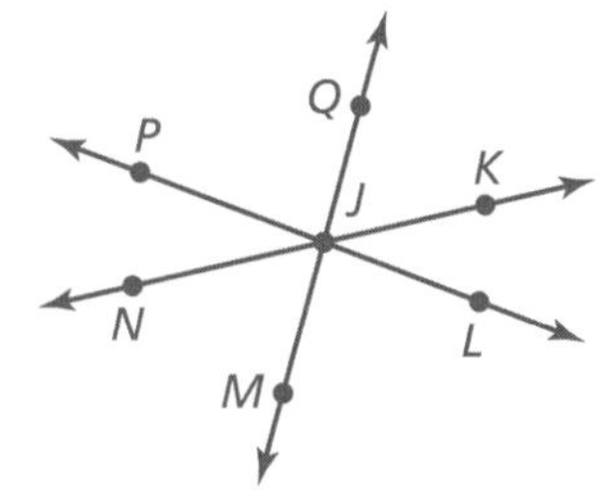

Multi-Language Glossary at BigIdeasMath.com

EXAMPLE 2 Using Adjacent and Vertical Angles

Tell whether the angles are *adjacent* or *vertical*. Then find the value of x.

a.

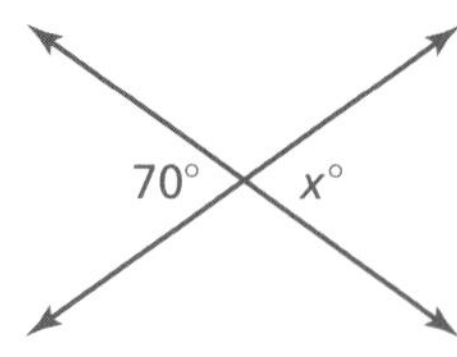

The angles are vertical angles. Because vertical angles are congruent, the angles have the same measure.

So, the value of x is 70.

Remember

You can add angle measures. When two or more adjacent angles form a larger angle, the sum of the measures of the smaller angles is equal to the measure of the larger angle.

b.

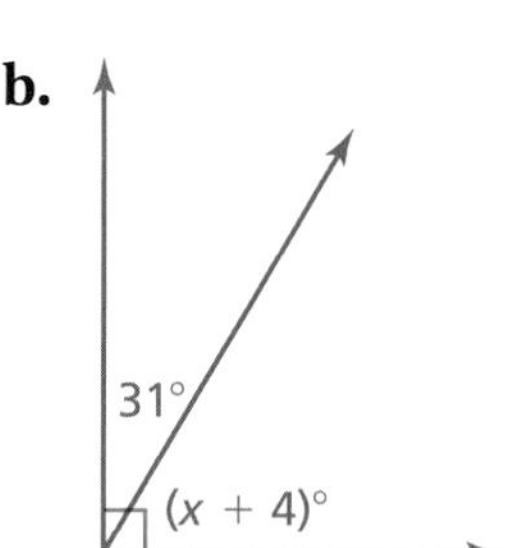

The angles are adjacent angles. Because the angles make up a right angle, the sum of their measures is 90°.

$(x + 4) + 31 = 90$ Write equation.

$x + 35 = 90$ Combine like terms.

$x = 55$ Subtract 35 from each side.

So, the value of x is 55.

EXAMPLE 3 Constructing Angles

Draw a pair of vertical angles with a measure of 40°.

Step 1: Use a protractor to draw a 40° angle.

Step 2: Use a straightedge to extend the sides to form two intersecting lines.

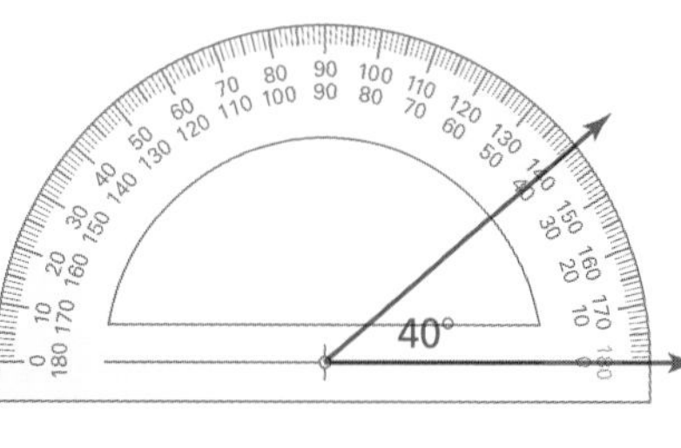

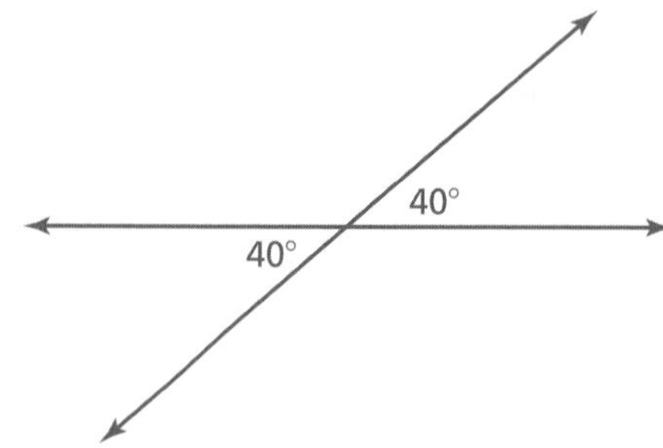

On Your Own

Now You're Ready
Exercises 8–17

Tell whether the angles are *adjacent* or *vertical*. Then find the value of x.

3. **4.**

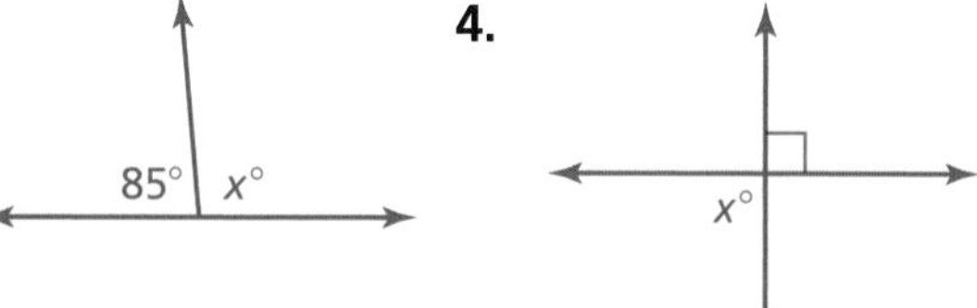

5.

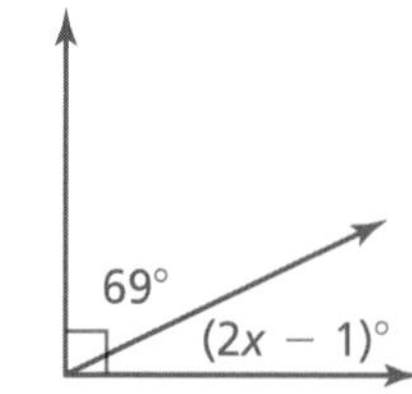

6. Draw a pair of vertical angles with a measure of 75°.

12.1 Exercises

Vocabulary and Concept Check

1. **VOCABULARY** When two lines intersect, how many pairs of vertical angles are formed? How many pairs of adjacent angles are formed?

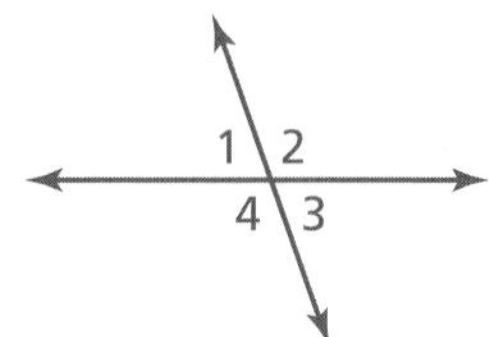

2. **REASONING** Identify the congruent angles in the figure. Explain your reasoning.

Practice and Problem Solving

Use the figure at the right.

3. Measure each angle formed by the intersecting lines.
4. Name two angles that are adjacent to $\angle ABC$.

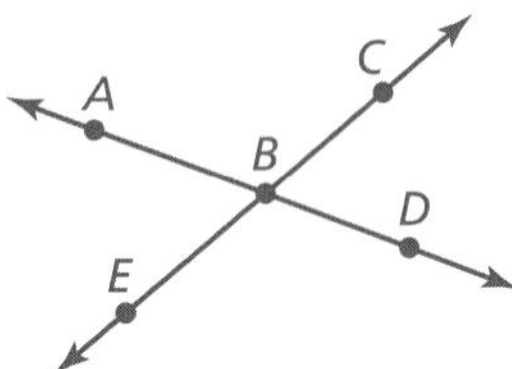

Name two pairs of adjacent angles and two pairs of vertical angles in the figure.

1 5.

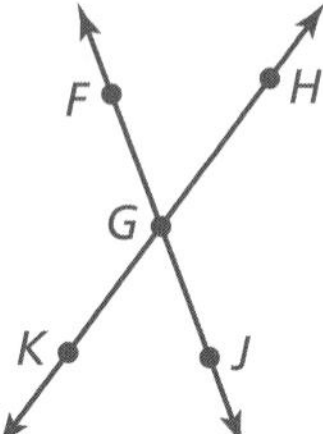

6.

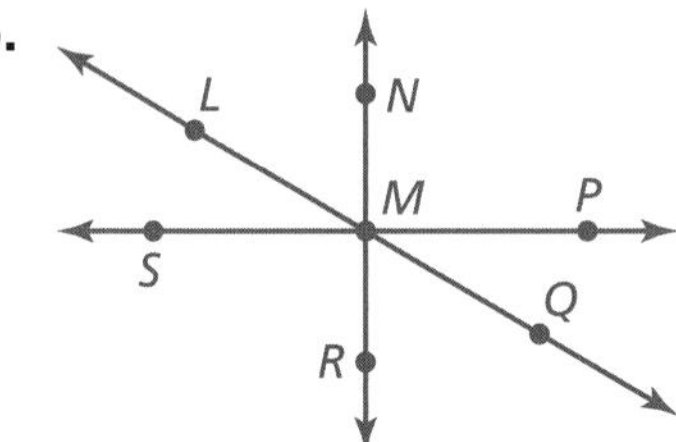

7. **ERROR ANALYSIS** Describe and correct the error in naming a pair of vertical angles.

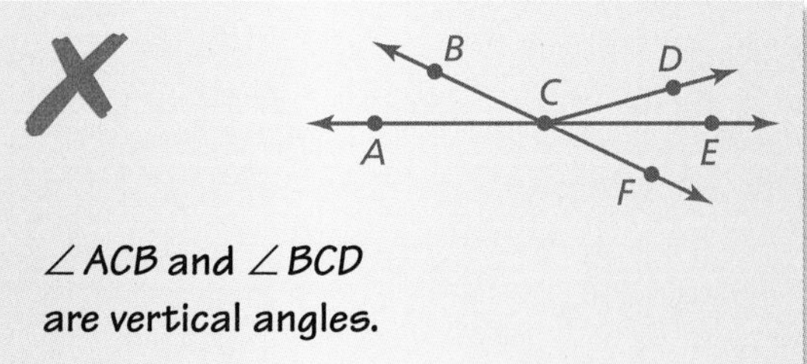

Tell whether the angles are *adjacent* or *vertical*. Then find the value of *x*.

2 8.

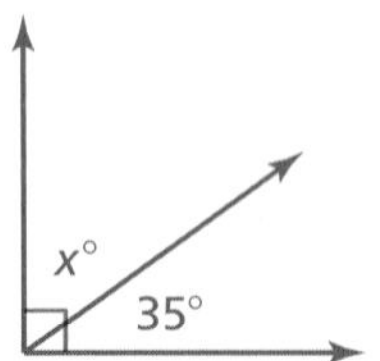

9.

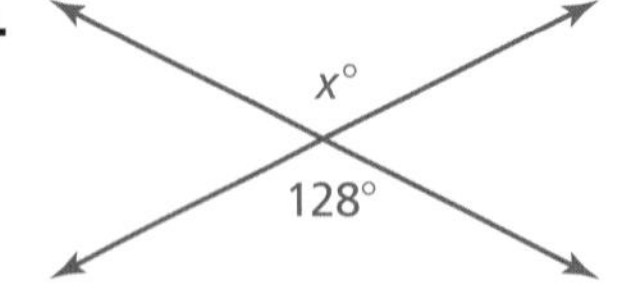

10.

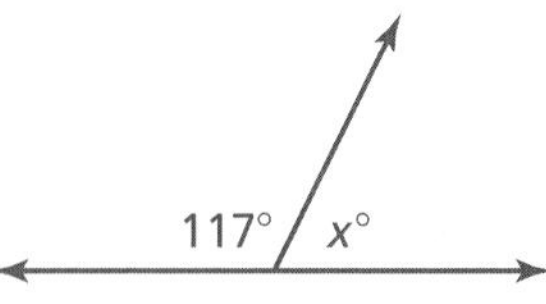

11.

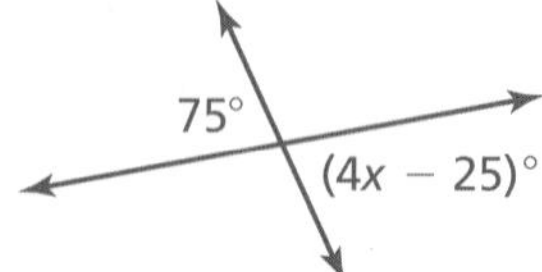

12.

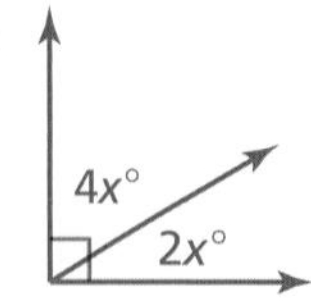

13.

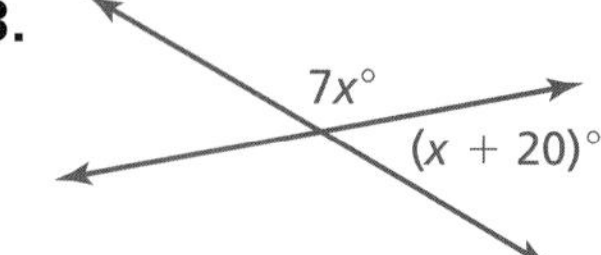

Draw a pair of vertical angles with the given measure.

3 **14.** 25° **15.** 85° **16.** 110° **17.** 135°

18. IRON CROSS The iron cross is a skiing trick in which the tips of the skis are crossed while the skier is airborne. Find the value of x in the iron cross shown.

19. OPEN-ENDED Draw a pair of adjacent angles with the given description.

a. Both angles are acute.

b. One angle is acute, and one is obtuse.

c. The sum of the angle measures is 135°.

20. PRECISION Explain two procedures that you can use to draw adjacent angles with given measures.

Determine whether the statement is *always, sometimes,* or *never* true.

21. When the measure of $\angle 1$ is 70°, the measure of $\angle 3$ is 110°.

22. When the measure of $\angle 4$ is 120°, the measure of $\angle 1$ is 60°.

23. $\angle 2$ and $\angle 3$ are congruent.

24. The measure of $\angle 1$ plus the measure of $\angle 2$ equals the measure of $\angle 3$ plus the measure of $\angle 4$.

25. REASONING Draw a figure in which $\angle 1$ and $\angle 2$ are acute vertical angles, $\angle 3$ is a right angle adjacent to $\angle 2$, and the sum of the measure of $\angle 1$ and the measure of $\angle 4$ is 180°.

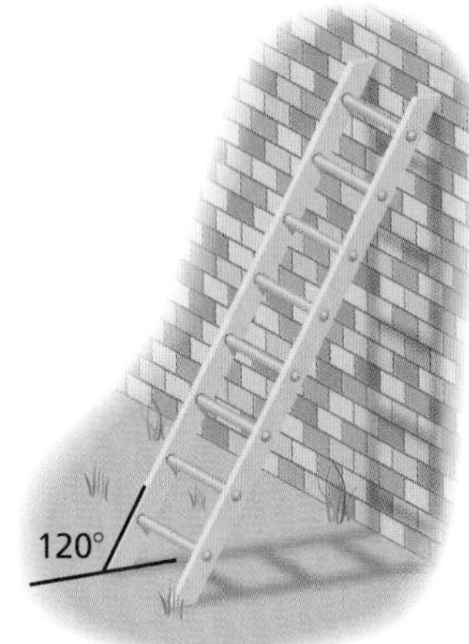

26. Structure For safety reasons, a ladder should make a 15° angle with a wall. Is the ladder shown leaning at a safe angle? Explain.

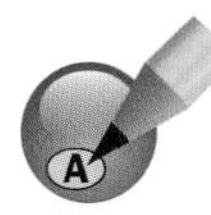

Fair Game Review What you learned in previous grades & lessons

Solve the inequality. Graph the solution. *(Section 11.3)*

27. $-6n > 54$ **28.** $-\frac{1}{2}x \le 17$ **29.** $-1.6 < \frac{m}{-2.5}$

30. MULTIPLE CHOICE What is the slope of the line that passes through the points (2, 3) and (6, 8)? *(Skills Review Handbook)*

Ⓐ $\frac{4}{5}$ Ⓑ $\frac{5}{4}$ Ⓒ $\frac{4}{3}$ Ⓓ $\frac{3}{2}$

12.2 Complementary and Supplementary Angles

Essential Question How can you classify two angles as complementary or supplementary?

1 ACTIVITY: Complementary and Supplementary Angles

Work with a partner.

a. The graph represents the measures of *complementary angles*. Use the graph to complete the table.

x		20°		30°	45°		75°
y	80°		65°	60°		40°	

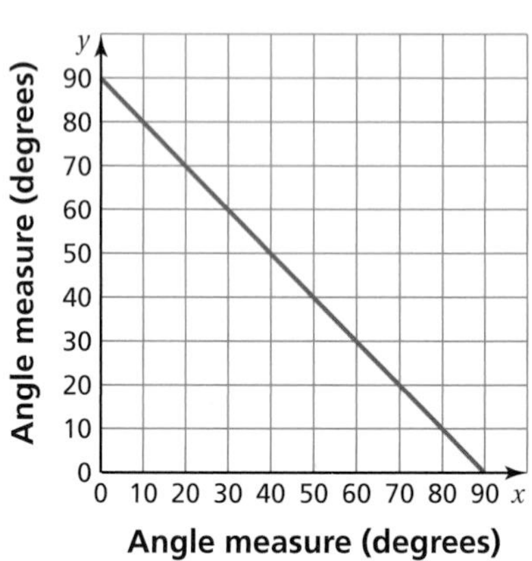

b. How do you know when two angles are complementary? Explain.

c. The graph represents the measures of *supplementary angles*. Use the graph to complete the table.

x	20°		60°	90°		140°	
y		150°		90°	50°		30°

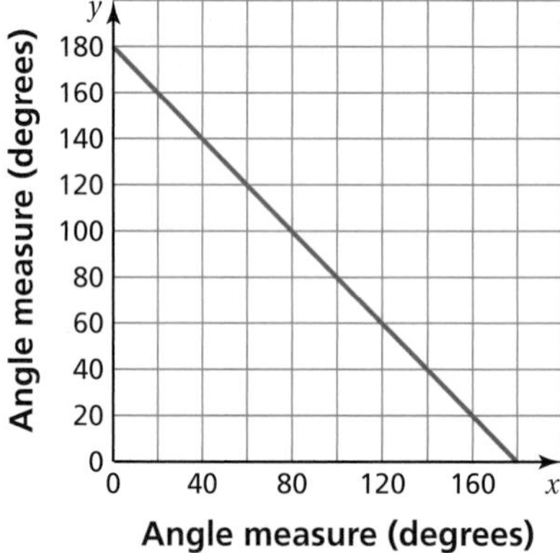

d. How do you know when two angles are supplementary? Explain.

Geometry
In this lesson, you will
- classify pairs of angles as complementary, supplementary, or neither.
- find angle measures using complementary and supplementary angles.

2 ACTIVITY: Exploring Rules About Angles

Work with a partner. Copy and complete each sentence with *always*, *sometimes*, or *never*.

a. If x and y are complementary angles, then both x and y are ________ acute.

b. If x and y are supplementary angles, then x is ________ acute.

c. If x is a right angle, then x is ________ acute.

d. If x and y are complementary angles, then x and y are ________ adjacent.

e. If x and y are supplementary angles, then x and y are ________ vertical.

3 ACTIVITY: Classifying Pairs of Angles

Work with a partner. Tell whether the two angles shown on the clocks are *complementary*, *supplementary*, or *neither*. Explain your reasoning.

a. 

b. 

c. 

d. 

4 ACTIVITY: Identifying Angles

Work with a partner. Use a protractor and the figure shown.

a. Name four pairs of complementary angles and four pairs of supplementary angles.

> **Math Practice**
>
> **Use Definitions**
>
> How can you use the definitions of *complementary*, *supplementary*, and *vertical angles* to answer the questions?

b. Name two pairs of vertical angles.

What Is Your Answer?

5. IN YOUR OWN WORDS How can you classify two angles as complementary or supplementary? Give examples of each type.

Practice Use what you learned about complementary and supplementary angles to complete Exercises 3–5 on page 512.

12.2 Lesson

Key Vocabulary
complementary angles, *p. 510*
supplementary angles, *p. 510*

Complementary Angles

Words Two angles are **complementary angles** when the sum of their measures is 90°.

Examples

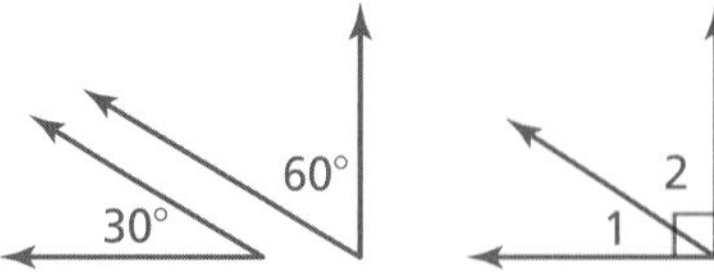

$\angle 1$ and $\angle 2$ are complementary angles.

Supplementary Angles

Words Two angles are **supplementary angles** when the sum of their measures is 180°.

Examples

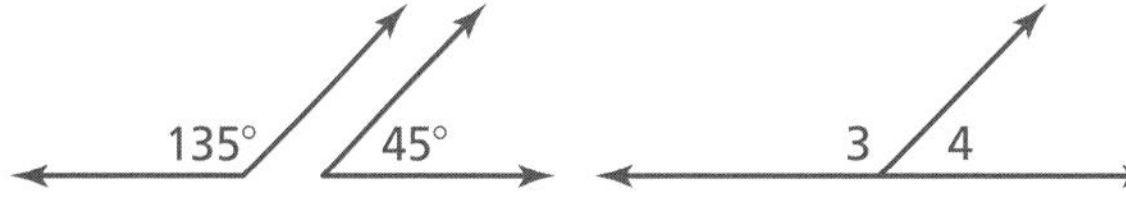

$\angle 3$ and $\angle 4$ are supplementary angles.

EXAMPLE 1 Classifying Pairs of Angles

Tell whether the angles are *complementary*, *supplementary*, or *neither*.

a. 70° 110°

$70^\circ + 110^\circ = 180^\circ$

So, the angles are supplementary.

b.

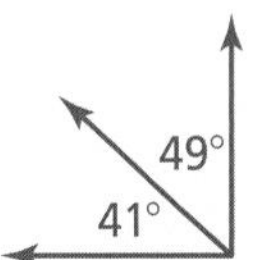

$41^\circ + 49^\circ = 90^\circ$

So, the angles are complementary.

c. 128° 62°

$128^\circ + 62^\circ = 190^\circ$

So, the angles are *neither* complementary nor supplementary.

On Your Own

Now You're Ready
Exercises 6–11

Tell whether the angles are *complementary*, *supplementary*, or *neither*.

1. 26° 64°

2.

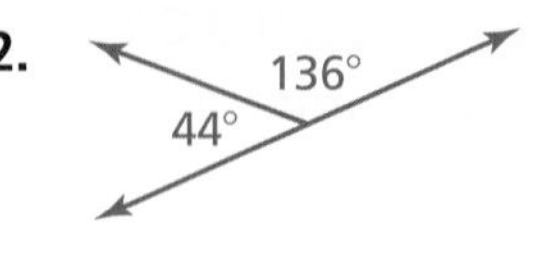

3.

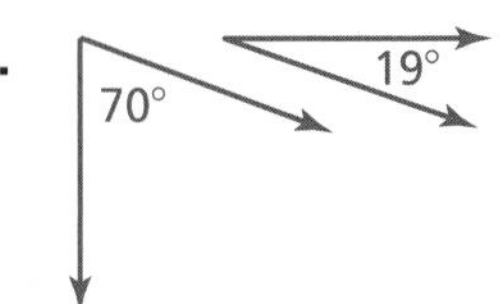

EXAMPLE 2 Using Complementary and Supplementary Angles

Tell whether the angles are *complementary* or *supplementary*. Then find the value of x.

a.

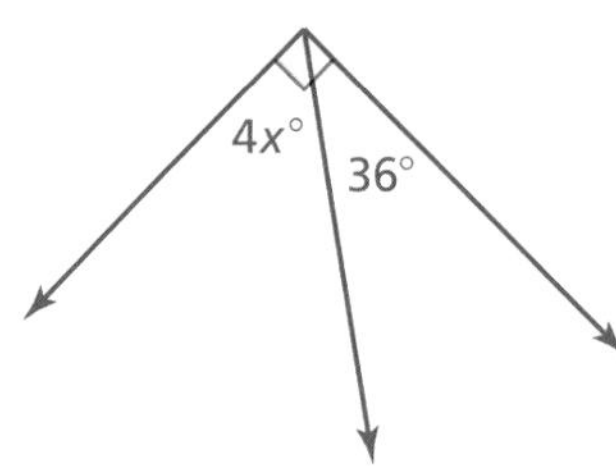

The two angles make up a right angle. So, the angles are complementary angles, and the sum of their measures is 90°.

$4x + 36 = 90$	Write equation.
$4x = 54$	Subtract 36 from each side.
$x = 13.5$	Divide each side by 4.

b.

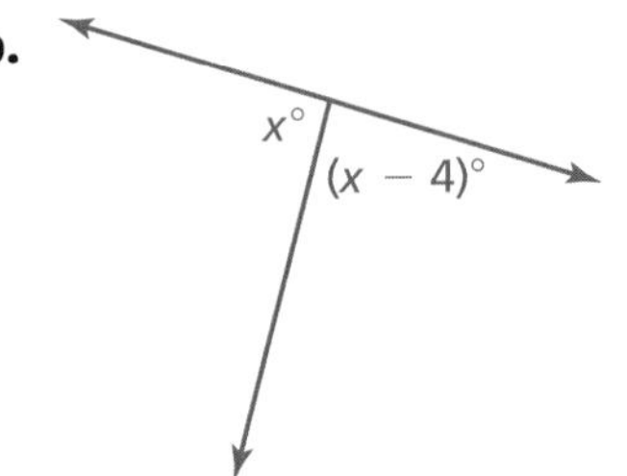

The two angles make up a straight angle. So, the angles are supplementary angles, and the sum of their measures is 180°.

$x + (x - 4) = 180$	Write equation.
$2x - 4 = 180$	Combine like terms.
$2x = 184$	Add 4 to each side.
$x = 92$	Divide each side by 2.

EXAMPLE 3 Constructing Angles

Draw a pair of adjacent supplementary angles so that one angle has a measure of 60°.

Step 1: Use a protractor to draw a 60° angle.

Step 2: Extend one of the sides to form a line.

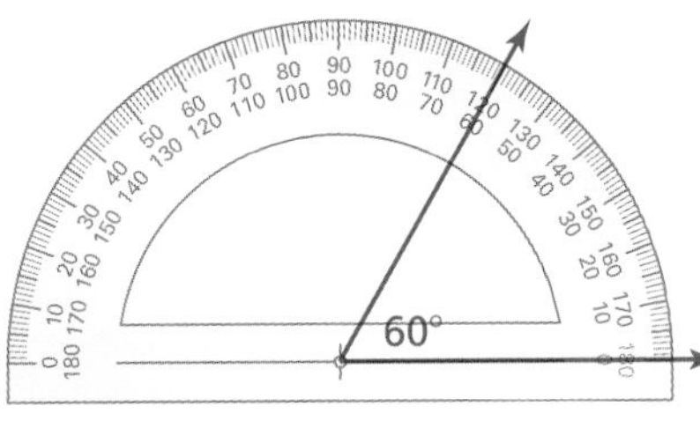

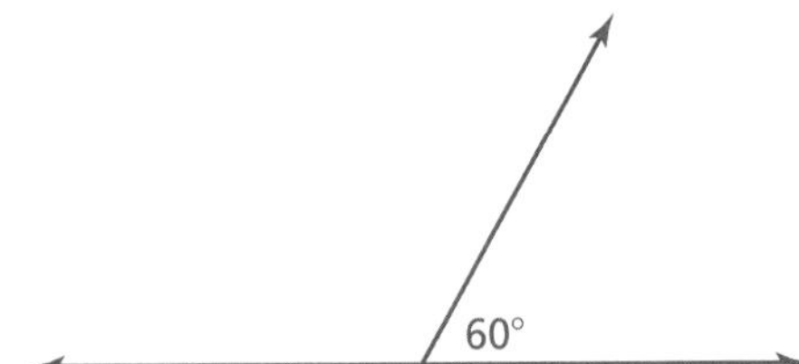

On Your Own

Now You're Ready
Exercises 12–14 and 17–20

Tell whether the angles are *complementary* or *supplementary*. Then find the value of x.

4.

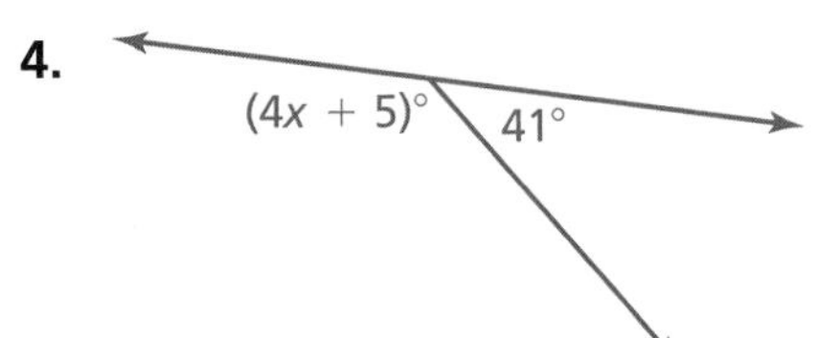

5.

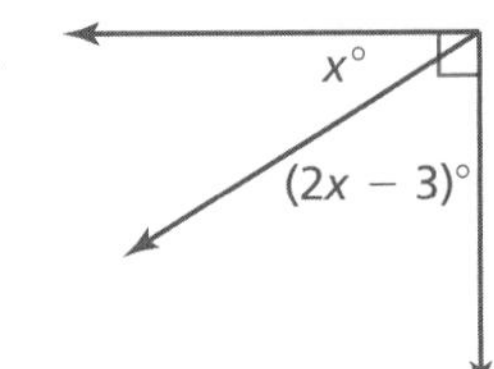

6. Draw a pair of adjacent supplementary angles so that one angle has a measure of 15°.

12.2 Exercises

Vocabulary and Concept Check

1. **VOCABULARY** Explain how complementary angles and supplementary angles are different.

2. **REASONING** Can adjacent angles be supplementary? complementary? neither? Explain.

Practice and Problem Solving

Tell whether the statement is *always, sometimes,* or *never* true. Explain.

3. If x and y are supplementary angles, then x is obtuse.

4. If x and y are right angles, then x and y are supplementary angles.

5. If x and y are complementary angles, then y is a right angle.

Tell whether the angles are *complementary, supplementary,* or *neither.*

(1) 6.

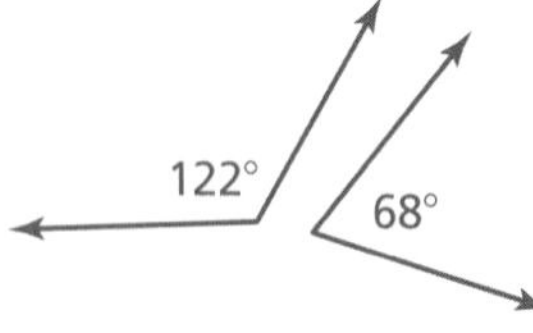

7.

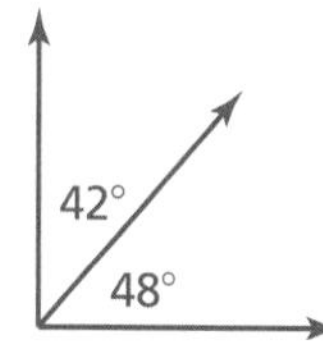

8.

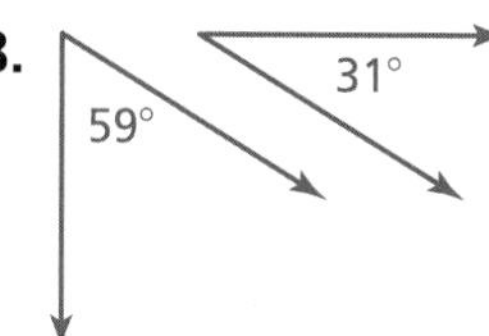

9.

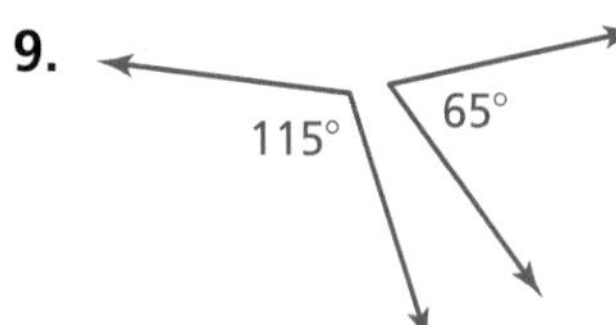

10.

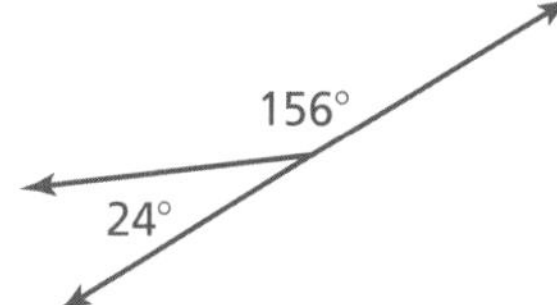

11.

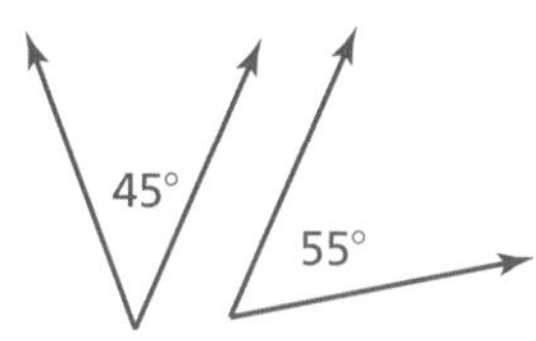

Tell whether the angles are *complementary* or *supplementary.* Then find the value of x.

(2) 12.

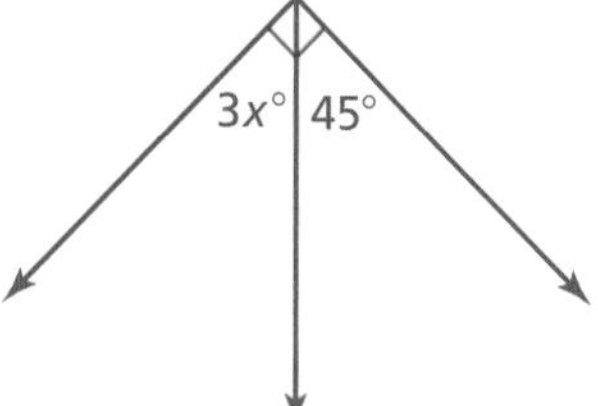

13.

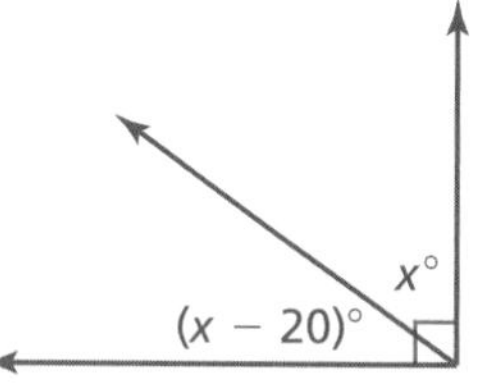

14.

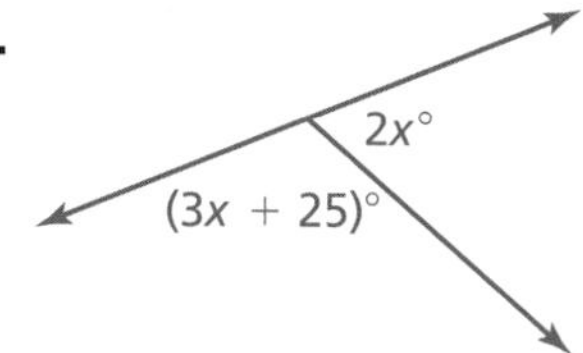

15. **INTERSECTION** What are the measures of the other three angles formed by the intersection?

50°
1
2
3

16. **TRIBUTARY** A tributary joins a river at an angle. Find the value of x.

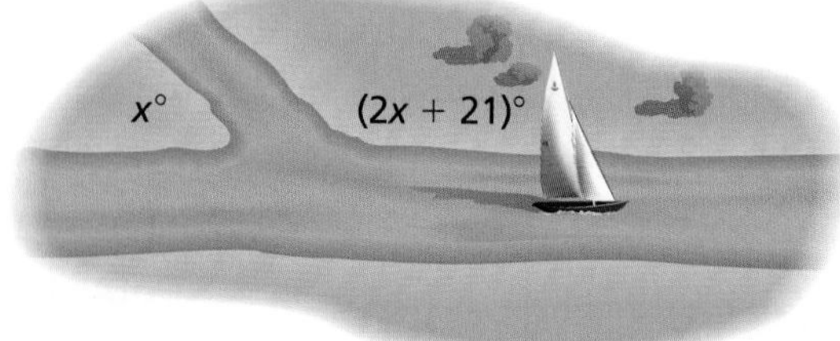

Draw a pair of adjacent supplementary angles so that one angle has the given measure.

3 **17.** 20° **18.** 35° **19.** 80° **20.** 130°

21. **PRECISION** Explain two procedures that you can use to draw two adjacent complementary angles. Then draw a pair of adjacent complementary angles so that one angle has a measure of 30°.

22. **OPEN-ENDED** Give an example of an angle that can be a supplementary angle but cannot be a complementary angle. Explain.

23. **VANISHING POINT** The vanishing point of the picture is represented by point *B*.

a. The measure of $\angle ABD$ is 6.2 times greater than the measure of $\angle CBD$. Find the measure of $\angle CBD$.

b. $\angle FBE$ and $\angle EBD$ are congruent. Find the measure of $\angle FBE$.

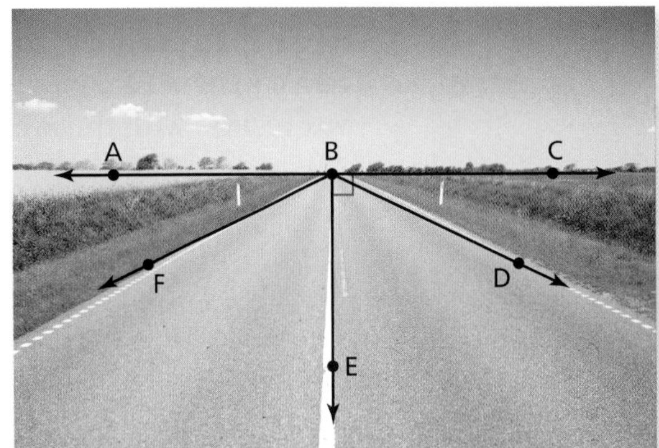

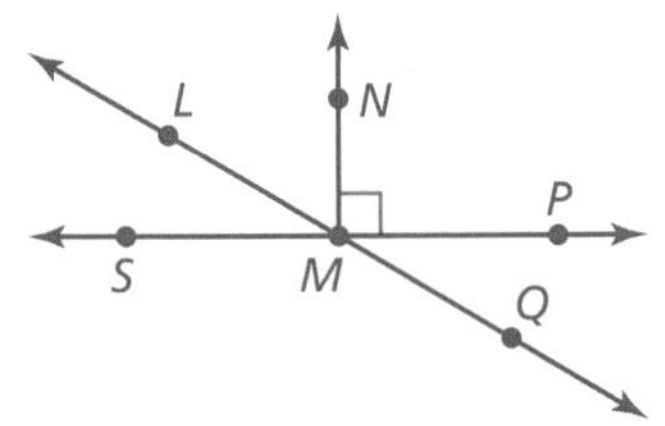

24. **LOGIC** Your friend says that $\angle LMN$ and $\angle PMQ$ are complementary angles. Is she correct? Explain.

25. **RATIO** The measures of two complementary angles have a ratio of 3 : 2. What is the measure of the larger angle?

26. **REASONING** Two angles are vertical angles. What are their measures if they are also complementary angles? supplementary angles?

27. Problem Solving Find the values of x and y.

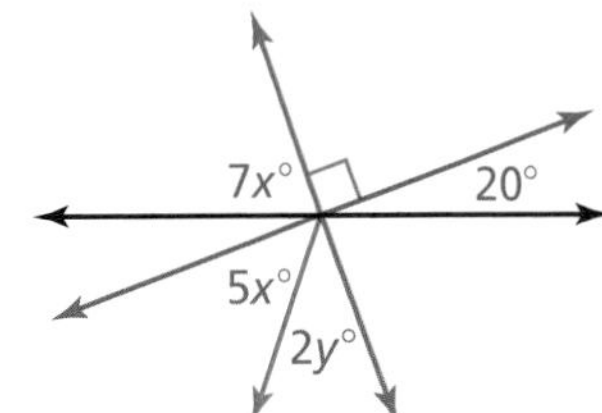

Fair Game Review What you learned in previous grades & lessons

Solve the equation. Check your solution. *(Section 1.1)*

28. $x + 7 = -8$ **29.** $\frac{1}{3} = n + \frac{3}{4}$ **30.** $-12.7 = y - 3.4$

31. **MULTIPLE CHOICE** Which decimal is equal to 3.7%? *(Skills Review Handbook)*

Ⓐ 0.0037 Ⓑ 0.037 Ⓒ 0.37 Ⓓ 3.7

12.3 Triangles

Essential Question How can you construct triangles?

1 ACTIVITY: Constructing Triangles Using Side Lengths

Work with a partner. Cut different-colored straws to the lengths shown. Then construct a triangle with the specified straws if possible. Compare your results with those of others in your class.

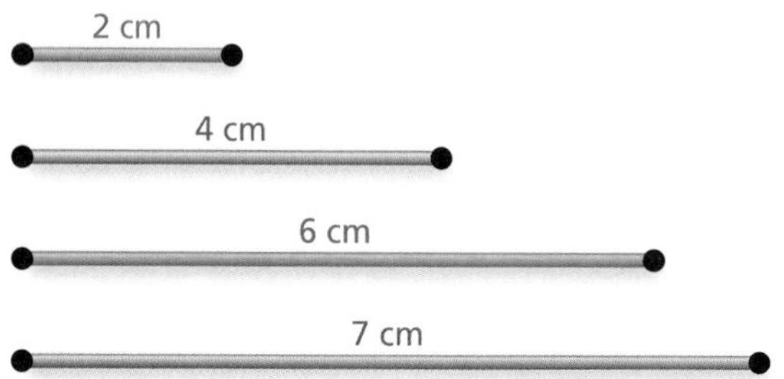

a. blue, green, purple

b. red, green, purple

c. red, blue, purple

d. red, blue, green

2 ACTIVITY: Using Technology to Draw Triangles (Side Lengths)

Work with a partner. Use geometry software to draw a triangle with the two given side lengths. What is the length of the third side of your triangle? Compare your results with those of others in your class.

a. 4 units, 7 units

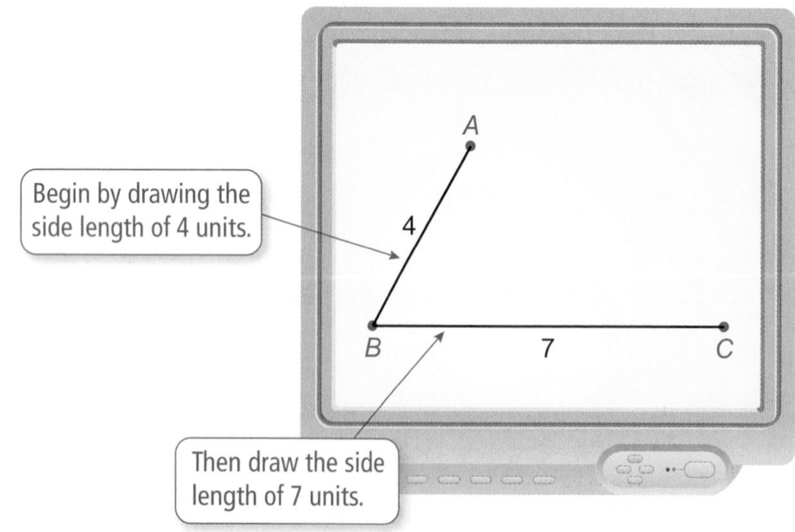

b. 3 units, 5 units

c. 2 units, 8 units

d. 1 unit, 1 unit

Geometry

In this lesson, you will

- construct triangles with given angle measures.
- construct triangles with given side lengths.

3 ACTIVITY: Constructing Triangles Using Angle Measures

Work with a partner. Two angle measures of a triangle are given. Draw the triangle. What is the measure of the third angle? Compare your results with those of others in your class.

a. 40°, 70°

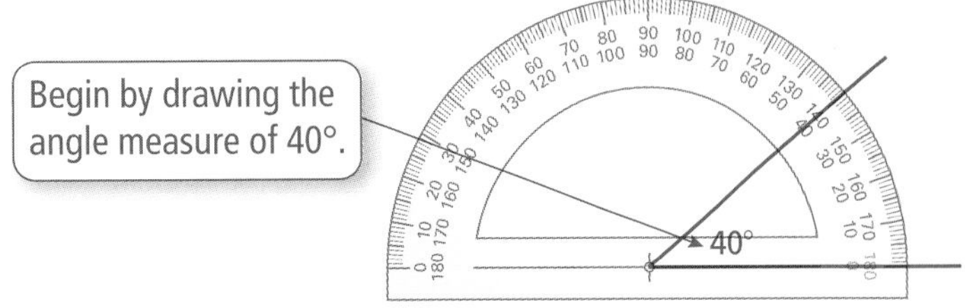

Begin by drawing the angle measure of 40°.
40°

b. 60°, 75° **c.** 90°, 30° **d.** 100°, 40°

4 ACTIVITY: Using Technology to Draw Triangles (Angle Measures)

Math Practice

Recognize Usefulness of Tools

What are some advantages and disadvantages of using geometry software to draw a triangle?

Work with a partner. Use geometry software to draw a triangle with the two given angle measures. What is the measure of the third angle? Compare your results with those of others in your class.

a. 45°, 55°

b. 50°, 40°

c. 110°, 35°

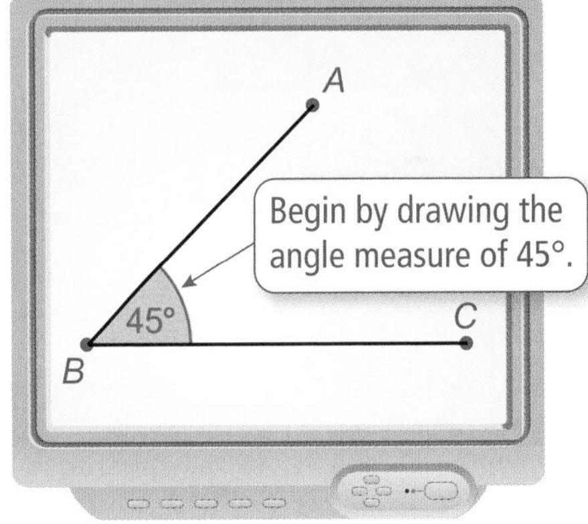

A
Begin by drawing the angle measure of 45°.
45°
B
C

What Is Your Answer?

5. **IN YOUR OWN WORDS** How can you construct triangles?

6. **REASONING** Complete the table below for each set of side lengths in Activity 2. Write a rule that compares the sum of any two side lengths to the third side length.

Side Length			
Sum of Other Two Side Lengths			

7. **REASONING** Use a table to organize the angle measures of each triangle you formed in Activity 3. Include the sum of the angle measures. Then describe the pattern in the table and write a conclusion based on the pattern.

Practice Use what you learned about constructing triangles to complete Exercises 3–5 on page 518.

12.3 Lesson

Key Vocabulary
congruent sides, *p. 516*

You can use side lengths and angle measures to classify triangles.

Classifying Triangles Using Angles

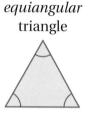

acute triangle	*obtuse* triangle	*right* triangle	*equiangular* triangle

all acute angles	1 obtuse angle	1 right angle	3 congruent angles

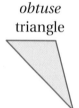

Classifying Triangles Using Sides

Congruent sides have the same length.

scalene triangle	*isosceles* triangle	*equilateral* triangle
	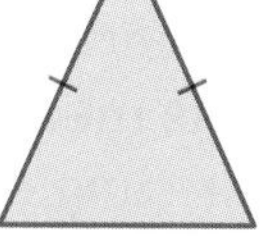	
no congruent sides	at least 2 congruent sides	3 congruent sides

Reading

Red arcs indicate congruent angles.
Red tick marks indicate congruent sides.

EXAMPLE 1 Classifying Triangles

Classify each triangle.

a.

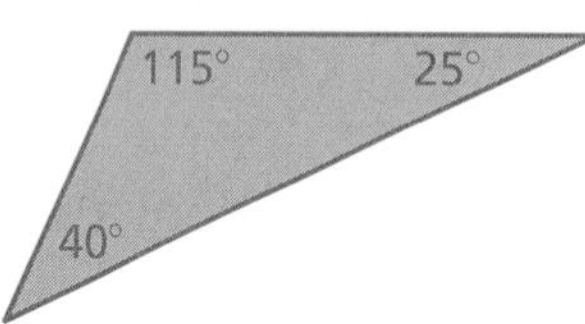

The triangle has one obtuse angle and no congruent sides.

So, the triangle is an obtuse scalene triangle.

b.

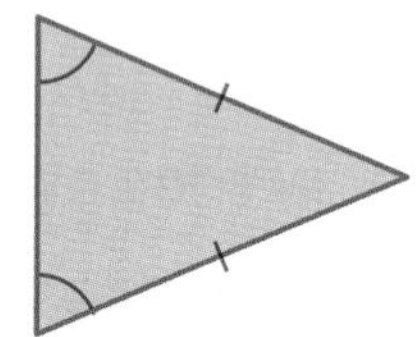

The triangle has all acute angles and two congruent sides.

So, the triangle is an acute isosceles triangle.

On Your Own

Now You're Ready
Exercises 6–11

Classify the triangle.

1.

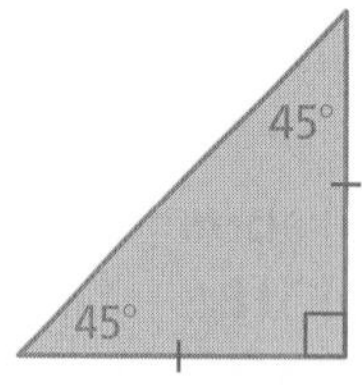

2.

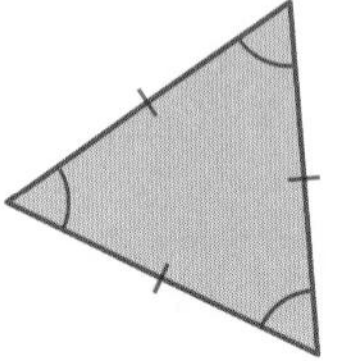

EXAMPLE 2 Constructing a Triangle Using Angle Measures

Draw a triangle with angle measures of 30°, 60°, and 90°. Then classify the triangle.

Step 1: Use a protractor to draw the 30° angle.

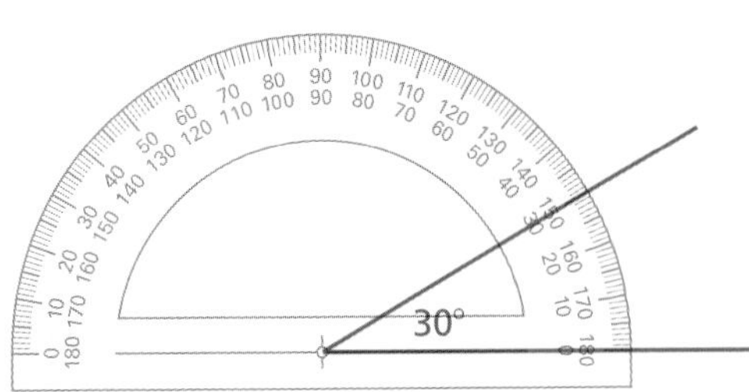

Step 2: Use a protractor to draw the 60° angle.

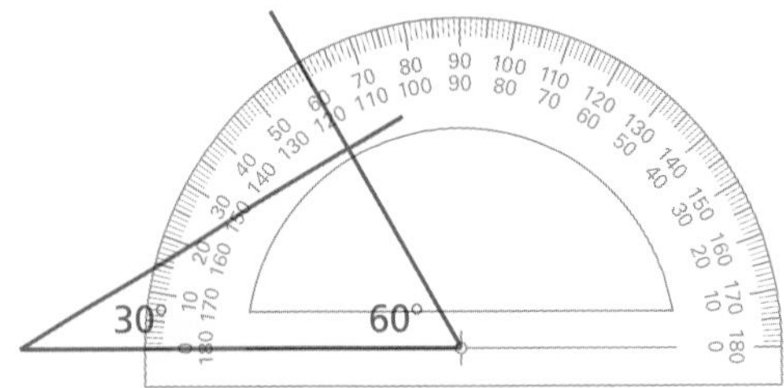

Step 3: The protractor shows that the measure of the remaining angle is 90°.

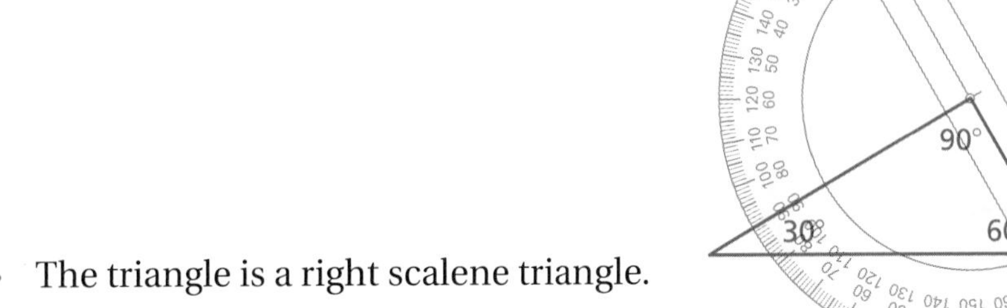

Study Tip

After drawing the first two angles, make sure you check the remaining angle.

∴ The triangle is a right scalene triangle.

EXAMPLE 3 Constructing a Triangle Using Side Lengths

Draw a triangle with a 3-centimeter side and a 4-centimeter side that meet at a 20° angle. Then classify the triangle.

Step 1: Use a protractor to draw a 20° angle.

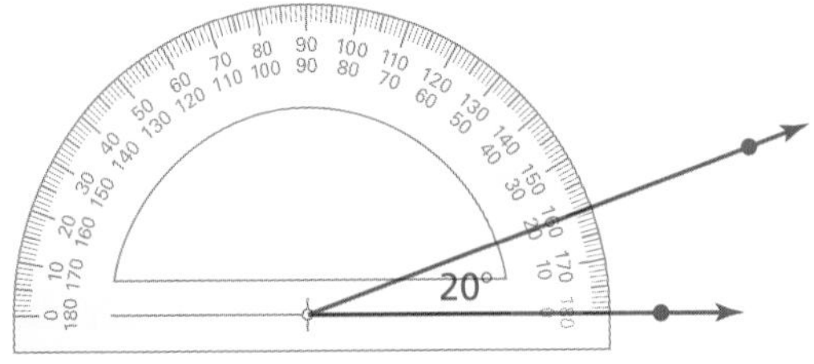

Step 2: Use a ruler to mark 3 centimeters on one ray and 4 centimeters on the other ray.

Step 3: Draw the third side to form the triangle.

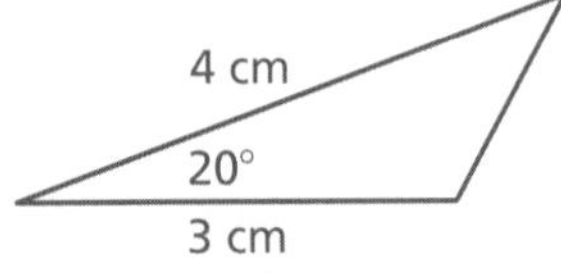

∴ The triangle is an obtuse scalene triangle.

On Your Own

3. Draw a triangle with angle measures of 45°, 45°, and 90°. Then classify the triangle.
4. Draw a triangle with a 1-inch side and a 2-inch side that meet at a 60° angle. Then classify the triangle.

12.3 Exercises

Vocabulary and Concept Check

1. **WRITING** How can you classify triangles using angles? using sides?
2. **DIFFERENT WORDS, SAME QUESTION** Which is different? Find "both" answers.

Construct an equilateral triangle.	Construct a triangle with 3 congruent sides.
Construct an equiangular triangle.	Construct a triangle with no congruent sides.

Practice and Problem Solving

Construct a triangle with the given description.

3. side lengths: 4 cm, 6 cm
4. side lengths: 5 cm, 12 cm
5. angles: 65°, 55°

Classify the triangle.

1 6.

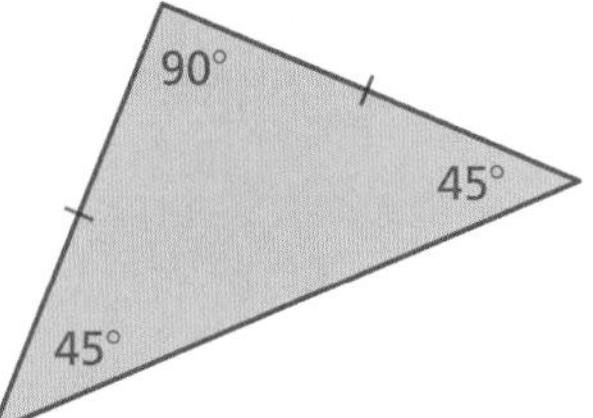

7.

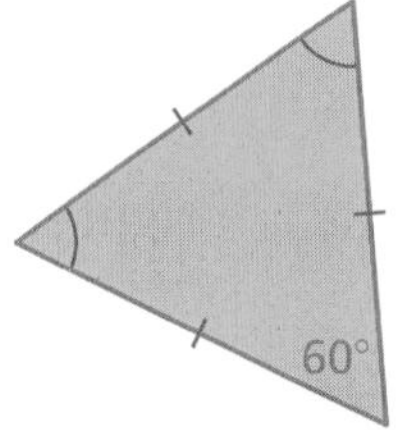

8.

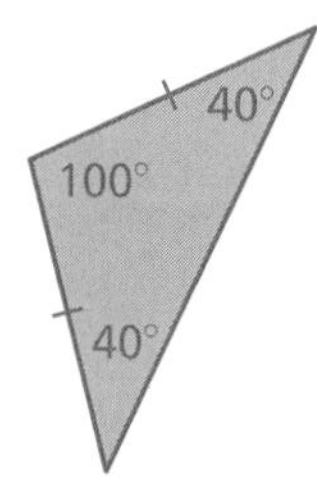

9.

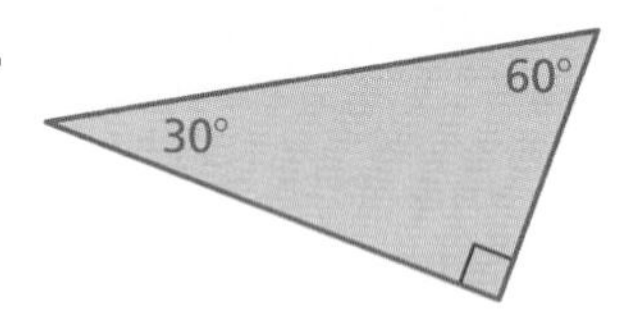

10.

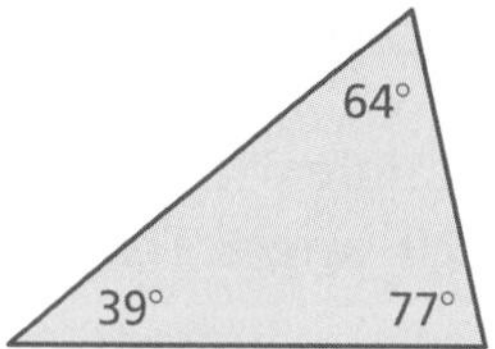

11.

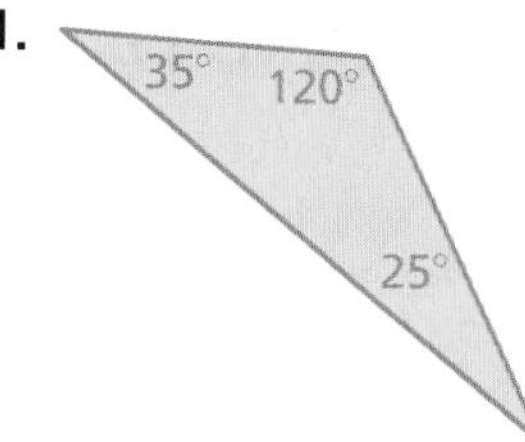

12. **ERROR ANALYSIS** Describe and correct the error in classifying the triangle.

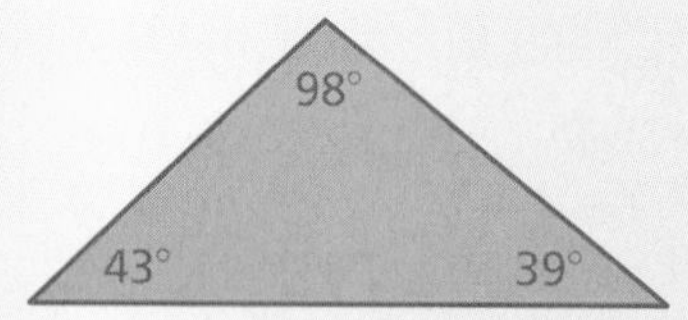

The triangle is acute and scalene because it has two acute angles and no congruent sides.

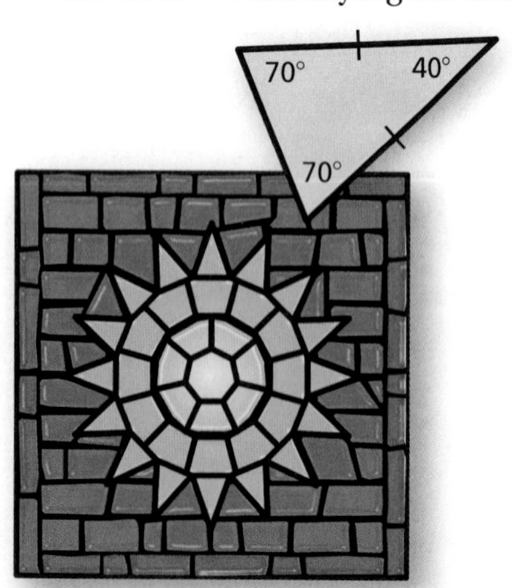

13. **MOSAIC TILE** A mosaic is a pattern or picture made of small pieces of colored material. Classify the yellow triangle used in the mosaic.

Draw a triangle with the given angle measures. Then classify the triangle.

2 **14.** 15°, 75°, 90° **15.** 20°, 60°, 100° **16.** 30°, 30°, 120°

Draw a triangle with the given description.

3 **17.** a triangle with a 2-inch side and a 3-inch side that meet at a 40° angle

18. a triangle with a 45° angle connected to a 60° angle by an 8-centimeter side

19. an acute scalene triangle

20. **LOGIC** You are constructing a triangle. You draw the first angle, as shown. Your friend says that you must be constructing an acute triangle. Is your friend correct? Explain your reasoning.

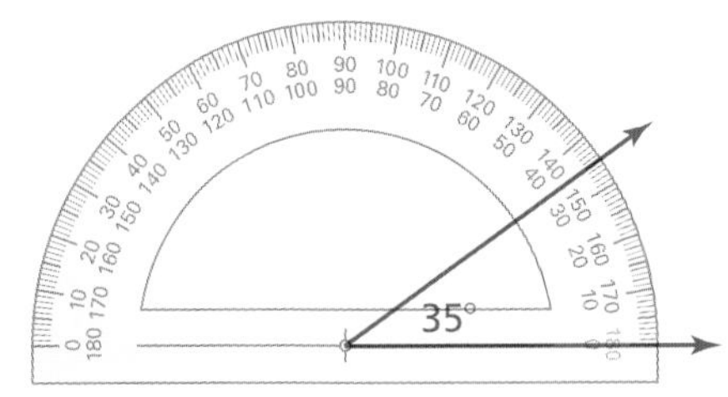

Determine whether you can construct *many, one,* or *no* triangle(s) with the given description. Explain your reasoning.

21. a triangle with angle measures of 50°, 70°, and 100°

22. a triangle with one angle measure of 60° and one 4-centimeter side

23. a scalene triangle with a 3-centimeter side and a 7-centimeter side

24. an isosceles triangle with two 4-inch sides that meet at an 80° angle

25. an isosceles triangle with two 2-inch sides and one 5-inch side

26. a right triangle with three congruent sides

27. Critical Thinking Consider the three isosceles triangles.

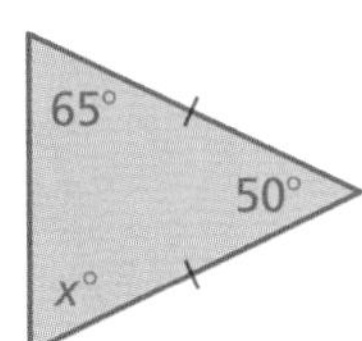

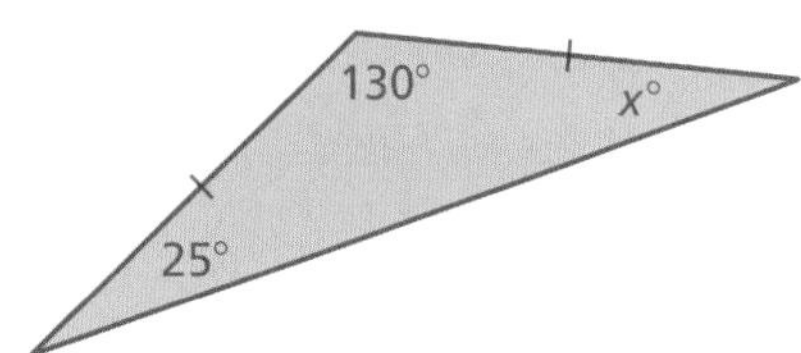

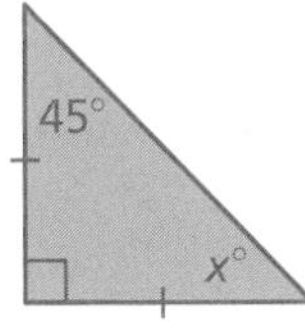

a. Find the value of x for each triangle.

b. What do you notice about the angle measures of each triangle?

c. Write a rule about the angle measures of an isosceles triangle.

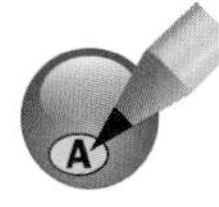

Fair Game Review What you learned in previous grades & lessons

Tell whether x and y show direct variation. Explain your reasoning. If so, find the constant of proportionality. *(Skills Review Handbook)*

28. $x = 2y$ **29.** $y - x = 6$ **30.** $xy = 5$

31. **MULTIPLE CHOICE** A savings account earns 6% simple interest per year. The principal is $800. What is the balance after 18 months? *(Skills Review Handbook)*

Ⓐ $864 Ⓑ $872 Ⓒ $1664 Ⓓ $7200

Extension 12.3 Angle Measures of Triangles

Sum of the Angle Measures of a Triangle

Words The sum of the angle measures of a triangle is 180°.

Algebra $x + y + z = 180$

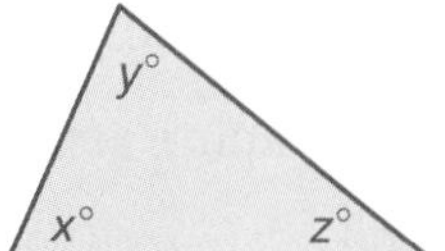

EXAMPLE 1 Finding Angle Measures

Geometry

In this extension, you will

- understand that the sum of the angle measures of any triangle is 180°.
- find missing angle measures in triangles.

Find each value of x. Then classify each triangle.

a.

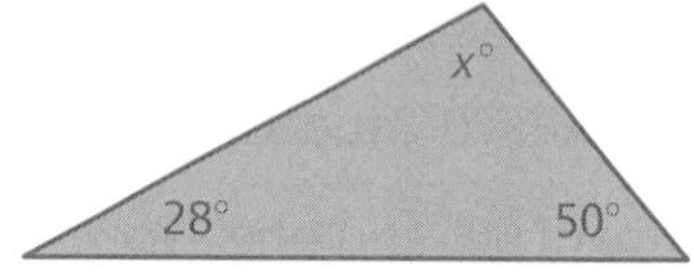

$$x + 28 + 50 = 180$$
$$x + 78 = 180$$
$$x = 102$$

The value of x is 102. The triangle has one obtuse angle and no congruent sides. So, it is an obtuse scalene triangle.

b.

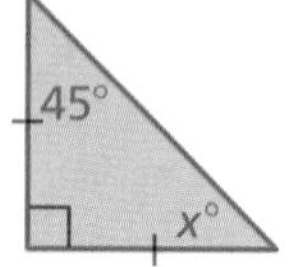

$$x + 45 + 90 = 180$$
$$x + 135 = 180$$
$$x = 45$$

The value of x is 45. The triangle has a right angle and two congruent sides. So, it is a right isosceles triangle.

Practice

Find the value of x. Then classify the triangle.

1.

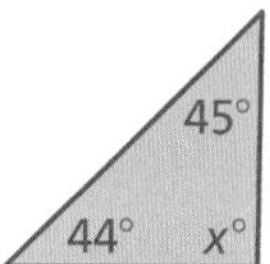

2.

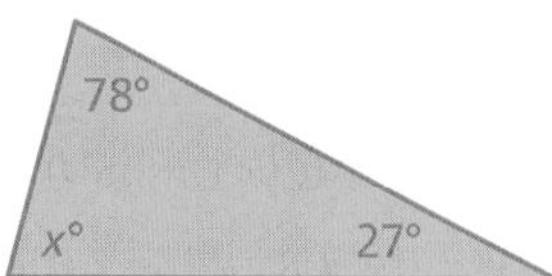

3.

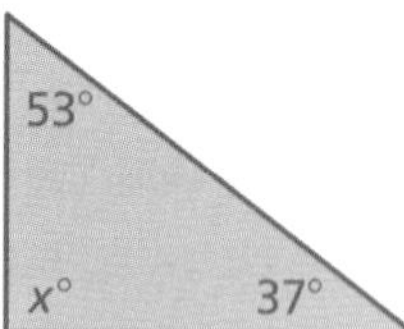

4.

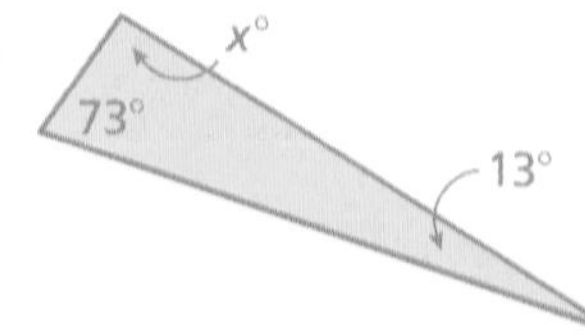

5.

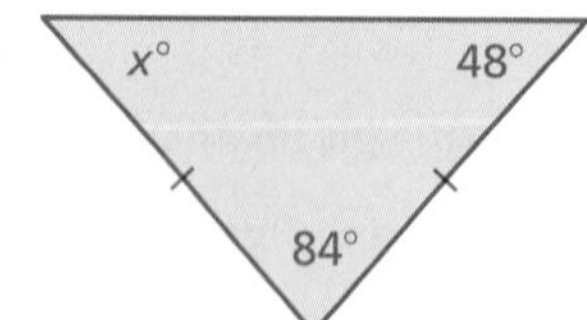

6.

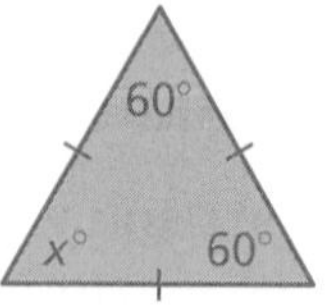

Tell whether a triangle can have the given angle measures. If not, change the first angle measure so that the angle measures form a triangle.

7. $76.2°$, $81.7°$, $22.1°$

8. $115.1°$, $47.5°$, $93°$

9. $5\frac{2}{3}°$, $64\frac{1}{3}°$, $87°$

10. $31\frac{3}{4}°$, $53\frac{1}{2}°$, $94\frac{3}{4}°$

EXAMPLE 2 Finding Angle Measures

Math Practice

Analyze Givens
What information is given in the problem? How can you use this information to answer the question?

Find each value of x. Then classify each triangle.

a. Flag of Jamaica

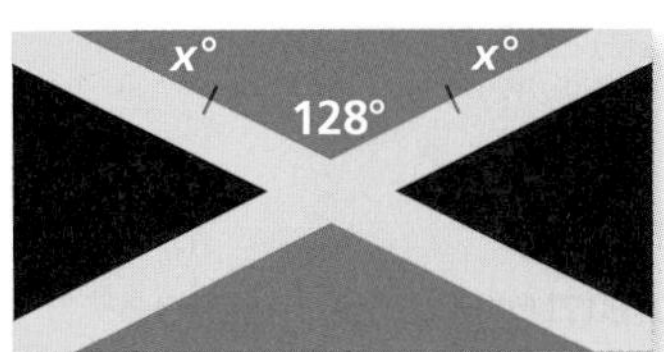

$$x + x + 128 = 180$$
$$2x + 128 = 180$$
$$2x = 52$$
$$x = 26$$

The value of x is 26. The triangle has one obtuse angle and two congruent sides. So, it is an obtuse isosceles triangle.

b. Flag of Cuba

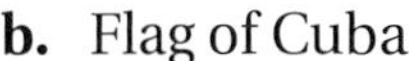

$$x + x + 60 = 180$$
$$2x + 60 = 180$$
$$2x = 120$$
$$x = 60$$

The value of x is 60. All three angles are congruent. So, it is an equilateral and equiangular triangle.

Practice

Find the value of x. Then classify the triangle.

11.

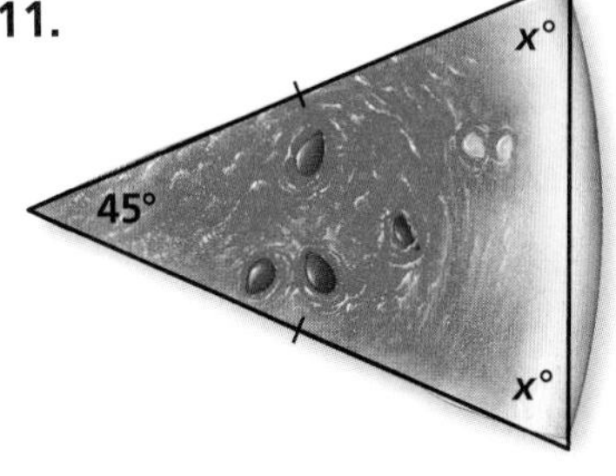

12.

13.

14.

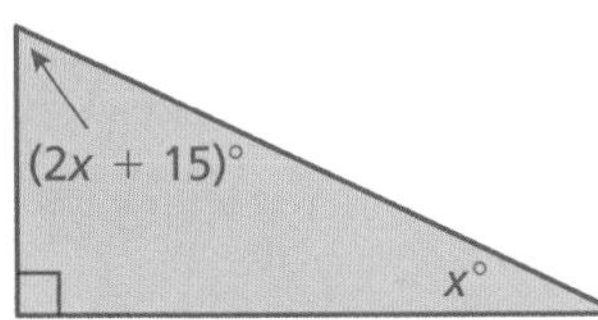

15.

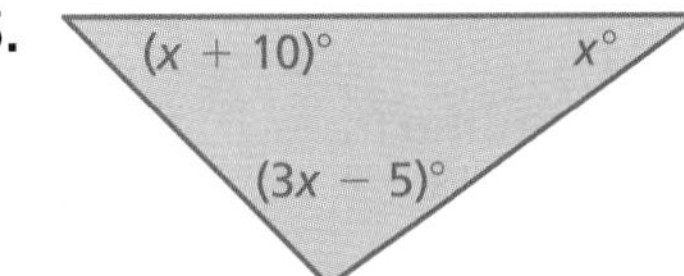

16. REASONING Explain why all triangles have at least two acute angles.

17. CARDS One method of stacking cards is shown.

a. Find the value of x.

b. Describe how to stack the cards with different angles. Is the value of x limited? If so, what are the limitations? Explain your reasoning.

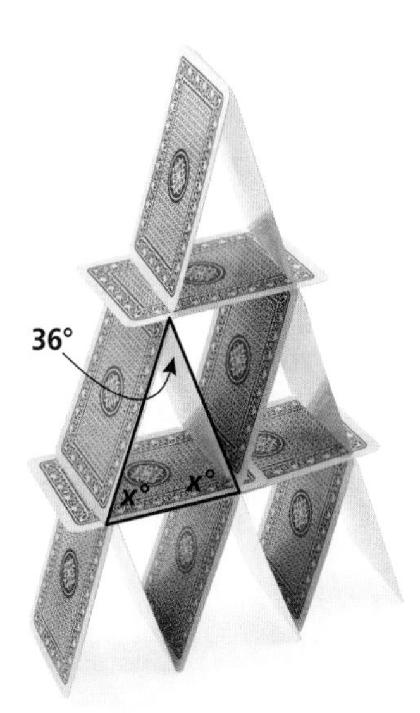

12 Study Help

You can use an **example and non-example chart** to list examples and non-examples of a vocabulary word or item. Here is an example and non-example chart for complementary angles.

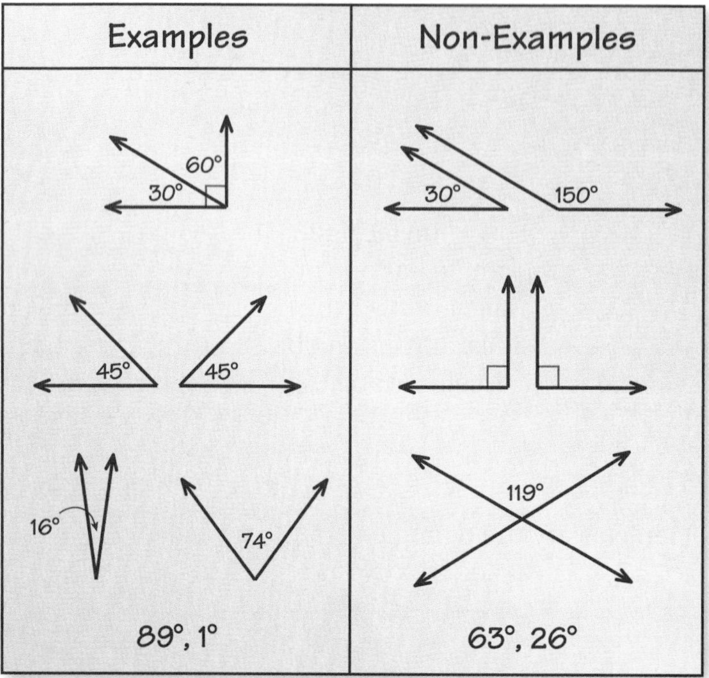

On Your Own

Make example and non-example charts to help you study these topics.

1. adjacent angles
2. vertical angles
3. supplementary angles

After you complete this chapter, make example and non-example charts for the following topics.

4. quadrilaterals
5. scale factor

"What do you think of my example & non-example chart for popular cat toys?"

12.1–12.3 Quiz

Name two pairs of adjacent angles and two pairs of vertical angles in the figure. *(Section 12.1)*

1. 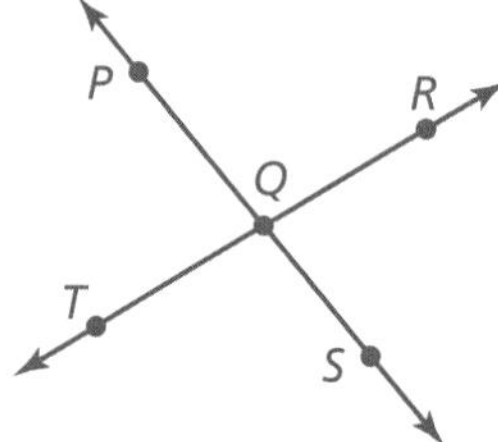

2. 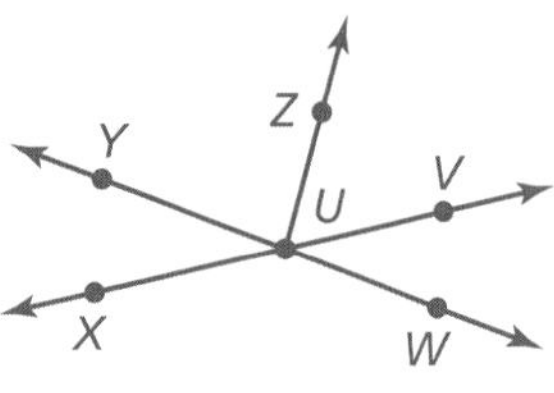

Tell whether the angles are *adjacent* or *vertical*. Then find the value of *x*. *(Section 12.1)*

3. 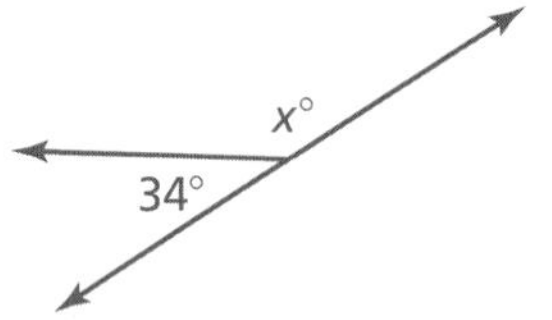

4. 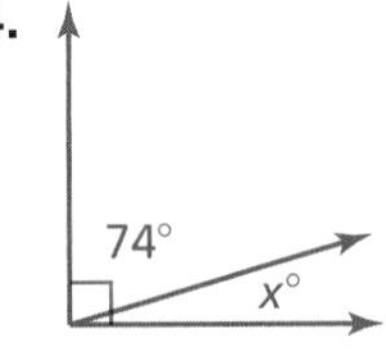

5. 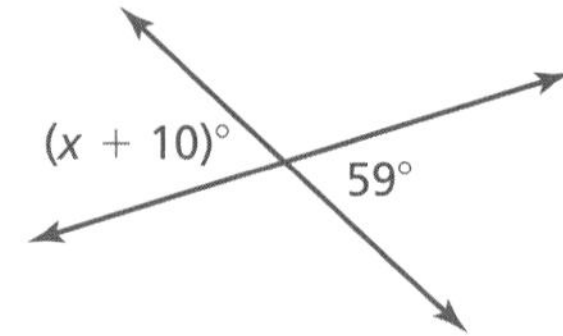

Tell whether the angles are *complementary* or *supplementary*. Then find the value of *x*. *(Section 12.2)*

6. 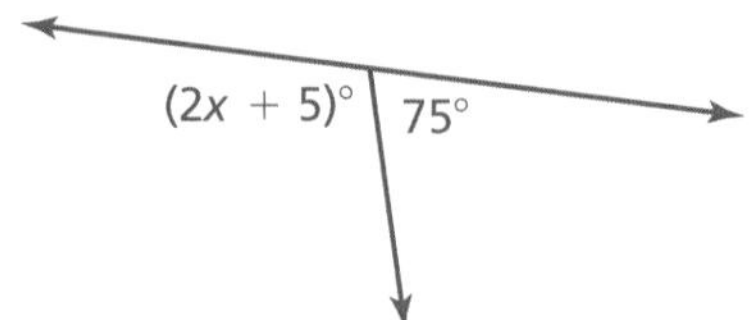

7. 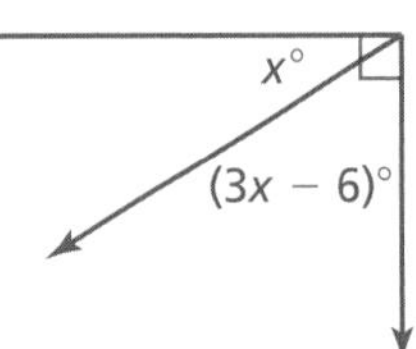

Draw a triangle with the given description. *(Section 12.3)*

8. a triangle with angle measures of 35°, 65°, and 80°
9. a triangle with a 5-centimeter side and a 7-centimeter side that meet at a 70° angle
10. an obtuse scalene triangle

Find the value of *x*. Then classify the triangle. *(Section 12.3)*

11. 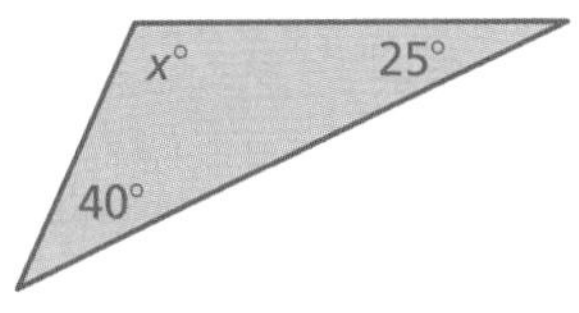

12. 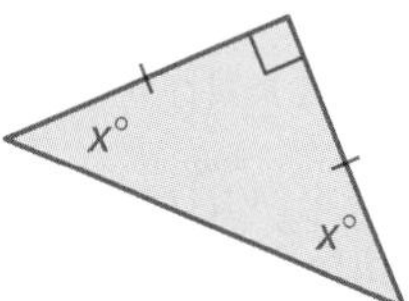

13. 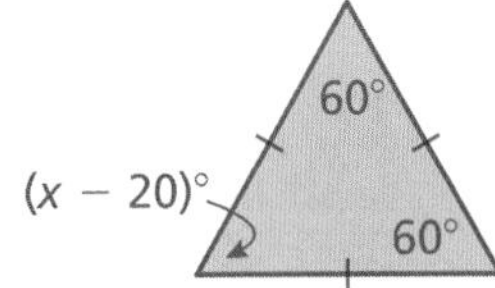

14. **RAILROAD CROSSING** Describe two ways to find the measure of $\angle 2$. *(Section 12.1 and Section 12.2)*

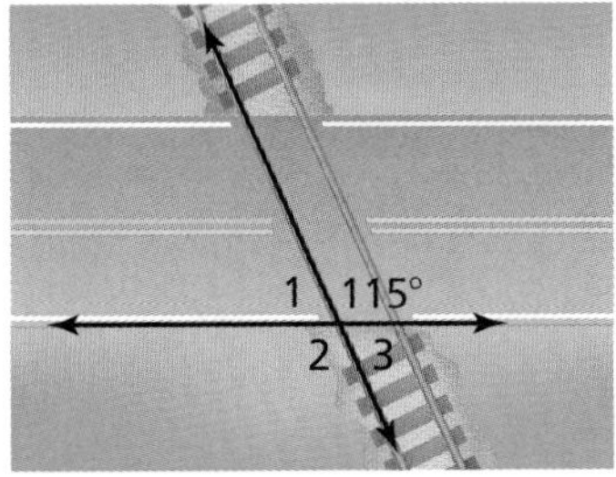

12.4 Quadrilaterals

Essential Question How can you classify quadrilaterals?

Quad means *four* and *lateral* means *side*. So, *quadrilateral* means a polygon with *four sides*.

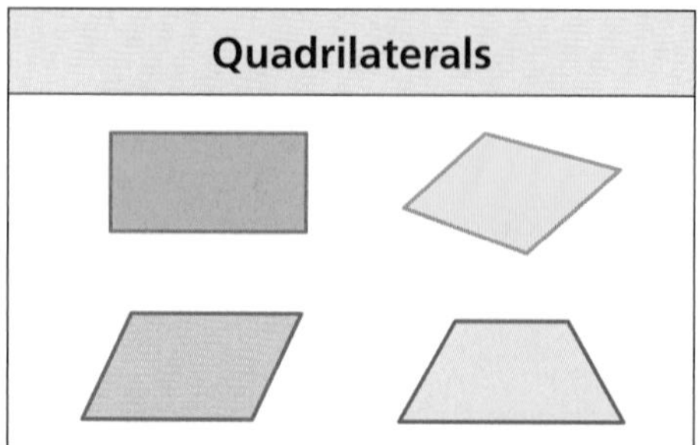
Quadrilaterals

1 ACTIVITY: Using Descriptions to Form Quadrilaterals

Work with a partner. Use a geoboard to form a quadrilateral that fits the given description. Record your results on geoboard dot paper.

a. Form a quadrilateral with exactly one pair of parallel sides.

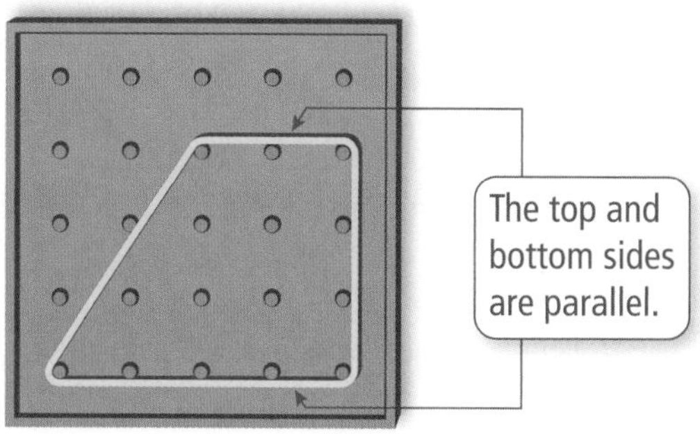
The top and bottom sides are parallel.

b. Form a quadrilateral with four congruent sides and four right angles.

c. Form a quadrilateral with four right angles that is *not* a square.

d. Form a quadrilateral with four congruent sides that is *not* a square.

e. Form a quadrilateral with two pairs of congruent adjacent sides and whose opposite sides are *not* congruent.

f. Form a quadrilateral with congruent and parallel opposite sides that is *not* a rectangle.

Geometry

In this lesson, you will

- understand that the sum of the angle measures of any quadrilateral is 360°.
- find missing angle measures in quadrilaterals.
- construct quadrilaterals.

2 ACTIVITY: Naming Quadrilaterals

Work with a partner. Match the names *square, rectangle, rhombus, parallelogram, trapezoid,* and *kite* with your 6 drawings in Activity 1.

3 ACTIVITY: Forming Quadrilaterals

Work with a partner. Form each quadrilateral on your geoboard. Then move *only one* vertex to create the new type of quadrilateral. Record your results on geoboard dot paper.

a. Trapezoid ⇨ Kite

b. Kite ⇨ Rhombus (*not* a square)

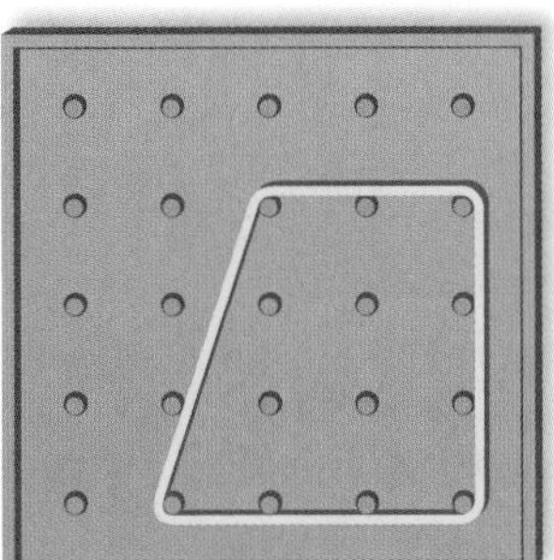

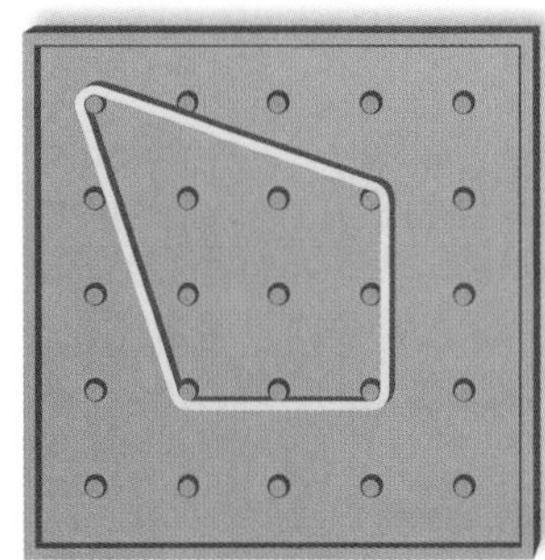

4 ACTIVITY: Using Technology to Draw Quadrilaterals

Math Practice

Use Technology to Explore

How does geometry software help you learn about the characteristics of a quadrilateral?

Work with a partner. Use geometry software to draw a quadrilateral that fits the given description.

a. a square with a side length of 3 units

b. a rectangle with a width of 2 units and a length of 5 units

c. a parallelogram with side lengths of 6 units and 1 unit

d. a rhombus with a side length of 4 units

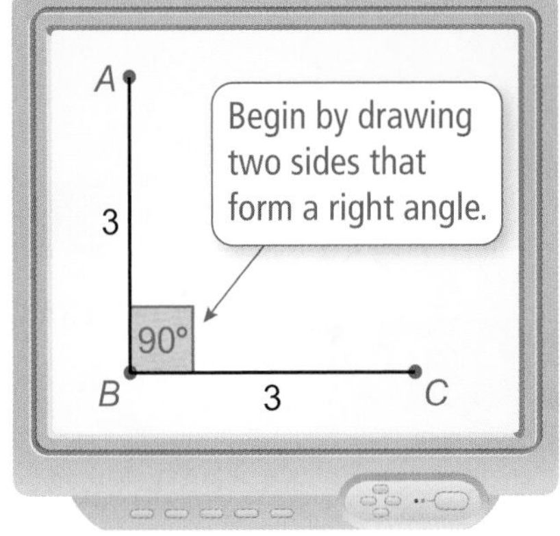

What Is Your Answer?

5. REASONING Measure the angles of each quadrilateral you formed in Activity 1. Record your results in a table. Include the sum of the angle measures. Then describe the pattern in the table and write a conclusion based on the pattern.

6. IN YOUR OWN WORDS How can you classify quadrilaterals? Explain using properties of sides and angles.

Practice Use what you learned about quadrilaterals to complete Exercises 4–6 on page 528.

12.4 Lesson

Check It Out
Lesson Tutorials
BigIdeasMath.com

Key Vocabulary
kite, *p. 526*

A quadrilateral is a polygon with four sides. The diagram shows properties of different types of quadrilaterals and how they are related. When identifying a quadrilateral, use the name that is most specific.

Reading
Red arrows indicate parallel sides.

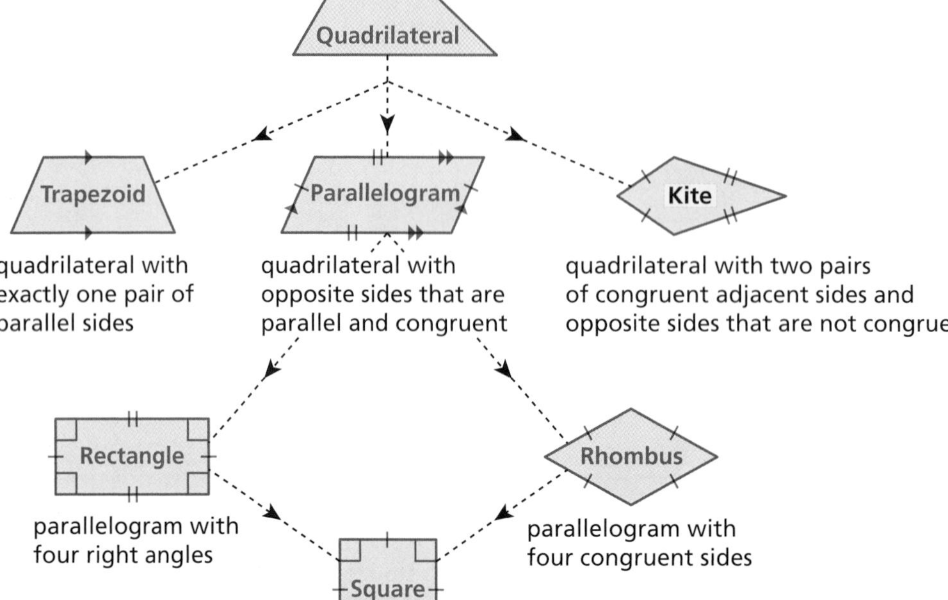
Quadrilateral
Trapezoid
Parallelogram
Kite
quadrilateral with exactly one pair of parallel sides
quadrilateral with opposite sides that are parallel and congruent
quadrilateral with two pairs of congruent adjacent sides and opposite sides that are not congruent
Rectangle
Rhombus
parallelogram with four right angles
parallelogram with four congruent sides
Square
parallelogram with four congruent sides and four right angles

EXAMPLE 1 Classifying Quadrilaterals

Classify the quadrilateral.

a.

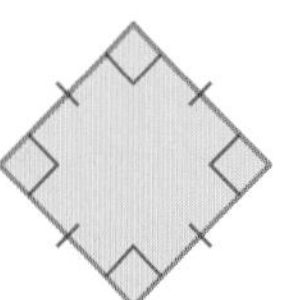

The quadrilateral has four congruent sides and four right angles.

So, the quadrilateral is a square.

b.

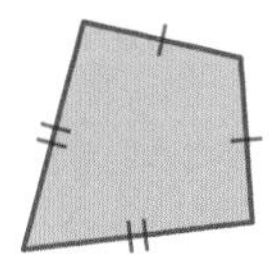

The quadrilateral has two pairs of congruent adjacent sides and opposite sides that are not congruent.

So, the quadrilateral is a kite.

Study Tip
In Example 1(a), the square is also a parallelogram, a rectangle, and a rhombus. Square is the most specific name.

On Your Own

Now You're Ready
Exercises 4–9

Classify the quadrilateral.

1.

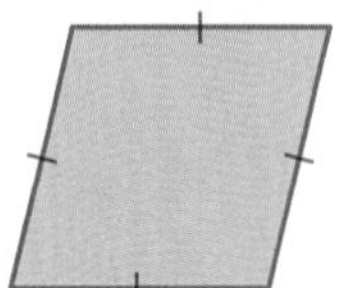

2.

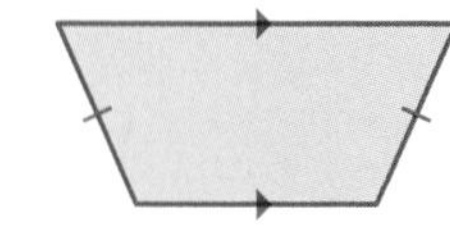

3.

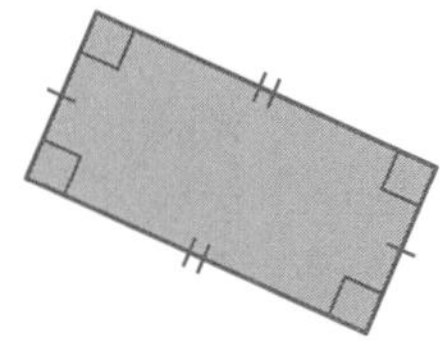

Sum of the Angle Measures of a Quadrilateral

Words The sum of the angle measures of a quadrilateral is 360°.

Algebra $w + x + y + z = 360$

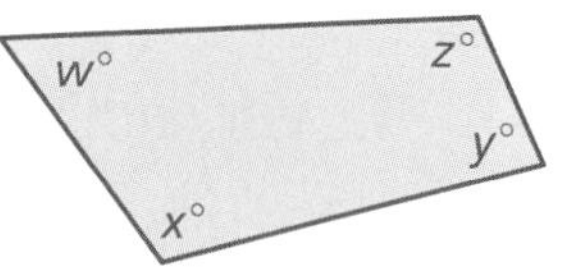

EXAMPLE 2 Finding an Angle Measure of a Quadrilateral

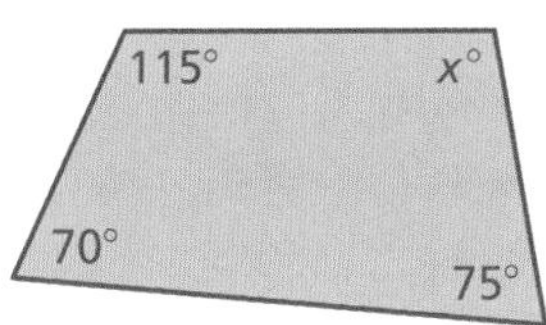

Find the value of x.

$70 + 75 + 115 + x = 360$		Write an equation.
$260 + x = 360$		Combine like terms.
$-260 \quad -260$		Subtraction Property of Equality
$x = 100$		Simplify.

The value of x is 100.

EXAMPLE 3 Constructing a Quadrilateral

Draw a parallelogram with a 60° angle and a 120° angle.

Step 1: Draw a line.

Step 2: Draw a 60° angle and a 120° angle that each have one side on the line.

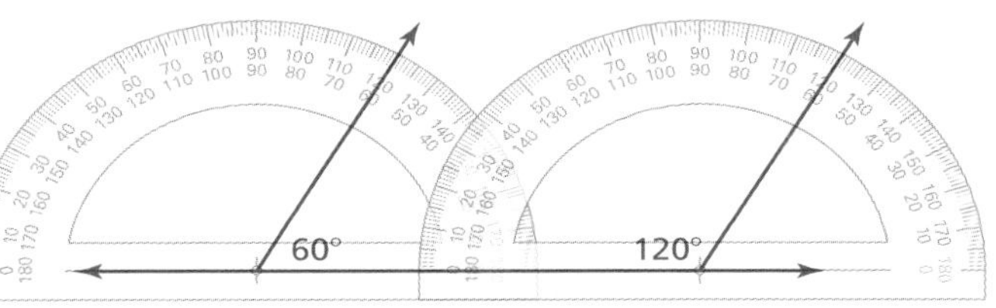

Step 3: Draw the remaining side. Make sure that both pairs of opposite sides are parallel and congruent.

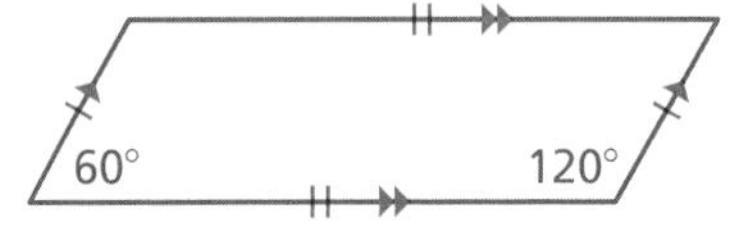

On Your Own

Now You're Ready
Exercises 10–12 and 14–17

Find the value of x.

4.

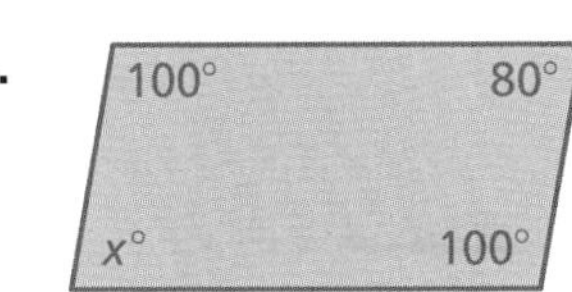

5.

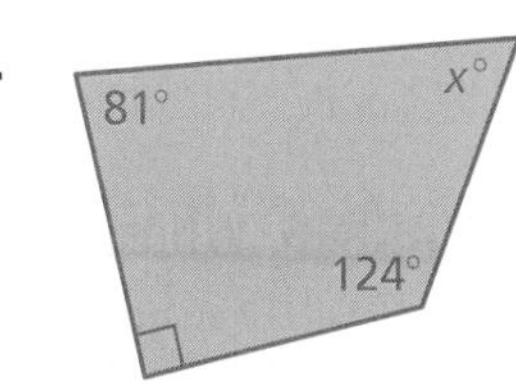

6. Draw a right trapezoid whose parallel sides have lengths of 3 centimeters and 5 centimeters.

12.4 Exercises

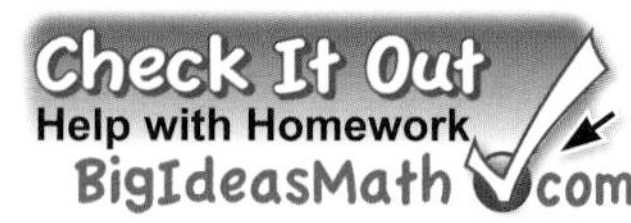

Vocabulary and Concept Check

1. **VOCABULARY** Which statements are true?

   a. All squares are rectangles.
   b. All squares are parallelograms.
   c. All rectangles are parallelograms.
   d. All squares are rhombuses.
   e. All rhombuses are parallelograms.

2. **REASONING** Name two types of quadrilaterals with four right angles.

3. **WHICH ONE DOESN'T BELONG?** Which type of quadrilateral does *not* belong with the other three? Explain your reasoning.

   rectangle | parallelogram | square | kite

Practice and Problem Solving

Classify the quadrilateral.

4. 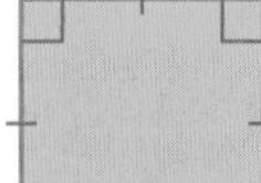

5. 

6. 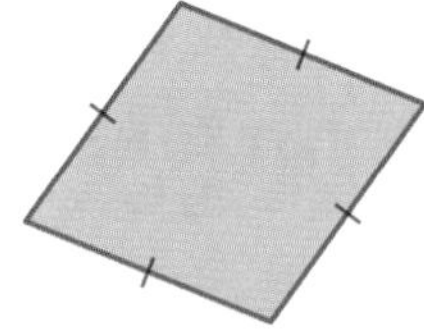

7. 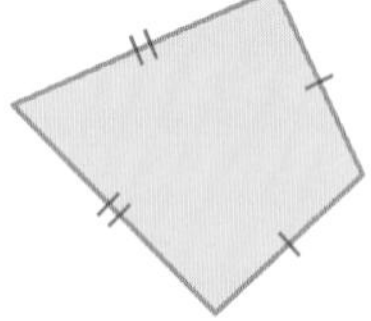

8. 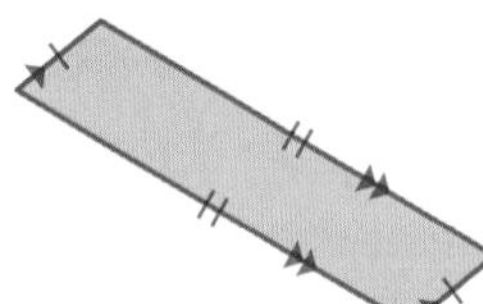

9. 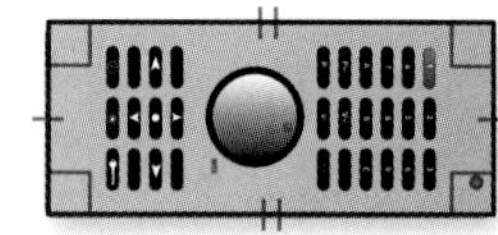

Find the value of x.

10. 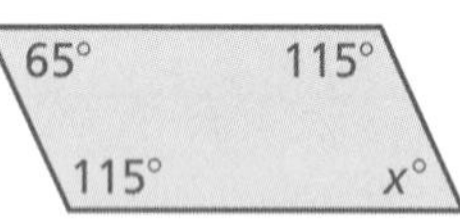

11. 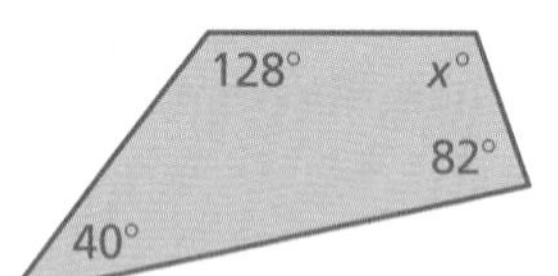

12. 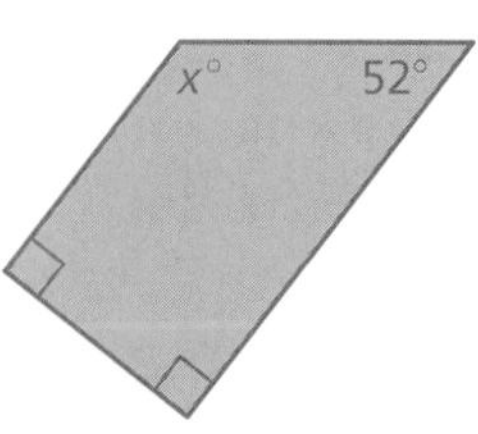

13. **KITE MAKING** What is the measure of the angle at the tail end of the kite?

Draw a quadrilateral with the given description.

3 **14.** a trapezoid with a pair of congruent, nonparallel sides

15. a rhombus with 3-centimeter sides and two 100° angles

16. a parallelogram with a 45° angle and a 135° angle

17. a parallelogram with a 75° angle and a 4-centimeter side

Copy and complete using *always, sometimes,* or *never.*

18. A square is __?__ a rectangle.

19. A square is __?__ a rhombus.

20. A rhombus is __?__ a square.

21. A parallelogram is __?__ a trapezoid.

22. A trapezoid is __?__ a kite.

23. A rhombus is __?__ a rectangle.

24. **DOOR** The dashed line shows how you cut the bottom of a rectangular door so it opens more easily.

a. Identify the new shape of the door. Explain.

b. What is the new angle at the bottom left side of the door? Explain.

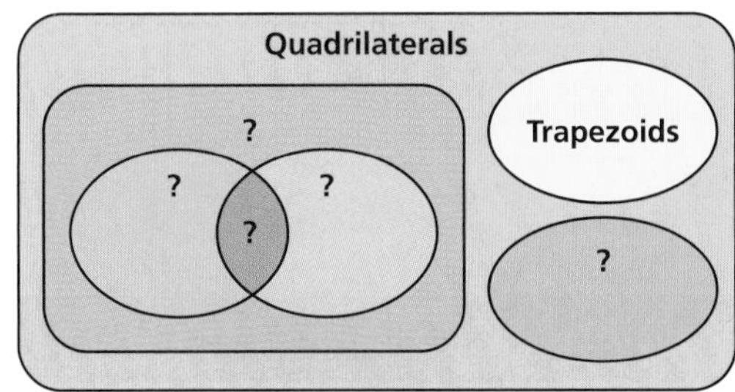

25. **VENN DIAGRAM** The diagram shows that some quadrilaterals are trapezoids, and all trapezoids are quadrilaterals. Copy the diagram. Fill in the names of the types of quadrilaterals to show their relationships.

26. Structure Consider the parallelogram.

a. Find the values of x and y.

b. Make a conjecture about opposite angles in a parallelogram.

c. In polygons, consecutive interior angles share a common side. Make a conjecture about consecutive interior angles in a parallelogram.

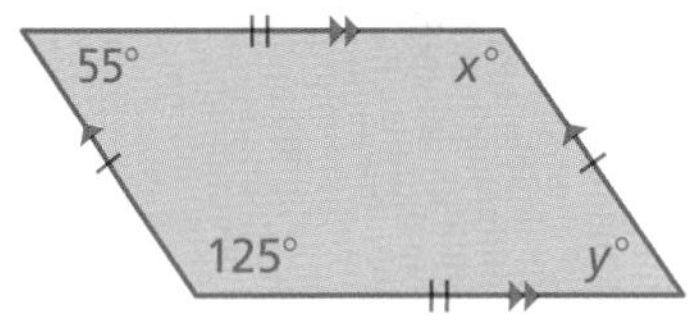

Fair Game Review What you learned in previous grades & lessons

Write the ratio as a fraction in simplest form. *(Skills Review Handbook)*

27. 3 turnovers : 12 assists

28. 18 girls to 27 boys

29. 42 pens : 35 pencils

30. **MULTIPLE CHOICE** Computer sales decreased from 40 to 32. What is the percent of decrease? *(Skills Review Handbook)*

Ⓐ 8% Ⓑ 20% Ⓒ 25% Ⓓ 80%

12.5 Scale Drawings

Essential Question How can you enlarge or reduce a drawing proportionally?

1 ACTIVITY: Comparing Measurements

Work with a partner. The diagram shows a food court at a shopping mall. Each centimeter in the diagram represents 40 meters.

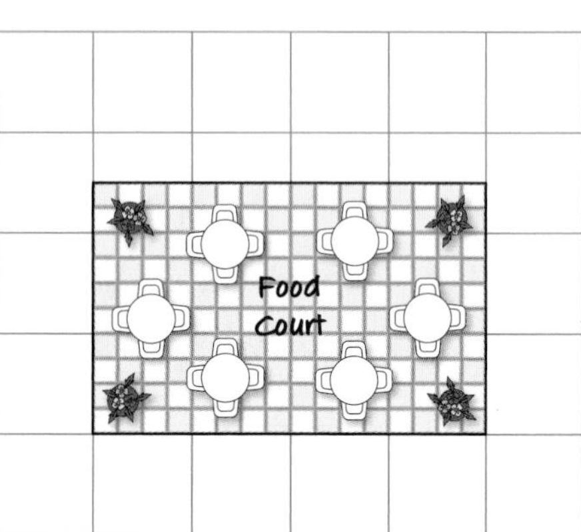

a. Find the length and the width of the drawing of the food court.

length: ___ cm width: ___ cm

b. Find the actual length and width of the food court. Explain how you found your answers.

length: ___ m width: ___ m

c. Find the ratios $\frac{\text{drawing length}}{\text{actual length}}$ and $\frac{\text{drawing width}}{\text{actual width}}$. What do you notice?

2 ACTIVITY: Recreating a Drawing

Work with a partner. Draw the food court in Activity 1 on the grid paper so that each centimeter represents 20 meters.

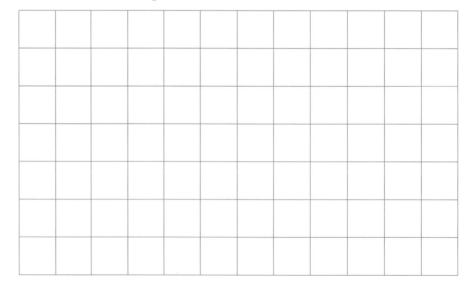

a. What happens to the size of the drawing?

b. Find the length and the width of your drawing. Compare these dimensions to the dimensions of the original drawing in Activity 1.

Geometry

In this lesson, you will

- use scale drawings to find actual distances.
- find scale factors.
- use scale drawings to find actual perimeters and areas.
- recreate scale drawings at a different scale.

3 ACTIVITY: Comparing Measurements

Work with a partner. The diagram shows a sketch of a painting. Each unit in the sketch represents 8 inches.

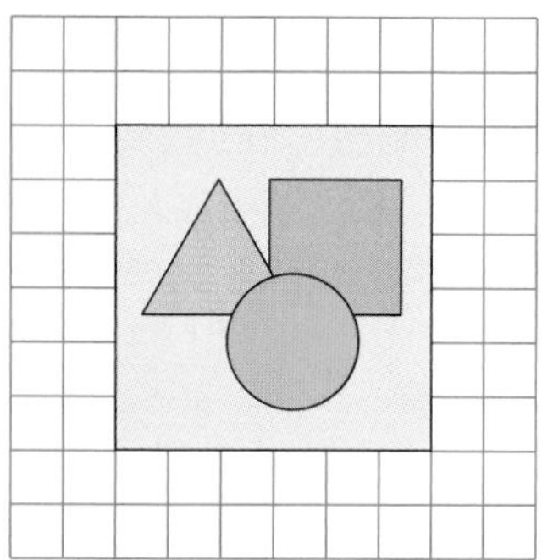

a. Find the length and the width of the sketch.

length: ____ units width: ____ units

b. Find the actual length and width of the painting. Explain how you found your answers.

length: ____ in. width: ____ in.

c. Find the ratios $\frac{\text{sketch length}}{\text{actual length}}$ and $\frac{\text{sketch width}}{\text{actual width}}$.

What do you notice?

4 ACTIVITY: Recreating a Drawing

Math Practice

Specify Units

How do you know whether to use feet or units for each measurement?

Work with a partner. Let each unit in the grid paper represent 2 feet. Now sketch the painting in Activity 3 onto the grid paper.

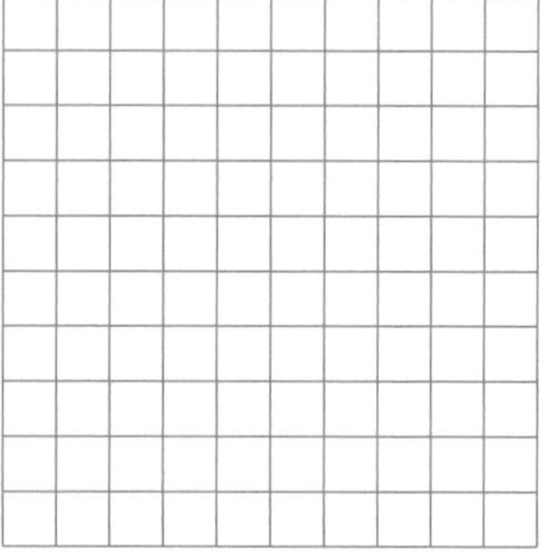

a. What happens to the size of the sketch?

b. Find the length and the width of your sketch. Compare these dimensions to the dimensions of the original sketch in Activity 3.

What Is Your Answer?

5. **IN YOUR OWN WORDS** How can you enlarge or reduce a drawing proportionally?

6. Complete the table for both the food court and the painting.

	Actual Object	Original Drawing	Your Drawing
Perimeter			
Area			

Compare the measurements in each table. What conclusions can you make?

7. **RESEARCH** Look at some maps in your school library or on the Internet. Make a list of the different scales used on the maps.

8. When you view a map on the Internet, how does the scale change when you zoom out? How does the scale change when you zoom in?

Practice Use what you learned about enlarging or reducing drawings to complete Exercises 4–7 on page 535.

12.5 Lesson

Key Vocabulary
scale drawing, *p. 532*
scale model, *p. 532*
scale, *p. 532*
scale factor, *p. 533*

Scale Drawings and Models

A **scale drawing** is a proportional, two-dimensional drawing of an object. A **scale model** is a proportional, three-dimensional model of an object.

Scale

The measurements in scale drawings and models are proportional to the measurements of the actual object. The **scale** gives the ratio that compares the measurements of the drawing or model with the actual measurements.

$\frac{1 \text{ in.}}{10 \text{ mi}}$ ← drawing distance / ← actual distance

1 in. : 10 mi (drawing : actual)

Study Tip

Scales are written so that the drawing distance comes first in the ratio.

EXAMPLE 1 Finding an Actual Distance

What is the actual distance *d* between Cadillac and Detroit?

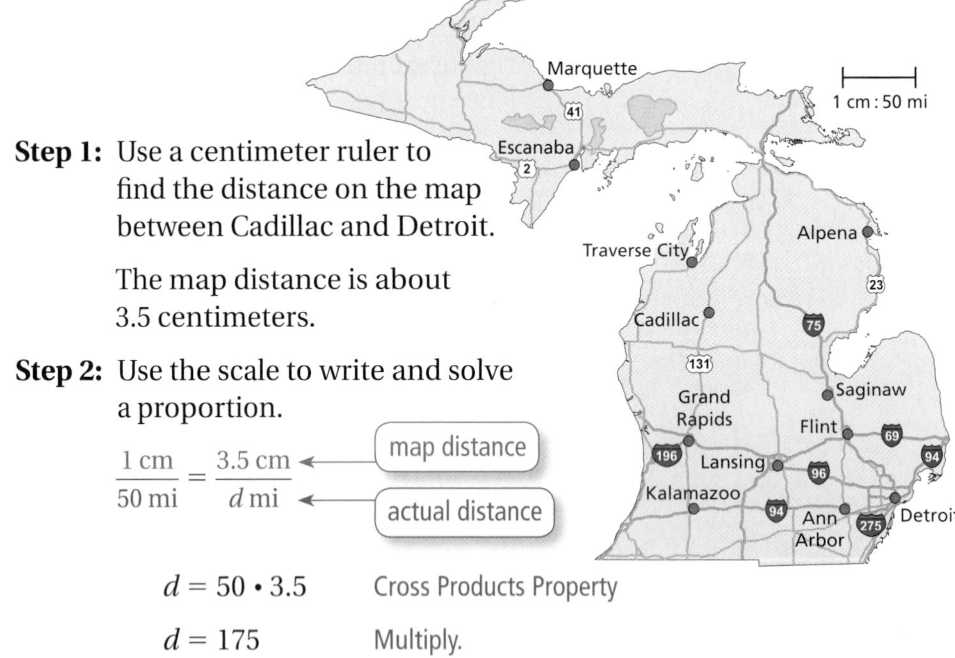

Step 1: Use a centimeter ruler to find the distance on the map between Cadillac and Detroit.

The map distance is about 3.5 centimeters.

Step 2: Use the scale to write and solve a proportion.

$$\frac{1 \text{ cm}}{50 \text{ mi}} = \frac{3.5 \text{ cm}}{d \text{ mi}}$$ ← map distance / ← actual distance

$d = 50 \cdot 3.5$ Cross Products Property

$d = 175$ Multiply.

So, the distance between Cadillac and Detroit is about 175 miles.

On Your Own

1. What is the actual distance between Traverse City and Marquette?

EXAMPLE 2 Finding a Distance in a Model

The liquid outer core of Earth is 2300 kilometers thick. A scale model of the layers of Earth has a scale of 1 in. : 500 km. How thick is the liquid outer core of the model?

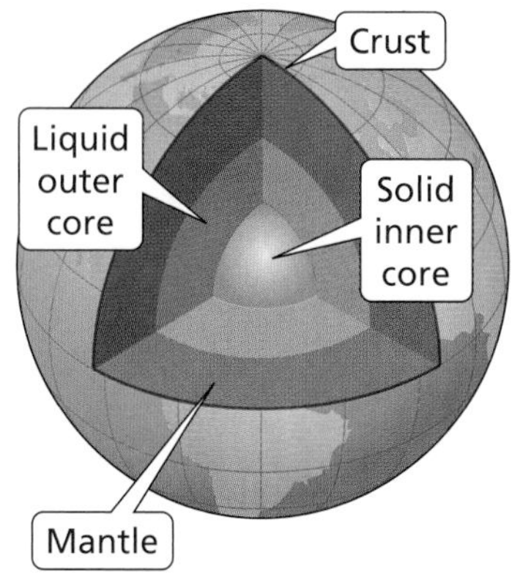

Ⓐ 0.2 in. Ⓑ 4.6 in. Ⓒ 0.2 km Ⓓ 4.6 km

$$\frac{1 \text{ in.}}{500 \text{ km}} = \frac{x \text{ in.}}{2300 \text{ km}}$$ ← model thickness, ← actual thickness

$$\frac{1 \text{ in.}}{500 \text{ km}} \cdot 2300 \text{ km} = \frac{x \text{ in.}}{2300 \text{ km}} \cdot 2300 \text{ km}$$ Multiplication Property of Equality

$$4.6 = x$$ Simplify.

So, the liquid outer core of the model is 4.6 inches thick. The correct answer is Ⓑ.

On Your Own

2. The mantle of Earth is 2900 kilometers thick. How thick is the mantle of the model?

A scale can be written without units when the units are the same. A scale without units is called a **scale factor**.

EXAMPLE 3 Finding a Scale Factor

A scale model of the Sergeant Floyd Monument is 10 inches tall. The actual monument is 100 feet tall.

a. What is the scale of the model?

$$\frac{\text{model height}}{\text{actual height}} = \frac{10 \text{ in.}}{100 \text{ ft}} = \frac{1 \text{ in.}}{10 \text{ ft}}$$

The scale is 1 in. : 10 ft.

b. What is the scale factor of the model?

Write the scale with the same units. Use the fact that 1 ft = 12 in.

$$\text{scale factor} = \frac{1 \text{ in.}}{10 \text{ ft}} = \frac{1 \text{ in.}}{120 \text{ in.}} = \frac{1}{120}$$

The scale factor is 1 : 120.

On Your Own

Now You're Ready Exercises 12–16

3. A drawing has a scale of 1 mm : 20 cm. What is the scale factor of the drawing?

EXAMPLE 4 Finding an Actual Perimeter and Area

The scale drawing of a computer chip helps you see the individual components on the chip.

a. Find the perimeter and the area of the computer chip in the scale drawing.

When measured using a centimeter ruler, the scale drawing of the computer chip has a side length of 4 centimeters.

So, the perimeter of the computer chip in the scale drawing is $4(4) = 16$ centimeters, and the area is $4^2 = 16$ square centimeters.

b. Find the actual perimeter and area of the computer chip.

$$\frac{1 \text{ cm}}{2 \text{ mm}} = \frac{4 \text{ cm}}{s \text{ mm}}$$ ← drawing distance (numerators), actual distance (denominators)

$s = 2 \cdot 4$ Cross Products Property

$s = 8$ Multiply.

The side length of the actual computer chip is 8 millimeters.

So, the actual perimeter of the computer chip is $4(8) = 32$ millimeters, and the actual area is $8^2 = 64$ square millimeters.

c. Compare the ratios $\frac{\text{drawing perimeter}}{\text{actual perimeter}}$ and $\frac{\text{drawing area}}{\text{actual area}}$ to the scale factor.

Use the fact that 1 cm = 10 mm.

$$\text{scale factor} = \frac{1 \text{ cm}}{2 \text{ mm}} = \frac{10 \text{ mm}}{2 \text{ mm}} = \frac{5}{1}$$

$$\frac{\text{drawing perimeter}}{\text{actual perimeter}} = \frac{16 \text{ cm}}{32 \text{ mm}} = \frac{1 \text{ cm}}{2 \text{ mm}} = \frac{5}{1}$$

$$\frac{\text{drawing area}}{\text{actual area}} = \frac{16 \text{ cm}^2}{64 \text{ mm}^2} = \frac{1 \text{ cm}^2}{4 \text{ mm}^2} = \left(\frac{1 \text{ cm}}{2 \text{ mm}}\right)^2 = \left(\frac{5}{1}\right)^2$$

So, the ratio of the perimeters is equal to the scale factor, and the ratio of the areas is equal to the square of the scale factor.

Study Tip

The ratios tell you that the perimeter of the drawing is 5 times the actual perimeter, and the area of the drawing is $5^2 = 25$ times the actual area.

On Your Own

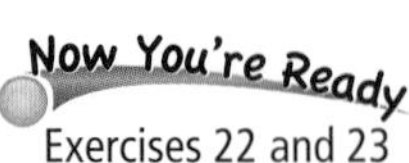

Exercises 22 and 23

4. **WHAT IF?** The scale of the drawing of the computer chip is 1 cm : 3 mm. How do the answers in parts (a)–(c) change? Justify your answer.

12.5 Exercises

Vocabulary and Concept Check

1. **VOCABULARY** Compare and contrast the terms *scale* and *scale factor*.
2. **CRITICAL THINKING** The scale of a drawing is 2 cm : 1 mm. Is the scale drawing *larger* or *smaller* than the actual object? Explain.
3. **REASONING** How would you find the scale factor of a drawing that shows a length of 4 inches when the actual object is 8 feet long?

Practice and Problem Solving

Use the drawing and a centimeter ruler. Each centimeter in the drawing represents 5 feet.

4. What is the actual length of the flower garden?
5. What are the actual dimensions of the rose bed?
6. What are the actual perimeters of the perennial beds?
7. The area of the tulip bed is what percent of the area of the rose bed?

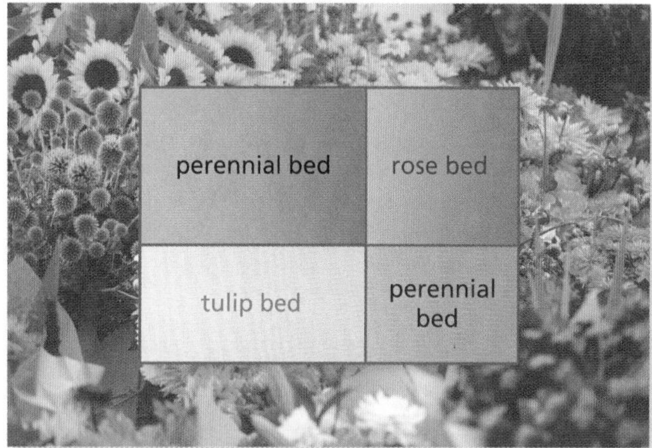

Use the map in Example 1 to find the actual distance between the cities.

① 8. Kalamazoo and Ann Arbor
9. Lansing and Flint
10. Grand Rapids and Escanaba
11. Saginaw and Alpena

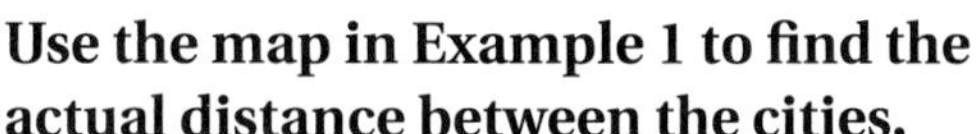

Find the missing dimension. Use the scale factor 1 : 12.

	Item	Model	Actual
② ③ 12.	Mattress	Length: 6.25 in.	Length: ___ in.
13.	Corvette	Length: ___ in.	Length: 15 ft
14.	Water tower	Depth: 32 cm	Depth: ___ m
15.	Wingspan	Width: 5.4 ft	Width: ___ yd
16.	Football helmet	Diameter: ___ mm	Diameter: 21 cm

17. **ERROR ANALYSIS** A scale is 1 cm : 20 m. Describe and correct the error in finding the actual distance that corresponds to 5 centimeters.

✗ $\frac{1 \text{ cm}}{20 \text{ m}} = \frac{x \text{ m}}{5 \text{ cm}}$

$x = 0.25$ m

Use a centimeter ruler to measure the segment shown. Find the scale of the drawing.

18. 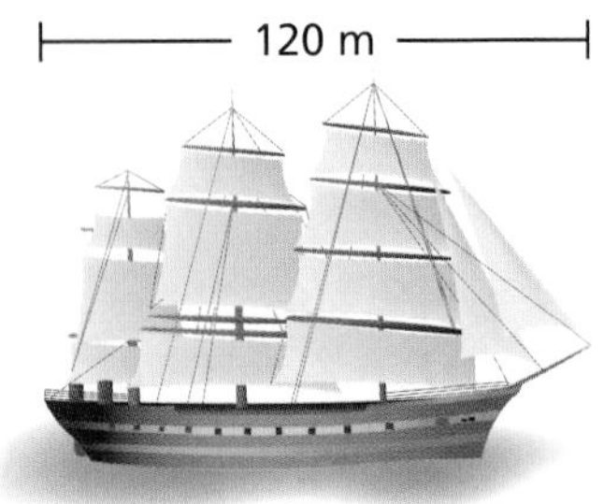

19. 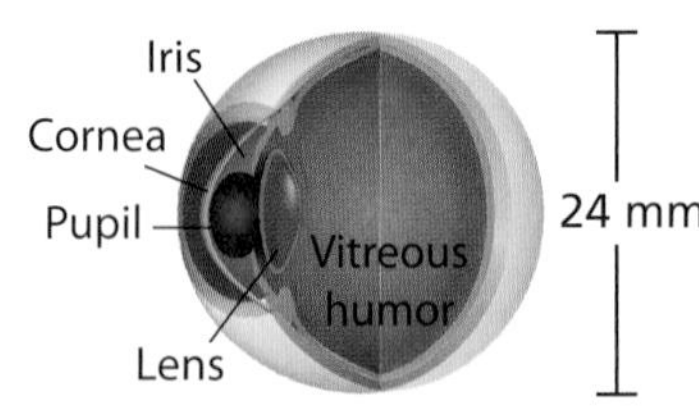

20. REASONING You know the length and the width of a scale model. What additional information do you need to know to find the scale of the model?

21. OPEN-ENDED You are in charge of creating a billboard advertisement with the dimensions shown.

a. Choose a product. Then design the billboard using words and a picture.

b. What is the scale factor of your design?

4 **22. CENTRAL PARK** Central Park is a rectangular park in New York City.

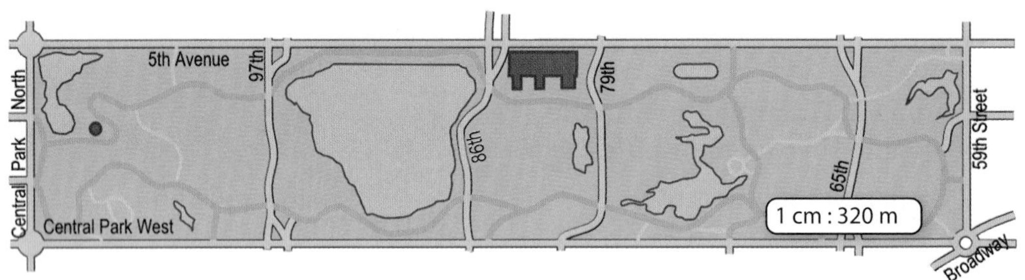

a. Find the perimeter and the area of Central Park in the scale drawing.

b. Find the actual perimeter and area of Central Park.

23. ICON You are designing an icon for a mobile app.

a. Find the perimeter and the area of the icon in the scale drawing.

b. Find the actual perimeter and area of the icon.

24. CRITICAL THINKING Use the results of Exercises 22 and 23 to make a conjecture about the relationship between the scale factor of a drawing and the ratios $\frac{\text{drawing perimeter}}{\text{actual perimeter}}$ and $\frac{\text{drawing area}}{\text{actual area}}$.

Recreate the scale drawing so that it has a scale of 1 cm : 4 m.

25. 

1 cm : 8 m

26. 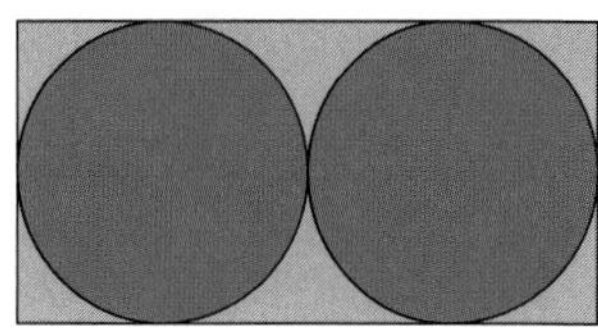

1 cm : 2 m

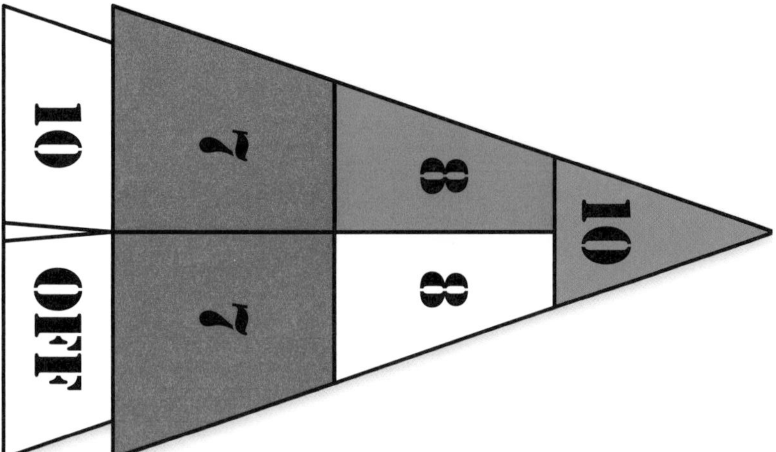

The shuffleboard diagram has a scale of 1 cm : 1 ft. Find the actual area of the region.

27. red region

28. blue region

29. green region

Reduced Drawing of Blueprint

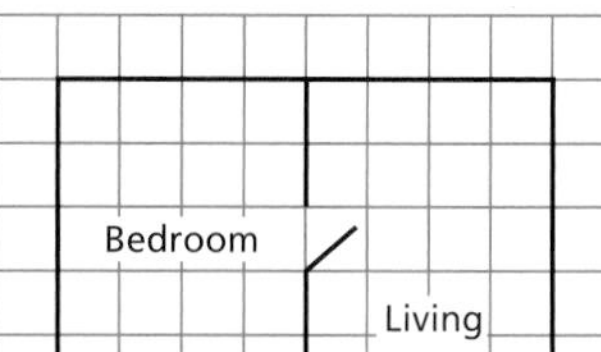

30. **BLUEPRINT** In a blueprint, each square has a side length of $\frac{1}{4}$ inch.

  a. Ceramic tile costs \$5 per square foot. How much would it cost to tile the bathroom?

  b. Carpet costs \$18 per square yard. How much would it cost to carpet the bedroom and living room?

  c. Which has a greater unit cost, the tile or the carpet? Explain.

31. **Modeling** You are making a scale model of the solar system. The radius of Earth is 6378 kilometers. The radius of the Sun is 695,500 kilometers. Is it reasonable to choose a baseball as a model of Earth? Explain your reasoning.

Fair Game Review What you learned in previous grades & lessons

Plot and label the ordered pair in a coordinate plane. *(Skills Review Handbook)*

32. $A(-4, 3)$ 33. $B(2, -6)$ 34. $C(5, 1)$ 35. $D(-3, -7)$

36. **MULTIPLE CHOICE** Which set of numbers is ordered from least to greatest? *(Skills Review Handbook)*

Ⓐ $\frac{7}{20}$, 32%, 0.45 Ⓑ 17%, 0.21, $\frac{3}{25}$ Ⓒ 0.88, $\frac{7}{8}$, 93% Ⓓ 57%, $\frac{11}{16}$, 5.7

12.4–12.5 Quiz

Classify the quadrilateral. *(Section 12.4)*

1. 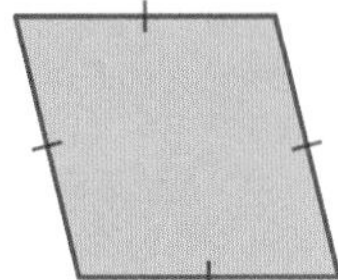

2. 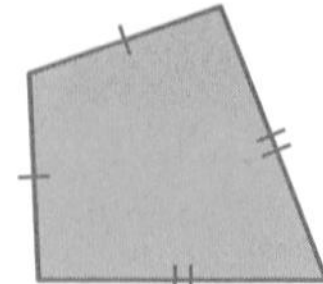

Find the value of *x*. *(Section 12.4)*

3. 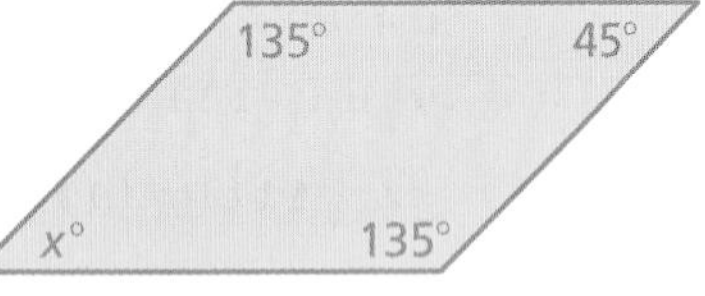

4. 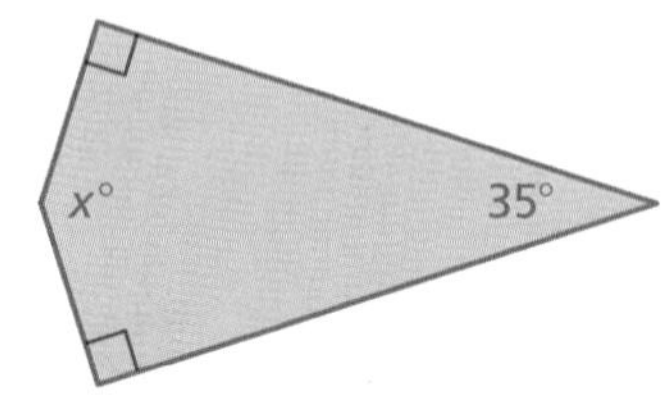

Draw a quadrilateral with the given description. *(Section 12.4)*

5. a rhombus with 2-centimeter sides and two 50° angles

6. a parallelogram with a 65° angle and a 5-centimeter side

Find the missing dimension. Use the scale factor 1 : 20. *(Section 12.5)*

	Item	Model	Actual
7.	Basketball player	Height: ☐ in.	Height: 90 in.
8.	Dinosaur	Length: 3.75 ft	Length: ☐ ft

9. SHED The side of the storage shed is in the shape of a trapezoid. Find the value of *x*. *(Section 12.4)*

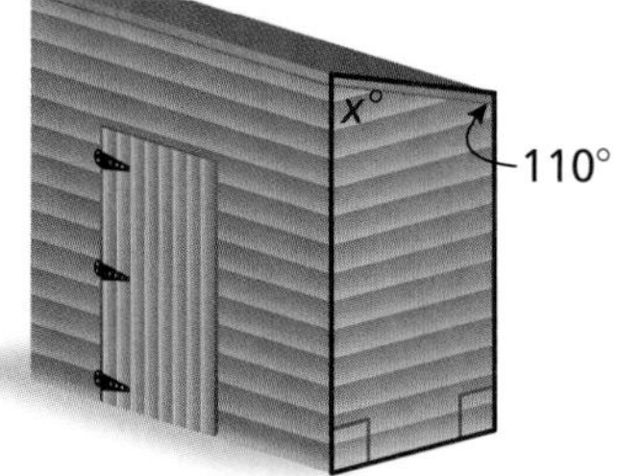

10. DOLPHIN A dolphin in an aquarium is 12 feet long. A scale model of the dolphin is $3\frac{1}{2}$ inches long. What is the scale factor of the model? *(Section 12.5)*

11. SOCCER A scale drawing of a soccer field is shown. The actual soccer field is 300 feet long. *(Section 12.5)*

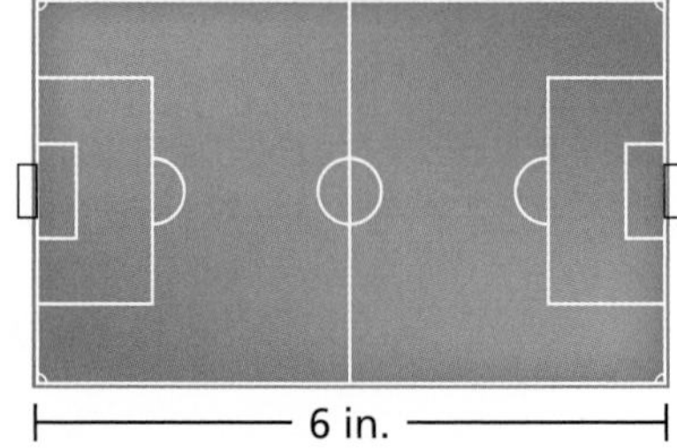

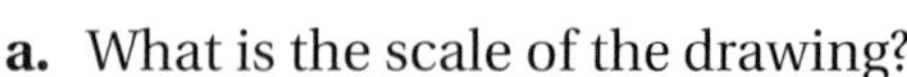

a. What is the scale of the drawing?

b. What is the scale factor of the drawing?

12 Chapter Review

Check It Out
Vocabulary Help
BigIdeasMath.com

Review Key Vocabulary

- adjacent angles, *p. 504*
- vertical angles, *p. 504*
- congruent angles, *p. 504*
- complementary angles, *p. 510*
- supplementary angles, *p. 510*
- congruent sides, *p. 516*
- kite, *p. 526*
- scale drawing, *p. 532*
- scale model, *p. 532*
- scale, *p. 532*
- scale factor, *p. 533*

Review Examples and Exercises

12.1 Adjacent and Vertical Angles *(pp. 502–507)*

Tell whether the angles are *adjacent* or *vertical*. Then find the value of *x*.

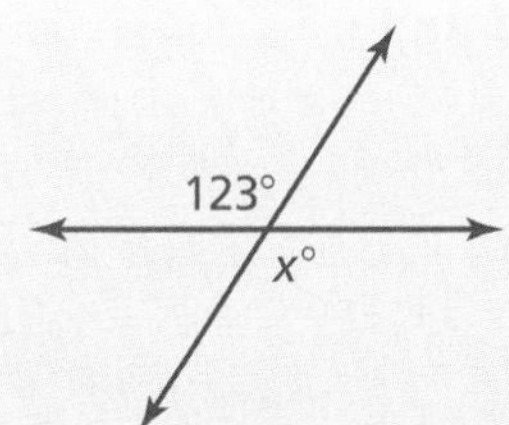

The angles are vertical angles. Because vertical angles are congruent, the angles have the same measure.

So, the value of x is 123.

Exercises

Tell whether the angles are *adjacent* or *vertical*. Then find the value of *x*.

1.

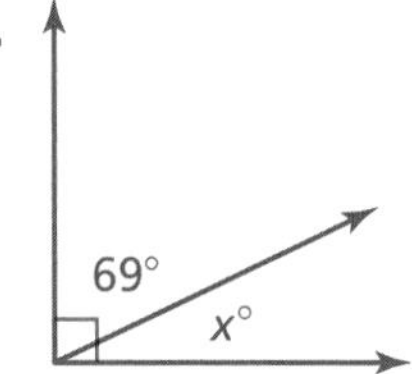

2.

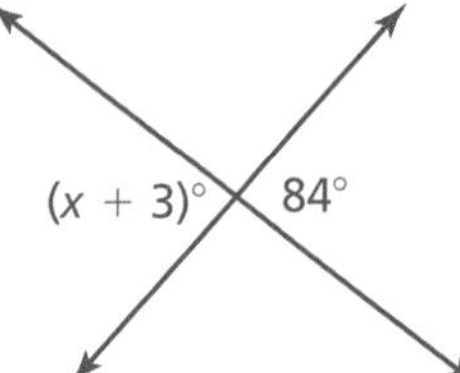

12.2 Complementary and Supplementary Angles *(pp. 508–513)*

Tell whether the angles are *complementary* or *supplementary*. Then find the value of *x*.

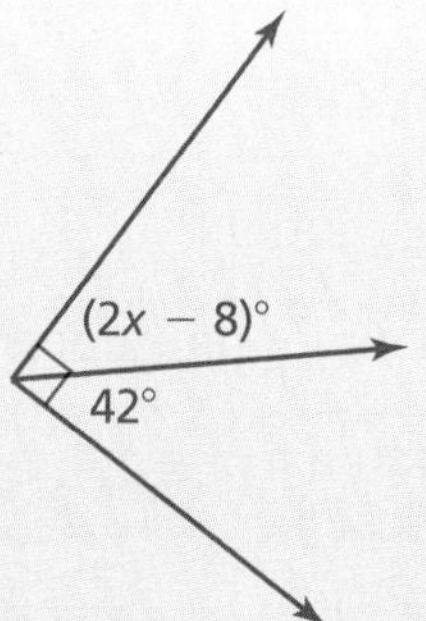

The two angles make up a right angle. So, the angles are complementary angles, and the sum of their measures is 90°.

$(2x - 8) + 42 = 90$	Write equation.
$2x + 34 = 90$	Combine like terms.
$2x = 56$	Subtract 34 from each side.
$x = 28$	Divide each side by 2.

So, the value of x is 28.

Exercises

Tell whether the angles are *complementary* or *supplementary*. Then find the value of *x*.

3.

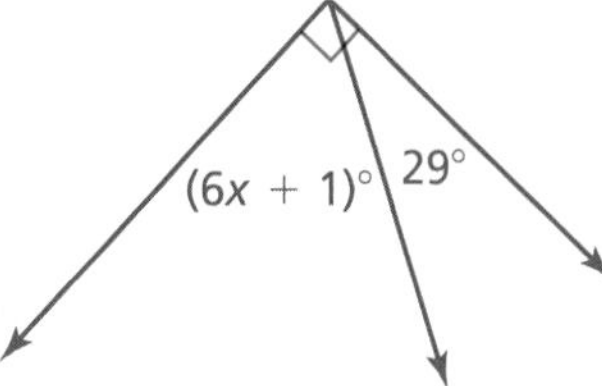

4.

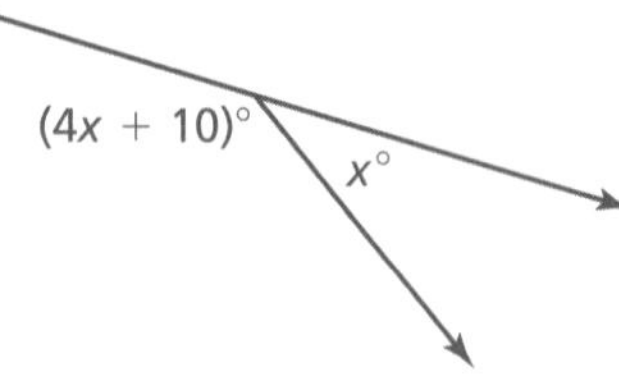

12.3 Triangles (pp. 514–521)

Draw a triangle with a 3.5-centimeter side and a 4-centimeter side that meet at a 25° angle. Then classify the triangle.

Step 1: Use a protractor to draw a 25° angle.

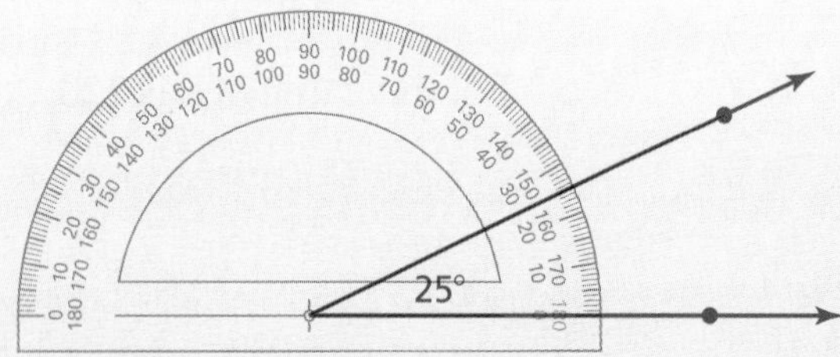

Step 2: Use a ruler to mark 3.5 centimeters on one ray and 4 centimeters on the other ray.

Step 3: Draw the third side to form the triangle.

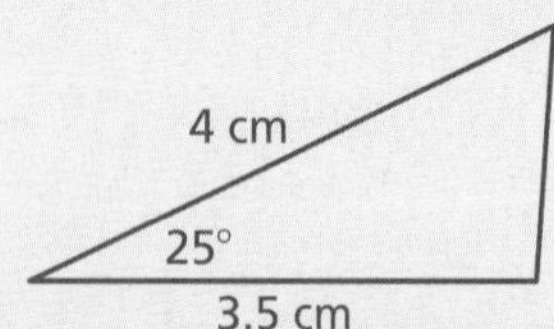

∴ The triangle is an obtuse scalene triangle.

Exercises

Draw a triangle with the given description.

5. a triangle with angle measures of 40°, 50°, and 90°
6. a triangle with a 3-inch side and a 4-inch side that meet at a 30° angle

Find the value of *x*. Then classify the triangle.

7.

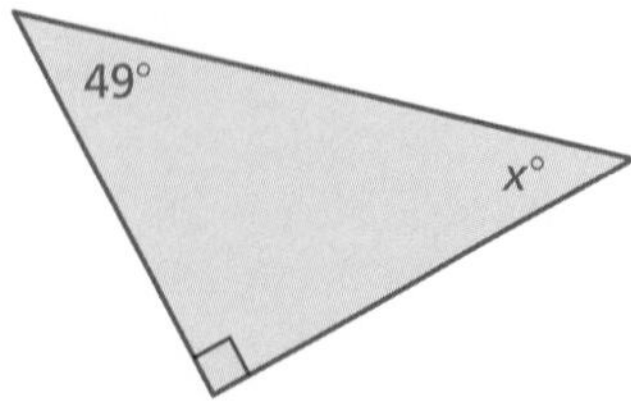

8.

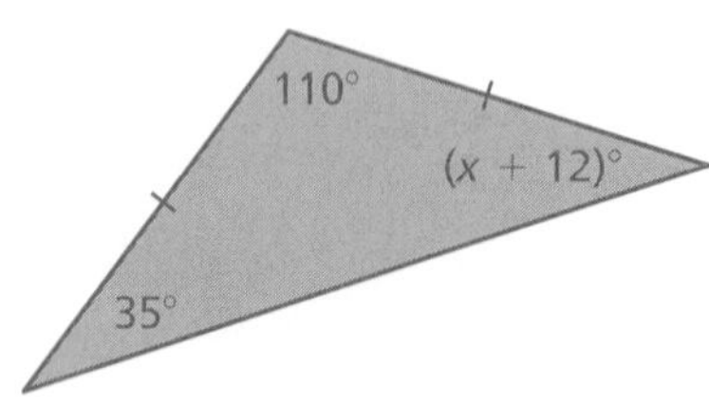

12.4 Quadrilaterals (pp. 524–529)

Draw a parallelogram with a 50° angle and a 130° angle.

Step 1: Draw a line.

Step 2: Draw a 50° angle and a 130° angle that each have one side on the line.

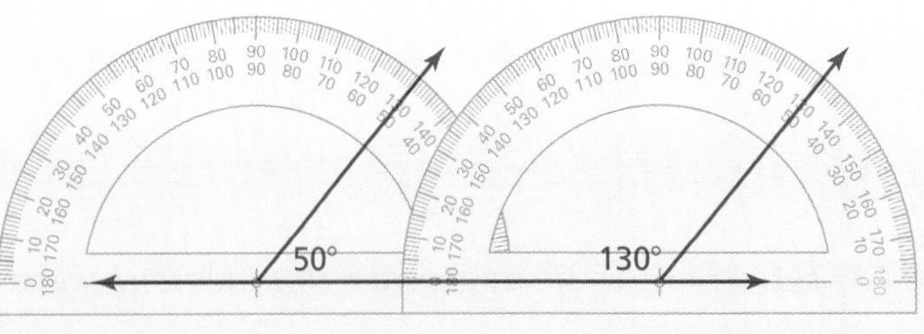

Step 3: Draw the remaining side. Make sure that both pairs of opposite sides are parallel and congruent.

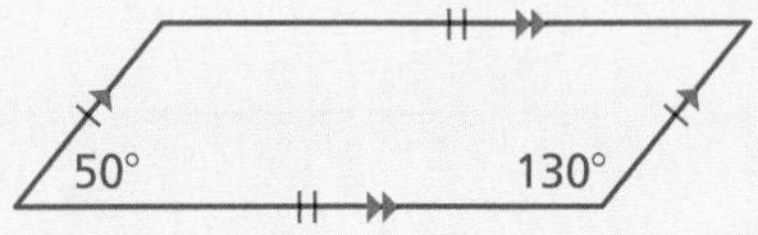

Exercises

Find the value of *x*.

9. 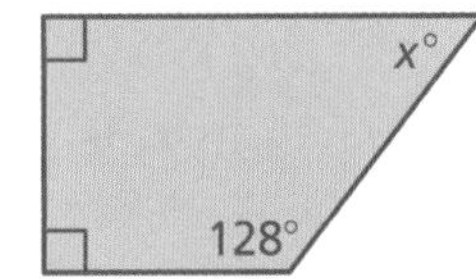

10. 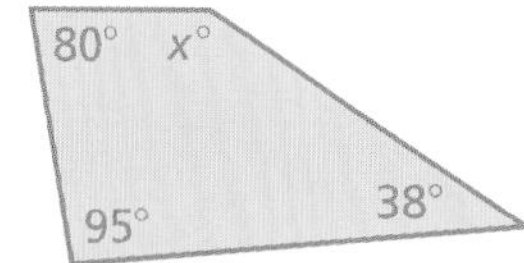

11. Draw a rhombus with 5-centimeter sides and two 120° angles.

12.5 Scale Drawings (pp. 530–537)

A lighthouse is 160 feet tall. A scale model of the lighthouse has a scale of 1 in. : 8 ft. How tall is the model of the lighthouse?

$$\frac{1 \text{ in.}}{8 \text{ ft}} = \frac{x \text{ in.}}{160 \text{ ft}}$$ ← model height (numerator), ← actual height (denominator)

$$\frac{1 \text{ in.}}{8 \cancel{\text{ft}}} \cdot 160 \cancel{\text{ft}} = \frac{x \text{ in.}}{160 \text{ ft}} \cdot 160 \text{ ft}$$ Multiplication Property of Equality

$$20 = x$$ Simplify.

So, the model of the lighthouse is 20 inches tall.

Exercises

Use a centimeter ruler to measure the segment shown. Find the scale of the drawing.

12. 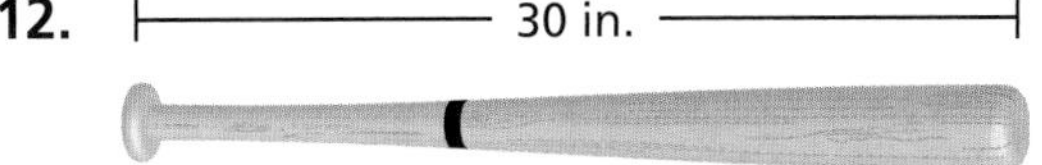

13. 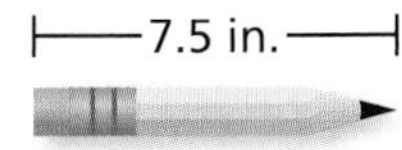

12 Chapter Test

Tell whether the angles are *adjacent* or *vertical*. Then find the value of x.

1. 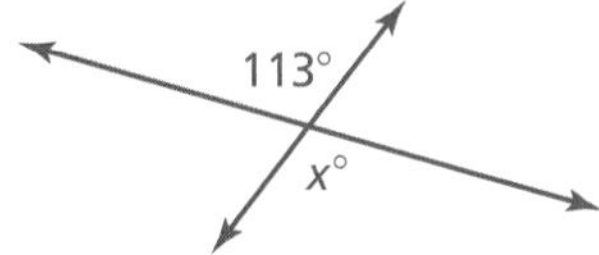

2. 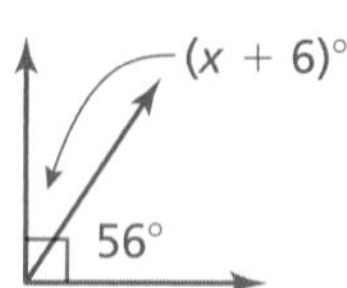

Tell whether the angles are *complementary* or *supplementary*. Then find the value of x.

3. 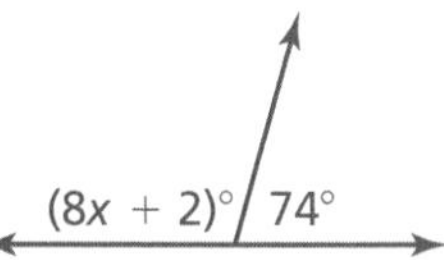

4. 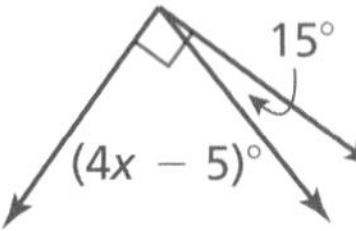

Draw a triangle with the given angle measures. Then classify the triangle.

5. 10°, 80°, 90°

6. 30°, 40°, 110°

Draw a triangle with the given description.

7. a triangle with a 5-inch side and a 6-inch side that meet at a 50° angle

8. a right isosceles triangle

Find the value of x. Then classify the triangle.

9. 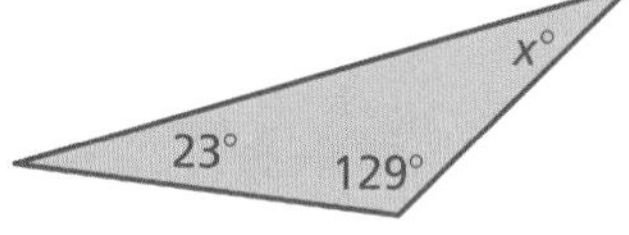

10. 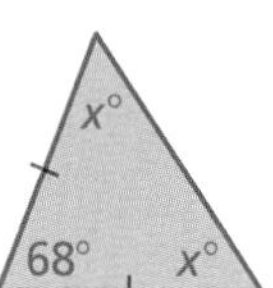

11. 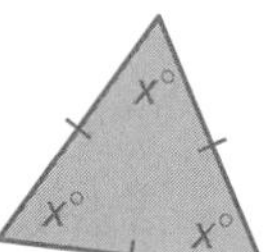

Find the value of x.

12. 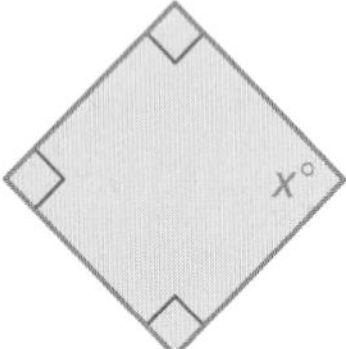

13. 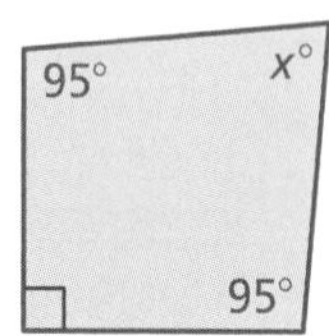

14. 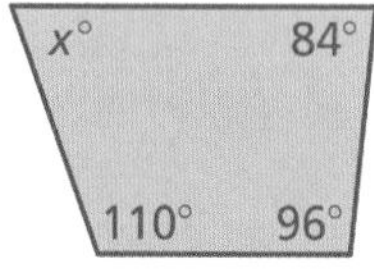

Draw a quadrilateral with the given description.

15. a rhombus with 6-centimeter sides and two 80° angles

16. a parallelogram with a 20° angle and a 160° angle

17. **FISH** Use a centimeter ruler to measure the fish. Find the scale factor of the drawing.

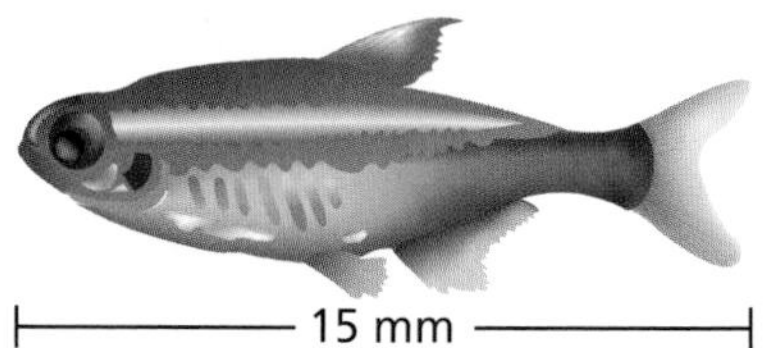

18. **CAD** An engineer is using computer-aided design (CAD) software to design a component for a space shuttle. The scale of the drawing is 1 cm : 60 in. The actual length of the component is 12.5 feet. What is the length of the component in the drawing?

12 Cumulative Assessment

1. The number of calories you burn by playing basketball is proportional to the number of minutes you play. Which of the following is a valid interpretation of the graph below?

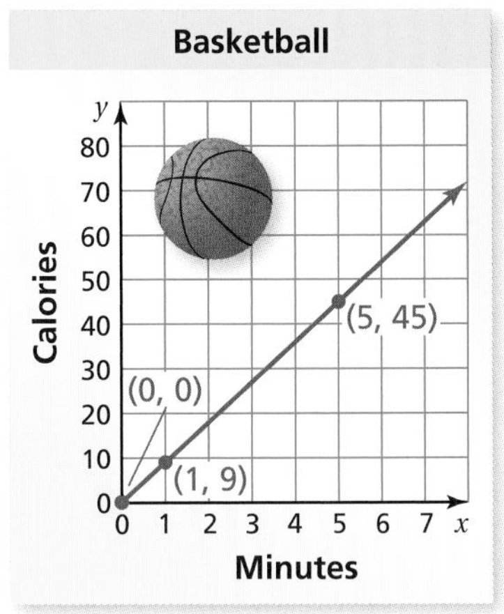

Test-Taking Strategy
Solve Problem Before Looking at Choices
Your paw has an area of 2 in.² A hyena's paw is twice as long. What is its area?
(A) 4 in.² (B) 6 in.² (C) 8 in.² (D) 10 in.²
4 times more area! Help!
CAT
HYENA
"Solve the problem before looking at the choices. You know area increases as the square of the scale. So, it's 8 in.²"

A. The unit rate is $\frac{1}{9}$ calorie per minute.

B. You burn 5 calories by playing basketball for 45 minutes.

C. You do not burn any calories if you do not play basketball for at least 1 minute.

D. You burn an additional 9 calories for each minute of basketball you play.

2. A lighting store is holding a clearance sale. The store is offering discounts on all the lamps it sells. As the sale progresses, the store will increase the percent of discount it is offering.

You want to buy a lamp that has an original price of \$40. You will buy the lamp when its price is marked down to \$10. What percent discount will you have received?

3. What is the value of the expression below?

$$2 - 6 - (-9)$$

F. -13

G. -5

H. 5

I. 13

4. What is the solution to the proportion below?

$$\frac{8}{12} = \frac{x}{18}$$

5. Which graph represents the inequality below?

$$-5 - 6x \leq -23$$

A. 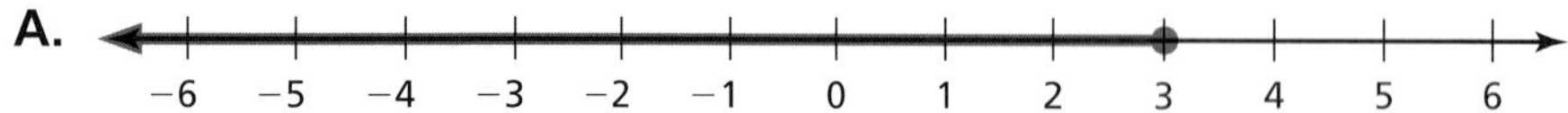

B. 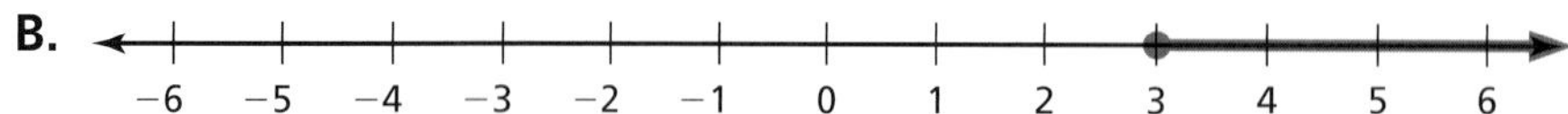

C. −6 −5 −4 −3 −2 −1 0 1 2 3 4 5 6

D. −6 −5 −4 −3 −2 −1 0 1 2 3 4 5 6

6. You are building a scale model of a park that is planned for a city. The model uses the scale below.

1 centimeter = 2 meters

The park will have a rectangular reflecting pool with a length of 20 meters and a width of 12 meters. In your scale model, what will be the area of the reflecting pool?

F. 60 cm^2

G. 120 cm^2

H. 480 cm^2

I. 960 cm^2

7. The quantities x and y are proportional. What is the missing value in the table?

x	y
$\frac{5}{7}$	10
$\frac{9}{7}$	18
$\frac{15}{7}$	30
4	

A. 38

B. 42

C. 46

D. 56

8. $\angle 1$ and $\angle 2$ form a straight angle. $\angle 1$ has a measure of $28°$. What is the measure of $\angle 2$?

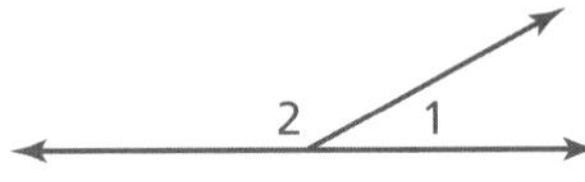

F. $62°$

G. $118°$

H. $152°$

I. $208°$

9. Brett solved the equation in the box below.

$$\frac{c}{5} - (-15) = -35$$
$$\frac{c}{5} + 15 = -35$$
$$\frac{c}{5} + 15 - 15 = -35 - 15$$
$$\frac{c}{5} = -50$$
$$\frac{c}{5} = \frac{-50}{5}$$
$$c = -10$$

What should Brett do to correct the error that he made?

A. Subtract 15 from -35 to get -20.

B. Rewrite $\frac{c}{5} - (-15)$ as $\frac{c}{5} - 15$.

C. Multiply each side of the equation by 5 to get $c = -250$.

D. Multiply each side of the equation by -5 to get $c = 250$.

10. A map of the state where Donna lives has the scale shown below.

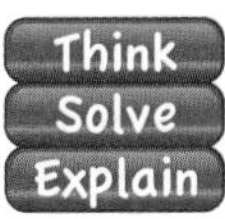

$$\frac{1}{2} \text{ inch} = 10 \text{ miles}$$

Part A Donna measured the distance between her town and the state capital on the map. Her measurement was $4\frac{1}{2}$ inches. Based on Donna's measurement, what is the actual distance, in miles, between her town and the state capital? Show your work and explain your reasoning.

Part B Donna wants to mark her favorite campsite on the map. She knows that the campsite is 65 miles north of her town. What distance on the map, in inches, represents an actual distance of 65 miles? Show your work and explain your reasoning.

13 Circles and Area

"Think of any number between 1 and 9."

"Okay, now add 4 to the number, multiply by 3, subtract 12, and divide by your original number."

"You end up with 3, don't you?"

"What do you get when you divide the circumference of a jack-o-lantern by its diameter?"

"Pumpkin pi, HE HE HE."

What You Learned Before

"The area of the circle is pi *r* squared. The area of the triangle is one-half *bh*."

Classifying Figures

Identify the basic shapes in the figure.

Example 1

Rectangle, right triangle

Example 2

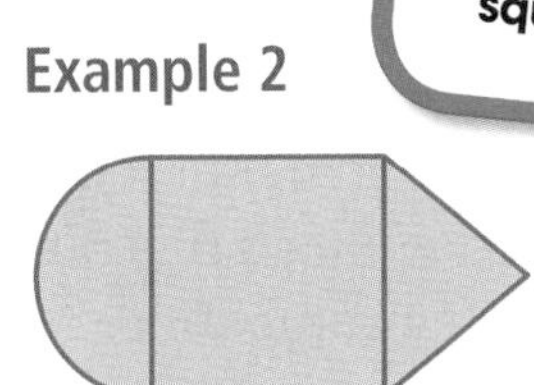

Semicircle, square, and triangle

Try It Yourself

Identify the basic shapes in the figure.

1. 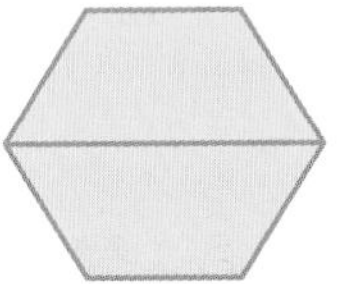

2. 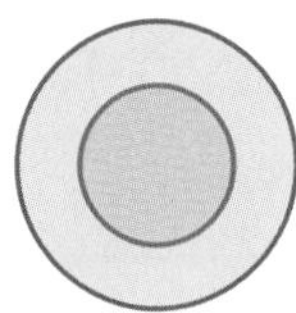

3. 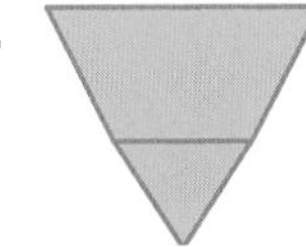

4. 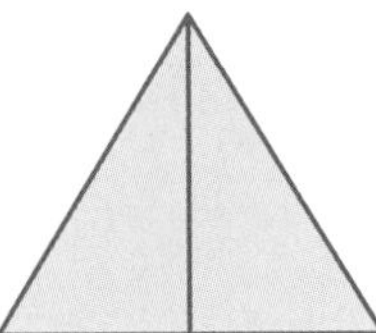

5. 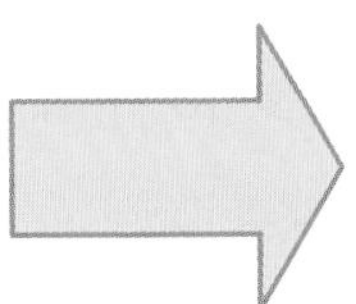

6. 

Squaring Numbers and Using Order of Operations

Example 3 **Evaluate 4^2.**

$$4^2 = 4 \cdot 4 = 16$$

4^2 means to multiply 4 by itself.

Example 4 **Evaluate $3 \cdot 6^2$.**

$$3 \cdot 6^2 = 3 \cdot (6 \cdot 6) = 3 \cdot 36 = 108$$

Use order of operations. Evaluate the exponent, and then multiply.

Try It Yourself

Evaluate the expression.

7. 5^2

8. 12^2

9. $3 \cdot 2^2$

10. $4 \cdot 7^2$

11. $3(1 + 8)^2$

12. $2(3 + 7)^2 - 3 \cdot 4$

13.1 Circles and Circumference

Essential Question How can you find the circumference of a circle?

Archimedes was a Greek mathematician, physicist, engineer, and astronomer.

Archimedes discovered that in any circle the ratio of circumference to diameter is always the same. Archimedes called this ratio pi, or π (a letter from the Greek alphabet).

$$\pi = \frac{\text{circumference}}{\text{diameter}}$$

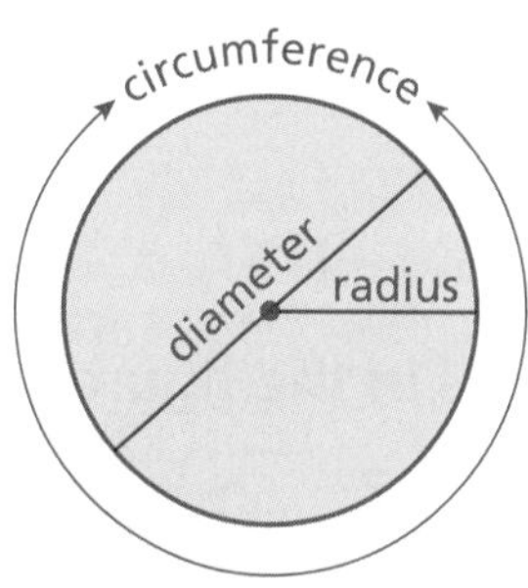

In Activities 1 and 2, you will use the same strategy Archimedes used to approximate π.

1 ACTIVITY: Approximating Pi

Work with a partner. Copy the table. Record your results in the table.

- **Measure the perimeter of the large square in millimeters.**
- **Measure the diameter of the circle in millimeters.**
- **Measure the perimeter of the small square in millimeters.**
- **Calculate the ratios of the two perimeters to the diameter.**
- **The average of these two ratios is an approximation of π.**

Large Square
Small Square

Sides	Large Perimeter	Diameter of Circle	Small Perimeter	$\frac{\text{Large Perimeter}}{\text{Diameter}}$	$\frac{\text{Small Perimeter}}{\text{Diameter}}$	Average of Ratios
4						
6						
8						
10						

Geometry
In this lesson, you will
- describe a circle in terms of radius and diameter.
- understand the concept of pi.
- find circumferences of circles and perimeters of semicircles.

The knights rode as fast as they could to the King's castle. Sir Cumference lived nearby, so his family came with him. Sir Cumference was married to Lady Di, who came from the town of Ameter.
They had a son named Radius. Radius was very small and quite young, but his keen mind and boundless energy more than made up for what he lacked in height and age.

A page from *Sir Cumference and the First Round Table* by Cindy Neuschwander

2 ACTIVITY: Approximating Pi

Continue your approximation of pi. Complete the table from Activity 1 using a hexagon (6 sides), an octagon (8 sides), and a decagon (10 sides).

Math Practice

Make Conjectures

How can you use the results of the activity to find an approximation of pi?

a.

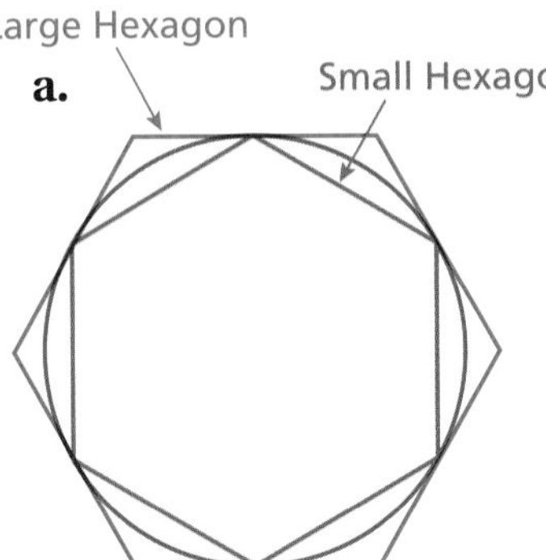

b.

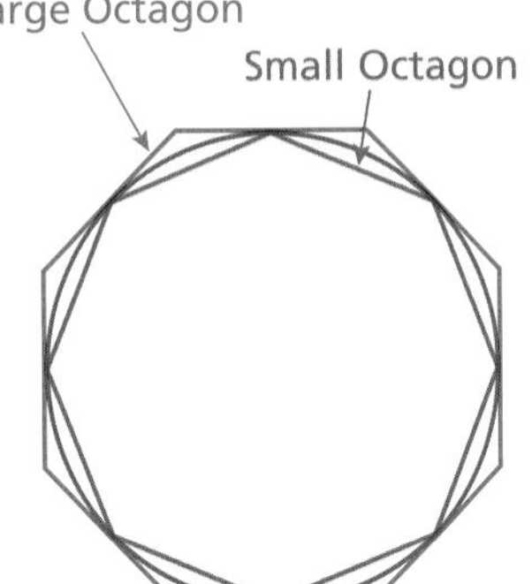

c.

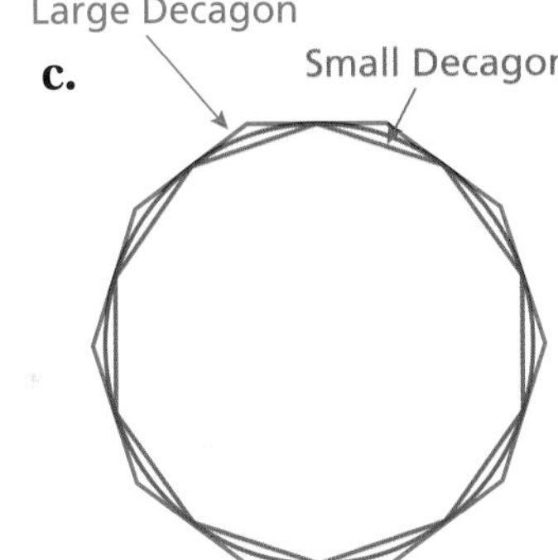

d. From the table, what can you conclude about the value of π? Explain your reasoning.

e. Archimedes calculated the value of π using polygons with 96 sides. Do you think his calculations were more or less accurate than yours?

What Is Your Answer?

3. **IN YOUR OWN WORDS** Now that you know an approximation for pi, explain how you can use it to find the circumference of a circle. Write a formula for the circumference C of a circle whose diameter is d.

4. **CONSTRUCTION** Use a compass to draw three circles. Use your formula from Question 3 to find the circumference of each circle.

Practice Use what you learned about circles and circumference to complete Exercises 9–11 on page 553.

13.1 Lesson

Check It Out
Lesson Tutorials
BigIdeasMath.com

Key Vocabulary
circle, *p. 550*
center, *p. 550*
radius, *p. 550*
diameter, *p. 550*
circumference, *p. 551*
pi, *p. 551*
semicircle, *p. 552*

A **circle** is the set of all points in a plane that are the same distance from a point called the **center**.

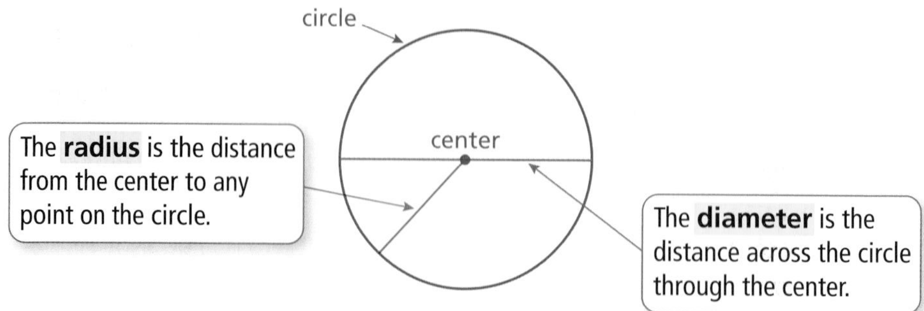
circle
center
The radius is the distance from the center to any point on the circle.
The diameter is the distance across the circle through the center.

Key Idea

Radius and Diameter

Words The diameter d of a circle is twice the radius r. The radius r of a circle is one-half the diameter d.

Algebra **Diameter:** $d = 2r$ **Radius:** $r = \frac{d}{2}$

EXAMPLE 1 Finding a Radius and a Diameter

a. The diameter of a circle is 20 feet. Find the radius.

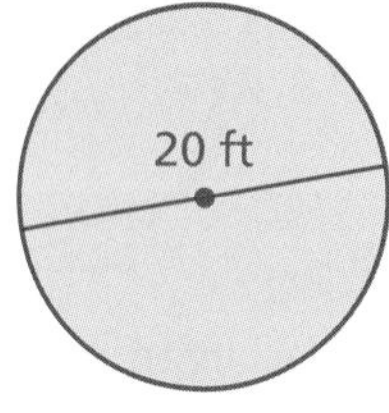
20 ft

$r = \frac{d}{2}$ Radius of a circle

$= \frac{20}{2}$ Substitute 20 for *d*.

$= 10$ Divide.

The radius is 10 feet.

b. The radius of a circle is 7 meters. Find the diameter.

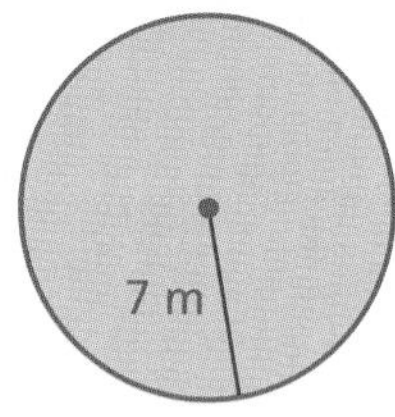
7 m

$d = 2r$ Diameter of a circle

$= 2(7)$ Substitute 7 for *r*.

$= 14$ Multiply.

The diameter is 14 meters.

On Your Own

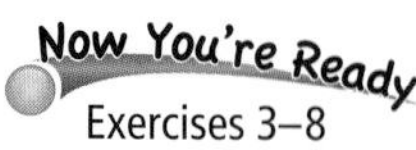
Now You're Ready
Exercises 3–8

1. The diameter of a circle is 16 centimeters. Find the radius.
2. The radius of a circle is 9 yards. Find the diameter.

Multi-Language Glossary at BigIdeasMath.com

The distance around a circle is called the **circumference**. The ratio $\frac{\text{circumference}}{\text{diameter}}$ is the same for *every* circle and is represented by the Greek letter π, called **pi**. The value of π can be approximated as 3.14 or $\frac{22}{7}$.

When the radius or diameter is a multiple of 7, it is easier to use $\frac{22}{7}$ as the estimate of π.

Circumference of a Circle

Words The circumference C of a circle is equal to π times the diameter d or π times twice the radius r.

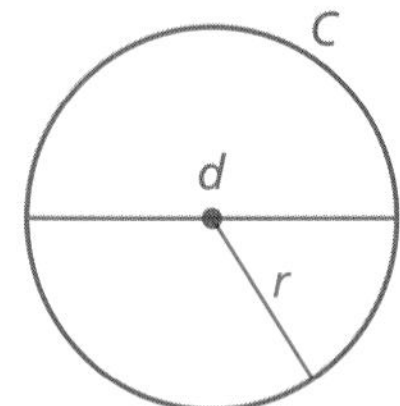

Algebra $C = \pi d$ or $C = 2\pi r$

EXAMPLE

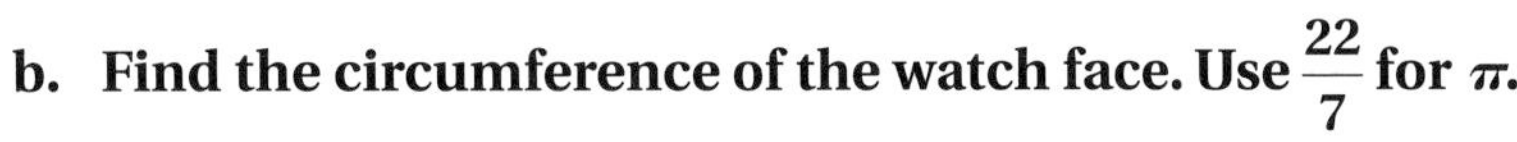

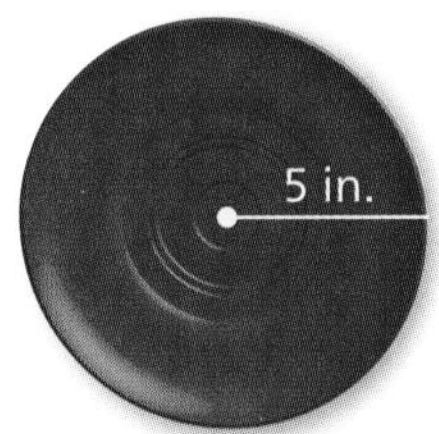

a. Find the circumference of the flying disc. Use 3.14 for π.

$C = 2\pi r$	Write formula for circumference.
$\approx 2 \cdot 3.14 \cdot 5$	Substitute 3.14 for π and 5 for r.
$= 31.4$	Multiply.

The circumference is about 31.4 inches.

b. Find the circumference of the watch face. Use $\frac{22}{7}$ for π.

$C = \pi d$	Write formula for circumference.
$\approx \frac{22}{7} \cdot 28$	Substitute $\frac{22}{7}$ for π and 28 for d.
$= 88$	Multiply.

The circumference is about 88 millimeters.

On Your Own

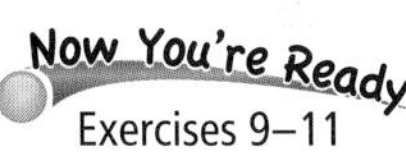

Find the circumference of the object. Use 3.14 or $\frac{22}{7}$ for π.

3. 2 cm

4.

5.

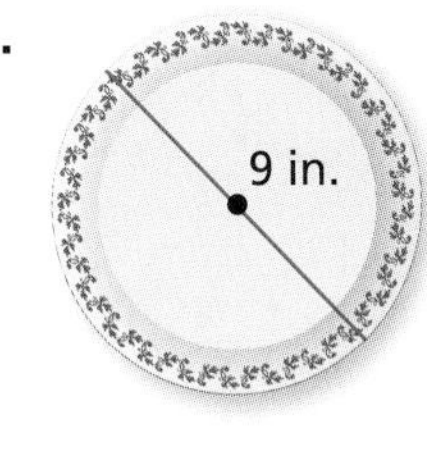

EXAMPLE 3 Estimating a Diameter

$C = 31.4$ in.

The circumference of the roll of caution tape decreases 10.5 inches after a construction worker uses some of the tape. Which is the best estimate of the diameter of the roll after the decrease?

(A) 5 inches (B) 7 inches (C) 10 inches (D) 12 inches

After the decrease, the circumference of the roll is $31.4 - 10.5 = 20.9$ inches.

$C = \pi d$	Write formula for circumference.
$20.9 \approx 3.14 \cdot d$	Substitute 20.9 for C and 3.14 for π.
$21 \approx 3d$	Round 20.9 up to 21. Round 3.14 down to 3.
$7 = d$	Divide each side by 3.

∴ The correct answer is (B).

On Your Own

6. **WHAT IF?** The circumference of the roll of tape decreases 5.25 inches. Estimate the diameter of the roll after the decrease.

EXAMPLE 4 Finding the Perimeter of a Semicircular Region

A semicircle is one-half of a circle. Find the perimeter of the semicircular region.

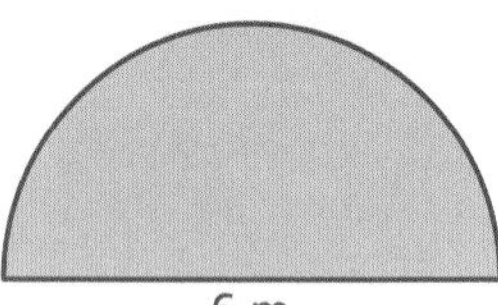

The straight side is 6 meters long. The distance around the curved part is one-half the circumference of a circle with a diameter of 6 meters.

$\frac{C}{2} = \frac{\pi d}{2}$	Divide the circumference by 2.
$\approx \frac{3.14 \cdot 6}{2}$	Substitute 3.14 for π and 6 for d.
$= 9.42$	Simplify.

∴ So, the perimeter is about $6 + 9.42 = 15.42$ meters.

On Your Own

Now You're Ready
Exercises 15 and 16

Find the perimeter of the semicircular region.

7.

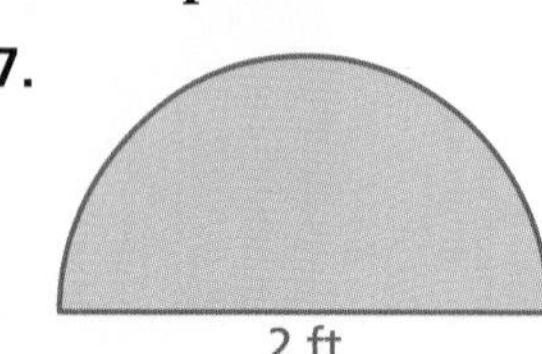

8.

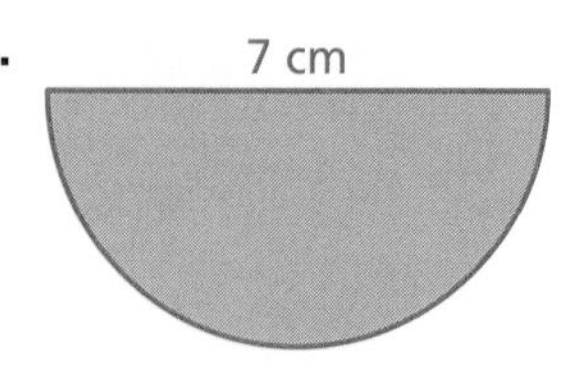

9.

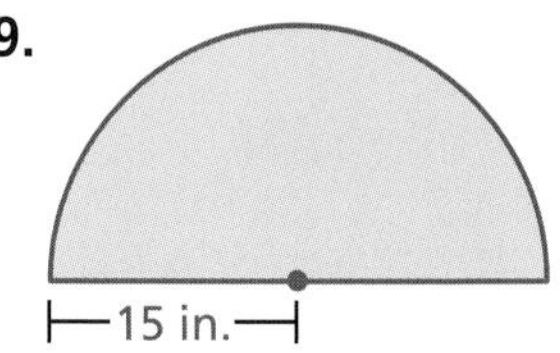

13.1 Exercises

Vocabulary and Concept Check

1. **VOCABULARY** What is the relationship between the radius and the diameter of a circle?

2. **WHICH ONE DOESN'T BELONG?** Which phrase does *not* belong with the other three? Explain your reasoning.

the distance around a circle	π times twice the radius
π times the diameter	the distance from the center to any point on the circle

Practice and Problem Solving

Find the radius of the button.

1 3.

4.

5.

Find the diameter of the object.

6.

7.

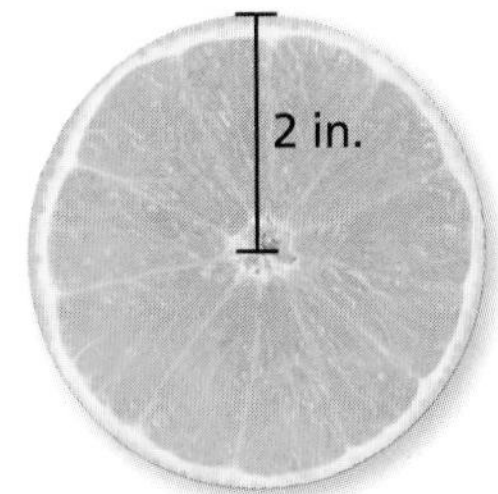

8.

Find the circumference of the pizza. Use 3.14 or $\frac{22}{7}$ for π.

2 9.

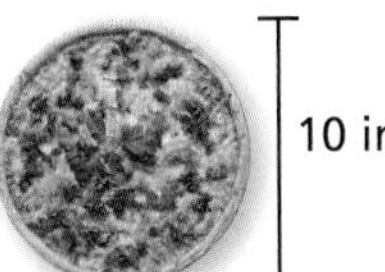

10.

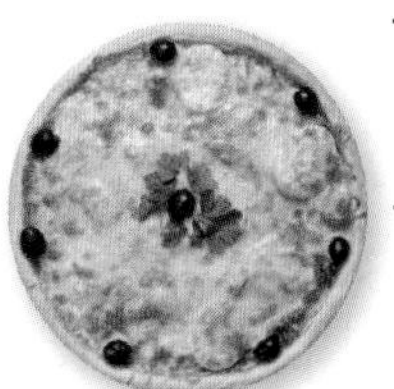

11.

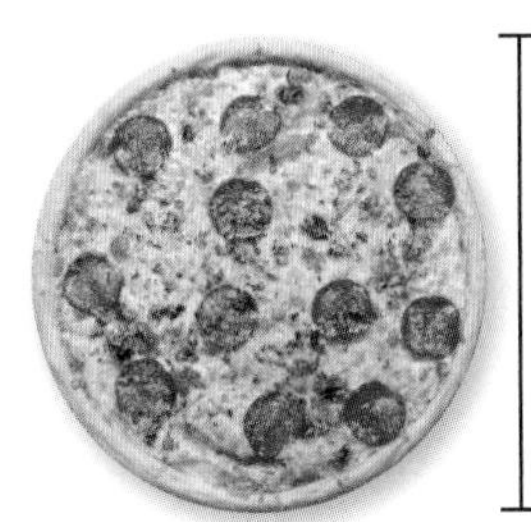

12. **CHOOSE TOOLS** Choose a real-life circular object. Explain why you might need to know its circumference. Then find the circumference.

13. **SINKHOLE** A circular sinkhole has a circumference of 75.36 meters. A week later, it has a circumference of 150.42 meters.

  a. Estimate the diameter of the sinkhole each week.

  b. How many times greater is the diameter of the sinkhole now compared to the previous week?

14. **REASONING** Consider the circles A, B, C, and D.

A 8 ft

B

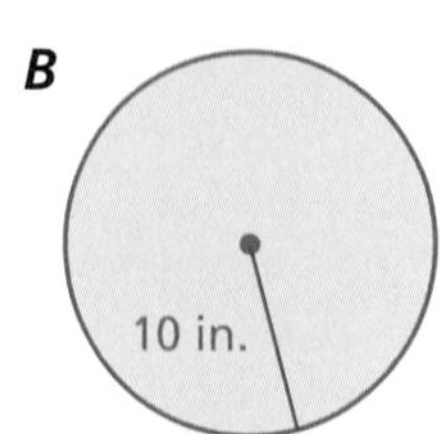

C

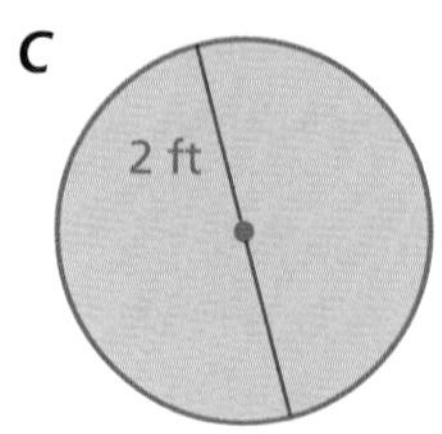

D

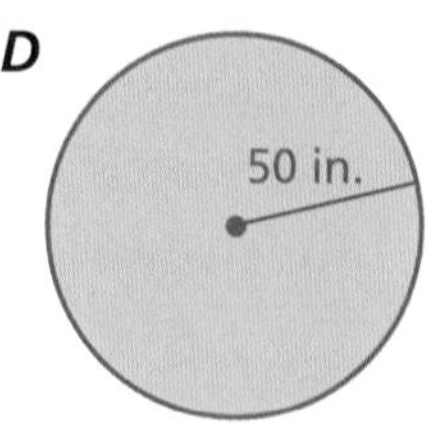

  a. Without calculating, which circle has the greatest circumference?

  b. Without calculating, which circle has the least circumference?

Find the perimeter of the window.

4 15.

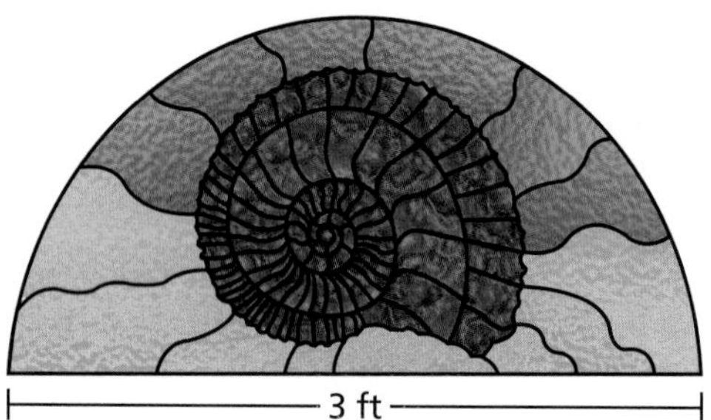

16.

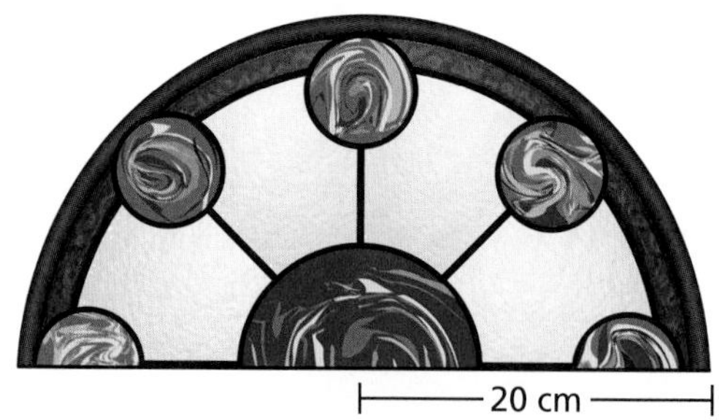

Find the circumferences of both circles.

17.

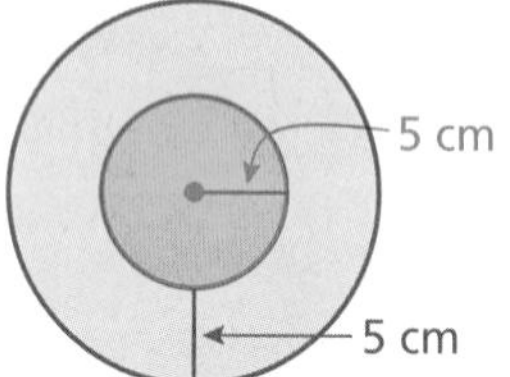

18.

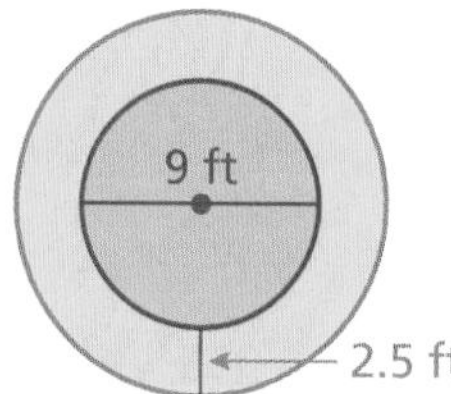

19.

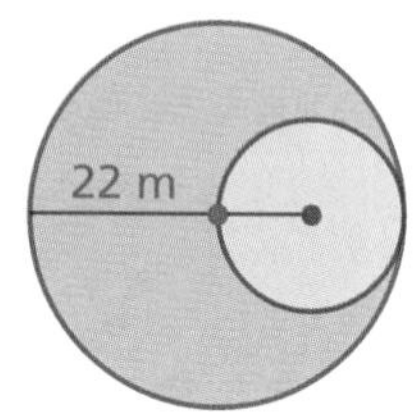

20. **STRUCTURE** Because the ratio $\frac{\text{circumference}}{\text{diameter}}$ is the same for every circle, is the ratio $\frac{\text{circumference}}{\text{radius}}$ the same for every circle? Explain.

21. **WIRE** A wire is bent to form four semicircles. How long is the wire?

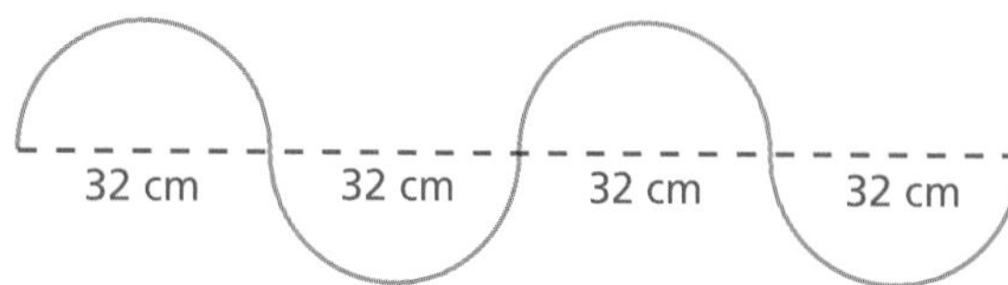

22. **CRITICAL THINKING** Explain how to draw a circle with a circumference of π^2 inches. Then draw the circle.

23. **AROUND THE WORLD** "Lines" of latitude on Earth are actually circles. The Tropic of Cancer is the northernmost line of latitude at which the Sun appears directly overhead at noon. The Tropic of Cancer has a radius of 5854 kilometers.

To qualify for an around-the-world speed record, a pilot must cover a distance no less than the circumference of the Tropic of Cancer, cross all meridians, and land on the same airfield where he started.

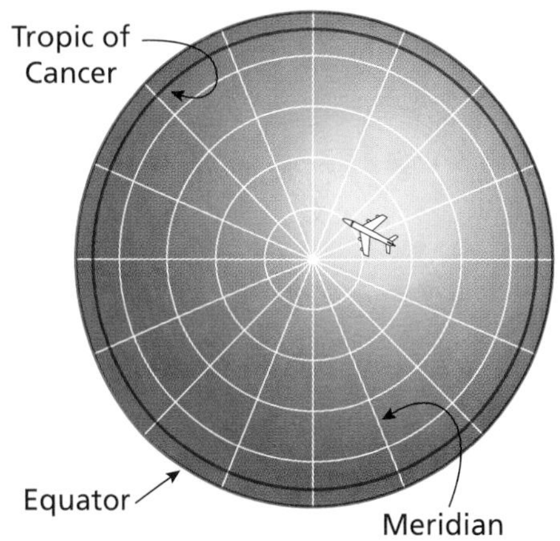

a. What is the minimum distance that a pilot must fly to qualify for an around-the-world speed record?

b. **RESEARCH** Estimate the time it would take for a pilot to qualify for the speed record.

24. **PROBLEM SOLVING** Bicycles in the late 1800s looked very different than they do today.

a. How many rotations does each tire make after traveling 600 feet? Round your answers to the nearest whole number.

b. Would you rather ride a bicycle made with two large wheels or two small wheels? Explain.

25. **Logic** The length of the minute hand is 150% of the length of the hour hand.

a. What distance will the tip of the minute hand move in 45 minutes? Explain how you found your answer.

b. In 1 hour, how much farther does the tip of the minute hand move than the tip of the hour hand? Explain how you found your answer.

Fair Game Review What you learned in previous grades & lessons

Find the perimeter of the polygon. *(Skills Review Handbook)*

26.

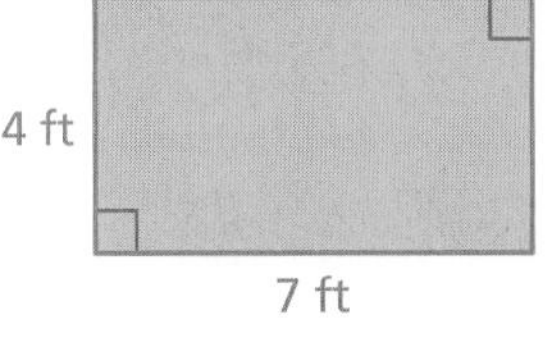

27.

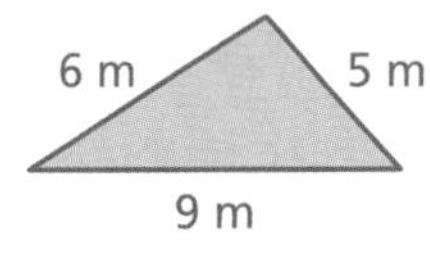

28.

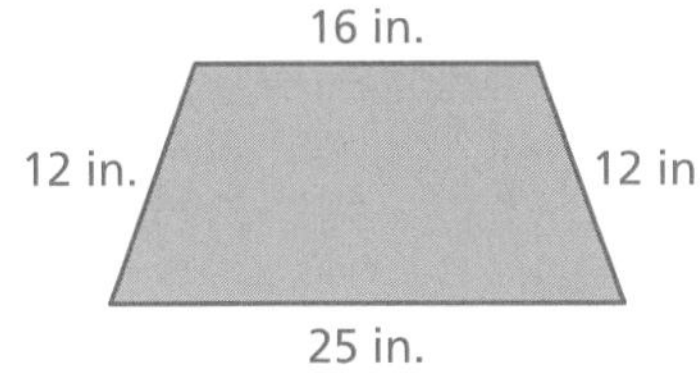

29. **MULTIPLE CHOICE** What is the median of the data set? *(Skills Review Handbook)*

12, 25, 16, 9, 5, 22, 27, 20

Ⓐ 7 Ⓑ 16 Ⓒ 17 Ⓓ 18

13.2 Perimeters of Composite Figures

Essential Question How can you find the perimeter of a composite figure?

1 ACTIVITY: Finding a Pattern

Work with a partner. Describe the pattern of the perimeters. Use your pattern to find the perimeter of the tenth figure in the sequence. (Each small square has a perimeter of 4.)

a.

b.

c.

2 ACTIVITY: Combining Figures

Work with a partner.

a. A rancher is constructing a rectangular corral and a trapezoidal corral, as shown. How much fencing does the rancher need to construct both corrals?

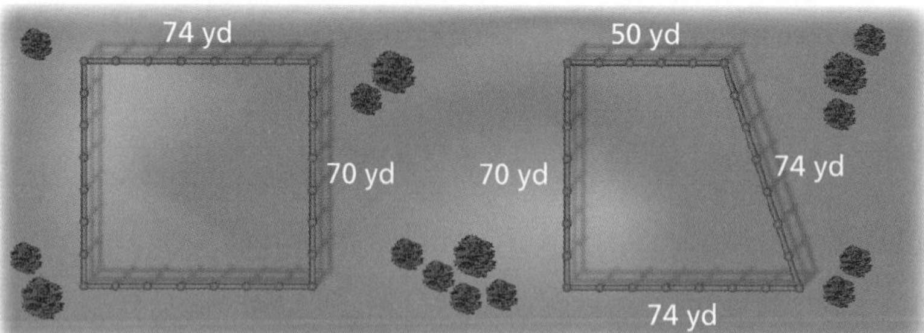

b. Another rancher is constructing one corral by combining the two corrals above, as shown. Does this rancher need more or less fencing? Explain your reasoning.

c. How can the rancher in part (b) combine the two corrals to use even less fencing?

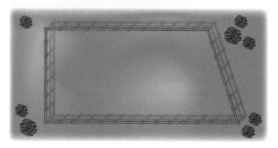

Geometry
In this lesson, you will
- find perimeters of composite figures.

3 ACTIVITY: Submitting a Bid

Work with a partner. You want to bid on a tiling contract. You will be supplying and installing the brown tile that borders the swimming pool. In the figure, each grid square represents 1 square foot.

- **Your cost for the tile is $4 per linear foot.**
- **It takes about 15 minutes to prepare, install, and clean each foot of tile.**

a. How many brown tiles do you need for the border?

b. Write a bid for how much you will charge to supply and install the tile. Include what you want to charge as an hourly wage. Estimate what you think your profit will be.

Math Practice

Communicate Precisely

What do you need to include to create an accurate bid? Explain.

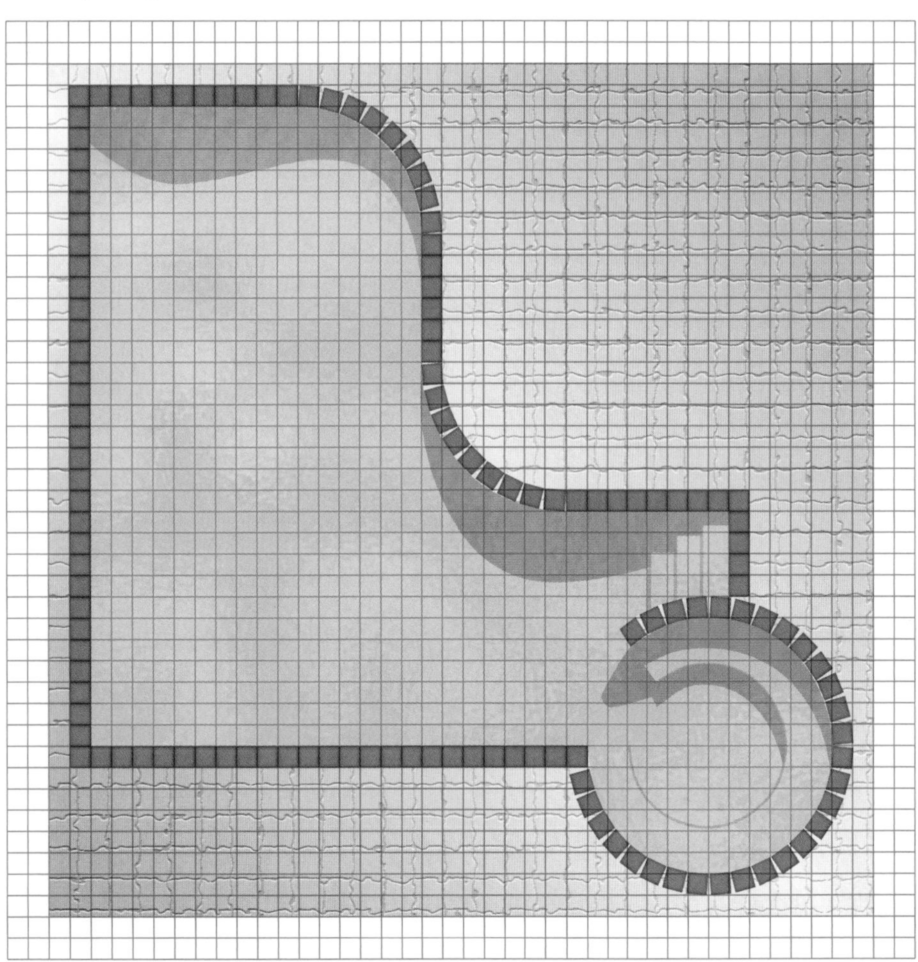

What Is Your Answer?

4. **IN YOUR OWN WORDS** How can you find the perimeter of a composite figure? Use a semicircle, a triangle, and a parallelogram to draw a composite figure. Label the dimensions. Find the perimeter of the figure.

Practice Use what you learned about perimeters of composite figures to complete Exercises 3–5 on page 560.

13.2 Lesson

Check It Out
Lesson Tutorials
BigIdeasMath.com

Key Vocabulary
composite figure, *p. 558*

A **composite figure** is made up of triangles, squares, rectangles, semicircles, and other two-dimensional figures. Here are two examples.

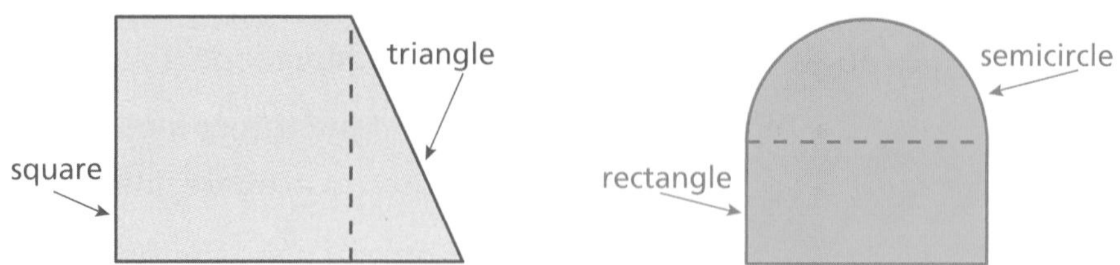
triangle
square
semicircle
rectangle

To find the perimeter of a composite figure, find the distance around the figure.

EXAMPLE 1 Estimating a Perimeter Using Grid Paper

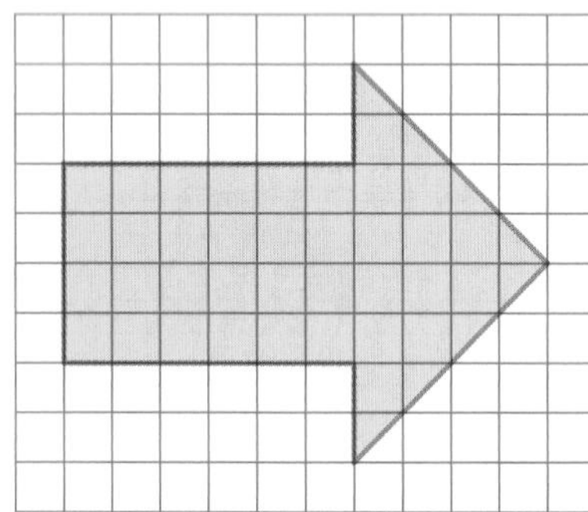

Estimate the perimeter of the arrow.

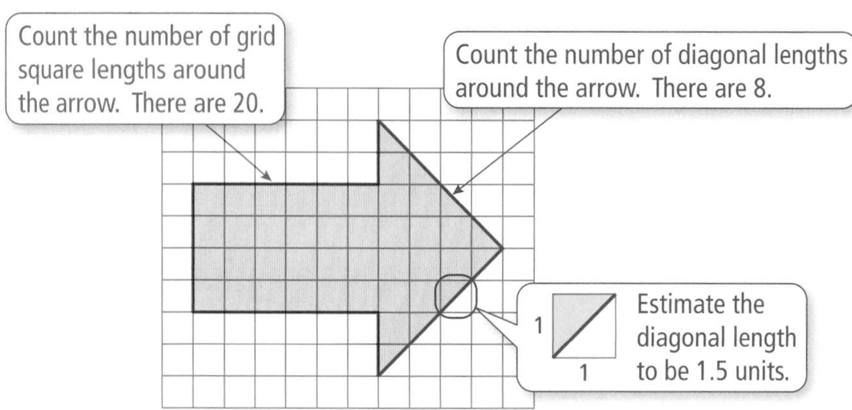
Count the number of grid square lengths around the arrow. There are 20.
Count the number of diagonal lengths around the arrow. There are 8.
1
1
Estimate the diagonal length to be 1.5 units.

Length of 20 grid square lengths: $20 \times 1 = 20$ units

Length of 8 diagonal lengths: $8 \times 1.5 = 12$ units

So, the perimeter is about $20 + 12 = 32$ units.

On Your Own

Now You're Ready
Exercises 3–8

Estimate the perimeter of the figure.

1. 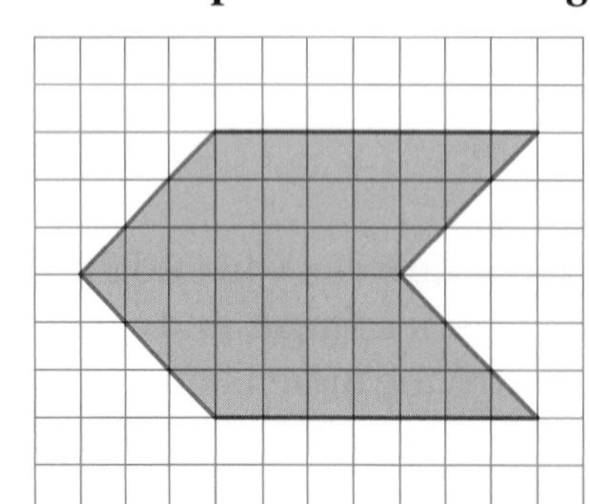

2. 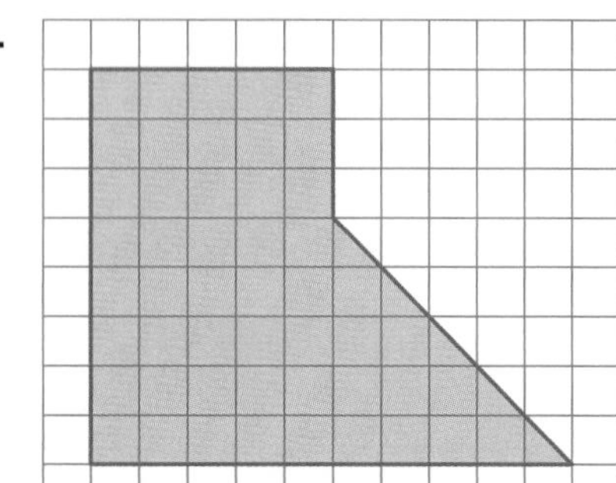

EXAMPLE 2 Finding a Perimeter

The figure is made up of a semicircle and a triangle. Find the perimeter.

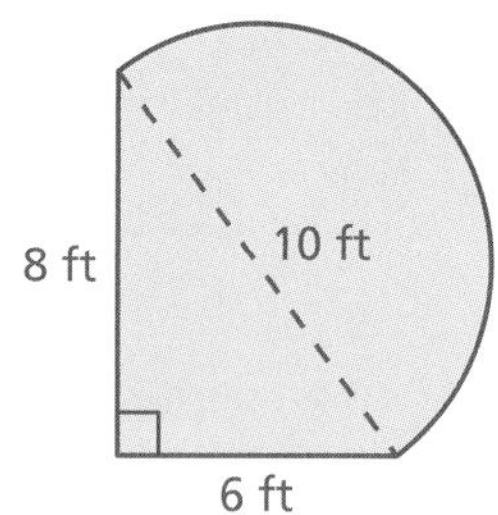

The distance around the triangular part of the figure is $6 + 8 = 14$ feet.

The distance around the semicircle is one-half the circumference of a circle with a diameter of 10 feet.

$$\frac{C}{2} = \frac{\pi d}{2}$$ Divide the circumference by 2.

$$\approx \frac{3.14 \cdot 10}{2}$$ Substitute 3.14 for π and 10 for d.

$$= 15.7$$ Simplify.

So, the perimeter is about $14 + 15.7 = 29.7$ feet.

EXAMPLE 3 Finding a Perimeter

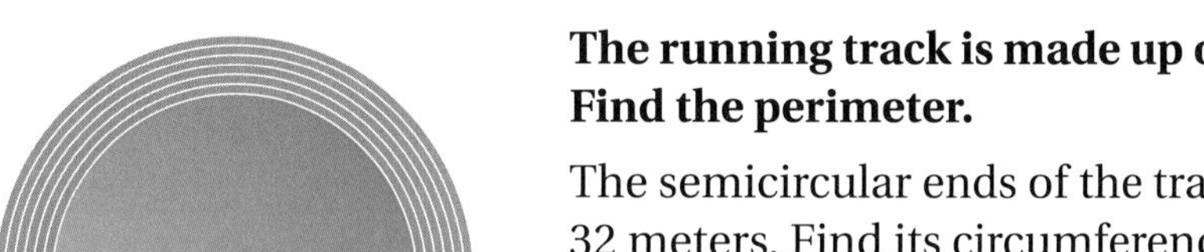

The running track is made up of a rectangle and two semicircles. Find the perimeter.

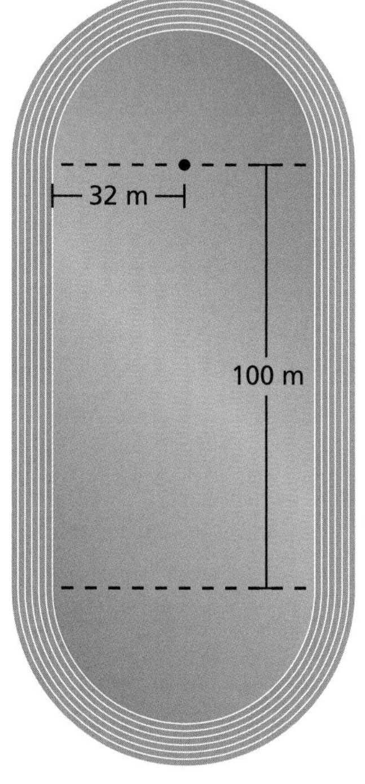

The semicircular ends of the track form a circle with a radius of 32 meters. Find its circumference.

$$C = 2\pi r$$ Write formula for circumference.

$$\approx 2 \cdot 3.14 \cdot 32$$ Substitute 3.14 for π and 32 for r.

$$= 200.96$$ Multiply.

So, the perimeter is about $100 + 100 + 200.96 = 400.96$ meters.

Now You're Ready
Exercises 9–11

On Your Own

3. The figure is made up of a semicircle and a triangle. Find the perimeter.

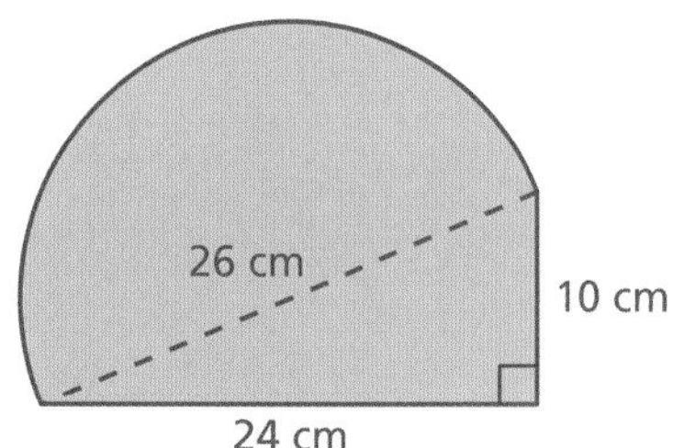

4. The figure is made up of a square and two semicircles. Find the perimeter.

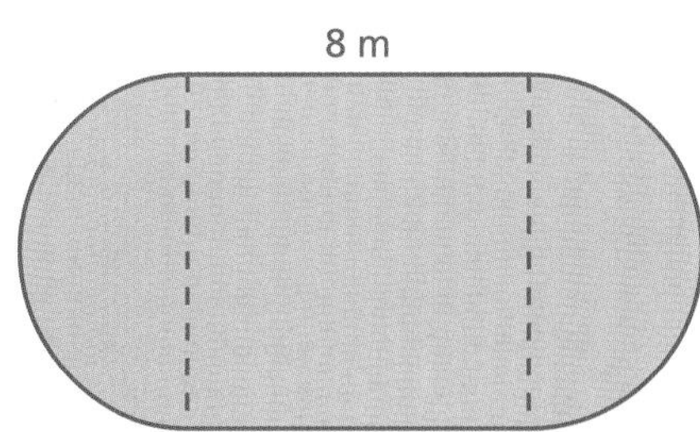

13.2 Exercises

Vocabulary and Concept Check

1. **REASONING** Is the perimeter of the composite figure equal to the sum of the perimeters of the individual figures? Explain.

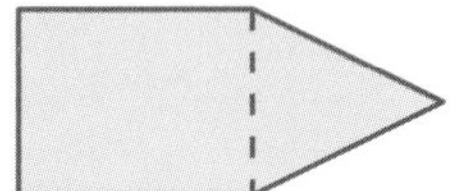

2. **OPEN-ENDED** Draw a composite figure formed by a parallelogram and a trapezoid.

Practice and Problem Solving

Estimate the perimeter of the figure.

3. 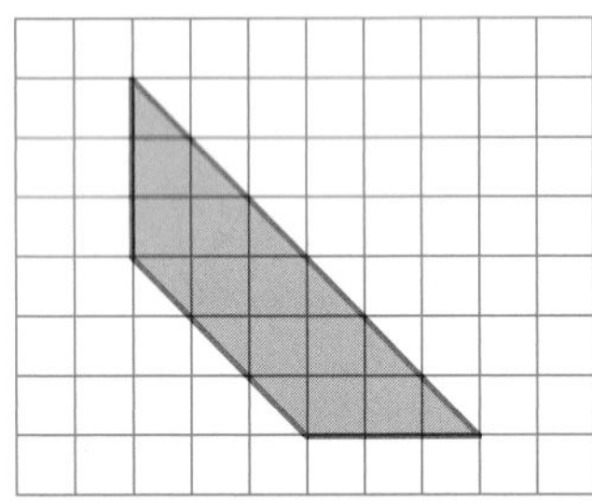

4. 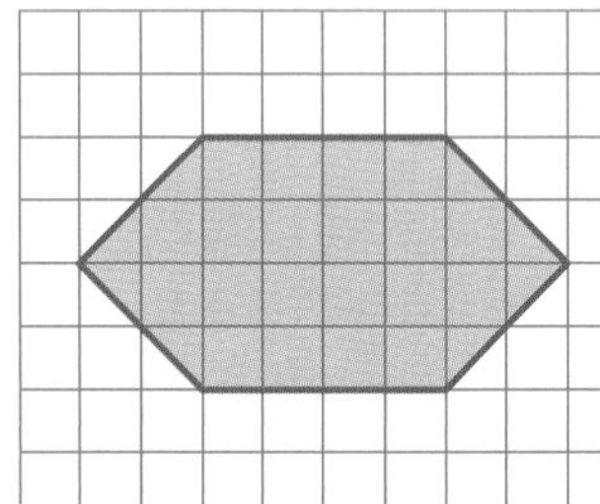

5. 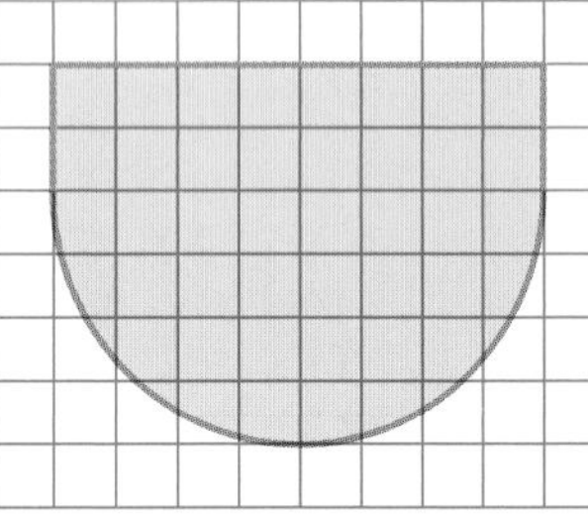

6. 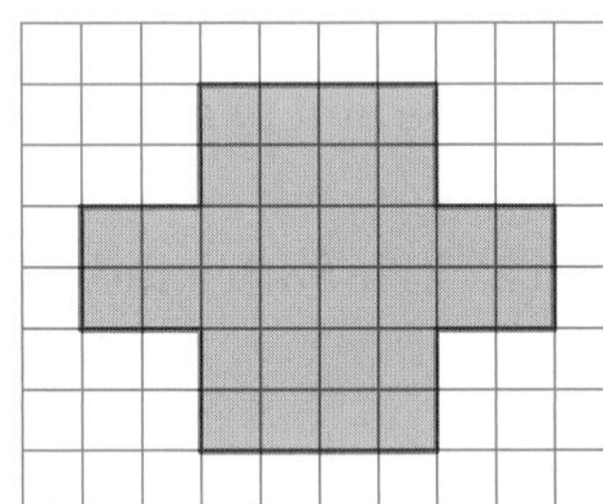

7. 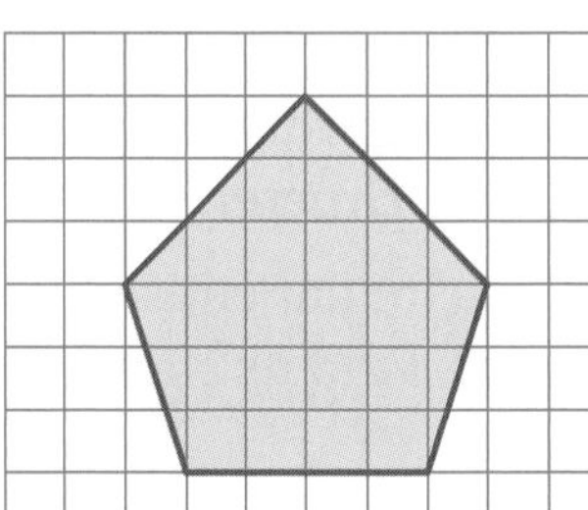

8. 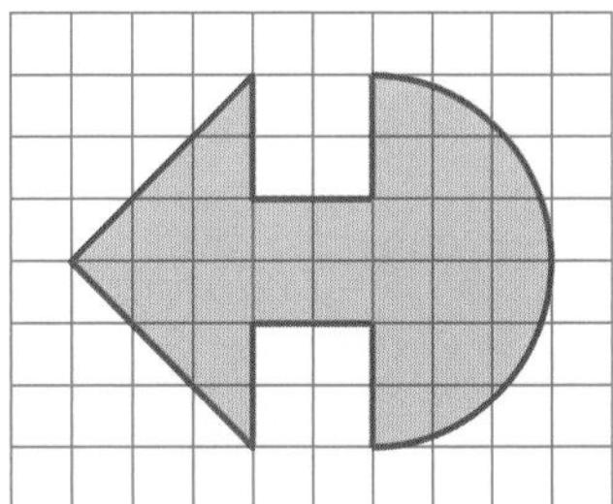

Find the perimeter of the figure.

2

9. 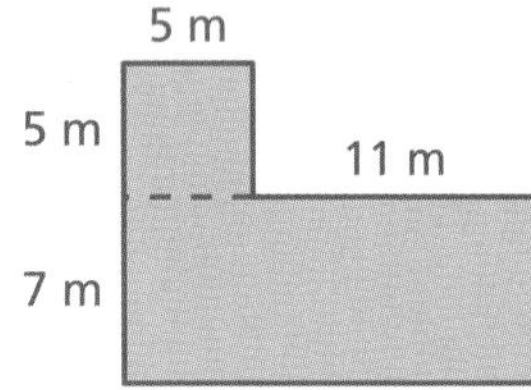

10. 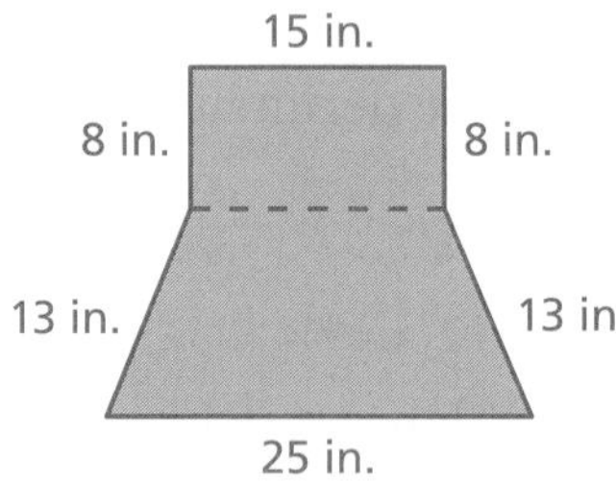

11. 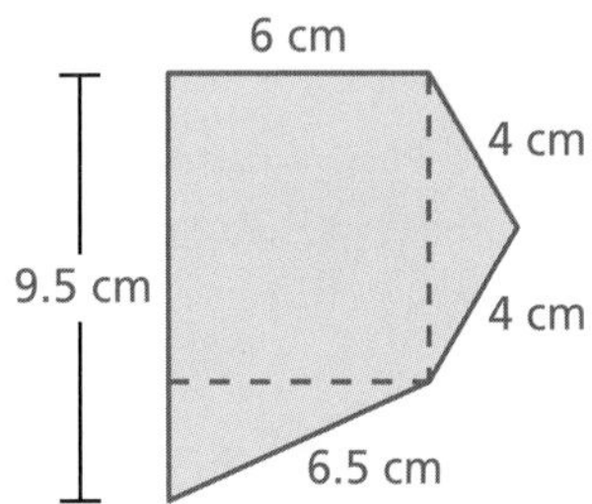

12. **ERROR ANALYSIS** Describe and correct the error in finding the perimeter of the figure.

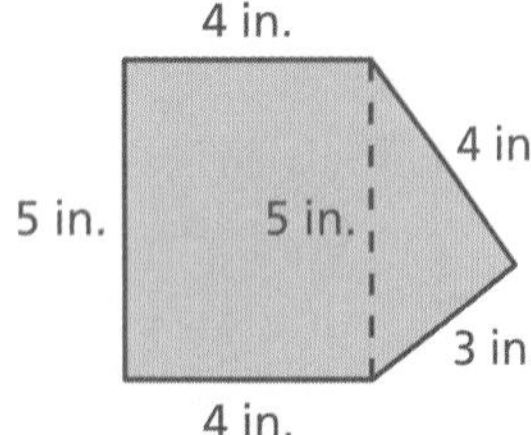

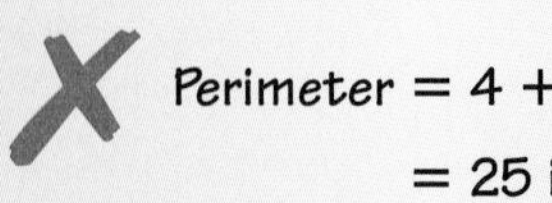

Find the perimeter of the figure.

13.

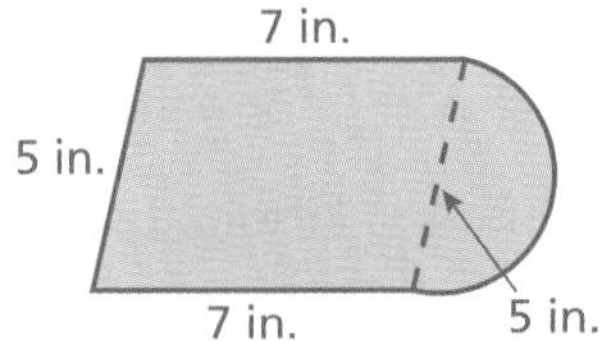

14.

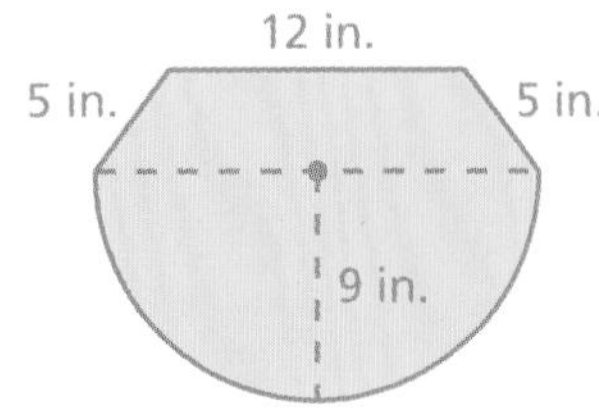

15.

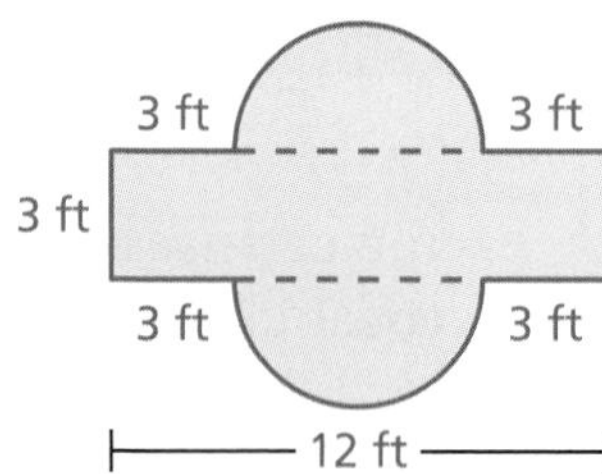

240 ft 285 ft

450 ft 450 ft

450 ft

16. PASTURE A farmer wants to fence a section of land for a horse pasture. Fencing costs $27 per yard. How much will it cost to fence the pasture?

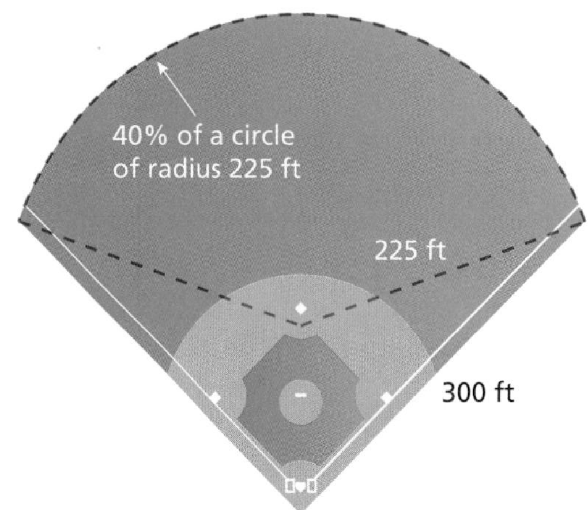

17. BASEBALL You run around the perimeter of the baseball field at a rate of 9 feet per second. How long does it take you to run around the baseball field?

18. TRACK In Example 3, the running track has six lanes. Explain why the starting points for the six runners are staggered. Draw a diagram as part of your explanation.

19. Critical Thinking How can you add a figure to a composite figure without increasing its perimeter? Draw a diagram to support your answer.

Fair Game Review What you learned in previous grades & lessons

Evaluate the expression. *(Skills Review Handbook)*

20. $2.15(3)^2$ **21.** $4.37(8)^2$ **22.** $3.14(7)^2$ **23.** $8.2(5)^2$

24. MULTIPLE CHOICE Which expression is equivalent to $(5y + 4) - 2(7 - 2y)$? *(Skills Review Handbook)*

Ⓐ $y - 10$ Ⓑ $9y + 18$ Ⓒ $3y - 10$ Ⓓ $9y - 10$

13 Study Help

You can use a **word magnet** to organize formulas or phrases that are associated with a vocabulary word or term. Here is an example of a word magnet for circle.

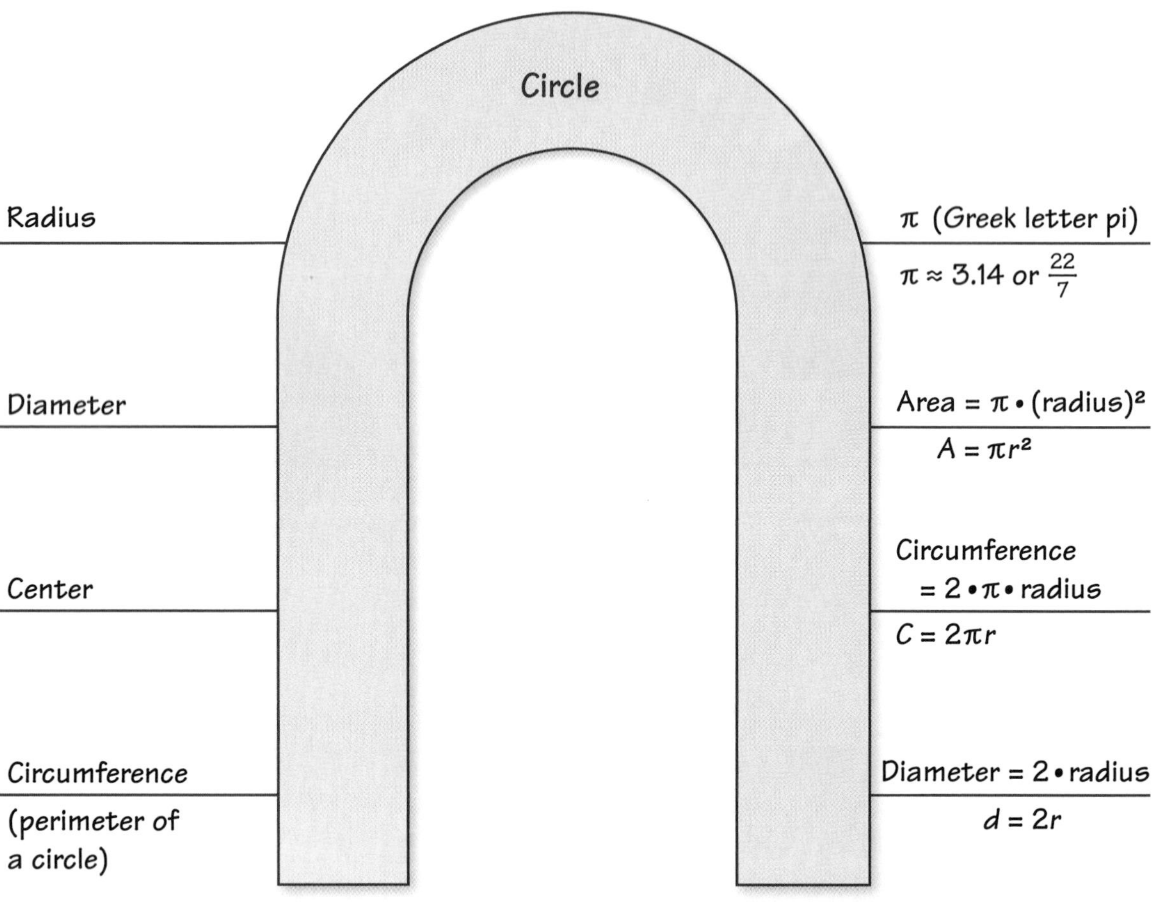

On Your Own

Make word magnets to help you study these topics.

1. semicircle
2. composite figure
3. perimeter

After you complete this chapter, make word magnets for the following topics.

4. area of a circle
5. area of a composite figure

"I'm trying to make a word magnet for happiness, but I can only think of two words."

13.1–13.2 Quiz

1. The diameter of a circle is 36 centimeters. Find the radius. *(Section 13.1)*

2. The radius of a circle is 11 inches. Find the diameter. *(Section 13.1)*

Estimate the perimeter of the figure. *(Section 13.2)*

3. 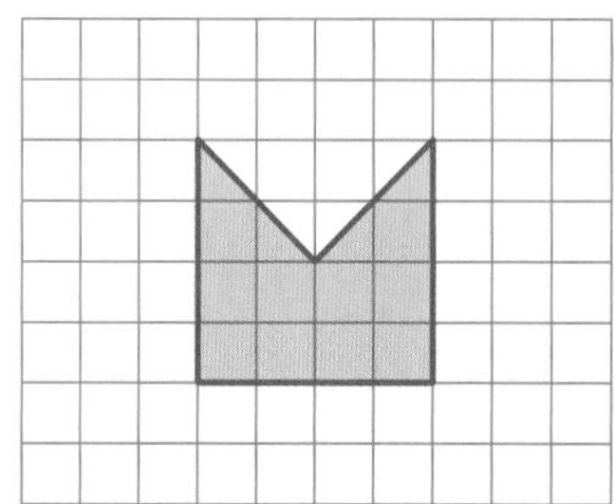

4. 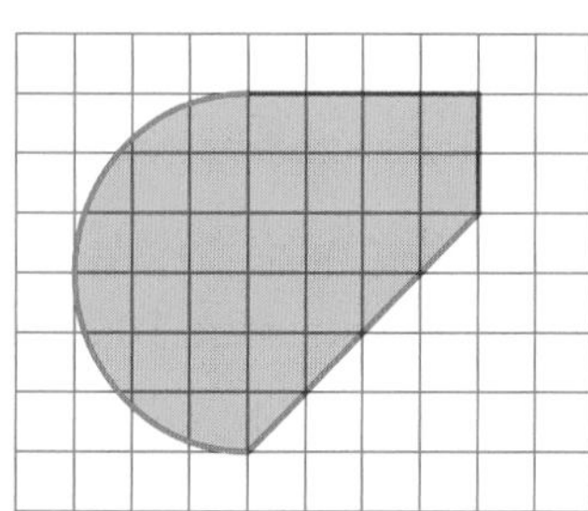

5. 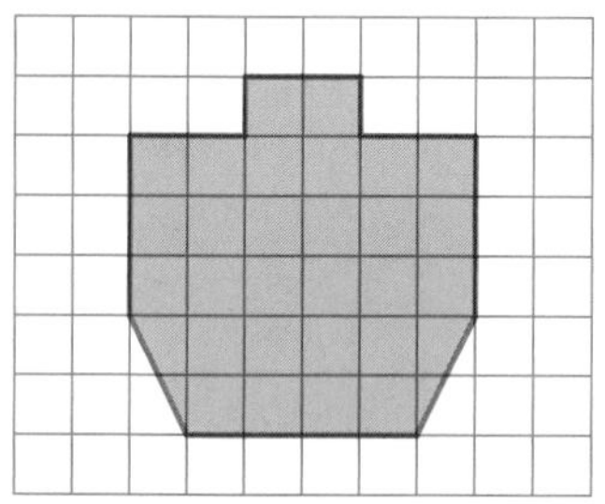

Find the circumference of the circle. Use 3.14 or $\frac{22}{7}$ for π. *(Section 13.1)*

6. 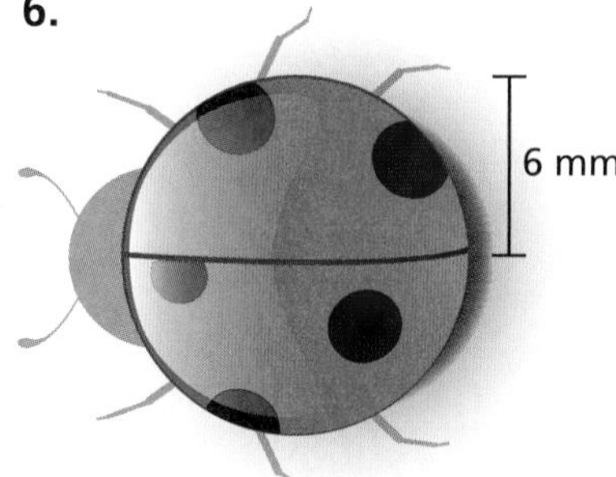

7. 

8. 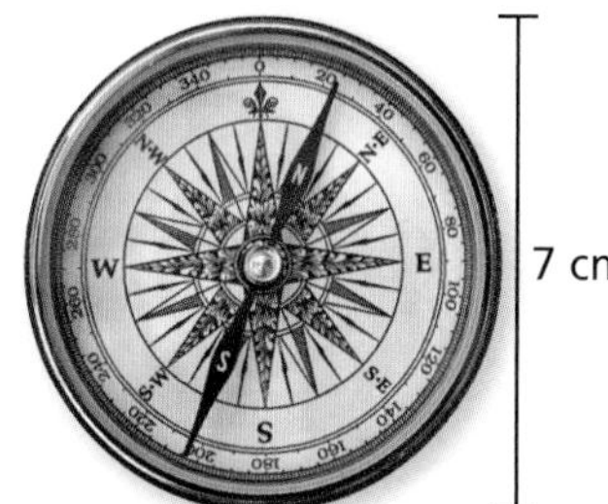

Find the perimeter of the figure. *(Section 13.1 and Section 13.2)*

9. 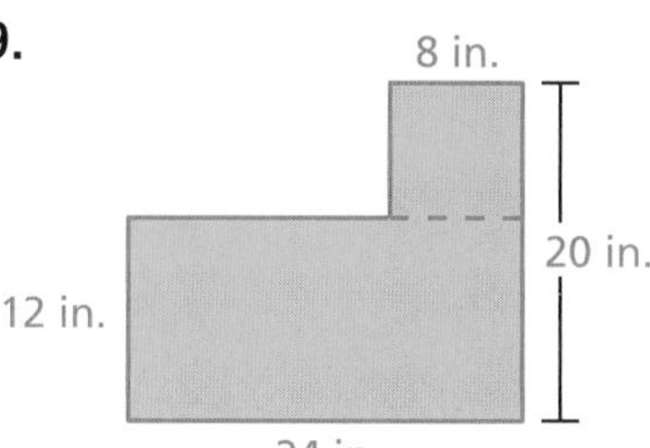

10. 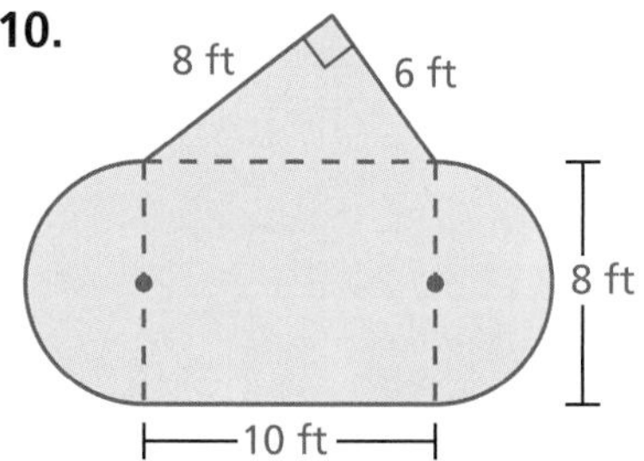

11. 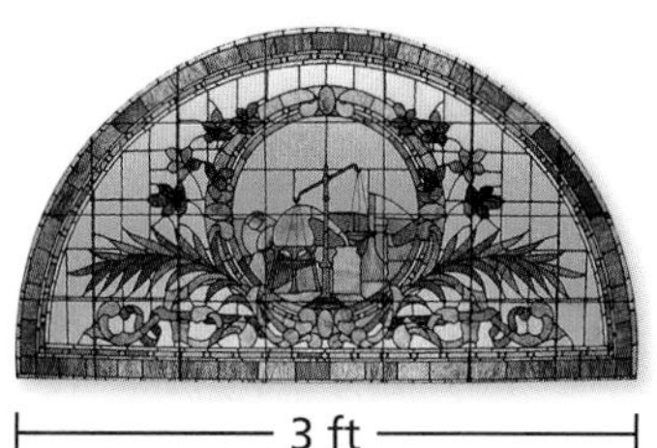

12. **BUTTON** What is the circumference of a circular button with a diameter of 8 millimeters? *(Section 13.1)*

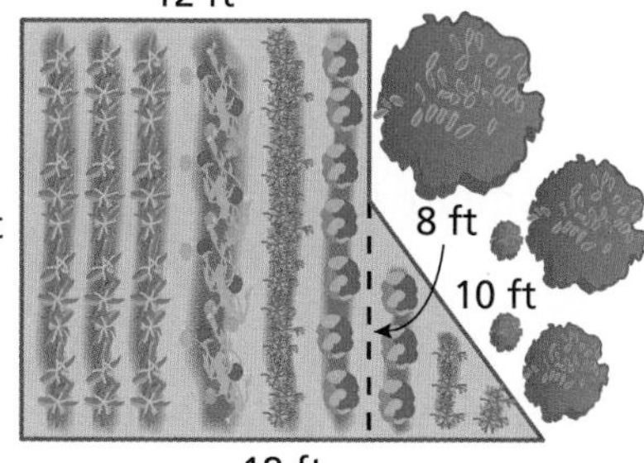

13. **GARDEN** You want to fence part of a yard to make a vegetable garden. How many feet of fencing do you need to surround the garden? *(Section 13.2)*

14. **BAKING** A baker is using two circular pans. The larger pan has a diameter of 12 inches. The smaller pan has a diameter of 7 inches. How much greater is the circumference of the larger pan than that of the smaller pan? *(Section 13.1)*

13.3 Areas of Circles

Essential Question How can you find the area of a circle?

1 ACTIVITY: Estimating the Area of a Circle

Work with a partner. Each square in the grid is 1 unit by 1 unit.

a. Find the area of the large 10-by-10 square.

b. Copy and complete the table.

Region	(square)	(triangle)	(square and triangle)
Area (square units)			

c. Use your results to estimate the area of the circle. Explain your reasoning.

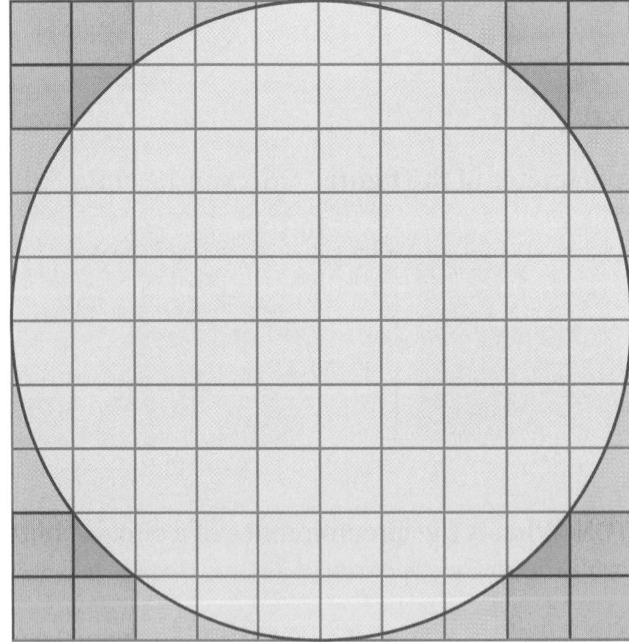

d. Fill in the blanks. Explain your reasoning.

Area of large square = ▭ • 5^2 square units

Area of circle ≈ ▭ • 5^2 square units

e. What dimension of the circle does 5 represent? What can you conclude?

Geometry

In this lesson, you will

- find areas of circles and semicircles.

2 ACTIVITY: Approximating the Area of a Circle

Work with a partner.

a. Draw a circle. Label the radius as r.

b. Divide the circle into 24 equal sections.

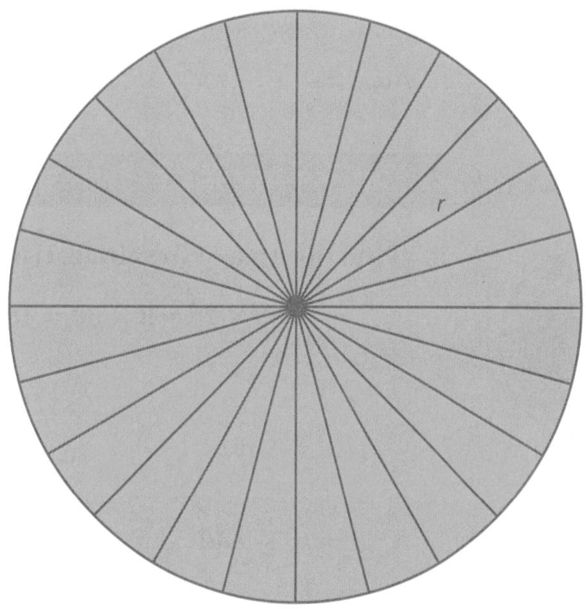

c. Cut the sections apart. Then arrange them to approximate a parallelogram.

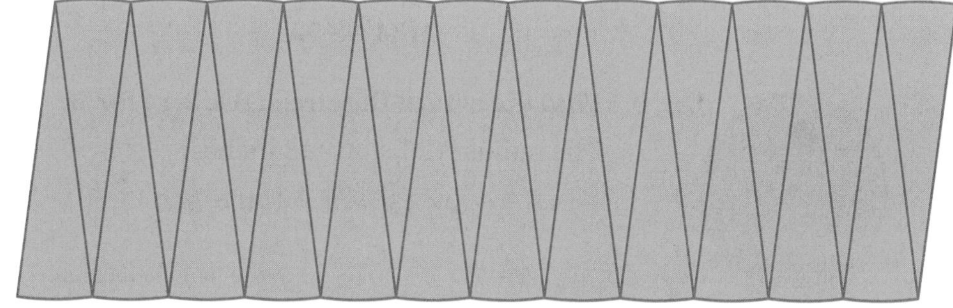

d. What is the approximate height and base of the parallelogram?

e. Find the area of the parallelogram. What can you conclude?

Math Practice

Interpret a Solution

What does the area of the parallelogram represent? Explain.

What Is Your Answer?

3. **IN YOUR OWN WORDS** How can you find the area of a circle?

4. Write a formula for the area of a circle with radius r. Find an object that is circular. Use your formula to find the area.

Practice Use what you learned about areas of circles to complete Exercises 3–5 on page 568.

13.3 Lesson

Area of a Circle

Words The area A of a circle is the product of π and the square of the radius.

Algebra $A = \pi r^2$

EXAMPLE 1 Finding Areas of Circles

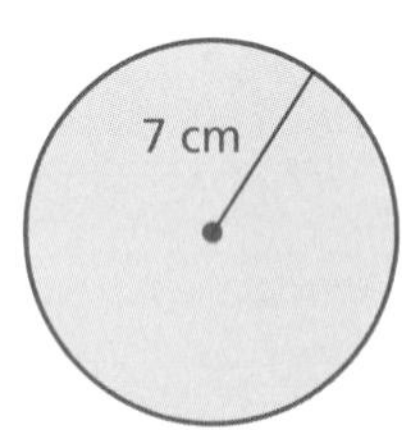

a. Find the area of the circle. Use $\frac{22}{7}$ for π.

Estimate $3 \times 7^2 \approx 3 \times 50 = 150$

$A = \pi r^2$ Write formula for area.

$\approx \frac{22}{7} \cdot 7^2$ Substitute $\frac{22}{7}$ for π and 7 for r.

$= \frac{22}{\cancel{7}_1} \cdot \cancel{49}^{7}$ Evaluate 7^2. Divide out the common factor.

$= 154$ Multiply.

The area is about 154 square centimeters.

Reasonable? $154 \approx 150$ ✓

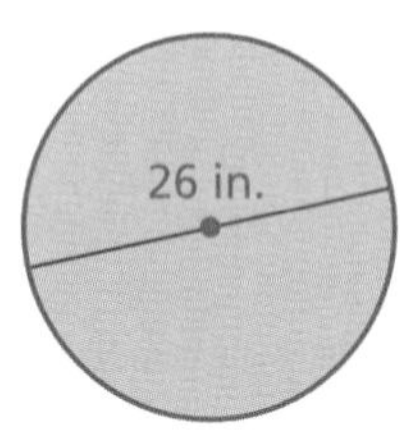

b. Find the area of the circle. Use 3.14 for π.

The radius is $26 \div 2 = 13$ inches.

Estimate $3 \times 13^2 \approx 3 \times 170 = 510$

$A = \pi r^2$ Write formula for area.

$\approx 3.14 \cdot 13^2$ Substitute 3.14 for π and 13 for r.

$= 3.14 \cdot 169$ Evaluate 13^2.

$= 530.66$ Multiply.

The area is about 530.66 square inches.

Reasonable? $530.66 \approx 510$ ✓

On Your Own

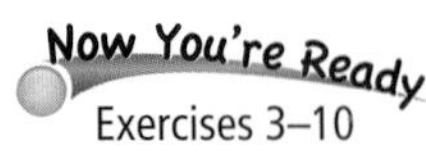

1. Find the area of a circle with a radius of 6 feet. Use 3.14 for π.
2. Find the area of a circle with a diameter of 28 meters. Use $\frac{22}{7}$ for π.

EXAMPLE 2 Describing a Distance

You want to find the distance the monster truck travels when the tires make one 360-degree rotation. Which best describes this distance?

Ⓐ the radius of the tire

Ⓑ the diameter of the tire

Ⓒ the circumference of the tire

Ⓓ the area of the tire

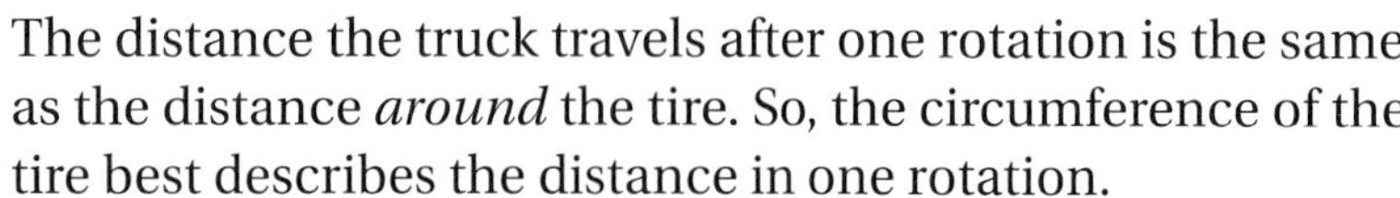

The distance the truck travels after one rotation is the same as the distance *around* the tire. So, the circumference of the tire best describes the distance in one rotation.

The correct answer is Ⓒ.

On Your Own

3. You want to find the height of one of the tires. Which measurement would best describe the height?

EXAMPLE 3 Finding the Area of a Semicircle

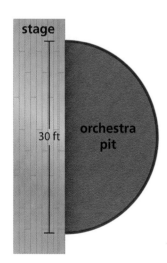

Find the area of the semicircular orchestra pit.

The area of the orchestra pit is one-half the area of a circle with a diameter of 30 feet.

The radius of the circle is 30 ÷ 2 = 15 feet.

$\frac{A}{2} = \frac{\pi r^2}{2}$	Divide the area by 2.
$\approx \frac{3.14 \cdot 15^2}{2}$	Substitute 3.14 for π and 15 for r.
$= \frac{3.14 \cdot 225}{2}$	Evaluate 15^2.
$= 353.25$	Simplify.

stage

30 ft

orchestra pit

So, the area of the orchestra pit is about 353.25 square feet.

On Your Own

Now You're Ready
Exercises 13–15

Find the area of the semicircle.

4. 8 m

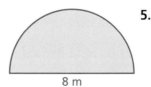

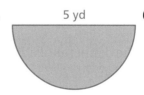

5. 5 yd

6. 11 cm

13.3 Exercises

Vocabulary and Concept Check

1. **VOCABULARY** Explain how to find the area of a circle given its diameter.
2. **DIFFERENT WORDS, SAME QUESTION** Which is different? Find "both" answers.

What is the area of a circle with a diameter of 1 m?	What is the area of a circle with a diameter of 100 cm?
What is the area of a circle with a radius of 100 cm?	What is the area of a circle with a radius of 500 mm?

Practice and Problem Solving

Find the area of the circle. Use 3.14 or $\frac{22}{7}$ for π.

1 3.

4.

5.

6.

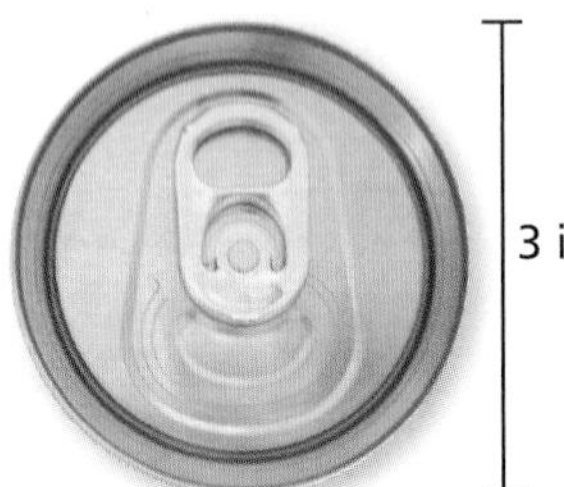

7.

8.

9. Find the area of a circle with a diameter of 56 millimeters.
10. Find the area of a circle with a radius of 5 feet.
11. **TORTILLA** The diameter of a flour tortilla is 12 inches. What is the area?
12. **LIGHTHOUSE** The Hillsboro Inlet Lighthouse lights up how much more area than the Jupiter Inlet Lighthouse?

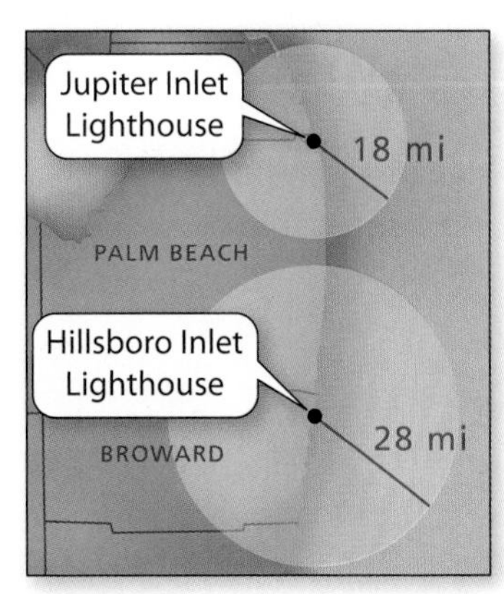

Find the area of the semicircle.

③ **13.**

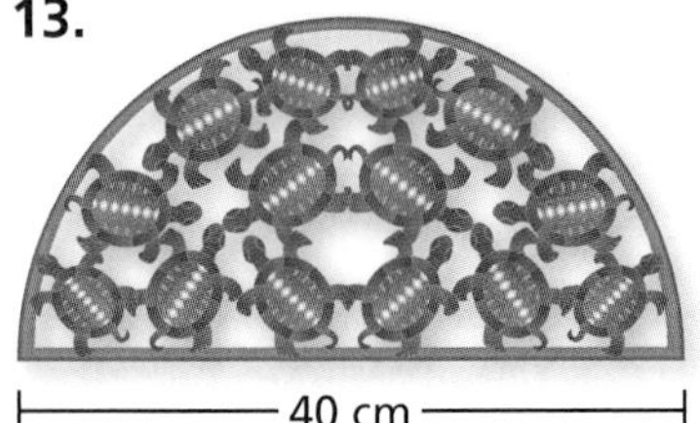

14.

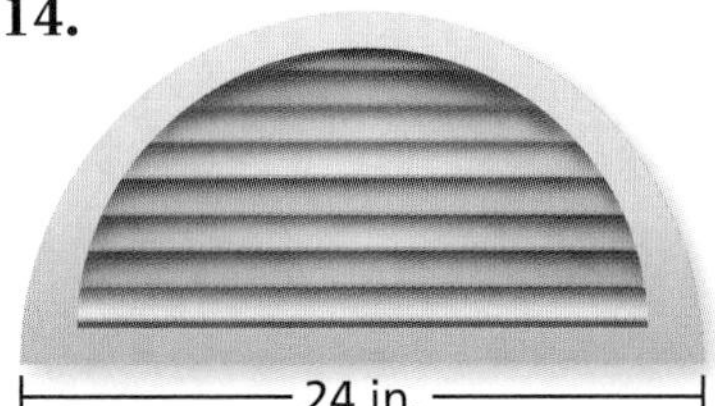

15.

16. REPEATED REASONING Consider five circles with radii of 1, 2, 4, 8, and 16 inches.

a. Copy and complete the table. Write your answers in terms of π.

b. Compare the areas and circumferences. What happens to the circumference of a circle when you double the radius? What happens to the area?

c. What happens when you triple the radius?

Radius	Circumference	Area
1	2π in.	π in.^{2}
2		
4		
8		
16		

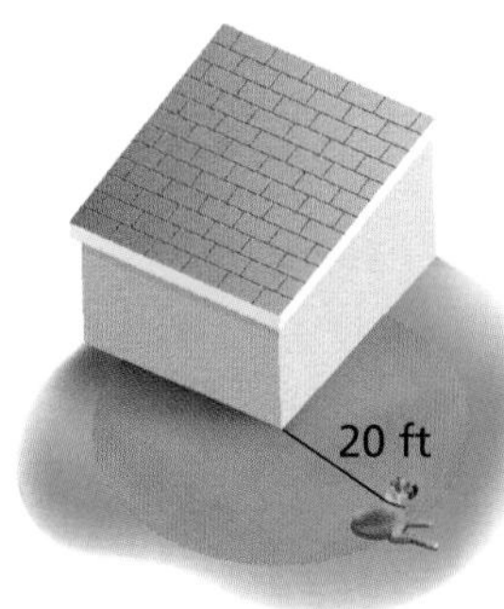

17. DOG A dog is leashed to the corner of a house. How much running area does the dog have? Explain how you found your answer.

18. CRITICAL THINKING Is the area of a semicircle with a diameter of x *greater than*, *less than*, or *equal to* the area of a circle with a diameter of $\frac{1}{2}x$? Explain.

Find the area of the shaded region. Explain how you found your answer.

19.

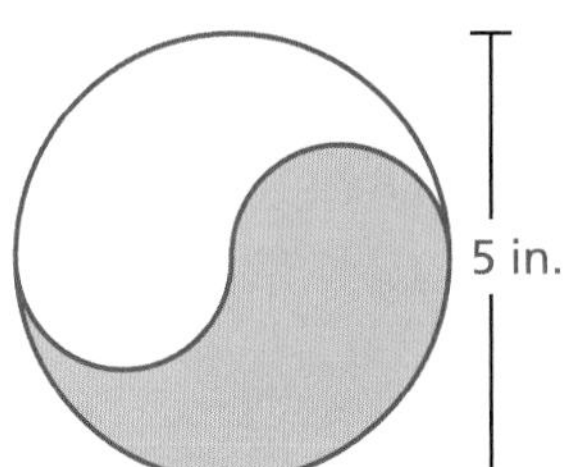

20.

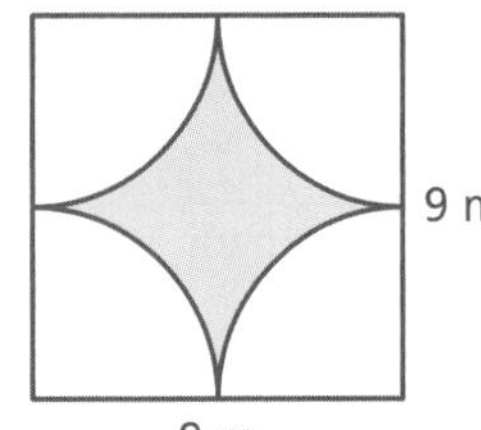

21.

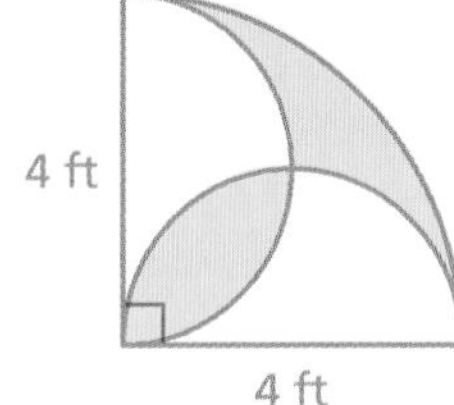

Fair Game Review What you learned in previous grades & lessons

Evaluate the expression. *(Skills Review Handbook)*

22. $\frac{1}{2}(7)(4) + 6(5)$

23. $\frac{1}{2} \cdot 8^2 + 3(7)$

24. $12(6) + \frac{1}{4} \cdot 2^2$

25. MULTIPLE CHOICE What is the product of $-8\frac{1}{3}$ and $3\frac{2}{5}$? *(Skills Review Handbook)*

Ⓐ $-28\frac{1}{3}$ Ⓑ $-24\frac{2}{15}$ Ⓒ $24\frac{2}{15}$ Ⓓ $28\frac{1}{3}$

13.4 Areas of Composite Figures

Essential Question How can you find the area of a composite figure?

1 ACTIVITY: Estimating Area

Work with a partner.

a. Choose a state. On grid paper, draw a larger outline of the state.

b. Use your drawing to estimate the area (in square miles) of the state.

c. Which state areas are easy to find? Which are difficult? Why?

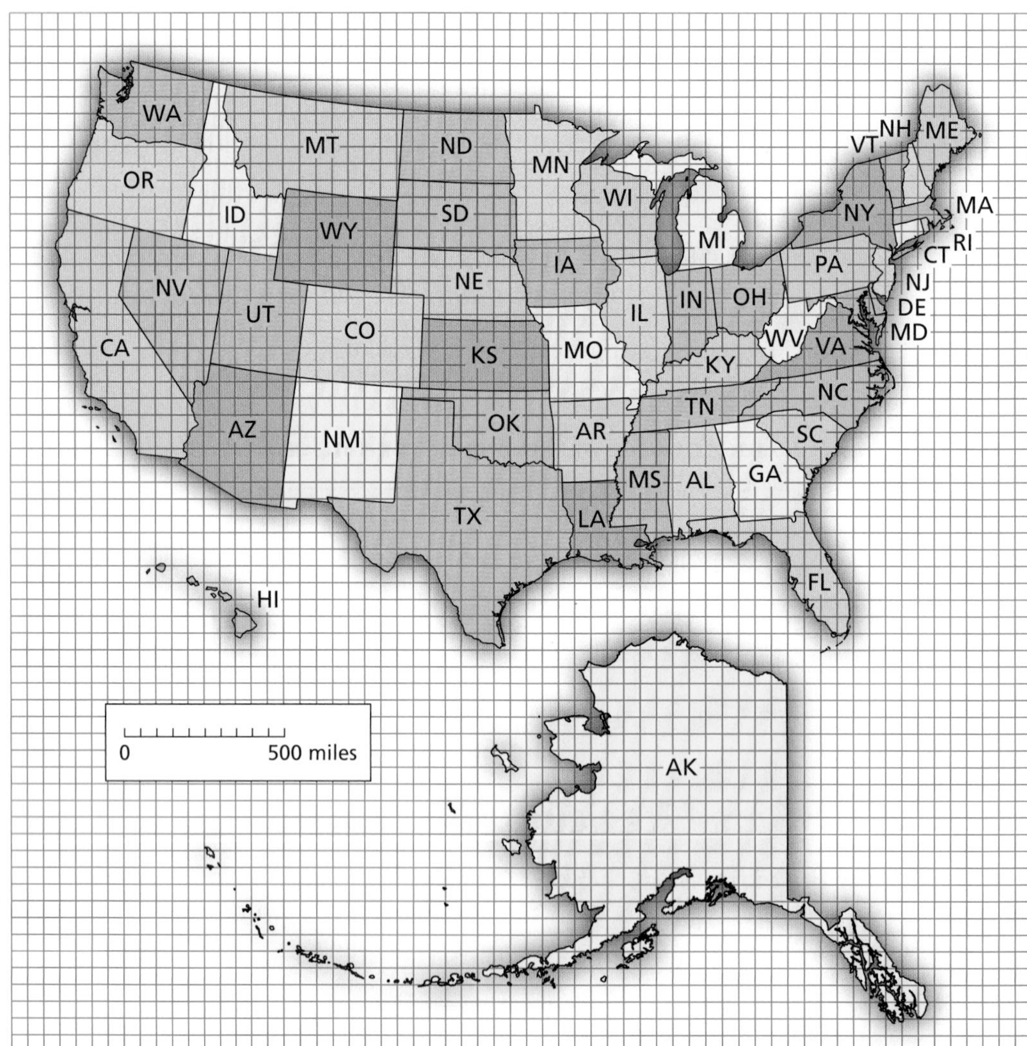

Geometry

In this lesson, you will

- find areas of composite figures by separating them into familiar figures.
- solve real-life problems.

2 ACTIVITY: Estimating Areas

Work with a partner. The completed puzzle has an area of 150 square centimeters.

a. Estimate the area of each puzzle piece.

b. Check your work by adding the six areas. Why is this a check?

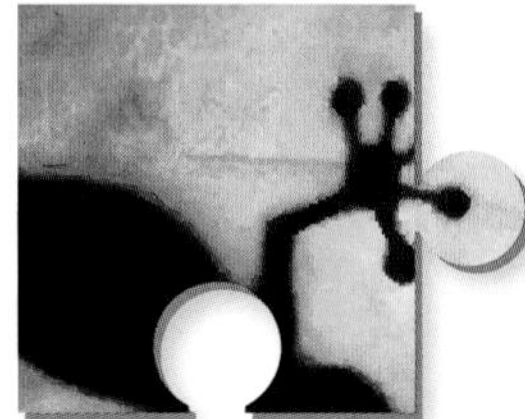

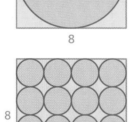

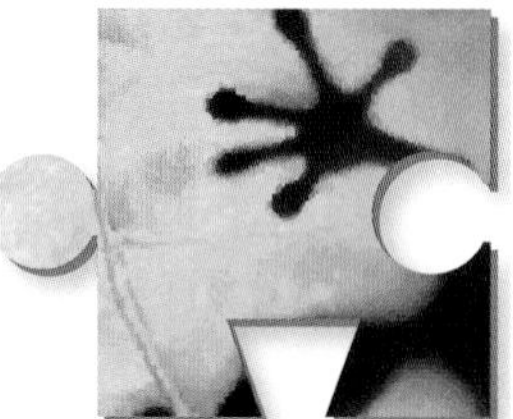

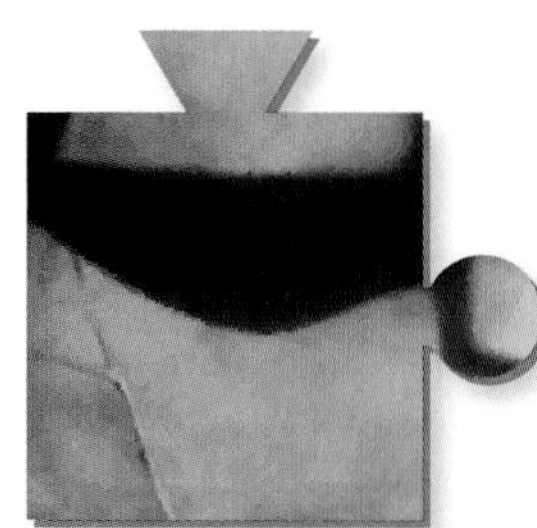

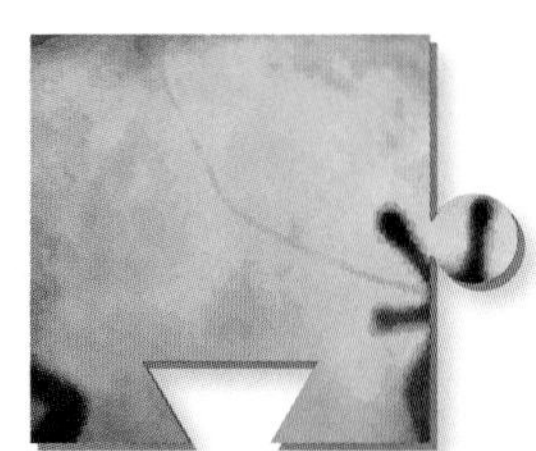

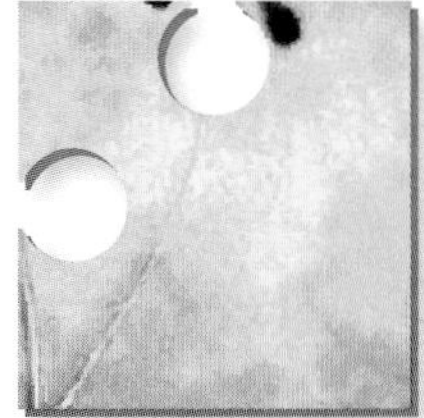

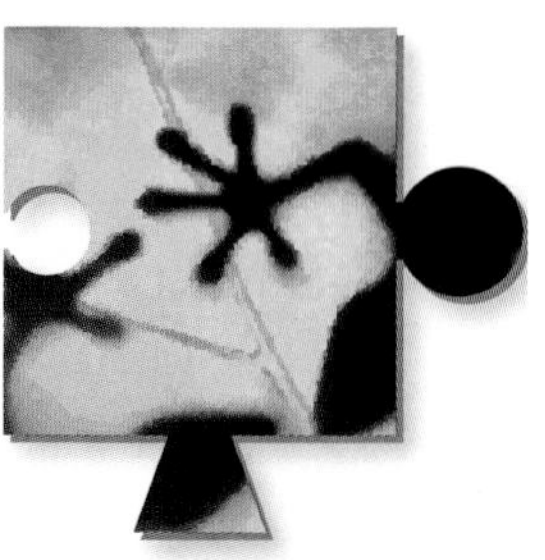

3 ACTIVITY: Filling a Square with Circles

Math Practice

Make a Plan

What steps will you use to solve this problem?

Work with a partner. Which pattern fills more of the square with circles? Explain.

a.

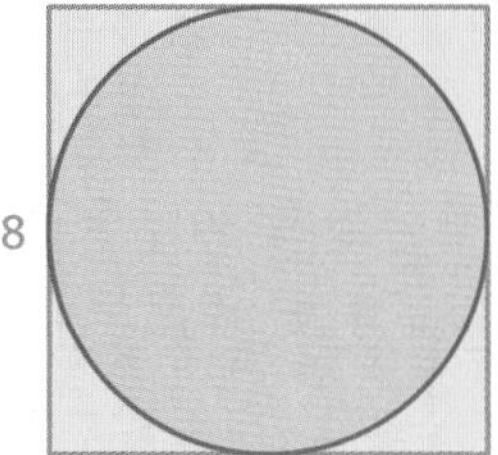

b.

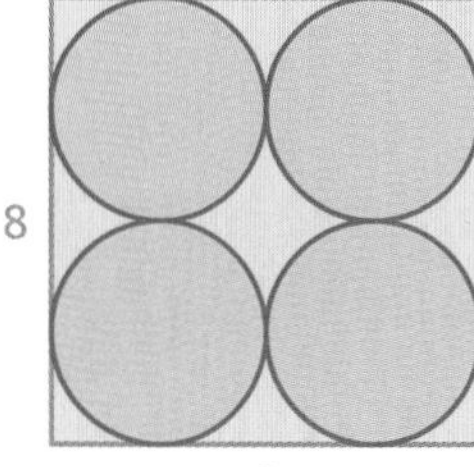

c.

8

8

d.

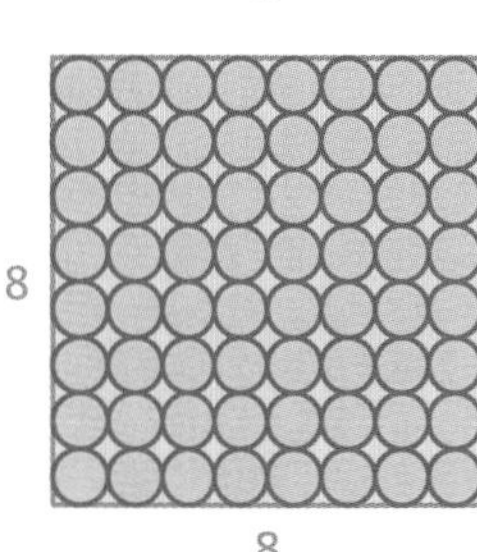

What Is Your Answer?

4. **IN YOUR OWN WORDS** How can you find the area of a composite figure?

5. Summarize the area formulas for all the basic figures you have studied. Draw a single composite figure that has each type of basic figure. Label the dimensions and find the total area.

Use what you learned about areas of composite figures to complete Exercises 3–5 on page 574.

13.4 Lesson

To find the area of a composite figure, separate it into figures with areas you know how to find. Then find the sum of the areas of those figures.

EXAMPLE 1 Finding an Area Using Grid Paper

Find the area of the yellow figure.

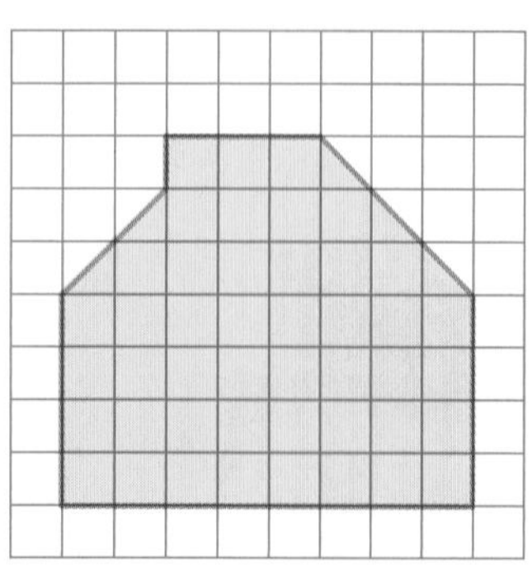

Count the number of squares that lie entirely in the figure. There are 45.

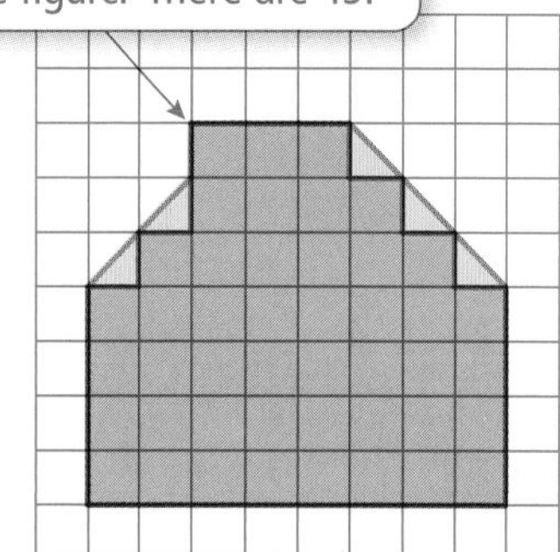

Count the number of half squares in the figure. There are 5.

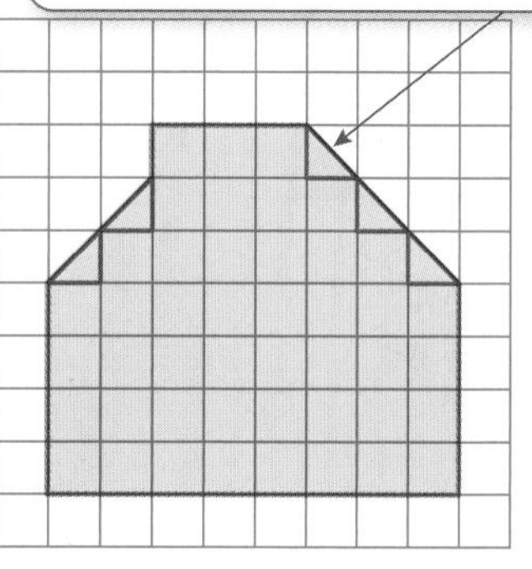

The area of a half square is $1 \div 2 = 0.5$ square unit.

Area of 45 squares: $45 \times 1 = 45$ square units

Area of 5 half squares: $5 \times 0.5 = 2.5$ square units

So, the area is $45 + 2.5 = 47.5$ square units.

On Your Own

Now You're Ready
Exercises 3–8

Find the area of the shaded figure.

1.

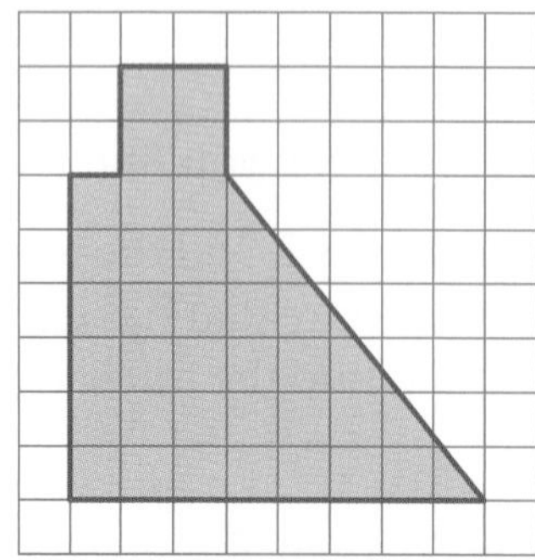

2.

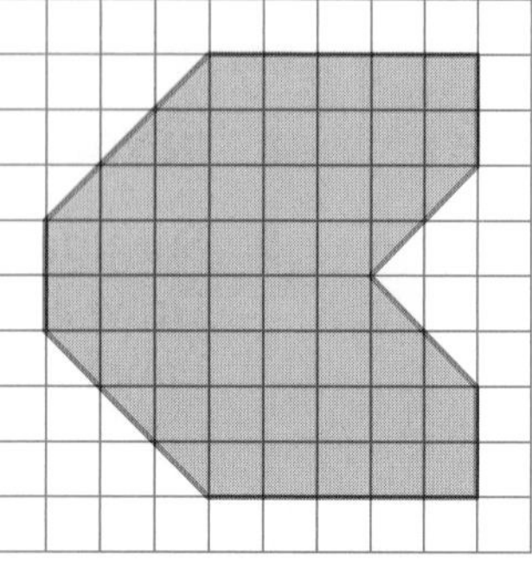

EXAMPLE 2 Finding an Area

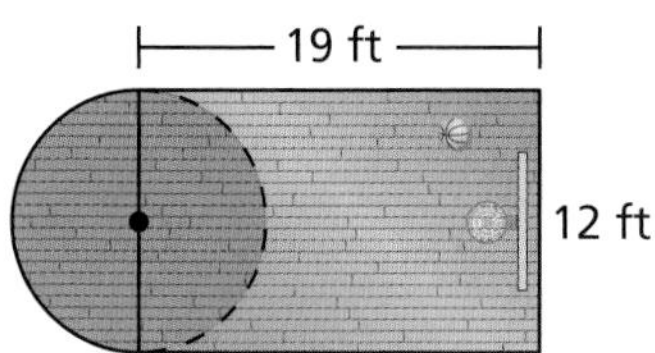

Find the area of the portion of the basketball court shown.

The figure is made up of a rectangle and a semicircle. Find the area of each figure.

Area of Rectangle

$$A = \ell w$$
$$= 19(12)$$
$$= 228$$

Area of Semicircle

$$A = \frac{\pi r^2}{2}$$
$$\approx \frac{3.14 \cdot 6^2}{2}$$
$$= 56.52$$

The semicircle has a radius of $\frac{12}{2} = 6$ feet.

So, the area is about $228 + 56.52 = 284.52$ square feet.

EXAMPLE 3 Finding an Area

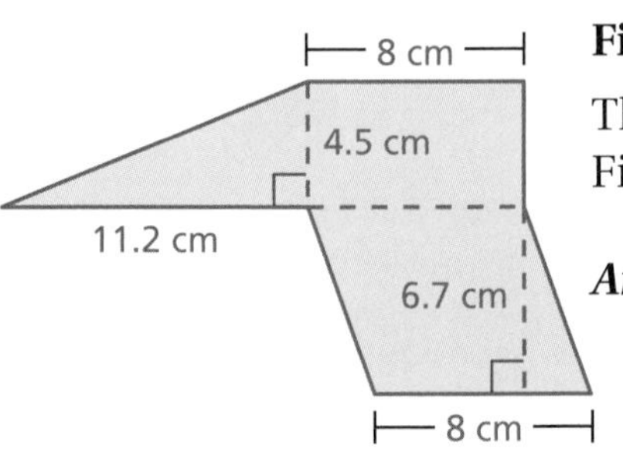

Find the area of the figure.

The figure is made up of a triangle, a rectangle, and a parallelogram. Find the area of each figure.

Area of Triangle

$$A = \frac{1}{2}bh$$
$$= \frac{1}{2}(11.2)(4.5)$$
$$= 25.2$$

Area of Rectangle

$$A = \ell w$$
$$= 8(4.5)$$
$$= 36$$

Area of Parallelogram

$$A = bh$$
$$= 8(6.7)$$
$$= 53.6$$

So, the area is $25.2 + 36 + 53.6 = 114.8$ square centimeters.

On Your Own

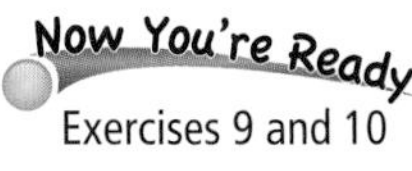

Exercises 9 and 10

Find the area of the figure.

3.

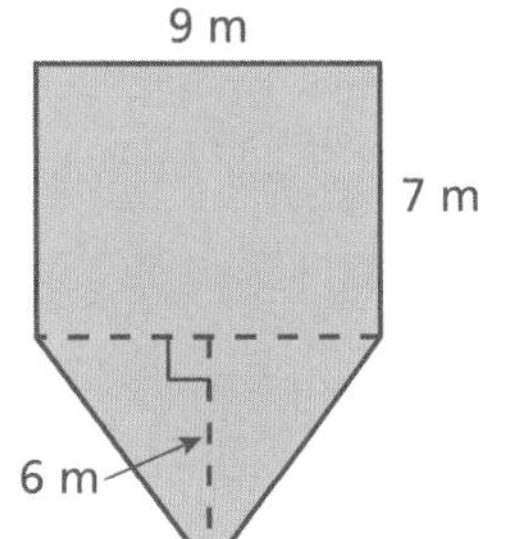

4.

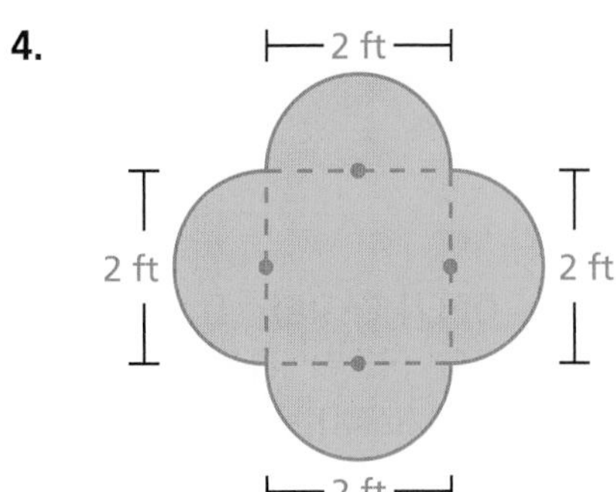

13.4 Exercises

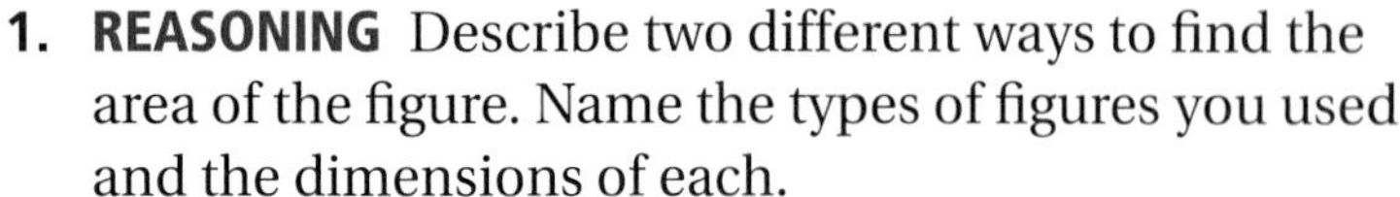

Vocabulary and Concept Check

1. **REASONING** Describe two different ways to find the area of the figure. Name the types of figures you used and the dimensions of each.

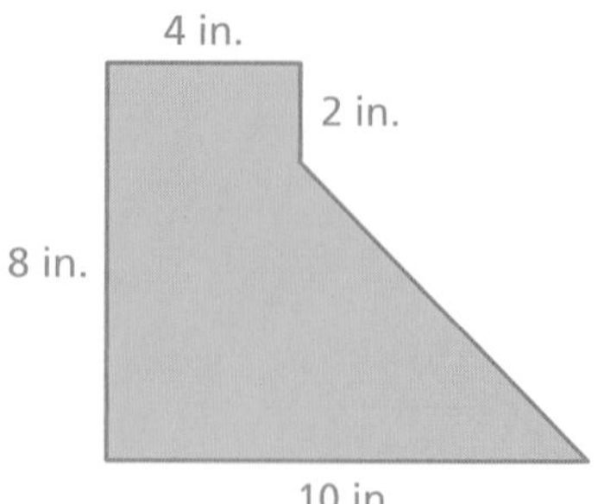

2. **REASONING** Draw a trapezoid. Explain how you can think of the trapezoid as a composite figure to find its area.

Practice and Problem Solving

Find the area of the figure.

1 3. 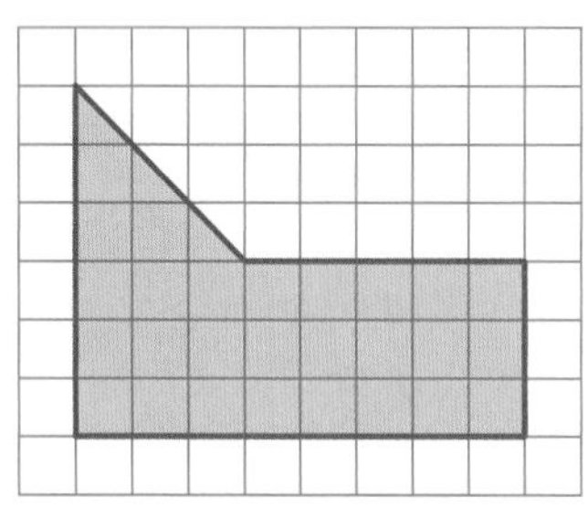

4. 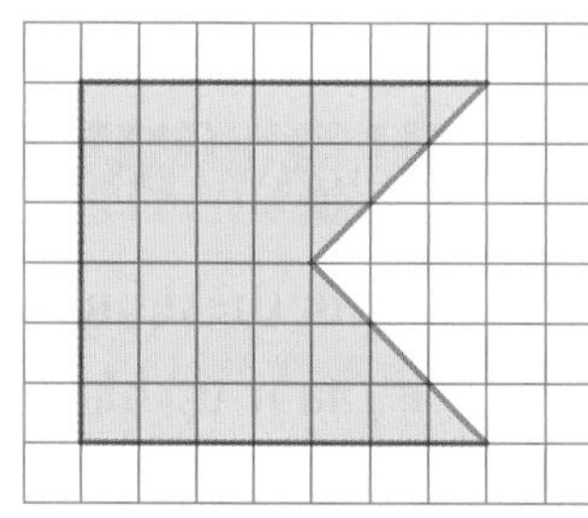

5. 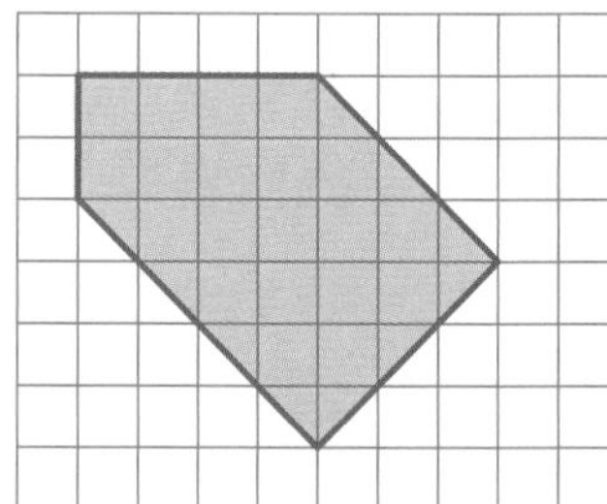

6. 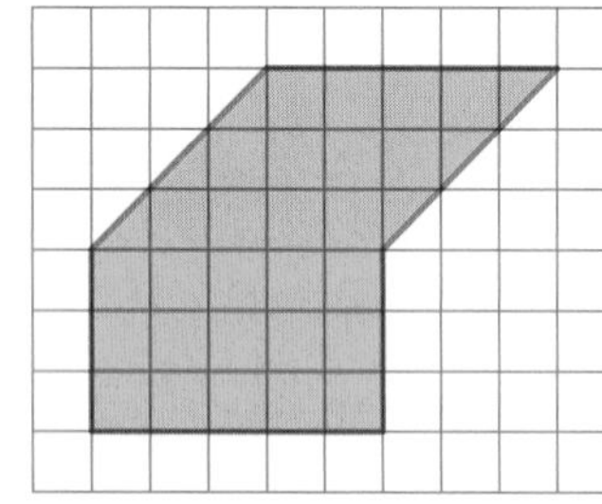

7. 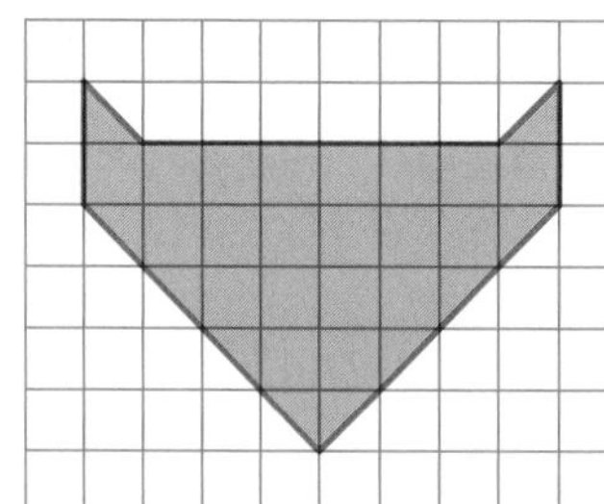

8. 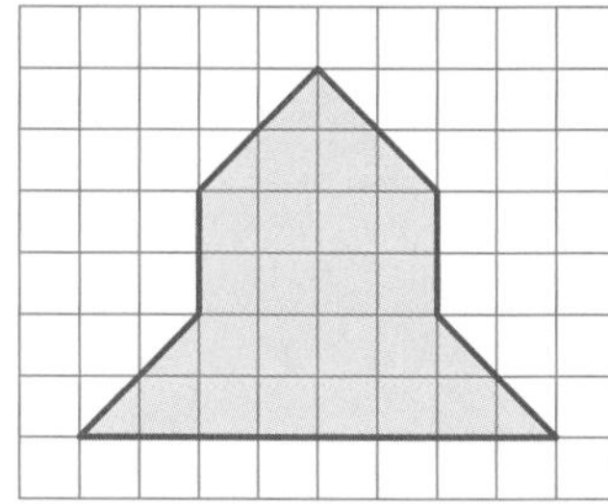

Find the area of the figure.

2 3 9. 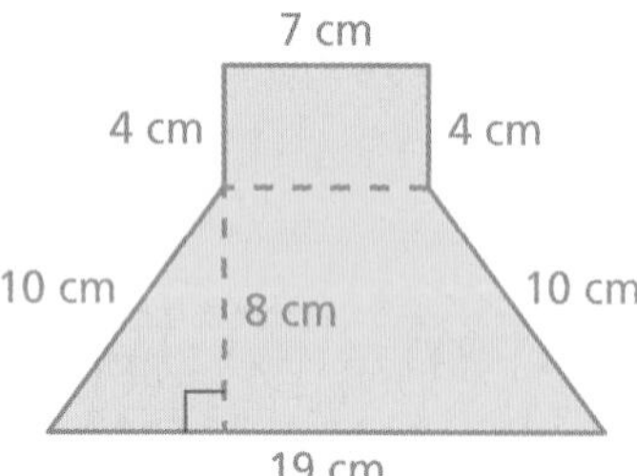

10. 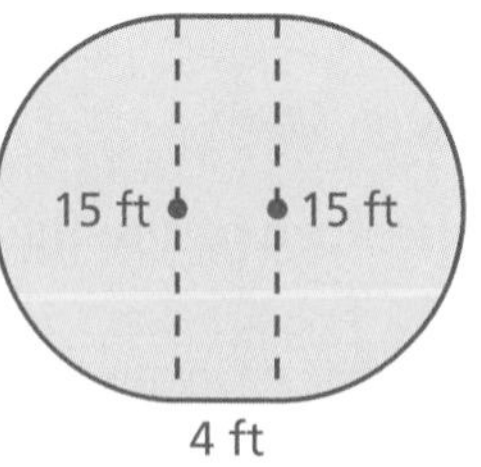

11. **OPEN-ENDED** Trace your hand and your foot on grid paper. Then estimate the area of each. Which one has the greater area?

Find the area of the figure.

12.

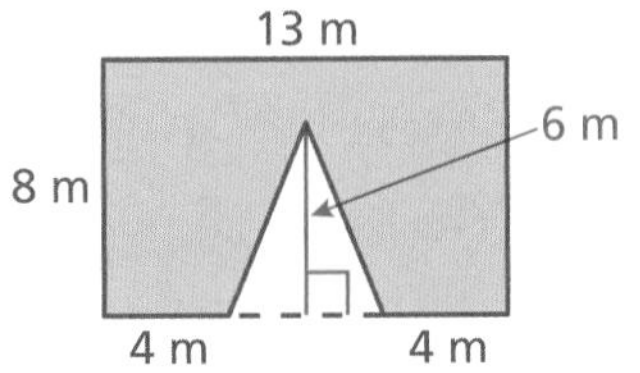

13.

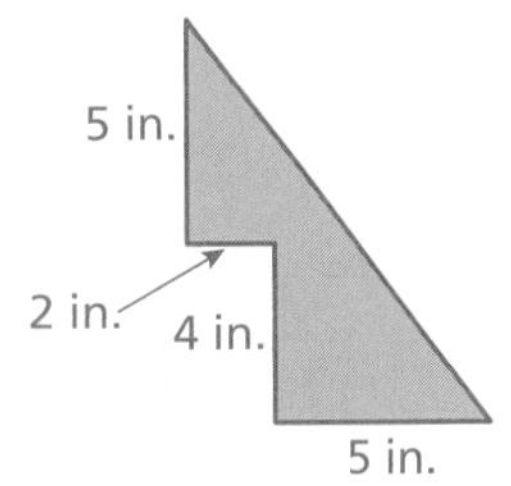

14.

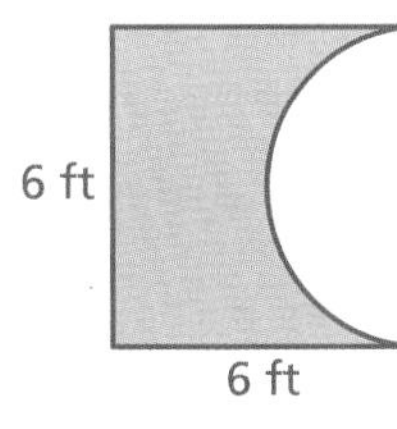

15. STRUCTURE The figure is made up of a square and a rectangle. Find the area of the shaded region.

7 m
3 m
16 m

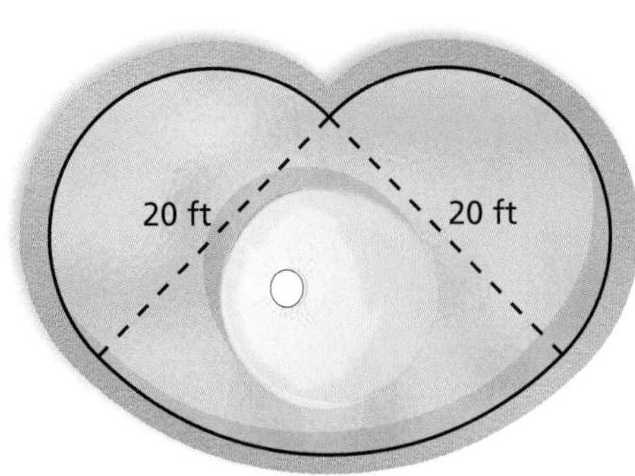

16. FOUNTAIN The fountain is made up of two semicircles and a quarter circle. Find the perimeter and the area of the fountain.

17. Critical Thinking You are deciding on two different designs for envelopes.

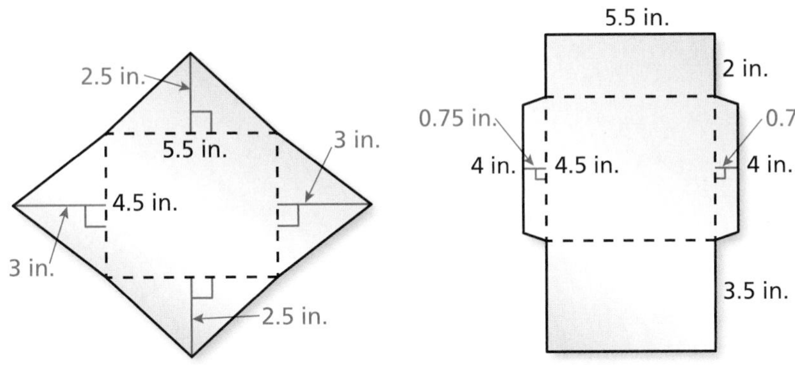

a. Which design has the greater area?

b. You make 500 envelopes using the design with the greater area. Using the same amount of paper, how many more envelopes can you make with the other design?

Fair Game Review What you learned in previous grades & lessons

Write the phrase as an expression. *(Skills Review Handbook)*

18. 12 less than a number x

19. a number y divided by 6

20. a number b increased by 3

21. the product of 7 and a number w

22. MULTIPLE CHOICE What number is 0.02% of 50? *(Skills Review Handbook)*

Ⓐ 0.01 Ⓑ 0.1 Ⓒ 1 Ⓓ 100

13.3–13.4 Quiz

Check It Out
Progress Check
BigIdeasMath.com

Find the area of the figure. *(Section 13.4)*

1. 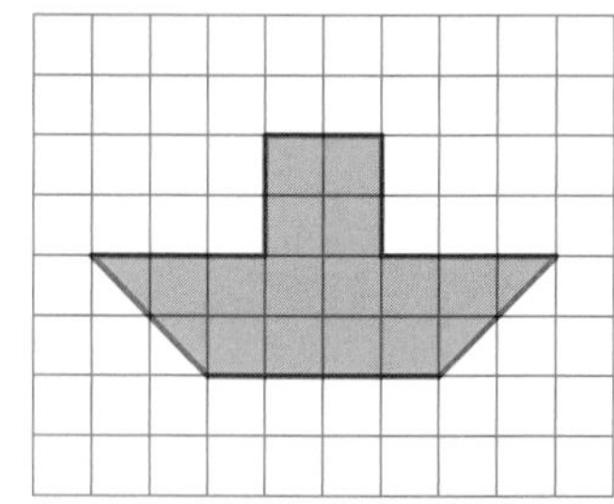

2. 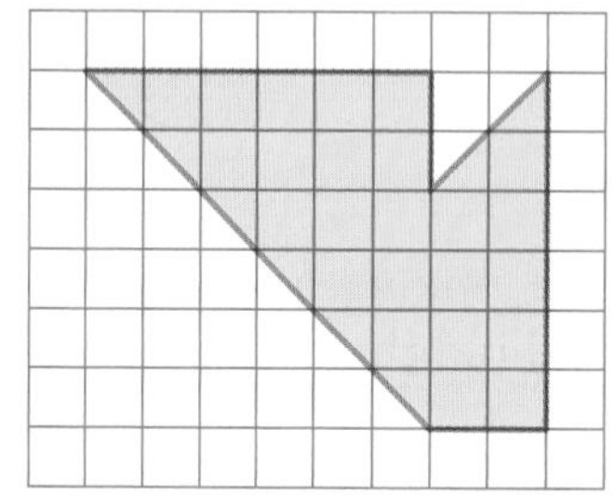

3. 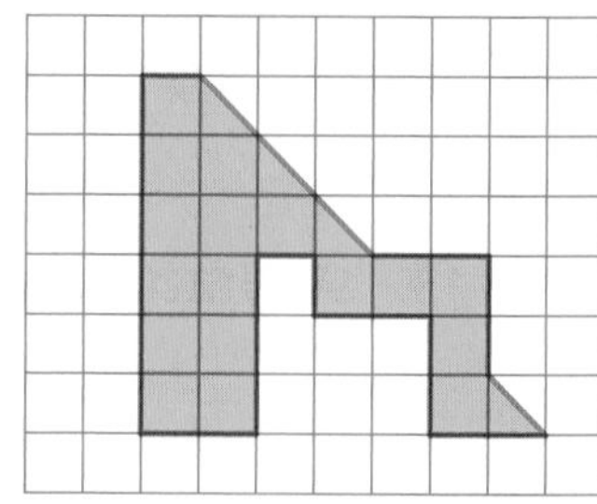

Find the area of the circle. Use 3.14 or $\frac{22}{7}$ for π. *(Section 13.3)*

4. 

5. 

6. 

Find the area of the figure. *(Section 13.4)*

7. 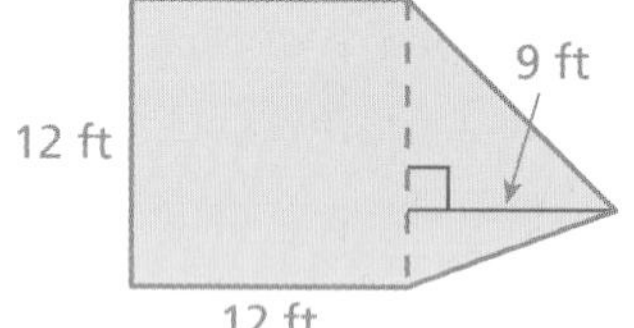

8. 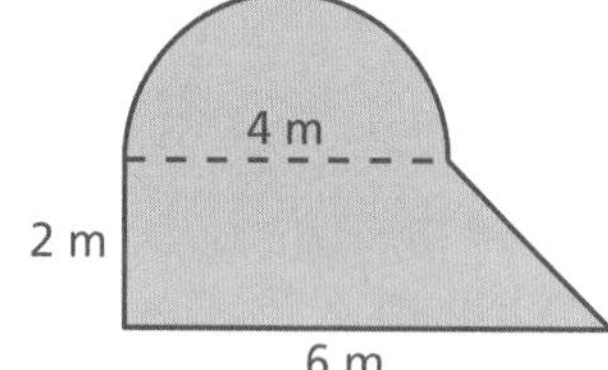

9. 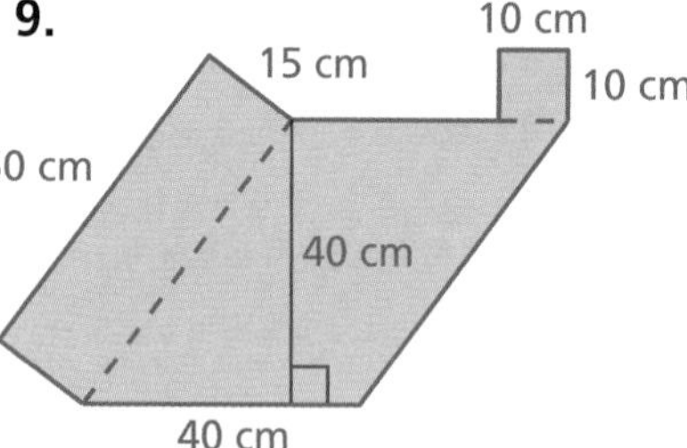

10. POT HOLDER A knitted pot holder is shaped like a circle. Its radius is 3.5 inches. What is its area? *(Section 13.3)*

11. CARD The heart-shaped card is made up of a square and two semicircles. What is the area of the card? *(Section 13.4)*

12. DESK A desktop is shaped like a semicircle with a diameter of 28 inches. What is the area of the desktop? *(Section 13.3)*

13. RUG The circular rug is placed on a square floor. The rug touches all four walls. How much of the floor space is *not* covered by the rug? *(Section 13.4)*

13 Chapter Review

Review Key Vocabulary

circle, *p. 550*
center, *p. 550*
radius, *p. 550*
diameter, *p. 550*
circumference, *p. 551*
pi, *p. 551*
semicircle, *p. 552*
composite figure, *p. 558*

Review Examples and Exercises

13.1 Circles and Circumference *(pp. 548–555)*

Find the circumference of the circle. Use 3.14 for π.

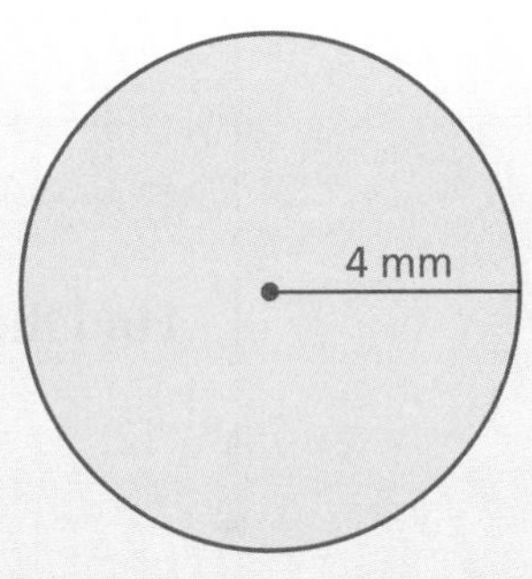

The radius is 4 millimeters.

$C = 2\pi r$	Write formula for circumference.
$\approx 2 \cdot 3.14 \cdot 4$	Substitute 3.14 for π and 4 for r.
$= 25.12$	Multiply.

∴ The circumference is about 25.12 millimeters.

Exercises

Find the radius of the circle with the given diameter.

1. 8 inches
2. 60 millimeters
3. 100 meters
4. 3 yards

Find the diameter of the circle with the given radius.

5. 20 feet
6. 5 meters
7. 1 inch
8. 25 millimeters

Find the circumference of the circle. Use 3.14 or $\frac{22}{7}$ for π.

9. 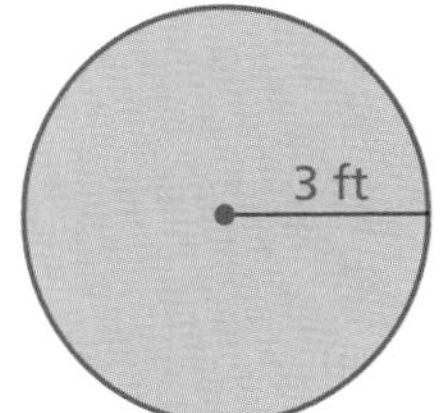

10. 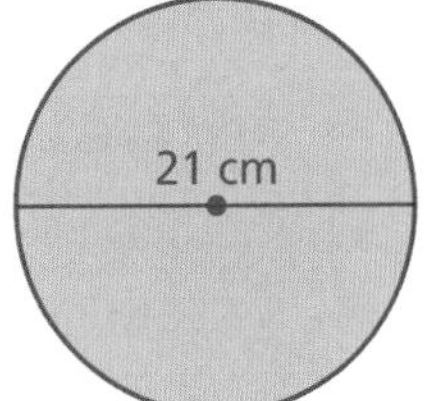

11. 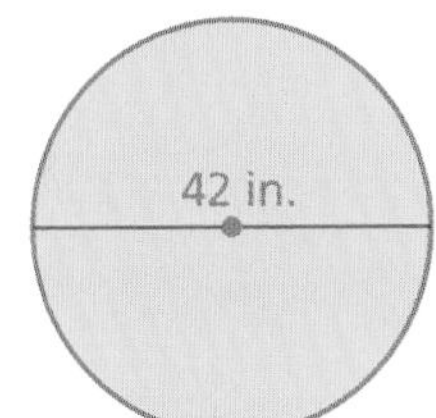

13.2 Perimeters of Composite Figures (pp. 556–561)

The figure is made up of a semicircle and a square. Find the perimeter.

The distance around the square part is $6 + 6 + 6 = 18$ meters. The distance around the semicircle is one-half the circumference of a circle with $d = 6$ meters.

$\frac{C}{2} = \frac{\pi d}{2}$ Divide the circumference by 2.

$\approx \frac{3.14 \cdot 6}{2}$ Substitute 3.14 for π and 6 for d.

$= 9.42$ Simplify.

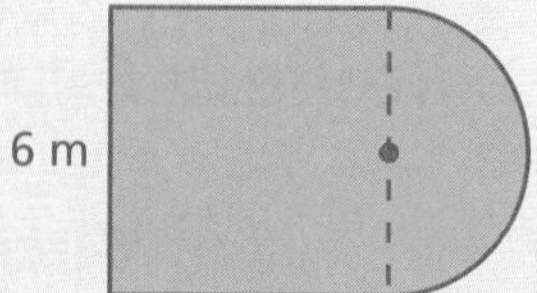

So, the perimeter is about $18 + 9.42 = 27.42$ meters.

Exercises

Find the perimeter of the figure.

12.

5 in.
4 in.
3 in.
5 in.
9 in.

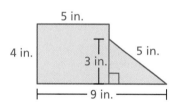

13.

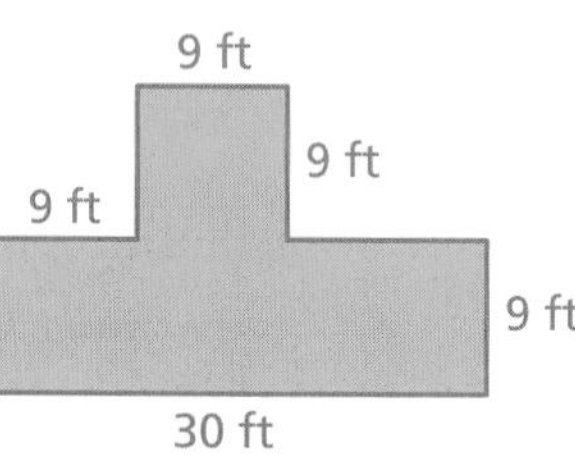

14.

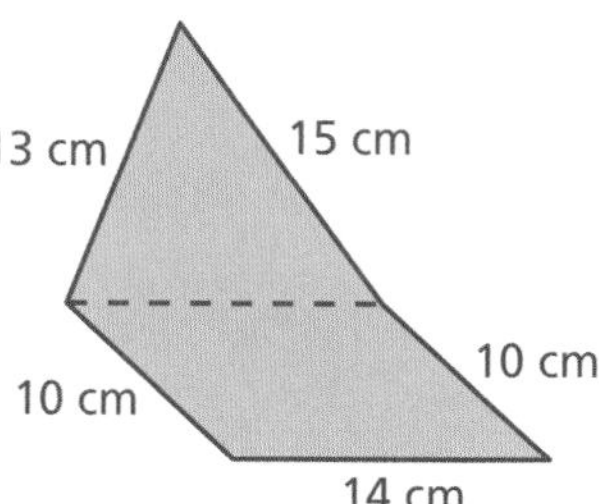

15.

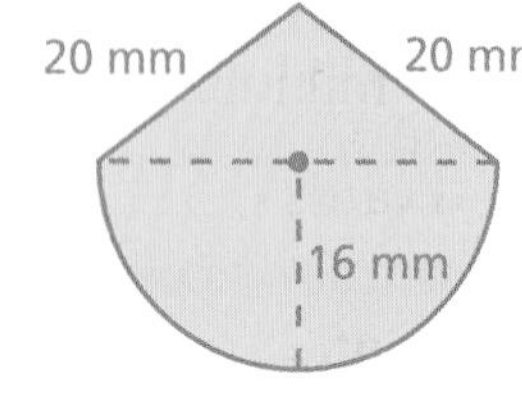

16.

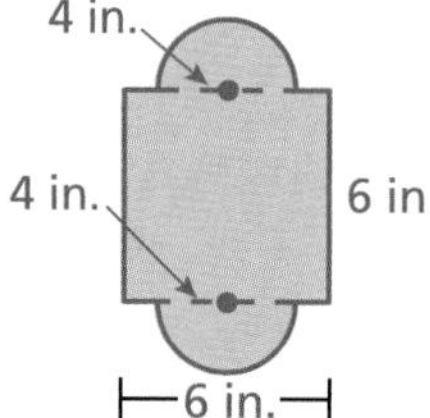

17.

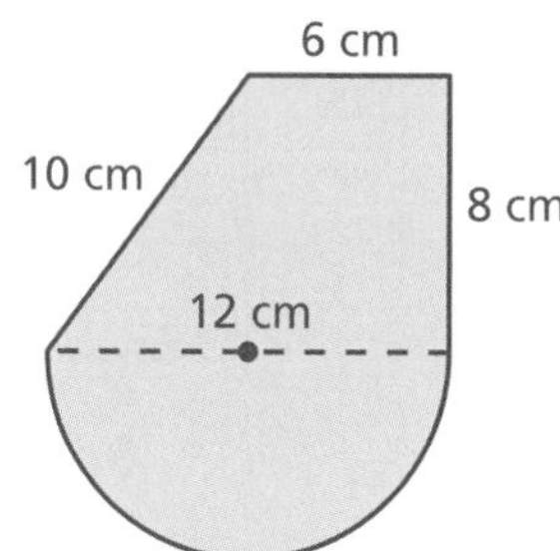

13.3 Areas of Circles (pp. 564–569)

Find the area of the circle. Use 3.14 for π.

$A = \pi r^2$ Write formula for area.

$\approx 3.14 \cdot 20^2$ Substitute 3.14 for π and 20 for r.

$= 1256$ Multiply.

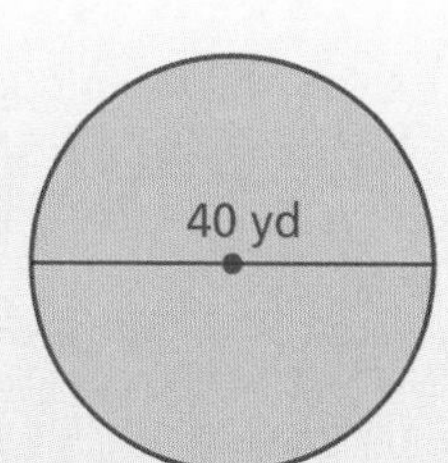

The area is about 1256 square yards.

Exercises

Find the area of the circle. Use 3.14 or $\frac{22}{7}$ for π.

18.

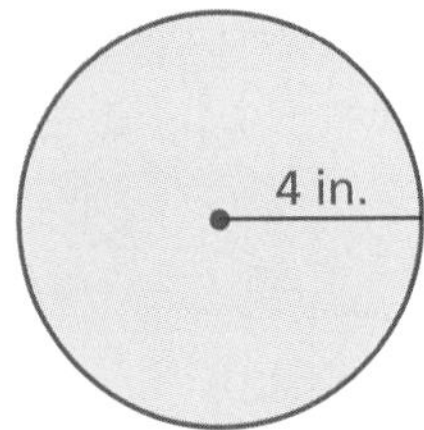

19.

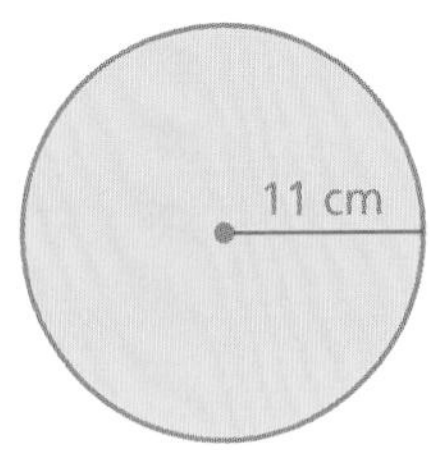

20.

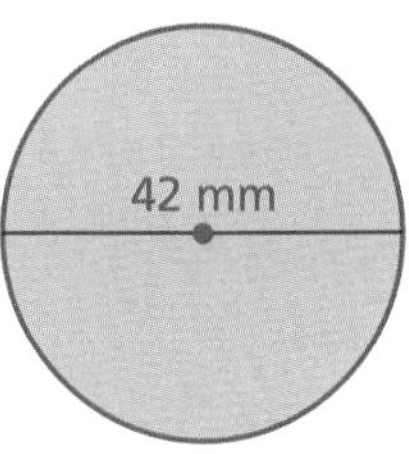

13.4 Areas of Composite Figures (pp. 570–575)

Find the area of the figure.

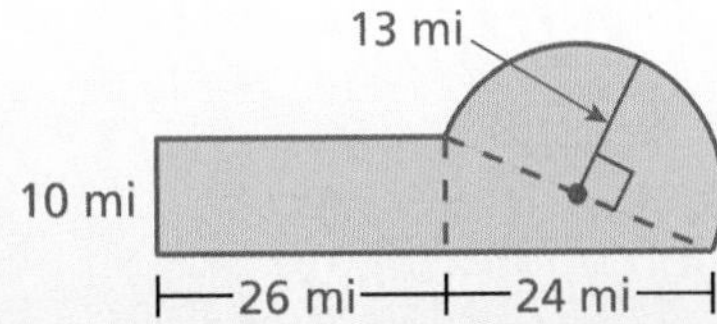

The figure is made up of a rectangle, a triangle and a semicircle. Find the area of each figure.

Area of Rectangle

$$A = \ell w$$
$$= 26(10)$$
$$= 260$$

Area of Triangle

$$A = \frac{1}{2}bh$$
$$= \frac{1}{2}(10)(24)$$
$$= 120$$

Area of Semicircle

$$A = \frac{\pi r^2}{2}$$
$$\approx \frac{3.14 \cdot 13^2}{2}$$
$$= 265.33$$

So, the area is about $260 + 120 + 265.33 = 645.33$ square miles.

Exercises

Find the area of the figure.

21.

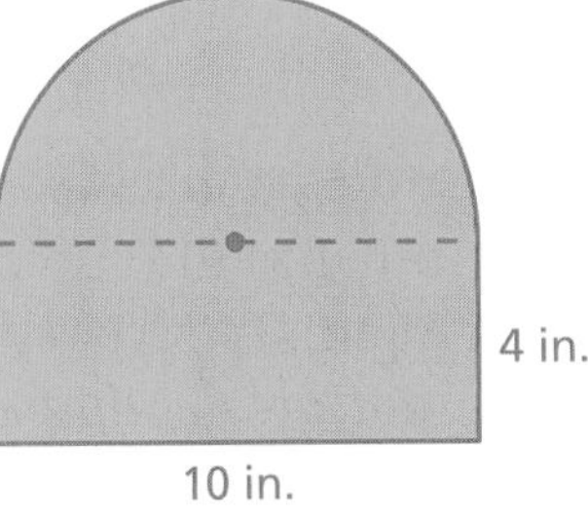

22.

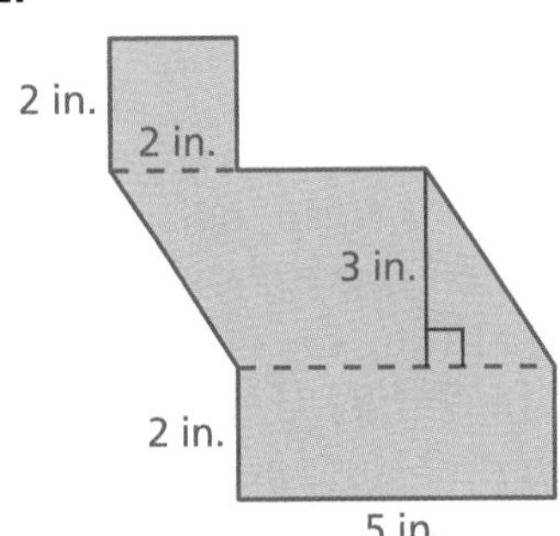

23.

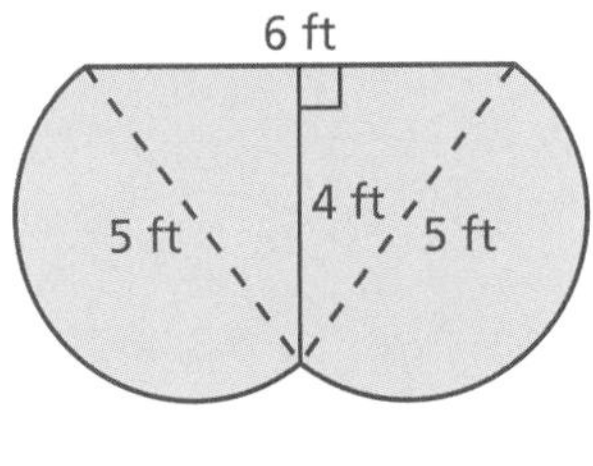

13 Chapter Test

Find the radius of the circle with the given diameter.

1. 10 inches
2. 5 yards

Find the diameter of the circle with the given radius.

3. 34 feet
4. 19 meters

Find the circumference and the area of the circle. Use 3.14 or $\frac{22}{7}$ for π.

5. 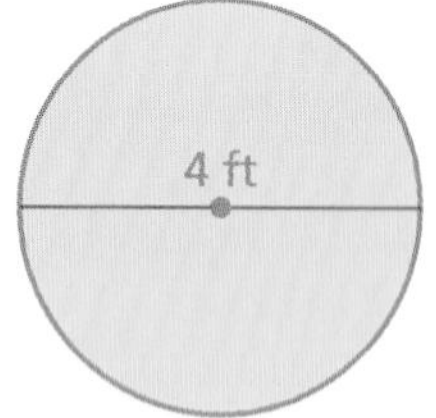

6. 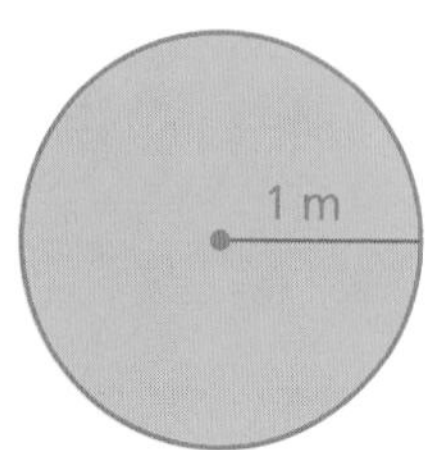

7. 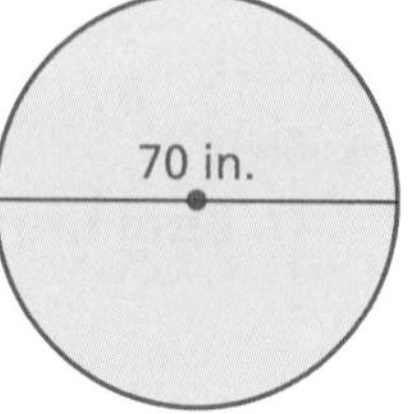

8. Estimate the perimeter of the figure. Then find the area.

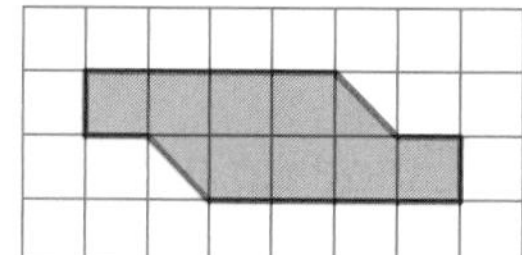

Find the perimeter and the area of the figure. Use 3.14 or $\frac{22}{7}$ for π.

9. 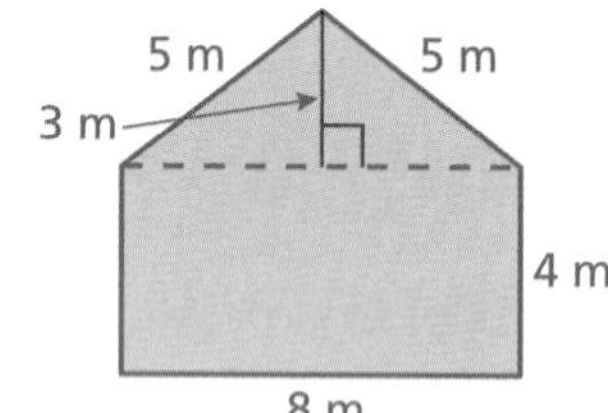

10. 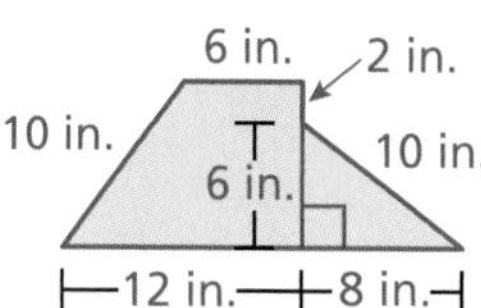

11. 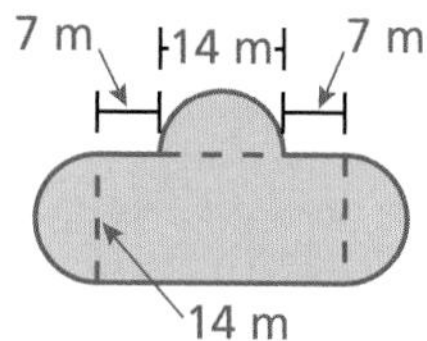

12. **MUSEUM** A museum plans to rope off the perimeter of the L-shaped exhibit. How much rope does it need?

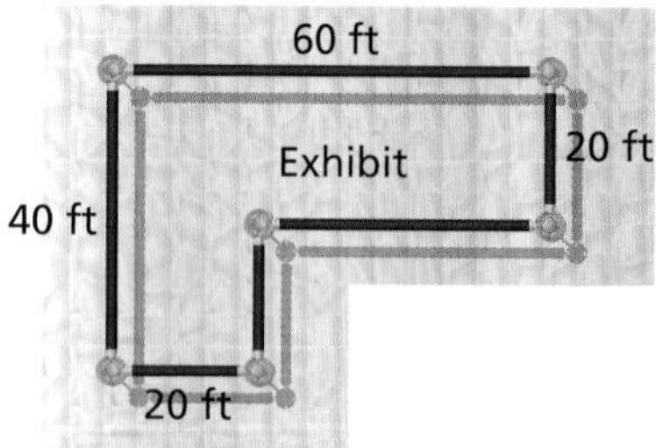

13. **ANIMAL PEN** You unfold chicken wire to make a circular pen with a diameter of 2.9 meters. How many meters of chicken wire do you need?

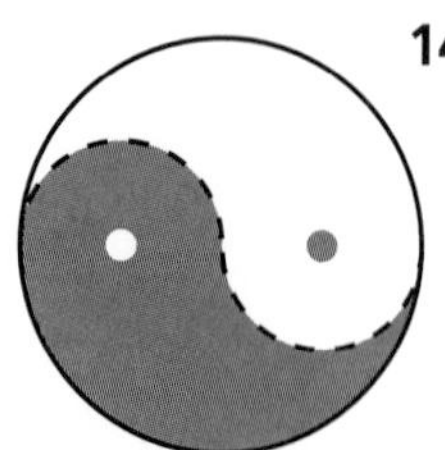

14. **YIN AND YANG** In the Chinese symbol for yin and yang, the dashed curve shows two semicircles formed by the curve separating the yin (dark) and the yang (light). Is the circumference of the entire yin and yang symbol *less than*, *greater than*, or *equal to* the perimeter of the yin?

13 Cumulative Assessment

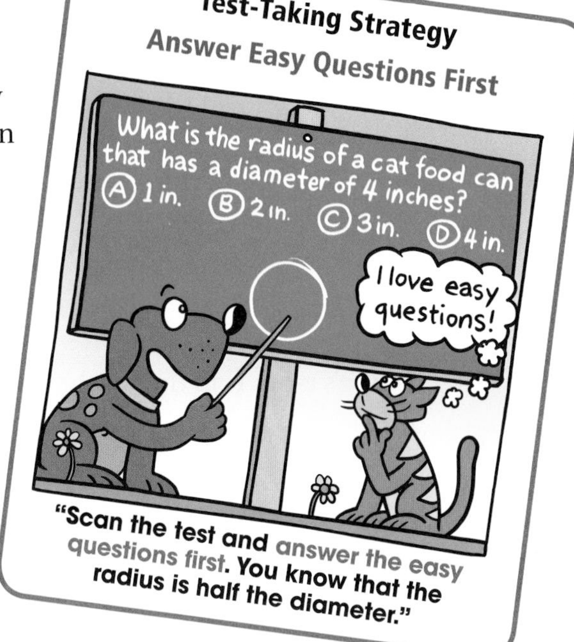

Test-Taking Strategy
Answer Easy Questions First
What is the radius of a cat food can that has a diameter of 4 inches?
Ⓐ 1 in. Ⓑ 2 in. Ⓒ 3 in. Ⓓ 4 in.
I love easy questions!
"Scan the test and answer the easy questions first. You know that the radius is half the diameter."

1. To make 6 servings of soup, you need 5 cups of chicken broth. You want to know how many servings you can make with 2 quarts of chicken broth. Which proportion should you use?

   A. $\frac{6}{5} = \frac{2}{x}$ **C.** $\frac{6}{5} = \frac{x}{8}$

   B. $\frac{6}{5} = \frac{x}{2}$ **D.** $\frac{5}{6} = \frac{x}{8}$

2. What is the value of x?

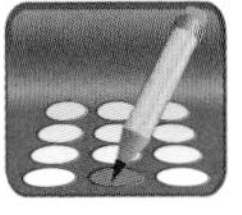

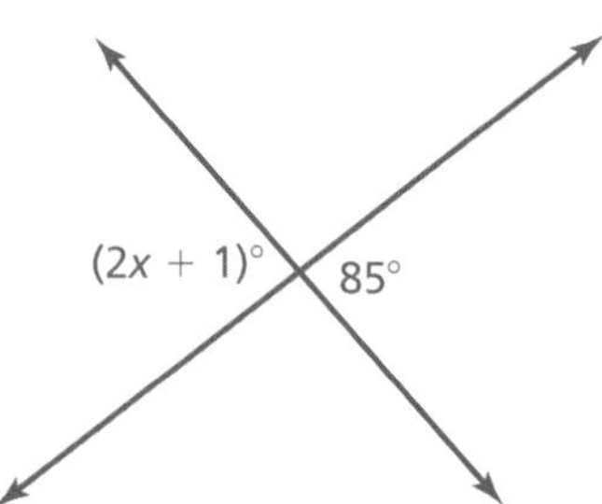

3. Your mathematics teacher described an equation in words. Her description is in the box below.

   "5 less than the product of 7 and an unknown number is equal to 42."

   Which equation matches your mathematics teacher's description?

   F. $(5 - 7)n = 42$ **H.** $5 - 7n = 42$

   G. $(7 - 5)n = 42$ **I.** $7n - 5 = 42$

4. What is the area of the circle below? $\left(\text{Use } \frac{22}{7} \text{ for } \pi.\right)$

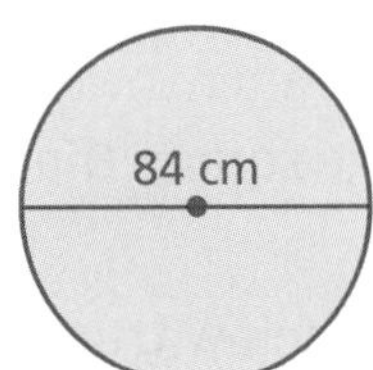

   A. 132 cm^2 **C.** 5544 cm^2

   B. 264 cm^2 **D.** $22{,}176 \text{ cm}^2$

5. John was finding the area of the figure below.

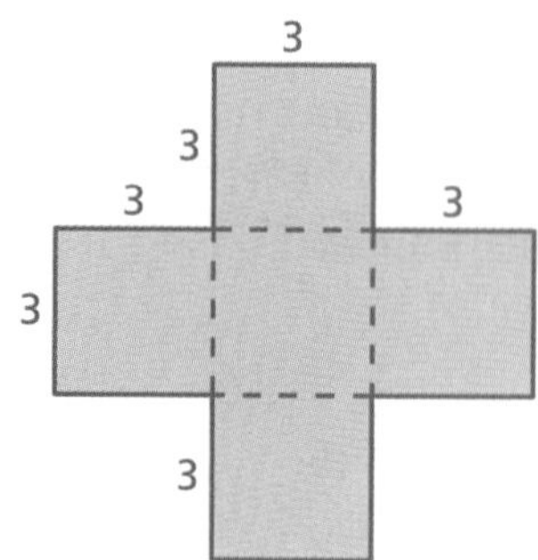

John's work is in the box below.

area of horizontal rectangle

$A = 3 \times (3 + 3 + 3)$

$= 3 \times 9$

$= 27$ square units

area of vertical rectangle

$A = (3 + 3 + 3) \times 3$

$= 9 \times 3$

$= 27$ square units

total area of figure

$A = 27 + 27$

$= 54$ square units

What should John do to correct the error that he made?

F. Add the area of the center square to the 54 square units.

G. Find the area of one square and multiply this number by 4.

H. Subtract the area of the center square from the 54 square units.

I. Subtract 54 from the area of a large square that is 9 units on each side.

6. Which value of x makes the equation below true?

$$5x - 3 = 11$$

A. 1.6

B. 2.8

C. 40

D. 70

7. What is the perimeter of the figure below? (Use 3.14 for π.)

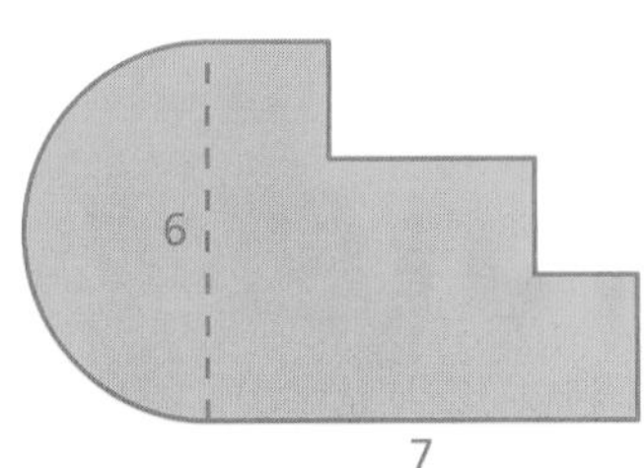

8. Which inequality has 5 in its solution set?

F. $5 - 2x \geq 3$

G. $3x - 4 \geq 8$

H. $8 - 3x > -7$

I. $4 - 2x < -6$

9. Four jewelry stores are selling an identical pair of earrings.

- Store A: original price of \$75; 20% off during sale
- Store B: original price of \$100; 35% off during sale
- Store C: original price of \$70; 10% off during sale
- Store D: original price of \$95; 30% off during sale

Which store has the least sale price for the pair of earrings?

A. Store A

B. Store B

C. Store C

D. Store D

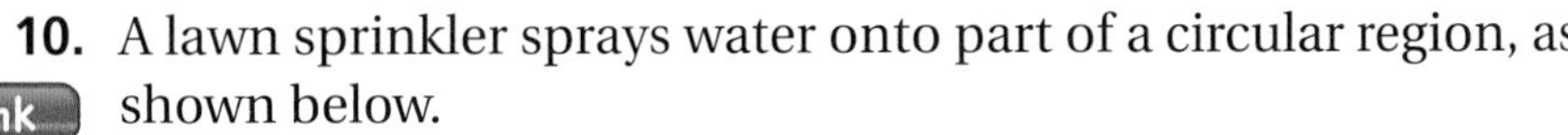

10. A lawn sprinkler sprays water onto part of a circular region, as shown below.

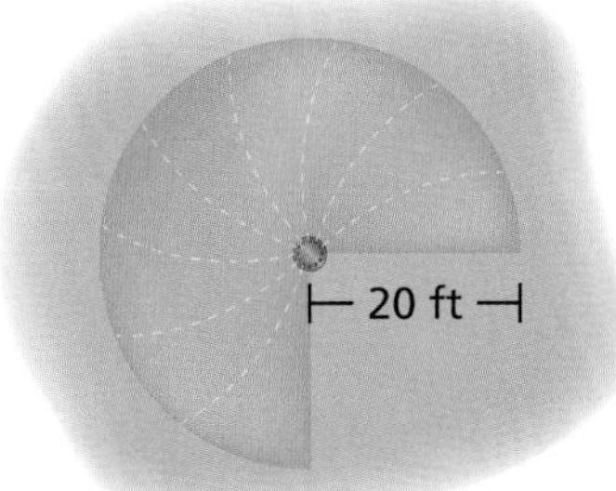

Part A What is the area, in square feet, of the region that the sprinkler sprays with water? Show your work and explain your reasoning. (Use 3.14 for π.)

Part B What is the perimeter, in feet, of the region that the sprinkler sprays with water? Show your work and explain your reasoning. (Use 3.14 for π.)

14 Surface Area and Volume

"I was thinking that I want the Pagodal roof instead of the Swiss chalet roof for my new doghouse."

"Because PAGODAL rearranges to spell 'A DOG PAL.' "

"Take a deep breath and hold it."

"Now, do you feel like your surface area or your volume is increasing more?"

What You Learned Before

"Descartes, how would you like it if I could double the height of your cat food can?"

Finding Areas of Squares and Rectangles

Example 1 **Find the area of the rectangle.**

Area $= \ell w$ Write formula for area.

$= 7(3)$ Substitute 7 for ℓ and 3 for w.

$= 21$ Multiply.

∴ The area of the rectangle is 21 square millimeters.

Try It Yourself

Find the area of the square or rectangle.

1. 

2. 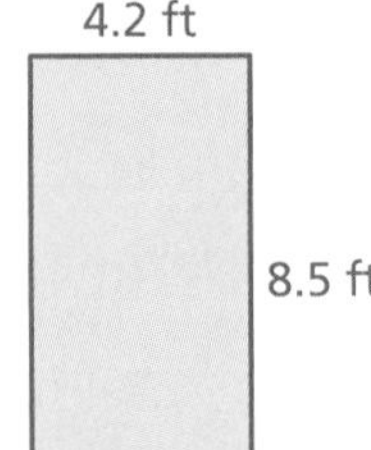

3. 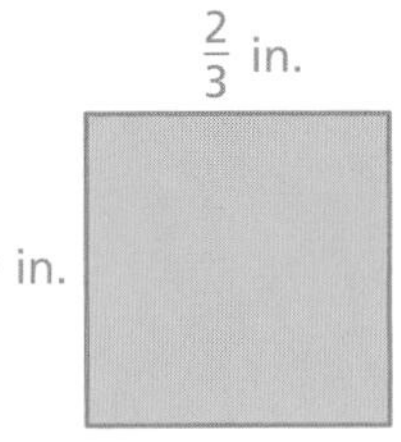

Finding Areas of Triangles

Example 2 **Find the area of the triangle.**

$A = \frac{1}{2}bh$ Write formula.

$= \frac{1}{2}(6)(7)$ Substitute 6 for b and 7 for h.

$= \frac{1}{2}(42)$ Multiply 6 and 7.

$= 21$ Multiply $\frac{1}{2}$ and 42.

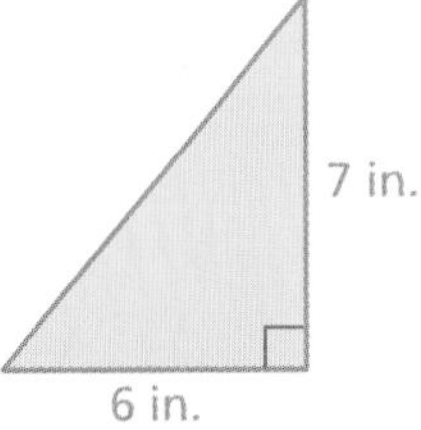

∴ The area of the triangle is 21 square inches.

Try It Yourself

Find the area of the triangle.

4. 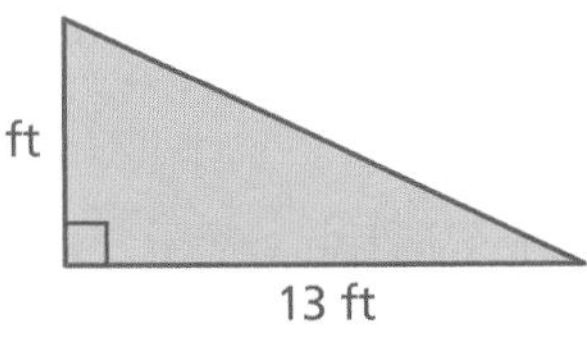

5. 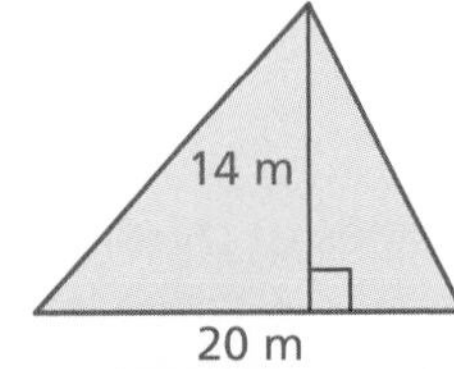

6. 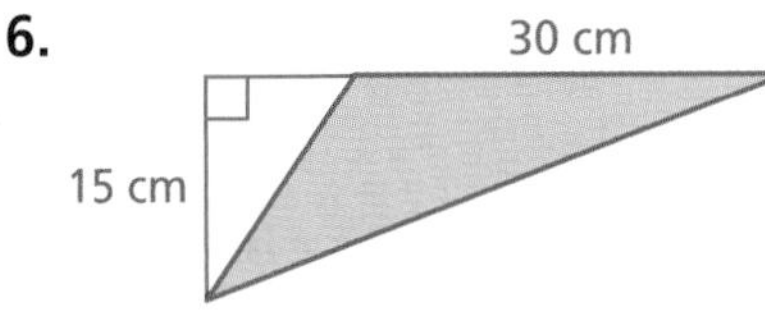

14.1 Surface Areas of Prisms

Essential Question How can you find the surface area of a prism?

1 ACTIVITY: Surface Area of a Rectangular Prism

Work with a partner. Copy the net for a rectangular prism. Label each side as *h*, *w*, or *ℓ*. Then use your drawing to write a formula for the surface area of a rectangular prism.

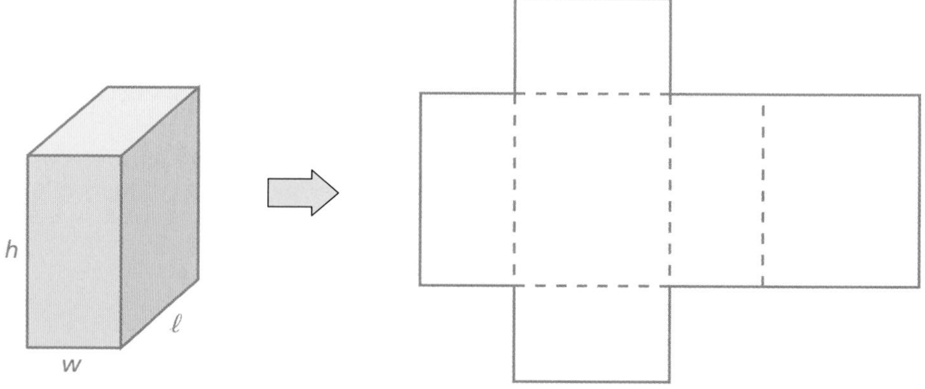

2 ACTIVITY: Surface Area of a Triangular Prism

Work with a partner.

a. Find the surface area of the solid shown by the net. Copy the net, cut it out, and fold it to form a solid. Identify the solid.

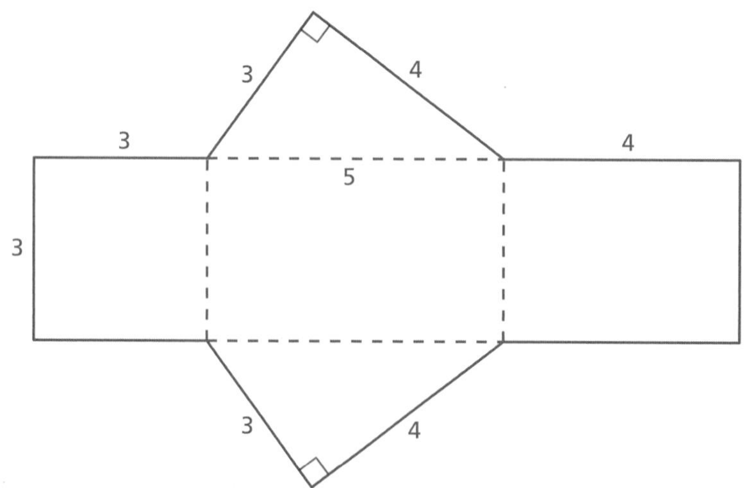

b. Which of the surfaces of the solid are bases? Why?

Geometry
In this lesson, you will
- use two-dimensional nets to represent three-dimensional solids.
- find surface areas of rectangular and triangular prisms.
- solve real-life problems.

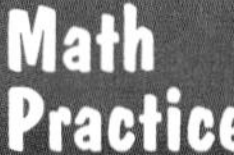

Math Practice

Construct Arguments

What method did you use to find the surface area of the rectangular prism? Explain.

3 ACTIVITY: Forming Rectangular Prisms

Work with a partner.

- **Use 24 one-inch cubes to form a rectangular prism that has the given dimensions.**
- **Draw each prism.**
- **Find the surface area of each prism.**

	Drawing	*Surface Area*
a. $4 \times 3 \times 2$	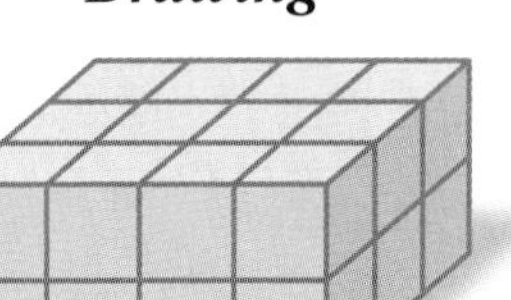	▭ in.^{2}

b. $1 \times 1 \times 24$ **c.** $1 \times 2 \times 12$ **d.** $1 \times 3 \times 8$

e. $1 \times 4 \times 6$ **f.** $2 \times 2 \times 6$ **g.** $2 \times 4 \times 3$

What Is Your Answer?

4. Use your formula from Activity 1 to verify your results in Activity 3.
5. **IN YOUR OWN WORDS** How can you find the surface area of a prism?
6. **REASONING** When comparing ice blocks with the same volume, the ice with the greater surface area will melt faster. Which will melt faster, the bigger block or the three smaller blocks? Explain your reasoning.

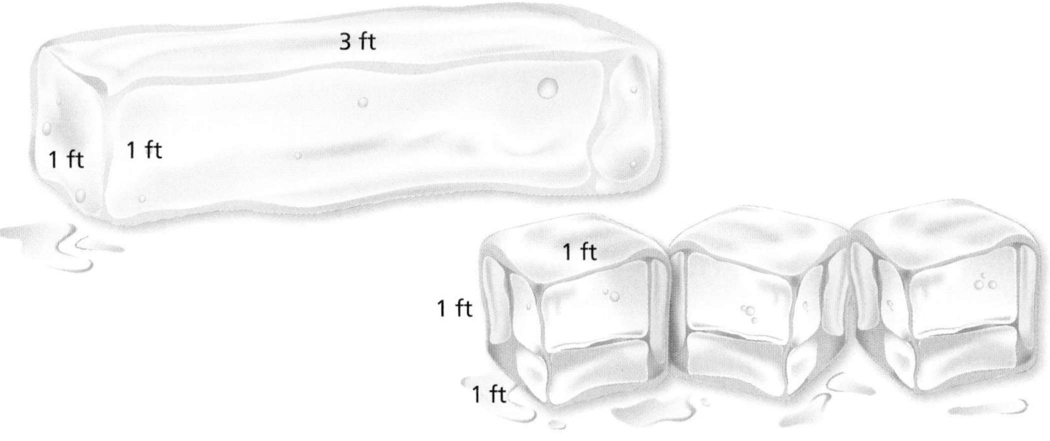

Practice Use what you learned about the surface areas of rectangular prisms to complete Exercises 4–6 on page 591.

Key Vocabulary
lateral surface area, *p. 590*

Key Idea

Surface Area of a Rectangular Prism

Words The surface area S of a rectangular prism is the sum of the areas of the bases and the lateral faces.

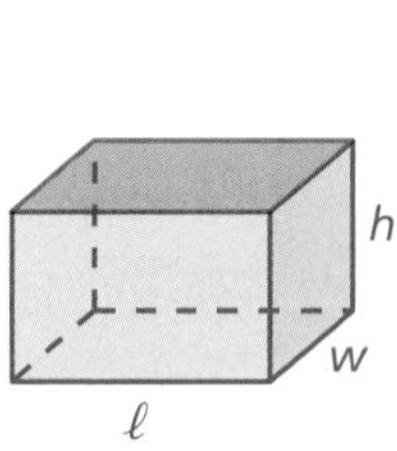

Algebra $S = 2\ell w + 2\ell h + 2wh$

Areas of bases ($2\ell w$); Areas of lateral faces ($2\ell h + 2wh$)

EXAMPLE 1 Finding the Surface Area of a Rectangular Prism

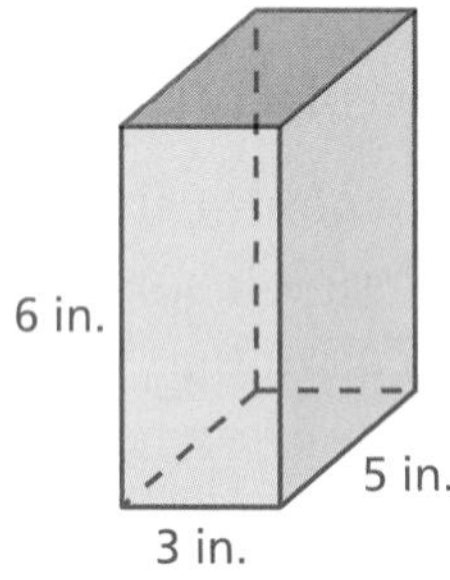

Find the surface area of the prism.

Draw a net.

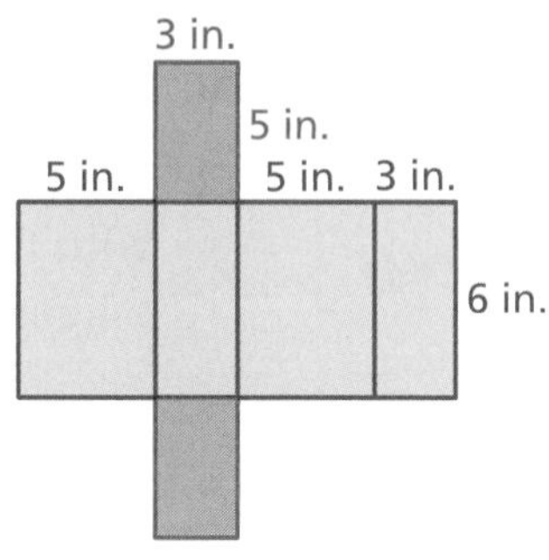

$$S = 2\ell w + 2\ell h + 2wh$$
$$= 2(3)(5) + 2(3)(6) + 2(5)(6)$$
$$= 30 + 36 + 60$$
$$= 126$$

The surface area is 126 square inches.

On Your Own

Now You're Ready
Exercises 7–9

Find the surface area of the prism.

1. 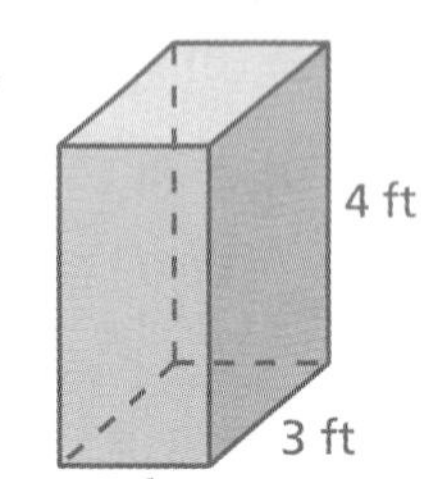

2. 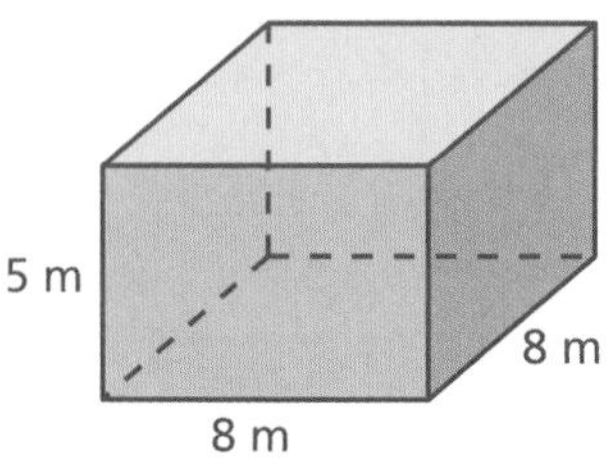

Key Idea

Surface Area of a Prism

The surface area S of any prism is the sum of the areas of the bases and the lateral faces.

$$S = \text{areas of bases} + \text{areas of lateral faces}$$

EXAMPLE 2 Finding the Surface Area of a Triangular Prism

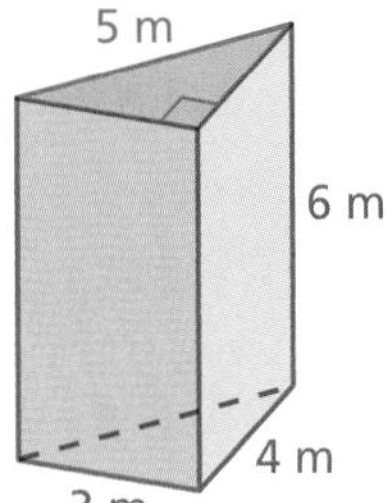

Find the surface area of the prism.

Draw a net.

3 m, 4 m, 3 m, 4 m, 5 m, 6 m

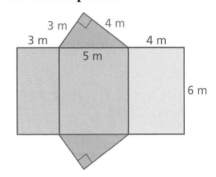

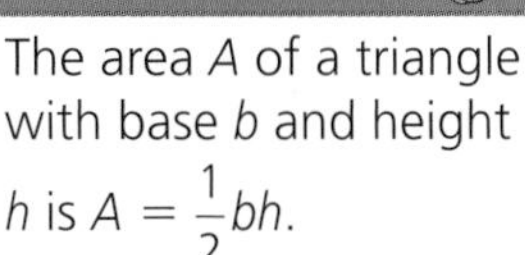

Remember

The area A of a triangle with base b and height h is $A = \frac{1}{2}bh$.

Area of a Base	***Areas of Lateral Faces***
Red base: $\frac{1}{2} \cdot 3 \cdot 4 = 6$	Green lateral face: $3 \cdot 6 = 18$
	Purple lateral face: $5 \cdot 6 = 30$
	Blue lateral face: $4 \cdot 6 = 24$

Add the areas of the bases and the lateral faces.

S = areas of bases + areas of lateral faces

$= 6 + 6 + 18 + 30 + 24$

There are two identical bases. Count the area twice.

$= 84$

∴ The surface area is 84 square meters.

On Your Own

Now You're Ready
Exercises 10–12

Find the surface area of the prism.

3.

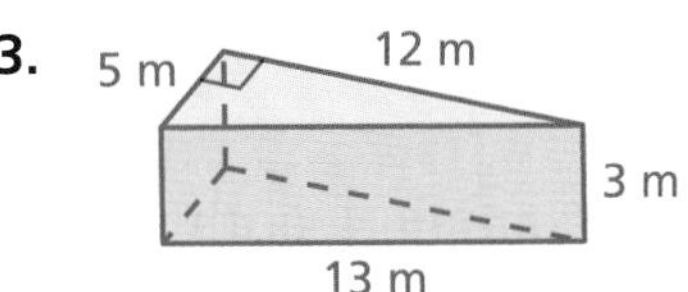

4.

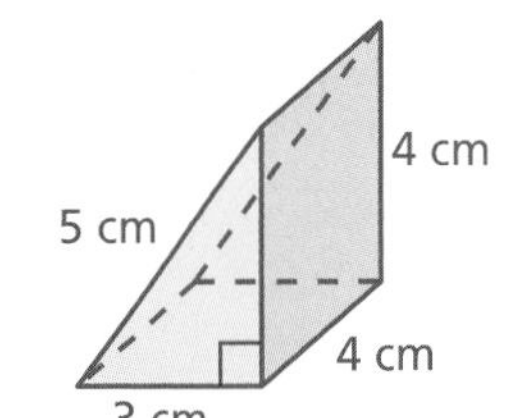

Remember

A cube has 6 congruent square faces.

When all the edges of a rectangular prism have the same length s, the rectangular prism is a cube. The formula for the surface area of a cube is

$S = 6s^2$. Formula for surface area of a cube

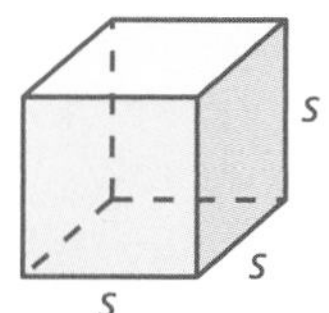

EXAMPLE 3 Finding the Surface Area of a Cube

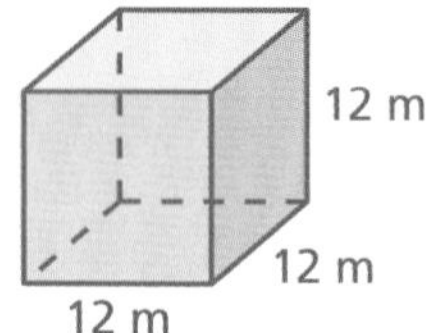

Find the surface area of the cube.

$S = 6s^2$	Write formula for surface area of a cube.
$= 6(12)^2$	Substitute 12 for s.
$= 864$	Simplify.

The surface area of the cube is 864 square meters.

The **lateral surface area** of a prism is the sum of the areas of the lateral faces.

EXAMPLE 4 Real-Life Application

The outsides of purple traps are coated with glue to catch emerald ash borers. You make your own trap in the shape of a rectangular prism with an open top and bottom. What is the surface area that you need to coat with glue?

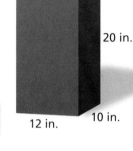

20 in.

12 in. 10 in.

Find the lateral surface area.

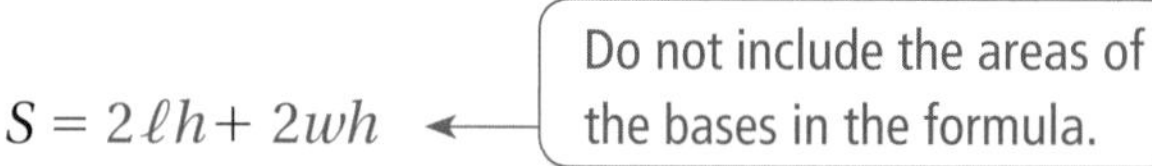

$S = 2\ell h + 2wh$ ← Do not include the areas of the bases in the formula.

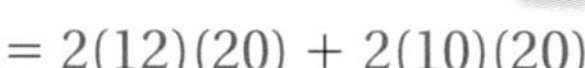

$= 2(12)(20) + 2(10)(20)$	Substitute.
$= 480 + 400$	Multiply.
$= 880$	Add.

So, you need to coat 880 square inches with glue.

On Your Own

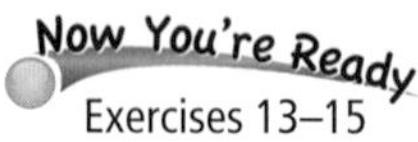

Now You're Ready
Exercises 13–15

5. Which prism has the greater surface area?

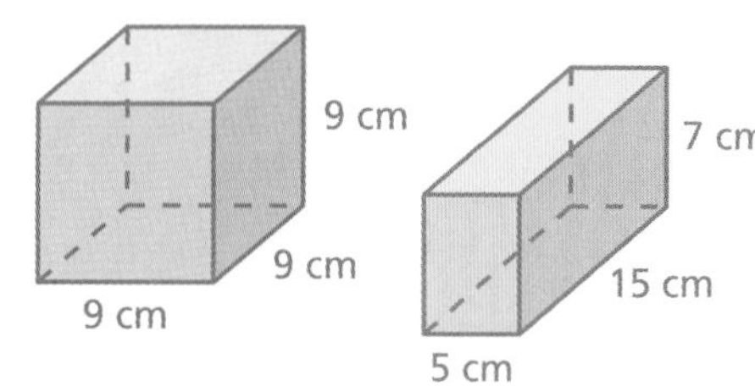

6. **WHAT IF?** In Example 4, both the length and the width of your trap are 12 inches. What is the surface area that you need to coat with glue?

14.1 Exercises

Vocabulary and Concept Check

1. **VOCABULARY** Describe two ways to find the surface area of a rectangular prism.
2. **WRITING** Compare and contrast a rectangular prism to a cube.
3. **DIFFERENT WORDS, SAME QUESTION** Which is different? Find "both" answers.

Find the surface area of the prism.	Find the area of the bases of the prism.
Find the area of the net of the prism.	Find the sum of the areas of the bases and the lateral faces of the prism.

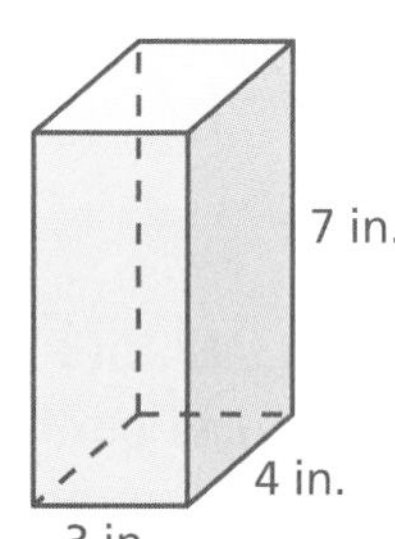

Practice and Problem Solving

Use one-inch cubes to form a rectangular prism that has the given dimensions. Then find the surface area of the prism.

4. $1 \times 2 \times 3$
5. $3 \times 4 \times 1$
6. $2 \times 3 \times 2$

Find the surface area of the prism.

1 7. 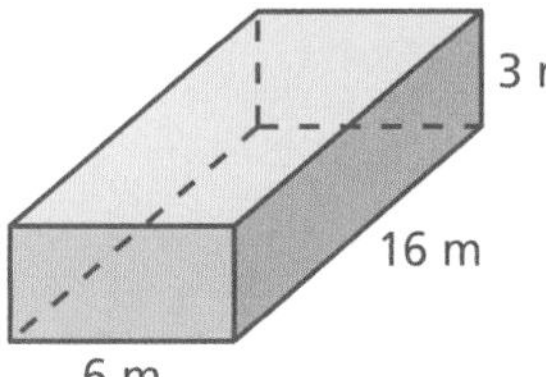

8. 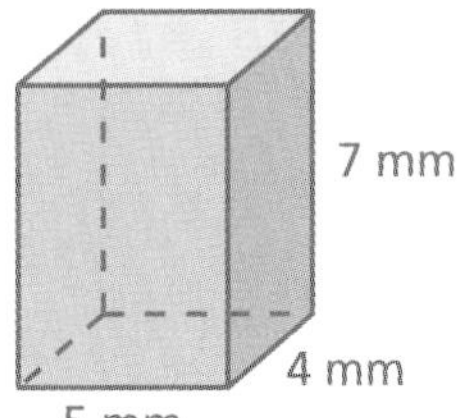

9. 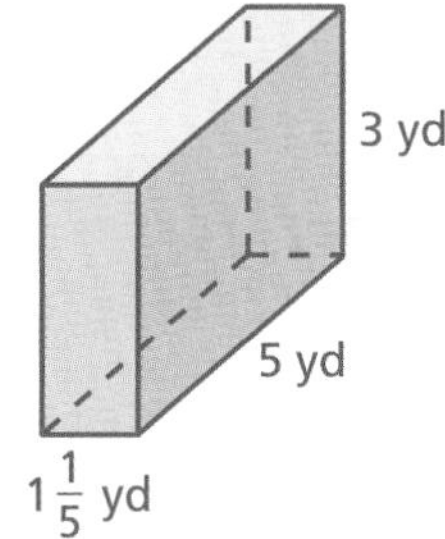

2 10. 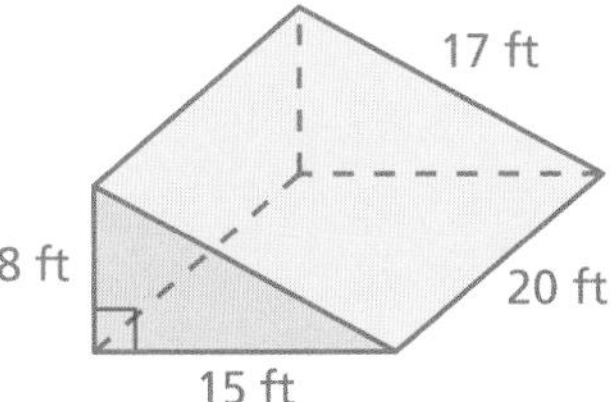

11. 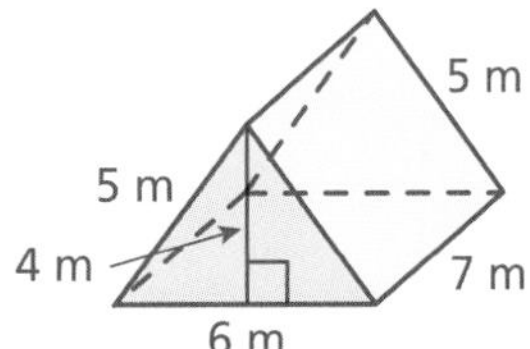

12. 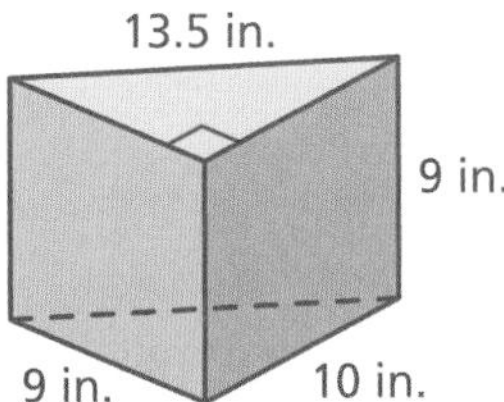

3 13. 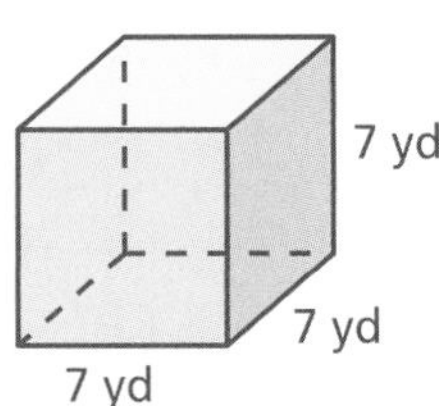

14. 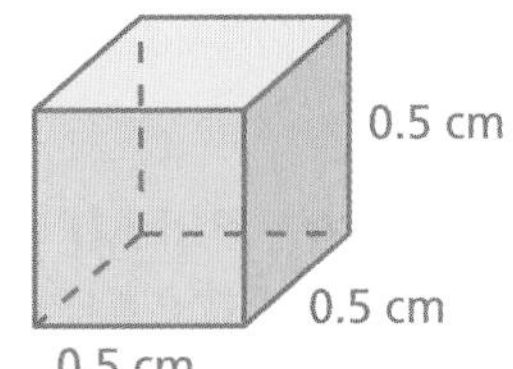

15. 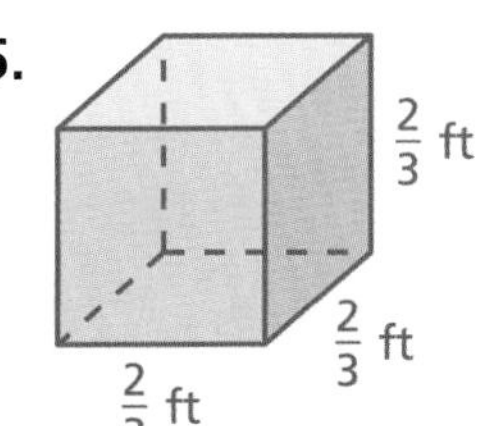

16. ERROR ANALYSIS Describe and correct the error in finding the surface area of the prism.

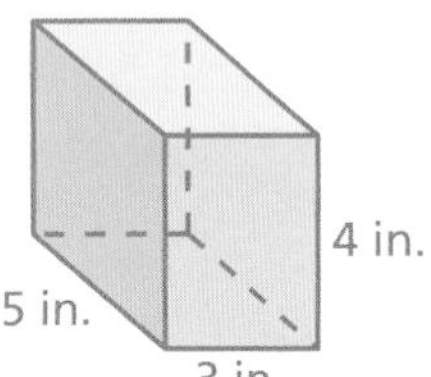

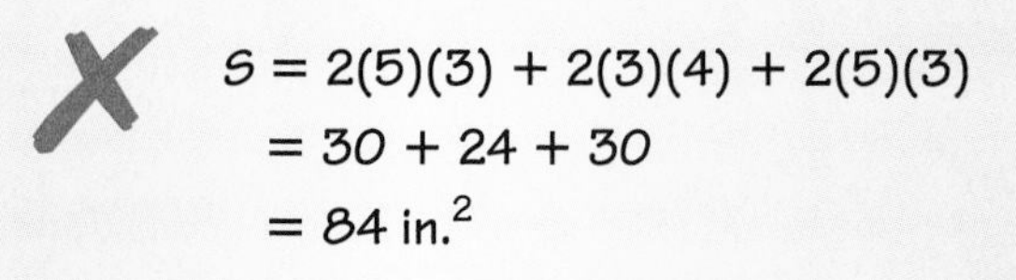

17. GAME Find the surface area of the tin game case.

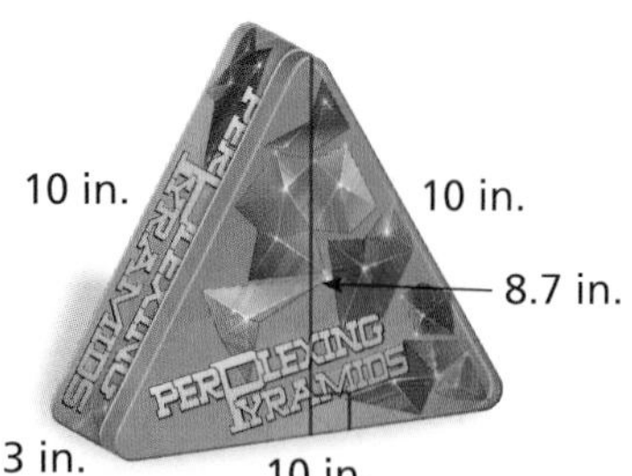

18. WRAPPING PAPER A cube-shaped gift is 11 centimeters long. What is the least amount of wrapping paper you need to wrap the gift?

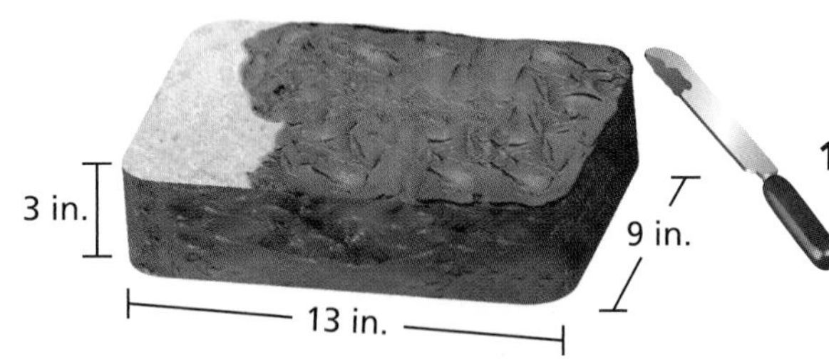

19. FROSTING One can of frosting covers about 280 square inches. Is one can of frosting enough to frost the cake? Explain.

Find the surface area of the prism.

20.

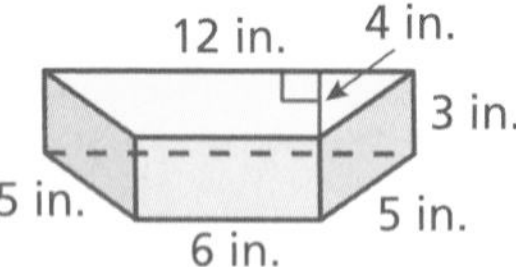

21.

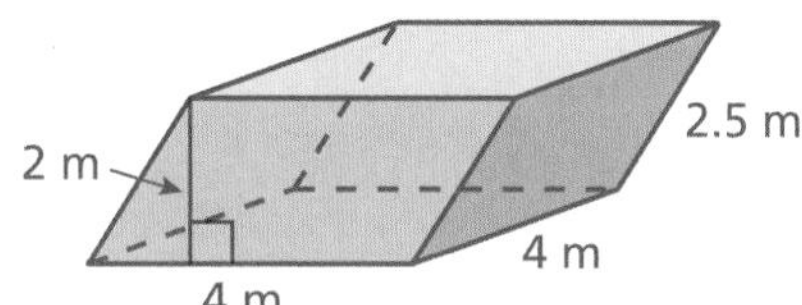

22. OPEN-ENDED Draw and label a rectangular prism that has a surface area of 158 square yards.

23. LABEL A label that wraps around a box of golf balls covers 75% of its lateral surface area. What is the value of x?

24. BREAD Fifty percent of the surface area of the bread is crust. What is the height h?

Compare the dimensions of the prisms. How many times greater is the surface area of the red prism than the surface area of the blue prism?

25.

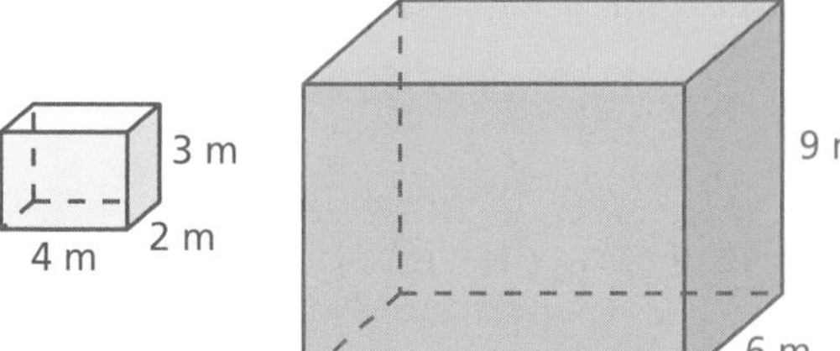

26.

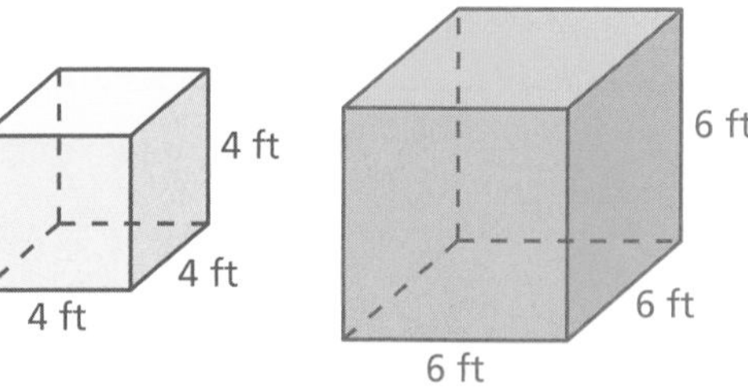

27. **STRUCTURE** You are painting the prize pedestals shown (including the bottoms). You need 0.5 pint of paint to paint the red pedestal.

   a. The side lengths of the green pedestal are one-half the side lengths of the red pedestal. How much paint do you need to paint the green pedestal?

   b. The side lengths of the blue pedestal are triple the side lengths of the green pedestal. How much paint do you need to paint the blue pedestal?

   c. Compare the ratio of paint amounts to the ratio of side lengths for the green and red pedestals. Repeat for the green and blue pedestals. What do you notice?

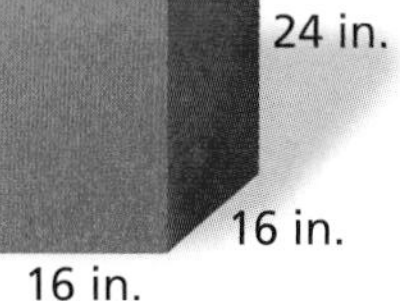

28. Number Sense A keychain-sized Rubik's Cube® is made up of small cubes. Each small cube has a surface area of 1.5 square inches.

   a. What is the side length of each small cube?

   b. What is the surface area of the entire Rubik's Cube®?

Fair Game Review What you learned in previous grades & lessons

Find the area of the triangle *(Skills Review Handbook)*

29.

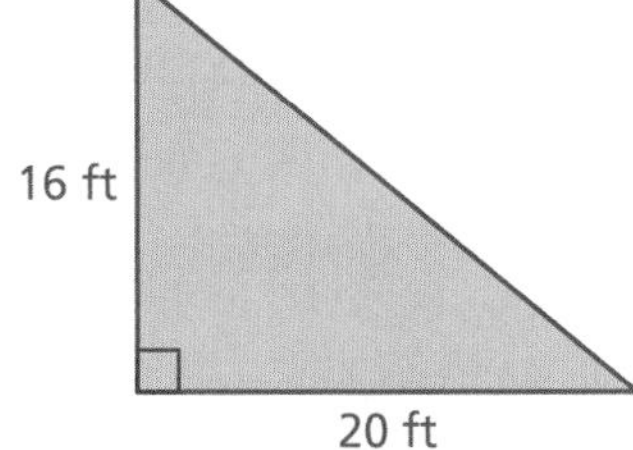

30.

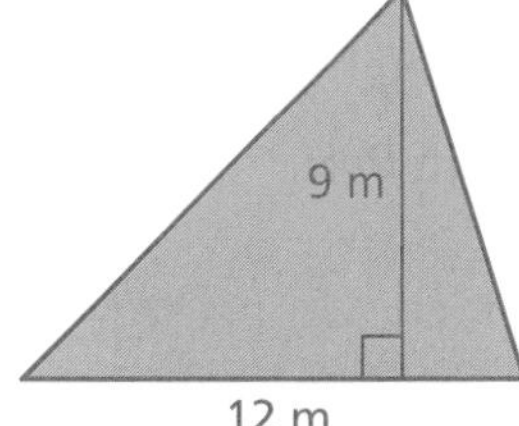

31.

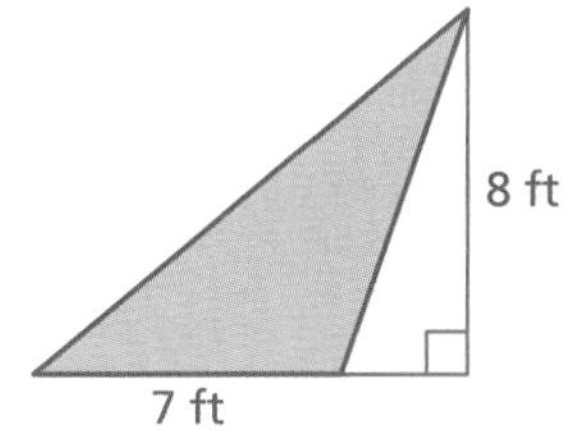

32. **MULTIPLE CHOICE** What is the circumference of the basketball? Use 3.14 for π. *(Section 13.1)*

   Ⓐ 14.13 in. Ⓑ 28.26 in. Ⓒ 56.52 in. Ⓓ 254.34 in.

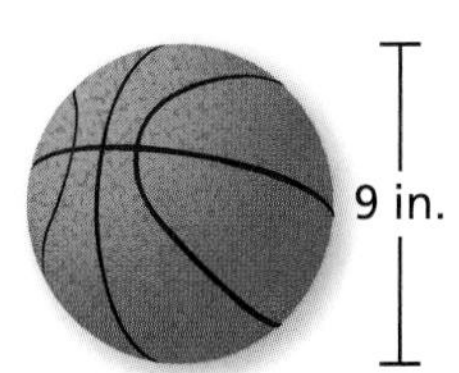

14.2 Surface Areas of Pyramids

Essential Question How can you find the surface area of a pyramid?

Even though many well-known pyramids have square bases, the base of a pyramid can be any polygon.

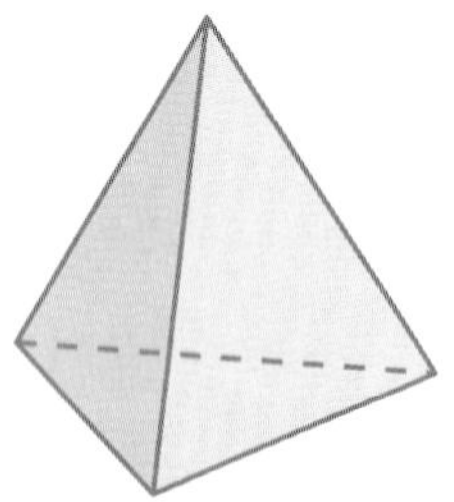

Triangular Base

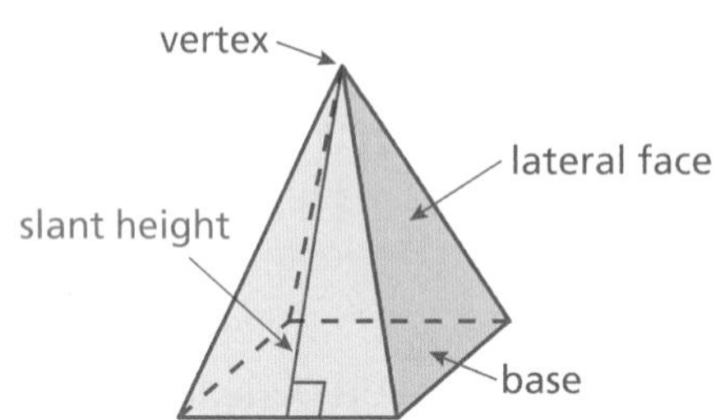

Square Base

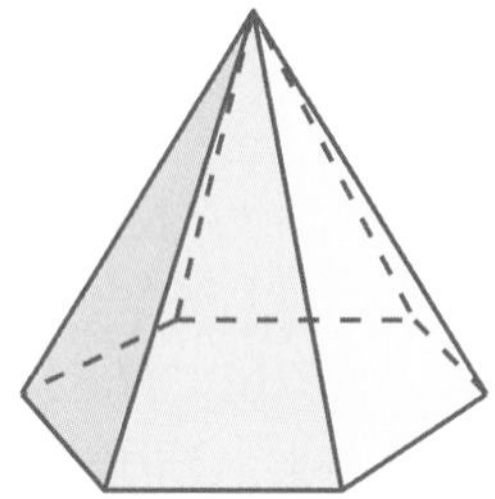

Hexagonal Base

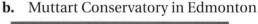

1 ACTIVITY: Making a Scale Model

Work with a partner. Each pyramid has a square base.

- **Draw a net for a scale model of one of the pyramids. Describe your scale.**
- **Cut out the net and fold it to form a pyramid.**
- **Find the lateral surface area of the real-life pyramid.**

a. Cheops Pyramid in Egypt

Side = 230 m, Slant height ≈ 186 m

b. Muttart Conservatory in Edmonton

Side = 26 m, Slant height ≈ 27 m

c. Louvre Pyramid in Paris

Side = 35 m, Slant height ≈ 28 m

d. Pyramid of Caius Cestius in Rome

Side = 22 m, Slant height ≈ 29 m

Geometry

In this lesson, you will
- find surface areas of regular pyramids.
- solve real-life problems.

2 ACTIVITY: Estimation

Math Practice

Calculate Accurately

How can you verify that you have calculated the lateral surface area accurately?

Work with a partner. There are many different types of gemstone cuts. Here is one called a brilliant cut.

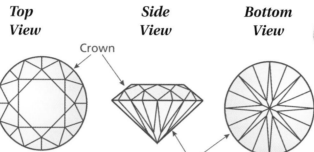

Top View ***Side View*** ***Bottom View***

Crown

Pavilion

The size and shape of the pavilion can be approximated by an octagonal pyramid.

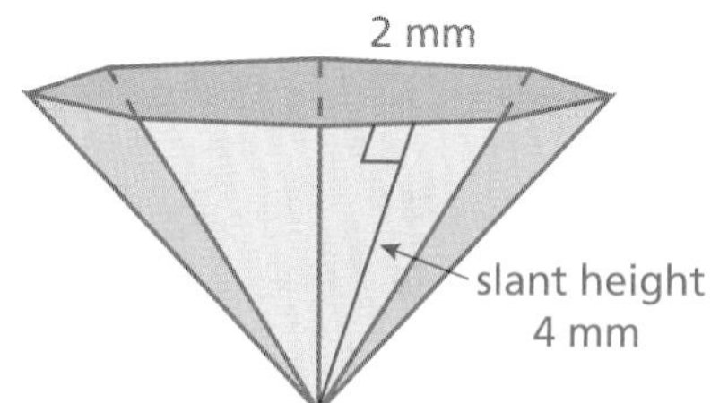

a. What does *octagonal* mean?

b. Draw a net for the pyramid.

c. Find the lateral surface area of the pyramid.

3 ACTIVITY: Comparing Surface Areas

Work with a partner. Both pyramids have the same side lengths of the base and the same slant heights.

a. **REASONING** Without calculating, which pyramid has the greater surface area? Explain.

b. Verify your answer to part (a) by finding the surface area of each pyramid.

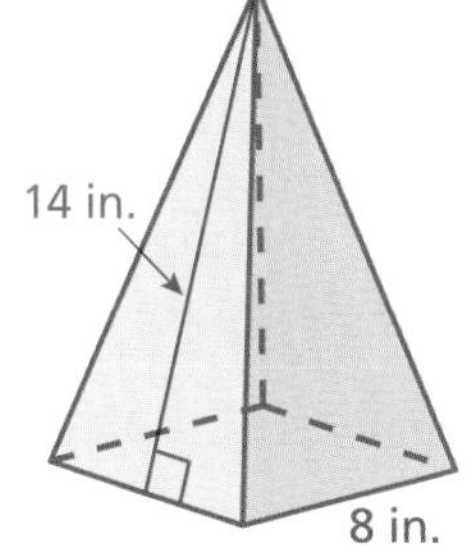

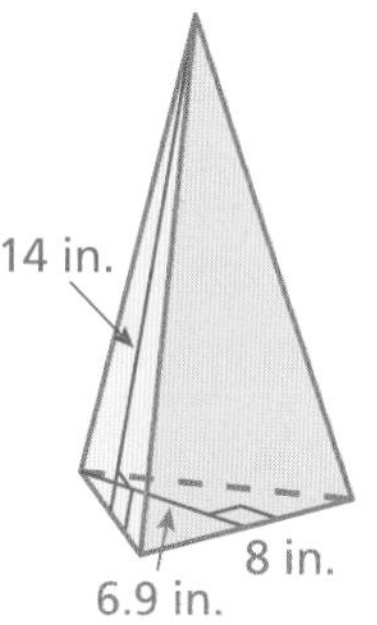

What Is Your Answer?

4. **IN YOUR OWN WORDS** How can you find the surface area of a pyramid? Draw a diagram with your explanation.

Practice Use what you learned about the surface area of a pyramid to complete Exercises 4–6 on page 598.

14.2 Lesson

Key Vocabulary
regular pyramid, *p. 596*
slant height, *p. 596*

A **regular pyramid** is a pyramid whose base is a regular polygon. The lateral faces are triangles. The height of each triangle is the **slant height** of the pyramid.

Remember

In a regular polygon, all the sides are congruent and all the angles are congruent.

Surface Area of a Pyramid

The surface area S of a pyramid is the sum of the areas of the base and the lateral faces.

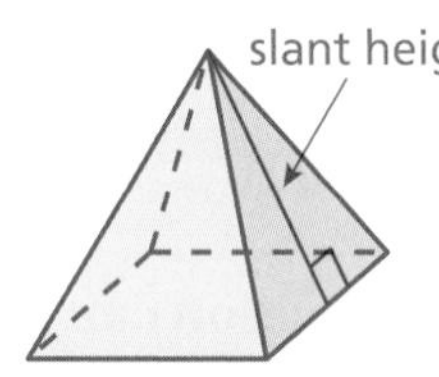

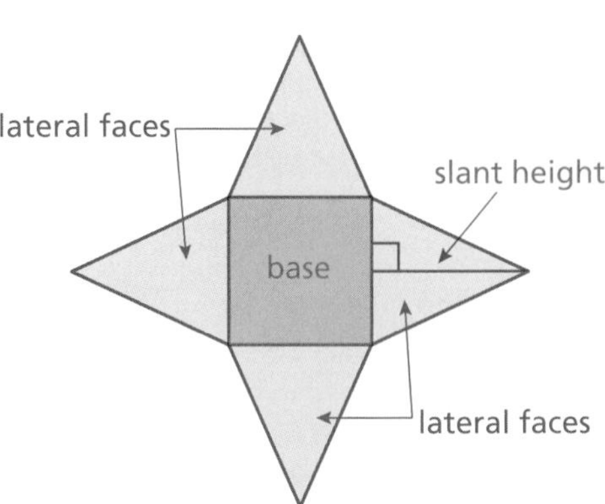

S = area of base + areas of lateral faces

EXAMPLE 1 Finding the Surface Area of a Square Pyramid

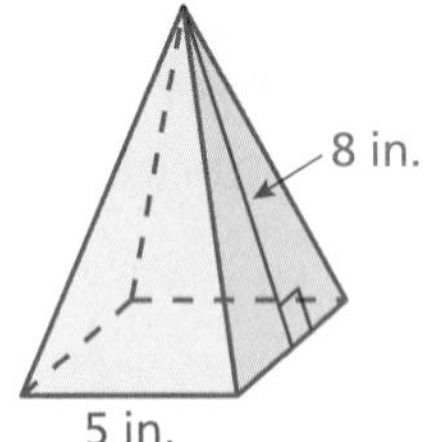

Find the surface area of the regular pyramid.

Draw a net.

Area of Base	*Area of a Lateral Face*
$5 \cdot 5 = 25$	$\frac{1}{2} \cdot 5 \cdot 8 = 20$

5 in.
8 in.
5 in.

Find the sum of the areas of the base and the lateral faces.

S = area of base + areas of lateral faces

$= 25 + 20 + 20 + 20 + 20$

$= 105$

There are 4 identical lateral faces. Count the area 4 times.

The surface area is 105 square inches.

On Your Own

1. What is the surface area of a square pyramid with a base side length of 9 centimeters and a slant height of 7 centimeters?

Multi-Language Glossary at BigIdeasMath.com

EXAMPLE 2 Finding the Surface Area of a Triangular Pyramid

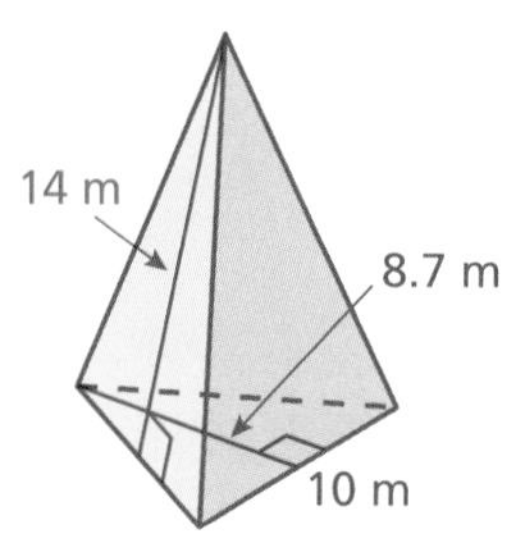

Find the surface area of the regular pyramid.

Draw a net.

Area of Base

$\frac{1}{2} \cdot 10 \cdot 8.7 = 43.5$

Area of a Lateral Face

$\frac{1}{2} \cdot 10 \cdot 14 = 70$

Find the sum of the areas of the base and the lateral faces.

$$S = \text{area of base} + \text{areas of lateral faces}$$
$$= 43.5 + \underbrace{70 + 70 + 70}$$
$$= 253.5$$

There are 3 identical lateral faces. Count the area 3 times.

The surface area is 253.5 square meters.

EXAMPLE 3 Real-Life Application

A roof is shaped like a square pyramid. One bundle of shingles covers 25 square feet. How many bundles should you buy to cover the roof?

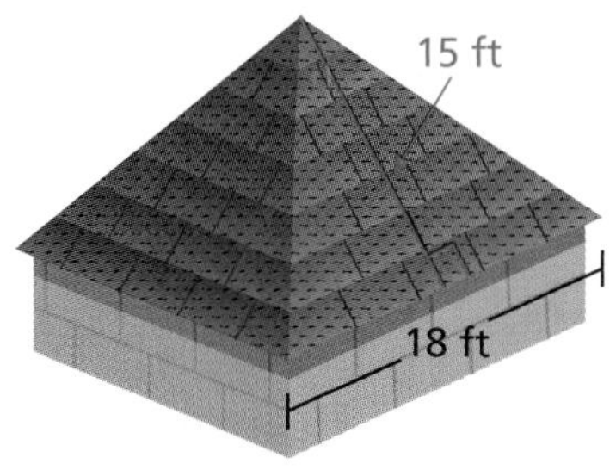

The base of the roof does not need shingles. So, find the sum of the areas of the lateral faces of the pyramid.

Area of a Lateral Face

$\frac{1}{2} \cdot 18 \cdot 15 = 135$

There are four identical lateral faces. So, the lateral surface area is

$$135 + 135 + 135 + 135 = 540.$$

Because one bundle of shingles covers 25 square feet, it will take $540 \div 25 = 21.6$ bundles to cover the roof.

So, you should buy 22 bundles of shingles.

On Your Own

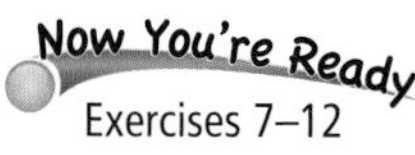

2. What is the surface area of the regular pyramid at the right?
3. **WHAT IF?** In Example 3, one bundle of shingles covers 32 square feet. How many bundles should you buy to cover the roof?

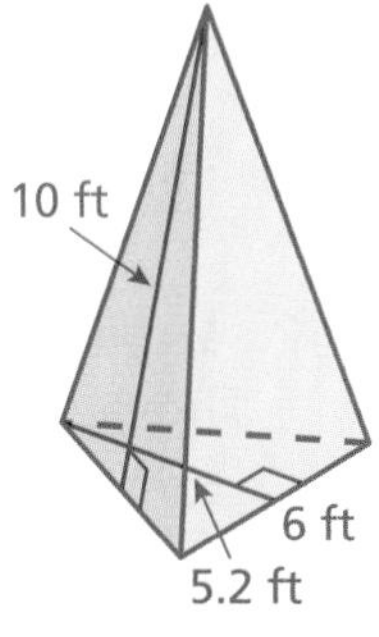

14.2 Exercises

Vocabulary and Concept Check

1. **VOCABULARY** Can a pyramid have rectangles as lateral faces? Explain.
2. **CRITICAL THINKING** Why is it helpful to know the slant height of a pyramid to find its surface area?
3. **WHICH ONE DOESN'T BELONG?** Which description of the solid does *not* belong with the other three? Explain your answer.

square pyramid	regular pyramid
rectangular pyramid	triangular pyramid

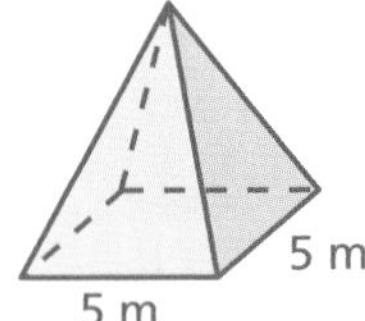

Practice and Problem Solving

Use the net to find the surface area of the regular pyramid.

4. 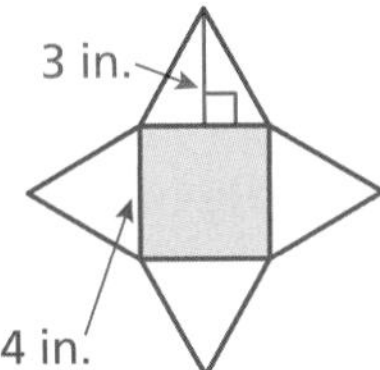

5. 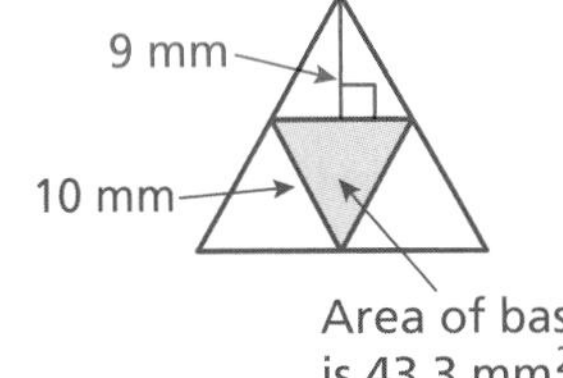

6. 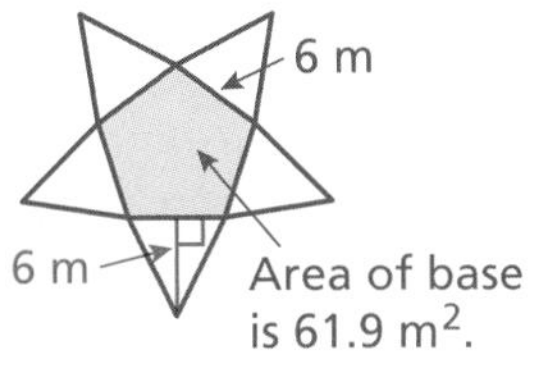

In Exercises 7–11, find the surface area of the regular pyramid.

7. 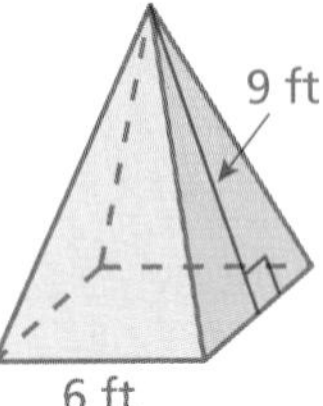

8. 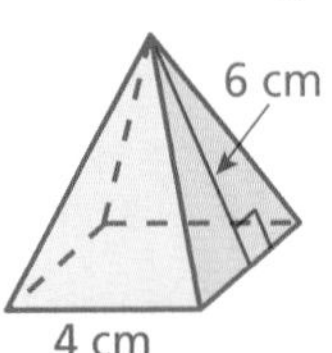

9. 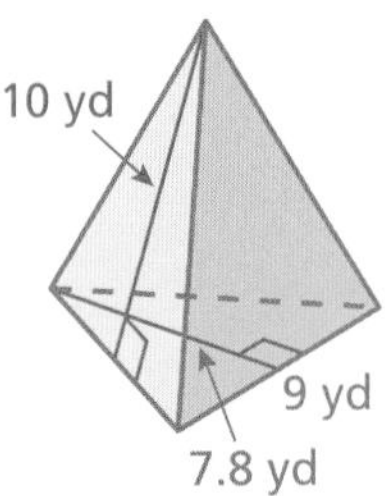

10. 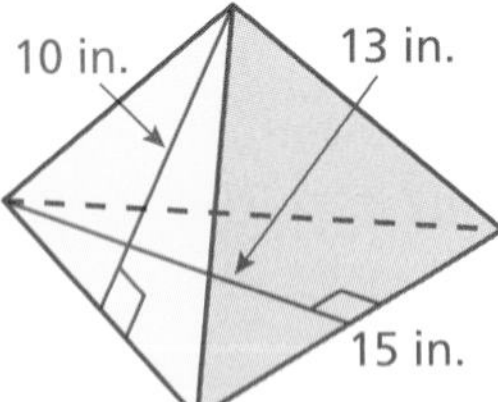

11. 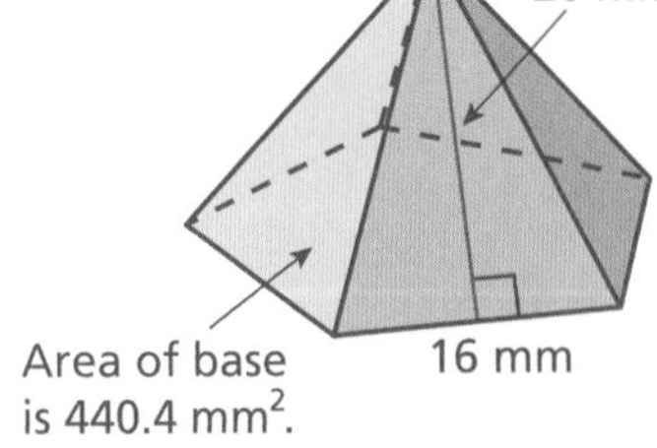
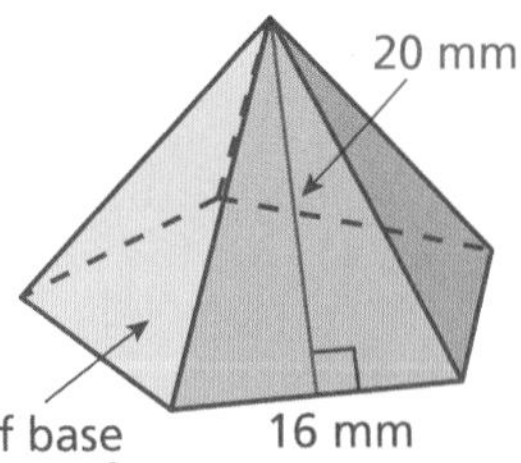

3 12. **LAMPSHADE** The base of the lampshade is a regular hexagon with a side length of 8 inches. Estimate the amount of glass needed to make the lampshade.

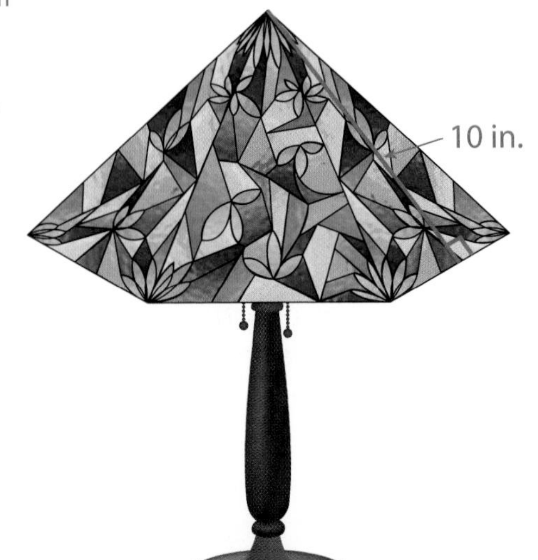

13. **GEOMETRY** The surface area of a square pyramid is 85 square meters. The base length is 5 meters. What is the slant height?

Find the surface area of the composite solid.

14.

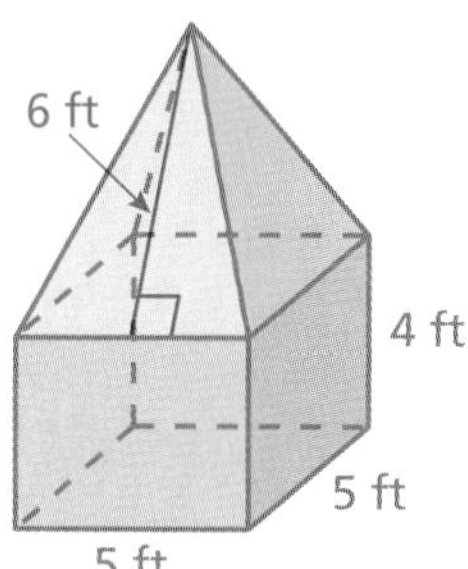

15.

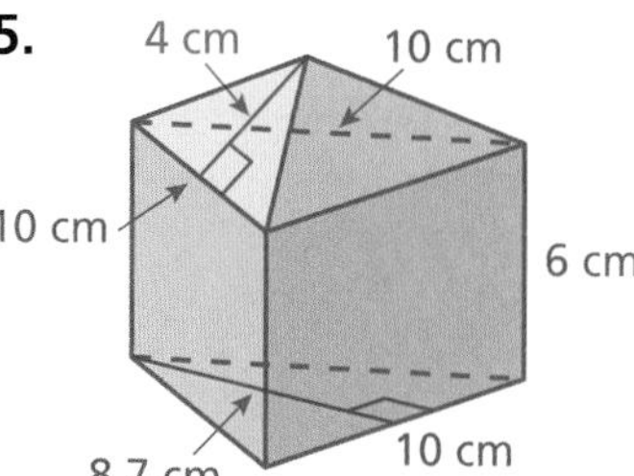

16.

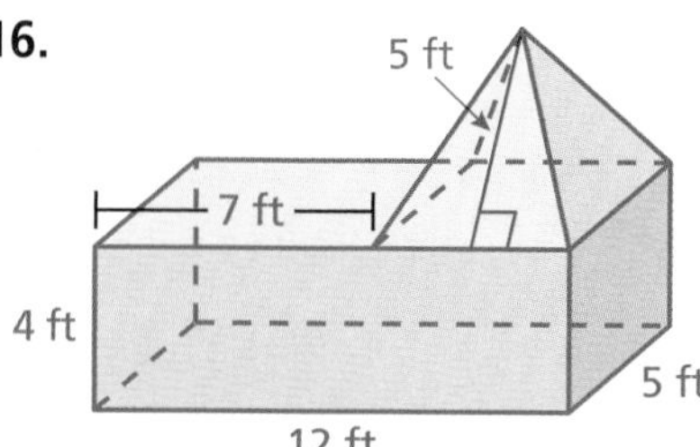

5 ft

4 ft

17. PROBLEM SOLVING You are making an umbrella that is shaped like a regular octagonal pyramid.

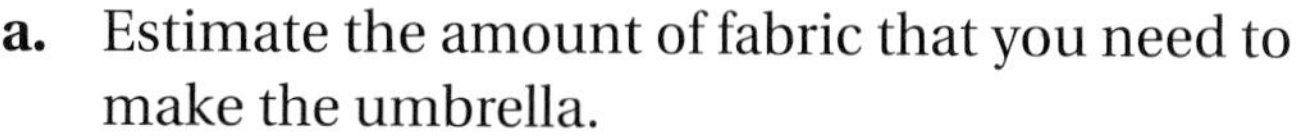

a. Estimate the amount of fabric that you need to make the umbrella.

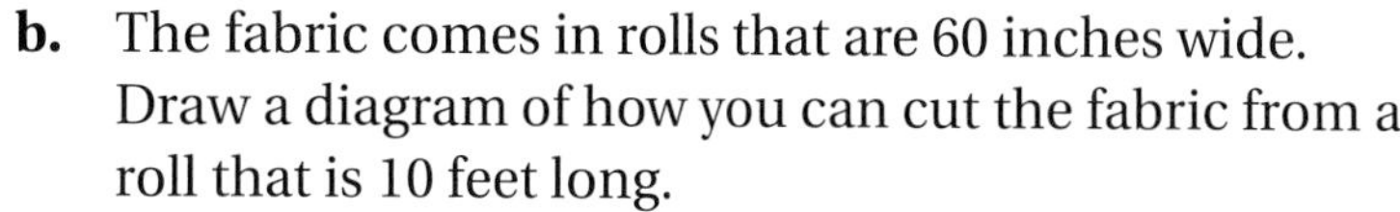

b. The fabric comes in rolls that are 60 inches wide. Draw a diagram of how you can cut the fabric from a roll that is 10 feet long.

c. How much fabric is wasted?

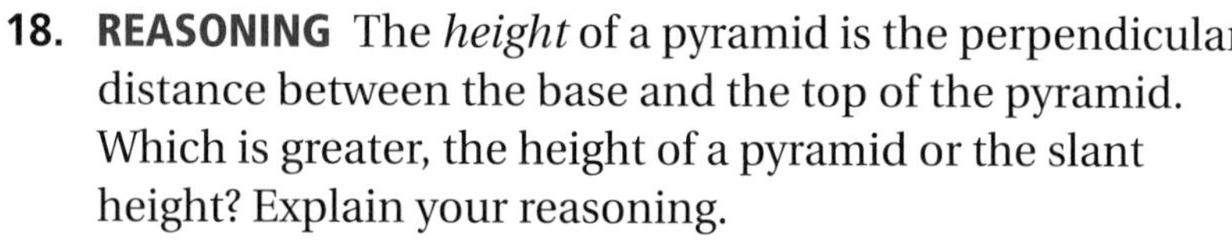

18. REASONING The *height* of a pyramid is the perpendicular distance between the base and the top of the pyramid. Which is greater, the height of a pyramid or the slant height? Explain your reasoning.

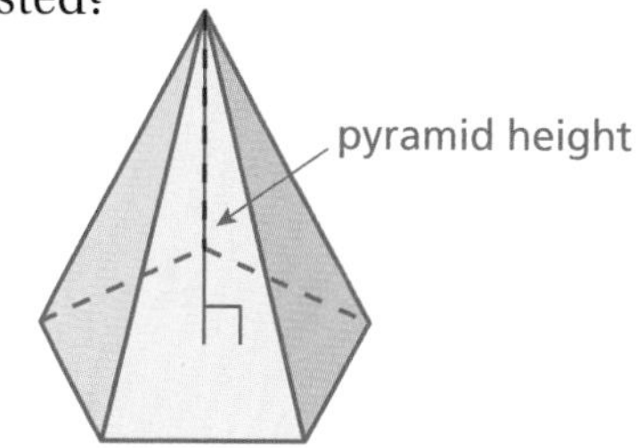

19. TETRAHEDRON A tetrahedron is a triangular pyramid whose four faces are identical equilateral triangles. The total lateral surface area is 93 square centimeters. Find the surface area of the tetrahedron.

20. Reasoning Is the total area of the lateral faces of a pyramid *greater than*, *less than*, or *equal to* the area of the base? Explain.

Fair Game Review What you learned in previous grades & lessons

Find the area and the circumference of the circle. Use 3.14 for π.
(Section 13.1 and Section 13.3)

21.

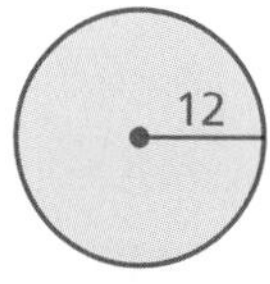

22.

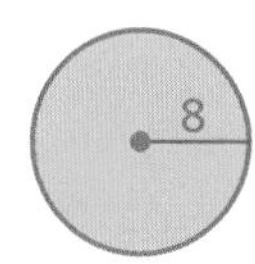

23.

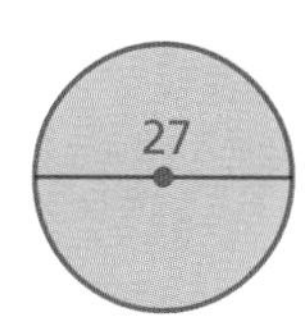

24. MULTIPLE CHOICE The distance between bases on a youth baseball field is proportional to the distance between bases on a professional baseball field. The ratio of the youth distance to the professional distance is 2 : 3. Bases on a youth baseball field are 60 feet apart. What is the distance between bases on a professional baseball field? *(Skills Review Handbook)*

Ⓐ 40 ft Ⓑ 90 ft Ⓒ 120 ft Ⓓ 180 ft

14.3 Surface Areas of Cylinders

Essential Question How can you find the surface area of a cylinder?

A *cylinder* is a solid that has two parallel, identical circular bases.

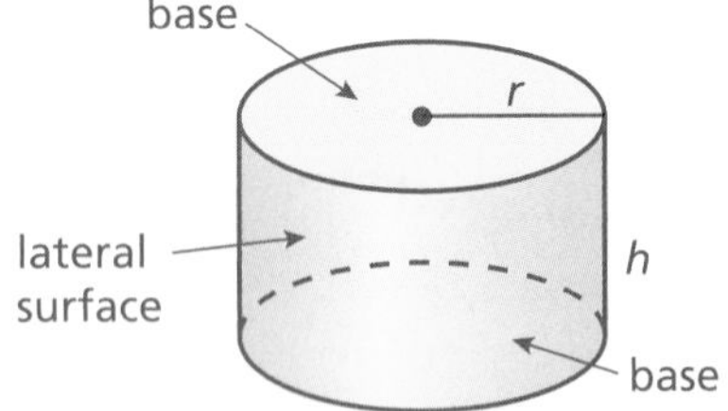

1 ACTIVITY: Finding Area

Work with a partner. Use a cardboard cylinder.

- **Talk about how you can find the area of the outside of the roll.**
- **Estimate the area using the methods you discussed.**
- **Use the roll and the scissors to find the actual area of the cardboard.**
- **Compare the actual area to your estimates.**

2 ACTIVITY: Finding Surface Area

Work with a partner.

- **Make a net for the can. Name the shapes in the net.**
- **Find the surface area of the can.**
- **How are the dimensions of the rectangle related to the dimensions of the can?**

Geometry
In this lesson, you will
- find surface areas of cylinders.

3 ACTIVITY: Estimation

Math Practice

View as Components

How can you use the results of Activity 2 to help you identify the components of the surface area?

Work with a partner. From memory, estimate the dimensions of the real-life item in inches. Then use the dimensions to estimate the surface area of the item in square inches.

a.

b.

c.

d.

What Is Your Answer?

4. IN YOUR OWN WORDS How can you find the surface area of a cylinder? Give an example with your description. Include a drawing of the cylinder.

5. To eight decimal places, $\pi \approx 3.14159265$. Which of the following is closest to π?

a. 3.14 **b.** $\frac{22}{7}$ **c.** $\frac{355}{113}$

"To approximate $\pi \approx 3.141593$, I simply remember 1, 1, 3, 3, 5, 5."

"Then I compute $\frac{355}{113} \approx 3.141593$."

Practice Use what you learned about the surface area of a cylinder to complete Exercises 3–5 on page 604.

14.3 Lesson

Surface Area of a Cylinder

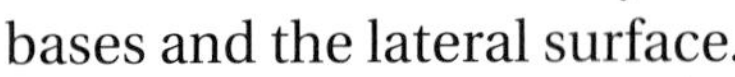

Words The surface area S of a cylinder is the sum of the areas of the bases and the lateral surface.

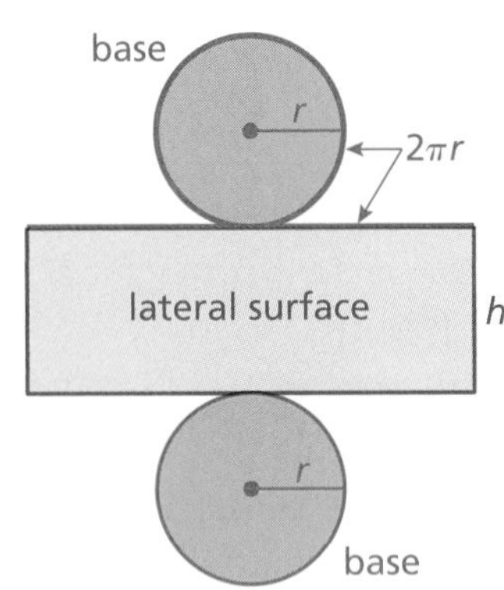

Algebra $S = 2\pi r^2 + 2\pi rh$

Areas of bases — Area of lateral surface

Remember

Pi can be approximated as 3.14 or $\frac{22}{7}$.

EXAMPLE 1 Finding the Surface Area of a Cylinder

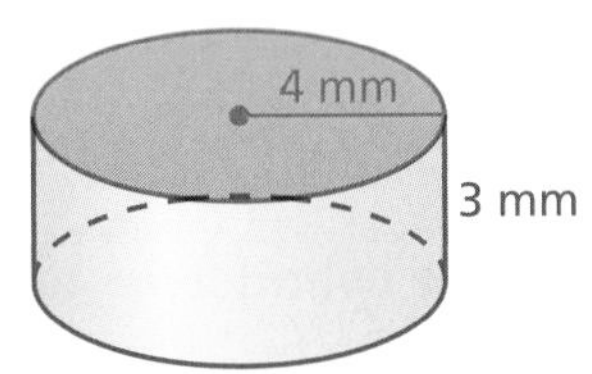

Find the surface area of the cylinder. Round your answer to the nearest tenth.

Draw a net.

$$S = 2\pi r^2 + 2\pi rh$$
$$= 2\pi(4)^2 + 2\pi(4)(3)$$
$$= 32\pi + 24\pi$$
$$= 56\pi$$
$$\approx 175.8$$

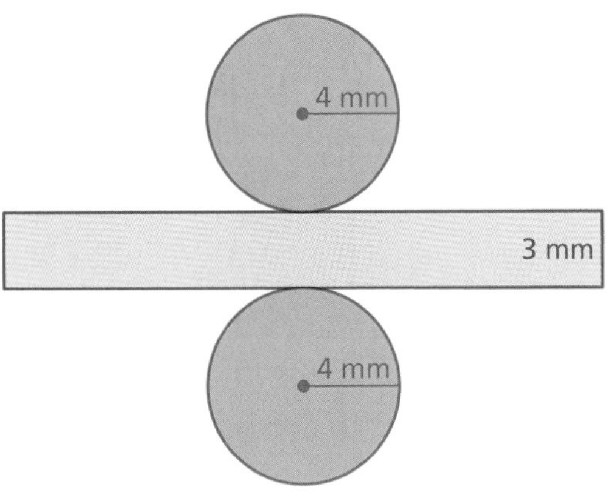

The surface area is about 175.8 square millimeters.

On Your Own

Find the surface area of the cylinder. Round your answer to the nearest tenth.

1. 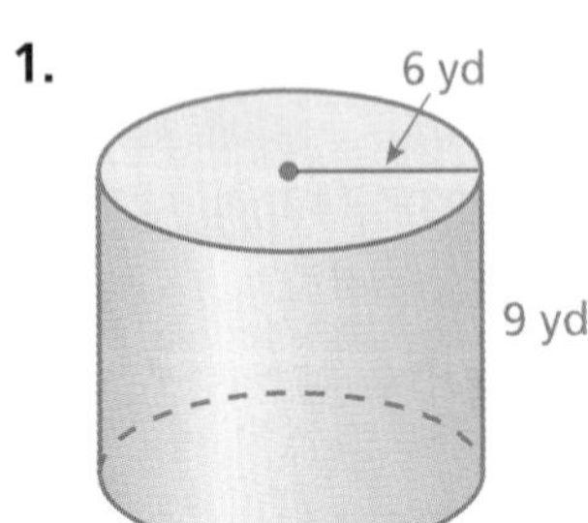

2. 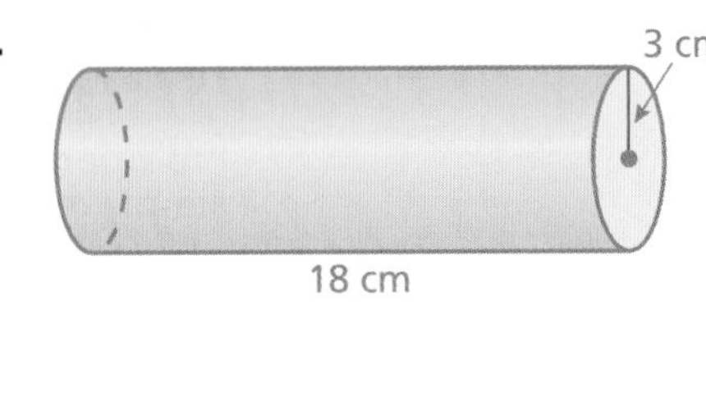

EXAMPLE 2 Finding Surface Area

How much paper is used for the label on the can of peas?

Find the lateral surface area of the cylinder.

$S = 2\pi rh$

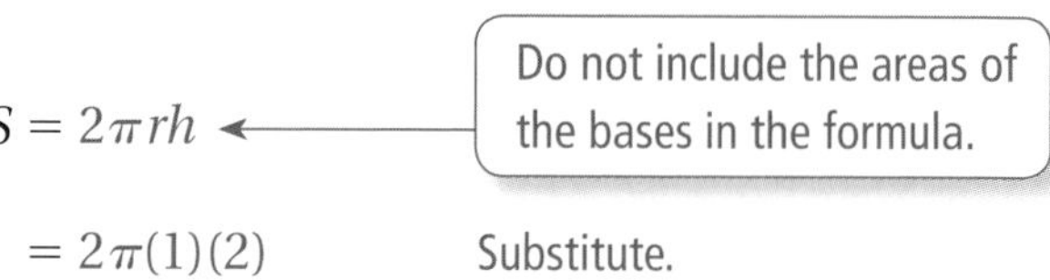

Do not include the areas of the bases in the formula.

$= 2\pi(1)(2)$ Substitute.

$= 4\pi \approx 12.56$ Multiply.

∴ About 12.56 square inches of paper is used for the label.

EXAMPLE 3 Real-Life Application

You earn \$0.01 for recycling the can in Example 2. How much can you expect to earn for recycling the tomato can? Assume that the recycle value is proportional to the surface area.

Find the surface area of each can.

Tomatoes	***Peas***
$S = 2\pi r^2 + 2\pi rh$	$S = 2\pi r^2 + 2\pi rh$
$= 2\pi(2)^2 + 2\pi(2)(5.5)$	$= 2\pi(1)^2 + 2\pi(1)(2)$
$= 8\pi + 22\pi$	$= 2\pi + 4\pi$
$= 30\pi$	$= 6\pi$

Use a proportion to find the recycle value x of the tomato can.

$$\frac{30\pi \text{ in.}^2}{x} = \frac{6\pi \text{ in.}^2}{\$0.01}$$

surface area

recycle value

$30\pi \cdot 0.01 = x \cdot 6\pi$ Cross Products Property

$5 \cdot 0.01 = x$ Divide each side by 6π.

$0.05 = x$ Simplify.

∴ You can expect to earn \$0.05 for recycling the tomato can.

On Your Own

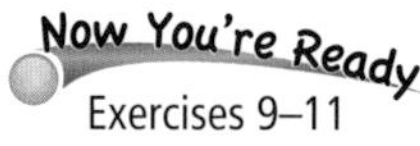

3. **WHAT IF?** In Examples 2 and 3, the height of the can of peas is doubled.

   a. Does the amount of paper used in the label double?

   b. Does the recycle value double? Explain.

14.3 Exercises

Vocabulary and Concept Check

1. **CRITICAL THINKING** Which part of the formula $S = 2\pi r^2 + 2\pi rh$ represents the lateral surface area of a cylinder?
2. **CRITICAL THINKING** You are given the height and the circumference of the base of a cylinder. Describe how to find the surface area of the entire cylinder.

Practice and Problem Solving

Make a net for the cylinder. Then find the surface area of the cylinder. Round your answer to the nearest tenth.

3. 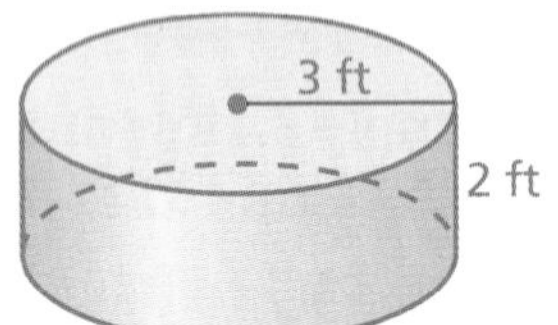

4. 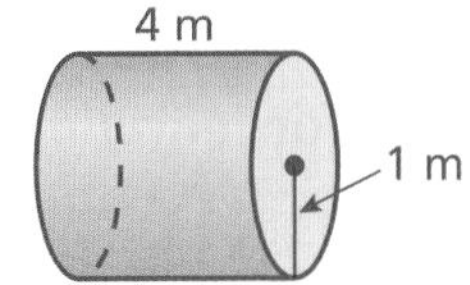

5. 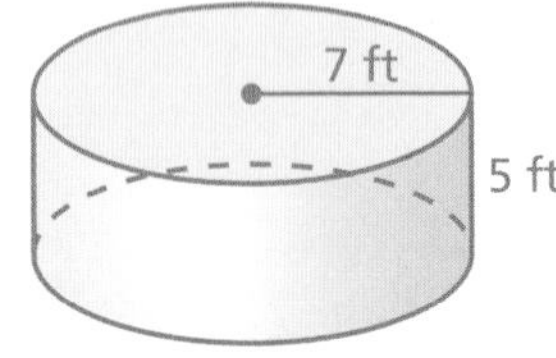

Find the surface area of the cylinder. Round your answer to the nearest tenth.

1 6. 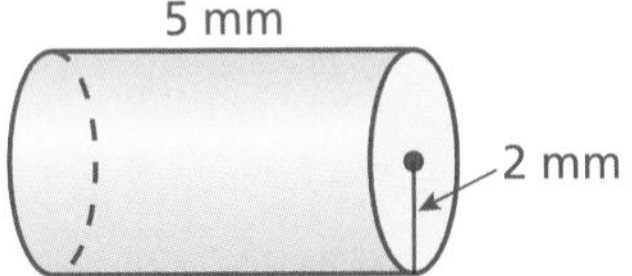

7. 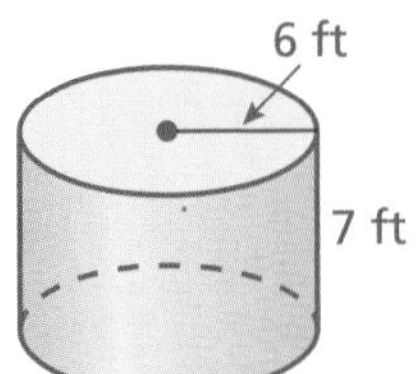

8. 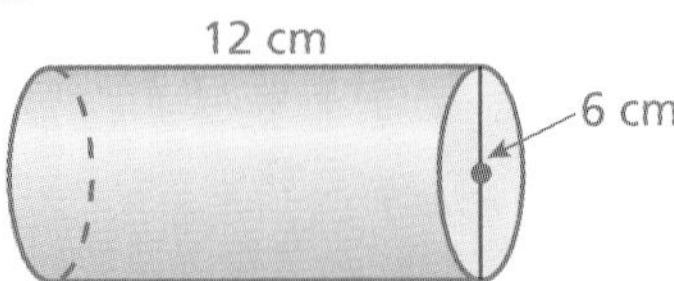

Find the lateral surface area of the cylinder. Round your answer to the nearest tenth.

2 9. 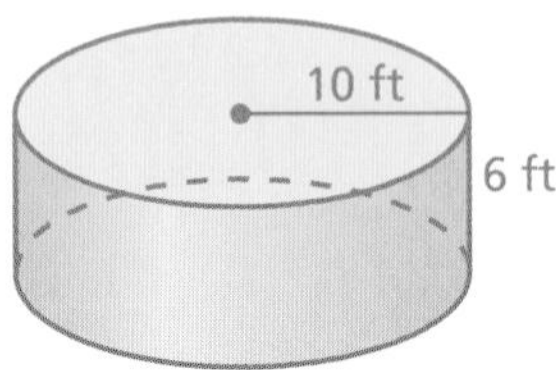

10. 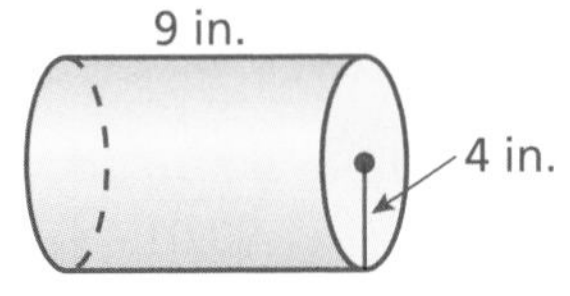

11. 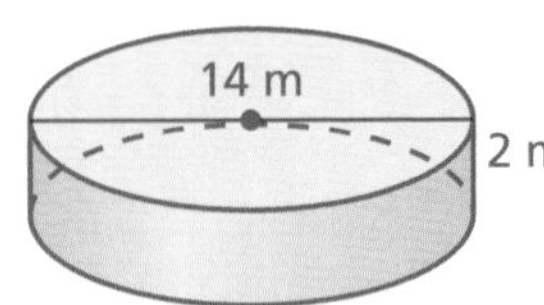

12. **ERROR ANALYSIS** Describe and correct the error in finding the surface area of the cylinder.

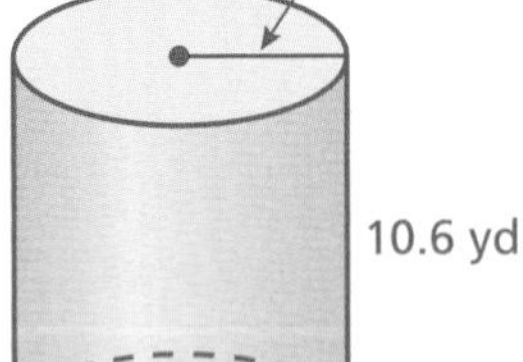

$$S = \pi r^2 + 2\pi rh$$
$$= \pi(5)^2 + 2\pi(5)(10.6)$$
$$= 25\pi + 106\pi$$
$$= 131\pi \approx 411.3 \text{ yd}^2$$

13. **TANKER** The truck's tank is a stainless steel cylinder. Find the surface area of the tank.

14. **OTTOMAN** What percent of the surface area of the ottoman is green (not including the bottom)?

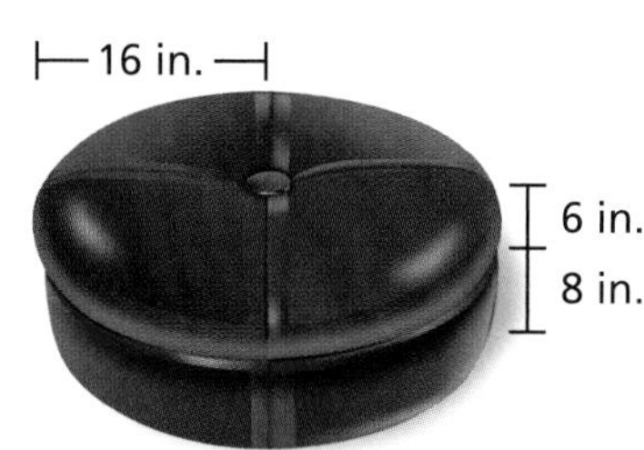

15. **REASONING** You make two cylinders using 8.5-by-11-inch pieces of paper. One has a height of 8.5 inches, and the other has a height of 11 inches. Without calculating, compare the surface areas of the cylinders.

16. **INSTRUMENT** A *ganza* is a percussion instrument used in samba music.

    a. Find the surface area of each of the two labeled ganzas.

    b. The weight of the smaller ganza is 1.1 pounds. Assume that the surface area is proportional to the weight. What is the weight of the larger ganza?

17. **BRIE CHEESE** The cut wedge represents one-eighth of the cheese.

    a. Find the surface area of the cheese before it is cut.

    b. Find the surface area of the remaining cheese after the wedge is removed. Did the surface area increase, decrease, or remain the same?

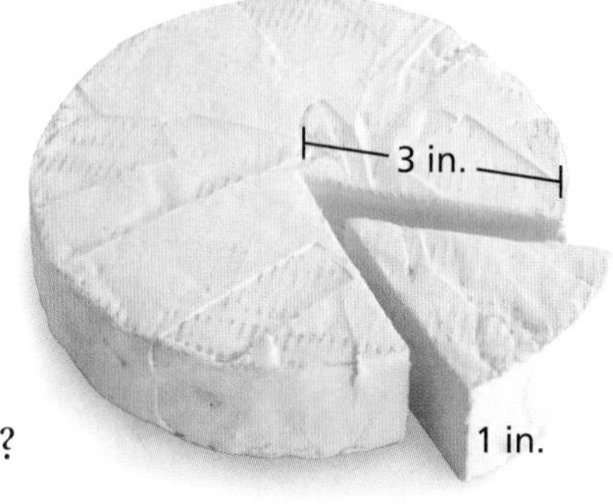

18. 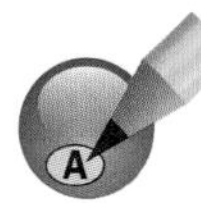

A cylinder has radius r and height h.

    a. How many times greater is the surface area of a cylinder when both dimensions are multiplied by a factor of 2? 3? 5? 10?

    b. Describe the pattern in part (a). How many times greater is the surface area of a cylinder when both dimensions are multiplied by a factor of 20?

Fair Game Review What you learned in previous grades & lessons

Find the area. *(Skills Review Handbook)*

19. 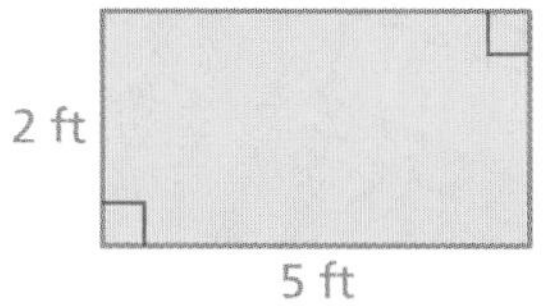

20. 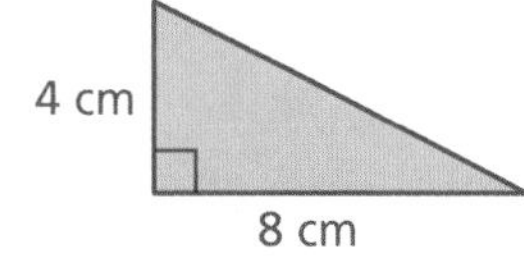

21. 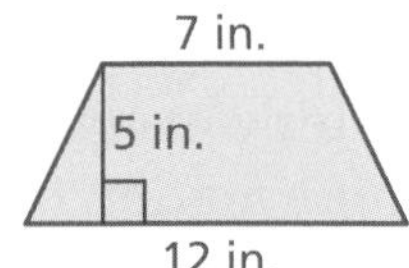

22. **MULTIPLE CHOICE** 40% of what number is 80? *(Skills Review Handbook)*

    (A) 32 (B) 48 (C) 200 (D) 320

14 Study Help

You can use an **information frame** to help you organize and remember concepts. Here is an example of an information frame for surface areas of rectangular prisms.

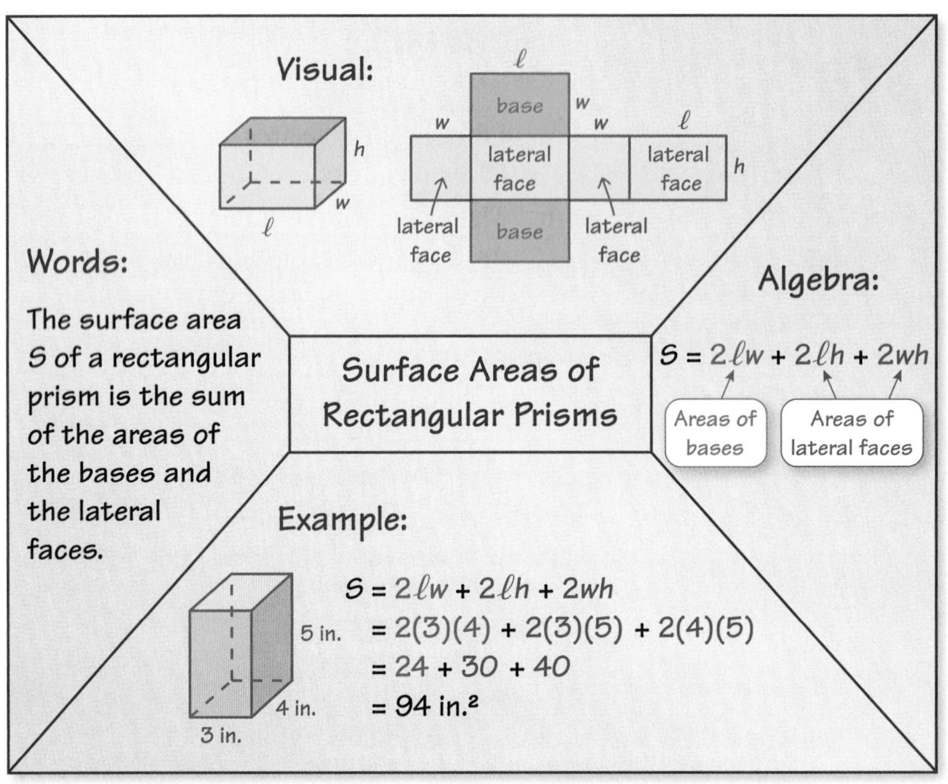

On Your Own

Make information frames to help you study the topics.

1. surface areas of prisms
2. surface areas of pyramids
3. surface areas of cylinders

After you complete this chapter, make information frames for the following topics.

4. volumes of prisms
5. volumes of pyramids

"I'm having trouble thinking of a good title for my information frame."

14.1–14.3 Quiz

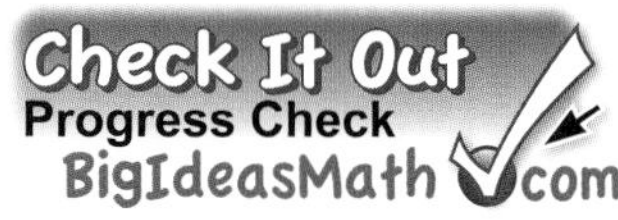

Find the surface area of the prism. *(Section 14.1)*

1. 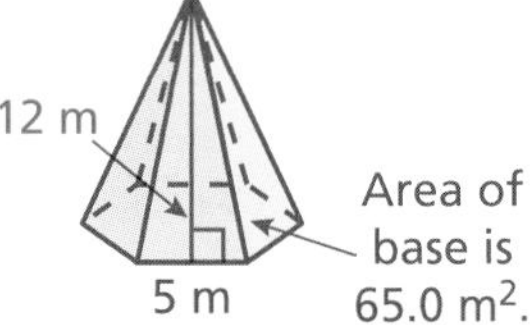

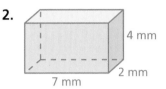

2. 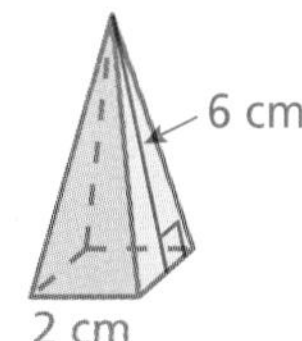

Find the surface area of the regular pyramid. *(Section 14.2)*

3. 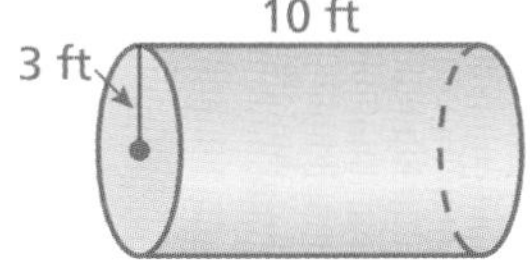

4. 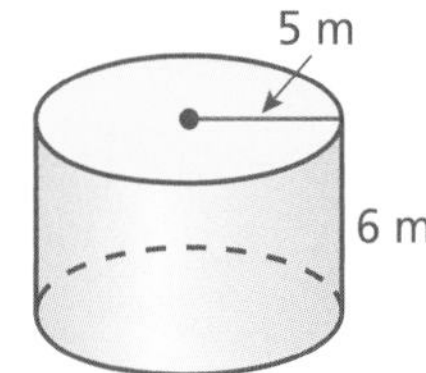

Find the surface area of the cylinder. Round your answer to the nearest tenth. *(Section 14.3)*

5. 3 ft, 10 ft

6. 5 m, 6 m

Find the lateral surface area of the cylinder. Round your answer to the nearest tenth. *(Section 14.3)*

7. 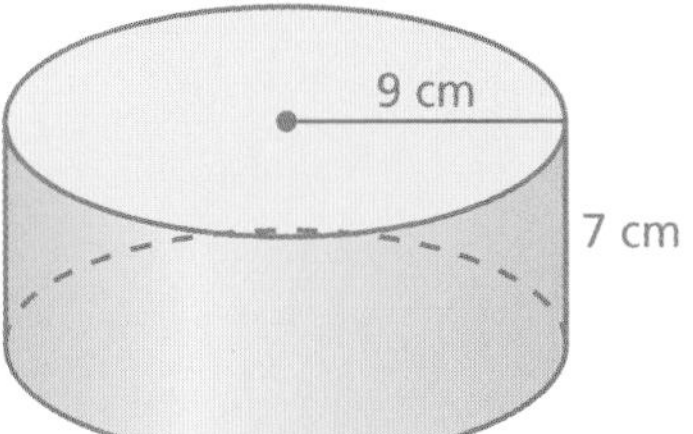

8. 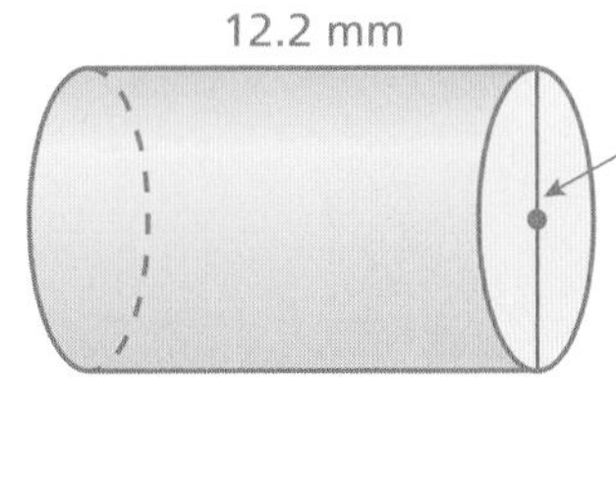

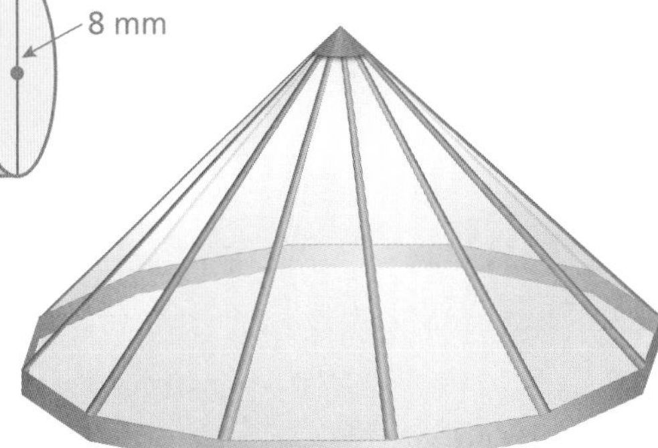

9. **SKYLIGHT** You are making a skylight that has 12 triangular pieces of glass and a slant height of 3 feet. Each triangular piece has a base of 1 foot. *(Section 14.2)*

   a. How much glass will you need to make the skylight?

   b. Can you cut the 12 glass triangles from a sheet of glass that is 4 feet by 8 feet? If so, draw a diagram showing how this can be done.

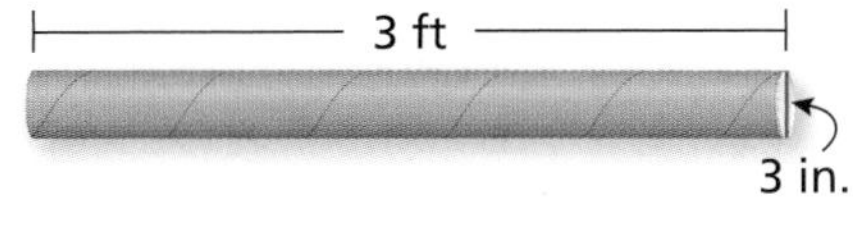

10. **MAILING TUBE** What is the least amount of material needed to make the mailing tube? *(Section 14.3)*

11. **WOODEN CHEST** All the faces of the wooden chest will be painted except for the bottom. Find the area to be painted, in *square inches*. *(Section 14.1)*

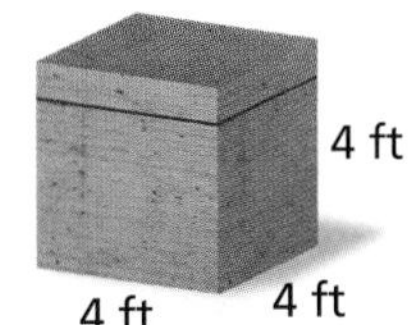

14.4 Volumes of Prisms

Essential Question How can you find the volume of a prism?

1 ACTIVITY: Pearls in a Treasure Chest

Work with a partner. A treasure chest is filled with valuable pearls. Each pearl is about 1 centimeter in diameter and is worth about $80.

Use the diagrams below to describe two ways that you can estimate the number of pearls in the treasure chest.

a.

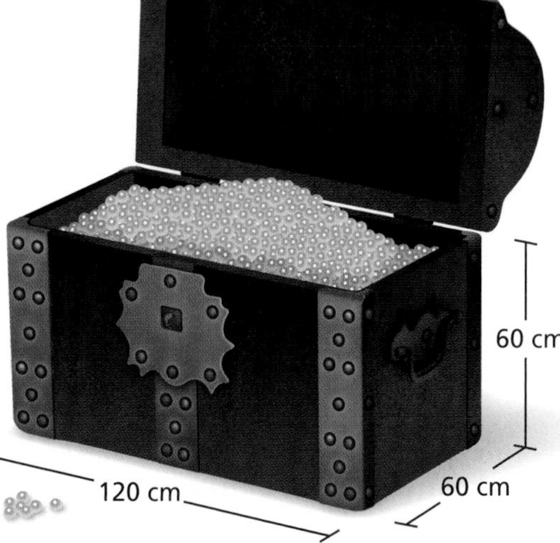

b.

c. Use the method in part (a) to estimate the value of the pearls in the chest.

Geometry
In this lesson, you will
- find volumes of prisms.
- solve real-life problems.

2 ACTIVITY: Finding a Formula for Volume

Work with a partner. You know that the formula for the volume of a rectangular prism is $V = \ell wh$.

a. Write a formula that gives the volume in terms of the area of the base B and the height h.

b. Use both formulas to find the volume of each prism. Do both formulas give you the same volume?

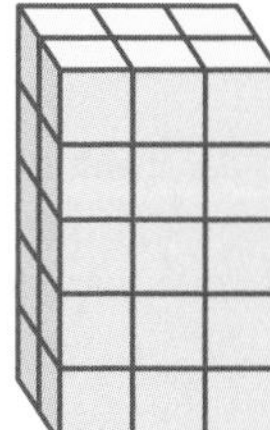

3 ACTIVITY: Finding a Formula for Volume

Math Practice

Use a Formula

What are the given quantities? How can you use the quantities to write a formula?

Work with a partner. Use the concept in Activity 2 to find a formula that gives the volume of any prism.

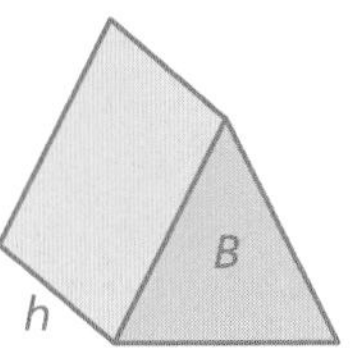

Triangular Prism

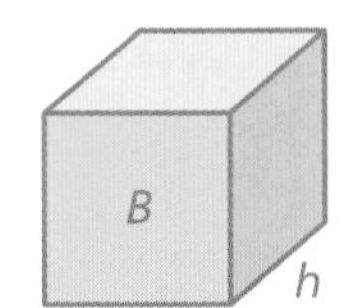

Rectangular Prism

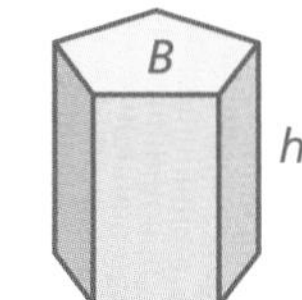

Pentagonal Prism

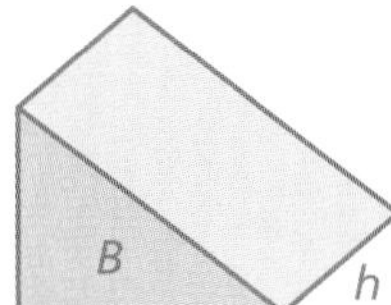

Triangular Prism

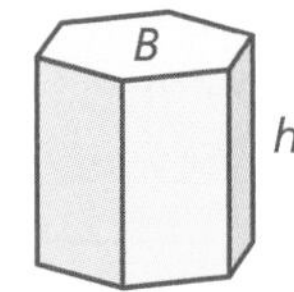

Hexagonal Prism

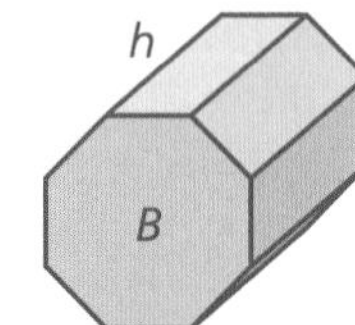

Octagonal Prism

4 ACTIVITY: Using a Formula

Work with a partner. A ream of paper has 500 sheets.

a. Does a single sheet of paper have a volume? Why or why not?

b. If so, explain how you can find the volume of a single sheet of paper.

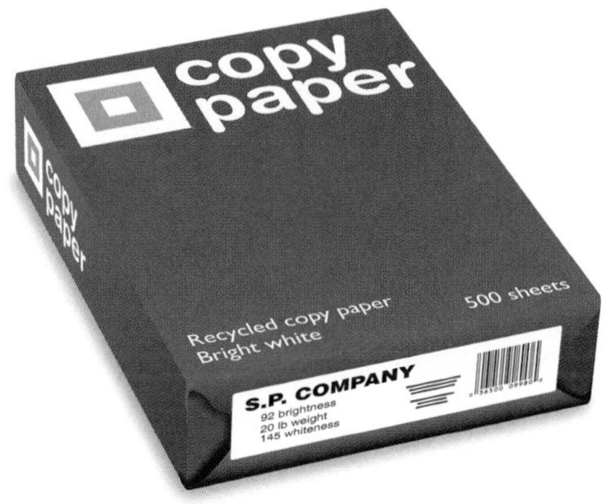

What Is Your Answer?

5. **IN YOUR OWN WORDS** How can you find the volume of a prism?

6. **STRUCTURE** Draw a prism that has a trapezoid as its base. Use your formula to find the volume of the prism.

Use what you learned about the volumes of prisms to complete Exercises 4–6 on page 612.

14.4 Lesson

The *volume* of a three-dimensional figure is a measure of the amount of space that it occupies. Volume is measured in cubic units.

Volume of a Prism

Words The volume V of a prism is the product of the area of the base and the height of the prism.

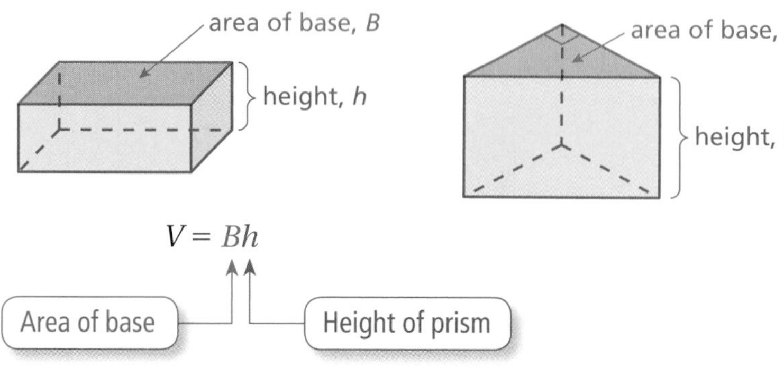

Algebra $V = Bh$

Area of base — Height of prism

Remember

The volume V of a cube with an edge length of s is $V = s^3$.

EXAMPLE 1 Finding the Volume of a Prism

Find the volume of the prism.

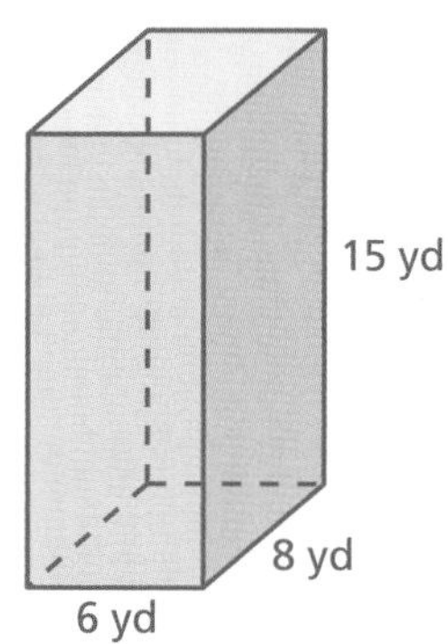

$V = Bh$	Write formula for volume.
$= 6(8) \cdot 15$	Substitute.
$= 48 \cdot 15$	Simplify.
$= 720$	Multiply.

The volume is 720 cubic yards.

Study Tip

The area of the base of a rectangular prism is the product of the length ℓ and the width w.

You can use $V = \ell wh$ to find the volume of a rectangular prism.

EXAMPLE 2 Finding the Volume of a Prism

Find the volume of the prism.

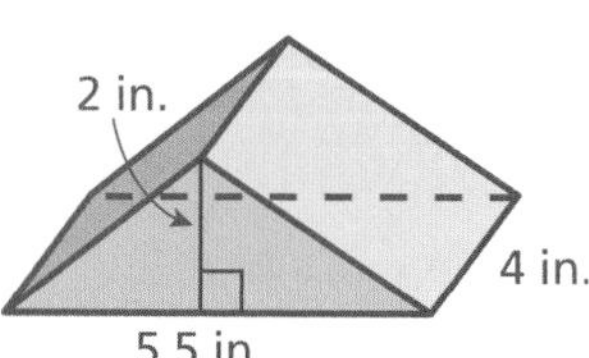

$V = Bh$	Write formula for volume.
$= \frac{1}{2}(5.5)(2) \cdot 4$	Substitute.
$= 5.5 \cdot 4$	Simplify.
$= 22$	Multiply.

The volume is 22 cubic inches.

On Your Own

Find the volume of the prism.

1. 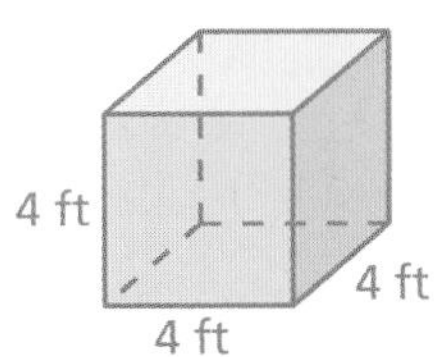

2. 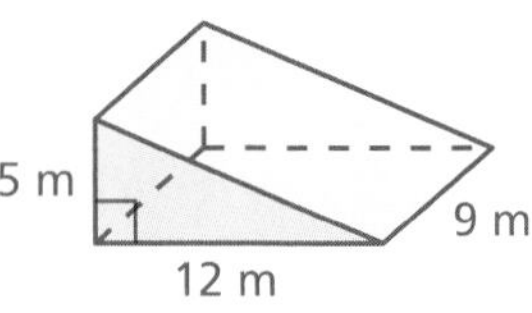

EXAMPLE 3 Real-Life Application

A movie theater designs two bags to hold 96 cubic inches of popcorn. (a) Find the height of each bag. (b) Which bag should the theater choose to reduce the amount of paper needed? Explain.

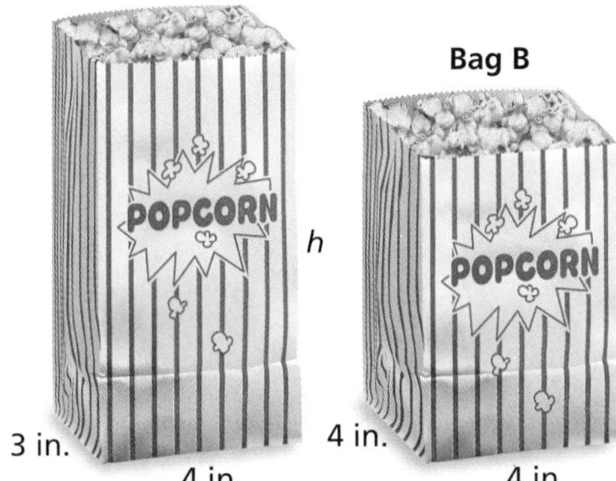

a. Find the height of each bag.

Bag A

$V = Bh$

$96 = 4(3)(h)$

$96 = 12h$

$8 = h$

The height is 8 inches.

Bag B

$V = Bh$

$96 = 4(4)(h)$

$96 = 16h$

$6 = h$

The height is 6 inches.

b. To determine the amount of paper needed, find the surface area of each bag. Do not include the top base.

Bag A

$$\begin{aligned} S &= \ell w + 2\ell h + 2wh \\ &= 4(3) + 2(4)(8) + 2(3)(8) \\ &= 12 + 64 + 48 \\ &= 124 \text{ in.}^2 \end{aligned}$$

Bag B

$$\begin{aligned} S &= \ell w + 2\ell h + 2wh \\ &= 4(4) + 2(4)(6) + 2(4)(6) \\ &= 16 + 48 + 48 \\ &= 112 \text{ in.}^2 \end{aligned}$$

The surface area of Bag B is less than the surface area of Bag A. So, the theater should choose Bag B.

On Your Own

3. You design Bag C that has a volume of 96 cubic inches. Should the theater in Example 3 choose your bag? Explain.

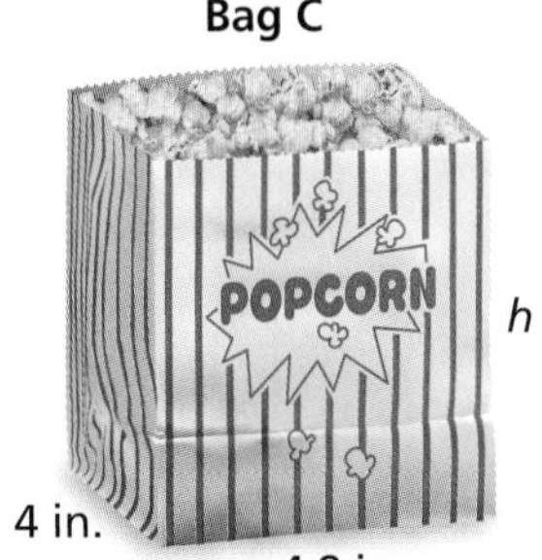

14.4 Exercises

Vocabulary and Concept Check

1. **VOCABULARY** What types of units are used to describe volume?
2. **VOCABULARY** Explain how to find the volume of a prism.
3. **CRITICAL THINKING** How are volume and surface area different?

Practice and Problem Solving

Find the volume of the prism.

4. 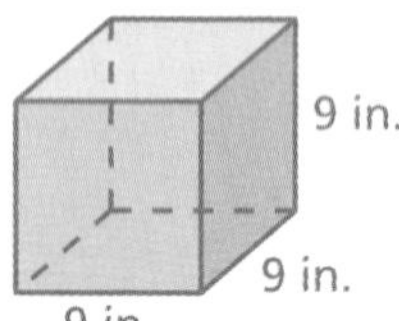

5. 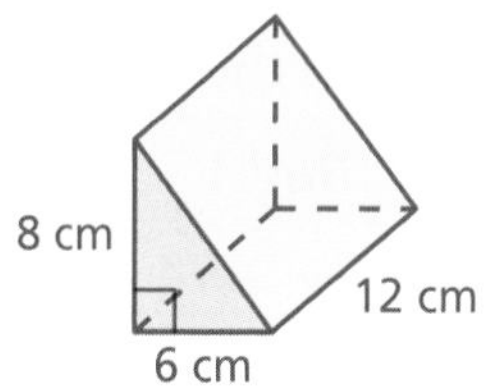

6. 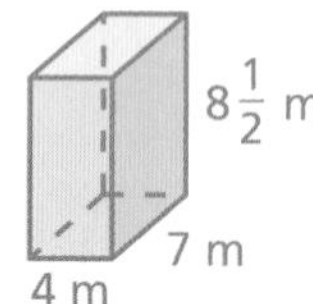

7. 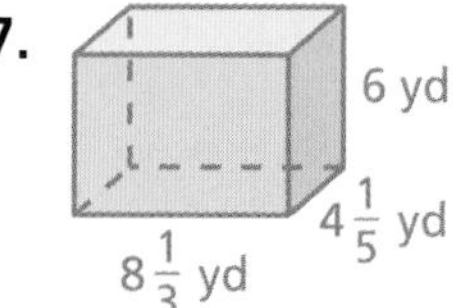

8. 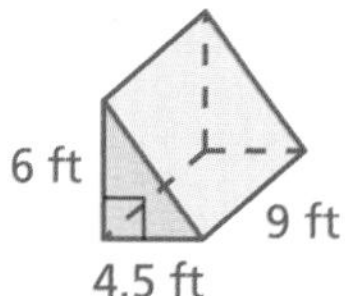

9. 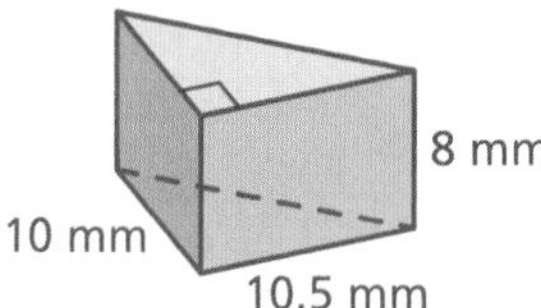

10. 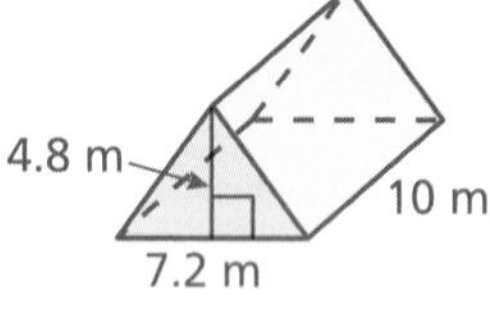

11. 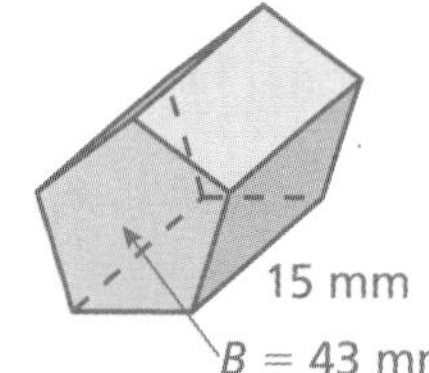

12. 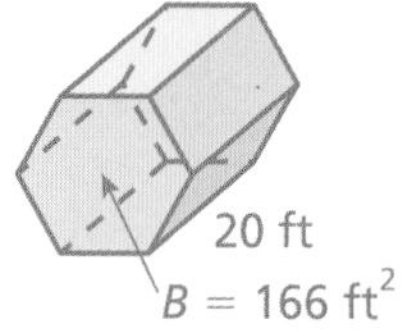

13. **ERROR ANALYSIS** Describe and correct the error in finding the volume of the triangular prism.

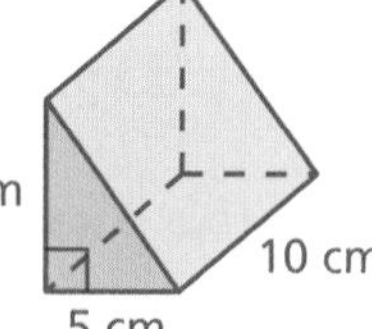

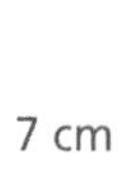

$$V = Bh$$
$$= 10(5)(7)$$
$$= 50 \cdot 7$$
$$= 350 \text{ cm}^3$$

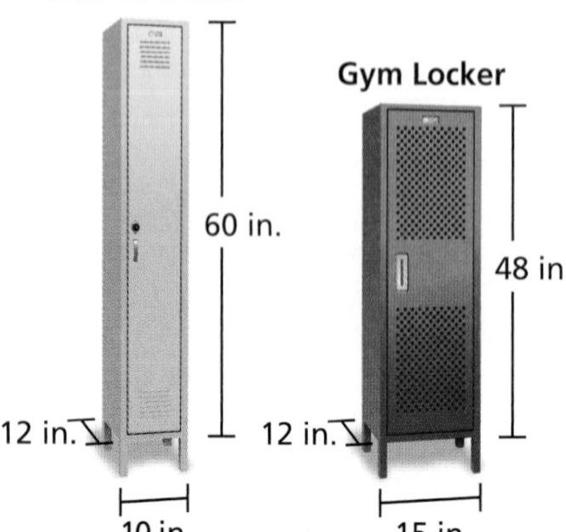

14. **LOCKER** Each locker is shaped like a rectangular prism. Which has more storage space? Explain.

15. **CEREAL BOX** A cereal box is 9 inches by 2.5 inches by 10 inches. What is the volume of the box?

Find the volume of the prism.

16.

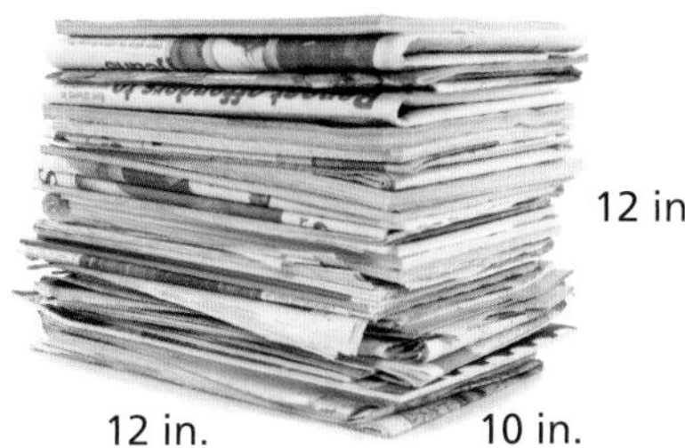

17.

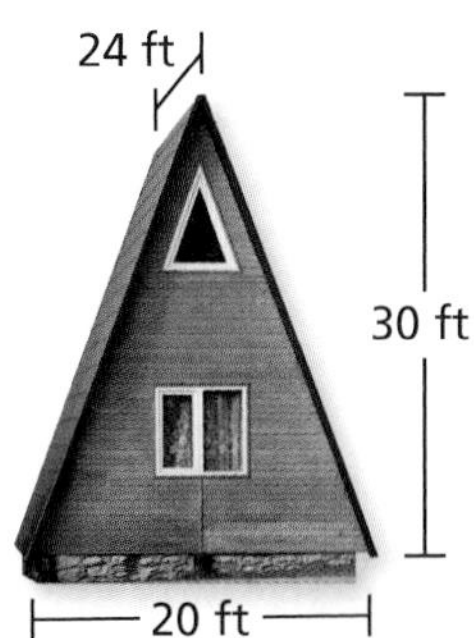

18. LOGIC Two prisms have the same volume. Do they *always, sometimes,* or *never* have the same surface area? Explain.

19. CUBIC UNITS How many cubic inches are in a cubic foot? Use a sketch to explain your reasoning.

20. CAPACITY As a gift, you fill the calendar with packets of chocolate candy. Each packet has a volume of 2 cubic inches. Find the maximum number of packets you can fit inside the calendar.

21. PRECISION Two liters of water are poured into an empty vase shaped like an octagonal prism. The base area is 100 square centimeters. What is the height of the water? (1 L = 1000 cm^3)

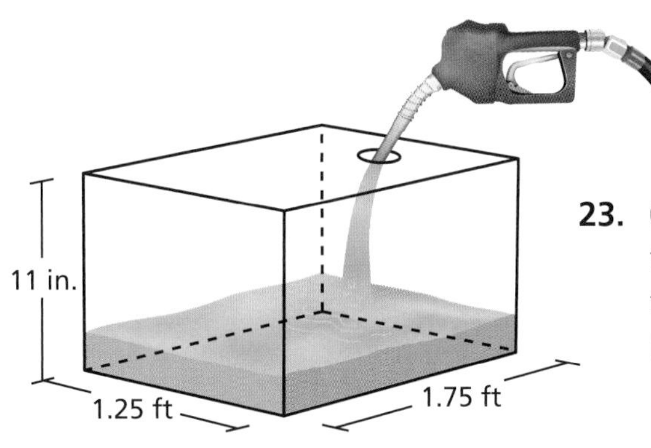

22. GAS TANK The gas tank is 20% full. Use the current price of regular gasoline in your community to find the cost to fill the tank. (1 gal = 231 in.^{3})

23. OPEN-ENDED You visit an aquarium. One of the tanks at the aquarium holds 450 gallons of water. Draw a diagram to show one possible set of dimensions of the tank. (1 gal = 231 in.^{3})

24. Critical Thinking How many times greater is the volume of a triangular prism when one of its dimensions is doubled? when all three dimensions are doubled?

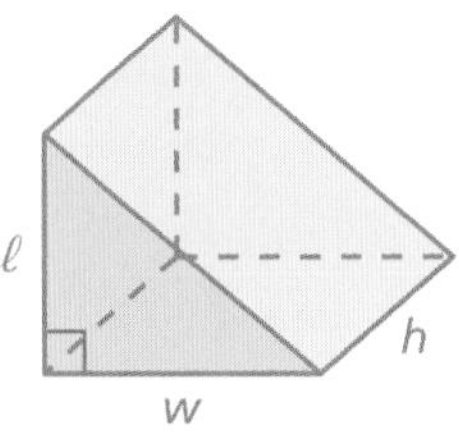

Fair Game Review What you learned in previous grades & lessons

Find the selling price. *(Skills Review Handbook)*

25. Cost to store: \$75
Markup: 20%

26. Cost to store: \$90
Markup: 60%

27. Cost to store: \$130
Markup: 85%

28. MULTIPLE CHOICE What is the approximate surface area of a cylinder with a radius of 3 inches and a height of 10 inches? *(Section 14.3)*

Ⓐ 30 in.^{2} Ⓑ 87 in.^{2} Ⓒ 217 in.^{2} Ⓓ 245 in.^{2}

14.5 Volumes of Pyramids

Essential Question How can you find the volume of a pyramid?

1 ACTIVITY: Finding a Formula Experimentally

Work with a partner.

- **Draw the two nets on cardboard and cut them out.**

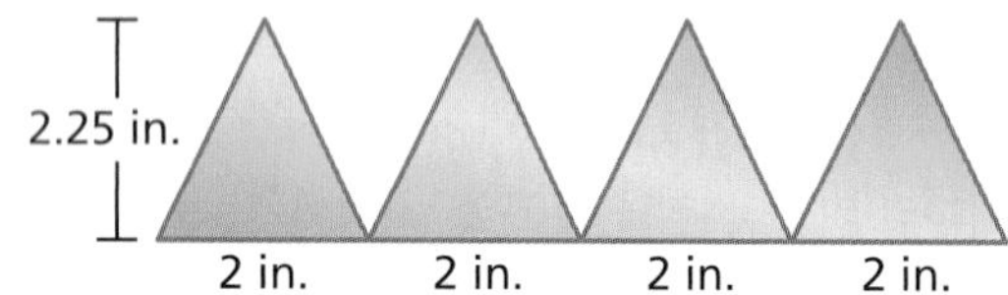

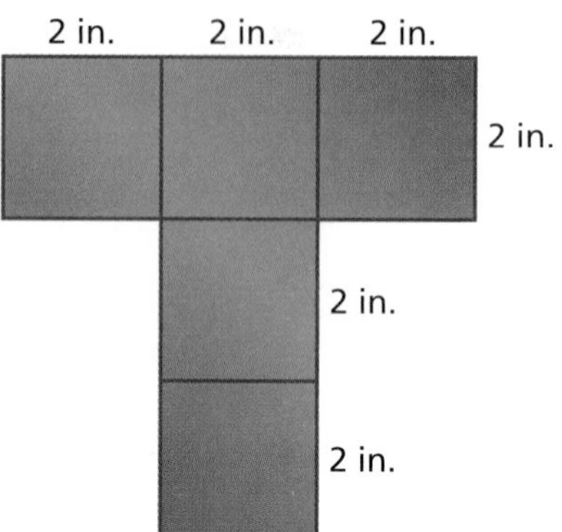

- **Fold and tape the nets to form an open square box and an open pyramid.**
- **Both figures should have the same size square base and the same height.**
- **Fill the pyramid with pebbles. Then pour the pebbles into the box. Repeat this until the box is full. How many pyramids does it take to fill the box?**
- **Use your result to find a formula for the volume of a pyramid.**

2 ACTIVITY: Comparing Volumes

Work with a partner. You are an archaeologist studying two ancient pyramids. What factors would affect how long it took to build each pyramid? Given similar conditions, which pyramid took longer to build? Explain your reasoning.

Geometry
In this lesson, you will
- find volumes of pyramids.
- solve real-life problems.

The Sun Pyramid in Mexico
Height: about 246 ft
Base: about 738 ft by 738 ft

Cheops Pyramid in Egypt
Height: about 480 ft
Base: about 755 ft by 755 ft

3 ACTIVITY: Finding and Using a Pattern

Math Practice

Look for Patterns

As the height and the base lengths increase, how does this pattern affect the volume? Explain.

Work with a partner.

- **Find the volumes of the pyramids.**
- **Organize your results in a table.**
- **Describe the pattern.**
- **Use your pattern to find the volume of a pyramid with a base length and a height of 20.**

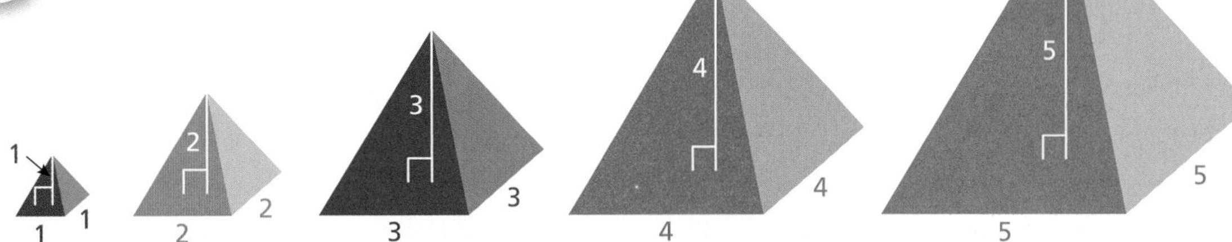

4 ACTIVITY: Breaking a Prism into Pyramids

Work with a partner. The rectangular prism can be cut to form three pyramids. Show that the sum of the volumes of the three pyramids is equal to the volume of the prism.

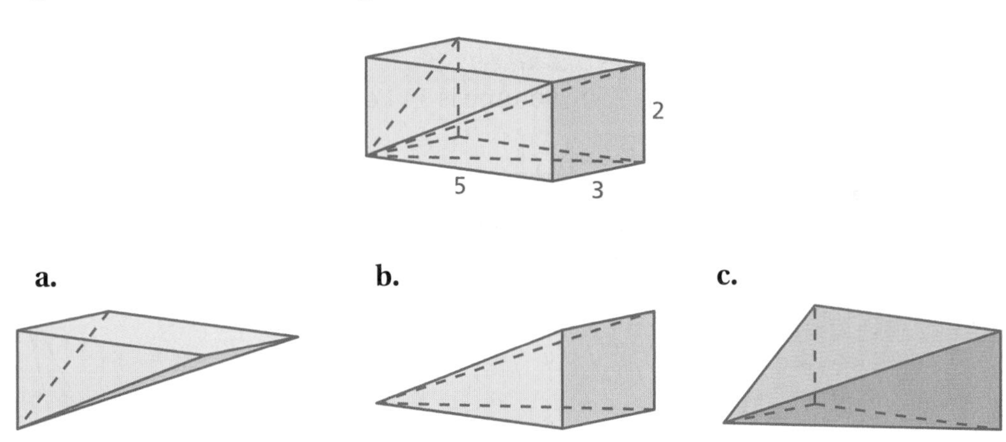

What Is Your Answer?

5. **IN YOUR OWN WORDS** How can you find the volume of a pyramid?

6. **STRUCTURE** Write a general formula for the volume of a pyramid.

Use what you learned about the volumes of pyramids to complete Exercises 4–6 on page 618.

14.5 Lesson

Key Idea

Volume of a Pyramid

Words The volume V of a pyramid is one-third the product of the area of the base and the height of the pyramid.

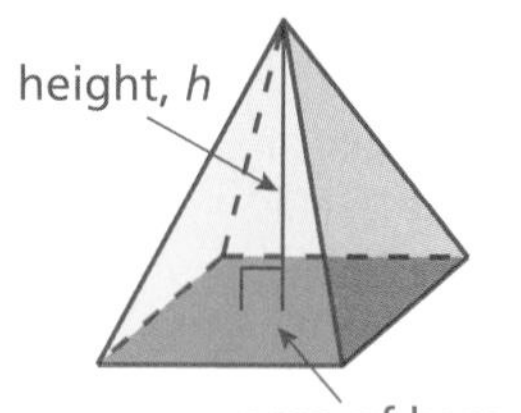

Algebra $V = \frac{1}{3}Bh$ (Area of base: B; Height of pyramid: h)

Study Tip

The *height* of a pyramid is the perpendicular distance from the base to the vertex.

EXAMPLE 1 Finding the Volume of a Pyramid

Find the volume of the pyramid.

9 mm

$B = 48\text{ mm}^2$

$V = \frac{1}{3}Bh$ Write formula for volume.

$= \frac{1}{3}(48)(9)$ Substitute.

$= 144$ Multiply.

The volume is 144 cubic millimeters.

EXAMPLE 2 Finding the Volume of a Pyramid

Study Tip

The area of the base of a rectangular pyramid is the product of the length ℓ and the width w.

You can use $V = \frac{1}{3}\ell wh$ to find the volume of a rectangular pyramid.

Find the volume of the pyramid.

a.

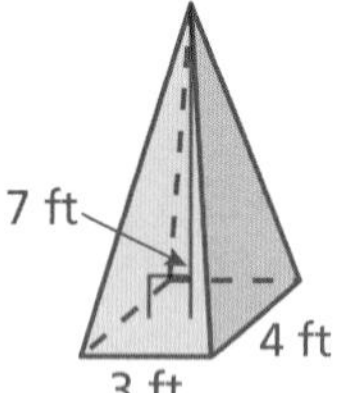

$V = \frac{1}{3}Bh$

$= \frac{1}{3}(4)(3)(7)$

$= 28$

The volume is 28 cubic feet.

b.

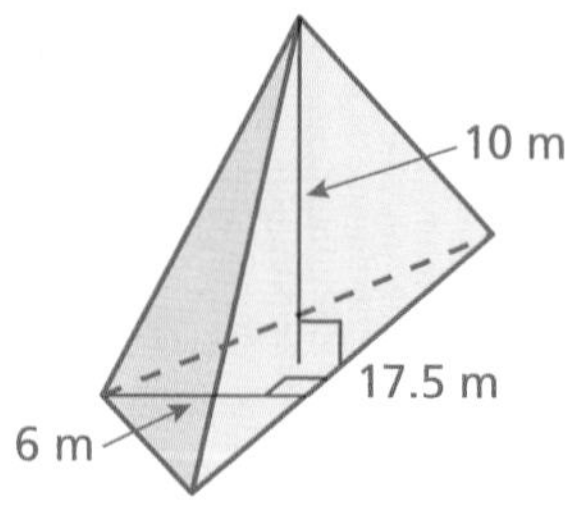

$V = \frac{1}{3}Bh$

$= \frac{1}{3}\left(\frac{1}{2}\right)(17.5)(6)(10)$

$= 175$

The volume is 175 cubic meters.

On Your Own

Now You're Ready
Exercises 4–11

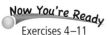

Find the volume of the pyramid.

1. 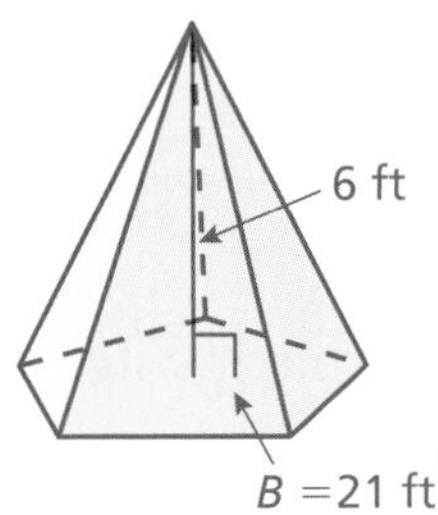

2. 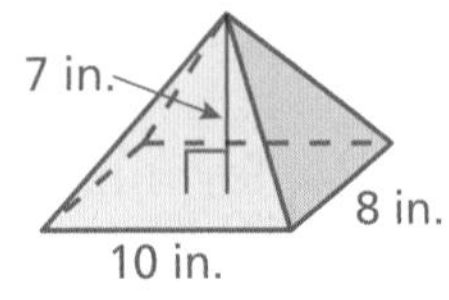

3. 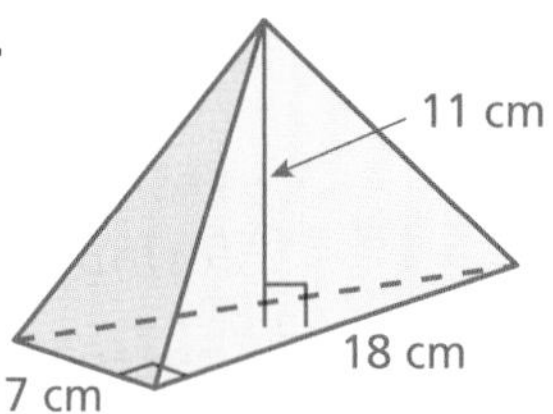

EXAMPLE 3 Real-Life Application

a. The volume of sunscreen in Bottle B is about how many times the volume in Bottle A?

b. Which is the better buy?

a. Use the formula for the volume of a pyramid to estimate the amount of sunscreen in each bottle.

Bottle A	***Bottle B***
$V = \frac{1}{3}Bh$	$V = \frac{1}{3}Bh$
$= \frac{1}{3}(2)(1)(6)$	$= \frac{1}{3}(3)(1.5)(4)$
$= 4 \text{ in.}^3$	$= 6 \text{ in.}^3$

So, the volume of sunscreen in Bottle B is about $\frac{6}{4} = 1.5$ times the volume in Bottle A.

b. Find the unit cost for each bottle.

Bottle A	***Bottle B***
$\frac{\text{cost}}{\text{volume}} = \frac{\$9.96}{4 \text{ in.}^3}$	$\frac{\text{cost}}{\text{volume}} = \frac{\$14.40}{6 \text{ in.}^3}$
$= \frac{\$2.49}{1 \text{ in.}^3}$	$= \frac{\$2.40}{1 \text{ in.}^3}$

The unit cost of Bottle B is less than the unit cost of Bottle A. So, Bottle B is the better buy.

On Your Own

Now You're Ready
Exercise 16

4. Bottle C is on sale for $13.20. Is Bottle C a better buy than Bottle B in Example 3? Explain.

14.5 Exercises

Vocabulary and Concept Check

1. **WRITING** How is the formula for the volume of a pyramid different from the formula for the volume of a prism?
2. **OPEN-ENDED** Describe a real-life situation that involves finding the volume of a pyramid.
3. **REASONING** A triangular pyramid and a triangular prism have the same base and height. The volume of the prism is how many times the volume of the pyramid?

Practice and Problem Solving

Find the volume of the pyramid.

4. 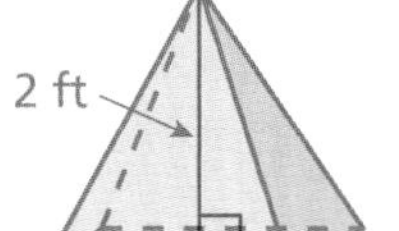

5. 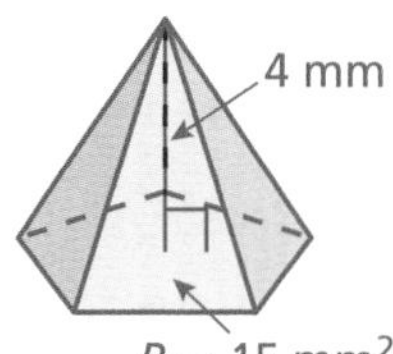

6. 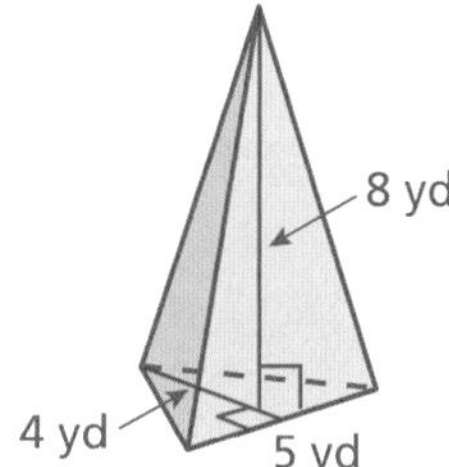

7. 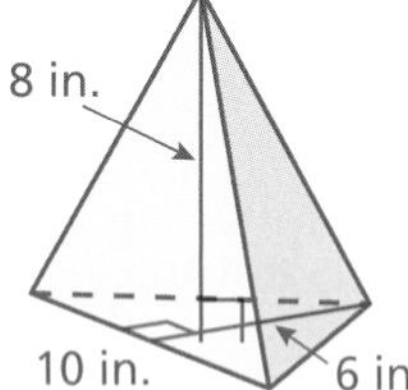

8. 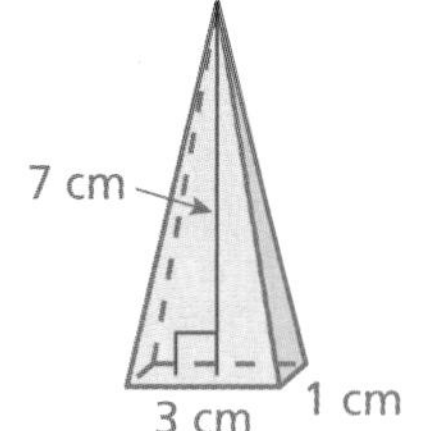

9. 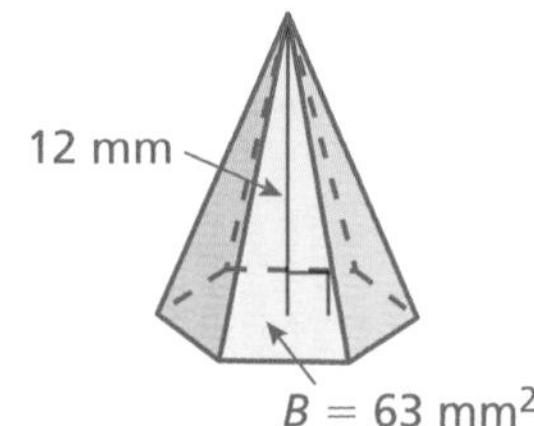

10. 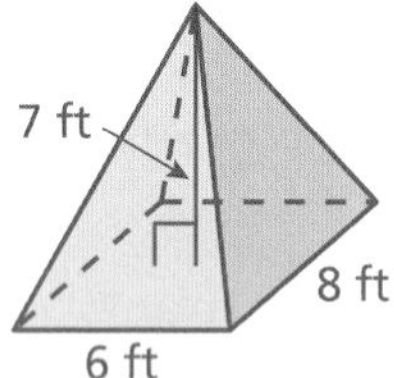

11. 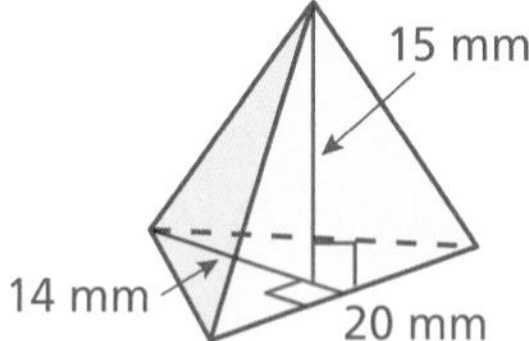

12. **PARACHUTE** In 1483, Leonardo da Vinci designed a parachute. It is believed that this was the first parachute ever designed. In a notebook, he wrote, "If a man is provided with a length of gummed linen cloth with a length of 12 yards on each side and 12 yards high, he can jump from any great height whatsoever without injury." Find the volume of air inside Leonardo's parachute.

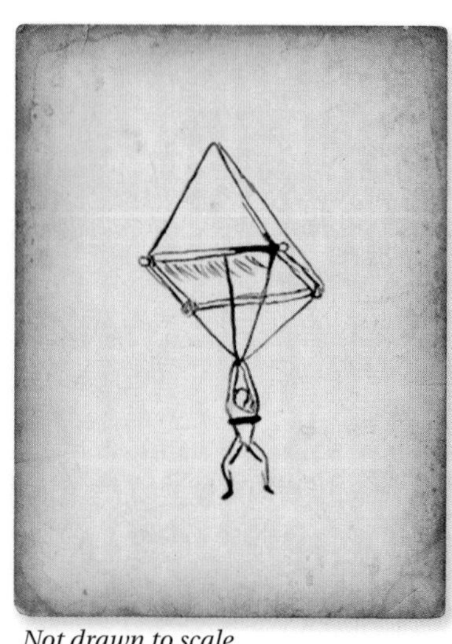

Not drawn to scale

Find the volume of the composite solid.

13.

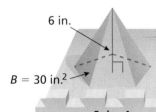

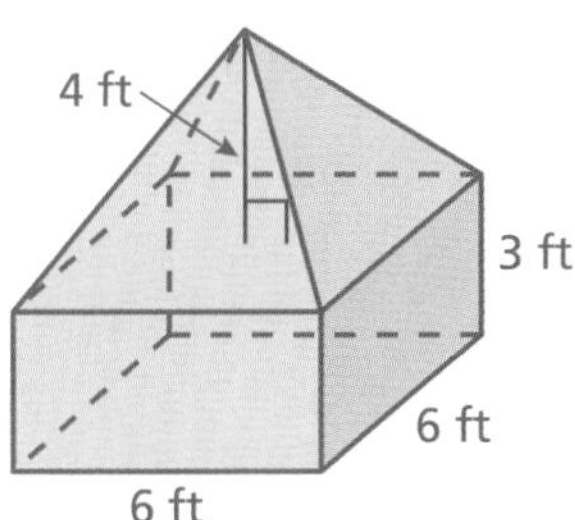

14.

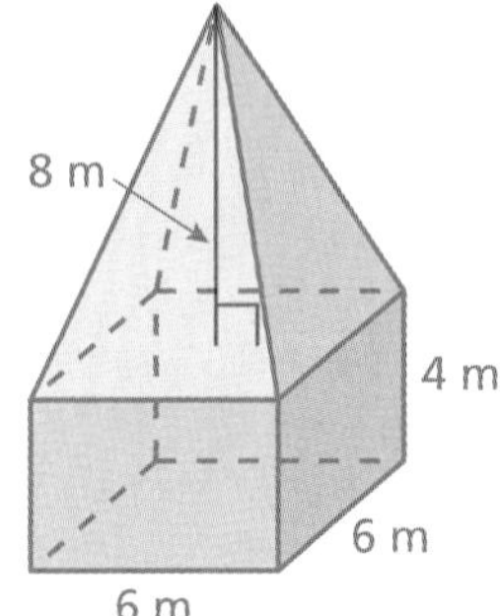

15.

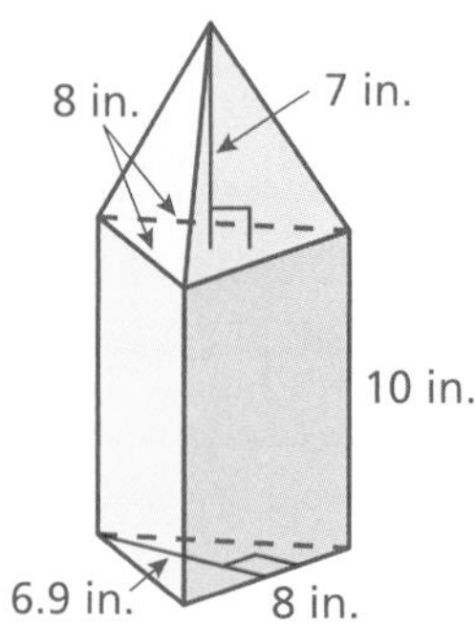

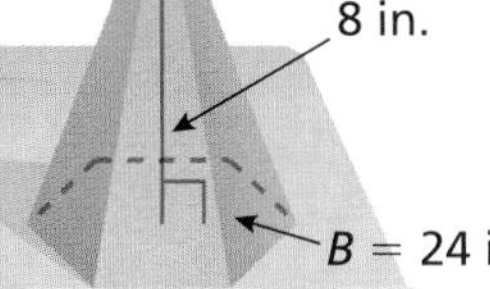

16. **SPIRE** Which sand-castle spire has a greater volume? How much more sand do you need to make the spire with the greater volume?

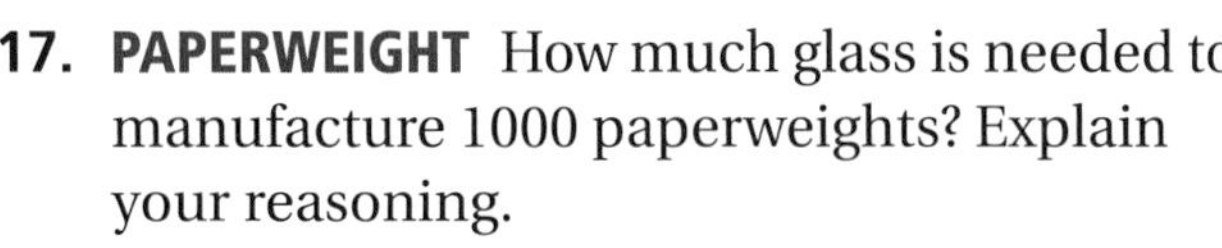

17. **PAPERWEIGHT** How much glass is needed to manufacture 1000 paperweights? Explain your reasoning.

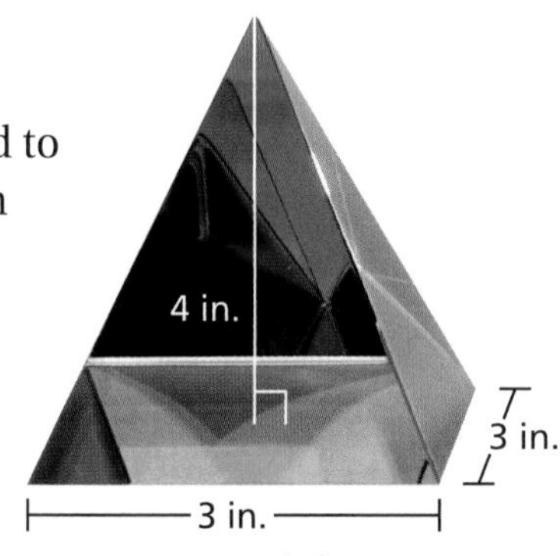

Paperweight

18. **PROBLEM SOLVING** Use the photo of the tepee.

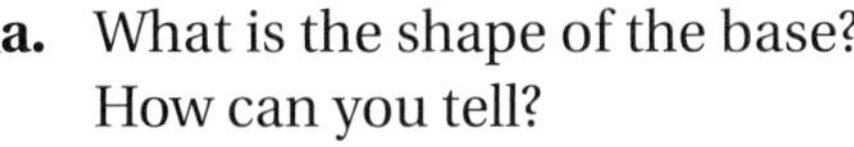

a. What is the shape of the base? How can you tell?

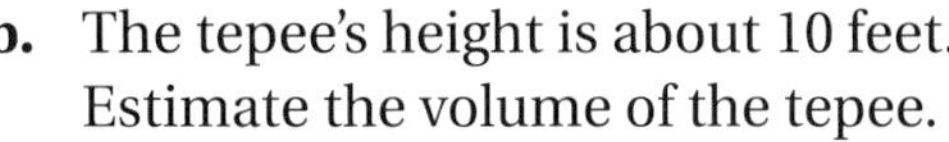

b. The tepee's height is about 10 feet. Estimate the volume of the tepee.

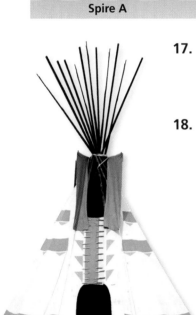

19. **OPEN-ENDED** A pyramid has a volume of 40 cubic feet and a height of 6 feet. Find one possible set of dimensions of the rectangular base.

20. **Reasoning** Do the two solids have the same volume? Explain.

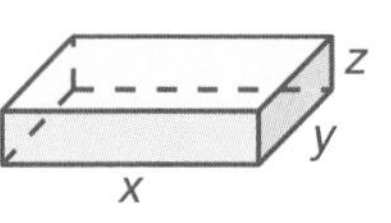

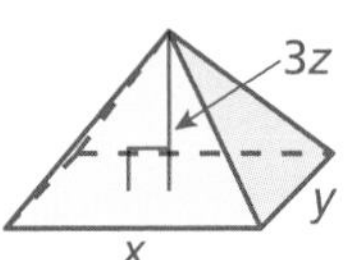

Fair Game Review What you learned in previous grades & lessons

For the given angle measure, find the measure of a supplementary angle and the measure of a complementary angle, if possible. *(Section 12.2)*

21. 27°

22. 82°

23. 120°

24. **MULTIPLE CHOICE** The circumference of a circle is 44 inches. Which estimate is closest to the area of the circle? *(Section 13.3)*

Ⓐ 7 in.^{2} Ⓑ 14 in.^{2} Ⓒ 154 in.^{2} Ⓓ 484 in.^{2}

Extension 14.5 Cross Sections of Three-Dimensional Figures

Key Vocabulary
cross section, *p. 620*

Consider a plane "slicing" through a solid. The intersection of the plane and the solid is a two-dimensional shape called a **cross section**. For example, the diagram shows that the intersection of the plane and the rectangular prism is a rectangle.

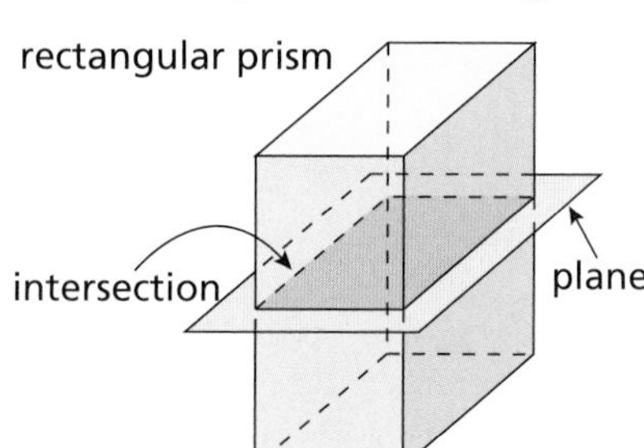

EXAMPLE  1 Describing the Intersection of a Plane and a Solid

Geometry
In this extension, you will
- describe the intersections of planes and solids.

Describe the intersection of the plane and the solid.

a.

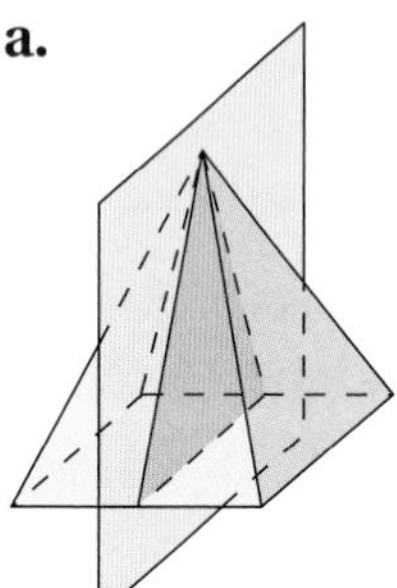

b.

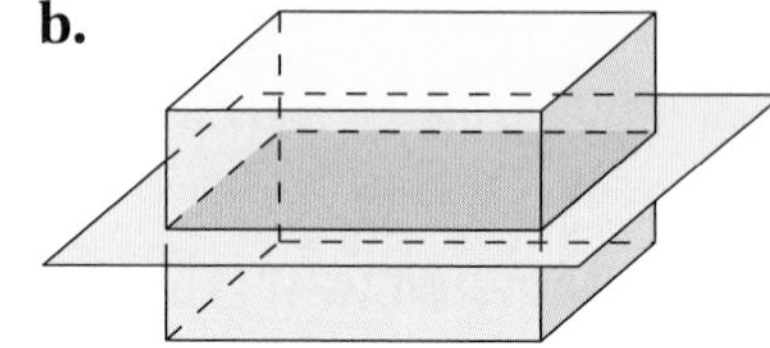

c.

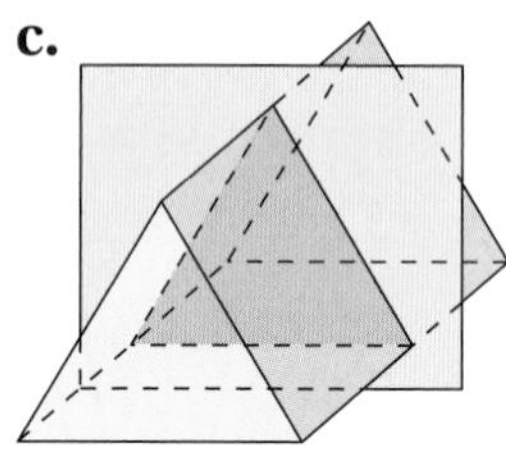

a. The intersection is a triangle.

b. The intersection is a rectangle.

c. The intersection is a triangle.

Practice

Describe the intersection of the plane and the solid.

1.

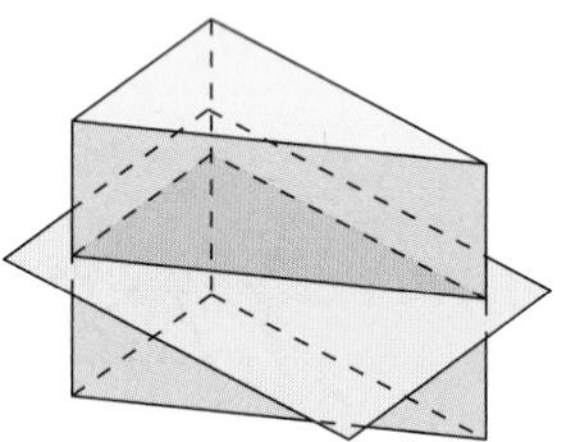

2.

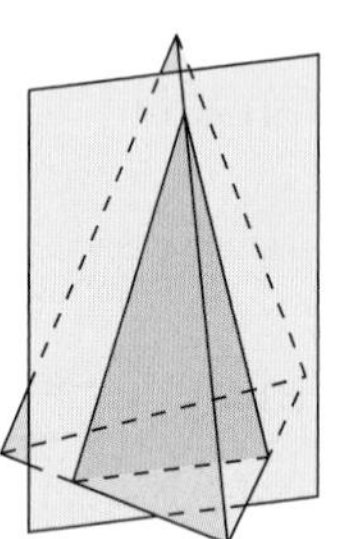

3.

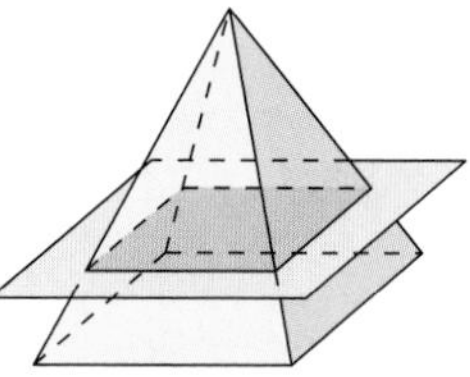

4.

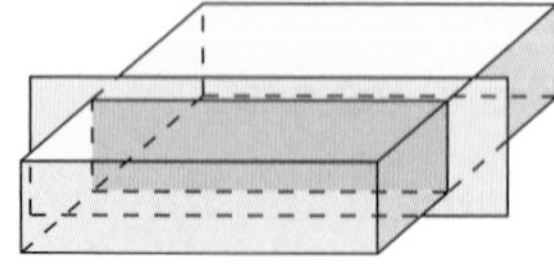

5.

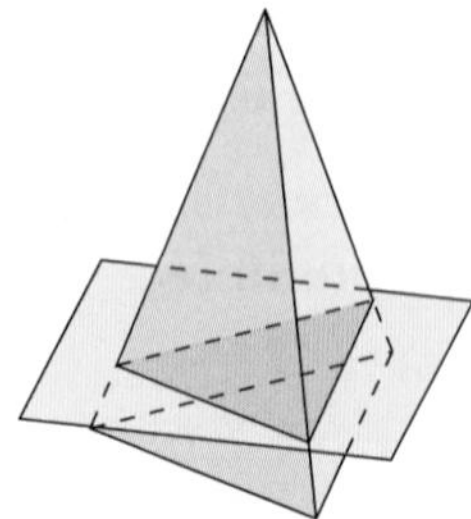

6.

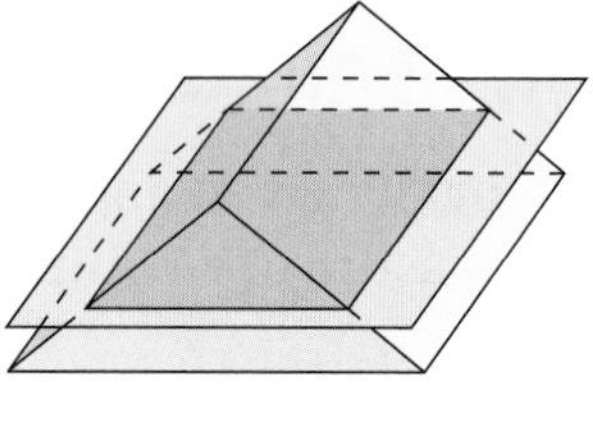

7. REASONING A plane that intersects a prism is parallel to the bases of the prism. Describe the intersection of the plane and the prism.

Example 1 shows how a plane intersects a polyhedron. Now consider the intersection of a plane and a solid having a curved surface, such as a cylinder or cone. As shown, a *cone* is a solid that has one circular base and one vertex.

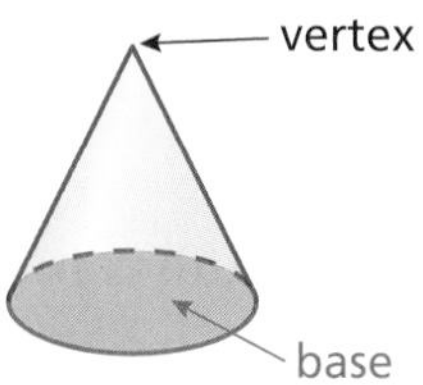

EXAMPLE 2 Describing the Intersection of a Plane and a Solid

Math Practice

Analyze Givens
What solid is shown? What are you trying to find? Explain.

Describe the intersection of the plane and the solid.

a.

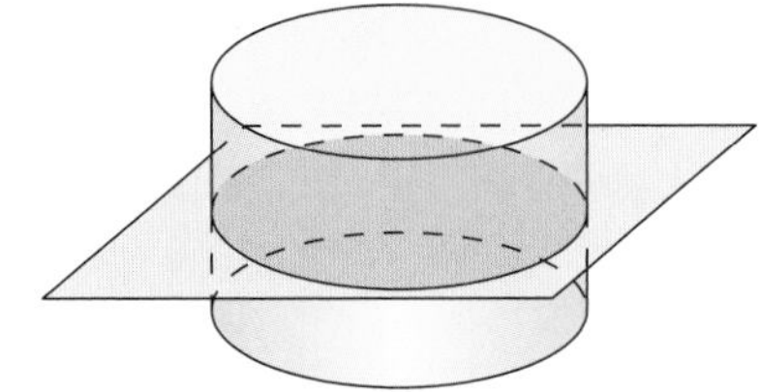

b.

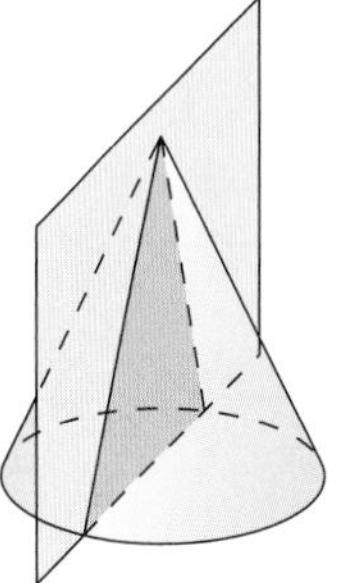

a. The intersection is a circle.

b. The intersection is a triangle.

Practice

Describe the intersection of the plane and the solid.

8.

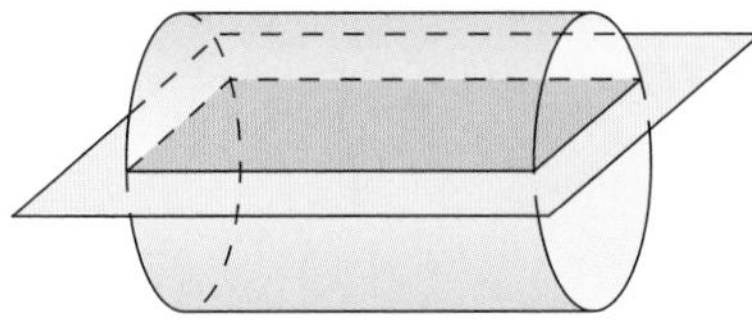

9.

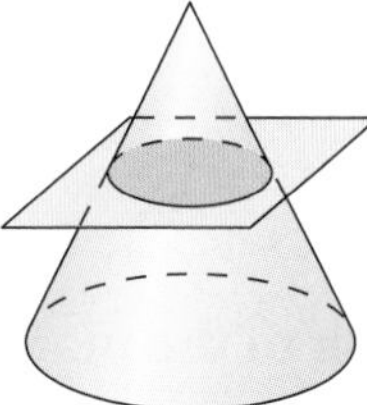

10.

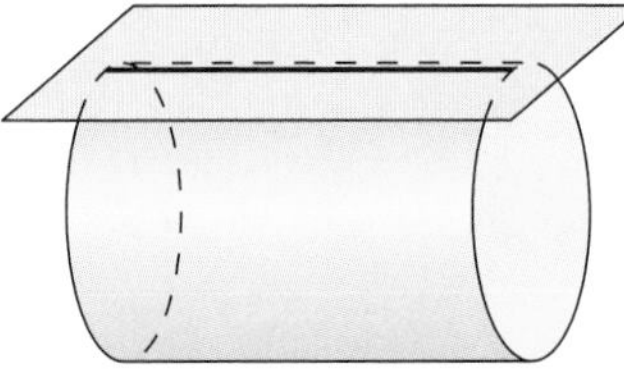

11.

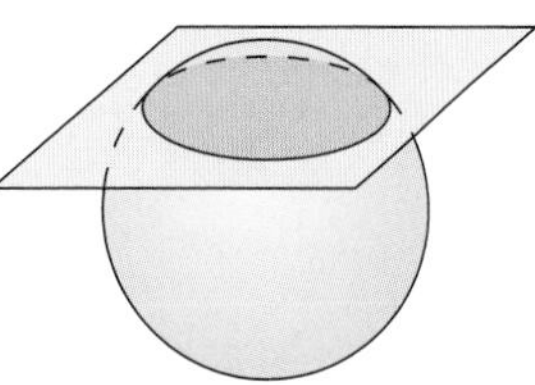

Describe the shape that is formed by the cut made in the food shown.

12.

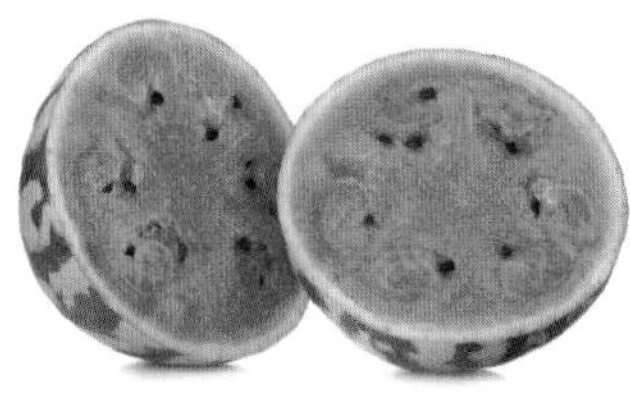

13.

14.

15. **REASONING** Explain how a plane can be parallel to the base of a cone and intersect the cone at exactly one point.

14.4–14.5 Quiz

Check It Out
Progress Check
BigIdeasMath.com

Find the volume of the prism. *(Section 14.4)*

1.

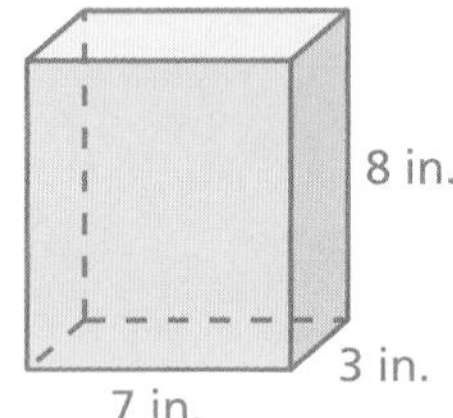

2.

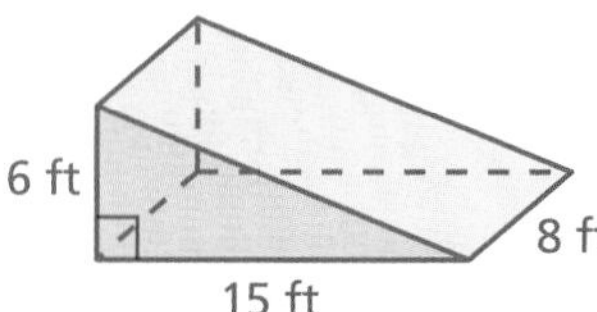

3.

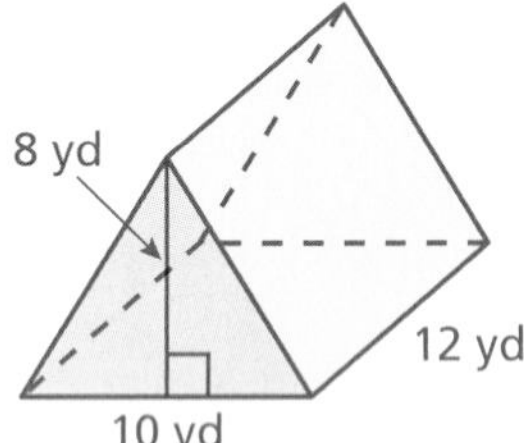

4.

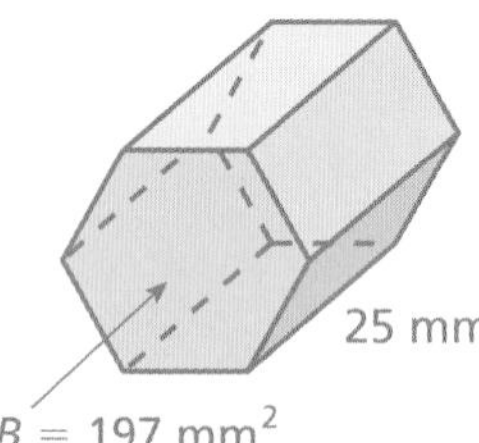

Find the volume of the solid. Round your answer to the nearest tenth if necessary. *(Section 14.5)*

5.

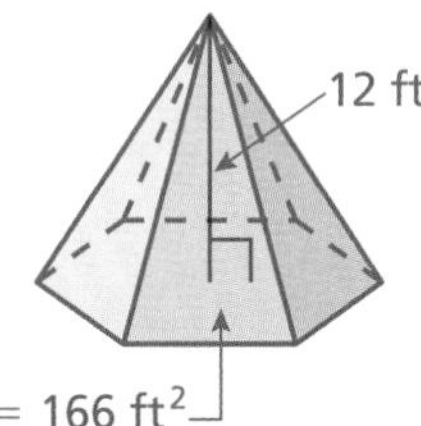

6.

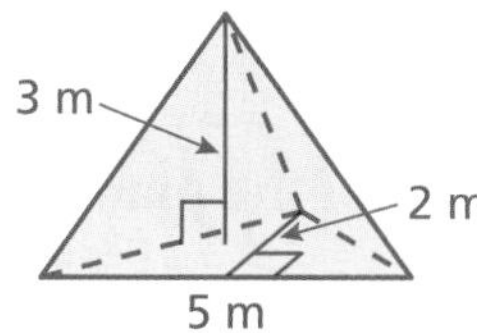

Describe the intersection of the plane and the solid. *(Section 14.5)*

7.

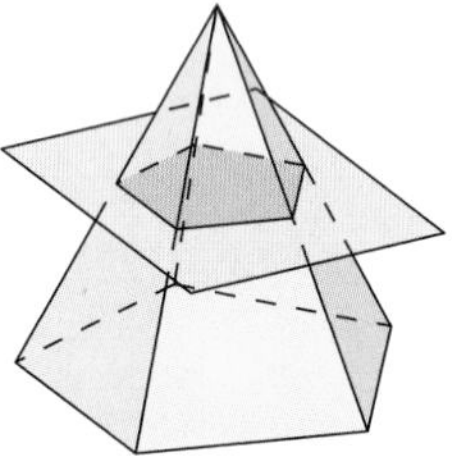

8.

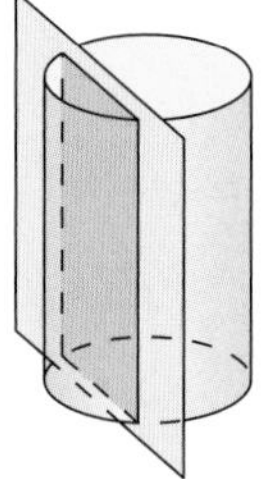

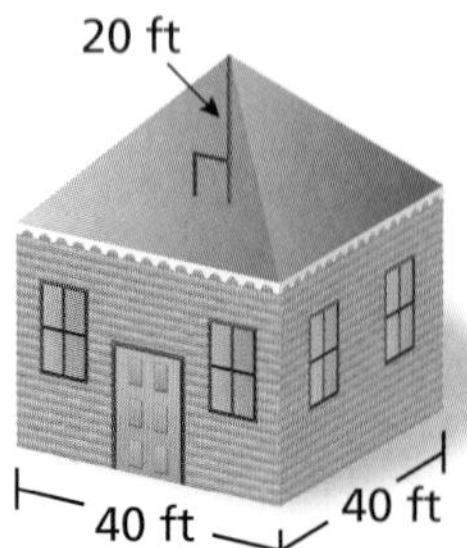

9. **ROOF** A pyramid hip roof is a good choice for a house in a hurricane area. What is the volume of the roof to the nearest tenth? *(Section 14.5)*

10. **CUBIC UNITS** How many cubic feet are in a cubic yard? Use a sketch to explain your reasoning. *(Section 14.4)*

14 Chapter Review

Check It Out
Vocabulary Help
BigIdeasMath.com

Review Key Vocabulary

lateral surface area, *p. 590*
regular pyramid, *p. 596*
slant height, *p. 596*
cross section, *p. 620*

Review Examples and Exercises

14.1 Surface Areas of Prisms *(pp. 586–593)*

Find the surface area of the prism.

Draw a net.

$$S = 2\ell w + 2\ell h + 2wh$$
$$= 2(6)(4) + 2(6)(5) + 2(4)(5)$$
$$= 48 + 60 + 40$$
$$= 148$$

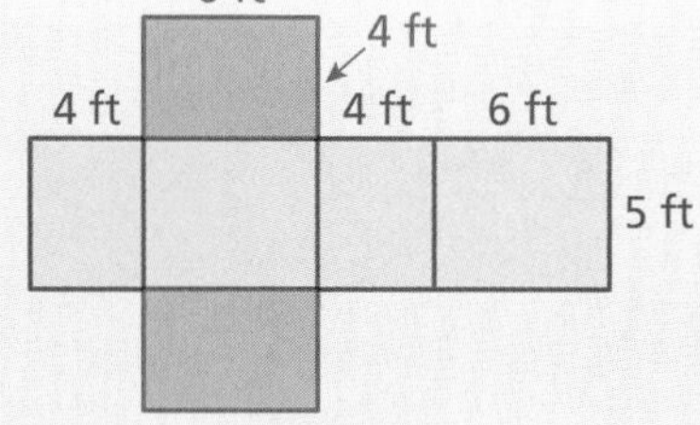

The surface area is 148 square feet.

Exercises

Find the surface area of the prism.

1. 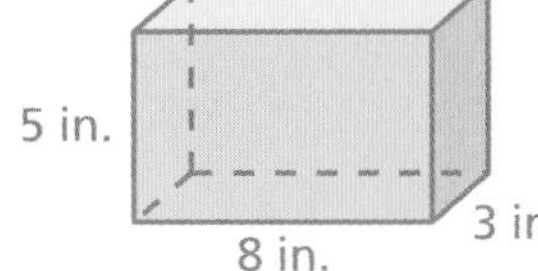

2. 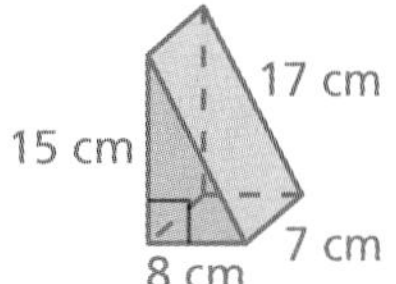

3. 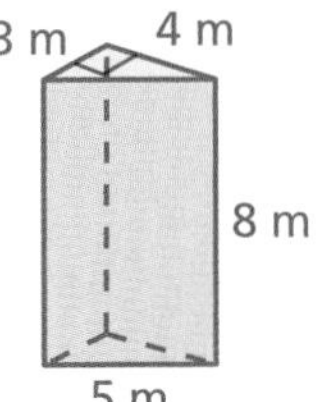

14.2 Surface Areas of Pyramids *(pp. 594–599)*

Find the surface area of the regular pyramid.

Draw a net.

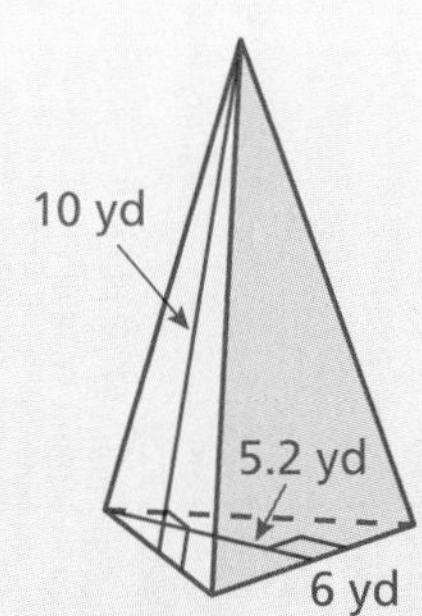

Area of Base

$\frac{1}{2} \cdot 6 \cdot 5.2 = 15.6$

Area of a Lateral Face

$\frac{1}{2} \cdot 6 \cdot 10 = 30$

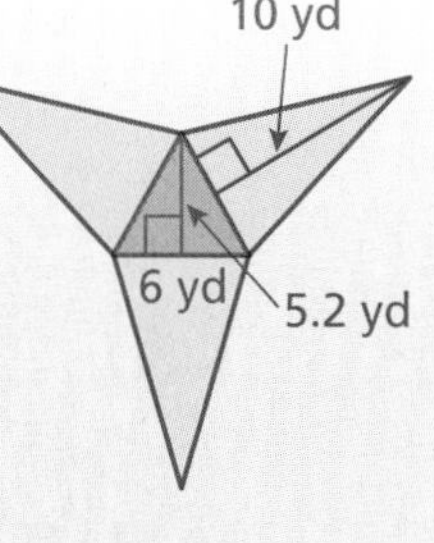

Find the sum of the areas of the base and all three lateral faces.

$$S = 15.6 + \underbrace{30 + 30 + 30}$$
$$= 105.6$$

There are 3 identical lateral faces. Count the area 3 times.

The surface area is 105.6 square yards.

Exercises

Find the surface area of the regular pyramid.

4.

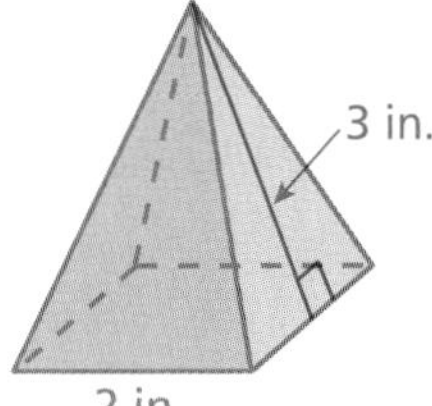

5.

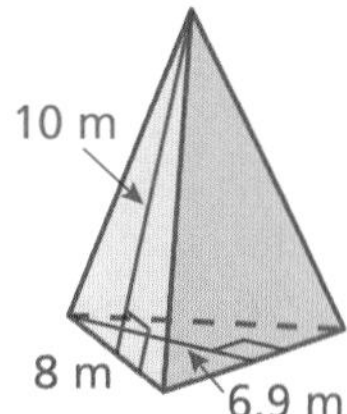

6.

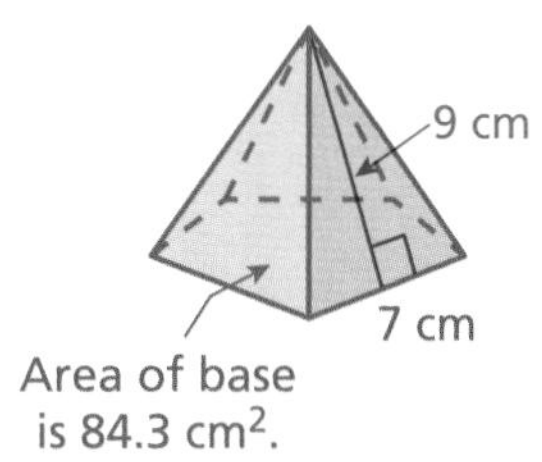

14.3 Surface Areas of Cylinders *(pp. 600–605)*

Find the surface area of the cylinder. Round your answer to the nearest tenth.

Draw a net.

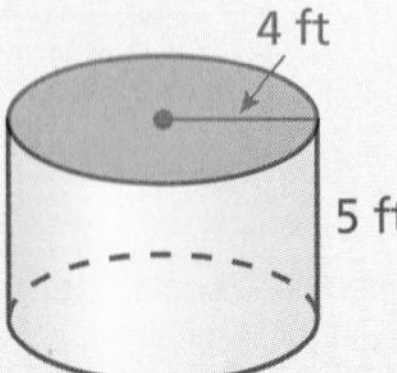

$$S = 2\pi r^2 + 2\pi rh$$
$$= 2\pi(4)^2 + 2\pi(4)(5)$$
$$= 32\pi + 40\pi$$
$$= 72\pi \approx 226.1$$

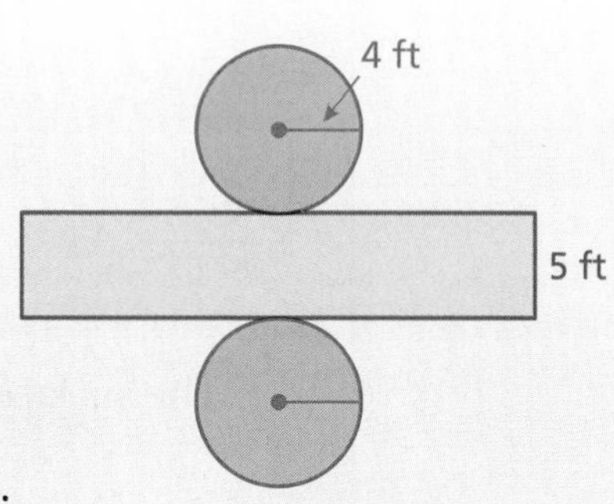

∴ The surface area is about 226.1 square millimeters.

Exercises

Find the surface area of the cylinder. Round your answer to the nearest tenth.

7.

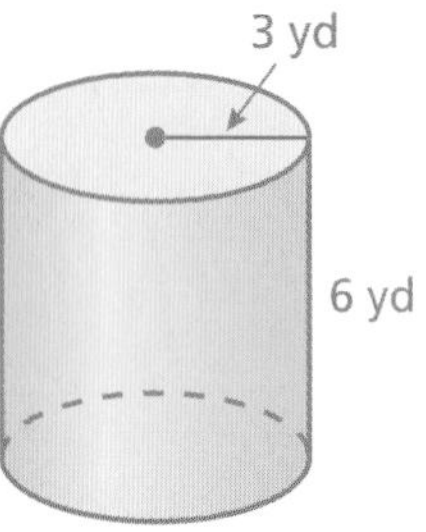

8.

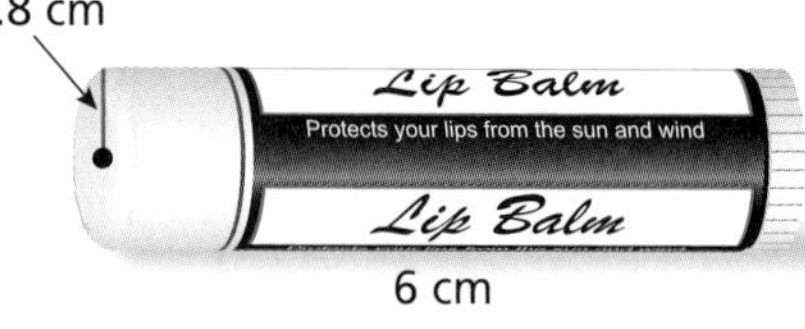

9. ORANGES Find the lateral surface area of the can of mandarin oranges.

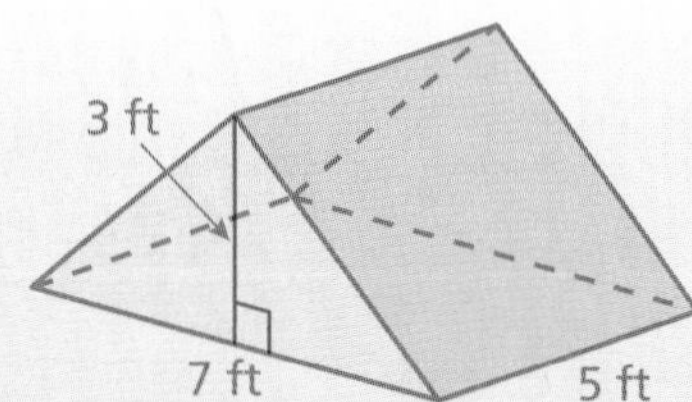

14.4 Volumes of Prisms *(pp. 608–613)*

Find the volume of the prism.

$V = Bh$	Write formula for volume.
$= \frac{1}{2}(7)(3) \cdot 5$	Substitute.
$= 52.5$	Multiply.

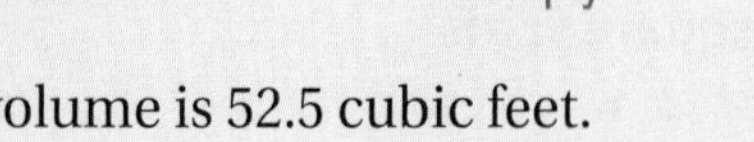

∴ The volume is 52.5 cubic feet.

Exercises

Find the volume of the prism.

10.

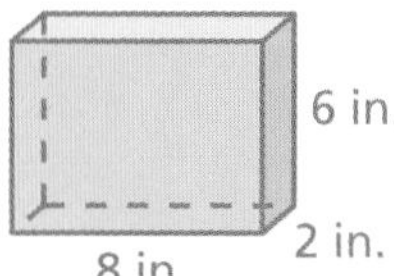

11.

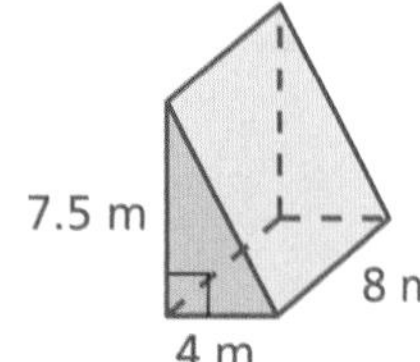

12.

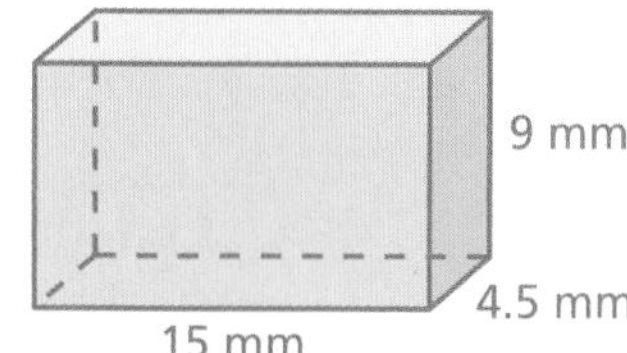

14.5 Volumes of Pyramids *(pp. 614–621)*

a. Find the volume of the pyramid.

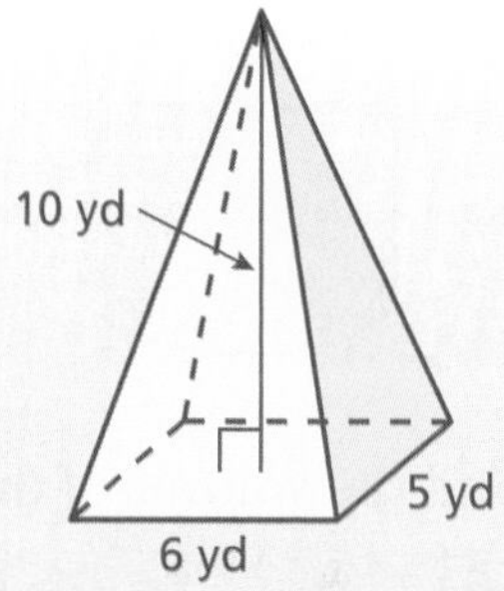

$V = \frac{1}{3}Bh$ Write formula for volume.

$= \frac{1}{3}(6)(5)(10)$ Substitute.

$= 100$ Multiply.

The volume is 100 cubic yards.

b. Describe the intersection of the plane and the solid.

i.

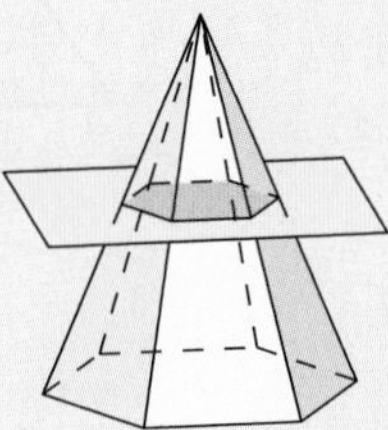

The intersection is a hexagon.

ii.

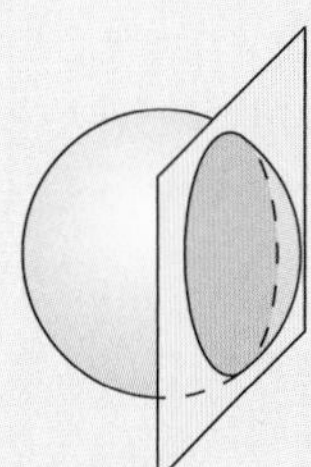

The intersection is a circle.

Exercises

Find the volume of the pyramid.

13.

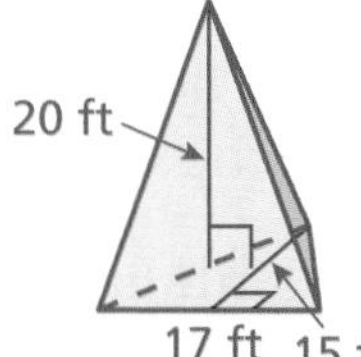

14.

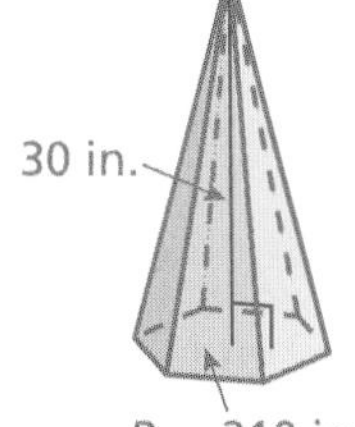

15.

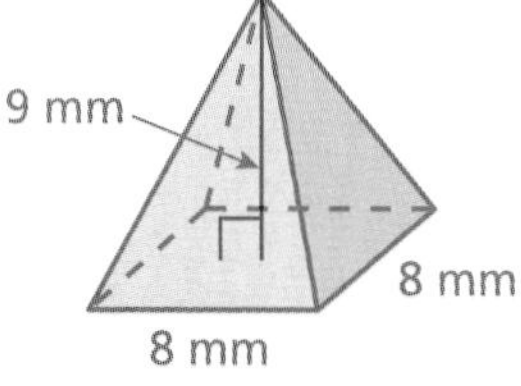

Describe the intersection of the plane and the solid.

16.

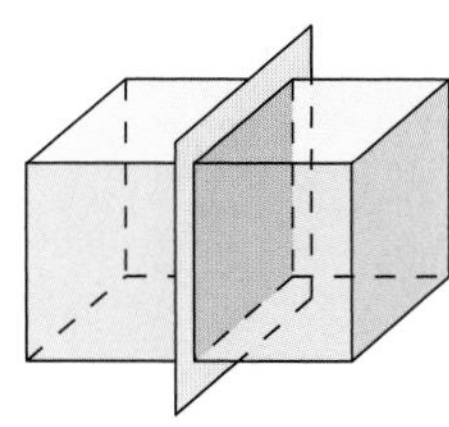

17.

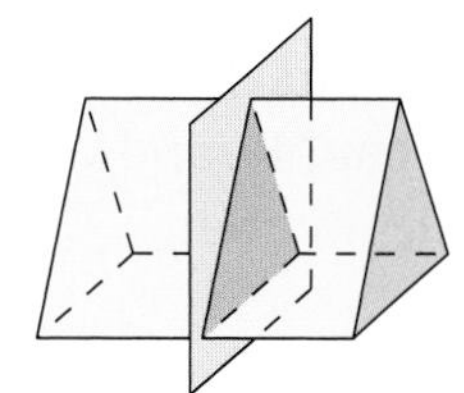

14 Chapter Test

Find the surface area of the prism or regular pyramid.

1. 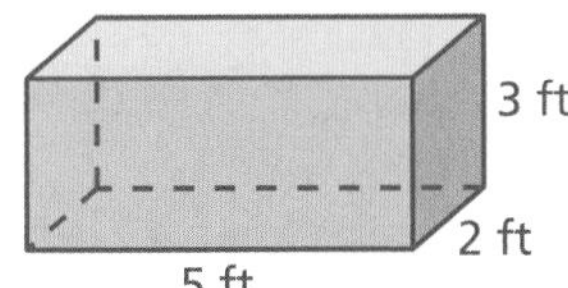

2. 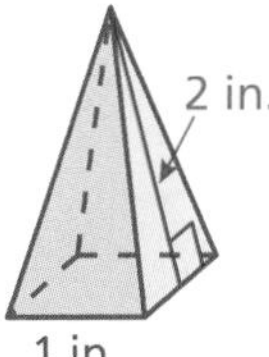

3. 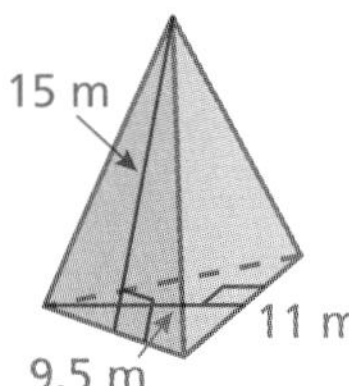

Find the surface area of the cylinder. Round your answer to the nearest tenth.

4. 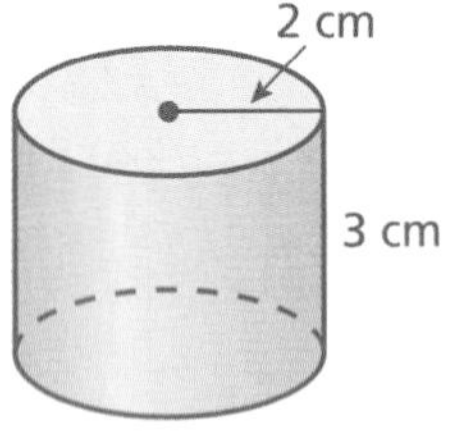

5. 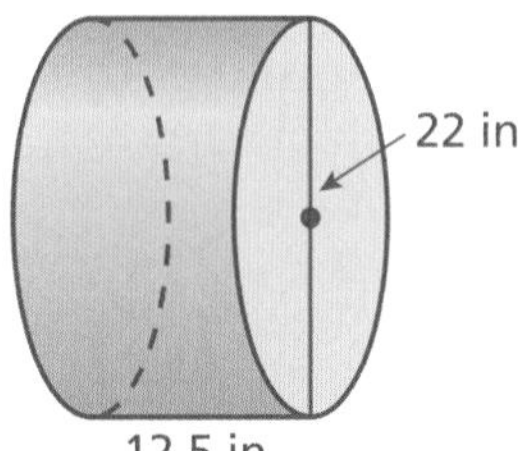

Find the volume of the solid.

6. 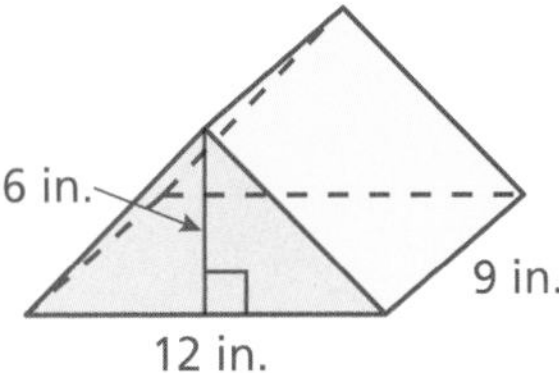

7. 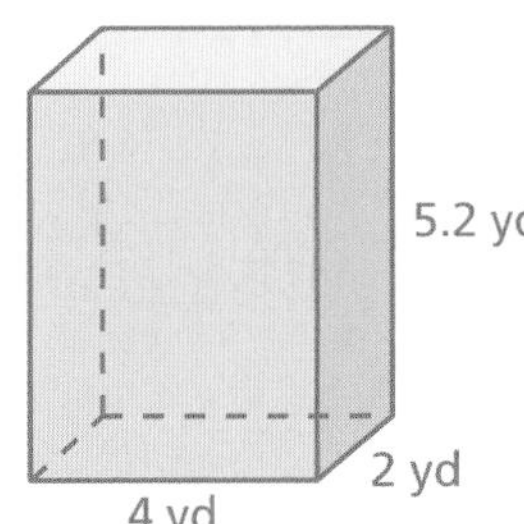

8. 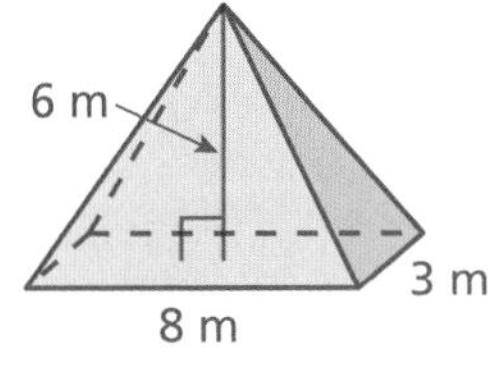

9. **SKATEBOARD RAMP** A quart of paint covers 80 square feet. How many quarts should you buy to paint the ramp with two coats? (Assume you will not paint the bottom of the ramp.)

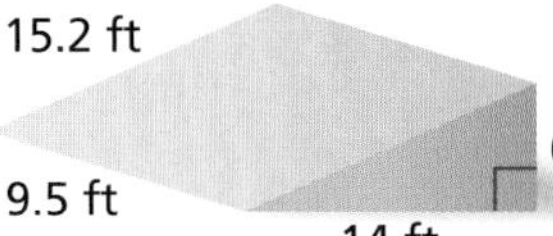

10. **GRAHAM CRACKERS** A manufacturer wants to double the volume of the graham cracker box. The manufacturer will either double the height or double the width.

    a. Which option uses less cardboard? Justify your answer.

    b. What is the volume of the new graham cracker box?

11. **SOUP** The label on the can of soup covers about 354.2 square centimeters. What is the height of the can? Round your answer to the nearest whole number.

14 Cumulative Assessment

1. A gift box and its dimensions are shown below.

What is the least amount of wrapping paper that you could have used to wrap the box?

A. 20 in.^{2}

B. 56 in.^{2}

C. 64 in.^{2}

D. 112 in.^{2}

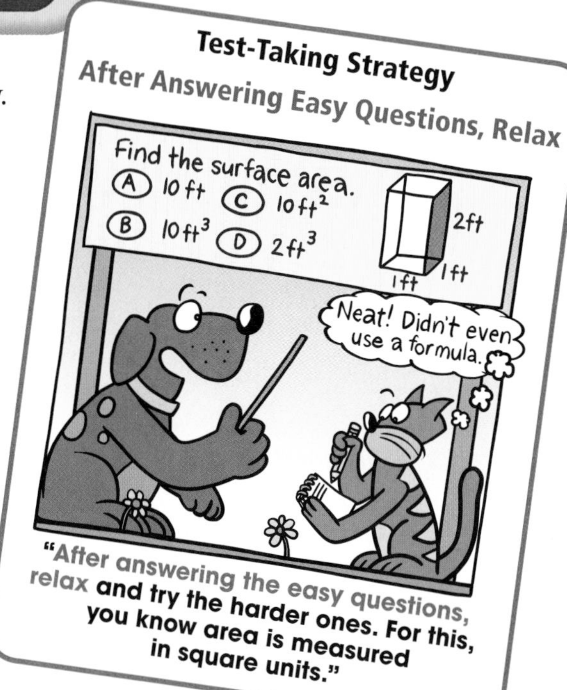

2. A student scored 600 the first time she took the mathematics portion of her college entrance exam. The next time she took the exam, she scored 660. Her second score represents what percent increase over her first score?

F. 9.1%

G. 10%

H. 39.6%

I. 60%

3. Raj was solving the proportion in the box below.

$$\frac{3}{8} = \frac{x-3}{24}$$

$$3 \cdot 24 = (x-3) \cdot 8$$

$$72 = x - 24$$

$$96 = x$$

What should Raj do to correct the error that he made?

A. Set the product of the numerators equal to the product of the denominators.

B. Distribute 8 to get $8x - 24$.

C. Add 3 to each side to get $\frac{3}{8} + 3 = \frac{x}{24}$.

D. Divide both sides by 24 to get $\frac{3}{8} \div 24 = x - 3$.

4. A line contains the two points plotted in the coordinate plane below.

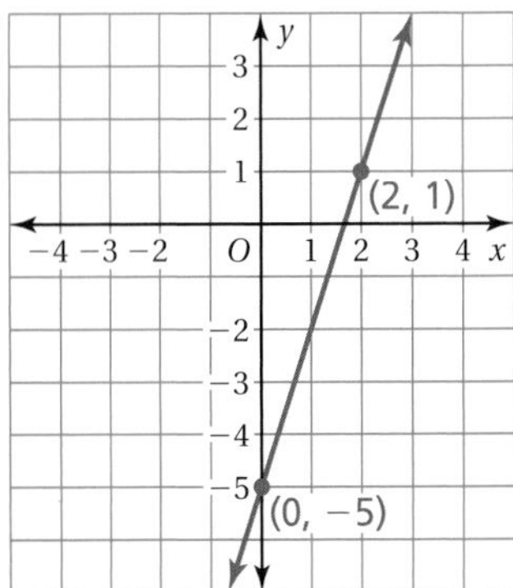

What is the slope of the line?

F. $\frac{1}{3}$

G. 2

H. 3

I. 6

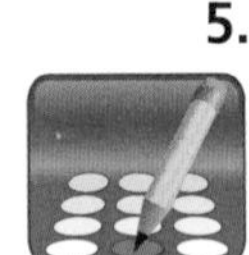

5. James is getting ready for wrestling season. As part of his preparation, he plans to lose 5% of his body weight. James currently weighs 160 pounds. How much will he weigh, in pounds, after he loses 5% of his weight?

6. How much material is needed to make the popcorn container?

A. 76π in.^{2}

B. 84π in.^{2}

C. 92π in.^{2}

D. 108π in.^{2}

7. To make 10 servings of soup you need 4 cups of broth. You want to know how many servings you can make with 8 pints of broth. Which proportion should you use?

F. $\frac{10}{4} = \frac{x}{8}$

G. $\frac{4}{10} = \frac{x}{16}$

H. $\frac{10}{4} = \frac{8}{x}$

I. $\frac{10}{4} = \frac{x}{16}$

8. A rectangular prism and its dimensions are shown below.

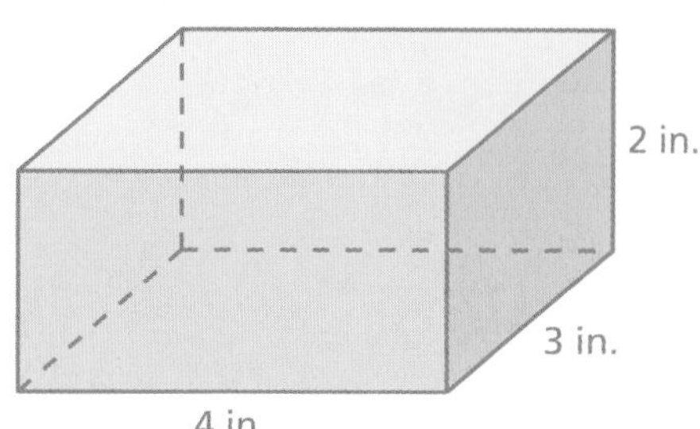

What is the volume, in cubic inches, of a rectangular prism whose dimensions are three times greater?

9. What is the value of x?

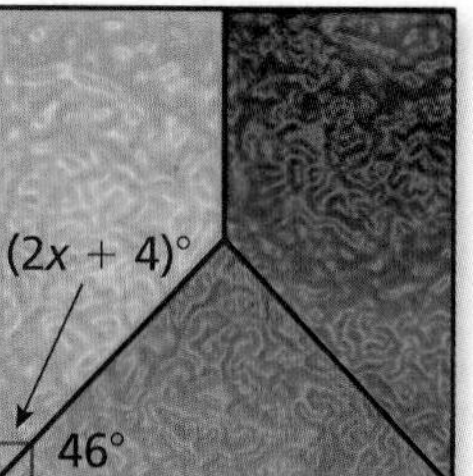

A. 20

B. 43

C. 44

D. 65

10. Which of the following could be the angle measures of a triangle?

F. 60°, 50°, 20°

G. 40°, 80°, 90°

H. 30°, 60°, 90°

I. 0°, 90°, 90°

11. The table below shows the costs of buying matinee movie tickets.

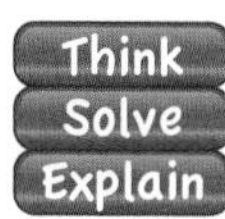

Matinee Tickets, x	2	3	4	5
Cost, y	\$9	\$13.50	\$18	\$22.50

Part A Graph the data.

Part B Find and interpret the slope of the line through the points.

Part C How much does it cost to buy 8 matinee movie tickets?

15 Probability and Statistics

RAWR!

"If there are 7 cats in a sack and I draw one at random,..."

It's ZERO! Because I'm not getting in a sack with 6 other cats!

"... what is the probability that I will draw you?"

CUBE CONSTRUCTION PLANS

"I'm just about finished making my two number cubes."

"Now, here's how the game works. You toss the two cubes."

I've got a better idea. Let's just toss one number cube.

"If the sum is even, I win. If it's odd, you win."

What You Learned Before

M = Mouse
B = Dog Biscuit
Why do we always have to use this spinner?

"Let's spin to decide what we will have for lunch."

Writing Ratios

Example 1 **There are 32 football players and 16 cheerleaders at your school. Write the ratio of cheerleaders to football players.**

cheerleaders → 16, football players → 32

$$\frac{16}{32} = \frac{1}{2}$$ Write in simplest form.

So, the ratio of cheerleaders to football players is $\frac{1}{2}$.

Example 2

a. Write the ratio of girls to boys in Classroom A.

	Boys	Girls
Classroom A	14	11
Classroom B	12	8

$$\frac{\text{Girls in Classroom A}}{\text{Boys in Classroom A}} = \frac{11}{14}$$

So, the ratio of girls to boys in Classroom A is $\frac{11}{14}$.

b. Write the ratio of boys in Classroom B to the total number of students in both classes.

$$\frac{\text{Boys in Classroom B}}{\text{Total number of students}} = \frac{12}{14 + 11 + 12 + 8} = \frac{12}{45} = \frac{4}{15}$$ Write in simplest form.

So, the ratio of boys in Classroom B to the total number of students is $\frac{4}{15}$.

Try It Yourself

Write the ratio in simplest form.

1. baseballs to footballs
2. footballs to total pieces of equipment
3. sneakers to ballet slippers
4. sneakers to total number of shoes

5. green beads to blue beads
6. red beads : green beads
7. green beads : total number of beads

15.1 Outcomes and Events

Essential Question In an experiment, how can you determine the number of possible results?

An *experiment* is an investigation or a procedure that has varying results. Flipping a coin, rolling a number cube, and spinning a spinner are all examples of experiments.

1 ACTIVITY: Conducting Experiments

Work with a partner.

a. You flip a dime.

There are ☐ possible results.

Out of 20 flips, you think you will flip heads ☐ times.

Flip a dime 20 times. Tally your results in a table. How close was your guess?

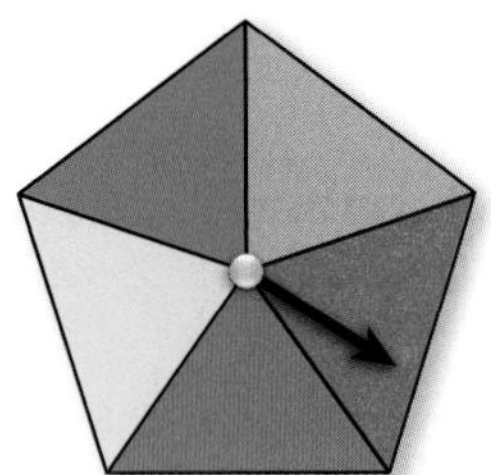

b. You spin the spinner shown.

There are ☐ possible results.

Out of 20 spins, you think you will spin orange ☐ times.

Spin the spinner 20 times. Tally your results in a table. How close was your guess?

c. You spin the spinner shown.

There are ☐ possible results.

Out of 20 spins, you think you will spin a 4 ☐ times.

Spin the spinner 20 times. Tally your results in a table. How close was your guess?

Probability and Statistics

In this lesson, you will

- identify and count the outcomes of experiments.

2 ACTIVITY: Comparing Different Results

Work with a partner. Use the spinner in Activity 1 (c).

a. Do you have a better chance of spinning an even number or a multiple of 4? Explain your reasoning.

b. Do you have a better chance of spinning an even number or an odd number? Explain your reasoning.

3 ACTIVITY: Rock Paper Scissors

Work with a partner.

Math Practice

Interpret a Solution

How do your results compare to the possible results? Explain.

a. Play Rock Paper Scissors 30 times. Tally your results in the table.

b. How many possible results are there?

c. Of the possible results, in how many ways can Player A win? Player B win? the players tie?

d. Does one of the players have a better chance of winning than the other player? Explain your reasoning.

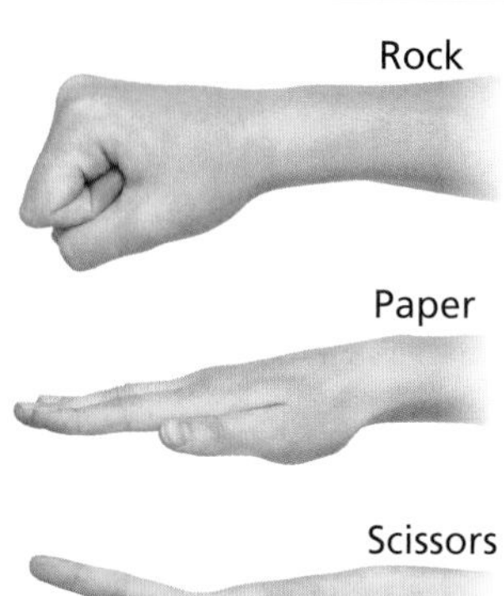

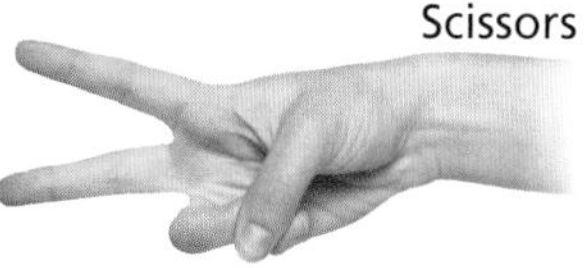

Rock *breaks* scissors.
Paper *covers* rock.
Scissors *cut* paper.

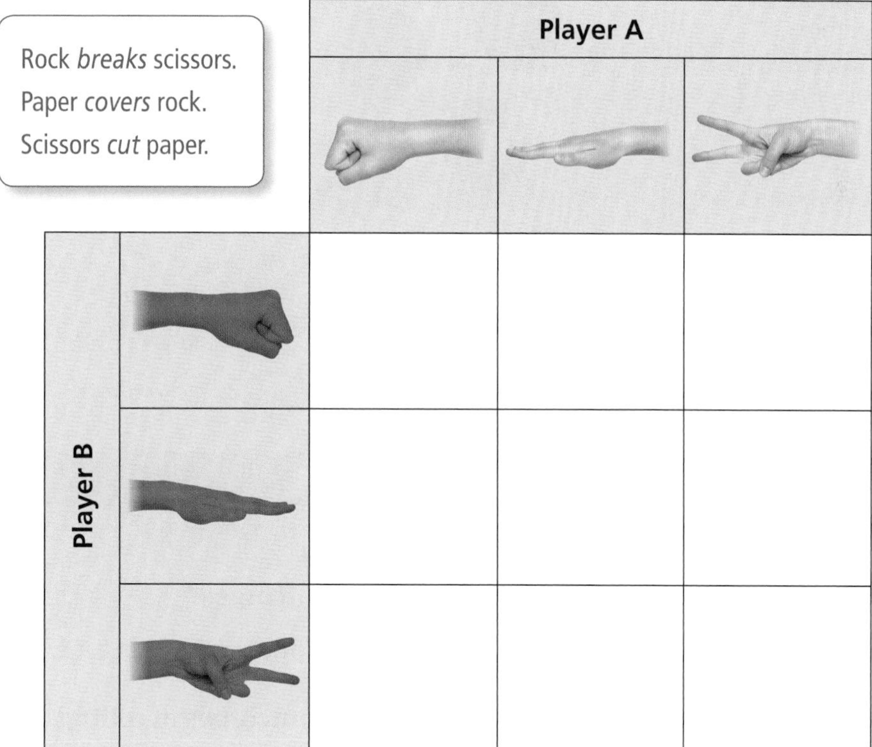

What Is Your Answer?

4. **IN YOUR OWN WORDS** In an experiment, how can you determine the number of possible results?

Use what you learned about experiments to complete Exercises 3 and 4 on page 636.

15.1 Lesson

Key Vocabulary
experiment, *p. 634*
outcomes, *p. 634*
event, *p. 634*
favorable outcomes, *p. 634*

Reading

When an experiment is performed *at random* or *randomly*, all of the possible outcomes are equally likely.

Key Ideas

Outcomes and Events

An **experiment** is an investigation or a procedure that has varying results. The possible results of an experiment are called **outcomes**. A collection of one or more outcomes is an **event**. The outcomes of a specific event are called **favorable outcomes**.

For example, randomly selecting a marble from a group of marbles is an experiment. Each marble in the group is an outcome. Selecting a green marble from the group is an event.

Possible outcomes

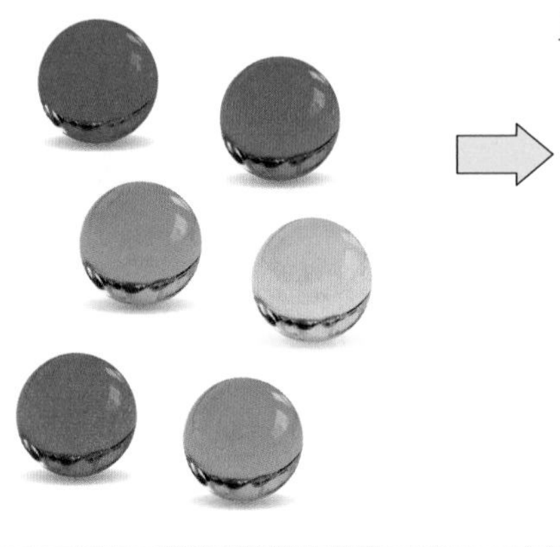

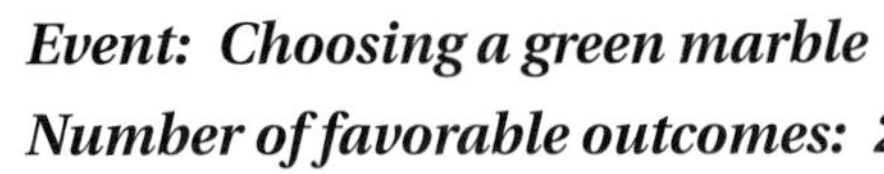

Event: Choosing a green marble

Number of favorable outcomes: 2

EXAMPLE 1 Identifying Outcomes

You roll the number cube.

a. What are the possible outcomes?

The six possible outcomes are rolling a 1, 2, 3, 4, 5, and 6.

b. What are the favorable outcomes of rolling an even number?

even	*not* even
2, 4, 6	1, 3, 5

The favorable outcomes of the event are rolling a 2, 4, and 6.

c. What are the favorable outcomes of rolling a number greater than 5?

greater than 5	*not* greater than 5
6	1, 2, 3, 4, 5

The favorable outcome of the event is rolling a 6.

On Your Own

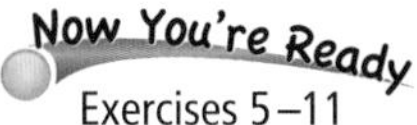

1. You randomly choose a letter from a hat that contains the letters A through K.
   a. What are the possible outcomes?
   b. What are the favorable outcomes of choosing a vowel?

EXAMPLE 2 Counting Outcomes

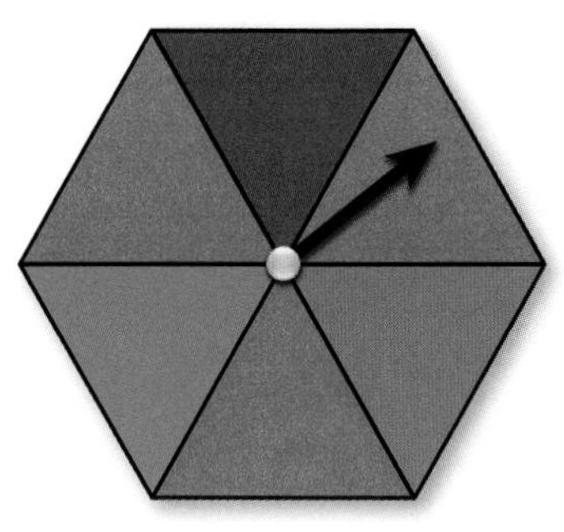

You spin the spinner.

a. How many possible outcomes are there?

The spinner has 6 sections. So, there are 6 possible outcomes.

b. In how many ways can spinning red occur?

The spinner has 3 red sections. So, spinning red can occur in 3 ways.

c. In how many ways can spinning *not* purple occur? What are the favorable outcomes of spinning *not* purple?

The spinner has 5 sections that are *not* purple. So, spinning *not* purple can occur in 5 ways.

purple	*not* purple
purple	red, red, red, green, blue

The favorable outcomes of the event are red, red, red, green, and blue.

On Your Own

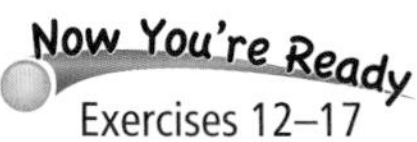

2. You randomly choose a marble.

   a. How many possible outcomes are there?
   b. In how many ways can choosing blue occur?
   c. In how many ways can choosing *not* yellow occur? What are the favorable outcomes of choosing *not* yellow?

15.1 Exercises

Vocabulary and Concept Check

1. **VOCABULARY** Is rolling an even number on a number cube an *outcome* or an *event*? Explain.
2. **WRITING** Describe how an outcome and a favorable outcome are different.

Practice and Problem Solving

You spin the spinner shown.

3. How many possible results are there?
4. Of the possible results, in how many ways can you spin an even number? an odd number?

① 5. **TILES** What are the possible outcomes of randomly choosing one of the tiles shown?

You randomly choose one of the tiles shown above. Find the favorable outcomes of the event.

6. Choosing a 6
7. Choosing an odd number
8. Choosing a number greater than 5
9. Choosing an odd number less than 5
10. Choosing a number less than 3
11. Choosing a number divisible by 3

You randomly choose one marble from the bag. (a) Find the number of ways the event can occur. (b) Find the favorable outcomes of the event.

② 12. Choosing blue
13. Choosing green
14. Choosing purple
15. Choosing yellow
16. Choosing *not* red
17. Choosing *not* blue
18. **ERROR ANALYSIS** Describe and correct the error in finding the number of ways that choosing *not* purple can occur.

✗

purple	*not* purple
purple	red, blue, green, yellow

Choosing *not* purple can occur in 4 ways.

19. **COINS** You have 10 coins in your pocket. Five are Susan B. Anthony dollars, two are Kennedy half-dollars, and three are presidential dollars. You randomly choose a coin. In how many ways can choosing *not* a presidential dollar occur?

Susan B. Anthony dollar

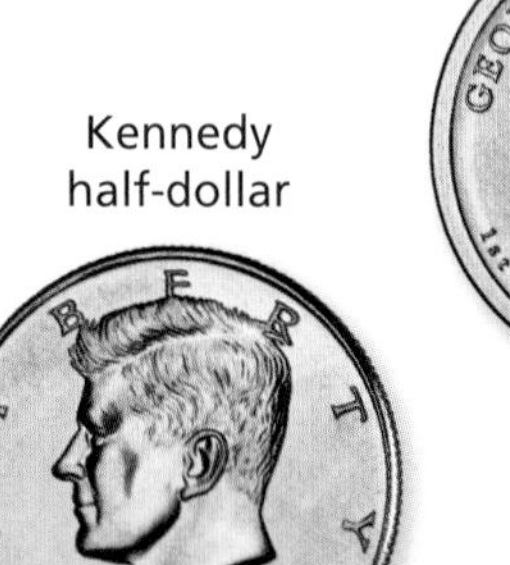

Kennedy half-dollar

Presidential dollar

Tell whether the statement is *true* or *false*. If it is false, change the italicized word to make the statement true.

Spinner A

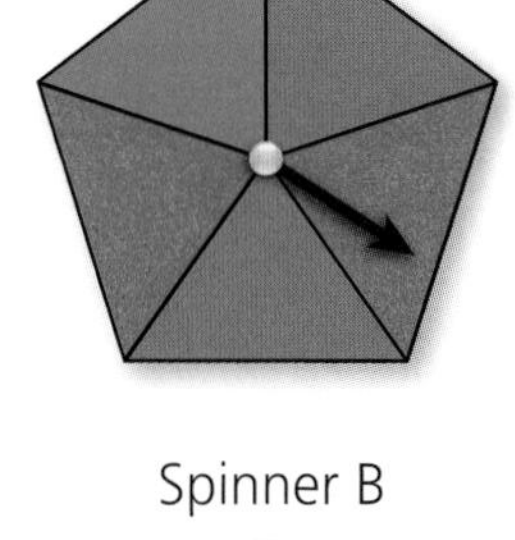

20. Spinning blue and spinning *green* have the same number of favorable outcomes on Spinner A.

21. Spinning blue has one *more* favorable outcome than spinning green on Spinner B.

22. There are *three* possible outcomes of spinning Spinner A.

23. Spinning *red* can occur in four ways on Spinner B.

24. Spinning not green can occur in *three* ways on Spinner B.

Spinner B

25. **MUSIC** A bargain bin contains classical and rock CDs. There are 60 CDs in the bin. Choosing a rock CD and *not* choosing a rock CD have the same number of favorable outcomes. How many rock CDs are in the bin?

26. **Precision** You randomly choose one of the cards and set it aside. Then you randomly choose a second card. Describe how the number of possible outcomes changes after the first card is chosen.

Fair Game Review What you learned in previous grades & lessons

Solve the proportion. *(Skills Review Handbook)*

27. $\frac{x}{10} = \frac{1}{5}$

28. $\frac{60}{n} = \frac{20}{7}$

29. $\frac{1}{3} = \frac{w}{36}$

30. $\frac{25}{17} = \frac{100}{b}$

31. **MULTIPLE CHOICE** What is the surface area of the rectangular prism? *(Section 14.1)*

Ⓐ 162 in.^{2} Ⓑ 264 in.^{2}

Ⓒ 324 in.^{2} Ⓓ 360 in.^{2}

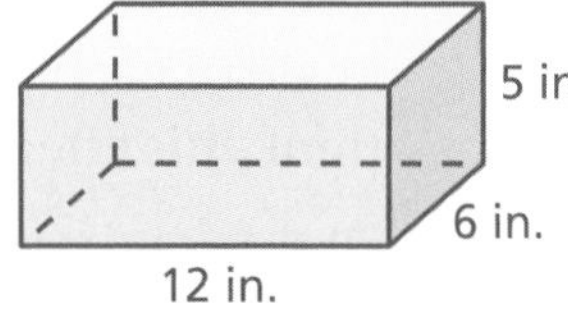

15.2 Probability

Essential Question How can you describe the likelihood of an event?

1 ACTIVITY: Black-and-White Spinner Game

Work with a partner. You work for a game company. You need to create a game that uses the spinner below.

a. Write rules for a game that uses the spinner. Then play it.

b. After playing the game, do you want to revise the rules? Explain.

Probability and Statistics

In this lesson, you will

- understand the concept of probability and the relationship between probability and likelihood.
- find probabilities of events.

c. **CHOOSE TOOLS** Using the center of the spinner as the vertex, measure the angle of each pie-shaped section. Is each section the same size? How do you think this affects the likelihood of spinning a given number?

d. Your friend is about to spin the spinner and wants to know how likely it is to spin a 3. How would you describe the likelihood of this event to your friend?

2 ACTIVITY: Changing the Spinner

Work with a partner. For each spinner, do the following.

- **Measure the angle of each pie-shaped section.**
- **Tell whether you are more likely to spin a particular number. Explain your reasoning.**
- **Tell whether your rules from Activity 1 make sense for these spinners. Explain your reasoning.**

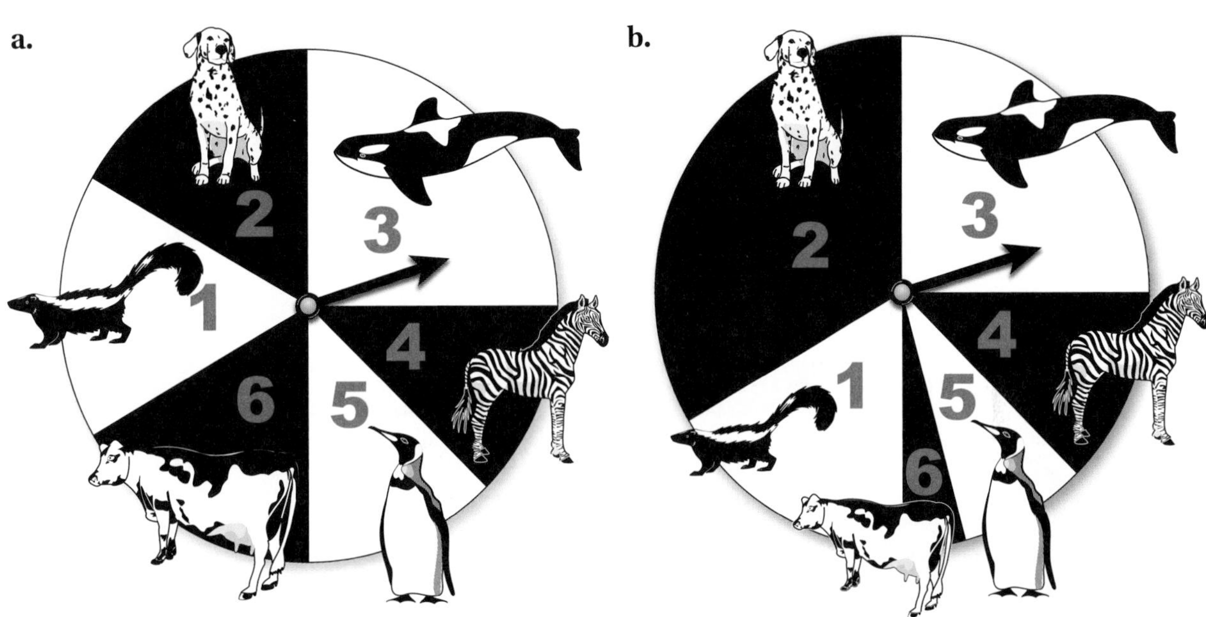

3 ACTIVITY: Is This Game Fair?

Math Practice

Use Prior Results

How can you use the results of the previous activities to determine whether the game is fair?

Work with a partner. Apply the following rules to each spinner in Activities 1 and 2. Is the game fair? Why or why not? If not, who has the better chance of winning?

- **Take turns spinning the spinner.**
- **If you spin an odd number, Player 1 wins.**
- **If you spin an even number, Player 2 wins.**

What Is Your Answer?

4. **IN YOUR OWN WORDS** How can you describe the likelihood of an event?
5. Describe the likelihood of spinning an 8 in Activity 1.
6. Describe a career in which it is important to know the likelihood of an event.

Practice Use what you learned about the likelihood of an event to complete Exercises 4 and 5 on page 642.

15.2 Lesson

Key Vocabulary
probability, *p. 640*

Probability

The **probability** of an event is a number that measures the likelihood that the event will occur. Probabilities are between 0 and 1, including 0 and 1. The diagram relates likelihoods (above the diagram) and probabilities (below the diagram).

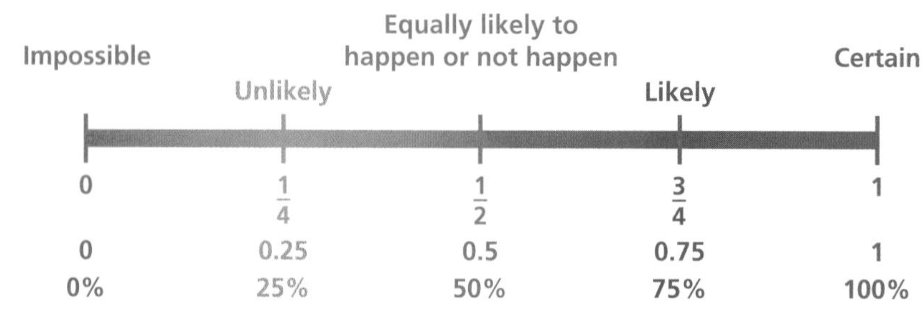

Study Tip

Probabilities can be written as fractions, decimals, or percents.

EXAMPLE 1 Describing the Likelihood of an Event

There is an 80% chance of thunderstorms tomorrow. Describe the likelihood of the event.

The probability of thunderstorms tomorrow is 80%.

∴ Because 80% is close to 75%, it is *likely* that there will be thunderstorms tomorrow.

On Your Own

Now You're Ready
Exercises 6–9

Describe the likelihood of the event given its probability.

1. The probability that you land a jump on a snowboard is $\frac{1}{2}$.
2. There is a 100% chance that the temperature will be less than 120°F tomorrow.

Key Idea

Finding the Probability of an Event

When all possible outcomes are equally likely, the probability of an event is the ratio of the number of favorable outcomes to the number of possible outcomes. The probability of an event is written as $P(\text{event})$.

$$P(\text{event}) = \frac{\text{number of favorable outcomes}}{\text{number of possible outcomes}}$$

EXAMPLE 2 Finding a Probability

You roll the number cube. What is the probability of rolling an odd number?

$$P(\text{event}) = \frac{\text{number of favorable outcomes}}{\text{number of possible outcomes}}$$

$$P(\text{odd}) = \frac{3}{6}$$

There are 3 odd numbers (1, 3, and 5).

There is a total of 6 numbers.

$$= \frac{1}{2}$$ Simplify.

The probability of rolling an odd number is $\frac{1}{2}$, or 50%.

EXAMPLE 3 Using a Probability

The probability that you randomly draw a short straw from a group of 40 straws is $\frac{3}{20}$. How many are short straws?

Ⓐ 4 Ⓑ 6

Ⓒ 15 Ⓓ 34

$$P(\text{short}) = \frac{\text{number of short straws}}{\text{total number of straws}}$$

$$\frac{3}{20} = \frac{n}{40}$$ Substitute. Let n be the number of short straws.

$$6 = n$$ Solve for n.

There are 6 short straws.

So, the correct answer is Ⓑ.

On Your Own

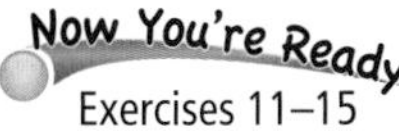

Now You're Ready
Exercises 11–15

3. In Example 2, what is the probability of rolling a number greater than 2?
4. In Example 2, what is the probability of rolling a 7?
5. The probability that you randomly draw a short straw from a group of 75 straws is $\frac{1}{15}$. How many are short straws?

15.2 Exercises

Vocabulary and Concept Check

1. **VOCABULARY** Explain how to find the probability of an event.
2. **REASONING** Can the probability of an event be 1.5? Explain.
3. **OPEN-ENDED** Give a real-life example of an event that is impossible. Give a real-life example of an event that is certain.

Practice and Problem Solving

You are playing a game using the spinners shown.

4. You want to move down. On which spinner are you more likely to spin "Down"? Explain.
5. You want to move forward. Which spinner would you spin? Explain.

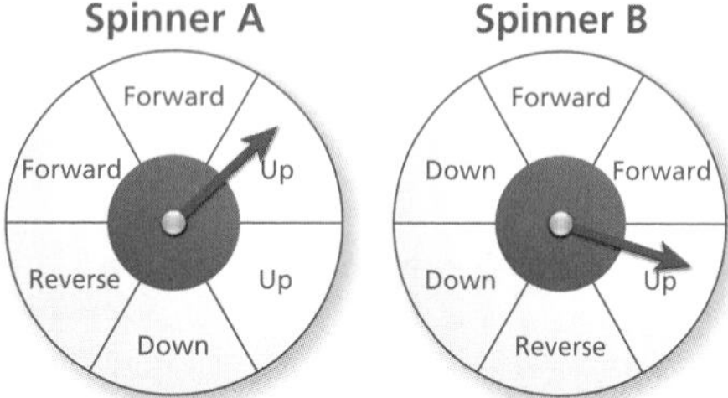

Describe the likelihood of the event given its probability.

(1) 6. Your soccer team wins $\frac{3}{4}$ of the time.

7. There is a 0% chance that you will grow 12 more feet.
8. The probability that the sun rises tomorrow is 1.
9. It rains on $\frac{1}{5}$ of the days in July.

10. **VIOLIN** You have a 50% chance of playing the correct note on a violin. Describe the likelihood of playing the correct note.

You randomly choose one shirt from the shelves. Find the probability of the event.

(2) 11. Choosing a red shirt

12. Choosing a green shirt
13. *Not* choosing a white shirt
14. *Not* choosing a black shirt
15. Choosing an orange shirt

16. **ERROR ANALYSIS** Describe and correct the error in finding the probability of *not* choosing a blue shirt from the shelves above.

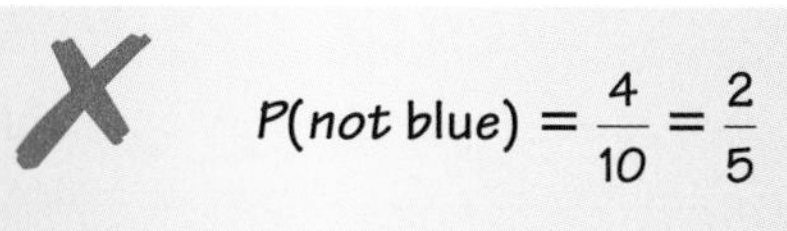

17. **CONTEST** The rules of a contest say that there is a 5% chance of winning a prize. Four hundred people enter the contest. Predict how many people will win a prize.

18. **RUBBER DUCKS** At a carnival, the probability that you choose a winning rubber duck from 25 ducks is 0.24.

    a. How many are *not* winning ducks?

    b. Describe the likelihood of *not* choosing a winning duck.

19. **DODECAHEDRON** A dodecahedron has twelve sides numbered 1 through 12. Find the probability and describe the likelihood of each event.

    a. Rolling a number less than 9

    b. Rolling a multiple of 3

    c. Rolling a number greater than 6

A Punnett square is a grid used to show possible gene combinations for the offspring of two parents. In the Punnett square shown, a boy is represented by *XY*. A girl is represented by *XX*.

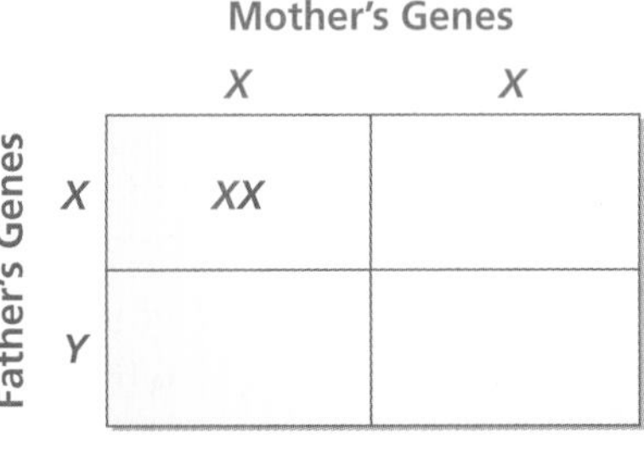

20. Complete the Punnett square.

21. Explain why the probability of two parents having a boy or having a girl is equally likely.

22. Critical Thinking Two parents each have the gene combination *Cs*. The gene *C* is for curly hair. The gene *s* is for straight hair.

    a. Make a Punnett square for the two parents. When all outcomes are equally likely, what is the probability of a child having the gene combination *CC*?

    b. Any gene combination that includes a *C* results in curly hair. When all outcomes are equally likely, what is the probability of a child having curly hair?

Fair Game Review What you learned in previous grades & lessons

Solve the inequality. Graph the solution. *(Section 11.2 and Section 11.3)*

23. $x + 5 < 9$

24. $b - 2 \geq -7$

25. $1 > -\frac{w}{3}$

26. $6 \leq -2g$

27. **MULTIPLE CHOICE** Find the value of x. *(Section 12.4)*

Ⓐ 85 Ⓑ 90

Ⓒ 93 Ⓓ 102

135°
86°
$x°$
46°

15.3 Experimental and Theoretical Probability

Essential Question How can you use relative frequencies to find probabilities?

When you conduct an experiment, the **relative frequency** of an event is the fraction or percent of the time that the event occurs.

$$\text{relative frequency} = \frac{\text{number of times the event occurs}}{\text{total number of times you conduct the experiment}}$$

1 ACTIVITY: Finding Relative Frequencies

Work with a partner.

a. Flip a quarter 20 times and record your results. Then complete the table. Are the relative frequencies the same as the probability of flipping heads or tails? Explain.

	Flipping Heads	Flipping Tails
Relative Frequency		

b. Compare your results with those of other students in your class. Are the relative frequencies the same? If not, why do you think they differ?

c. Combine all of the results in your class. Then complete the table again. Did the relative frequencies change? What do you notice? Explain.

d. Suppose everyone in your school conducts this experiment and you combine the results. How do you think the relative frequencies will change?

2 ACTIVITY: Using Relative Frequencies

Probability and Statistics
In this lesson, you will
- find relative frequencies.
- use experimental probabilities to make predictions.
- use theoretical probabilities to find quantities.
- compare experimental and theoretical probabilities.

Work with a partner. You have a bag of colored chips. You randomly select a chip from the bag and replace it. The table shows the number of times you select each color.

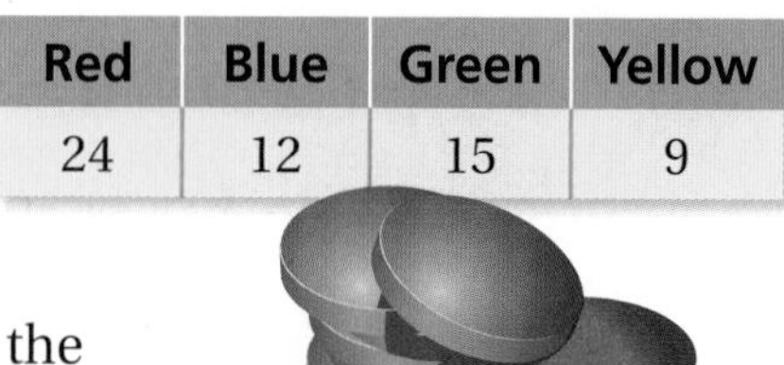

Red	Blue	Green	Yellow
24	12	15	9

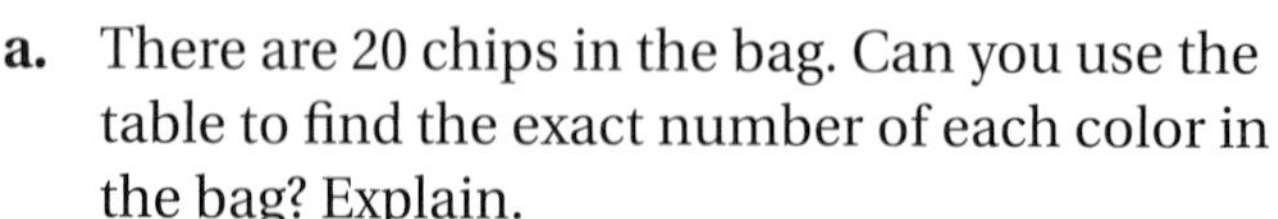

a. There are 20 chips in the bag. Can you use the table to find the exact number of each color in the bag? Explain.

b. You randomly select a chip from the bag and replace it. You do this 50 times, then 100 times, and you calculate the relative frequencies after each experiment. Which experiment do you think gives a better approximation of the exact number of each color in the bag? Explain.

3 ACTIVITY: Conducting an Experiment

Work with a partner. You toss a thumbtack onto a table. There are two ways the thumbtack can land.

a. Your friend says that because there are two outcomes, the probability of the thumbtack landing point up must be $\frac{1}{2}$. Do you think this conclusion is true? Explain.

Point up

On its side

b. Toss a thumbtack onto a table 50 times and record your results. In a *uniform probability model,* each outcome is equally likely to occur. Do you think this experiment represents a uniform probability model? Explain.

Use the relative frequencies to complete the following.

P(point up) = ▭ P(on its side) = ▭

> **Math Practice**
>
> **Analyze Relationships**
>
> How can you use the results of your experiment to determine whether this is a uniform probability model?

What Is Your Answer?

4. **IN YOUR OWN WORDS** How can you use relative frequencies to find probabilities? Give an example.
5. Your friend rolls a number cube 500 times. How many times do you think your friend will roll an odd number? Explain your reasoning.
6. In Activity 2, your friend says, "There are no orange-colored chips in the bag." Do you think this conclusion is true? Explain.
7. Give an example of an experiment that represents a uniform probability model.
8. Tell whether you can use each spinner to represent a uniform probability model. Explain your reasoning.

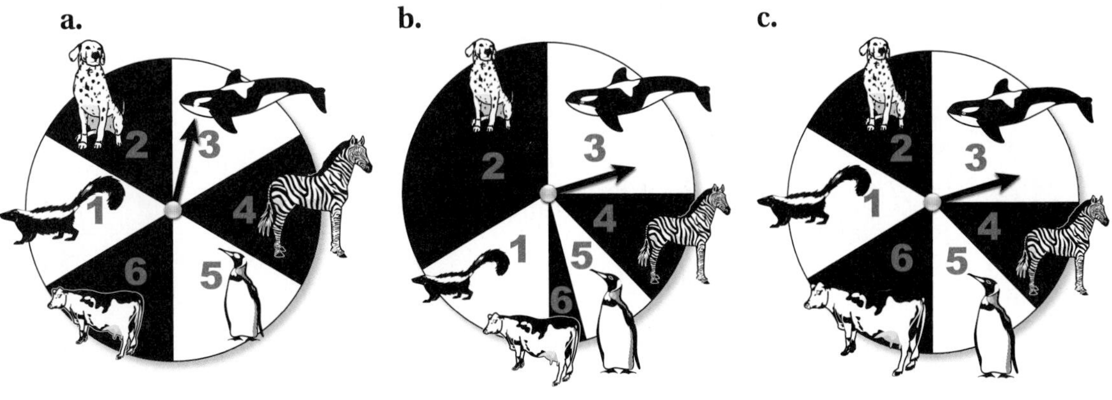

Practice Use what you learned about relative frequencies to complete Exercises 6 and 7 on page 649.

15.3 Lesson

Key Vocabulary
relative frequency, *p. 644*
experimental probability, *p. 646*
theoretical probability, *p. 647*

Key Idea

Experimental Probability

Probability that is based on repeated trials of an experiment is called **experimental probability**.

$$P(\text{event}) = \frac{\text{number of times the event occurs}}{\text{total number of trials}}$$

EXAMPLE 1 Finding an Experimental Probability

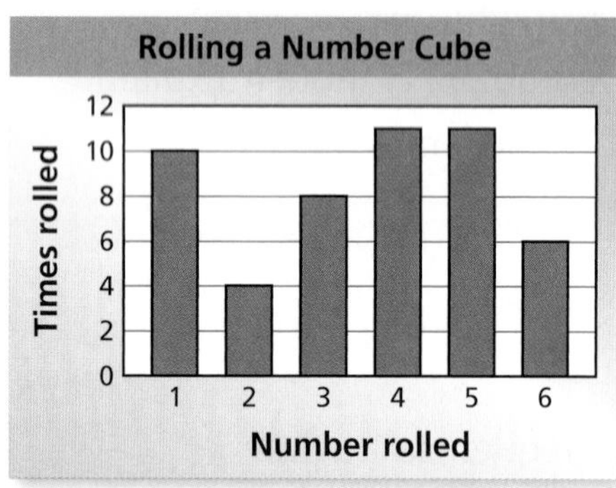

The bar graph shows the results of rolling a number cube 50 times. What is the experimental probability of rolling an odd number?

The bar graph shows 10 ones, 8 threes, and 11 fives. So, an odd number was rolled $10 + 8 + 11 = 29$ times in a total of 50 rolls.

$$P(\text{event}) = \frac{\text{number of times the event occurs}}{\text{total number of trials}}$$

$$P(\text{odd}) = \frac{29}{50}$$

An odd number was rolled 29 times.

There was a total of 50 rolls.

The experimental probability is $\frac{29}{50}$, 0.58, or 58%.

EXAMPLE 2 Making a Prediction

"April showers bring May flowers." Old Proverb, 1557

It rains 2 out of the last 12 days in March. If this trend continues, how many rainy days would you expect in April?

Find the experimental probability of a rainy day.

$$P(\text{event}) = \frac{\text{number of times the event occurs}}{\text{total number of trials}}$$

It rains 2 days.

$$P(\text{rain}) = \frac{2}{12} = \frac{1}{6}$$

There is a total of 12 days.

To make a prediction, multiply the probability of a rainy day by the number of days in April.

$$\frac{1}{6} \cdot 30 = 5$$

So, you can predict that there will be 5 rainy days in April.

On Your Own

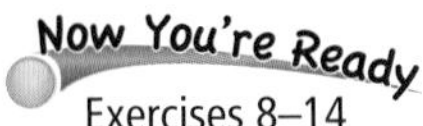

Now You're Ready
Exercises 8–14

1. In Example 1, what is the experimental probability of rolling an even number?
2. At a clothing company, an inspector finds 5 defective pairs of jeans in a shipment of 200. If this trend continues, about how many pairs of jeans would you expect to be defective in a shipment of 5000?

Key Idea

Theoretical Probability

When all possible outcomes are equally likely, the **theoretical probability** of an event is the ratio of the number of favorable outcomes to the number of possible outcomes.

$$P(\text{event}) = \frac{\text{number of favorable outcomes}}{\text{number of possible outcomes}}$$

EXAMPLE 3 Finding a Theoretical Probability

You randomly choose one of the letters shown. What is the theoretical probability of choosing a vowel?

$$P(\text{event}) = \frac{\text{number of favorable outcomes}}{\text{number of possible outcomes}}$$

$$P(\text{vowel}) = \frac{3}{7}$$

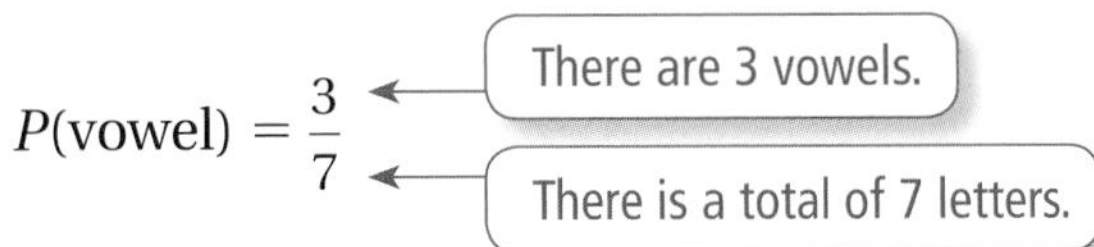

There are 3 vowels.
There is a total of 7 letters.

The probability of choosing a vowel is $\frac{3}{7}$, or about 43%.

EXAMPLE 4 Using a Theoretical Probability

The theoretical probability of winning a bobblehead when spinning a prize wheel is $\frac{1}{6}$. The wheel has 3 bobblehead sections. How many sections are on the wheel?

$$P(\text{bobblehead}) = \frac{\text{number of bobblehead sections}}{\text{total number of sections}}$$

$$\frac{1}{6} = \frac{3}{s}$$ Substitute. Let s be the total number of sections.

$$s = 18$$ Cross Products Property

So, there are 18 sections on the wheel.

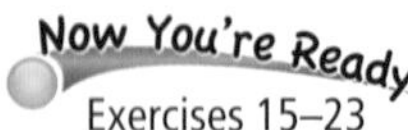

3. In Example 3, what is the theoretical probability of choosing an X?

4. The theoretical probability of spinning an odd number on a spinner is 0.6. The spinner has 10 sections. How many sections have odd numbers?

5. The prize wheel in Example 4 was spun 540 times at a baseball game. About how many bobbleheads would you expect were won?

EXAMPLE 5 Comparing Experimental and Theoretical Probability

The bar graph shows the results of rolling a number cube 300 times.

a. What is the experimental probability of rolling an odd number?

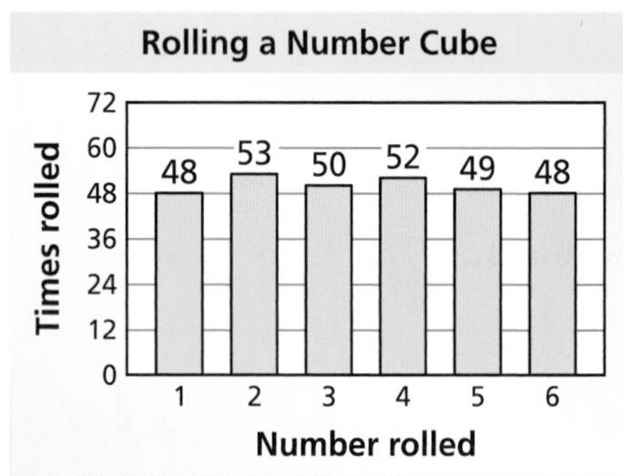

The bar graph shows 48 ones, 50 threes, and 49 fives. So, an odd number was rolled $48 + 50 + 49 = 147$ times in a total of 300 rolls.

$$P(\text{event}) = \frac{\text{number of times the event occurs}}{\text{total number of trials}}$$

$$P(\text{odd}) = \frac{147}{300}$$

An odd number was rolled 147 times.

There was a total of 300 rolls.

$$= \frac{49}{100}, \text{ or } 49\%$$

b. How does the experimental probability compare with the theoretical probability of rolling an odd number?

In Section 15.2, Example 2, you found that the theoretical probability of rolling an odd number is 50%. The experimental probability, 49%, is close to the theoretical probability.

c. Compare the experimental probability in part (a) to the experimental probability in Example 1.

As the number of trials increased from 50 to 300, the experimental probability decreased from 58% to 49%. So, it became closer to the theoretical probability of 50%.

On Your Own

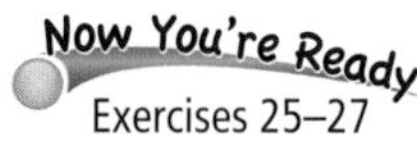

6. Use the bar graph in Example 5 to find the experimental probability of rolling a number greater than 1. Compare the experimental probability to the theoretical probability of rolling a number greater than 1.

15.3 Exercises

Vocabulary and Concept Check

1. **VOCABULARY** Describe how to find the experimental probability of an event.
2. **REASONING** You flip a coin 10 times and find the experimental probability of flipping tails to be 0.7. Does this seem reasonable? Explain.
3. **VOCABULARY** An event has a theoretical probability of 0.5. What does this mean?
4. **OPEN-ENDED** Describe an event that has a theoretical probability of $\frac{1}{4}$.
5. **LOGIC** A pollster surveys randomly selected individuals about an upcoming election. Do you think the pollster will use experimental probability or theoretical probability to make predictions? Explain.

Practice and Problem Solving

Use the bar graph to find the relative frequency of the event.

6. Spinning a 6
7. Spinning an even number

Use the bar graph to find the experimental probability of the event.

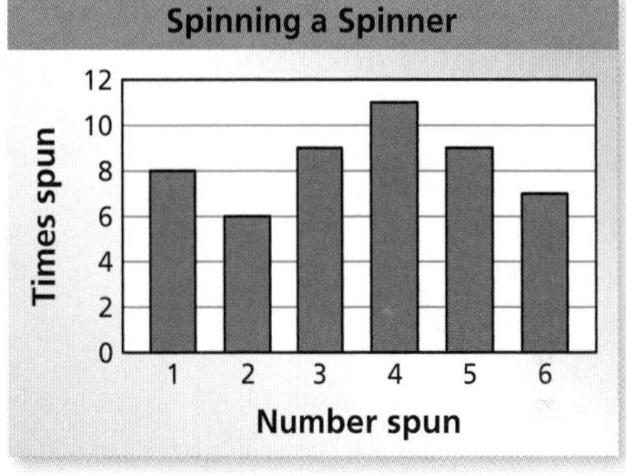

1 8. Spinning a number less than 3
9. *Not* spinning a 1
10. Spinning a 1 or a 3
11. Spinning a 7

12. **EGGS** You check 20 cartons of eggs. Three of the cartons have at least one cracked egg. What is the experimental probability that a carton of eggs has at least one cracked egg?

2 13. **BOARD GAME** There are 105 lettered tiles in a board game. You choose the tiles shown. How many of the 105 tiles would you expect to be vowels?

14. **CARDS** You have a package of 20 assorted thank-you cards. You pick the four cards shown. How many of the 20 cards would you expect to have flowers on them?

Use the spinner to find the theoretical probability of the event.

③ **15.** Spinning red

16. Spinning a 1

17. Spinning an odd number

18. Spinning a multiple of 2

19. Spinning a number less than 7

20. Spinning a 9

21. LETTERS Each letter of the alphabet is printed on an index card. What is the theoretical probability of randomly choosing any letter except Z?

④ **22. GAME SHOW** On a game show, a contestant randomly chooses a chip from a bag that contains numbers and strikes. The theoretical probability of choosing a strike is $\frac{3}{10}$. The bag contains 9 strikes. How many chips are in the bag?

23. MUSIC The theoretical probability that a pop song plays on your MP3 player is 0.45. There are 80 songs on your MP3 player. How many of the songs are pop songs?

24. MODELING There are 16 females and 20 males in a class.

a. What is the theoretical probability that a randomly chosen student is female?

b. One week later, there are 45 students in the class. The theoretical probability that a randomly chosen student is a female is the same as last week. How many males joined the class?

The bar graph shows the results of spinning the spinner 200 times. Compare the theoretical and experimental probabilities of the event.

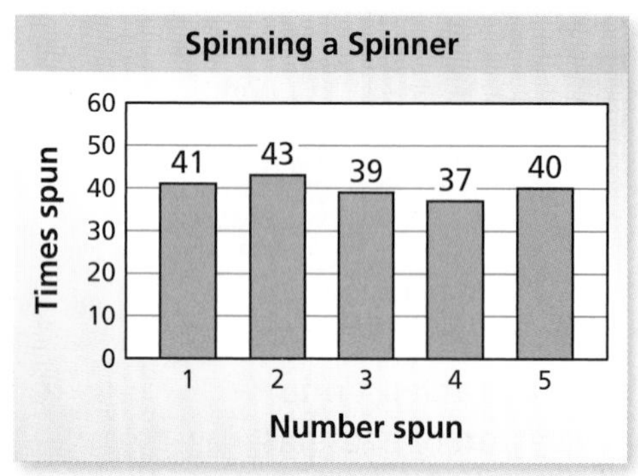

⑤ **25.** Spinning a 4

26. Spinning a 3

27. Spinning a number greater than 4

28. Should you use *theoretical* or *experimental* probability to predict the number of times you will spin a 3 in 10,000 spins?

29. NUMBER SENSE The table at the right shows the results of flipping two coins 12 times each.

HH	HT	TH	TT
2	6	3	1

a. What is the experimental probability of flipping two tails? Using this probability, how many times can you expect to flip two tails in 600 trials?

HH	HT	TH	TT
23	29	26	22

b. The table at the left shows the results of flipping the same two coins 100 times each. What is the experimental probability of flipping two tails? Using this probability, how many times can you expect to flip two tails in 600 trials?

c. Why is it important to use a large number of trials when using experimental probability to predict results?

You roll a pair of number cubes 60 times. You record your results in the bar graph shown.

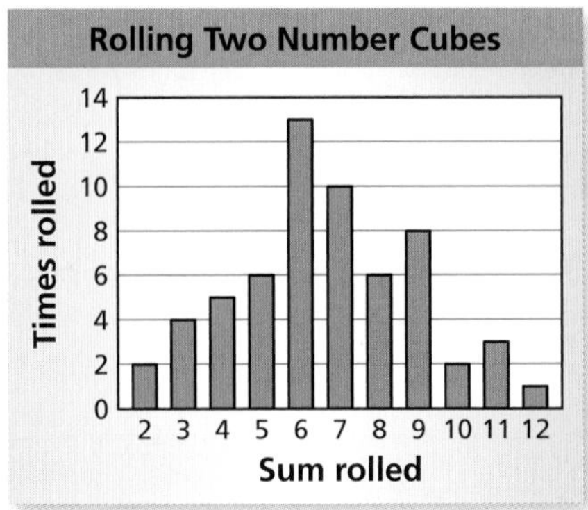

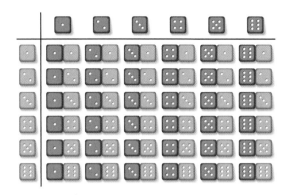

30. Use the bar graph to find the experimental probability of rolling each sum. Is each sum equally likely? Explain. If not, which is most likely?

31. Use the table to find the theoretical probability of rolling each sum. Is each sum equally likely? Explain. If not, which is most likely?

32. **PROBABILITIES** Compare the probabilities you found in Exercises 30 and 31.

33. **REASONING** Consider the results of Exercises 30 and 31.

a. Which sum would you expect to be most likely after 500 trials? 1000 trials? 10,000 trials?

b. Explain how experimental probability is related to theoretical probability as the number of trials increases.

34. **Project** When you toss a paper cup into the air, there are three ways for the cup to land: *open-end up*, *open-end down*, or *on its side*.

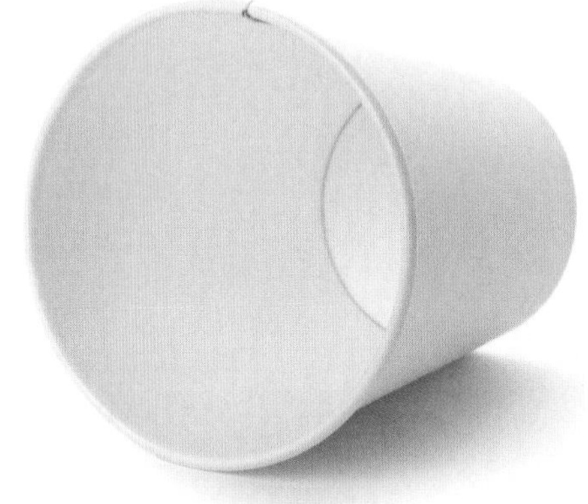

a. Toss a paper cup 100 times and record your results. Do the outcomes for tossing the cup appear to be equally likely? Explain.

b. What is the probability of the cup landing open-end up? open-end down? on its side?

c. Use your results to predict the number of times the cup lands on its side in 1000 tosses.

d. Suppose you tape a quarter to the bottom of the cup. Do you think the cup will be *more likely* or *less likely* to land open-end up? Justify your answer.

Fair Game Review What you learned in previous grades & lessons

Find the annual interest rate. *(Skills Review Handbook)*

35. $I = \$16$, $P = \$200$, $t = 2$ years

36. $I = \$26.25$, $P = \$500$, $t = 18$ months

37. **MULTIPLE CHOICE** The volume of a prism is 9 cubic yards. What is its volume in cubic feet? *(Section 14.4)*

Ⓐ 3 ft^3 Ⓑ 27 ft^3 Ⓒ 81 ft^3 Ⓓ 243 ft^3

15.4 Compound Events

Essential Question How can you find the number of possible outcomes of one or more events?

1 ACTIVITY: Comparing Combination Locks

Work with a partner. You are buying a combination lock. You have three choices.

a. This lock has 3 wheels. Each wheel is numbered from 0 to 9.

The least three-digit combination possible is ____.

The greatest three-digit combination possible is ____.

How many possible combinations are there?

b. Use the lock in part (a).

There are ____ possible outcomes for the first wheel.

There are ____ possible outcomes for the second wheel.

There are ____ possible outcomes for the third wheel.

How can you use multiplication to determine the number of possible combinations?

c. This lock is numbered from 0 to 39. Each combination uses three numbers in a right, left, right pattern. How many possible combinations are there?

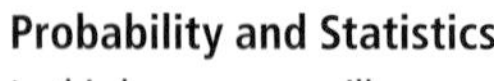

Probability and Statistics

In this lesson, you will
- use tree diagrams, tables, or a formula to find the number of possible outcomes.
- find probabilities of compound events.

d. This lock has 4 wheels.

Wheel 1: 0–9
Wheel 2: A–J
Wheel 3: K–T
Wheel 4: 0–9

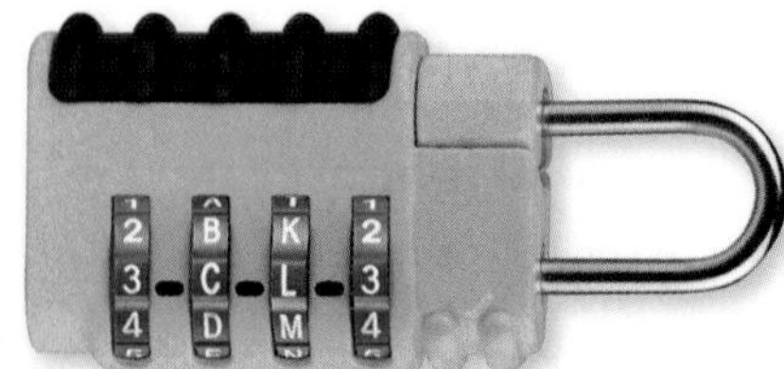
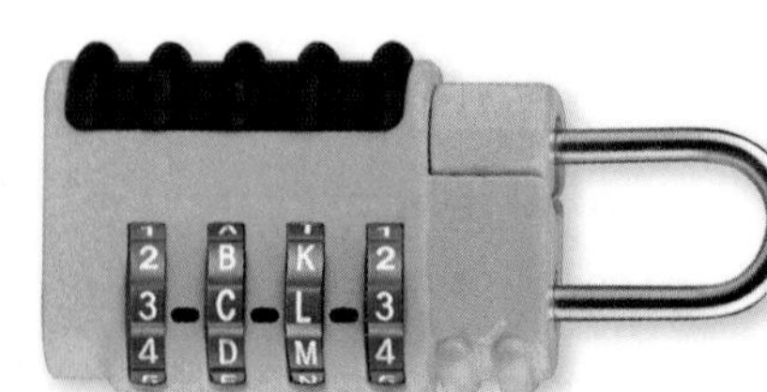

How many possible combinations are there?

e. For which lock is it most difficult to guess the combination? Why?

2 ACTIVITY: Comparing Password Security

Work with a partner. Which password requirement is most secure? Explain your reasoning. Include the number of different passwords that are possible for each requirement.

a. The password must have four digits.

Username: funnydog
Password: 2335
Sign in

b. The password must have five digits.

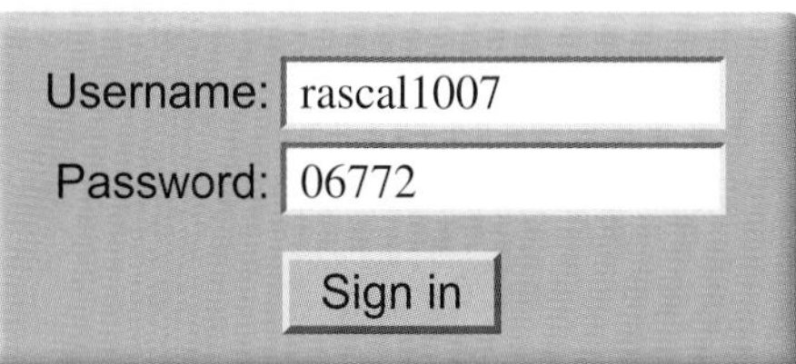

c. The password must have six letters.

Username: supergrowl
Password: AFYYWP
Sign in

d. The password must have eight digits or letters.

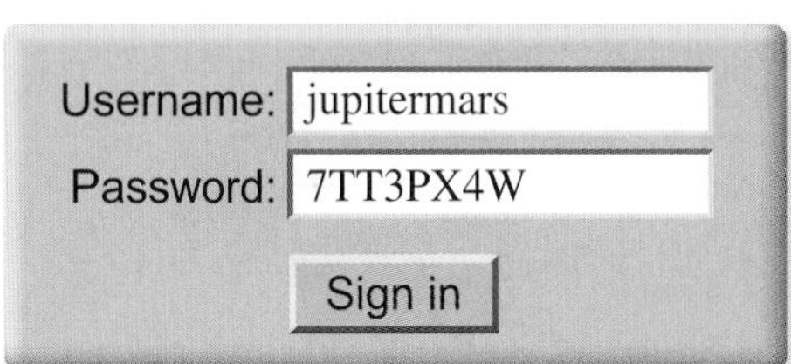

Math Practice

View as Components

What is the number of possible outcomes for each character of the password? Explain.

What Is Your Answer?

3. **IN YOUR OWN WORDS** How can you find the number of possible outcomes of one or more events?

4. **SECURITY** A hacker uses a software program to guess the passwords in Activity 2. The program checks 600 passwords per minute. What is the greatest amount of time it will take the program to guess each of the four types of passwords?

Practice Use what you learned about the total number of possible outcomes of one or more events to complete Exercise 5 on page 657.

15.4 Lesson

Key Vocabulary
sample space, *p. 654*
Fundamental Counting Principle, *p. 654*
compound event, *p. 656*

The set of all possible outcomes of one or more events is called the **sample space**.

You can use tables and tree diagrams to find the sample space of two or more events.

EXAMPLE 1 Finding a Sample Space

Crust

- Thin Crust
- Stuffed Crust

Style

- Hawaiian
- Mexican
- Pepperoni
- Veggie

You randomly choose a crust and style of pizza. Find the sample space. How many different pizzas are possible?

Use a tree diagram to find the sample space.

Crust	*Style*	***Outcome***
Thin	Hawaiian	Thin Crust Hawaiian
	Mexican	Thin Crust Mexican
	Pepperoni	Thin Crust Pepperoni
	Veggie	Thin Crust Veggie
Stuffed	Hawaiian	Stuffed Crust Hawaiian
	Mexican	Stuffed Crust Mexican
	Pepperoni	Stuffed Crust Pepperoni
	Veggie	Stuffed Crust Veggie

There are 8 different outcomes in the sample space. So, there are 8 different pizzas possible.

On Your Own

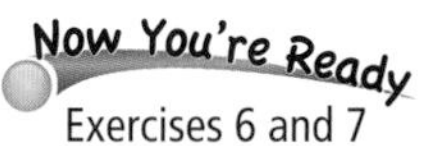

Exercises 6 and 7

1. **WHAT IF?** The pizza shop adds a deep dish crust. Find the sample space. How many pizzas are possible?

Another way to find the total number of possible outcomes is to use the **Fundamental Counting Principle**.

Study Tip
The Fundamental Counting Principle can be extended to more than two events.

Fundamental Counting Principle

An event M has m possible outcomes. An event N has n possible outcomes. The total number of outcomes of event M followed by event N is $m \times n$.

EXAMPLE 2 Finding the Total Number of Possible Outcomes

Find the total number of possible outcomes of rolling a number cube and flipping a coin.

Method 1: Use a table to find the sample space. Let H = heads and T = tails.

	1	2	3	4	5	6
Heads	1H	2H	3H	4H	5H	6H
Tails	1T	2T	3T	4T	5T	6T

∴ There are 12 possible outcomes.

Method 2: Use the Fundamental Counting Principle. Identify the number of possible outcomes of each event.

Event 1: Rolling a number cube has 6 possible outcomes.

Event 2: Flipping a coin has 2 possible outcomes.

$6 \times 2 = 12$ Fundamental Counting Principle

∴ There are 12 possible outcomes.

EXAMPLE 3 Finding the Total Number of Possible Outcomes

How many different outfits can you make from the T-shirts, jeans, and shoes in the closet?

Use the Fundamental Counting Principle. Identify the number of possible outcomes for each event.

Event 1: Choosing a T-shirt has 7 possible outcomes.

Event 2: Choosing jeans has 4 possible outcomes.

Event 3: Choosing shoes has 3 possible outcomes.

$7 \times 4 \times 3 = 84$ Fundamental Counting Principle

∴ So, you can make 84 different outfits.

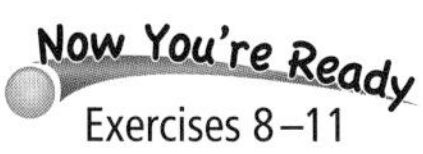

2. Find the total number of possible outcomes of spinning the spinner and choosing a number from 1 to 5.
3. How many different outfits can you make from 4 T-shirts, 5 pairs of jeans, and 5 pairs of shoes?

A **compound event** consists of two or more events. As with a single event, the probability of a compound event is the ratio of the number of favorable outcomes to the number of possible outcomes.

EXAMPLE 4 Finding the Probability of a Compound Event

In Example 2, what is the probability of rolling a number greater than 4 and flipping tails?

There are two favorable outcomes in the sample space for rolling a number greater than 4 and flipping tails: 5T and 6T.

$$P(\text{event}) = \frac{\text{number of favorable outcomes}}{\text{number of possible outcomes}}$$

$$P(\text{greater than 4 and tails}) = \frac{2}{12} \qquad \text{Substitute.}$$

$$= \frac{1}{6} \qquad \text{Simplify.}$$

∴ The probability is $\frac{1}{6}$, or $16\frac{2}{3}\%$.

EXAMPLE 5 Finding the Probability of a Compound Event

You flip three nickels. What is the probability of flipping two heads and one tails?

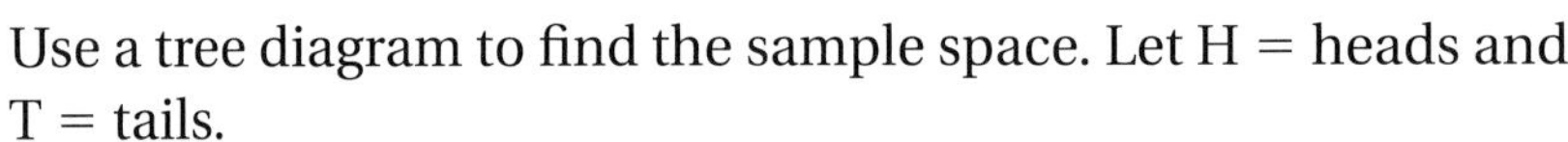

Use a tree diagram to find the sample space. Let H = heads and T = tails.

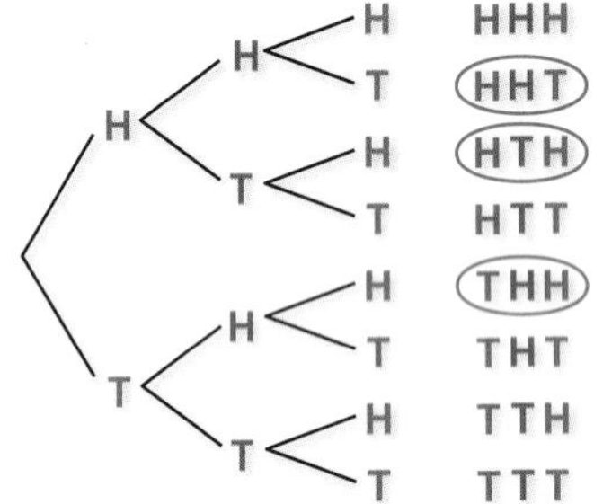

There are three favorable outcomes in the sample space for flipping two heads and one tails: HHT, HTH, and THH.

$$P(\text{event}) = \frac{\text{number of favorable outcomes}}{\text{number of possible outcomes}}$$

$$P(\text{2 heads and 1 tails}) = \frac{3}{8} \qquad \text{Substitute.}$$

∴ The probability is $\frac{3}{8}$, or 37.5%.

On Your Own

4. In Example 2, what is the probability of rolling at most 4 and flipping heads?
5. In Example 5, what is the probability of flipping at least two tails?
6. You roll two number cubes. What is the probability of rolling double threes?
7. In Example 1, what is the probability of choosing a stuffed crust Hawaiian pizza?

15.4 Exercises

Vocabulary and Concept Check

1. **VOCABULARY** What is the sample space of an event? How can you find the sample space of two or more events?

2. **WRITING** Explain how to use the Fundamental Counting Principle.

3. **WRITING** Describe two ways to find the total number of possible outcomes of spinning the spinner and rolling the number cube.

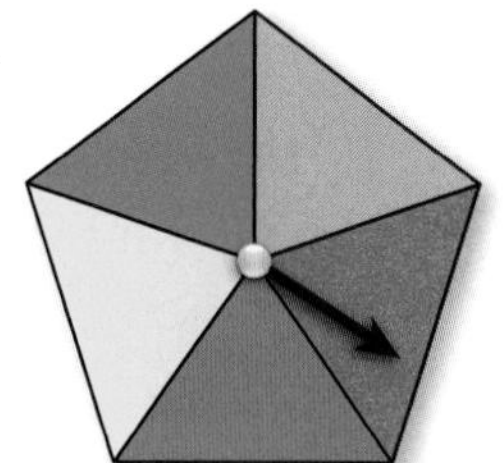

4. **OPEN-ENDED** Give a real-life example of a compound event.

Practice and Problem Solving

5. **COMBINATIONS** The lock is numbered from 0 to 49. Each combination uses three numbers in a right, left, right pattern. Find the total number of possible combinations for the lock.

Use a tree diagram to find the sample space and the total number of possible outcomes.

① 6.

Birthday Party	
Event	Miniature golf, Laser tag, Roller skating
Time	1:00 P.M.–3:00 P.M., 6:00 P.M.–8:00 P.M.

7.

New School Mascot	
Type	Lion, Bear, Hawk, Dragon
Style	Realistic, Cartoon

Use the Fundamental Counting Principle to find the total number of possible outcomes.

② 8.

Beverage	
Size	Small, Medium, Large
Flavor	Root beer, Cola, Diet cola, Iced tea, Lemonade, Water, Coffee

9.

MP3 Player	
Memory	2 GB, 4 GB, 8 GB, 16 GB
Color	Silver, Green, Blue, Pink, Black

③ 10.

Clown	
Suit	Dots, Stripes, Checkers board
Wig	One color, Multicolor
Talent	Balloon animals, Juggling, Unicycle, Magic

11.

Meal	
Appetizer	Nachos, Soup, Spinach dip, Salad, Fruit
Entrée	Chicken, Beef, Spaghetti, Fish
Dessert	Cake, Cookies, Ice cream

12. **NOTE CARDS** A store sells three types of note cards. There are three sizes of each type. Show two ways to find the total number of note cards the store sells.

13. **ERROR ANALYSIS** A true-false quiz has five questions. Describe and correct the error in using the Fundamental Counting Principle to find the total number of ways that you can answer the quiz.

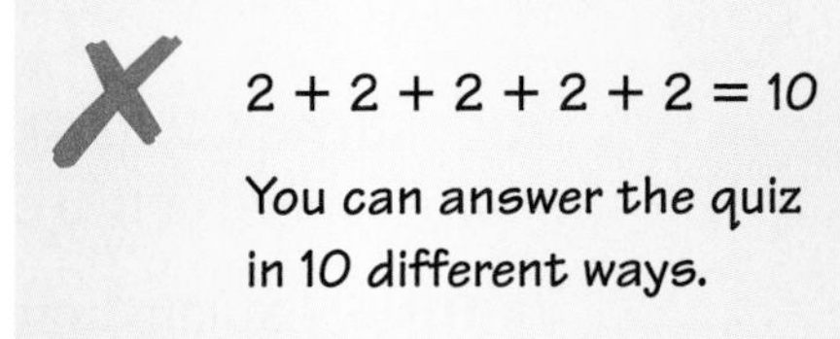

2 + 2 + 2 + 2 + 2 = 10
You can answer the quiz in 10 different ways.

14. **CHOOSE TOOLS** You randomly choose one of the marbles. Without replacing the first marble, you choose a second marble.

    a. Name two ways you can find the total number of possible outcomes.

    b. Find the total number of possible outcomes.

You spin the spinner and flip a coin. Find the probability of the compound event.

(4) 15. Spinning a 1 and flipping heads

16. Spinning an even number and flipping heads

17. Spinning a number less than 3 and flipping tails

18. Spinning a 6 and flipping tails

19. *Not* spinning a 5 and flipping heads

20. Spinning a prime number and *not* flipping heads

You spin the spinner, flip a coin, then spin the spinner again. Find the probability of the compound event.

(5) 21. Spinning blue, flipping heads, then spinning a 1

22. Spinning an odd number, flipping heads, then spinning yellow

23. Spinning an even number, flipping tails, then spinning an odd number

24. *Not* spinning red, flipping tails, then *not* spinning an even number

25. **TAKING A TEST** You randomly guess the answers to two questions on a multiple-choice test. Each question has three choices: A, B, and C.

    a. What is the probability that you guess the correct answers to both questions?

    b. Suppose you can eliminate one of the choices for each question. How does this change the probability that your guesses are correct?

26. **PASSWORD** You forget the last two digits of your password for a website.

   a. What is the probability that you randomly choose the correct digits?

   b. Suppose you remember that both digits are even. How does this change the probability that your choices are correct?

27. **COMBINATION LOCK** The combination lock has 3 wheels, each numbered from 0 to 9.

   a. What is the probability that someone randomly guesses the correct combination in one attempt?

   b. You try to guess the combination by writing five different numbers from 0 to 999 on a piece of paper. Explain how to find the probability that the correct combination is written on the paper.

28. **TRAINS** Your model train has one engine and eight train cars. Find the total number of ways you can arrange the train. (The engine must be first.)

29. **REPEATED REASONING** You have been assigned a 9-digit identification number.

   a. Why should you use the Fundamental Counting Principle instead of a tree diagram to find the total number of possible identification numbers?

   b. How many identification numbers are possible?

   c. **RESEARCH** Use the Internet to find out why the possible number of Social Security numbers is not the same as your answer to part (b).

30. Problem Solving From a group of 5 candidates, a committee of 3 people is selected. In how many different ways can the committee be selected?

Fair Game Review What you learned in previous grades & lessons

Name two pairs of adjacent angles and two pairs of vertical angles in the figure. *(Section 12.1)*

31.

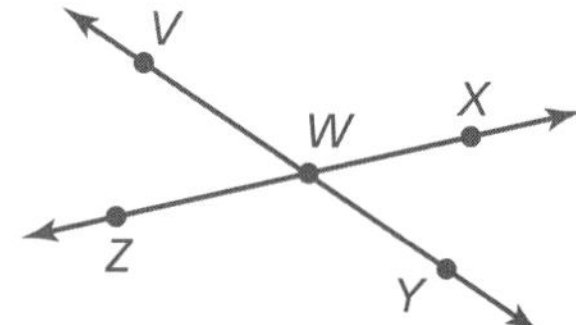

32.

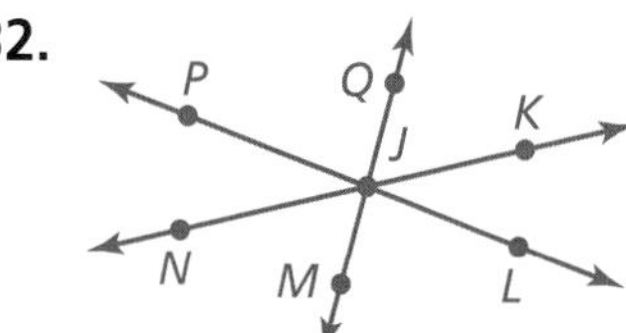

33. **MULTIPLE CHOICE** A drawing has a scale of 1 cm : 1 m. What is the scale factor of the drawing? *(Section 12.5)*

   Ⓐ 1 : 1   Ⓑ 1 : 100   Ⓒ 10 : 1   Ⓓ 100 : 1

15.5 Independent and Dependent Events

Essential Question What is the difference between dependent and independent events?

1 ACTIVITY: Drawing Marbles from a Bag (With Replacement)

Work with a partner. You have three marbles in a bag. There are two green marbles and one purple marble. Randomly draw a marble from the bag. Then put the marble back in the bag and draw a second marble.

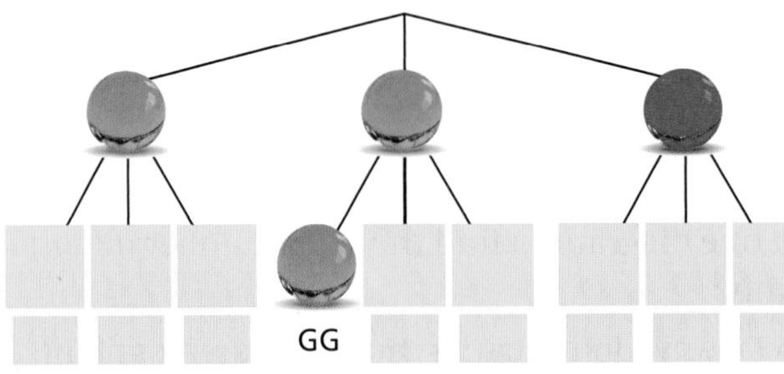

a. Complete the tree diagram. Let G = green and P = purple. Find the probability that both marbles are green.

First draw:

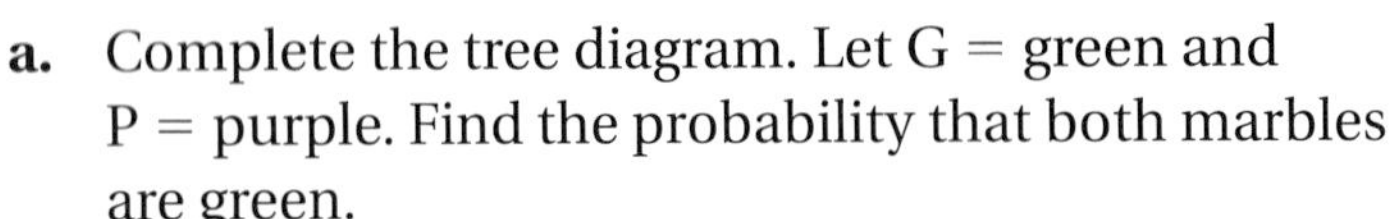

Second draw:

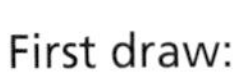

GG

b. Does the probability of getting a green marble on the second draw *depend* on the color of the first marble? Explain.

2 ACTIVITY: Drawing Marbles from a Bag (Without Replacement)

Work with a partner. Using the same marbles from Activity 1, randomly draw two marbles from the bag.

a. Complete the tree diagram. Let G = green and P = purple. Find the probability that both marbles are green.

First draw:

Second draw:

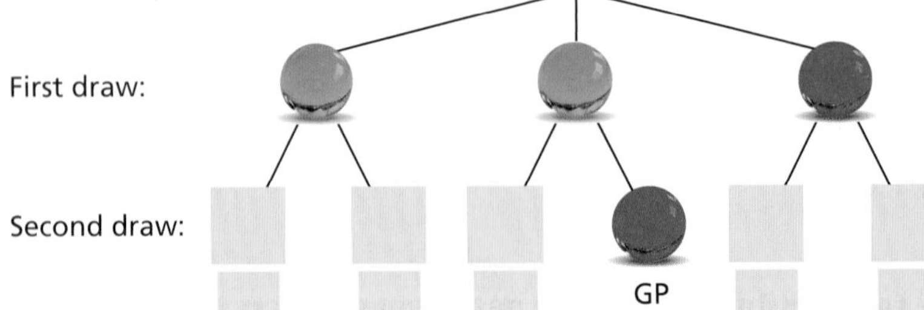

GP

Is this event more likely than the event in Activity 1? Explain.

b. Does the probability of getting a green marble on the second draw *depend* on the color of the first marble? Explain.

Probability and Statistics

In this lesson, you will

- identify independent and dependent events.
- use formulas to find probabilities of independent and dependent events.

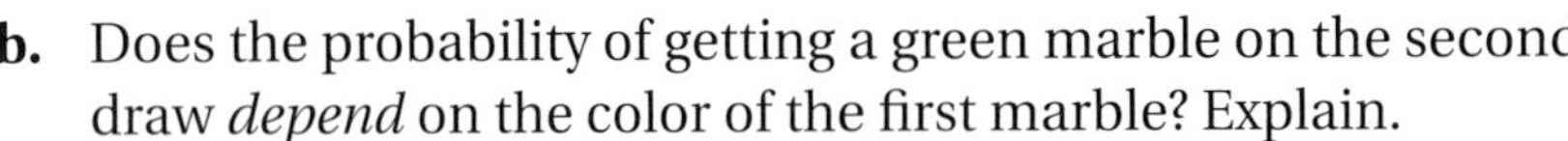

3 ACTIVITY: Conducting an Experiment

Work with a partner. Conduct two experiments.

a. In the first experiment, randomly draw one marble from the bag. Put it back. Draw a second marble. Repeat this 36 times. Record each result. Make a bar graph of your results.

b. In the second experiment, randomly draw two marbles from the bag 36 times. Record each result. Make a bar graph of your results.

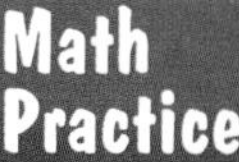

Use Definitions

In what other mathematical context have you seen the terms *independent* and *dependent*? How does knowing these definitions help you answer the questions in part (d)?

First Experiment Results

Frequency: 0, 5, 10, 15, 20, 25, 30, 35, 40

Result: GG, GP, PP

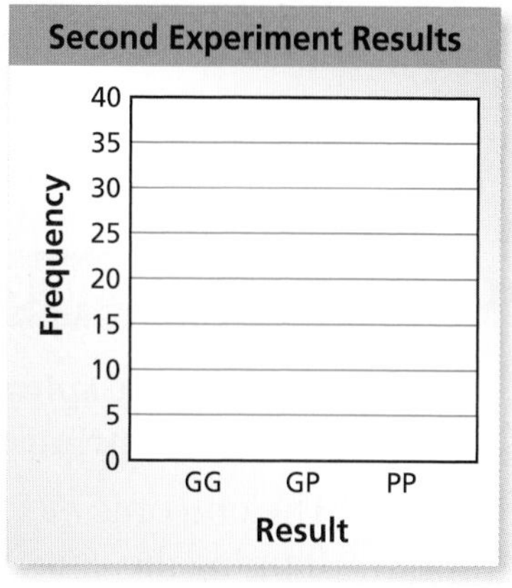

c. For each experiment, estimate the probability of drawing two green marbles.

d. Which experiment do you think represents *dependent events*? Which represents *independent events*? Explain your reasoning.

What Is Your Answer?

4. IN YOUR OWN WORDS What is the difference between dependent and independent events? Describe a real-life example of each.

In Questions 5–7, tell whether the events are *independent* or *dependent*. Explain your reasoning.

5. You roll a 5 on a number cube and spin blue on a spinner.

6. Your teacher chooses one student to lead a group, and then chooses another student to lead another group.

7. You spin red on one spinner and green on another spinner.

8. In Activities 1 and 2, what is the probability of drawing a green marble on the first draw? on the second draw? How do you think you can use these two probabilities to find the probability of drawing two green marbles?

Practice Use what you learned about independent and dependent events to complete Exercises 3 and 4 on page 665.

15.5 Lesson

Key Vocabulary
independent events, p. 662
dependent events, p. 663

Compound events may be *independent events* or *dependent events*. Events are **independent events** if the occurrence of one event *does not* affect the likelihood that the other event(s) will occur.

Probability of Independent Events

Words The probability of two or more independent events is the product of the probabilities of the events.

Symbols

$$P(A \text{ and } B) = P(A) \cdot P(B)$$

$$P(A \text{ and } B \text{ and } C) = P(A) \cdot P(B) \cdot P(C)$$

EXAMPLE 1 Finding the Probability of Independent Events

You spin the spinner and flip the coin. What is the probability of spinning a prime number and flipping tails?

The outcome of spinning the spinner does not affect the outcome of flipping the coin. So, the events are independent.

$P(\text{prime}) = \frac{3}{5}$ ← There are 3 prime numbers (2, 3, and 5). / There is a total of 5 numbers.

$P(\text{tails}) = \frac{1}{2}$ ← There is 1 tails side. / There is a total of 2 sides.

Use the formula for the probability of independent events.

$$P(A \text{ and } B) = P(A) \cdot P(B)$$

$$P(\text{prime and tails}) = P(\text{prime}) \cdot P(\text{tails})$$

$$= \frac{3}{5} \cdot \frac{1}{2} \quad \text{Substitute.}$$

$$= \frac{3}{10} \quad \text{Multiply.}$$

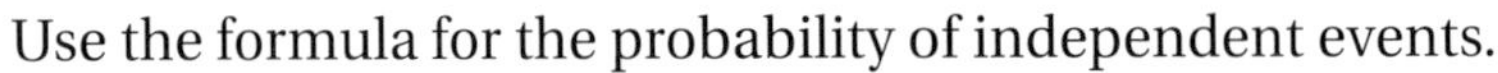

The probability of spinning a prime number and flipping tails is $\frac{3}{10}$, or 30%.

On Your Own

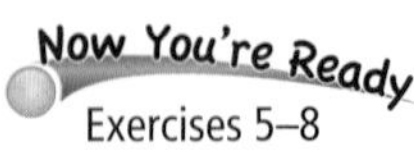

1. What is the probability of spinning a multiple of 2 and flipping heads?

Events are **dependent events** if the occurrence of one event *does* affect the likelihood that the other event(s) will occur.

Key Idea

Probability of Dependent Events

Words The probability of two dependent events A and B is the probability of A times the probability of B after A occurs.

Symbols $P(A \text{ and } B) = P(A) \cdot P(B \text{ after } A)$

EXAMPLE 2 Finding the Probability of Dependent Events

People are randomly chosen to be game show contestants from an audience of 100 people. You are with 5 of your relatives and 6 other friends. What is the probability that one of your relatives is chosen first, and then one of your friends is chosen second?

Choosing an audience member changes the number of audience members left. So, the events are dependent.

$$P(\text{relative}) = \frac{5}{100} = \frac{1}{20}$$

There are 5 relatives.

There is a total of 100 audience members.

$$P(\text{friend}) = \frac{6}{99} = \frac{2}{33}$$

There are 6 friends.

There is a total of 99 audience members left.

Use the formula for the probability of dependent events.

$$P(A \text{ and } B) = P(A) \cdot P(B \text{ after } A)$$

$$P(\text{relative and friend}) = P(\text{relative}) \cdot P(\text{friend after relative})$$

$$= \frac{1}{20} \cdot \frac{2}{33} \quad \text{Substitute.}$$

$$= \frac{1}{330} \quad \text{Simplify.}$$

∴ The probability is $\frac{1}{330}$, or about 0.3%.

On Your Own

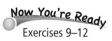

Now You're Ready
Exercises 9–12

2. What is the probability that you, your relatives, and your friends are *not* chosen to be either of the first two contestants?

EXAMPLE 3 Finding the Probability of a Compound Event

A student randomly guesses the answer for each of the multiple-choice questions. What is the probability of answering all three questions correctly?

1. In what year did the United States gain independence from Britain?

A. 1492 **B.** 1776 **C.** 1788 **D.** 1795 **E.** 2000

2. Which amendment to the Constitution grants citizenship to all persons born in the United States and guarantees them equal protection under the law?

A. 1st **B.** 5th **C.** 12th **D.** 13th **E.** 14th

3. In what year did the Boston Tea Party occur?

A. 1607 **B.** 1773 **C.** 1776 **D.** 1780 **E.** 1812

Choosing the answer for one question does not affect the choice for the other questions. So, the events are independent.

Method 1: Use the formula for the probability of independent events.

$$P(\#1 \text{ and } \#2 \text{ and } \#3 \text{ correct}) = P(\#1 \text{ correct}) \cdot P(\#2 \text{ correct}) \cdot P(\#3 \text{ correct})$$

$$= \frac{1}{5} \cdot \frac{1}{5} \cdot \frac{1}{5} \quad \text{Substitute.}$$

$$= \frac{1}{125} \quad \text{Multiply.}$$

∴ The probability of answering all three questions correctly is $\frac{1}{125}$, or 0.8%.

Method 2: Use the Fundamental Counting Principle.

There are 5 choices for each question, so there are $5 \cdot 5 \cdot 5 = 125$ possible outcomes. There is only 1 way to answer all three questions correctly.

$$P(\#1 \text{ and } \#2 \text{ and } \#3 \text{ correct}) = \frac{1}{125}$$

∴ The probability of answering all three questions correctly is $\frac{1}{125}$, or 0.8%.

On Your Own

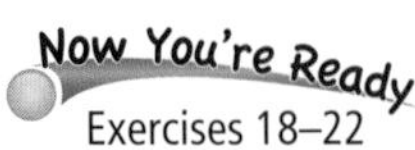

3. The student can eliminate Choice A for all three questions. What is the probability of answering all three questions correctly? Compare this probability with the probability in Example 3. What do you notice?

15.5 Exercises

Vocabulary and Concept Check

1. **DIFFERENT WORDS, SAME QUESTION** You randomly choose one of the chips. Without replacing the first chip, you choose a second chip. Which question is different? Find "both" answers.

   What is the probability of choosing a 1 and then a blue chip?

   What is the probability of choosing a 1 and then an even number?

   What is the probability of choosing a green chip and then a chip that is *not* red?

   What is the probability of choosing a number less than 2 and then an even number?

2. **WRITING** How do you find the probability of two events A and B when A and B are independent? dependent?

Practice and Problem Solving

Tell whether the events are *independent* or *dependent*. Explain.

3. You roll a 4 on a number cube. Then you roll an even number on a different number cube.
4. You randomly draw a lane number for a 100-meter race. Then your friend randomly draws a lane number for the same race.

You spin the spinner and flip a coin. Find the probability of the compound event.

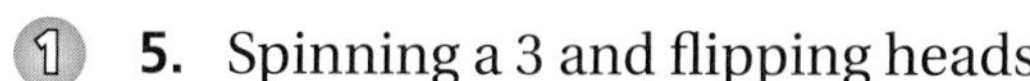

① 5. Spinning a 3 and flipping heads

6. Spinning an even number and flipping tails
7. Spinning a number greater than 1 and flipping tails
8. *Not* spinning a 2 and flipping heads

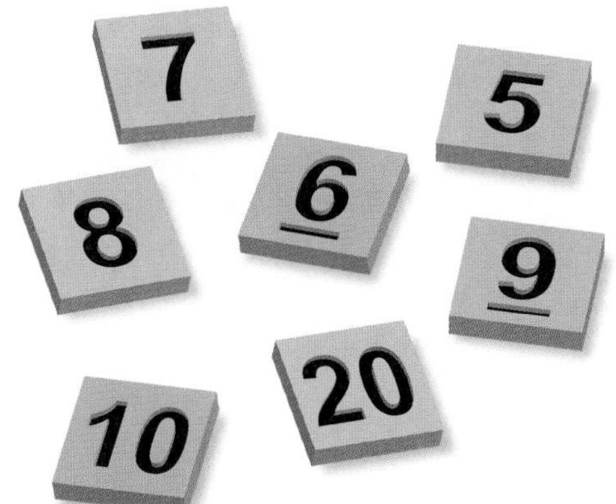

You randomly choose one of the tiles. Without replacing the first tile, you choose a second tile. Find the probability of the compound event.

② 9. Choosing a 5 and then a 6

10. Choosing an odd number and then a 20
11. Choosing a number less than 7 and then a multiple of 4
12. Choosing two even numbers

13. ERROR ANALYSIS Describe and correct the error in finding the probability.

You randomly choose one of the marbles. Without replacing the first marble, you choose a second marble. What is the probability of choosing red and then green?

$P(\text{red and green}) = \frac{1}{4} \cdot \frac{1}{4} = \frac{1}{16}$

14. LOGIC A bag contains three marbles. Does the tree diagram show the outcomes for *independent* or *dependent* events? Explain.

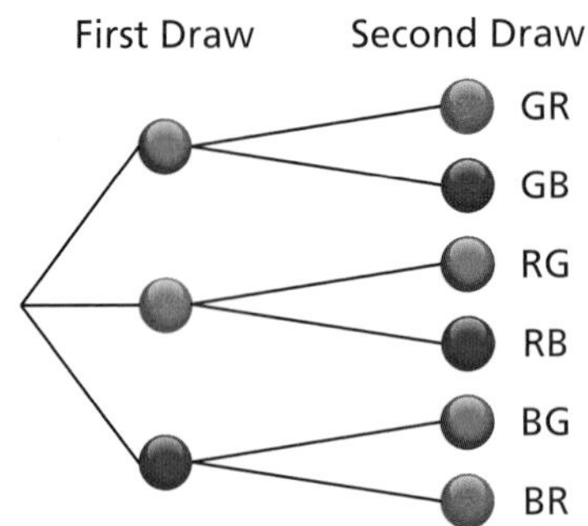

15. EARRINGS A jewelry box contains two gold hoop earrings and two silver hoop earrings. You randomly choose two earrings. What is the probability that both are silver hoop earrings?

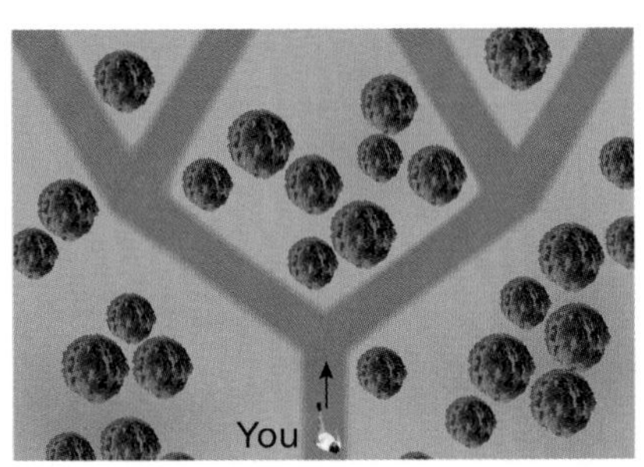

16. HIKING You are hiking to a ranger station. There is one correct path. You come to a fork and randomly take the path on the left. You come to another fork and randomly take the path on the right. What is the probability that you are still on the correct path?

17. CARNIVAL At a carnival game, you randomly throw two darts at the board and break two balloons. What is the probability that both of the balloons you break are purple?

You spin the spinner, flip a coin, then spin the spinner again. Find the probability of the compound event.

3 **18.** Spinning a 4, flipping heads, then spinning a 7

19. Spinning an odd number, flipping heads, then spinning a 3

20. Spinning an even number, flipping tails, then spinning an odd number

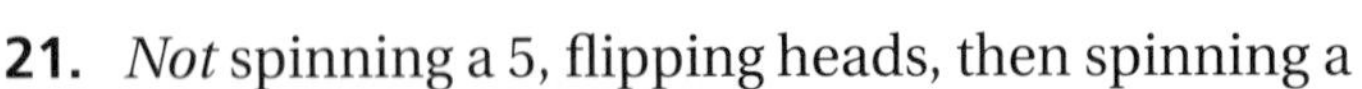

21. *Not* spinning a 5, flipping heads, then spinning a 1

22. Spinning an odd number, *not* flipping heads, then *not* spinning a 6

23. LANGUAGES There are 16 students in your Spanish class. Your teacher randomly chooses one student at a time to take a verbal exam. What is the probability that you are *not* one of the first four students chosen?

24. SHOES Twenty percent of the shoes in a factory are black. One shoe is chosen and replaced. A second shoe is chosen and replaced. Then a third shoe is chosen. What is the probability that *none* of the shoes are black?

25. PROBLEM SOLVING Your teacher divides your class into two groups, and then randomly chooses a leader for each group. The probability that you are chosen to be a leader is $\frac{1}{12}$. The probability that both you and your best friend are chosen is $\frac{1}{132}$.

a. Is your best friend in your group? Explain.

b. What is the probability that your best friend is chosen as a group leader?

c. How many students are in the class?

26. Structure After ruling out some of the answer choices, you randomly guess the answer for each of the story questions below.

1. Who was the oldest?

A. Ned **B.** Yvonne **C.** Sun Li **D.** Angel **E.** Dusty

2. What city was Stacey from?

A. Raleigh **B.** New York **C.** Roanoke **D.** Dallas **E.** San Diego

a. How can the probability of getting both answers correct be 25%?

b. How can the probability of getting both answers correct be $8\frac{1}{3}\%$?

Fair Game Review What you learned in previous grades & lessons

Draw a triangle with the given angle measures. Then classify the triangle. *(Section 12.3)*

27. 30°, 60°, 90° **28.** 20°, 50°, 110° **29.** 50°, 50°, 80°

30. MULTIPLE CHOICE Which set of numbers is in order from least to greatest? *(Skills Review Handbook)*

Ⓐ $\frac{2}{3}$, 0.6, 67%

Ⓑ 44.5%, $\frac{4}{9}$, $0.4\overline{6}$

Ⓒ 0.269, 27%, $\frac{3}{11}$

Ⓓ $2\frac{1}{7}$, 214%, $2.\overline{14}$

Extension 15.5 Simulations

Key Vocabulary
simulation, *p. 668*

A **simulation** is an experiment that is designed to reproduce the conditions of a situation or process. Simulations allow you to study situations that are impractical to create in real life.

EXAMPLE 1 Simulating Outcomes That Are Equally Likely

HTH	HTT
HTT	HTH
HTT	TTT
(HHH)	HTT
HTT	TTT
HTT	HTH
HTH	(HHH)
HTT	HTT
TTT	HTH
HTH	HTT

A couple plans on having three children. The gender of each child is equally likely. (a) Design a simulation involving 20 trials that you can use to model the genders of the children. (b) Use your simulation to find the experimental probability that all three children are boys.

a. Choose an experiment that has two equally likely outcomes for each event (gender), such as tossing three coins. Let heads (H) represent a boy and tails (T) represent a girl.

b. To find the experimental probability, you need repeated trials of the simulation. The table shows 20 trials.

$$P(\text{three boys}) = \frac{2}{20} = \frac{1}{10}$$

HHH occurred 2 times.

There is a total of 20 trials.

∴ The experimental probability is $\frac{1}{10}$, 0.1, or 10%.

EXAMPLE 2 Simulating Outcomes That Are Not Equally Likely

Study Tip

In Example 2, the digits 1 through 6 represent 60% of the possible digits (0 through 9) in the tens place. Likewise, the digits 1 and 2 represent 20% of the possible digits in the ones place.

There is a 60% chance of rain on Monday and a 20% chance of rain on Tuesday. Design and use a simulation involving 50 randomly generated numbers to find the experimental probability that it will rain on both days.

Use the random number generator on a graphing calculator. Randomly generate 50 numbers from 0 to 99. The table below shows the results.

Let the digits 1 through 6 in the tens place represent rain on Monday. Let digits 1 and 2 in the ones place represent rain on Tuesday. Any number that meets these criteria represents rain on both days.

```
randInt(0,99,50)
{52 66 73 68 75...
```

(52)	66	73	68	75	28	35	47	48	2
16	68	49	3	77	35	92	78	6	6
58	18	89	39	24	80	(32)	(41)	77	(21)
(32)	40	96	59	86	1	(12)	0	94	73
40	71	28	(61)	1	24	37	25	3	25

$$P(\text{rain both days}) = \frac{7}{50}$$

7 numbers meet the criteria.

There is a total of 50 trials.

∴ The experimental probability is $\frac{7}{50}$, 0.14, or 14%.

EXAMPLE 3 Using a Spreadsheet to Simulate Outcomes

Probability and Statistics
In this extension, you will
- use simulations to find experimental probabilities.

Each school year, there is a 50% chance that weather causes one or more days of school to be canceled. Design and use a simulation involving 50 randomly generated numbers to find the experimental probability that weather will cause school to be canceled in at least three of the next four school years.

Use a random number table in a spreadsheet. Randomly generate 50 four-digit whole numbers. The spreadsheet below shows the results.

Let the digits 1 through 5 represent school years with a cancellation. The numbers in the spreadsheet that contain at least three digits from 1 through 5 represent four school years in which at least three of the years have a cancellation.

Study Tip

To create a four-digit random number table in a spreadsheet, follow these steps.
1. Highlight the group of cells to use for your table.
2. Format the cells to display four-digit whole numbers.
3. Enter the formula RAND()*10000 into each cell.

	A	B	C	D	E	F
1	7584	3974	8614	2500	4629	
2	3762	3805	2725	7320	6487	
3	3024	1554	2708	1126	9395	
4	4547	6220	9497	7530	3036	
5	1719	0662	1814	6218	2766	
6	7938	9551	8552	4321	8043	
7	6951	0578	5560	0740	4479	
8	4714	4511	5115	6952	5609	
9	0797	3022	9067	2193	6553	
10	3300	5454	5351	6319	0387	
11						

$$P\left(\begin{array}{l}\text{cancellation in at least three}\\ \text{of the next four school years}\end{array}\right) = \frac{17}{50}$$

17 numbers contain at least three digits from 1 to 5.

There is a total of 50 trials.

The experimental probability is $\frac{17}{50}$, 0.34, or 34%.

Practice

1. **QUIZ** You randomly guess the answers to four true-false questions. (a) Design a simulation that you can use to model the answers. (b) Use your simulation to find the experimental probability that you answer all four questions correctly.
2. **BASEBALL** A baseball team wins 70% of its games. Assuming this trend continues, design and use a simulation to find the experimental probability that the team wins the next three games.
3. **WHAT IF?** In Example 3, there is a 40% chance that weather causes one or more days of school to be canceled each school year. Find the experimental probability that weather will cause school to be canceled in at least three of the next four school years.
4. **REASONING** In Examples 1–3 and Exercises 1–3, try to find the theoretical probability of the event. What do you think happens to the experimental probability when you increase the number of trials in the simulation?

15 Study Help

You can use a **notetaking organizer** to write notes, vocabulary, and questions about a topic. Here is an example of a notetaking organizer for probability.

Write important vocabulary or formulas in this space. →

If *P*(event) = 0, the event is *impossible*. If *P*(event) = 0.25, the event is *unlikely*. If *P*(event) = 0.5, the event is *equally likely* to happen or not happen. If *P*(event) = 0.75, the event is *likely*. If *P*(event) = 1, the event is *certain*.	Probability A number that measures the likelihood that an event will occur Can be written as a fraction, decimal, or percent Always between 0 and 1, inclusive
How do you find the probability of two or more events?	

← Write your notes about the topic in this space.

← Write your questions about the topic in this space.

On Your Own

Make notetaking organizers to help you study these topics.

1. experimental probability
2. theoretical probability
3. Fundamental Counting Principle
4. independent events
5. dependent events

After you complete this chapter, make notetaking organizers for the following topics.

6. sample
7. population

Formulas: Newton = beagle
Notes: Likes dog biscuits
Questions: What is my greatest accomplishment?
I hope it's better than your 200-page essay on bacon.

"I am using a notetaking organizer to plan my autobiography."

15.1–15.5 Quiz

You randomly choose one butterfly. Find the number of ways the event can occur. *(Section 15.1)*

1. Choosing red
2. Choosing brown
3. Choosing *not* blue

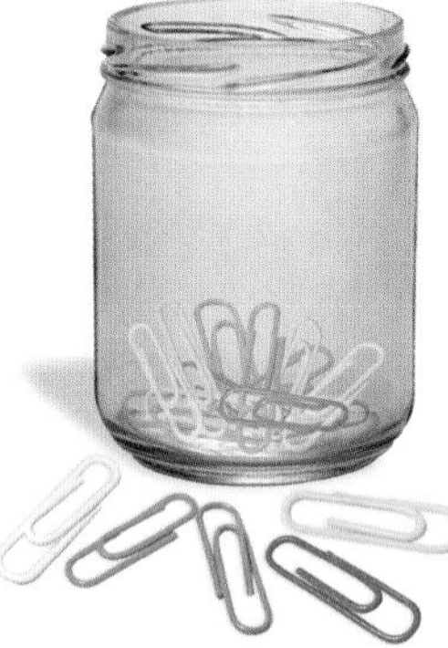

6 Green
3 White
4 Red
2 Blue
5 Yellow

You randomly choose one paper clip from the jar. Find the probability of the event. *(Section 15.2)*

4. Choosing a green paper clip
5. Choosing a yellow paper clip
6. *Not* choosing a yellow paper clip
7. Choosing a purple paper clip

Use the bar graph to find the experimental probability of the event. *(Section 15.3)*

8. Rolling a 4
9. Rolling a multiple of 3
10. Rolling a 2 or a 3
11. Rolling a number less than 7

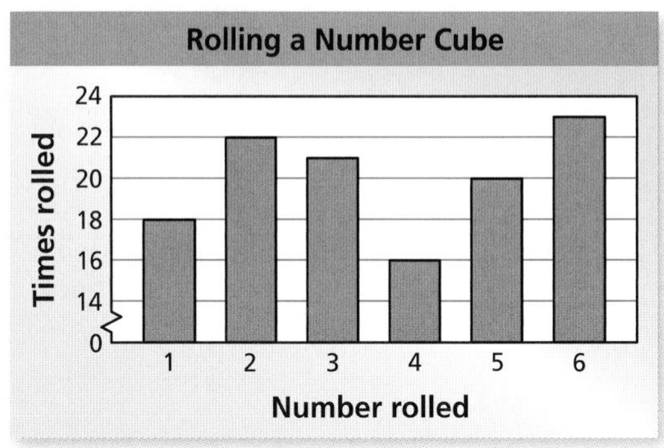

Use the Fundamental Counting Principle to find the total number of possible outcomes. *(Section 15.4)*

12.

Calculator	
Type	Basic display, Scientific, Graphing, Financial
Color	Black, White, Silver

13.

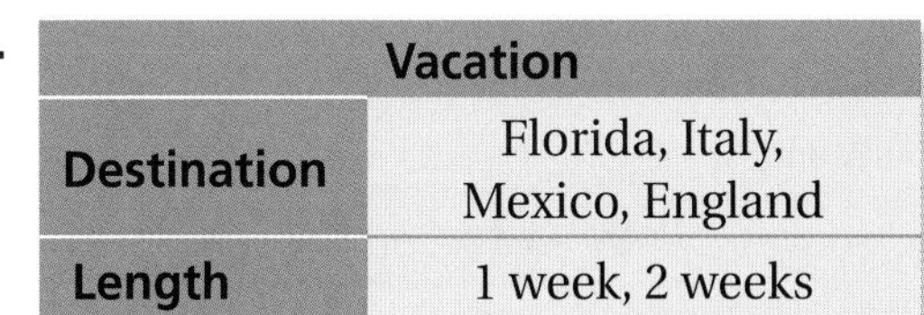

Vacation	
Destination	Florida, Italy, Mexico, England
Length	1 week, 2 weeks

14. **BLACK PENS** You randomly choose one of the pens shown. What is the theoretical probability of choosing a black pen? *(Section 15.3)*

15. **BLUE PENS** You randomly choose one of the five pens shown. Your friend randomly chooses one of the remaining pens. What is the probability that you and your friend both choose a blue pen? *(Section 15.5)*

15.6 Samples and Populations

Essential Question How can you determine whether a sample accurately represents a population?

A **population** is an entire group of people or objects. A **sample** is a part of the population. You can use a sample to make an *inference*, or conclusion, about a population.

Identify a population.	Select a sample.	Interpret the data in the sample.	Make an inference about the population.
Population →	Sample →	Interpretation →	Inference

1 ACTIVITY: Identifying Populations and Samples

Work with a partner. Identify the population and the sample.

a.

The students in a school

The students in a math class

b.

The grizzly bears with GPS collars in a park

The grizzly bears in a park

c.

150 quarters

All quarters in circulation

d.

All books in a library

10 fiction books in a library

Probability and Statistics

In this lesson, you will

- determine when samples are representative of populations.
- use data from random samples to make predictions about populations.

2 ACTIVITY: Identifying Random Samples

Work with a partner. When a sample is selected at random, each member of the population is equally likely to be selected. You want to know the favorite extracurricular activity of students at your school. Determine whether each method will result in a random sample. Explain your reasoning.

a. You ask members of the school band.

b. You publish a survey in the school newspaper.

c. You ask every eighth student who enters the school in the morning.

d. You ask students in your class.

There are many different ways to select a sample from a population. To make valid inferences about a population, you must choose a random sample very carefully so that it accurately represents the population.

3 ACTIVITY: Identifying Representative Samples

Work with a partner. A new power plant is being built outside a town. In each situation below, residents of the town are asked how they feel about the new power plant. Determine whether each conclusion is valid. Explain your reasoning.

a. A local radio show takes calls from 500 residents. The table shows the results. The radio station concludes that most of the residents of the town oppose the new power plant.

New Power Plant	
For	70
Against	425
Don't know	5

Math Practice

Understand Quantities

Can the size of a sample affect the validity of a conclusion about a population?

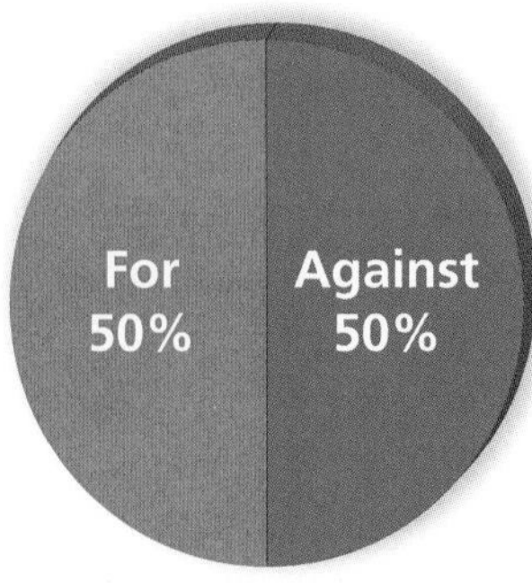

b. A news reporter randomly surveys 2 residents outside a supermarket. The graph shows the results. The reporter concludes that the residents of the town are evenly divided on the new power plant.

c. You randomly survey 250 residents at a shopping mall. The table shows the results. You conclude that there are about twice as many residents of the town against the new power plant than for the new power plant.

New Power Plant	
For	32%
Against	62%
Don't know	6%

What Is Your Answer?

4. **IN YOUR OWN WORDS** How can you determine whether a sample accurately represents a population?

5. **RESEARCH** Choose a topic that you would like to ask people's opinions about, and then write a survey question. How would you choose people to survey so that your sample is random? How many people would you survey? Conduct your survey and display your results. Would you change any part of your survey to make it more accurate? Explain.

6. Does increasing the size of a sample necessarily make the sample representative of a population? Give an example to support your explanation.

Practice Use what you learned about populations and samples to complete Exercises 3 and 4 on page 676.

15.6 Lesson

Key Vocabulary
population, *p. 672*
sample, *p. 672*
unbiased sample, *p. 674*
biased sample, *p. 674*

An **unbiased sample** is representative of a population. It is selected at random and is large enough to provide accurate data.

A **biased sample** is not representative of a population. One or more parts of the population are favored over others.

EXAMPLE 1 Identifying an Unbiased Sample

You want to estimate the number of students in a high school who ride the school bus. Which sample is unbiased?

Ⓐ 4 students in the hallway

Ⓑ all students in the marching band

Ⓒ 50 seniors at random

Ⓓ 100 students at random during lunch

Choice A is not large enough to provide accurate data.

Choice B is not selected at random.

Choice C is not representative of the population because seniors are more likely to drive to school than other students.

Choice D is representative of the population, selected at random, and large enough to provide accurate data.

So, the correct answer is Ⓓ.

On Your Own

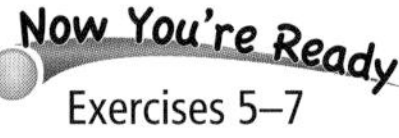

1. **WHAT IF?** You want to estimate the number of seniors in a high school who ride the school bus. Which sample is unbiased? Explain.

2. You want to estimate the number of eighth-grade students in your school who consider it relaxing to listen to music. You randomly survey 15 members of the band. Your friend surveys every fifth student whose name appears on an alphabetical list of eighth graders. Which sample is unbiased? Explain.

The results of an unbiased sample are proportional to the results of the population. So, you can use unbiased samples to make predictions about the population.

Biased samples are not representative of the population. So, you should not use them to make predictions about the population because the predictions may not be valid.

EXAMPLE 2 Determining Whether Conclusions Are Valid

You want to know how the residents of your town feel about adding a new stop sign. Determine whether each conclusion is valid.

a. You survey the 20 residents who live closest to the new sign. Fifteen support the sign, and five do not. So, you conclude that 75% of the residents of your town support the new sign.

The sample is not representative of the population because residents who live close to the sign are more likely to support it.

∴ So, the sample is biased, and the conclusion is not valid.

b. You survey 100 residents at random. Forty support the new sign, and sixty do not. So, you conclude that 40% of the residents of your town support the new sign.

The sample is representative of the population, selected at random, and large enough to provide accurate data.

∴ So, the sample is unbiased, and the conclusion is valid.

EXAMPLE 3 Making Predictions

Movies per Week

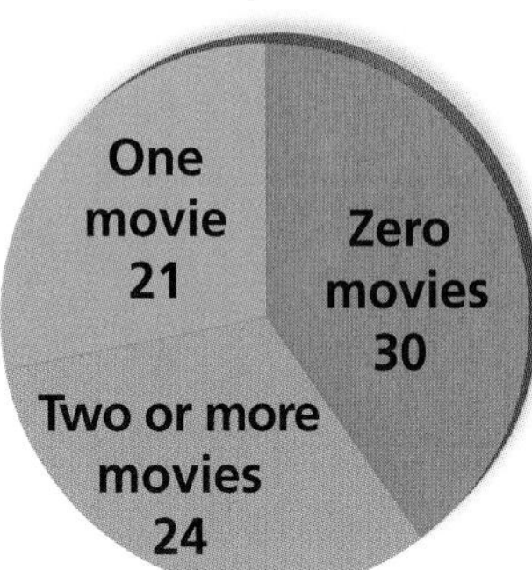

You ask 75 randomly chosen students how many movies they watch each week. There are 1200 students in the school. Predict the number *n* of students in the school who watch one movie each week.

The sample is representative of the population, selected at random, and large enough to provide accurate data. So, the sample is unbiased, and you can use it to make a prediction about the population.

Write and solve a proportion to find n.

$$\overset{\textbf{Sample}}{\frac{\text{students in survey (one movie)}}{\text{number of students in survey}}} = \overset{\textbf{Population}}{\frac{\text{students in school (one movie)}}{\text{number of students in school}}}$$

$$\frac{21}{75} = \frac{n}{1200} \qquad \text{Substitute.}$$

$$336 = n \qquad \text{Solve for } n.$$

∴ So, about 336 students in the school watch one movie each week.

On Your Own

Now You're Ready
Exercises 8, 9, and 12

3. In Example 2, each of 25 randomly chosen firefighters supports the new sign. So, you conclude that 100% of the residents of your town support the new sign. Is the conclusion valid? Explain.

4. In Example 3, predict the number of students in the school who watch two or more movies each week.

15.6 Exercises

Vocabulary and Concept Check

1. **VOCABULARY** Why would you survey a sample instead of a population?
2. **CRITICAL THINKING** What should you consider when conducting a survey?

Practice and Problem Solving

Identify the population and the sample.

3. Residents of New Jersey   Residents of Ocean County

4.

4 cards

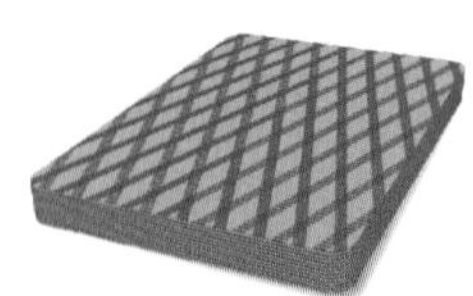

All cards in a deck

Determine whether the sample is *biased* or *unbiased*. Explain.

1 5. You want to estimate the number of students in your school who play a musical instrument. You survey the first 15 students who arrive at a band class.

6. You want to estimate the number of books students in your school read over the summer. You survey every fourth student who enters the school.

7. You want to estimate the number of people in a town who think that a park needs to be remodeled. You survey every 10th person who enters the park.

Determine whether the conclusion is valid. Explain.

2 8. You want to determine the number of students in your school who have visited a science museum. You survey 50 students at random. Twenty have visited a science museum, and thirty have not. So, you conclude that 40% of the students in your school have visited a science museum.

9. You want to know how the residents of your town feel about building a new baseball stadium. You randomly survey 100 people who enter the current stadium. Eighty support building a new stadium, and twenty do not. So, you conclude that 80% of the residents of your town support building a new baseball stadium.

Which sample is better for making a prediction? Explain.

10.

Predict the number of students in a school who like gym class.	
Sample A	A random sample of 8 students from the yearbook
Sample B	A random sample of 80 students from the yearbook

11.

Predict the number of defective pencils produced per day.	
Sample A	A random sample of 500 pencils from 20 machines
Sample B	A random sample of 500 pencils from 1 machine

3 **12.** **FOOD** You ask 125 randomly chosen students to name their favorite food. There are 1500 students in the school. Predict the number of students in the school whose favorite food is pizza.

Favorite Food	
Pizza	58
Hamburger	36
Pasta	14
Other	17

Determine whether you would survey the population or a sample. Explain.

13. You want to know the average height of seventh graders in the United States.

14. You want to know the favorite types of music of students in your homeroom.

15. You want to know the number of students in your state who have summer jobs.

Theater Ticket Sales	
Adults	**Students**
522	210

16. **THEATER** You su0.rvey 72 randomly chosen students about whether they are going to attend the school play. Twelve say yes. Predict the number of students who attend the school.

17. **CRITICAL THINKING** Explain why 200 people with email addresses may not be a random sample. When might it be a random sample?

18. **LOGIC** A person surveys residents of a town to determine whether a skateboarding ban should be overturned.

a. Describe how the person could conduct the survey so that the sample is biased toward overturning the ban.

b. Describe how the person could conduct the survey so that the sample is biased toward keeping the ban.

19. Reasoning A guidance counselor surveys a random sample of 60 out of 900 high school students. Using the survey results, the counselor predicts that approximately 720 students plan to attend college. Do you agree with her prediction? Explain.

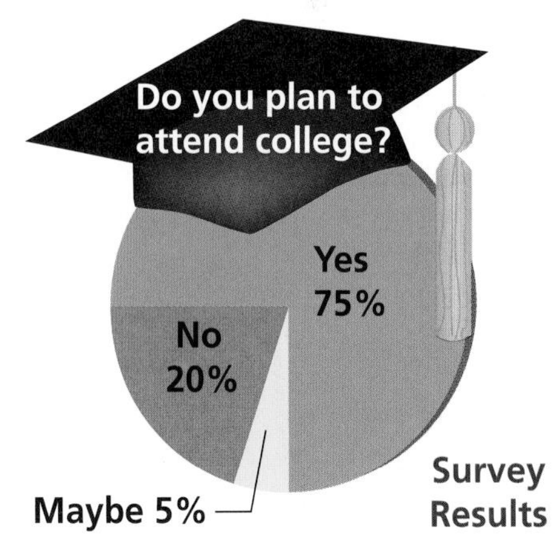

Fair Game Review What you learned in previous grades & lessons

Write and solve a proportion to answer the question. *(Skills Review Handbook)*

20. What percent of 60 is 18?

21. 70% of what number is 98?

22. 30 is 15% of what number?

23. What number is 0.6% of 500?

24. **MULTIPLE CHOICE** What is the volume of the pyramid? *(Section 14.5)*

Ⓐ 40 cm^3 Ⓑ 50 cm^3

Ⓒ 100 cm^3 Ⓓ 120 cm^3

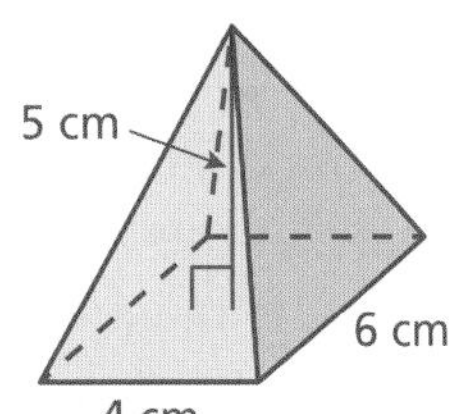

Extension 15.6 Generating Multiple Samples

You have already used unbiased samples to make inferences about a population. In some cases, making an inference about a population from only one sample is not as precise as using multiple samples.

1 ACTIVITY: Using Multiple Random Samples

Work with a partner. You and a group of friends want to know how many students in your school listen to pop music. There are 840 students in your school. Each person in the group randomly surveys 20 students.

Step 1: The table shows your results. Make an inference about the number of students in your school who prefer pop music.

Favorite Type of Music			
Country	Pop	Rock	Rap
4	10	5	1

Step 2: The table shows Kevin's results. Use these results to make another inference about the number of students in your school who prefer pop music.

Favorite Type of Music			
Country	Pop	Rock	Rap
2	13	4	1

Compare the results of Steps 1 and 2.

Probability and Statistics
In this extension, you will
- use multiple samples to make predictions about populations.

Step 3: The table shows the results of three other friends. Use these results to make three more inferences about the number of students in your school who prefer pop music.

	Favorite Type of Music			
	Country	Pop	Rock	Rap
Steve	3	8	7	2
Laura	5	10	4	1
Ming	5	9	3	3

Step 4: Describe the variation of the five inferences. Which one would you use to describe the number of students in your school who prefer pop music? Explain your reasoning.

Step 5: Show how you can use all five samples to make an inference.

Practice

1. **PACKING PEANUTS** Work with a partner. Mark 24 packing peanuts with either a red or a black marker. Put the peanuts into a paper bag. Trade bags with other students in the class.

   a. Generate a sample by choosing a peanut from your bag six times, replacing the peanut each time. Record the number of times you choose each color. Repeat this process to generate four more samples. Organize the results in a table.

   b. Use each sample to make an inference about the number of red peanuts in the bag. Then describe the variation of the five inferences. Make inferences about the numbers of red and black peanuts in the bag based on all the samples.

   c. Take the peanuts out of the bag. How do your inferences compare to the population? Do you think you can make a more accurate prediction? If so, explain how.

2 ACTIVITY: Using Measures from Multiple Random Samples

Hours Worked Each Week
1: 6, 8, 6, 6, 7, 4, 10, 8, 7, 8
2: 10, 4, 4, 6, 8, 6, 7, 12, 8, 8
3: 10, 9, 8, 6, 5, 8, 6, 6, 9, 10
4: 4, 8, 4, 4, 5, 4, 4, 6, 5, 6
5: 6, 8, 8, 6, 12, 4, 10, 8, 6, 12
6: 10, 10, 8, 9, 16, 8, 7, 12, 16, 14
7: 4, 5, 6, 6, 4, 5, 6, 6, 4, 4
8: 16, 20, 8, 12, 10, 8, 8, 14, 16, 8

Work with a partner. You want to know the mean number of hours students with part-time jobs work each week. You go to 8 different schools. At each school, you randomly survey 10 students with part-time jobs. Your results are shown at the left.

Step 1: Find the mean of each sample.

Step 2: Make a box-and-whisker plot of the sample means.

Step 3: Use the box-and-whisker plot to estimate the actual mean number of hours students with part-time jobs work each week.

How does your estimate compare to the mean of the entire data set?

3 ACTIVITY: Using a Simulation

Work with a partner. Another way to generate multiple samples of data is to use a simulation. Suppose 70% of all seventh graders watch reality shows on television.

Step 1: Design a simulation involving 50 packing peanuts by marking 70% of the peanuts with a certain color. Put the peanuts into a paper bag.

Step 2: Simulate choosing a sample of 30 students by choosing peanuts from the bag, replacing the peanut each time. Record the results. Repeat this process to generate eight more samples. How much variation do you expect among the samples? Explain.

Step 3: Display your results.

Practice

2. **SPORTS DRINKS** You want to know whether student-athletes prefer water or sports drinks during games. You go to 10 different schools. At each school, you randomly survey 10 student-athletes. The percents of student-athletes who prefer water are shown.

   60% 70% 60% 50% 80% 70% 30% 70% 80% 40%

   a. Make a box-and-whisker plot of the data.

   b. Use the box-and-whisker plot to estimate the actual percent of student-athletes who prefer water. How does your estimate compare to the mean of the data?

3. **PART-TIME JOBS** Repeat Activity 2 using the medians of the samples.

4. **TELEVISION** In Activity 3, how do the percents in your samples compare to the given percent of seventh graders who watch reality shows on television?

5. **REASONING** Why is it better to make inferences about a population based on multiple samples instead of only one sample? What additional information do you gain by taking multiple random samples? Explain.

15.7 Comparing Populations

Essential Question How can you compare data sets that represent two populations?

1 ACTIVITY: Comparing Two Data Distributions

Work with a partner. You want to compare the shoe sizes of male students in two classes. You collect the data shown in the table.

Male Students in Eighth-Grade Class														
7	9	8	$7\frac{1}{2}$	$8\frac{1}{2}$	10	6	$6\frac{1}{2}$	8	8	$8\frac{1}{2}$	9	11	$7\frac{1}{2}$	$8\frac{1}{2}$
Male Students in Sixth-Grade Class														
6	$5\frac{1}{2}$	6	$6\frac{1}{2}$	$7\frac{1}{2}$	$8\frac{1}{2}$	7	$5\frac{1}{2}$	5	$5\frac{1}{2}$	$6\frac{1}{2}$	7	$4\frac{1}{2}$	6	6

a. How can you display both data sets so that you can visually compare the measures of center and variation? Make the data display you chose.

b. Describe the shape of each distribution.

c. Complete the table.

	Mean	Median	Mode	Range	Interquartile Range (IQR)	Mean Absolute Deviation (MAD)
Male Students in Eighth-Grade Class						
Male Students in Sixth-Grade Class						

d. Compare the measures of center for the data sets.

e. Compare the measures of variation for the data sets. Does one data set show more variation than the other? Explain.

f. Do the distributions overlap? How can you tell using the data display you chose in part (a)?

g. The double box-and-whisker plot below shows the shoe sizes of the members of two girls basketball teams. Can you conclude that at least one girl from each team has the same shoe size? Can you conclude that at least one girl from the Bobcats has a larger shoe size than one of the girls from the Tigers? Explain your reasoning.

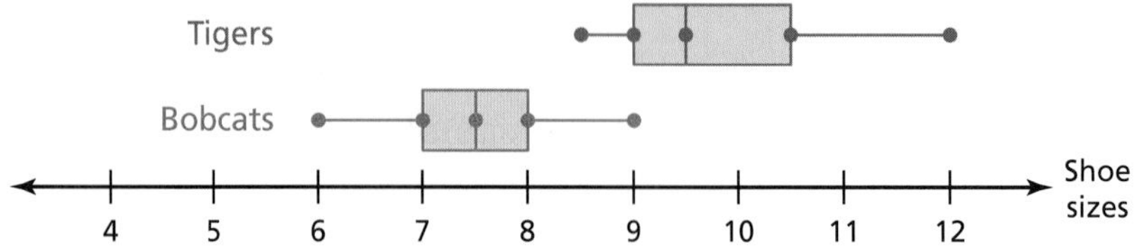

Probability and Statistics

In this lesson, you will

- use measures of center and variation to compare populations.
- use random samples to compare populations.

2 ACTIVITY: Comparing Two Data Distributions

Work with a partner. Compare the shapes of the distributions. Do the two data sets overlap? Explain. If so, use measures of center and the least and the greatest values to describe the overlap between the two data sets.

a.

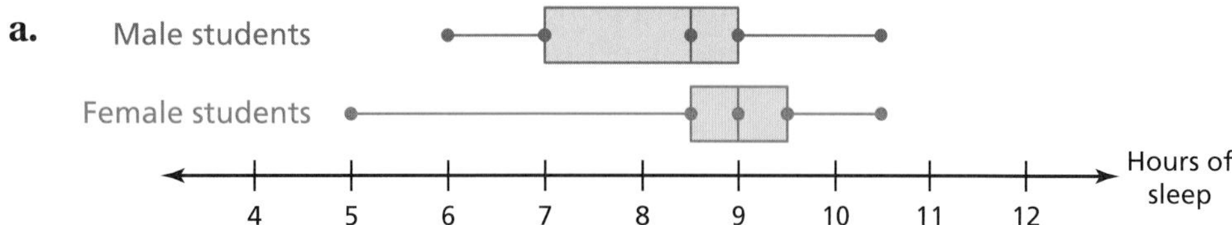

b.

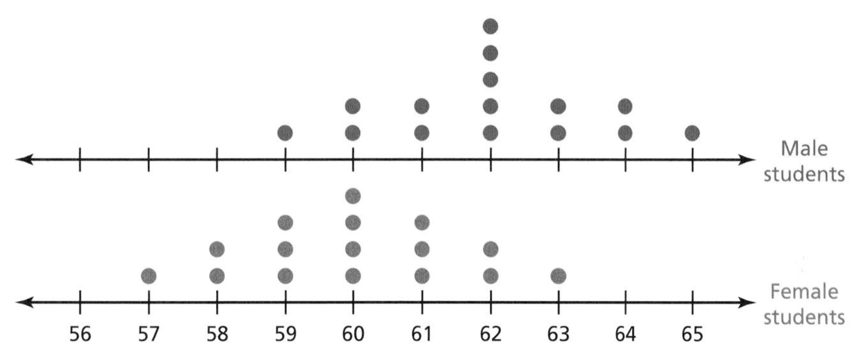

c. **Ages of People in Two Exercise Classes**

10:00 A.M. Class		8:00 P.M. Class
	1	8 9
	2	1 2 2 7 9 9
	3	0 3 4 5 7
9 7 3 2 2 2	4	0
7 5 4 3 1	5	
7 0 0	6	
0	7	

Key: 1 | 8 = 18

Math Practice

Recognize Usefulness of Tools

How is each type of data display useful? Which do you prefer? Explain.

What Is Your Answer?

3. **IN YOUR OWN WORDS** How can you compare data sets that represent two populations?

Practice Use what you learned about comparing data sets to complete Exercise 3 on page 684.

15.7 Lesson

Recall that you use the mean and the mean absolute deviation (MAD) to describe symmetric distributions of data. You use the median and the interquartile range (IQR) to describe skewed distributions of data.

To compare two populations, use the mean and the MAD when both distributions are symmetric. Use the median and the IQR when either one or both distributions are skewed.

EXAMPLE 1 Comparing Populations

The double dot plot shows the time that each candidate in a debate spent answering each of 15 questions.

Study Tip

You can more easily see the visual overlap of dot plots that are aligned vertically.

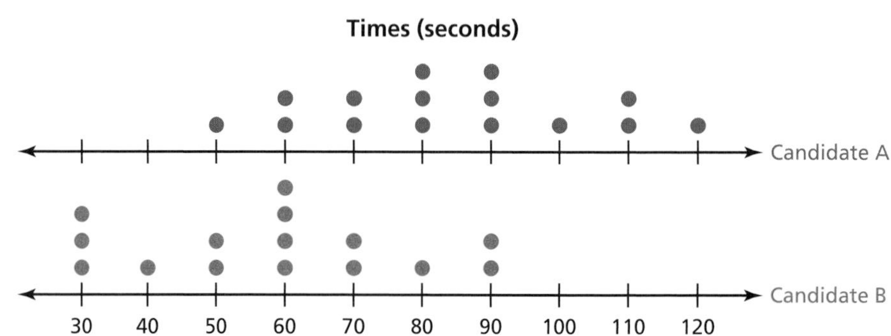

a. Compare the populations using measures of center and variation.

Both distributions are approximately symmetric, so use the mean and the MAD.

Candidate A

$$\text{Mean} = \frac{1260}{15} = 84$$

$$\text{MAD} = \frac{244}{15} \approx 16$$

Candidate B

$$\text{Mean} = \frac{870}{15} = 58$$

$$\text{MAD} = \frac{236}{15} \approx 16$$

So, the variation in the times was about the same, but Candidate A had a greater mean time.

Study Tip

When two populations have similar variabilities, the value in part (b) describes the visual overlap between the data. In general, the greater the value, the less the overlap.

b. Express the difference in the measures of center as a multiple of the measure of variation.

$$\frac{\text{mean for Candidate A} - \text{mean for Candidate B}}{\text{MAD}} = \frac{26}{16} \approx 1.6$$

So, the difference in the means is about 1.6 times the MAD.

On Your Own

Now You're Ready
Exercises 4–6

1. **WHAT IF?** Each value in the dot plot for Candidate A increases by 30 seconds. How does this affect the answers in Example 1? Explain.

You do not need to have all the data from two populations to make comparisons. You can use random samples to make comparisons.

EXAMPLE 2 Using Random Samples to Compare Populations

You want to compare the costs of speeding tickets in two states.

a. The double box-and-whisker plot shows a random sample of 10 speeding tickets issued in two states. Compare the samples using measures of center and variation. Can you use this to make a valid comparison about speeding tickets in the two states? Explain.

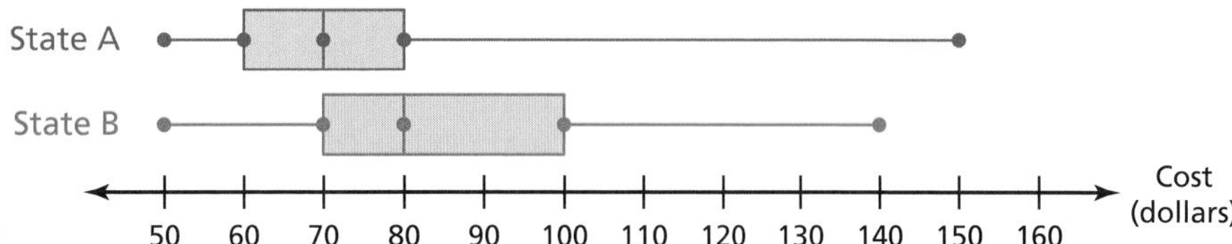

Both distributions are skewed right, so use the median and the IQR.

The median and the IQR for State A, 70 and 20, are less than the median and the IQR for State B, 80 and 30. However, the sample size is too small and the variability is too great to conclude that speeding tickets generally cost more in State B.

b. The double box-and-whisker plot shows the medians of 100 random samples of 10 speeding tickets for each state. Compare the variability of the sample medians to the variability of the sample costs in part (a).

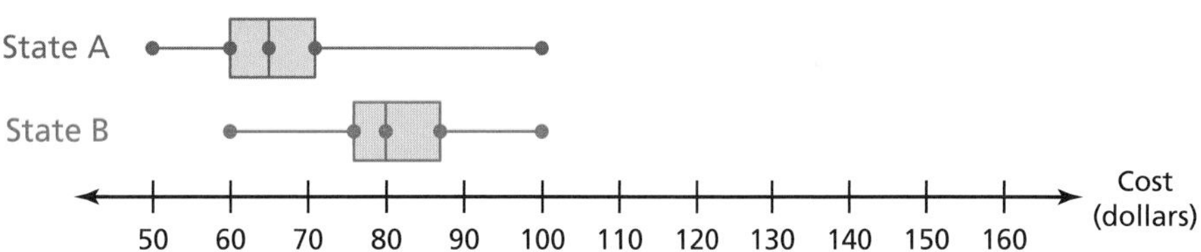

The IQR of the sample medians for each state is about 10.

So, the sample medians vary much less than the sample costs.

c. Make a conclusion about the costs of speeding tickets in the two states.

The sample medians show less variability. Most of the sample medians for State B are greater than the sample medians for State A.

So, speeding tickets generally cost more in State B than in State A.

2. **WHAT IF?** A random sample of 8 speeding tickets issued in State C has a median of $120. Can you conclude that a speeding ticket in State C costs more than in States A and B? Explain.

15.7 Exercises

Vocabulary and Concept Check

1. **REASONING** When comparing two populations, when should you use the mean and the MAD? the median and the IQR?

2. **WRITING** Two data sets have similar variabilities. Suppose the measures of center of the data sets differ by 4 times the measure of variation. Describe the visual overlap of the data.

Practice and Problem Solving

3. **SNAKES** The tables show the lengths of two types of snakes at an animal store.

Garter Snake Lengths (inches)					
26	30	22	15	21	24
28	32	24	25	18	35

Water Snake Lengths (inches)					
34	25	24	35	40	32
41	27	37	32	21	30

a. Find the mean, median, mode, range, interquartile range, and mean absolute deviation for each data set.

b. Compare the data sets.

1 4. **HOCKEY** The double box-and-whisker plot shows the goals scored per game by two hockey teams during a 20-game season.

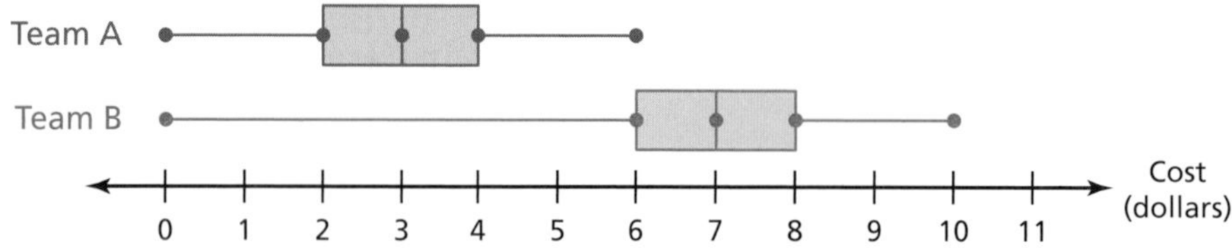

a. Compare the populations using measures of center and variation.

b. Express the difference in the measures of center as a multiple of the measure of variation.

5. **TEST SCORES** The dot plots show the test scores for two classes taught by the same teacher.

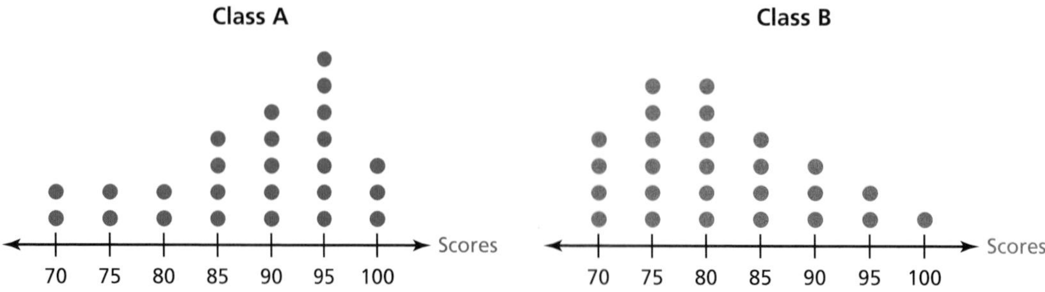

a. Compare the populations using measures of center and variation.

b. Express the difference in the measures of center as a multiple of each measure of variation.

6. **ATTENDANCE** The tables show the attendances at volleyball games and basketball games at a school during the year.

Volleyball Game Attendance						
112	95	84	106	62	68	53
75	88	93	127	98	117	60
49	54	85	74	88	132	

Basketball Game Attendance						
202	190	173	155	169	188	195
176	141	152	181	198	214	179
163	186	184	207	219	228	

   a. Compare the populations using measures of center and variation.
   b. Express the difference in the measures of center as a multiple of each measure of variation.

7. **NUMBER SENSE** Compare the answers to Exercises 4(b), 5(b), and 6(b). Which value is the greatest? What does this mean?

2 8. **MAGAZINES** You want to compare the number of words per sentence in a sports magazine to the number of words per sentence in a political magazine.

   a. The data represent random samples of 10 sentences in each magazine. Compare the samples using measures of center and variation. Can you use this to make a valid comparison about the magazines? Explain.

      Sports magazine: 9, 21, 15, 14, 25, 26, 9, 19, 22, 30

      Political magazine: 31, 22, 17, 5, 23, 15, 10, 20, 20, 17

   b. The double box-and-whisker plot shows the means of 200 random samples of 20 sentences. Compare the variability of the sample means to the variability of the sample numbers of words in part (a).

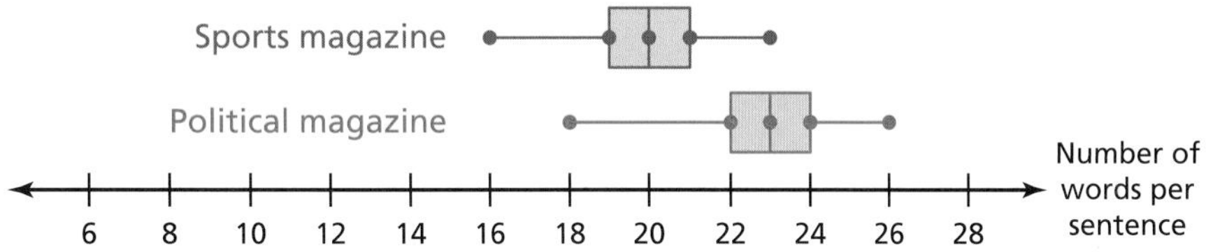

   c. Make a conclusion about the numbers of words per sentence in the magazines.

9. **Project** You want to compare the average amounts of time students in sixth, seventh, and eighth grade spend on homework each week.

   a. Design an experiment involving random sampling that can help you make a comparison.
   b. Perform the experiment. Can you make a conclusion about which students spend the most time on homework? Explain your reasoning.

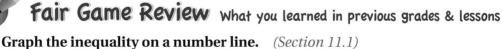

Fair Game Review What you learned in previous grades & lessons

Graph the inequality on a number line. *(Section 11.1)*

10. $x > 5$ **11.** $b \leq -3$ **12.** $n < -1.6$ **13.** $p \geq 2.5$

14. **MULTIPLE CHOICE** The number of students in the marching band increased from 100 to 125. What is the percent of increase? *(Skills Review Handbook)*

Ⓐ 20% Ⓑ 25% Ⓒ 80% Ⓓ 500%

15.6–15.7 Quiz

1. Which sample is better for making a prediction? Explain. *(Section 15.6)*

Predict the number of students in your school who play at least one sport.	
Sample A	A random sample of 10 students from the school student roster
Sample B	A random sample of 80 students from the school student roster

2. **GYMNASIUM** You want to estimate the number of students in your school who think the gymnasium should be remodeled. You survey 12 students on the basketball team. Determine whether the sample is *biased* or *unbiased*. Explain. *(Section 15.6)*

3. **TOWN COUNCIL** You want to know how the residents of your town feel about a recent town council decision. You survey 100 residents at random. Sixty-five support the decision, and thirty-five do not. So, you conclude that 65% of the residents of your town support the decision. Determine whether the conclusion is valid. Explain. *(Section 15.6)*

4. **FIELD TRIP** Of 60 randomly chosen students surveyed, 16 chose the aquarium as their favorite field trip. There are 720 students in the school. Predict the number of students in the school who would choose the aquarium as their favorite field trip. *(Section 15.6)*

5. **FOOTBALL** The double box-and-whisker plot shows the points scored per game by two football teams during the regular season. *(Section 15.7)*

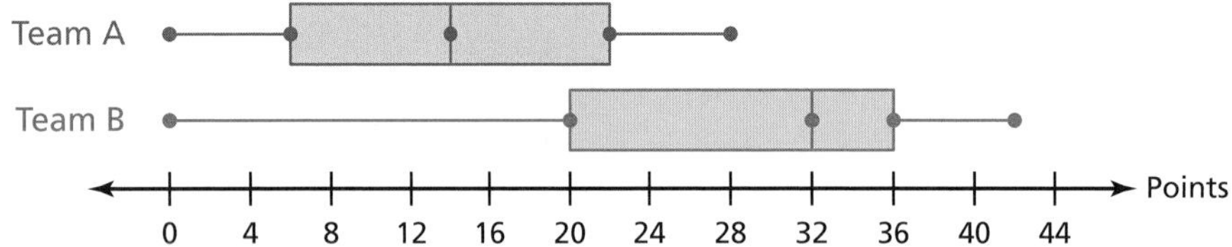

a. Compare the populations using measures of center and variation.

b. Express the difference in the measures of center as a multiple of the measure of variation.

6. **SUMMER CAMP** The dot plots show the ages of campers at two summer camps. *(Section 15.7)*

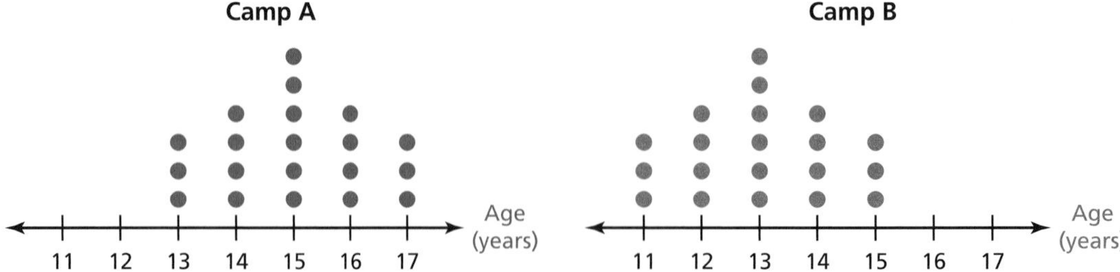

a. Compare the populations using measures of center and variation.

b. Express the difference in the measures of center as a multiple of the measure of variation.

15 Chapter Review

Check It Out
Vocabulary Help
BigIdeasMath.com

Review Key Vocabulary

experiment, *p. 634*
outcomes, *p. 634*
event, *p. 634*
favorable outcomes, *p. 634*
probability, *p. 640*
relative frequency, *p. 644*
experimental probability, *p. 646*
theoretical probability, *p. 647*
sample space, *p. 654*
Fundamental Counting Principle, *p. 654*
compound event, *p. 656*
independent events, *p. 662*
dependent events, *p. 663*
simulation, *p. 668*
population, *p. 672*
sample, *p. 672*
unbiased sample, *p. 674*
biased sample, *p. 674*

Review Examples and Exercises

15.1 Outcomes and Events *(pp. 632–637)*

You randomly choose one toy race car.

a. In how many ways can choosing a green car occur?

b. In how many ways can choosing a car that is *not* green occur? What are the favorable outcomes of choosing a car that is *not* green?

a. There are 5 green cars. So, choosing a green car can occur in 5 ways.

b. There are 2 cars that are *not* green. So, choosing a car that is *not* green can occur in 2 ways.

green	*not* green
green, green, green, green, green	blue, red

The favorable outcomes of the event are blue and red.

Exercises

You spin the spinner. (a) Find the number of ways the event can occur. (b) Find the favorable outcomes of the event.

1. Spinning a 1
2. Spinning a 3
3. Spinning an odd number
4. Spinning an even number
5. Spinning a number greater than 0
6. Spinning a number less than 3

15.2 Probability *(pp. 638–643)*

You flip a coin. What is the probability of flipping tails?

$$P(\text{event}) = \frac{\text{number of favorable outcomes}}{\text{number of possible outcomes}}$$

$$P(\text{tails}) = \frac{1}{2}$$

There is 1 tails.

There is a total of 2 sides.

The probability of flipping tails is $\frac{1}{2}$, or 50%.

Exercises

7. You roll a number cube. Find the probability of rolling an even number.

15.3 Experimental and Theoretical Probability *(pp. 644–651)*

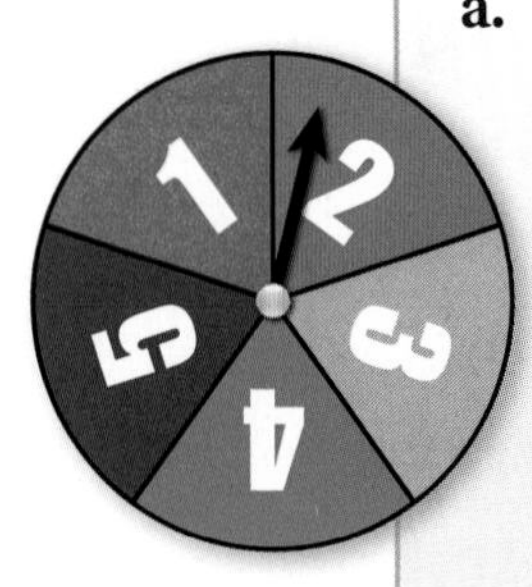

a. The bar graph shows the results of spinning the spinner 70 times. What is the experimental probability of spinning a 2?

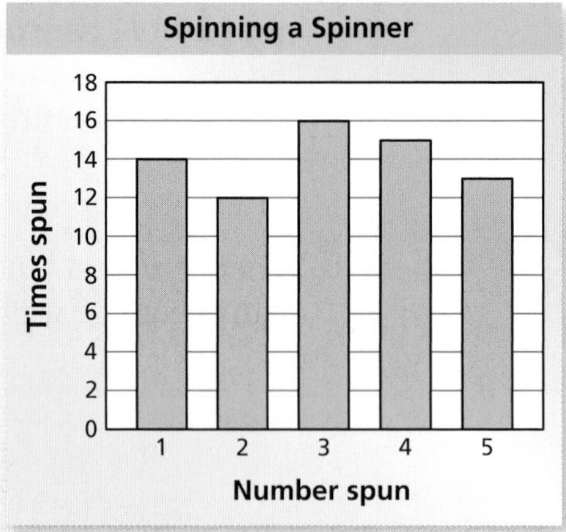

The bar graph shows 12 twos. So, the spinner landed on two 12 times in a total of 70 spins.

$$P(\text{event}) = \frac{\text{number of times the event occurs}}{\text{total number of trials}}$$

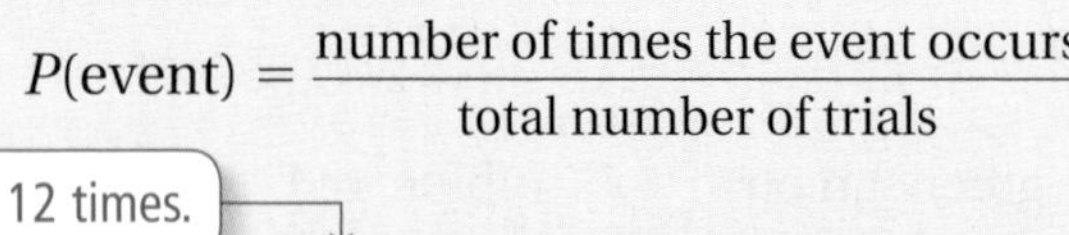

$$P(2) = \frac{12}{70} = \frac{6}{35}$$

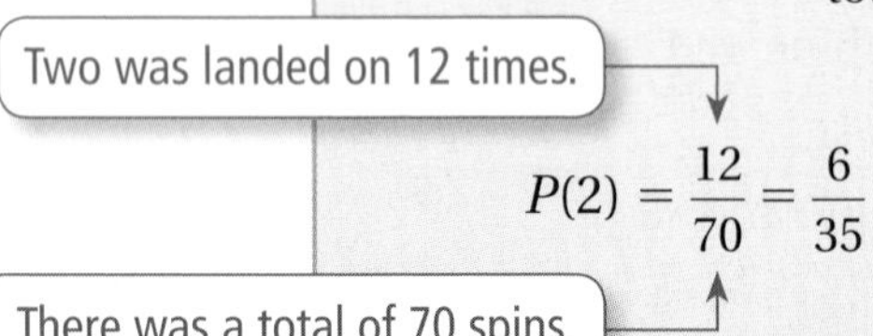

The experimental probability is $\frac{6}{35}$, or about 17%.

b. The theoretical probability of choosing a purple grape from a bag is $\frac{2}{9}$. There are 8 purple grapes in the bag. How many grapes are in the bag?

$$P(\text{purple}) = \frac{\text{number of purple grapes}}{\text{total number of grapes}}$$

$\frac{2}{9} = \frac{8}{g}$ Substitute. Let g be the total number of grapes.

$g = 36$ Solve for g.

So, there are 36 grapes in the bag.

Exercises

Use the bar graph on page 456 to find the experimental probability of the event.

8. Spinning a 3

9. Spinning an odd number

10. *Not* spinning a 5

11. Spinning a number greater than 3

Use the spinner to find the theoretical probability of the event.

12. Spinning blue

13. Spinning a 1

14. Spinning an even number

15. Spinning a 4

16. The theoretical probability of spinning an even number on a spinner is $\frac{2}{3}$. The spinner has 8 even-numbered sections. How many sections are on the spinner?

15.4 Compound Events (pp. 652–659)

a. How many different home theater systems can you make from 6 DVD players, 8 TVs, and 3 brands of speakers?

$6 \times 8 \times 3 = 144$ Fundamental Counting Principle

∴ So, you can make 144 different home theater systems.

b. You flip two pennies. What is the probability of flipping two heads?

Use a tree diagram to find the probability. Let H = heads and T = tails.

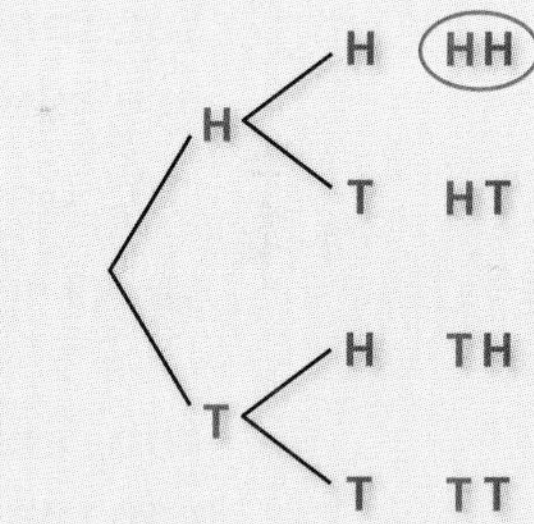

There is one favorable outcome in the sample space for flipping two heads: HH.

$$P(\text{event}) = \frac{\text{number of favorable outcomes}}{\text{number of possible outcomes}}$$

$$P(2 \text{ heads}) = \frac{1}{4}$$ Substitute.

∴ The probability is $\frac{1}{4}$, or 25%.

Exercises

17. You have 6 bracelets and 15 necklaces. Find the number of ways you can wear one bracelet and one necklace.

18. You flip two coins and roll a number cube. What is the probability of flipping two tails and rolling an even number?

15.5 Independent and Dependent Events (pp. 660–669)

You randomly choose one of the tiles and flip the coin. What is the probability of choosing a vowel and flipping heads?

Choosing one of the tiles does not affect the outcome of flipping the coin. So, the events are independent.

$P(\text{vowel}) = \frac{2}{7}$ ← There are 2 vowels (A and E). ← There is a total of 7 tiles.

$P(\text{tails}) = \frac{1}{2}$ ← There is 1 tails side. ← There is a total of 2 sides.

Use the formula for the probability of independent events.

$$P(A \text{ and } B) = P(A) \cdot P(B)$$

$$= \frac{2}{7} \cdot \frac{1}{2} = \frac{1}{7}$$

∴ The probability of choosing a vowel and flipping heads is $\frac{1}{7}$, or about 14%.

Exercises

You randomly choose one of the tiles above and flip the coin. Find the probability of the compound event.

19. Choosing a blue tile and flipping tails

20. Choosing the letter G and flipping tails

You randomly choose one of the tiles above. Without replacing the first tile, you randomly choose a second tile. Find the probability of the compound event.

21. Choosing a green tile and then a blue tile

22. Choosing a red tile and then a vowel

15.6 Samples and Populations (pp. 672–679)

You want to estimate the number of students in your school whose favorite subject is math. You survey every third student who leaves the school. Determine whether the sample is *biased* or *unbiased*.

The sample is representative of the population, selected at random, and large enough to provide accurate data.

∴ So, the sample is unbiased.

Exercises

23. You want to estimate the number of students in your school whose favorite subject is biology. You survey the first 10 students who arrive at biology club. Determine whether the sample is *biased* or *unbiased*. Explain.

15.7 Comparing Populations *(pp. 680–685)*

The double box-and-whisker plot shows the test scores for two French classes taught by the same teacher.

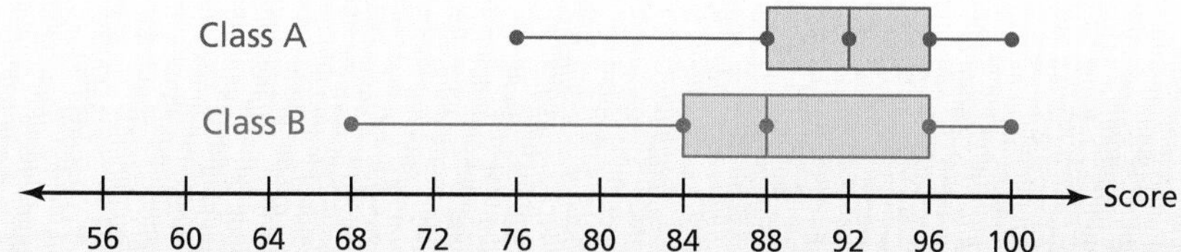

a. Compare the populations using measures of center and variation.

Both distributions are skewed left, so use the median and the IQR.

The median for Class A, 92, is greater than the median for Class B, 88. The IQR for Class B, 12, is greater than the IQR for Class A, 8. The scores in Class A are generally greater and have less variability than the scores in Class B.

b. Express the difference in the measures of center as a multiple of each measure of variation.

$$\frac{\text{median for Class A} - \text{median for Class B}}{\text{IQR for Class A}} = \frac{4}{8} = 0.5$$

$$\frac{\text{median for Class A} - \text{median for Class B}}{\text{IQR for Class B}} = \frac{4}{12} = 0.3$$

So, the difference in the medians is about 0.3 to 0.5 times the IQR.

Exercises

24. SPANISH TEST The double box-and-whisker plot shows the test scores of two Spanish classes taught by the same teacher.

a. Compare the populations using measures of center and variation.

b. Express the difference in the measures of center as a multiple of each measure of variation.

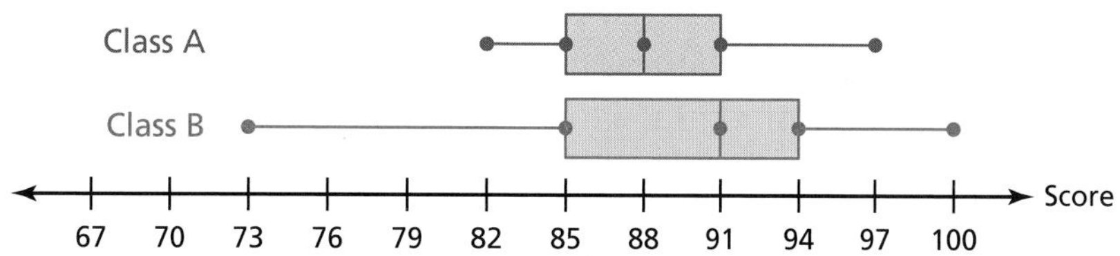

15 Chapter Test

You randomly choose one game piece. (a) Find the number of ways the event can occur. (b) Find the favorable outcomes of the event.

1. Choosing green
2. Choosing *not* yellow

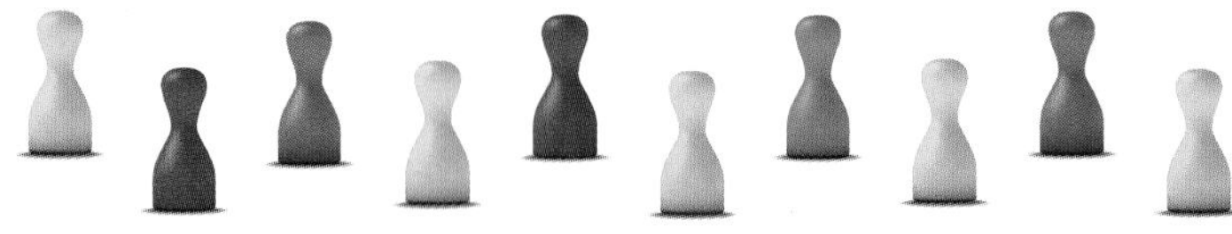

3. Use the Fundamental Counting Principle to find the total number of different sunscreens possible.

Sunscreen	
SPF	10, 15, 30, 45, 50
Type	Lotion, Spray, Gel

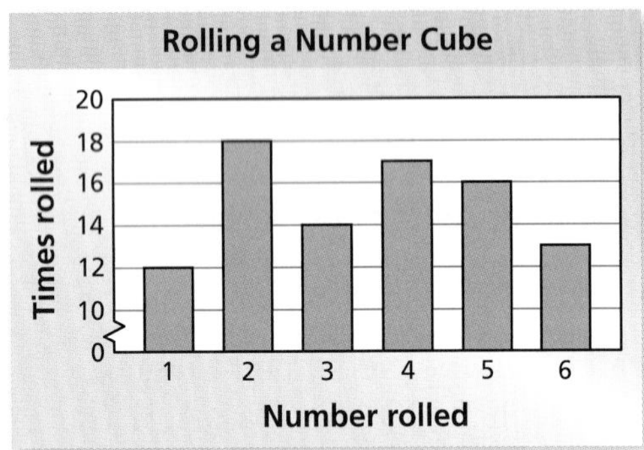

Use the bar graph to find the experimental probability of the event.

4. Rolling a 1 or a 2
5. Rolling an odd number
6. *Not* rolling a 5

Use the spinner to find the theoretical probability of the event(s).

7. Spinning an even number
8. Spinning a 1 and then a 2

You randomly choose one chess piece. Without replacing the first piece, you randomly choose a second piece. Find the probability of choosing the first piece, then the second piece.

9. Bishop and bishop
10. King and queen

11. **LUNCH** You want to estimate the number of students in your school who prefer to bring a lunch from home rather than buy one at school. You survey five students who are standing in the lunch line. Determine whether the sample is *biased* or *unbiased*. Explain.

12. **AGES** The double box-and-whisker plot shows the ages of the viewers of two television shows in a small town.

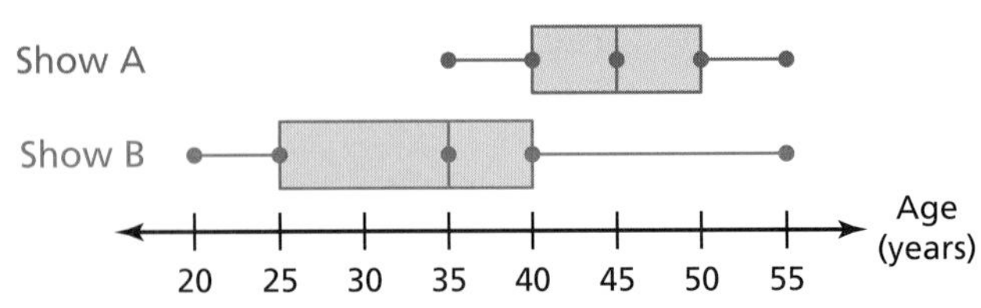

   a. Compare the populations using measures of center and variation.

   b. Express the difference in the measures of center as a multiple of each measure of variation.

15 Cumulative Assessment

Test-Taking Strategy
Use Intelligent Guessing
What's the probability of drawing 1 hyena out of a bag with 2 hyenas and 3 mice?
(A) -10% (B) 40% (C) 60% (D) 500%
40% < 60% I'm hoping 40%.
"You know it can't be -10% or 500%. So, you can intelligently guess between 40% and 60%."

1. A school athletic director asked each athletic team member to name his or her favorite professional sports team. The results are below:

   - D.C. United: 3
   - Florida Panthers: 8
   - Jacksonville Jaguars: 26
   - Jacksonville Sharks: 7
   - Miami Dolphins: 22
   - Miami Heat: 15
   - Miami Marlins: 20
   - Minnesota Lynx: 4
   - New York Knicks: 5
   - Orlando Magic: 18
   - Tampa Bay Buccaneers: 17
   - Tampa Bay Lightning: 12
   - Tampa Bay Rays: 28
   - Other: 6

   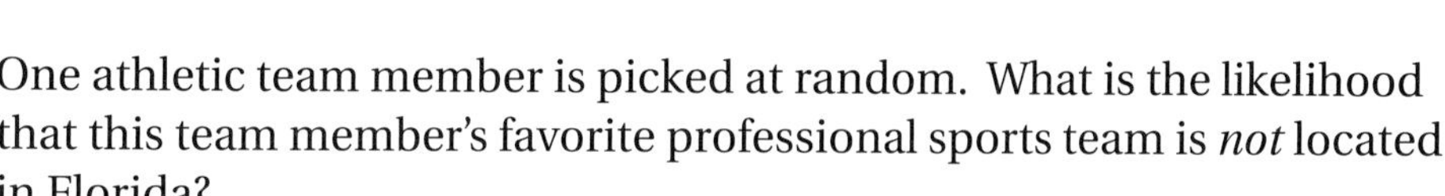

   One athletic team member is picked at random. What is the likelihood that this team member's favorite professional sports team is *not* located in Florida?

   A. certain
   B. likely, but not certain
   C. unlikely, but not impossible
   D. impossible

2. Each student in your class voted for his or her favorite day of the week. Their votes are shown below:

Favorite Day of the Week

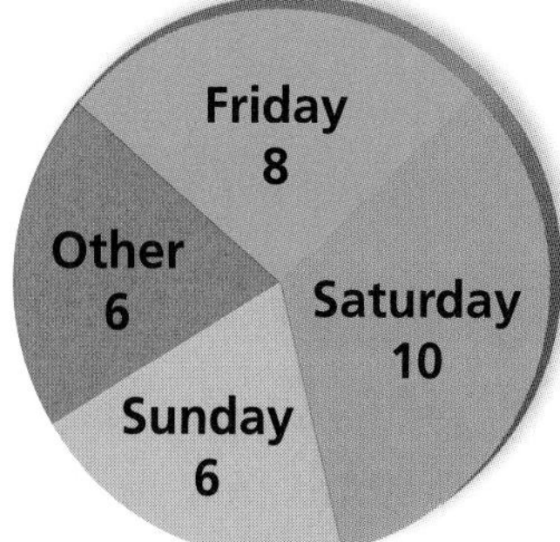

A student from your class is picked at random. What is the probability that this student's favorite day of the week is Sunday?

3. How far, in millimeters, will the tip of the hour hand of the clock travel in 2 hours? (Use $\frac{22}{7}$ for π.)

F. 44 mm

G. 88 mm

H. 264 mm

I. 528 mm

4. Nathaniel solved the proportion in the box below.

$$\frac{16}{40} = \frac{p}{27}$$

$$16 \cdot p = 40 \cdot 27$$

$$16p = 1080$$

$$\frac{16p}{16} = \frac{1080}{16}$$

$$p = 67.5$$

What should Nathaniel do to correct the error that he made?

A. Add 40 to 16 and 27 to p.

B. Subtract 16 from 40 and 27 from p.

C. Multiply 16 by 27 and p by 40.

D. Divide 16 by 27 and p by 40.

5. A North American hockey rink contains 5 face-off circles. Each of these circles has a radius of 15 feet. What is the total area, in square feet, of all the face-off circles? (Use 3.14 for π.)

F. 706.5 ft^2

G. 2826 ft^2

H. 3532.5 ft^2

I. 14,130 ft^2

6. A spinner is divided into eight congruent sections, as shown below.

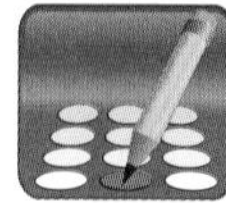

You spin the spinner twice. What is the probability that the arrow will stop in a yellow section both times?

7. What is the surface area, in square inches, of the square pyramid?

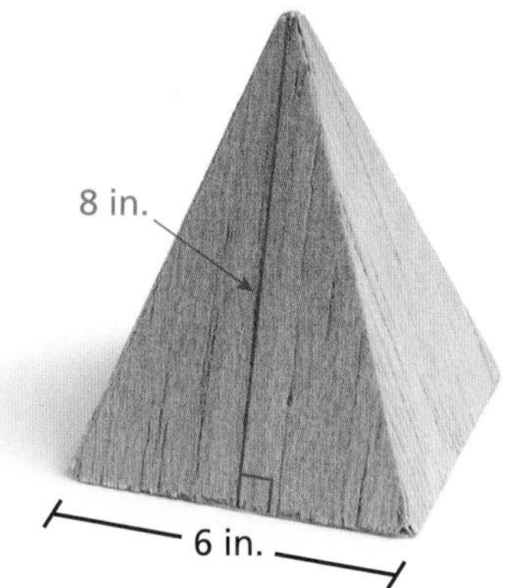

A. 24 in.^{2}

B. 96 in.^{2}

C. 132 in.^{2}

D. 228 in.^{2}

8. The value of one of Kevin's baseball cards was $6.00 when he first got it. The value of this card is now $15.00. What is the percent increase in the value of the card?

F. 40%

G. 90%

H. 150%

I. 250%

9. You roll a number cube twice. You want to roll two even numbers.

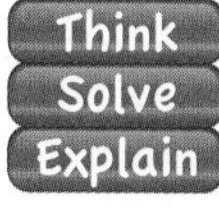

Part A Determine whether the events are independent or dependent.

Part B Find the number of favorable outcomes and the number of possible outcomes of each roll.

Part C Find the probability of rolling two even numbers. Explain your reasoning.

Appendix A
My Big Ideas Projects

My Big Ideas Projects

A.1 Literature Project

Swiss Family Robinson

1 Getting Started

Swiss Family Robinson is a novel about a Swiss family who was shipwrecked in the East Indies. The story was written by Johann David Wyss, and was first published in 1812.

Essential Question How does the knowledge of mathematics provide you and your family with survival tools?

Read *Swiss Family Robinson*. As you read the exciting adventures, think about the mathematics the family knew and used to survive.

Sample: The tree house built by the family was accessed by a long rope ladder. The ladder was about 30 feet long with a rung every 10 inches. To make the ladder, the family had to plan how many rungs were needed. They decided the number was $1 + 12(30) \div 10$. Why?

2 Things to Include

- Suppose you lived in the 18th century. Plan a trip from Switzerland to Australia. Describe your route. Estimate the length of the route and the number of miles you will travel each day. About how many days will the entire trip take?
- Suppose that your family is shipwrecked on an island that has no other people. What do you need to do to survive? What types of tools do you hope to salvage from the ship? Describe how mathematics could help you survive.
- Suppose that you are the oldest of four children in a shipwrecked family. Your parents have made you responsible for the education of your younger siblings. What type of mathematics would you teach them? Explain your reasoning.

3 Things to Remember

- You can download each part of the book at *BigIdeasMath.com*.
- Add your own illustrations to your project.
- Organize your math stories in a folder, and think of a title for your report.

A.2 History Project

Mathematics in Ancient China

1 Getting Started

Mathematics was developed in China independently of the mathematics that was developed in Europe and the Middle East. For example, the Pythagorean Theorem and the computation of pi were used in China prior to the time when China and Europe began communicating with each other.

Essential Question How have tools and knowledge from the past influenced modern day mathematics?

Sample: Here are the names and symbols that were used in ancient China to represent the digits from 1 through 10.

1	yi	一
2	er	二
3	san	三
4	si	四
5	wu	五
6	liu	六
7	qi	七
8	ba	八
9	jiu	九
10	shi	十

Life-size Terra-cotta Warriors

A Chinese Abacus

2 Things to Include

- Describe the ancient Chinese book *The Nine Chapters on the Mathematical Art* (c. 100 B.C.). What types of mathematics are contained in this book?
- How did the ancient Chinese use the abacus to add and subtract numbers? How is the abacus related to base 10?
- How did the ancient Chinese use mathematics to build large structures, such as the Great Wall and the Forbidden City?
- How did the ancient Chinese write numbers that are greater than 10?
- Describe how the ancient Chinese used mathematics. How does this compare with the ways in which mathematics is used today?

Ancient Chinese Teapot

The Great Wall of China

3 Things to Remember

- Add your own illustrations to your project.
- Organize your math stories in a folder, and think of a title for your report.

Chinese Guardian Fu Lions

A.3 Art Project

Building a Kaleidoscope

1 Getting Started

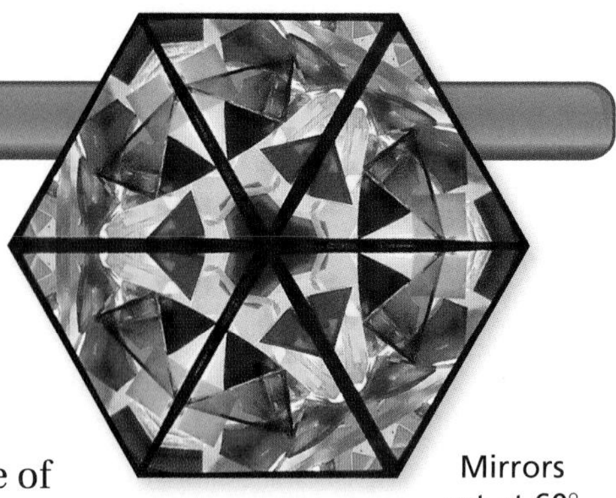
Mirrors set at 60°

A kaleidoscope is a tube of mirrors containing loose colored beads, pebbles, or other small colored objects. You look in one end and light enters the other end, reflecting off the mirrors.

Essential Question How does the knowledge of mathematics help you create a kaleidoscope?

If the angle between the mirrors is 45°, you see 8 duplicate images. If the angle is 60°, you see 6 duplicate images. If the angle is 90°, you see 4 duplicate images. As the tube is rotated, the colored objects tumble, creating various patterns.

Write a report about kaleidoscopes. Discuss the mathematics you need to know in order to build a kaleidoscope.

Sample: A kaleidoscope whose mirrors meet at 60° angles has reflective symmetry and rotational symmetry.

Reflect

Rotate 120°

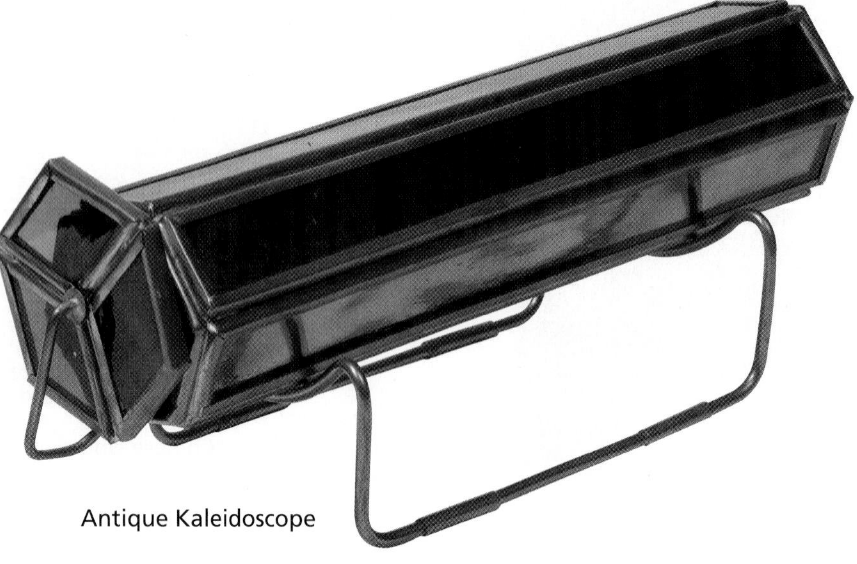
Antique Kaleidoscope

2 Things to Include

- How does the angle at which the mirrors meet affect the number of duplicate images that you see?
- What angles can you use other than 45°, 60°, and 90°? Explain your reasoning.
- Research the history of kaleidoscopes. Can you find examples of kaleidoscopes being used before they were patented by David Brewster in 1816?
- Make your own kaleidoscope.
- Describe the mathematics you used to create your kaleidoscope.

Mirrors set at 90°

Mirrors set at 60°

3 Things to Think About

- Add your own drawings and pattern creations to your project.
- Organize your report in a folder, and think of a title for your report.

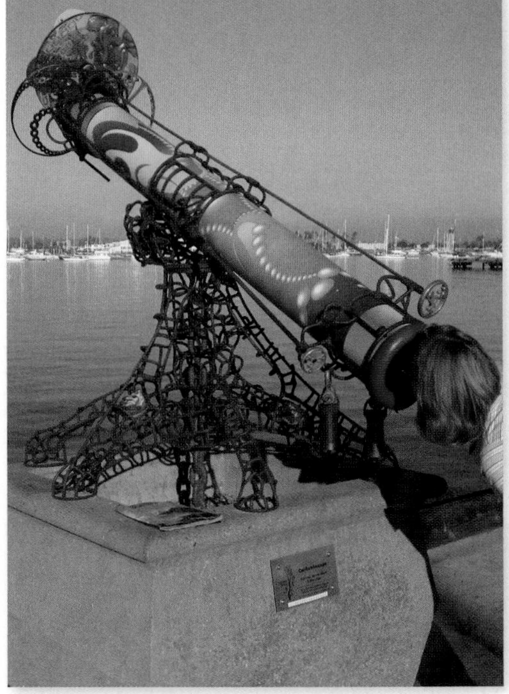

Giant Kaleidoscope, San Diego harbor

Mirrors set at 45°

A.4 Science Project

Our Solar System

1 Getting Started

Our solar system consists of four inner planets, four outer planets, dwarf planets such as Pluto, several moons, and many asteroids and comets.

Essential Question How do the characteristics of a planet influence whether or not it can sustain life?

Sample: The average temperatures of the eight planets in our solar system are shown in the graph.

The average temperature tends to drop as the distance between the Sun and the planet increases.

An exception to this rule is Venus. It has a higher average temperature than Mercury, even though Mercury is closer to the Sun.

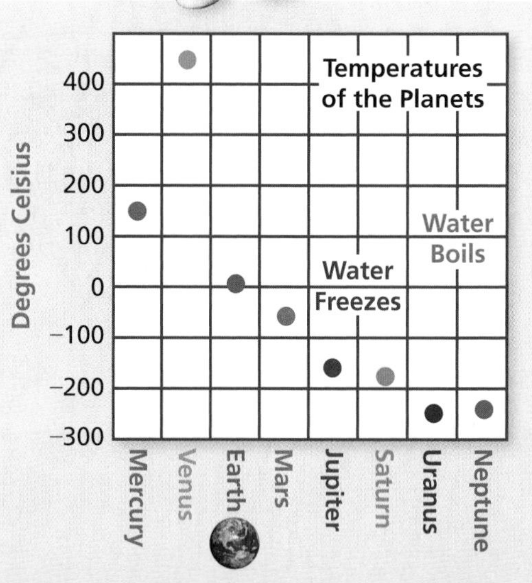

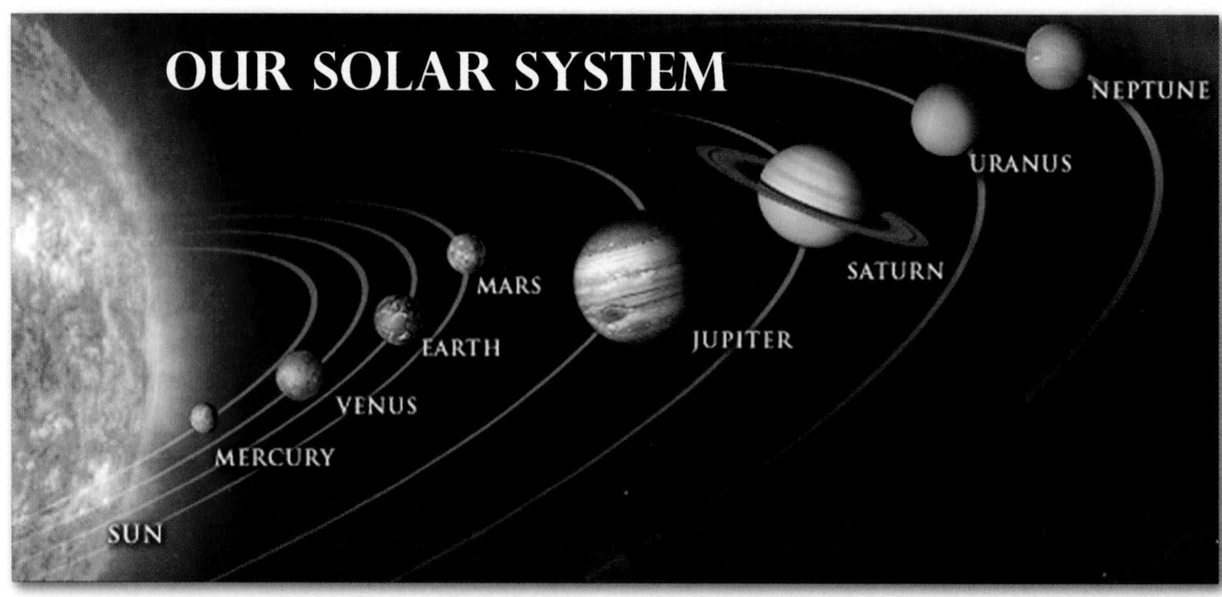

2 Things to Include

- Compare the masses of the planets.
- Compare the gravitational forces of the planets.
- How long is a "day" on each planet? Why?
- How long is a "year" on each planet? Why?
- Which planets or moons have humans explored?
- Which planets or moons could support human life? Explain your reasoning.

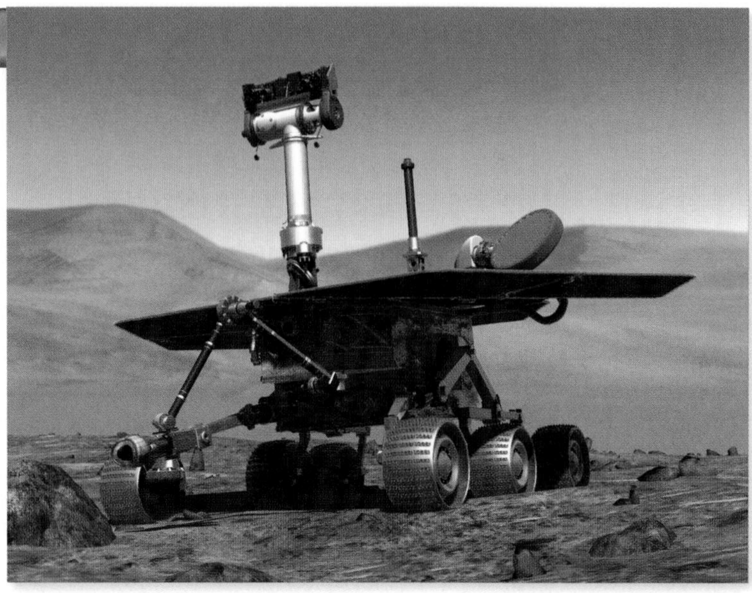

Mars Rover

3 Things to Remember

- Add your own drawings or photographs to your report. You can download photographs of the solar system and space travel at *NASA.gov*.
- Organize your report in a folder, and think of a title for your report.

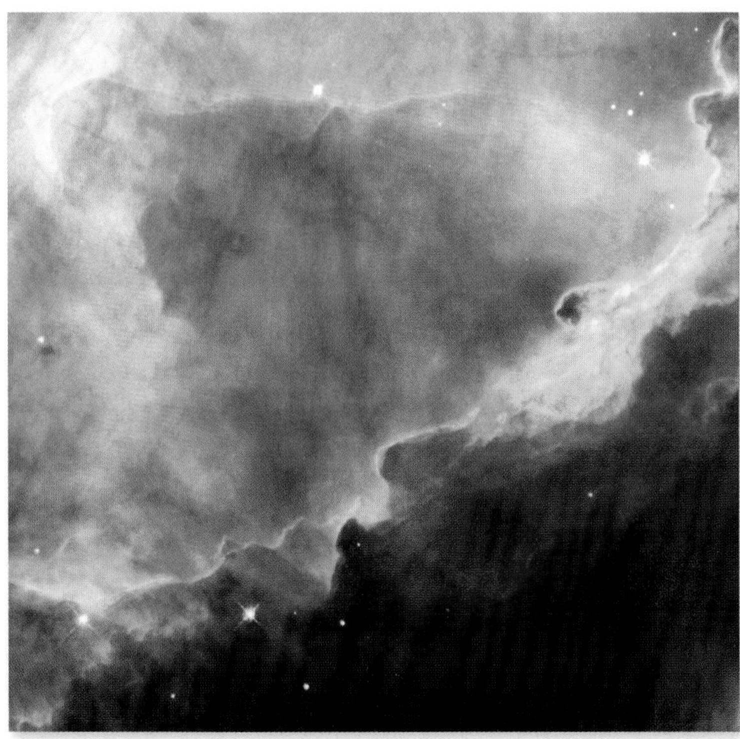

Hubble Image of Space

Hubble Spacecraft

Selected Answers

Section 1.1 Solving Simple Equations (pages 7–9)

1. + and − are inverses. × and ÷ are inverses.

3. $x - 3 = 6$; It is the only equation that does not have $x = 6$ as a solution.

5. $x = 57$ 7. $x = -5$ 9. $p = 21$ 11. $x = 9\pi$ 13. $d = \frac{1}{2}$ 15. $n = -4.9$

17. **a.** $105 = x + 14$; $x = 91$

b. no; Because $82 + 9 = 91$, you did not knock down the last pin with the second ball of the frame.

19. $n = -5$ 21. $m = 7.3\pi$ 23. $k = 1\frac{2}{3}$ 25. $p = -2\frac{1}{3}$

27. They should have added 1.5 to each side.

$$-1.5 + k = 8.2$$
$$k = 8.2 + 1.5$$
$$k = 9.7$$

29. $6.5x = 42.25$; \$6.50 per hour

31. $420 = \frac{7}{6}b$, $b = 360$; \$60

33. $h = -7$ 35. $q = 3.2$ 37. $x = -1\frac{4}{9}$

39. greater than; Because a negative number divided by a negative number is a positive number.

41. 3 mg 43. 12 in. 45. $7x - 4$ 47. $\frac{25}{4}g - \frac{2}{3}$

Section 1.2 Solving Multi-Step Equations (pages 14 and 15)

1. $2 + 3x = 17$; $x = 5$ 3. $k = 45$; 45°, 45°, 90° 5. $b = 90$; 90°, 135°, 90°, 90°, 135°

7. $c = 0.5$ 9. $h = -9$ 11. $x = -\frac{2}{9}$ 13. 20 watches

15. $4(b + 3) = 24$; 3 in. 17. $\frac{2580 + 2920 + x}{3} = 3000$; 3500 people

19. $<$ 21. $>$

Section 1.3 Solving Equations with Variables on Both Sides (pages 23–25)

1. no; When 3 is substituted for x, the left side simplifies to 4 and the right side simplifies to 3.

3. $x = 13.2$ in. 5. $x = 7.5$ in. 7. $k = -0.75$

9. $p = -48$ 11. $n = -3.5$ 13. $x = -4$

15. The 4 should have been added to the right side.

$$\begin{aligned} 3x - 4 &= 2x + 1 \\ 3x - 2x - 4 &= 2x + 1 - 2x \\ x - 4 &= 1 \\ x - 4 + 4 &= 1 + 4 \\ x &= 5 \end{aligned}$$

17. $15 + 0.5m = 25 + 0.25m$; 40 mi

19. $x = \frac{1}{3}$

21. no solution

23. infinitely many solutions

25. $x = 2$

27. no solution

29. infinitely many solutions

31. *Sample answer:* $8x + 2 = 8x$; The number $8x$ cannot be equal to 2 more than itself.

33. It's never the same. Your neighbor's total cost will always be \$75 more than your total cost.

35. no; $2x + 5.2$ can never equal $2x + 6.2$.

37. 7.5 units

39. Remember that the box is with priority mail and the envelope is with express mail.

41. 10 mL

43. **a.** 40 ft

b. no;

$$\begin{aligned} 2(\text{white area}) &= \text{black area} \\ 2[5(6x)] &= 4[6(x + 1)] \\ 60x &= 24x + 24 \\ 36x &= 24 \\ x &= \frac{2}{3} \end{aligned}$$

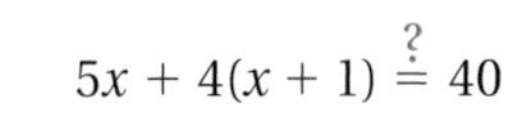

$$5x + 4(x + 1) \stackrel{?}{=} 40$$

Length of hallway is $5\left(\frac{2}{3}\right) + 4\left(\frac{2}{3} + 1\right) \stackrel{?}{=} 40$

$$10 \neq 40$$

45. 15.75 cm^3

47. C

Section 1.4 Rewriting Equations and Formulas

(pages 30 and 31)

1. no; The equation only contains one variable.

3. **a.** $A = \frac{1}{2}bh$ **b.** $b = \frac{2A}{h}$ **c.** $b = 12$ mm

5. $y = 4 - \frac{1}{3}x$

7. $y = \frac{2}{3} - \frac{4}{9}x$

9. $y = 3x - 1.5$

11. The y should have a negative sign in front of it.

$$\begin{aligned} 2x - y &= 5 \\ -y &= -2x + 5 \\ y &= 2x - 5 \end{aligned}$$

13. **a.** $t = \frac{I}{Pr}$

b. $t = 3$ yr

15. $m = \frac{e}{c^2}$

17. $\ell = \dfrac{A - \frac{1}{2}\pi w^2}{2w}$

19. $w = 6g - 40$

21. **a.** $F = 32 + \frac{9}{5}(K - 273.15)$

b. 32°F

c. liquid nitrogen

23. $r^3 = \frac{3V}{4\pi}$; $r = 4.5$ in.

25. $-5\frac{1}{3}$

27. $1\frac{1}{4}$

Section 2.1 Congruent Figures
(pages 46 and 47)

1. **a.** $\angle A$ and $\angle D$, $\angle B$ and $\angle E$, $\angle C$ and $\angle F$

   b. Side AB and Side DE, Side BC and Side EF, Side AC and Side DF

3. $\angle V$ does not belong. The other three angles are congruent to each other, but not to $\angle V$.

5. congruent

7. $\angle P$ and $\angle W$, $\angle Q$ and $\angle V$, $\angle R$ and $\angle Z$, $\angle S$ and $\angle Y$, $\angle T$ and $\angle X$;
   Side PQ and Side WV, Side QR and Side VZ, Side RS and Side ZY,
   Side ST and Side YX, Side TP and Side XW

9. not congruent; Corresponding side lengths are not congruent.

11. The corresponding angles are not congruent, so the two figures are not congruent.

13. What figures have you seen in this section that have at least one right angle?

15. **a.** true; Side AB corresponds to Side YZ.

    b. true; $\angle A$ and $\angle X$ have the same measure.

    c. false; $\angle A$ corresponds to $\angle Y$.

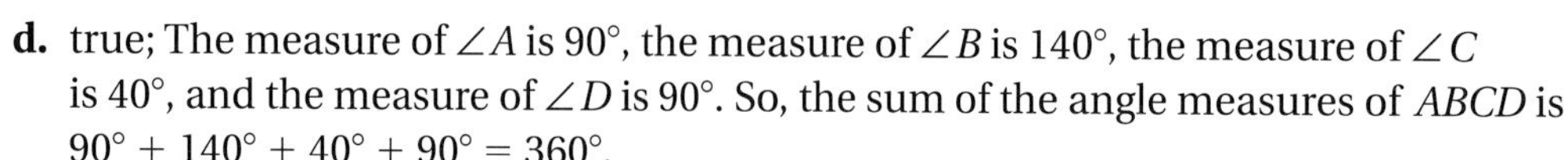

    d. true; The measure of $\angle A$ is 90°, the measure of $\angle B$ is 140°, the measure of $\angle C$ is 40°, and the measure of $\angle D$ is 90°. So, the sum of the angle measures of $ABCD$ is $90° + 140° + 40° + 90° = 360°$.

17 and 19.

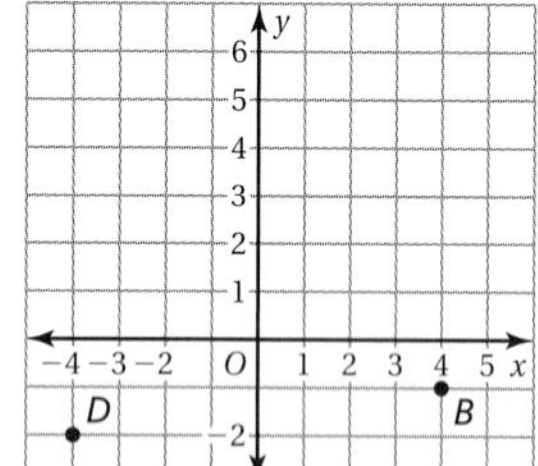

Section 2.2 Translations
(pages 52 and 53)

1. A

3. yes; Translate the letters T and O to the end.

5. no

7. yes

9. no

11. $A'(-3, 0)$, $B'(0, -1)$, $C'(1, -4)$, $D'(-3, -5)$

13.

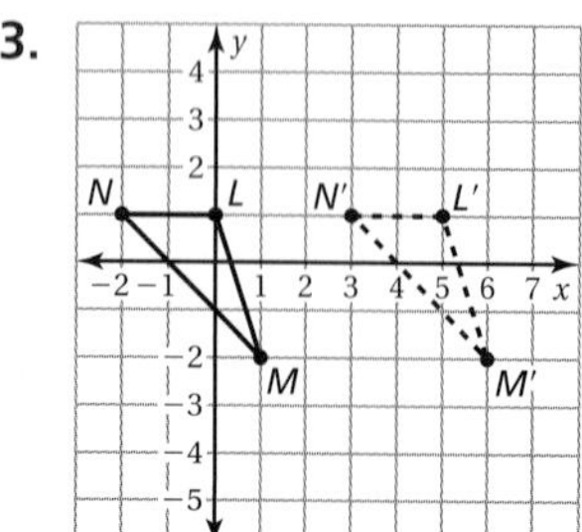

15.

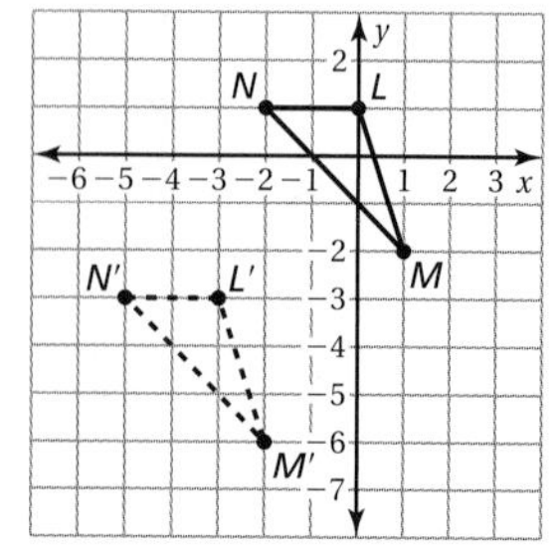

17. 2 units left and 2 units up

19. 6 units right and 3 units down

21. a. 5 units right and 1 unit up
 b. no; It would hit the island.
 c. 4 units up and 4 units right

23. If you are doing more than 10 moves and have not moved the knight to g5, you might want to start over.

25. no

27. yes

Section 2.3 Reflections
(pages 58 and 59)

1. The third one because it is not a reflection.

3. Quadrant IV

5. yes

7. no

9. no

11. $M'(-2, -1)$, $N'(0, -3)$, $P'(2, -2)$

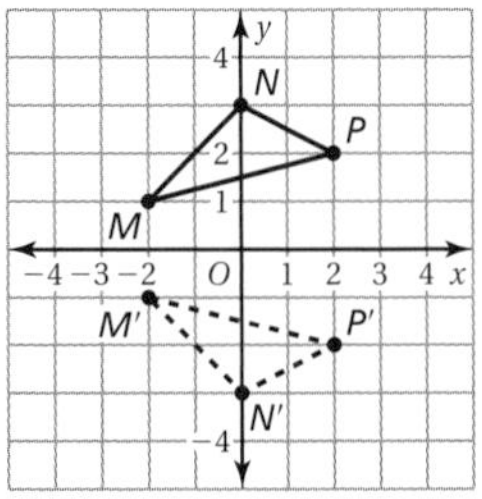

13. $D'(-2, 1)$, $E'(0, 1)$, $F'(0, 5)$, $G'(-2, 5)$

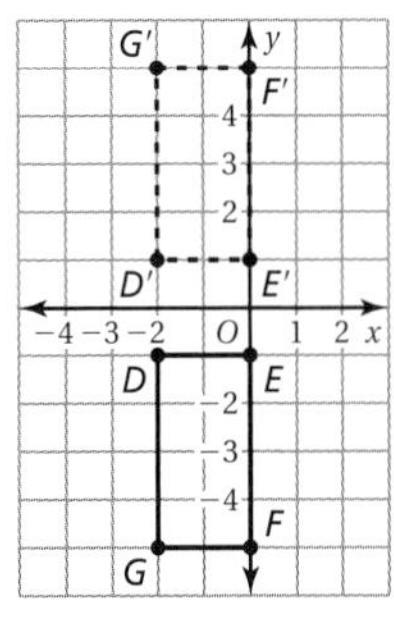

15. $T'(-4, -2)$, $U'(-4, 2)$, $V'(-6, -2)$

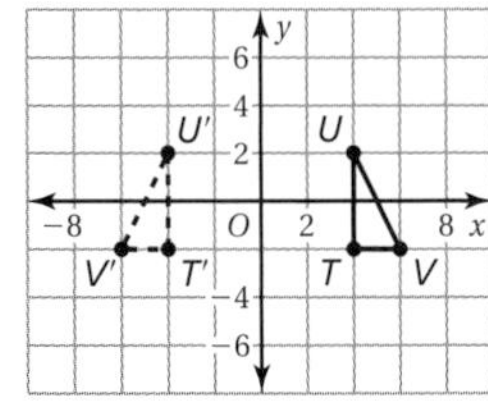

17. $J'(-2, 2)$, $K'(-7, 4)$, $L'(-9, -2)$, $M'(-3, -1)$

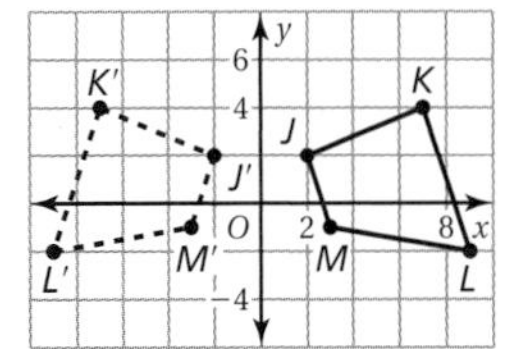

19. x-axis

21. y-axis

23. $R'(3, -4)$, $S'(3, -1)$, $T'(1, -4)$

25. yes; Translations and reflections produce images that are congruent to the original figure.

27. If you are driving a vehicle and want to see who is following you, where would you look?

29. obtuse

31. right

33. B

Section 2.4 Rotations
(pages 65–67)

1. (0, 0); (1, −3)

3. Quadrant IV

5. Quadrant II

7. reflection

9. translation

11. yes; 90° counterclockwise

13. $A'(2, 2)$, $B'(1, 4)$, $C'(3, 4)$, $D'(4, 2)$

15. $J'(0, -3)$, $K'(0, -5)$, $L'(-4, -3)$

17. $W'(-2, 6)$, $X'(-2, 2)$, $Y'(-6, 2)$, $Z'(-6, 5)$

Section 2.4 Rotations *(continued)* *(pages 65–67)*

19. It only needs to rotate 120° to produce an identical image.

21. It only needs to rotate 180° to produce an identical image.

23. $J''(4, 4)$, $K''(3, 4)$, $L''(1, 1)$, $M''(4, 1)$

25. *Sample answer:* Rotate 180° about the origin and then rotate 90° clockwise about vertex $(-1, 0)$; Rotate 90° counterclockwise about the origin and then translate 1 unit left and 1 unit down.

27. Use Guess, Check, and Revise to solve this problem.

29. (2, 4), (4, 1), (1, 1)

31. yes

33. no

Section 2.5 Similar Figures *(pages 74 and 75)*

1. They are congruent.

3. Yes, because the angles are congruent and the side lengths are proportional.

5. not similar; Corresponding side lengths are not proportional.

7.

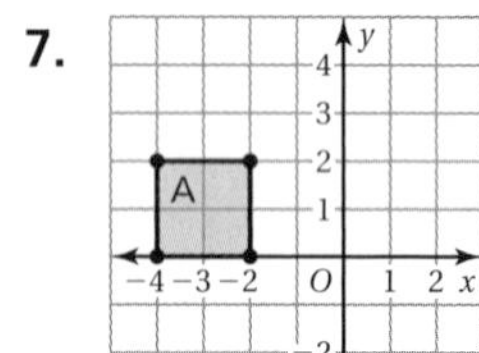

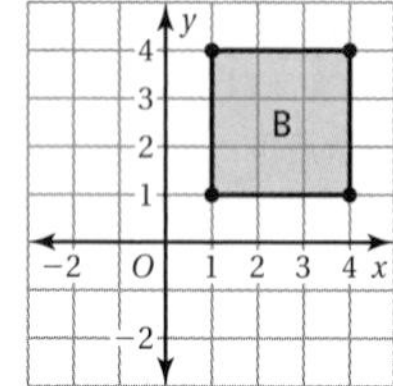

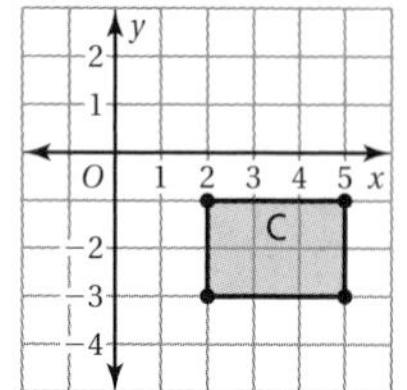

A and B; Corresponding side lengths are proportional and corresponding angles are congruent.

9. $6\frac{2}{3}$

11. 14

13. 30 in.

15. What types of quadrilaterals can have the given angle measures?

Hmmm.

17. 3 times

19. **a.** yes

b. yes; It represents the fact that the sides are proportional because you can split the isosceles triangles into smaller right triangles that will be similar.

21. $\frac{16}{81}$

23. $\frac{49}{16}$

25. C

Section 2.6 Perimeters and Areas of Similar Figures
(pages 80 and 81)

1. The ratio of the perimeters is equal to the ratio of the corresponding side lengths.

3. Because the ratio of the corresponding side lengths is $\frac{1}{2}$, the ratio of the areas is equal to $\left(\frac{1}{2}\right)^2$. To find the area, solve the proportion $\frac{30}{x} = \frac{1}{4}$ to get $x = 120$ square inches.

5. $\frac{5}{8}; \frac{25}{64}$

7. $\frac{14}{9}; \frac{196}{81}$

9. The area is 9 times larger.

11. 25.6

13. 39 in.; 93.5 in.^{2}

15. 108 yd

17. **a.** 400 times greater; The ratio of the corresponding lengths is $\frac{120 \text{ in.}}{6 \text{ in.}} = \frac{20}{1}$. So, the ratio of the areas is $\left(\frac{20}{1}\right)^2 = \frac{400}{1}$.

b. 1250 ft^2

19. 15 m

21. $x = -2$

23. $n = -4$

Section 2.7 Dilations
(pages 87–89)

1. A dilation changes the size of a figure. The image is similar, not congruent, to the original figure.

3. The middle red figure is not a dilation of the blue figure because the height is half of the blue figure and the base is the same. The left red figure is a reduction of the blue figure and the right red figure is an enlargement of the blue figure.

5.

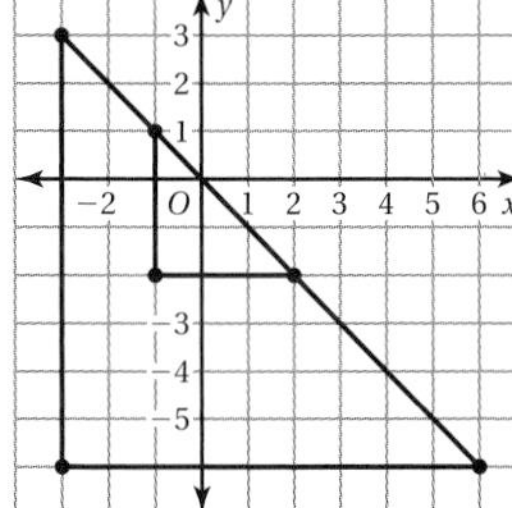

The triangles are similar.

7. yes

9. no

11. yes

13.

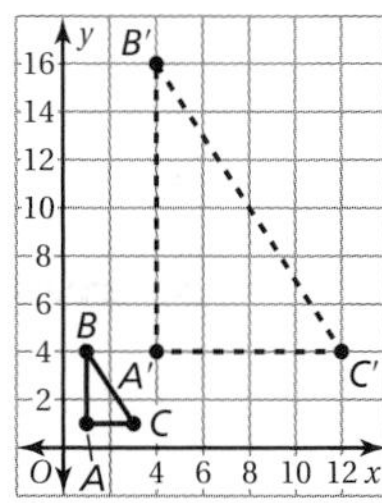

enlargement

15.

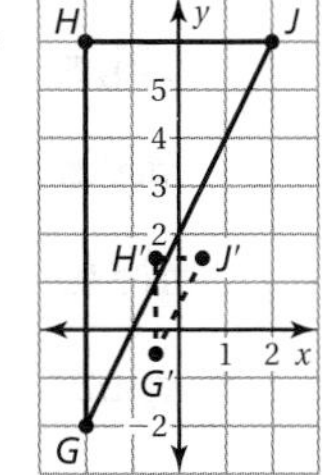

reduction

17.

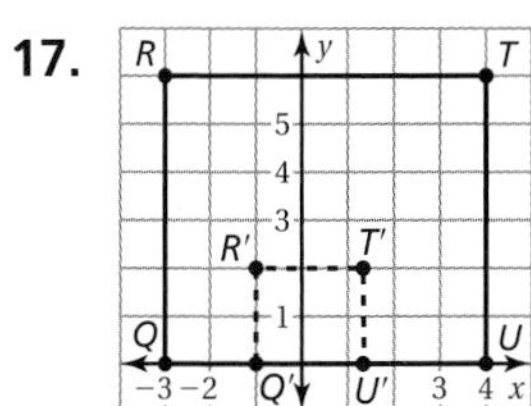

reduction

19. Each coordinate was multiplied by 2 instead of divided by 2. The coordinates should be $A'(1, 2.5)$, $B'(1, 0)$, and $C'(2, 0)$.

Section 2.7 Dilations *(continued)* *(pages 87–89)*

21. reduction; $\frac{1}{4}$

23. $A''(10, 6)$, $B''(4, 6)$, $C''(4, 2)$, $D''(10, 2)$

25. $J''(3, -3)$, $K''(12, -9)$, $L''(3, -15)$

27. *Sample answer:* Rotate 90° counterclockwise about the origin and then dilate with respect to the origin using a scale factor of 2

29. Exercise 27: yes; Exercise 28: no; Explanations will vary based on sequences chosen in Exercises 27 and 28.

31. **a.** enlargement

b. center of dilation

c. $\frac{4}{3}$

d. The shadow on the wall becomes larger. The scale factor will become larger.

33. The transformations are a dilation using a scale factor of 2 and then a translation of 4 units right and 3 units down; similar; A dilation produces a similar figure and a translation produces a congruent figure, so the final image is similar.

35. The transformations are a dilation using a scale factor of $\frac{1}{3}$ and then a reflection in the x-axis; similar; A dilation produces a similar figure and a reflection produces a congruent figure, so the final image is similar.

37. $A'(-2, 3)$, $B'(6, 3)$, $C'(12, -7)$, $D'(-2, -7)$; Methods will vary.

39. supplementary; $x = 16$

41. B

Section 3.1 Parallel Lines and Transversals *(pages 107–109)*

1. *Sample answer:*

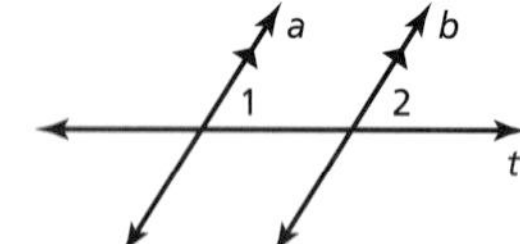

3. m and n

5. 8

7. $\angle 1 = 107°$, $\angle 2 = 73°$

9. $\angle 5 = 49°$, $\angle 6 = 131°$

11. 60°; Corresponding angles are congruent.

13. *Sample answer:* rotate 180° and translate down

15. $\angle 6 = 61°$; $\angle 6$ and the given angle are vertical angles.
$\angle 5 = 119°$ and $\angle 7 = 119°$; $\angle 5$ and $\angle 7$ are supplementary to the given angle.
$\angle 1 = 61°$; $\angle 1$ and the given angle are corresponding angles.
$\angle 3 = 61°$; $\angle 1$ and $\angle 3$ are vertical angles.
$\angle 2 = 119°$ and $\angle 4 = 119°$; $\angle 2$ and $\angle 4$ are supplementary to $\angle 1$.

17. $\angle 2 = 90°$; $\angle 2$ and the given angle are vertical angles.
$\angle 1 = 90°$ and $\angle 3 = 90°$; $\angle 1$ and $\angle 3$ are supplementary to the given angle.
$\angle 4 = 90°$; $\angle 4$ and the given angle are corresponding angles.
$\angle 6 = 90°$; $\angle 4$ and $\angle 6$ are vertical angles.
$\angle 5 = 90°$ and $\angle 7 = 90°$; $\angle 5$ and $\angle 7$ are supplementary to $\angle 4$.

19. 132°; *Sample answer:* $\angle 2$ and $\angle 4$ are alternate interior angles and $\angle 4$ and $\angle 3$ are supplementary.

21. 120°; *Sample answer:* $\angle 6$ and $\angle 8$ are alternate exterior angles.

23. 61.3°; *Sample answer:* $\angle 3$ and $\angle 1$ are alternate interior angles and $\angle 1$ and $\angle 2$ are supplementary.

25. They are all right angles because perpendicular lines form 90° angles.

27. 130

29. **a.** no; They look like they are spreading apart. **b.** Check students' work.

31. 13

33. 51

35. B

Section 3.2 Angles of Triangles

(pages 114 and 115)

1. Subtract the sum of the given measures from 180°.

3. 115°, 120°, 125°

5. 40°, 65°, 75°

7. 25°, 45°, 110°

9. 48°, 59°, 73°

11. 45

13. 140°

15. The measure of the exterior angle is equal to the sum of the measures of the two nonadjacent interior angles. The sum of all three angles is not 180°;

$$(2x - 12) = x + 30$$

$$x = 42$$

The exterior angle is $(2(42) - 12)° = 72°$.

17. 126°

19. sometimes; The sum of the angle measures must equal 180°.

21. never; If a triangle had more than one vertex with an acute exterior angle, then it would have to have more than one obtuse interior angle which is impossible.

23. $x = -4$

25. $n = -3$

Section 3.3 Angles of Polygons

(pages 123–125)

1. *Sample answer:*

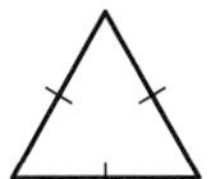

3. What is the measure of an interior angle of a regular pentagon?; 108°; 540°

5. 1260°

7. 360°

9. 1260°

Section 3.3 Angles of Polygons *(continued)* *(pages 123–125)*

11. no; The interior angle measures given add up to 535°, but the sum of the interior angle measures of a pentagon is 540°.

13. 90°, 135°, 135°, 135°, 135°, 90°

15. 140°

17. 140°

19. The sum of the interior angle measures should have been divided by the number of angles, 20. $3240° \div 20 = 162°$; The measure of each interior angle is 162°.

21. 24 sides

23. 75°, 93°, 85°, 107°

25. 60°; The sum of the interior angle measures of a hexagon is 720°. Because it is regular, each angle has the same measure. So, each interior angle is $720° \div 6 = 120°$ and each exterior angle is 60°.

27. 120°, 120°, 120°

29. interior: 135°; exterior: 45°

31. 120°

33. **a.** *Sample answer:*

b. *Sample answer:*
square, regular hexagon

c. *Sample answer:*

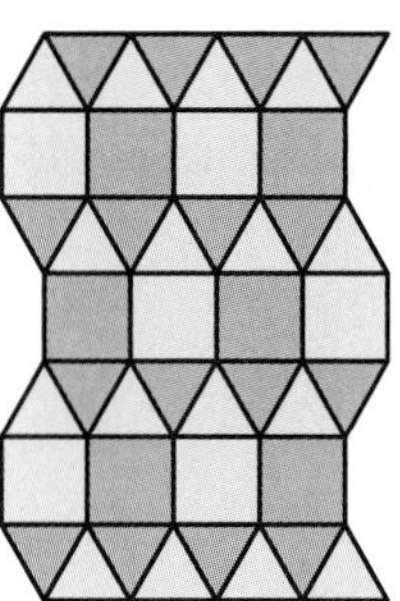

d. *Answer should include, but is not limited to:* a discussion of the interior and exterior angles of the polygons in the tessellation and how they combine to add up to 360° where the vertices meet.

35. 2

37. 6

Section 3.4 Using Similar Triangles *(pages 130 and 131)*

1. Write a proportion that uses the missing measurement because the ratios of corresponding side lengths are equal.

3. *Sample answer:* Two of the angles are congruent, so they have the same sum. When you subtract this from 180°, you will get the same third angle.

5. Student should draw a triangle with the same angle measures as the ones given in the textbook.

If the student's triangle is larger than the one given, then the ratio of the corresponding side lengths, $\frac{\text{student's triangle length}}{\text{book's triangle length}}$, should be greater than 1. If the student's triangle is smaller than the one given, then the ratio of the corresponding side lengths, $\frac{\text{student's triangle length}}{\text{book's triangle length}}$, should be less than 1.

7. no; The triangles do not have two pairs of congruent angles.

9. yes; The triangles have the same angle measures, 81°, 51°, and 48°.

11. yes; The triangles have two pairs of congruent angles.

13. Think of the different ways that you can show that two triangles are similar.

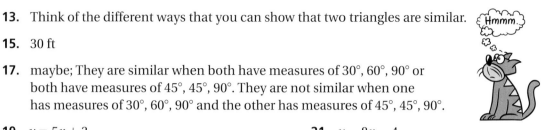

15. 30 ft

17. maybe; They are similar when both have measures of 30°, 60°, 90° or both have measures of 45°, 45°, 90°. They are not similar when one has measures of 30°, 60°, 90° and the other has measures of 45°, 45°, 90°.

19. 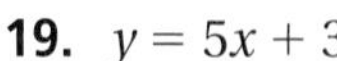 $y = 5x + 3$

21. $y = 8x - 4$

Section 4.1 Graphing Linear Equations
(pages 146 and 147)

1. a line

3.  *Sample answer:*

x	0	1
$y = 3x - 1$	-1	2

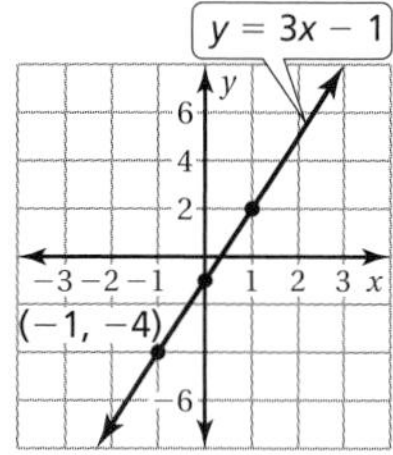

5. 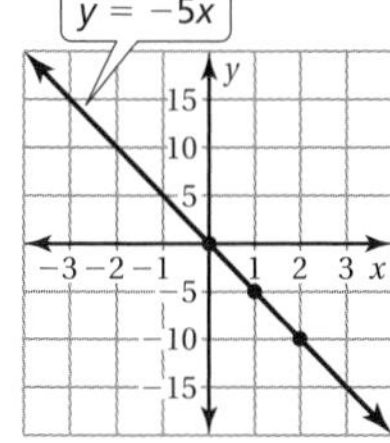

7. 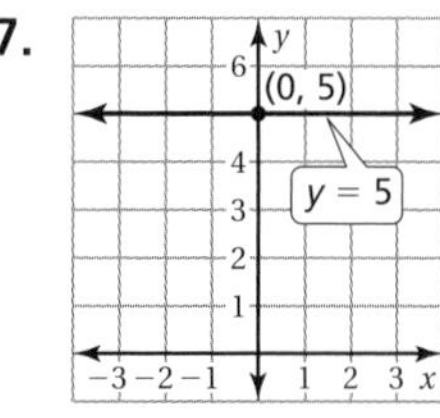

9. 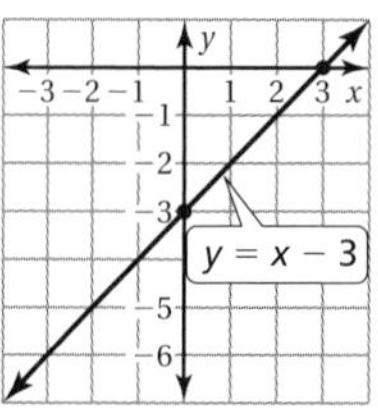

11. 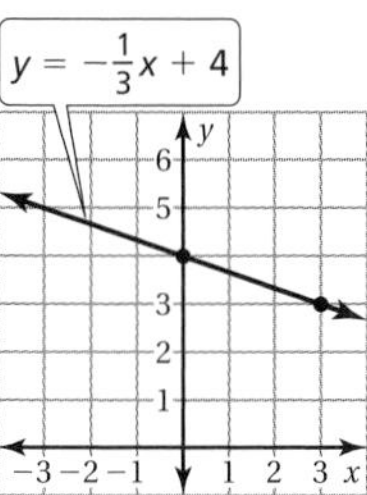

13. 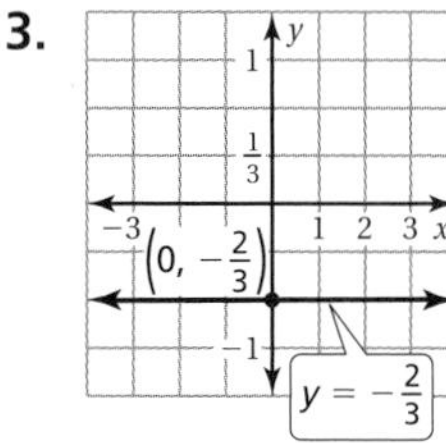

15. 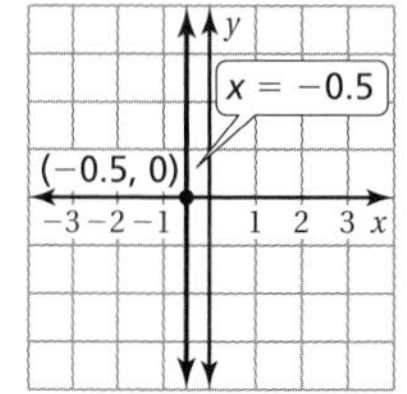

17. The equation $x = 4$ is graphed, not $y = 4$.

19. a. 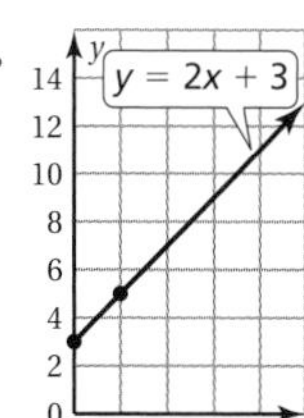

b. about \$5

c. \$5.25

21. $y = -\frac{5}{2}x + 2$

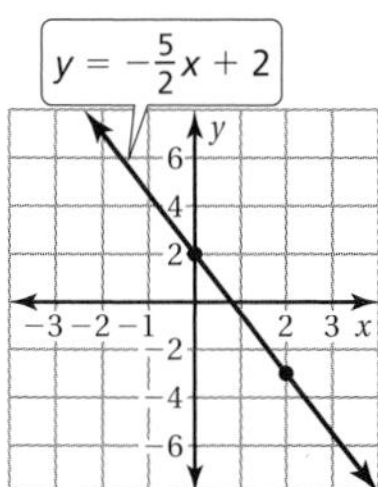

23. $y = -2x + 3$

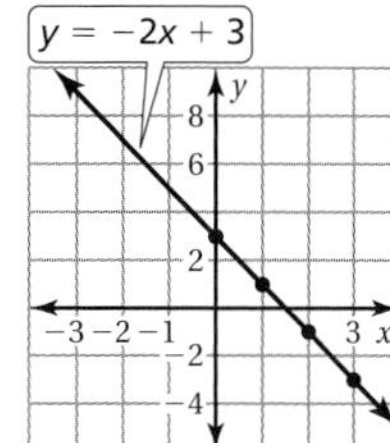

Selected Answers

Section 4.1 Graphing Linear Equations *(continued)*

(pages 146 and 147)

25. **a.** *Sample answer:*

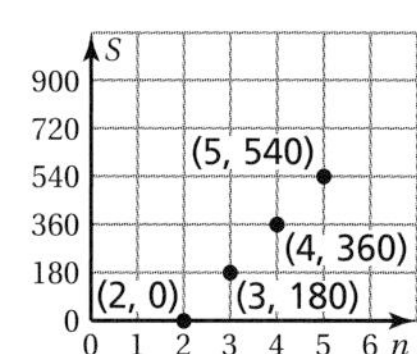

Yes; The graph of the equation is a line.

b. No, $n = 3.5$ does not make sense because a polygon cannot have half a side.

27. Begin this exercise by listing all of the given information.

29. $(-6, 6)$

31. $(-4, -3)$

Section 4.2 Slope of a Line

(pages 153–155)

1. **a.** B and C

b. A

c. no; None of the lines are vertical.

3. The line is horizontal.

5.

The lines are parallel.

7. $\frac{3}{4}$

9. $-\frac{3}{5}$

11. 0

13. 0

15. undefined

17. $-\frac{11}{6}$

19. The denominator should be $2 - 4$.

$m = -1$

21. 4

23. $-\frac{3}{4}$

25. $\frac{1}{3}$

27. $k = 11$

29. $k = -5$

31. **a.** $\frac{3}{40}$

b. The cost increases by \$3 for every 40 miles you drive, or the cost increases by \$0.075 for every mile you drive.

33. yes; The slopes are the same between the points.

35. When you switch the coordinates, the differences in the numerator and denominator are the opposite of the numbers when using the slope formula. You still get the same slope.

37. $b = 25$

39. $x = 7.5$

Extension 4.2 Slopes of Parallel and Perpendicular Lines
(pages 156 and 157)

1. blue and red; They both have a slope of -3.

3. yes; Both lines are horizontal and have a slope of 0.

5. yes; Both lines are vertical and have an undefined slope.

7. blue and green; The blue line has a slope of 6. The green line has a slope of $-\frac{1}{6}$. The product of their slopes is $6 \cdot \left(-\frac{1}{6}\right) = -1$.

9. yes; The line $x = -2$ is vertical. The line $y = 8$ is horizontal. A vertical line is perpendicular to a horizontal line.

11. yes; The line $x = 0$ is vertical. The line $y = 0$ is horizontal. A vertical line is perpendicular to a horizontal line.

Section 4.3 Graphing Proportional Relationships
(pages 162 and 163)

1. (0, 0)

3. no; *Sample answer:* The graph of the equation does not pass through the origin.

5. yes; $y = \frac{1}{3}x$; *Sample answer:* The rate of change in the table is constant.

7. 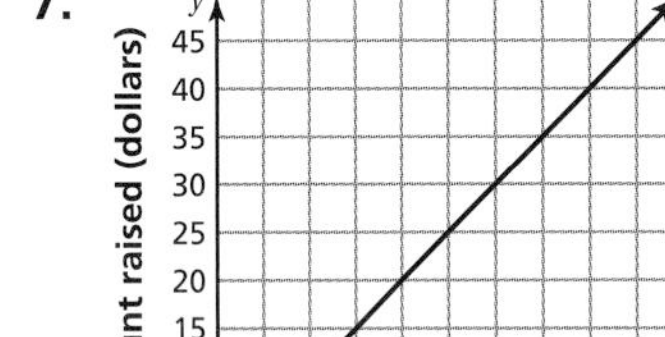

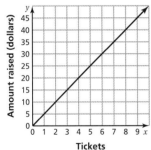

Each ticket costs $5.

9. **a.** the car; *Sample answer:* The equation for the car is $y = 25x$. Because 25 is greater than 18, the car gets better gas mileage.

   b. 56 miles

11. Consider the direct variation equation and that the graph passes through the origin.

Selected Answers

Section 4.3 Graphing Proportional Relationships *(continued)* *(pages 162 and 163)*

13. a. yes; The equation is $d = 6t$, which represents a proportional relationship.

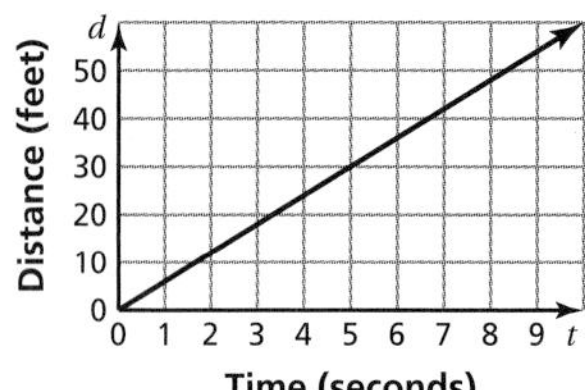

b. yes; The equation is $d = 50r$, which represents a proportional relationship.

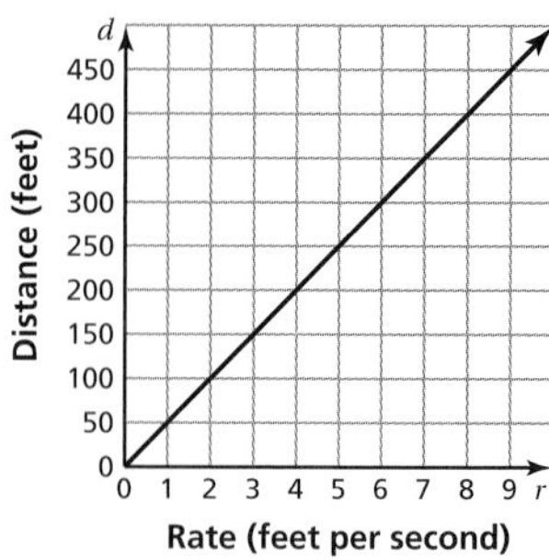

c. no; The equation is $t = \frac{300}{r}$, which does not represent a proportional relationship.

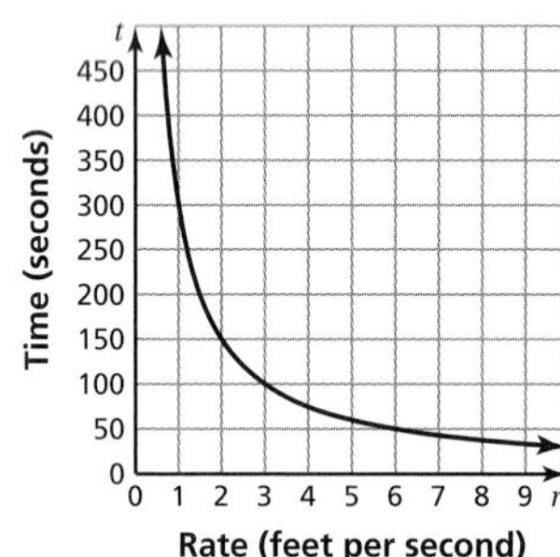

d. part c; It is called inverse variation because when the rate increases, the time decreases, and when the rate decreases, the time increases.

15.

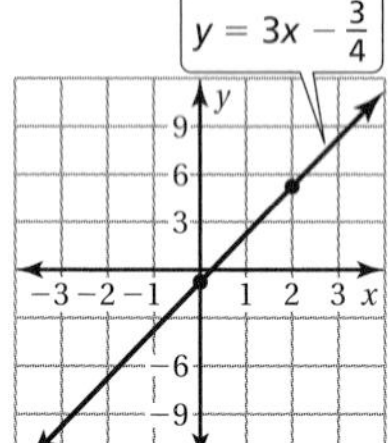

17. B

Section 4.4 Graphing Linear Equations in Slope-Intercept Form *(pages 170 and 171)*

1. Find the x-coordinate of the point where the graph crosses the x-axis.

3. *Sample answer:* The amount of gasoline y (in gallons) left in your tank after you travel x miles is $y = -\frac{1}{20}x + 20$. The slope of $-\frac{1}{20}$ means the car uses 1 gallon of gas for every 20 miles driven. The y-intercept of 20 means there is originally 20 gallons of gas in the tank.

5. A; slope: $\frac{1}{3}$; y-intercept: -2

7. slope: 4; y-intercept: -5

9. slope: $-\frac{4}{5}$; y-intercept: -2

11. slope: $\frac{4}{3}$; y-intercept: -1

13. slope: -2; y-intercept: 3.5

15. slope: 1.5; y-intercept: 11

17. a.

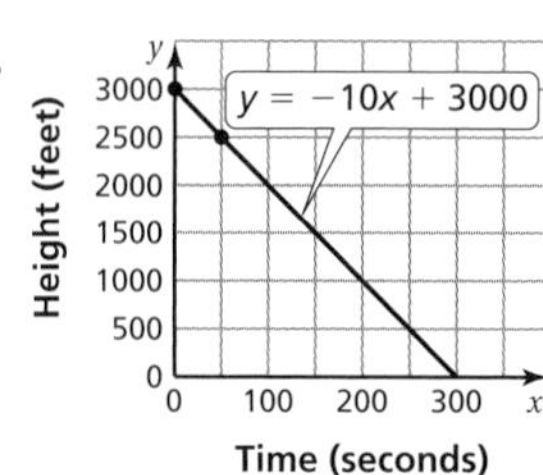

b. The x-intercept of 300 means the skydiver lands on the ground after 300 seconds. The slope of -10 means that the skydiver falls to the ground at a rate of 10 feet per second.

19.

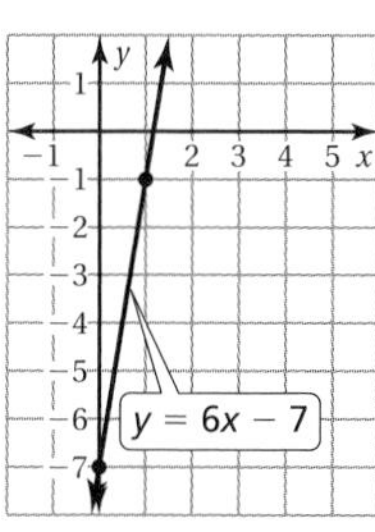

x-intercept: $\frac{7}{6}$

21.

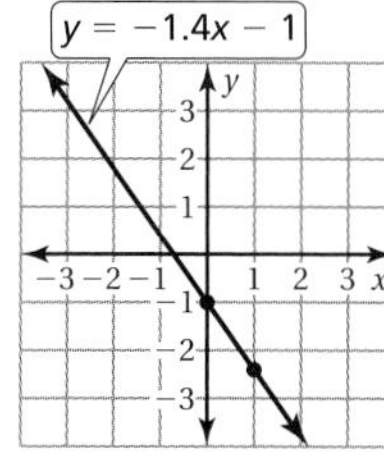

x-intercept: $-\frac{5}{7}$

23.

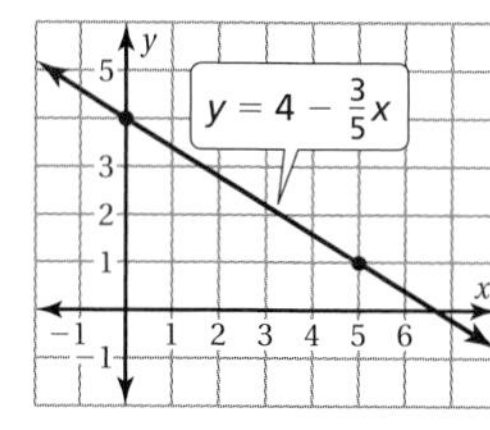

x-intercept: $\frac{20}{3}$

25. a. $y = 2x + 4$ and $y = 2x - 3$ are parallel because the slope of each line is 2; $y = -3x - 2$ and $y = -3x + 5$ are parallel because the slope of each line is -3.

b. $y = 2x + 4$ and $y = -\frac{1}{2}x + 2$ are perpendicular because the product of their slopes is -1;

$y = 2x - 3$ and $y = -\frac{1}{2}x + 2$ are perpendicular because the product of their slopes is -1;

$y = -\frac{1}{3}x - 1$ and $y = 3x + 3$ are perpendicular because the product of their slopes is -1.

27. $y = 2x + 3$

29. $y = \frac{2}{3}x - 2$

31. B

Section 4.5 Graphing Linear Equations in Standard Form

(pages 176 and 177)

1. no; The equation is in slope-intercept form.

3. x = pounds of peaches

y = pounds of apples

$y = -\frac{4}{3}x + 10$

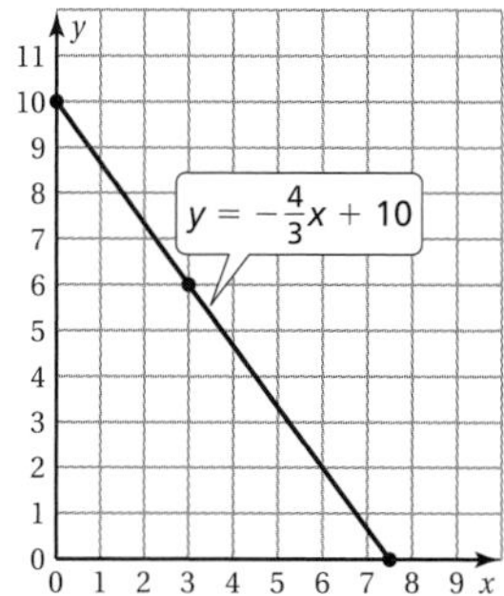

5. $y = -2x + 17$

7. $y = \frac{1}{2}x + 10$

9.

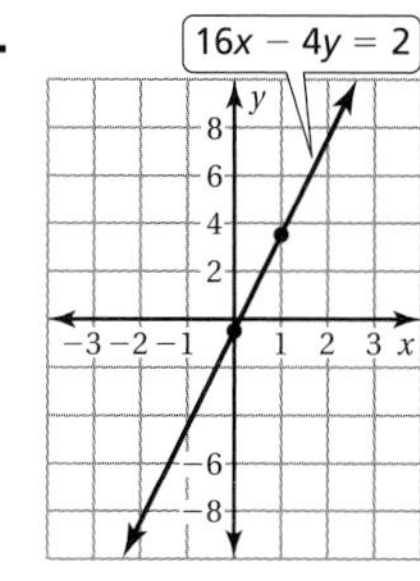

11. B

13. C

Section 4.5 Graphing Linear Equations in Standard Form *(continued)* *(pages 176 and 177)*

15. a. 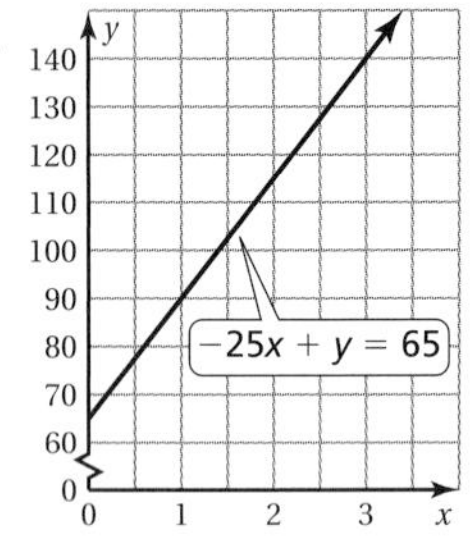

b. \$390

17. 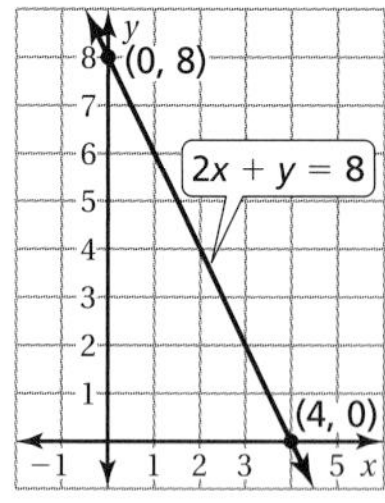

19. x-intercept: 9

y-intercept: 7

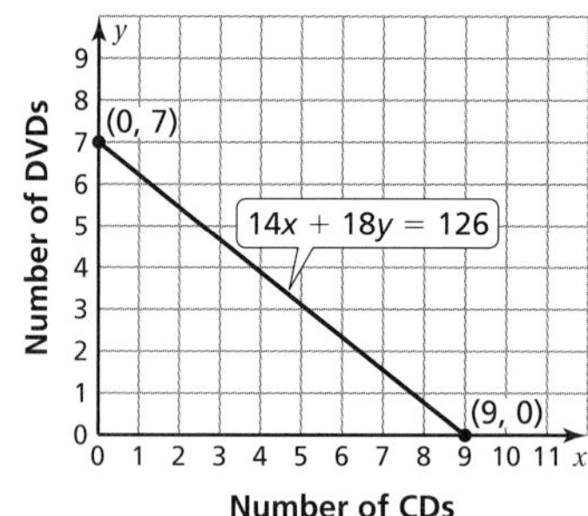

21. a. $9.45x + 7.65y = 160.65$

b. 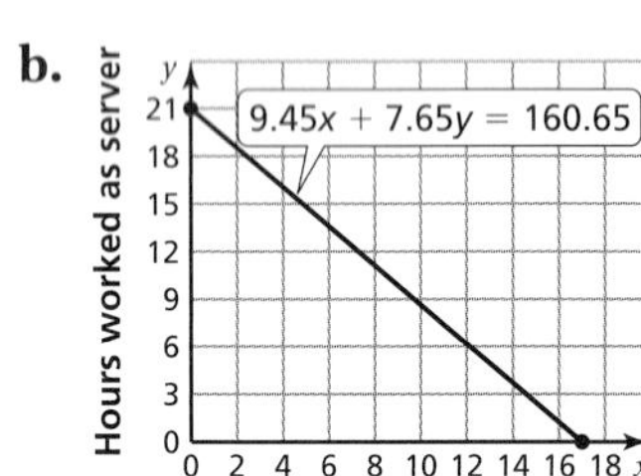

23. a. $y = 40x + 70$

b. x-intercept: $-\frac{7}{4}$; no;

You cannot have a negative time.

c. 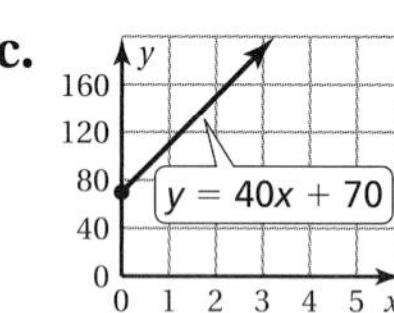

25. $\frac{1}{2}$

Section 4.6 Writing Equations in Slope-Intercept Form *(pages 182 and 183)*

1. *Sample answer:* Find the ratio of the rise to the run between the intercepts.

3. $y = 3x + 2$; $y = 3x - 10$; $y = 5$; $y = -1$

5. $y = x + 4$

7. $y = \frac{1}{4}x + 1$

9. $y = \frac{1}{3}x - 3$

11. The x-intercept was used instead of the y-intercept. $y = \frac{1}{2}x - 2$

13. $y = 5$

15. $y = -2$

17. a–b. 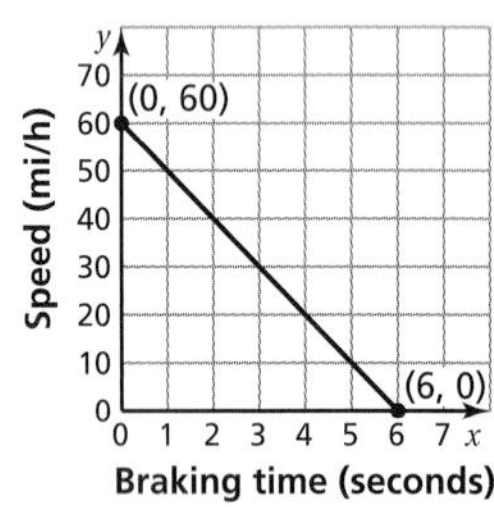

(0, 60) represents the speed of the automobile before braking. (6, 0) represents the amount of time it takes to stop. The line represents the speed y of the automobile after x seconds of braking.

c. $y = -10x + 60$

19. Be sure to check that your rate of growth will not lead to a 0-year-old tree with a negative height.

21 and 23.

Section 4.7 Writing Equations in Point-Slope Form *(pages 188 and 189)*

1. $m = -2$; $(-1, 3)$

3. $y - 0 = \frac{1}{2}(x + 2)$

5. $y + 1 = -3(x - 3)$

7. $y - 8 = \frac{3}{4}(x - 4)$

9. $y + 5 = -\frac{1}{7}(x - 7)$

11. $y + 4 = -2(x + 1)$

13. $y = 2x$

15. $y = \frac{1}{4}x$

17. $y = x + 1$

19. a. $V = -4000x + 30,000$

    b. $30,000

21. The rate of change is 0.25 degree per chirp.

23. a. $y = 14x - 108.5$

    b. 4 meters

25. 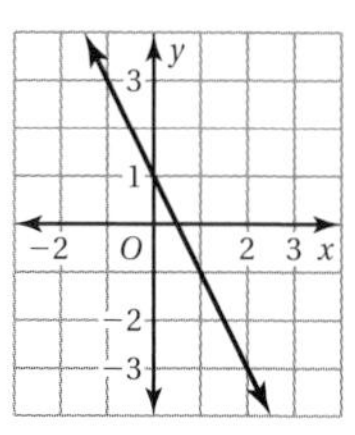

27. D

Section 5.1 Solving Systems of Linear Equations by Graphing *(pages 206 and 207)*

1. yes; The equations are linear and in the same variables.

3. Check whether (3, 4) is a solution of each equation.

5. (4, 176)

7. B; (6, 7)

9. C; (3, −1)

11. (−5, 1)

13. (12, 15)

15. (8, 1)

17. (5, 1.5)

19. (−6, 2)

21. no; Two lines cannot intersect in exactly two points.

23. Make a table to compare your distance to your friend's distance.

25. $c = 8$

27. $x = 11$

Section 5.2 Solving Systems of Linear Equations by Substitution *(pages 212 and 213)*

1. **Step 1:** Solve one of the equations for one of the variables.

   Step 2: Substitute the expression from Step 1 into the other equation and solve.

   Step 3: Substitute the value from Step 2 into one of the original equations and solve.

3. sometimes; A solution obtained by graphing may not be exact.

5. *Sample answer:* $x + 2y = 6$

   $x - y = 3$

7. $4x - y = 3$; The coefficient of y is -1.

9. $2x + 10y = 14$; Dividing by 2 to solve for x yields integers.

11. $(6, 17)$

13. $(4, 1)$

15. $\left(\frac{1}{4}, 6\right)$

17. **a.** $x = 2y$

    $64x + 132y = 1040$

    b. adult tickets: \$8; student tickets: \$4

19. $(-2, 4)$

21. The expression for y was substituted back into the same equation; solution: $(2, 1)$

23. 30 cats, 35 dogs

25. Make a diagram to help visualize the problem.

27. $2x - 5y = -8$

29. B

Section 5.3 Solving Systems of Linear Equations by Elimination *(pages 221–223)*

1. **Step 1:** Multiply, if necessary, one or both equations by a constant so at least one pair of like terms has the same or opposite coefficients.

   Step 2: Add or subtract the equations to eliminate one of the variables.

   Step 3: Solve the resulting equation for the remaining variable.

   Step 4: Substitute the value from Step 3 into one of the original equations and solve.

3. $2x + 3y = 11$

   $3x - 2y = 10$;

   You have to use multiplication to solve the system by elimination.

5. $(6, 2)$

7. $(2, 1)$

9. $(1, -3)$

11. $(3, 2)$

13. The student added y-terms, but subtracted x-terms and constants; solution $(1, 2)$

15. **a.** $2x + y = 10$

    $2x + 3y = 22$

    b. 6 minutes

17. $(5, -1)$

19. $(-2, -1)$

21. $(4, 3)$

23. **a.** ± 4

    b. ± 7

25. yes; The lines are perpendicular.

27. **a.** $23x + 10y = 86$

$28x + 5y = 76$

b. Multiple choice: 2 points each Short response: 4 points each

29. \$95

31. 5 grams of 90% gold alloy, 3 grams of 50% gold alloy

33. $(-1, 2, 1)$

35. yes

37. D

Section 5.4 Solving Special Systems of Linear Equations
(pages 228 and 229)

1. The graph of a system with no solution is two parallel lines, and the graph of a system with infinitely many solutions is one line.

3. infinitely many solutions; all points on the line $y = 4x + \frac{1}{3}$

5. no solution; The lines have the same slope and different y-intercepts.

7. infinitely many solutions; The lines are identical.

9. $(-1, -2)$

11. infinitely many solutions; all points on the line $y = -\frac{1}{6}x + 5$

13. $(-2.4, -3.5)$

15. no; because they are running at the same speed and your pig had a head start

17. When the slopes are different, there is one solution. When the slopes are the same, there is no solution if the y-intercepts are different and infinitely many solutions if the y-intercepts are the same.

19. $y = 0.99x + 10$

$y = 0.99x$

no; Because you paid \$10 before buying the same number of songs at the same price, you spend \$10 more.

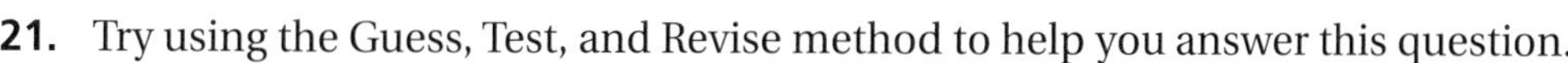

21. Try using the Guess, Test, and Revise method to help you answer this question.

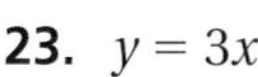

23. $y = 3x$

25. $y = -\frac{1}{2}x + 2$

Extension 5.4 Solving Linear Equations by Graphing
(pages 230 and 231)

1. $x = \frac{1}{2}$

3. no solution

5. $x = 2$

7. *Sample answer:* $6x - 3 = 6x$; Subtract 3 from the right side.

9. $x = \frac{21}{2}$

11. 6 mo

Section 6.1 Relations and Functions
(pages 246 and 247)

1. the first number; the second number

3. As each input increases by 1, the output increases by 4.

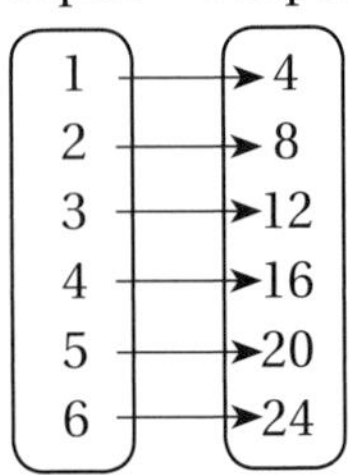

5. As each input increases by 1, the output increases by 5.

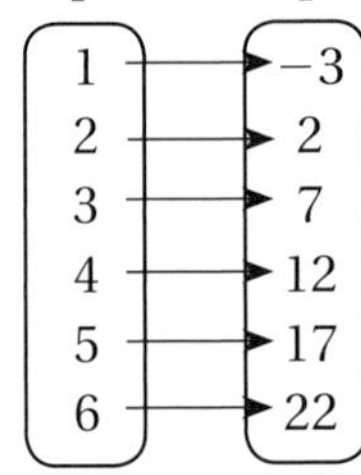

7. (1, 8), (3, 8), (3, 4), (5, 6), (7, 2)

9. no

11. yes

13. 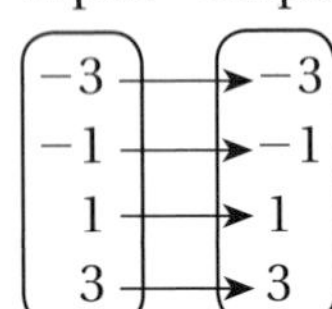

As each input increases by 2, the output increases by 2.

15. 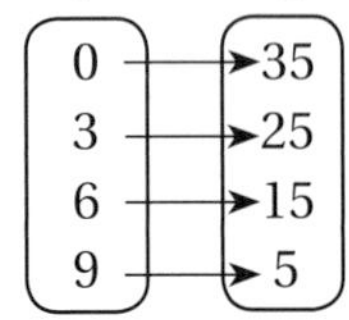

As each input increases by 3, the output decreases by 10.

17. **a.** 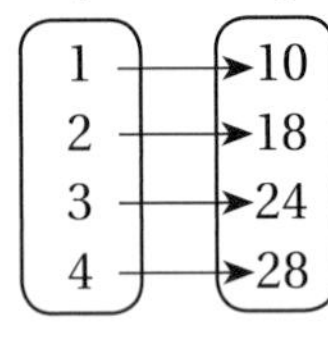

b. yes; Each input has exactly one output.

c. The pattern is that for each input increase of 1, the output increases by $2 less than the previous increase. For each additional movie you buy, your cost per movie decreases by $1.

19. y-axis

21. x-axis

Section 6.2 Representations of Functions
(pages 253–255)

1. input variable: x; output variable: y

3. What output is twice the sum of the input 3 and 4?; $2(3 + 4) = 14$; $2(3) + 4 = 10$

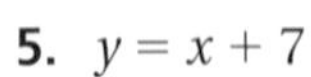

5. $y = x + 7$

7. $y = \frac{1}{2}x$

9. $y = x - 3$

11. $y = 6x$

13. 8

15. −17

17. 54

19. 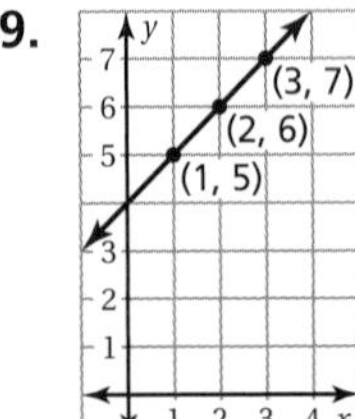

21. 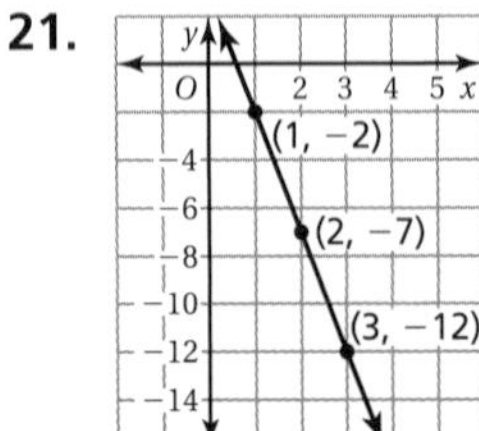

23. 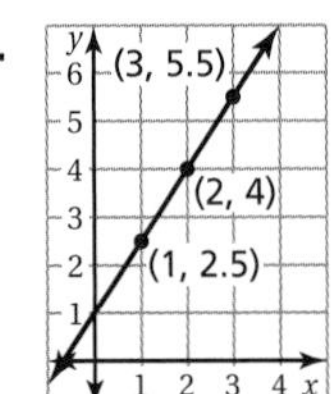

25. The order of the x- and y-coordinates is reversed in each coordinate pair.

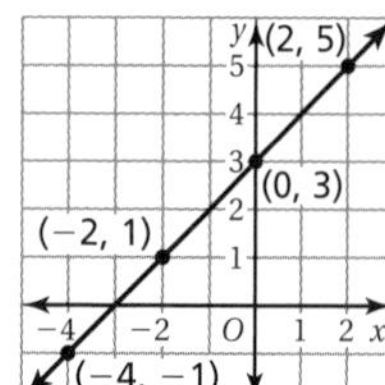

27. B

29. A

31. -4

33. **a.** $P = 3.50b - 84$

b. independent variable: b; dependent variable: P; The profit depends on the number of bracelets sold.

c. 24 bracelets

35. **a.** $G = 35 + 10h$

b. $S = 25h$

c. Snake Tours; For 2 hours, Gator Tours cost \$55 and Snake Tours cost \$50.

37. *Sample answer:*

Side Length	1	2	3	4	5
Perimeter	4	8	12	16	20

Side Length	1	2	3	4	5
Area	1	4	9	16	25

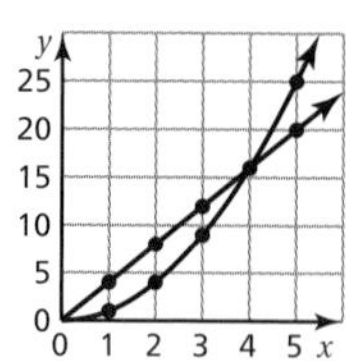

Sample answer: The perimeter function appears to form a line, and the area function appears to form a curve. When the side length is less than 4, the perimeter function is greater. When the side length is greater than 4, the area function is greater. When the side length is 4, the two functions are equal.

39. 1

41. $\frac{1}{3}$

Section 6.3 Linear Functions
(pages 261–263)

1. yes; The graph of $y = mx$ is a nonvertical line, so it is a linear function.

3. $y = \pi x$; x is the diameter; y is the circumference.

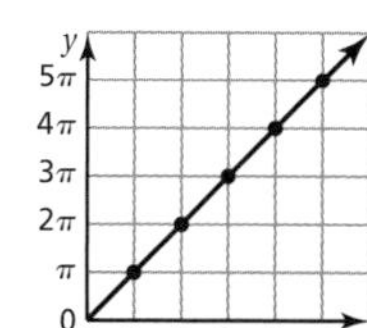

5. $y = \frac{4}{3}x + 2$

7. $y = 3$

9. $y = -\frac{1}{4}x$

11. **a.** independent variable: x; dependent variable: y

b. $y = 3x$; It costs \$3 to rent one movie.

c.

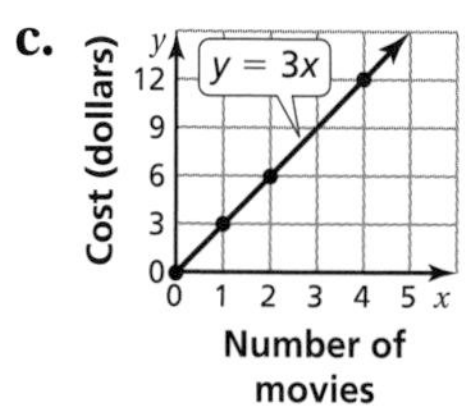

d. \$9

Section 6.3 Linear Functions *(continued)*

(pages 261–263)

13. **a.** $y = -0.2x + 1$

b. The slope indicates that the power decreases by 20% per hour. The x-intercept indicates that the battery lasts 5 hours. The y-intercept indicates that the battery power is at 100% when you turn on the laptop.

c. 1.25 hours

15. **a.** hiking

b. 67.5 calories

17. yes; A horizontal line is a nonvertical line.

19. **a.**

Temperature (°F), t	94	95	96	97	98
Heat Index (°F), H	122	126	130	134	138

b. independent variable: t; dependent variable: H

c. $H = 4t - 254$

d. 146°F

21. $w = 1.5$

23. C

Section 6.4 Comparing Linear and Nonlinear Functions

(pages 270 and 271)

1. A linear function has a constant rate of change. A nonlinear function does not have a constant rate of change.

3. 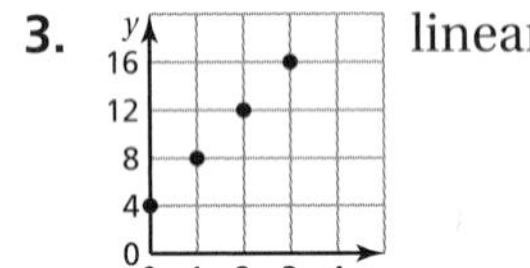

linear

5. nonlinear

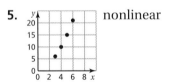

7. linear; The graph is a line.

9. linear; As x increases by 6, y increases by 4.

11. nonlinear; As x increases by 1, V increases by different amounts.

13. linear; You can rewrite the equation in slope-intercept form.

15. nonlinear; As x decreases by 65, y increases by different amounts.

17. **a.** nonlinear; When graphing the points, they do not lie on a line.

b. Tree B; After ten years, the height of Tree A is 20 feet and the height of Tree B is at least 23 feet.

19. **a.** 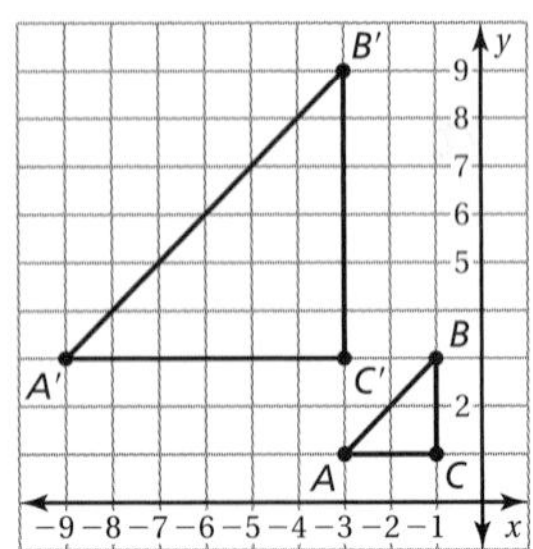

enlargement

21. C

Section 6.5 Analyzing and Sketching Graphs
(pages 276 and 277)

1. F
3. A
5. D
7. The volume of the balloon increases at a constant rate, then stays constant, then increases at a constant rate, then stays constant, and then increases at a constant rate.
9. Horsepower increases at an increasing rate and then increases at a decreasing rate.
11. The hair length increases at a constant rate, then decreases instantly, then increases at a constant rate, then decreases instantly, and then increases at a constant rate.
13. **a.** The usage decreases at an increasing rate.

    b. The usage decreases at a decreasing rate.
15. 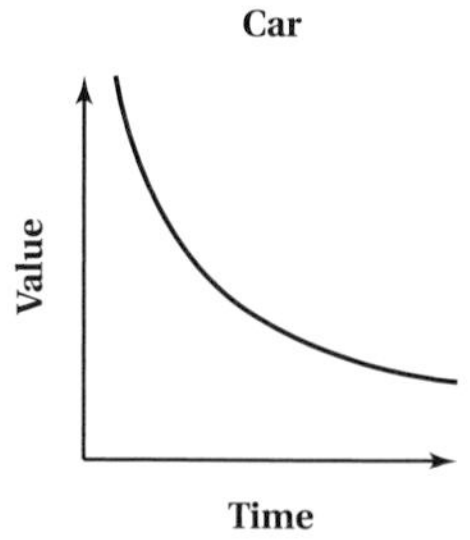

17. 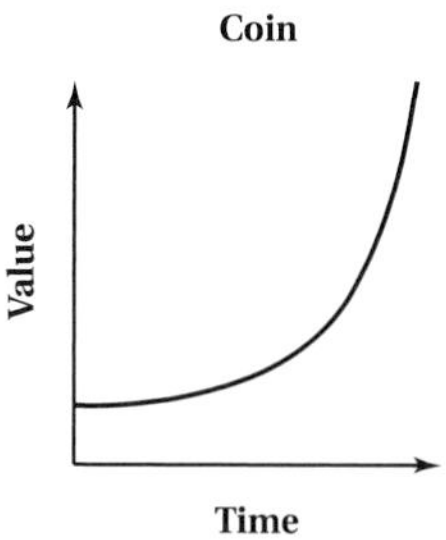

19. Think about the real-life meanings of the words "surplus" and "shortage."

21. (2, −1)
23. C

Section 7.1 Finding Square Roots
(pages 292 and 293)

1. no; There is no integer whose square is 26.
3. $\sqrt{256}$ represents the positive square root because there is not a − or a ± in front.
5. $s = 1.3$ km
7. 3 and −3
9. 2 and −2
11. 25
13. $\frac{1}{31}$ and $-\frac{1}{31}$
15. 2.2 and −2.2
17. −19
19. The positive and negative square roots should have been given.

    $\pm\sqrt{\frac{1}{4}} = \frac{1}{2}$ and $-\frac{1}{2}$
21. −116
23. 9
25. 25
27. 40
29. because a negative radius does not make sense
31. =
33. 9 ft
35. 8 m/sec
37. 2.5 ft
39. $y = 3x - 2$
41. $y = \frac{3}{5}x + 1$

Section 7.2 Finding Cube Roots
(pages 298 and 299)

1. no; There is no integer that equals 25 when cubed.

3. 50 in. **5.** 0.4 m **7.** -5

9. 12 **11.** $\frac{7}{4}$ **13.** $3\frac{5}{8}$

15. $\frac{7}{12}$ **17.** 74 **19.** -276

21. 30 cm **23.** $>$ **25.** $<$

27. $-1, 0, 1$

29. The side length of the square base is 18 inches and the height of the pyramid is 9 inches.

31. $x = 3$ **33.** $x = 4$

35. 289 **37.** 49

Section 7.3 The Pythagorean Theorem
(pages 304 and 305)

1. The hypotenuse is the longest side and the legs are the other two sides.

3. 29 km **5.** 9 in. **7.** 24 cm

9. The length of the hypotenuse was substituted for the wrong variable.

$$\begin{aligned} a^2 + b^2 &= c^2 \\ 7^2 + b^2 &= 25^2 \\ 49 + b^2 &= 625 \\ b^2 &= 576 \\ b &= 24 \end{aligned}$$

11. 16 cm **13.** Use a right triangle to find the distance.

15. *Sample answer:* length = 20 ft, width = 48 ft, height = 10 ft; $BC = 52$ ft, $AB = \sqrt{2804}$ ft

17. a. *Sample answer:* **b.** 45 ft

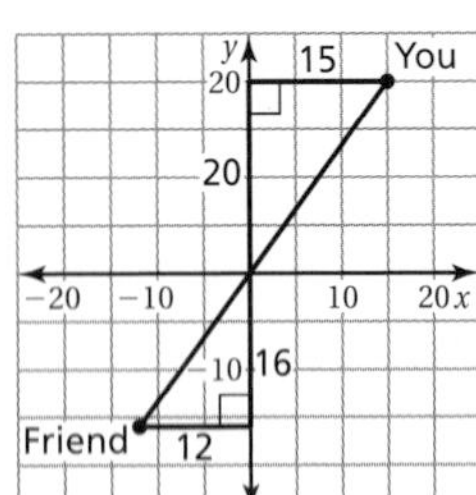

19. 6 and -6 **21.** 13 **23.** C

Section 7.4 Approximating Square Roots
(pages 313–315)

1. A rational number can be written as the ratio of two integers. An irrational number cannot be written as the ratio of two integers.

3. all rational and irrational numbers; *Sample answer:* $-2, \frac{1}{8}, \sqrt{7}$

5. yes
7. no
9. whole, integer, rational
11. irrational
13. rational
15. irrational
17. 144 is a perfect square. So, $\sqrt{144}$ is rational.
19. **a.** If the last digit is 0, it is a whole number. Otherwise, it is a natural number.

    b. irrational number **c.** irrational number

21. **a.** 26 **b.** 26.2
23. **a.** -10 **b.** -10.2
25. **a.** -13 **b.** -12.9
27. $\sqrt{15}$; $\sqrt{15}$ is positive and -3.5 is negative.
29. $\frac{2}{3}$; $\frac{2}{3}$ is to the right of $\sqrt{\frac{16}{81}}$.
31. $-\sqrt{182}$; $-\sqrt{182}$ is to the right of $-\sqrt{192}$.
33. true
35. 8.1 ft
37. 8.5 ft
39. 20.6 in.
41. Create a table of integers whose cubes are close to the radicand. Determine which two integers the cube root is between. Then create another table of numbers between those two integers whose cubes are close to the radicand. Determine which cube is closest to the radicand; 2.4
43. *Sample answer:* $a = 82$, $b = 97$
45. 1.1
47. 30.1 m/sec
49. Falling objects do not fall at a linear rate. Their speed increases with each second they are falling.

51. 40 m
53. 9 cm

Extension 7.4 Repeating Decimals
(pages 316 and 317)

1. $\frac{1}{9}$
3. $-1\frac{2}{9}$
5. Because the solution does not change when adding/subtracting two equivalent equations; Multiply by 10 so that when you subtract the original equation, the repeating part is removed.
7. $-\frac{13}{30}$
9. $\frac{3}{11}$
11. Pattern: Digits that repeat are in the numerator and 99 is in the denominator; Use 9 as the integer part, 4 as the numerator, and 99 as the denominator of the fractional part.

Selected Answers

Section 7.5 Using the Pythagorean Theorem
(pages 322 and 323)

1. the Pythagorean Theorem and the distance formula

3. If a^2 is odd, then a is an odd number; true when a is an integer; A product of two integers is odd only when each integer is odd.

5. yes 7. no 9. yes 11. $\sqrt{52}$ 13. $\sqrt{29}$ 15. $\sqrt{85}$

17. The squared quantities under the radical should be added not subtracted; $\sqrt{136}$

19. yes 21. yes

23. no; The measures of the side lengths are $\sqrt{5000}$, $\sqrt{3700}$, and $\sqrt{8500}$ and $(\sqrt{5000})^2 + (\sqrt{3700})^2 \neq (\sqrt{8500})^2$.

25. Notice that the picture is not drawn to scale. Use right triangles.

27. mean: 13; median: 12.5; mode: 12

29. mean: 58; median: 59; mode: 59

Section 8.1 Volumes of Cylinders
(pages 338 and 339)

1. How much does it take to cover the cylinder?; $170\pi \approx 534.1$ cm^2; $300\pi \approx 942.5$ cm^3

3. $486\pi \approx 1526.8$ ft^3 5. $245\pi \approx 769.7$ ft^3

7. $90\pi \approx 282.7$ mm^3 9. $252\pi \approx 791.7$ in.^{3}

11. $256\pi \approx 804.2$ cm^3 13. $\dfrac{125}{8\pi} \approx 5$ ft

15. 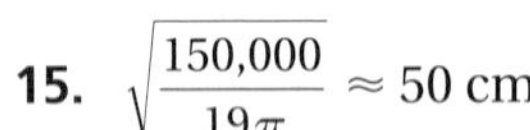 $\sqrt{\dfrac{150{,}000}{19\pi}} \approx 50$ cm

17. Divide the volume of one round bale by the volume of one square bale.

19. $8325 - 729\pi \approx 6035$ m^3 21. yes

23. no

Section 8.2 Volumes of Cones
(pages 344 and 345)

1. The height of a cone is the perpendicular distance from the base to the vertex.

3. Divide by 3. 5. $9\pi \approx 28.3$ m^3

7. $\dfrac{2\pi}{3} \approx 2.1$ ft^3 9. $\dfrac{147\pi}{4} \approx 115.5$ yd^3 11. $\dfrac{125\pi}{6} \approx 65.4$ in.^{3}

13. The diameter was used instead of the radius;

$$V = \frac{1}{3}(\pi)(1)^2(3) = \pi \text{ m}^3$$

15. 1.5 ft

17. $2\sqrt{\dfrac{10.8}{4.2\pi}} \approx 1.8$ in.

19. 24.1 min

21. $3y$

23. $A'(-1, 1), B'(-3, 4), C'(-1, 4)$

25. D

Section 8.3 Volumes of Spheres
(pages 352 and 353)

1. A hemisphere is one-half of a sphere.

3. $\dfrac{500\pi}{3} \approx 523.6 \text{ in.}^3$

5. $972\pi \approx 3053.6 \text{ mm}^3$

7. $36\pi \approx 113.1 \text{ cm}^3$

9. 9 mm

11. 4.5 ft

13. 2.5 in.

15. $256\pi + 128\pi = 384\pi \approx 1206.4 \text{ ft}^3$

17. $r = \dfrac{3}{4}h$

19. 5400 in.^2; $27{,}000 \text{ in.}^3$

21. enlargement; 2

23. A

Section 8.4 Surface Areas and Volumes of Similar Solids
(pages 359–361)

1. Similar solids are solids of the same type that have proportional corresponding linear measures.

3. a. $\dfrac{9}{4}$; because $\left(\dfrac{3}{2}\right)^2 = \dfrac{9}{4}$

   b. $\dfrac{27}{8}$; because $\left(\dfrac{3}{2}\right)^3 = \dfrac{27}{8}$

5. no

7. no

9. $b = 18$ m; $c = 19.5$ m; $h = 9$ m

11. 1012.5 in.^2

13. $13{,}564.8 \text{ ft}^3$

15. 673.75 cm^2

17. a. 9483 pounds; The ratio of the height of the original statue to the height of the small statue is 8.4 : 1. So, the ratio of the weights, or volumes is $\left(\dfrac{8.4}{1}\right)^3$.

    b. 221,184 lb

19. a. yes; Because all circles are similar, the slant height and the circumference of the base of the cones are proportional.

    b. no; because the ratio of the volumes of similar solids is equal to the cube of the ratio of their corresponding linear measures

21.

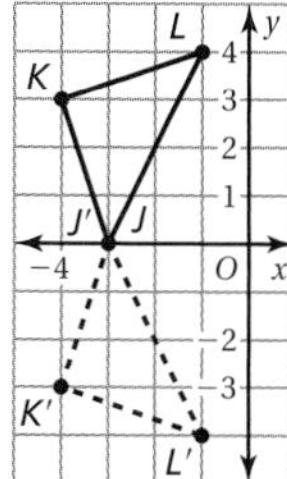

$J'(-3, 0), K'(-4, -3), L'(-1, -4)$

Section 9.1 Scatter Plots

(pages 376 and 377)

1. They must be ordered pairs so there are equal amounts of x- and y-values.

3. no relationship; A student's shoe size is not related to his or her IQ.

5. nonlinear relationship; On each successive bounce, the ball rebounds to a height less than its previous bounce.

7. **a.** (22, 152), (40, 94), (28, 134), (35, 110), (46, 81)

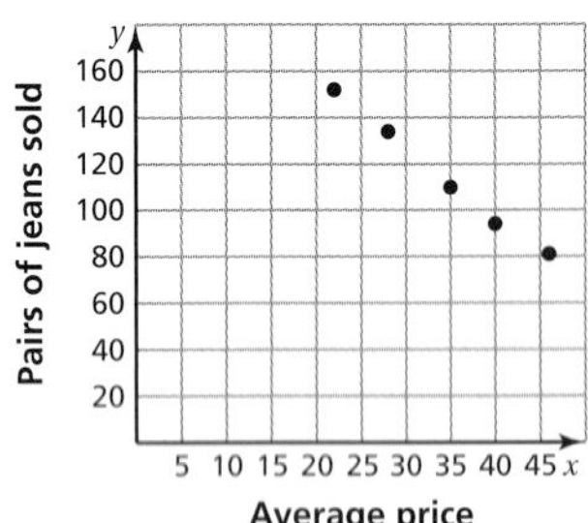

   b. As the average price of jeans increases, the number of pairs of jeans sold decreases.

9. **a.** 3.5 h **b.** $85

   c. There is a positive linear relationship between hours worked and earnings.

11. nonlinear relationship; no outliers, gaps, or clusters

13. positive linear relationship

15. *Sample answer:* bank account balance during a shopping spree

17. Could there be another event that is causing the sales of both items to increase?

19. 8

21. B

Section 9.2 Lines of Fit

(pages 382 and 383)

1. You can estimate and predict values.

3. -0.98, because it is closer to -1 than 0.91 is to 1. $(|-0.98| > |0.91|)$

5. **a.**

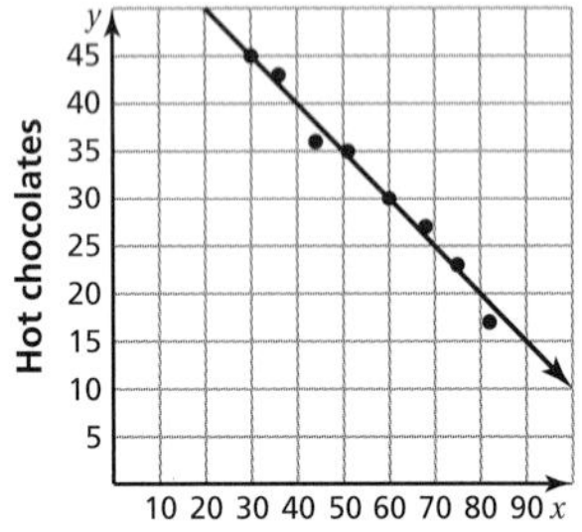

   b. *Sample answer:* $y = -0.5x + 60$

   c. *Sample answer:* The slope is -0.5 and the y-intercept is 60. So, you could predict that 60 hot chocolates are sold when the temperature is 0°F, and the sales decrease by about 1 hot chocolate for every 2°F increase in temperature.

   d. 50 hot chocolates

7. no; There is no line that lies close to most of the points.

9. $y = 0.9x + 4$; $r \approx 0.999$; The relationship between x and y is a strong positive correlation and the equation closely models the data; 4 in.

11. **a.** $y = 48x + 11$; $r \approx 0.98$; The relationship between x and y is a strong positive correlation and the equation closely models the data.

    b. 251 ft

    c. The height of a hit baseball is not linear. The best fit line from part (a) only models a small part of the data.

13. $-2\frac{7}{9}$

15. $\frac{9}{11}$

Section 9.3 Two-Way Tables
(pages 390 and 391)

1. The joint frequencies are the entries in the two-way table that differentiate the two categories of data collected. The marginal frequencies are the sums of the rows and columns of the two-way table.

3. total of females surveyed: 73;
   total of males surveyed: 59

5. 51

7. 71 students are juniors.
   75 students are seniors.
   93 students are attending the school play.
   53 students are not attending the school play.

9. **a.** 19; 42

   b. 72 6th-graders were surveyed.
   74 7th-graders were surveyed.
   65 8th-graders were surveyed.
   112 students chose grades.
   40 students chose popularity.
   59 students chose sports.

   c. about 8.5%

11. **a.**

Gender		Eye Color			
		Green	Blue	Brown	Total
	Male	5	16	27	48
	Female	3	19	18	40
	Total	8	35	45	88

b. 48 males were surveyed.
40 females were surveyed.
8 students have green eyes.
35 students have blue eyes.
45 students have brown eyes.

c.

Gender		Eye Color		
		Green	Blue	Brown
	Male	63%	46%	60%
	Female	38%	54%	40%

Sample answer: About 63% of the students with green eyes are male. 40% of the students with brown eyes are female.

13. Be careful not to count the females with green eyes twice.

15. $y = 5x - 2$

17. B

Section 9.4 Choosing a Data Display
(pages 397–399)

1. yes; Different displays may show different aspects of the data.

3. *Sample answer:*

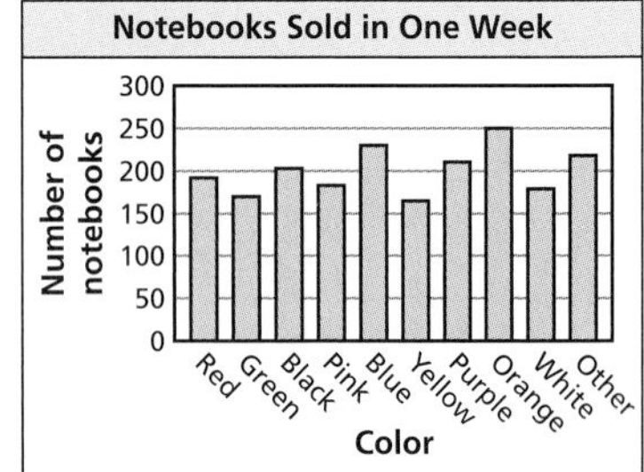

A bar graph shows the data in different color categories.

5. *Sample answer:* line graph; shows changes over time

7. *Sample answer:* line graph; shows changes over time

9. **a.** yes; The circle graph shows the data as parts of the whole.

   b. no; The bar graph shows the number of students, not the portion of students.

11. The pictures of the bikes are the largest on Monday and the smallest on Wednesday, which makes it seem like the distance is the same each day.

13. The intervals are not the same size.

15. *Sample answer:* bar graph; Each bar can represent a different vegetable.

17. *Sample answer:* dot plot

19. Does one display better show the differences in digits?

21. **a.** -9

    b. -8.6

23. A

Section 10.1 Exponents
(pages 414 and 415)

1. -3^4 is the negative of 3^4, so the base is 3, the exponent is 4, and its value is -81. $(-3)^4$ has a base of -3, an exponent of 4, and a value of 81.

3. 3^4

5. $\left(-\frac{1}{2}\right)^3$

7. π^3x^4

9. $(6.4)^4b^3$

11. 25

13. 1

15. $\frac{1}{144}$

17. The negative sign is not part of the base; $-6^2 = -(6 \cdot 6) = -36$.

19. $-\left(\frac{1}{4}\right)^4$

21. 29

23. 5

25. 66

27.

h	1	2	3	4	5
$2^h - 1$	1	3	7	15	31
2^{h-1}	1	2	4	8	16

$2^h - 1$; The option $2^h - 1$ pays you more money when $h > 1$.

29. Remember to add the black keys when finding how many notes you travel.

31. Associative Property of Multiplication

33. B

Section 10.2 Product of Powers Property
(pages 420 and 421)

1. when multiplying powers with the same base

3. 3^4 5. $(-4)^{12}$ 7. h^7

9. $\left(-\frac{5}{7}\right)^{17}$ 11. 5^{12} 13. 3.8^{12}

15. The bases should not be multiplied. $5^2 \cdot 5^9 = 5^{2+9} = 5^{11}$

17. $216g^3$ 19. $\frac{1}{25}k^2$ 21. $r^{12}t^{12}$

23. no; $3^2 + 3^3 = 9 + 27 = 36$ and $3^5 = 243$

25. 496 27. 78,125

29. a. $16\pi \approx 50.27$ in.^{3}

 b. $192\pi \approx 603.19$ in.^{3} Squaring each of the dimensions causes the volume to be 12 times larger.

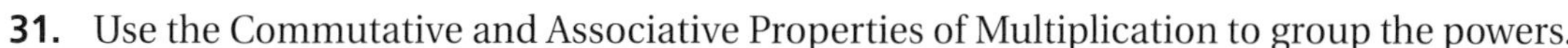

31. Use the Commutative and Associative Properties of Multiplication to group the powers.

33. 4 35. 3

37. B

Section 10.3 Quotient of Powers Property
(pages 426 and 427)

1. To divide powers means to divide out the common factors of the numerator and denominator. To divide powers with the same base, write the power with the common base and an exponent found by subtracting the exponent in the denominator from the exponent in the numerator.

3. 6^6 5. $(-3)^3$ 7. 5^6

9. $(-17)^3$ 11. $(-6.4)^2$ 13. b^{13}

15. You should subtract the exponents instead of dividing them. $\frac{6^{15}}{6^5} = 6^{15-5} = 6^{10}$

17. 2^9 19. π^8 21. k^{14}

23. $64x$ 25. $125a^3b^2$ 27. x^7y^6

29. You are checking to see if there is a linear relationship between memory and price, not if the change in price is constant for consecutive sizes of MP3 players.

31. 10^{13} galaxies 33. -9

35. 61 37. B

Section 10.4 Zero and Negative Exponents

(pages 432 and 433)

1. no; Any nonzero base raised to a zero exponent is always 1.

3. $5^{-5}, 5^0, 5^4$

5. 1

7. 1

9. $\frac{1}{36}$

11. $\frac{1}{16}$

13. $5\frac{1}{4}$

15. $\frac{1}{125}$

17. The negative sign goes with the exponent, not the base. $(4)^{-3} = \frac{1}{4^3} = \frac{1}{64}$

19. 2^0; 10^0

21. $\frac{a^7}{64}$

23. $5b$

25. 12

27. $\frac{w^6}{9}$

29. 100 mm

31. 1,000,000 nanometers

33. **a.** 10^{-9} m **b.** equal to

35. Write the power as 1 divided by the power and use a negative exponent. Justifications will vary.

37. 10^9

39. 10^4

Section 10.5 Reading Scientific Notation

(pages 440 and 441)

1. Scientific notation uses a factor greater than or equal to 1 but less than 10 multiplied by a power of 10. A number in standard form is written out with all the zeros and place values included.

3. 5,600,000,000,000

5. 87,300,000,000,000,000

7. yes; The factor is greater than or equal to 1 and less than 10. The power of 10 has an integer exponent.

9. no; The factor is greater than 10.

11. yes; The factor is greater than or equal to 1 and less than 10. The power of 10 has an integer exponent.

13. no; The factor is less than 1.

15. 70,000,000

17. 500

19. 0.000044

21. 1,660,000,000

23. 9,725,000

25. **a.** 810,000,000 platelets
    b. 1,350,000,000,000 platelets

27. **a.** Bellatrix
    b. Betelgeuse

29. 1555.2 km^2

31. 35,000,000 km^3

33. 4^5

35. $(-2)^3$

Section 10.6 Writing Scientific Notation
(pages 446 and 447)

1. If the number is greater than or equal to 10, the exponent will be positive. If the number is less than 1 and greater than 0, the exponent will be negative.

3. 2.1×10^{-3}

5. 3.21×10^{8}

7. 4×10^{-5}

9. 4.56×10^{10}

11. 8.4×10^{5}

13. 72.5 is not less than 10. The decimal point needs to move one more place to the left. 7.25×10^{7}

15. 6.09×10^{-5}, 6.78×10^{-5}, 6.8×10^{-5}

17. 4.8×10^{-8}, 4.8×10^{-6}, 4.8×10^{-5}

19. 6.88×10^{-23}, 5.78×10^{23}, 5.82×10^{23}

21. 4.01×10^{7} m

23. 680, 6.8×10^{3}, $\frac{68{,}500}{10}$

25. 6.25×10^{-3}, 6.3%, 0.625, $6\frac{1}{4}$

27. 1.99×10^{9} watts

29. carat; Because 1 carat $= 1.2 \times 10^{23}$ atomic mass units and 1 milligram $= 6.02 \times 10^{20}$ atomic mass units, and $1.2 \times 10^{23} > 6.02 \times 10^{20}$.

31. natural, whole, integer, rational

33. irrational

Selected Answers

Section 10.7 Operations in Scientific Notation
(pages 452 and 453)

1. Use the Distributive Property to group the factors together. Then subtract the factors and write it with the power of 10. The number may need to be rewritten so that it is still in scientific notation.

3. 8.34×10^{7}

5. 4.947×10^{11}

7. 5.8×10^{5}

9. 5.2×10^{8}

11. 7.555×10^{7}

13. 1.037×10^{7}

15. You have to rewrite the numbers so they have the same power of 10 before adding; 3.03×10^{9}

17. 2.9×10^{-3}

19. 1.5×10^{0}

21. 2.88×10^{-7}

23. 1.12×10^{-2}

25. 4.006×10^{9}

27. 1.962×10^{8} cm

29. First find the total length of the ridges and valleys.

31. 3×10^{8} m/sec

33. $\frac{1}{8}$

35. C

Section 11.1 Writing and Graphing Inequalities (pages 468 and 469)

1. A closed circle would be used because -42 is a solution.

3. no; $x < 5$ is all values of x less than 5. $5 < x$ is all values of x greater than 5.

5. $x \le -4$; all values of x less than or equal to -4

7. $w + 2.3 > 18$

9. $b - 4.2 < -7.5$

11. yes

13. no

15. yes

17.

19.

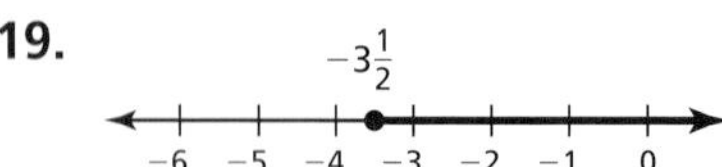

21. $p \ge 53$

23. yes

25. yes

27. **a.** any value that is greater than -2

b. any value that is less than or equal to -2; $b \le -2$

c. They represent the entire set of real numbers; yes

29. $p = 11$

31. $x = -7$

Section 11.2 Solving Inequalities Using Addition or Subtraction (pages 474 and 475)

1. Yes, because of the Subtraction Property of Inequality.

3. $x \ge 11$;

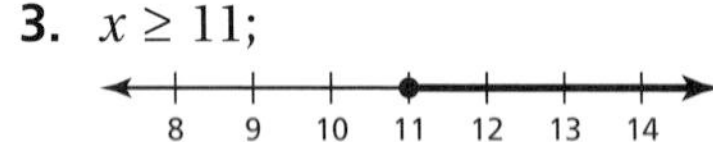

5. $-4 \le g$;

7. $-6 < y$;

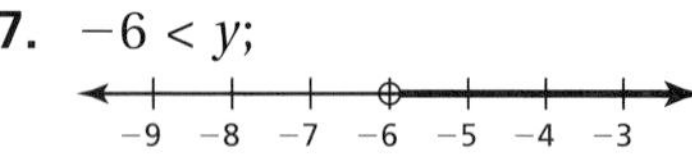

9. $t \le -2$;

11. $-\frac{3}{7} > b$;

13. $-2.8 < d$;

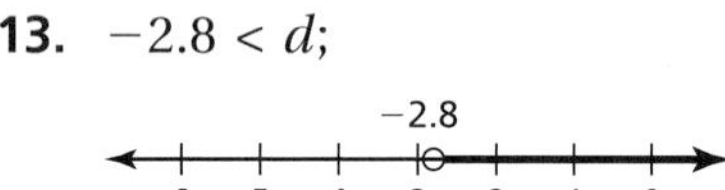

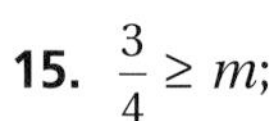

15. $\frac{3}{4} \ge m$;

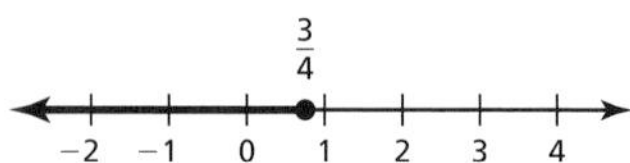

17. $h \le -2.4$;

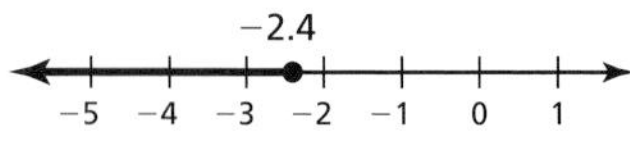

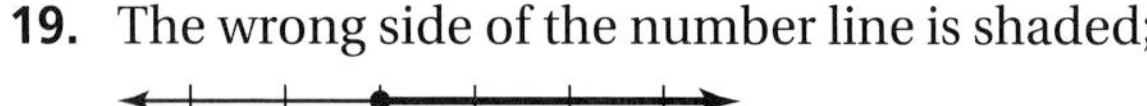

19. The wrong side of the number line is shaded;

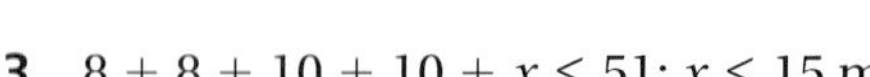

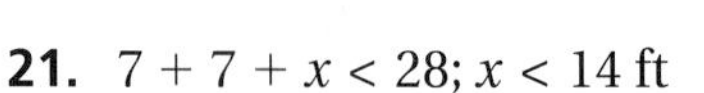

21. $7 + 7 + x < 28$; $x < 14$ ft

23. $8 + 8 + 10 + 10 + x \le 51$; $x \le 15$ m

25. $x - 3 \ge 5$; $x \ge 8$ ft

27. The three items must use less than 2400 watts total.

29. $x = 9$

31. $b = -22$

33. A

Section 11.3 Solving Inequalities Using Multiplication or Division *(pages 483–485)*

1. Multiply each side by 3.

3. *Sample answer:* $-4x < 16$

5. $x \geq -1$

7. $x \leq -35$

9. $x \leq \frac{3}{2}$

11. $c \leq -36$;

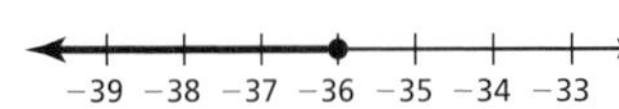

13. $x < -32$;

−35 −34 −33 −32 −31 −30 −29

15. $k > 2$;

−1 0 1 2 3 4 5

17. $y \leq -3$;

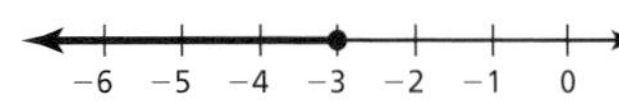

19. The inequality sign should not have been reversed.

$$\frac{x}{3} < -9$$
$$3 \cdot \frac{x}{3} < 3 \cdot (-9)$$
$$x < -27$$

21. $\frac{x}{7} < -3$; $x < -21$

23. $-2x > 30$; $x < -15$

25. **a.** $2.40x \leq 9.60$; $x \leq 4$ avocados

b. no; You must buy a whole number of avocados.

27. $n \geq -3$;

−6 −5 −4 −3 −2 −1 0

29. $h \leq -24$;

−27 −26 −25 −24 −23 −22 −21

31. $y > \frac{14}{3}$;

$\frac{14}{3}$

1 2 3 4 5 6 7

33. $m > -27$;

−30 −29 −28 −27 −26 −25 −24

35. $b > 6$;

3 4 5 6 7 8 9

37. $-2.5x < -20$; $x > 8$ h

39. $10x \geq 120$; $x \geq 12$ cm

41. $\frac{x}{5} < 100$; $x < \$500$

43. *Answer should include, but is not limited to:* Use the correct number of months that the novel has been out.

45. $n \geq -12$ and $n \leq -5$;

−12 −11 −10 −9 −8 −7 −6 −5

47. $s < 14$;

49. $v = 45$

51. $m = 4$

Section 11.4 Solving Two-Step Inequalities

(pages 490 and 491)

1. *Sample answer:* They use the same techniques, but when solving an inequality, you must be careful to reverse the inequality symbol when you multiply or divide by a negative number.

3. C

5. $y < 1$;

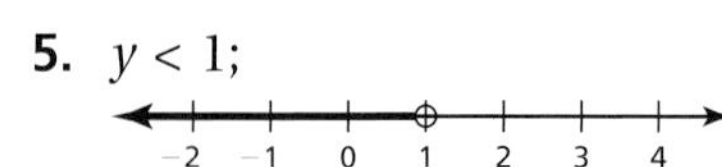

7. $h > \frac{9}{2}$;

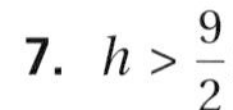

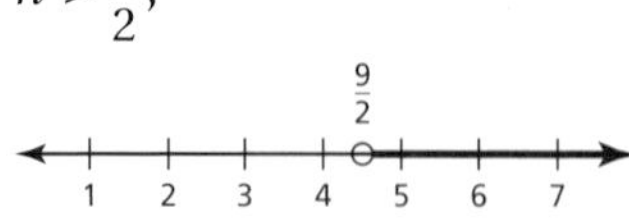

9. $b \le -6$;

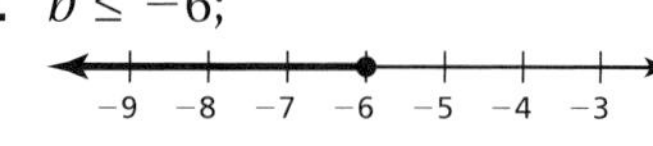

11. They did not perform the operations in the proper order.

$$\frac{x}{3} + 4 < 6$$
$$\frac{x}{3} < 2$$
$$x < 6$$

13. $w \le 3$;

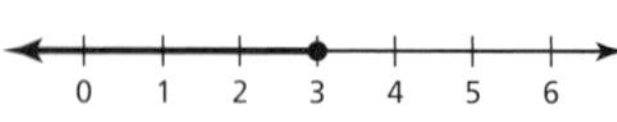

15. $d > -9$;

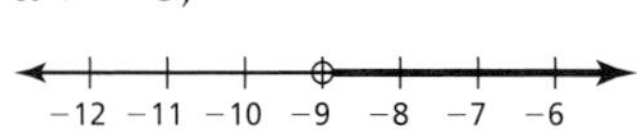

17. $c \ge -1.95$;

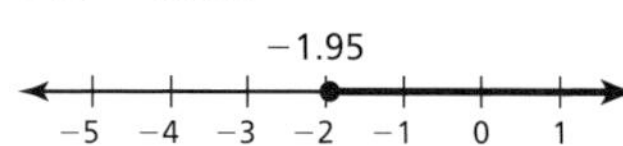

19. $x \ge 4$;

21. $-12x - 38 < -200$; $x > 13.5$ min

23. **a.** $9.5(70 + x) \ge 1000$; $x \ge 35\frac{5}{19}$, which means that at least 36 more tickets must be sold.

b. Because each ticket costs $1 more, fewer tickets will be needed for the theater to earn $1000.

25.

Flutes	7	21	28
Clarinets	4	12	16

7:4, 21:12, and 28:16

27. A

Section 12.1 Adjacent and Vertical Angles

(pages 506 and 507)

1. two pairs; four pairs

3. $\angle ABC = 120°$, $\angle CBD = 60°$, $\angle DBE = 120°$, $\angle ABE = 60°$

5. *Sample answer:* adjacent: $\angle FGH$ and $\angle HGJ$, $\angle FGK$ and $\angle KGJ$;
vertical: $\angle FGH$ and $\angle JGK$, $\angle FGK$ and $\angle JGH$

7. $\angle ACB$ and $\angle BCD$ are adjacent angles, not vertical angles.

9. vertical; 128

11. vertical; 25

13. adjacent; 20

15.

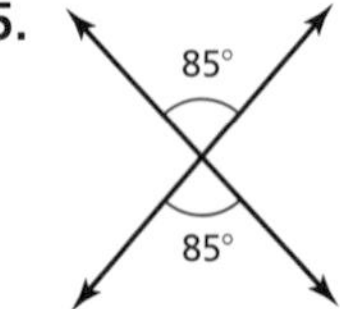

17.

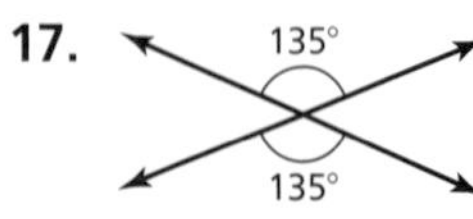

19. **a.** *Sample answer:*

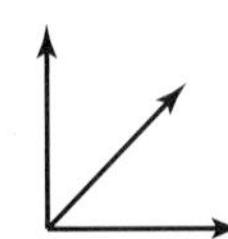

b. *Sample answer:*

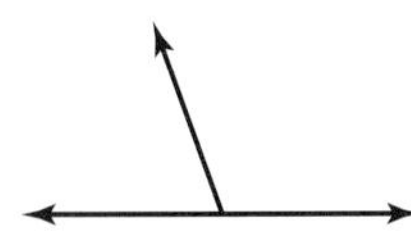

c. *Sample answer:*

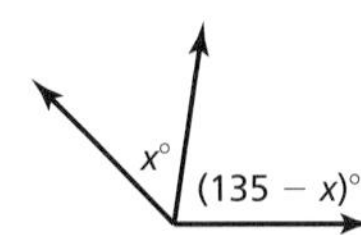

21. never

23. sometimes

25.

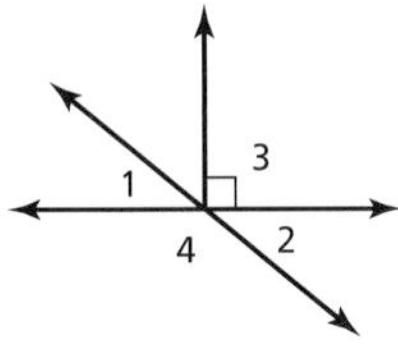

27. $n < -9$

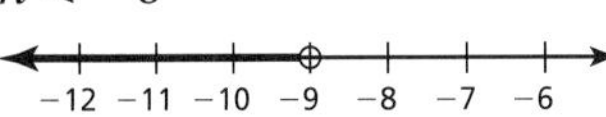

29. $m < 4$

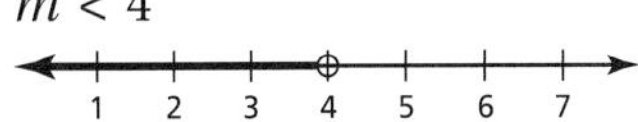

Section 12.2 Complementary and Supplementary Angles
(pages 512 and 513)

1. The sum of the measures of two complementary angles is 90°. The sum of the measures of two supplementary angles is 180°.

3. sometimes; Either x or y may be obtuse.

5. never; Because x and y must both be less than 90° and greater than 0°.

7. complementary

9. supplementary

11. neither

13. complementary; 55

15. $\angle 1 = 130°$, $\angle 2 = 50°$, $\angle 3 = 130°$

17. 20°

19. 80°

21. *Sample answer:* 1) Draw one angle, then draw the other using a side of the first angle; 2) Draw a right angle, then draw the shared side.

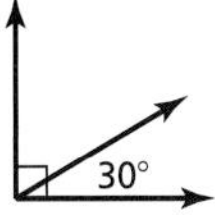

23. **a.** 25° **b.** 65°

25. 54°

27. $x = 10$; $y = 20$

29. $n = -\frac{5}{12}$

31. B

Section 12.3 Triangles
(pages 518 and 519)

1. *Angles:* When a triangle has 3 acute angles, it is an acute triangle. When a triangle has 1 obtuse angle, it is an obtuse triangle. When a triangle has 1 right angle, it is a right triangle. When a triangle has 3 congruent angles, it is an equiangular triangle.

   Sides: When a triangle has no congruent sides, it is a scalene triangle. When a triangle has 2 congruent sides, it is an isosceles triangle. When a triangle has 3 congruent sides, it is an equilateral triangle.

3. *Sample answer:*

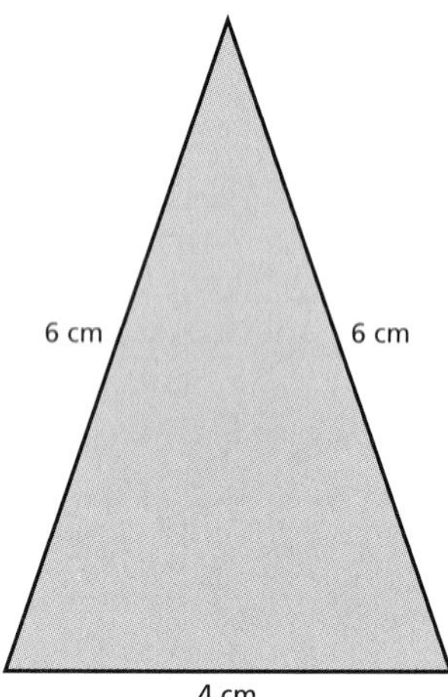

5. *Sample answer:*

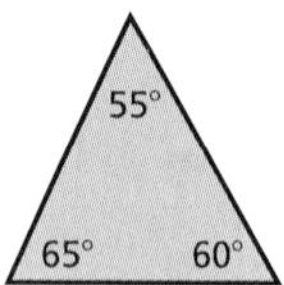

7. equilateral equiangular

9. right scalene

11. obtuse scalene

13. acute isosceles

15.

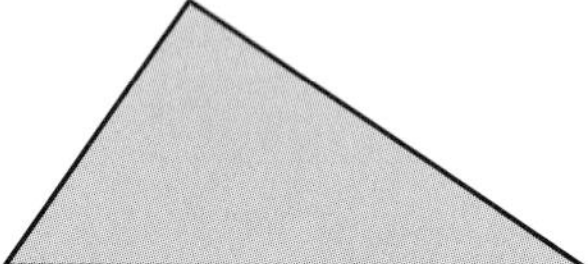

obtuse scalene triangle

17.

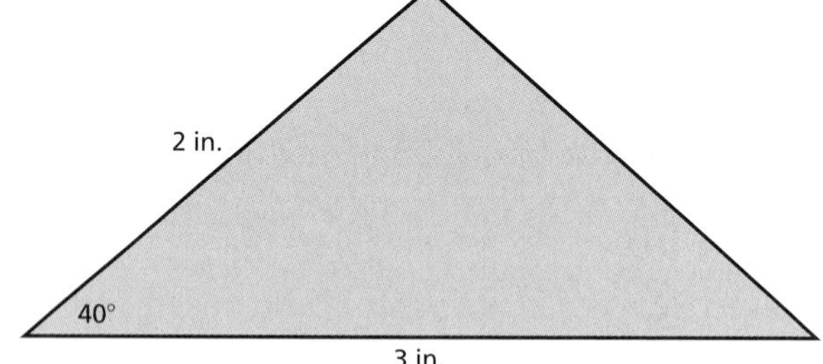

19.

21. no; The sum of the angle measures must be 180°.

23. many; You can change the angle formed by the two given sides to create many triangles.

25. no; The sum of any two side lengths must be greater than the remaining length.

27. **a.** green: 65; purple: 25; red: 45

    b. The angles opposite the congruent sides are congruent.

    c. An isosceles triangle has at least two angles that are congruent.

29. no; The equation cannot be written as $y = kx$.

31. B

Extension 12.3 Angle Measures of Triangles
(pages 520 and 521)

1. 91; obtuse scalene triangle
3. 90; right scalene triangle
5. 48; acute isosceles triangle
7. yes
9. no; $28\frac{2}{3}$
11. 67.5; acute isosceles triangle
13. 24; obtuse isosceles triangle
15. 35; obtuse scalene triangle
17. **a.** 72

    b. You can change the distance between the bottoms of the two upright cards; yes; x must be greater than 60 and less than 90; If x were less than or equal to 60, the two upright cards would have to be exactly on the edges of the base card or off the base card. It is not possible to stack cards at these angles. If x were equal to 90, then the two upright cards would be vertical, which is not possible. The card structure would not be stable. In practice, the limits on x are probably closer to $70 < x < 80$.

Section 12.4 Quadrilaterals
(pages 528 and 529)

1. all of them
3. kite; It is the only type of quadrilateral listed that does not have opposite sides that are parallel and congruent.
5. trapezoid
7. kite
9. rectangle
11. 110
13. 58°
15. 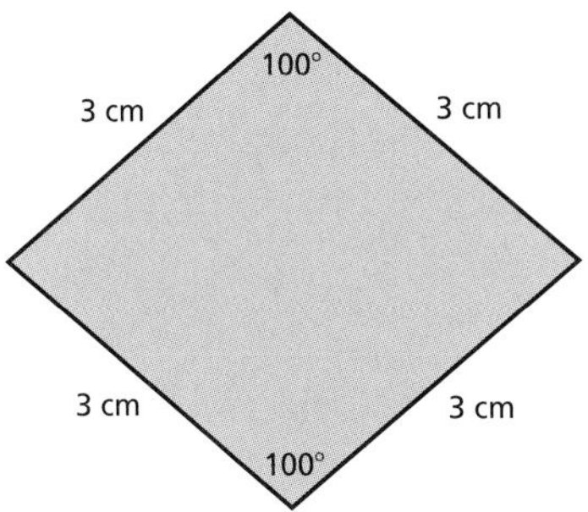

17. 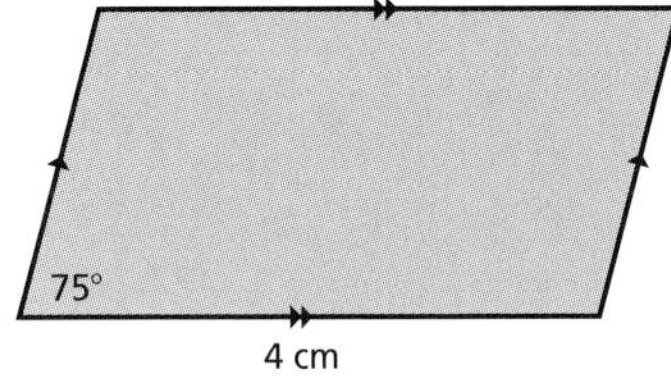

19. always
21. never
23. sometimes
25. 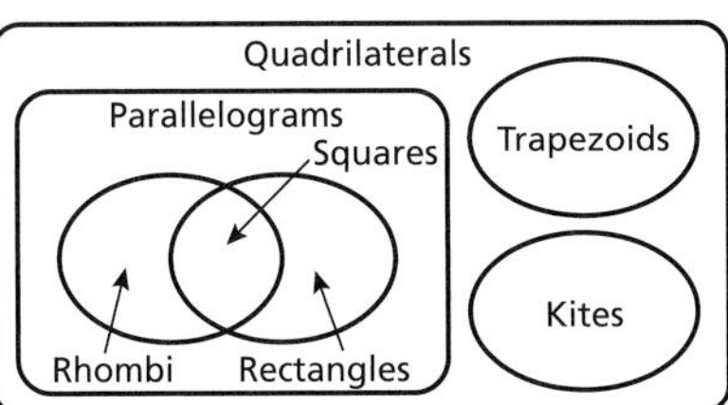

27. $\frac{1}{4}$
29. $\frac{6}{5}$

Selected Answers

Section 12.5 Scale Drawings
(pages 535–537)

1. A scale is the ratio that compares the measurements of the drawing or model with the actual measurements. A scale factor is a scale without any units.

3. Convert one of the lengths into the same units as the other length. Then, form the scale and simplify.

5. 10 ft by 10 ft
7. 112.5%
9. 50 mi
11. 110 mi
13. 15 in.
15. 21.6 yd

17. The 5 cm should be in the numerator.

$$\frac{1 \text{ cm}}{20 \text{ m}} = \frac{5 \text{ cm}}{x \text{ m}}$$

$$x = 100 \text{ m}$$

19. 2.4 cm; 1 cm : 10 mm

21. **a.** *Answer should include, but is not limited to:* Make sure words and picture match the product.

    b. Answers will vary.

23. **a.** 16 cm; 16 cm^2 **b.** 40 mm; 100 mm^2

25.

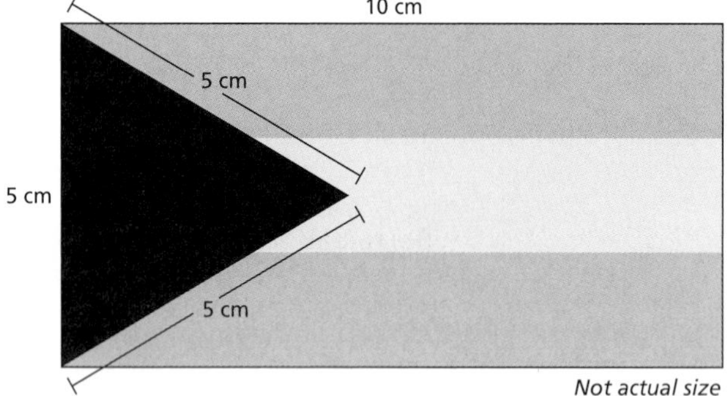

27. 15 ft^2

29. 3 ft^2

31. Find the size of the object that would represent the model of the Sun.

33 and 35.

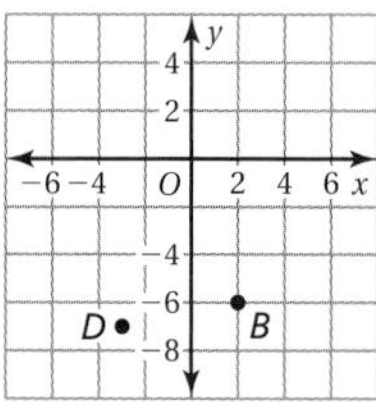

Section 13.1 Circles and Circumference
(pages 553–555)

1. The radius is one-half the diameter.

3. 2.5 cm
5. $1\frac{3}{4}$ in.
7. 4 in.

9. about 31.4 in.
11. about 56.52 in.

13. **a.** about 25 m; about 50 m

    b. about 2 times greater

15. about 7.71 ft

17. about 31.4 cm; about 62.8 cm

19. about 69.08 m; about 138.16 m

21. about 200.96 cm

23. Draw a diagram of the given information.

25. **a.** about 254.34 mm; First find the length of the minute hand. Then find $\frac{3}{4}$ of the circumference of a circle whose radius is the length of the minute hand.

b. about 320.28 mm; Subtract $\frac{1}{12}$ of the circumference of a circle whose radius is the length of the hour hand from the circumference of a circle whose radius is the length of the minute hand.

27. 20 m

29. D

Section 13.2 Perimeters of Composite Figures

(pages 560 and 561)

1. no; The perimeter of the composite figure does not include the measure of the shared side.

3. 19.5 units

5. 25.5 units

7. 19 units

9. 56 m

11. 30 cm

13. about 26.85 in.

15. about 36.84 ft

17. First find the total distance you run.

19. *Sample answer:* By adding the triangle shown by the dashed line to the L-shaped figure, you *reduce* the perimeter.

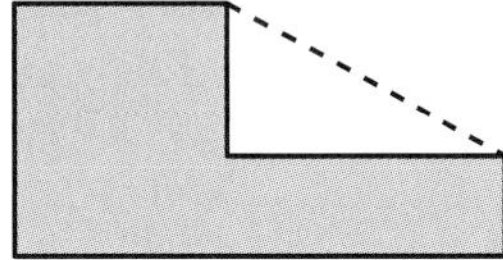

21. 279.68

23. 205

Section 13.3 Areas of Circles

(pages 568 and 569)

1. Divide the diameter by 2 to get the radius. Then use the formula $A = \pi r^2$ to find the area.

3. about 254.34 mm^2

5. about 314 in.^{2}

7. about 3.14 cm^2

9. about 2461.76 mm^2

11. about 113.04 in.^{2}

13. about 628 cm^2

15. about 1.57 ft^2

Section 13.3 Areas of Circles *(continued)* *(pages 568 and 569)*

17. What fraction of the circle is the dog's running area?

19. about 9.8125 in.^{2}; The two regions are identical, so find one-half the area of the circle.

21. about 4.56 ft^2; Find the area of the shaded regions by subtracting the areas of both unshaded regions from the area of the quarter-circle containing them. The area of each unshaded region can be found by subtracting the area of the smaller shaded region from the semicircle. The area of the smaller shaded region can be found by drawing a square about the region.

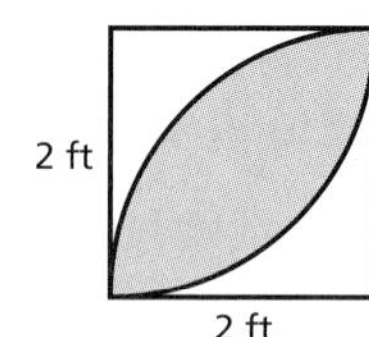

Subtract the area of a quarter-circle from the area of the square to find an unshaded area. Then subtract both unshaded areas from the square's area to find the shaded region's area.

23. 53

25. A

Section 13.4 Areas of Composite Figures *(pages 574 and 575)*

1. *Sample answer:* You could add the areas of an 8-inch × 4-inch rectangle and a triangle with a base of 6 inches and a height of 6 inches. Also you could add the area of a 2-inch × 4-inch rectangle to the area of a trapezoid with a height of 6 inches, and base lengths of 4 inches and 10 inches.

3. 28.5 units^2

5. 25 units^2

7. 25 units^2

9. 132 cm^2

11. *Answer will include but is not limited to:* Tracings of a hand and foot on grid paper, estimates of the areas, and a statement of which is greater.

13. 23.5 in.^{2}

15. 24 m^2

17. Each envelope can be broken up into 5 smaller figures to find the area.

19. $y \div 6$

21. $7w$

Section 14.1 Surface Areas of Prisms *(pages 591–593)*

1. *Sample answer:* 1) Use a net. 2) Use the formula $S = 2\ell w + 2\ell h + 2wh$.

3. Find the area of the bases of the prism; 24 in.^{2}; 122 in.^{2}

5.

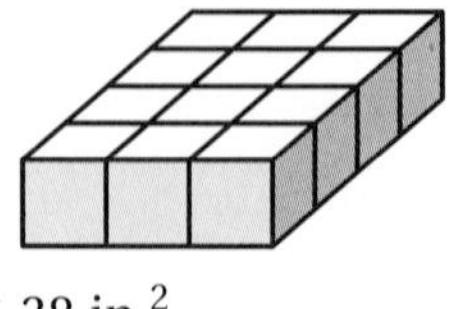

38 in.^{2}

7. 324 m^2

9. 49.2 yd^2

11. $136\ m^2$ 13. $294\ yd^2$ 15. $2\frac{2}{3}\ ft^2$ 17. $177\ in.^2$

19. yes; Because you do not need to frost the bottom of the cake, you only need 249 square inches of frosting.

21. $68\ m^2$ 23. $x = 4$ in.

25. The dimensions of the red prism are three times the dimensions of the blue prism. The surface area of the red prism is 9 times greater than the surface area of the blue prism.

27. **a.** 0.125 pint **b.** 1.125 pints

c. red and green: The ratio of the paint amounts (red to green) is 4 : 1 and the ratio of the side lengths is 2 : 1.

green and blue: The ratio of the paint amounts (blue to green) is 9 : 1 and the ratio of the side lengths is 3 : 1.

The ratio of the paint amounts is the square of the ratio of the side lengths.

29. $160\ ft^2$ 31. $28\ ft^2$

Section 14.2 Surface Areas of Pyramids

(pages 598 and 599)

1. no; The lateral faces of a pyramid are triangles.

3. triangular pyramid; The other three are names for the pyramid.

5. $178.3\ mm^2$ 7. $144\ ft^2$ 9. $170.1\ yd^2$

11. $1240.4\ mm^2$ 13. 6 m 15. $283.5\ cm^2$

17. How many green triangles and how many blue triangles are there in the umbrella?

19. $124\ cm^2$

21. $A \approx 452.16\ units^2$; $C \approx 75.36$ units

23. $A \approx 572.265\ units^2$; $C \approx 84.78$ units

Section 14.3 Surface Areas of Cylinders

(pages 604 and 605)

1. $2\pi rh$

3.

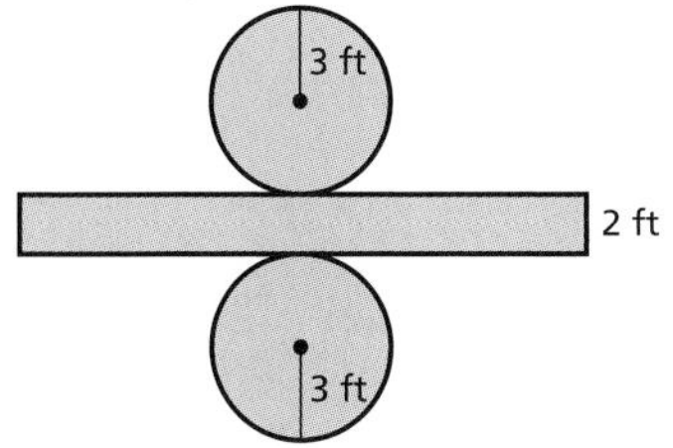

$30\pi \approx 94.2\ ft^2$

5.

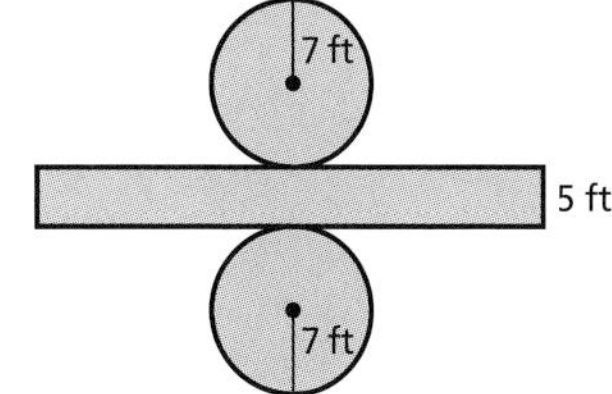

$168\pi \approx 527.5\ ft^2$

7. $156\pi \approx 489.8\ ft^2$ 9. $120\pi \approx 376.8\ ft^2$ 11. $28\pi \approx 87.9\ m^2$

Section 14.3 Surface Areas of Cylinders *(continued)*
(pages 604 and 605)

13. $432\pi \approx 1356.48 \text{ ft}^2$

15. The surface area of the cylinder with the height of 8.5 inches is greater than the surface area of the cylinder with the height of 11 inches.

17. After removing the wedge, is there any new surface area added?

19. 10 ft^2

21. 47.5 in.^2

Section 14.4 Volumes of Prisms
(pages 612 and 613)

1. cubic units

3. The volume of an object is the amount of space it occupies. The surface area of an object is the sum of the areas of all its faces.

5. 288 cm^3

7. 210 yd^3

9. 420 mm^3

11. 645 mm^3

13. The area of the base is wrong.

$$V = \frac{1}{2}(7)(5) \cdot 10$$
$$= 175 \text{ cm}^3$$

15. 225 in.^3

17. 7200 ft^3

19. 1728 in.^3

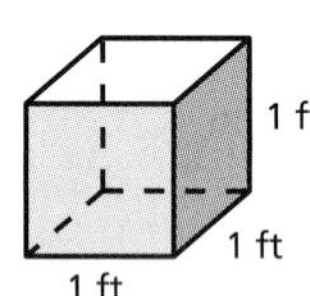

$1 \times 1 \times 1 = 1 \text{ ft}^3$

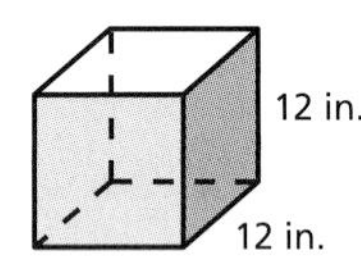

$12 \times 12 \times 12 = 1728 \text{ in.}^3$

21. 20 cm

23. You can write the volume in cubic inches and use prime factorization to find the dimensions.

25. \$90

27. \$240.50

Section 14.5 Volumes of Pyramids
(pages 618 and 619)

1. The volume of a pyramid is $\frac{1}{3}$ times the area of the base times the height. The volume of a prism is the area of the base times the height.

3. 3 times

5. 20 mm^3

7. 80 in.^3

9. 252 mm^3

11. 700 mm^3

13. 156 ft^3

15. 340.4 in.^3

17. $12{,}000 \text{ in.}^3$; The volume of one paperweight is 12 cubic inches. So, 12 cubic inches of glass is needed to make one paperweight. So, it takes $12 \times 1000 = 12{,}000$ cubic inches to make 1000 paperweights.

19. *Sample answer:* 5 ft by 4 ft

21. 153°; 63°

23. 60°; none

Extension 14.5 Cross Sections of Three-Dimensional Figures
(pages 620 and 621)

1. triangle **3.** rectangle **5.** triangle

7. The intersection is the shape of the base. **9.** circle

11. circle **13.** rectangle

15. The intersection occurs at the vertex of the cone.

Section 15.1 Outcomes and Events
(pages 636 and 637)

1. event; It is a collection of several outcomes.

3. 8 **5.** 1, 2, 3, 4, 5, 6, 7, 8, 9 **7.** 1, 3, 5, 7, 9

9. 1, 3 **11.** 3, 6, 9

13. **a.** 1 way **b.** green **15.** **a.** 1 way **b.** yellow

17. **a.** 7 ways

b. red, red, red, purple, purple, green, yellow

19. 7 ways **21.** true **23.** true **25.** 30 rock CDs

27. $x = 2$ **29.** $w = 12$ **31.** C

Section 15.2 Probability
(pages 642 and 643)

1. The probability of an event is the ratio of the number of favorable outcomes to the number of possible outcomes.

3. *Sample answer:* You will not have any homework this week.; You will fall asleep tonight.

5. either; Both spinners have the same number of chances to land on "Forward."

7. impossible **9.** unlikely

11. $\frac{1}{10}$ **13.** $\frac{9}{10}$ **15.** 0 **17.** 20

19. **a.** $\frac{2}{3}$; likely **b.** $\frac{1}{3}$; unlikely

c. $\frac{1}{2}$; equally likely to happen or not happen

21. There are 2 combinations for each.

23. $x < 4$;

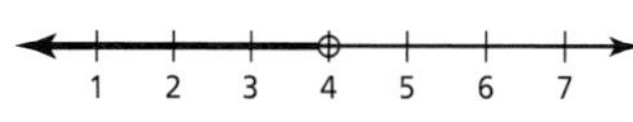

25. $w > -3$;

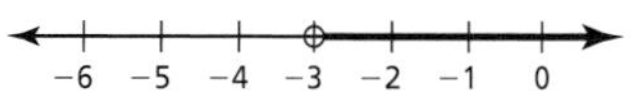

27. C

Selected Answers

Section 15.3 Experimental and Theoretical Probability
(pages 649–651)

1. Perform an experiment several times. Count how often the event occurs and divide by the number of trials.

3. There is a 50% chance you will get a favorable outcome.

5. experimental probability; The population is too large to survey every person, so a sample will be used to predict the outcome.

7. $\frac{12}{25}$, or 48%

9. $\frac{7}{25}$, or 28%

11. 0, or 0%

13. 45 tiles

15. $\frac{1}{3}$, or about 33.3%

17. $\frac{1}{2}$, or 50%

19. 1, or 100%

21. $\frac{25}{26}$, or about 96.2%

23. 36 songs

25. theoretical: $\frac{1}{5}$, or 20%;

    experimental: $\frac{37}{200}$, or 18.5%

    The experimental probability is close to the theoretical probability.

27. theoretical: $\frac{1}{5}$, or 20%;

    experimental: $\frac{1}{5}$, or 20%

    The probabilities are equal.

29. **a.** $\frac{1}{12}$; 50 times **b.** $\frac{11}{50}$; 132 times

    c. A larger number of trials should result in a more accurate probability, which gives a more accurate prediction.

31. Make a list of all possible ways to get each sum.

35. 4%

37. D

33. **a.** As a number of trials increases, the most likely sum will change from 6 to 7.

    b. As an experiment is repeated over and over, the experimental probability of an event approaches the theoretical probability of the event.

Section 15.4 Compound Events
(pages 657–659)

1. A sample space is the set of all possible outcomes of an event. Use a table or tree diagram to list all the possible outcomes.

3. You could use a tree diagram or the Fundamental Counting Principle. Either way, the total number of possible outcomes is 30.

5. 125,000

7. Sample space: Realistic Lion, Realistic Bear, Realistic Hawk, Realistic Dragon, Cartoon Lion, Cartoon Bear, Cartoon Hawk, Cartoon Dragon; 8 possible outcomes

9. 20

11. 60

13. The possible outcomes of each question should be multiplied, not added. The correct answer is $2 \times 2 \times 2 \times 2 \times 2 = 32$.

15. $\frac{1}{10}$, or 10%

17. $\frac{1}{5}$, or 20%

19. $\frac{2}{5}$, or 40%

21. $\frac{1}{18}$, or $5\frac{5}{9}$%

23. $\frac{1}{9}$, or $11\frac{1}{9}$%

25. **a.** $\frac{1}{9}$, or about 11.1%

b. It increases the probability that your guesses are correct to $\frac{1}{4}$, or 25%, because you are only choosing between 2 choices for each question.

27. **a.** $\frac{1}{1000}$, or 0.1%

b. There are 1000 possible combinations. With 5 tries, someone would guess 5 out of the 1000 possibilities. So, the probability of getting the correct combination is $\frac{5}{1000}$, or 0.5%.

29. **a.** The Fundamental Counting Principle is more efficient. A tree diagram would be too large.

b. 1,000,000,000 or one billion

c. *Sample answer:* Not all possible number combinations are used for Social Security Numbers (SSN). SSNs are coded into geographical, group, and serial numbers. Some SSNs are reserved for commercial use and some are forbidden for various reasons.

31. *Sample answer:* adjacent: $\angle XWY$ and $\angle ZWY$, $\angle XWY$ and $\angle XWV$; vertical: $\angle VWX$ and $\angle YWZ$, $\angle YWX$ and $\angle VWZ$

33. B

Section 15.5 Independent and Dependent Events (pages 665–667)

1. What is the probability of choosing a 1 and then a blue chip?; $\frac{1}{15}$; $\frac{1}{10}$

3. independent; The outcome of the first roll does not affect the outcome of the second roll.

5. $\frac{1}{8}$

7. $\frac{3}{8}$

9. $\frac{1}{42}$

11. $\frac{2}{21}$

13. The two events are dependent, so the probability of the second event is $\frac{1}{3}$.

$P(\text{red and green}) = \frac{1}{4} \cdot \frac{1}{3} = \frac{1}{12}$

15. $\frac{1}{6}$, or about 16.7%

17. $\frac{2}{35}$

19. $\frac{5}{162}$, or about 3.1%

21. $\frac{4}{81}$, or about 4.9%

23. $\frac{3}{4}$

25. **a.** Because the probability that both you and your best friend are chosen is $\frac{1}{132}$, you and your best friend are not in the same group. The probability that you both are chosen would be 0 because only one leader is chosen from each group.

b. $\frac{1}{11}$ **c.** 23

Selected Answers

Section 15.5 Independent and Dependent Events *(continued)* *(pages 665–667)*

27.

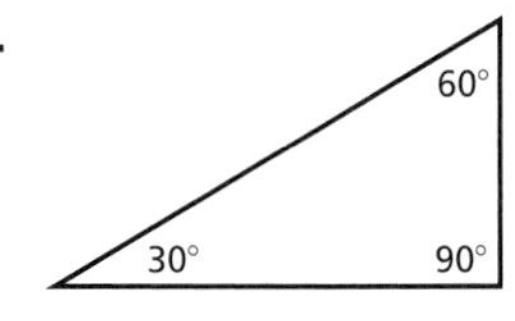

right scalene

29.

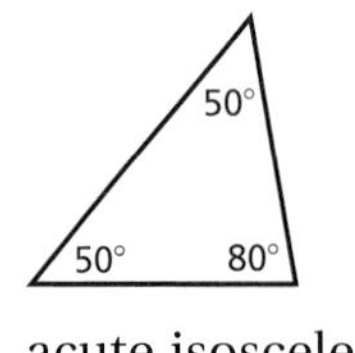

acute isosceles

Extension 15.5 Simulations *(page 669)*

1. **a.** *Sample answer:* Roll four number cubes. Let an odd number represent a correct answer and an even number represent an incorrect answer. Run 40 trials.

b. Check students' work. The probability should be "close" to 6.25% (depending on the number of trials, because that is the theoretical probability).

3. *Sample answer:* Using the spreadsheet in Example 3 and using digits 1–4 as successes, the experimental probability is 16%.

Section 15.6 Samples and Populations *(pages 676 and 677)*

1. Samples are easier to obtain.

3. Population: Residents of New Jersey
Sample: Residents of Ocean County

5. biased; The sample is not selected at random and is not representative of the population because students in a band class play a musical instrument.

7. biased; The sample is not representative of the population because people who go to a park are more likely to think that the park needs to be remodeled.

9. no; the sample is not representative of the population because people going to the baseball stadium are more likely to support building a new baseball stadium. So, the sample is biased and the conclusion is not valid.

11. Sample A; It is representative of the population.

13. a sample; It is much easier to collect sample data in this situation.

15. a sample; It is much easier to collect sample data in this situation.

17. Not everyone has an email address, so the sample may not be representative of the entire population. *Sample answer:* When the survey question is about technology or which email service you use, the sample may be representative of the entire population.

19. Use the survey results to find the number of students in the school who plan to attend college.

21. 140

23. 3

Extension 15.6 Generating Multiple Samples

(pages 678 and 679)

1. **a.** Check students' work. **b.** Check students' work.

   c. Check students' work. *Sample answer:* yes; Increase the number of random samples.

3. **Step 1:** 7, 7.5, 8, 4.5, 8, 10, 5, 11

   Step 2:

   4.5 6 7.75 9 11

   Median hours worked each week

   3 4 5 6 7 8 9 10 11 12 13

   Step 3: *Sample answer:* The actual median number of hours probably lies within the interval 6 to 9 hours (the box). So, about 7.5 is a good estimate.

   The median of the data is 8. So, the estimate is close.

5. The more samples you have, the more accurate your inferences will be. By taking multiple random samples, you can find an interval where the actual measurement of a population may lie.

Section 15.7 Comparing Populations

(pages 684 and 685)

1. When comparing two populations, use the mean and the MAD when each distribution is symmetric. Use the median and the IQR when either one or both distributions are skewed.

3. **a.** garter snake: mean = 25, median = 24.5, mode = 24, range = 20, IQR = 7.5, MAD ≈ 4.33
   water snake: mean = 31.5, median = 32, mode = 32, range = 20, IQR = 10, MAD ≈ 5.08

   b. The water snakes have greater measures of center because the mean, median, and mode are greater. The water snakes also have greater measures of variation because the interquartile range and mean absolute deviation are greater.

5. **a.** Class A: median = 90, IQR = 12.5
   Class B: median = 80, IQR = 10

   The variation in the test scores is about the same, but Class A has greater test scores.

   b. The difference in the medians is 0.8 to 1 times the IQR.

7. Arrange the dot plots in Exercise 5 vertically and construct a double box-and-whisker plot in Exercise 6 to help you visualize the distributions.

9. **a.** Check students' work. Experiments should include taking many samples of a manageable size from each grade level. This will be more doable if the work of sampling is divided among the whole class, and the results are pooled together.

   b. Check students' work. The data may or may not support a conclusion.

11. 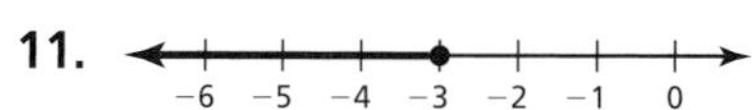

13. 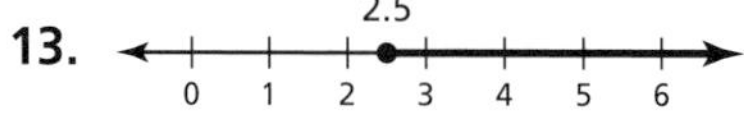

Selected Answers

Key Vocabulary Index

Mathematical terms are best understood when you see them used and defined *in context.* This index lists where you will find key vocabulary. A full glossary is available in your Record and Practice Journal and at *BigIdeasMath.com.*

Student Index

This student-friendly index will help you find vocabulary, key ideas, and concepts. It is easily accessible and designed to be a reference for you whether you are looking for a definition, real-life application, or help with avoiding common errors.

D

Student Index

E

F

G

M

N

O

Student Index

P

S

T

Photo Credits

Front matter

iv Big Ideas Learning, LLC; **viii** *top* ©iStockphoto.com/Lisa Thornberg, ©iStockphoto.com/Ann Marie Kurtz; *bottom* ©iStockphoto.com/Jane norton; **ix** *top* Kasiap/Shutterstock.com, ©iStockphoto.com/Ann Marie Kurtz; *bottom* wavebreakmedia ltd/Shutterstock.com; **x** *top* ©iStockphoto.com/sumnersgraphicsinc, ©iStockphoto.com/Ann Marie Kurtz; *bottom* Odua Images/Shutterstock.com; **xi** *top* ©iStockhphoto.com/Jonathan Larsen; *bottom* James Flint/ Shutterstock.com; **xii** *top* stephan kerkhofs/Shutterstock.com, Cigdem Sean Cooper/Shutterstock.com, ©iStockphoto.com/Andreas Gradin; *bottom* william casey/Shutterstock.com; **xiii** *top* ©iStockphoto.com/ ALEAIMAGE, ©iStockphoto.com/Ann Marie Kurtz; *bottom* Edyta Pawlowska/Shutterstock.com; **xiv** *top* ©iStockphoto/Michael Flippo, ©iStockphoto.com/Ann Marie Kurtz; *bottom* PETER CLOSE/ Shutterstock.com; **xv** *top* ©iStockphoto.com/ALEAIMAGE, ©iStockphoto.com/Ann Marie Kurtz; *bottom* ©iStockphoto.com/ ranplett; **xvi** *top* ©iStockphoto.com/Alistair Cotton; *bottom* ©iStockphoto.com/Noraznen Azit; **xvii** *top* Varina and Jay Patel/ Shutterstock.com, ©iStockphoto.com/Ann Marie Kurtz; *bottom* ©iStockphoto.com/Thomas Perkins; **xviii** *top* ©iStockhphoto.com/ Jonathan Larsen; *bottom* Evok20/Shutterstock.com; **xix** *top* ©iStockphoto.com/Lisa Thornberg, ©iStockphoto.com/Ann Marie Kurtz; *bottom* Sinisa Bobic/Shutterstock.com; **xx** *top* ©iStockphoto/ Michael Flippo, ©iStockphoto.com/Ann Marie Kurtz; *bottom* Artpose Adam Borkowski/Shutterstock.com; **xxi** *top* stephan kerkhofs/ Shutterstock.com, Cigdem Sean Cooper/Shutterstock.com, ©iStockphoto.com/Andreas Gradin; *bottom* ©iStockphoto.com/ Aldo Murillo; **xxii** *top* infografick/Shutterstock.com, ©iStockphoto.com/ Ann Marie Kurtz; *bottom* ©iStockphoto.com/kate_sept2004; **xxiii** *top* ©iStockphoto.com/Björn Kindler; *bottom* ©iStockphoto.com/ Ferli Achirulli; **xxiv** Ljupco Smokovski/Shutterstock.com

Chapter 1

1 ©iStockphoto.com/Lisa Thornberg, ©iStockphoto.com/Ann Marie Kurtz; **6** ©iStockphoto.com/David Freund; **7** ©iStockphoto.com/nicolas hansen; **8** amskad/Shutterstock.com; **9** ©iStockphoto.com/Ryan Lane; **12** ©iStockphoto.com/Harley McCabe; **13** ©iStockphoto.com/Jacom Stephens; **14** ©iStockphoto.com/Harry Hu; **15** ©iStockphoto.com/ Ralf Hettler, Vibrant Image Studio/Shutterstock.com; **23** ©iStockphoto.com/Andrey Krasnov; **24** Shawn Hempel/ Shutterstock.com; **31** *top right* ©iStockphoto.com/Alan Crawford; *center left* ©iStockphoto.com/Julio Yeste; *bottom right* ©iStockphoto.com/Mark Stay; **36** *center right* Ljupco Smokovski/ Shutterstock.com; *bottom left* emel82/Shutterstock.com

Chapter 2

40 Kasiap/Shutterstock.com, ©iStockphoto.com/Ann Marie Kurtz; **48** Azat1976/Shutterstock.com; **52** ©iStockphoto.com/Er Ten Hong; **53** *center left* ©iStockphoto.com/Sergey Galushko; *center right* ©iStockphoto.com/Tryfonov levgenii; **54** ©iStockphoto.com/ingmar wesemann; **59** ©iStockphoto.com/Hazlan Abdul Hakim; **67** ©iStockphoto.com/Maksim Shmeljov; **70** *top* ©iStockphoto.com/ Viatcheslav Dusaleev; *bottom left* ©iStockphoto.com/Jason Mooy; *bottom right* ©iStockphoto.com/Felix Möckel; **73** gary718/ Shutterstock.com; **83** Diego Cervo/Shutterstock.com; **90** *center left* Antonio Jorge Nunes/Shutterstock.com, Tom C Amon/ Shutterstock.com; *center right* ©iStockphoto.com/Alex Slobodkin

Chapter 3

100 ©iStockphoto.com/sumnersgraphicsinc, ©iStockphoto.com/ Ann Marie Kurtz; **102** PILart/Shutterstock.com, Wildstyle/ Shutterstock.com; **103** Estate Craft Homes, Inc.; **114** Marc Dietrich/ Shutterstock.com; **120** *bottom left* ©iStockphoto.com/Evgeny Terentev; *bottom right* ©iStockphoto.com/Vadym Volodin; **121** NASA; **124** iStockphoto.com/Evelyn Peyton; **125** *top right* ©iStockphoto.com/ Terraxplorer; *top left* ©iStockphoto.com/Lora Clark; *center right* ©iStockphoto.com/Jennifer Morgan

Chapter 4

140 ©iStockphoto.com/Jonathan Larsen; **145** NASA; **146** ©iStockphoto.com/David Morgan; **147** *top right* NASA; *center left* ©iStockphoto.com/jsemeniuk; **154** ©iStockphoto.com/Amanda Rohde; **155** Julian Rovagnati/Shutterstock.com; **159** RyFlip/Shutterstock.com; **162** Luke Wein/Shutterstock.com; **165** AVAVA/Shutterstock.com; **170** ©iStockphoto.com/Dreamframer; **171** *top right* Jerry Horbert/ Shutterstock.com; *center left* ©iStockphoto.com/Chris Schmidt; **173** ©iStockphoto.com/biffspandex; **176** ©iStockphoto.com/ Stephen Pothier; **177** *top left* Gina Smith/Shutterstock.com; *center left* Dewayne Flowers/Shutterstock.com; **181** Herrenknecht AG; **182** ©iStockphoto.com/Adam Mattel; **183** *top left* ©iStockphoto.com/ Gene Chutka; *center right* ©iStockphoto.com/marcellus2070, ©iStockphoto.com/beetle8; **187** ©iStockphoto.com/Connie Maher; **188** ©iStockphoto.com/Jacom Stephens; **189** *top right* ©iStockphoto.com/Petr Podzemny; *bottom left* ©iStockphoto.com/ adrian beesley; **190** Richard Goldberg/Shutterstock.com; **196** Thomas M Perkins/Shutterstock.com

Chapter 5

200 stephan kerkhofs/Shutterstock.com, Cigdem Sean Cooper/ Shutterstock.com, ©iStockphoto.com/Andreas Gradin; **202** Howard Sandler/Shutterstock.com, ©iStockphoto.com/Dori OConnell; **205** Richard Paul Kane/Shutterstock.com; **206** ©iStockphoto.com/ Kathy Hicks; **208** *top right* YuriyZhuravov/Shutterstock.com; *bottom right* Talvi/Shutterstock.com; **211** aguilarphoto/Shutterstock.com; **212** Kiselev Andrey Valerevich/Shutterstock.com; **213** *center left* Susan Schmitz/ Shutterstock.com; *center right* akva/Shutterstock.com; **215** Andrey Yurlov/ Shutterstock.com; **216** Steve Cukrov/Shutterstock.com; **220** Le Do/ Shutterstock.com, Quang Ho/Shutterstock.com, SergeyIT/ Shutterstock.com, jon Le-Bon/Shutterstock.com; **221** Ariwasabi/ Shutterstock.com; **222** Ewa/Shutterstock.com; **223** *top left* Gordana Sermek/Shutterstock.com; *center right* Rashevskyi Viacheslav/ Shutterstock.com; **224** ©iStockphoto.com/walik; **228** ©iStockphoto.com/ Corina Estepa; **229** ©iStockphoto.com/Tomislav Forgo; **231** Kateryna Larina/Shutterstock.com; **232** Selena/Shutterstock.com; **236** kostudio/Shutterstock.com

Chapter 6

240 ©iStockphoto.com/ALEAIMAGE, ©iStockphoto.com/Ann Marie Kurtz; **247** ©iStockphoto.com/Kevin Panizza; **249** ©iStockphoto.com/ Jacom Stephens; **252** ©iStockphoto.com/DivaNir4a; **254** *top left* ©iStockphoto.com/Manuel Angel Diaz Blanco; *bottom right* ©iStockphoto.com/Sergey Lemeshencko; **255** ©iStockphoto.com/ Robert Rushton; **259** General Atomics Aeronautical Systems, Inc.; **262** ©iStockphoto.com/Mlenny Photography; **263** ©iStockphoto.com/ medobear; **267** ©iStockphoto.com/PeskyMonkey; **271** ©iStockphoto.com/Tom Buttle; **278** gillmar/Shutterstock.com

Chapter 7

286 ©iStockphoto/Michael Flippo, ©iStockphoto.com/Ann Marie Kurtz; **291** Perfectblue97; **292** ©iStockphoto.com/Benjamin Lazare; **293** *top right* ©iStockphoto.com/iShootPhotos, LLC; *center left* ©iStockphoto.com/Jill Chen, Oleksiy Mark/Shutterstock.com; **298** Gary Whitton/Shutterstock.com; **299** Michael Stokes/ Shutterstock.com; **300** ©Oxford Science Archive/Heritage Images/ Imagestate; **304** ©iStockphoto.com/Melissa Carroll; **307** *center left* ©iStockphoto.com/Yvan Dubé; *bottom right* Snvv/Shutterstock.com; **308** ©iStockphoto.com/Kais Tolmats; **312** *top left* ©iStockphoto.com/ Don Bayley; *center left* ©iStockphoto.com/iLexx; **315** ©iStockphoto.com/Marcio Silva; **319** Monkey Business Images/ Shutterstock.com; **327** LoopAll/Shutterstock.com; **328** CD Lanzen/ Shutterstock.com

Chapter 8

332 ©iStockphoto.com/ALEAIMAGE, ©iStockphoto.com/Ann Marie Kurtz; **334** ©iStockphoto.com/Jill Chen; **337** ©iStockphoto.com/ camilla wisbauer; **339** *Exercises 13 and 14* ©iStockphoto.com/Prill Mediendesigns & Fotografie; *Exercise 15* ©iStockphoto.com/subjug; *center left* ©iStockphoto.com/Matthew Dixon; *center right* ©iStockphoto.com/nilgun bostanci; **345** ©iStockphoto.com/Stefano Tiraboschi; **351** Donald Joski/Shutterstock.com; **352** ©iStockphoto.com/ Yury Kosourov; **353** Carlos Caetano/Shutterstock.com; **360** Courtesy of Green Light Collectibles; **361** *top right* ©iStockphoto.com/wrangel; *center left* ©iStockphoto.com/ivanastar; *bottom left* ©iStockphoto.com/ Daniel Cardiff; **362** Eric Isselée/Shutterstock.com; **366** ©iStockphoto.com/Daniel Loiselle

Chapter 9

370 ©iStockphoto.com/Alistair Cotton; **372** *baseball* Kittisak/Shutterstock.com; *golf ball* tezzstock/Shutterstock.com; *basketball* vasosh/Shutterstock.com; *tennis ball* UKRID/Shutterstock.com; *water polo ball* John Kasawa/Shutterstock.com; *softball* Ra Studio/Shutterstock.com; *volleyball* vberla/Shutterstock.com; **376** ©iStockphoto.com/Jill Fromer; **377** ©iStockphoto.com/Janis Litavnieks; **378** Gina Brockett; **379** ©iStockphoto.com/Craig Dingle; **381** Sashkin/Shutterstock.com; **382** ©iStockphoto.com/Brian McEntire; **385** Dwight Smith/Shutterstock.com; **386** Aptyp_koK/Shutterstock.com; **391** Alberto Zornetta/Shutterstock.com; **392** *center left* ©iStockphoto.com/Tony Campbell; *bottom right* Eric Isselee/Shutterstock.com; **393** *top right* Larry Korhnak; *bottom right* Photo by Andy Newman; **399** *top left* ©iStockphoto.com/Jane norton; *bottom right* ©iStockphoto.com/Krzysztof Zmij; **400** IrinaK/Shutterstock.com; **404** Lim Yong Hian/Shutterstock.com

Chapter 10

408 Varina and Jay Patel/Shutterstock.com, ©iStockphoto.com/Ann Marie Kurtz; **410** ©iStockphoto.com/Franck Boston; **411** *Activity 3a* ©iStockphoto.com/Manfred Konrad; *Activity 3b* NASA/JPL-Caltech/R.Hurt (SSC); *Activity 3c and d* NASA; *bottom right* Stevyn Colgan; **413** ©iStockphoto.com/Philippa Banks; **414** ©iStockphoto.com/clotilde hulin; **415** ©iStockphoto.com/Boris Yankov; **420** ©iStockphoto.com/VIKTORIIA KULISH; **421** *top right* ©iStockphoto.com/Paul Tessier; *center left* ©iStockphoto.com/subjug, ©iStockphoto.com/Valerie Loiseleux, ©iStockphoto.com/Linda Steward; **426** ©iStockphoto.com/Petrovich9; **427** *top right* Dash/Shutterstock.com; *center left* NASA/JPL-Caltech/L.Cieza (UT Austin); **431** ©iStockphoto.com/Aliaksandr Autayeu; **432** EugeneF/Shutterstock.com; **433** ©iStockphoto.com/Nancy Louie; **435** ©iStockphoto.com/Dan Moore; **436** ©iStockphoto.com/Kais Tolmats; **437** *Activity 3a and d* Tom C Amon/Shutterstock.com; *Activity 3b* Olga Gabay/Shutterstock.com; *Activity 3c* NASA/MODIS Rapid Response/Jeff Schmaltz; *Activity 3f* HuHu/Shutterstock.com; *Activity 4a* PILart/Shutterstock.com; *Activity 4b* Matthew Cole/Shutterstock.com; *Activity 4c* Yanas/Shutterstock.com; *Activity 4e* unkreativ/Shutterstock.com; **439** *top left* ©iStockphoto.com/Mark Stay; *top center* ©iStockphoto.com/Frank Wright; *top right* ©iStockphoto.com/Evgeniy Ivanov; *bottom left* ©iStockphoto.com/Oliver Sun Kim; **440** ©iStockphoto.com/Christian Jasiuk; **441** Microgen/Shutterstock.com; **442** *Activity 1a* ©iStockphoto.com/Susan Trigg; *Activity 1b* ©iStockphoto.com/subjug; *Activity 1c* ©iStockphoto.com/camilla wisbauer; *Activity 1d* ©iStockphoto.com/Joe Belanger; *Activity 1e* ©iStockphoto.com/thumb; *Activity 1f* ©iStockphoto.com/David Freund; **443** NASA; **444** *center* Google and YouTube logos are registered trademarks of Google Inc., used with permission.; **445** *top left* Elaine Barker/Shutterstock.com; *center right* ©iStockphoto.com/breckeni; **446** *bottom left* ©iStockphoto.com/Max Delson Martins Santos; *bottom right* ©iStockphoto.com/Jan Rysavy; **447** *top right* BORTEL Pavel/Shutterstock.com; *center right* ©iStockphoto.com/breckeni; **451** *center left* Sebastian Kaulitzki/Shutterstock.com; *center right* ©iStockphoto.com/Jan Rysavy; **453** ©iStockphoto.com/Boris Yankov; **454** mmutlu/Shutterstock.com; **458** *bottom right* ©iStockphoto.com/Eric Holsinger; *bottom left* TranceDrumer/Shutterstock.com

Chapter 11

462 ©iStockphoto.com/Jonathan Larsen; **464** *Activity 1a* Filaphoto/Shutterstock.com; *Activity 1b* Palmer Kane LLC/Shutterstock.com; *Activity 1c* bitt24/Shutterstock.com; *Activity 1d* kropic1/Shutterstock.com, ©iStockphoto.com/Chris Schmidt, ©iStockphoto.com/Jane norton; **469** Gregory James Van Raalte/Shutterstock.com; **470** *top left* Boy Scouts of America; *bottom right* William Seng/CC BY ND; **471** PILart/Shutterstock.com; **473** eddtoro/Shutterstock.com; **474** Khafizov Ivan Harisovich/Shutterstock.com; **475** molchunya/Shutterstock.com; **483** Oleksiy Mark/Shutterstock.com; **484** LockStockBob/Shutterstock.com; **485** Jacek Chabraszewski/Shutterstock.com; **490** ©Keddie. Image from BigStockPhoto.com; **491** ©Lars Christenen/123RF; **492** *center left* Nattika/Shutterstock.com; *bottom right* Svetlana Mihailova/Shutterstock.com; **496** *center right* ©iStockphoto.com/Jack Puccio; *bottom left* ©iStockphoto.com/itographer

Chapter 12

500 ©iStockphoto.com/Lisa Thornberg, ©iStockphoto.com/Ann Marie Kurtz; **507** mountainpix/Shutterstock.com; **509** Scott J. Carson/Shutterstock.com; **513** ©iStockphoto.com/Jorgen Jacobsen; **521** *Exercise 11* ©iStockphoto.com/Chih-Feng Chen; *Exercise 12* ©iStockphoto.com/Andreas Gradin; *Exercise 13* ©iStockphoto.com/Jim Lopes; *bottom right* ©iStockphoto.com/Zoran Kolundzija; **533** Andrew B Hall/Shutterstock.com; **534** Serg64/Shutterstock.com; **535** ©iStockphoto.com/Dan Moore; **536** *center right* ©iStockphoto.com/Aldo Murillo; *center left* MapMaster / CC-BY-SA-3.0; *bottom right* VectorWeb/Shutterstock.com

Chapter 13

546 ©iStockphoto.com/Michael Flippo, ©iStockphoto.com/Ann Marie Kurtz; **548** ©iStockphoto.com/HultonArchive; **549** Sir Cumference and the First Round Table; text copyright © 1997 by Cindy Neuschwander; illustrations copyright © 1997 by Wayne Geehan. Used with permission by Charlesbridge Publishing, Inc., 85 Main Street, Watertown, MA 02472; 617-926-0329; www.charlesbridge.com. All rights reserved.; **553** *Exercises 3 and 4* ©iStockphoto.com/zentilia; *Exercise 5* ©iStockphoto.com/Dave Hopkins; *Exercise 6* ©iStockphoto.com/Boris Yankov; *Exercise 7* ©iStockphoto.com/ALEAIMAGE; *Exercise 8* ©iStockphoto.com/iLexx; *Exercise 9–11* ©iStockphoto.com/alexander mychko; **555** *center left* ©iStockphoto.com/HultonArchive; *center right* ©iStockphoto.com/winterling; **561** ©iStockphoto.com/Scott Slattery; **563** *Exercise 6* ©iStockphoto.com/DivaNir4A; *Exercise 7* ©iStockphoto.com/Stacey Walker; *Exercise 8* ©iStockphoto.com/Creativeye99; *Exercise 11* Kalamazoo (Michigan) Public Library; **567** ©iStockphoto.com/Brian Sullivan; **568** *Exercise 3* ©iStockphoto.com/zentilia; *Exercises 4 and 5* ©iStockphoto.com/PhotoHamster; *Exercise 6* ©iStockphoto.com/subjug; *Exercise 7* ©iStockphoto.com/Dave Hopkins; *Exercise 8* ©iStockphoto.com/7nuit; **574** Big Ideas Learning, LLC; **576** *Exercise 4* ©iStockphoto.com/Mr_Vector; *Exercises 5 and 6* ©iStockphoto.com/sndr

Chapter 14

584 stephan kerkhofs/Shutterstock.com, Cigdem Sean Cooper/Shutterstock.com, ©iStockphoto.com/Andreas Gradin; **587** ©iStockphoto.com/Remigiusz Załucki; **590** *center left* Bob the Wikipedian / CC-BY-SA-3.0; *center right* ©iStockphoto.com/Sherwin McGehee; **592** Yasonya/Shutterstock.com; **593** Stankevich/Shutterstock.com; **594** *top left* ©iStockphoto.com/Luke Daniek; *top right* ©iStockphoto.com/Jeff Whyte; *bottom left* ©Michael Mattox. Image from BigStockPhoto.com; *bottom right* ©iStockphoto.com/Hedda Gjerpen; **595** ©iStockphoto.com/josh webb; **604** ©iStockphoto.com/Tomasz Pietryszek; **605** *center left* Newcastle Drum Centre; *center right* ©iStockphoto.com/scol22; **613** *top left* ©iStockphoto.com/david franklin; *top right* ©Ruslan Kokarev. Image from BigStockPhoto.com; *center right* ©iStockphoto.com/Lev Mel, ©iStockphoto.com/Ebru Baraz; **614** *bottom left* ©iStockphoto.com/Jiri Vatka; *bottom right* Patryk Kosmider/Shutterstock.com; **618** ©iStockphoto.com/ranplett, Image © Courtesy of Museum of Science, Boston; **619** *center right* ©iStockphoto.com/James Kingman; *center left* ©iStockphoto.com/Yails; **621** *Exercise 12* ©iStockphoto.com/AlexStar; *Exercise 13* Knartz/Shutterstock.com; *Exercise 14* SOMMAI/Shutterstock.com

Chapter 15

630 infografick/Shutterstock.com, ©iStockphoto.com/Ann Marie Kurtz; **633** ryasick photography/Shutterstock.com; **634** *center* ©iStockphoto.com/sweetym; *bottom left* Big Ideas Learning, LLC; **635** ©iStockphoto.com/sweetym; **636** ©iStockphoto.com/Joe Potato Photo, ©iStockphoto.com/sweetym; **637** *top left* ©iStockphoto.com/Jennifer Morgan; *top center* United States coin image from the United States Mint; **641** *top left* design56/Shutterstock.com; *center right* ©iStockphoto.com/spxChrome; **642** *center right* Daniel Skorodyelov/Shutterstock.com; *bottom left* Mitrofanova/Shutterstock.com; **643** Jamie Wilson/Shutterstock.com; **644** *top right* James Steidl/Shutterstock.com; *bottom right* traudl/Shutterstock.com; **645** Warren Goldswain/Shutterstock.com; **646** ©iStockphoto.com/Frank van de Bergh; **647** ©iStockphoto.com/Eric Ferguson; **651** Feng Yu/Shutterstock.com; **652** *top right* John McLaird/Shutterstock.com; *center right* Robert Asento/Shutterstock.com; *bottom right* Mark Aplet/Shutterstock.com; **657** ©iStockphoto.com/Justin Horrocks; **658** ©iStockphoto.com/sweetym; **659** *top right* John McLaird/Shutterstock.com; *center* tele52/Shutterstock.com; **660 and 661** ©iStockphoto.com/Joe Potato Photo, ©iStockphoto.com/sweetym; **663** Univega/Shutterstock.com; **664** James Steidl/Shutterstock.com; **666** *top left* ©iStockphoto.com/sweetym; *center right* ©iStockphoto.com/Andy Cook; **667** -Albachiaraa-/Shutterstock.com; **671** ©iStockphoto.com/Doug Cannell; **672** *Activity 1a left* ©iStockphoto.com/Shannon Keegan; *Activity 1a right* ©iStockphoto.com/Lorelyn Medina; *Activity 1b left* Joel Sartore/joelsartore.com; *Activity 1b right* Feng Yu/Shutterstock.com; *Activity 1c left* ©iStockphoto.com/kledge; *Activity 1c right* ©iStockphoto.com/spxChrome; *Activity 1d* ©iStockphoto.com/Alex Slobodkin; **674** ©iStockphoto.com/Philip Lange; **676** ©iStockphoto.com/blackwaterimages, ©iStockphoto.com/Rodrigo Blanco, ©iStockphoto.com/7nuit; **679** Feng Yu/Shutterstock.com; **683** artis777/Shutterstock.com; **689** GoodMood Photo/Shutterstock.com; **690** Asaf Eliason/Shutterstock.com; **692** *top right* ©iStockphoto.com/Frank Ramspott; *bottom left* ©iStockphoto.com/7nuit; **694** mylisa/Shutterstock.com; **695** ra-design/Shutterstock.com

Appendix A

A0 *background* ©iStockphoto.com/Björn Kindler; *top left* ©iStockphoto.com/mika makkonen; *top right* ©iStockphoto.com/Hsing-Wen Hsu; **A1** *top right* ©iStockphoto.com/toddmedia; *bottom left* ©iStockphoto.com/Loretta Hostettler; *bottom right* NASA; **A4** *top right* ©iStockphoto.com/Hsing-Wen Hsu; *bottom left* ©iStockphoto.com/Thomas Kuest; *bottom right* Lim ChewHow/Shutterstock.com; **A5** *top right* ©iStockphoto.com/Richard Cano; *bottom left* ©iStockphoto.com/best-photo; *bottom right* ©iStockphoto.com/mika makkonen; **A6** *top right* ©iStockphoto.com/Loretta Hostettler; *bottom* ©iStockphoto.com/toddmedia; **A7** *top right* LudmilaM/Shutterstock.com; *center left and bottom right* ©iStockphoto.com/Clayton Hansen; *bottom left* Billwhittaker at en.wikipedia; **A8 and A9** NASA

Cartoon illustrations Tyler Stout

Learning Progression

Kindergarten

Counting and Cardinality	– Count to 100 by Ones and Tens; Compare Numbers
Operations and Algebraic Thinking	– Understand and Model Addition and Subtraction
Number and Operations in Base Ten	– Work with Numbers 11–19 to Gain Foundations for Place Value
Measurement and Data	– Describe and Compare Measurable Attributes; Classify Objects into Categories
Geometry	– Identify and Describe Shapes

Grade 1

Operations and Algebraic Thinking	– Represent and Solve Addition and Subtraction Problems
Number and Operations in Base Ten	– Understand Place Value for Two-Digit Numbers; Use Place Value and Properties to Add and Subtract
Measurement and Data	– Measure Lengths Indirectly; Write and Tell Time; Represent and Interpret Data
Geometry	– Draw Shapes; Partition Circles and Rectangles into Two and Four Equal Shares

Grade 2

Operations and Algebraic Thinking	– Solve One- and Two-Step Problems Involving Addition and Subtraction; Build a Foundation for Multiplication
Number and Operations in Base Ten	– Understand Place Value for Three-Digit Numbers; Use Place Value and Properties to Add and Subtract
Measurement and Data	– Measure and Estimate Lengths in Standard Units; Work with Time and Money
Geometry	– Draw and Identify Shapes; Partition Circles and Rectangles into Two, Three, and Four Equal Shares

Grade 3

Operations and Algebraic Thinking	– Represent and Solve Problems Involving Multiplication and Division; Solve Two-Step Problems Involving Four Operations
Number and Operations in Base Ten	– Round Whole Numbers; Add, Subtract, and Multiply Multi-Digit Whole Numbers
Number and Operations—Fractions	– Understand Fractions as Numbers
Measurement and Data	– Solve Time, Liquid Volume, and Mass Problems; Understand Perimeter and Area
Geometry	– Reason with Shapes and Their Attributes

Grade 4

Operations and Algebraic Thinking	– Use the Four Operations with Whole Numbers to Solve Problems; Understand Factors and Multiples
Number and Operations in Base Ten	– Generalize Place Value Understanding; Perform Multi-Digit Arithmetic
Number and Operations—Fractions	– Build Fractions from Unit Fractions; Understand Decimal Notation for Fractions
Measurement and Data	– Convert Measurements; Understand and Measure Angles
Geometry	– Draw and Identify Lines and Angles; Classify Shapes

Grade 5

Operations and Algebraic Thinking	– Write and Interpret Numerical Expressions
Number and Operations in Base Ten	– Perform Operations with Multi-Digit Numbers and Decimals to Hundredths
Number and Operations—Fractions	– Add, Subtract, Multiply, and Divide Fractions
Measurement and Data	– Convert Measurements within a Measurement System; Understand Volume
Geometry	– Graph Points in the First Quadrant of the Coordinate Plane; Classify Two-Dimensional Figures

Mathematics Reference Sheet

Conversions

U.S. Customary
1 foot = 12 inches
1 yard = 3 feet
1 mile = 5280 feet
1 acre = 43,560 square feet
1 cup = 8 fluid ounces
1 pint = 2 cups
1 quart = 2 pints
1 gallon = 4 quarts
1 gallon = 231 cubic inches
1 pound = 16 ounces
1 ton = 2000 pounds
1 cubic foot ≈ 7.5 gallons

U.S. Customary to Metric
1 inch = 2.54 centimeters
1 foot ≈ 0.3 meter
1 mile ≈ 1.61 kilometers
1 quart ≈ 0.95 liter
1 gallon ≈ 3.79 liters
1 cup ≈ 237 milliliters
1 pound ≈ 0.45 kilogram
1 ounce ≈ 28.3 grams
1 gallon ≈ 3785 cubic centimeters

Time
1 minute = 60 seconds
1 hour = 60 minutes
1 hour = 3600 seconds
1 year = 52 weeks

Temperature
$C = \frac{5}{9}(F - 32)$

$F = \frac{9}{5}C + 32$

Metric
1 centimeter = 10 millimeters
1 meter = 100 centimeters
1 kilometer = 1000 meters
1 liter = 1000 milliliters
1 kiloliter = 1000 liters
1 milliliter = 1 cubic centimeter
1 liter = 1000 cubic centimeters
1 cubic millimeter = 0.001 milliliter
1 gram = 1000 milligrams
1 kilogram = 1000 grams

Metric to U.S. Customary
1 centimeter ≈ 0.39 inch
1 meter ≈ 3.28 feet
1 kilometer ≈ 0.62 mile
1 liter ≈ 1.06 quarts
1 liter ≈ 0.26 gallon
1 kilogram ≈ 2.2 pounds
1 gram ≈ 0.035 ounce
1 cubic meter ≈ 264 gallons

Number Properties

Commutative Properties of Addition and Multiplication

$a + b = b + a$

$a \cdot b = b \cdot a$

Associative Properties of Addition and Multiplication

$(a + b) + c = a + (b + c)$

$(a \cdot b) \cdot c = a \cdot (b \cdot c)$

Addition Property of Zero

$a + 0 = a$

Multiplication Properties of Zero and One

$a \cdot 0 = 0$

$a \cdot 1 = a$

Distributive Property:

$a(b + c) = ab + ac$

$a(b - c) = ab - ac$

Properties of Equality

Addition Property of Equality

If $a = b$, then $a + c = b + c$.

Subtraction Property of Equality

If $a = b$, then $a - c = b - c$.

Multiplication Property of Equality

If $a = b$, then $a \cdot c = b \cdot c$.

Multiplicative Inverse Property

$n \cdot \frac{1}{n} = \frac{1}{n} \cdot n = 1, n \neq 0$

Division Property of Equality

If $a = b$, then $a \div c = b \div c, c \neq 0$.

Squaring both sides of an equation

If $a = b$, then $a^2 = b^2$.

Cubing both sides of an equation

If $a = b$, then $a^3 = b^3$.

Properties of Inequality

Addition Property of Inequality

If $a > b$, then $a + c > b + c$.

Subtraction Property of Inequality

If $a > b$, then $a - c > b - c$.

Multiplication Property of Inequality

If $a > b$ and c is positive, then $a \cdot c > b \cdot c$.

If $a > b$ and c is negative, then $a \cdot c < b \cdot c$.

Division Property of Inequality

If $a > b$ and c is positive, then $a \div c > b \div c$.

If $a > b$ and c is negative, then $a \div c < b \div c$.

Properties of Exponents

Product of Powers Property: $a^m \cdot a^n = a^{m+n}$

Quotient of Powers Property: $\dfrac{a^m}{a^n} = a^{m-n}, a \neq 0$

Power of a Power Property: $(a^m)^n = a^{mn}$

Power of a Product Property: $(ab)^m = a^m b^m$

Zero Exponents: $a^0 = 1, a \neq 0$

Negative Exponents: $a^{-n} = \dfrac{1}{a^n}, a \neq 0$

Slope

$$m = \frac{\text{rise}}{\text{run}} = \frac{\text{change in } y}{\text{change in } x} = \frac{y_2 - y_1}{x_2 - x_1}$$

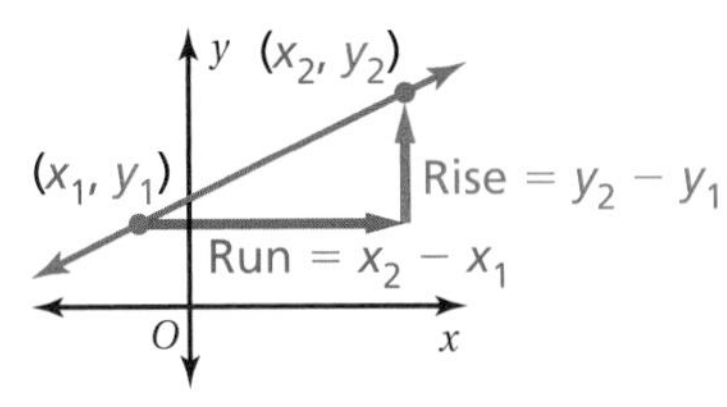

Equations of Lines

Slope-intercept form

$y = mx + b$

Standard form

$ax + by = c, a, b \neq 0$

Point-slope form

$y - y_1 = m(x - x_1)$

Pythagorean Theorem

$a^2 + b^2 = c^2$

Converse of the Pythagorean Theorem

If the equation $a^2 + b^2 = c^2$ is true for the side lengths of a triangle, then the triangle is a right triangle.

Distance Formula

$$d = \sqrt{(x_2 - x_1)^2 + (y_2 - y_1)^2}$$

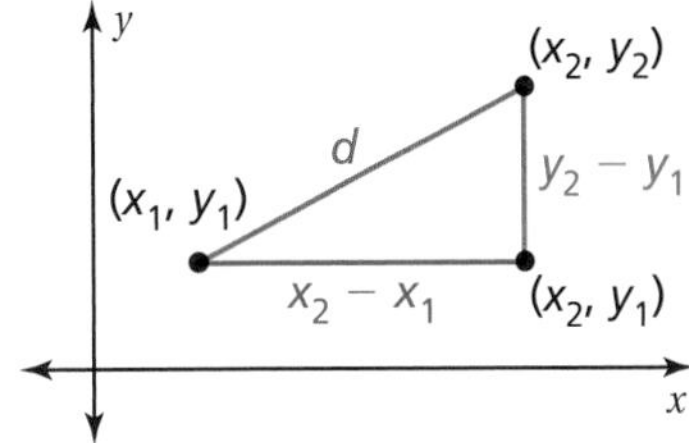

Formulas in Geometry

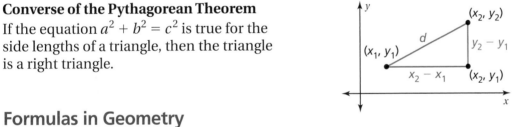

Prism

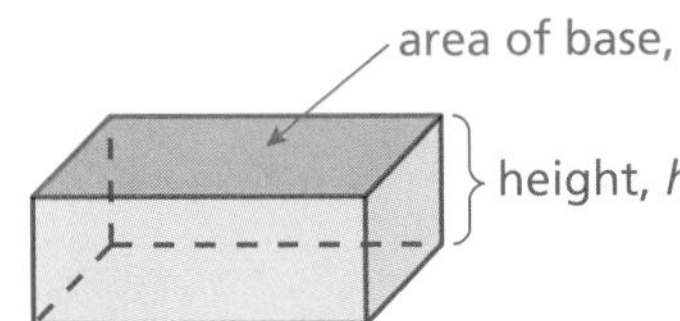

$S =$ areas of bases + areas of lateral faces

$V = Bh$

Pyramid

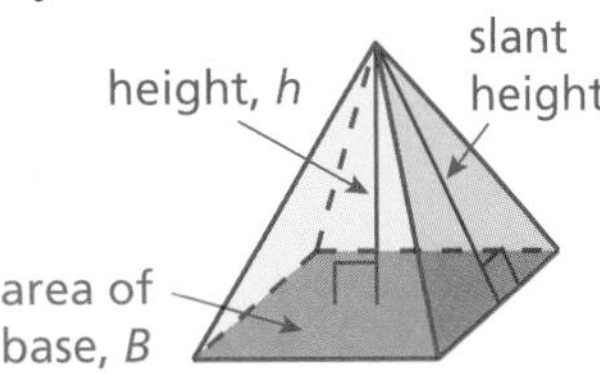

$S =$ area of base + areas of lateral faces

$V = \frac{1}{3}Bh$

Circle

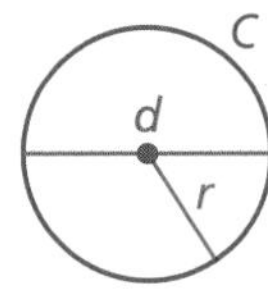

$C = \pi d$ or $C = 2\pi r$

$A = \pi r^2$

$\pi \approx \frac{22}{7}$, or 3.14

Cylinder

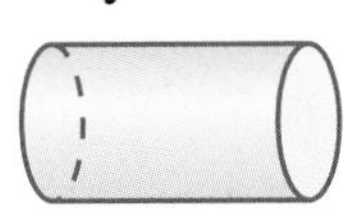

$V = Bh = \pi r^2 h$

$S = 2\pi r^2 + 2\pi rh$

Cone

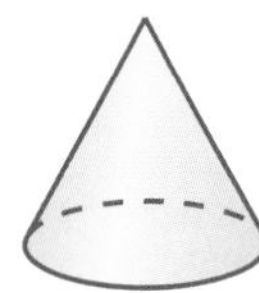

$V = \frac{1}{3}Bh = \frac{1}{3}\pi r^2 h$

Sphere

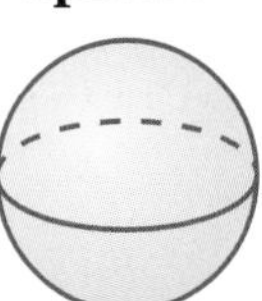

$V = \frac{4}{3}\pi r^3$